Air Quality

Science, Impacts, and Management

Ranjeet S. Sokhi
Centre for Climate Change Research (C3R), School of Physics, Engineering and Computer Science,
University of Hertfordshire, Hatfield, United Kingdom

ELSEVIER

Elsevier
Radarweg 29, PO Box 211, 1000 AE Amsterdam, Netherlands
125 London Wall, London EC2Y 5AS, United Kingdom
50 Hampshire Street, 5th Floor, Cambridge, MA 02139, United States

ISBN: 978-0-12-822591-2

For Information on all Elsevier publications
visit our website at https://www.elsevier.com/books-and-journals

Publisher: Candice Janco
Acquisitions Editor: Jennette McClain
Editorial Project Manager: Sara Valentino
Production Project Manager: Paul Prasad Chandramohan
Cover Designer: Greg Harris

Typeset by MPS Limited, Chennai, India

Air Quality
Science, Impacts, and Management

Dedication

For those who stood before us and those who continue in their endeavors to understand our atmosphere.

Contents

List of contributors

Ummugulsum Alyuz Centre for Climate Change Research (C3R), School of Physics, Engineering and Computer Science, University of Hertfordshire, Hatfield, United Kingdom

Zorana Jovanovic Andersen Department of Public Health, Section of Environmental Health, University of Copenhagen, Copenhagen, Denmark

Maria de Fatima Andrade Institute of Astronomy, Geophysics and Atmospheric Sciences, University of São Paulo, São Paulo, Brazil

Somayyeh Arghavani Centre for Climate Change Research (C3R), School of Physics, Engineering and Computer Science, University of Hertfordshire, Hatfield, United Kingdom

Yusuf Aydin Department of Climate and Marine Sciences, Eurasia Institute of Earth Sciences, Istanbul Technical University, Maslak, Istanbul, Turkey

Jaana Bäck Institute for Atmospheric and Earth System Research, University of Helsinki, Helsinki, Finland

Evangelos Bagkis Environmental Informatics Research Group, School of Mechanical Engineering, Aristotle University of Thessaloniki, Thessaloniki, Greece

Alexander Baklanov Climate and Geophysics, Niels Bohr Institute, University of Copenhagen, Copenhagen, Denmark

Jo Barnes University of the West of England, Bristol, United Kingdom

Mario Gavidia Calderon Institute of Astronomy, Geophysics and Atmospheric Sciences, University of São Paulo, São Paulo, Brazil

Gregory R. Carmichael Department of Chemical and Biochemical Engineering, The University of Iowa, Iowa City, IA, United States

Timothy Chaston Environment Protection Authority Victoria, Australia

Jason Choi Environment Protection Authority Victoria, Australia

Isabelle Coll Univ Paris Est Creteil and Université Paris Cité, CNRS, LISA, Créteil, France

Helen F. Dacre Department of Meteorology, University of Reading, Reading, United Kingdom

Edmilson Dias de Freitas Institute of Astronomy, Geophysics and Atmospheric Sciences, University of São Paulo, São Paulo, Brazil

Goksel Demir Hamidiye Faculty of Sciences, Department of Occupational Health, and Safety, University of Health Sciences Turkey, Uskudar, Istanbul, Turkey

Ali Deniz Department of Climate and Marine Sciences, Eurasia Institute of Earth Sciences, Istanbul Technical University, Maslak, Istanbul, Turkey

Anna Di Leo ARPA Lombardia (Environmental Protection Agency of Lombardia Region), Milan

Sergio Ibarra Espinosa Cooperative Institute for Research in Environmental Sciences, University of Colorado Boulder, Boulder, CO, United States; NOAA Global Monitoring Laboratory, Boulder, CO, United States

Sandro Finardi ARIANET, Milan

Evangelia Fragkou Department of Mechanical Engineering, Aristotle University of Thessaloniki, Thessaloniki, Greece

Meng Gao Department of Geography, Hong Kong Baptist University, Hong Kong SAR, P.R. China

Camilla Geels Department of Environmental Science, Aarhus University, Denmark

Akshi Goyal Department of Environment Studies, Panjab University, Chandigarh, India

Yasemin Guler Faculty of Engineering and Physical Sciences, Global Centre for Clean Air Research (GCARE), School of Sustainability, Civil and Environmental Engineering, University of Surrey, Guildford, United Kingdom

Enda Hayes University of the West of England, Bristol, United Kingdom

Ulas Im Department of Environmental Science, Aarhus University, Denmark

Leonardo Yoshiaki Kamigauti Faculty of Engineering and Physical Sciences, Global Centre for Clean Air Research (GCARE), School of Sustainability, Civil and Environmental Engineering, University of Surrey, Guildford, United Kingdom

Kostas Karatzas Environmental Informatics Research Group, School of Mechanical Engineering, Aristotle University of Thessaloniki, Thessaloniki, Greece

Theodosios Kassandros Environmental Informatics Research Group, School of Mechanical Engineering, Aristotle University of Thessaloniki, Thessaloniki, Greece

Veli-Matti Kerminen Institute for Atmospheric and Earth System Research, University of Helsinki, Helsinki, Finland

Kenza Khomsi Mohammed 6 University of Sciences and Health, Casablanca, Morocco

Tom Kokkonen Institute for Atmospheric and Earth System Research, University of Helsinki, Helsinki, Finland

Soma Sekhara Rao Kolluru Faculty of Engineering and Physical Sciences, Global Centre for Clean Air Research (GCARE), School of Sustainability, Civil and Environmental Engineering, University of Surrey, Guildford, United Kingdom

Shaofei Kong Department of Atmospheric Sciences, School of Environmental Studies, China University of Geosciences, Wuhan, P.R. China

Patricia Krecl Federal University of Technology (UTFPR), Graduate Program in Environmental Engineering, Londrina, Brazil

Markku Kulmala Institute for Atmospheric and Earth System Research, University of Helsinki, Helsinki, Finland

Prashant Kumar Faculty of Engineering and Physical Sciences, Global Centre for Clean Air Research (GCARE), School of Sustainability, Civil and Environmental Engineering, University of Surrey, Guildford, United Kingdom

Saurabh Kumar Centre for Climate Change Research (C3R), University of Hertfordshire, Hatfield, United Kingdom

Abidemi Kuye Faculty of Engineering and Physical Sciences, Global Centre for Clean Air Research (GCARE), School of Sustainability, Civil and Environmental Engineering, University of Surrey, Guildford, United Kingdom

Guido Lanzani ARPA Lombardia (Environmental Protection Agency of Lombardia Region), Milan

Alexander Mahura Faculty of Science, Institute for Atmospheric and Earth System Research, University of Helsinki, Helsinki, Finland

Jorge Martins Graduate Program in Environmental Engineering, Federal University of Technology, Paraná, Brazil

Leila Droprinchinski Martins Graduate Program in Environmental Engineering, Federal University of Technology, Paraná, Brazil

Brian McDonald NOAA Chemical Sciences Laboratory, Boulder, CO, United States

Ana Paula Mendes Emygdio University of Surrey, Surrey, United Kingdom

Regina Miranda School of Arts, Sciences and Humanities, University of São Paulo, São Paulo, Brazil

Luisa T. Molina Molina Center for Energy and the Environment, Boston, MA, United States

Suman Mor Department of Environment Studies, Panjab University, Chandigarh, India

Nicolas Moussiopoulos Aristotle University of Thessaloniki, Thessaloniki, Greece

Anil Namdeo Department of Geography and Environmental Sciences, Northumbria University, Newcastle upon Tyne, United Kingdom

Thiago Nogueira Public Health Faculty, University of São Paulo, São Paulo, Brazil

Roman Nuterman Niels Bohr Institute, University of Copenhagen, Copenhagen, Denmark

Hamid Omidvarborna Faculty of Engineering and Physical Sciences, Global Centre for Clean Air Research (GCARE), School of Sustainability, Civil and Environmental Engineering, University of Surrey, Guildford, United Kingdom

Gabriel Yoshikazu Oukawa Federal University of Technology (UTFPR), Graduate Program in Environmental Engineering, Londrina, Brazil

Guilherme Pereira Institute of Astronomy, Geophysics and Atmospheric Sciences, University of São Paulo, São Paulo, Brazil

Pedro Jose Perez Martinez Faculty of Engineering, Architecture and Urbanism, University of Campinas, Campinas, São Paulo, Brazil

Tuukka Petäjä Institute for Atmospheric and Earth System Research, University of Helsinki, Helsinki, Finland

Paola Radice ARIANET, Milan

Khaiwal Ravindra Department of Community Medicine and School of Public Health, Post Graduate Institute of Medical Education and Research (PGIMER), Chandigarh, India

Armando Retama Molina Center for Energy and the Environment, Boston, MA, United States

Pauliina Schiestl-Aalto Institute for Atmospheric and Earth System Research, University of Helsinki, Helsinki, Finland

Imane Sekmoudi Independent Researcher, Casablanca, Morocco

Arushi Sharma Interdisciplinary Program in Climate Studies, Indian Institute of Technology Bombay, Mumbai, Maharashtra, India

Vikas Singh National Atmospheric Research Laboratory (NARL), Gadanki, Andhra Pradesh, India

Ranjeet S. Sokhi Centre for Climate Change Research (C3R), School of Physics, Engineering and Computer Science, University of Hertfordshire, Hatfield, United Kingdom

Carlos Eduardo Souto-Oliveira Institute of Astronomy, Geophysics and Atmospheric Sciences, University of São Paulo, São Paulo, Brazil

Nigel Tapper School of Earth, Atmosphere and Environment, Monash University, Australia

Admir Créso Targino Federal University of Technology (UTFPR), Graduate Program in Environmental Engineering, Londrina, Brazil

Pushp Raj Tiwari Centre for Atmospheric and Climate Physics Research, University of Hertfordshire, Hatfield, United Kingdom; Centre for Climate Change Research (C3R), University of Hertfordshire, Hatfield, United Kingdom; Centre for Future Society Research, University of Hertfordshire, Hatfield, United Kingdom

George Tsegas Department of Mechanical Engineering, Aristotle University of Thessaloniki, Thessaloniki, Greece

Bhishma Tyagi Department of Earth and Atmospheric Sciences, National Institute of Technology Rourkela, Rourkela, India

Alper Unal Department of Climate and Marine Sciences, Eurasia Institute of Earth Sciences, Istanbul Technical University, Maslak, Istanbul, Turkey

Perola Vasconcellos Institute of Chemistry, University of São Paulo, São Paulo, Brazil

Chandra Venkataraman Interdisciplinary Program in Climate Studies, Indian Institute of Technology Bombay, Mumbai, Maharashtra, India; Department of Chemical Engineering, Indian Institute of Technology Bombay, Mumbai, Maharashtra, India

Fan Wang Department of Geography, Hong Kong Baptist University, Hong Kong SAR, P.R. China

Oliver Wild Lancaster Environment Centre, Lancaster University, Lancaster, United Kingdom

Huai-Wen Wu Faculty of Engineering and Physical Sciences, Global Centre for Clean Air Research (GCARE), School of Sustainability, Civil and Environmental Engineering, University of Surrey, Guildford, United Kingdom

Ilona Ylivinkka Institute for Atmospheric and Earth System Research, University of Helsinki, Helsinki, Finland

Huang Zheng Department of Atmospheric Sciences, School of Environmental Studies, China University of Geosciences, Wuhan, P.R. China

Preface

When we look at the sky, we assume that the atmosphere is a vast layer of gas that encircles our planet. This is of course true, but what is not so obvious is that the atmosphere is a relatively thin layer, which not only provides us with the oxygen to live but also shields us from hazardous solar radiation. As an interlinked component of the holistic Earth system, it is responsible for creating our climate and weather and nourishes the complex ecosystems of our planet. Not surprisingly, we are becoming increasingly concerned with the state of our atmosphere and, in particular, the quality of the air we breathe. Although air pollution has been a historical issue, there has been a growing public awareness of poor air quality over the past few decades. Despite wide interest, however, considerable progress is still needed to make air safe for everyone.

In most cases, we relate to air pollution through our own experiences or because it affects other people. This usually happens because we live near a major road, or we are in the vicinity of industry or are affected by hazards such as dust and wildfires. What is clear is that the quality of air affects us every time we breathe. While exposure to poor local air pollution is an unfortunate part of our everyday lives in many towns and cities, regional and global contributions are also important and can exacerbate local air quality. As our understanding of air quality has progressed, so has our appreciation of its impacts on health and the environment, which has stimulated public and policy concerns in many parts of the world.

The science and management of air quality are complex and involve multiple aspects. These include fundamental processes that control the dynamics and chemistry of air pollutant species, quantifying their emissions and developing approaches to monitor, manage, and control air pollution. All these factors are essential for understanding the resulting impacts on our environment and health.

Although weather plays a key role in determining the short- and medium-term variations in air pollutant concentrations at any given location, longer-term interactions with climate are also important and can influence the state of air quality. The prevailing processes, emissions, and interactions are all spatially and temporally dependent, which leads to significant differences in the extent and severity of poor air quality between regions on top of the local-scale gradients that are evident in our towns and cities.

The subject of air quality has been the focus of several textbooks over the years. Inevitably, these texts have adopted their own approach and preferences for the topics that they include. There are also several related books on the broader subject of atmospheric science and climate, which are complementary to the texts on air quality and air pollution. This book differs from the other texts concerning air quality in several ways.

The overall approach of this book relies on providing the main fundamental concepts of atmospheric science relevant to air quality. It then progresses to different but interlinked aspects of air quality science. Given that air quality science is interdisciplinary, this book consists of interconnected chapters written by international experts from several disciplines. In most chapters, the concepts are supported with worked examples and supplemented with further questions to test and challenge the reader regarding their understanding of the topics. Each chapter adopts a textbook style but also contains a significant number of references that highlight some of the historical and recent advances in the subject through a brief state-of-the-art review of that topic.

With respect to the specific chapters and the structure of the book, there is a logical progression beginning with fundamental knowledge of the main concepts, processes, and mechanisms that determine and influence air quality at multiple scales. The chapters focus on the main sources and emissions of air pollution and then build on aspects such as atmospheric chemistry, dynamics, and boundary layer physics, which are essential for understanding the next two chapters on the theory of atmospheric dispersion and air quality models that are used to predict the spatial and temporal distributions of air pollution.

Models rely on observations; hence, some of the main methods for measuring air quality are described and discussed in the next chapter. After discussing regional air quality, the book examines the key interactions between air quality and climate, which also operate on large spatial scales and can significantly influence local- and city-scale air quality.

Poor air quality is generally perceived to be an outdoor problem, but indoor environments heavily determine peoples' overall exposure to air pollution and hence are covered in the next chapter. The following chapter provides an overview of key methods for quantitatively analyzing complex air pollution-related datasets, which are essential for monitoring and assessing air quality. One of the main motivations for studying air quality is to better understand the impacts of air quality on human health and ecosystems, which forms the focus of the next two chapters before moving to the chapter on air quality management and policy approaches for improving air quality. The last chapter includes case studies of air quality trends in selected major cities and provides readers with the current state of play of air pollution in the main regions of the world.

The subject matter of the book will be suitable for undergraduate and postgraduate courses as well as for informed readers working in the environmental field, such as consultants and those active in air quality research, or for those who wish to gain an understanding of the scientific intricacies of the subject of air quality. Although most chapters have an introductory component, to appreciate the more advanced aspects, ideally, it would be necessary for the reader to have an equivalent of first- or second-year university-level background in a relevant discipline such as physics, chemistry, meteorology, or environmental science with a mathematical slant. Whatever the level of educational background of the reader, we hope that they will find this textbook interesting and informative and will not only enhance but also stimulate their curiosity in the field. If we succeed even partially in this endeavor, our efforts would be worthwhile.

Ranjeet S. Sokhi

Acknowledgments

This book would not have been possible if had not been for the generosity of the following people and organizations to whom we are deeply grateful.

The financial support has been provided through multiple projects related to the Pan-Eurasian EXperiment (PEEX) program (coordinated by the University of Helsinki, Institute for Atmospheric and Earth System Research, Helsinki, Finland) including Academy of Finland project ACCC (grant #337549), European Union's Horizon 2020 Programme project RI-URBANS (grant #101036245), and Horizon Europe project FOCI (grant #101056783) including funding from Innovate UK (UKRI Reference Number 10040465).

Authors thank the support from Innovate UK funded project "MyGlobalHome" (Technology Strategy Board File Reference of 106168), EPSRC-funded CO-TRACE (EP/W001411/1), COVAIR (EP/V052462/1), INHALE (EP/T003189/1), and the University of Surrey's HEIF funded RealAir project.

We acknowledge the support from the nextGEMS project, which received funding from the European Union's Horizon 2020 research and innovation program under the grant agreement No. 101003470.

The authors express sincere thanks to Giovanni and NASA GEC DISC (https://giovanni.gsfc.nasa.gov/giovanni/; https://disc.gsfc.nasa.gov/) for providing the datasets and tool/method screenshots related to remote sensing and satellite observations.

We extend our gratitude to the Environmental Company of the State of São Paulo (CETESB) for generously providing the air pollution data, and to the Institute of Astronomy, Geophysics and Atmospheric Science of the University of São Paulo (IAG/USP) for supplying the meteorological data.

The authors are grateful to Mexico City Secretariat of Environment (SEDEMA) for providing access to the data used in our analysis.

In the case of the Indian city case studies, we acknowledge the Indian Megacity project PROMOTE, part of the Air Pollution and Human Health programme, funded by the UK's Natural Environment Research Council (NERC) and the Indian Ministry of Earth Sciences (MOES), Grant reference number NE/P016391/1 and NE/R001782/1. We also acknowledge the NCAP-COALESCE project funded by the Ministry of Environment Forests and Climate Change through grant NO.14/10/2014-CC (Vol. II). The $PM_{2.5}$ concentration data has been obtained from an open dataset provided by the Central Pollution Control Board (CPCB).

For the London case study, we are grateful to the Department of Environment, Food and Rural Affairs (DEFRA, UK) for making the air quality data publicly available.

For the Paris case study, the air quality analysis relies on measurement data as well as cartographic representations produced by the AIRPARIF air quality monitoring network.

We would like to thank the reviewers for their very useful comments that helped to improve the content and structure of the book.

Thanks to John Methven at the University of Reading for useful discussions about Chapter 5.

Finally, we are grateful to the editorial staff at Elsevier for their assistance in guiding us through the publication process. In particular, our thanks go to Sara Valentino for her support during the manuscript preparation phase and to Paul Prasad Chandramohan during the production stages of the book.

General note:

The views expressed in this document are solely those of the authors and do not necessarily reflect those of any of the organizations or governmental departments or ministries mentioned here. These organizations and government departments and ministries do not endorse any products or commercial services mentioned or implied in this publication.

Useful constants and other data

Parameter	Symbol	Value and units
Boltzmann constant	k_B	1.38×10^{-23} J K^{-1}
Planck constant	h	6.626×10^{-34} J s^{-1}
Stefan–Boltzmann constant	σ_B	5.670×10^{-8} W m^{-2} K^{-4}
Wien's displacement constant	B	2897.8 μmK or 2897.8×10^{-6} m·K
Solar constant	S	1365 W m^{-2} (mean value at the top of the Earth's atmosphere)
Avogadro number	N_A	$6.02214076 \times 10^{23}$ mol^{-1}
Molecular weight of dry air	M	28.965 g mol^{-1}
Mean air temperature of the lower troposphere	T_a	288 K
Acceleration due to gravity on Earth	g	9.81 m s^{-2}
Universal gas constant	R^*	8.314 J K^{-1} mol^{-1} or 0.08205 L atm mol^{-1} K^{-1}
Specific gas constant for dry air	R_d	287 J K^{-1} kg^{-1}
Specific heat of pure water	C	4186 J K^{-1} kg^{-1}
Specific heat of dry air at sea level at constant pressure	c_p	1004 J K^{-1} kg^{-1}
Standard atmospheric pressure	P	1.013×10^5 Pa or 1013.25 hPa
Environmental lapse rate	G_e	6.5°C km^{-1} (typical value)
Dry atmospheric lapse rate	G_d	9.8°C km^{-1}
Global mean radius of the Earth	R_{Earth}	6.37×10^6 m

Chapter 1

Introduction—why does air quality matter?

Ranjeet S. Sokhi
Centre for Climate Change Research (C3R), School of Physics, Engineering and Computer Science, University of Hertfordshire, Hatfield, United Kingdom

... air pollution has become recognized as the single biggest environmental threat to human health based on its notable contribution to disease burden.

(WHO, 2021a)

This introductory chapter begins our journey to understanding air quality. While we usually take the quality of the air we breathe for granted, there has been a universal recognition that air quality is the biggest environmental risk to human health (WHO, 2021a). Some of the key concepts that will be necessary to understand and navigate through the rest of the book are presented here. After a short summary of the book, a brief historical perspective is presented before explaining some of the key terms associated with air quality. A short survey of the main properties and impacts of the key air pollutants is given. Some of these are at the core of the air quality debate. A glimpse of the current state of air quality around the world is given before outlining the structure of the book.

As we delve into this chapter, readers with a background in the field of air quality science may also find it useful to consult other comprehensive reviews of the topic, such as HEI (2020), Molina (2021), Morawska et al. (2021), Sicard et al. (2021), and Sokhi et al. (2022). These provide a research synthesis of the state of the art in this field to various degrees.

1.1 The book in a nutshell

Over recent years, air quality has become a household phrase and is routinely discussed on television, in the general media, in our everyday discourse and even has been the subject of legal cases (Kenis & Barratt, 2022). When one looks for information on air quality on the many apps, internet or national and international media news outlets, more often than not, the information is expressed as simple, easy-to-follow indices which can even be color coded to show the severity of the air quality at a particular location. However, the state of air quality at any given location, while being expressed as this simple color-coded information, is dependent on complex interactions and on a large number of factors. We can group these factors into (i) emissions of air pollutants, (ii) chemical and physical changes that can influence the air pollutants in our atmosphere and (iii) dynamical processes that mix, transport and spread air pollution from the point of emission to towns and cities, countries, continents and even the whole globe. Many of these factors are directly or indirectly influenced by weather conditions which itself are changeable on space and time scales. In the longer term, climate is changing and will also affect the quality of the air that we breathe.

Through this book, you will gain knowledge of the main concepts, processes, and mechanisms that determine air quality, how it affects our environment and health and how we manage it. The book also explains how air pollution is measured and predicted which are essential tools to monitor and understand how air pollution spreads, how it changes in our atmosphere and how it affects us and our environment.

Air Quality. DOI: https://doi.org/10.1016/B978-0-12-822591-2.00001-9

1.2 What do we mean by air quality and related terms?

Before we delve into the science of air quality, we need to understand some key terms. **Air** is usually defined as a naturally occurring mixture of gases, mainly nitrogen and oxygen with small amounts of argon, carbon dioxide, water vapor, and various air pollutants (a definition adopted by the Royal Meteorological Society, UK and similar to that from the American Meteorological Society).

From a science perspective, **air quality** is considered to be the state of the lower atmosphere in terms of its chemical constituents at a given time and space. However, such a definition is incomplete as the term has broader connotations.

For example, the World Health Organization (WHO) defines air quality as "the chemical state of the atmosphere at a given time and place and how it affects people and ecosystems" (WHO, 2006, 2021a).

While this definition is accurate, air quality also has a policy relevance, especially, when comparing the quality of air that we breathe at any given period of time and type of location with metrics established by organizations such as WHO itself (air quality guidelines, AQGs), the European Union (EU) (air quality limit values) and standards set by governments like the UK Government (see Chapter 16 for more details). These are often legally binding, as in the case of national standards and EU limit values which define threshold values for several air pollutants. These thresholds are assigned specific values along with how they are defined and should not be exceeded to provide a level of protection to avoid or reduce detrimental health and environmental impacts.

In atmospheric and air quality science, the term **chemical state** can also describe a broader set of chemical species as these vary in the atmosphere and are not all included in the list of policy standards and limit values. Furthermore, air quality is usually related to near-ground airborne concentrations of specific chemical species.

Hence, an alternative definition of air quality can be "the chemical state of the lower atmosphere expressed as airborne concentrations of gaseous and particulate matter (PM) air pollutant species according to environment and health-related metrics specified in national and international policy frameworks."

Using this definition, air pollutants relevant to air quality, include PM ($PM_{2.5}$ and PM_{10}), nitrogen dioxide (NO_2), carbon monoxide (CO), ozone (O_3), sulfur dioxide (SO_2) and PM along with several other chemical species. Chemical components such as carbon dioxide (CO_2), methane (CH_4) and nitrous oxide (N_2O) are not considered to be air quality-related air pollutants as they are not included in the policy-established standards and limit values even though they play a critical role in the state of our atmosphere and associated impacts such as climate change.

Several other terms are often used interchangeably with air quality. However, they are different in important aspects. For example, a related term to air quality is **air pollution** which is associated with harmful effects on human health and damage to the environment. WHO defines air pollution as "contamination of the indoor or outdoor environment by any chemical, physical or biological agent that modifies the natural characteristics of the atmosphere."

Another relatively recent term that is often used in the field of air quality forecasting, and can seem confusing, is that of **chemical weather**. We are accustomed to talking about the weather in our everyday conversations and it must be one of the most popular topics of social and general discourse. In the field of meteorology, **weather** is defined as the state or condition of the atmosphere in terms of its properties including wind strength and direction, temperature, precipitation and pressure at a specific time and place (adapted from the Royal Meteorological Society, UK). In everyday life, we think of weather in terms of, for example, how hot or cold the temperature will be, how much and for how long it will rain, will the day be calm or stormy, or if the day will be clear or cloudy. This takes us to the term chemical weather which is understood as the condition of ambient air quality (or even composition of the lower atmosphere) at a given place and time. For the interested reader, this term is discussed further by Yang and Baklanov (2020).

1.3 A short historical perspective of air pollution

This section provides a short historical perspective on air pollution to provide context to the state of the air quality that we face in current times. For a deeper treatment of the history of air pollution, the reader is directed to references such as Jacobson (2002), Bachmann (2007), Brimblecombe (1987, 2011, 2012), Fowler et al. (2020), and EPA (2020).

1.3.1 Early history

Air pollution is not a modern problem. While natural emissions, such as eruptions of volcanoes, marine spray, biogenic species and airborne dust have been a part of the earth's history, these were subsequently followed by anthropogenic sources such as the human discovery of fire. As human society has developed, there have been numerous examples of documented evidence of incidents of air pollution and its impact on human health. Much of the direct evidence dates

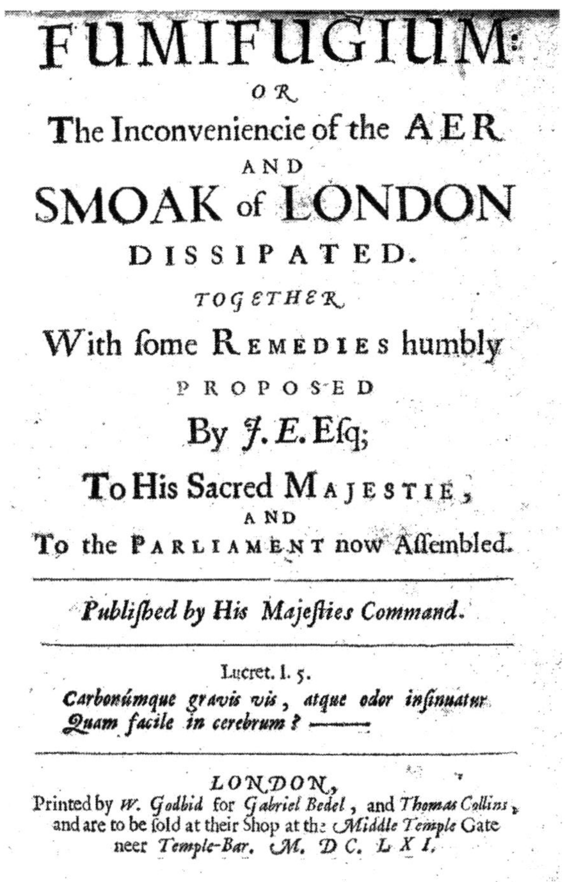

FUMIFUGIUM:
OR,
The Inconveniencie of the AER
AND
SMOAK of LONDON
DISSIPATED.
TOGETHER
With ſome REMEDIES humbly
PROPOSED
By *J. E.* Eſq;
To His Sacred MAJESTIE,
AND
To the PARLIAMENT now Aſſembled.

Publiſhed by His Majeſties Command.

Lucret. l. 5.
Carbonúmque gravis vis, atque odor inſinuatur
Quam facile in cerebrum? ——

LONDON,
Printed by *W. Godbid* for *Gabriel Bedel*, and *Thomas Collins*, and are to be ſold at their Shop at the *Middle Temple* Gate neer *Temple-Bar*. *M. D C. L X I.*

FIGURE 1.1 Image of the front cover of Fumifugium (Evelyn, 1661). *From: https://commons.wikimedia.org/wiki/File:%22Fumifugium%22,_Evelyn_Wellcome_L0009664.jpg.*

back to medieval times as reported in some of the references cited above. Notable examples have included build-up of air pollution (e.g., smoke from wood burning for cooking and oil lamps) in dwellings such as huts leading to respiratory symptoms, foul odor and concentration of air pollutants in crowded areas of towns. There is also evidence that in classical Greece times, air quality management practices were being implemented to ensure industrial processes were located outside the city boundaries as they relied on wood for fuel (Brimblecombe, 2011).

It is not only residential air pollution that affects the health of people but also industrial activities such as metal production and mining of coal as it increasingly replaced wood for fuel. The poisonous nature of coal smoke was recognized as far back as 1300s (Haagen-Smit, 1950) and has been documented in the essay Fumifugium by John Evelyn (Evelyn, 1661), one of the earliest works on air pollution (Fig. 1.1) dating back to the 17th century (see views and analysis of this work by Brimblecombe, 1987; Jenner, 1995 and Cavert, 2016).

1.3.2 Role of the industrial revolution

The period referred to as the Industrial Revolution is commonly considered to have started around 1760 to about 1830–40 and was largely driven by the rapidly increasing use of coal to produce energy and the introduction of machinery in industrial workplaces. However, Wrigley (2018) argues that in Britain at least, this process started much sooner in 1600s with England having an earlier use of coal and a higher rate of population increase than in other European countries. This was a key reason why Britain is generally considered to be the birthplace of the Industrial

Revolution before it spread to the rest of Europe and North America. With greater industrial productivity and economic development, urban populations started to grow with people moving from rural areas into cities where they often worked in crowded conditions in factories and other industrial workplaces with increasing use of machinery. However, the rapidly industrialized society concentrated in urban areas also faced several environmental pressures and greater risks from air pollution primarily from SO_2 and PM. While we will not discuss the history of climate change in detail, it is worth noting that analysis of global warming using increases in concentrations of greenhouse gases such as carbon dioxide (CO_2), shows that significant anthropogenic contributions leading to global warming started around the late 19th century. However, there is evidence that it may have started even earlier around 1830 in the midst of the industrial revolution (Abram et al., 2016). Environmental risks from both air pollution and climate change continued into the next transition phase of society.

1.3.3 Air quality after the industrial revolution

From the late 1800s and into the 1900s, there was greater industrialization (sometimes known as the Second Industrial Revolution) driven by developments in metal industries (e.g., steel production), electricity generation and the rise of road vehicles. With the increased use of cleaner coal and desulfurization technologies, concentrations of SO_2 started to decline and road traffic rose as an increasingly important source of air pollution along with greater recognition of gaseous air pollutants as well as PM.

Hugh Miller, a geologist, journalist and newspaper editor, graphically describes the conditions of pungent air pollution (Miller, 1889):

"Manchester I found as true a representative of the great manufacturing town of modern England, as York of the old English ecclesiastical city. One receives one's first intimation of its existence from the lurid gloom of the atmosphere that overhangs it. There is a murky blot in one section of the sky, however clear the weather, which broadens and heightens as we approach, until at length it seems to spread over half the firmament. And now the innumerable chimneys come in view, tall and dim in the dun haze, each bearing a-top its own troubled pennon of darkness."

The evolution of air pollution since the first industrial revolution has been examined by Ritchie (2017), for example, by using data on concentrations of PM and economic metrics such as gross domestic product (GDP) per capita (Fouquet, 2011). The trends show that overall PM concentrations peaked around 1890–1900 and reduced relatively quickly from 1950 onwards. However, evidence over the last decade is increasingly demonstrating, that even lower concentrations of air pollutants such as $PM_{2.5}$ still exhibit adverse health impacts (e.g., Chen & Hoek, 2020; WHO, 2021a).

While the general concentrations of air pollutants such as SO_2, were beginning to decline, for example, due to the use of low sulfur coal, there have been a number of notable incidents over the 1900s of high air pollution events that again brought the issue back into the limelight and in the public radar (see Fig. 1.2 showing an image of air pollution commonly witnessed over Widnes, Cheshire, England).

The previous high air pollution events were driven primarily by the excessive use of low-grade coal, but a new form of episodes was emerging that occurred during summertime and was driven by a mixture of chemicals with their reactions mediated by sunlight and hence was most serious during daylight hours. The word "smog" which is a portmanteau combining "smoke" and "fog," was routinely used to describe high air pollution conditions occurring during stagnant atmospheric conditions. It was now also being used for describing photochemically driven air pollution episodes such as those affecting Los Angeles. Some of the key high air pollution incidents occurring in the first half of the 20th century are summarized in Table 1.1 shown below.

FIGURE 1.2 Image of air pollution over Widnes (Cheshire, England) (Hardie, 1950). *From https://commons.wikimedia.org/wiki/File:Widnes_Smoke.jpg.*

TABLE 1.1 Notable historical incidents of high air pollution over the first half of the 20th century.

Air pollution incident	Nature of air pollution incident	Health impacts	References
Meuse Valley, Belgium, First week of December 1930	Foggy conditions with a low level inversion leading to a stable atmosphere trapping air pollution (e.g., SO_2, fine particulate matter, fluoride) from factories in the valley	60 people died with thousands affected by acute respiratory symptoms, and worsening of pre-existing cardiorespiratory problems	Firket (1936), Nemery et al. (2001), Anderson (2009), Brimblecombe (2011)
Los Angeles, California, USA, 1940–50 such as smog episode on July 26, 1943 (in the midst of World War 2)	"LA smog" associated with summer season and daytime due to photochemistry and formation of ozone with road traffic being a key emissions source	Severe eye irritation, sharp odor (similar to bleach), plant damage	Haagen-Smit (1952), Brimblecombe (2011), Hoekman and Welstand (2021)
Donora, Pennsylvania, USA, October 1948	Prevalent stable and foggy conditions led to high concentrations of particulate matter and SO_2 arising from zinc works emissions	Even pets (e.g., dogs) became ill and died	Schrenk et al. (1949), Helfand et al. (2001), Brimblecombe (2011)
London, UK, December 5–9, 1952	"Great London Smog" was due to emissions from industry and burning low-grade sulfur-rich coal for home heating during stagnant, stable atmospheric conditions	Acute exposure led to about 4000 excess deaths with persistent effects giving a total of 12,000 deaths	Stone (2002), Hunt et al. (2003), Anderson (2009), Brimblecombe (2011)

FIGURE 1.3 Arie Jan Haagen-Smit (1900–77), often known as the "father" of air pollution control. *Courtesy of Caltech Archives and Special Collections.*

Although some improvements in reducing air pollution emissions have taken place over the 1900s, high air pollution events have continued and affected many parts of the world (e.g., Kukkonen et al., 2005; Morawska et al., 2021). Coal was progressively replaced by gas and road traffic volumes and related emissions were increasing in urban areas while regulation to control air pollution was being implemented in some parts of the world. Despite these controls, a large number of air pollution incidents continued to have major negative impacts on health and ecosystems. These included chemical accidents (e.g., Serveso in 1976 and Bhopal in 1984), natural emissions from volcanoes and seasonal episodes driven by the accumulation of air pollutants, for example, during winter stagnation conditions and those associated with summertime photochemical events. For more information on other air pollution events, the reader is referred to sources such as Brimblecombe (2017), Morawska et al. (2021), and Ohja et al. (2022).

Science has also continually developed to better understand air pollution and its impacts. The severe photochemical smog affecting Los Angeles was highlighted by Haagen-Smit (1949, 1950, 1952) who is often referred to as the "father" of air pollution control (see Fig. 1.3). He highlighted the role of photochemistry involving emissions from road traffic and sunlight which led to controls of emissions from road transport. Relatively recent advances in the understanding of photochemical reaction pathways leading to ozone formation in the atmosphere were recognized through the Nobel Prize of Chemistry awarded to Paul Crutzen, Mario J. Molina and F. Sherwood Roland in 1995. A review of tropospheric ozone and its chemistry can be found in Monks et al. (2015).

1.3.4 Air quality in recent times

There are many emerging pressures that directly or indirectly affect air quality. In this section, we mention a few that have been topical and provide a wider perspective on air quality as one of the greatest environmental risks to people across the world.

1.3.4.1 Role of climate change and short-lived climate pollutants

Over the last couple of decades, pressures from climate change have been increasingly recognized and felt across the world. These go beyond sea level rise and flooding and some of these changes have a direct connection with air quality, see for example, Hong et al. (2019) and Sokhi et al. (2022). A particular group of air pollutants, that have significant importance for climate change and have increasingly received attention are termed short-lived climate pollutants (SCLPs) (Szopa et al., 2021). The key SLCP species include PM components such as sulfate, nitrate, ammonium, carbonaceous aerosols, mineral dust and sea spray, and reactive gases such as methane, ozone, some halogenated compounds, nitrogen oxides, carbon monoxide, nonmethane volatile organic compounds (VOCs), sulfur dioxide and ammonia (from Szopa et al., 2021). These are important as they have relatively short lifetimes in the atmosphere from a few hours to months to about a decade. As a consequence, many of these are heterogeneously distributed in the atmosphere affecting both air quality (either as primary or secondary species) and contribute to global warming either by cooling (e.g., sulfates and nitrates compounds) or warming (e.g., black carbon, BC). *Primary* refers to those air pollutants that are emitted directly into the atmosphere and *secondary* are those that have formed through chemical formation.

Climate change is likely to increase concentrations of O_3, (for example, Stowell et al., 2017) and have complex influences on $PM_{2.5}$ (see for example, Park et al., 2020) although uncertainties still exist in future projections. Our understanding of how aerosols influence climate is far from complete and large uncertainties still exist. For example, interactions and feedbacks between air pollutants such as aerosols affect the climate system and its components, as discussed by Thornhill et al. (2021). These interactions are not only important for anthropogenic emissions of greenhouse gases and air pollutants but also between the atmosphere and the biosphere where climate-driven global warming can lead to increased natural emissions of biogenic volatile organic compounds (BVOCs). These BVOCs are involved in the formation of ozone and secondary organic aerosols (SOA) which then can modulate interactions between aerosols and clouds affecting the radiation balance of the Earth (Yli-Juuti et al., 2021). High aerosol loading may also affect extreme weather events such as the occurrence of heatwaves and their increasing severity in the future as the climate changes (Zhao et al., 2019).

1.3.4.2 Wildfires—an increasing threat to air quality

Over the past few years, increasing incidents of wildfires have been observed (UNEP, 2022; WMO, 2022). Wildfires (see Fig. 1.4) are considered to be unplanned burning of natural vegetation areas (e.g., grasslands and forests). While their initiation can be natural, such as lightning strikes, or set alight deliberately, their impacts can be devastating to people and animals, properties, the environment and the local economy (UNEP, 2022). Risks from wildfires increase during dry periods, high temperatures and high winds. Many regions of the world are affected each year including Australia, Brazil, China, Canada, the USA, and many parts of Europe. Emissions from these fires consist of CO, BC, and other components of PM, NO_x, and VOCs amongst other species and consequently can dramatically affect air quality and hence health. With the increasing occurrence of wildfire incidents, there is a serious concern about how their frequency and intensity are likely to be affected by changes in climate in the future (e.g., Yang et al., 2022).

1.3.4.3 Reflections on the impact of COVID-19 pandemic on air quality

Most of the world literally came to a standstill in the early part of 2020 with the emergence and rapid spread of COVID-19 (the disease caused by SARS-CoV-2). In order to curtail the spread of COVID-19, governments across the globe severely limited the movement of people, for several months at a time. One significant impact of these reduced mobility measures was to change concentrations of air pollutants but, as many studies showed, the responses of air

FIGURE 1.4 Image of wildfires showing the devastation that can be cause by the fires and often lead to poor air quality in urban areas (Elsevier stock).

quality to these measures were complex and only partially beneficial (e.g., Sokhi et al., 2021 and references therein). While there were noticeable reductions in concentrations of air pollutants such as NO_2, CO, SO_2, and other species, $PM_{2.5}$ and O_3, showed complex behavior depending on the prevailing meteorology, urban or rural locations and different regions of the world. Some clear conclusions, however, can be made and are briefly summarised below.

The complex nature of air pollutants such as $PM_{2.5}$ and O_3 and their precursors underlined that our understanding of responses of secondary air pollutants to emission changes, varying meteorological conditions and a changing climate needs to be strengthened. For NO_2, universal reductions were observed but in order to benefit from these, measures to limit emissions would need to be implemented on a long-term basis which will have socioeconomical implications.

Taking a wider context, studies on how COVID-19 lockdown measures affected air quality, clearly showed that in many regions of the world, the dramatic reductions in air pollution emissions were not in itself sufficient to meet the WHO 2005AQG (WHO, 2006) values which implied that air quality policies will need to be more stringent and regionally tailored.

1.4 Air quality policy frameworks for reducing air pollution

Agreed national and regional policies along with global treaties serve a critical function to ensure consistency in the processes for regulation and setting targets to protect the environment and human health. In the context of air pollution, this process has evolved over decades and continues as scientific understanding develops and new evidence emerges. Its main elements in general consist of identifying sources of air pollutants and quantifying their emissions, measuring and monitoring regulated air pollutant concentrations, modeling predictions of air pollution for scenarios, estimating future projections and ensuring reporting obligations are met.

The first Air Pollution Control Act in the USA was introduced in Federal legislation in 1955 and provided funding for research into air pollution. The Clean Air Act in 1963 was the first one that specifically aimed at controlling air pollution and introduced a US Public Health Service with a remit to research into monitoring and controlling air pollution. The first (1965) and second (1967) amendments set standards for controlling emissions for specified vehicles, and introduced powers for enforcement and extended monitoring and inspections of stationary sources of air pollution. In the 1970 Clean Air Act, a number of air quality management measures were introduced such as the National Ambient Air Quality Standards (NAAQS). The 1990 amendment tackled emerging air pollution-related problems of acid rain and toxic compounds as well as ozone depletion.

In current times, air quality regulation and policy have continued to develop across the world but heterogeneously. A comprehensive assessment has been conducted by UNEP (2021a) which has revealed that about a third of the countries do not have a legislative framework to regulate ambient air quality, with most of these having no formal air quality standards and the rest only having guidelines. More details are given in Chapter 16, but here the air quality policy framework adopted by the EU is discussed as an example before outlining the efforts being made by the WHO to achieve better air quality.

In the EU, the Air Quality Framework Directive 96/62/EC introduced in legislation in 1996 set limit values that had to be met by all member states for ozone, PM_{10}, and nitrogen dioxide (NO_2) for a period up to 2004. The next 2004 Ambient Air Quality Directive (AAQD) and daughter Directive (2004/107/EC) were focused on controlling emissions from mobile sources, improving fuel quality, and ensuring that the transport and energy sectors met environmental protection requirements. The Thematic Strategy on Air Pollution led to a consolidation of the directives into the Clean Air For Europe 2008/50/EC AAQD which also set objectives for $PM_{2.5}$. There have been other updates since the 2008 Directive. The air quality directives as part of the overall air quality policy have undergone reviews to assess their fitness in light of new knowledge and evidence (Karamfilova, 2022).

The air quality directives of the EU are but one strand of the overall air quality policy framework. The other two are reducing air pollution emissions through the National Emissions Ceiling Directive (NECD) and setting emissions standards for key sources of pollution. The NECD introduced in 2016 (EU Directive 2016/2284) sets national reduction commitments for SO_2, NO_x, NMVOC, NH_3 and $PM_{2.5}$ all of which have negative impacts on human health and the environment. The longer-term ambitions of the EU are set in the "zero pollution vision for 2050" for air, water and soil pollution. The aim is to reduce emissions so that concentrations of pollutants are no longer considered to be harmful to health or the natural ecosystems. The third element of the air quality policy is a set of a large number of directives targeted at a range of emissions sectors.

As mentioned earlier, air pollution is not necessarily a local problem and cities are often subject to contributions from long-range transport of air pollutants. The cooperation between different countries to control long-range transport of air pollution is enshrined in the United Nations Economic Commission for Europe Convention on Long-Range Transboundary Air Pollution (often referred to as The Air Convention). Since its adoption in 1979, the Air Convention now includes eight Protocols, the Gothenburg Protocol being of key importance. The Gothenburg Protocol was agreed in 1999 followed by an amendment in 2012 and set emissions reduction commitments by member nations for 2020 and beyond. These commitments for the EU were established in legislation in 2017 as part of the wider clean air program that describes the overall air quality policy framework (EC, 2013).

In addition to regional and country-specific air quality policies and regulations supported by standards and limit values, the WHO publishes AQGs that act as recommendations to stimulate local, national and regional action to protect public health. The first release of AQG for Europe was in 1987 with a revision in 2000 (WHO, 1987, 2000). A major milestone was the global update for 2005 (WHO, 2006) which released health-based AQGs for $PM_{2.5}$, PM_{10}, O_3, NO_2, and SO_2. An evolution of WHO AQGs until this point can be found in WHO (2017). Recently, stricter AQGs were published in WHO (2021a) which proposed significantly lower AQG values for $PM_{2.5}$, PM_{10} and NO_2. It also introduced new additional AQGs for O_3, NO_2, SO_2, and CO.

An important observation reported by UNEP (2021a) is that the alignment of the national standards with the WHO AQG values is weak. While there may be many regionally specific technical, governance and socio-economic factors that may hinder such an alignment, it is however necessary to achieve further improvement beyond national standards that tend to be less stringent than the WHO AQGs (WHO, 2021a).

The situation relating to indoor air quality (IAQ) is less defined in legislation across the world. Although, WHO has established guidelines for IAQ (WHO, 2010), IAQ standards have not been established in most countries and hence they do not have any legal status. For example, UNEP (2021a) has estimated that only 7% of countries have relevant IAQ standards. Given that population exposure is largely determined by the time spent indoors, this is an area that requires urgent development.

1.5 A brief survey of key air pollutants

We survey here briefly some of the key air pollutants that are closely associated with air quality. These are often referred to *as regulated* or *criteria* air pollutants. A more detailed treatment of the sources and emissions of air pollutants and chemical reactions that occur in the atmosphere leading to their formation and removal are covered in Chapters 3 and 4 respectively. The health impacts of air pollutants are covered in Chapter 14. Important characteristics of these air pollutants are summarized in Table 1.2.

1.5.1 Particulate matter

PM represents a group of airborne particles that are emitted into the air through a diverse range of ubiquitous sources and represent a complex group of chemical compounds and other materials, many of which are hazardous to health (HEI, 2020).

Before proceeding with this section, it is important to clarify the term "aerosol" which is often used interchangeably with PM. However, strictly these two terms are not interchangeable or equivalent. Aerosols, whose sizes range typically from 0.01 to 10 μm, can be considered to be a *system* consisting of solid and/or liquid particles suspended and dispersed

TABLE 1.2 Properties of key air pollutants associated with air quality. The World Health Organization (WHO) air quality guideline values (AQG) published in 2021 (WHO, 2021a) are also given along with EU air quality limit values where indicated. AQG or limit values, like national standards, are usually expressed for specific time averages (e.g. annual, 24 hour, 8 hour) and percentile scores.

Air pollutant	Primary or secondary (main mode)	Main sources	Associated health hazards (see Chapter 14)	WHO air quality guidelines (2021) or EU Air Quality Limit Values where indicated
Particulate matter: These constitute a complex mixture of inorganic and organic compounds	Primary and secondary	Anthropogenic: road traffic (exhaust and nonexhaust emissions), industry, construction activities and wood stoves Natural: natural origin sources such as sea salt, dust and pollen	Cardiovascular disease (e.g., heart attacks and strokes), respiratory illness (e.g., exacerbation of asthma, COPD), increased risk of cancer, damage to the nervous system and reproductive system	$PM_{2.5}$: 5 $\mu g\,m^{-3}$ (annual) 15 $\mu g\,m^{-3}$ (24 h)[a] PM_{10}: 15 $\mu g\,m^{-3}$ (Annual) 45 $\mu g\,m^{-3}$ (24 h)[a]
Nitrogen oxides: Nitrogen dioxide (NO_2), Nitrogen oxide (NO), $NO_x = NO_2 + NO$	Primary and secondary	Anthropogenic: road traffic (mainly coal, oil, gas or diesel), cooking Natural: volcanoes, biological decay, lightning strikes and naturally occurring wildfires	Chronic obstructive pulmonary disease (COPD), inflammation of the lungs, exacerbation of coughs and wheezing, asthma, allergies diabetes and heart disease	NO_2: 10 $\mu g\,m^{-3}$ (annual) 25 $\mu g\,m^{-3}$ (24 h)[a]
Ozone (O_3)	Secondary	Mainly photochemical reactions of precursors and sunlight in the troposphere	Exacerbation of underlying respiratory conditions such as asthma Damage to ecosystems (vegetation, crops)	60 $\mu g\,m^{-3}$ (peak season)[b] 100 $\mu g\,m^{-3}$ (8 hour)[a]
Carbon monoxide (CO)	Primary	Anthropogenic: Incomplete combustion of carbonaceous fuels (e.g., diesel and petroleum products, gas or solid fuels), main source is road transport other sources include, industrial processes, waste incinerators and coal fired power plants. Indoors sources include faulty boilers and heaters Natural: volcanoes, naturally occurring wildfires and oceans	High exposure affects critical organs such as the heart and brain, dizziness, confusion, unconsciousness and death where even short term high exposure can be dangerous especially to vulnerable people with predisposed conditions	4 mg m^{-3} (24 hour)[a]
Sulfur dioxide (SO_2)	Primary	Anthropogenic: Combustion of fossil fuel burning by power plants, residential homes and industrial and commercial facilities. Other sources include industrial processes (e.g., metal ore extraction), manufacturing industry and shipping Natural: Volcanoes	Respiratory system, e.g., worsening of breathing, asthma	40 $\mu g\,m^{-3}$ (24 h)[a]
Organic compounds with benzo[a]pyrene as an example of a polycyclic aromatic hydrocarbon (PAH)	Primary	Anthropogenic: incomplete combustion, e.g., use of coal as fuel, vehicle exhausts incinerators, and domestic cooking as well as tobacco smoke Natural: Seepage from coal, oil and gas deposits, vegetation and wood fires and volcanic eruptions	Lung function damage, worsening of asthma, obstructive lung diseases, cardiovascular disease, impaired cognitive function in children, carcinogenic effects from long term exposure	EU limit value for BaP 1 ng m^{-3} (annual), measured in PM_{10}
Metals (Lead as an example)	Primary	Anthropogenic: Combustion of coal, mining, industrial production (e.g., smelting), batteries, paint and previously road transport Natural: mainly volcanic eruptions, rock weathering, soil erosion, re-suspension of dust and sediments, evaporation from water bodies	Infections, brain function, kidney damage, high blood pressure, hazard during pregnancy	EU Limit values: Lead (Pb) 0.5 $\mu g\,m^{-3}$ (annual) measured in PM_{10}

[a] *99th percentile (i.e. 3–4 exceedance days per year).*
[b] *Average of daily maximum 8-hour mean O_3 concentration in the six consecutive months with the highest 6-month running-average O_3 concentration.*

within a gas medium. In relation to the present context, the gas is always air. Smog and haze are examples of aerosol systems which consist of solid particles (e.g., smoke aerosols, dust) and often are mixed with condensed liquid particles depending on meteorological conditions. Fog and mist are examples of suspended condensed liquid particles in air. The origin of aerosols can either be natural (e.g., fog, mist, sea spray, volcano emissions, windblown dust) or anthropogenic (e.g., vehicle exhausts, industry and power generation combustion processes).

In relation to air quality, the term "*particulate matter*" is very important. All suspended airborne particles, usually referred to as PM, can also be considered to be aerosols. There are many components of PM including elemental carbon (EC), trace metals, mineral material, sulfate and nitrate compounds, organic matter, SOA, and compounds such as polycyclic aromatic hydrocarbons (PAHs).

In addition to the composition of PM, one property that is closely linked to health effects is the aerodynamic size of the particles. While PM exhibits a distribution of sizes in the atmosphere, for air quality and health purposes and for regulation, they are usually expressed as $PM_{2.5}$ and PM_{10}. $PM_{2.5}$ represents particles of aerodynamic size of 2.5 μm or less and PM_{10} represents particles of aerodynamic size of 10 μm or less. Strictly, the size is not a precise physical upper cut-off but is defined according to airborne particles sampled with an inlet with a given collection efficiency. $PM_{2.5}$ is hence defined as particles of a size that pass through a size-selective inlet with a 50% efficiency cut-off at 2.5 μm aerodynamic diameter (e.g., ISO 7708:1995). Similarly, PM_{10} is defined as particles of a size that pass through a size selective inlet with a 50% efficiency cut off at 10 μm aerodynamic diameter (ISO 7708:1995) (ISO, 2017). Fig. 1.5 depicts both fractions ($PM_{2.5}$ and PM_{10}) in comparison to other common entities.

PM can be introduced into the atmosphere directly (primary origin), e.g., EC, dust from rock or surface erosion and those formed through chemical reactions (secondary origin), e.g., secondary organic aerosols and compounds of sulfates, nitrates and ammonium.

Sources of PM can be human-made (anthropogenic origin) such as road traffic (exhaust and nonexhaust emissions), industry, construction activities and wood stoves and of natural origin such as sea salt, marine spray, dust and vegetation (e.g., pollen). Components such as EC and OC have primary sources (e.g., road traffic and industrial combustion processes) and components such as sulfate and nitrate compounds have secondary origin. Particles that originate from combustion sources (e.g., vehicle exhausts, domestic cooking and heating, power generation, industry, agricultural burning and wildfires) are mostly of submicron size (less than 1 μm) and hence are part of the $PM_{2.5}$, which is also known as the fine fraction. Particles that are formed in the air through gas-particle (heterogeneous) reactions begin their lifetime also in the submicron range and often in the ultrafine size of less than 100 nm. Depending on the nature of particles and meteorological conditions (e.g., relative humidity and temperature), the ultrafine particles can grow into larger size fractions as they travel away from their immediate sources.

A term that is often used interchangeably with elemental carbon (EC) is black carbon (BC). BC results from combustion sources such as agricultural burning, wildfires, road transport, industry and power generation, residential heating and cooking activities and hence is a component of $PM_{2.5}$ (Briggs & Long, 2016; Singh et al., 2018). Although

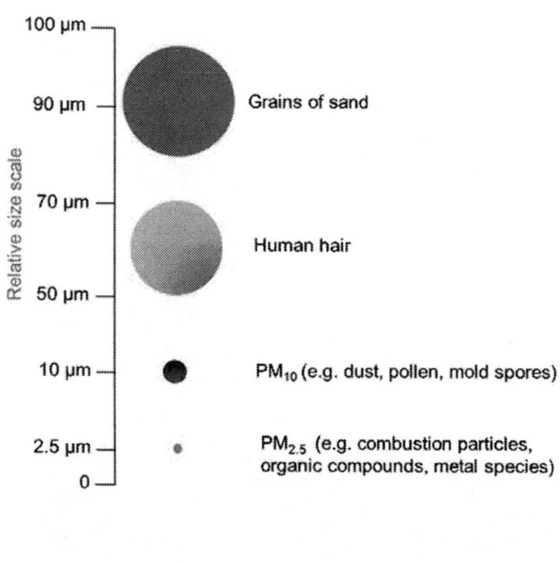

FIGURE 1.5 Relative size comparison of particulate matter to human hair strand and grain of sand.

correlation has been observed between measurements of EC and BC, the two components are not strictly identical as they are also dependent on the measurement technique (Briggs & Long, 2016). BC is an important component of PM as it is relevant for air quality and for climate change and is associated with a net warming effect on our climate.

The coarse fraction of PM has aerodynamic particle sizes between 10 and 2.5 μm (denoted as $PM_{2.5-10}$). It usually consists of particles that have been formed through mechanical abrasion-type activities such as quarrying, construction and even demolition work in addition to resuspended dust and dust that can be transported by wind (e.g., Saharan dust). These coarse components consist largely of aluminum, silicon, iron and calcium fractions (e.g., see AQEG, 2012).

Secondary components of PM consist of two main components (i) secondary inorganic aerosols (SIAs) and (ii) secondary organic aerosols (SOAs). A large proportion of SIA consists of ammonium salts of sulfate and nitrate which are formed by oxidation reactions involving precursors of SO_2 and NO_x leading to the formation of sulfuric and nitric acid respectively and then neutralization with ammonia to form mainly ammonium sulfate ($(NH_4)_2SO_4$) and ammonium nitrate (NH_4NO_3) (e.g., McFiggans et al., 2015; Seinfeld & Pandis, 2016).

SOA is also an important component of $PM_{2.5}$ and, like SIA, exhibits considerable variability depending on meteorological conditions and location along with distribution of natural (biogenic) and anthropogenic emissions of VOC precursors. Natural sources include vegetation and forested areas and anthropogenic sources such as combustion processes, transport, solvent evaporation and wildfires (see also Section 1.5.7). SOA component is highly relevant for both air quality and climate change and displays a complex composition profile (e.g., McFiggans et al., 2015; Shrivastava et al., 2017; Srivastava et al., 2022; Zhu et al., 2017). The chemistry of SOA is also complex and involves many volatile and semivolatile compounds undergoing oxidation with O_3, hydroxyl (OH) and other radicals as well as other chemical species through a cascade of reactions to partition from the gas phase into the solid phase and either form new particles or undergo condensation on preexisting particles.

PM, especially the fine fraction, $PM_{2.5}$, can penetrate deep into the lungs and is associated with serious health effects (e.g., USEPA, 2019). These include reduction in life expectance through short-term and long-term exposure, cardiovascular disease (e.g., heart attacks and strokes), respiratory illness (e.g., exacerbation of asthma, chronic obstructive pulmonary disease (COPD)), increased risk of cancer, and damage to the nervous system and reproductive system.

1.5.2 Nitrogen oxides

In terms of air quality, nitrogen oxides, namely nitric oxide (NO) and nitrogen dioxide (NO_2), are two important gaseous species. Within the atmospheric and air quality science communities, the sum of NO and NO_2 is denoted by NO_x. However, within the air quality management and policy regulation framework, the key component is NO_2 ambient concentrations which are subject to strict national and EU standards and WHO AQGs (see Table 1.1). In terms of air quality science, other oxides of nitrogen play an important role in chemical reactions that can lead to the formation of pollutants such as $PM_{2.5}$ and ozone (see Chapter 4 for more details).

Much of the emissions of NO_x is in the form of NO, for example, from the combustion of fossil fuels (coal, oil, gas, or diesel) and hence road traffic and energy generation sectors are both important emitters. Although one important pathway for the formation of NO_2 in the atmosphere is through the chemical oxidation of NO by ozone, there are significant primary emissions of NO_2 associated with diesel vehicles as noted by Carslaw (2005), which was one of the first studies to highlight this issue. As oxides of nitrogen form through combustion processes, NO_2 can also be emitted indoors during cooking when using fossil fuels such as wood or natural gas. In addition to anthropogenic sources, there are natural sources of NO_x and these include volcanoes, biological decay, lightning strikes and naturally occurring wildfires.

Near combustion sources, where concentrations of NO_x are high, as in the case of urban areas with high road traffic, NO reacts quickly with O_3 to form NO_2. This process is known as titration and reduces O_3 concentrations (or we can consider this reaction to be a sink for O_3). During the daytime, when solar radiation is high, photolysis of NO_2 occurs where the molecule spits into NO and O and the oxygen atom reacts with an oxygen molecule to form O_3. Because of the $NO-NO_2-O_3$ chemical reaction cycle, the sum of NO and NO_2 (NO_x) is considered an inert quantity. During nighttime, with no solar radiation and hence no photolysis, O_3 concentrations drop due to titration and NO_2 starts to build up because of primary emissions especially when the surface layer of the atmosphere is stable such as during winter seasons (see Chapter 6).

As NO_2 is largely associated with road traffic, some of the highest concentrations are observed near road locations with heavy traffic (e.g., GLA, 2020). There are important implications for exposure to people in urban areas as its concentrations drop significantly away from the road due to increased dilution and reach background levels after about 500 m. Within an urban area, its spatial distribution, therefore, can show significant spatial heterogeneities depending on the distance from major roads. With road transport being a common source, some studies have examined the correlation observed between NO_2 and other road traffic-related air pollutants such as EC, BC, CO, and $PM_{2.5}$ (Beckerman et al., 2008; Olstrup et al., 2021).

NO_2 is not only associated with a number of health hazards related to the respiratory system including inflammation of the lungs, exacerbation of coughs and wheezing, asthma and allergies but also with other health effects such as diabetes and heart disease (e.g., USEPA, 2016).

1.5.3 Ozone

Ozone (O_3) which is a highly reactive, oxidizing gas occurs naturally in our stratosphere where is it commonly referred to as the ozone layer. This protective stratospheric ozone is vital to reducing harmful ultraviolet radiation from the sun reaching lower levels of the atmosphere (see Chapter 2). Ozone in the lower troposphere near the surface (often referred to as ground-level ozone), however, is an important air pollutant which is produced through a complex set of chemical reactions involving nitrogen oxides (NO_x), nonmethanic volatile organic compounds (NMVOCs) and other chemical species such as hydroxyl radical (OH), carbon monoxide (CO) and methane (CH_4) in the presence of sunlight. As a consequence, it is often referred to as a "photochemical" air pollutant and is a key component of photochemical smog that affects many cities across the world (Sicard et al., 2020; Sillman, 2014). As ozone in the lower atmosphere is not emitted directly and is formed through chemical reactions, it is a secondary air pollutant. We refer to directly emitted chemicals that react to form another air pollutant as precursors of that secondary air pollutant. In this case NO_x and VOCs are precursors of O_3. It affects human health (e.g., the respiratory system) as well as food crops, forests, and vegetation.

The chemical reactions that lead to the formation of O_3 are complex starting with emissions of NO_x and VOCs precursors which are emitted by sources such as road transport and industrial processes as well as natural sources such as biogenic emissions from vegetation (e.g., isoprene). The natural emissions will be dependent on land cover and vegetation type and NO_x produced from microbial activities (nitrification and denitrification) in soil. O_3 is formed as a result of the interplay between sunlight and fresh emissions of these precursors, other reactive chemicals, and remanent background ozone which all mix and disperse due to the prevailing meteorology (e.g., wind flows and turbulence). Conditions that lead to particularly high concentrations that can be harmful to health are when the precursor chemicals are present with lighter wind flows, lower vertical ventilation in the atmosphere with strong sunshine and high temperatures. This is the key reason why high ozone is associated with the summer season. People with underlying respiratory conditions such as asthma, children and older populations are at particular risk. Ground-level O_3 also affects vegetation, crops and ecosystems in general. Major impacts have been recorded where high levels of O_3 have been damaging to crop yields, forests, and biodiversity (Agathokleous et al., 2020; Zhang et al., 2019).

Because of this interplay between complex chemistry and prevailing meteorology, high levels of O_3 can result in cities but even higher concentrations can be observed outside the urban areas in suburban locations as precursors get transported by wind flows while ozone formation is taking place. Consequently, the spatial spread of O_3 can leave a much larger footprint (several hundred kilometers) compared to say, NO_2.

Ozone concentrations in the lower troposphere are not only influenced by photochemical reactions of precursor emissions but also by large-scale global drivers of stratosphere-troposphere exchange (STE) and circulation patterns including El Niño–Southern Oscillation (ENSO), Atlantic Multidecadal Oscillation (AMO) and changes in the Hadley circulation (Boothe & Homeyer, 2017; Holton et al., 1995; Stohl et al., 2003). About 10% of the global tropospheric ozone can be attributed to STE intrusion of O_3 (e.g., Lu et al., 2019a; 2019b). Concentrations of O_3 in the lower troposphere, have also been shown to be sensitive to other large-scale circulations, such as the Arctic Oscillation (AO) and North Atlantic Oscillation (NAO) (Hess & Lamarque, 2007; Liu & Hu, 2021).

1.5.4 Methane

Methane (CH_4) is also an important precursor that contributes to the formation of O_3. CH_4 is the second most important greenhouse gas (after carbon dioxide) that affects the radiation balance of our planet and contributes to global warming (UNEP, 2021b). It has important anthropogenic and natural sources. Anthropogenic sources include livestock and agricultural activities (e.g., use of manure and enteric fermentation), wastewater treatment, landfills (e.g., decay of organic waste), oil and natural gas production and processing systems and coal mining. Natural sources of CH_4 include natural wetlands (e.g., decay of plant material), permafrost, water systems (e.g., marine, lakes), seepage near coal and oil deposits, and emissions from volcanoes (e.g., Lu et al., 2019a).

1.5.5 Carbon monoxide

As CO is a colorless and odorless gas, it can be difficult to notice and can be hazardous when concentrations are high. Outdoor (ambient) concentrations of carbon monoxide (CO) have reduced dramatically in many parts of the world (e.g., over Europe) where road transport emissions have decreased. For example, in the UK, CO emissions from road transport have reduced significantly as a result of the implementation of EU-wide emission standards (e.g., DEFRA, 2022).

CO is released whenever there is incomplete combustion of carbonaceous fuels such as diesel and petroleum products, gas or solid fuels. Road transport is still a major source in many areas of the world. In addition to road transport, sources include, industrial processes, waste incinerators and coal-fired power plants. Because of the high use of solid fuels (e.g., wood, coal, and charcoal) for cooking and heating, concentrations of CO along with other air pollutants can be high in regions of Asia and the Pacific region (IEA, 2016; UNEP, 2019).

For indoor environments, people will be most familiar with CO monitors to ensure there are no leakages from heating boilers and other gas-fueled devices. However, indoor concentrations can increase due to unvented use of cooking and heating fuels (e.g., gas stoves and wood burning) and fugitive (leaking or escaping) emissions such as from chimneys.

CO is present in the environment through natural sources too, including emissions from volcanoes, naturally occurring wildfires and oceans (Conte et al., 2019). It also plays an important role in the formation of O_3 through photochemical reactions involving NO_x, VOCs, and CH_4.

As the immediate effects of CO can go unnoticed, the hazard can be missed till the exposure becomes seriously high. While small amounts of CO are produced naturally inside the body, exposure to CO can reduce the absorption of oxygen in the lungs. This can happen when breathing air with high concentrations of CO which then affects critical organs such as the heart and brain (Levy, 2015). High exposure can lead to dizziness, confusion, unconsciousness and death whereas even short-term high exposure can be dangerous, especially to vulnerable people with predisposed conditions. In regions such as Europe, it is unlikely to have high exposure to CO from ambient air, but concentrations can be much higher in enclosed spaces due to faulty boilers and heaters.

1.5.6 Sulfur dioxide

As a corrosive and acidic gas which is water-soluble and colorless, sulfur dioxide (SO_2) is harmful to human health and ecosystems (USEPA, 2020). The major sources of atmospheric SO_2 include fossil fuel burning by power plants, residential homes and industrial and commercial facilities. Other sources include industrial processes (e.g., metal ore extraction), manufacturing industry and shipping. Since the 1990s, SO_2 emissions in Europe have significantly reduced from the energy production sector, due to the increasing use of low sulfur fuels and the ceasing of coal-fired power plants and they are continuing to fall (e.g., EEA (2020)). The move away from high sulfur fuels (e.g. coal), toward lower sulfur content fuels along with an increase in the use of gas and biomass fuels has made a major difference in reducing SO_2 emissions. This has also been aided by the implementation of flue gas desulfurization technologies for coal-fired power plants. Much of these developments have been largely stimulated by policy developments such as the Convention on Long Range Transboundary Air Pollution (CLRTAP) and National Emissions Ceiling Regulations (NECR).

Historically, emissions of SO_2, along with NO_x, and their transport over large distances (known as long-range transport) have been responsible for the acidification of precipitation, commonly referred to as *acid rain* which has caused significant damage to ecosystem habitats and buildings (Grennfelt et al., 2020). As these gases are emitted into the atmosphere, they react with water vapor and other atmospheric species including oxygen to form sulfuric acid in the case of SO_2, and nitric acid in the case of NO_x. In the 1970s and 1980s, acid rain caused severe damage to habitats and ecosystems in Scandinavia. As mentioned above in regions such as North America and Europe, acid rain has been largely eliminated as emissions and hence concentrations of SO_2 have been reduced dramatically. However, the problem persists in other regions such as Southeast Asia (Payus et al., 2020).

SO_2, along with NO_x, ammonia (NH_3) from farming activity emissions, and organic compounds from biogenic and anthropogenic sources, is a major precursor to ambient secondary components of $PM_{2.5}$ concentrations. Reactions of these species in the atmosphere, lead to the formation of components such as sulfate (SO_4^{2-}), nitrate (NO_3^-) and ammonium (NH_4^+) compounds and secondary organic aerosols (see also Section 1.5.1).

SO_2 can penetrate the lungs and adversely affects human health through irritation and constriction of the airways especially showing rapid effects for people with conditions such as asthma. Health effects can occur very rapidly, making short-term exposure to peak concentrations important (DEFRA, 2022; Health Canada, 2016).

1.5.7 Other regulated air pollutants

In addition to the air pollutants mentioned above, regions such as the EU have designated limit values for other airborne chemical species of metals and organic compounds, namely lead, arsenic, cadmium, nickel, benzene, and benzo[a]pyrene. Other regions also include some of these such as lead which is one of the criteria pollutants of the North American NAAQS. There are many sources responsible for emissions of metals into the atmosphere with the main ones being combustion of coal, mining, industrial production of metals and road transport. In the case of lead, while the emissions from vehicles have decreased dramatically as the lead was phased out of petrol, there is evidence that the resuspension of historic deposits in soil is a significant contributor to airborne concentrations (e.g., Resongles et al., 2021). Airborne metals are driven by anthropogenic emissions of PM and is a serious issue for many cities across the globe (e.g., McNeill et al., 2020).

The properties of PAHs of being chemically stable, with low water solubility and high sorption capacity makes them particularly persistent in the environment and hence are found in air, as well as in water and soil (Brandt & Einhenkel-Arle, 2016). For this reason, PAH are classified as persistent organic pollutant (POP) compounds. As there are numerous PAH compounds, Benzo[a]pyrene is often used as a marker for this group of pollutants (EA, 2019). Their emissions are mainly from incomplete combustion (e.g., coal, vehicle exhausts incinerators, and domestic cooking as well as tobacco smoke) and hence make them important to both outdoor and IAQ. Naturally, they can arise from vegetation wood fires and volcanic eruptions. Long-term exposure to airborne PAHs is considered to be carcinogenic (see WHO, 2021b for a full review). There are also several other health effects associated with PAH and these include impairing lung function leading to worsening of asthma, increased morbidity and mortality from obstructive lung diseases as well as being linked to cardiovascular disease and adverse impacts on cognitive function in children (WHO, 2021b).

Another group of air pollutants that are ubiquitous are VOCs which are emitted from biogenic sources (e.g., vegetation, peatlands and forests) and from anthropogenic sources such as combustion processes, building construction materials, paints and solvents during wildfires. Because of the diversity of sources, VOCs affect both outdoor (ambient) and indoor air quality. As mentioned previously, they are one of the important precursors for O_3 and SOA.

Formaldehyde is a commonly found indoor VOC and is emitted from paints, wood preservatives, pesticides, and aerosol sprays, amongst many other sources. VOCs form a group of a large number of compounds and generally are associated with a wide range of adverse health effects including risk of irritation (e.g., eye, nose, and throat), respiratory diseases (e.g., worsening of asthma), diseases of the nervous system, gene mutation and different types of cancer (Cheng et al., 2019). Another important VOC compound is benzene, which is regulated in the EU, North America and many other parts of the world as it is associated with the risk of leukemia, skin cancer and other tumors.

There are a large group of toxic air pollutants termed hazardous air pollutants (HAPs) which are regulated in North America as they are associated with adverse health effects including cancer, birth defects as well as environmental impacts. Some of the air pollutants already mentioned above such as VOCs and PAHs as well as metal compounds are listed in the HAP register which includes benzene and lead compounds. Air pollutants may interact with other entities in the atmosphere such as biological materials like pollen and have a synergistic effect on human health (Anenberg et al., 2020). Studies of synergistic effects during the COVID-19 pandemic have revealed possible mechanisms where air pollutants such as HAP could be associated with increased mortality from infections from the COVID-19 virus (Petroni et al., 2020).

1.6 Air quality around the world

High air pollution episodes where concentrations of air pollutants can be significantly higher than normal levels for a few days to a few weeks, often exceed legislative limits and standards. Studies have highlighted that air pollution concentrations and their impacts are not uniformly felt by people and that the environmental, socioeconomic, and other regional influences are evident which means that strategies for controlling and managing air pollution have to be tailored to take account of these factors (Morawska et al., 2021; Sokhi et al., 2021). Consequently, regional inequalities can have significant impacts on people's exposure to air pollution. For example, Shaddick et al. (2020) highlight that trends of annual average population-weighted concentrations in North America and Europe are decreasing but are increasing for central and southern Asia.

In their 2019 analysis, HEI (2020) shows that there was a small decrease in the global average of $PM_{2.5}$ exposures between 2010 and 2019. However, although some regions in south Asia showed an improvement overall (e.g., southeast

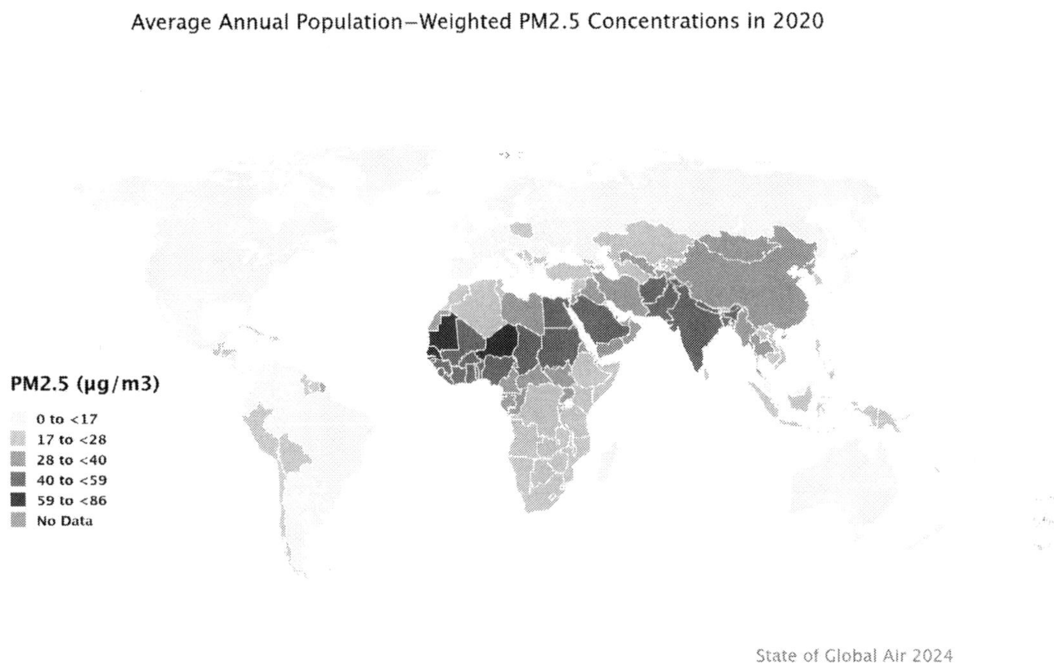

FIGURE 1.6 Average annual population weighted $PM_{2.5}$ concentrations in 2020. Health Effects Institute. 2024. State of Global Air 2024. Available: www.stateofglobalair.org [accessed 24/02/2025]. Data source: Global Burden of Disease Study 2021. IHME, 2024, https://www.stateofglobalair.org/data/#/air/plot.

and east Asia, including China, Vietnam, and Thailand), annual $PM_{2.5}$ levels in other regions such as the Indian subcontinent, north Africa, the Middle East, and sub-Saharan Africa, showed very little decline or even an increase.

Population-weighted average values of the annual mean for $PM_{2.5}$ are shown for 2019 in Fig. 1.6. The distribution starkly shows that the highest values are found in some parts of Latin America, north and central Africa and over large parts of Asia. When compared to WHO (2006) annual AQG for $PM_{2.5}$ of 10 $\mu g\,m^{-3}$, all regions other than higher-income ones, exceeded this value. However, when compared to the stricter 2021 WHO AQG (see Table 1.2), most regions of the world would exceed the equivalent annual $PM_{2.5}$ AQG value of 5 $\mu g\,m^{-3}$. Chapter 17 gives examples of the status and trends of air quality for selected major cities across the world.

1.7 Scope and structure of the book

This book has been organized to provide a logical progression for the reader. The Introduction provides a nonmathematical summary of some of the main concepts related to air quality to enable the reader to better understand the more detailed chapters. This is followed up in Chapter 2 which considers necessary fundamental scientific concepts and the key principles that underpin air quality science. Sources and emissions of air pollutants are detailed in Chapter 3 before moving to the main chemical transformations that are relevant to air quality in Chapter 4. The more dynamical concepts and processes that control the larger circulation patterns and behavior of the lower atmospheric boundary layer, both of which are fundamental to understanding and predicting air quality, are covered in Chapters 5 and 6. The theory of the development and application of air quality models is described in Chapters 7 and 8 before moving to discussing the main methods and techniques for measuring air quality in Chapter 9. These observations are also critical for improving and evaluating air quality models. The next two chapters adopt a more regional to global scale approach and consider air quality at larger spatial scales which can significantly influence local and urban scale air quality (Chapter 10) and examine the interactions of how climate change can influence atmospheric composition and hence air quality (Chapter 11). While we often consider air quality to be solely an outdoor environmental issue, peoples' overall exposure to air pollution is inextricably determined by IAQ which is discussed in Chapter 12. With the increasing complexity of air quality and related data, Chapter 13 provides an overview of methods to analyze this data quantitatively. The next two chapters provide an understanding of the impacts of air quality on human health (Chapter 14) and on our ecosystems (Chapter 15) and not surprisingly, they form the bedrock for the rationale and motivation for pursuing air quality science. Chapter 16 describes and examines different levels of air quality management practices and policy tools to control and reduce air pollution and hence improve air quality. Finally, in Chapter 17, we consider air quality trends in selected cities across the world which provides a further appreciation of the challenges and outlook facing people across the world.

It is important to note that the field of air quality science is a continually evolving area of study. In many respects, the material covered in this book also has a deep association with our own personal lives as each and every moment of our existence critically depends on the quality of the air that we breathe.

References

Abram, N. J., McGregor, H. V., Tierney, J. E., Evans, M. N., McKay, N. P., & Kaufman, D. S.Consortium, a. t. (2016). Early onset of industrial-era warming across the oceans and continents. *Nature*, *536*, 411–418. Available from https://doi.org/10.1038/nature19082.

Agathokleous, E., Feng, Z., Oksanen, E., Sicard, P., Wang, Q., Saitanis, C. J., & P. (2020). Ozone affects plant, insect, and soil microbial communities: A threat to terrestrial ecosystems and biodiversity. *Science Advances*, *6*, eabc1176.

Anderson, H. A. (2009). Air pollution and mortality: A history. *Atmosphere Environment*, *43*, 142–152.

Anenberg, S. C., Haines, S., Wang, E., Nassikas, N., & Kinney, P. L. (2020). Synergistic health effects of air pollution, temperature, and pollen exposure: a systematic review of epidemiological evidence. *Environmental Health*, *19*, 130.

AQEG. (2012). Fine Particulate Matter (PM2.5) in the United Kingdom. A report from the Air Quality Expert Group (AQEG) to the Department for Environment, Food and Rural Affairs; Scottish Government; Welsh Government; and Department of the Environment in Northern Ireland, UK, Crown copyright.

Bachmann, J. (2007). Critical review: Will the circle be unbroken: A history of the U.S. National Ambient Air Quality Standards. *Journal of the Air & Waste Management Association*, *S7*, 652–697.

Beckerman, B., Jerrett, M., Brook, J. R., Verma, D. K., Arain, M. A., & Finkelstein, M. M. (2008). Correlation of nitrogen dioxide with other traffic pollutants near a major expressway. *Atmospheric Environment*, *42*, 275–290.

Boothe, A. C., & Homeyer, C. R. (2017). Global large-scale stratosphere–troposphere exchange in modern reanalyses. *Atmospheric Chemistry and Physics*, *17*, 5537–5559. Available from https://doi.org/10.5194/acp-17-5537-2017.

Brandt, M., & Einhenkel-Arle, D. (2016). Polycyclic aromatic hydrocarbons harmful to the environment toxic! inevitable? German Environment Agency.

Briggs, N. L., & Long, C. M. (2016). Critical review of black carbon and elemental carbon source apportionment in Europe and the United States. *Atmospheric Environment*, *144*, 409–427.

Brimblecombe, P. (1987). *The big smoke: A history of air pollution in London since medieval times*. Routledge.

Brimblecombe, P. (2011). Air pollution history. Chapter 1: World Atlas of atmospheric pollution. In R. S. Sokhi (Ed.), *Anthem advances in atmospheric environment science*. UK: Anthem Press.

Brimblecombe, P. (2012). *The big smoke: A history of air pollution in London since medieval times*. Routledge Revivals.

Brimblecombe, P. (2017). In P. Brimblecombe (Ed.), *Air pollution episodes (Vols. air pollution reviews* (6). World Scientific Publishing Europe Ltd.

Carslaw, D. C. (2005). Evidence of an increasing NO_2/NO_X emissions ratio from road traffic emissions. *Atmospheric Environment*, *39*, 4793–4802. Available from https://doi.org/10.1016/j.atmosenv.2005.06.023.

Cavert, W. M. (2016). *The smoke of London, energy and environment in the early modern city*. Cambridge University Press. Available from https://doi.org/10.1017/CBO9781139680967.013.

Chen, J., & Hoek, G. (2020). Long-term exposure to PM and all-cause and cause-specific mortality: A systematic review and meta-analysis. *Environment International*, *143*105974. Available from https://doi.org/10.1016/j.envint.2020.105974.

Cheng, S., Zhang, J., Wang, Y., Zhang, D., Teng, G., Chang-Chien., & Yan, P. (2019). Global research trends in health effects of volatile organic compounds during the last 16 years: A bibliometric analysis. *Aerosol and Air Quality Research*, *19*, 1834–1843. Available from https://doi.org/10.4209/aaqr.2019.06.0327.

Conte, L., Szopa, S., Séférian, R., & Bopp, L. (2019). The oceanic cycle of carbon monoxide and its emissions to the atmosphere. *Biogeosciences*, *16*, 881–902. Available from https://doi.org/10.5194/bg-16-881-2019.

DEFRA. (2022). Air pollution in the UK 2021. Department for Environment, Food and Rural Affairs, UK.

EA. (2019). Polycyclic aromatic hydrocarbons (PAHs): Sources, pathways and environmental data. Report by the Environment Agency, UK.

EC. (2013). COM(2013) 918 final. https://eur-lex.europa.eu/LexUriServ/LexUriServ.do?uri = COM:2013:0918:FIN:EN:PDF (accessed 15.11.22).

EEA. (2020). Air quality in Europe—2020 report. European Environment Agency.

EPA. (2020). Cleaning the air we breathe, a half century of progress. EPA Alumni Association, USA.

Evelyn, J. (1661). Fumifugium, or, the inconveniencie of the aer and smoak of London dissipated. The Rota, University of Exeter. Reproduced 1976.

Firket, J. (1936). Fog along the Meuse Valley. *Transactions of the Faraday Society*, *32*, 1192–1197.

Fouquet, R. (2011). Long run trends in energy-related external costs. *Ecological Economics*, *70*, 2380–2389.

Fowler, D., Brimblecombe, P., Burrows, J., Heal, M. R., Stevenson, D. S., & Vieno, M. (2020). A chronology of global air quality. *Philosphical Transactions of Royal Society*, *A378*20190314. Available from https://doi.org/10.1098/rsta.2019.0314.

GLA. (2020). Air quality in London 2016-2020 London environment strategy: Air quality impact evaluation. Greater London Authority, London, UK.

Grennfelt, P., Engleryd, A., Forsius, M., Hov, Ø., Rodhe, H., & Cowling, E. (2020). Acid rain and air pollution: 50 years of progress in environmental science and policy. *Ambio*, *49*, 849–864. Available from https://doi.org/10.1007/s13280-019-01244-4.

Haagen-Smit, A. (1952). Chemistry and physiology of Los Angeles smog. *Industrial and Engineering Chemistry Research*, *44*, 1342–1346.

Haagen-Smit, A. J. (1949). *The analysis of air contaminants*. Los Angeles: Report to the Los Angeles County Air Pollution Control District.

Haagen-Smit, A. J. (1950). The air pollution problem in Los Angeles. *Engineering and Science, XIV*, 7–13.

Hardie, D.W. F. 1950. A history of the chemical industry in Widnes. Imperial Chemical Industries Limited. ASIN: B0007JBZRQ, ICI.

Health Canada. (2016). Human health risk assessment for sulphur dioxide: Analysis of Ambient exposure to and health effects of sulphur dioxide in the Canadian population. Her Majesty the Queen in Right of Canada as represented by the Minister of Health Canada, Cat.: H144-29/2016E-PDF, Pub: 150211.

HEI. (2020). State of global air 2020. Special Report. Health Effects Institute, Boston, MA, USA. Data source: Global Burden of Disease Study 2019. IHME, 2020.

Helfand, W. H., Lazarus, J., & Theerman, P. (2001). Donora, Pennsylvania: An environmental disaster of the 20th century. *American Journal of Public Health*, *91*, 553–554.

Hess, P. G., & Lamarque, J.-F. (2007). Ozone source attribution and its modulation by the Arctic oscillation during the spring months. *Journal of Geophysics Research*, *112*D11303. Available from https://doi.org/10.1029/2006jd007557.

Hoekman, S., & Welstand, J. (2021). Vehicle emissions and air quality: The early years (1940s–1950s). *Atmosphere*, *12*, 1354. Available from https://doi.org/10.3390/atmos12101354.

Holton, J. R., Haynes, H. P., Mcintyre, M. E., Douglass, A. R., Rood, R. B., & Pfister, L. (1995). Stratosphere-troposphere exchange. *Reviews of Geophysics*, *33*, 403–439.

Hong, C., Zhang, Q., Zhang, Y., Davis, S. J., Tong, D., Zheng, Y., Liu, Z., Guan, D., He, K., & Schellnhuber, H. J. (2019). Impacts of climate change on future air quality and human health in China. *Proceedings of the National Academy of Sciences USA.*, *116*(35), 17193–17200. Available from https://doi.org/10.1073/pnas.1812881116. Epub 2019 Aug 12. PMID: 31405979.

Hunt, A., Abraham, J. L., Judson, B., & Berry, C. L. (2003). Toxicologic and epidemiologic clues from the characterization of the 1952 London Smog fine particulate matter in archival autopsy lung tissues. *Environmental Health Perspectives*, *111*, 1209–1212.

IEA. (2016). World Energy Outlook 2016 Special Report: Energy and Air Pollution. Paris: International Energy Agency, France.

ISO. (2017). <https://www.iso.org/standard/14534.html>.

Jacobson, M. Z. (2002). *Atmospheric pollution: History, science, and regulation*. Cambridge University Press.

Jenner, M. (1995). The politics of London air John Evelyn's Fumifugium and the restoration. *The Historical Journal*, *38*, 535–551.

Karamfilova, E. (2022). Revision of the EU ambient air quality directives. European Parliamentary Research Service (EPRS), PE 734.679.

Kenis, A., & Barratt, B. (2022). The role of the media in staging air pollution: The controversy on extreme air pollution along Oxford Street and other debates on poor air quality in London. *Politics and Space*, *40*, 611–628. Available from https://doi.org/10.1177/2399654420981607.

Kukkonen, J., Pohjola, M., Sokhi, R. S., Luhana, L., Kitwiroon, N., Fragkou, L., & Finardi, S. (2005). Analysis and evaluation of selected PM10 air pollution episodes affecting four European cities: Helsinki, London, Milan and Oslo. *Atmospheric Environment*, *39*, 2759–2774.

Levy, R. L. (2015). Carbon monoxide pollution and neurodevelopment: A public health concern. *Neurotoxicology and Teratology*, *49*, 31–40.

Liu, M., & Hu, D. (2021). Different relationships between Arctic oscillation and ozone in the stratosphere over the Arctic in January and February. *Atmosphere*, *12*, 129. Available from https://doi.org/10.3390/atmos12020129.

Lu, X., Zhang, L., & Shen, L. (2019a). Meteorology and climate influences on tropospheric ozone: A review of natural sources, chemistry, and transport patterns. *Current Pollution Reports*, *5*, 238–260. Available from https://doi.org/10.1007/s40726-019-00118-3.

Lu, X., Zhang, L., Zhao, Y., Jacob, D. J., Hu, Y., Hu, L., & Querel, R. (2019b). Surface and tropospheric ozone trends in the Southern Hemisphere since 1990: Possible linkages to poleward expansion of the Hadley Circulation. *Science Bulletin*, *64*(6), 400–409. Available from https://doi.org/10.1016/j.scib.2018.12.021.

McFiggans, G., Alfarra, M. R., Allan, J. D., Coe, H., Hamilton, J. F., Harrison, R. M., Jenkin, M. E., Lewis, A. C., Moller, S. J., Topping, D. O., & Williams, P. I. (2015). A review of the state-of-the science relating to secondary particulate matter of relevance to the composition of the UK atmosphere. Full technical report to Defra, project AQ0732.

McNeill, J., Snider, G., Weagle, C. L., Walsh, B., Bissonnette, P., Stone, E., & Martin, R. V. (2020). Large global variations in measured airborne metal concentrations driven by anthropogenic sources. *Science Reports*, *10*, 21817. Available from https://doi.org/10.1038/s41598-020-78789-y.

Miller, H. (1889). First impressions of England and its people. W. P. Nimmo, Hay and Mitchell. <https://minorvictorianwriters.org.uk/miller/b_impressions.htm> and <https://minorvictorianwriters.org.uk/miller/c_impressions_1.htm#III>.

Molina, L. T. (2021). Introductory lecture: Air quality in megacities. *Faraday Discussions*, *226*, 9. Available from https://doi.org/10.1039/D0FD00123F.

Monks, P., Archibald, A., Colette, A., Cooper, O., Coyle, M., Derwent, R., & Williams, M. (2015). Tropospheric ozone and its precursors from the urban to the global scale from air quality to short-lived climate forcer. *Atmospheric Chemistry and Physics*, *15*, 8889–8973.

Morawska, L., Zhu, T., Liu, N., Torkmahalleh, M. A., Andrade, M. D., Barratt, B., & Ye, C. (2021). The state of science on severe air pollution episodes: Quantitative and qualitative analysis. *Environment International*, *156*106732. Available from https://doi.org/10.1016/j.envint.2021.106732.

Nemery, B., Hoet, P. H., & Nemmar, A. (2001). The Meuse Valley fog of 1930: An air pollution disaster. *Lancet*, *357*, 704–708. Available from https://doi.org/10.1016/s0140-6736(00)04135-0.

Ojha, N., Soni, M., Kumar, M., Girach, I., Sharma, S., & Gunthe, S. (2022). Air pollution episodes: Brief history, mechanisms and outlook. In P. Saxena, A. Shukla, & A. K. Gupta (Eds.), *Extremes in atmospheric processes and phenomenon: Assessment, impacts and mitigation. disaster resilience and green growth.* Singapore: Springer. Available from https://doi.org/10.1007/978-981-16-7727-4_12.

Olstrup, H., Johansson, C., Forsberg, B., Åström, C., & Orru, H. (2021). Seasonal variations in the daily mortality associated with exposure to particles, nitrogen dioxide, and ozone in Stockholm, Sweden, from 2000 to 2016. *Atmosphere, 12*, 1481.

Park, S., Allen, R. J., & Lim, C. H. (2020). A likely increase in fine particulate matter and premature mortality under future climate change. *Air Quality, Atmosphere & Health, 13*, 143−151. Available from https://doi.org/10.1007/s11869-019-00785-7.

Payus, C. M., Jikilim, C., & Sentian, J. (2020). Rainwater chemistry of acid precipitation occurrences due to long-range transboundary haze pollution and prolonged drought events during southwest monsoon season: Climate change driven. *Heliyon, 6*. Available from https://doi.org/10.1016/j.heliyon.2020.e04997, e04997.

Petroni, M., Hill, D., Younes, L., Barkman, L., Howard, S., Howell, I. B., & Collins, M. B. (2020). Hazardous air pollutant exposure as a contributing factor to COVID-19 mortality in the United States. *Environmental Research Letters, 15*, 0940a9.

Resongles, E., Dietze, V., Green, D., Harrison, R. M., Ochoa-Gonzalez, R., Tremper, A. H., & Weiss, D. J. (2021). Strong evidence for the continued contribution of lead deposited during the 20th century to the atmospheric environment in London of today. *PNAS, 118*e2102791118. Available from https://doi.org/10.1073/pnas.2102791118.

Ritchie, H. (2017). What the history of London's air pollution can tell us about the future of today's growing megacities. *Our world in data.* <https://ourworldindata.org/london-air-pollution> (accessed 10.11.22).

Schrenk, H., Heimann, H., Clayton, G., Gafafer, W., & Wexler, H. (1949). Air pollution in Donora PA, epidemiology of the unusual smog episode of October 1948: Preliminary report. Public Health Bull Number 306, Federal Security Agency, Public Health Service, Bureau of State Services, Division of Industrial Hygiene, Washington, D.C.

Seinfeld, J. H., & Pandis, S. N. (2016). *Atmospheric chemistry and physics: From air pollution to climate change.* Wiley, Publisher.

Shaddick, G., Thomas, M. L., Mudu, P., Ruggeri, G., & Gumy, S. (2020). Half the world's population are exposed to increasing air pollution. *NPJ Climate and Atmospheric Science, 3*, 23. Available from https://doi.org/10.1038/s41612-020-0124-2, 2020.

Shrivastava, M., Cappa, C. D., Fan, J., Goldstein, A. H., Guenther, A. B., Jimenez, J. L., & Zhang, Q. (2017). Recent advances in understanding secondary organic aerosol: Implications for global climate forcing. *Reviews of Geophysics (Washington, D.C.: 1985), 55*, 509−559. Available from https://doi.org/10.1002/2016RG000540.

Sicard, P., Agathokleous, E., De Marco, A., Paoletti, E., & Calatayud, V. (2021). Urban population exposure to air pollution in Europe over the last decades. *Environmental Science Europe, 33*, 28.

Sicard, P., Paoletti, E., Agathokleous, E., Araminien, V., Proietti, C., Coulibaly, F., & De Marco, A. (2020). Ozone weekend effect in cities: Deep insights for urban air pollution control. *Environmental Research, 191*110193. Available from https://doi.org/10.1016/j.envres.2020.110193.

Sillman, S. (2014). Tropospheric ozone and photochemical smog. In *Treatise on geochemistry* (Second Edition) (Vol. 11, pp. 415−437). Available from https://doi.org/10.1016/B978-0-08-095975-7.00911-6.

Singh, V., Khaiwal, R., Sahu, L., & Sokhi, R. (2018). Trends of atmospheric black carbon concentration over United Kingdom. *Atmospheric Environment, 178*, 148−157.

Sokhi, R. S., Moussiopoulos, N., Baklanov, A., Bartzis, J., Coll, I., Finardi, S., & Kukkonen, J. (2022). Advances in air quality research—Current and emerging challenges. *Atmospheric Chemistry & Physics, 22*, 4615−4703. Available from https://doi.org/10.5194/acp-22-4615-2022.

Sokhi, R. S., Singh, V., Querol, X., Finardi, S., C, T. A., & Zavala, M. (2021). A global observational analysis to understand changes in air quality during exceptionally low anthropogenic emission conditions. *Environment International, 157*106818. Available from https://doi.org/10.1016/j.envint.2021.106818.

Srivastava, D., Vu, T. V., Tong, S., Shi, Z., & Harrison, R. M. (2022). Formation of secondary organic aerosols from anthropogenic precursors in laboratory studies. *NPJ Climate Atmospheric Science, 5*, 22. Available from https://doi.org/10.1038/s41612-022-00238-6.

Stohl, A., Bonasoni, P., Cristofanelli, P., Collins, W., Feichter, J., Frank, A., & Zerefos, C. (2003). Stratosphere-troposphere exchange: A review, and what we have learned from STACCATO. *Journal of Geophysics Reseaerch - Atmosphere., 108*(D12), 8516. Available from https://doi.org/10.1029/2002jd002490.

Stone, R. (2002). Counting the cost of London's Killer Smog. *Science (New York, N.Y.), 298*, 2106−2107. Available from https://doi.org/10.1126/science.298.5601.2106b.

Stowell, J. D., Kim, Y.-M., Gao, Y., Fu, J. S., Chang, H. H., & Y, L. (2017). The impact of climate change and emissions control on future ozone levels: Implications for human health. *Environment International, 108*, 41−50. Available from https://doi.org/10.1016/j.envint.2017.08.001.

Szopa, S., Naik, V., Adhikary, B., Artaxo, P., Berntsen, T., Collins, W. D., & Zanis, P. (2021). Short-lived climate forcers. In *Climate change 2021: The physical science basis.* Contribution of Working Group I to the Sixth Assessment Report of the Intergovernmental Panel on Climate Change. University Press, Cambridge, United Kingdom and New York, NY, USA. Available from https://doi.org/10.1017/9781009157896.008.

Thornhill, G., Collins, W., Olivié, D., Skeie, R. B., Archibald, A., Bauer, S., & Weber, J. (2021). Climate-driven chemistry and aerosol feedbacks in CMIP6 Earth system models. *Atmospheric Chemistry and Physics, 21*, 1105−1126. Available from https://doi.org/10.5194/acp-21-1105-2021.

UNEP. (2019). UNEP 2019. Air pollution in Asia and the Pacific: Science-based solutions. United Nations Environment Programme (UNEP).

UNEP. (2021a) Regulating air quality: The first global assessment of air pollution legislation. United Nations Environment Programme. ISBN No: 978-92-807-3872-8.

UNEP. (2021b). Global methane assessment: Benefits and costs of mitigating methane emissions. United Nations Environment Programme and Climate and Clean Air Coalition.

UNEP. (2022). Spreading like wildfire—The rising threat of extraordinary landscape fires. Nairobi: United Nations Environment Programme. A UNEP Rapid Response Assessment.

USEPA. (2016). Integrated science assessments (ISAs). <http://www.epa.gov/ncea/isa> (accessed 15.11.22).

USEPA. (2019). Integrated science assessment for particulate matter. U.S. Environmental Protection Agency, EPA 600/R-19/188.

USEPA. (2020). Integrated science assessment for oxides of nitrogen, oxides of sulfur, and particulate matter - Ecological criteria. US Environmental Protection Agency, EPA/600/R-20/278. <http://www.epa.gov/ncea/isa>.

WHO. (1987). Air quality guidelines for Europe. WHO regional publications. European series: No.23, Copenhagen.

WHO. (2000). Air quality guidelines for Europe, European Series, No. 91, 2nd Edition. WHO Regional Publications, Copenhagen.

WHO. (2006). WHO air quality guidelines for particulate matter, ozone, nitrogen dioxide and sulfur dioxide. Global update 2005, Summary of risk assessment. WHO Regional Office for Europe.

WHO. (2010). WHO guidelines for indoor air quality: Selected pollutants. World Health Organization. WHO Regional Office for Europe. ISBN 978 92 890 0213 4.

WHO. (2017). Evolution of WHO air quality guidelines: Past, present and future. WHO Regional Office for Europe, Copenhagen.

WHO. (2021a). WHO global air quality guidelines. Particulate matter (PM2.5 and PM10), ozone, nitrogen dioxide, sulfur dioxide and carbon monoxide. World Health Organization.

WHO. (2021b). Human health effects of polycyclic aromatic hydrocarbons as ambient air pollutants: Report of the working group on polycyclic aromatic hydrocarbons of the joint task force on the health aspects of air pollution. WHO Regional Office for Europe. Licence: CC BY-NC-SA 3.0 IGO.

WMO. (2022). WMO air quality and climate bulletin. No. 2. Publication by the World Meteorological Organization.

Wrigley, E. A. (2018). Reconsidering the industrial revolution: England and Wales. *Journal of Interdisciplinary History*, *49*(01), 9–42.

Yang, C., Fu, J. S., Liu, Y., Dong, X., & Liu, Y. (2022). Projections of future wildfires impacts on air pollutants and air toxics in a changing climate over the western United States. *Environmental Pollution*, *304*119213. Available from https://doi.org/10.1016/j.envpol.2022.119213.

Yang, Y., & Baklanov, A. (2020). Training materials and best practices for chemical weather/air quality forecasting. Report ETR-26, World Meteorological Organization.

Yli-Juuti, T., Mielonen, T., Heikkinen, L., Arola, A., Ehn, M., Isokääntä, S., & Virtanen, A. (2021). Significance of the organic aerosol driven climate feedback in the boreal area. *Nature Communications*, *12*, 5637. Available from https://doi.org/10.1038/s41467-021-25850-7.

Zhang, J., Wei, Y., & Fang, Z. (2019). Ozone pollution: A major health hazard worldwide. *Frontiers in Immunology*. Available from https://doi.org/10.3389/fimmu.2019.02518, Article 2518.

Zhao, A., Bollasina, M. A., & Stevenson, D. S. (2019). Strong influence of aerosol reductions on future heatwaves. *Geophysical Research Letters*, *46*, 4913–4923. Available from https://doi.org/10.1029/2019GL082269.

Zhu, J., Penner, J. E., Lin, G., Zhou, C., Xu, L., & Zhuang, B. (2017). Mechanism of SOA formation determines magnitude of radiative effects. *PNAS*, *114*(48), 12685–12690. Available from https://doi.org/10.1073/pnas.1712273114.

Chapter 2

The atmosphere—fundamental concepts

Ranjeet S. Sokhi, Ummugulsum Alyuz and Somayyeh Arghavani
Centre for Climate Change Research (C3R), School of Physics, Engineering and Computer Science, University of Hertfordshire, Hatfield, United Kingdom

2.1 Introduction

In this chapter, we cover the main fundamentals of atmospheric science, which provide the understanding of basic physics, chemistry, and meteorology concepts needed for subsequent chapters. It lays the foundation by first considering the energy from the Sun, and then moving to the transfer and distribution of solar energy, the thermal structure, and pressure variations within the lower atmosphere. To understand air quality and its impacts, we also need knowledge of the composition of the atmosphere in terms of key species and their sources.

The principles explained in this chapter provide the basis to understand the next chapters, namely, Chapter 3 (Sources and emissions), Chapter 4 (Chemistry of the atmosphere) as well as Chapter 5 (Dynamics of the atmosphere) and Chapter 6 (Physics of the planetary boundary layer). There are a number of relevant reviews that the reader will find useful to expand their knowledge in this area, such as Molina (2021) and Sokhi et al. (2022).

As a vital organ of our planet, the Earth's *atmosphere* provides a critical function for everything we do. Not only does it provide the air that we breathe, it also acts as a protective barrier against the dangerous radiation and heat arriving from the Sun. Although invisible to the eye, *air* is actually a mixture of solid, liquid, and gaseous components and forms what we generally call the *atmosphere*. We *feel* the air when the wind flows at a discernible rate and directly affects our state of wellbeing. Without the atmosphere, as in the case of the moon, numerous processes and interactions that lead to weather features would not occur. Without air, almost all life on the planet would cease to exist.

2.2 Energy and solar radiation

The energy of the Earth comes from two sources, namely geothermal energy from the Earth's interior and radiation from our nearest star, the Sun. Internal processes of the Earth that produce volcanoes, earthquakes, and mountains are driven by the heat remaining from the formation of the Earth and the radiogenic heat that is continuously generated by radioactive decay. The Sun is the most important source of energy for life on Earth, and its vast amount of energy is produced by nuclear fusion with a core temperature of more than 15,000,000°C and a surface temperature of about 5500°C.

When we speak of ***temperature*** we refer to the mean energy of a system. Temperature is commonly expressed in units of Kelvin (K) and/or Celsius (°C). These Celsius and Kelvin temperature scales are related through a simple relationship given below in Eq. (2.1).

$$[^\circ C] = [K] - 273.15 \tag{2.1}$$

2.2.1 Solar radiation incident on the Earth's surface

The sun is the source of more than 99.9% of the energy that heats Earth's surface (Kren et al., 2017; Sellers, 1965). Incoming solar radiation arriving at the earth's surface is termed ***insolation***. A related term is ***total solar irradiance*** (TSI), previously known as the solar constant, and is one of the fundamental parameters in atmospheric physics. The main contribution of energy at the Earth comes from TSI, which represents the solar power per unit area (or equivalent to surface power density) calculated at the mean distance of the Earth from the Sun and has units of Watts per unit area

Air Quality. DOI: https://doi.org/10.1016/B978-0-12-822591-2.00002-0

($W\ m^{-2}$). Its mean value at the top of the Earth's atmosphere is 1365 $W\ m^{-2}$, however, the value at ground level can vary depending on factors such as location, time of day, and atmospheric conditions.

This energy from the Sun is not distributed equally over the surface of the Earth, with much of the energy absorbed by the oceans, mainly in tropical waters at the equator. This unequal heating over the Earth's surface drives the ocean's currents and creates global winds in the atmosphere. To understand how the atmosphere works, we need to understand the reasons for the unequal distribution of energy from the Sun and what causes the variations in the amount of solar energy that reaches Earth.

The first factor that affects the sunlight that is coming to Earth is the motion of the Earth itself. Two principal motions of Earth are rotation around its axis and revolution of the orbit around the Sun. The other factor affecting the solar radiation received by Earth is the tilt of the Earth's axis. In climate physics, the cycles of these motions are important on long time scales. Their importance for affecting solar insolation and hence the natural causes of climate change variations were recognized by the Serbian scientist Milutin Milankovitch in early the 20th century. In the context of air quality, we usually consider short time scales of less than a few years, and hence these long-term cycles are not discussed further, but the basic description of the Earth's motion is given below.

2.2.1.1 Earth's rotation

Earth spins about its axis, taking 24 hours to complete one rotation. Usually in air quality and meteorology, we refer to this 24-hour cycle as a ***diurnal cycle***. For example, we can consider diurnal cycles of air quality and meteorological parameters such as temperature, insolation, and concentrations of air pollutants. The ***axis of the Earth*** is the imaginary line through the Earth that extends from the North Pole to the South Pole. The spinning of Earth about its axis causes changes in solar radiation from the Sun falling on the Earth's surface. The portion of the Earth that is away from the Sun experiences night while it is the daylight in the other half of the Earth that faces the Sun, as shown in Fig. 2.1. As the Earth rotates about its axis, we see a "moving" Sun across the sky from our position where we stand.

2.2.1.2 Earth's revolution

The other motion, ***revolution***, or ***orbit*** of the Earth around the Sun occurs in a counterclockwise path that appears fixed but varies on long timescales of tens of thousands of years (one of the Milankovitch cycles). The reason for the Earth orbiting around the Sun is the gravity pull of the Sun on Earth, which is also affected by the other solar system planets, mainly Jupiter and Saturn, because of their large masses. The gravitational force between the Sun and the Earth is strong because the mass of the Sun is 300,000 times greater than Earth's mass, and the distance between Earth and the Sun is adequate for this force to have a significant influence on the Earth. Although the Earth travels about 113,000 $km\ h^{-1}$ in its orbit, we don't perceive this force, that is, feel the air rushing past us, because the atmosphere is also bound by these forces, and it is carried along at the same speed as the planet.

The motion of Earth around the Sun takes approximately 365 days and 6 hours to revolve around the sun. However, for convenience, we consider a year as consisting of 365 days, and one additional day is added to February every four years, which is called a ***leap year***.

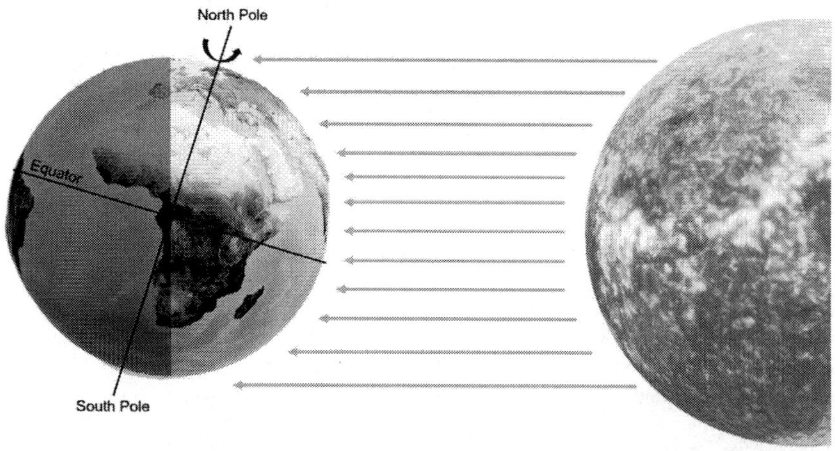

FIGURE 2.1 Day and night on the Earth due to its rotation with incoming solar radiation (not to scale). Map lines delineate study areas and do not necessarily depict accepted national boundaries.

The distance of the Earth from the Sun, known as the solar distance or one astronomical unit (AU), is 150 million km on average and varies as the Earth's orbit around the Sun is not perfectly circular. As illustrated in Fig. 2.2, the Earth is 147 km from the Sun on about 3rd of January, which is the closest distance of Earth to the Sun and is called ***perihelion***. The solar distance of the Earth is 152 km on around 4th of July, the farthest distance away from the Sun, which is known as the ***aphelion***.

In the northern hemisphere, it is winter during January and summer during July. You may wonder why January is in the winter season when the solar distance is the smallest and why July is summer when the distance is the largest. The sections below explain the other factors that lead to seasonal weather conditions.

2.2.1.3 The axial tilt of Earth

The axis of the Earth is an imaginary line through the poles, as illustrated in Fig. 2.3, and is currently tilted by 23.4 degrees from the perpendicular orbital axis, which is termed the ***inclination of the axis***. This tilt angle of Earth's rotational axis from its orbital axis is called the ***axial tilt of Earth*** or ***obliquity***. The reason for this tilt is thought to be caused by a giant object, as large as Mars, which scientists call Theia, that crashed on the Earth when the Earth was young and caused the Earth to lean away from its orbital axis (Herwartz et al., 2014). The Earth's axial tilt oscillates between 22.1 and 24.5 degrees in 41,000 years. These changes are explained by the Milankovitch cycles mentioned earlier.

Due to the axial tilt of the Earth, the Earth receives the Sun's direct rays over different parts of its surface throughout the year and hence experiences different seasons. Receiving direct light from the Sun means that the Sun is overhead of the region of the Earth, and solar energy arrives on the Earth's surface at 90 degrees. As shown on the right-hand side of Fig. 2.4, during one-half of Earth's orbit around the Sun, the rotational axis is toward the Sun from the

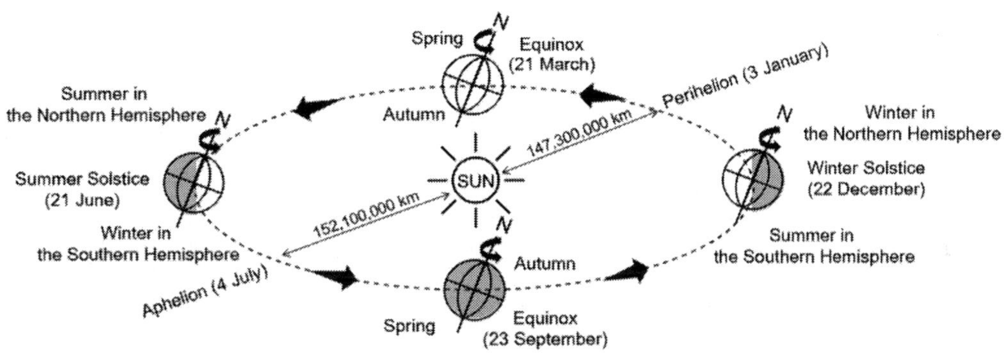

FIGURE 2.2 Solar distance of the Earth with the perihelion (closest) and aphelion (farthest) positions from the Sun.

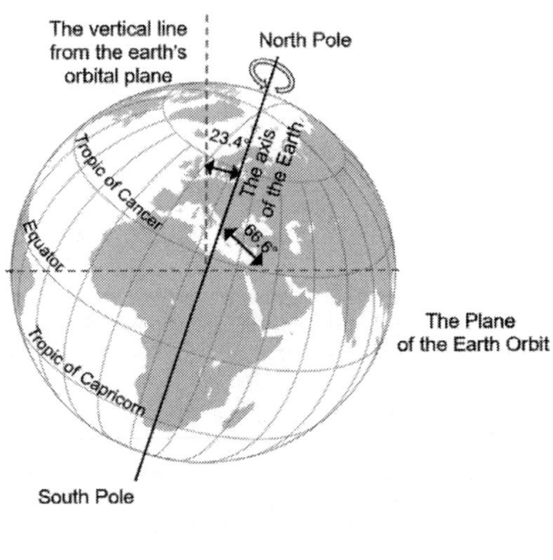

FIGURE 2.3 Axial tilt of the Earth. Map lines delineate study areas and do not necessarily depict accepted national boundaries.

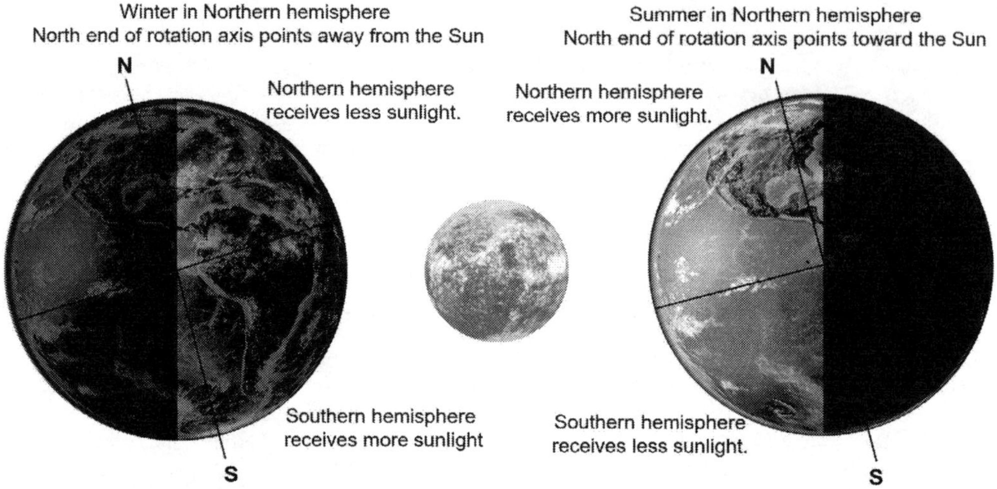

FIGURE 2.4 Change of incoming solar radiation between Northern and Southern Hemispheres depending on the angle of the sunlight (not to scale).

North. Thus, more energy reaches the Northern hemisphere than the Southern hemisphere. In this case, the Northern Hemisphere experiences spring and summer. Conversely, the north end of the rotation axis is further away from the Sun during the other half of Earth's orbit, as shown on the left-hand side of Fig. 2.4, and hence, the Southern hemisphere receives more energy from the Sun than the Northern hemisphere. The season now is fall and winter in the Northern hemisphere, while it is spring and summer in the Southern hemisphere.

The Northern Hemisphere is tilted toward the Sun around the 21st of June when rays from the Sun fall directly on the ***Tropic of Cancer*** (the most Northern circle of latitude on Earth at which the Sun is directly overhead, also referred to as the ***Northern Tropic***) (see Fig. 2.3). As a result, these areas receive more solar radiation, with the longest day and the shortest night occurring at these places. The position of the Earth at this time is called the **June or *Summer Solstice*** for the Northern hemisphere and is taken to be the first day of summer. Since the North Pole is inclined toward the sun, the places around the Arctic Circle experience continuous daylight for about six months, while the reverse occurs in the Southern Hemisphere, where it is the winter season, and the nights are longer than the days.

Around the 22nd of December, the Tropic of Capricorn (the most Southern latitude on Earth) at which the Sun is directly overhead (see Fig. 2.3) receives rays directly from the Sun as the South Pole tilts toward the Sun. As the sun's rays fall vertically, a more significant portion of the Southern Hemisphere is exposed to the Sun's rays, and it is summer season in the Southern Hemisphere, with longer days and shorter nights. The reverse happens in the Northern Hemisphere. This position of the Earth is called the **December or *Winter Solstice*** for the Northern hemisphere and is taken as the first day of Winter.

On around the 21st of March and 23rd of September, the vertical rays of the Sun come over 0 degree latitude, the equator. In this position, called an **equinox**, neither of the poles is tilted toward the sun, and days and nights are equal on the whole earth. On around 23rd September, it is the astronomical autumn season in the Northern Hemisphere and spring season in the Southern Hemisphere. The opposite happens around the 21st of March, when it is spring in the Northern Hemisphere and autumn in the Southern Hemisphere.

2.2.2 Mechanisms of heat transfer

Atoms and molecules constantly vibrate and start moving faster when they receive heat energy, and consequently the temperature of the substance rises. Similarly, when heat energy is removed, the atoms of the substance move slower, and its temperature decreases. In thermodynamics, we understand ***heat*** as the transfer of thermal energy from one system to another due to a difference in temperature between the systems. The ***kinetic energy*** of a system or an object is its possessed energy of motion. Heat represents the ***thermal energy*** contained within a system or an object by virtue of the moving atoms and molecules that make up the system or object. For the Earth, sunlight is the primary source of heat. Heat is transferred throughout the Earth by three mechanisms: conduction, convection, and radiation.

2.2.2.1 Conduction

Conduction is the transfer of heat (thermal energy) in a substance from the higher-temperature particles to those at the lower temperature through intermolecular collisions. The ability of matter to conduct heat depends on the distance between the molecules and atoms and hence varies between solids, liquids, and gases. In general, solids are better conductors of heat than liquids, which in turn are better than gases.

Since air is a poor conductor, transfer of thermal energy in the atmosphere through conduction occurs mostly near Earth's surface. As shown in Fig. 2.5, sunlight heats the ground on clear, sunny days with little or no wind, which in turn heats the air directly above it via conduction. Conversely, the ground cools at night, and the heat flows from the warmer air to the cooler ground via conduction. So, conduction modulates air temperature only within a few centimeters above the surface of the Earth. In terms of heat transfer in the atmosphere, conduction is considered the least significant compared to convection and radiation.

2.2.2.2 Convection

Convection is the transfer of heat by moving particles or circulation within fluids. Unlike conduction, atoms and molecules are more free to move in convection, so convection will only occur in air, the ocean, or any fluid. Fig. 2.5 illustrates the rising of warm air from the surface of the Earth by convection on a warm and sunny day. Most of the heat is acquired in the atmosphere's lowest layer through radiation (next section) and conduction. The air parcel adjacent to the surface is warm and less dense compared to upper levels. This parcel rises upward, and heat is transferred aloft by the ***convective flow***. As a result of unequally distributed heat, the warm air will continue to turn over, producing a ***convective circulation***. This type of convection transfers heat and moisture in the air. Columns of warm air rising upwards, or ***thermals***, aid the formation of clouds because of the water vapor that they transport vertically. These thermals also play an essential role in the vertical mixing of pollutants in the atmosphere. When warm air thermals rise since they are less dense than the surrounding air, the rising air is called ***positively buoyant***. As air moves both vertically and horizontally in convective circulation, the term "convection" is usually used to describe vertical heat transfer. By contrast, the term ***advection*** is used to refer to the horizontal component of convective flow, which is wind.

2.2.2.3 Radiation

The Earth receives heat energy from the Sun (solar energy), which warms the Earth, causes wind and weather, and sustains plant and animal life on the planet. Solar energy reaches Earth with one of the heat transfer mechanisms known as ***radiation***. ***Radiant energy*** from the Sun, also known as ***radiation***, ***electromagnetic radiation***, or ***electromagnetic waves***, travels through space at the speed of light at different wavelengths and frequencies. A representation of the waves is shown in Fig. 2.6. For our purposes, the length of the wave is the essential characteristic of radiant energy, called ***wavelength***. Wavelength is the distance from one crest to the next in a wave, as shown in Fig. 2.6, while ***spectral radiance*** describes the amount of energy a body emits as radiation of different frequencies. Radio waves have the longest wavelengths (lowest frequency of the waves), ranging tens to kilometers. In contrast, gamma waves are the shortest (highest frequency of the waves) and are less than a billionth of a centimeter long. The unit of wavelengths can be in meters (m), centimeters (cm), micrometers (1 μm $= 10^{-6}$ m), nanometers (1 nm $= 10^{-9}$ m), or Angstrom (1 A $= 10^{-10}$ m).

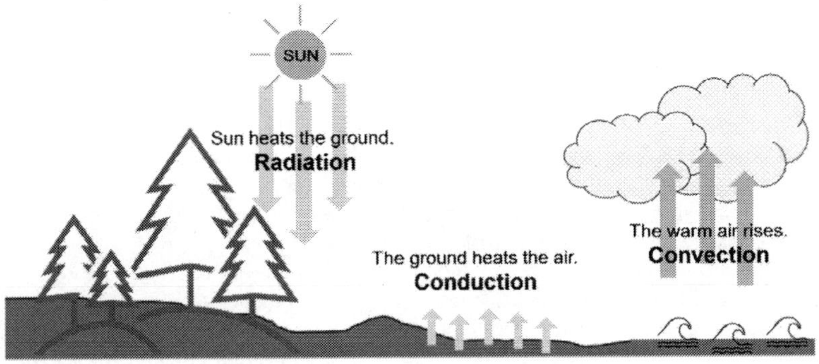

FIGURE 2.5 The main mechanisms of heat transfer.

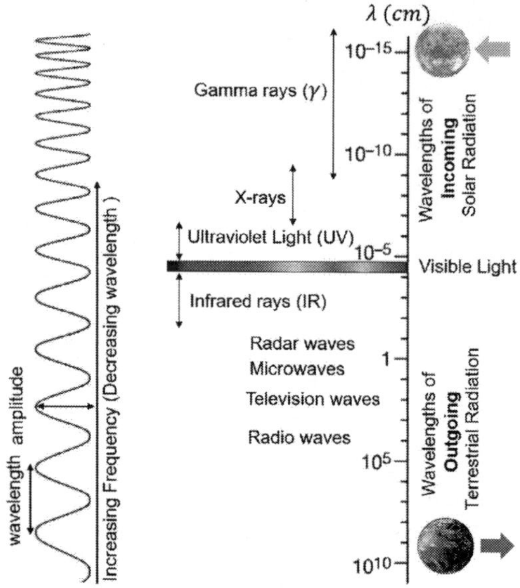

FIGURE 2.6 Wavelengths of incoming radiation from the Sun and outgoing radiation from the Earth.

The wavelength of an electromagnetic wave is given by Eq. (2.2).

$$\lambda = \frac{c}{\nu} = \frac{1}{\tilde{\nu}} \tag{2.2}$$

where λ is the wavelength (m), c is the speed (m s^{-1}), and ν is the frequency of the wave's oscillation, which is the number of wavelengths that pass through a point per unit of time (s^{-1}). $\tilde{\nu}$ is defined as the wavenumber, the number of the wavelengths per unit distance (and equals $1/\lambda$, usually with unit, m^{-1}).

The electromagnetic (EM) spectrum describes the distribution of electromagnetic radiation and their respective energies and wavelengths. The electromagnetic spectrum has seven regions: gamma-rays, X-rays, ultraviolet light, visible light, infrared, microwaves, and radio waves in order of decreasing energy, which means decreasing frequency and increasing wavelength. A brief summary of the main characteristics of these seven EM radiations is given below:

Gamma rays are produced by the universe's most energetic, hottest objects and some radioactive materials. The frequency of gamma-rays is the highest, while they have the shortest wavelengths in the spectrum. Gamma rays arriving on the Earth's atmosphere are usually detected with high-altitude balloons and orbiting telescopes, as most are absorbed by the atmosphere.

X-rays are widely used for their ability to penetrate tissues such as the skin. As they arrive on the Earth, they are mostly blocked by the upper atmosphere and don't reach the surface.

Ultraviolet (UV) radiation has a wavelength of 10–400 nm in the electromagnetic spectrum. UV radiation is commonly subdivided into UV-A, UV-B, and UV-C. Our atmosphere absorbs almost all UV-C rays, which are the most harmful UV rays. The ozone layer protects us from exposure to UV-B rays, which cause sunburn and may also cause cellular damage in living organisms, leading to skin cancer.

Visible light has a wavelength between about 400 and 750 nm and is the only portion of the wavelengths that the human eye can perceive. Collectively, all the wavelengths in the visible spectrum are defined as white light. However, white light is an array of colors, each representing a specific wavelength. This white light can be divided into rainbow colors using a prism, and we perceive red light at a wavelength of around 740 nm, green at around 500 nm, and violet at around 380 nm. All other colors that the human eye can see are mixtures of these three colors. Most of the radiant energy from the Sun is concentrated in the visible and near-visible parts of the electromagnetic spectrum. As the atmosphere is almost transparent in the visible spectrum, most of the visible light reaches Earth's surface.

Infrared (IR) radiation is both received and emitted by Earth. The solar energy radiated back from the Earth, called ***terrestrial radiation***, has longer wavelengths than the incident shortwave solar radiation. Most terrestrial radiation has wavelengths between 1 and 30 μm, placing it well within the infrared range. Outgoing infrared radiation from Earth's surface is absorbed by water vapor, carbon dioxide, methane, and other trace gases in the atmosphere. These atmospheric gases then scatter and reemit infrared radiation in all directions, both outward toward space and downward toward Earth, causing the greenhouse effect on the Earth, as discussed in the upcoming sections. Aerosols also play a key role in determining the radiation balance of the Earth, as they are critically involved in absorbing and scattering solar radiation as it travels in the atmosphere.

Microwaves have wavelengths ranging from 1 mm to around 30 cm, and the wavelengths of ***radio waves*** can reach hundreds of meters.

As the radiation arrives on the Earth, the narrow band of visible light represents over 43% of the total emitted radiation, and the bulk of the remainder (49%), lies in the infrared zone nearest to the visible band, and 7% is from the UV range. Less than 1% of solar radiation is emitted as radio waves, gamma waves, and x-rays (Gueymard, 2004; ASTM, 2020).

2.2.3 Global distribution of radiation

Irradiating energy in the Earth's atmosphere is emitted from three sources: the Sun, the surface of the Earth, and the atmosphere itself. Upon reaching the receiving body, be it a surface or a particle, the radiation can be subject to several processes such as scattering, reflection, absorption, refraction, dispersion, or transmission.

2.2.3.1 Incoming solar radiation

As discussed previously, sunlight is the primary source of energy for the Earth, which includes waves infrared, visible light, and ultraviolet radiation. ***Solar radiation*** is the term given to the energy emitted by the Sun.

Solar flux (or ***solar constant***) is the amount of energy per unit area that is incident just above the top of Earth's atmosphere. Assuming an annual mean distance between the Sun and the Earth, solar flux is approximately 1365 W m^{-2}. As a comparison, the solar flux for Venus is larger at 2641 W m^{-2}, since it is closer to the Sun, while it is 576 W m^{-2} for Mars, which is farther from the Sun than the Earth.

The amount of sunlight, however, is not equal throughout the day in all places over the Earth because of Earth's rotation and shape. Instead, the Earth intercepts the incoming solar flux with its cross-sectional area that faces the Sun. The cross-sectional surface area of the Earth that receives sunlight is πR_{Earth}^2 (R_{Earth} is the radius of the Earth). By considering the radius (R_{Earth}) of the Earth as 6.4×10^6 m and the solar constant as 1365 W m^{-2} (e.g., Haigh, 2011, Hartmann, 2016; Kopp & Lean, 2011), the incoming solar radiation falling per m^2 on the Earth is calculated to give approximately 341 W m^{-2} as follows:

$$\begin{matrix}\text{The incoming solar radiation}\\ \text{falling on the Earth per } m^2\end{matrix}\left(\frac{W}{m^2}\right) = \text{Solar constant} \times \frac{\begin{matrix}\text{The amount of sunlight}\\ \text{falling on the Earth }(W)\end{matrix}}{\text{Earth surface area }(m^2)}$$

$$\text{Solar constant} \times \frac{\pi R^2}{4\pi R^2} = 1365\frac{\text{W}}{\text{m}^2} \times \frac{\pi \times (6.4 \times 10^6\text{m})^2}{4\pi \times (6.4 \times 10^6\text{m})^2} = 341 \text{ W m}^{-2} \tag{2.3}$$

This value from Eq. (2.3) is an average number to ease the calculations since the equator receives more and the poles receive less energy during daylight.

Once solar radiation reaches the Earth, it may be scattered, reflected, or absorbed, as illustrated in Fig. 2.7 and discussed by Wild et al. (2015) and Forster et al. (2021). Here, the critical question is what determines whether solar radiation will be reflected, scattered, or absorbed, and this mainly depends on the wavelength of the energy and properties of the reflecting substance.

2.2.3.1.1 Scattering

The gases and dust particles in the atmosphere can change the direction of solar radiation that travels in a straight line. This process is called ***atmospheric scattering***. The light is diffused with atmospheric scattering, making our sky bright during the daytime. If there were no atmosphere on Earth, the sky would be dark even during daylight hours, as on the Moon. Because of this scattering process, the light diffuses and reaches shaded areas, forming the colors we observe in the sky.

The degree of scattering is determined by several factors, including the wavelength of the radiation, the size of particles and gas molecules, the distance of the radiation that travels through the atmosphere, and the number of suspended particles and gas molecules. When these particles, especially gas molecules, scatter light, it is scattered in all directions,

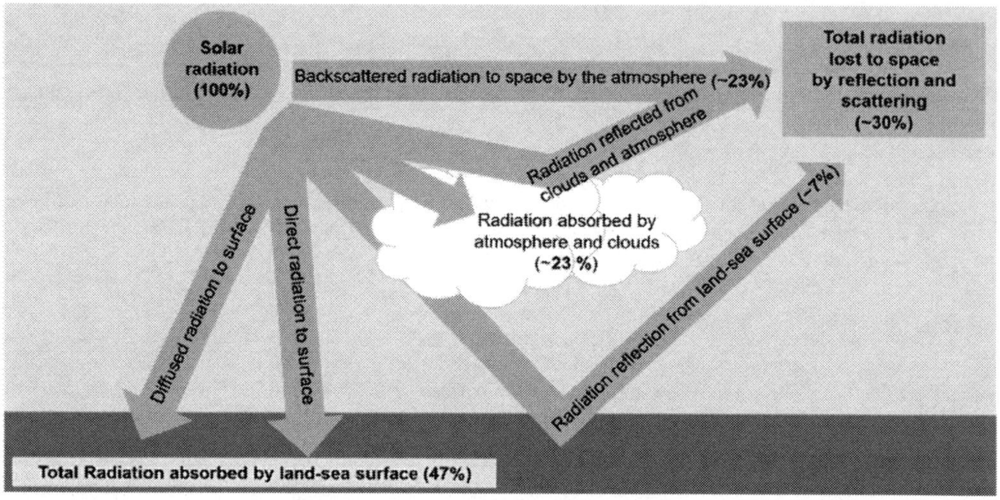

FIGURE 2.7 Average distribution of incoming solar radiation by percentage. *Data from Wild, M., Folini, D., Hakuba, M. Z., Schär, C., Seneviratne, S. L., Kato, S., Rutan, D., Ammann, C., Wood, E. F., & König-Langlo, G. (2015). The energy balance over land and oceans: An assessment based on direct observations and CMIP5 climate models.* Climate Dynamics, *44(11–12), 3393–3429. https://doi.org/10.1007/s00382-014-2430-z and Forster, P., Storelvmo, T., Armour, K., Collins, W., Dufresne, J. -L., Frame, D., Lunt, D. J., Mauritsen, T., Palmer, M. D., Watanabe, M., Wild, M. & H. Zhang, H. (2021). The Earth's energy budget, climate feedbacks, and climate sensitivity. In V. Masson-Delmotte, P. Zhai, A. Pirani, S. L. Connors, C. Pean, S. Berger, N. Caud, Y. Chen, L. Goldfarb, M. I. Gomis, M. Huang, K. Leitzell, E. Lonnoy, J. B. R. Matthews, T. K. Maycock, T. Waterfield, O. Yelekci, R. Yu, B. Zhou (Eds.),* Climate change 2021: The physical science basis. contribution of working group I to the sixth assessment report of the intergovernmental panel on climate change. *Cambridge University Press, pp. 923–1054, https://doi.org/10.1017/9781009157896.009.*

with some of the light returning toward space and the remainder continuing to travel toward Earth's surface but in different directions. In this process, the wavelength of the light doesn't change.

There are three main types of scattering in the atmosphere: Rayleigh scattering, Mie scattering, and nonselective scattering.

Rayleigh scattering (named after Lord Rayleigh, a British physicist) occurs when electromagnetic radiation interacts with particles that are smaller than the wavelengths of light. Smaller dust particles, nitrogen (N_2) molecules, and oxygen (O_2) are examples of such "particles." Gas molecules scatter the shorter wavelength (blue and violet) of visible light more effectively than the longer wavelengths (red and orange). As mentioned above, scattering also produces ***diffused light***, which accounts for the brightness of the daytime sky.

At sunset and sunrise, the traveling distance of sunlight to the Earth's surface is longer as the sun is relatively lower in the horizon. Longer traveling distance leads to more scattering of shorter (blue) wavelengths of radiation, and hence the Earth's surface receives longer (orange and red) wavelengths. We see an orange or red sky instead of blue, as the shorter blue wavelength light is scattered by the particles in the atmosphere, leaving red and orange colors to pass through. However, solar radiation travels the shortest distance through the atmosphere during the day; more shortwave radiation reaches the surface, and gas molecules scatter these shorter wavelengths. As a result of Rayleigh scattering, a clear sky is observed as blue.

Mie scattering (named after a German physicist, Gustav Mie) occurs when the wavelength of electromagnetic radiation is similar in size to particles in the atmosphere and tends to scatter all wavelengths of light equally. This type of scattering is important for larger particles in the air, such as dust, pollen, smoke, and water vapor. Since these large particles are more abundant in the lower atmosphere, Mie scattering generally occurs in the lower layers of the troposphere.

Non-selective scattering occurs when the particle sizes are much larger than the wavelength of electromagnetic radiation. Light incident on water droplets and larger dust particles experiences this type of scattering. The most prominent example of nonselective scattering is the color of clouds, since the clouds consist of water droplets. Clouds scatter light of all wavelengths, and hence they appear white. The colors making up the sky color indicate the number and size of the particles present in the atmosphere, with red sunsets occurring when there are many small particles and a whiter sky when there are larger particles (e.g., water droplets) in the atmosphere.

Scattering of sunlight by airborne particles can result in major impacts on weather. The famous example of the Indonesian volcano Krakatau, which erupted in 1883, was followed by cooler summers in Europe for three years. A more recent eruption of Krakatau is shown in Fig. 2.8. The cooling effect results from significant radiation loss caused by backscattering from atmospheric particles remaining from the volcanic eruption. Simultaneously, spectacular sunsets occurred worldwide, attributed to the large quantities of fine dust or smoke particles that penetrated the stratosphere.

FIGURE 2.8 Krakatau volcano in Indonesia. *From Elsevier stock, https://www.shutterstock.com/image-photo/volcano-eruption-anak-krakatau-indonesia-72070270.*

2.2.3.1.2 Reflection

The fraction of the radiation that is diffusely reflected by a surface is called **albedo**. It is usually expressed on a scale from 0 for a completely black surface where all the light is absorbed to 1 where the surface is completely reflecting and all of the light is reflected back. Albedo values are often expressed as percentages. The albedo of various surfaces is given in Table 2.1. As the values in Table 2.1 show, dark-colored surfaces tend to be less reflective than light-colored surfaces as they absorb most of the received radiation and thus have lower albedo.

For a given location on the Earth, the albedo can vary considerably depending on the amount of cloud cover, the particulate matter concentration in the air, and the angle at which the Sun's rays strike the Earth's surface. In a cloudy atmosphere, solar radiation can be reflected and absorbed depending on the cloud thickness, sizes, and concentration of the water droplets making up the clouds. Clouds significantly reflect solar radiation, and because of this reason, clouds are responsible for much of the brightness in satellite images, which is a source of interference when resolving surface characteristics and detecting environmental variables (see Fig. 2.7).

In Fig. 2.7, ~7% of the incoming solar radiation is reflected from Earth's land-sea surface, while ~23% is backscattered, reflected to space, or reflected by clouds. The sum of these values is 30%, which is approximately the planetary albedo of the Earth.

Although the Moon has an albedo of 7% (see Table 2.1), it can still provide significant illumination on a clear night, as seen in Fig. 2.9. The reason we perceive bright light from the Moon despite its low albedo is because the Moon lacks clouds or an atmosphere that would absorb the radiation.

FIGURE 2.9 Moonlight reflection of the sea surface. From Flicker photo archive (Mrazek, 2022).

TABLE 2.1 The albedo of selected surfaces expressed as percentage (Stull, 2011; Hungate & Koch, 2023).

Surface	Typical albedos (%)
Water, deep	5–20
Snow, fresh	75–95
Snow, old	35–70
Cloud, thick	70–95
Cloud, thin	20–65
Green grass[a]	22–28
Crops[a]	15–26
Forest[a]	5–18
Forest, coniferous	5–15
Forest, deciduous	10–25
Soil, wet	6–8
Soil, dry	16–18
Stones	20–30
Urban, mean	15
Buildings	9
Asphalt	5–15
Concrete	15–37
Moon	7

[a]*Data taken from Hungate and Koch (2023).*
Source: Data from Stull, R. B. (1988). *An introduction to boundary layer meteorology*. Kluwer Academic Publishers. p. 670; Hungate, B.A. & Koch, G.W. (2023). GLOBAL CHANGE, Biospheric impacts and feedbacks, *Encyclopaedia of atmospheric sciences*, Academic Press.

2.2.3.1.3 Absorption

Approximately 70% of the solar radiation incident on the Earth is absorbed by the atmosphere and the surface, with about 30% being reflected back. As illustrated in Fig. 2.7, nearly a fifth of the incoming solar radiation is absorbed by clouds and gases in the atmosphere. Absorption of solar radiation plays a vital role in determining the temperature of the atmosphere as it leads to an increase in the energy of the gas molecules.

Absorptivity is a measure of the fraction of how much a body absorbs radiation. A blackbody has zero reflectance and transmittance and hence 100% absorptivity as it absorbs all the incident radiation internally. Such an ideal body is called a ***blackbody***. In real life, there is no ideal blackbody, but this concept is useful to ease calculations.

It should now be apparent that not all wavelengths of electromagnetic radiation emitted from the Sun reach the Earth since some of the incoming energy is absorbed by the atmosphere. However, most of the earth's infrared energy is absorbed. Fig. 2.10 illustrates the absorptivity of selected primary gases in the atmosphere as a function of wavelength in the lower part of the figure. The radiation intensity of the solar and terrestrial radiation (see Section 2.2.3.2) is given on the top of the figure with wavelength.

From Fig. 2.7, you will recall that the atmosphere absorbs 19% of the incoming solar radiation, primarily by water vapor, ozone, and oxygen. While water vapor (H_2O) is an effective absorber of incoming solar radiation, nitrogen in the form of nitrous oxide (N_2O) is a relatively poor absorber. However, water vapor is also the most abundant constituent in the atmosphere. Shorter wavelength ultraviolet radiation from the Sun is efficiently absorbed by oxygen (O_2) in the troposphere, mostly above an altitude of about 50 km. This absorption is stronger for wavelengths between 100 and 200 nm and weaker between 200 and 245 nm. In comparison, ozone (O_3) absorbs longer wavelength (between 200 and 300 nm) ultraviolet radiation (harmful to the skin) in the stratospheric region of the atmosphere (see ozone layer in Section 2.3.1 for further information).

Although oxygen and ozone also absorb between 300 and 800 nm in the stratosphere, their absorptivity is around 10%, as seen in Fig. 2.10, and hence most of the solar radiation in these wavelengths is transmitted to the troposphere region. Wavelengths

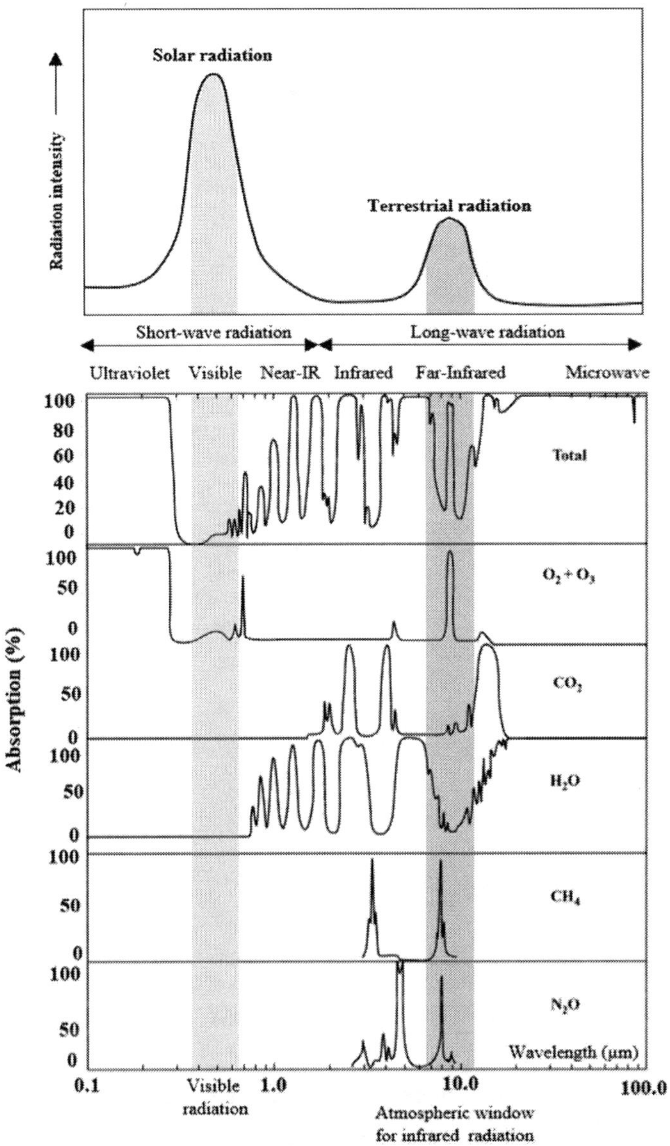

FIGURE 2.10 The absorptivity of the total atmosphere and selected gases (Lutgens & Tarbuck, 1997; Lazaridis, 2011). *Adapted from Lutgens, F. & Tarbuck, E. (1997). The atmosphere. Prentice Hall Inc.; Lazaridis, M. (2011). First principles of meteorology and air pollution.* Environmental Pollution, *Springer.*

between this range of 300 and 800 nm correspond to the visible range of the electromagnetic spectrum (see Fig. 2.6). which explains why a significant fraction of solar radiation reaching Earth's surface is mainly in the visible part of the electromagnetic spectrum. In other words, the atmosphere is transparent to incoming solar radiation if it is in the visible part of the spectrum.

This can easily be confused with the concept of "atmospheric windows." As explained, the atmosphere blocks certain frequency bands of light from the sun reaching the Earth's surface. This implies that for a specific range of frequencies there is little or no absorption and the light can pass through the atmosphere. This concept is referred to as "atmospheric windows" and can be thought of as frequencies for which the atmosphere is optically thin or transparent. These windows are used to peer into the atmosphere using remote sensing technology (Emery & Camps, 2017).

2.2.3.2 Terrestrial radiation

Just over half of the solar energy reaches Earth's surface directly or indirectly (scattered) and is absorbed by land and sea (see Fig. 2.7), and most of this energy is re-radiated upward. As discussed in Section 2.2.2.3, solar energy reaching the Earth is shortwave visible light. By contrast, the outgoing energy from Earth's surface, referred to as ***terrestrial radiation***, is longwave infrared (heat) radiation because Earth has a much lower surface temperature than the Sun.

The wavelengths of most of the terrestrial radiation are between 1 and 30 μm, corresponding to the infrared range of the electromagnetic spectrum. From Fig. 2.10, we see that in the 1–30 μm wavelength range, the atmosphere is an efficient absorber of radiation, mainly by water vapor and carbon dioxide. Particularly, water vapor absorbs most of the radiation in this range compared to other gases.

Once the gases in the atmosphere absorb radiation, they become warmer and reradiate this energy. Some of this reradiated energy travels toward the surface of the Earth and is again absorbed by the surface, while the rest goes skyward, where other gas molecules may absorb it. This cycle and the continuing Sun's incoming solar radiation keep the average temperature of the Earth warmer than it would otherwise, making it habitable for humans and other species. This phenomenon is referred to as the ***greenhouse effect*** (see Section 2.9 and Chapter 11 for further details).

Clouds are absorbers of terrestrial radiation, and they play essential roles in modulating the temperature of the Earth's atmosphere, including keeping the surface warmer at night. For example, the Earth's surface is warmer on cloudy or humid nights than it would be on a clear and dry night, as cloud cover absorbs most of the terrestrial radiation and reradiates it back to the surface. Clouds can absorb up to 25% of the incoming solar radiation and reflect 70%–80% depending on droplet size and concentration, as well as the thickness of the clouds.

Although terrestrial radiation is absorbed well by the atmosphere, it is absorbed very little between wavelengths of 8–11 μm. In the case of carbon dioxide and water vapor, which do not absorb in these wavelengths, this band is an ***atmospheric window***, as radiation with wavelengths in the 8–11 μm range is not absorbed by the atmosphere and escapes to outer space. However, the amount of this "lost" terrestrial radiation through this window is comparatively small.

2.2.3.3 Global heat budget

Although seasonal cold spells and heat waves occur on Earth on relatively short time scales, the average annual temperature remains relatively constant. The longer-term effects of climate change are the subject of Chapter 11. Conceptually, we can consider ***the heat budget*** of our planet to understand the annual balance of incoming solar radiation and outgoing terrestrial radiation.

Fig. 2.11 illustrates Earth's heat budget in terms of energy flux by using 340 W m^{-2} as an average solar constant (similar value to that calculated with Eq. 2.3). The values indicated represent the conditions averaged over all latitudes, seasons, and for all clear and cloudy days. Further information can be found in Trenberth et al. (2009), Wild et al. (2015), and Forster et al. (2021). With new data from observations and climate models emerging, the distribution of Earth's energy budget values according to different pathways has been updated, and hence Fig. 2.11 shows data from Trenberth et al. (2019) and the more recent estimates from Wild et al. (2015), which have been used by Forster et al. (2021).

As Earth's planetary albedo is about 30%, hence nearly this amount (~100 W m^{-2}) of the incoming solar radiation (340 W m^{-2}) is directly reflected to space by the Earth. For the remaining 70% (240 W m^{-2}) of the radiation, the atmosphere absorbs ~23% (80 W m^{-2}), while Earth's land-sea surface absorbs about 47% (160 W m^{-2}).

After absorption of a fraction of incoming shortwave solar radiation by the Earth, much of this fraction is lost in the form of longwave radiation. A small amount, about 13% (21 W m^{-2}) of the absorbed radiation by Earth's land-sea surfaces, is transferred to the atmosphere mostly by convection but also partly by conduction. On the global scale, atmospheric convection promotes worldwide air circulation, which helps the redistribution of heat between the poles and equatorial regions.

2.2.3.4 Surface energy exchange

In atmospheric science, ***sensible heat flux*** refers to the transported energy from the surface of the Earth to the atmosphere by conduction, convection, and then advection described in the previous sections. Note that the term sensible heat refers to the exchange of heat energy between a surface and the atmosphere in this case and hence results in a change of temperature between the two. When this energy exchange takes place, the state of the surface or body does not change, and other variables such as volume and temperature also do not change. By examining Fig. 2.11, the reader will notice that ~51% (82 W m^{-2}) of absorbed radiation by Earth's land-sea surface (160 W m^{-2}) is transported to the atmosphere by evaporation and transpiration of water at the surface and subsequent condensation of water vapor in the

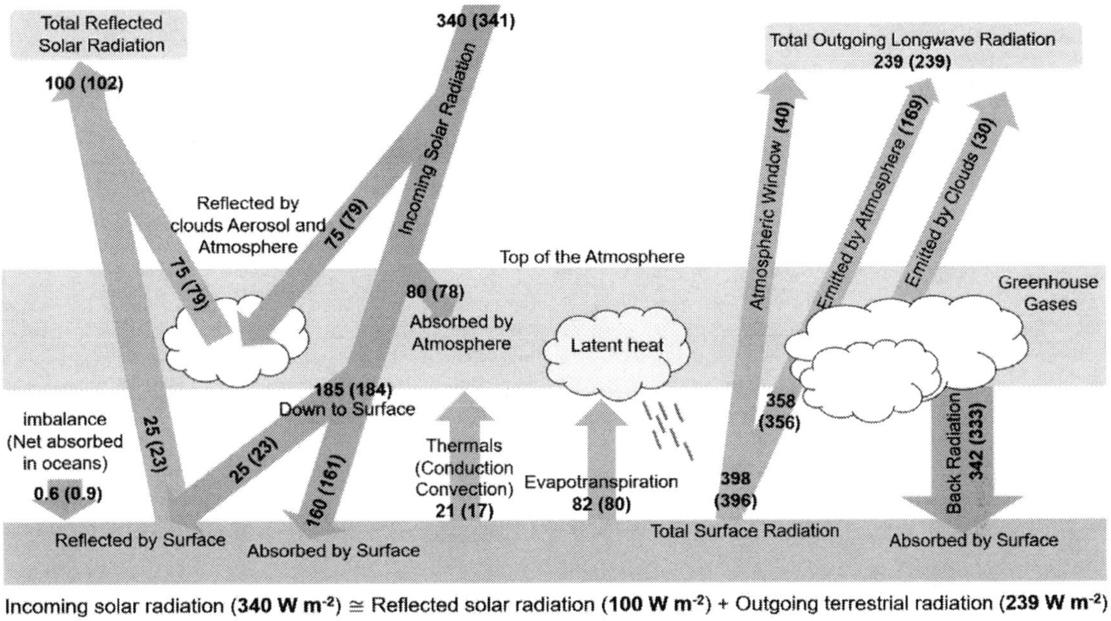

FIGURE 2.11 The Earth's annual global mean energy budget (in W m^{-2}) for the present-day climate conditions at the beginning of the twenty-first century (Wild et al., 2015) and, for comparison, values given in parenthesis for the period from March 2000 to May 2004 (Trenberth et al., 2009). Incoming radiation is depicted in saffron, while outgoing radiation is represented in blue. The residual heat flux (top of the atmosphere energy imbalance) is estimated as 0.6 ($\sim$0.9) W m^{-2} and is mostly absorbed by the oceans. *Data from Wild, M., Folini, D., Hakuba, M. Z., Schär, C., Seneviratne, S. L., Kato, S., Rutan, D., Ammann, C., Wood, E. F., & König-Langlo, G. (2015). The energy balance over land and oceans: An assessment based on direct observations and CMIP5 climate models.* Climate Dynamics, *44(11–12), 3393–3429. https://doi.org/10.1007/s00382-014-2430-z; Trenberth, K. E., Fasullo, J. T., & Kiehl, J. (2009). Earth's global energy budget.* American Meteorological Society, *90(3), 311–324.*

troposphere (lowest 8–18 km of atmosphere; see Section 2.3.1 for details). The name given to these heat fluxes is ***latent heat flux*** (literally meaning hidden heat).

The right-hand side of Fig. 2.11 shows the longwave radiation exchange between the Earth's surface and the atmosphere. Notice that on average, Earth's surface loses 398 W m^{-2} of radiation while it receives about 340 W m^{-2}. This exchange between Earth's radiation and atmospheric radiation makes a net loss of about 17% from Earth's surface. Notice $\sim$10% (40 W m^{-2}) (on the right side of Fig. 2.11) passes directly through the atmosphere as a consequence of atmospheric windows and represents the fraction of net energy loss to space with wavelength in the 8–11 μm range (see Section 2.2.3.2). The amount of longwave radiation emitted into space by the atmosphere totals to 50% (sum of 169 W m^{-2} and 30 W m^{-2}, making 199 W m^{-2}).

An overall radiation balance between incoming and outgoing solar radiation is achieved because the sum of 100 W m^{-2} (reflected and backscattered radiation lost to space), 40 W m^{-2} (lost longwave radiation from the atmospheric window), and 199 W m^{-2} (emitted longwave radiation by the atmosphere itself) makes nearly 340 W m^{-2}, which is equal to the energy the Earth receives as incoming solar radiation.

2.2.4 Laws of radiation

Four fundamental laws of radiation essentially govern how the Sun's energy interacts with Earth's surface and atmosphere. These radiation laws describe the amount and type of energy emitted by an object and are Planck's Law, Stefan–Boltzmann law, Kirchhoff's law, and Wien's displacement law, which are briefly described below.

2.2.4.1 Planck's law

Put simply, Planck's law states that ***all objects emit radiation all the time and at all wavelengths, whatever their temperature***. So, energy is not only emitted by hot objects like the Sun but also by the Earth, including the colder polar regions, and all other objects, including us. The distribution of the emitted radiation from different objects will be different according to the predictions of Planck's law discussed below.

Planck's law gives a distribution that peaks with wavelength. As the temperature increases, the peak shifts to shorter wavelengths and the area under the curve increases, as shown in Fig. 2.12. Blackbodies absorb all radiation and emit the maximum possible intensity of radiant energy at a given wavelength and temperature. Eqs. (2.4a and b) is the mathematical expression for ***Planck's law*** at a given wavelength (λ),

$$B_{\lambda,T} = \frac{2hc^2}{\lambda^5\left[\exp\left(\frac{hc}{\lambda k_B T}\right) - 1\right]} \tag{2.4a}$$

where $B_{\lambda,T}$ is spectral radiance as a function of λ and T, k_B is the Boltzmann's constant (1.38×10^{-23} J K^{-1} or W s K^{-1}), h is the Planck constant (6.626×10^{-34} J s^{-1}), c is the speed of light in a medium such as air or vacuum (m s^{-1}), and T is the absolute temperature of the body in units of Kelvin (K). In the units of $B_{\lambda,T}$, W m^{-3} sr^{-1}, radiance is reflected as the energy emitted per unit area per unit time per unit wavelength per incremental solid angle (units of steradians, given as "sr"). Note, steradian (or solid radian) is a unit used in 3D geometry and represents a solid angle subtended at a point by an area projected on the surface of a sphere. As $4\pi r^2$ is the surface area of a sphere, it follows that the maximum steradian that can be subtended at any point is 4π.

Given that $\lambda = c/\nu$, Eq. (2.4a) can also be rewritten as;

$$B_{\nu,T} = \frac{8\pi h\nu^3}{c^3\left[\exp\left(\frac{h\nu}{k_B T}\right) - 1\right]} \tag{2.4b}$$

where $B_{\nu,T}$ is the spectral radiance at frequency ν and temperature T.

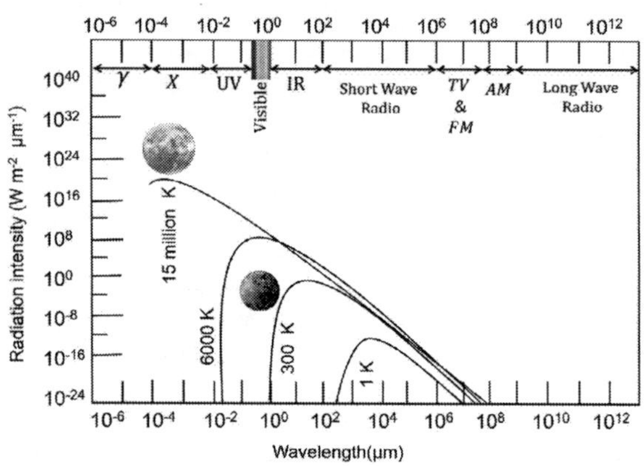

FIGURE 2.12 Schematic to illustrate the spectral irradiance (using Planck's and Stefan Boltzmann laws) as a function of wavelength and temperature resulting from emission by a blackbody. *Adapted from Lazaridis, M. (2011). First principles of meteorology and air pollution. Environmental Pollution, Springer.*

2.2.4.2 Stefan–Boltzmann law

Stefan-Boltzmann law states simply that ***hotter objects radiate more total energy per unit area than colder objects***. More specifically, the law states that the total energy radiated per unit surface area of a black body per unit time from all wavelengths is proportional to the fourth power of its thermodynamic temperature. Mathematically, the Stefan–Boltzmann law is expressed by integrating the spectral irradiance over all wavelengths given by Eq. (2.5),

$$E = \pi\int_0^\infty B_{\lambda,T} d\lambda = \sigma_B T^4 \tag{2.5}$$

where E is the rate of the radiation emitted by a body and T is the blackbody's thermodynamic temperature (K). σ_B is a derived constant and is given as,

$$\sigma_B = \frac{2\pi^5 {k_B}^4}{15c^2h^3} \tag{2.6}$$

Here, σ_B is Stefan–Boltzmann constant (5.6704×10^{-8} W m^{-2} K^{-4}), k_B is the Boltzmann constant (1.38×10^{-23} J K^{-1} or W s K^{-1}), c is the light speed in a vacuum (m s^{-1}), and h is Planck's constant (6.626×10^{-34} J s^{-1}).

The Planck distribution function for the Sun at a surface temperature of 15 million K is shown in Fig. 2.12. As shown in Fig. 2.12, much more energy is emitted from the hotter bodies than cooler bodies (see the radiation intensity axis). Note the rapid rise of solar radiation intensity as wavelengths decrease and a peak value at the shorter wavelengths, which are primarily gamma rays and X-rays.

2.2.4.3 Kirchhoff's law

Simply expressed, Kirchhoff's law states that **if an object is a good absorber of radiation, it is also a good emitter at the same wavelengths**. We can also say that **if a body emits energy efficiently at specific wavelengths, then it will also absorb well at those wavelengths**.

As a body absorbs energy, its temperature increases, and the emitted energy by the body also increases. Eventually, a thermal equilibrium is reached over time where the body's temperature no longer rises. In this thermodynamic equilibrium, the absorbed radiant energy (absorbed radiant power $\acute{E}$) must therefore be equal to the emitted radiant energy (emitted radiant power E) within a specific period:

$$\acute{E} = \alpha . E \tag{2.7}$$

where α is the absorption coefficient. When a body has similar characteristics to those of a blackbody with an emissive coefficient (or emissivity) ε:

$$\varepsilon = \acute{E}/E \tag{2.8}$$

and then $\alpha = E$ at the same temperature, which is also the mathematical expression of Kirchhoff's law. In other words, when a body is at thermal equilibrium, its emissivity is equal to its absorption coefficient. For a black body the value of ϵ is unity and for real objects the value is between 0 and 1.

2.2.4.4 Wien's displacement law

Wien's law states that ***when the temperature of a radiating black body increases, the wavelength of maximum radiated energy gets shorter or in order words, is inversely proportional to its absolute temperature***. Notice in Fig. 2.12 that the peak of hotter objects moves toward shorter wavelengths, which is consistent with Planck's radiation law, where the intensity of black-body radiation is a function of radiation wavelength at any given temperature.

Wien's displacement law can be expressed mathematically through Eq. (2.9), which shows the inverse relationship between the maximum wavelength (λ_m) of the emitted radiation and the absolute temperature of a black body:

$$\lambda_{\max} = \frac{b}{T} \tag{2.9}$$

where b is Wien's displacement constant (2897.8 μm K), and T is the temperature of the radiating body.

Worked example 2.1

Estimate the peak wavelength of radiation emitted from the Sun by assuming that the temperature of the Sun's outer layer is 5778 K.

Solution:

To calculate peak wavelength of Sun's radiation we use Wien's displacement law given in Eq. (2.9).

$$\lambda_{\max} = \frac{C}{T} = \frac{2897.8\ \mu\text{m K}}{5778\ \text{K}} = 0.5015\ \mu\text{m}$$

So, for the Sun, with a temperature of approximately 5778K, the peak wavelength of its emitted radiation is estimated to be around 0.5015 μm. This wavelength corresponds to the greenish-yellow part of the electromagnetic spectrum given in Fig. 2.12.

2.3 Thermal stratification and vertical structure

Nowadays, balloon measurements (***radiosondes***) are equipped with high-technology devices. At specific worldwide locations, they have produced two measurements per day since the late 1920s. For an example, see a database of the measurements from the Wyoming University web page (Wyoming University, 2022). With technological progress, rockets and airplanes are being used to collect atmospheric data at different layers of the atmosphere. The best-known platforms by the public are weather radar and satellites, which are often mentioned with colorful images of clouds, for example, during television weather forecasts. In this section, we introduce properties of the layers of the atmosphere based on our understanding of the data derived from these types of instruments and platforms.

2.3.1 Layers of the atmosphere

2.3.1.1 Temperature stratification of the atmosphere

As the atmosphere is variable in its composition and structure, we can examine its properties by adopting the concept of the term "***standard atmosphere***," which is often used in atmospheric sciences to standardize values and ease comparison and calculations. Standard atmosphere represents the horizontally and temporally averaged structure of the atmosphere as a function of height. For example, the vertical temperature distribution is shown in Fig. 2.13 for a standard atmosphere and is typically used for classifying the different atmospheric layers according to thermal stratification.

The bottom layer of the atmosphere, where we live and have near immediate impact, is called the ***troposphere***. The temperature profile of the troposphere is determined by many factors, including vertical mixing processes, incoming solar radiation, amount of moisture (that is, humidity of the atmosphere), and altitude. Due to warm surface temperatures and strong thermal mixing, the troposphere extends up to about 16–20 km above the Equator while it is about 7–9 km above the polar regions. The troposphere accounts for around 80% of the mass of the atmosphere and almost all-weather phenomena, such as clouds, precipitation, and storms, which occur in the troposphere because of strong mixing and transport of heat and moisture.

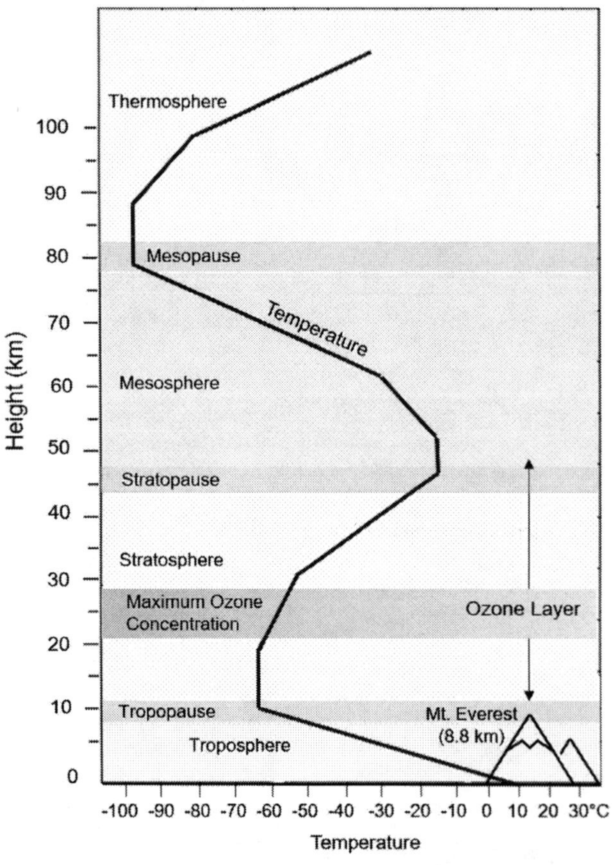

FIGURE 2.13 The vertical distribution of atmospheric layers according to temperature in a standard atmosphere.

In this layer, the temperature decreases as a function of altitude, to a minimum of about −50°C at midlatitudes, to about −80°C above the Equator, and down to about −40 to − 50°C above the polar regions. We use the term ***lapse rate*** to represent the rate at which temperature changes with altitude. Overall, the temperature in the troposphere decreases with increasing altitude. This decrease in temperature is termed the ***environmental (or normal) lapse rate***. The average value of the environmental lapse rate is about 6.5°C km^{-1}, which can vary considerably with diurnal and seasonal changes in the weather and from place to place. Hence, the environmental lapse rate is specified for a given place and time.

While there is a general decrease in air temperature with height within the troposphere, there are embedded thin sublayers of ***temperature inversions*** in which temperature increases with altitude instead of decreasing. When temperature inversion exists, warmer air from the surface cannot rise, and therefore these layers can strongly inhibit vertical mixing of air pollution and hence are of much importance for air quality. Further details on atmospheric stability are given in Section 2.3.3.

The boundary between the troposphere and stratosphere is known as the ***tropopause***. While large-scale turbulence and mixing drive atmospheric properties below the tropopause, they are not maintained above it. There are three "pauses" in the atmosphere: tropopause, stratopause, and mesopause. Temperature variation in these "pauses" is only slight, as shown in Fig. 2.13.

The ***stratosphere*** is above the tropopause and extends to heights up to 50 km. As shown in Fig. 2.13, the temperature just above the tropopause remains nearly constant to a height of approximately 20 km. Incidental solar radiation is absorbed by ozone in this layer and causes an increase in the temperature of the atmosphere as a function of altitude up to about 50 km. This ozone layer concentrated in the stratosphere serves as a protective layer from the harmful effects of the incident UV radiation. Absorption of the ultraviolet solar radiation (UV) by ozone (O_3) and by molecular oxygen (O_2) causes higher observed temperatures. Heating by ozone molecules is responsible for the temperature maximum at around 50 km that defines the stratopause. ***Stratopause*** is the boundary between the stratosphere and mesosphere. Stratosphere behaves as a strong thermal inversion, and hence vertical mixing is strongly inhibited in this layer.

In the middle atmosphere, the layer above the stratosphere is the ***mesosphere***, which extends up to 80–90 km from the Earth's surface. Temperatures decrease throughout the layer as a function of altitude up to −90°C or −100°C (the coldest atmospheric temperature) until the ***mesopause***. The decrease in temperature is caused by reducing concentrations of gases such as ozone and oxygen as we go to higher altitudes.

Above the mesopause, the temperature rises and is typically between 200°C and 500°C in the ***thermosphere***. At higher altitudes, temperatures rise to very high values of more than 1000°C in the upper thermosphere. The air mass in the thermosphere is extremely rare; however, the temperature increases with altitude because of the absorption of high-energy shortwave solar radiation and the dissociation of diatomic nitrogen and oxygen molecules along with ionization of atoms. These processes are referred to as ***photodissociation*** and ***photoionization***, respectively. Although the small number of gas molecules in this layer have very high temperatures, there would be minimal heat transfer because of their sparseness, and therefore, an astronaut would not feel hot in the thermosphere.

2.3.1.2 Compositional stratification of the atmosphere

The atmospheric layers can also be classified with respect to other properties, such as atmospheric composition. The main constituents of the atmosphere are molecular nitrogen, which makes about 78%, and molecular oxygen, which makes around 21% of the composition. Together with argon (around 1% of the composition), these gases make up most of the atmosphere's compositional mass of around 99.9%. Atmospheric constituents that are important for the radiation balance of the atmosphere include water vapor, carbon dioxide, ozone, methane, nitrous oxide, and other species that make up the remaining atmosphere's mass (see Section 2.6 for further information).

There is also stratification of the atmosphere in terms of its major passive constituents, where it is often divided into two layers called the ***homosphere*** and the ***heterosphere***. The composition of the atmosphere is generally uniform from the surface of the Earth to an altitude of about 80 to 100 km, a layer referred to as the ***homosphere***, because of large-scale vertical movement and turbulence (irregular motion) of air. The homosphere layer includes the thermally stratified layers discussed earlier of the troposphere, stratosphere, mesosphere, and the lower thermosphere. Above this altitude, the atmosphere is heterogeneous, and gases are stratified according to their molecular weight. In this ***heterosphere*** layer, heavier atoms of oxygen and nitrogen are more abundant in the lower part of the layer, followed by helium (He), with hydrogen being mostly present in the uppermost part of the layer. Hence, in the heterosphere, the chemical composition of the atmosphere depends on the altitude, with the stratification being driven mainly by molecular diffusion.

The ***ionosphere*** is in the altitude range between about 80 and 1000 km. This layer is composed of electrically charged ionized molecules of atoms and molecules such as nitrogen and oxygen. Ionization of molecules by highly energetic ultraviolet radiation and x-rays create a blanket of electrons. There can be recombination and neutralisation of the ions leading to a plasma of neutral atoms and molecules and ions. This layer contains the auroral zone which leads to the colorful auroral borealis (northern lights) and aurora australis (southern lights). The ***exosphere*** extends from above about 500 km to the outermost parts of the atmosphere merging into outer space. This layer consists of hydrogen and helium and other molecules which experience no or negligible collisions and can escape into space.

2.3.2 Spatial variation of temperature

In the atmosphere, the temperature changes vertically and spatially. Spatial distribution of global average 2 m temperature is shown in Fig. 2.14 for January over the period 2000–2023 extracted from monthly averaged observation assimilated model data from Hersbach et al. (2023). Temperatures over the South Pole (−90 degree latitude) are warmer than those over the North Pole (90 degree latitude) since the map is for January. In January, the average temperature in the Northern Hemisphere is lower than that in the Southern Hemisphere. This is mainly because the Northern Hemisphere has larger land masses, which results in faster heat loss. Conversely, the Southern Hemisphere's average temperature is relatively higher during this time, as its oceans retain heat more effectively due to their higher heat capacity compared to land masses. However, this situation is reversed in July.

2.3.3 Stability of the atmosphere

Most air pollutants are emitted close to the ground and disperse vertically and horizontally in the atmosphere. Although the amount of air into which pollutants are initially mixed is mainly governed by wind speed, atmospheric stability determines the vertical motions that mix the ground-level pollutants with cleaner air above. In stable atmospheres, air resists vertical displacement, so the pollutants cannot move vertically. The categorization of the stability of the atmosphere helps to understand how mixing in the air affects air pollution and hence the quality of the air that we breathe. The three main types of atmospheric stability categories are ***stable***, ***unstable***, and ***neutral***. Vertical movement of air is suppressed in a stable atmosphere, enhanced in an unstable atmosphere, and neither suppressed nor enhanced in a neutral atmosphere.

The atmosphere's stability is largely determined by what happens to a gas as it rises and how it is affected by the surrounding atmosphere. The ***air parcel*** term is a useful theoretical concept in atmospheric sciences that is used to

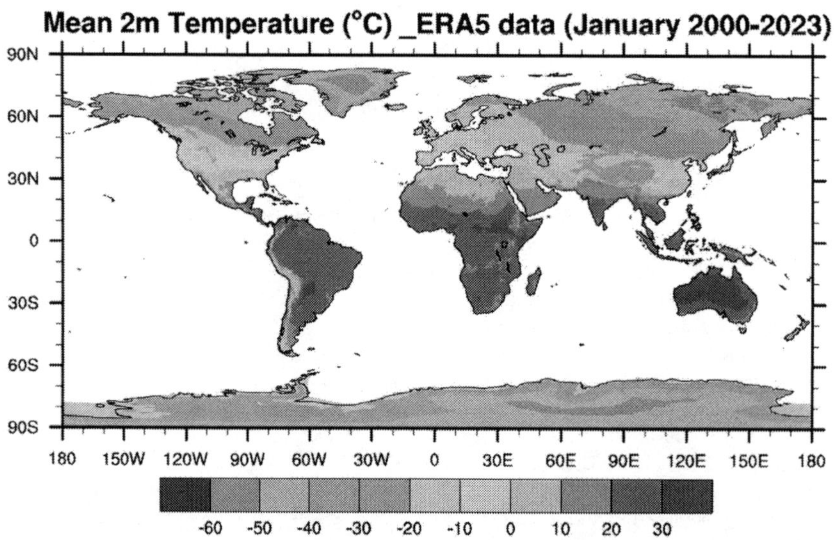

FIGURE 2.14 Global average temperature distribution for January over 2000–2023 (mapped with the observation assimilated model). *Data from Hersbach, H., Bell, B., Berrisford, P., Biavati, G., Horányi, A., Muñoz Sabater, J., Nicolas, J., Peubey, C., Radu, R., Rozum, I., Schepers, D., Simmons, A., Soci, C., Dee, D., & Thépaut, J-N. (2023): ERA5 monthly averaged data on single levels from 1940 to present.* Copernicus Climate Change Service (C3S) Climate Data Store (CDS), *https://doi.org/10.24381/cds.f17050d7 (accessed on 27-07-2023).*

define a moving air volume with uniform temperature that can expand and/or compress but not mix with the surrounding air. When we refer to it as an ***adiabatic air parcel***, we mean that it does not exchange heat with its surroundings. For the purposes of air quality science, the vertical mixing in the atmosphere can be approximated as involving many rising and falling air parcels. A rising air parcel expands and cools, and a sinking air parcel shrinks and warms.

Atmospheric stability is determined through several factors, but the key is the variation of temperature with altitude. The air parcel concept facilitates the description of atmospheric stability by comparing the temperature of a rising air parcel with that of the surrounding air at the same level. A useful parameter to describe atmospheric stability in terms of vertical temperature variation is ***lapse rate***, which is defined as the rate of change of air temperature with height.

The ***environmental lapse rate (ELR)*** describes the temperature variation of the air that surrounds air parcels. As described in Section 2.3.1, temperature generally decreases with height in the troposphere. Typically, ELR is about $-6.5°C\ km^{-1}$. On the other hand, if the temperature of an atmospheric layer increases with altitude, it is called a ***temperature inversion layer***. An inversion occurs when warm air lies above cool air, and this condition is highly stable. These conditions severely inhibit air pollution from mixing and dispersing and hence lead to high air pollutant concentrations. Environmental lapse rate (Γ_e) is calculated from:

$$\Gamma_e = -\frac{dT}{dz} = \frac{T_2 - T_1}{Z_2 - Z_1} \tag{2.10}$$

where T_2 is upper layer temperature, T_1 is lower layer temperature, Z_2 is the height at which the air temperature is T_2, and Z_1 is the height at which the air temperature is T_1.

The temperature of an air parcel can be defined in two ways according to its moisture content. ***Dry adiabatic lapse rate*** (DALR) is the rate at which the temperature of dry air (e.g., a dry air parcel) will change as it moves up and down without exchanging heat with the surrounding air. The average dry adiabatic lapse rate equals $-9.8°C\ km^{-1}$ for an ascending air parcel through the Earth's atmosphere. ***Wet adiabatic lapse rate***, or moist adiabatic lapse rate, is used when the water content is available in the adiabatic air parcel. Air parcel continues to rise and cool at the DALR rate till it reaches the dew point of the air. At this point, the water vapor starts to condense and releases ***latent heat***, which slows the rate of cooling of the air. Latent heat (meaning hidden heat) refers to the energy released or absorbed by a body when it undergoes a change of state, such as gas to liquid (or the other way around) without a change in its temperature and pressure (e.g., water vapor condensing into liquid water or vice versa).

Unlike DALR, the wet adiabatic lapse rate depends on the temperature and pressure of the atmosphere and ranges approximately between 3.6°C and $5.5°C\ km^{-1}$. Details on calculating dry and wet adiabatic lapse rates are given in Sections 2.5.9 and 2.5.10.

Fig. 2.15 illustrates the three main stability conditions by comparing temperature changes with height for both the air parcel (DALR) and the surrounding air (ELR). When the air parcel rises from a to b, the temperature change of the air parcel follows the DALR line. In Fig. 2.15A, the ELR line is steeper than the DALR (***superadiabatic conditions***). This means that the temperature of the surrounding atmosphere drops faster than that of the air parcel moving upwards

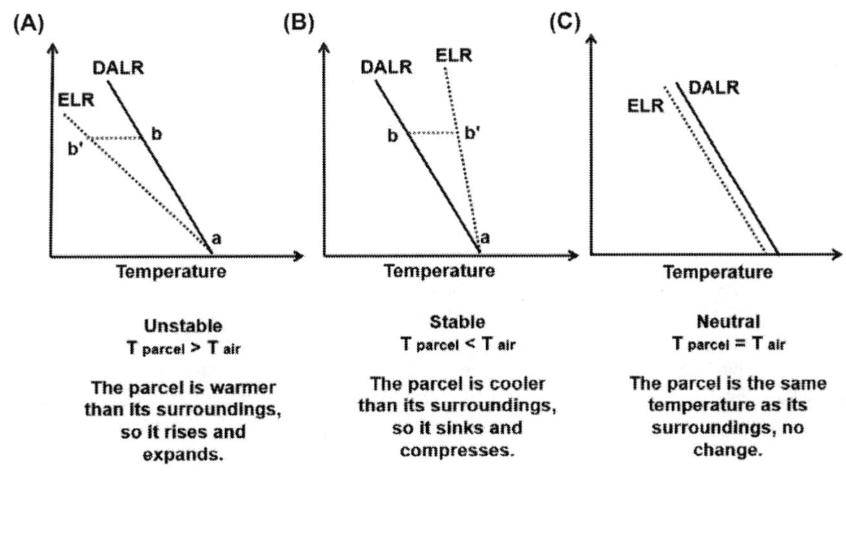

FIGURE 2.15 Comparing the temperature profiles of an ideal air parcel to that of the surrounding environment in unstable, stable and neutral atmospheres.

at DALR. Therefore, the parcel's temperature at point b is warmer than the environment (point b'). Thus, the resulting acceleration of the air parcel is upwards, and it will rise. That is an example of an unstable atmosphere that enhances vertical motion and mixing of the air, which usually occur on sunny days and when wind speed is low.

In the case of a stable atmosphere illustrated in Fig. 2.15B, the ELR line is not as steep as the DALR (***subadiabatic***). When we force the air parcel to rise from a to b, it cools less than the DALR. Therefore, the ideal air parcel is cooler than the surrounding atmosphere at point b'. In this case, it will sink back to its original level. This type of atmosphere resists vertical motion and mixing and hence is usually associated with higher air pollution concentrations. These conditions often occur during nighttime and early winter mornings before sunrise when there is no or light wind.

The ELR in the neutral atmosphere (see Fig. 2.15C) is the same as DALR (adiabatic). Therefore, the vertical movement of the air parcel is neither enhanced nor suppressed. These neutral stability conditions usually prevail when the wind speed is higher and strong heating or cooling of the Earth's surface is inhibited, such as when the sky is cloudy.

Worked example 2.2

Over a mountain, what will be the temperature of a rising air parcel from sea level to 1500 m if the lapse rate is 6.5°C km^{-1} and the temperature at sea level is 25°C?

Solution:

Since the lapse rate is 6.5°C km^{-1} we can express lapse rate using Eq. (2.10).

$$\Gamma_e = -\frac{dT}{dz} = -\frac{6.5°C}{1\ km}$$

Which means that when we move to an altitude of 1000 m upwards, the temperature will drop by 6.5°C. Hence, the temperature at 1500 m height can be written as follows by considering see level as 0 m.

$$\Gamma_e = -\frac{6.5°C}{1000\ m} = \frac{T_2 - 25}{1500\ m}$$

So, T_2 will be 15.25°C.

2.3.4 Atmospheric boundary layer

The lower troposphere exhibits distinctive behavior as it is directly influenced by the Earth's surface. The part of the troposphere near the Earth's surface is called the boundary layer (BL) and is also known as the atmospheric boundary layer (ABL) or planetary boundary layer (PBL). The layer of the atmosphere above the PBL where the influence of the Earth's surface is negligible is referred to as the ***free atmosphere***. See Fig. 2.16 for an illustration of the boundary layer and its diurnal evolution.

The boundary layer height changes mostly according to surface forcings (a term often used to describe the perturbations caused by the Earth's surface), including heat transfer, airflow modification due to terrain properties, frictional drag, emitted emissions from the surface, evaporation, and transpiration. Frictional drag is one of the forcings that affect the air flowing over the ground and causes wind shears that frequently become turbulent. Obstacles like buildings and trees also change the flow direction and cause deflection of turbulent winds around obstacles. Convection, explained previously, is another essential surface forcing due to heat exchange leading to rising warmer air, which in turn causes swirling of the air flow (called ***eddies***). These surface forcings cause irregular motions of the air, which is termed ***turbulence***.

In the lowest layer of the atmospheric boundary layer, the ***surface layer*** or ***constant flux layer***, the change of turbulent fluxes and stress is less than about 10% of their magnitude (see Fig. 2.16). The lowest few centimeters of air are dominated by molecular transport compared to turbulent transport, the so-called ***microlayer*** or ***interfacial layer***.

Three major components of this boundary layer are the stable boundary layer (usually at nighttime, i.e., nocturnal and/or early mornings before sunrise on cold days), residual layer, and mixed layer. On a clear, sunny day with little or no clouds, soon after sunrise, solar heating of the ground leads warmer air to rise (thermals or convective flows, see Section 2.2.2.2) from the ground. These thermals cause a turbulent mixing layer growing in depth called a ***mixed layer*** or ***convective mixed layer***. The maximum depth of the mixing layer is reached by late afternoon. Due to daytime turbulent mixing, air pollutants emitted near the surface are mixed and dispersed, reducing ambient concentrations of these species.

There is intense mixing in this layer because of air exchange between the ground and upper levels. However, this mixing is not extended throughout the whole atmosphere because of the entrainment zone that can act as a cap on the top of the mixing layer, which restricts the dispersion of air pollutants emitted in this layer. By the end of the day, this layer is well mixed.

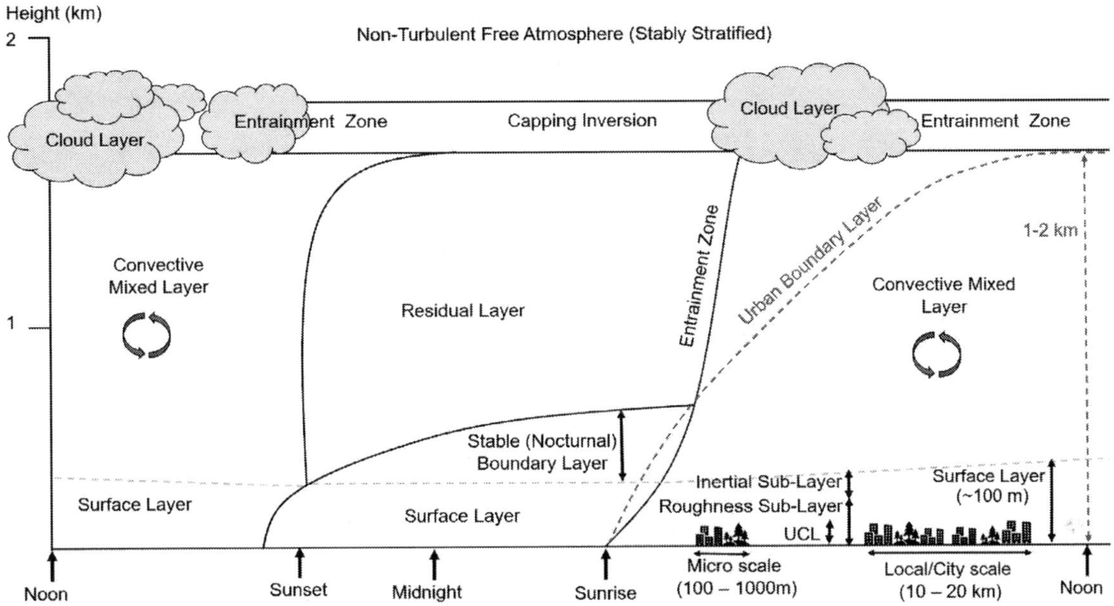

FIGURE 2.16 The evolution of atmospheric boundary layer height through a diurnal cycle and daytime evolution of urban boundary layer (Stull, 1988; Barlow, 2014). The red dashed line shows the top of the urban boundary layer. The urban canopy layer (UCL) is usually taken to represent the mean height up to the roof tops, and the roughness sub-layer (RSL) height is typically 2–5 times the mean rooftop height (UCL height). *Adapted from Stull, R. B. (1988). An introduction to boundary layer meteorology. Kluwer Academic Publishers. p. 670; Barlow, J. F. (2014). Progress in observing and modelling the urban boundary layer.* Urban Climate, 10, *216–240.*

As sunset is about to begin, turbulence decreases due to the reduction in solar heating of the ground and the reduced thermals (in the absence of cold air advection). The resulting air, which is mixed well throughout the day and no longer mixes, is called the ***residual layer***. This layer is neutrally stratified, meaning that the turbulence's intensity is approximately equal in all directions. In these conditions, a plume emitted from a factory stack will travel in a cone-like shape since the air of the residual layer is spread at almost the same rate in all directions. Air pollutants that are present in this residual layer may undergo chemical transformations, and photochemical reactions may also occur due to incoming solar radiation before the residual layer is entrained into the new mixing layer in the early morning.

The ***nocturnal boundary layer*** is a ***stable layer*** which forms from the bottom part of the residual layer as the night progresses when the surface is cooler than the above air. When the sky is clear at night, the land surface cools, which in turn cools the adjacent air through heat transfer via conduction, convection, and radiation. The depth of this layer may grow up to a few hundred meters depending on the season and is statistically stable with weaker turbulence. At the top of this stable boundary layer, a maximum in the wind speed often develops, called a ***low-level jet*** or ***nocturnal jet***. Emitted air pollutants into the nocturnal boundary layer will spread out horizontally in thin layers since there is little vertical mixing.

From the air quality perspective, the height of the mixing layer is an essential parameter since most pollutants are emitted near the Earth's surface. In shallower mixing depths, air pollutants are trapped closer to the surface, which means we are exposed to poorer air quality. Concepts on the boundary layer are developed further in Chapter 6.

2.4 Pressure and winds

The atmosphere exerts a pressure force on all objects within it, that is, the Earth's surface, people, and animals, as it has weight due to the force of gravity. ***Air pressure*** is the weight (force) of air exerted on a surface. Atmospheric pressure is a function of a number of parameters, including the mass of the planet (which determines gravitational force), composition and vertical distribution of the gases, density variations in the air (e.g., due to temperature changes), and the Earth's rotation (leading to the Coriolis effect).

Hydrostatic air pressure is the weight of air in a column above a surface at a given altitude. However, air pressure is ***hydrostatic*** only if the air is static or has a constant vertical speed where the downward and upward forces on a slab of air are in balance of motion. When these forces are not balanced, air pressure is termed ***nonhydrostatic***. We can assume that in many cases hydrostatic pressure applies when air pressure is averaged over larger (few kilometers) horizontal areas.

"Barometer" is an instrument used for the measurement of atmospheric pressure, and hence atmospheric pressure is also referred to as **barometric pressure**. At sea level, the average atmospheric pressure is 1013.25 hectopascal (hPa), which is taken as ***standard sea-level surface pressure***. At a given location and time, sea level pressure varies from −20 to +10 mb from standard sea level pressure.

In the realm of pressure measurement, various units are commonly employed. In meteorology, the International System of Units (SI) prevails, with the Pascal (Pa) serving as a standard unit for pressure. One Pascal is equivalent to one Newton per square meter ($1\ Pa = 1\ N\ m^{-2}$). Additionally, the hectopascal (hPa) is a widely used subunit, where 1 hPa equals 100 Pa. An important benchmark for atmospheric pressure is the standard atmospheric pressure, denoted as 1 atm, which corresponds to 101,325 Pa, or 1.013×10^5 Pa, or 1013.25 hPa.

In the past, the unit "bar" was commonly utilized, with 1 bar equaling 100,000 Pa or 0.987 atmospheres (atm). Another common subunit is the millibar (mbar), which is equivalent to 1 hPa. For certain applications, millimeters of mercury (mmHg) are employed to express pressure. Notably, 1013.25 mbars are equivalent to 1013.25 hPa or 760 mmHg.

Atmospheric pressure is generally used to characterize weather systems affecting a region. Simply, if the area receives low-pressure systems, then the air moves upward, usually leading to cloudiness and wind, accompanied by precipitation. Fair and calm weather is usually experienced over the regions when high-pressure systems are prevailing.

2.4.1 Pressure variation with height

Most of the mass of the gases that make up the atmosphere is near the Earth's surface because of gravity. Roughly about half of the mass of the air is between the Earth's surface and the 500 mb pressure level, which is at about 5.6 km altitude (about two-thirds of the height of Mount Everest). Approximately 90% of the air mass is contained in the lower 16 km of the atmosphere. For example, fewer air molecules exist in the stratosphere compared to the troposphere. So, it is not surprising that at the surface of the Earth, pressure should be highest due to the highest number of molecules exerting force on the Earth's surface. Conversely, for the upper levels, a decrease in pressure is expected since there is a reduction in the number of air molecules with altitude. The vertical variation of pressure in the atmosphere is illustrated in Fig. 2.17.

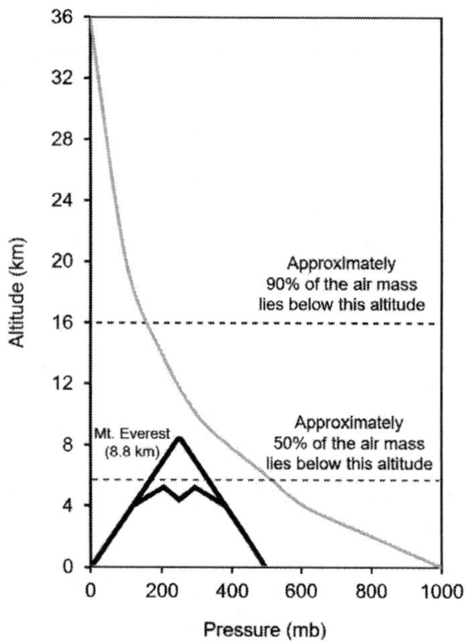

FIGURE 2.17 Pressure variation in the atmosphere with altitude (orange line). Mount Everest is shown schematically to provide a height guide.

2.4.2 Pressure variation horizontally

Although the vertical movement of air is due to temperature differences in the atmosphere, the horizontal movement of air (***wind***) is due to horizontal pressure differences. In this section, we consider vertical pressure variations and the main forces that control the flow of wind, that is, the pressure gradient, Coriolis and friction forces.

2.4.2.1 Pressure gradient force

An object moves because there is an unbalanced (unequal) force exerted on it in one direction. The pressure difference (from high to low), therefore, across a surface that results in movement is called the ***pressure gradient force***. Pressure gradient is defined as the rate of change in air pressure as a function of horizontal distance at a given time and hence has units of pascal per meter (Pa m^{-1}).

The unbalanced force that makes wind flow results because of the horizontal pressure differences. The air moves from the high-pressure region to the low-pressure region, causing the wind flow. The greater the difference between high- and low-pressure regions, the greater the wind speed. Simply put, the primary reason for observed pressure differences on the Earth's surface is the unequal heating of the Earth's land-sea surface.

It is common to see pressure maps, which are produced using barometric pressure readings taken from weather stations and then shown graphically using ***isobars***. Isobars are the lines connecting points that are of equal pressure at a given time interval and shown on a spatial map to illustrate the distribution of pressure over a region. An example is given in Fig. 2.18, which shows closely packed isobars around a low-pressure center, indicating strong pressure and high wind speeds. Widely spaced isobars indicate a weak pressure gradient and low wind speeds. The pressure gradient force is always directed at right angles to the isobars.

In Fig. 2.19, red half circles represent warm fronts, where blue triangles show cold fronts. When cold air collides with warm air, the denser and colder cold air slides underneath the warmer air, causing it to lift. This interaction initiates the expansion and cooling of the warm air, leading to the release of moisture and the formation of clouds and precipitation along the ***cold front***. The blue triangles indicate the direction of the cold air moving beneath the warm air. Cold fronts are commonly associated with local and intense thunderstorms and rainstorms, which tend to be short-lived, dissipating as the front progresses. Conversely, in the case of a ***warm front***, the warmer air pushes the colder air forward. As the warm air pushes, it also rises. During this ascent, the warm air undergoes cooling, resulting in the development of clouds and precipitation. Unlike cold fronts, warm fronts generate more widespread rainstorms rather than localized, intense ones. The red half moons on the map illustrate the direction in which warm air pushes the colder air. ***Occluded front***s occur when the cold front "catches up" with the warm front forcing the warm air aloft.

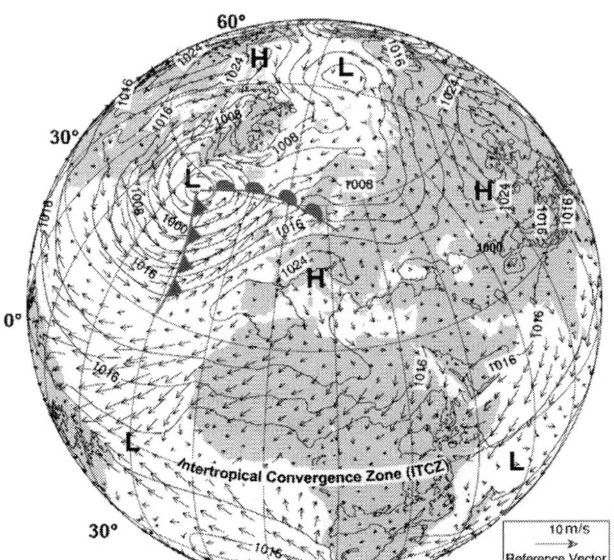

FIGURE 2.18 Pressure gradient force shown on a map derived from observations assimilated model data for February 2019 (Hersbach et al., 2023). Lines show isobars, and arrows show winds. Cold and warm fronts are shown as a blue line with triangles and a red line with semicircles, respectively. Closely spaced isobars indicate a strong pressure gradient and high wind speeds, whereas widely spaced isobars indicate a weak pressure gradient and low wind speed. Anticlockwise strong winds around the low pressure (L) and clockwise weak winds around the high pressure (H) systems in the Northern hemisphere are clear on the figure. *Data from Hersbach, H., Bell, B., Berrisford, P., Biavati, G., Horányi, A., Muñoz Sabater, J., Nicolas, J., Peubey, C., Radu, R., Rozum, I., Schepers, D., Simmons, A., Soci, C., Dee, D., & Thépaut, J-N. (2023): ERA5 monthly averaged data on single levels from 1940 to present.* Copernicus Climate Change Service (C3S) Climate Data Store (CDS), *https://doi.org/10.24381/cds.f17050d7 (accessed on 27-07-2023).*

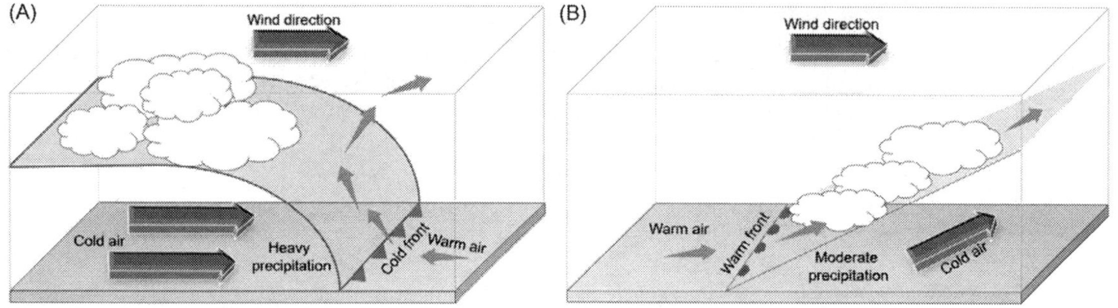

FIGURE 2.19 Formation of (A) cold front where cold air pushes warm air and cause heavy preciptation and (B) warm front where warm air pushes cold air and cause moderate preciptation.

2.4.2.2 Coriolis effect

All free-moving objects experience a deflection in their motion. This deflection occurs because of Earth's rotation and is named the ***Coriolis effect*** after the French mathematician Gaspard Gustave de Coriolis (1792–1843). The Coriolis effect is strongest at the poles and weakens at the equator, where it is almost nonexistent.

As the shape of the Earth is a near spheroid, different parts of the Earth rotate at different speeds depending on latitude. This difference in rotational speed causes the deflection of moving objects, including air masses. In everyday instances, the Coriolis effect is imperceptible, but for our purposes, it is important when air moves over large distances. As a result of the Coriolis effect, circulating air masses do not simply move from high to low pressure regions but are deflected toward the right in the Northern Hemisphere and toward the left in the Southern Hemisphere, resulting in curved trajectories. Consequently, air circulates (such as in hurricanes) anticlockwise in the Northern Hemisphere and clockwise in the Southern Hemisphere. Sometimes, the term "Coriolis force" is used, and it should be noted that this force is not real but virtual or fictitious in nature, as the "effect" it causes is due to a rotating frame of reference (that is, the Earth in this case) relative to an inertial frame.

2.4.2.3 Friction

Friction is a force that slows moving objects and always acts in the opposite direction of the motion. Friction significantly influences airflow near Earth's surface, but its effect reduces as altitude increases and eventually becomes negligible above a few kilometers. Hence, the wind flows aloft (that is, above the atmospheric boundary layer) are not affected by friction, while friction is significant for wind flows closer to the surface.

2.4.3 Formation of wind flows

As wind flows are influenced by frictional forces, the formation of these flows can be considered under two main categories: winds aloft and surface winds.

At higher altitudes, as the wind speed of the air parcel increases, deflection due to Coriolis force increases, and the winds turn away from their high-pressure to low-pressure trajectories. When the wind flow direction is parallel to the isobars (lines of constant pressure at a given height), the opposing Coriolis force balances the pressure gradient force. Under this condition, the wind will continue to flow parallel to the isobars at a constant speed as long as this balance continues. This balance is referred to as ***geostrophic balance* or *equilibrium***, and the idealized wind is called ***geostrophic wind***. In reality, this condition is rarely reached, as it occurs when there is no friction and isobars are straight. However, isobars are mostly curved, as in Fig. 2.18. Therefore, geostrophic balance does not hold, but winds still flow parallel to the isobars while following curved paths around low- and high-pressure cells. In these conditions, there is a third force, the centrifugal force, which leads to a gradient wind balance of the three forces instead of the geostrophic balance between the pressure gradient and the Coriolis forces alone. Winds that flow parallel to curved isobars with a constant speed are called ***gradient winds***.

As mentioned previously, air will move from high to low pressure regions, and the speed of flow increases when the pressure difference is high. In the case of high-pressure systems (anticyclones), which are associated with calmer winds and clearer skies, the circulation of air is clockwise in the northern hemisphere as the Coriolis effect deflects the wind toward the right as it moves away from the center. The opposite happens in the Southern Hemisphere, resulting in counterclockwise circulations for high-pressure systems. In the case of low-pressure systems (cyclones), the winds circulate in a counterclockwise direction as Coriolis effect deflects the wind towards the left as it moves away from the centre. Again, the wind circulates in the opposite direction (clockwise) in the Southern Hemisphere. In reality, the precise characteristics of wind circulation are determined by a large number of factors, including the effects of surface frictions as well as the balance between the pressure gradient, Coriolis, and centrifugal forces. More details can be found in Chapter 5, which deals with atmospheric dynamics and air pollution transport.

In Fig. 2.20A, following the explanation from above, as long as the flow due to pressure gradient force is initiated, it is deflected by the Coriolis force leading to a gradient wind flow. Since the Coriolis force deflects the flow to the right in the Northern Hemisphere, the resulting wind blows counterclockwise around the low-pressure zone. Low-pressure centres are called ***cyclones***, and their ***cyclonic*** flows rotate counterclockwise in the Northern Hemisphere and clockwise in the Southern Hemisphere.

Conversely, as shown in Fig. 2.20B, the outward-directed pressure gradient force is balanced with the inward-directed Coriolis force around a high-pressure cell. Finally, a clockwise flow occurs. High-pressure centers are called ***anticyclones***, and the flow around them is called ***anticyclonic***.

Frictional forces cause a net inflow around a cyclone called ***convergence*** and a net outflow, ***divergence***, around an anticyclone in either hemisphere, as illustrated in Fig. 2.21. The low-pressure system (cyclones) has converging surface

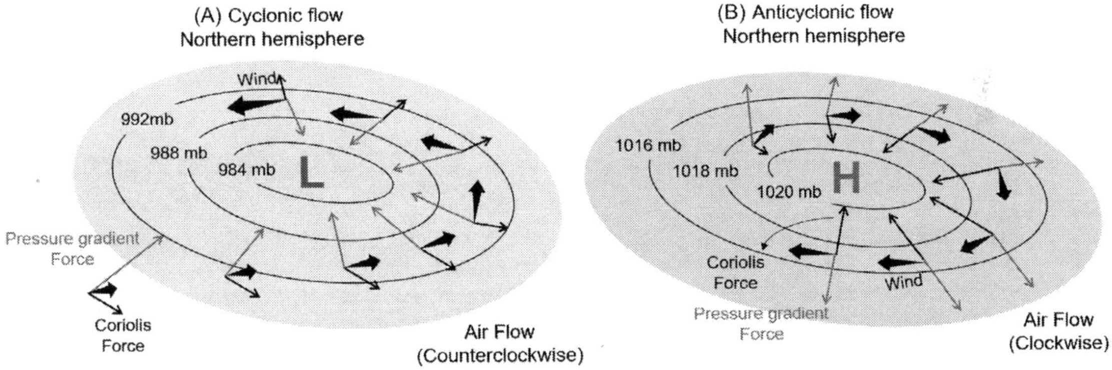

FIGURE 2.20 Idealized illustration depicting expected airflow aloft around (A) a low-pressure zone (L) and (B) a high-pressure zone (H). The circles represent isobars, with black arrows indicating winds deflected by the Coriolis force. In a cyclonic flow around L, the wind is deflected counterclockwise, while in an anticyclonic flow, it is deflected clockwise in the Northern Hemisphere.

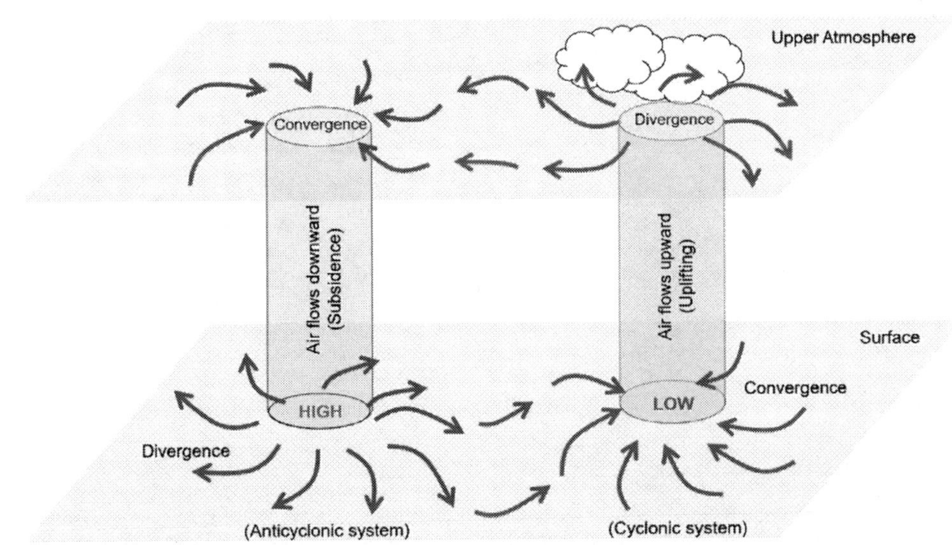

FIGURE 2.21 Relationship between converging and diverging air in cyclones and anticyclones and vertical air motion.

winds and rising air with an uplift of moisture, causing cloudy conditions. In contrast, high-pressure systems (anticyclones) have diverging surface winds and descending dryer air, leading to clear skies, calm weather in summer, and clear and cold days in winter.

From an air quality perspective, when an anticyclone settles over a region, this descending air may cause temperature inversions (see Section 2.3.3 for further information), which limits dispersion of air, and air pollutants may get trapped close to the ground layer. Conversely, a cyclone (low-pressure system) brings wet and windy conditions to a region. In such a case, the storms may wash air pollutants or transport them horizontally to other areas (see also Chapter 5 for further information on atmospheric transport).

Worked example 2.3

The synoptic chart for the North Atlantic Ocean and Europe on July 20, 2023, is provided in Fig. 2.22. Identify isobars, high- and low-pressure systems, as well as cold and warm fronts.

Solution:

On Fig. 2.22, black lines connect areas of equal atmospheric pressure, known as isobars (see Section 2.4.2.1 for definition). Isobars are crucial on such maps as they reveal the wind's direction and strength. They often form circular patterns, indicating areas of high or low pressure. Identifying an "H" or "L" with an associated number helps determine whether it's a high- or low-pressure system. The closer the isobars (black lines) are, the windier the conditions.

Cold fronts are denoted by blue semi triangles, warm fronts by red half circles, and an occluded front, where a cold front catches up with a warm front (see Section 2.4.2 for definitions), is represented by purple lines containing both triangles and semicircles.

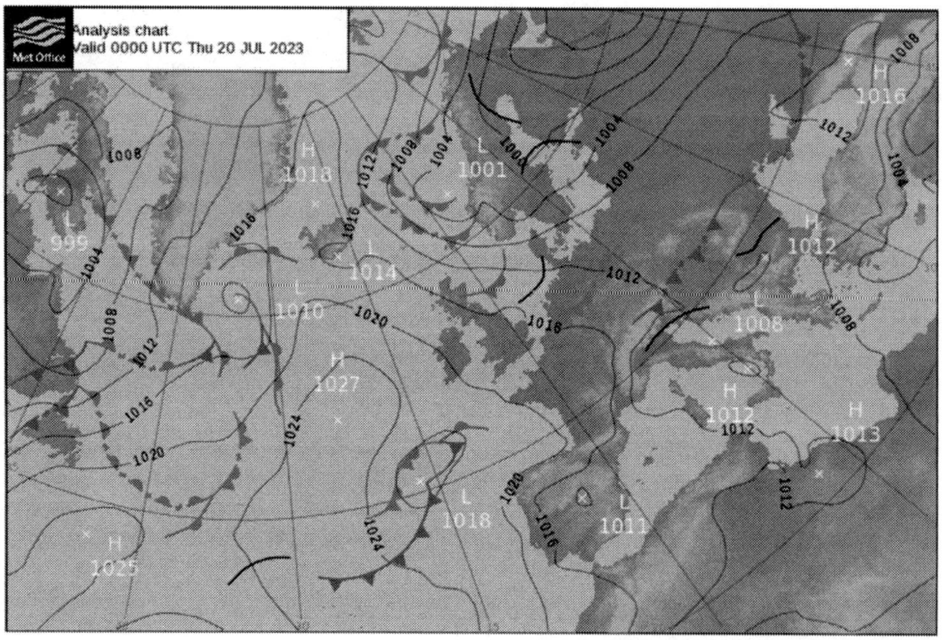

FIGURE 2.22 High (H) and low (L) pressure systems around the UK and Europe (Met Office, 2023). *Courtesy of Met Office. (2023).* UK daily temperature data, part of the Met Office Integrated Data Archive System (MIDAS). *Available at: https://www.metoffice.gov.uk/ (accessed July 20, 2023).*

2.4.4 Global circulation

Global circulation of air is fundamentally driven by uneven heating of the Earth's surface by solar radiation. Without an understanding of global circulation, we cannot understand the different wind flow patterns that occur on our planet. As we discussed in Section 2.2.3.3, the total energy that Earth receives from the Sun is balanced by the total energy radiated back to space. However, because of the tilt of the planet and its axial rotation, this radiation balance varies from location to location. For example, polar regions receive less solar energy but emit more than they receive (negative or deficit balance), while the reverse is valid for the tropics (positive or surplus balance).

Global atmospheric circulation (large-scale movement of air) along with ocean circulation (large-scale movement of ocean water) plays an essential role in driving the weather and climate of our planet. Importantly, large-scale movement of air governs transport of heat and moisture from topical to polar regions. In the 1920s, a ***three-cell circulation model*** (Bjerknes, 1921) was proposed for each hemisphere. In this section, the three-cell global circulation model is described by considering pressure zones on Earth.

According to this atmospheric model, three circulation cells operate in each hemisphere and circulate the air through the entire troposphere: Hadley, Ferrel, and Polar cells. With the effect of the Coriolis force, the circulation from three circulation cells results in global circulation with the net effect of transferring thermal energy and moisture from equatorial regions to polar regions.

Due to this uneven heating on the earth's surface, seven distinctly identifiable ***pressure zones*** form where pressure regimes are homogeneous horizontally. These pressure zones are also called ***pressure belts***. These seven pressure belts are two polar highs, two subpolar lows, two subtropical highs, and an equatorial low. Except for the equator low, all pressure belts have matching pairs in each hemisphere, as shown in Fig. 2.23.

Hadley cells extend from the equator to 30 and 40 degree latitudes in both hemispheres. In those regions, the circulation closely resembles the convective model used by Hadley. The circulation is driven by the surplus heat of the equatorial region due to incoming vertical rays of solar radiation (see Section 2.2.3) throughout the year. Intensely heated air near the equator rises and causes a low-pressure region and is referred to as ***equatorial low***. As the Hadley cell is driven by heat, it is referred to as a ***thermally direct*** cell. The rising moist air releases latent heat and forms cumulus towers near the equator that provide rainfalls that maintain the vegetation of tropical rainforests. The equatorial low is also called the ***intertropical convergence zone*** (ITCZ) due to the converging northeast and the southeast trade winds over the equatorial region. Trade winds are the winds that flow east to west (easterly) near above and below the equator (see below).

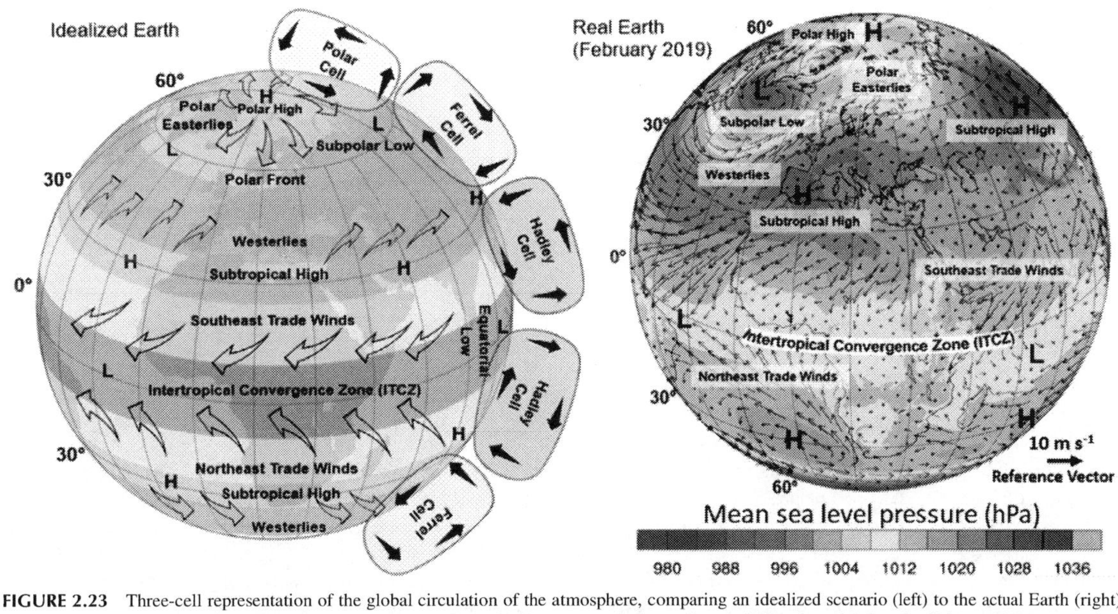

FIGURE 2.23 Three-cell representation of the global circulation of the atmosphere, comparing an idealized scenario (left) to the actual Earth (right). The real Earth figure on the right-hand side was plotted using monthly averaged observation assimilated model data for February 2019. *Data from Hersbach, H., Bell, B., Berrisford, P., Biavati, G., Horányi, A., Muñoz Sabater, J., Nicolas, J., Peubey, C., Radu, R., Rozum, I., Schepers, D., Simmons, A., Soci, C., Dee, D., & Thépaut, J-N. (2023). ERA5 monthly averaged data on single levels from 1940 to present.* Copernicus Climate Change Service (C3S) Climate Data Store (CDS), *https://doi.org/10.24381/cds.f17050d7 (accessed on 27-07-2023).*

In the band of latitudes of about 20–35 degrees (popularly called ***horse latitudes***) on both sides of the equator, ascending equatorial air currents start to descend. Descending air leads to high pressure on the surface called ***subtropical highs*** and is associated with calm and sunny weather with generally dry air flow.

Since upper-level winds flow from high-pressure to low-pressure regions, the descending winds in the subtropical region are divided into two parts. One part blows to equatorial lows, and the other blows to subpolar lows. The first part of the winds blowing from the subtropical highs to the equator is deflected due to Coriolis force and forms prevailing easterly ***trade winds***. The trade winds and associated ocean currents enabled early sailing ships to conduct commercial trade between European and African ports (hence the name trade winds or simply trades). ITCZ is formed within the Hadley cells, where trade winds blow toward the equator and ascend and sink as they reach higher latitudes. This sinking air in the high-pressure regions is mostly over the world's hot deserts, such as the Sahara Desert in North Africa. Trade winds are important for air quality as they help to transport dust to different regions of the world, such as Latin America and the Caribbean region. The second part of the air that blows from subtropical highs to subpolar lows has a strong westerly component due to the Coriolis force. South-westerlies are stronger and more persistent in the southern hemisphere than the north-westerlies.

Ferrel cells, located between 60 and 70 degree latitudes, are between Hadley and Polar cells and are dependent on their existence (Ferrel, 1889). They are associated with strong temperature gradients and eddies (turbulent movement or air) that mix heat in mid-latitudes by transporting thermal energy between polar and subtropical regions. This mixing happens because eddies play a key role in transporting warmer air toward the polar regions as it rises and cooler air moving toward the equatorial regions as the air sinks. As a result of the deflection from the Coriolis effect, the prevailing westerly wind in the Ferrel cells gathers moisture when traveling over the oceans and transports warm air to meet cold polar air. The uplift of warm air causes unstable weather conditions over mid-latitudes. As a result, ***baroclinic instability*** can occur because of strong flows and temperature gradients at midlatitudes leading to the formation of eddies near the surface of the Earth. Because the flow is in the opposite sense, with cold air rising and warm air sinking, Ferrel cells are referred to as ***thermally indirect*** cells.

Between the Ferrel and Polar cells, cold polar easterlies and warmer westerlies converge and make a low-pressure zone called ***subpolar lows*** around 60–65 degree latitudes of both hemispheres. The ***polar front*** is the boundary between these two air masses, with the region usually experiencing violent storms in winter seasons. The ***polar cells*** are from 60 to 70 degree latitudes to the poles in both the South and North hemispheres. Polar cells are the smallest and weakest cells compared to Hadley and Ferrel cells.

2.5 Atmospheric thermodynamics

Atmospheric thermodynamics is an essential theoretical basis for explaining much of the dynamical processes occurring in the troposphere. It deals with the transformations of the energy that take place in the earth's atmosphere and govern our weather and climate. These transformations can act on multiple scales, ranging from large-scale general circulation to the microphysical processes leading to the formation of clouds, including interactions with chemical species such as aerosols. Without an understanding of the basic thermodynamic principles as applied to atmospheric processes, we are unable to explain or predict the fate of air pollutants that are responsible for the quality of air that we all breathe. This section surveys some of the main principles of atmospheric thermodynamics. More details can be found, for example, in Wallace and Hobbs (2006) and Ambaum (2010).

2.5.1 Ideal gas law

The atmosphere is a mixture of gases that can expand or compress, and under certain conditions, the molecules can interact with each other. It is not practical to consider all the interactions and changes that may occur, and hence the ideal gas concept is used as an approximation of real atmospheric conditions. The main assumptions of an ***ideal gas*** are: (1) gases consist of point-like molecules therefore do not have volume themselves; (2) there are no attraction or repulsion forces between the molecules; and (3) fully elastic molecular collisions occur between these molecules.

Pressure (p), temperature (T), mass (m), and volume (V) are some of the properties of the gases that describe the state of the gas or a mixture of gases, as in the case of air. An equation for the state of the gas can be written by considering observations and the relationship between these parameters. In the mid-1600s, Sir Robert Boyle observed that for a fixed mass of gas, at constant temperature, the volume of the gas is inversely proportional to its pressure. So, the product of pressure and volume is an exact constant for an ideal gas. This relationship is referred to as ***Boyle's Law*** as follows:

$$pV = \text{constant} \tag{2.11}$$

where, p is pressure and V is the volume.

In the early 1800s, Charles and Gay-Lussac observed that for a given mass, at constant pressure, the volume (V) of a given mass of gas is directly proportional to the temperature (T). So, the ratio of V and temperature is constant, as in Eq. (2.12), and called as the ***Charles and Gay-Lussac law***.

$$\frac{V}{T} = \text{constant} \tag{2.12}$$

Ideal gas law is expressed in Eq. (2.13) and combines Boyle and Charles and Gay-Lussac laws in a single ***equation of state*** and can be used for gases such as air for most applications.

$$\frac{pV}{T} = nR^* \tag{2.13}$$

where p is the air pressure (in the unit of Pa where 1 Pa is equal to 1 kg m^{-1} s^{-2}), V is the volume (m^3) occupied by the gas, n is the number of moles of gas, R^* is the universal gas constant for the ideal gas (8.314 J K^{-1} mol^{-1}), and T is the temperature (K).

In a more familiar form:

$$pV = nR^*T \tag{2.14}$$

Mole number is calculated as follows:

$$n = \frac{m}{M} \tag{2.15}$$

where m is the mass of the gas (g), and M is the molecular weight of the gas (g mol^{-1}). And density of the air is (ρ) expressed with the ratio of mass (m) and volume (V) of the air expressed as:

$$\rho = \frac{m}{V} \tag{2.16}$$

Another form of the gas constant for a particular gas or a mixture (e.g., air), the specific gas constant (R), can be calculated as

$$R = \frac{R^*}{M} \tag{2.17}$$

We can derive the molar form of Eq. (2.14) by substituting Eqs. (2.15–2.17) into Eq. (2.13),

$$p = \rho RT \tag{2.18}$$

Note that here we use grams instead of kilograms as it is more generally common in chemistry. If kg is used (to be consistent with SI units) then a factor of 1000 has to be accounted for in the calculations.

Worked example 2.4

Calculate the specific gas constant for one gram of dry air (R_d). Molecular weight of dry air (M) is 28.97 g mol^{-1}.

Solution:

The specific gas constant for 1 g of dry air, R_d, can be calculated from Eq. (2.17)

$$R_d = \frac{R^*}{M} = \frac{8.314\ \text{J K}^{-1}\,\text{mol}^{-1}}{28.97\ \text{g mol}^{-1}} = 0.2869\ \text{J K}^{-1}\,\text{g}^{-1} = 287\ \text{J K}^{-1}\,\text{kg}^{-1}$$

Note that for a mixture of gases, as in the case of air, the weighted mean of the molecular weights of the individual gases is used.

A basic concept of chemistry is the ***Avogadro's hypothesis*** (named after Amedeo Avogadro, 1776–1856), which states that equal volumes of gases at the same temperature and pressure contain the same number of molecules. It follows that one mol of any substance will contain the same number of particles or constituents (e.g., atoms or molecules) at standard atmospheric temperature (273.15 K) and pressure (1×10^5 Pa). This is because the masses contained in one mol of different substances are the same as the ratios of their molecular weights (see, e.g., Wallace & Hobbs, 2006 for further details). The number of particles in one mol of any substance is called the Avogadro number or constant and has a value of 6.022×10^{23}. It is a universal constant and is equal to the number of atoms contained in 12 g of the carbon isotope ^{12}C (12 g C = 1 mol).

The equation of state can be written for per molecule using the Boltzmann's constant, k, (named after Ludwig Boltzmann, 1844–1906), which is defined as:

$$k = \frac{R^*}{N_A} \tag{2.19}$$

where N_A is the Avogadro's constant. Hence the ideal gas law can be rewritten for a gas containing n_0 molecules per unit volume,

$$p = n_0 kT \tag{2.20}$$

We can define α_d as the specific volume of dry air where $\alpha_d = 1/\rho$ and ρ is the density of dry air. Specific volume is the volume occupied by 1 kg of the gas at a given pressure p and temperature T. Ideal gas law for dry air can now be written as in Eq. (2.21):

$$p_d \alpha_d = R_d T \tag{2.21}$$

where p_d is the pressure of dry air, and the specific gas constant of dry air is R_d.

By considering the mole number given in Eq. (2.15), the apparent molecular weight M_d of dry air can be rewritten as the ratio of the total mass (in grams) of the constituent gases in dry air to the total number of moles of the constituent gases as follows:

$$M_d = \frac{\sum_i m_i}{\sum_i \frac{m_i}{M_i}} \tag{2.22}$$

where m_i is the mass (in grams), and M_i is the molecular weight of the substance for the ith constituent of the air mixture. As stated earlier, the apparent molecular weight of dry air is 28.97 g mol^{-1}.

2.5.2 Virtual temperature

Virtual temperature is the temperature of a moist air parcel at which a theoretical dry air parcel would have a total pressure and density equal to that of the moist air parcel. The apparent molecular weight of the moist air is less than that of dry air, and hence (see Eq. 2.21) the value for R_d is larger than that for 1 kg of dry air. As the value for R_d is dependent on moisture content in the atmosphere, which is highly variable, the use of the theoretical virtual temperature makes the calculations of the ideal gas equation more convenient in practice.

We discuss the main steps to derive an expression for the virtual temperature, but more details can be found in works such as Wallace and Hobbs (2006) and Ambaum (2010). To derive an expression for the virtual temperature, we assume a volume V of moist air at temperature T and pressure p and hold a mass of dry air of m_d and a mass of water vapor of m_v in that volume. The expression for density (ρ) of the moist air in this volume V is given by,

$$\rho = \frac{m_d + m_v}{V} = \rho'_d + \rho'_v \tag{2.23}$$

where ρ'_d is the density of dry air of the same mass occupying the whole of the same volume V alone and ρ'_v is the density of water vapor of the same mass occupying the whole of the same volume alone. The densities of dry air and water vapor can be considered partial densities as they represent densities of the components of moist air.

Following on from above, we can apply the ideal gas law (with the form including air density as in Eq. 2.18) to moist air. When we rewrite Eq. (2.18) for the water vapor we get,

$$e = \rho'_v R_v T \tag{2.24}$$

where e is partial pressure exerted by water vapor. Then, Eq. (2.18) is rewritten for the dry air:

$$p'_d = \rho'_d R_d T \tag{2.25}$$

where p'_d is the partial pressure exerted by dry air. By rewriting Eq. (2.24),

$$\rho'_v = \frac{e}{R_v T} \tag{2.26}$$

by rewriting Eq. (2.25)

$$\rho'_d = \frac{p'_d}{R_d T} \tag{2.27}$$

From Dalton's law of partial pressures, we have

$$p = p'_d + e \tag{2.28}$$

Hence by rewriting Eq. (2.28),

$$p'_d = p - e \tag{2.29}$$

Substituting Eq. (2.29) into Eq. (2.27), we get

$$\rho'_d = \frac{(p-e)}{R_d T} \tag{2.30}$$

Using Eq. (2.23) for partial densities, and Eqs. (2.26 and 2.30), we get

$$\rho = \frac{p-e}{R_d T} + \frac{e}{R_v T} \tag{2.31}$$

Eq. (2.31) can be written as, that is, removing $p/R_d T$ outside the bracket and defining $\varepsilon = R_d/R_v$;

$$\rho = \frac{p-e}{R_d T} + \frac{e}{\frac{R_d}{\varepsilon} T} = \frac{p}{R_d T} - \frac{e}{R_d T} + \frac{e\varepsilon}{R_d T} = \frac{p}{R_d T}\left[1 - \frac{e}{p} + \frac{e\varepsilon}{p}\right] \tag{2.32}$$

In the simplified form we get,

$$\rho = \frac{p}{R_d T}\left[1 - \frac{e}{p}(1-\varepsilon)\right] \tag{2.33}$$

where e/p represents the ratio of partial pressure exerted by water vapor to the total air pressure.

We can rearrange Eq. (2.33) to give an expression for pressure (p):

$$p = \frac{\rho R_d T}{\left[1 - \frac{e}{p}(1-\varepsilon)\right]} \tag{2.34}$$

We can now define the virtual temperature term T_v as

$$T_v = \frac{T}{1 - \frac{e}{p}(1-\varepsilon)} \tag{2.35}$$

By substituting T_v in Eq. (2.35) into Eq. (2.34), we get

$$p = \rho R_d T_v \tag{2.36}$$

Note that Eq. (2.36) is a moist air form of the ideal gas equation for dry air. Virtual temperature can also be calculated from Eq. (2.36) as follows:

$$T_v = \frac{p}{\rho R_d} \tag{2.37}$$

A more complete treatment and derivation of virtual temperature is given in Wallace and Hobbs (2006).

Worked example 2.5

At 0°C, the density of dry air alone is 1.275 kg m^{-3}, and the density of water vapor alone is 4.770×10^{-3} kg m^{-3}. What would be the total pressure exerted by the mixture of dry air and water vapor at 0°C? The specific gas constant for 1 kg of dry air (R_d) is 287.0 J K^{-1} kg^{-1}, the density of the water vapor is 4.770×10^{-3} kg m^{-3} at 273K, and the gas constant for 1 kg of water vapor is (R_v) 461.5 J K^{-1} kg^{-1}.

Solution:

The partial pressure exerted by dry air is calculated by using Eq. (2.36),

$$p = \rho R_d T_v = \left(1.275 \frac{\text{kg}}{\text{m}^3}\right) \times \left(287 \frac{\text{J}}{\text{K} \cdot \text{kg}}\right) \times (0 + 273.15 \text{ K}) = 99952 \text{ Pa} = 999.52 \text{ hP}a$$

The partial pressure exerted by water vapor is calculated by using Eq. (2.24) as

$$e = \rho'_v R_v T = 4.770 \times 10^{-3} \frac{\text{kg}}{\text{m}^3} \times 461.5 \text{ J K}^{-1} \text{ kg}^{-1} \times (0 + 273.15)\text{K} = 601.67 \text{ Pa} = 6.02 \text{ hpa}$$

The total pressure exerted by the mixture of dry air and water vapor:

Total pressure $= p'_d + e = 999.52$ hpa $+ 6.02$ hpa $= 1005.54$ (to two decimal places)

2.5.3 Hydrostatic equation

From basic physics, you will recall that pressure is defined as forces exerted per unit area. It follows that the pressure at a given altitude of the atmosphere is determined by the force exerted by the part of the atmosphere above that altitude. In other words, as the mass of the atmosphere above a particular altitude decreases with altitude, the pressure will decrease as height increases, with the highest pressure being felt at the Earth's surface.

We can consider a thin layer of the atmosphere with thickness, δz, at a height, z, above the Earth's surface. The forces acting on such a layer are shown in Fig. 2.24. Note that for a unit area, pressure is numerically equal to force. The downward force is due to gravity and hence due to the weight of the air above the layer, and the upward force is due to pressure exerted by the air below as the layer experiences a decrease in pressure with height (vertical pressure gradient force). The third force is due to the weight of the air within this layer acting downwards. When the upward and downward forces are equal, that is, there is no vertical acceleration (or negligible compared to acceleration due to gravity), the atmosphere is in ***hydrostatic equilibrium***. On average and in the absence of atmospheric motion, the downward gravitational and upward pressure gradient forces are usually almost in balance in the atmosphere.

If we consider the column of air with the unit horizontal cross-sectional area shown in Fig. 2.24, we can deduce the following by assuming that there is no vertical acceleration and we have hydrostatic balance as explained below.

p is the upward pressure acting on a thin layer from the air below the layer.

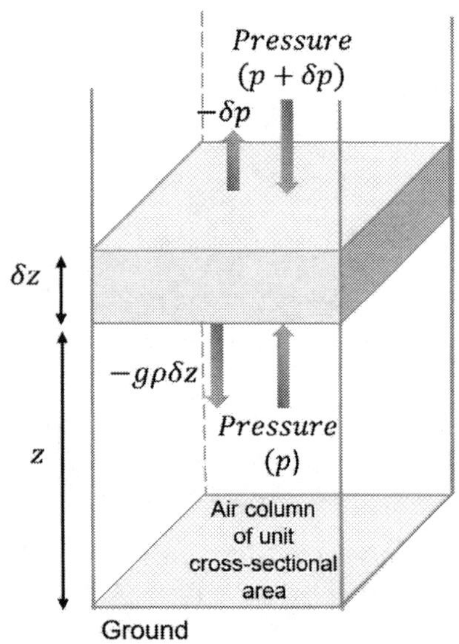

FIGURE 2.24 Forces acting on an idealized layer within an air column.

$p + \delta p$ is the downward pressure on this layer, where δp is the change in pressure from height z to $z + \delta z$, where δz is the thickness of the layer.

We can assume that each layer has a unit area, and hence pressure is numerically equal to force.

The upward pressure on the layer will be greater than the downward pressure as pressure decreases with height, and hence the net upward pressure on the layer of air or force for the unit area layer is given by:

$$p - (p + \delta p) = -\delta p \tag{2.38}$$

As air pressure drops from height z to $z + \delta z$, the change in pressure, δp, in the layer will be numerically negative, and hence the numerical value of $-\delta p$ will be positive. In other words, the upward pressure, p, at height z is greater than the upward pressure at height $z + \delta z$ by δp, and hence the net vertical force on the air layer due to the vertical pressure gradient is upward.

The mass of the air layer will be given by ρV, where ρ is the density of air at height z (assumed to be incompressible) and V is the volume of the layer equal to δz times the area of the layer. As we are assuming that the layer has unit area, the mass of a unit area of the layer with thickness δz in the column is given by $\rho\delta z$.

The downward gravitational force due to the weight of this unit area mass of air layer is:

$$= g\rho\delta z \tag{2.39}$$

where g is the acceleration due to gravity at height z.

The condition for hydrostatic balance (or equilibrium) to exist is:

Upwards pressure gradient force = downward gravitational force (note force is equal to pressure per unit area)

$$-\delta p = g\rho\delta z \tag{2.40}$$

or

$$-\frac{\delta p}{\delta z} = -g\rho \tag{2.41}$$

By setting the limit $\delta z \to 0$, we arrive at the simple differential form of the ***hydrostatic equation***:

$$\frac{\partial p}{\partial z} = -g\rho \tag{2.42}$$

The partial differential notation on the left-hand side for the variation of air pressure p with height z assumes that this pressure variation is independent of the other variables, which are invariant. The negative sign in Eq. (2.42) implies that the pressure decreases as we go higher in height. The hydrostatic equation is a valid assumption mainly for average larger scale atmospheric processes and breaks down for mesoscale and finer scale processes where the air motion is rapid and changeable where the vertical acceleration of the air may not be negligible compared to g and hence the balance between pressure gradient and gravitational forces is not maintained.

We can derive an expression for pressure at a given height, $p(z)$, by rearranging Eq. (2.41) in its differential form:

$$-dp = g\rho dz \tag{2.43}$$

We now integrate from $p\ (z)$ to $p\ (\infty)$ for atmospheric pressure and from z to ∞ for height.

$$-\int_{p(z)}^{p(\infty)} dp = \int_{z}^{\infty} g\rho dz \tag{2.44}$$

At height of infinity, pressure is zero, and changing the integration limits on the right-hand side removes the negative sign to give:

$$\int_{0}^{p(z)} dp = \int_{z}^{\infty} g\rho dz \tag{2.45}$$

Which gives the expression to calculate pressure at a height z above the Earth's surface:

$$p(z) = \int_{z}^{\infty} g\rho dz \tag{2.46}$$

Eq. (2.46) means that the pressure at z *is equal to the weight of the air in a vertical column* of unit cross-sectional area above this height.

If the column of air started at the sea level, the pressure would be 1013 hPa which is called *one standard atmosphere* (usually written as 1 atm).

Worked example 2.6

Estimate total mass of the atmosphere by considering the globally averaged surface pressure to be 1000 mb. The acceleration due to gravity is 9.807 m s^{-2}, radius of Earth is 6.37×10^6 m and the atmosphere can be assumed to be in hydrostatic equilibrium.

Solution:

According to Eq. (2.16),

$$\rho = \frac{m}{V}$$

So, the mass (m) equals to

$$m = \rho V$$

Assuming constant density throughout the atmosphere, the volume can be expressed as the product of the surface area (A) and the height (h), and the equation becomes:

$$m = \rho A h \tag{2.47}$$

In order to calculate the density (ρ) of the air in the atmosphere, Eq. (2.46) can be rewritten simply as by neglecting the minor variations of g with latitude, longitude, and height and instead considering its average value over the Earth's surface. As gravity is also accepted as a constant value for this example, all the terms are taken outside of the integral, and we get the following formula, which is also called the **barometric formula**.

$$p = \rho g h \tag{2.48}$$

So, the density (ρ) equals to

$$\rho = \frac{p}{gh} \tag{2.49}$$

Then Eq. (2.47) can be rewritten for the mass of the atmosphere as by considering the area of the Earth (A_{Earth}),

$$m = \rho A h = \frac{p}{gh} A_{\text{Earth}} h \tag{2.50}$$

The height (h) cancels out, and we are left with:

$$\text{m} = \frac{pA_{Earth}}{g} \tag{2.51}$$

The surface area of the Earth is calculated by considering the radius of the Earth as,

$$A_{\text{Earth}} = 4\pi R_{\text{Earth}}^2 = 4\pi (6.37 \times 10^6 \text{m})^2$$

Now, let's use the globally averaged surface pressure ($P = 1000$ mb) and the approximate acceleration due to gravity ($g = 9.807$ m s^{-2}) to calculate the mass of the air:

$$m = \frac{p \times A_{\text{Earth}}}{g} = \frac{1000 \text{ mb} \times 4\pi \times (6.37 \times 10^6 \text{m})^2}{9.807 \text{ ms}^{-1}} = 5.19 \times 10^{18} \text{ kg}$$

Keep in mind that this is a rough estimate, and actual values may vary based on atmospheric conditions and regional variations.

2.5.4 Geopotential and geopotential height

The ***geopotential*** is a very useful concept employed in meteorology for interpreting and analyzing weather behavior such as cyclones and anticyclones. As an example, a related parameter of ***geopotential height*** is commonly calculated at a constant pressure level such as 500 hPa (equivalent to an altitude of about 5.6 km above sea level; see Section 2.4.1) for synoptic analysis. By relating to mean sea level pressure, geopotential height anomalies can be displayed on synoptic charts to show areas of high pressure (anticyclones) and low pressure (depressions and cyclones).

When defining geopotential, Φ, at a specific point, as the name implies, the height of this point in the atmosphere is related to its potential energy. The geopotential at any point can be defined as the amount of work that must be done against gravitational force to raise a mass of 1 kg from sea level to that point. For this reason, geopotential is often referred to as ***the gravitational potential per unit mass*** (e.g., see Wallace & Hobbs, 2006). The main expression that mathematically defines the geopotential at a height z above the surface of the Earth, Φ (z), is:

$$\Phi(z) = \int_0^z g dz \tag{2.52}$$

where g is the acceleration due to gravity. This expression represents the work that would be needed to lift 1 km of mass of air from the surface ($z = 0$) to a height z. It follows that the geopotential at the surface, Φ (0), is taken as zero.

A related parameter is that of the geopotential height (Z), which is defined by the expression:

$$Z \equiv \frac{\Phi(z)}{g_0} = \frac{1}{g_0}\int_0^z g dz \tag{2.53}$$

where g_0 is the mean acceleration due to gravity at the surface of the Earth taken to be 9.81 m s^{-2}. Geopotential height is used as a vertical reference coordinate system to take account of gravitational dependence on altitude and latitude and is often used for large-scale atmospheric circulation charts. As mentioned earlier, it enables height at any point to be referenced to the mean sea level, where geopotential is assumed to be zero. The value of Z is either equal to (at the Earth's surface) or smaller (as height increases) than the geometric height. Its value is very close to the actual geometric height near the surface, and, even for much of the troposphere, the difference is very small.

2.5.5 Scale height

Another important and useful parameter to define the thickness of the atmosphere is the ***scale height*** (H). This is defined as the height over which a characteristic parameter of the atmosphere (such as pressure or density) will decrease by a factor of $1/e$ (e is the exponential with a value of 2.7183 to four decimal places). In other words, pressure or density will decrease exponentially with height, and hence H is sometimes referred to as the ***e-folding height or depth***. For an isothermal atmosphere (where temperature remains constant), the value of H remains constant. This will mean that as the height above the surface increases by $1H$, the pressure (as an example) will decrease by a factor of $1/e$, that is, by 36.8%. As height increases to $2H$, the pressure will reduce by a further 13.5%. The exponential dependence for air pressure on height, z, and scale height, H, can be written as:

$$p(z) = p_0 e^{-\frac{z}{H}} \tag{2.54}$$

where the scale height is given by:

$$H = \frac{kT}{mg} = R^*T/M_d g \tag{2.55}$$

where k is the Boltzmann's constant (1.3806×10^{-23} J K^{-1}), T is the air temperature (in units of K), m is the mean molecular mass of the gas (in units of kg), g is the acceleration due to gravity (9.81 m s^{-2}), R^* is the universal gas constant (8.3145 J K^{-1} mol^{-1}), and M_d is the mean mass of one mol of air or apparent molecular weight (0.02897 kg for dry air).

Note:

$R^* = k N_A$ from Eq. (2.17)

$N_A = M_d/m$, as the definition of N_A is the number of atoms or molecules in one mole of a substance.

Often in books on atmospheric science you will find the gas constant for 1 g of dry air (R_d) stated, which is equal to R^*/M_d, where M_d is expressed in grams (28.97 g in the case for dry air). Hence, for 1 kg of dry air, R_d is 8.3145/0.02897, which equals 287.0 J K^{-1} kg^{-1}. The above scale height Eq. (2.55) can be written as:

$$H = \frac{R_d T}{g} \tag{2.56}$$

Substituting in the values for R_d (287.0 J K^{-1} kg^{-1}) and g (9.81 m s^{-2}) we get:

$$H = 29.3T \tag{2.57}$$

If we assume a typical mean temperature of the troposphere and stratosphere is 255K, we find that the scale height is about 7.5 km, and for the lower troposphere, where the mean air temperature is around 288K, the value for H is about 8.4 km. As we move higher from the surface by each scale height, properties of the atmosphere such as pressure and density reduce by a factor of $1/e$.

2.5.6 The first law of thermodynamics

When an air parcel warms, either the density of the air parcel must decrease to keep the pressure constant or the pressure will increase. In the atmosphere, minor pressure differences are equalized as the air parcel warms and its pressure and density decrease, and hence it begins to rise and expand in volume. Expansion occurs because the pressure of the air parcel is higher than the surrounding air and eventually to maintain equilibrium. Not all of the thermal energy is used to warm the air parcel, with some being used for work needed for expansion of the parcel. Hence, as heat is added to the air parcel and its temperature increases, it expands while maintaining constant pressure. Under these conditions, a part of the heat added is needed to do work to enable the air parcel to expand while maintaining constant pressure of the surrounding air.

In terms of the simple form of the first law of thermodynamics, ***energy is neither created nor destroyed*** and the total amount of energy must be conserved. In the atmosphere, this can be understood by considering that when an amount of heat is applied to a mass of air, it must be equal to the total sum of the warming (or internal energy) of the air and the amount of work done per unit mass of air. This can be expressed as

$$dQ = dU + dW \tag{2.58}$$

where dQ is the added heat, dU is the warming or internal energy, and dW is work done by the expanding air.

From Eq. (2.36), we can consider dU to be the difference ($dQ - dW$) between the heat added to the air parcel and the work done by the expanding air against the surrounding environment. Recall from basic thermodynamics that the change (increase in this case) in work done (dW) when an air parcel undergoes an incremental expansion by an increase in volume (V) by dV is given by the pressure (p) (remaining constant) and the increment change in its volume, that is:

$$dW = pdV \tag{2.59}$$

Hence we can represent dU as:

$$dU = dQ - pdV \tag{2.60}$$

For a unit mass of air, we can use a specific volume (α) where

$$\alpha = 1/\rho \tag{2.61}$$

and a small change in volume (dV) is equivalent to a small change in the specific volume of the air parcel ($d\alpha$). We can then write Eq. (2.36) as:

$$dQ = dU + pd\alpha \tag{2.62}$$

Which is one form of the expression for the first law of thermodynamics.

If we assume that the volume of the air parcel is kept constant, we can define the *specific heat of air at constant volume (c_v) as*

$$c_v = dU/dT \; (V = \text{constant}) \tag{2.63}$$

Therefore, we can write

$$dU = c_v dT \tag{2.64}$$

We can then write an alternative expression of the first law of thermodynamics:

$$dQ = c_v dT + pd\alpha \tag{2.65}$$

As a reminder from basic thermodynamics, ***specific heat*** is the amount of heat needed to raise the temperature of one gram (g) of a substance by one degree Celsius, or Kelvin. The specific heat of pure water is 4186 J K^{-1} kg^{-1}, while the specific heat of dry air at sea level is about 1004 J K^{-1} kg^{-1}. Approximately four times the energy is needed to raise the temperature of 1 kg of water by 1°C than is needed to do the same for 1 kg of dry air.

Often for atmospheric calculations, change in ***enthalpy*** is used instead of the warming term. Enthalpy can be considered to refer to quantifying the total heat content of an air parcel. The change in enthalpy (*dh*) for a unit m of material (or air in this case) can be written in the form:

$$dh = c_p \Delta T \tag{2.66}$$

where c_p is the specific heat or heat capacity at constant pressure, and T is the temperature (K).

The integral form of Eq. (2.66) is:

$$h = c_p T \tag{2.67}$$

The change in enthalpy for a parcel of mass m is also given as a function of the change in temperature (T) as;

$$dh = mc_p dT \tag{2.68}$$

where dh is the change in enthalpy.

Yet another alternate form of the first law of thermodynamics can be written as:

$$dQ = c_p dT - \alpha dp \tag{2.69}$$

where α is the specific volume as we are considering unit mass of air parcel, or in the enthalpy form:

$$dQ = dh - \alpha dp \tag{2.70}$$

This equation can be thought of as

$$\text{Heat transferred (dQ)} = \text{Enthalpy change} - \text{Pressure change per unit mass} \tag{2.71}$$

When the pressure change is zero, the equation (use the number from the above dQ equation) becomes:

$$dQ = dh = c_p dT \tag{2.72}$$

2.5.7 The second law of thermodynamics

When considering the first law of thermodynamics, we discussed that some work has to be done by the air parcel to expand. The second law of thermodynamics evolved with the works of Nicholas Leonard Sadi Carnot (1796–1832), Rudolf Clausius (1822–1888), and William Thomson, 1st Baron Kelvin (1824–1907) and relates to the maximum proportion of heat energy that can be used for work. There are number of versions of this law. Some of the common ones include that for heat to be converted to do work in a cyclic process, it can only be transferred from a warmer to a colder body; the reverse of this version is that in the absence of work done by an external influence, heat cannot of itself transfer from a colder to a warmer body in a cyclic process.

Another form states that there exists a useful state variable called ***entropy***. Entropy is a measure of the disorder and randomness that exist in a thermodynamical system and can also be considered to impose a limit on how much energy is available—in other words, how much energy is not available to perform useful work. For a ***reversible*** process (i.e., which can revert to a previous state if the surrounding state is changed), the second law states that for a reversible transformation there is no change in the entropy of the universe. In other words, the increase in entropy of a system due to the input of heat will be equal to the decrease in entropy of the surrounding. Note that the word "universe" refers collectively to the system and its surrounding environment. In reality, most thermodynamical processes are ***irreversible***; that is, the system will not revert to its previous state even if the surroundings are changed. This leads to the corollary statement for the second law as that for irreversible transformation, the entropy of the universe will always increase, which means that in real situations the disorder within systems will increase over time if there is no external influence. Encapsulating these ideas, a common form of the second law of thermodynamics states that *the system's entropy cannot diminish; it can only either remain constant or increase.*

We can express the above concepts mathematically as follows: we assume that the change in entropy from the initial state (1) is given by δS_1 and for the final state (2) is given by δS_2. We can also assume that the change in entropy in the "universe" (system + surrounding environment) is δS_{uni}, in the surrounding environment is δS_{env}, and in the system is δS_{sys}. We can now state:

$$\delta S_{uni} = \delta S_{sys} + \delta S_{env} \tag{2.73}$$

For reversible processes or those in equilibrium with their environment, we have:

$$\delta S_{\text{uni}} = 0 \text{ or } \delta\, S_2 = \delta S_1 \tag{2.74}$$

For irreversible (referred to as spontaneous) processes:

$$\delta S_{\text{uni}} > 0 \text{ or } \delta\, S_2 > \delta S_1 \tag{2.75}$$

For a reversible transformation, if a certain small amount of heat dQ is supplied quasistatically to a system with an absolute temperature of T, then the entropy of the system increases by:

$$TdS - dQ = 0 \tag{2.76}$$

By rewriting Eq. (2.76), to define entropy, we get:

$$dS = \frac{dQ}{T} \tag{2.77}$$

where S is the entropy and dQ/T is the differential of the state function S. It follows that for an adiabatic transformation ($dQ = 0$), the entropy is invariant of the system (i.e., the mean value does not change).

For an *irreversible* transformation

$$TdS - dQ' = 0 \tag{2.78}$$

where dQ' has been introduced and is the Clausius noncompensated heat. It follows that for an irreversible adiabatic ($dQ = 0$) transformation, the entropy produced by the system is

$$dS = \frac{dQ'}{T} \tag{2.79}$$

It should be appreciated that S is a state variable, which means that it describes the property of the state of a system and does not depend on how the system reached that state. W, which is work done on the system, is not a state variable. Although S is a state variable, the product of TdS is not a state variable. Further discussion can be found in Wallace and Hobbs (2006) and Ambaum (2010).

2.5.8 Potential temperature

The stability of the troposphere depends on both how the temperature and pressure of air change with altitude (see Sections 2.3.1 and 2.3.3). Pressure and temperature of the air can be combined into a single potential temperature, which is an important variable in atmospheric science and provides a robust way of classifying atmospheric stability. It is a more useful variable than temperature as, under dry adiabatic conditions, it is not influenced by air ascending or descending by, for example, flowing over obstacles such as hills, whereas air temperature will change. In fact, potential temperature is conserved when an air parcel moves under adiabatic conditions and can be considered to remain constant.

As mentioned before in Section 2.3.3, an adiabatic process is where no heat is transferred between the air parcel and the environment, namely $dQ = 0$. Hence, we get

$$dh = Vdp \tag{2.80}$$

and by replacing the enthalpy term (dh) with Eq. (2.68), we arrive at the first law of thermodynamics expressed as

$$mc_p dT = Vdp \tag{2.81}$$

Dividing by V on both sides, we have

$$\frac{m}{V} c_p dT = dp \tag{2.82}$$

which can be written as

$$\rho c_p dT = dp \tag{2.83}$$

Substituting $\rho = p/RT$ from the ideal gas law (see Eq. 2.18), we have

$$pc_p dT/RT = dp \tag{2.84}$$

rearranging we get

$$\frac{c_p}{R}\frac{dT}{T} = \frac{dp}{p} \tag{2.85}$$

This equation can also be derived from a slightly different point (see Wallace & Hobbs, 2006) starting from the form of the first law of thermodynamics (see Eq. 2.69).

$$dQ = c_p dT - \alpha dp \tag{2.86}$$

For adiabatic conditions, this equation can be written as:

$$0 = c_p dT - \alpha dp \text{ or } c_p dT = \alpha dp \tag{2.87}$$

Recall that the specific volume $(\alpha) = 1/\rho$, where ρ is the density of the air parcel, we get:

$$c_p dT = \frac{dp}{\rho} \tag{2.88}$$

As above, substituting for $\rho = p/RT$ into Eq. (2.88), we get

$$c_p dT = RT\frac{dp}{p} \tag{2.89}$$

We can then write

$$\frac{c_p}{R}\frac{dT}{T} = \frac{dp}{p} \tag{2.90}$$

which is the same form above in Eq. (2.85).

Integrating from the standard pressure (p_0) where the air temperature is θ to pressure p where the air temperature is T, the potential temperature, θ, is then defined as the temperature that an air parcel would have if it were brought from a pressure (p) to a standard pressure (p_0) through adiabatic expansion or compression. It is commonly assumed that the standard pressure is 1000 hPa. An important property of potential temperature is that it is conserved as a dry air parcel moves adiabatically and remains close to being invariant, that is, it remains constant.

The temperature that the air parcel at the ground should have after an adiabatic motion from a thermodynamic state (T, p). Let p_s be the ground pressure (or a reference pressure, 1000 hPa). We obtain the so-called Poisson equation,

$$\theta(\mathrm{p},\mathrm{T}) = T\left(\frac{p_s}{p}\right)^{R/c_p} \tag{2.91}$$

For the case of dry air, where R_d is used instead of R, we calculate $R_d/c_p = 0.286$ since $c_p = 1004\ \mathrm{J\ K^{-1}\ kg^{-1}}$ and R_d is 287 $\mathrm{J\ K^{-1}kg^{-1}}$.

2.5.9 Potential temperature and atmospheric stability

Having defined potential temperature, we can now consider how potential temperature helps us to characterize the stability of the atmosphere. The profile of potential temperature followed by an air parcel as it moves vertically can be used to ascertain the stability of the atmosphere. Essentially, if the potential temperature of the air parcel increases with height, the atmosphere can be described as being generally stable, and if it decreases with height, the atmosphere is generally unstable. The atmosphere is neutral when the potential temperature remains constant. We can illustrate this as follows.

As mentioned in Section 2.5.8, the potential temperature is conserved during an adiabatic process. The potential temperature has, therefore, a constant vertical profile if the atmosphere behaves adiabatically. We assume that the atmosphere is dry or unsaturated and is in hydrostatic equilibrium (see Section 2.5.3) T is the temperature of the air (K), p_0 is the standard pressure (Pa), p is the pressure of the air, R_d is the gas constant for dry air (287 $\mathrm{J\ K^{-1}\ kg^{-1}}$) and c_p is the specific heat for air at a pressure p ($\mathrm{J\ K^{-1}\ kg^{-1}}$).

Take natural logs of Eq. (2.91) to get

$$\ln\theta = \ln T - \frac{R_d}{c_p}\ln p + \text{constant} \tag{2.92}$$

Note that $p_0^{(R/cp)}$ is a constant, and we have $p^{-(R/cp)}$ as p was in the denominator, and hence the second term is negative when you take the log of Eq. (2.91).

Multiply by c_p and get

$$c_p\ln\theta = c_p\ln T - R_d\ln p + \text{constant} \tag{2.93}$$

We can now differentiate, noting that c_p and R_d are both constants and derivatives of $\ln x = 1/x$, and get

$$c_p\frac{d\theta}{\theta} = c_p\frac{dT}{T} - R_d\frac{dp}{p} \tag{2.94}$$

Multiplying by T, we get

$$c_pT\frac{d\theta}{\theta} = c_pdT - R_dT\frac{dp}{p} \tag{2.95}$$

From the ideal gas law (see Eq. 2.21) we know p α = RT, and hence we can replace R_dT by $p\ \alpha$ (remember α is the specific volume of the air parcel).

$$c_pT\frac{d\theta}{\theta} = c_pdT - p\alpha\frac{dp}{p} \tag{2.96}$$

(p cancels) giving

$$c_pT\frac{d\theta}{\theta} = c_pdT - \alpha dp \tag{2.97}$$

From the hydrostatic balance or equilibrium, we know $gdz = -\alpha dp$, or $\alpha dp = -gdz$ (see Section 2.5.3 and Chapter 5), and therefore

$$c_pT\frac{d\theta}{\theta} = c_pdT + gdz \tag{2.98}$$

We can rearrange Eq. (2.98) by letting

$$d\theta = \left(\frac{\partial\theta}{\partial z}\right)dz \text{ and } dT = \left(\frac{\partial T}{\partial z}\right)dz \tag{2.99}$$

To get

$$c_pT\left(\frac{\partial\theta}{\partial z}\right)\frac{dz}{\theta} = c_p\left(\frac{\partial T}{\partial z}\right)dz + gdz \tag{2.100}$$

and then dividing by c_pTdz which gives

$$\frac{1}{\theta}\frac{\partial\theta}{\partial z} = \frac{1}{T}\left(\frac{\partial T}{\partial z} + \frac{g}{c_p}\right) \tag{2.101}$$

However, we know from Section 2.3.3 that

$-dT/dz = \Gamma$ (actual lapse rate) and $g/c_p = \Gamma_d$ (dry adiabatic lapse rate), and hence we arrive at

$$\frac{1}{\theta}\frac{\partial\theta}{\partial z} = \frac{1}{T}(\Gamma_d - \Gamma) \tag{2.102}$$

This equation neatly relates potential temperature and lapse rate terms. When $\Gamma < \Gamma_d$, we note that the left-hand side term is positive, and hence $\partial\theta/\partial z$ is positive for statically stable atmospheric conditions where vertical movement of air is limited. It follows that under these conditions, potential temperature must increase with height. When the opposite is true, then convective conditions prevail in the atmosphere and there is greater mixing of air pollutants that are emitted into the atmosphere. We can summarize the relationships between conditions of stability of an unsaturated atmosphere, lapse rate, and potential temperature in Table 2.2.

TABLE 2.2 Characterization of the stability of unsaturated atmosphere according to lapse rate and potential temperature.

Condition of the atmosphere	Lapse rate	Potential temperature variation with height
Stable	$\Gamma < \Gamma_d$	$\partial\theta/\partial z$ is increasing, >0
Unstable	$\Gamma > \Gamma_d$	$\partial\theta/\partial z$ is decreasing, <0
Neutral	$\Gamma = \Gamma_d$	$\partial\theta/\partial z$ is constant, =0

2.5.10 The moist adiabatic lapse rate

A rising moist air parcel will cool with height at the dry adiabatic lapse rate (see Section 2.3.3). It will continue to rise adiabatically till it becomes saturated with water vapor, after which point the moisture will condense (e.g., as liquid water) as it continues to rise while releasing latent heat of condensation. Because latent heat is released, the rate at which the temperature of the rising air parcel cools reduces. As long as the condensation is not lost from the air parcel (that is, it is conserved) and the latent heat is not transferred to the surrounding air, the process is considered to be reversible and can be described as **saturated adiabatic**. Under these conditions, the temperature of the air parcel will decrease at the ***saturated adiabatic lapse rate***. On the other hand, if the condensation (e.g., liquid water) is lost from the air parcel, the process is considered irreversible and is now classified as ***pseudoadiabatic***. As noted by Wallace and Hobbs (2006), the amount of heat lost through the removal of condensation product is small, and hence the saturated adiabatic and the pseudoadiabatic lapse rates are similar. Whereas the dry adiabatic lapse rate is a fixed number, the saturated lapse rate is a function of temperature and pressure of the air and ranges from 4 K km^{-1} (surface) to 7 K km^{-1} (at the midlevel of the troposphere).

In the case of moist air where latent heat of condensation is released as the air parcel rises, the term ***equivalent potential temperature*** is introduced, and a similar version of the Eq. (2.102) can be derived (for further discussions and derivation, see Wallace & Hobbs, 2006).

$$\theta_e \cong \theta \exp\left(\frac{L_v w_s}{c_p T}\right) \tag{2.103}$$

where θ_e is the equivalent potential temperature for saturated air, θ is the potential temperature of unsaturated air, L_v is the latent heat of vaporization or evaporation, w_s is the saturation mixing ratio, c_p is the specific heat of air at constant pressure (p), and T is the temperature of the air parcel. Note that w_s is a function of state and represents the maximum amount of water that can be held in a given parcel of air for a specific temperature and pressure. It is defined as the ratio of the mass of water vapor needed for a volume of air to be saturated (relative humidity of 100%) to the mass of the dry air parcel of the same volume. θ_e can be used to characterize stability of moist (i.e., unsaturated) air in an analogous manner as discussed above (Section 2.5.9) for dry air stability and summarized in Table 2.2.

2.6 Atmospheric composition

The atmosphere that surrounds our planet is a mixture of many gases. Table 2.3 summarizes the approximate mean composition of the gases in the atmosphere (NOAA, 2024) in two units of mixing ratio for volume relative to dry air. The mixing ratio is the relative fraction of a given component relative to the sum of the other components (see Section 2.6.3).

The composition of air varies from place to place and from time to time. However, two gases, nitrogen (N_2) and oxygen (O_2), make up approximately 99% of the volume of clean, dry air in the bottom 100 km of the Earth's atmosphere. Argon (Ar), a chemically inert gas, makes up most of the remaining 1%, but there are many other species also in smaller concentrations, which will be discussed in the next sections.

2.6.1 Fixed gases

Nitrogen, oxygen (O_2), argon (Ar), neon (Ne), helium (He), krypton (Kr), and xenon (Xe) are fixed gases. The volume mixing ratio of these ***fixed gases*** does not change substantially in time or space.

TABLE 2.3 Main atmospheric gaseous species (concentration of variable gases as of November 2022 from Mauna Loa observatory, Hawaii measurement data).

	Name	Formula	Volume mixing ratio (for dry air)	
			%	ppmv
Fixed gases	Nitrogen	N_2	78.08	780,900
	Oxygen	O_2	20.95	209,360
	Argon	Ar	0.93	9300
	Neon	Ne	0.0018	18
	Helium	He	0.0005	5
	Krypton	Kr	0.0001	1
	Xenon	Xe	0.000005	0.05
Variable gases	Carbon dioxide	CO_2	0.0417	417
	Methane	CH_4	0.00019	1.9
	Ozone	O_3	0.0000003–0.001	0.03–10
	Water vapor	H_2O	0.00001–4.0 in wet air	0.1–40,000 in wet air

Mixing ratios are relative to dry air except for water vapor.
Source: NOAA. (2024). *Data provided by carbon cycle and greenhouse gases group*, 325 Broadway R/GML, Boulder, CO 80305 (https://gml.noaa.gov/ccgg/).

The most abundant gas in Earth's atmosphere is ***nitrogen*** in its diatomic form, with a mean concentration of about 78%, which plays a vital role in the nitrogen cycle. As part of this cycle, nitrogen constantly exchanges between living organisms and the atmosphere. In addition to its importance for ecosystems, nitrogen also plays a role in climate change and air quality. Nitrogen compounds contribute to aerosol formation and to ozone production and, together with nitrogen dioxide, these species all have serious impacts on human health (see Chapter 14). A major source of nitrogen into the environment is agricultural use associated with emissions of ammonia (NH_3), which are associated with damage to the natural ecosystems and crop yields. Nitrogen molecules in the atmosphere are affected by high temperatures during lightning strikes, which lead to the formation of oxides of nitrogen that are involved in ozone chemistry (see Chapter 4).

About 21% of Earth's atmosphere is ***oxygen***, but this was not the case in the early history of our planet, which had almost no oxygen (for further information see Rosen & Egger, 2014). However, ancient organisms (called cyanobacteria or blue-green algae) are believed to have led to the oxygenation of the Earth's atmosphere. They were able to conduct oxygenic ***photosynthesis***, using sunshine, water, and carbon dioxide to produce carbohydrates and oxygen as a by-product through a reaction:

$$6CO_2 + 6H_2O \overset{hv}{\rightarrow} C_6H_{12}O_6 + 6O_2 \tag{2.104}$$

where hv is a photon of radiation (sunlight), and $C_6H_{12}O_6$ is glucose (carbohydrates). Absorption of light (usually sunlight) is performed by chlorophyll in plants, trees, blue-green algae, and certain bacteria.

Oxygen critically plays a role in processes such as respiration, which is a necessary function of almost all life on Earth. Plants use oxygen during respiration but also produce oxygen through photosynthesis.

Oxygen is a very reactive atom and is found as part of many compounds that are important for air quality, including ozone (O_3), nitrogen dioxide (NO_2), nitric oxide (NO), carbon monoxide (CO), as well as greenhouse gases such as carbon dioxide (CO_2), nitrous oxide (N_2O), and water vapor (H_2O). ***Greenhouse gas***, in the context of this book, is a gas that can absorb infrared radiation and hence trap heat in the atmosphere. O_3 is a significant greenhouse gas as well as a major air pollutant.

Argon, neon, helium, krypton, and xenon are the other fixed gases abundant in the atmosphere. They are all inert gases and hence do not react chemically. Their main source is the radioactive decay of elements in the Earth's crust. Argon is the third most abundant gas in the atmosphere after nitrogen and oxygen.

2.6.2 Variable gases

There are many gases in the atmosphere for which their concentration varies in space and time. Some of the important ***variable gases*** that are essential for air quality and climate are water vapor, aerosols, carbon dioxide, methane, and ozone.

The percentage of ***water vapor*** in the atmosphere varies considerably from almost zero to approximately 4%. Although it is a small fraction of the air, it is a very significant constituent of air and has major impacts on the behavior of our weather and climate. Besides being a source of precipitation and clouds, it can absorb terrestrial radiation and some incoming solar radiation, as explained in Section 2.2.3, making it an important greenhouse gas.

Most of the water vapor in the atmosphere (about 85%) comes from ocean surface evaporation, followed by transpiration from plants and vegetation, glaciers, sea ice soils, other water bodies, fuel combustion, and many gas-phase chemical reactions.

In low temperatures, as in the North and South Poles, water vapor readily condenses as a liquid or deposits as ice. Under these conditions, the mixing ratio of water vapor in the atmosphere is almost zero. At high temperatures, water evaporates, and ice sublimates to the gas phase, making the water vapor mixing ratio high in air. Over water bodies, the mixing ratio may be as high as 4% or more, especially in equatorial regions where the temperatures are high and evaporation occurs readily over the surface of water. Water absorbs or releases latent heat when it changes phase from one state to another without a change in its temperature.

In terms of abundance, ***carbon dioxide (CO_2)*** is the most important greenhouse gas. To maintain the habitability of our planet, the Earth's radiation balance is essential, and the role of CO_2 (along with other greenhouse gases) in ensuring a habitable climate is pivotal. Consequently, CO_2 is the subject of global treaties and wide discourse on climate change and how to achieve a carbon-neutral and sustainable future. It is a relatively inert and well-mixed gas and has a long life in the environment. Atmospheric lifetimes range from 5 to 200 years depending on the different uptake and sink pathways (e.g., absorption in oceans and uptake by plants and vegetation for photosynthesis) (IPCC, 2021a). It is produced during cellular respiration by trees and plants, by evaporation from the oceans, released from volcanic eruptions, fossil fuel combustion, and biological decomposition of dead organic matter. Cellular respiration is essentially the reverse of Eq. (2.104).

Due to fossil fuel combustion, global CO_2 mixing ratios are continually increasing. In spring and summer, an important removal pathway for CO_2 from the air is by photosynthesis (Eq. 2.104). In the autumn and winter seasons, plants decompose biologically and release CO_2 into the air. Despite the seasonal cycle of CO_2, its mixing ratio is increasing, as illustrated in Fig. 2.25.

Methane (CH_4) is a major component of natural gas and has an atmospheric lifetime of around 10–12 years. It is emitted from many anthropogenic and natural sources, such as landfills, wetlands, rice paddies, farming activities, and coal mining. In terms of anthropogenic sources, agricultural emissions are a predominant contributor. In terms of natural sources, wetlands are the most important source. In environments where oxygen is missing, such as rice fields and swamp lands, methane-producing bacteria (methanogens) consume organic material and produce methane. It is also produced from decomposing fossilized carbon under the ground, burning biomass, and chemical reactions in the atmosphere.

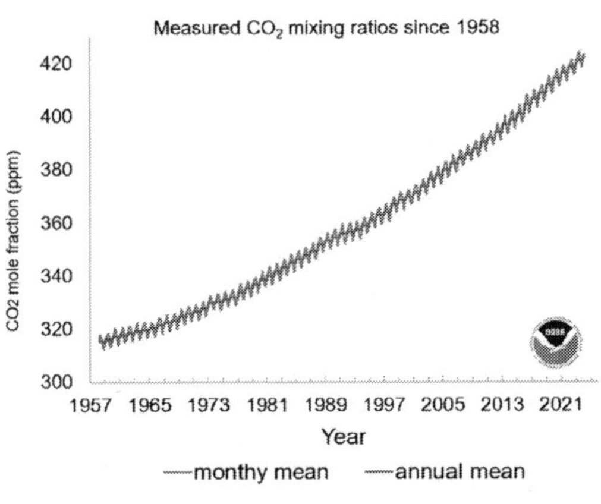

FIGURE 2.25 Annual and seasonal fluctuations in CO_2 mixing ratios from 1958 to 2022. Measurements were taken at Mauna Loa Observatory. *Data source: NOAA. (2024). Data provided by carbon cycle and greenhouse gases group, 325 Broadway R/GML, Boulder, CO 80305 (https://gml.noaa.gov/ccgg/).*

The 2021 AR6 report of the IPCC (2021a) gives two values for the global warming potential for methane as 29.8 for fossil fuel origin and 27.2 for nonfossil fuel origin for 100 years, while for CO_2, it is accepted as one given that it is the reference gas against which other greenhouse gases are compared.

Methane has an important role in ozone production chemistry along with other precursors, oxides of nitrogen (NO_x), and volatile organic compounds (VOCs) (Seinfeld & Pandis, 2016). It therefore has a dual role in affecting our climate and the quality of air that we breathe.

Aerosols are tiny and often invisible particles that can originate from many sources, including breaking waves via sea salts, fine soil blown into the air, fires, transport combustion, pollen, microorganisms lifted by the wind, and volcanic eruptions. Although aerosols are most abundant on the near surface layer of the Earth, they can transport to the upper levels of the atmosphere.

Many aerosols act as surfaces on which water vapor may condense. They also absorb and reflect incoming solar radiation depending on their physical and chemical properties and hence play a crucial role in determining the radiation balance of the atmosphere (IPCC, 2021a). Fine particulate matter ($PM_{2.5}$ with an aerodynamic diameter of equal or less than 2.5 μm) is especially one of the most hazardous air pollutants in terms of their health impacts (WHO, 2021). In addition, they are associated with microphysical and radiative interactions influencing clouds. A more detailed discussion on aerosols is given in Section 2.7.

Ozone consists of three oxygen atoms per molecule (O_3). Most of the ozone is present in the stratosphere, where it is naturally produced through the photolysis of molecular oxygen. The lower part of the stratosphere, which contains most of the naturally occurring ozone (referred to as the ozone layer in Section 2.3.1.1), protects us from the harmful UV-B and UV-C ultraviolet radiation bands, and hence in common language it is labeled as "good ozone." Compounds containing chlorine and bromine are particularly efficient at destroying ozone and are termed ozone depleting substances (ODS). These include chlorofluorocarbons (CFCs), hydrochlorofluorocarbons (HCFCs), hydrobromofluorocarbons (HBFCs), and methyl bromide.

The presence of ozone in the lower troposphere is generally harmful to humans, animals, and crops and plants as it is chemically reactive. As a consequence, ozone is one of the major air pollutants. Concentrations of ozone are at their highest level during the warmer summer months, when its production is strong following photochemical reactions involving sunlight and precursors such as NO_x, VOC, and methane. Stratospheric photolysis of oxygen does not take place in the lower troposphere, as the wavelength of radiation required for this reaction is absorbed by the diatomic molecular oxygen in the upper atmosphere. The net ozone present is the difference between the production and the loss of ozone. Typical summer mixing ratios of ozone in the stratosphere are 5–10 ppmv, whereas in the lower troposphere or ground level they are 60–70 ppbv in urban background areas and around 100 ppbv or more in rural areas (expressed as daily maximum eight-hour mean concentrations).

2.6.3 Expressing the amount of species in air

There are several expressions for describing the amount of a substance in a fluid, such as a liquid solution or air, as in our case. It can be expressed as ***mole*** (n), or mol, in the SI unit system.

Avogadro number is a constant number of atoms or molecules in 1 mol of the species and expressed as $N_A = 6.02214076 \times 10^{23}\ \mathrm{mol}^{-1}$ as described in Section 2.5.1. Therefore, a ***mole*** can be defined as the amount of a substance containing the same number of basic chemical entities (i.e., atoms, molecules, and ions) as the Avogadro's number ($6.02214076 \times 10^{23}$). Previously, Avogadro's number was defined in relation to the number of atoms contained in exactly 12 g of the C-12 isotope of carbon (or ^{12}C). For most calculations, Avogadro's number can be approximated to 6.022×10^{23}. It follows that the molecular weight or molar mass of a substance in grams contains exactly the Avogadro's number of basic entities.

Taking H_2O as a simple example:

Molecular weight or molar mass of

hydrogen = 1.008 g

oxygen = 15.998 g

Molar mass of H_2O is $2 \times 1.008 + 15.999 = 18.015\ \mathrm{g\ mol}^{-1}$

Similarly, for CO_2, the molas mass $= 1 \times 12.011 + 2 \times 15.999 = 44.009\ \mathrm{g\ mol}^{-1}$ (in this calculation we assume that carbon consists of 98.9% ^{12}C and 1.1% ^{13}C).

Usually, calculations are conducted in grams, but if a kilogram is used, then 1 kg molar mass of a substance would contain 6.022×10^{26} entities, that is 1000 times the Avogadro's number. ***Mass concentration*** (C_i) is an important variable in air quality. Consider an atmospheric species, *i*. If we want to express the amount of species in a volume, we use mass concentration, which is calculated as follows:

$$C_i = \frac{m_i}{V_{\text{air}}} \tag{2.105}$$

where m_i is the amount (or mass) of species and V_{air} is the volume of the air (e.g., cubic meter of air). If we want to express, for example, the amount of CO_2 occupying the air, we can say g CO_2 per m^3 of air (g CO_2 m^{-3} air). In the case of particulate matter, the most common unit for expressing mass concentration in air is $\mu g\ m^{-3}$.

In the case of the ***molar concentration*** (*often called* ***molarity***) (c_i) mass is replaced with number of moles (usually in grams as explained above) as follows,

$$c_i = \frac{n_i}{V_{\text{air}}} \tag{2.106}$$

where n_i is the number of moles of species *i*. The amount of CO_2 in the molar concentration unit is expressed as mole CO_2 per m^3 of air (mole CO_2 m^{-3} air).

The ***Ideal gas law*** is given in Eq. (2.14). Often pressure, *P*, is expressed in units of atmosphere, atm and the volume of air, *V*, is given in the unit Litre, L. The mole of a particular species is expressed as a number, *n*, and the temperature in a unit of Kelvin, K.

In this case the universal gas constant, *R*, for one mole, temperature 273K and pressure 1 atm at sea level, can be calculated as:

$$R = pV/nT = 1\ (\text{atm}) \times 22.4\ (\text{L})\ /\ 1\ (\text{mole}) \times 273\ (\text{K}) = 0.08205\ \text{L atm mol}^{-1}\ \text{K}^{-1}$$

Generally, *R* is used as 0.082 L atm mol^{-1} K^{-1}. Note from basic chemistry, one mole of gas occupies a volume of 22.4 L or 0.0224 m^3 for pressure of air is 1 atm and air temperature of 273 K. Also note that 1 atm is equal to 101325 Pa or 1013.25 hPa.

The mixing ratio is the molar fraction of the species and is very commonly used in atmospheric science and especially in air quality science to express the concentration (C_i) of a specific air pollutant, *i*. For the purposes of this book, the mixing ratio for a species, r_i, is usually calculated relative to an equivalent measure of air using:

$$r_i = \frac{c_i}{c_{\text{air}}} \tag{2.107}$$

where c_i is the molar concentration of *i*, and c_{air} is the total molar concentration of air. For water vapor, the mixing ratio is the ratio of the mass of water vapor divided by the mass of *dry* air for a given volume.

For gas species, ***mixing ratio by volume*** is used in atmospheric science. For example, in the case of nitrogen, which is the most abundant species in the atmosphere, the mixing ratio by volume is 0.7808 mol mol^{-1} or 78.08%. This can be written in ***part per million volume (ppmv)*** as 780,900 ppmv, meaning that there are 780,900 moles of nitrogen per 1 million moles of air. Another similar unit for trace species is that of **parts per billion by volume *(ppbv)*** and is regularly used for lower tropospheric or ground level concentrations of air pollutants such as ozone and nitrogen dioxide. Unless noted otherwise, we will use the units for mixing ratio as ppbv for trace gaseous species except for carbon monoxide, which is found in higher concentrations near the surface and is expressed in ppmv.

In the case of particulate matter ($PM_{2.5}$ and PM_{10}), the concentrations are expressed as mass per volume of air (usually $\mu g\ m^{-3}$) instead of volume mixing ratio (as ppmv or ppbv). However, for gaseous trace species in air quality science, there are numerous instances when we need to convert between units for mixing ratios expressed as ppb or ppm (by volume) and mass per unit volume ($\mu g\ m^{-3}$).

To establish a conversion formula, we'll employ the ideal gas law given in Eq. (2.14), where *P* is atmospheric pressure (in atm), *V* is volume (in litres), *n* is the mole number, *R* is the universal gas constant (0.082 L atm mol^{-1} K^{-1} as explained earlier), and *T* is the temperature (in Kelvin). Here we can replace the number of moles *n* with the mass of the air pollutant (*m*) divided by its molecular weight (M_w) to give the equivalent number of moles:

$$p.V = \frac{m}{M_w}.R.T \tag{2.108}$$

Rearranging Eq. (2.108), we arrive at

$$\frac{m}{V} = \frac{p.M_w}{R.T} \tag{2.109}$$

The left-hand side of the above equation represents the concentration (C) of an air pollutant in terms of mass per unit volume of air. Breaking down the units in Eq. (2.109), we get:

$$C\left(\frac{\mathrm{g}}{\mathrm{L}}\right) = \frac{m(\mathrm{g})}{V(\mathrm{L})} = \frac{p(\mathrm{atm}) \times M_w\left(\frac{\mathrm{g}}{\mathrm{mol}}\right)}{R\left(\frac{\mathrm{L.atm}}{\mathrm{mol.K}}\right) \times T(\mathrm{K})} \tag{2.110}$$

Noting that $1\mathrm{g} = 10^6 \mu\mathrm{g}$ and $1\mathrm{L} = 10^{-3}\ \mathrm{m}^3$, left hand side of the Eq. (2.110) which shows the amount of the specie in the air can be rewritten as:

$$C\left(\frac{\mu\mathrm{g}}{\mathrm{m}^3}\right) = \frac{m(10^6 \mu\mathrm{g})}{V(10^{-3}\mathrm{m}^3)} = \frac{p(\mathrm{atm}) \times M_w\left(\frac{\mathrm{g}}{\mathrm{mol}}\right)}{R\left(\frac{\mathrm{L.atm}}{\mathrm{mol.K}}\right) \times T(\mathrm{K})} \tag{2.111}$$

Simplifying further:

$$C\left(\frac{\mu\mathrm{g}}{\mathrm{m}^3}\right) = \frac{m(\mu\mathrm{g})}{V(\mathrm{m}^3)} \times 10^9 = \frac{p(\mathrm{atm}) \times M_w\left(\frac{\mathrm{g}}{\mathrm{mol}}\right)}{R\left(\frac{\mathrm{L.atm}}{\mathrm{mol.K}}\right) \times T(\mathrm{K})} \tag{2.112}$$

Dividing both side by 10^9, we get

$$C\left(\frac{\mu\mathrm{g}}{\mathrm{m}^3}\right) = \frac{m(\mu\mathrm{g})}{V(\mathrm{m}^3)} = \frac{1}{10^9} \times \frac{p(\mathrm{atm}) \times M_w\left(\frac{\mathrm{g}}{\mathrm{mol}}\right)}{R\left(\frac{\mathrm{L.atm}}{\mathrm{mol.K}}\right) \times T(\mathrm{K})} \tag{2.113}$$

Recognizing that $1/10^9$ is parts per billion (ppbv), which is also a concentration unit, we arrive at the conversion formula to convert ppbv to $\mu\mathrm{g\ m}^{-3}$:

$$C\left(\frac{\mu\mathrm{g}}{\mathrm{m}^3}\right) = \frac{m(\mu\mathrm{g})}{V(\mathrm{m}^3)} = C(\mathrm{ppbv}) \times \frac{p(\mathrm{atm}) \times M_w\left(\frac{\mathrm{g}}{\mathrm{mol}}\right)}{R\left(\frac{\mathrm{L.atm}}{\mathrm{mol.K}}\right) \times T(\mathrm{K})} \tag{2.114}$$

Similarly, the conversion formula from ppmv to $\mu\mathrm{g\ m}^{-3}$; is:

$$C\left(\frac{\mu\mathrm{g}}{\mathrm{m}^3}\right) = \frac{m(\mu\mathrm{g})}{V(\mathrm{m}^3)} = C(\mathrm{ppmv}) \times 1000 \times \frac{p(\mathrm{atm}) \times M_w\left(\frac{\mathrm{g}}{\mathrm{mol}}\right)}{R\left(\frac{\mathrm{L.atm}}{\mathrm{mol.K}}\right) \times T(\mathrm{K})} \tag{2.115}$$

Frequently, these expressions are simplified for Normal Temperature and Pressure (NTP) conditions, which are defined as having an air temperature of 25°C (298K) and an air pressure of 1 atm, which gives the expression for converting concentrations from ppbv to $\mu\mathrm{g\ m}^{-3}$:

$$C\left(\frac{\mu\mathrm{g}}{\mathrm{m}^3}\right) = \frac{\mathrm{m}(\mu\mathrm{g})}{\mathrm{V}(\mathrm{m}^3)} = C(\mathrm{ppbv}) \times \frac{\mathrm{M_w}\left(\frac{\mathrm{g}}{\mathrm{mol}}\right)}{24.45} \tag{2.116}$$

Note that the volume occupied at NTP by one mole of gas is 24.45 L. As mentioned previously, at an air pressure of 1 atm and an air temperature of 273K, one mole of gas occupies a volume of 22.4 L.

Similarly, for ppmv conversion to $\mu\mathrm{g\ m}^{-3}$:

$$C\left(\frac{\mu\mathrm{g}}{\mathrm{m}^3}\right) = \frac{\mathrm{m}(\mu\mathrm{g})}{\mathrm{V}(\mathrm{m}^3)} = C(\mathrm{ppmv}) \times 1000 \times \frac{M_w\left(\frac{\mathrm{g}}{\mathrm{mol}}\right)}{24.45} \tag{2.117}$$

which gives

$$C\left(\frac{\mu\mathrm{g}}{\mathrm{m}^3}\right) = C(\mathrm{ppmv}) \times M_w\left(\frac{\mathrm{g}}{\mathrm{mol}}\right) \times 40.9 \tag{2.118}$$

Worked example 2.7

Calculate 50 ppbv O_3 and NO_2 concentrations in the unit of $\mu\mathrm{g\ m}^{-3}$ for normal temperature and pressure conditions (25°C and 1 atm). Universal gas constant R is 0.082 L atm $\mathrm{mol}^{-1}\ \mathrm{K}^{-1}$, molecular weight of O_3 is 48.00 g mol^{-1}, and NO_2 is 46.01 g mol^{-1}.

Solution:

The conversion formula from ppbv to $\mu\mathrm{g\ m}^{-3}$ is given in Eq. (2.114) as

$$C\left(\frac{\mu\mathrm{g}}{\mathrm{m}^3}\right) = C(\mathrm{ppbv}) \times \frac{p(\mathrm{atm}) \times \mathrm{M_w}\left(\frac{\mathrm{g}}{\mathrm{mol}}\right)}{R\left(\frac{\mathrm{L.atm}}{\mathrm{mol.K}}\right) \times \mathrm{T(K)}}$$

50 ppbv O_3 in $\mu g\ m^{-3}$ is calculated as:

$$C\left(\frac{\mu g}{m^3}\right) = 50\ \text{ppbv}\ O_3 \times \frac{1\text{atm} \times \left(48\frac{g}{mol}\right)}{\left(0.082\frac{L.atm}{mol.K}\right) \times (25 + 273.15)K} = 98.17\ \mu g\ m^{-3}$$

50 ppbv NO_2 in $\mu g\ m^{-3}$ is calculated as:

$$C\left(\frac{\mu g}{m^3}\right) = 50\ \text{ppbv}\ NO_2 \times \frac{1\text{atm} \times \left(46.01\frac{g}{mol}\right)}{\left(0.082\frac{L.atm}{mol.K}\right) \times (25 + 273.15)K} = 94.09\ \mu g\ m^{-3}$$

These calculations will give the same result with the simplified Eq. (2.116).

2.6.4 Timescales of atmospheric transport and residence time

The average time that a species resides in the atmosphere before removal is the ***lifetime*** of the species. This ***residence time*** is important to understand to assess the impact of human-made emissions on the atmosphere and the general environment.

Fig. 2.26 illustrates the various lifetimes of chemical species within the atmosphere, categorizing them as short-lived, moderately long-lived, and long-lived. The figure also shows indicative temporal and spatial scales of variability for atmospheric constituents. Examining the figure, we observe that chemical species exhibit diverse residence times in the troposphere—ranging from seconds (e.g., hydroxyl radical, OH), days to weeks (e.g., ozone, O_3; ammonia, NH_3), and years (e.g., methane, CH_4, approximately 10 years; carbon dioxide, CO_2, lasting from a few years to around 200 years) to even thousands or millions of years (e.g., oxygen, O_2; nitrogen, N_2), depending on the effectiveness of the removal processes by sinks. CO_2, for example, can be absorbed from the lower atmosphere into leafy plants and into oceans in less than a few years, but as it is being added into the atmosphere through anthropogenic emissions, it remains in the air for a long time.

Production and removal of the gases occur simultaneously in the atmosphere. Chemical processes, biological activity, volcanic eruptions, human activities, and radioactive decay are the sources of the gases in the atmosphere. Removal mechanisms from the atmosphere are physical processes (such as state changes or particle formations), chemical transformations, uptake and emissions by the oceans, and deposition on the water bodies (wet deposition) and land masses (dry deposition),

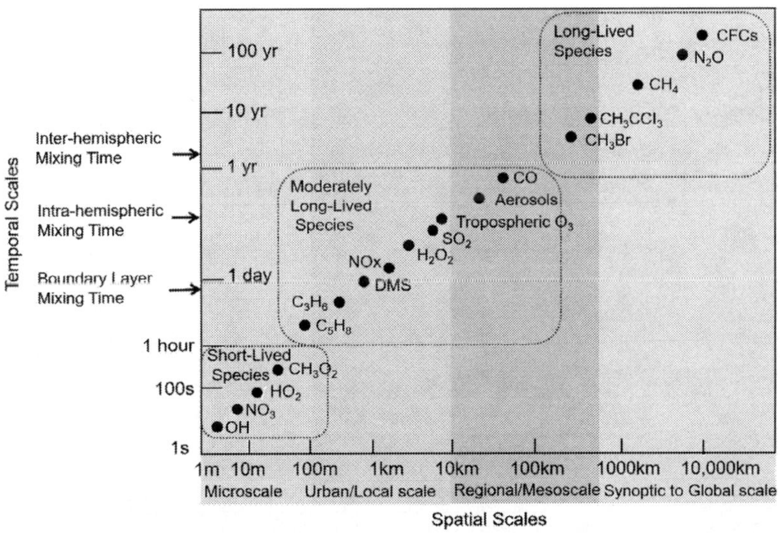

FIGURE 2.26 Temporal and spatial scales of variability for atmospheric constituents. The vertical axis signifies the time required for changes in atmospheric species, while the horizontal axis denotes the distance these species can travel before changing. Categorization into short-lived, moderately long-lived, and long-lived groups is determined by their positions on these axes. *Adapted from Seinfeld, J. H., & Pandis, S. N. (2016). Atmospheric chemistry and physics: From air pollution to climate change. Wiley.*

as well as exchanges with ecosystems. The average lifetime of a species depends on the effectiveness of these types of removal mechanisms as well as their production rates.

Wind fields are the primary factor determining horizontal transport of airborne species. In the west–east direction (or vice versa), the characteristic wind velocity is referred to as the **zonal flow** (U). In the south–north direction (or vice versa), the wind velocity is termed **meridional flow**, typically having a lower value compared to zonal flow (NOAA, 2023). When a trace species is emitted, the characteristic timescale (τ) can be calculated for the transport of the emitted trace species:

$$\tau = \frac{L}{U} \tag{2.119}$$

where L is the characteristic spatial scale in the horizontal direction, and U is the wind speed.

Two mechanisms contribute to interhemispheric mass exchange: large-scale eddy transfer and mean meridional circulations. As discussed in previous sections, dynamical and transport processes take place on a range of spatial (micro to global) and temporal (seconds to decades) scales. Spatially, the scales of motion in the atmosphere vary from small eddies of a centimeter or less in size to substantial airmass movements of over continental dimensions. Large-scale movement of air masses and associated species involving interhemispheric exchange is influenced by complex circulations, including large-scale eddy transfer and mean meridional circulations, the Inter-Tropical Convergence Zone (ITCZ), and trade winds (see Section 2.4.4). In the ITCZ region, for example, surface low pressure develops with strong updrafts of warm and wet air masses, which play an essential role in the equatorial region to move air masses between hemispheres. The combined effect of these influences is that the typical timescale for the interhemispheric exchange is the order of about one year. The timescale for mixing to occur within the troposphere up to the tropopause is about a few months, whereas transport exchange between the troposphere and the stratosphere can take a few years.

For the lower part of the atmosphere, which is critical for air quality, the vertical distribution of temperature governs the vertical motion (see section 2.3). The strength of the vertical movement is determined by a vertical turbulent diffusion coefficient, K_z, which has a wide range of values from 1 to about 30 $m^2\ s^{-1}$ depending on different meteorological conditions. Vertical diffusion can have a significant impact on the concentration of air pollutants as it affects how vertical exchanges take place (see, e.g., Huszar et al., 2020). In the vertical motion, the molecular diffusion can be neglected, except in a thin layer just above the Earth's surface.

The main air pollution and global pressures are shown schematically in Fig. 2.27 according to temporal and spatial scales. Depending on the nature of the episodic air pollution events, their duration tends to be from a few to several

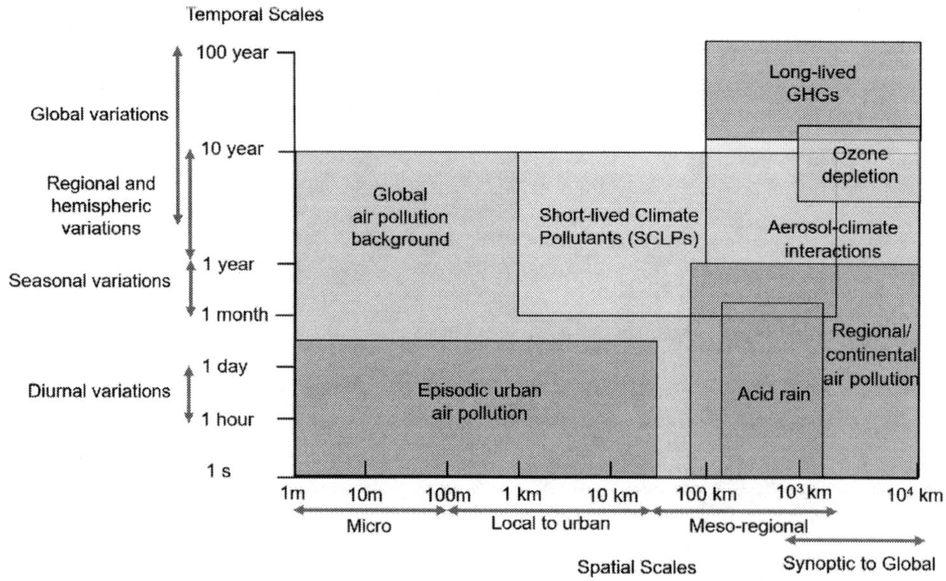

FIGURE 2.27 A schematic illustration of the main air pollution-related events according to temporal and spatial scales. Note that the scales are neither linear nor definitive and merely serve to be indicative in terms of space and time.

days, whereas regional and global air pollution background can be spread over large spatial scales and persist for months to years, especially without appropriate mitigation. Transport distances for acid rain are usually up to a few thousand kilometers, and stratospheric ozone depletion impacts spatial scales of typically 1000s of kilometers. Climate-related effects manifest themselves from months, as in the case of SLCPs, to many years in the case of longer-lived GHGs.

As indicated in Figs. 2.26 and 2.27, atmospheric scales of motion can be classified into four general categories: microscale, local to urban scale, meso and regional scale, synoptic to global scales. ***Global scale*** atmospheric transport occurs on spatial scales from about a few thousand kilometers. Global air pollution problems such as stratospheric ozone depletion and the increase in greenhouse gases are examples of global-scale changes. ***The synoptic*** scale includes motions of whole weather systems on several hundred kilometers to thousands of kilometers, which overlaps with the ***mesoscale*** transport taking place over a few hundred kilometers to typically a few thousand kilometers. Examples of meteorological phenomena taking place over synoptic and mesoscales include land-sea breezes, mountain-valley winds, and migratory high- and low-pressure fronts.

Atmospheric variations occurring on scales of up to a few 100 m to a few tens of kilometers, such as the meandering and dispersion of a chimney plume, occur over city or urban and finer scales. While the impact of long-lived GHGs (e.g., CO_2) can take the order of 10s to around 100 years if not longer, short-lived climate pollutants (SLCPs) or forcers (SLCFs) can have an impact on much shorter spatial and temporal scales. Acid rain impacts, while not strictly considered as an air quality issue, take place over 100 to a few thousand kilometers. On the other end of the scale, turbulent flow regimes, for example around buildings and street canyons, usually take place over ***micro- to local- and urban scales***. When considering spatial scales of relevance to air quality, these range from meters to a few 10s of kilometers (local and urban air pollution) to around 100 km to a few 10^4 km for regional or continental air pollution.

2.7 Atmospheric aerosols

Particles in the atmosphere arise from natural sources, such as windborne dust, volcanoes, and many anthropogenic combustion activities. Although aerosols can be considered to be a suspension of solid or liquid particles in air, in our context, we are usually interested in the solid particulate matter components. For a broader review of atmospheric aerosols, the reader is directed to Carslaw (2022), McCoy (2019), Ramachandran (2018), and Boucher (2015). Table 2.4 shows the main sources and annual emission fluxes of natural and anthropogenic aerosols.

2.7.1 Classification of aerosols

For air quality purposes, particulate matter is usually classified in terms of their aerodynamic size. Particles with an aerodynamic diameter of equal to or less than 2.5 μm ($PM_{2.5}$) are called ***fine particles***, and those with an aerodynamic diameter between 2.5 and 10 μm are referred to as ***coarse particles***. A common fraction is that of PM_{10}, which represents a fraction of particles with an aerodynamic diameter of equal to or less than 10 μm and hence contains both the fine ($PM_{2.5}$) and coarse fractions ($PM_{2.5-10}$). The distinction between fine and coarse particles is fundamental since their origin, transformation mechanisms, deposition patterns, chemical composition, radiative optical properties, and their health impact differ significantly.

Natural aerosols include those emitted by volcanoes, fog, sea spray, vegetation, pollen, and windblown mineral dust (i.e., dust lifted from deserts by wind). These are characterized as ***primary aerosols*** as they are emitted directly from their sources, which in this case are all natural. Aerosols that are emitted directly as a consequence of human activity are termed ***primary anthropogenic aerosols*** (e.g., industrial processes, construction, road transport).

Secondary aerosols are those that are formed through chemical reactions in the atmosphere. Transformation from the gas phase takes place through gas-particle conversion processes (usually referred to as heterogeneous reactions as two phases are involved). Once emitted, their size and composition can change through further chemical reactions or processes such as coagulation and accumulation with other particles or becoming seeds (condensation nuclei) for fog and cloud droplets in the presence of water.

Aerosols are further classified as either ***inorganic or organic aerosols*** depending on their chemical composition. ***Primary inorganic aerosols*** include black carbon, mineral dust, and sea salt. ***Secondary inorganic aerosols*** are formed with reactions involving SO_2, NO_x, and NH_3 and produce compounds including sulfates (SO_4^{2-}), nitrates (NO_3^-), and ammonium (NH_4^+). The predominant fraction of secondary inorganic aerosols is found in the form of NH_4NO_3 and $(NH_4)_2SO_4$ as well as $NaSO_4$.

A major fraction of fine aerosols consists of ***organic aerosols (OA)***, sometimes referred to as ***particulate organic material***. In addition to being an important component of PM2.5, they also play a key role in influencing climate forcing (see, e.g., Shrivastava et al., 2017). As with inorganic aerosols, organic aerosols can also be classified as primary and secondary according to their source. Sources of ***primary organic aerosols*** (POA) include biomass burning (e.g., agricultural fires or wildfires), fossil fuel combustion (e.g., energy production), road transport, industrial emissions, volcanic eruptions, and sea spray. Secondary organic aerosols (SOA) are formed through photochemical oxidation of primary organic matter and gas-phased volatile organic compounds (VOCs). The resulting oxidized compounds undergo nucleation or condensation onto the surfaces of preexisting particles or form new particles. A multitude of volatile organic compounds (VOCs) and their variations present in the atmosphere undergo complex reactions with hydroxyl radicals (OH), ozone (O_3), and nitrate radicals (NO_3), leading to organic products that can partition onto preexisting particles or nucleate, producing new SOA particles. Both biogenic VOCs (naturally emitted from vegetation, BVOCs, such as isoprene and terpenes) and anthropogenic VOCs (emitted from human activities, AVOCs) are involved in these reactions, which take place in the gas-to-gas and gas-to-particle phases (heterogeneous reactions) (see Shrivastava et al., 2017 for more details).

Another major category of natural aerosols is ***sand and dust***, as shown in Table 2.4, which shows that dust is the largest emissions of natural aerosols (e.g., Boloorani et al., 2023; Boucher, 2015). Sand and dust are largely emitted from dry, arid, and semiarid regions since these regions cover one-third of the global land area on the Earth. The Sahara/Sahal in North Africa, Middle East, and Central Asia regions, for example, are some of the largest sand and mineral dust sources (see, e.g., Zender et al., 2004; Kok et al., 2021). Natural dust emissions cause a large uncertainty

TABLE 2.4 Sources of aerosols and annual emissions.

Aerosol type	Emissions (per year)
Natural primary aerosols	
Desert sand and dust	1000–3000 Tg
Sea spray	1000–6000 Tg
Biomass burning aerosols	20–35 Tg
Terrestrial primary biogenic aerosols	Order of 1000 Tg
Including bacteria	40–1800 Gg
Including spores	30 Tg
Precursors of natural secondary aerosols	
Dimethylsulphide (DMS)	20–40 Tg S
Volcanic SO_2	6–20 Tg S
Terpenes	40–400 Tg
Anthropogenic primary aerosols	
Industrial dust	40–130 Tg
Biomass burning aerosols	50–90 Tg
Black carbon (from fossil fuel)	6–10 Tg
Organic carbon (from fossil fuel)	20–30 Tg
Anthropogenic secondary aerosols	
SO_2	70–90 Tg S
Volatile organic compounds (VOCs)	100–560 Tg C
NH_3	20–50 Tg N
NO_x	30–40 Tg N

Note: Tg = 10^{12} g = 1 million tons. Gg = 10^9 g = 1 thousand tons,
C, carbon; *S*, sulfur; *N*, nitrogen.
Source: Courtesy of Boucher, O. (2015). Atmospheric aerosols, properties and climate impacts. Springer. ISBN: 978-94-017-9648-4.

in the air quality and climate model estimates because of the complexity of the estimation of processes that raise dust from their sources (e.g., Arias et al., 2021; Pai et al., 2022). The studies by Sokhi et al. (2021) and Pai et al. (2022), as examples, illustrate that reductions in anthropogenic sources of air pollutants such as $PM_{2.5}$ may not be enough to meet the current air quality guidelines issued by the World Health Organization (WHO, 2021).

Anthropogenic aerosols are produced by human activities such as fossil fuel combustion, road transport, biomass burning, and industrial production. Although less abundant in the atmosphere than natural aerosols (see Table 2.4), anthropogenic aerosols are a significant hazard to human health (WHO, 2021). Fig. 2.28 shows the global distribution of aerosols shown as monthly averaged values of aerosol optical depth (AOD) derived from satellite observations (NASA, 2023) (see Chapter 9, Section 9.6 for an explanation of AOD). The figure clearly shows the heterogeneity of aerosol loadings, with high AOD values indicating high amounts of aerosols and hazy conditions in the atmosphere and low values depicting clear sky conditions. The heterogeneities in aerosol loading arise, for example, in parts of South America from emissions related to land clearing and agricultural fires during dry conditions. Similar agricultural fires are prevalent in parts of Africa and regions of Southeast Asia. As mentioned above, sources such as dust storms can also increase aerosols in the atmosphere dramatically (Kok et al., 2021).

2.7.2 Processes affecting aerosols

The processes that influence particle sizes in the atmosphere are shown in an idealized schematic in Fig. 2.29, which depicts a typical distribution of atmospheric aerosols in terms of their surface area (Whitby and Cantrell, 1976).

For most air quality applications, the size of atmospheric aerosols can often be divided into three modes. The first mode can be considered to include particles with diameters up to 10 nm that are accounted for in ***nucleation (or nuclei) mode*** and those with a size range between 10 and 100 nm (0.1 μm), which is referred to as the ***Aitken mode***. These are also often referred to as **ultrafine aerosols** and exhibit a very high number concentration but low mass concentration in the atmosphere. The second mode is the ***accumulation mode***, which extends from 100 nm to 2.5 μm, while the third model is the ***coarse mode*** and includes particles of diameter greater than 2.5 μm.

Fig. 2.29 also illustrates the corresponding formation and removal mechanisms relevant for each mode. Aerosols in the nucleation model are primarily formed by condensation of low-volatility compounds and heterogeneous gas-particle reactions leading to either new particle formation or growth on surfaces of existing aerosols. As mentioned earlier, many atmospheric species are involved in the early stages of aerosol formation, such as the nucleation of sulfuric acid, water, and ammonia (Almeida et al., 2013; Schobesberger et al., 2013; Semeniuk & Dastoor, 2018).

These are then undergoing ***coagulation***, which is the result of particles colliding and aggregating into larger particles and moving them into the ***accumulation mode***. Through this process, the number concentration of the ultrafine aerosols rapidly decreases while the mass concentration increases (Seinfeld & Pandis, 2016).

A substantial part of the aerosol mass (with number concentration dropping) belongs to the accumulation mode, and a large proportion of aerosols in the accumulation mode result from direct emissions as well (e.g., primary emissions from combustion).

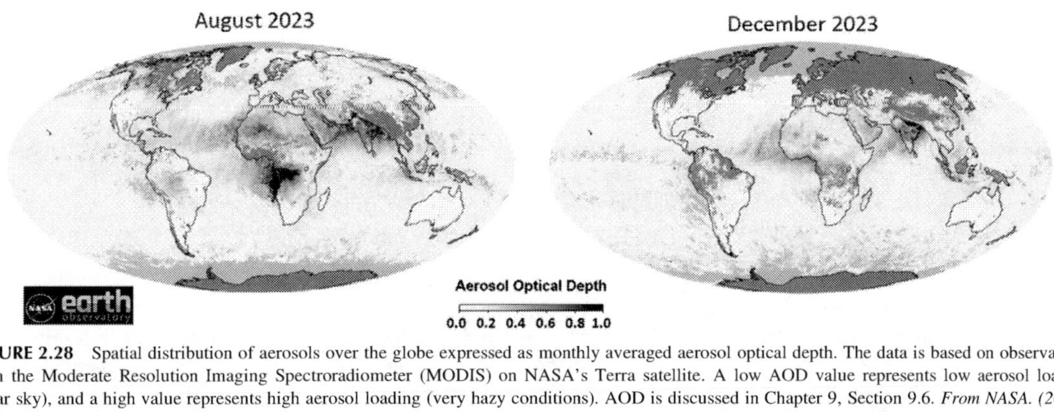

FIGURE 2.28 Spatial distribution of aerosols over the globe expressed as monthly averaged aerosol optical depth. The data is based on observations from the Moderate Resolution Imaging Spectroradiometer (MODIS) on NASA's Terra satellite. A low AOD value represents low aerosol loading (clear sky), and a high value represents high aerosol loading (very hazy conditions). AOD is discussed in Chapter 9, Section 9.6. *From NASA. (2023). NASA Earth Observatory.* Aerosol optical depth. *Retrieved from https://earthobservatory.nasa.gov/global-maps/MODAL2_M_AER_OD. (accessed in 2024).*

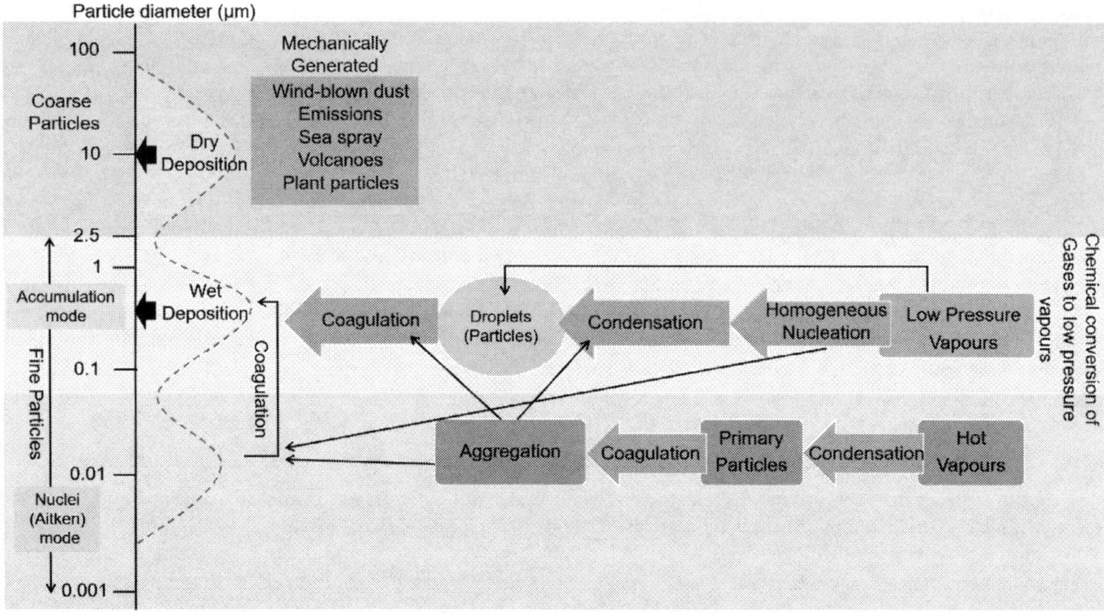

FIGURE 2.29 Idealized schematic of the distribution of particle surface area of atmospheric aerosols. Principal modes, sources, and particle formation and removal mechanisms are indicated. Note, that nuclei mode is sometimes used to refer to nucleation mode. *Adapted from Whitby, K. T., and Cantrell, B. (1976). Fine particles, Proc. Int. Conf. Environmental Sensing and Assessment, Las Vegas, NV, Institute of Electrical and Electronic Engineers; Lazaridis, M. (2011). First principles of meteorology and air pollution. Environmental Pollution, Springer.*

In terms of air quality applications, these modes form part of the $PM_{2.5}$ size fraction of atmospheric particulate matter along with other directly emitted and secondary aerosol contributions.

In the coarse mode, particle sizes between 2.5 and 10 μm, particles are generated mainly by mechanical processes with sources such as dust, abrasion particles, and construction. Since their diameter is large, their number concentration is very low but is responsible for a large fraction of the mass concentrations. Their lifetime in the atmosphere is low compared to finer aerosols as they have significantly larger mass, which results in their removal by gravitational sedimentation and impaction. Generally, $PM_{2.5}$ and PM_{10} are removed through wet (through precipitation) and dry (settling) deposition.

As mentioned previously, for air quality purposes, we are concerned with $PM_{2.5}$ and PM_{10} fractions. However, the role of ultrafine aerosols in climate interactions is very important, as already discussed, and while more research is needed, their impact on health is also being recognized (e.g., Schraufnagel, 2020).

2.8 Ozone depletion

One of the most important human-induced global-scale hazards is that of ***ozone depletion***, which is strongest over the polar regions, especially the Antarctic. Usually, it is not treated as an air pollution issue, although many of the chemical species that are responsible for depleting stratospheric ozone originate from ground-level emission sources. We discuss it briefly here to place it in the broader context of anthropogenically driven atmospheric hazards. As discussed in Sections 2.2.2.3 and 2.6.2, the ozone layer protects life from the harmful part of solar radiation. Some ultraviolet wavelengths destroy microorganisms on the ground and cause human skin cancer.

Ozone-depleting substances (ODS) emitted into the atmosphere, especially the group of chlorofluorocarbon (CFC) compounds, chemically destroy ozone in the stratosphere. These gases are used as coolants for air conditioning, in refrigerators, as propellants for aerosol sprays, and in some industrial processes. Ozone reduction in the stratosphere directly increases the intensity of ultraviolet radiation from the Sun reaching the Earth's surface. Colloquially, this phenomenon has been called the ***ozone hole***, but it is not a "hole" but a large area where stratospheric ozone concentrations significantly diminish for several months (e.g., over the south pole) compared to typical levels.

Ozone depletion was first highlighted in 1985 by Farman et al. (1985), who observed a large, continuous decline in ozone concentrations in the Antarctic since the 1970s. This and other confirmatory studies led to the adoption of the widely regarded successful Montreal Protocol that banned ODSs in 1987. Some of the most important ODSs are the chlorofluorocarbon compounds (CFC) mentioned above. Chlorine plays a fundamental role in destroying ozone, as explained below. CFCs include many compounds such as trichlorofluoromethane, $CFCl_3$ (CFC-11), dichlorodifluoromethane, CF_2Cl_2 (CFC-12), and 1,1,2 trichlorotrifluoroethane, $CF_2ClCFCl_2$ (CFC-113).

Sharp drops in ozone concentrations were observed over Antarctica during the southern hemispheric spring, and this is why this phenomenon is sometimes called ***Antarctic ozone depletion***.

The main mechanisms that lead to this reduction in stratospheric ozone over the Antarctic are complex but can be summarized as follows: Emissions of anthropogenic ODS chemicals (e.g., propellants and refrigerants) are released into the atmosphere and are eventually transported into the stratosphere as they are relatively inert and persistent over decadal timescales. During the winter months of the Southern Hemisphere (July and August), low-pressure systems associated with strong winds circulate in the region and form the polar vortex. Once formed, the polar vortex inhibits mixing and transport of air masses containing ODS species with the surrounding air. The vortex is stronger and lasts several months up to November in the southern polar region compared to the north pole, where it is weaker, has a shorter duration, and breaks up because of greater regional mixing. Because of dark winter months, temperatures reach extremely low values (typically ~ -75 to $-80°C$). Under these conditions, polar stratospheric clouds (PSC) form as water and soluble acids (nitric and sulfuric) condense. These clouds can persist for several months in the Antarctic compared to a couple of months in the Arctic, as temperatures remain lower longer in the southern pole.

The usually inert ODSs (e.g., CFC mentioned above) react chemically on the surface of PSC liquid and solid ice particles, releasing chlorine (Cl_2). As solar radiation builds up from early spring (from September in the Antarctic), chlorine molecules split photochemically into highly reactive free chlorine atoms, which then chemically react with ozone, which is also a reactive molecule, essentially destroying ozone.

There are other halogen-containing compounds in the stratosphere that also play a significant role in the ozone depletion chemistry. The most important in the stratosphere are hydrogen chloride (HCl), chlorine nitrate ($ClONO_2$), and bromine nitrate ($BrONO_2$), which act as reservoirs for chlorine and bromine in the stratosphere (see, e.g., Salawitch et al., 2019). Although they do not react directly with ozone, they can be converted to reactive species such as chlorine monoxide (ClO) and similarly to bromine monoxide (BrO) along with chlorine and bromine atoms. Almost all of HCl and $ClONO_2$, for example, are converted to ClO on PSC particles. A typical but simple example of a photochemical reaction catalytic cycle involving CFC, chlorine, and ozone molecules is:

$$CFCl_3 + h\nu \rightarrow Cl + CFCl_2 \tag{2.120}$$

Since both chlorine atoms and ozone are highly reactive, they react, destroying the ozone molecule to form chlorine monoxide and an oxygen molecule:

$$Cl + O_3 \rightarrow ClO + O_2 \tag{2.121}$$

Chlorine monoxide, reacts with another ozone molecule:

$$ClO + O_3 \rightarrow Cl + 2O_2 \tag{2.122}$$

This second reaction produces more chlorine atoms, which then continue the catalytic cycle to destroy ozone. ClO also reacts with oxygen atoms, which form when O_3 and O_2 molecules split through photolysis:

$$ClO + O \rightarrow Cl + O_2 \tag{2.123}$$

A net reaction can be written as:

$$O_3 + O \rightarrow 2O_2 \tag{2.124}$$

Chlorine atoms act as catalysts, and hence a single free Cl atom can destroy thousands of O_3 molecules. Compounds containing bromine (Br) also play a large role to destroy ozone. During this spring season over the Antarctic, the stratospheric ozone increasingly depletes, and the area affected over the south pole grows. Eventually, as the solar radiation increases and hence temperatures rise during the spring months, the polar vortex subsides and mixing of air masses strengthens, leading to dispersion of the harmful ODSs and any free chlorine atoms. In the Antarctic region, ozone depletion is observed till around November, after which concentrations recover. The whole cycle repeats in the following year. Fortunately, as a consequence of the banning of ODSs following the Montreal Protocol, there are positive signs that there is a slow recovery of the ozone layer over these regions, as has been reported in the United Nations assessment report (WMO, 2022). Further details on ozone depletion and its impacts can be found in Muller (2012) and Reimann et al. (2018).

2.9 Global climate change

Chapter 11 provides a detailed treatment of the interactions between air quality and climate change. This section introduces climate change briefly to provide a broader context. ***Global climate change*** generally refers to the long-term changes in global air temperature, sea level rise, and weather patterns. These changes are driven by natural and anthropogenic causes. However, since the 1800s, human activities such as combustion of fossil fuels like coal, gas, and oil have been the major cause of anthropogenic global climate change. These combustion activities generate greenhouse gas emissions, such as CO_2, that absorb infrared radiation and are raising the air temperatures above preindustrial levels. Some of the significant negative consequences of climate change include melting polar ice, rising sea levels, water scarcity in some regions, intense droughts, severe wildfires, flooding, catastrophic storms, and declining biodiversity (IPCC, 2021a).

Greenhouse gas (GHG) concentrations have been at their highest levels since the last million years and are still increasing (e.g., Lüthi et al., 2008). Consequently, the Earth is about 1.1°C warmer than it was in the late 1800s. The warmest decade on record was 2011–2020 (IPCC, 2021a).

2.9.1 Key greenhouse gases and sources

Carbon dioxide (CO_2), methane (CH_4), nitrous oxide (N_2O), fluorinated gases (F-gases), and aerosol components such as black carbon (BC), sulfates (SO_4), and nitrates (NO_3) are all radiative forcers, or sometimes called climate forcers (species that influence the radiation balance of the planet) and influence the Earth's greenhouse effect. The relative proportion of these anthropogenic emissions of global greenhouse species is illustrated in Fig. 2.30. Other important greenhouse gases are vapor (H_2O) that is present naturally and ozone (O_3), which resides in the stratosphere in large amounts as the protective ozone layer.

CO_2 is a necessary GHG species that is critical for the radiation balance of the planet and is naturally present as part of the Earth's carbon cycle. Compared to all anthropogenic GHG species, CO_2 makes about 65% of the emissions. Its main emission sources are fossil fuel combustion for energy and heat production (i.e., use of coal, natural gas, and oil), transport, some industrial processes (e.g., manufacturing), construction, and land-use changes (e.g., deforestation and soil degradation). Methane (CH_4) makes 16% of the GHG emissions and is emitted by leakages from all stages of natural gas production, storage and distribution, coal mining, agricultural activities (e.g., livestock production), waste management, and biomass burning.

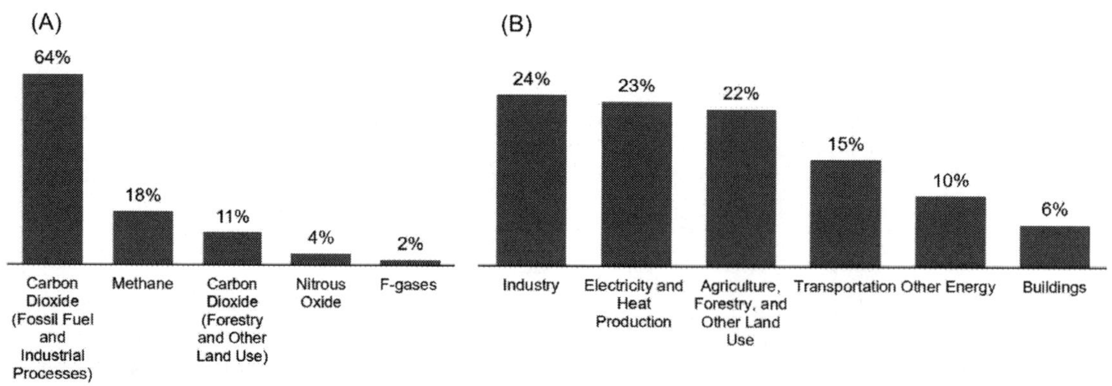

FIGURE 2.30 Proportion of (A) global greenhouse gas emissions (B) economic sectors contributing to the total of global greenhouse gas emissions for 2019 (IPCC, 2022). *Data from IPCC. (2022).* Climate change 2022: Mitigation of climate change. Contribution of working group III to the sixth assessment report of the intergovernmental panel on climate change. *In P. R. Shukla, J. Skea, R. Slade, A. Al Khourdajie, R. van Diemen, D. McCollum, M. Pathak, S. Some, P. Vyas, R. Fradera, M. Belkacemi, A. Hasija, G. Lisboa, S. Luz, J. Malley (Eds.). Cambridge University Press. doi: 10.1017/9781009157926.*

The main source of nitrous oxide is agricultural activities, especially fertilizer usage and fuel combustion. Other sources include fuel combustion, wastewater management, and some industrial processes. Fluorinated gases are mainly synthetic in origin and originate from industrial activities, refrigeration, and consumer cosmetic products such as perfume sprays. They were developed mainly as alternatives to ozone-depleting substances. Fluorinated gases include hydrofluorocarbons (HFCs), perfluorocarbons (PFCs), and sulfur hexafluoride (SF6). While their emissions are relatively small compared to other GHGs, they have a very large global warming potential.

2.9.2 Climate responses from different greenhouse gas species

Responses from different atmospheric GHG species depend on a number of factors, including their radiative forcing properties, their abundance, and their lifetimes. In the latest IPCC report on the Physical Basis of Climate Change (IPCC, 2021a), the ***effective radiative forcing (ERF)*** metric is used as the measure of energy imbalance to a radiative perturbation from a climate forcing species (those that cause changes in the radiation balance of the Earth) (e.g., Smith et al., 2020). It is a useful metric as it includes adjustments and feedback from the atmosphere and the Earth's surface. A positive value of ERF indicates a warming effect, and a negative value indicates a cooling effect.

Fig. 2.31 shows the effective radiative forcing and the resultant contribution to global surface temperatures for each of the key components over the period 1750–2019 based on the latest model simulations from the Coupled Model Intercomparison Project Phase 6 (From IPCC WG1 Technical Summary report Arias et al., 2021). The figure shows the radiative forcing and global surface temperature changes due to long lifetime (CO_2), medium lifetime (e.g., CH_4), and short-lived climate forcers (SLCFs) (e.g., NO_x, black carbon, and ozone) GHG components.

SLCFs (also called short-lived climate pollutants, SLCPs) include methane (CH_4), nitrogen oxides (NO_x), sulfate (SO_4), sulfur dioxide (SO_2), nitrate (NO_3), ammonia (NH_3), organic carbon (OC), black carbon (BC), and ozone (O_3). The life-time of SLCFs (apart from CH_4, whose life-time is about 10 years in the atmosphere) is short, from a few hours to a couple of months. These species are particularly important for air quality, as species SO_4, NO_3, OC, and BC are all components of particulate matter. GHG components such as NH_3 are involved in the formation of secondary particulate matter, and CH_4, along with water vapor, is involved in ozone chemistry. Nitrogen dioxide (NO_2), one of the oxides of nitrogen, is a key air pollutant, especially in cities and towns. Ozone is formed through precursors such as NO_x and volatile organic compounds (VOCs) under conditions of sunlight and is a significant air pollutant and harmful to health (e.g., under photochemical episodes) and ecosystems. Mitigation strategies that control the SLCFs provide an important ***cobenefit*** to improve air quality as well as reduce the impacts from climate change. However, the situation is complicated as SLCFs such as sulfates and nitrates have a cooling effect, whereas BC has a warming effect on climate-induced temperature changes (see, e.g., Szopa et al., 2021).

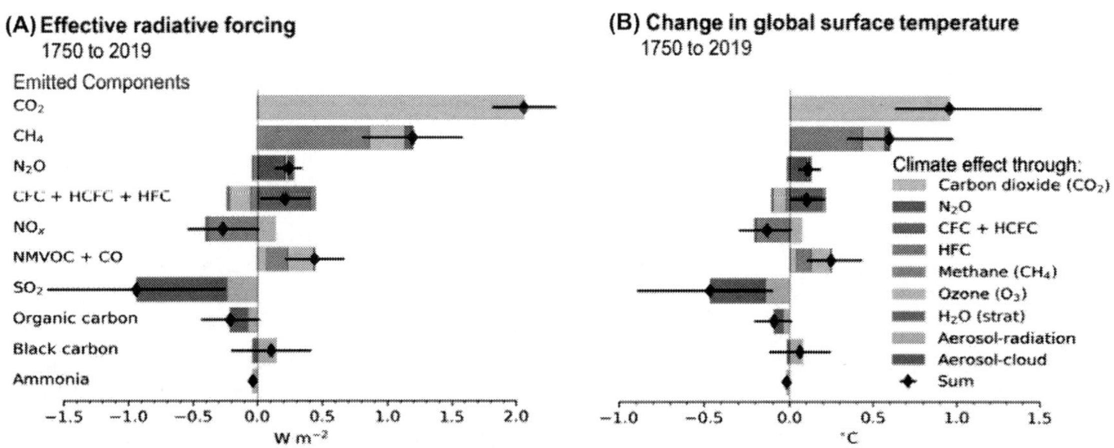

FIGURE 2.31 Contributions from GHG species to (A) effective radiative forcing (ERF) and (B) global surface temperature change for the period 1750–2019. This data is based on the latest Coupled Model Intercomparison Project Phase 6 (CMIP6) model outputs. This figure is taken from Figure TS.15, panels A and B, in Arias et al., (2021), with the courtesy of the IPCC. The error bars represent the 5%–95% confidence intervals, which indicate the uncertainties in the mean values. The methodology for deriving this data is described in Vautard et al. (2020).

2.9.2.1 Climate feedback mechanisms

There are many outcomes possible when a climate system component is altered. These arise from processes that cause a perturbation to one component affecting another and vice versa and are referred to as ***climate feedback***. In our context, we are concerned with feedback that alters the energy imbalance of the Earth. When a feedback process results in warming of the atmosphere, it is ***positive*** and amplifies the initial imbalance; when it results in cooling, it is ***negative*** and dampens the initial imbalance.

The term **positive feedback mechanisms** is used when the alterations reinforce the initial change. The most important example is the increase in evaporation with warmer surface temperatures, which leads to water vapor increases in the atmosphere. Water vapor is an important absorber of terrestrial radiation, and as it increases in the atmosphere with increasing temperatures, it further exacerbates the rise in temperature. There are other important positive climate feedbacks, including the release of methane in the arctic as a consequence of thawing permafrost peat bogs, rainforest drying, which reduced their uptake of CO_2, forest fires, which release CO_2 and air pollutants, changes in land-use/land-cover practices and increasing desertification, snow/ice-albedo feedback, and cloud feedback, which is still associated with significant uncertainties when quantifying the impacts of climate change (Vautard et al., 2020).

Negative feedback mechanisms act to reduce or dampen initial change and tend to offset it. For example, cloud cover may increase due to high moisture levels in the atmosphere resulting from higher temperatures. These additional clouds will reflect more shortwave solar radiation and thus diminish the amount of solar energy leading to a cooling of the atmosphere. However, clouds also have positive feedback by absorbing and emitting more longwave terrestrial radiation simultaneously and preventing it from being lost into the Earth's atmosphere. The situation is complex as it is dependent on cloud type, seasons, and regions in addition to being influenced by thermodynamical temperature-humidity effects, large-scale circulation, and microphysical (small-scale) processes that take place within the clouds. Aerosols add another layer of complexity as they can act as condensation cloud nuclei or, more simply, act as seeds for the formation of clouds. As mentioned previously, there is less confidence in understanding the role of cloud and aerosol-cloud feedback on climate (see, e.g., Seinfeld et al., 2016).

The fundamental radiation Stefan–Boltzmann law introduces an important negative feedback. The law states that the radiation emitted by a black body depends on the fourth power of its temperature (see Section 2.2.4.2). This implies that as the Earth's temperature increases, so does the amount of longwave radiation that is lost according to the fourth power of the temperature, which has a cooling effect on the planet. This feedback, which has a stabilizing effect on climate, is referred to as Stefan-Boltzmann feedback, Planck feedback, or longwave feedback. Another negative feedback is that related to the carbon cycle, where the uptake of carbon (e.g., CO_2 and CH_4) increases as their concentrations increase, for example, to oceans and through the photosynthesis process. The relative importance of the various positive and negative feedbacks can be found in IPCC (2021b).

2.10 Questions

2.1 Indicate the wind direction and any other feature you consider relevant on the plot for both warm and cool air as the movement of air evolves. Warm air originates from the east of the mountain, while the direction of the cooler air is upward from the surface of the valley.

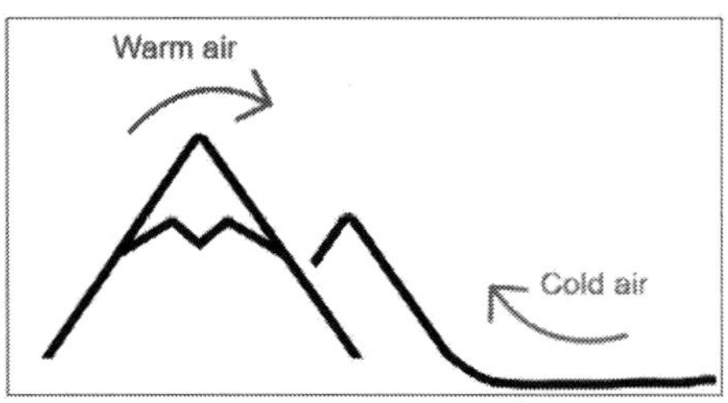

Indicate features as requested in question 2.1.

2.2 What is the concentration of carbon dioxide (CO_2) in parts per million by volume (ppmv) in a $3 \times 3 \times 3$ m test room at 25°C and 1 atm, given that 500 mg of carbon dioxide has been injected into the air of the test room in an experiment? The molecular weight of carbon dioxide is 44.01 g mol^{-1}.

2.3 Consider the Earth as a black body with a temperature of 25°C. Calculate the total electromagnetic flux emitted from the Earth's surface. Stefan Boltzmann constant is 5.67×10^{-8} Wm^{-2} K^{-4}. Consider the radius of the Earth as 6.4×10^6 m.

2.4 Approximately what height above sea level corresponds to half of the mass of the atmosphere? Consider an exponential pressure dependence with H = 8 km and neglecting the small vertical variation of gravitational force with height.

2.5 Convert 2 ppmv CO and 6 ppmv SO_2 concentrations in the unit of $\mu g\ m^{-3}$ for normal temperature and pressure conditions (25°C and 1 atm). Universal gas constant R is 0.082 L atm mol^{-1} K^{-1}, molecular weight of CO is 28.01 g mol^{-1}, and SO_2 is 64.06 g mol^{-1}.

Solutions to the end-of-chapter questions will be available at: https://www.elsevier.com/books-and-journals/book-companion/9780128225912.

References

Almeida, J., Schobesberger, S., Kürten, A., Ortega, I. K., Kupiainen-Määttä, O., Praplan, A. P., Adamov, A., Amorim, A., Bianchi, F., Breitenlechner, M., David, A., Dommen, J., Donahue, N. M., Downard, A., Dunne, E., Duplissy, J., Ehrhart, S., Flagan, R. C., Franchin, A., … Kirkby, J. (2013). Molecular understanding of sulphuric acid–amine particle nucleation in the atmosphere. *Nature*, *502*(7471), 359–363. Available from https://doi.org/10.1038/nature12663.

Ambaum, M. H. P. (2010). *Thermal physics of the atmosphere. Advancing weather and climate science* (p. 240) Wiley-Blackwell, ISBN 978-0-470-74515-1.

Arias, P. A., Bellouin, N., Coppola, E., Jones, R. G., Krinner, G., Marotzke, J., Naik, V., Palmer, M. D., Plattner, G.-K., Rogelj, J., Rojas, M., Sillmann, J., Storelvmo, T., Thorne, P. W., Trewin, B., Achuta Rao, K., Adhikary, B., Allan, R. P., Armour, K., … Zickfeld, K. (2021). Technical summary. In V. Masson-Delmotte, P. Zhai, A. Pirani, S. L. Connors, C. Péan, S. Berger, N. Caud, Y. Chen, L. Goldfarb, M. I. Gomis, M. Huang, K. Leitzell, E. Lonnoy, J. B. R. Matthews, T. K. Maycock, T. Waterfield, O. Yelekçi, R. Yu, & B. Zhou (Eds.), *Climate change 2021: The physical science basis. Contribution of working group I to the sixth assessment report of the intergovernmental panel on climate change* (pp. 33–144). Cambridge University Press. Available from http://doi.org/10.1017/9781009157896.002.

ASTM International. (2020). ASTM G173-03(2020), Standard Tables for Reference Solar Spectral Irradiances: Direct Normal and Hemispherical on 37° Tilted Surface. ASTM International. Available from https://doi.org/10.1520/G0173-03R20.

Barlow, J. F. (2014). Progress in observing and modelling the urban boundary layer. *Urban Climate*, *10*(2014), 216–240.

Bjerknes, V. (1921). *On the dynamics of the circular vortex: With applications to the atmosphere and atmospheric vortex and wave motions*. Grøndahl & Søns Boktrykkeri.

Boloorani, A. D., Soleimani, M., Papi, R., Neysani Samany, N., Teymouri, P., & Soleimani, Z. (2023). Sources, drivers, and impacts of sand and dust storms: A global view. In A. Al-Dousari, & M. Z. Hashmi (Eds.), *Dust and health. Emerging contaminants and associated treatment technologies*. Springer. Available from https://doi.org/10.1007/978-3-031-21209-3_3.

Boucher, O. (2015). *Atmospheric aerosols, properties and climate impacts*. Springer, ISBN: 978-94-017-9648-4.

Carslaw, K. S. (2022). *Aerosols and climate* (1st ed.). Elsevier, ISBN: 9780128197660.

Emery, W., & Camps, A. (2017). *Introduction to satellite remote sensing. Chapter 2 – Basic electromagnetic concepts and applications to optical sensors* (pp. 43–83). Elsevier ISBN 9780128092545. Available from https://doi.org/10.1016/B978-0-12-809254-5.00002-6.

Farman, J. C., Gardiner, B. G., & Shanklin, J. D. (1985). Large losses of total ozone in Antarctica reveal seasonal ClO_x/NO_x interaction. *Nature*, *315* (6016), 207–210.

Ferrel, W. (1889). *Popular treatise on the winds: Comprising the general motions of the atmosphere, monsoons, cyclones, tornadoes, waterspouts, hail-storms, etc* (p. 536) John Wiley & Sons and Chapman & Hall Limited.

Forster, P., Storelvmo, T., Armour, K., Collins, W., Dufresne, J.-L., Frame, D., Lunt, D. J., Mauritsen, T., Palmer, M. D., Watanabe, M., Wild, M., & Zhang, H. (2021). The Earth's energy budget, climate feedbacks, and climate sensitivity. In V. Masson-Delmotte, P. Zhai, A. Pirani, S. L. Connors, C. Pean, S. Berger, N. Caud, Y. Chen, L. Goldfarb, M. I. Gomis, M. Huang, K. Leitzell, E. Lonnoy, J. B. R. Matthews, T. K. Maycock, T. Waterfield, O. Yelekci, R. Yu, & B. Zhou (Eds.), *Climate change 2021: The physical science basis. contribution of working group I to the sixth assessment report of the intergovernmental panel on climate change* (pp. 923–1054). Cambridge University Press. Available from https://doi.org/10.1017/9781009157896.009.

Gueymard, C. A. (2004). The sun's total and spectral irradiance for solar energy applications and solar radiation models. Solar Energy, 76(4), 423–453. Available from https://doi.org/10.1016/j.solener.2003.08.039.

Haigh, J. (2011). Solar influences on climate (Grantham Institute for Climate Change Briefing Paper No. 5). Imperial College London.

Hartmann, D. L. (2016). *Global physical climatology* (2nd ed.). Elsevier Science. Available from https://doi.org/10.1016/C2009-0-00030-0, ISBN: 978-0-12-328531-7.

Hersbach, H., Bell, B., Berrisford, P., Biavati, G., Horányi, A., Muñoz Sabater, J., Nicolas, J., Peubey, C., Radu, R., Rozum, I., Schepers, D., Simmons, A., Soci, C., Dee, D., & Thépaut, J.-N. (2023). ERA5 monthly averaged data on single levels from 1940 to present. *Copernicus Climate Change Service (C3S) Climate Data Store (CDS)*. Available from https://doi.org/10.24381/cds.f17050d7, accessed on 27-07-2023.

Herwartz, D., Pack, A., Friedrichs, B., & Bischoff, A. (2014). Identification of the giant impactor Theia in lunar rocks. *Science*, *344*(6188), 1146–1150. Available from https://doi.org/10.1126/science.1251117.

Hungate, B. A., & Koch, G. W. (2023). *GLOBAL CHANGE, Biospheric impacts and feedbacks. Encyclopaedia of atmospheric sciences*. Academic Press.

Huszar, P., Karlický, J., Ďoubalová, J., Šindelářová, K., Nováková, T., Belda, M., Halenka, T., Žák, M., & Pišoft, P. (2020). Urban canopy meteorological forcing and its impact on ozone and PM2.5: Role of vertical turbulent transport. Atmospheric Chemistry and Physics, 20(4), 1977–2016. Available from https://doi.org/10.5194/acp-20-1977-2020.

IPCC. (2021a). Climate change 2021: The physical science basis. In: Masson-Delmotte, V., P. Zhai, A. Pirani, S.L. Connors, C. Péan, S. Berger, N. Caud, Y. Chen, L. Goldfarb, M.I. Gomis, M. Huang, K. Leitzell, E. Lonnoy, J.B.R. Matthews, T.K. Maycock, T. Waterfield, O. Yelekçi, R. Yu, B. Zhou (eds.). Contribution of working group I to the sixth assessment report. Cambridge University Press. Available from https://doi.org/10.1017/9781009157896.

IPCC. (2021b). Summary for policymakers. In: V. Masson-Delmotte, P. Zhai, A. Pirani, S. L. Connors, C. Péan, S. Berger, N. Caud, Y. Chen, L. Goldfarb, M. I. Gomis, M. Huang, K. Leitzell, E. Lonnoy, J.B.R. Matthews, T. K. Maycock, T. Waterfield, O. Yelekçi, R. Yu, B. Zhou. Climate change 2021: The physical science basis. Contribution of working group I to the sixth assessment report of the intergovernmental panel on climate change, Cambridge University Press.

IPCC. (2022). Climate change 2022: Mitigation of climate change. In P. R. Shukla, J. Skea, R. Slade, A. Al Khourdajie, R. van Diemen, D. McCollum, M. Pathak, S. Some, P. Vyas, R. Fradera, M. Belkacemi, A. Hasija, G. Lisboa, S. Luz, & J. Malley (Eds.), *Contribution of working group III to the sixth assessment report of the intergovernmental panel on climate change*. Cambridge University Press. Available from http://doi.org/10.1017/9781009157926.

Kok, J. F., Adebiyi, A. A., Albani, S., Balkanski, Y., Checa-Garcia, R., Chin, M., Colarco, P. R., Hamilton, D. S., Huang, Y., Ito, A., Klose, M., Li, L., Mahowald, N. M., Miller, R. L., Obiso, V., Pérez García-Pando, C., Rocha-Lima, A., & Wan, J. S. (2021). Contribution of the world's main dust source regions to the global cycle of desert dust. *Atmospheric Chemistry and Physics*, *21*, 8169–8193. Available from https://doi.org/10.5194/acp-21-8169-2021.

Kopp, G., & Lean, J. L. (2011). A new, lower value of total solar irradiance: Evidence and climate significance. Geophysical Research Letters, 38(1). Available from https://doi.org/10.1029/2010GL045777.

Kren, A. C., Pilewskie, P., & Coddington, O. (2017). Where does Earth's atmosphere get its energy? *Journal of Space Weather and Space Climate*, *7*, A10. Available from https://doi.org/10.1051/swsc/2017007.

Lazaridis, M. (2011). *First principles of meteorology and air pollution. Environmental Pollution* (1st ed.). Springer.

Lutgens, F., & Tarbuck, E. (1997). *The atmosphere* (2nd ed.). Prentice Hall Inc.

Lüthi, D., Le Floch, M., Bereiter, B., Blunier, T., Barnola, J.-M., Siegenthaler, U., Raynaud, D., Jouzel, J., Fischer, H., Kawamura, K., & Stocker, T. F. (2008). High-resolution carbon dioxide concentration record 650,000-800,000 years before present. *Nature*, *453*, 379–382. Available from https://doi.org/10.1038/nature06949.

McCoy, L. (2019). Atmospheric aerosols: Characterization, properties and climate impacts (p. 217) Syrawood Publishing House.

Met Office. (2023). *UK daily temperature data, part of the Met Office Integrated Data Archive System (MIDAS)*. Available at: https://www.metoffice.gov.uk/ (accessed July 20, 2023).

Molina, T. L. (2021). Introductory lecture: Air quality in megacities. *Faraday Discussions*, *226*, 9. Available from https://doi.org/10.1039/d0fd00123f.

Mrazek, T. 2022. *Flicker photo archive*. Available at: https://www.flickr.com/photos/tommrazek/19059453883 (accessed Novemver 2, 2022). Licensed under CC BY 2.0

Muller, R. (Ed.), (2012). *Stratospheric ozone depletion and climate change*. Royal Society of Chemistry. Available from https://www.rsc.org.

NASA. (2023). NASA Earth Observatory. *Aerosol optical depth*. Retrieved from https://earthobservatory.nasa.gov/global-maps/MODAL2_M_AER_OD. (accessed in 2024).

NOAA. (2023). *Glossary of NOAA's National Weather Service*. https://w1.weather.gov/glossary/

NOAA. (2024). Data provided by carbon cycle and greenhouse gases group, 325 Broadway R/GML, Boulder, CO 80305 (https://gml.noaa.gov/ccgg/).

Pai, S. J., Carter, T. S., Heald, C. L., & Kroll, J. H. (2022). Updated world health organization air quality guidelines highlight the importance of non-anthropogenic PM2.5. *Environmental Science & Technology Letters*, *9*(6), 501–506. Available from https://doi.org/10.1021/acs.estlett.2c00203.

Ramachandran, S. (2018). *Atmospheric aerosols*. Taylor & Francis INC International Concepts.

Reimann, S., Elkins, J. W., Fraser, P. J., Hall, B. D., Kurylo, M. J., Mahieu, E., Montzka, S. A., Prinn, R. G., Rigby, M., Simmonds, P. G., & Weiss, R. F. (2018). Observing the atmospheric evolution of ozone-depleting substances. *Comptes Rendus Geoscience*, *350*(7), 384–392. Available from https://doi.org/10.1016/j.crte.2018.08.008.

Rosen, J., & Egger, A. E. (2014). History of Earth's atmosphere, I. The origin of the modern atmosphere. *Visionlearning*, *2*.

Salawitch, R.J., Fahey, D.W., Hegglin, M.I., McBride, L.A., Tribett, W.R., & Doherty, S.J. (2019). Twenty questions and answers about the ozone layer: 2018 update. *Scientific assessment of ozone depletion: 2018*. World Meteorological Organization. Geneva, Switzerland. ISBN: 978-1-7329317-2-5

Schobesberger, S., Junninen, H., Bianchi, F., Lönn, G., Ehn, M., Lehtipalo, K., Dommen, J., Ehrhart, S., Ortega, I. K., Franchin, A., Nieminen, T., Riccobono, F., Hutterli, M., Duplissy, J., Almeida, J., Amorim, A., Breitenlechner, M., Downard, A. J., Dunne, E. M., … Worsnop, D. R. (2013). Molecular understanding of atmospheric particle formation from sulfuric acid and large oxidized organic molecules. *Proceedings of the National Academy of Sciences*, *110*(43), 17223–17228. Available from https://doi.org/10.1073/pnas.1306973110.

Schraufnagel, D. E. (2020). The health effects of ultrafine particles. Experimental & Molecular Medicine, 52(3), 311–317. Available from https://doi.org/10.1038/s12276-020-0403-3.

Seinfeld, J. H., Bretherton, C., Carslaw, K. S., Wood, R., et al. (2016). Improving our fundamental understanding of the role of aerosol – cloud interactions in the climate system. *Proceedings of the National Academy of Sciences (PNAS)*, *113*(21), 5781–5790. Available from https://doi.org/10.1073/pnas.1514043113.

Seinfeld, J. H., & Pandis, S. N. (2016). *Atmospheric chemistry and physics: From air pollution to climate change* (3rd ed.). Wiley.

Sellers, W. D. (1965). *Physical climatology*. University of Chicago Press, ISBN-10: 0226746992 / ISBN-13: 978-0226746999.

Semeniuk, K., & Dastoor, A. (2018). Current state of aerosol nucleation parameterizations for air-quality and climate modeling. *Atmospheric Environment*, *179*, 77–106. Available from https://doi.org/10.1016/j.atmosenv.2018.01.039.

Shrivastava, M., Cappa, C. D., Fan, J., Goldstein, A. H., Guenther, A. B., Jimenez, J. L., Kuang, C., Laskin, A., Martin, S. T., Ng, N. L., Petaja, T., Pierce, J. R., Rasch, P. J., Roldin, P., Seinfeld, J. H., Shilling, J., Smith, J. N., Thornton, J. A., Volkamer, R., Wang, J., Worsnop, D. R., Zaveri, R. A., Zelenyuk, A., Zhang, Q., & others. (2017). Recent advances in understanding secondary organic aerosol: Implications for global climate forcing. *Reviews of Geophysics*, *55*(2), 509–559. Available from https://doi.org/10.1002/2016RG000540.

Smith, C. J., Kramer, R. J., Myhre, G., Alterskjær, K., Collins, W., Sima, A., Boucher, O., Dufresne, J.-L., Nabat, P., Michou, M., Yukimoto, S., Cole, J., Paynter, D., Shiogama, H., O'Connor, F. M., Robertson, E., Wiltshire, A., Andrews, T., Hannay, C., . . . Forster, P. M. (2020). Effective radiative forcing and adjustments in CMIP6 models. *Atmospheric Chemistry and Physics*, *20*, 9591–9618. Available from https://doi.org/10.5194/acp-20-9591-2020.

Sokhi, R. S., Moussiopoulos, N., Baklanov, A., Bartzis, J., Coll, I., Finardi, S., Friedrich, R., Geels, C., Grönholm, T., Halenka, T., Ketzel, M., Maragkidou, A., Matthias, V., Moldanova, J., Ntziachristos, L., Schäfer, K., Suppan, P., Tsegas, G., Carmichael, G., . . . Kukkonen, J. (2022). Advances in air quality research – Current and emerging challenges. *Atmospheric Chemistry and Physics*, *22*, 4615–4703. Available from https://doi.org/10.5194/acp-22-4615-2022.

Sokhi, R. S., Singh, V., Querol, X., Finardi, S., Targino, A. C., Andrade, M. F., Pavlovic, R., Garland, R. M., Massagué, J., Kong, S., Baklanov, A., Ren, L., Tarasova, O., Carmichael, G., Peuch, V.-H., Anand, V., Arbilla, G., Badali, K., Beig, G., . . . Zavala, M. (2021). A global observational analysis to understand changes in air quality during exceptionally low anthropogenic emission conditions. *Environment International*, *157*, 106818. Available from https://doi.org/10.1016/j.envint.2021.106818.

Stull, R. (2011). *Meteorology for scientists & engineers* (3rd Edition, p. 938)Univ. of British Columbia.

Stull, R. B. (1988). *An introduction to boundary layer meteorology* (p. 670) Kluwer Academic Publishers.

Szopa, S., Naik, V., Adhikary, B., Artaxo, P., Berntsen, T., Collins, W. D., Fuzzi, S., Gallardo, L., Kiendler-Scharr, A., Klimont, Z., Liao, H., Unger, N., & Zanis, P. (2021). Short-lived climate forcers. In V. Masson-Delmotte, P. Zhai, A. Pirani, S. L. Connors, C. Péan, S. Berger, N. Caud, Y. Chen, L. Goldfarb, M. I. Gomis, M. Huang, K. Leitzell, E. Lonnoy, J. B. R. Matthews, T. K. Maycock, T. Waterfield, O. Yelekçi, R. Yu, & B. Zhou (Eds.), Climate change 2021: The physical science basis. contribution of working group I to the sixth assessment report of the intergovernmental panel on climate change (pp. 817–922). Cambridge University Press. Available from https://doi.org/10.1017/9781009157896.008.

Trenberth, K. E., Fasullo, J. T., & Kiehl, J. (2009). Earth's global energy budget. *American Meteorological Society*, *90*(3), 311–324.

Vautard, R., Kadygrov, N., Iles, C., Boberg, F., Buonomo, E., Bülow, K., Coppola, E., Corre, L., van Meijgaard, E., Nogherotto, R., Sandstad, M., Schwingshackl, C., Somot, S., Aalbers, E., Christensen, O. B., Ciarlo, J. M., Demory, M.-E., Giorgi, F., Jacob, D., ... Wulfmeyer, V. (2020). Evaluation of the large EURO-CORDEX regional climate model ensemble. Journal of Geophysical Research: Atmospheres, 125(22). Available from https://doi.org/10.1029/2019JD032344.

Wallace, J.M., & Hobbs, P.V. (2006). *Atmospheric science: An introductory survey*. Elsevier. ISBN 978-0-12-732951-2. Available from https://doi.org/10.1016/C2009-0-00034-8.

Whitby, K. T., & Cantrell, B. (1976). Fine particles, Proc. Int. Conf. Environmental Sensing and Assessment, Las Vegas, NV, Institute of Electrical and Electronic Engineers.

Wild, M., Folini, D., Hakuba, M. Z., Schär, C., Seneviratne, S. l, Kato, S., Rutan, D., Ammann, C., Wood, E. F., & König-Langlo, G. (2015). The energy balance over land and oceans: An assessment based on direct observations and CMIP5 climate models. *Climate Dynamics*, *44*(11–12), 3393–3429. Available from https://doi.org/10.1007/s00382-014-2430-z.

World Health Organization. (2021). *WHO global air quality guidelines. Particulate matter (PM2.5 and PM10), ozone, nitrogen dioxide, sulfur dioxide and carbon monoxide*. Geneva: World Health Organization. Licence: CC BY-NC-SA 3.0 IGO.

World Meteorological Organization (WMO). (2022). *Scientific Assessment of Ozone Depletion* (GAW Report No. 278, Executive Summary). Geneva. 56 pp. ISBN 978-9914-733-99-0.

Wyoming University. (2022). *Radiosonde soundings*. Available at: https://weather.uwyo.edu/upperair/sounding.html (accessed Novemver 10, 2022).

Zender, C. S., Miller, R. L., & Tegen, I. (2004). Quantifying mineral dust mass budgets: Terminology, constraints, and current estimates. *Eos, Transactions American Geophysical Union*, *85*(48), 509–512.

Chapter 3

Sources and emissions of air pollution

Sergio Ibarra Espinosa[1,2], Mario Gavidia Calderon[3], Pedro Jose Perez Martinez[4], Leila Droprinchinski Martins[5], Ana Paula Mendes Emygdio[6], Carlos Eduardo Souto-Oliveira[3], Regina Miranda[7], Perola Vasconcellos[8], Edmilson Dias de Freitas[3] and Maria de Fatima Andrade[3]

[1]Cooperative Institute for Research in Environmental Sciences, University of Colorado Boulder, Boulder, CO, United States, [2]NOAA Global Monitoring Laboratory, Boulder, CO, United States, [3]Institute of Astronomy, Geophysics and Atmospheric Sciences, University of São Paulo, São Paulo, Brazil, [4]Faculty of Engineering, Architecture and Urbanism, University of Campinas, Campinas, São Paulo, Brazil, [5]Graduate Program in Environmental Engineering, Federal University of Technology, Paraná, Brazil, [6]University of Surrey, Surrey, United Kingdom, [7]School of Arts, Sciences and Humanities, University of São Paulo, São Paulo, Brazil, [8]Institute of Chemistry, University of São Paulo, São Paulo, Brazil

3.1 Introduction

The natural atmosphere is composed of many gases and particles, which are present in background (i.e., in trace) concentrations. However, natural and anthropogenic sources can emit compounds to the atmosphere, producing an increase in concentration levels when the emission is not in equilibrium with sinking processes. Besides, emission sources—from natural or anthropogenic sources—can increase the diversity of substances in the atmosphere. Atmospheric concentration of compounds is determined by emission levels and dynamic processes. This includes atmospheric conditions, which can vary with time and space, ultimately impacting concentrations.

In this chapter, the main characteristics and emissions of natural and anthropogenic sources to atmospheric composition will be presented, including their role as drivers of air quality on regional and global scales. Section 3.2 provides relevant contributions of natural processes to air quality and ecosystem services. For instance, several soil, biogenic, and marine systems regulate nutrient cycles and act as a sink of gaseous and particles from the air. On the other hand, volcano eruptions, forest wildfires, and natural dust can act as important pollutant sources in the atmosphere. Subsequently, in the following sections, anthropogenic sources are introduced, and their amounts emitted into the atmosphere are presented.

The organization of this chapter and respective subsections is schematically presented in Fig. 3.1. In addition, relevant references are provided for those looking for more detailed information.

3.2 Natural sources—biogenic, bioaerosols, volcanoes, and others

Particles and gaseous compounds in the atmosphere come from various natural sources such as windborne dust, vegetation, sea spray, and volcanoes, as well as anthropogenic origins such as those associated with energy generation, industries, and vehicles. Atmospheric aerosol is defined as a set of solid and liquid particles suspended in the atmosphere. Primary aerosol is the one emitted directly by sources, while that formed by gas-to-particle conversion in the atmosphere is denoted secondary. Particles display different sizes and chemical compositions, having significant effects on health and climate, contributing to the scattering and absorption of radiation (direct effect) and also interacting with the microphysical process of cloud formation (indirect effect). Due to these numerous effects, it is key that the main sources of aerosols and their characteristics are known. The sources of gas and particles involved in several physical and chemical processes in the atmosphere will be presented in the next sections of this chapter.

3.2.1 Soil and vegetation

Soil and vegetation interact directly with the air and, consequently, play important ecosystem services through filtration, fixation, degradation, and storage of pollutants. These factors directly impact the air quality. Biogeochemical cycles of carbon, oxygen, nitrogen, sulfur, and water, regulate the balance of these fundamental species among earth's system

Air Quality. DOI: https://doi.org/10.1016/B978-0-12-822591-2.00003-2

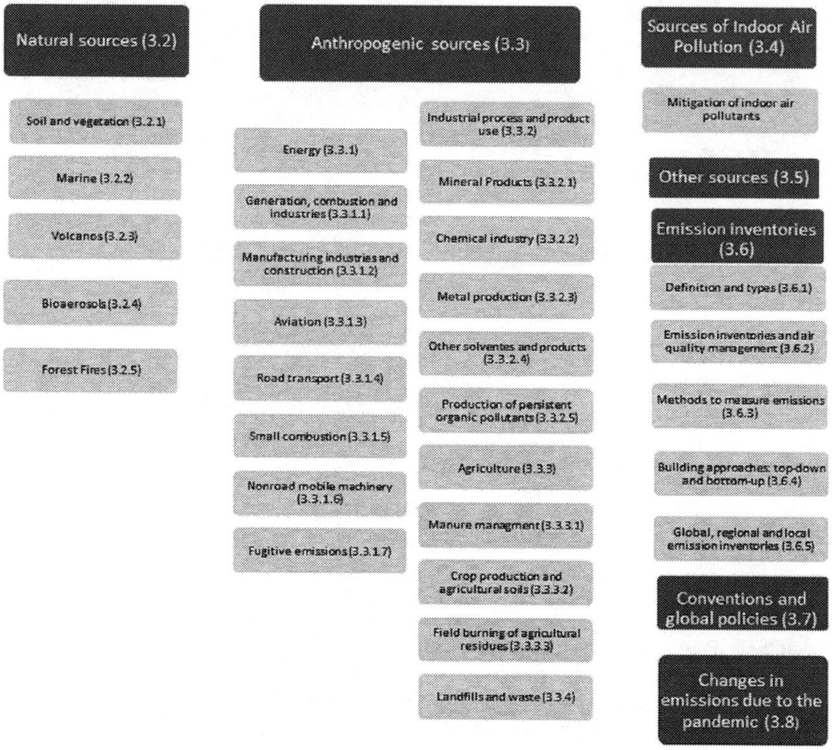

FIGURE 3.1 Diagram of the main sessions and subsections of Chapter 3.

compartments. For instance, the main carbon dioxide (CO_2) exchange between the atmosphere and biosphere is via photosynthesis and respiration. The atmosphere is the main reservoir of nitrogen; however, biological processes in the soil and vegetation are essential to fix nitrogen and make it available for living beings (Nguyen et al., 2019). In this line, the Intergovernmental Science-Policy Platform on Biodiversity and Ecosystem Services (IPBES) classified 18 categories of nature's contributions to the population, one of them applicable to the regulation of air quality (IPBES, 2018).

Despite their positive function as a sink of gases and organic compounds, soils can act negatively as a source of particles and other pollutants in the atmosphere. Sand and dust storms are well known as natural sources affecting air quality on a global scale due to high emissions and long-range transportation. On regional or local scales, natural dust emissions, land management practices, or dust from construction/demolition activities, further discussed in the next sections, are important drivers that worsen air quality (Giltrap et al., 2021). In addition, the increase in the magnitude of extreme events such as heatwaves, wildfires, and drought, possibly associated with climate changes, has been linked to negative impacts on ecosystem services as presented in the 2022 IPCC (Pörtner et al., 2022).

3.2.2 Marine (including the emissions by the biosphere in the ocean)

Seas all over the world, which comprise three-quarters of the planet's total, constitute the major source of aerosols (particles and gases) in the atmosphere, known as Sea-Spray Aerosols (SSA). SSA are emitted from the ocean due to the bursting of air bubbles through the action of waves generated by winds. The droplets produced have radii ranging from 10 nm to 2 μm. The exchange between the ocean surface and the atmosphere occurs through the sea-surface microlayer interface, as described by Poschl and Shiraiwa (2015).

Usually, more attention is paid to the inorganic emission (NaCl salt, bromine, and sulfur, among other compounds), but the emissions associated with biological activity (including proteins, enzymes, bacteria, viruses, and other organic compounds, such as fatty acids and sugars) by the oceans have been under attention due to uncertainties in the estimations, as reviewed by (Schiffer et al., 2018). Particles smaller than 1 μm emitted in the marine boundary layer have high participation of organic compounds, such as fatty acids, hydroxy acids, polysaccharides, proteins, and organosulfur compounds, in their composition (Poschl & Shiraiwa, 2015). A large fraction of small particles has its origin not only from primary emissions but also from the secondary formation of Sulfate and secondary organic aerosols (SOAs). Secondary inorganic sulfate and SOA are formed from the oxidation of biogenic dimethyl sulfide and volatile organic compound (VOC) emitted by phytoplankton (Quigg et al., 2021). The impacts of ocean acidification on processes that control the production of these species are potentially important but still not fully determined (Gruber, 2011). Forestieri et al. (2018), performing experiments with a tank, a reference for marine aerosol production simulations, showed that the composition of seawater may have the same impact as water temperature in SSA production.

Although these emissions have direct and indirect effects on radiative balance, through the formation of cloud condensation nuclei and ice nuclei, uncertainties in numbers regarding primary emissions and secondary formation due to marine aerosols are very high. Large uncertainties in determinations of organic species emission by the marine layer are related to the lack of measurements of reactive organic species (Carpenter & Nightingale, 2015; Shaw et al., 2010). These emissions can occur in remote regions, with fewer measurements. Thus, albeit possibly having strong impacts on aerosol formation, they may be underrepresented in the description of the physical and chemistry processes globally, bringing more uncertainties to climate models. However, this emission is quite important, having a great impact on climate change. Meanwhile, the ocean is also a source of greenhouse gases (GHGs), such as nitrous oxide (N_2O), halocarbons, VOCs, primary organic carbon, and particles, which have an important role in radiative balance (Quinn et al., 2015).

Sea salt aerosols are the main source of reactive species bromine, chlorine, and iodine (Horowitz et al., 2020), which have a significant impact on the chemistry of ozone and methane. Using the Harvard-NASA Emissions Component (HEMCO) for the 0.5 degrees lat. x0.625 degrees lon., considering the accumulation and coarse modes (from 0.5 to 8 μm), Weng et al. (2020) estimated the total amount of particles from oceanic origin. Between 2014 and 2017, estimations reached the value of 3860 Tg $year^{-1}$.

The latest IPCC report (AR6 WGIII) underscores the urgent need to address the impacts of climate change on marine ecosystems, emphasizing the growing challenges of rising temperatures, ocean acidification, pollutant sedimentation, and resource extraction. These factors must be evaluated within an evolving environmental landscape to effectively manage and protect marine biodiversity (Bindoff et al., 2019).

3.2.3 Volcanoes

In volcanoes, magma, which is less dense than surrounding rocks, rises through ruptures in the crust, giving rise to an eruption. At the surface, with lower pressure, gases dissolved in the magma expand, producing incandescent rocky materials, such as rock fragments, lava, ash, and gases. These gases, when released into the atmosphere, can be transported by the wind, and chemically turn into toxic acids. Volcanic eruptions can then cause global climate variations. Gases released by the magma are called fumaroles, and these exhalations are composed mainly of a mixture of sulfur dioxide (SO_2), hydrochloric acid (HCl), hydrofluoric acid (HF), and silicon tetrafluoride (SiF_4). Compositions vary according to temperature and to the amount of dissolved minerals. Particles include silicates, halide salts, and sulfates. After the eruption, sulfates formation continues, as gaseous SO_2 is oxidized to sulfuric acid and sulfates (Andres et al., 1991; Oppenheimer et al., 2011).

The eruption of large volcanoes can inject SO_2 into the stratosphere, increasing the concentration of stratospheric sulfate aerosols, providing surfaces for heterogeneous chemistry, or serving as nuclei for polar stratospheric cloud formation. Sulfate particles can scatter incoming solar radiation, causing a surface cooling effect. These particles can also contribute to the warming of the stratosphere, as they absorb infrared radiation. However, this warming effect is of lesser magnitude, and the net effect of a volcanic eruption is a cooling result. The eruption of the Pinatubo Volcano in June 1991 and its effects on the atmosphere have been extensively studied. The optical depth of aerosol due to the eruption influenced the atmosphere for up to three years (Niranjan et al., 1999; Olmo et al., 1999), decreasing in late 1993. Global temperatures decreased by up to 0.5°C at the surface (Parker et al., 1996). Recently, in January 2022, an undersea volcano, located in the South Pacific Island nation of Tonga, erupted and its effects are still being studied (Carvajal et al., 2022). Scientists believe that the global average surface temperature will decrease by a few tenths of a degree in the next two years due to volcanic eruptions (Adam, 2022; Zhang et al., 2022).

Worked Example 3.1

Calculating the SO_2 Emissions from a Volcanic Eruption

A volcanic eruption releases 50,000 tonnes of sulfur dioxide (SO_2) per day into the atmosphere. Calculate the total amount of SO_2 emissions from the eruption over a 30-day period.

Solution:

Calculate the total amount of SO_2 emissions per day: 50,000 tonnes of SO_2 per day and then calculate the total amount of SO_2 emissions over a 30-day period: 50,000 tonnes of $SO_2\,day^{-1} \times 30$ days $=$ 1.5 million tonnes of SO_2

emissions over a 30-day period. Therefore the total amount of SO_2 emissions from the volcanic eruption over a 30-day period is 1.5 million tonnes of SO_2.

3.2.4 Bioaerosols

Bioaerosols are very heterogeneous atmospheric particles, which include organisms, such as bacteria, algae, archaea, and lichens, which can be alive or dead; fragments and excrement, such as plant debris; and dispersal units, such as pollens and fungal spores (definition presented by Fröhlich-Nowoisky et al., 2016 and Després et al., 2012). They range from nanometers to more than 100 μm (Després et al., 2012). However, these particles can display different number and concentrations, depending on size, urbanization rate, weather conditions, land use, and other anthropogenic factors (Després et al., 2012; Fröhlich-Nowoisky et al., 2012, 2016; Matthias-maser & Jaenicke, 2000). For instance, estimates point out that at a continental site the average number concentration of bacteria can be around 1×10^4 cells m^{-3} during out-of-cloud conditions (Bauer et al., 2002), while fungal spores number concentration can be around $10^3-10^4\,m^{-3}$ in continental boundary layer air (Després et al., 2012; Elbert et al., 2007).

There are several sources of bioaerosols, which include both natural and anthropogenic sources. In general, the atmosphere acts as a temporary home for these particles, as reviewed by Xie et al. (2021). These bioaerosols can agglomerate and adhere to other particles, and be transported over long distances, allowing the dispersal of reproductive material and/or pathogens that may cause disease and infections (Burrows et al., 2009; Fröhlich-Nowoisky et al., 2016).

Among natural sources are the soil, water, and vegetation (Bauer et al., 2002). Microorganisms require nutrients to grow and multiply. The soil is a rich source of these nutrients, and thus an important source of microorganisms, which can be transported to the atmosphere. Bioaerosols found in soil can include a variety of bacteria, fungi, archaea, lichens, and bryophytes. In addition, other biological fragments or dispersal units can also be transported to the atmosphere when deposited in the soil (reviewed by Després et al., 2012; Xie et al., 2021). Soil bacteria can be transported to the atmosphere through rain formation, with the vertical transport, which can function as a disperser of bacteria (Joung et al., 2017). Water bodies (e.g., rivers, lakes, oceans) are also an important source of bioaerosols, including bacteria, viruses, cyanobacteria algae. As reviewed by Després et al. (2012), the bubble-bursting mechanism, for example, can release various microorganisms such as bacteria into the atmosphere. Finally, plants and animals are also important sources of bioaerosols and can be origin of various types of debris (including leaf, skin, dander, eggs, and insect excrements), microorganisms (e.g., bacteria, viruses, and fungi), and pollen and fungal spores, which play an important role in allergic diseases (Després et al., 2012; Xie et al., 2021).

Bioaerosols can also be emitted from anthropogenic sources, which include both the human body itself and its daily activities, as well as other environments created by human action, such as farms, wastewater treatment plants, and landfills. Different types of bacteria and viruses live inside and outside the human body (Xie et al., 2021) and can be transported to the atmosphere, influencing the concentration of bioaerosols. Farms and agriculture can be artificial sources of bioaerosols, such as bacteria, viruses, fungi, endotoxins, and mycotoxins, and can ultimately influence their diversity in the atmosphere (Martin et al., 2010; White et al., 2019; Xie et al., 2021). Landfills and wastewater treatment plants also become sources of microorganisms, metabolites, and toxins, with a special impact on the emission of pathogenic microorganisms. In a landfill, several pathogenic species of bioaerosols have been identified. However, concentration and type of bioaerosol vary depending on factors such as the type of waste, weather conditions, and the nature of the microorganism present (Nair, 2021, Wikuats et al., 2020). Bioaerosols emitted by agriculture, landfill, and wastewater treatment plants can create health problems for workers and adjoining public (Martin et al., 2010; Nair, 2021; Sánchez-Monedero et al., 2008).

3.2.5 Forest fires

Forest fires are an important source of air pollution and GHGs globally. Recently, it was detected that despite global burned area is declining, emissions are increasing (Zheng et al., 2021). It is important to differentiate these emissions from those from agricultural management and crop residues (see Section 3.3.3.3). Hence, monitoring of these sources is imperative to comprehensively grasp their impact and accurately project future emissions trends. Studies and reports indicate that forest fires, particularly under the influence of climate change, have the potential to elevate PM_{10} concentrations by as much as 400% (Liu et al., 2016). The Global Fire Emissions Database (GFED)[1] is a system that analyses biomass burning globally, providing maps and emissions estimates (Giglio et al., 2013). Brazil is severely affected by

1. https://globalfiredata.org/pages/amazon-dashboard/

biomass burning, including crop residues (Junior et al., 2020). The Brazilian Biomass Burning Emissions Model (3BEM) (Longo et al., 2010) was developed to estimate biomass burning emissions based on remote sensing satellites data. Furthermore, the National Institute for Space Research in Brazil (INPE in Portuguese) maintains the website https://queimadas.dgi.inpe.br/queimadas/portal to provide real-time biomass burning maps from South America. Emission factors (EF) are usually reported as mass per burned area. Hence, the European Emissions Guidelines (San-Miguel-Ayanz & Steinbrecher, 2019) present the following average EF for burned areas for nitrogen oxides (NO_X), carbon monoxide (CO), and nonmethane volatile organic compound (NMVOC): 190 (kg $NO_X\,ha^{-1}$ burned), 5400 (g $CO\,ha^{-1}$ burned), and 500 (g $NMVOC\,ha^{-1}$ burned) for temperate forest.

3.3 Anthropogenic sources

Anthropogenic emissions, that is, the air pollutant released into the atmosphere due to human activity, have impacts on local, regional, and global scales, such as climate change (Stocker, 2014). Crutzen (2006) defined that the earth system has entered a new period called the "Anthropocene" due to the impact of human activities. Therefore it is crucial to understand the sources and release of pollutants considering space and time resolution. The identification of sources from human origin is difficult due to the complexity of human activities. However, significant efforts from the last decades have resulted in guidelines to estimate emissions, such as the AP-42: Compilation of Air Emissions Factors from the U.S. Environmental Protection Agency (EPA)[2] or the European Emissions Guidelines (European Union (EU)/ European Environmental Agency (EEA), Amon et al., 2019).[3,4]

According to the U.S. EPA[5], the basic equation to estimate emissions is given as follows:

$$E = A \times \mathrm{EF} \times \left(1 - \frac{\mathrm{ER}}{100}\right), \tag{3.1}$$

where E denotes the emissions, A is the activity rate, EF is the emission factor (amount of pollutant released by the activity), and ER is the overall emission reduction efficiency (in %). This is a basic concept that can be expanded to account for the complexities in emissions sources which depend on activity data. To illustrate this equation, one can take as an example the vehicular emissions, where activity rate is the distance driven; for instance, considering 10 km in 1 hour, EFr as 1 g of CO emitted per km, and finally, let us suppose that ER equals to 0%, meaning no reduction efficiency. Then, the emissions in this simplistic example will be $E = 10\ (\mathrm{km\,h^{-1}}) \times 1\ (\mathrm{g\ CO\,km^{-1}}) = 10\ (\mathrm{g\ CO\,h^{-1}})$ during that run.

One important aspect to consider is the existence of "key" sources. These types of sources are characterized by a higher amount of emissions in comparison with others for the same pollutant. Hence, complete and detailed estimation of key sources require greater efforts (Pulles & Heslinga, 2007). Another important concept is the EF "tiers," which consist of a hierarchical classification of EF and activity data based on complexity level. Tier 1 groups emission factors are recommended for national level and lower complexity, Tier 2 is recommended for subnational level with medium complexity, and Tier 3 is recommended for city level with a high complexity. However, it is also possible to apply Tier 3 emission factor to a region with any extension if the emission source is a key category. This makes sense since emissions from the most important sources need to be estimated with a higher level of detail. In the following sections, all figures presented to illustrate emissions have their respective codes available at https://github.com/ibarraespinosa/whobook.

3.3.1 Energy

Emissions related to the energy sector consider not only that derived from fuel consumption but also the fugitive emissions from fuel use. This section was based on the European Emission Guidelines[6], as it provides a comprehensive framework to identify the sources of air pollution. Furthermore, we also included recent research to highlight some of the last advances in each emission characterization by sectors.

2. https://www.epa.gov/air-emissions-factors-and-quantification/ap-42-compilation-air-emissions-factors
3. https://www.eea.europa.eu/publications/emep-eea-guidebook-2019
4. https://www.eea.europa.eu/themes/air/air-pollution-sources-1/emep-eea-air-pollutant-emission-inventory-guidebook
5. https://www.epa.gov/air-emissions-factors-and-quantification/basic-information-air-emissions-factors-and-quantification
6. https://www.eea.europa.eu/publications/emep-eea-guidebook-2019

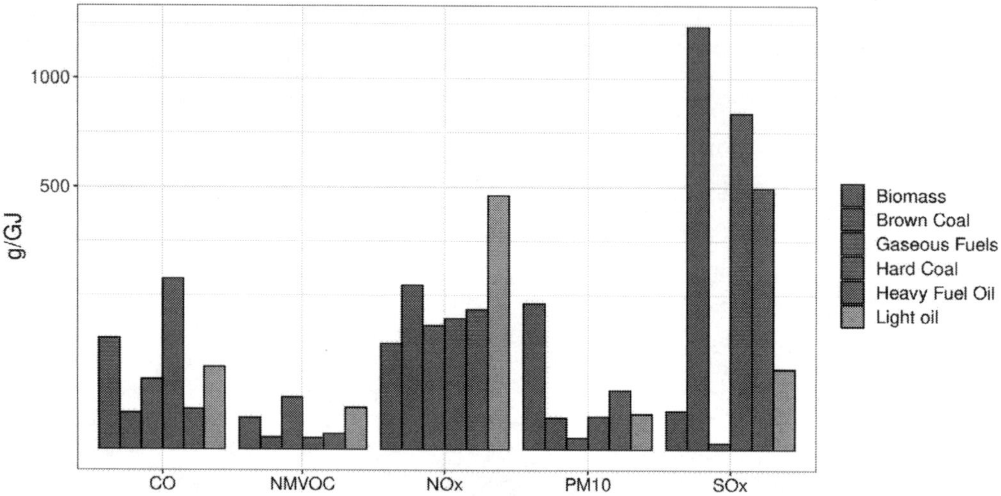

FIGURE 3.2 Average emission factors ($g\,GJ^{-1}$) during the combustion of different fuel types.

3.3.1.1 Energy generation, combustion, and industries

Sources from the power plant sector are constituted mainly by (1) public electricity and heat production, (2) petroleum refining associated with energy generation, and (3) manufacture of solid fuels. The subactivities in this category include boilers, turbines, engines, and ovens. The amount of pollutant released and its composition depend on the type of fuel used and other parameters inherent to the activity, such as temperature of combustion and after-treatment devices (Fig. 3.2).

Many types of fuels can be used to generate electricity from combustion. For instance, coal combustion reaches a temperature of about 900°C–1400°C emitting great quantities of CO_2, CO, and NO_X, among other pollutants. The general types of fuels considered for this activity are shown in Fig. 3.2, which was based on the Table 1 of the European Emissions Guidelines, Chapter 1.A.1: Energy Industries (Trozzi & Kuenen, 2019). The process of refining petroleum returns several daily-use products such as gasoline and diesel. There are many processes associated with this activity that emit large amounts of pollutants (e.g., flaring activities). The main pollutant in this context is NO_X, followed by CO, with respective values of 63 and 12.1 $g\,GJ^{-1}$ according to the Tier 1 European Emissions Guidelines (Trozzi & Kuenen, 2019).

Other activity is the use of solid fuels such as coke. Coke is a hard fuel with a high concentration of carbon which results from the anaerobic process during coal heating. Coke is mostly associated with the steel industry. The main pollutants during this process are sulfur oxides (SO_X) and particulate matter with an aerodynamic diameter smaller than 10 μm (PM_{10}) with 91 and 79 $g\,GJ^{-1}$ of emission, respectively.

The inundation of land for hydropower projects implies the conversion of terrestrial areas into water reservoirs for hydroelectric energy generation. Following the flooding process, the release of GHGs such as carbon dioxide (CO_2), methane (CH_4), and nitrous oxide (N_2O) is anticipated. In tropical regions, CH_4 emissions assume particular significance, exacerbated by factors including reservoir age, depth, and extent, as delineated in the 2006 IPCC emissions guidelines (Eggleston et al., 2006).

As mentioned above, tropical wetlands emit a large amount of CH_4, and since the Amazon basin is the most important tropical ecosystem, many authors have studied CH_4 emissions in this landscape. For instance, Wilson et al. (2021) applied an inversion modeling to estimate posterior emissions in South America, finding that average emissions between the periods of 2010–13 and 2014–17 had an increment from 49.8 to 55.6 ($Tg\,year^{-1}$), being the largest increases observed during the wet seasons. Furthermore, higher emissions were found during the exceptional 2013–14[7] flooding season in the amazon basin.

Worked Example 3.2

Calculating Emissions from a Power Plant

7. Consider looking this YouTube video: https://www.youtube.com/watch?v = Y-roT6uT9go

A power plant burns coal to generate electricity. The coal has a carbon content of 80% by weight. The plant generates 1000 MW of electricity and operates for 8000 hours per year. Calculate the annual carbon dioxide (CO_2) emissions from the power plant.

Solution:

Calculate the amount of coal burned per year: 1000 MW $\times$ 8000 hours year^{-1} $\times$ 1 tonne coal per MW hour = 8 million tonnes of coal per year and then calculate the amount of carbon in the coal burned per year: 8 million tonnes of coal $\times$ 0.8 (80% carbon content) = 6.4 million tonnes of carbon per year.

Convert the amount of carbon to CO_2 emissions: 1 tonne of carbon produces 3.67 tonnes of CO_2 and 6.4 million tonnes of carbon $\times$ 3.67 tonnes of CO_2 per tonne of carbon = 23.5 million tonnes of CO_2 emissions per year.

Therefore the power plant emits 23.5 Thousand tonnes of CO_2 per year.

3.3.1.2 Combustion in manufacturing industries and construction

Here we review the combustion process inside the manufacturing and construction industries. Subsectors included are as follows:

- Iron and steel: including flat furnaces, sintering and pelletizing plants, reheating furnaces, and gray iron foundries;
- Nonferrous metals: includes primary and secondary metal production. For more information, see Trozzi & Kuenen (2019), Section 1.A.2;
- Chemicals;
- Pulp, paper, and print;
- Food processing and tobacco;
- Nonmetallic minerals. Include cement, lime and glass manufacture, mineral wool, bricks, tiles and ceramics, and enamel manufacture. Modern cement plants accept different fuels, including waste such as tires, recovered liquid fuel and plastic.

These activities cover combustion in boilers, gas turbines, stationary engines, and furnaces. Each one of these sources has its own subprocesses and related emissions. Therefore it is up to the researcher/technician to define the importance of the source to account for the complete emissions processing. EFs are essentially the same, with the difference found in the type of fuel used, since it is covering the combustion process. Fig. 3.3 shows the average EF (g GJ^{-1}) by fuel type. Higher emissions are found in the combustion sector of Solid Fuels for CO and SO_X. In the case of NO_X, Liquid Fuels are the main responsible. Finally, Biomass emits large amounts of CO and NMVOC.

3.3.1.3 Aviation

Aviation is a sector that can be an important contributor to air pollution at local and regional scales. Most aircrafts can be individually identified in open platforms such as Flight Radar (https://www.flightradar24.com/40.04,-105.28/6), providing information such as origin, destination, aircraft type, altitude, and ground speed. To estimate emissions from aviation, an important concept to consider is the LTO, which stands for Landing and Taking Off. Emissions related to LTO have been well studied and there is available software to assess its environmental impact, such as the Aviation Environmental Design Tool (AEDT), developed by the Federal Aviation Administration/U.S. Department of Transportation (https://aedt.faa.gov/3c_information.aspx). Previous to AEDT, Federal Aviation Administration (FAA)

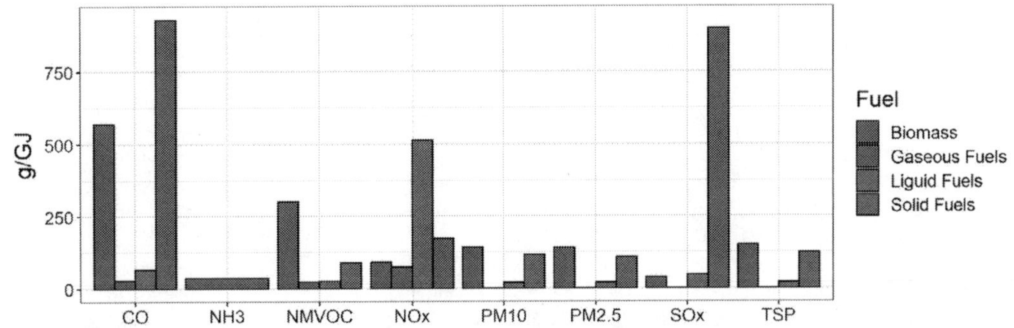

FIGURE 3.3 Average emission factors (g GJ^{-1}) during the combustion in manufacturing industries.

developed the Emissions Dispersion and Modeling System to estimate aircraft emissions. In this software, the user basically has to enter the number of LTO cycles by type of aircraft during the desired period. The European Emissions Guidelines provides an Excel spreadsheet to estimate these emissions as well (https://www.eea.europa.eu/publications/emep-eea-guidebook-2019/part-b-sectoral-guidance-chapters/1-energy/1-a-combustion/1-a-3-a-aviation-1-annex5-LTO/view).

During the coronavirus disease (COVID-19) pandemic, many countries implemented lockdowns, which heavily affected the aviation sector and resulted in a substantial drop in emissions, including of CO_2 (Liu et al., 2020). In Brazil, it was found that the COVID-19-associated measures resulted in a 62% drop in CO_2 emissions just in the aviation sector (Bazzo Vieira et al., 2022), representing -4.7 million metric tonnes $CO_2\,year^{-1}$. This reduction was directly associated with the diminished demand for passenger air transportation.

3.3.1.4 *Road transport*

Road transportation has been considered the main source of air pollution in megacities for more than 30 years (Andrade et al., 2017; Chow et al., 2004). This makes sense since urbanization is directly associated with economic growth and demand for vehicles. Nevertheless, the importance of the sources has a temporal and geographical component. For instance, regarding spatial dependence, the main source of air pollution in Northeast China during the haze episodes is residential coal and agricultural burning (Chen et al., 2018; Gao et al., 2018). However, the transportation sector is still the most important source of regulated air pollutants in urban centers across the globe.

The estimation of vehicular emissions is complex because it must represent a spatial and temporal process and it is difficult to obtain representative data with appropriate resolution. Several approaches have been developed, ranging from simple (top-down) to more detailed (bottom-up) alternatives. In the next section, some differences among top-down, bottom-up, and fuel calibration approaches are presented.

There are many emission processes in vehicles that need to be considered. According to the U.S. EPA emission model Motor Vehicle Emission Simulator (MOVES) (Koupal et al., 2008), the official approach to estimate vehicular emissions in the United States and assess the environmental policies, emission processes are exhaust running, exhaust start, exhaust extended idle emissions (long-haul combination trucks only), crankcase running exhaust, crankcase start exhaust, crankcase exhaust extended idle emissions (long-haul combination trucks only), brake wear (particulate matter (PM) only), tire wear (PM only), evaporative permeation (hydrocarbons [HC] only), evaporative fuel vapor venting (HC only), refueling displacement vapor loss (HC only), refueling spillage loss (HC only), and auxiliary power exhaust (long-haul combination trucks only). A comprehensive revision of the associated technical documentation can be found at https://www.epa.gov/moves/moves-onroad-technical-reports. As we can see, the representation of emission processes is complex and relies on numerous parameters, which are not always available. Models for estimation of the emissions consider the following categories:

- Exhaust: These emissions come from combustion inside the pistons, which go through the control devices such as catalytic converter and then released into the atmosphere. These emissions depend on many mass and kinematic parameters. One classic approach has been to estimate emissions as speed functions. In general, emissions follow a polynomial adjustment based on measurements, which have higher values both at low and higher speeds. For instance, the script "fig_vehicular_ef.R:," available in the repository from this Chapter, produces a figure which shows the average EF of EU/EEA, Brazil, and China. EU/EEA emission factors have a minimum value at $100\,km\,h^{-1}$ (Ntziachristos & Samaras, 2016). In China values have a discrete distribution (Ministry of Ecology and Environment MEE, 2018). Emission factors in Brazil depend on the age of use, but are based on an average speed (CETESB, 2020). Readers are encouraged to run the "fig_vehicular_ef.R" script for further visualization.
- Exhaust start (cold start) emissions occur when the temperature of the engine and the catalysts have not reached the optimal temperature for correct functioning, which is around 300°C (Weilenmann et al., 2009). In fact, these emissions can be largely responsible for most pollutants in environments where cold temperatures are typical and impact all vehicles with an Otto cycle. Furthermore, these emissions also occur in compression engines that consume diesel, but they are significantly lower.
- Tire and brake wear are significant sources of PM (Ntziachristos & Boulter, 2003). These emissions are also affected by electric cars and subway trains. In this sense, there are reports of high PM concentrations in subways due to brake and tire wear (Martins et al., 2015).
- The evaporation of fuels is an important source of VOCs. This process occurs due to the temperature differential between the ambient environment and the fuel tank, causing the fuel to expand and evaporate. In vehicles without canisters, the excess evaporation is released directly into the atmosphere. However, even in vehicles equipped with canisters, these emissions can be released when the canister becomes saturated (Mellios & Ntziachristos, 2016).

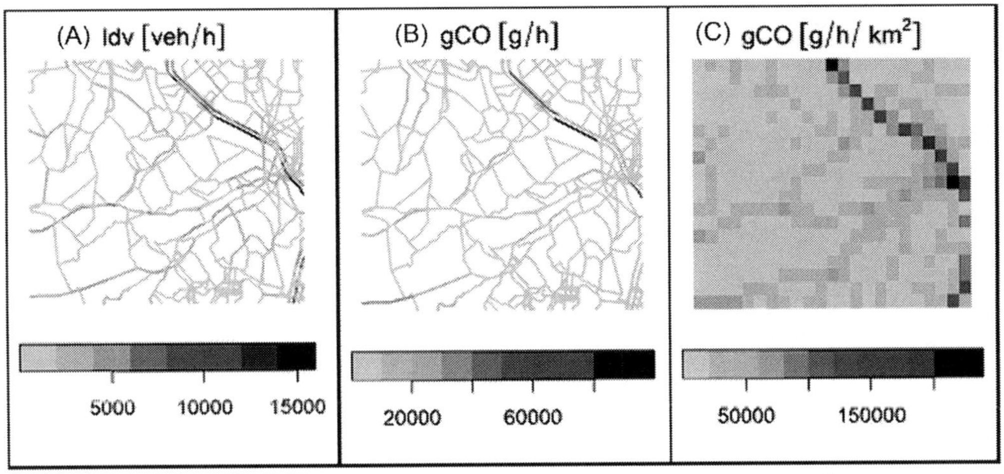

FIGURE 3.4 Emission estimating for vehicles. (A) Morning rush volume of traffic for light duty vehicles (ldv, veh h^{-1}). (B) Street emissions ($g h^{-1}$) of carbon monoxide. (C) Gridded emissions of CO ($g h^{-1} km^{-2}$), with a spatial gridding of approximately 500 m, using VEIN (Ibarra-Espinosa et al., 2018).

Evaporation also takes place during refueling. At fuel stations, gasoline vapors inside the tank are released into the atmosphere during this process. These emissions can be controlled through the use of Onboard Refueling Vapor Recovery (ORVR), a system designed to continuously recover gasoline vapors, and Stage 2, a system implemented in fuel stations to prevent vapors from being released into the atmosphere (Fung & Maxwell, 2011). Currently in Brazil, these systems are not being utilized. Simulations using Chemical Transport Models have indicated that the implementation of Stage 2 and ORVR could lead to a similar reduction in tropospheric ozone production (Ibarra-Espinosa et al., 2022).

In Brazil there are many studies about the emissions estimation of vehicles. Recently, a manuscript was published which presents a top-down estimation for Brazil at the municipality level, including NMVOC speciation (Hoinaski et al., 2022). For example, in Fig. 3.4, we can see on panel A the morning rush hour volume of traffic for west São Paulo, Brazil; on panel B the estimation of emissions at each street; and on panel C the gridded emissions of CO that shows the visual representation of the street emissions at each pixel. Please note that Fig. 3.4, panel C, expresses the emissions as mass flux (mass/time/area). Emissions inventory for mobile vehicles considers not only the regulated pollutants but also CO_2, due to its importance on climate and use as indicative of engine efficiency. As an example, there is the work of Perez-Martinez et al. (2020), who estimated high-resolution CO_2 emissions from fuel-based vehicles in the metropolitan area of São Paulo.

Similarly to navigation and shipping, emissions from railways and trains depend on the type of fuel. In most cases, it consists of a diesel engine. However, many marine vessels use heavy diesel, which emits large amounts of pollution, resulting in severe air pollution in coastal cities with ports. Furthermore, activities such as shipping, hoteling, and maneuvering at ports contribute to more pollution and are responsible for almost 3% of global GHG emissions, according to the International Maritime Organization (IMO, Faber et al., 2021). The most important pollutant for railways is NO_X, and for shipping SO_X and NO_X (De Lauretis et al., 2021). Diesel vessels also have emissions standards, and in the case of the United States, since 1999, these standards started as 0.4–0.14 $g k^{-1}$ Wh for PM_{10}, as informed in the website: transportpolicy.net.[8]

Worked Example 3.3

A consumer is thinking of buying a new car and is considering three options. One option is a car with an engine <1400 cc, another is a medium car with an engine between 1400 and 2000 cc, and the third option is a bigger car with an engine >2000 cc. The consumer wants to take into account an option more friendly to the environment, considering the CO_2 emissions and also to consider if the speed of the vehicle affects the amount of emissions.

Solution:

To solve this problem, first of all, let us run the script https://github.com/ibarraespinosa/whobook/blob/main/fuel_consumption_CO2.R, which will provide the following figure showing the amount of CO_2 emitted associated to the

8. https://www.transportpolicy.net/standard/us-marine-emissions

speed of the car for the three categories of engines. The CO_2 emissions were estimated applying the European emission factors (Ntziachristos & Samaras, 2016) using the VEIN (Vehicular Emissions Model; Ibarra-Espinosa et al., 2018).

The consumer is interested in the following questions:

1. Does the engine size matter for CO_2 emissions?
2. Is there any difference in the average speed of driving in CO_2 emissions?

The figure clearly indicates that larger engine sizes correspond to higher CO_2 emissions. Additionally, there is a smaller difference between small and medium-sized vehicles, with average emissions of 176.94 and 206.63 $g\,km^{-1}$, respectively, while for larger vehicles the average CO_2 emission is 277.31 $g\,km^{-1}$. However, these averages are calculated across all speeds. Moving on to the second question, it is evident that emission factors vary depending on speed. Taking an average speed of 33 $km\,h^{-1}$, the emission factors for small, medium, and large cars are 174.56, 208.02, and 302.65 $g\,km^{-1}$, respectively. Lastly, we can examine at which speeds the lowest emission factors are observed. For small cars, the minimum emission factor is 139.86 $g\,km^{-1}$ at a speed of 66 $km\,h^{-1}$. Medium-sized cars show a minimum emission factor of 167.88 $g\,km^{-1}$ at a speed of 75 $km\,h^{-1}$. Finally, for large cars, the emission factor is 207.49 $g\,km^{-1}$ at a speed of 90 $km\,h^{-1}$. In recent years, the electrification of the fleet has been seen as a solution to decrease GHG emissions from road sector. However, detractors argued that the GHGs emitted during the production of batteries for electric vehicles are higher than emitted by a gasoline vehicle. To clarify this concerns, the U.S EPA launched the site "Electric Vehicle Myths" where it can be seen clearly how the environmental footprint and use of the electric vehicle is lower than gasoline cars (https://www.epa.gov/greenvehicles/electric-vehicle-myths).

3.3.1.5 Small combustion

Here the categories include combustion with thermal capacity <50 MW and heating boilers >50 kW. Sectors considered are commercial/institutional, residential, agriculture/forestry, and other small stationary combustion. In the category of small combustion, there is a wide range of technologies and fuels utilized, including hard coals used in boilers and steam boilers, solid biomass, liquid fuel, gaseous fuels (hot and water boilers, stream boilers, gas turbines, ovens, etc.). Emissions of pollutants by energy generated are presented in Fig. 3.5.

3.3.1.6 Nonroad mobile machinery

This category covers all mobile sources that are not restricted to roads, including construction, agricultural, residential, and military machinery. The emissions depend mainly on the fuel used, size, and power. Fuel includes diesel, gasoline, and liquefied petroleum gas. The European Emissions Guidelines also provides an Excel file to help in the estimation of these sources (https://www.eea.europa.eu/publications/emep-eea-guidebook-2019/part-b-sectoral-guidance-chapters/

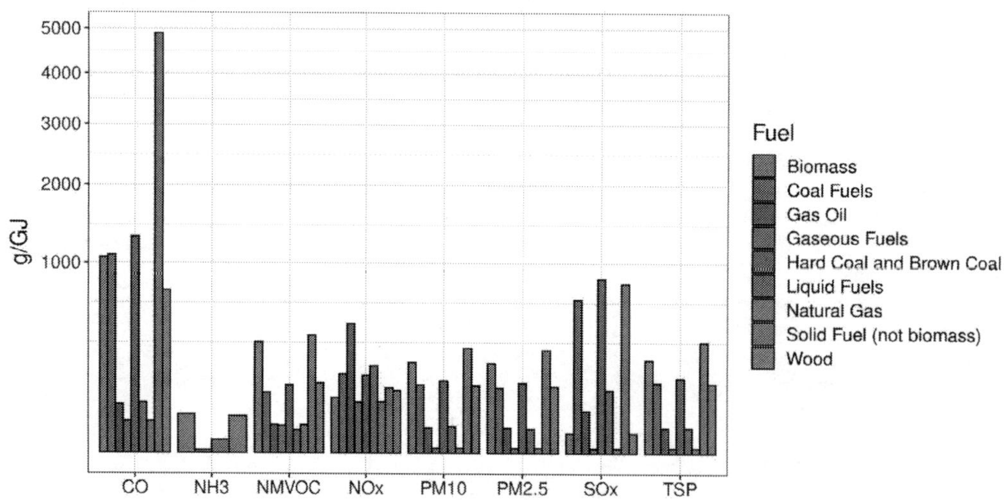

FIGURE 3.5 Average emission factors in small combustion process by type of fuel ($g\,GJ^{-1}$).

1-energy/1-a-combustion/1-a-4-non-road/view). These emissions can be very relevant at regions where the agricultural activity demands the intensive use of machinery, which is not subject to the control of emissions at many countries in the world (Crippa et al., 2020; Lang et al., 2018). In the United States, nonroad mobile sources emissions can be estimated using the model MOVES (Koupal et al., 2002). MOVES covers the sectors of agriculture, construction, lawn, garden, among others.

3.3.1.7 Fugitive emissions from fuels

This category covers solid fuels, such as coal mining and handling, and solid fuel transformation. Coal mining covers extraction, treatment, and developing access to the coal deposit. Most emissions of coal mining involve NMVOC and particulate matter emissions. The transformation of solid fuel refers to fugitive and leakage emissions from coke ovens. Most emissions are CO, black carbon (BC), NMVOC, and particulate matter. For more information, readers are encouraged to check Section 1.B of European Emissions Guidelines. Recently, Tu et al. (2022) used Tropomi[9] to detect coal mine emissions in Poland.

This class of emission also covers exploration, production, transport, refining, storage, distribution of oil products (nonsolid), venting and flaring, and other sources from energy production. There are both onshore and offshore sources, including fugitive emissions from production platforms, crude oil stabilization, glycerol regeneration, tanker loading and losses during transport in marine tanker and gas pipelines, and more. The main emissions are NMVOC and CH_4, which can be detected from space using Tropomi in accident investigations (Pandey et al., 2019). For instance, on August 30, 2019, a natural gas well blowout (uncontrolled release of gas) occurred south of Shreveport (Louisiana, United States), and it took 38 days to have the wells closed. Then, Maasakkers et al. (2022) applied an inverse modeling approach using WRF-Chem and Tropomi data to estimate emissions during this incident.

3.3.2 Industrial process and product use

3.3.2.1 Mineral products

This section covers the production of cement, lime, glass, quarrying, and minerals other than coal, construction and demolition, and storage, handling, and transport of mineral products. In the manufacture of cement, natural raw materials are heated to temperatures between 600°C and 1500°C to produce cement clinker, an intermediate product (Wikipedia Contributors, 2022a). Emissions in cement manufacturing are produced by the physical and chemical reactions of raw material and fuels, with carbon dioxide, nitrogen, dust, sulfur dioxide, and carbon monoxide being the major emitted compounds (Kuenen, Berdowski, et al., 2019). Tier 1 emission factors for cement production are 234 and 130 ($g\,Mg^{-1}$ clinker) of PM_{10} and particulate matter with an aerodynamic diameter smaller than 2.5 μm ($PM_{2.5}$), respectively. In general, there are four stages involved:

1. Extraction and preprocessing of raw materials such as calcium oxide, silicon oxide, aluminum oxide, and ferrous oxide.
2. Pyroprocessing to produce clinker involves a kiln system, a type of high-temperature oven. In this process, raw materials are transformed into new substances through high temperatures. Initially, calcium carbonate is heated to approximately 900°C, resulting in the production of calcium oxide and the emission of CO_2. Subsequently, at around 1400°C, calcium oxide reacts with silica, alumina, and ferrous oxide, ultimately yielding clinker.
3. Blending and grinding of clinker to cement, which consists in intergridding clinker and sulfates.
4. Storage, packing, and delivery of cement.

Lime (CaO) is the result of the calcination of limestone, a carbonate sedimentary rock, at high temperatures. Calcium limestone is composed of 97% − 98% of carbonate, but may contain between 35% − 45% of magnesium carbonate. Main emissions during lime manufacture are particulate matter, specifically 3500 and 700 ($g\,Mg^{-1}$ lime) for PM_{10} and $PM_{2.5}$, respectively (Kuenen, Pacyna, et al., 2019).

Glass is made from natural and abundant raw materials (sand, soda ash, and limestone) that are melted at very high temperatures to form a new glass material. The main pollutant emitted is CO_2 and others from the combustion process. Recycled glass is also used as input in glass manufacture. Emissions also depend on the type of glass produced because they require different energies. Nevertheless, Tier 1 emission factors for glass production are 270 and 240 ($g\,Mg^{-1}$ glass) of PM_{10} and $PM_{2.5}$, respectively (Kuenen, van der Most, et al., 2019).

9. http://www.tropomi.eu/

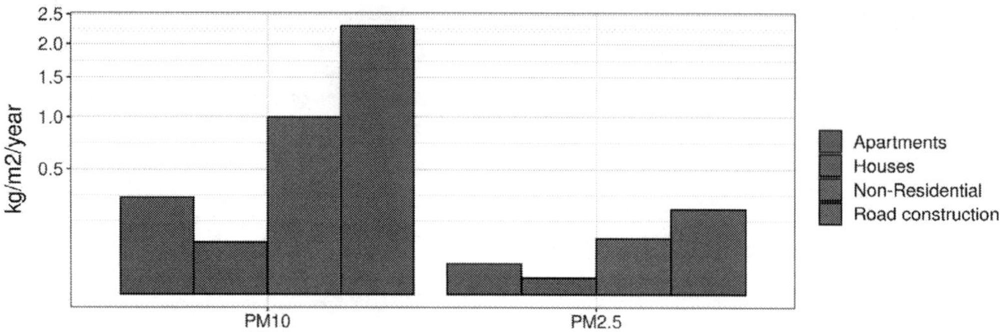

FIGURE 3.6 Emission factors for construction and demolition (kg m^{-2} $year^{-1}$) (Kuenen, Coenen, et al., 2019).

Quarry is a large deep pit from which stones are extracted. Hence, quarrying consists of the extraction of stone or other materials from a quarry. Noncoal quarrying activity emits large amounts of particulate matter. Emission factors are 50 and 5 (g Mg^{-1} mineral) of PM_{10} and $PM_{2.5}$, respectively. The European Emissions Guidelines also provide an Excel file to assist in the emissions estimation for this activity (https://www.eea.europa.eu/publications/emep-eea-guidebook-2019/part-b-sectoral-guidance-chapters/2-industrial-processes/2-a-mineral-products/2-a-5-a-quarrying-1/view).

Construction and demolition activities also produce a large amount of particulate matter. Activity sites can have a significant impact on air pollution on a local scale. Many subactivities are associated, such as land clearing, soil and equipment moving, mobile debris, vehicular transport, and loading of hauling material, building debris such as concrete and drilling, and windblown dust from unpaved roads. As an example, Tier 1 emission factors are shown in Fig. 3.6.

Worked Example 3.4

Calculating the CO_2 Emissions from Cement Production

A cement plant produces 1 million tonnes of cement per year. The plant uses a dry process to produce cement, which involves heating limestone and other raw materials in a kiln to produce clinker. The clinker is then ground into a fine powder to produce cement. The process emits CO_2 due to the calcination of limestone and the combustion of fuels in the kiln. The plant uses coal as the primary fuel for the kiln and produces 0.6 tonnes of CO_2 per tonne of cement produced. Additionally, the plant uses electricity from the grid, which is generated from a mix of sources that emit 0.5 tonnes of CO_2 per MWh of electricity generated. The plant consumes 10 MWh of electricity per tonne of cement produced. Calculate the total annual CO_2 emissions from the cement plant.

Solution:

Calculating the total amount of CO_2 emissions from the combustion of coal: 0.6 tonnes of CO_2 per tonne of cement $\times$ 1 million tonnes of cement $=$ 600,000 tonnes of CO_2 emissions per year, and then calculating the total amount of CO_2 emissions from electricity consumption: 10 MWh of electricity per tonne of cement $\times$ 1 million tonnes of cement $=$ 10 million MWh of electricity per year:

10 million MWh of electricity $\times$ 0.5 tonnes of CO_2 per MWh $=$ 5 million tonnes of CO_2 emissions per year.

Calculating the total annual CO_2 emissions from the cement plant:

Total CO_2 emissions $=$ CO_2 emissions from coal combustion $+$ CO_2 emissions from electricity consumption

Total CO_2 emissions $=$ 600,000 tonnes $+$ 5 million tonnes $=$ 5.6 million tonnes of CO_2 emissions per year.

Therefore the cement plant emits 5.6 million tonnes of CO_2 per year.

3.3.2.2 Chemical industry

Chemical industry covers the production of compounds such as ammonia, nitric acid, adipic acid, and carbide. Each industry is constituted by particular complex processes; thus it is recommended that the readers consult the Specific Emission Guidelines for further details of each context. The main pollutant in this category is NO_X. In general, most emissions of ammonia production are 1 (kg $NO_X t^{-1}$ NH_3), 0.1 (kg CO t^{-1} NH_3), and 0.01 (kg $NH_3 t^{-1}$ NH_3). Nitric acid emits 10 (kg $NO_X t^{-1}$ NH_3). Adipic acid production emits 8 (kg $NO_X Mg^{-1}$) and 0.4 (kg CO Mg^{-1}). Calcium carbide production emits 0.1 (kg TSP Mg^{-1} product).

3.3.2.3 *Metal production*

This section covers the production of iron and steel, ferroalloys, aluminum, magnesium, lead, zinc, copper, nickel, and others. There are primary and secondary productions of materials according to the technology. To provide guidance, Fig. 3.7 displays the average EF by production type as $g\,Mg^{-1}$ of produced materials.

3.3.2.4 *Other solvent and products*

Solvents is a type of source that recently has gained more importance. The subactivities covered are domestic solvent use, road paving with asphalt, asphalt roofing, coating applications, degreasing, dry cleaning, printing, and others. The major pollutant is the NMVOC and default emission factor is normalized per capita. Main categories are cosmetics, household products, construction products such as adhesives and paint removers, car care products, and pesticides. Tier 1 emission factors for NMVOC are $1.8\,kg\,capita^{-1}$ for western Europe and $1.2\,kg\,capita^{-1}$ for other countries.

3.3.2.5 *Production of persistent organic pollutants*

Persistent organic pollutants (POPs), known as forever chemicals, are toxic pollutants that resist and remain in the environment (Wikipedia Contributors, 2022b). Many POPs can bioaccumulate on animals, like whales and other marine fauna, presenting a high risk for human health through consumption of these animals. In 2001 the Stockholm Convention on Persistent Organic Pollutants aimed to eradicate the emissions of POPs among their parties (http://www.pops.int/). Due to their toxicity, these pollutants have been thoroughly studied (Breivik et al., 2007). The available EFs for these pollutants are included in the European Emissions Guidelines, in the vehicular emissions section for Polychlorinated Dibenzodioxins (PCDD or just dioxins), Polychlorinated Dibenzofurans (PCDF), and Polychlorinated Biphenyl (PCB). Fig. 3.8 presents the vehicular EF for PCB, PCDD, and PCDF by type of vehicle,

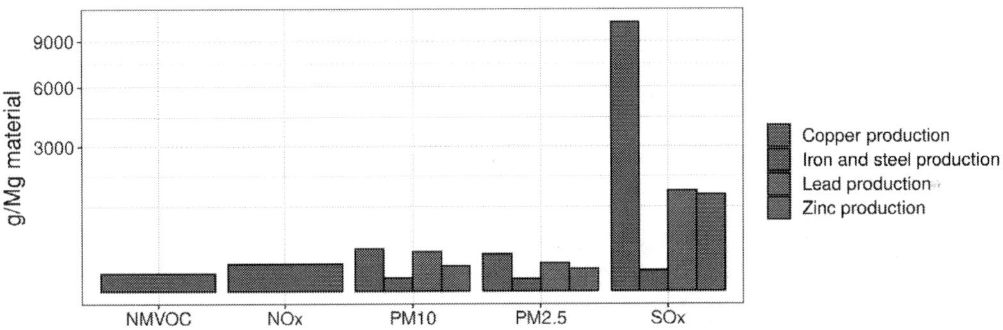

FIGURE 3.7 Emission factors from metal production in $g\,Mg^{-1}$ material produced. These emission factors have been averaged by sector.

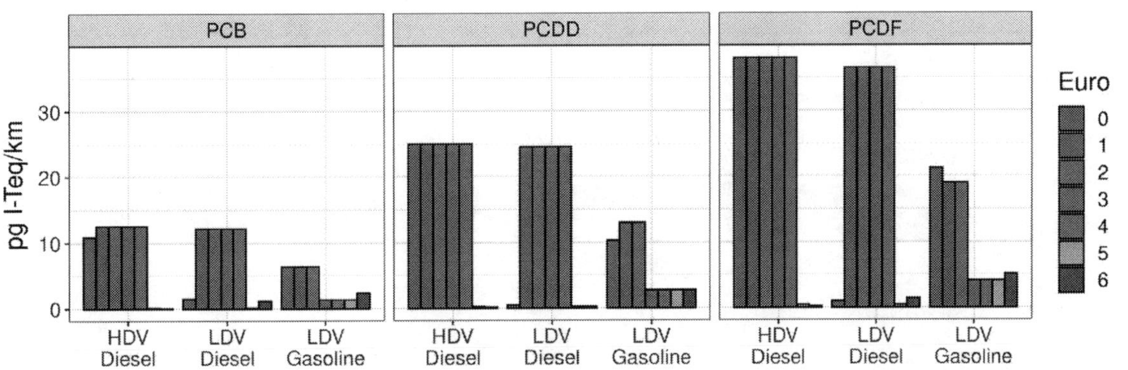

FIGURE 3.8 Vehicular emission factors of persistent organic pollutants for high-duty vehicles (HDV) and light-duty vehicles (LDV) according to the Euro emissions control phases (pg $I\text{-}Teq\,km^{-1}$).

fuel, and Euro standard expressed as picograms in International Toxic Equivalent Quantity per km (pg I-Teq km^{-1}). Vehicles are divided by type: high-duty vehicles (HDV) and light-duty vehicles (LDV). In general, newer standards have lower POP emissions.

3.3.3 Agriculture

Agriculture is an important category, covering manure management, crop production, pesticides, and field burning.

3.3.3.1 Manure management

According to the Cambridge dictionary, manure is the solid waste from animals, especially horses, that is spread on the land in order to make plants grow well.[10]

In other words, manure is the material that fertilizes land. Main pollutants emitted are NH_3 and N_2O. The source of NH_3 is the nitrogen excreted by livestock. Thus these emissions also vary by animal species, agricultural practice, and climate. NO is formed by the nitrification process on the surface. NMVOC and CH_4 are also emitted by livestock production. The average emission factor by type of animal, expressed as kg per average annual population (AAP) is shown in Fig. 3.9. It can be observed that dairy cattle and buffalo are the main categories for all pollutants. Note that the European Emissions Guidelines also provide software to estimate these emissions (https://www.eea.europa.eu/publications/emep-eea-guidebook-2019/part-b-sectoral-guidance-chapters/4-agriculture/manure-management-n-flow-tool/view).

3.3.3.2 Crop production and agricultural soils

This section covers the emission of NH_3, NO, NMVOC, and particulate matter from crop production and agricultural soils. The NH_3 is released from cultivated crops and crop residues. This category includes the following sources: inorganic nitrogen fertilizers including urea, livestock mature in soils, sewage sludge in soils, other organic fertilizers in soils, urine and dung deposit by livestock, crop residues applied to soils, indirect emissions, farm-level operations, and

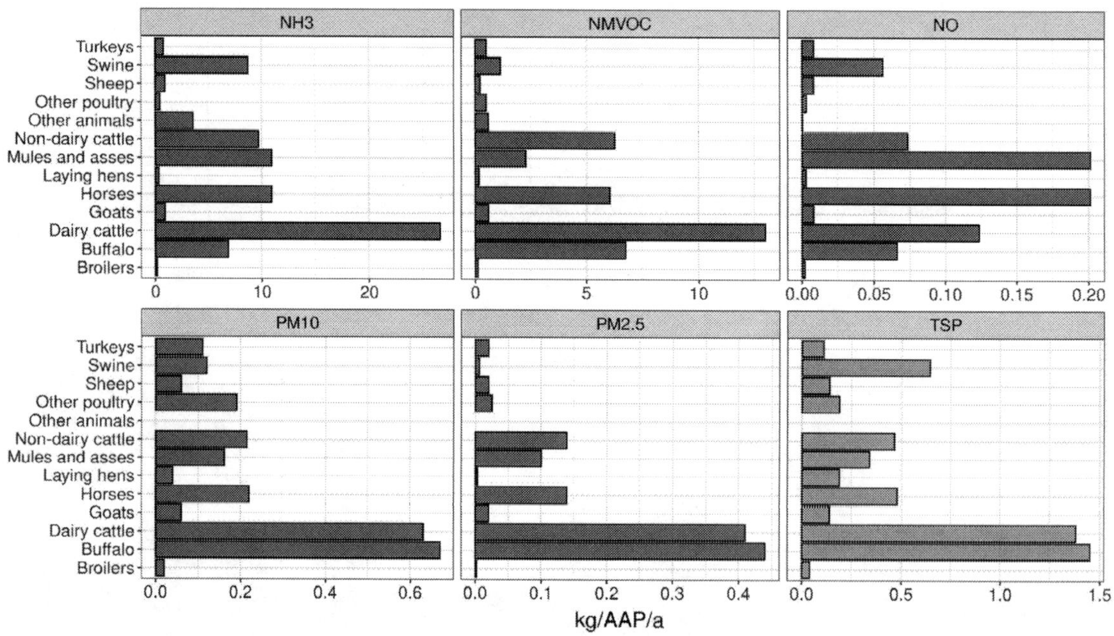

FIGURE 3.9 Average emission by agriculture manure by type of animal (kg AAP a^{-1}).

10. https://dictionary.cambridge.org/us/dictionary/english/manure

cultivated crops. NH_3 emissions occur 7–10 days after nitrogen fertilizer application. The average emission factors are 1.9, and 2.24 (kg NH_3 AAP $year^{-1}$) for excreta deposited by grazing livestock and manure applied to soils, respectively. The postharvest process (cleaning, drying, classification, etc.) of grains is also an emission source, mainly of particles and CO. The drying process involves furnaces and wood to provide heat to dry the grains, and emissions are usually not considered in the inventories. as the same applies to the cleaning of grains, which mainly produces particulate matter.

3.3.3.3 Field burning of agricultural residues

This section refers specifically to clearing land and rapidly burning crop residues. Note that this activity is different from burning waste. It is an important source of emissions in agricultural countries. Many factors influence the amount of pollutants released, such as the combustion temperature, climate, moisture, oxygen, and more. The European Emissions Guidelines report emission factors of 0.0054 (kg PM2.5 kg^{-1}-dry-matter) and 0.0005 (kg BC kg^{-1}-dry-matter). Emission related to this activity can be estimated as follows:

$$AR_{ResidueBurnt} = A \times Y \times s \times d \times p_b \times Cf, \tag{3.2}$$

where A (ha) is the area of land with crops residue burnt, Y (kg ha^{-1} fresh-weight) is the average yield of those crops, s is the ratio between the mass of crop and residues and the crop yield, d is the dry matter content of the yield, pb is the proportion of residues that are burned, and Cf is the combustion factor (Webb et al., 2019).

3.3.4 Landfills and waste

The major compounds emitted from landfills and waste disposal are GHGs, mainly CH_4 (Crippa et al., 2020). Regarding air pollutants, main emissions are of NMVOC. Specifically, Tier 1 emission factors are 1500 (g NMVOC Mg^{-1}), 0.219 (g PM_{10} Mg^{-1}), and 0.033 (g $PM_{2.5}$ Mg^{-1}). Composting is a process in which the organic matter is recycled, fertilizing the land, with little emissions but from which odors can be perceived. Emission factor is 0.24 (kg NH_3 Mg^{-1} organic waste). According to the U.S. EPA, "Biogas is generated during anaerobic digestion when microorganisms break down (eat) organic materials in the absence of air (or oxygen). Biogas is mostly CH_4 and carbon CO_2, with very small amounts of water vapor and other gases. The CO_2 and other gases can be removed, leaving only the CH_4. Methane is the primary component of natural gas."[11]

There are mainly three technologies for municipal waste incineration: (1) mass burn units, in which municipal wastes are directly incinerated, with an average of 45–900 t day^{-1}; (2) modular combustors, that incinerate not preprocessed but shop-fabricated wastes, resulting in smaller wastes; and (3) fluidized bed combustors, which have a bed where materials are agitated. In the latter, preprocessing is needed to obtain smaller sizes. The average emission factors are 1071 (g NO_X Mg^{-1}), 87 (g SO_2 Mg^{-1}), 41 (g CO Mg^{-1}), and 5.9 (g NMVOC Mg^{-1}). Industrial waste incineration follows the same principle, but types of residues are different; hence the emission factors of 870 (g NO_X Mg^{-1}), 47 (g SO_2 Mg^{-1}), 70 (g CO Mg^{-1}), and 7400 (g NMVOC Mg^{-1}). Hospital waste is toxic and dangerous and is required to be incinerated. Average emission factors are 2300 (g NO_X Mg^{-1}), 540 (g SO_2 Mg^{-1}), 190 (g CO Mg^{-1}), and 700 (g NMVOC Mg^{-1}). Finally, cremation is the incineration of corpses and carcasses, which emits particulate matter and other pollutants. In India, cremation can cause high levels of air pollution (Bhadauria et al., 2022). In addition, treatment of wastewater can generate emissions as well, especially of NMVOC in urban centers. In this case, emission factor is 15 (mg NMVOC m^{-3} wastewater handled).

3.4 Sources of indoor air pollution

Many chemical compounds are introduced and formed indoors. VOC are ubiquitous both indoors and outdoors. Furnishing and office products are inherent to the modern lifestyle, and together with paint, solvent, and cleaning products, VOC can be present at a higher concentration. Other human activities, such as cooking and tobacco smoke, can enormously affect the levels of indoor VOC. Together with the introduction of outdoor pollutants, frequently indoor environments become much more harmful than outdoors. Solid fuels are extensively used in developing countries and are considered the main sources of indoor pollution. Approximately 220 tons of solid fuels are consumed in these

11. https://www.epa.gov/anaerobic-digestion/basic-information-about-anaerobic-digestion-ad#:~:text=Biogas%20is%20generated%20during%20anaerobic,water%20vapor%20and%20other%20gases.

countries. Incomplete burning produces BC, organic gases and particles, and inorganic species like SO_X and NO_X, metals, and ions (Mohammad et al., 2021).

World Health Organization (2018) reported that over 3 billion people worldwide used kerosene, biomass, and coal for cooking, and nearly 4 million people died due to indoor pollution by cooking with solid fuel and kerosene. A large study in China involving over 30,000 people, showed the association between the use of solid fuel for cooking and heating with depressive symptoms (Zhang et al., 2022). Some VOCs can react in indoor ambient. The compound 4-phenyl cyclohexane, present in carpets, reacts with latex paints, and secondary aldehydes can be formed (Morrison & Nazaroff, 2002).

The literature reports a large list of sources of indoor pollution and the most used chemicals. Aliphatic and aromatic hydrocarbons, ketones, and aldehydes are often linked to vehicular emissions and some of these are emitted indoors by furniture, paints, and insulation. Alcohols, halocarbons, and esters are emitted from personal care products (Miguel et al., 1995; Yue et al., 2021). The large use of computers and printers is associated with the presence of organic compounds and, at poor ventilated environment, concentrations of aromatic and aliphatic hydrocarbons, as well as ketones and aldehydes, can increase, affecting indoor air quality (Ayoko et al., 2004). A recent study (Seo et al., 2022) reports the emissions of polycyclic aromatic hydrocarbons (PAH) using computers. The most abundant compounds were naphthalene and phenanthrene, and the concentrations of PAH were 2–5 times higher than in the outdoor atmosphere. Other minor contaminants can also be found indoors, such as esters, ethers, terpenes, and amines coming from candles, consumer products, and air-conditioning systems. Many allergens come from plants, pollen, and mold. In developing countries where the burning of biomass for cooking is routine, health problems are reported. World Health Organization (2021) reports that around 2.6 billion people in the world use open fires and stoves fueled by kerosene, coal, and biomass to cook their foods, and over 4 million people die of illnesses attributable to these activities

3.4.1 Mitigating indoor air pollutants

Yue et al. (2021) reviewed adsorption material and catalytic oxidation processes to mitigate indoor pollution problems. Due to the complexity of indoor pollutants and sources, improving air quality becomes a great challenge. To decrease the concentration of VOCs and $PM_{2.5}$, technologies using functional molecular materials seem promising, since they present large specific surface area and pore size overcoming the use of zeolites and activated carbon materials. Acetone, benzene, toluene, and other VOCs are easily absorbed by different biochar produced by raw materials, and the adsorption capacity depends mainly on the surface area and its noncarbonized organic matter. Many parameters have to be considered, including type of biochar, pyrolysis temperature, and surface modification. Other materials of interest for adsorption are those with metal-organic framework (MOFs), formed with organic ligands and metal ions. They have a specific surface area and pore volume with high adsorption capacity for VOCs. Due to their excellent properties, MOFs are an interesting choice for environmental mitigation. Morawska et al. (2024) discussed the need to be create an Indoor Air Quality index to guarantee the health of the population.

3.5 Other sources

Microplastics have already been found not only in the air we breathe, in terrestrial, marine, and freshwater environments, but also in tap and bottled water, sea salt, foods, seafood, and in fish consumed by humans. A large proportion of the plastic produced in the world accumulates in the environment as waste, and its fragmentation can lead to the presence of secondary plastic in the atmospheric, aquatic, and terrestrial environments. Primary microplastics are those that can be seen at the size they were manufactured. Secondary microplastics are those that have suffered abrasion or deterioration by ultraviolet (UV) radiation, temperature, and humidity, among other factors. A study carried out in the United States analyzed dry and wet atmospheric deposition in conservation areas, showing that microplastics were present in 98% of the samples. Microplastics deposited in wet conditions are larger in size and lower in number. In dry depositions, plastics are smaller in size and can be subject to global dispersion (Brahney et al., 2020). Microplastics in the atmospheric environment are still poorly studied and can also be found indoors (Akdogan & Guyen, 2019; Dris et al., 2017). Microplastics can also enter the human body through the process of food digestion, absorption, and metabolism through the digestive system, as well as by inhalation through the respiratory system (Zhang et al., 2021; Cox et al., 2019).

Worked Example 3.5

Emissions of CO by an Industrial Boiler

An industrial boiler burns 80,000 L of fuel oil per day. Estimate the CO emissions (kg day^{-1}) considering that there are no emission control devices. *Note*: EF (emission factor) for CO = 0.6 kg per 1000 L oil burned.

Solution:

$E = 80{,}000\ \text{L oil day}^{-1} \times 0.6\ \text{kg/1000 L} = 48\ \text{kg CO day}^{-1}$

3.6 Emission inventories—top down, bottom up, global, regional, and city scales

3.6.1 Definition and types

In the previous sections we identified the sources of air pollutants. Now, we will present how to quantify the emissions of these sources. As it is not possible to manage the unknown, emission inventories are quite important in the context of air quality management and air quality forecasts. Emission inventories are defined as databases that describe the quantity of emitted pollutants by type of pollutant source, for a defined time and space (Vallero, 2014). This definition can help us determine the classification of emission inventories. For example, vehicular emission inventories only quantify pollutants emitted by vehicular fleets; regarding spatial coverage, there are global and regional emission inventories; and about type of pollutant, there are greenhouse emission inventories. When building the emission inventory, more resources need to be spent on the calculation of the most important source.

3.6.2 Emission inventories and air quality management

In the context of air quality management, emission inventories allow us to evaluate whether mitigation measures are being effective. In the context of air quality forecasts, they are an important input for air quality models, frequently pointed out as the reason when a given air quality model does not reach the performance benchmark (Pulles & Heslinga, 2007). According to Thunis et al. (2022), errors in spatial and temporal allocation can greatly affect model results in urban scales modeling. Countries are compelled to estimate their national emissions through many international treaties and conventions. One of the oldest is "The Convention on Long-Range Transboundary Air Pollution (CLRTAP)" (Matthias et al., 2018). For example, the Pollutant Release and Transfer Registers (PRTR) are applied to the country's members of the Organisation for Economic Co-operation and Development (OECD). The IPCC also helps countries estimate their greenhouse emissions inventories with methodologies and protocols. Likewise, states and cities are compelled to measure or estimate their emissions by national laws or state mandates.

3.6.3 Methods to measure emissions

Quantifying emissions is not a simple task, and there are many ways to proceed with it. We can directly measure emissions by using a gas analyzer to measure emissions from a chimney, or, alternatively, when trying to measure emissions from vehicles, by using laboratory measurements from dynamometers. On the other hand, we can also estimate emissions using the EF. To this, it is possible to use the Eq. (3.1) discussed in Section 3.2. Emission factors are values which relate emissions with an activity factor (grams of emitted pollutant by kg of burned fuel). Most important EF databases are the AP42[12] from the U.S. EPA and the EMEP/EEA air pollutant inventory guidebook from the European Environmental Agency.[13]

3.6.4 Building approaches: top-down and bottom-up

There are two approaches to calculate emission inventories: bottom-up and top-down. This classification is based on the direction and detail of information. For example, using measurements from the source level (like plants or vehicles) in an emission estimation is a bottom-up approach. On the other hand, using national inventory statistics is a top-down approach (Pulles & Heslinga, 2007). Theoretically, these approaches should produce similar results (i.e., emission reconciliation). Consider the following example: calculating vehicular emission by using activity data from a traffic model (traffic count), vehicle composition, speed recording, and length of roads, and EF that depend on speed and fuel, would be considered a bottom-up approach. However, if vehicular fleet number and composition, fuel consumption, mean speeds, and country balance are used, it would be a top-down approach. Ultimately, we could use the total fuel consumption to reconcile both emission inventories.

12. In: https://www.epa.gov/air-emissions-factors-and-quantification/ap-42-compilation-air-emissions-factors

13. In: https://www.eea.europa.eu/themes/air/air-pollution-sourcs-1/emep-eea-air-pollutant-emission-inventory-guidebook

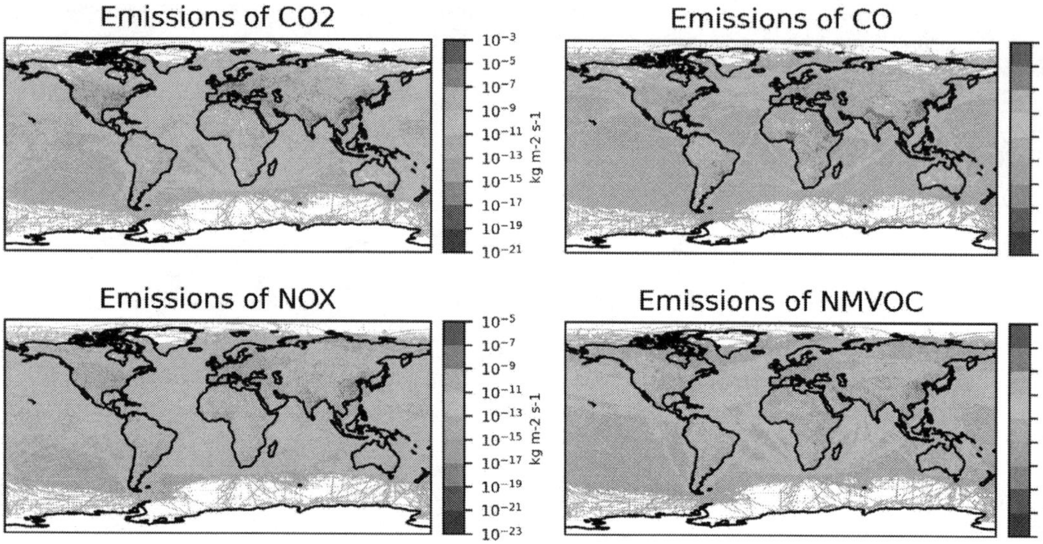

FIGURE 3.10 EDGARv6.0 total emission rate (in $kg\,m^{-2}\,s^{-1}$) for 2018 (all sources) for CO_2, CO, NO_X, and nonmethane volatile organic compound (NMVOC). *Based on https://edgar.jrc.ec.europa.eu.*

3.6.5 Global, regional, and local emission inventories

Global emissions inventories are available in the Emissions of Atmospheric Compounds and Compilation of Ancillary Data (ECCAD) site (https://eccad.aeris-data.fr/). It hosts different spatially disaggregated emissions for different sources, including anthropogenic, biomass burning, and biogenic emission species, for greenhouse, reactive, and VOCs gases. Well-known emission inventories, like the Emissions Database for Global Atmospheric Research (EDGAR), REanalysis of the TROpospheric chemical composition (RETRO), and the Evaluating the Climate and Air Quality Impacts of Short-Lived Pollutants (ECLIPSE) project are also publicly available (Matthias et al., 2018). As global inventories, they usually have a spatial resolution of $1°-0.1°$, and a temporal resolution of yearly and monthly periods. Fig. 3.10 displays the emission rate in $kg\,m^{-2}\,s^{-1}$ of different pollutants from EDGARv6.0 for all the sources available for 2018. EDGARv6.0 is a global inventory with 0.1° of spatial resolution and with a monthly temporal resolution that also has emission rates for different sources, as detailed in the previous subsections.

3.7 Conventions and global policies

Energy consumption in industrialized countries is based primarily on fossil fuels and is mainly associated with the main negative impacts of diffuse sectors such as transportation: climate change, air pollution, congestion, and accidents (Ewbank et al., 2020). Although the emissions of many pollutants are being reduced due to improvements in engines and fuels, the consequences for human health are a growing concern, particularly the risks imposed by NO_X and PM. Emissions of CO_2, the gas considered the major responsible for the greenhouse effect, are also increasing, especially regarding the transportation sector.

The world's developing economies are involved in an ecological transition toward reducing the consumption of fossil fuels and related emissions. For instance, the European Commission's White Papers on Transport concluded that this sector will be based on energy-efficient clean modes, which will also help to reduce the demand and end-use energy. The diffuse sectors present opportunities for establishing effective strategy acts. These plans require a greater commitment to the processes of economic deregulation due to involving several factors: consumer awareness of price considerations; the establishment of mechanisms to ensure that these prices reflect actual costs, according to the "polluter pays" principle; the promotion of energy savings; and the prevention of health effects and economical costs (Pimpin et al., 2018). This principle was underlined in the Green Paper on energy end-use efficiency and services, which suggests that overall consumption in the European Union can be reduced by up to 20% until 2040 without compromising economic profitability. Policy measures require time to take effect and must be supported by changes in lifestyle (Rodenburg et al., 2002). In the transportation sector, a reduction in emissions can be also achieved by balancing modal distribution (Schipper et al., 2001; Steenhof et al., 2006).

3.8 Changes in emissions due to the pandemic

During the COVID-19 pandemic, many studies around the world focused on the impacts of restrictions on mobility in urban areas. This was associated with a reduction in exhaust and nonexhaust emissions (brakes and tires). In general, a decrease in some primary pollutants, such as CO and NO_X, was observed, but PM and ozone presented different behaviors depending on the site of origin (Sokhi et al., 2021). It was also observed that other sources have great participation for some pollutants. For instance, the biomass burning on CO and particles, and rural emissions for NH_3, influencing the secondary formation of particles and ozone. Fig. 3.11 shows the monthly deviations of bus passenger trips, NO_X concentrations, LDV and HDV flow, congested kilometers, and the number of COVID-19 deaths in São Paulo, Brazil. During the initial months of the epidemic, bus passenger flows decreased by ~70%, with 42%, 55%, 18%, and 94% reductions in NO_X, LDVs, HDVs, and congestion, respectively. These reductions were the direct consequence of mobility restrictions established in São Paulo to battle COVID-19, which showed an initial peak in June 2020, with an average of 96 deaths per day. The isolation rate increased from 25% to 52% during this period—with more than half of the citizens experiencing restricted mobility. Transit vehicle occupation rates decreased to a minimum of 25 passengers per operated bus, down from the normal rate of ~34 passengers (decrease of 25%), influencing transport policies toward

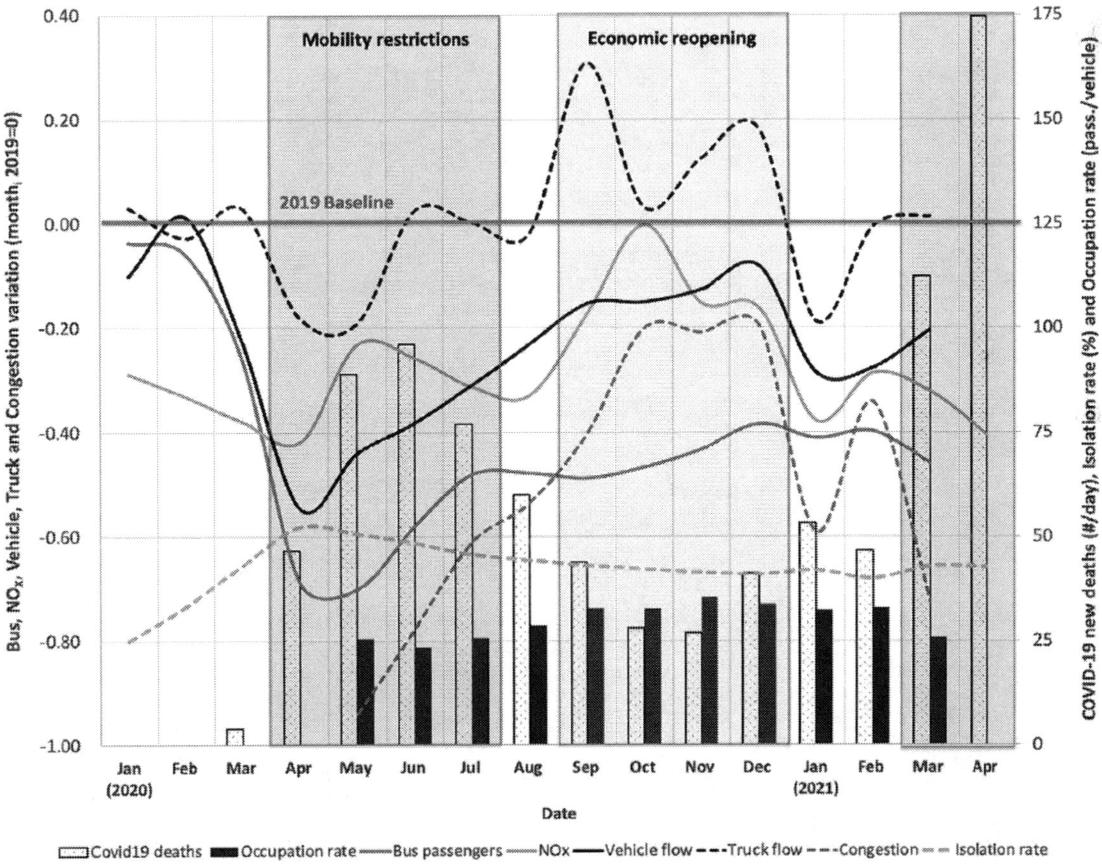

FIGURE 3.11 Monthly deviations of bus passenger trips, NO_X concentrations, vehicle and truck flows, and congested kilometers compared with COVID-19 deaths, bus occupation, and isolation rates during the months since the beginning of the epidemic in São Paulo city, Brazil. *Notes*: deviations of the month *i* (January 2020–April 2021) are estimated from 2019 baseline records (deviation month = (month − months, 2019)/months, 2019); red and yellow vertical shaded bars represent the months with some mobility restrictions and reopening of the economic activities, respectively. *With permission from Journal of Transport Geography, 101 (2022) 103349 (Pérez-Martínez et al., 2022).*

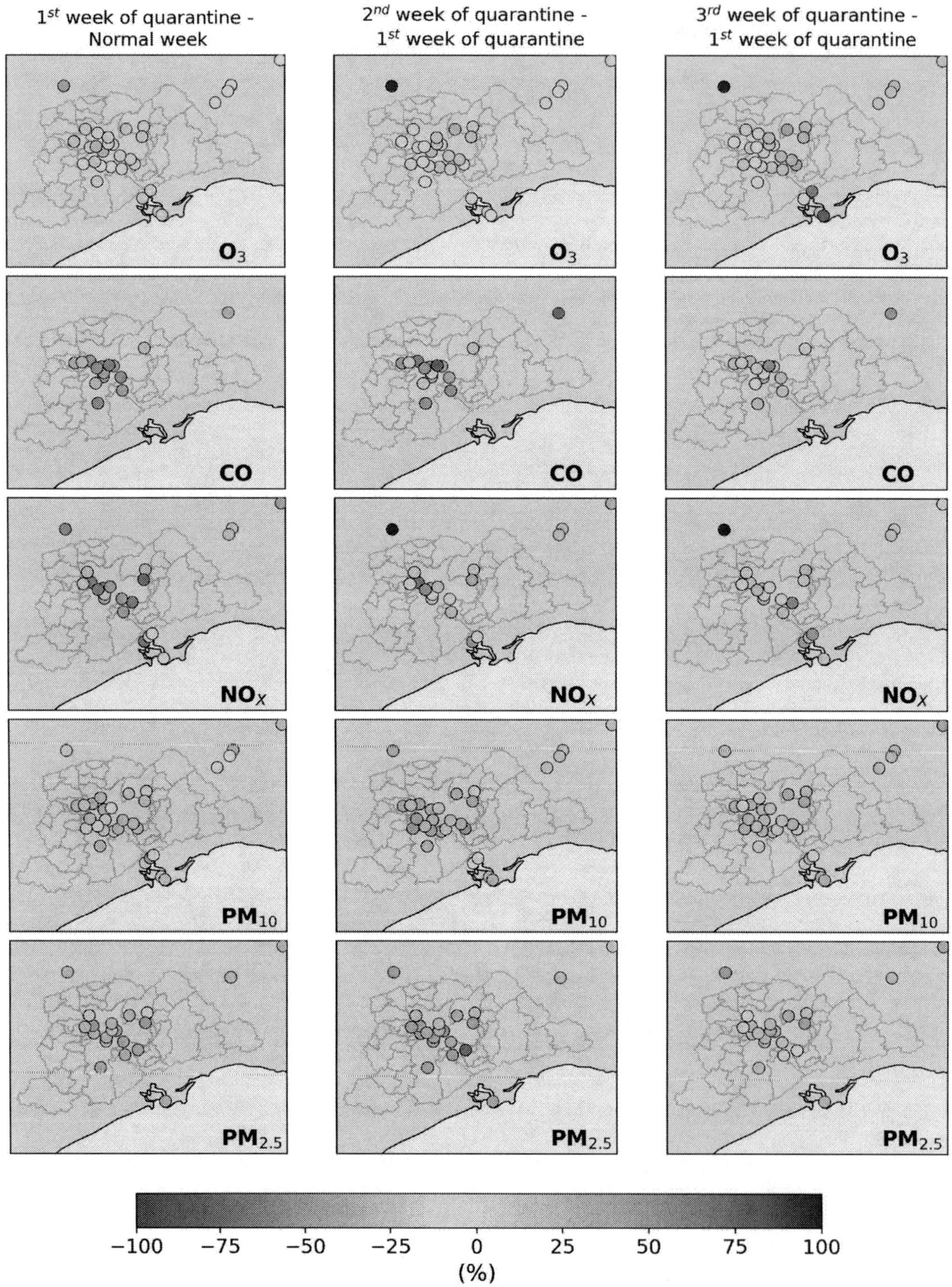

FIGURE 3.12 Difference in pollutant concentration between a baseline, normal week (March 15–21, 2020), first week of quarantine (March 22–23, 2020), second week of quarantine (March 2–April 4, 2020), and third week of quarantine (April 5–11, 2020) in the metropolitan area of Sao Paulo, Brazil, for O_3, CO, NO_X, PM_{10}, and $PM_{2.5}$.

social distancing and the influence of the new Delta variant in March and April 2021. The reduction of COVID-19-associated mortality to ~25 deaths per day led to a reopening of economic activities by the end of 2020. NO_X concentrations and LDV flows almost recovered to prior prepandemic values (HDV flows even increased in comparison to 2019 by 20%). Bus passengers and congestion values recovered from the previous 40% and 20%, respectively, reductions regarding the baseline period. Transit occupation values increased to normal operating values and isolation rates remained at a constant value of 42% (an increment of 68% compared to the previous 25%, corresponding to restricted mobility under normal conditions). We can observe that, in October 2020, NO_X concentrations returned to near-baseline levels. In São Paulo, this pollutant predominately results from heavy truck and bus emissions (CETESB, 2020b). Although bus circulation rates did not recover to baseline levels, the circulation of trucks increased during this economic reopening period. There was an overall reduction of the mobility patterns corresponding to new mobility restrictions linked to different peaks of COVID-19 deaths at the beginning of 2021. In April 2021, the epidemic reached a maximum peak of 175 new deaths per day. In response to new peaks, different mobility restriction policies were applied from January 2021 onwards and deviations of NO_X concentrations, vehicle and truck flow, and congestion fell to 38%, 29%, 19%, and 59%, respectively in relation to 2019. Bus passengers slightly decreased their activities and deviations fell under 40% (corresponding to a new decrease in the occupation rates to 26 passengers per vehicle).

In Fig. 3.12, difference in pollutant concentrations of O_3, CO, NO_X, PM_{10}, and $PM_{2.5}$ are presented between what was considered as a normal week (March 15–21, 2020, before the full lockdown), the first week of quarantine (March 22–23, 2020), second week of quarantine (March 2 to April 4, 2020), and third week of quarantine (April 5–11, 2020), in the metropolitan area of São Paulo, Brazil. This figure is a good example of the complexity of ozone formation. Many authors reported that ozone increased during the lockdown due to the decrease in NO_X. But when analyzing in more detail urban and suburban air quality stations, it was observed that more detailed analysis is needed for interpreting ozone concentrations close to roads or distant from emissions sources, within the meteorological conditions of each specific period of emission reductions of primary pollutants.

3.9 Questions

1. Using R, install the package VEIN and run the following script presented in Box 3.1. The idea is to compare how the emission varies depending on speed. What can be concluded? Do the emission factors converge to some speed? Analyze different pollutants. The resulting emission factors give units of $g\,km^{-1}$. Then, multiply by different distances (your house to school/work, for instance) and estimate emissions.
2. With the help of the recent literature, present an estimation of the global emissions of SO_2, CO, NO_X, and CO_2 due to anthropogenic and natural processes, for the year 2020.
3. Discuss the primary source of pollutants in your region (State or National Level) for PM and regulated pollutants.
4. Enumerate sources of metal production in your country.
5. Due to the control legislation of sulfur emission, concentrations have decreased, mainly in the Northern Hemisphere. Discuss the consequences to the radiative balance in the atmosphere.
6. Methane is the most abundant hydrocarbon in the atmosphere, displaying a long lifetime. Discuss its main sources in natural and urban areas.

Solutions to the end-of-chapter questions will be available at https://www.elsevier.com/books-and-journals/book-companion/9780128225912.

BOX 3.1 Application using VEIN (Vehicular Emission Model).

```
# install.packages("vein")# if not installed library(vein)
ef1 <- ef_eea(category = "PC," fuel = "G," segment = "Small," euro = "III," tech = "GDI," pol = "NOX," mode = NA, slope = 0, load = 0)
ef2 <- ef_eea(category = "PC," fuel = "G," segment = "Small," euro = "I," tech = NA, pol = "NOX," mode = NA, slope = 0, load = 0)
df <- data.frame(EF_Euro_III = ef1(0:100), EF_Euro_I = ef2(0:100))
colplot(df, main = "NOX (g km-1)," xlab = "Speed (km h-1)")
```

References

Adam, D. (2022). Tonga volcano created puzzling atmospheric ripples. *Nature*, *602*, 497–498.

Akdogan, Z., & Guyen, B. (2019). Microplastics in the environment: A critical review of current understanding and identification of future research needs. *Environmental Pollution*, *254*, 113011. Available from https://doi.org/10.1016/j.envpol.2019.113011.

Amon, B., Hutchings, N., Dämmgen, U., Sommer, S., & Webb, J. (2019). EMEP/EEA air pollutant emission inventory Guidebook 2019. *Technical Guidance to Prepare National Emission Inventories*, *26*.

Andrade, M., Kumar, P., de Freitas, E. D., Ynoue, R. Y., Martins, J., Martins, L. D., Nogueira, T., Perez-Martinez, P., de Miranda, R. M., Albuquerque, T., Gonçalves, F. L. T., Oyama, B., & Zhang, Y. (2017). Air quality in the megacity of São Paulo: Evolution over the last 30 years and future perspectives. *Atmospheric Environment*, *159*, 66–82. Available from https://doi.org/10.1016/j.atmosenv.2017.03.051.

Andres, R. J., Rose, W. I., Kyle, P. R., DeSilva, S., Francis, P., Gardeweg, M., & Moreno Roa, H. (1991). Excessive sulfur dioxide emissions from Chilean volcanoes. *Journal of Volcanology and Geothermal Research*, *46*(3–4), 323–329. Available from https://doi.org/10.1016/0377-0273(91)90091-D.

Ayoko, G. A., Morawska, L., Kokot, S., & Gilbert, D. (2004). Application of multicriteria decision making methods to air quality in the microenvironments of residential houses in Brisbane, Australia. *Environmental Science & Technology*, *38*(9), 2609–2616.

Bauer, H., Kasper-Giebl, A., Löflund, M., Giebl, H., Hitzenberger, R., Zibuschka, F., & Puxbaum, H. (2002). The contribution of bacterial and fungal spores to the organic carbon content of cloud water, precipitation and aerosols. *Atmospheric Research*, *64*(1–4). Available from https://doi.org/10.1016/S0169-8095(02)00084-4.

Bazzo Vieira, J. P., Vieira Braga, C. K., & Pereira, R. H. M. (2022). The impact of COVID-19 on air passenger demand and CO_2 emissions in Brazil. *Energy Policy*, *164*. Available from https://doi.org/10.1016/j.enpol.2022.112906.

Bhadauria, V., Parmar, D., Ganguly, R., Rathi, A. K., & Kumar, P. (2022). Exposure assessment of $PM_{2.5}$ in temple premises and crematoriums in Kanpur, India. *Environmental Science and Pollution Research*, *29*, 38374–38384.

Bindoff, N. L., Cheung, W. L., Kairo, J. G., Javier, A., Guinder, V. A., Hallberg, R., H, N. J. M., Jiao, N., Levin, L., O'Donoghue, S., Purca Cuicapusa, S. R., Rinkevich, B., Suga, T., Tagliabue, A., & Williamson, P. (2019). Changing ocean, marine ecosystems, and dependent communities. In H.-O. Pörtner, D. C. Roberts, V. Masson-Delmotte, P. Zhai, M. Tignor, E. Poloczanska, et al. (Eds.), *IPCC special report on the ocean and cryosphere in a changing climate* (pp. 477–587). Intergovernmental Panel on Climate Change, Switzerland.

Brahney, J., Hallerud, M., Heim, E., Hahnenberger, M., & Sukumaran, S. (2020). Plastic rain in protected areas of the United States. *Science (New York, N.Y.)*, *368*(6496), 1257–1260. Available from https://doi.org/10.1126/science.aaz5819.

Breivik, K., Sweetman, A., Pacyna, J. M., & Jones, K. C. (2007). Towards a global historical emission inventory for selected PCB congeners—A mass balance approach: 3. An update. *Science of the Total Environment*, *377*(2–3), 296–307.

Burrows, S. M., Elbert, W., Lawrence, M. G., & Pöschl, U. (2009). Bacteria in the global atmosphere – Part 1: Review and synthesis of literature data for different ecosystems. *Atmospheric Chemistry and Physics*, *9*, 9263–9280.

Carpenter, L., & Nightingale, P. D. (2015). Chemistry and release of gases from the surface ocean. *Chemical Reviews*, *115*(10), 4015–4034. Available from https://doi.org/10.1021/cr5007123.

Carvajal, M., Sepúlveda, I., Gubler, A., & Garreaud, R. (2022). Worldwide signature of the 2022 Tonga volcanic tsunami. *Geophysical Research Letters*, *49*, e2022GL098153. Available from https://doi.org/10.1029/2022GL098153.

CETESB: Emissões Veiculares no Estado de São Paulo. (2020). Available from <http://veicular.cetesb.sp.gov.br/relatorios-e-publicacoes/> Accessed 30.04.22.

Chen, W., Zhang, S., Tong, Q., Zhang, X., Zhao, H., Ma, S., Xiu, A., & He, Y. (2018). Regional characteristics and causes of haze events in Northeast China. *Chinese Geographical Science*, *28*(5), 836–850. Available from https://doi.org/10.1007/s11769-018-0965-3.

Chow, J. C., Watson, J. G., Shah, J. J., Kiang, C. S., Loh, C., Lev-On, M., Lents, J. M., Molina, M. J., & Molina, L. T. (2004). Megacities and atmospheric pollution. *Journal of the Air and Waste Management Association*, *54*(10), 1226–1235. Available from https://doi.org/10.1080/10473289.2004.10470995.

Cox, K. D., Covernton, G. A., Davies, H. L., Dower, J. F., Juanes, F., & Dudas, S. E. (2019). Human consumption of microplastics. *Environmental Science & Technology*, *53*(12), 7068–7074. Available from https://doi.org/10.1021/acs.est.9b01517.

Crippa, M., Solazzo, E., Huang, G., Guizzardi, D., Koffi, E., Muntean, M., & Janssens-Maenhout, G. (2020). High resolution temporal profiles in the emissions database for global atmospheric research. *Scientific Data*, *7*(1), 1–17.

Crutzen, P. J. (2006). The "Anthropocene.". In E. Ehlers, & T. Krafft (Eds.), *Earth system science in the Anthropocene*. Berlin, Heidelberg: Springer. Available from https://doi.org/10.1007/3-540-26590-2 3.

Després, V. R., Alex Huffman, J., Burrows, S. M., Hoose, C., Safatov, A. S., Buryak, G., et al. (2012). Primary biological aerosol particles in the atmosphere: A review. *Tellus, Series B: Chemical and Physical Meteorology*, *64*(1). Available from https://doi.org/10.3402/tellusb.v64i0.15598.

Dris, R., Gasperi, J., Mirande, C., Mandin, C., Guerrouache, M., Langlois, V., & Tassin, B. (2017). A first overview of textile fibers, including microplastics, in indoor and outdoor environment. *Environmental Pollution*. Available from https://doi.org/10.1016/j.envpol.2016.12.013.

Eggleston, H.S., Buendia, L., Miwa, K., Ngara, T., & Tanabe, K. (2006). *IPCC guidelines for national greenhouse gas inventories*.

Elbert, W., Taylor, P. E., Andreae, M. O., & Poschl, U. (2007). Contribution of fungi to primary biogenic aerosols in the atmosphere: Wet and dry discharged spores, carbohydrates, and inorganic ions. *Atmospheric Chemistry Physics*, *7*, 45694588. Available from https://doi.org/10.5194/acp-7-4569-2007.

Ewbank, H., Vieira, J. G. V., Fransoo, J., & Ferreira, M. A. (2020). The impact of urban freight transport and mobility on transport externalities in the SPMR. *Transportation Research Procedia*, *46*, 101–108.

Faber, S., Hanayama, S., Zhang, S., Pereda, P., Comer, B., Hauerhof, E., et al. (2021). *Fourth greenhouse gas study 2020*. London: International Maritime Organization (IMO). Available from https://www.imo.org/en/OurWork/Environment/Pages/Fourth-IMO-Greenhouse-Gas-Study-2020.aspx.

Forestieri, S. D., Moore, K. A., Martinez Borrero, R., Wang, A., Stokes, M. D., & Cappa, C. D. (2018). Temperature and composition dependence of sea spray aerosol production. *Geophysical Research Letters*, *45*(14), 7218–7225. Available from https://doi.org/10.1029/2018GL078193.

Fröhlich-Nowoisky, J., Burrows, S. M., Xie, Z., Engling, G., Solomon, P. A., Fraser, M. P., et al. (2012). Biogeography in the air: Fungal diversity over land and oceans. *Biogeosciences*, *9*(3), 1125–1136. Available from https://doi.org/10.5194/bg-9-1125-2012.

Fröhlich-Nowoisky, J., Kampf, C. J., Weber, B., Huffman, J. A., Pöhlker, C., Andreae, M. O., et al. (2016). Bioaerosols in the earth system: Climate, health, and ecosystem interactions. *Atmospheric Research*, *182*, 346–376. Available from https://doi.org/10.1016/j.atmosres.2016.07.018.

Fung, F., & Maxwell, B. (2011). *Onboard refueling vapor recovery: Evaluation of the ORVR program in the United States*. Washington, DC: ICCT.

Gao, J., Zhang, J., Li, H., Li, L., Xu, L., Zhang, Y., Wang, Z., et al. (2018). Comparative study of volatile organic compounds in ambient air using observed mixing ratios and initial mixing ratios taking chemical loss into account – A case study in a typical urban area in Beijing. *Science of the Total Environment*, *628–629*(8), 791–804. Available from https://doi.org/10.1016/j.scitotenv.2018.01.175.

Giglio, L., Randerson, J. T., & Van Der Werf, G. R. (2013). Analysis of daily, monthly, and annual burned area using the fourth-generation global fire emissions database (GFED4). *Journal of Geophysical Research: Biogeosciences*, *118*(1), 317–328.

Giltrap, D., Cavanagh, J., Stevenson, B., & Ausseil, A. (2021). The role of soils in the regulation of air quality. *Philosophical Transactions of the Royal Society A*, *376*, 20200172. Available from https://doi.org/10.1098/rstb.2020.0172.

Gruber, N. (2011). Warming up, turning sour, losing breath: Ocean biogeochemistry under global change. *Philosophical Transactions of the Royal Society A*, *369*, 1980–1996. Available from https://doi.org/10.1098/rsta.2011.0003.

Hoinaski, L., Vasques, T. V., Ribeiro, C. B., & Meotti, B. (2022). Multispecies and high spatiotemporal resolution database of vehicular emissions in Brazil. *Earth System Science Data*, *14*, 2939–2949.

Horowitz, H. M., Holmes, C., Wright, A., Sherwen, T., Wang, X., Evans, M., Huang, J., Jaeglé, L., Chen, Q., Zhai, S., & Alexander, B. (2020). Effects of sea salt aerosol emissions for marine cloud brightening on atmospheric chemistry: Implications for radiative forcing. *Geophysical Research Letters*, *47*(4). Available from https://doi.org/10.1029/2019GL085838.

Ibarra-Espinosa, S., Freitas, E. D. D., Andrade, M. D. F., & Landulfo, E. (2022). Effects of evaporative emissions control measurements on ozone concentrations in Brazil. *Atmosphere*, *13*(1), 82.

Ibarra-Espinosa, S., Ynoue, R., O'Sullivan, S., Pebesma, E., Andrade, M. F., & Osses, M. (2018). VEINv0. 2.2: An R package for bottom-up vehicular emissions inventories. *Geoscientific Model Development*, *11*(6), 2209–2229.

IPBES. (2018). The IPBES regional assessment report on biodiversity and ecosystem services for the Americas. In J. Rice, C. S. Seixas, M. E. Zaccagnini, M. Bedoya-Gaitán, & N. Valderrama (Eds.), *Secretariat of the intergovernmental science-policy platform on biodiversity and ecosystem services, Bonn, Germany*. IPBES, 656 pp. Available from https://doi.org/10.5281/zenodo.3236252.

Joung, Y. S., Ge, Z., & Buie, C. R. (2017). Bioaerosol generation by raindrops on soil. *Nature Communications*, *8*. Available from https://doi.org/10.1038/ncomms14668.

Junior, C. H. S., Celentano, D., Rousseau, G. X., de Moura, E. G., van Deursen Varga, I., Martinez, C., & Martins, M. B. (2020). Amazon forest on the edge of collapse in the Maranhão State, Brazil. *Land Use Policy*, *97*, 104806.

Koupal, J., Cumberworth, M., & Michaels, H. (2008). *Design and implementation of moves: EPA's new generation mobile source emission model*. U. S. Environmental Protection Agency.

Koupal, J., Michaels, H., Cumberworth, M., Bailey, C., & Brzezinski, D. (2002). EPA's plan for moves: A comprehensive mobile source emissions model. In *Proceedings of the 12th CRC on-road vehicle emissions workshop* (pp. 15–17). San Diego, CA.

Kuenen, J., Berdowski, J., Van der Most, P., Boer, R. W., Rentz, O., Oertel, D., ... Appelman, W. (2019). Cement Production EMEP/EEA Air Pollutant Emission Inventory Guidebook 2016: NFR 2. A. 1. *European Environment Agency*.

Kuenen, J., Pacyna, J. M., Rentz, O., Oertel, D., Trozzi, C., Pulles, T., & Appelman, W. (2019). EMEP/EEA emissions inventory guidebook. NFR: 2. A. 2 Lime production. Technical Report. Copenhagen. European Environmental Agency.

Kuenen, J., van der Most, P., Rentz, O., Nunge, S., Trozzi, C., Pulles, T., & Appelman, W. (2019). NFR: 2. A. 3 Glass production SNAP: 040613 Glass (decarbonizing) ISIC: 2610 Manufacture of glass and glass products Version Guidebook 2013.

Kuenen, J., Coenen, P., Hausmann, K., Kampffmeyer, T., Kludt, R., & Visschedijk, A. (2019). *EMEP/EEA emission inventory guidebook; construction and demolition (Tech. Rep.)*. Copenhagen: European Environment Agency.

Lang, J., Tian, J., Zhou, Y., Li, K., Chen, D., Huang, Q., et al. (2018). A high temporal-spatial resolution air pollutant emission inventory for agricultural machinery in China. *Journal of Cleaner Production*, *183*, 1110–1121.

De Lauretis, R., Ntziachristos, L., & Trozzi, C. (2021). *EMEP/EEA emission inventory guidebook; International maritime navigation, international inland navigation, national navigation (shipping), national fishing, military (shipping), and recreational boats (Tech. Rep.)*. Copenhagen: European Environment Agency.

Liu, J. C., Mickley, L. J., Sulprizio, M. P., Dominici, F., Yue, X., Ebisu, K., Anderson, G. B., Khan, R. F., Bravo, M. A., & Bell, M. L. (2016). Particulate air pollution from wildfires in the Western US under climate change. *Climatic Change*, *138*(3), 655–666.

Liu, Z., Ciais, P., Deng, Z., Lei, R., Davis, S. J., Feng, S., et al. (2020). Near-real-time monitoring of global CO_2 emissions reveals the effects of the COVID-19 pandemic. *Nature Communications*, *11*(1), 1–12.

Longo, K. M., Freitas, S. R., Andreae, M. O., Setzer, A., Prins, E., & Artaxo, P. (2010). The coupled aerosol and tracer transport model to the Brazilian developments on the regional atmospheric modeling system (CATT-BRAMS) – Part 2: Model sensitivity to the biomass burning inventories. *Atmospheric Chemistry and Physics*, *10*(13), 5785–5795.

Maasakkers, J. D., Omara, M., Gautam, R., Lorente, A., Pandey, S., Tol, P., Borsdorff, T., Houweling, S., & Aben, I. (2022). Reconstructing and quantifying methane emissions from the full duration of a 38-day natural gas well blowout using space-based observations. *Remote Sensing of Environment*, *270*, 112755.

Martin, E., Kämpfer, P., & Jäckel, U. (2010). Quantification and identification of culturable airborne bacteria from duck houses. *Annals of Occupational Hygiene*, *54*(2), 217−227. Available from https://doi.org/10.1093/annhyg/mep088.

Martins, V., Moreno, T., Minguillón, M. C., Amato, F., de Miguel, E., Capdevila, M., & Querol, X. (2015). Exposure to airborne particulate matter in the subway system. *Science of The Total Environment*, *511*, 711−722. Available from https://doi.org/10.1016/j.scitotenv.2014.12.013.

Matthias, V., Arndt, J. A., Aulinger, A., Bieser, J., Denier van der Gon, H., Kranenburg, R., et al. (2018). Modeling emissions for three-dimensional atmospheric chemistry transport models. *International Journal of the Air and Waste Management Association*, *68*(8), 763−800. Available from https://doi.org/10.1080/10962247.2018.1424057.

Matthias-maser, S., & Jaenicke, R. (2000). The size distribution of primary biological aerosol particles in the multiphase atmosphere. *Aerobiologia*, 207−210.

Mellios, G., & Ntziachristos, L. (2016). *EMEP/EEA emission inventory guidebook; gasoline evaporation from vehicles (Tech. Rep.)*. Copenhagen: European Environment Agency.

Miguel, A. H., de Neto, F. R., Cardoso, J. N., de Vasconcellos, P., Pereira, A. S., & Marquez, K. S. (1995). Characterization of indoor air quality in the cities of Sao Paulo and Rio de Janeiro, Brazil. *Environmental Science & Technology*, *29*(2), 338−345.

Ministry of Ecology and Environment (MEE). (2018). *Technical guidelines for road motor vehicle emission inventory of air pollutants*. MEE, China.

Mohammad, A. F., Ikegaya, N., Hikizu, R., & Zaki, S. A. (2021). Turbulence effect of urban-canopy flow on indoor velocity fields under sheltered and cross-ventilation conditions. *Sustainability*, *13*(2), 586. Available from https://doi.org/10.3390/su13020586.

Morawska, L., Allen, J., Bahnfleth, W., Bennett, B., Bluyssen, P. M., Boerstra, A., Buonanno, G., Cao, J., Dancer, S. J., Floto, A., Franchimon, F., Greenhalgh, T., Haworth, C., Hogeling, J., Isaxon, C., Jimenez, J. L., Kennedy, A., Kumar, P., Kurnitski, J., . . . Yao, M. (2024). Mandating indoor air quality for public buildings. *Science (New York, N.Y.)*, *383*(6690), 1418−1420. Available from https://doi.org/10.1126/science.adl0677.

Morrison, G. C., & Nazaroff, W. W. (2002). Ozone interactions with carpet: Secondary emissions of aldehydes. *Environmental Science & Technology*, *36*(10), 2185−2192.

Nair, A. T. (2021). Bioaerosols in the landfill environment: An overview of microbial diversity and potential health hazards. *Aerobiologia*, *37*(2), 185−203. Available from https://doi.org/10.1007/s10453-021-09693-9.

Nguyen, L. T., Broughton, K., Osanai, Y., Anderson, I. C., Bange, M. P., Tissue, D. T., & Singh, B. K. (2019). Effects of elevated temperature and elevated CO_2 on soil nitrification and ammonia-oxidizing microbial communities in field-grown crop. *Science of the Total Environment*, *675*, 81−89.

Niranjan, K., Thulasiraman, S., & Ramprasad, T. R. (1999). Pinatubo volcanic aerosol characteristics as observed from a low latitude location in India using a ground-based multiwavelength solar radiometer. *Journal of Aerosol Science*, *30*(9), 1181−1189. Available from https://doi.org/10.1016/S0021-8502(99)00019-1.

Ntziachristos, L., & Boulter, P. (2003). Road vehicle tyre and brake wear. Road surface wear. *Joint EMEP/CORINAIR emission inventory guidebook*, B770-1.

Ntziachristos, L., & Samaras, Z. (2016). *EMEP/EEA emission inventory guidebook; road transport: Passenger cars, light commercial trucks, heavy-duty vehicles including buses and motorcycles*. Copenhagen: European Environment Agency.

Olmo, F. J., Tovar, J., Alados-Arboledas, L., Okulov, O., & Ohvril, H. O. (1999). A comparison of ground level solar radiative effects of recent volcanic eruptions. *Atmospheric Environment*, *33*(28), 4589−4596. Available from https://doi.org/10.1016/S1352-2310(99)00271-X.

Oppenheimer, C., Scaillet, B., & Martin, R. S. (2011). Sulfur degassing from volcanoes: Source conditions, surveillance, plume chemistry and earth system impacts. *Reviews in Mineralogy and Geochemistry*, *73*, 363−421. Available from https://doi.org/10.2138/rmg.2011.73.13.

Pandey, S., Gautam, R., Houweling, S., Van Der Gon, H. D., Sadavarte, P., Borsdorff, T., Hasekamp, O., Landgraf, J., Tol, P., Van Kempen, T., & Hoogeveen, R. (2019). Satellite observations reveal extreme methane leakage from a natural gas well blowout. *Proceedings of the National Academy of Sciences*, *116*(52), 26376−26381.

Parker, D. E., Wilson, H., Jones, P. D., Christy, J. R., & Folland, C. K. (1996). The impact of Mount Pinatubo on world-wide temperatures. *International Journal of Climatology*, *16*, 487−497. Available from https://doi.org/10.1002/(SICI)1097-0088.

Perez-Martinez, P. J., Miranda, R. M. D., Andrade, M. D. F., & Kumar, P. (2020). Air quality and fossil fuel driven transportation in the Metropolitan Area of São Paulo. *Transportation Research Interdisciplinary Perspectives*, *5*, 100137. Available from https://doi.org/10.3390/atmos13010052.

Pérez-Martínez, P. J., Dunck, J. A., de Assunção, J. V., Connerton, P., Slovic, A. D., Ribeiro, H., & Miranda, R. M. (2022). Long-term commuting times and air quality relationship to COVID-19 in São Paulo. *Journal of Transport Geography*, *101*, 103349. ISSN 0966-6923. Available from https://doi.org/10.1016/j.jtrangeo.2022.103349. (https://www.sciencedirect.com/science/article/pii/S0966692322000722).

Pimpin, L., Retat, L., Fecht, D., de Preux, L., Sassi, F., Gulliver, J., et al. (2018). Estimating the costs of air pollution to the National Health Service and social care: An assessment and forecast up to 2035. *PLoS Medicine*, *15*(7), e1002602.

Poschl, U., & Shiraiwa, M. (2015). Multiphase chemistry at the atmosphere−biosphere interface influencing climate and public health in the Anthropocene. *Chemical Reviews. 2015*, *115*(10), 4440−4475. Available from https://doi.org/10.1021/cr500487s.

Pulles, T., & Heslinga, D. (2007). *The art of emission inventorying*. TNO. Available from https://doi.org/10.13140/RG.2.1.2082.8007.

Pörtner, H.-O., Roberts, D. C., Adams, H., Adelekan, I., Adler, C., Adrian, R., et al. (2022). *Climate change: Impacts, adaptation and vulnerability. contribution of working group II to the sixth assessment report of the intergovernmental panel on climate change* (pp. 37−118). Cambridge/New York, NY: Cambridge University Press. Available from http://doi.org/10.1017/9781009325844.002.

Quigg, A., Santschi, P. H., Burd, A., Chin, W. C., Kamalanathan, M., Xu, C., & Ziervogel, K. (2021). From nano-gels to marine snow: A synthesis of gel formation processes and modeling efforts involved with particle flux in the ocean. *Gels*, *7*(3). Available from https://doi.org/10.3390/gels7030114.

Quinn, P., Collins, D. B., Grassian, V. H., Prather, K. A., & Bates, T. (2015). Chemistry and related properties of freshly emitted sea spray aerosol. *Chemical Reviews*, *115*(10), 4383–4399. Available from https://doi.org/10.1021/cr500713g.

Rodenburg, C. A., Ubbels, B., & Nijkamp, P. (2002). Policy scenarios for achieving sustainable transportation in Europe. *Transport Reviews*, *22*(4), 449–472.

San-Miguel-Ayanz, J., & Steinbrecher, R. (2019). *EMEP/EEA emission inventory guidebook; forest and other vegetation fires (Tech. Rep.)*. Copenhagen: European Environment Agency.

Schiffer, J. M., Mael, L. E., Prather, K. A., Amaro, R. E., & Grassian, V. H. (2018). Sea spray aerosol: Where marine biology meets atmospheric chemistry. *ACS Central Science*, *4*(12), 1617–1623. Available from https://doi.org/10.1021/acscentsci.8b00674.

Schipper, L., Unander, F., Murtishaw, S., & Ting, M. (2001). Indicators of energy use and carbon emissions: Explaining the energy economy link. *Annual Review of Energy and the Environment*, *26*, 49–81.

Seo, S. H., Jung, K. S., Park, M. K., Kwon, H. O., & Choi, S. D. (2022). Indoor air pollution of polycyclic aromatic hydrocarbons emitted by computers. *Building and Environment*, *218*, 109107.

Shaw, S. L., Gantt, B., & Meskhidze, N. (2010). Production and emissions of marine isoprene and monoterpenes: A review. *Advances in Meteorology*, 24. Available from https://doi.org/10.1155/2010/408696, Article ID 408696.

Sokhi, R. S., Singh, V., Querol, X., Finardi, S., Targino, A. C., de Fatima Andrade, M., … Zavala, M. (2021). A global observational analysis to understand changes in air quality during exceptionally low anthropogenic emission conditions. *Environment international*, *157*, 106818. Available from https://doi.org/10.1016/j.envint.2021.106818.

Steenhof, P., Woudsma, C., & Sparling, E. (2006). Greenhouse gas emissions and the surface transport of freight in Canada. *Transportation Research Part D: Transport and Environment*, *11*(5), 369–376.

Stocker, T. (2014). *Climate change 2013: The physical science basis: Working group I contribution to the fifth assessment report of the intergovernmental panel on climate change*. Cambridge University Press.

Sánchez-Monedero, M. A., Aguilar, M. I., Fenoll, R., & Roig, A. (2008). Effect of the aeration system on the levels of airborne microorganisms generated at wastewater treatment plants. *Water Research*, *42*(14), 3739–3744. Available from https://doi.org/10.1016/j.watres.2008.06.028.

Thunis, P., Clappier, A., Pisoni, E., Bessagnet, B., Kuenen, J., Guevara, M., & Lopez-Aparicio, S. (2022). A multipollutant and multisectorial approach to screening the consistency of emission inventories. *Geoscientific Model Development*, *15*(13), 5271–5286.

Trozzi, C., & Kuenen, J. (2019). *EMEP/EEA emission inventory guidebook; Manufacturing industries and construction (combustion)*. Copenhagen: European Environment Agency.

Tu, Q., Schneider, M., Hase, F., Khosrawi, F., Ertl, B., Necki, J., … Fang, D. (2022). Quantifying CH 4 emissions in hard coal mines from TROPOMI and IASI observations using the wind-assigned anomaly method. *Atmospheric Chemistry and Physics*, *22*(15), 9747–9765. Available from https://doi.org/10.5194/acp-22-9747-2022.

Tu, Q., Schneider, M., Hase, F., Khosrawi, F., Ertl, B., Necki, J., Dubravica, D., Diekmann, C. J., Blumenstock, T., & Fang, D. (2022). Quantifying hard coal mines CH_4 emissions from TROPOMI and IASI observations using high-resolution CAMS forecast data and the wind-assigned anomaly method. *Atmospheric Chemistry and Physics Discussions*, 1–22.

Vallero, D. (2014). *Fundamentals of air pollution* (5th ed.). Academic Press.

Webb, J., Hutchings, N., & Amon, B. (2019). *EMEP/EEA emission inventory guidebook; field burning of agricultural residues (Tech. Rep.)*. Copenhagen: European Environment Agency.

Weilenmann, M., Favez, J.-Y., & Alvarez, R. (2009). Cold-start emissions of modern passenger cars at different low ambient temperatures and their evolution over vehicle legislation categories. *Atmospheric Environment*, *43*(15), 2419–2429. Available from https://doi.org/10.1016/j.atmosenv.2009.02.005.

Weng, H., Lin, J., Martin, R., Millet, D. B., Jaeglé, L., Ridley, D., Keller, C., Li, C., Du, M., & Meng, J. (2020). Global high-resolution emissions of soil NO_x, sea salt aerosols, and biogenic volatile organic compounds. *Scientific Data*, *7*(1). Available from https://doi.org/10.1038/s41597-020-0488-5.

White, J. K., Nielsen, J. L., & Madsen, A. M. (2019). Microbial species and biodiversity in settling dust within and between pig farms. *Environmental Research*, *171*, 558–567. Available from https://doi.org/10.1016/j.envres.2019.01.008.

Wikipedia Contributors. (2022a). Portland cement. In *Wikipedia, the free encyclopedia*. <https://en.wikipedia.org/w/index.php?title = Portland_cement&oldid = 1085850070> Accessed 10.05.22.

Wikipedia Contributors. (2022b). Persistent organic pollutant. In *Wikipedia, the free encyclopedia*. <https://en.wikipedia.org/w/index.php?title = Persistent_organic_pollutant&oldid = 1083832503> Accessed 10.05.22.

Wikuats, C. F. H., Duarte, E. H., Prates, K. V. M. C., Janiaski, L. L. L., de Oliveira Gabriel, B., da Cunha Molina, A., & Martins, L. D. (2020). Assessment of airborne particles and bioaerosols concentrations in a waste recycling environment in Brazil. *Science Reports*, *10*, 1–14. Available from https://doi.org/10.1038/s41598-020-71787-0.

Wilson, C., Chipperfield, M. P., Gloor, M., Parker, R. J., Boesch, H., McNorton, J., Gatti, L. V., Miller, J. B., Basso, L. S., & Monks, S. A. (2021). Large and increasing methane emissions from eastern Amazonia derived from satellite data, 2010–2018. *Atmospheric Chemistry and Physics*, *21*(13), 10643–10669. Available from https://doi.org/10.5194/acp-21-10643-2021.

World Health Organization. (2018). *Air pollution and child health: prescribing clean air: summary* (No. WHO/CED/PHE/18.01). World Health Organization.

World Health Organization. (2021). *Review of evidence on health aspects of air pollution: REVIHAAP project: technical report* (No. WHO/EURO: 2013-4101-43860-61757). World Health Organization. Regional Office for Europe.

Xie, W., Li, Y., Bai, W., Hou, J., Ma, T., Zeng, X., Zhang, L., & An, T. (2021). The source and transport of bioaerosols in the air: A review. *Frontiers of Environmental Science and Engineering*, *15*(3), 1−19. Available from https://doi.org/10.1007/s11783-020-1336-8.

Yue, X., Ma, N. L., Sonne, C., Guan, R., Lam, S. S., Van Le, Q., et al. (2021). Mitigation of indoor air pollution: A review of recent advances in adsorption materials and catalytic oxidation. *Journal of Hazardous Materials*, *405*, 124138.

Zhang, H., Wang, F., Li, J., Duan, Y., Zhu, C., & He, J. (2022). Potential impact of Tonga volcano eruption on global mean surface air temperature. *Journal of Meteorological Research*, *36*(1), 1−5. Available from https://doi.org/10.1007/s13351-022-2013-6.

Zhang, N., Li, Y. B., He, H. R., Zhang, J. F., & Ma, G. S. (2021). You are what you eat: Microplastics in the feces of young men living in Beijing. *Science of the Total Environment*, *767*, 144345. Available from https://doi.org/10.1016/j.scitotenv.2020.144345, ISSN 0048-9697.

Zheng, B., Ciais, P., Chevallier, F., Chuvieco, E., Chen, Y., & Yang, H. (2021). Increasing forest fire emissions despite the decline in global burned area. *Science Advances*, *7*(39), eabh2646.

Chapter 4

Chemistry of the atmosphere

Maria de Fatima Andrade[1], Leila Droprinchinski Martins[2], Edmilson Dias de Freitas[1], Regina Miranda[3], Perola Vasconcellos[4], Jorge Martins[2], Thiago Nogueira[5] and Guilherme Pereira[1]

[1]*Institute of Astronomy, Geophysics and Atmospheric Sciences, University of São Paulo, São Paulo, Brazil,* [2]*Graduate Program in Environmental Engineering, Federal University of Technology, Paraná, Brazil,* [3]*School of Arts, Sciences and Humanities, University of São Paulo, São Paulo, Brazil,* [4]*Institute of Chemistry, University of São Paulo, São Paulo, Brazil,* [5]*Public Health Faculty, University of São Paulo, São Paulo, Brazil*

4.1 Introduction

The designation "chemistry of the atmosphere," as the title of this chapter, is according to the Britannica Encyclopedia: "the science that deals with the properties, composition, and structure of substances (defined as elements and species), the transformations they undergo, and the energy that is released or absorbed during these processes." The kinetics and photochemical process are crucial to understand how the compounds emitted by sources interact in the atmosphere under different meteorological conditions, forming other compounds, and eventually being removed from the process.

The atmosphere consists of a thin layer formed by a mixture of gases that surrounds the Earth's surface. Disregarding water vapor, atmospheric air is composed of 78.1% (by volume) nitrogen, 21% oxygen, 0.9% argon, and 0.04% carbon dioxide. Under normal conditions, air contains between 1% and 3% water vapor by volume, plus a wide variety of trace-level gases below 0.002%, which include neon, helium, methane (CH_4), krypton, nitrous oxide, hydrogen, xenon, sulfur dioxide, ozone (O_3), nitrogen dioxide (NO_2), ammonia, and carbon monoxide (CO) (Seinfeld & Pandis, 2016). The behavior of the atmosphere is a consequence of the gases that form it, from natural or anthropogenic sources, in addition to the energy and physical forces acting on it. Emissions of air contaminants generated by human activity have changed the chemistry of Earth's atmosphere, with consequences for the health of people and the planet (UN Environment, 2019).

From a global public health perspective, especially considering urban areas, the two most important air pollutants are particulate matter (PM) and its components, mainly the fine particles, and tropospheric ozone (UN Environment, 2019). Many urban areas in the world experience air pollution episodes during which high levels of ozone and PM concentrations are present. The formation and accumulation of ozone is called photochemical smog (the historical perspective of air pollution is presented in Chapter 1) and is sometimes characterized as a misplaced ozone layer, in contrast to the stratospheric ozone depletion problem (discussed in Section 4.9).

Solar radiation is the engine for the chemistry of the atmosphere, dissociating molecules, and catalyzing many reactions. The actinic flux (that causes the photochemical reactions) is defined as the radiative flux from all directions on a volume of air (Demerjian et al., 1980). Many reactions are initiated by the dissociation of molecules induced by the actinic radiation in the atmosphere, as the UV-photolysis of ozone and nitrogen dioxide, among other reactions, presented in the following sections.

The photochemical process involves the dissociation of molecules due to the interaction with the shortwave solar radiation, most in the ultraviolet region of the spectra. The radiative balance of the Earth involves the income of solar radiation and the emission of infrared radiation by the Earth's surface. This terrestrial radiation is absorbed by gases in the atmosphere which are called greenhouse gases, as well as by other constituents, such as clouds. This absorbed radiation is then reemitted to the atmosphere. Part of the energy is also involved in sensible and latent heat. This balance is also a function of the concentration of the constituent gases, particles, and other physical processes, including surface absorption and scattering.

Air Quality. DOI: https://doi.org/10.1016/B978-0-12-822591-2.00004-4

4.2 Gaseous chemistry

Some specific gases are important because of their high concentrations, suggesting that they do not decay rapidly. Others are significant because they react swiftly to generate one or more dangerous or important compounds in the atmospheric process.

In polluted urban regions, ozone has a relatively short atmospheric lifetime (few hours), where concentrations of its precursors are high enough not only to be involved in the production but also in the consumption of this pollutant. On the other hand, ozone in the free troposphere has a lifetime of several weeks (Monks et al., 2015; Young et al., 2018), time enough to be transported over intercontinental scales. Hence, ozone has a role as a priority pollutant on an urban scale, and it may also influence air quality on a hemispheric scale (Akimoto et al., 2020; Monks et al., 2015). The process of ozone formation is initiated by the reaction of the $\cdot$OH radical with organic molecules. Hydroxyl ($\cdot$OH) is mainly produced by the UV-photolysis of ozone and is the most important radical because it reacts with most atmospheric species, leading to oxidation reactions that produce ozone and secondary aerosols (inorganic and organic particles, or SIA and SOA) in the troposphere. The subsequent reaction sequence of ozone formation is catalyzed by nitrogen oxides (NO_x) in an intricate network of free-radical reactions. Most of the direct emission of NO_x to the atmosphere is in the form of nitric oxide (NO). NO_2 is formed in the atmosphere by the oxidation conversion of NO. As in the stratosphere, the two oxides of nitrogen (NO and NO_2) are generally grouped together as NO_x because interconversion between NO and NO_2 is rapid (with a tropospheric time scale of $\sim$5 minutes) compared to the time scale for organic oxidation (1 to several hours).

NO_x mixing ratios range from 5 to 20 ppb in urban areas, about 1 ppb in rural areas during region-wide episodes, and from 10 to 100 ppt in the remote troposphere (Seinfeld & Pandis, 2016). In remote troposphere, ozone formation is sustained by the oxidation of CO and CH_4, each through reactions with $\cdot$OH. Both CO and CH_4 are long-lived species, with atmospheric lifetimes against $\cdot$OH, reaction of about 2 months and 9 years, respectively. Ozone formation in urban and regional atmospheres is driven by the oxidation of much shorter-lived volatile organic compounds (VOCs) emitted from anthropogenic and biogenic sources. Two classes of precursor compounds are involved in tropospheric ozone formation, VOCs and NO_x, arising a key question: how do changes in VOCs and NO_x levels affect the amount of ozone formed? (Seinfeld & Pandis, 2016).

Gases in the atmosphere have the tendency of being oxidized when reacting (Fig. 4.1), which is generally initiated by the action of solar radiation in presence of their fundamental constituents (O_2, H_2O, and N_2). Molecules under the action of radiation are dissociated into highly reactive fragments that promote a series of chemical reactions. This dissociation depends on the probability of occurrence of a collision between the photon radiative flux with a determined amount of energy and the molecule in a volume of air, that is, of the actinic flux (photons cm^{-2} s^{-1}) that can cause the photochemical reactions (Madronich & Flocke, 1999).

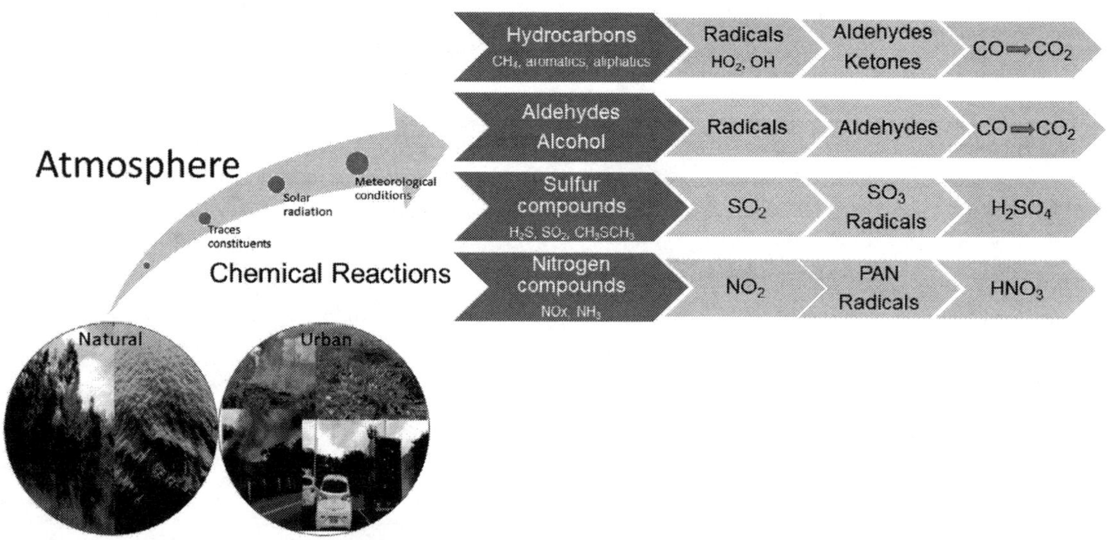

FIGURE 4.1 Schematic illustration of gaseous chemical reactions in the atmosphere.

According to Planck's law, the energy of a photon of frequency ν is $h\nu$. Thus considering the generic photolysis reaction $A + h\nu \rightarrow B1 + B2$, the velocity of reaction V is written as $V = J\,[A]$, and the rate of A as $d[A]/dt = J\,[A]$, where J is the photolysis constant for species A in s^{-1} and [A] is the concentration of molecules per volume of air.

Besides photolysis reactions, kinetic reactions are common in the atmosphere, which are bimolecular or trimolecular and can happen by thermal decomposition, isomerization, and collision, with the latter being the most common type. Many chemical reactions involve several intermediate reactive species, such as free radicals (an atom or group of atoms possessing an odd unpaired electron, which generally includes a dot to represent the odd electron). Because of their high reactivity, free radicals are virtually consumed as quickly as they are formed. They are usually treated assuming the validity of pseudo-steady-state approximation (PSSA) when deriving global rates of chemical reactions.

The atmospheric gaseous chemistry will be treated here under the view of the natural atmosphere, where there is a higher influence of natural emission sources, and of the urban atmosphere with expressive impacts of human activities. Further, several chemical reaction chains were summarized, with only the main reactions involving ubiquitous chemical species being illustrated to provide an overview of a complex gaseous atmospheric chemistry (Seinfeld & Pandis, 2016).

4.2.1 Natural atmosphere

Oxygen and ozone are the most abundant oxidants in the atmosphere. However, they are relatively unreactive due to the large bond energies, being $\cdot OH$ the key species in the oxidation of compounds in the atmosphere. There, $\cdot OH$ is originated via photolysis of O_3, which reacts with water vapor to form the radical:

$$O_3 + h\nu \rightarrow O^* + O_2 \tag{4.1}$$

$$O^* + H_2O \rightarrow 2\cdot OH \tag{4.2}$$

A small fraction of oxygen molecules in the excited state (O^*) reacts with water, providing enough $\cdot OH$ for oxidation processes in the atmosphere. The amount of $\cdot OH$ across the globe varies depending on region and meteorological conditions, such as temperature and humidity. More details can be found in Section 6.1 on Seinfeld and Pandis (2016).

Methane is largely emitted by natural processes. Its oxidation in the troposphere happens mainly with $\cdot OH$ (a small fraction can also react with nitrate radical—NO_3), constituting an important chemical species for chemistry in the natural atmosphere.

$$H_3C - H + \cdot OH \rightarrow H_3C\cdot + H_2O \tag{4.3}$$

The radical $\cdot CH_3$ reacts fast with oxygen to form alkyl peroxide radical (RO_2, $R = CH_3$)

$$CH_3\cdot + O_2 + M \rightarrow CH_3O_2\cdot + M \tag{4.4}$$

$$CH_3O_2\cdot + NO \rightarrow CH_3O\cdot + NO_2 \tag{4.5}$$

$$CH_3O\cdot + O_2 \rightarrow HCHO + HO_2\cdot \tag{4.6}$$

Alternatively, reactions (4.5) and (4.6) can be integrated, considering the great mass participation of O_2 in the atmosphere and that reaction (4.6) is fast:

$$CH_3O_2\cdot + NO \rightarrow HCHO + HO_2\cdot + NO_2 \text{(in presence of } O_2\text{)}.$$

Thus the following reactions are:

$$CH_3O_2\cdot + HO_2\cdot \rightarrow CH_3OOH + O_2 \tag{4.7}$$

$$CH_3OOH + h\nu \rightarrow HCHO + HO_2\cdot + \cdot OH \text{(in presence of } O_2\text{)} \tag{4.8}$$

$$\begin{aligned} CH_3OOH + \cdot OH &\rightarrow HCHO + H_2O + \cdot OH \\ &\rightarrow CH_3O_2\cdot + H_2O \end{aligned} \tag{4.9}$$

The formaldehyde (HCHO) formed in the system (reaction 4.6 and via reactions with methyl hydroperoxide, CH_3OOH) also dissociates by photolysis and reacts with $\cdot OH$ to form, finally, CO. Therefore the chemistry of methane is related to the CO and the oxidation has as net balance the formation of CO_2, O_3, and the regeneration of $\cdot OH$.

$$\begin{aligned} HCHO + h\nu &\rightarrow 2HO_2\cdot + CO \text{(in presence of } O_2\text{)} \\ &\rightarrow H_2 + CO \end{aligned} \tag{4.10}$$

$$HCHO + \cdot OH \rightarrow HO_2\cdot + CO + H_2O \tag{4.11}$$

$$CO + \cdot OH \rightarrow CO_2 + HO_2\cdot (\text{in presence of } O_2) \tag{4.12}$$

Besides methane, there are other biogenic-origin hydrocarbons relevant to atmospheric chemistry. These biogenic hydrocarbons are often primary compounds of the alkenes chemical family, which are very reactive in the atmosphere due to the presence of the double-bond (C=C). They react with O_3 and NO_3, besides with radical $\cdot OH$. Terpenes group $(C_5H_8)_n$, which includes isoprene (C_5H_8), are among the most abundant biogenic VOCs, considering the total emission globally, with significant contribution from tropical regions, which is yet relatively less studied and understood in comparison to other regions (Batista et al., 2019; Mu et al., 2022; see Chapter 3 for emissions details). The products of their oxidation are important as precursors for ozone and for SOA formation, thus influencing regional air quality and climate. Wells et al. (2020) showed, using satellite data, that the GEOS-Chem model underpredicts isoprene emissions in Southern Africa.

A recent review from Artaxo et al. (2022) addressed the effects of forests (tropical and bureau) on the atmosphere, pointing out that the aerosol composition and chemistry are similar, particularly concerning the main chemical components in the forest's atmospheres, being both dominated by an organic fraction. In both forests, isoprene and monoterpene are the main biogenic VOCs emitted and the major precursors of SOA.

Paulot et al. (2009) present the oxidation mechanism of isoprene by $\cdot OH$ under low NO_x conditions, which is typically found in the natural atmosphere or regions far from biomass and fossil fuel combustion. The oxidation is evidenced by the high yields of gas-phase formation of epoxydiols (IEPOX), a gas-phase precursor for SOA. Oxidation by $\cdot OH$ occurs in the double-bond by addition of $\cdot OH$ as follows:

$$H_2C_{\underline{1}} = C_{\underline{2}}(CH_3) - C_{\underline{3}}H = C_{\underline{4}}H_2 + \cdot OH + O_2 \rightarrow ISOPO_2, \tag{4.13}$$

or, in a simplified way, as ISOP + $\cdot OH + O_2 \rightarrow$ $ISOPO_2$, (ISOP equal $H_2C_{\underline{1}} = C_{\underline{2}}\ (CH_3) - C_{\underline{3}}H = C_{\underline{4}}H_2$; C_5H_8; the underlined numbers identify the double-bond in the molecule). Under low NO, $ISOPO_2$ reacts mainly with HO_2, and, afterwards, the product with $\cdot OH$, forming IEPOX.

$$ISOPO_2 + HO_2\cdot \rightarrow ISOPOOH + O_2 \tag{4.14}$$

$$ISOPOOH + \cdot OH \rightarrow IEPOX + \cdot OH \tag{4.15}$$

Addition of $\cdot OH$ can occur at any of the four carbons that, in presence of O_2, react quickly forming the peroxy radicals. The fate of peroxy radicals depends on the relative concentrations of NO and $HO_{\underline{2}}$. At low NO concentrations, the oxidation will produce isoprene hydroxyhydroperoxides (ISOPOOH) and further, as more recently evidenced, IEPOX.

Isoprene displays a high reactivity and can also react with O_3 and NO_3, inducing the oxidation of isoprene during the night. Isoprene is emitted during the day, and at night the reaction with $\cdot OH$ can be disregarded since $\cdot OH$ concentration is low. The reactions of isoprene with O_3 are by addition, forming two ozonides as in:

$$H_2C_{\underline{1}} = C_{\underline{2}}(CH_3) - C_{\underline{3}}H = C_{\underline{4}}H_2 + O_3 \rightarrow H_2C = C(CH_3) - CH - O_3 - CH_2 + H_2C - O_3 - C(CH_3) - CH = CH_2 \tag{4.16}$$

The two products decompose to mainly form formaldehyde, methacrolein (MACR), methyl vinyl ketone (MVK), and $\cdot OH$ radical. The reaction of isoprene with NO_3 is also by addition in the carbons 1 and 4 (double-bond) of isoprene, which quickly reacts with oxygen to form nitrooxy peroxy radical. The main products from the night-time reaction of isoprene with NO_3 are carbonyl nitrates, hydroxynitrates, and nitrooxyhydroperoxides (Seinfeld & Pandis, 2016). The chemistry of isoprene is complex and involves reactions in gas and particle phases and is yet not completely known and evidenced through observations (Claeys & Maenhaut, 2021).

Terpenes chemistry is also complex, and they react with $\cdot OH$, O_3, and NO_3 by addition. Pinene, a monoterpene, is the most abundant (isomers α and β pinene). Their oxidation readily leads to the formation of SOA due to oxidized products (radicals) that have low volatility, and condensate to form aerosol. The main products pointed out in the literature are pinonaldehyde, pinonic acid, pinic acid, acetone, formaldehyde, among others (Barreira et al., 2021; Seinfeld & Pandis, 2016).

Besides hydrocarbons, nitrogen compounds have an important role in tropospheric chemistry in natural and in urban atmospheres. Nitrogen oxides are precursors of tropospheric O_3, HNO_3, and many other secondary species. Besides NO_x, ammonia (NH_3) is an important gas-phase base for atmospheric chemistry. Ammonia, emitted by natural and anthropic activities (see Chapter 3), is the main species to neutralize acids in the atmosphere. Their reactions involve the aqueous and particulate phases with solubilization of gas NH_3 and reactions, mainly, with the inorganic acids in the atmosphere (Behera et al., 2013).

Worked example 4.1

It is estimated that $\cdot OH$ has a global concentration of 10^6 molecules cm^{-3}. The radical is the main reagent in the VOCs oxidation reactions in the atmosphere. Explain the primary formation of $\cdot OH$ and their recycling in the CO loss with O_3 formation.

Solution:

The primary formation of OH in the atmosphere is via photolysis of O_3 as:

$$O_3 + h\nu \rightarrow O(^1D) + O_2(\lambda < 340\ \text{nm})$$

$$O(^1D) + M \rightarrow O + M\ (\text{around } 97\%)$$

$$O(^1D) + H_2O \rightarrow 2\cdot OH\ (\text{around } 3\%)$$

The recycling in the CO loss is via the following reactions with the formation of one O_3 molecule:

$$CO + \cdot OH \rightarrow CO_2 + H$$

$$H + O_2 + M \rightarrow HO_2\cdot + M$$

$$HO_2 + NO \rightarrow \cdot OH + NO_2$$

$$NO_2 + h\nu \rightarrow NO + O\cdot$$

$$O\cdot + O_2 + M \rightarrow O_3 + M$$

$$CO + 2O_2 \rightarrow CO_2 + O_3(\text{net})$$

Then the OH is recycled feeding the oxidation process of VOCs and the O_3 formation.

4.2.2 Urban atmosphere

The urban atmosphere is influenced by several chemical species released by human activities and changes in land use. Meteorological and chemical conditions, as well as their interconnections, are significantly altered in relation to the natural atmosphere.

The gases and particles emitted change the chemical composition and are precursors of secondary species. A typical and recognized result is the photochemical smog.

Typically, ozone is present and formed due to several chemical reactions that occur in the urban atmosphere, with solar radiation having a fundamental role. The basic cycle of ozone formation is:

$$NO_2 + h\nu \rightarrow NO + O^* \quad (4.17)$$

$$O^* + O_2 + M \rightarrow O_3 + M \quad (4.18)$$

$$NO + O_3 \rightarrow NO_2 + O_2 \quad (4.19)$$

M being used to represent any third coreactant, that is, N_2. This cycle, composed by the three reactions mentioned above, indicates that ozone concentration is proportional to the concentration ratio of NO_2/NO, since ozone is consumed in the reaction with NO, and NO_2 should be photolyzed to provide an oxygen atom in the troposphere. However, the presence of VOC and CO alters the production of ozone in this cycle by providing the fast conversion of NO to NO_2 and regenerating the hydroxyl radical ($\cdot OH$). A simple set of reactions for CO describes this role (reactions 4.20–4.23):

$$CO + \cdot OH \rightarrow CO_2 + HO_2\cdot\ (\text{in presence of } O_2) \quad (4.20)$$

$$HO_2\cdot + NO \rightarrow NO_2 + \cdot OH \quad (4.21)$$

$$HO_2\cdot + HO_2\cdot \rightarrow H_2O_2 + O_2 \quad (4.22)$$

The hydrogen peroxide (H_2O_2) formed is a reservoir of radical $\cdot OH$ and HO_2, and the reaction (4.21) provides NO_2 to be photolyzed and form O_3.

$$NO_2 + h\nu \rightarrow NO + O^* \quad (4.23)$$

The termination chain occurs when $\cdot OH$ reacts with NO_2, since it removes the radicals ($\cdot OH$ and $HO_2\cdot$) and NO_x ($NO + NO_2$) from the system, as exemplified below by reaction (4.24):

$$\cdot OH + NO_2 + M \rightarrow HNO_3 + M \quad (4.24)$$

CO is a precursor of ozone, but relevant only at natural atmosphere, as it provides only one ozone molecule by each CO oxidation. VOCs are composed of several organic species, which vary significantly in reactivity, that is, kOH. They can also react with other chemical species, which are able to form radicals more efficiently and quickly. Radicals act to convert NO to NO_2 and, therefore, provide NO_2 to be photolyzed when in presence of solar radiation enough to break this molecule (reactions 4.17–4.19).

The amount of ozone in the troposphere is not explained by its transport from the stratosphere. However, it is a fundamental source of $\cdot OH$ radicals for tropospheric chemistry. Ozone molecules formed in the atmosphere are photolyzed and, together with water vapor, form $\cdot OH$ (reactions 4.1–4.2).

The chemistry during the day is dominated by the photochemical reactions with the interconversion of NO–NO_2 (as it can be observed in CO chain oxidation—reactions 4.20–4.23). However, during the night, the atmospheric chemistry does not stop, and other reactions become more relevant. At night, the photolysis of NO_2 does not occur and, therefore, the NO_x cycle is quite different. All NO present will quickly react with O_3 (reaction 4.19), converting NO to NO_2, and then at night, almost all NO_x is NO_2.

The reaction of NO_2 with O_3 at night provides radical nitrate for the atmosphere, being the only direct source in the atmosphere.

$$NO_2 + O_3 \rightarrow NO_3 + O_2 \tag{4.25}$$

Meanwhile, during the day, radical NO_3 is quickly photolyzed by visible radiation, forming NO_x, which comprises a group of key chemical species for ozone photochemical formation. Additionally, a dinitrogen pentoxide (N_2O_5) can also be formed in a reversing process, according to the reaction as follows:

$$NO_2 + NO_3 + M \rightarrow N_2O_5 + M \tag{4.26}$$

This N_2O_5 can later decompose back to NO_2 and NO_3.

Worked example 4.2

Converting concentration of CO from ppm to $\mu mol\ mol^{-1}$.

Solution: CO is a common pollutant in the atmosphere that can have harmful effects on human health. It is often measured in parts per million (ppm) or micromoles per mole ($\mu mol\ mol^{-1}$). To convert the concentration of CO from ppm to $\mu mol\ mol^{-1}$, we can use the following equation:

Concentration ($\mu mol\ mol^{-1}$) = Concentration (ppm) × (24.45/Molecular Weight)

where Molecular Weight of CO is 28.01 $g\ mol^{-1}$.

For example, let us say we have a measurement of the concentration of CO in the atmosphere with a value of 0.5 ppm. Using the above equation, we can convert this to a concentration in $\mu mol\ mol^{-1}$ as follows:

Concentration ($\mu mol\ mol^{-1}$) = 0.5 ppm × (24.45/28.01) = 0.435 $\mu mol\ mol^{-1}$

Therefore the concentration of CO in this example is 0.435 $\mu mol\ mol^{-1}$.

Oxidation of VOC

The VOC emitted to the atmosphere can be removed via deposition (dry and wet) and chemical reactions. Reactions can occur in the gaseous and aqueous phases by the dissolution of gases in water, following by reactions with other constituents present in this aqueous solution (addressed in Section 4.3.2). Several chemical reactions happen in the gas-phase with VOC in urban atmosphere, taking the oxidation of VOC to CO_2 (see Fig. 4.1). In the atmosphere, reactions of VOC occur preferably with $\cdot OH$, but also with radiation ($h\nu$), NO_3, and O_3, according to functional groups of VOC.

The oxidation process of VOC will be illustrated here only for major species representing the main groups of VOC which are found in the urban tropospheric environment. VOC oxidation initiated by $\cdot OH$ during daytime produces several organic products of oxygenated functional groups, such as aldehyde, ketone, alcohol, carboxylic acid, hydroperoxide, percarboxylic acid, and peroxyacyl nitrate groups (Seinfeld & Pandis, 2016).

The largest fraction of VOC in the urban atmosphere is of hydrocarbons, species constituted exclusively of hydrogen and carbon, which are divided into aliphatic and aromatics, based on their structures. The latter contain aromatic rings (e.g., benzene). Among them, toluene, xylenes, and benzene are present at higher amounts in the urban atmosphere, as presented in Table 4.1. Aliphatics are further divided into families: alkanes, alkenes, alkynes, and their cyclic analogs.

Alkanes are hydrocarbons with single common bonds C–H, such as methane. Typically, the chemistry of VOC treats methane oxidation separately due to low reactivity (only C–H bonds are available to react) in the atmosphere and due to the presence of other relevant hydrocarbons for chemistry in the urban atmosphere. Therefore usually, the tropospheric chemistry is represented by nonmethane volatile organic compounds (NMVOC).

TABLE 4.1 Overview of main nonmethane volatile organic compounds (NMVOC) found in urban areas and their typical concentrations ($\mu g\, m^{-3}$).

Site	Benzene	Toluene	Xylene[j]	Ethane	Propane	Butane[k]	Pentane[l]	Ethene	Acetylene	Formaldehyde	Acetaldehyde	Ethanol
RJ, BR[a]	2.7–4.7	9.3–46.8	2.3–14.9	–	–	31.2–39.3	7.1–53.1	–	–	–	–	–
MASP, BR[b]	0.13–17.6	0.38–65.5	0.17–33.8	0.34–17.0	0.85–83.8	0.26–40.1	0.12–26.5	0.23–41.6	0.10–12.6	–	–	–
São Paulo, BR[c]	2.9	9.9	3.7	4.1	15.8	8.3	35.7	4.9	2.6	8.1	8.4	52.2
Mexico City, MX[d]	1.7	10.7	1.2	6.5	37.5	28.6	13.4	7.8	10.2	–	–	–
Paris, EU[e]	0.3	2.7	1.8	1.6	1.0	2.7	2.9	0.7	0.5	7.3	1.9	–
London, UK[f]	0.5	0.8	0.5	4.1	2.3	3.3	2.0	0.6	0.3		6.5	9.4
Shenyang, CN[g]	1.4	1.6	0.48	6.2	8.7	5.7	3.3	3.7	–	4.8	2.9	–
Beijing, CN[h]	2.7	6.4	2.5	3.9	5.1	7.8	8.0	2.8	3.0	–	–	–
Delhi, IN[i]	12.8	39.4	12.2	23.5	56.8	87.9	44.6	14.8	6.8	–	–	79.0

[a] *RJ: Rio de Janeiro, Brazil (Da Silva et al., 2017).*
[b] *MASP: Metropolitan Area of São Paulo, Brazil (Dominutti et al., 2016).*
[c] *Dominutti et al. (2020).*
[d] *T0 day values (Apel et al., 2010).*
[e] *Borbon et al. (2013).*
[f] *DEFRA data (Dominutti et al., 2020).*
[g] *Center of Shenyang City, China (Ma et al., 2019).*
[h] *Wang et al. (2021).*
[i] *Average of pre- and postmonsoon (Stewart et al., 2021).*
[j–l] *i + n-butane, i + n-pentane, and m + p-xylene.*

Ethane is among the NMVOC displaying the highest amount in urban atmosphere, associated with vehicular emission. Its oxidation under tropospheric conditions occurs with ·OH and NO_3, together with methane and other aliphatic species. Aliphatic species such as alkanes, alkenes, and alkynes are compounds of relatively low polarity and are insoluble in water. They have similar physical properties, and their boiling points are very dependent on molecule size. These features influence the removal processes that occur in the atmosphere.

Consider the ethane oxidation reaction under a rich NO_x atmosphere to illustrate the chain of reactions of alkanes which is usually found in urban areas:

$$CH_3CH_3\cdot + \cdot OH \rightarrow CH_3CH_2OO\cdot + H_2O(\text{in presence of } O_2) \tag{4.27}$$

$$CH_3CH_2OO\cdot + NO \rightarrow CH_3CHO\cdot + NO_2 + HO_2\cdot(\text{in presence of } O_2), \tag{4.28}$$

or in a single overall reaction that can be written as:

$$CH_3CH_3 + \cdot OH + NO \rightarrow CH_3CHO\cdot + NO_2 + HO_2\cdot(\text{in presence of } O_2). \tag{4.29}$$

The products of these reactions are an aldehyde (in this case, CH_3CHO, i.e., acetaldehyde), radical hydroxyl, and NO_2. The NO produced is then converted to NO_2, providing the essential chemical species to the ozone cycle (reaction 4.17). In a generic way, the $RO_2\cdot$ radical (in enough amount of NO_x) reacts to convert NO to NO_2, and form $RONO_2$ and peroxynitrates.

$$RO_2\cdot + NO \rightarrow RO\cdot + NO_2 \tag{4.30}$$

$$RO_2\cdot + NO + M \rightarrow RONO_2 + M \tag{4.31}$$

$$RO_2\cdot + NO_2 + M \rightarrow ROONO_2 + M \tag{4.32}$$

Other important alkanes are propane, butane, and pentane, due to higher kOH. The amount found in urban atmosphere is associated with fuel combustion in power plants, vehicles engines, and industries. Table 4.1 presents an overview of the main NMVOC species, and their concentrations found in large urban areas in the world.

Worked example 4.3

The chemical species released into the atmosphere are usually oxidized. The NO_x emitted, for example, can form HNO_3, which is a strong acid that can decrease the pH of the liquid cloud water. Consider the natural pH of 5.6 and then calculate the concentration of H^+ and pH, supposing that the HNO_3 (5 ppm) was also present in that liquid cloud water.

Solution:

$HNO_3 + H_2O(l) \rightarrow H^+ + NO_3\cdot$

$[HNO_3] = [H^+] = [NO_3\cdot] = 5$ ppm since nitric acid is strong.

$[H^+] = 5\ mg\,L^{-1}$ or $7.9 \times 10^{-5}\ mol\,L^{-1}$ since 1 mol of HNO_3 has 63.01 $g\,mol^{-1}$.

The pH $= -\log\,[H^+]$, then the new pH of the water cloud will be 4.1.

The pH decreased from 5.6 to 4.1 when HNO_3 at 5 ppm was also present.

Alkenes, also known as olefins, are unsaturated hydrocarbons, since they contain less hydrogen than alkanes and have a carbon-carbon double-bond (C=C), a particular feature of alkenes structure. Ethene, C_2H_4 ($H_2C = CH_2$, also called ethylene) is the most important species, considering their amount and reactivity (kOH) in the atmosphere. The presence of a double-bond favors the reaction by breaking the weaker bond, which is the typical addition reaction. Due to this feature, alkenes can react with several reagents, as follows:

$$H_2C = CH_2 + \cdot OH \rightarrow HOH_2C - CH_2 \tag{4.33}$$

$$HOH_2C - CH_2\cdot + O_2 \rightarrow HOCH_2CH_2O_2\cdot \tag{4.34}$$

As mentioned above, it is a fast reaction. Thus the overall reaction is written as:

$$C_2H_4 + \cdot OH + O_2 \rightarrow HOCH_2CH_2O_2\cdot \tag{4.35}$$

$$HOCH_2CH_2O_2\cdot + NO \rightarrow HOCH_2CH_2O + NO_2 \tag{4.36}$$

The product is the radical $HOCH_2CH_2O$, which is unstable and decomposes by reacting with O_2. The products are mainly formaldehyde, $HO_2\cdot$, and NO_2, which was oxidized from NO. The overall reaction according to Seinfeld & Pandis (2016) is

$$C_2H_4 + \cdot OH + 2\,NO \rightarrow 2\,NO_2 + 1.44\,HCHO + 0.28\,HOCH_2CHO + \cdot OH(\text{glycolaldehyde, } HOCH_2CHO) \tag{4.37}$$

Relevant reactions with other alkenes involve dienes and trienes (see Seinfeld & Pandis, 2016 for more details). Alkenes can also react with nitrate radical, which is an important oxidant in the night-time period. The reaction is of addition in the double-bond, as follows:

$$H_2C = CH_2 + NO_3 \rightarrow H_2C(ONO_2) - COOH_2 \cdot (\text{in presence of } O_2) \tag{4.38}$$

In the same way, the radical reacts with NO converting it to NO_2.

Besides reacting with $\cdot$OH and NO_3, alkenes, due to their double bond, react with O_3, which during the day competes with OH and, during the night, with NO_3. Atkinson (2000) calculated the lifetime for some VOC submitted to photolysis, for the reaction with $\cdot$OH radical, the reaction with NO_3 radical, and for the reaction with O_3. For example, lifetime of ethene reaction with $\cdot$OH was calculated to be of 1.4 days, while isoprene (and diene) lifetime was of 1.4 hour with $\cdot$OH, 50 minutes with NO_3, and 2.1 hours with O_3.

As mentioned above, isoprene and terpenes are important hydrocarbons in atmospheric chemistry and, in the urban atmosphere, isoprene, which is also emitted by anthropogenic sources, can influence the oxidation rate of other species and contribute to ozone formation. In urban areas under sufficient NO_x levels, the product of isoprene oxidation by OH reacts almost exclusively with NO. The reactions are quite alike those presented above for alkenes, following the same sequence in a general way:

$$ISOP + \cdot OH \rightarrow ISOPO_2 (\text{in presence of } O_2) \tag{4.39}$$

$$ISOPO_2 + NO \rightarrow ISOPO + NO_2 \tag{4.40}$$

$$ISOPO \rightarrow OVOC + HO_2 \cdot, \ \text{OVOC is other VOC} \tag{4.41}$$

$$HO_2 \cdot + NO \rightarrow NO_2 + \cdot OH \tag{4.42}$$

The NO_2 formed then feeds the cycle of ozone (reactions 4.17–4.19), and the termination results in HNO_3 and H_2O_2 products as shown by reactions (4.22) and (4.24), respectively. Additional details of the oxidation process of isoprene in the urban atmosphere are addressed, for example, by Gu et al. (2022), which suggested that MVK and MACR are derived, mainly, from the oxidation of isoprene.

Alkynes are also an unsaturated and highly reactive family of species, mainly those with low molecular weight, and their feature is the presence of carbon-carbon triple bond (C≡C). Acetylene, C_2H_2, is the simplest and most abundant chemical species that has high reactivity in the atmosphere. Alkynes react usually via addition reactions and, in the atmosphere, the reaction for acetylene with $\cdot$OH (Senosiain et al., 2005) is as follows:

$$H - C \equiv C - H + \cdot OH + M \rightarrow HOH - C = C - H + M \text{ or} \tag{4.43}$$

$$H - C \equiv C - H + \cdot OH \rightarrow HOCHCHO_2 \cdot (\text{in presence of } O_2) \tag{4.44}$$

As discussed to the aliphatic hydrocarbons, the carbon-carbon single bond has low reactivity, while the carbon–carbon double and triple bonds display higher reactivity (present one and two π bonds, easier to break than σ bonds; $\sigma\pi$ bonds, easier to break than σ bonds) and, therefore, these bounds may also react with other species apart from $\cdot$OH.

Biogenic hydrocarbon oxidation is also important in urban atmosphere since there is the presence of urban green parks and trees near the streets. Besides that, the atmosphere is dynamic, and air masses are continually mixing and being transported at the same time as the chemical reactions are occurring. Another important observation is that isoprene, although having vegetation as its main source globally, is also primarily emitted by human activities in urban areas. This mixing of sources and physical–chemical processes poses more complexity in atmospheric chemistry within and around urban areas (Heald & Kroll, 2020).

Apart from hydrocarbons (including NMVOC), there are other important chemical structures such as aldehydes, ketones, and alcohols. Their importance relies on their reactivity and potential to form ozone and other secondary species, as well as on the considerable amount found in urban atmosphere.

Aldehydes have a general formula RCHO (R is H or CH_x, e.g., formaldehyde HCHO, acetaldehyde CH_3CHO) with functional carbonyl group C=O. Ketones also contain the same functional group and general formula RR′CO (R and R′ may be an aliphatic or aromatic group), which are often referred to as carbonyl compounds altogether. The main species, in terms of amount and reactivity, are formaldehyde, acetaldehyde, and acetone (propanone), which can react with different lifetimes, mainly with $\cdot$OH or/and in the presence of radiation. For example, formaldehyde lifetime due to reaction with $\cdot$OH is 1.2 day, while its photolysis is 4 hours. Although having the same functional group that widely defines their chemistry, aldehydes have a hydrogen atom bonded to the carbonyl group which makes them more reactive and easier to be oxidized than ketones (Seinfeld & Pandis, 2016).

To illustrate the chain reactions of aldehydes in the urban tropospheric atmosphere, consider acetaldehyde (CH_3CHO), which is the second more abundant aldehyde after formaldehyde (see reactions 4.10–4.11). Together, they are the most relevant aldehydes in urban atmosphere due to direct emissions by vehicular activity and secondary formation through oxidation of other NMVOC.

$$H_3CHC = O + \cdot OH \rightarrow CH_3CO\cdot + H_2O \tag{4.45}$$

$$CH_3CO\cdot + O_2 + M \rightarrow CH_3C(O)OO\cdot + M(\text{fast reaction}) \tag{4.46}$$

Then, the acyl peroxy radical $CH_3C(O)OO\cdot$ reacts with NO, forming NO_2 and peroxyacyl nitrate, represented as RC (O)$OONO_2$:

$$CH_3C(O)OO\cdot + NO \rightarrow CH_3C(O)O\cdot + NO_2 \tag{4.47}$$

$$CH_3C(O)O\cdot \rightarrow CH_3\cdot + CO_2 \tag{4.48}$$

$$CH_3C(O)O\cdot + NO_2 + M \leftrightarrow CH_3C(O)OONO_2 + M \tag{4.49}$$

Peroxyacetyl nitrate $CH_3C(O)OONO_2$, or PAN, is formed and, together with tropospheric ozone, is a typical chemical specie found in photochemical smog, and is a relevant reservoir of NO_2 after the decomposition reaction (4.49). Aldehydes, such as formaldehyde and acetaldehyde, are oxidized by photolysis

$$CH_3CHO + h\nu \rightarrow CH_4 + CO \rightarrow \cdot CH_3 + HCO\cdot \tag{4.50}$$

Alcohol is another NMVOC functional group (-OH), which has significant relevance in urban atmosphere. Methanol and ethanol are those found in higher amounts and react mainly with $\cdot$OH as follows:

$$CH_3OH + \cdot OH + O_2 \rightarrow H_2O + HO_2\cdot + HCHO\cdot \tag{4.51}$$

For instance, in Brazil, ethanol is widely used as a vehicular fuel (Dominutti et al., 2020; Nogueira et al., 2017), where the passenger car fleet is powered by hydrated ethanol or gasoline with ethanol blend (gasohol, with the addition of 27% ethanol in volume, or E27). The gasohol adopted in Brazil contains a relatively high content of ethanol. In European countries and the United States, gasoline can be blended with less ethanol (5%–15%), while in China E10 can be found in 12 provinces (Abel et al., 2021; Pacheco & Silva, 2019). Usually, in other countries, it is an additive in gasoline. Similarly to methanol, its oxidation is also via $\cdot$OH. Ethanol, although itself not presenting a high ozone potential, has a large influence on urban chemistry due to the large amount emitted and the consequent formation of secondary aldehydes—formaldehyde and, mainly, acetaldehyde, which have high ozone potentials. The overall reaction according to Seinfeld & Pandis (2016) is:

$$CH_3CH_2OH + OH + 0.05\ NO \rightarrow 0.05\ NO_2 + 0.014\ HOCH_2CHO + 0.072\ HCHO + 0.95\ CH_3CHO + HO_2 + H_2O \tag{4.52}$$

4.2.2.1 *Ozone production and the relation between NO_x and volatile organic compound*

The NMVOC and NO_x, as observed in the previous reactions, have a key role in urban tropospheric chemistry, especially in ozone formation. The concentration of NO_x in the atmosphere (depending on emission sources) will determine the availability of radical HO_x due to VOC and CO oxidation. Typically, the tropospheric urban atmosphere is rich in NO_x since there are several contributing sources of combustion, with a highlight for power vehicles and energy generation processes (refineries and thermal power) that usually are located nearby urban areas.

Radicals regenerated in the oxidation of VOCs and CO not only convert NO to NO_2 but also consume O_3 in an atmosphere with low NO concentrations, which are denoted by low-NO_x and high-NO_x limits.

$$\cdot OH + O_3 \rightarrow HO_2\cdot + O_2 \tag{4.53}$$

$$HO_2\cdot + O_3 \rightarrow 2\ O_2 + \cdot OH \tag{4.54}$$

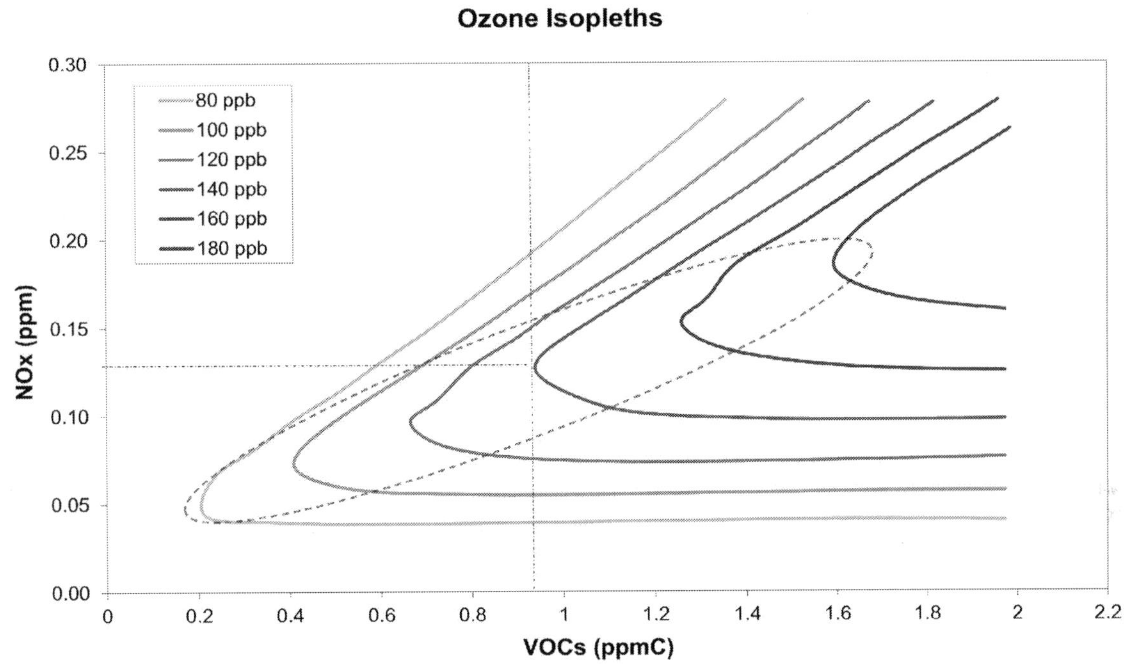

FIGURE 4.2 Nonlinear relationship between NO_x and volatile organic compounds (VOCs) for tropospheric urban ozone formation.

$$HO_2\cdot + NO \rightarrow NO_2 + \cdot OH \tag{4.55}$$

Therefore the net O_3 results from a competition between the reactions above with HO_x. This nonlinear and complex process, due to several chemical reactions, is typically illustrated by the ozone isopleths as shown in Fig. 4.2. Further, meteorological conditions are important to define the occurrence of O_3 episodes, which are characterized by high concentrations (above the air quality standards) during some days. High-pressure systems, calm winds, and relatively high temperatures are conditions that favor ozone formation.

Worked example 4.4

Converting mass mixing ratio of ozone from ppb to $\mu g\,kg^{-1}$.

Solution:

Ozone (O_3) is an important atmospheric gas that plays a role in both air quality and climate. It is often measured in parts per billion (ppb) or micrograms per kilogram ($\mu g\,kg^{-1}$). To convert the mass mixing ratio of O_3 from ppb to $\mu g\,kg^{-1}$, we can use the following equation:

Mass Mixing Ratio ($\mu g\,kg^{-1}$) = Concentration (ppb) × (Molecular Weight/24.45)

where Molecular Weight of O_3 is 48.0 $g\,mol^{-1}$.

For example, let us say we have a measurement of the mass mixing ratio of O_3 in the atmosphere with a value of 50 ppb. Using the above equation, we can convert this to a concentration in $\mu g\,kg^{-1}$ as follows:

Mass Mixing Ratio ($\mu g\,kg^{-1}$) = 50 ppb × (48.0 $g\,mol^{-1}$/24.45) = 97.8 $\mu g\,kg^{-1}$

Therefore the mass mixing ratio of O_3 in this example is 97.8 $\mu g\,kg^{-1}$.

4.2.2.2 Other important species

Besides organic compounds, nitrogen and carbon oxides, sulfur oxides (SOx) are relevant species for atmospheric chemistry in urban atmosphere due to the impact in the inorganic aerosol formation, and consequently direct and indirect effects on climate. Urban areas are impacted by the presence of sulfur due to the use of fossil fuels containing sulfur (see Chapter 3), although efforts have been made to reduce its concentrations in the last decades (Li et al., 2020). SO_2 can react as:

$$SO_2 + \cdot OH + M \rightarrow HOSO_2\cdot + M \tag{4.56}$$

$$HOSO_2\cdot + O_2 \rightarrow HO_2 + SO_3 \quad (4.57)$$

$$SO_3 + H_2O + M \rightarrow H_2SO_4 + M \quad (4.58)$$

Hydrogen sulfide (H_2S) is not expected to be found in high concentrations in urban environments. However, radicals produced in the reaction with $\cdot OH$ ($H_2S + \cdot OH \rightarrow SH\cdot + H_2O$) undergo a series of reactions, which ultimately result in the formation of SO_2. Thus sources around the urban areas, such as landfills, sewage treatment plants, and farm production of animals can contribute to SO_2 concentrations found in cities (Seinfeld & Pandis, 2016).

Ammonia, NH_3, is associated with emission by agricultural activity, mainly soil fertilization, decay of livestock waste, and biomass burning, and can impact urban areas through transport (Gu et al., 2014). In urban areas, the control of NO_x vehicular emissions, through three-way catalyst system in light-duty vehicles (LDVs) and selective catalytic reduction (SCR) in heavy-duty vehicles (HDVs), is associated with an increase in NH_3 emissions (Nevalainen et al., 2018). It is involved in many reactions producing ammonium (NH_4^+) in the particulate phase. Secondary inorganic aerosol (SIA), discussed in the next section, is mainly composed of ammonium sulfate and ammonium nitrate, responsible for a large part of the mass of fine particles in regions impacted by fossil fuels from the energy sector. Particulate ammonium is produced after the reaction of gaseous ammonia and acidic species in the atmosphere (nitric, sulfuric, and hydrochloric acids) or in the acidic surface of anthropogenic particles (Ianniello et al., 2011, Mouli et al., 2003, Wang et al., 2006) according to the following reactions:

$$H_2SO_4 + NH_3 \rightarrow NH_4HSO_4 \quad (4.59)$$

$$NH_4HSO_4 + NH_3 \rightarrow (NH_4)_2SO_4 \quad (4.60)$$

$$HNO_3 + NH_3 \rightarrow NH_4NO_3 \quad (4.61)$$

$$HCl + NH_3 \rightarrow NH_4Cl \quad (4.62)$$

Further, halogens have been recognized as important to tropospheric chemistry. Halogens are of special relevance because they are involved in many reaction cycles affecting the oxidation power of the atmosphere, indirectly by influencing the main oxidants, as O_3 and its photolysis product $\cdot OH$, and directly, for example, by reactions of the $Cl\cdot$ radical with hydrocarbons (e.g., CH_4) (Badia et al., 2021), that occurs in several urban areas around the world near the marine coast, the main global source of halogens.

4.3 Aerosols formation and composition

The study of aerosol is up-to-date research due to its role in almost all aspects of the biogeochemical cycles. Aerosol is ubiquitous in the physicochemical process at different compartments of soil, ocean, and atmosphere. Aerosol is a suspension of solid and liquid particles in a gas (being the gas the atmospheric air). PM is frequently used as synonymous for aerosol, but the former represents the particles in a gas phase. Not all sizes are relevant for the atmospheric process. It is important to consider particles that have a residence time long enough not to sediment fast, which means particle diameters in the range of nanometers (small clusters usually above 3 nm) to 100 μm (larger diameters are associated to a significant sedimentation process). Thus smaller particles can be transported to more distant sites, affecting the air quality in those places. Many sources emit particles or their precursors to the atmosphere, usually classified as natural or anthropogenic sources. Emissions of compounds are discriminated in Chapter 3 and the physical-chemistry process involving aerosols will be discussed in the present chapter.

Compounds generated by secondary processes or primarily emitted in the atmosphere can be inorganic and organic. The inorganic fraction has been studied and detailed by many authors since the late 1970s (Whitby, 1967). Whitby provided many basic definitions, including size distribution by number, mass, and surface of sulfur compounds. Metals, metalloids, and ionic species, such as sulfate, nitrate, and ammonium, explain most of the atmospheric PM in many cities in the world (Jimenez et al., 2009; Ng et al., 2010). Atmospheric metals and metalloids are present in the form of oxides, and these species can account for more than 30% of PM_{10} collected near roadways (Alves et al., 2015). In recent studies in São Paulo, Brazil, metal emissions were determined for LDVs and HDVs. Authors observed that copper, aluminum, barium, and iron were associated with ethanol and gasohol-powered LDVs (Pereira et al., 2023). The carbonaceous component of the PM (total carbon) is the sum of the organic (OC) and elemental (EC) carbon. OC, present in organic species, has its origin in direct emission from natural (biological material from plants and animals, resuspension of dust, sea spray, and natural fires) and anthropogenic sources (fossil fuel, biomass burning of forests induced by human activities) and is called primary organic carbon (POC) or indirect from the conversion of gas to particles, called secondary organic aerosol (SOA), accordingly with the volatility of the VOC. EC is directly emitted by the incomplete combustion of fossil fuels, and other types of organic matter.

The total carbon (TC) content of the PM is the sum of all carbon present in the particles (except the inorganic carbonates). Its measurement is usually done by thermochemical oxidation and posterior CO_2 analysis, given according to the temperature of volatilization of the organic and elemental fraction. But it is important to consider that there is no real sharp division between these fractions, but a continuous decrease of thermochemical refractiveness (Pöschl & Shirawa, 2015).

A significant fraction of the carbon content in the particles comes from the incomplete combustion of fuels. This carbon is released in the form of soot, composed mainly of elemental carbon, in addition to numerous other minor organic compounds (Tapia et al., 2018). Soot is an important constituent of air pollution and has importance over climate change and will be further discussed in more detail below, together with other carbon species.

The first important point is the terminology to describe the different processes of formation and composition of carbon compounds. Some differences in the classification of the carbon compounds are due to measurement differences.

Black carbon (BC), which can be considered mature soot particles, can be defined by its characteristics of strong absorption in the visible spectral region, graphitic structure, no solubility, and high sublimation temperature of approximately 4000 K. The measurement is usually based on a reflectometer and another denomination is refractory BC (Bond et al., 2013).

Another important category is brown carbon (BrC), formed by particles which absorb less than BC and the absorption is enhanced at short visible wavelengths relative to longer wavelengths (Morais et al., 2022). These particles are emitted during biomass and biofuel burning. Morais et al. (2022) evaluated the properties of BrC and BC in the Amazon region, assessing contributions to absorption at 440 nm, showing geographical and seasonal variability and that BrC radiative properties must not be neglected.

Michelsen et al. (2020) present a review of the definitions and properties of soot formation and terminology. These soot particles are generated by the combustion process, and size, morphology, and composition vary according to the fuel, temperature, and stage of the burning process. Soot is formed through incomplete combustion or pyrolysis. Precursors of soot include the polycyclic aromatic hydrocarbons (PAHs), radicals, and oxygenated hydrocarbons (Michelsen et al., 2020). According to this definition, BC is the most absorbing part of the soot, mainly in the visible spectral region with a graphitic structure and lack of solubility in polar and nonpolar solvents (Michelsen et al., 2020).

Soot particles can exhibit a huge surface area of $\sim 100\ m^2\,g^{-1}$ and are mostly constituted by elemental carbon, with a carbon skeleton embedded with several oxygen-containing groups and complex organic species (saturated and unsaturated hydrocarbons, alkanoic acids, aromatic species, including the PAHs and its derivatives, alcohols, ketones). The surface groups of soot contain aromatic (ring; $C{=}C-C$), aliphatic ($C{-}H$), and ($C{=}O$) groups in varied proportions, besides that, organic compounds (e.g., PAHs) can also be observed. Thus, these species can be very reactive, providing sites for heterogeneous reactions (Liu et al., 2022).

Soot can react with several gaseous species: O_2, NO_2, SO_2. Some reactions can occur more easily on the soot surface due to its redox and catalytic activity, not only promoting the removal of gas-phase species but also producing new species. The great importance of soot-NO_2 and soot-HNO_3 reactions as HONO sources has been pointed out in recent studies. Besides that, more advances in the comprehension of the reaction of soot with other species such as $\cdot OH$ radicals, PAN, N_2O_5, and others have been reached more recently (Liu et al., 2022).

Soot is linked with climate change due to its absorbing and scattering properties. When mixed with other heterogeneous products, soot absorption ability can be enhanced in the so-called "lensing effect" (Zhang et al., 2018). Another effect is that the originally emitted hydrophobic soot becomes more hydrophilic with the heterogeneous reactions and mixing, increasing the ability to act as cloud condensation nuclei (CCN).

4.3.1 Gas-to-particle conversion

As discussed in Chapter 3, the primary emission of PM is usually associated with natural sources (e.g., soil, marine spray, volcanoes, and vegetation). The secondary formation of particles in the atmosphere occurs through gas-to-particle conversion, when saturated vapor in the atmosphere condenses to form new particles (Zhang et al., 2012). This process is dependent on temperature, humidity, and the presence of different gases and particles in the atmosphere.

Formation of new particles can occur through homogeneous nucleation of gaseous species from natural and/or anthropogenic origin. Differently, when gases nucleate on a surface of a preexisting particle, the phenomenon is called heterogeneous nucleation. Nucleation can be homomolecular, when the nucleation occurs with only one gas compound, binary process, with two gases, and ternary nucleation, with three gases (Jacobson et al., 2000).

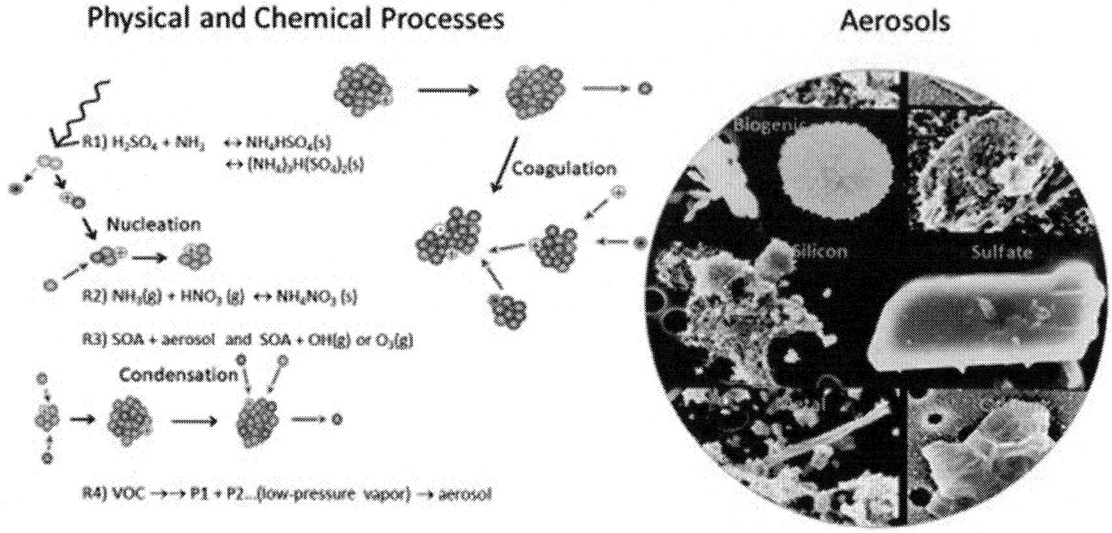

FIGURE 4.3 Schematic illustration of physical and chemical processes of aerosols and their morphological characteristics (P.S. biogenic, chloride, silicon, sulfate, metal, and calcium).

An important example of particle formation from gases is the binary homogeneous nucleation of sulfuric acid with water, generating particles in the 3–20 nm range. The homogeneous nucleation of water, for instance, does not occur under the Earth's atmospheric conditions, being always heterogeneous. Cloud droplets originate with the condensation of water vapor into existing PM. These hygroscopic particles are called CCN (Pöschl, 2005).

Volatility of organic compounds can be used to classify as described by Donahue et al. (2012), from extremely low volatility organic compounds to VOCs. Atmospheric oxidation can reduce the volatility of gaseous pollutants, leading to the formation of SOA (Fig. 4.3).

The main global source of SOA precursors is the vegetation, with the emissions of isoprene and terpenes, followed by anthropogenic emissions of species such as aromatic compounds emitted by fossil fuel burning (Kelly et al., 2018). Kelly et al. (2018) simulated the formation of SOA globally, finding a total of 73.6 Tg (SOA) per year. They simulated the formation of organic compounds using the database from the Aerosol Mass Spectrometry (AMS) global network website (https://sites.google.com/site/amsglobaldatabase/). At urban areas, VOC is mainly emitted by the evaporative process during refueling (Tipton et al., 2022) and at industrial plants. SOA is formed by the gas-to-particle conversion of VOC, and due to different characteristics of the volatility of the VOC. SVOC (semivolatile organic compounds) suffer nucleation and growth, and can also be adsorbed or absorbed by already existing aerosol or cloud droplets. Pöschl and Shiraiwa (2015) presented the components of organic aerosol and sources, distinguishing the great influence of biomass and fossil-fuel combustion emission in the secondary formation.

Sea spray is also responsible for a fraction of the VOC and Primary Organic Aerosol (POA) of biogenic origin emitted to the atmosphere (Quigg et al., 2021), but the estimates present great uncertainties due to the lack of measurements. As already mentioned, VOCs are ubiquitous in the atmosphere, and the description of the secondary process in the atmosphere is the subject of many different studies and an overview of the process is presented in the next sections.

4.3.2 Gas–particle partitioning

The gas–particle (G/P) partition affects chemical reactions, deposition, long-range transport, concentration of semivolatile organic species, and causes impact on human and ecosystem health. G/P partition of semivolatile organic species may involve simple adsorption when the PM is composed uniquely of mineral materials. However, as urban PM often contains a great amount of amorphous organic carbon, absorptive partitioning can play a role in urban air and sites affected by urban sources.

Acidic or basic gases can partition into atmospheric PM and fog/cloud droplets, influencing the condensed phase acidity. Acidity of the aqueous phase can also influence the partitioning of its species with the gas phase. This property influences the transfer or evaporation of dissociating species into the gaseous phase and acidity buffering process. The process of partitioning takes below 1 hour for fine mode aqueous particles and small cloud droplets. Equilibrium conditions are assumed for aqueous-phase concentrations estimation. The exceptions are larger droplets with higher pH, viscous particles or droplets, or the ones with coatings in the surface, and highly reactive dissolving species, where there are mass transfer limitations in gas or aqueous phase that can prevent attainment of equilibrium partitioning in convenient timescales (Tilgner et al., 2021).

In an ideal aqueous solution at equilibrium, factors such as mass transport limitations, chemical production, nonideal solution effects, and degradation processes are not considered. The aqueous phase concentration of a soluble species ($[A]_{aq}$) is proportional to its partial pressure in the gas phase ($pA_{(air)}$) and its Henry's law constant (H_A). Henry's law constant is defined by the following equation (mol L^{-1} atm^{-1}):

$$H_A = [A]_{aq}/pA_{(air)} \tag{4.1}$$

When an acid dissociates in an aqueous solution, it produces H^+ and A^- and is characterized by an equilibrium dissociation constant K_a and the solution acidity. The acidity of an aqueous solution, as well as the liquid water content (LWC), affect the partitioning of species that dissociate in aqueous particles or cloud droplets. The increasing acidity of a particle leads to a decrease in the partitioning of acid and an increase in the partitioning of bases. Nitrate in the particulate phase depends on the range of PM pH. With pH = 1, almost 100% of nitrate exists as HNO_3 in the gaseous phase, while at pH = 4 it is almost completely in the particulate phase. Aqueous atmospheric particles are often acidic, and bases can be predominantly found in the protonated form, not varying with pH variations. Therefore small biases in predicting PM pH can lead to over or underestimations of the $PM_{2.5}$ mass.

4.3.3 Coagulation

Particles suspended in the atmosphere collide with each other due to their random motion, coalescing and becoming larger-sized particles. This is an important phenomenon and helps to understand part of the dynamics involved in the size distribution of particles in the atmosphere.

Several factors, including the Brownian motion of particles, turbulence, presence of wind shear, and external forces, such as gravitational and electrical, can contribute to coagulation. Coagulation occurs when movement due to one of these factors forces two particles to adhere, forming an aggregate belonging to a larger size class. Among the various factors, Brownian coagulation has been more widely investigated (e.g., Sitarski, 2012). Although other factors contribute to this process, Brownian coagulation is that which effectively contributes to the submicron domain of particles. The rate of reduction in particle number, by Brownian coagulation, is proportional to the number of particles in a given volume of the atmosphere. The coagulation coefficient represents this constant of proportionality. Coagulation coefficient is calculated between two particles. Fig. 4.4 describes the Brownian coagulation coefficient for particles of different sizes. Note that efficiency of coagulation decreases substantially as particle size of a given particle pair approach, whatever the size of the particles. This indicated that the greatest coagulation efficiency occurs when very small particles, possibly from the gas–particle conversion, meet larger particles in their path, coalescing with them.

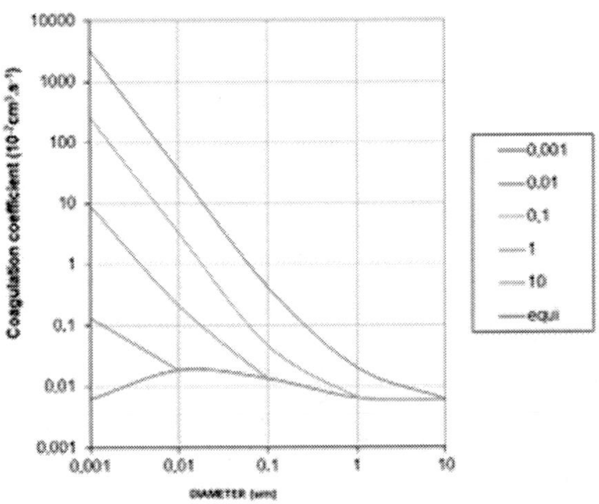

FIGURE 4.4 Brownian coagulation coefficient for typical aerosol size particles in the range 0.001–10 μm, under atmospheric conditions. The lower limit line (equi) means an equilibrium point where the constant undergoes inflection.

4.4 Particle size distribution

Aerosol particles occur at various diameters varying from 1 nm to 100 μm, and this distribution is at a continuous function which can be divided into fine (<2.5 μm) and coarse particles ($2.5 < d_a < 10$ μm) for regulatory purposes. Inhalable PM is the sum of fine and coarse particles. Fine particles are more related to the gas-to-particle conversion of compounds emitted primarily from combustion sources, while coarser particles are associated with mechanical processes (resuspension of soil dust, abrasion of tires and pavement, marine spray, and vegetation debris). The fine particle range includes accumulation mode (0.1–1.0 μm), Aitken mode (20–100 nm), and nucleation mode (3–20 nm). A common convention is to define "ultrafine particles" as the ones with a diameter below 100 nm (comprising the Aitken and nucleation modes). Mass concentrations can be reported in health-related PM sizes, such as PM_{10}, $PM_{2.5}$, and PM_1, which simulate their different depths of penetration in human airways (Colbeck & Lazaridis, 2014). Fig. 4.3 illustrates the formation and composition of particles from the gas-to-particle conversion.

The size distribution of particles is directly related to the formation process and many authors have presented the schematic representation of number, surface, and volume distribution with their characteristic modes (Seinfeld et al., 2016). The improvement of analytical procedures which provided the determination and quantification of particles just formed by the nucleation process (gas-to-particle conversion) allowed the discrimination of nanoparticles. Considering this evolution, size distribution is now divided from ultrafine (UFP) to coarse particles, ranging from 1–3 nm to 100 μm (Finlayson-Pitts & Pitts, 2000). Particles with sizes greater than 10 μm are not so important in atmospheric processes due to their limited lifetime.

Size distribution of particles is a product of many different processes, including emission, chemical reactions, transport, and sinks. UFP particles have very little mass but are formed in great quantities in urban areas due to the nucleation of gases emitted by combustion of fuels. Condensation of gaseous compounds in the surface of preexisting particles and coagulation of particles form the Aitken and accumulation mode (particles in the size range of 100 nm to 1 μm). The various roles that particles play in atmospheric processes, such as interaction with radiation or water uptake, are highly dependent on their size distribution.

Fig. 4.5 illustrates the mass size distribution of particles collected with a nano-MOUDI impactor. Distributions presented in this figure are a combination of different modes, characterizing size fractions that are associated with different sources and processes in the atmosphere of São Paulo, Brazil. In this distribution, mass of each stage is presented per logarithmic of particle diameter plotted against particle diameter. The distribution is commonly normalized by the log of the diameter due to the large interval of values covered by the aerosol particles. The total mass obtained during this sampling was approximately 30 $\mu g\ m^{-3}$.

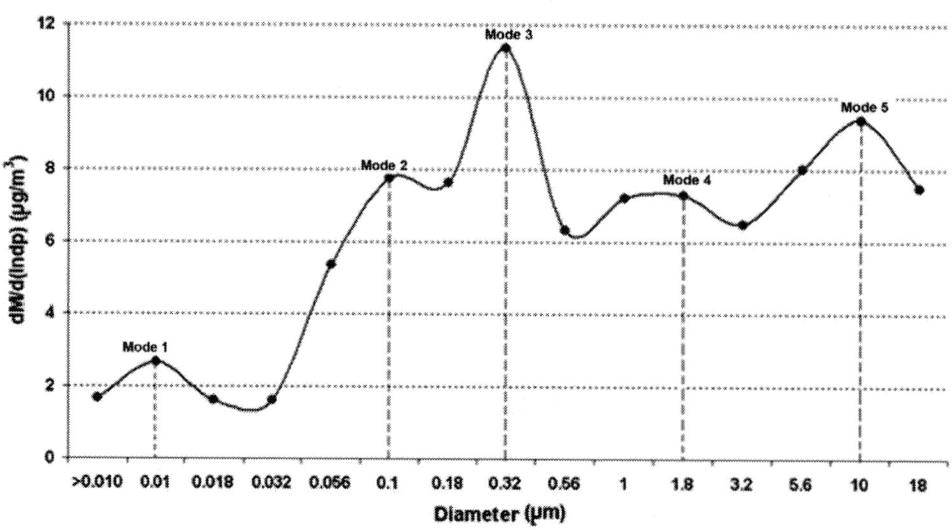

FIGURE 4.5 Mass size distribution of aerosol collected with a nano-MOUDI impactor in São Paulo, Brazil.

4.5 Particulate matter and air quality guidelines

The potential of a pollutant to affect human health is determined by both the degree of exposure and its toxicity. According to the World Health Organization (WHO), the main associations that affect the greatest number of people are pulmonary diseases (more than 50% of cases of pneumonia in children); cardiovascular diseases and strokes; cancer predisposition and diabetes; impaired cognitive development in children; and dementia in the elderly (WHO, 2018a,b). PM and toxic air compounds have entered the list of carcinogenic substances associated with lung and bladder cancer after prolonged exposure (Pope, 2002; IARC, 2013; Hamra et al., 2014). Poor air quality can harm health for life—even before birth (Veras et al., 2010). A study reveals that in cities with the highest concentration of pollutants, long-term exposure (over 15 years) to polluted air led to an increase of up to 15% in COVID-19 mortality rate (Wu et al., 2020). This work showed that the population exposed to air pollutants is more vulnerable to the SARS-CoV-2 virus.

The WHO established in 2021 (WHO, 2021) new guidelines for recommended concentrations of gases and particles in the atmosphere. Table 4.2 displays these new guideline values for PM, O_3, NO_2, SO_2, and CO. Guidelines are presented as interim targets and fine expected levels. In the previous air quality guidelines (AQG), the interim target 4 was the expected AQG level. Proposed levels for AQG are an indication of concentrations that have reduced health risks associated with pollutant concentrations.

Worked example 4.5

Calculating the mass concentration of atmospheric aerosols.

Solution: The mass concentration of atmospheric aerosols is an important parameter in atmospheric chemistry that is used to quantify the amount of PM in the air. The mass concentration is typically measured in micrograms per cubic meter ($\mu g\,m^{-3}$). To calculate the mass concentration of aerosols, we can use the following equation:

Mass Concentration = (Density × Volume Concentration)/1000

where Density is the density of the aerosol particles (in $g\,cm^{-3}$), Volume Concentration is the concentration of aerosol particles (in number per cubic centimeter), and the factor of 1000 is used to convert the result from $g\,m^{-3}$ to $\mu g\,m^{-3}$.

For example, let us say we have a sample of atmospheric aerosols with a density of $1.5\,g\,cm^{-3}$ and a volume concentration of 3.0×10^6 particles cm^{-3}. Using the above equation, we can calculate the mass concentration of the aerosols as:

Mass Concentration = $(1.5\,g\,cm^{-3} \times 3.0 \times 10^6$ particles $cm^{-3})\,1000 = 4.5\,\mu g\,m^{-3}$

Therefore the mass concentration of the atmospheric aerosols in this example is $4.5\,\mu g\,m^{-3}$.

TABLE 4.2 Recommended air quality guidelines levels and interim targets (WHO, 2021).

Air pollutants	Averaging time	Interim target				Final target
		1	2	3	4	
PM_{10} ($\mu g\,m^{-3}$)	Annual	70	50	30	20	15
	24-h	150	100	75	50	45
$PM_{2.5}$ ($\mu g\,m^{-3}$)	Annual	35	25	15	10	5
	24-h	75	50	37.5	25	15
NO_2 ($\mu g\,m^{-3}$)	Annual	40	30	20	–	10
	24-h	120	50	–	–	25
O_3 ($\mu g\,m^{-3}$)	Peak season[a]	100	70	–	–	60
	8-h	160	120	–	–	100
SO_2 ($\mu g\,m^{-3}$)	24-h	125	50	–	–	40
CO ($mg\,m^{-3}$)	24-h	7	–	–	–	4

[a]Average of daily maximum 8-h mean O_3 concentration during the six consecutive months with the highest six-month running average O_3.

4.6 Aerosol in the Earth system

4.6.1 Microphysical process: in-cloud processes

A very important climatic effect of PM is its role in the formation of cloud droplets and ice crystals. Water molecules interact with atmospheric particles and get adsorbed in their surfaces or absorbed in their bulk. Particles constituted of water-soluble material can uptake water vapor, forming an aqueous solution droplet and increasing particle diameter, in the process of hygroscopic growth. This growth process occurs even below water vapor supersaturation (relative humidity below 100%). When there is water vapor supersaturation (RH > 100%), these particles act as nuclei to form droplets inside clouds and act as CCN. Without these particles, the formation of water droplets from a homogeneous gas phase would take a very high supersaturation (much above 100%). In the atmosphere, water vapor saturation is rarely exceeded by more than 10% (when exceedance occurs, it is usually below 1%), since PM induces heterogeneous nucleation, condensation, and cloud formation. Formation of ice crystals in clouds can also be induced by particles. In this case, they are called ice-nuclei (IN). This process requires temperatures below 0°C, leading to a high-water vapor supersaturation concerning ice. Critical supersaturation is the minimum supersaturation in which aerosol particles are effective as CCN or IN and is related to their physical structure and chemical composition, as it decreases with the increase of the particles' size (Pöschl, 2005).

Water-soluble compounds favor the CCN effect of an atmospheric particle. These particles may present organic and inorganic species or a mixture of both. Physical properties are complex, including solubility, surface activity, hydrophobicity, and morphology. Sulfates are the most important inorganic CCNs (e.g., $(NH_4)_2SO_4$, ammonium sulfate) in continental areas, while sodium chloride (NaCl) plays an important role as CCN near marine areas. The fraction of organic species that are water soluble can range from 20% to 80%, with some groups presenting strong polarity, surface activity, solubility, and water affinity (Pöschl, 2005; Sun et al., 2006).

Different pathways or modes are linked to the nucleation of ice crystals: (1) deposition mode, where water vapor is adsorbed and converted to the IN surface; (2) condensation freezing mode, where the particles first act as CCN, inducing the formation of supercooled droplets; then freezing on (3) immersion mode, where IN is incorporated into aqueous droplets, inducing ice formation after cooling; and (4) contact mode, where freezing of the supercooled droplet initiates after the contact with IN surface.

Clouds have a key role in the planet's radiation budget, absorbing infrared radiation. Perturbations in the population of CCN can impact clouds optical properties, as well as their lifetime, size distribution, glaciation rates, and water mass distribution (Pöschl, 2005; Sun et al., 2006).

Water-soluble aerosol components are typically hygroscopic and can absorb water, forming aqueous solutions when RH is below 100%. Transition of a dry particle into saturated aqueous solutions is termed deliquescence, occurring when there is an exceedance of substance-specific RH threshold value, or the deliquescence relative humidity (DRH). The reverse transition (from saturated aqueous solutions to dry particles) is termed efflorescence and depends on the efflorescence relative humidity (ERH).

The Köhler theory is employed to model cloud nucleating potential of PM, where the equilibrium saturation vapor pressure ratio observed at the particle's surface relates to the size of the wet particle. Hygroscopic growth and CCN activation of the aqueous droplets are described by combining Raoult's law or other formulations for water's activity in aqueous solutions and Kelvin equation for the dependence of vapor pressure on the surface tension and curvature of a droplet (Lowe et al., 2016, Pöschl, 2005; Sun et al., 2006). The Köhler equations represent the vapor pressure over an aqueous solution droplet, the combination of the Kelvin effect (saturation over the droplet of pure water) and the solute effect (decrease in the vapor pressure due to the addition of a salt).

Vegetation can produce CCN from biogenic origin, and it is important in rural and forested areas, although biomass burning and vehicular anthropogenic sources can be also a great contributor to organic CCN (Sun et al., 2006).

Bacteria can act as CCN and they have been observed in the boundary layer, in upper troposphere, and, more rarely, in the stratosphere. Bacteria represent a small amount of OC (0.01%) in cloud water, snow, rain, and PM. There is a flux of bacteria from plant surfaces and soil into the atmosphere, including ice-nucleating species, that have been found in clouds, fog, raindrops, and hailstones. Bacteria can be effective as IN, as some studies found IN bacteria at 6 km of altitude. Strains of the plant pathogenic bacteria *Erwinia carotovora* and *E. carotovora atroseptica* were found to be active CCN, and authors have proposed that a significant number of tested species were actively involved in atmospheric precipitation processes and cloud formation. However, other authors suggested that bacteria may not be an important source of CCN, since they represent a low fraction of OC. Despite the low concentration, bacteria are among the most effective ice nucleators in clouds, allowing the nucleation of ice at temperatures very close to 0°C. This is the case of *Pseudomonas syringae*, whose scientific experiments have demonstrated nucleation at temperatures of up to $-2°C$ (Lukas et al., 2022). The role of other bioaerosols is still not well understood, including large molecules or volatile compounds, and remnants of biological activities, such as pollen, viruses, and fungi (Sun et al., 2006).

Organic aerosols can be important as CCN and their activation can be as effective as sulfate aerosols, especially in the case of monocarboxylic and dicarboxylic acids. Formic and acetic acids are dominant monocarboxylic acids in gaseous and particulate phases, and were observed in precipitation, cloud, and fog water. Dicarboxylic acids dominate in the particulate phase. Although it constitutes only a small fraction of water-soluble organic carbon (WSOC) in the atmosphere, the most abundant species are oxalic, malonic, and succinic acid, with concentrations higher in polluted urban areas (Sun et al., 2006).

Humic-like substances (HULIS) may also have an impact on aerosol's hygroscopicity and CCN formation. HULIS is composed of polysaccharide and aliphatic structures and are biogenic-originated major components of continental organic aerosols, comprising 20%–50% of WSOA in urban, rural, high Alpine sites, tropical rainforests (Amazonia), and fog samples. HULIS can be formed in agricultural burning and heterogeneous reactions as the oxidation of soot, which may be related to CCN effectiveness of soot aerosols (Sun et al., 2006).

A significant fraction of continental and marine inorganic aerosols can be coated by organic films, originating a mix of organic and inorganic species in the same particle. These organic coatings can modify particle properties, such as hygroscopicity, CCN formation and viability, and heterogeneous chemical reactivity. Reported organic coatings include fatty acids and inverted micelle structures, with short carbon chains in marine aerosols. Longer-chained fatty acids can display hydrophobic properties, which influences CCN ability. On the other hand, oxidation of these aerosols may lead to a propensity to water uptake and participation in cloud formation. An example is an increase in CCN ability after the oxidation of oleic acid aerosols by ozone. Therefore particle hygroscopicity and CCN effectiveness may increase as the aerosols are transported and aged. The self-assembled monolayer (SAM) is an organic layer adsorbed on an inorganic particle (e.g., sulfate aerosols), acting as a solid surface. The van der Waals interaction between SAM and ozone leads to the formation of polar species, which may affect its CCN effectiveness (Sun & Ariya, 2006).

4.6.2 Dry and wet depositions

Gaseous and particulate pollutants can be carried out by turbulent or molecular diffusion, or advection to buildings, trees, vegetation, and ocean surfaces in a process called dry deposition (Jacobson, 2002). Wet deposition refers to the process of removing airborne species by precipitation scavenging, including in-cloud and below-cloud scavenging, with the removal of vapor phase and particulate chemical species by rain and snow precipitation, although it may also happen during dew formation, mists, and fog. To illustrate this, Box 4.1 presents an unpredictable event of black rain that happened over the metropolitan area of São Paulo, Brazil, with wet deposition of particles due to intense biomass burning in South America at the time. The in-cloud scavenging is the process where atmospheric particles act as CCN or can be captured by cloud water (rainout). Below-cloud scavenging is the removal of these particles by raindrops and snow (washout).

Toxic nonessential metals and metalloids, such as Hg, Pb, Cd, As, and Al, as well as persistent organic pollutants (POPs), can be removed from the atmosphere through dry and wet deposition processes. Excess concentrations of metals and metalloids, such as Cu, Zn, Fe, Mn, and Ni, can also be toxic. These species can be present in soluble and insoluble fractions of wet depositions. The soluble fraction is bioaccessible and can deposit into the ecosystem, readily dissolved in rivers and lakes, absorbed by plants through leaves and roots and leached into the soil (Cheng et al., 2021; Vasseghian et al., 2021). On the other hand, precipitation can lead to the cleansing of vegetation, removing adhered particles on the surface. Thus rainfall may have an effect on the accumulation of POPs in the vegetation (Barber et al., 2004; Kinnersley et al., 1996).

The deposition of particulate species can be quantified by deposition fluxes. Dry deposition flux is a function of atmospheric concentrations and the velocity of the particle dry deposition of these airborne species to the Earth's surface. It depends on emission sources, meteorology, particle size distributions, and surface type. Wet deposition flux is a function of the precipitation concentration of a chemical species and the precipitation rate. The first parameter is affected by both in- and below-cloud scavenging; both processes are considered in calculations on wet deposition of chemical species (Cheng et al., 2021). In the wet deposition, the initial period in the precipitation is often accountable for most of the total deposition, due to the cleansing of particulate and gaseous POPs (Barber et al., 2004).

Although physical and chemical processes that control acidic deposition are well known, the emission and deposition data are mostly known for developed countries. This knowledge is very important since anthropogenic emissions

BOX 4.1 São Paulo (Brazil) - Dark skies and "black rain".

On August 19th, 2019, the population observed abnormally darkened skies in the early afternoon, preceded by a dark precipitation over the metropolitan area of São Paulo (Brazil), an unprecedented event termed "black rain". The event occurred after intense biomass-burning episodes in the South American hinterland. Satellite imagery, back trajectory analysis, and meteorological reports showed that air masses transported thick plumes of biomass-burning aerosols originated partly from the Amazon Basin and Bolivia. Rainwater samples collected in different sites of the metropolitan area during that event presented higher concentrations of a biomass-burning monosaccharide anhydride marker (levoglucosan), with deposition fluxes more than three times higher during the event, in comparison to control samples collected in "clean periods" (Pereira et al., 2021). Other monosaccharide markers suggested the presence of plant cell wall material derived from the partial combustion of wood and grass. Total polycyclic aromatic hydrocarbons (PAHs) concentrations were generally higher during the event than in post-event control samples collected.

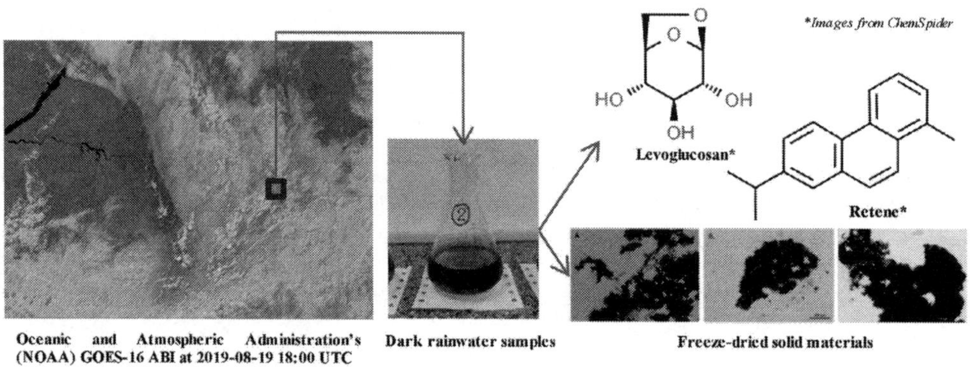

of sulfur and nitrogen oxides can exceed natural emissions on a global scale and influence the precipitation acidity. Sulfate can be a dominant contributor to acidic deposition in most regions, although nitrate can be important in others. Organic acids can be important contributors in the southern hemisphere, although not properly quantified. In the last decades, deposition of sulfur has decreased in developed countries, although in less developed areas deposition levels have become important. Measurements of dry deposition can be sparser than wet deposition (Vet et al., 2014).

On August 19, 2019, abnormally darkened skies were observed by the population in the early afternoon, preceded by a dark precipitation that was observed over the metropolitan area of São Paulo (Brazil), in an unprecedented event termed "black rain." The event took place after intense biomass burning episodes in the South American hinterland. Satellite imagery, back trajectory analysis, and meteorological reports illustrated air masses transporting thick plumes of biomass burning aerosols originated in part from the Amazon Basin and Bolivia. Rainwater samples collected in different sites of the metropolitan area during that event presented higher concentrations of a biomass burning monosaccharide anhydride marker (levoglucosan), with deposition fluxes more than three times higher during the event, in comparison to control samples collected in "clean periods" (Pereira et al., 2021). Other monosaccharide markers suggested the presence of plant cell wall material derived from the partial combustion of wood and grass. Total PAHs concentrations were, in general, higher during the event than in postevent control samples collected.

Dry deposition occurs in the absence of the effect of water droplets condensation when chemical species are deposited as gases or dry particles. It can be deposited as a gas (e.g., SO_2) or a fine and dry aerosol particle (e.g., ammonium sulfate). The collection and evaluation of wet deposition procedures are relatively uncomplicated, since it is easier to collect precipitation events, while the process of dry deposition is a continuous and slow procedure, and very difficult to measure since there are local factors that may alter wind conditions and seasonal factors, as well. On a regional basis, both types of deposition can be approximately equal. However, in more urbanized areas (or nearer to emission sources), dry deposition can be considered a major process (Hendrey, 2001).

4.6.3 Acid depositions

Acid rain is a popular term for the mix of precipitation (snow, sleet, hail, fog, and cloud water) acidified anthropogenically and the dry deposition of acidifying gases and particles. It is commonly defined that the pH of rainwater affected only by carbonic acid is approximately 5.6, and natural rainwater pH can range between 5 and 5.6, while acid rain and fog are between 2 and 5.6 (Jacobson, 2002). Anthropogenic emissions of NO_x and SO_2 began to increase mostly during the context of the Industrial Revolution, however, the acid-neutralizing capacity of the atmosphere, including dust particles, may have prevented major changes in the precipitation pH until around the mid-1950. Emission control of acid precursors, which includes SO_2, led to a reduction of acid precipitation in eastern North America, as was observed in samples collected recently. These were 80% less acid than those collected in the 1960s, although precipitation is still four or five times more acidic than a nonpolluted atmosphere would expected to be (Likens, 2021).

As observed in Chapter 3, SO_2 and NO_x can be emitted by stationary and mobile sources. These gases can be transported for thousands of kilometers, which can lead to acid rain phenomena, crossing political boundaries and different ecosystems. Thus it reinforces the importance to have regulations controlling these emissions in the developed world (the United States, Canada, and Europe). Recently, an increasing role of NO_x emissions to the contribution of nitric acid precipitation has been highlighted. It has been observed in areas of North America that nitric acid is becoming predominant in precipitation, due to changes in the controls of NO_x and SO_2 emissions (Likens, 2021).

Acid deposition has several environmental negative effects. Increased acidity is able to mobilize aluminum in the soil and this species, when dissolved, can be very toxic. Another effect is the leaching of magnesium and calcium in the soil, impacting its buffer capacity. Acidification of surface water may lead to stress and death of individual pH-sensitive organisms. Food chain can also be affected, as acidification may affect food webs (Likens, 2021). The lack of pollutant controls makes the issue of acid precipitation problematic in urban regions of developing countries. These countries may also be under the influence of crop and forest fires (Grennfelt et al., 2020). However, some advances have been observed in China, once very affected by it by the heavy releases of nitrogen and sulfur dioxides. Proportion of acid deposition-affected cities has then decreased from 52.8%, in 1998, to 37.6%, in 2018, reflecting an overall air quality improvement (Xuan et al., 2021).

Worked example 4.6

Scavenging of pollutants.

Imagine a severe case where precipitation is carrying particulate pollutants in an urban area, after a long dry period, on September 01, 2019. What may represent a worst case of polycyclic aromatic contamination in the soil and plants? Researchers found benzo(a)pyrene concentrations of $10\,ng\,mL^{-1}$ in rain samples collected in site A and $8\,ng\,mL^{-1}$ in samples collected in site B. They obtained a figure showing daily precipitation estimations from Climate Hazards Group InfraRed Precipitation with Station data (CHIRPS), Fig. 4.6. Is it possible to rely only on chemical species concentration in rainfall to assess how impacted a studied site is?

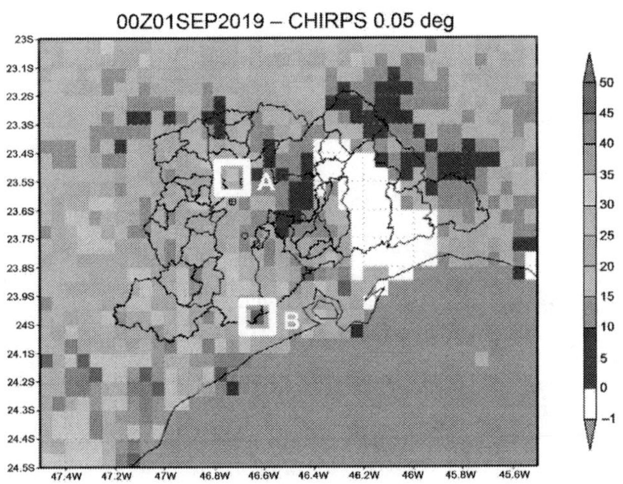

FIGURE 4.6 Daily precipitation estimations from Climate Hazards Group InfraRed Precipitation with Station data (CHIRPS). *Data from Climate Hazards Group InfraRed Precipitation with Station data (CHIRPS).*

Solution:

The pollutant concentration in rainwater itself is not enough to assess how impacted a site is by its precipitation with the rainfall or to compare rainfall events, as the amount of pollutants reaching the soil can be larger if there is a greater amount of rainfall.

The wet deposition flux can be easily calculated with the following equation, where C is the pollutant concentration ($mg\,m^{-3}$) and R is the rainfall amount (m).

Deposition flux ($mg\,m^{-2}$) $= C \times R$

Site A:

We can see that the precipitation in site A was between 30 and 35 mm.

The deposition flux of benzo(a)pyrene will be a value between $(10 \times 10^{-6}\,mg\,m^{-3}) \times (30 \times 10^{-3}\,m)$ and $(10 \times 10^{-6}\,mg\,m^{-3}) \times (35 \times 10^{-3}\,m)$, or 3.0×10^{-7} and $3.5 \times 10^{-7}\,mg\,m^{-2}$.

Site B:

We can see that the precipitation in site B was between 45 and 50 mm.

The deposition flux of benzo(a)pyrene will be between $(8 \times 10^{-6}\,mg\,m^{-3}) \times (45 \times 10^{-3}\,m)$ and $(8 \times 10^{-6}\,mg\,m^{-3}) \times (50 \times 10^{-3}\,m)$, or 3.6×10^{-7} and $4.0 \times 10^{-7}\,mg\,m^{-2}$.

Conclusion: Site B will be more impacted, with a higher deposition flux, besides the lower concentration found in the rainwater samples.

4.7 Aerosol light absorption and scattering

Aerosols suspended in the atmosphere influence solar radiation reaching the Earth's surface. They can scatter and absorb electromagnetic radiation, resulting in the redistribution of energy from the sun (shortwave radiation) or from Earth's surface (longwave radiation), causing warming or cooling of the atmosphere, depending on the physicochemical and optical properties of the particles. The mixture of absorbing and scattering particles presented in the atmosphere results in a change in Earth's energy budget.

There are two types of interactions between aerosol and radiation. The first type is called direct radiative effect, when particles can absorb and/or scatter radiation, reducing the amount of solar radiation reaching Earth's surface, which can cause surface cooling. Another direct effect occurs when particles (such as BC) cause local warming, contributing to changes in the stability of the atmosphere. The second type is the indirect radiative effect: aerosol particles acting as CCN, causing changes in cloud microphysics and optical properties (Seinfeld et al., 2016), ultimately leading to an increase in the albedo of the clouds and their lifetime.

Greenhouse gases such as CO_2, CH_4, N_2O, and hydrofluorocarbons have a positive net radiative effect, causing a warming of the atmosphere and Earth's surface, due to their properties of absorbing and emitting long-wave radiation. Some species found in PM, such as nitrates, sulfates, organic carbon, and mineral dust, tend to scatter solar radiation causing a cooling effect in the atmosphere. BC has a positive forcing, absorbing solar radiation and warming the atmosphere. Fig. 4.7 shows the values of the global average radiative forcing of the various compounds found in the atmosphere with their respective uncertainties, comparing them to values under preindustrial conditions, starting in 1750 until 2011, whose unit of measurement is the watt per square ($W\,m^{-2}$) (IPCC, 2013). Horizontal bars indicate the overall uncertainty, while vertical bars are for individual components (vertical bar lengths proportional to the relative uncertainty, with a total length equal to the bar width for a $\pm 50\%$ uncertainty). Best estimates for the totals and individual components (from left to right) of the response are given in the right column. Values are radiative forcing (RF) except for the effective radiative forcing (ERF) due to aerosol–cloud interactions (ERFaci) and rapid adjustment associated with the RF due to aerosol–radiation interaction (RFari Rapid Adjust).

The AR6 report from IPCC presents updated values for the radiative forcing, considering 1750 to 2019, and includes the global surface temperature change. Fig. 4.8, extracted from this report, illustrates the new evaluation of the estimative for radiative forcing and surface temperature from 1750 to 2019 (https://www.ipcc.ch/report/ar6/wg1/downloads/report/IPCC_AR6_WGI_TS.pdf).

Aerosol optical depth (AOD) is a measure of the amount of scattered and absorbed radiation by the aerosols distributed within a column of air from the surface to the top of the atmosphere. A map of AOD around the world can be found at https://earthobservatory.nasa.gov/global-maps/MODAL2_M_AER_OD. National Aeronautics and Space Administration (NASA) has a network of sun photometers around the world that provide long-term, continuous, and readily accessible public domain database of aerosol optical, microphysical, and radiative properties for aerosol research

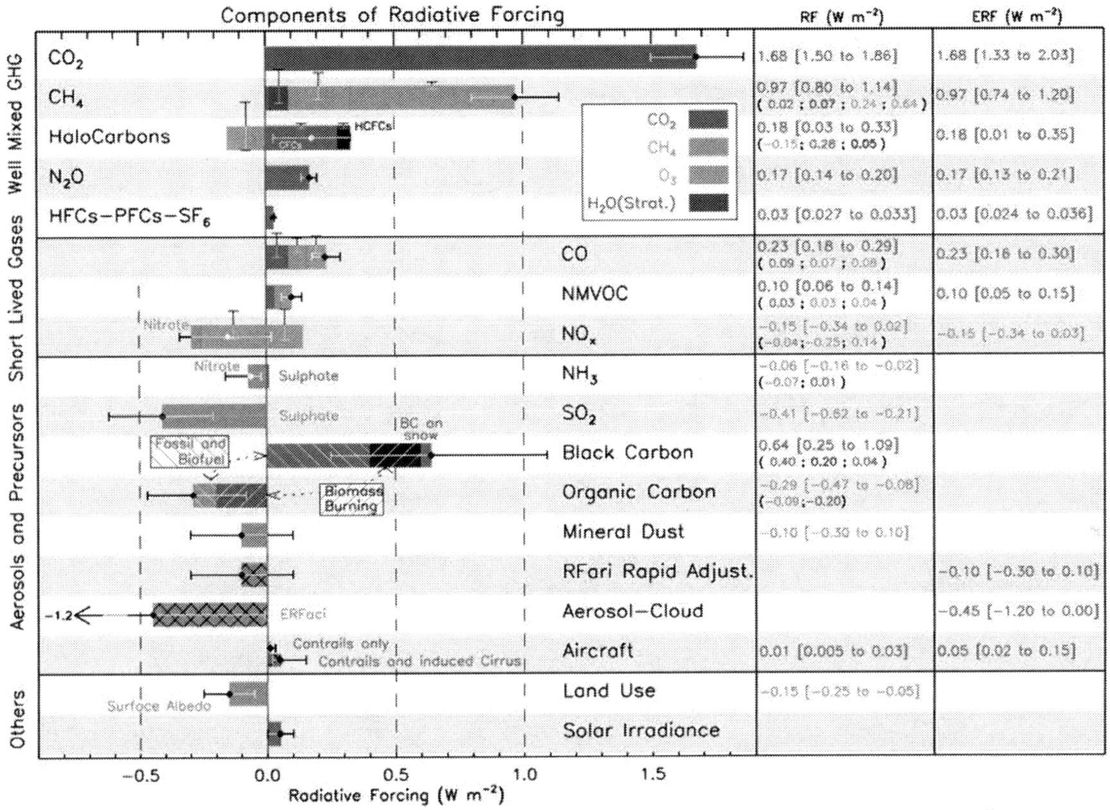

FIGURE 4.7 Radiative forcing (RF) of climate change during the Industrial Era shown by emitted components from 1750 to 2011. *Data from https://www.ipcc.ch/report/ar5/wg1/technical-summary/*

and characterization, validation of satellite retrievals, and synergism with other databases, including AOD (https://aeronet.gsfc.nasa.gov/) (Holben et al., 1998). Fig. 4.9 shows monthly AOD averages at three locations in Brazil for the year 2021: an urban region (SP-EACH), another in the middle of the Amazon rainforest (ATTO Tower), and a third in a city in western Brazil, heavily influenced by forest fires (Rio Branco). Despite the lack of data for the whole year, it is possible to observe an increase in the AOD in months when forest fires are more frequent in the country (between August and October), including in the city of São Paulo, which, besides local pollution, can also be influenced by pollutants coming from other places (de Miranda et al., 2017; Pereira et al., 2021).

Worked example 4.7

Calculating the optical depth of atmospheric aerosols.

The optical depth of atmospheric aerosols is a measure of how much light is scattered and/or absorbed by aerosol particles in the air. It is an important parameter for understanding the effects of aerosols on climate and air quality. The optical depth is typically calculated using the following equation:

$$\text{Optical Depth} = -\ln(T)$$

where T is the transmittance of light through the atmosphere. The transmittance is calculated by measuring the amount of light that passes through the atmosphere compared to the amount of light that would pass through a clear atmosphere.

For example, let us say we have a measurement of the transmittance of light through a polluted atmosphere with a value of 0.5. Using the above equation, we can calculate the optical depth of the atmospheric aerosols as:

$$\text{Optical Depth} = -\ln(T) = -\ln(0.5) = 0.693$$

Therefore the optical depth of the atmospheric aerosols in this example is 0.693.

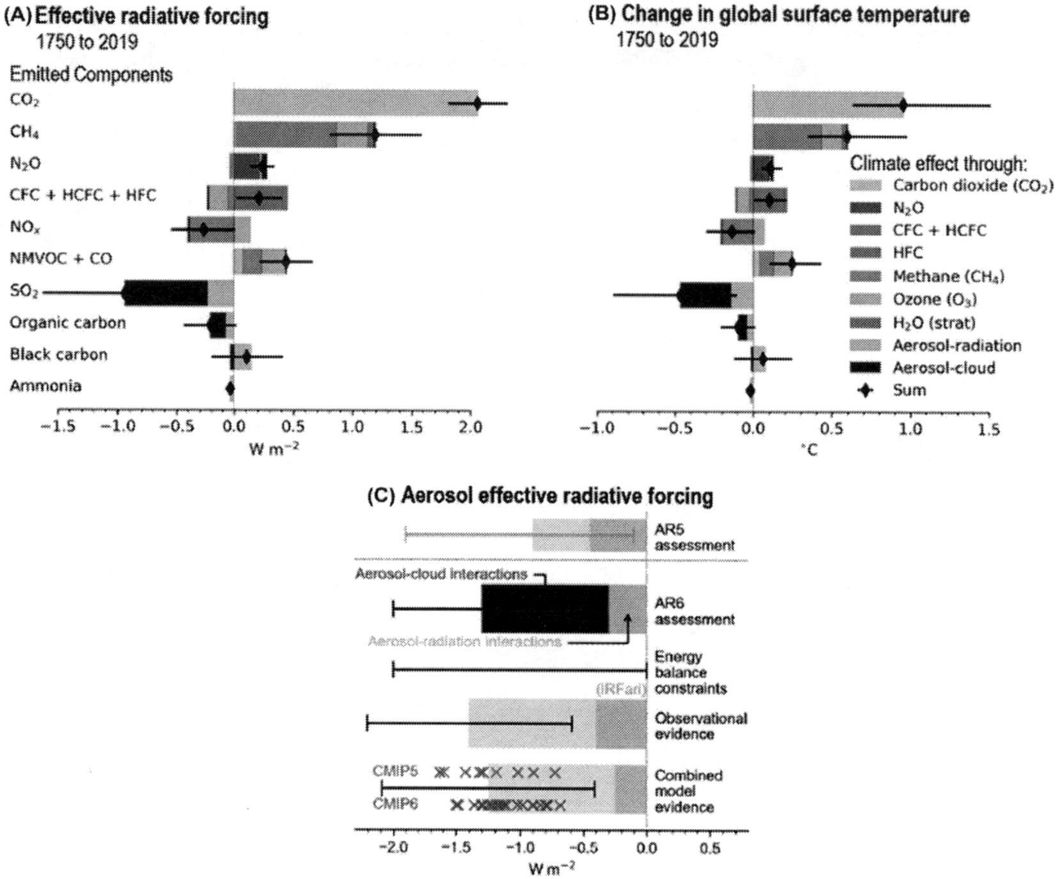

FIGURE 4.8 Effective radiative forcing (A), change in global surface temperature (B), and aerosol effective radiative forcing (C) presented in the last IPCC report. *Data from https://www.ipcc.ch/report/ar6/wg1/downloads/report/IPCC_AR6_WGI_TS.pdf*

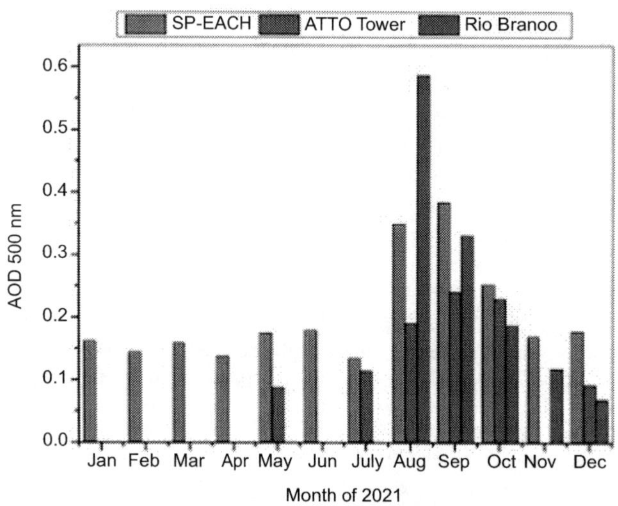

FIGURE 4.9 Time series of aerosol optical depth (AOD) 500 nm for the year 2021 in three different locations in Brazil. *Data from https://aeronet.gsfc.nasa.gov/*

4.8 Stratospheric chemistry

The chemistry of the stratosphere has a main chemical gas trace species concentration, the ozone expressed in Dobson Unit (DU = 2.60×10^{16} molecules cm^{-2}), which is formed from oxygen photolysis under the action of ultraviolet radiation (less than 242 nm) that arrives at the top of the atmosphere by solar radiation.

$$O = O + h\nu \rightarrow O + O(\lambda < 242 \text{ nm}) \tag{4.63}$$

$\rightarrow O + O^*$ ($\lambda < 175$ nm), O^* is the oxygen atom in an excited state.

$$O + O_2 + (M) \rightarrow O_3 + (M) \tag{4.64}$$

The ozone formed is destroyed by the interaction with ultraviolet radiation and other natural and anthropogenic stratospheric chemical species, being the balance between the formation and destruction in the stratosphere ozone layer (approximately 90% of the total in the atmosphere) around 25 km of altitude (Seinfeld & Pandis, 2016). The location of the ozone layer at the stratosphere is a result of the balance between the flux of photons that arrives to dissociate oxygen and the O_2 concentrations. While the latter decrease with altitude, the flux of photons increases with it. However, the distribution of O_3 is inequal around the globe because meteorological condition is not found at the same altitude at different altitudes.

The reaction with UV is fundamental for the existence of life on the Earth since ozone absorbs the more energetic UV radiation (220–280 nm and a large fraction until 320 nm) and dissociates it.

$$O_3 + h\nu \rightarrow O + O_2(\lambda < 320 \text{ nm}) \tag{4.65}$$

The higher production rates of ozone are found in the tropics (near the Equator), while the highest stratospheric ozone concentration is found in the middle-high latitudes due to the availability of radiation to photolyze O_2. Then, vertical and horizontal air masses transport redistributes these concentrations in air masses through the globe. Therefore there are asymmetries in relation to the location and time of ozone abundance and rate of production around the globe, with peaks in the north hemisphere at high latitudes in the winter, while in the south hemisphere it is at middle latitudes in springtime and around 15–20 km altitude (Seinfeld & Pandis, 2016).

Besides the destruction of ozone by interaction with UV radiation, hydrogen radicals, nitrogen oxides, and halogens are important ingredients in stratospheric chemistry and ozone chemistry. The reactions are summarized as:

$$X + O_3 \rightarrow XO + O_2 \tag{4.66}$$

$$XO + O \rightarrow X + O_2, \tag{4.67}$$

with X being HO_x, NO_x, Cl, and Br. The reactions are referred to as the cycle of HO_x ($OH + HO_2$), NO_x ($NO + NO_2$), ClO_x, and BrO_x, that are coupled to and destroy O_3 (Farman et al., 1985; Molina & Rowland, 1974). These cycles termination occurs when the reactive species participate in other reactions (competition reactions) and become relatively more stable. To illustrate the reaction, we can take the NO_x cycle as an example. The origin of NO_x in the stratosphere is mainly due to N_2O emission, which is stable enough to arrive until the stratosphere, and thereafter is photolyzed and destroyed, providing NO:

$$N_2O + h\nu \rightarrow N_2 + O^* \tag{4.68}$$

Around 90% is destroyed and the remainder reacts with excited oxygen.

$$N_2O + O^* \rightarrow 2\,NO \tag{4.69}$$

The NO provided by N_2O reacts with O_3 in the cycle as:

$$NO + O_3 \rightarrow NO_2 + O_2 \tag{4.70}$$

$$NO_2 + O \rightarrow NO + O_2 \tag{4.71}$$

The net result is the destruction of $O_3 + O \rightarrow 2O_2$. The action of NO reactive species finishes when NO_2 is photolyzed in the competition reaction with O (reaction 4.71). Then, if reaction (4.70) is followed by reaction (4.72), the action to destroy O_3 is null because the O atom formed in reaction (4.72) rapidly combines with O_2 to re-form O_3 in reaction (4.73) (Seinfeld & Pandis, 2016).

$$NO_2 + h\nu \rightarrow NO + O \tag{4.72}$$

$$O + O_2 + M \rightarrow O_3 + M \tag{4.73}$$

Another two examples of reactions that break the cycles are:

$$NO_2 + \cdot OH + M \rightarrow HNO_3 + M \tag{4.74}$$

$$Cl + CH_4 \rightarrow HCl + CH_3\cdot \tag{4.75}$$

The acids are then removed from the stratosphere by solubilization in water clouds and further precipitation.

The depletion of stratospheric ozone naturally occurs by reaction in the cycles of HO_x, NO_x, and ClO_x, since the species which are responsible for these in the stratosphere have natural sources on the surface. The significant issue involving the ozone layer is the large number of chemical species emitted by human activities that are sources of those reactive species that catalytically destroy ozone (Fig. 4.10).

Chipperfield et al. (2017) discussed the results of the Montreal Protocol and its amendments to the ozone layer. They pointed out that a decrease in anthropogenic contribution sources to ozone depletion may lead to an expected ozone recovery. However, the influence of other factors, mainly due to the increase of greenhouse gases that have been changing the atmosphere and climate, poses a complexity in this recovery estimation (Son et al., 2009; Yook et al., 2022). Another important aspect is that the chemical species responsible to destroy ozone have a long atmospheric lifetime, which implies a delay in the recovery of it.

4.9 Questions

1. Fig. 4.11 shows the concentrations of NO, NO_2, and O_3 on three consecutive days (Friday, Saturday, and Sunday) at a given location in the city of São Paulo, Brazil. Explain why the peaks occur at different times. Can the day of the week influence ozone concentrations? How? Data source: https://cetesb.sp.gov.br/ar/qualar/
2. At the link https://qualar.cetesb.sp.gov.br/qualar/home.do you can find air quality and meteorological data for several cities in Brazil. Fig. 4.12 shows PM_{10} concentrations and meteorological parameters obtained by an air quality station for June, July, and August 2019 for a region of the city of São Paulo (Congonhas): PM_{10} ($\mu g\, m^{-3}$), Precipitation—Precip (mm), Temperature—T (°C), Relative Humidity—RH (%), Wind Speed—WS ($m\, s^{-1}$). Notice graphically that some parameters are inversely proportional. Use the raw data and apply Pearson's Correlation to show these correlations statistically.
3. In some regions, there is the occurrence of an ozone nocturnal peak. Why does this occur?
4. Elaborate on a diagram the secondary formation of organic aerosols due to vegetation emissions of isoprene and terpenes.
5. Contributions to the carbon fraction in aerosol samples are very variable all over the world due to the specificities of each region. Enumerate the main sources of organic and elemental carbon on each continent.

Solutions to the end-of-chapter questions will be available at: https://www.elsevier.com/books-and-journals/book-companion/9780128225912.

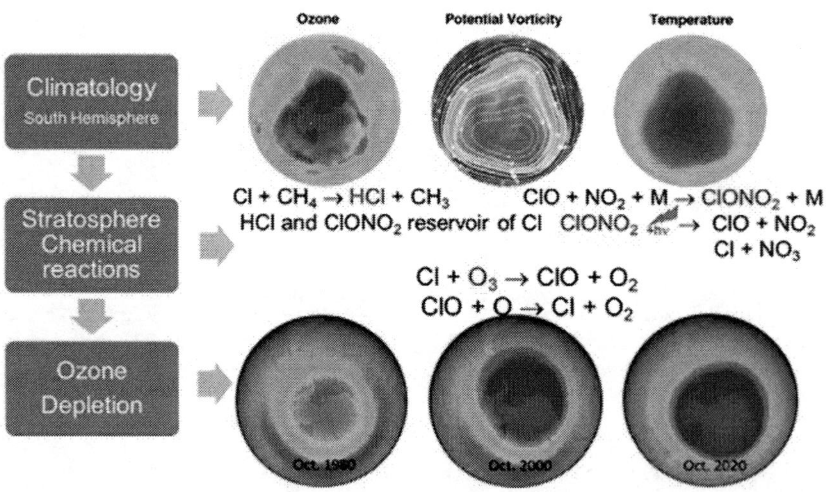

FIGURE 4.10 Schematic stratospheric ozone mechanism. *Ozone maps from Nasa Ozone Watch at https://ozonewatch.gsfc.nasa.gov/*

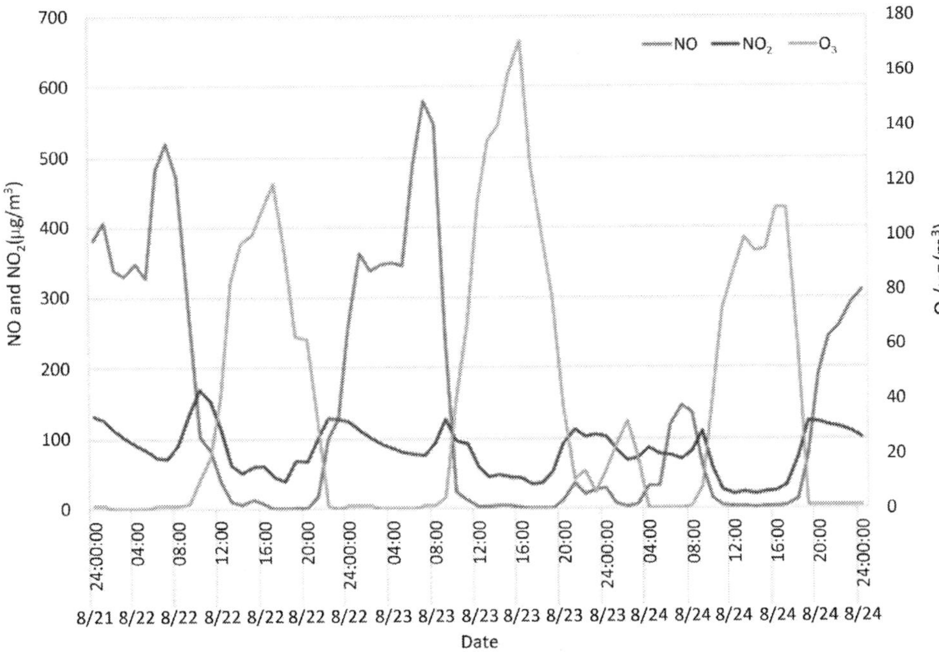

FIGURE 4.11 Concentrations of NO, NO_2, and O_3 on three consecutive days (Friday, Saturday, and Sunday) at a given location in the city of São Paulo. *Data from CETESB (Companhia de Tecnologia e Saneamento Ambiental do Estado de São Paulo) air quality station.*

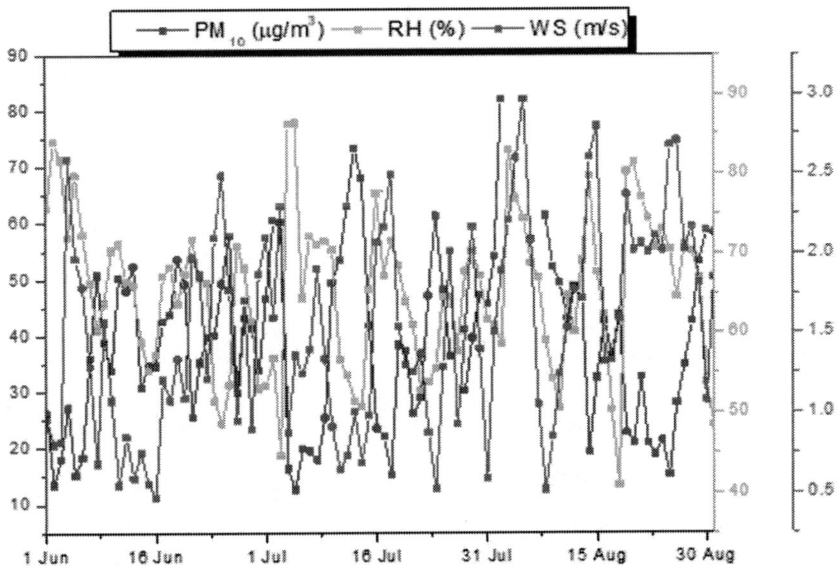

FIGURE 4.12 PM_{10} concentrations and meteorological parameters obtained by an air quality station for June, July, and August 2019 for a region of the city of São Paulo (Congonhas): PM_{10} ($\mu g\,m^{-3}$), Precipitation—Precip (mm), Temperature—T (°C), Relative Humidity—RH (%), Wind Speed—WS ($m\,s^{-1}$). *Data from CETESB (Companhia de Tecnologia e Saneamento Ambiental do Estado de São Paulo) air quality station.*

References

Abel, R.C., Coney, K., Johnson, C., Thornton, M.J., Zigler, B.T., & McCormick, R.L. (2021). *Global ethanol-blended-fuel vehicle compatibility study* (No. NREL/TP-5400-81252). National Renewable Energy Lab.(NREL), Golden, CO (United States). Available from: https://doi.org/10.2172/1832216.

Akimoto, H., Nagashima, T., Tanimoto, H., Klimont, Z., & Amann, M. (2020). An empirical approach toward the SLCP reduction targets in Asia for the mid-term climate change mitigation. *Progress in Earth and Planetary Science*, *7*(1). Available from https://doi.org/10.1186/S40645-020-00385-5.

Alves, C. A., Gomes, J., Nunes, T., Duarte, M., Calvo, A., Custódio, D., Pio, C., Karanasiou, A., & Querol, X. (2015). Size-segregated particulate matter and gaseous emissions from motor vehicles in a road tunnel. *Atmospheric Research*, *153*, 134–144. Available from https://doi.org/10.1016/j.atmosres.2014.08.002, http://www.elsevier.com/inca/publications/store/5/0/3/3/2/3/index.htt.

Apel, E. C., Emmons, L. K., Karl, T., Flocke, F., Hills, A. J., Madronich, S., Lee-Taylor, J., Fried, A., Weibring, P., Walega, J., Richter, D., Tie, X., Mauldin, L., Campos, T., Weinheimer, A., Knapp, D., Sive, B., Kleinman, L., Springston, S., ... Riemer, D. D. (2010). Chemical evolution of volatile organic compounds in the outflow of the Mexico City metropolitan area. *Atmospheric Chemistry and Physics*, *10*(5), 2353–2376. Available from https://doi.org/10.5194/acp-10-2353-2010, http://www.atmos-chem-phys.net/volumes_and_issues.html.

Artaxo, P., Hansson, H.-C., Andreae, M. O., Bäck, J., Alves, E. G., Barbosa, H. M. J., Bender, F., Bourtsoukidis, E., Carbone, S., Chi, J., Decesari, S., Després, V. R., Ditas, F., Ezhova, E., Fuzzi, S., Hasselquist, N. J., Heintzenberg, J., Holanda, B. A., Guenther, A., Hakola, H., Heikkinen, L., Kerminen, V.-M., Kontkanen, J., Krejci, R., Kulmala, M., Lavric, J. V., de Leeuw, G., Lehtipalo, K., Machado, L. A. T., McFiggans, G., Franco, M. A. M., Meller, B. B., Morais, F. G., Mohr, C., Morgan, W., Nilsson, M. B., Peichl, M., Petäjä, T., Praß, M., Pöhlker, C., Pöhlker, M. L., Pöschl, U., Von Randow, C., Riipinen, I., Rinne, J., Rizzo, L. V., Rosenfeld, D., Silva Dias, M. A. F., Sogacheva, L., Stier, P., Swietlicki, E., Sörgel, M., Tunved, P., Virkkula, A., Wang, J., Weber, B., Yáñez-Serrano, A. M., Zieger, P., Mikhailov, E., Smith, J. N., & Kesselmeier, J. (2022). Tropical and Boreal Forest – Atmosphere Interactions: A Review. *Tellus B: Chemical and Physical Meteorology*, *74*(2022), 24–163. Available from https://doi.org/10.16993/tellusb.34.

Atkinson, R. (2000). Atmospheric chemistry of VOCs and NO_x. *Atmospheric Environment*, *34*, 2063–2101. Available from https://doi.org/10.1016/S1352-2310(99)00460-4.

Badia, A., Iglesias-Suarez, F., Fernandez, R. P., Cuevas, C. A., Kinnison, D. E., Lamarque, J.-F., et al. (2021). The role of natural halogens in global tropospheric ozone chemistry and budget under different 21st century climate scenarios. *Journal of Geophysical Research: Atmospheres*, *126*e2021JD034859. Available from https://doi.org/10.1029/2021JD034859.

Barber, J. L., Thomas, G. O., Kerstiens, G., & Jones, K. C. (2004). Current issues and uncertainties in the measurement and modelling of air-vegetation exchange and within-plant processing of POPs. *Environmental Pollution*, *128*(1-2)), 99–138. Available from https://doi.org/10.1016/j.envpol.2003.08.024, http://www.elsevier.com/inca/publications/store/4/0/5/8/5/6.

Barreira, L. M. F., Ylisirniö, A., Pullinen, I., Buchholz, A., Li, Z., Lipp, H., et al. (2021). A synthesis review on atmospheric wet deposition of particulate elements: Scavenging ratios, solubility, and flux measurements. *Environmental Reviews*, *29*(3), 849–864. Available from https://doi.org/10.1007/s13280-019-01244-4.

Batista, C. E., Ye, J., Ribeiro, I. O., Guimarães, P. C., Medeiros, A. S. S., Barbosa, R. G., Oliveira, R. L., Duvoisin, S., Jardine, K. J., Gu, D., Guenther, A. B., McKinney, K. A., Martins, L. D., Souza, R. A. F., & Martinc, S. T. (2019). Intermediate-scale horizontal isoprene concentrations in the near-canopy forest atmosphere and implications for emission heterogeneity. *Proceedings of the National Academy of Sciences of the United States of America*, *116*(39), 19318–19323. Available from https://doi.org/10.1073/pnas.1904154116, https://www.pnas.org/content/pnas/116/39/19318.full.pdf.

Behera, S. N., Sharma, M., Aneja, V. P., & Balasubramanian, R. (2013). Ammonia in the atmosphere: A review on emission sources, atmospheric chemistry and deposition on terrestrial bodies. *Environmental Science and Pollution Research*, *20*(11), 8092–8131. Available from https://doi.org/10.1007/s11356-013-2051-9.

Bond, T. C., Doherty, S. J., Fahey, D. W., Forster, P. M., Berntsen, T., DeAngelo, B. J., ... Zender, C. S. (2013). Bounding the role of black carbon in the climate system: A scientific assessment. *Journal of geophysical research: Atmospheres*, *118*(11), 5380–5552. Available from https://doi.org/10.1002/jgrd.50171.

Borbon, A., Gilman, J. B., Kuster, W. C., Grand, N., Chevaillier, S., Colomb, A., Dolgorouky, C., Gros, V., Lopez, M., Sarda-Esteve, R., Holloway, J., Stutz, J., Petetin, H., McKeen, S., Beekmann, M., Warneke, C., Parrish, D. D., & Gouw, J. A. D. (2013). Emission ratios of anthropogenic volatile organic compounds in northern mid-latitude megacities: Observations versus emission inventories in Los Angeles and Paris. *Journal of Geophysical Research Atmospheres*, *118*(4), 2041–2057. Available from https://doi.org/10.1002/jgrd.50059, http://onlinelibrary.wiley.com/journal/10.1002/(ISSN)2169-8996.

Cheng, I., Al Mamun, A., & Zhang, L. (2021). A synthesis review on atmospheric wet deposition of particulate elements: Scavenging ratios, solubility, and flux measurements. *Environmental Reviews*, *29*(3), 340–353. Available from https://doi.org/10.1139/er-2020-0118, https://www.nrcresearchpress.com/loi/er.

Chipperfield, M. P., Bekki, S., Dhomse, S., Harris, N. R. P., Hassler, B., Hossaini, R., Steinbrecht, W., Thiéblemont, R., & Weber, M. (2017). Detecting recovery of the stratospheric ozone layer. *Nature*, *549*(7671), 211–218. Available from https://doi.org/10.1038/nature23681, http://www.nature.com/nature/index.html.

Claeys, M., & Maenhaut, W. (2021). Secondary Organic Aerosol Formation from Isoprene: Selected Research, Historic Account and State of the Art. *Atmosphere*, *12*, 728. Available from https://doi.org/10.3390/atmos12060728.

Colbeck, I., & Lazaridis, M. (2014). *Aerosol science. Technology and applications*. John Wiley & Sons.

Da Silva, C. M., Da Silva, L. L., Corrêa, S. M., & Arbilla, G. (2017). Speciation analysis of ozone precursor volatile organic compounds in the air basins of the Rio de Janeiro metropolitan area. *Revista Virtual de Quimica*, *9*(5), 1887–1909. Available from https://doi.org/10.21577/1984-6835.20170111, http://rvq.sbq.org.br/audiencia_pdf.asp?aid2 = 807&nomeArquivo = v9n5a08.pdf.

de Miranda, R. M., Lopes, F., do Rosário, N. É., Yamasoe, M. A., Landulfo, E., & de Fatima Andrade, M. (2017). The relationship between aerosol particles chemical composition and optical properties to identify the biomass burning contribution to fine particles concentration: A case study for São Paulo city, Brazil. *Environmental Monitoring and Assessment*, *189*(1). Available from https://doi.org/10.1007/s10661-016-5659-7.

Demerjian, K. L., Schere, K. L., & Peterson, J. (1980). *Theoretical estimates of actinic (spherically integrated) flux and photolytic rate constants of atmospheric species in the lower troposphere. Advances in environmental science and technology*. Wiley. Available from https://books.google.com.br/books?id=xUdQYAAACAAJ.

Dominutti, P. A., Nogueira, T., Borbon, A., Andrade, M. D. F., & Fornaro, A. (2016). One-year of NMHCs hourly observations in São Paulo megacity: Meteorological and traffic emissions effects in a large ethanol burning context. *Atmospheric Environment*, *142*, 371–382. Available from https://doi.org/10.1016/j.atmosenv.2016.08.008, http://www.elsevier.com/locate/atmosenv.

Dominutti, P., Nogueira, T., Fornaro, A., & Borbon, A. (2020). One decade of VOCs measurements in São Paulo megacity: Composition, variability, and emission evaluation in a biofuel usage context. *Science of the Total Environment*, *738*. Available from https://doi.org/10.1016/j.scitotenv.2020.139790, http://www.elsevier.com/locate/scitotenv.

Donahue, N. M., Henry, K. M., Mentel, T. F., Kiendler-Scharr, A., Spindler, C., Bohn, B., Brauers, T., Dorn, H. P., Fuchs, H., Tillmann, R., Wahner, A., Saathoff, H., Naumann, K. H., Möhler, O., Leisner, T., Müller üMüller, L., Reinnig, M. C., Hoffmann, T., Salo, K., … Baltensperger, U. (2012). Aging of biogenic secondary organic aerosol via gas-phase OH radical reactions. *Proceedings of the National Academy of Sciences of the United States of America*, *109*(34), 13503–13508. Available from https://doi.org/10.1073/pnas.1115186109, http://www.pnas.org/content/109/34/13503.full.pdf + html.

Farman, J. C., Gardiner, B. G., & Shanklin, J. D. (1985). Large losses of total ozone in Antarctica reveal seasonal ClOx/NOx interaction. *Nature*, *315* (6016), 207–210. Available from https://doi.org/10.1038/315207a0.

Finlayson-Pitts, B. J., & Pitts, J. N., Jr (1999). *Chemistry of the upper and lower atmosphere: theory, experiments, and applications*. Elsevier.

Finlayson-Pitts, B. J., & Pitts, J. N. (2000). *Chemistry of the Upper and Lower Atmosphere*. Academic Press ISBN 9780122570605. Available from. Available from https://doi.org/10.1016/B978-0-12-257060-5.50026-5.

Grennfelt, P., Engleryd, A., Forsius, M., Hov, Ø. Ø., Rodhe, H., & Cowling, E. (2020). Acid rain and air pollution: 50 years of progress in environmental science and policy. *Ambio*, *49*(4), 849–864. Available from https://doi.org/10.1007/s13280-019-01244-4, http://www.bioone.org/loi/ambi.

Gu, B., Sutton, M. A., Chang, S. X., Ge, Y., & Chang, J. (2014). Agricultural ammonia emissions contribute to China's urban air pollution. *Frontiers in Ecology and the Environment*, *12*(5), 265–266. Available from https://doi.org/10.1890/14.WB.007, http://www.esajournals.org/doi/pdf/10.1890/14.WB.007.

Gu, C., Wang, S., Zhu, J., Wu, S., Duan, Y., Gao, S., & Zhou, B. (2022). Investigation on the urban ambient isoprene and its oxidation processes. *Atmospheric Environment*, *270*, 118870. Available from https://doi.org/10.1016/j.atmosenv.2021.118870.

Hamra, G. B., Guha, N., Cohen, A., Laden, F., Raaschou-Nielsen, O., Samet, J. M., … Loomis, D. (2014). Outdoor particulate matter exposure and lung cancer: a systematic review and meta-analysis. *Environmental health perspectives*. Available from https://doi.org/10.1289/ehp.1408092.

Heald, C. L., & Kroll, J. H. (2020). The fuel of atmospheric chemistry: Toward a complete description of reactive organic carbon. *Science Advances*, *6*(6). Available from https://doi.org/10.1126/sciadv.aay8967, https://advances.sciencemag.org/content/6/6/EAAY8967/tab-pdf.

Hendrey, G. R. (2001). *Acid rain and deposition* (pp. 1–15). Elsevier BV. Available from 10.1016/b0-12-226865-2/00001-8.

Holben, B. N., Eck, T. F., Slutsker, I., Tanré, D., Buis, J. P., Setzer, A., Vermote, E., Reagan, J. A., Kaufman, Y. J., Nakajima, T., Lavenu, F., Jankowiak, I., & Smirnov, A. (1998). AERONET - a federated instrument network and data archive for aerosol characterization. *Remote Sensing of Environment*, *66*(1), 1–16. Available from https://doi.org/10.1016/S0034-4257(98)00031-5.

Ianniello, A., Spataro, F., Esposito, G., Allegrini, I., Hu, M., & Zhu, T. (2011). Chemical characteristics of inorganic ammonium salts in PM2.5 in the atmosphere of Beijing (China). *Atmospheric Chemistry and Physics*, *11*(21), 10803–10822. Available from https://doi.org/10.5194/acp-11-10803-2011.

IARC International Agency for Research on Cancer. (2016). *Outdoor Air Pollution*. Estados Unidos: International Agency for Research on Cancer, World Health Organization.

IPCC (2013). *The physical science basis. Contribution of Working Group I to the Fifth Assessment Report of the Intergovernmental Panel on Climate Change* [Stocker TF, Qin D, Plattner GK, Tignor M, Allen SK, Boschung J, Nauels A, Xia Y, Bex V, Midgley PM (eds)]. Cambridge University Press, Cambridge/New York. Available from: http://www.climatechange2013.org/images/report/WG1AR5_TS_FINAL.pdf.

Jacobson, M. C., Hansson, H. C., Noone, K. J., & Charlson, R. J. (2000). Organic atmospheric aerosols: Review and state of the science. *Reviews of Geophysics*, *38*(2), 267–294. Available from https://doi.org/10.1029/1998RG000045.

Jacobson, M. Z. (2002). *Atmospheric pollution: History, science, and regulation*. Cambridge University Press.

Jimenez, J. L., Canagaratna, M. R., Donahue, N. M., Prevot, A. S. H., Zhang, Q., Kroll, J. H., DeCarlo, P. F., Allan, J. D., Coe, H., Ng, N. L., Aiken, A. C., Docherty, K. S., Ulbrich, I. M., Grieshop, A. P., Robinson, A. L., Duplissy, J., Smith, J. D., Wilson, K. R., Lanz, V. A., … Worsnop, D. R. (2009). Evolution of organic aerosols in the atmosphere. *Science*, *326*(5959), 1525–1529. Available from https://doi.org/10.1126/science.1180353.

Kelly, J. M., Doherty, R. M., O'Connor, F. M., & Mann, G. W. (2018). The impact of biogenic, anthropogenic, and biomass burning volatile organic compound emissions on regional and seasonal variations in secondary organic aerosol. *Atmospheric Chemistry and Physics*, *18*(10), 7393–7422. Available from https://doi.org/10.5194/acp-18-7393-2018, http://www.atmos-chem-phys.net/volumes_and_issues.html.

Kinnersley, R. P., Shaw, G., Bell, J. N. B., Minski, M. J., & Goddard, A. J. H. (1996). Loss of particulate contaminants from plant canopies under wet and dry conditions. *Environmental Pollution*, *91*(2), 227–235. Available from https://doi.org/10.1016/0269-7491(95)00047-X, http://www.elsevier.com/inca/publications/store/4/0/5/8/5/6.

Li, X., Zhao, B., Zhou, W., Shi, H., Yin, R., Cai, R., Yang, D., Dällenbach, K., Deng, C., Fu, Y., Qiao, X., Wang, L., Liu, Y., Yan, C., Kulmala, M., Zheng, J., Hao, J., Wang, S., & Jiang, J. (2020). Responses of gaseous sulfuric acid and particulate sulfate to reduced SO2 concentration: A perspective from long-term measurements in Beijing. *Science of the Total Environment, 721*, 137700. Available from https://doi.org/10.1016/j.scitotenv.2020.137700.

Likens, G. E. (2021). *Acid rain. Fundamentals of ecosystem science* (2nd ed., pp. 293–299). United States: Elsevier. Available from https://www.sciencedirect.com/book/9780128127629, 10.1016/B978-0-12-812762-9.00016-2.

Liu, Y., He, G., Chu, B., Ma, Q., & He, H. J. (2023). Atmospheric heterogeneous reactions on soot: A review. *Fundamental Research, 3*, 579–591. Available from https://doi.org/10.1016/j.fmre.2022.02.012, ISSN 2667-3258.

Lowe, S., Partridge, D. G., Topping, D., & Stier, P. (2016). Inverse modelling of Köhler theory - part 1: A response surface analysis of CCN spectra with respect to surface-active organic species. *Atmospheric Chemistry and Physics, 16*(17), 10941–10963. Available from https://doi.org/10.5194/acp-16-10941-2016, http://www.atmos-chem-phys.net/volumes_and_issues.html.

Lukas, M., Schwidetzky, R., Eufemio, R. J., Bonn, M., & Meister, K. (2022). Toward understanding bacterial ice nucleation. *Journal of Physical Chemistry B, 126*(9), 1861–1867. Available from https://doi.org/10.1021/acs.jpcb.1c09342, http://pubs.acs.org/journal/jpcbfk.

Ma, Z., Liu, C., Zhang, C., Liu, P., Ye, C., Xue, C., Zhao, D., Sun, J., Du, Y., Chai, F., & Mu, Y. (2019). The levels, sources and reactivity of volatile organic compounds in a typical urban area of Northeast China. *Journal of Environmental Sciences, 79*, 121–134. Available from https://doi.org/10.1016/j.jes.2018.11.015.

Madronich, S., & Flocke, S. (1999). *The role of solar radiation in atmospheric chemistry* (pp. 1–26). Springer Science and Business Media LLC. Available from 10.1007/978-3-540-69044-3_1.

Michelsen, H. A., Colket, M. B., Bengtsson, P. E., D'Anna, A., Desgroux, P., Haynes, B. S., Miller, J. H., Nathan, G. J., Pitsch, H., & Wang, H. (2020). A review of terminology used to describe soot formation and evolution under combustion and pyrolytic conditions. *ACS Nano, 14*(10), 12470–12490. Available from https://doi.org/10.1021/acsnano.0c06226, http://pubs.acs.org/journal/ancac3.

Molina, M. J., & Rowland, F. S. (1974). Stratospheric sink for chlorofluoromethanes: Chlorine atomc-atalysed destruction of ozone. *Nature, 249* (5460), 810–812. Available from https://doi.org/10.1038/249810a0.

Monks, P. S., Archibald, A. T., Colette, A., Cooper, O., Coyle, M., Derwent, R., Fowler, D., Granier, C., Law, K. S., Mills, G. E., Stevenson, D. S., Tarasova, O., Thouret, V., von Schneidemesser, E., Sommariva, R., Wild, O., & Williams, M. L. (2015). Tropospheric ozone and its precursors from the urban to the global scale from air quality to short-lived climate forcer. *Atmospheric Chemistry and Physics, 15*(15), 8889–8973. Available from https://doi.org/10.5194/acp-15-8889-2015.

Morais, F. G., Franco, M. A., Palácios, R., Machado, L. A., Rizzo, L. V., Barbosa, H. M., … Artaxo, P. (2022). Relationship between land use and spatial variability of atmospheric brown carbon and black carbon aerosols in Amazonia. *Atmosphere, 13*(8), 1328. Available from https://doi.org/10.3390/atmos13081328.

Mouli, P. C., Mohan, S. V., & Reddy, S. J. (2003). A study on major inorganic ion composition of atmospheric aerosols at Tirupati. *Journal of Hazardous Materials, 96*(2-3), 217–228. Available from https://doi.org/10.1016/S0304-3894(02)00214-5.

Mu, Z., Llusià, J., Zeng, J., Zhang, Y., Asensio, D., Yang, K., Yi, Z., Wang, X., & Peñuelas, J. (2022). An overview of the isoprenoid emissions from tropical plant species. *Frontiers in Plant Science, 13*. Available from https://doi.org/10.3389/fpls.2022.833030, https://www.frontiersin.org/journals/plant-science.

Nevalainen, P., Kinnunen, N. M., Kirveslahti, A., Kallinen, K., Maunula, T., Keenan, M., & Suvanto, M. (2018). Formation of NH3 and N2O in a modern natural gas three-way catalyst designed for heavy-duty vehicles: The effects of simulated exhaust gas composition and ageing. *Applied Catalysis A: General, 552*, 30–37. Available from https://doi.org/10.1016/j.apcata.2017.12.017, http://www.sciencedirect.com/science/journal/0926860X.

Ng, N. L., Canagaratna, M. R., Zhang, Q., Jimenez, J. L., Tian, J., Ulbrich, I. M., Kroll, J. H., Docherty, K. S., Chhabra, P. S., Bahreini, R., Murphy, S. M., Seinfeld, J. H., Hildebrandt, L., Donahue, N. M., DeCarlo, P. F., Lanz, V. A., Prévôt, A. S. H., Dinar, E., Rudich, Y., & Worsnop, D. R. (2010). Organic aerosol components observed in northern hemispheric datasets from aerosol mass spectrometry. *Atmospheric Chemistry and Physics, 10*(10), 4625–4641. Available from https://doi.org/10.5194/acp-10-4625-2010.

Nogueira, T., Dominutti, P., Fornaro, A., & Andrade, M. (2017). Seasonal trends of formaldehyde and acetaldehyde in the megacity of São Paulo. *Atmosphere, 8*(8), 144. Available from https://doi.org/10.3390/atmos8080144.

Pacheco, R., & Silva, C. (2019). Global warming potential of biomass-to-ethanol: Review and sensitivity analysis through a case study. *Energies, 12* (13), 2535. Available from https://doi.org/10.3390/en12132535.

Paulot, F., Crounse, J., Kjaergaard, H. G., Kurten, A., Clair, J. M. S., Seinfeld, J. H., & Wennberg, P. O. (2009). Unexpected Epoxide Formation in the Gas-Phase Photooxidation of Isoprene. *Science, 325*, 730–733. Available from https://doi.org/10.1126/science.1172910.

Pereira, G. M., da Silva Caumo, S. E., Grandis, A., do Nascimento, E. Q. M., Correia, A. L., de Melo Jorge Barbosa, H., Marcondes, M. A., Buckeridge, M. S., & de Castro Vasconcellos, P. (2021). Physical and chemical characterization of the 2019 "black rain" event in the Metropolitan Area of São Paulo, Brazil. *Atmospheric Environment, 248*. Available from https://doi.org/10.1016/j.atmosenv.2021.118229, http://www.elsevier.com/locate/atmosenv.

Pereira, G. M., Nogueira, T., Kamigauti, L. Y., Monteiro dos Santos, D., Nascimento, E. Q. M., Martins, J. V., Vicente, A., Artaxo, P., Alves, C., de Castro Vasconcellos, P., & de Fatima Andrade, M. (2023). Particulate matter fingerprints in biofuel impacted tunnels in South America's largest metropolitan area. *Science of the Total Environment, 856*. Available from https://doi.org/10.1016/j.scitotenv.2022.159006, http://www.elsevier.com/locate/scitotenv.

Pope, C. A., III, Burnett, R. T., Thun, M. J., Calle, E. E., Krewski, D., Ito, K., & Thurston, G. D. (2002). Lung cancer, cardiopulmonary mortality, and long-term exposure to fine particulate air pollution. *Jama, 287*(9), 1132–1141. Available from https://doi.org/10.1001/jama.287.9.1132.

Pöschl, U. (2005). Atmospheric aerosols: Composition, transformation, climate and health effects. *Angewandte Chemie*, *44*(46), 7520–7540. Available from https://doi.org/10.1002/anie.200501122.

Pöschl, U., & Shiraiwa, M. (2015). Multiphase chemistry at the atmosphere-biosphere interface influencing climate and public health in the anthropocene. *Chemical Reviews*, *115*(10), 4440–4475. Available from https://doi.org/10.1021/cr500487s, http://pubs.acs.org/journal/chreay.

Quigg, A., Santschi, P. H., Burd, A., Chin, W. C., Kamalanathan, M., Xu, C., & Ziervogel, K. (2021). From nano-gels to marine snow: A synthesis of gel formation processes and modeling efforts involved with particle flux in the ocean. *Gels*, *7*(3). Available from https://doi.org/10.3390/gels7030114, https://www.mdpi.com/2310-2861/7/3/114/pdf.

Seinfeld, J. H., & Pandis, S. N. (2016). *Atmospheric chemistry and physics: from air pollution to climate change* (Third edition). Hoboken, New Jersey: John Wiley & Sons, Inc.

Seinfeld, J. H., Bretherton, C., Carslaw, K. S., Coe, H., DeMott, P. J., Dunlea, E. J., Feingold, G., Ghan, S., Guenther, A. B., Kahn, R., Kraucunas, I., Kreidenweis, S. M., Molina, M. J., Nenes, A., Penner, J. E., Prather, K. A., Ramanathan, V., Ramaswamy, V., Rasch, P. J., ... Wood, R. (2016). Improving our fundamental understanding of the role of aerosol-cloud interactions in the climate system. *Proceedings of the National Academy of Sciences of the United States of America*, *113*(21), 5781–5790. Available from https://doi.org/10.1073/pnas.1514043113, http://www.pnas.org/content/113/21/5781.full.pdf.

Senosiain, J. P., Klippenstein, S. J., & Miller, J. A. (2005). The reaction of acetylene with hydroxyl radicals. *J Phys Chem A*, *109*(27), 6045–6055. Available from https://doi.org/10.1021/jp050737g, PMID: 16833940.

Sitarski, M. A. (2012). Brownian coagulation of knudsen particles on evaporating/condensing aerosol droplet; modeling of capture of soot particles by Marine fog. *Aerosol and Air Quality Research*, *12*(6), 1189–1199. Available from https://doi.org/10.4209/aaqr.2012.05.0140, http://aaqr.org/VOL12_No6_December2012/15_AAQR-12-05-OA-0140_1189-1199.pdf.

Son, S. W., Tandon, N. F., Polvani, L. M., & Waugh, D. W. (2009). Ozone hole and Southern Hemisphere climate change. *Geophysical Research Letters*, *36*(15). Available from https://doi.org/10.1029/2009GL038671, http://onlinelibrary.wiley.com/journal/10.1002/(ISSN)1944-8007/issues?year = 2012.

Stewart, G. J., Nelson, B. S., Drysdale, W. S., Acton, W. J. F., Vaughan, A. R., Hopkins, J. R., Dunmore, R. E., Hewitt, C. N., Nemitz, E., Mullinger, N., Langford, B., Reyes-Villegas, S. E., Gadi, R., Rickard, A. R., Lee, J. D., & Hamilton, J. F. (2021). Sources of non-methane hydrocarbons in surface air in Delhi, India. *Faraday Discussions*, *226*, 409–431. Available from https://doi.org/10.1039/d0fd00087f, http://pubs.rsc.org/en/journals/journal/fd.

Sun, J., & Ariya, P. A. (2006). Atmospheric organic and bio-aerosols as cloud condensation nuclei (CCN): A review. *Atmospheric Environment*, *40*(5), 795–820. Available from https://doi.org/10.1016/j.atmosenv.2005.05.052.

Tapia, A., Salgado, S., Martín, P., Villanueva, F., García-Contreras, R., & Cabañas, B. (2018). Chemical composition and heterogeneous reactivity of soot generated in the combustion of diesel and GTL (Gas-to-Liquid) fuels and amorphous carbon Printex U with NO_2 and CF_3COOH gases. *Atmospheric Environment*, *177*, 214–221. Available from https://doi.org/10.1016/j.atmosenv.2018.01.028.

Tilgner, A., Schaefer, T., Alexander, B., Barth, M., Collett, J. L., Fahey, K. M., Nenes, A., Pye, H. O. T., Herrmann, H., & McNeill, V. F. (2021). Acidity and the multiphase chemistry of atmospheric aqueous particles and clouds. *Atmospheric Chemistry and Physics*, *21*(17), 13483–13536. Available from https://doi.org/10.5194/acp-21-13483-2021, http://www.atmos-chem-phys.net/volumes_and_issues.html.

Tipton, M. J., Lathem, T. L., Fu, J. S., & Tschantz, M. F. (2022). Effectiveness of emissions standards on automotive evaporative emissions in Europe under normal and extreme temperature conditions. *Environmental Research Communications*, *4*(8), 081003. Available from https://doi.org/10.1088/2515-7620/ac8b69.

UN Environment (2019). *Global Environment Outlook – GEO-6: Healthy Planet, Healthy People*. Nairobi. Available from: https://doi.org/10.1017/9781108627146.

Vasseghian, Y., Hosseinzadeh, S., Khataee, A., & Dragoi, E.-N. (2021). The concentration of persistent organic pollutants in water resources: A global systematic review, meta-analysis and probabilistic risk assessment. *Science of the Total Environment*, *796*, 149000. Available from https://doi.org/10.1016/j.scitotenv.2021.149000.

Veras, M. M., Caldini, E. G., Dolhnikoff, M., & Saldiva, P. H. N. (2010). Air pollution and effects on reproductive-system functions globally with particular emphasis on the Brazilian population. *Journal of Toxicology and Environmental Health, Part B*, *13*(1), 1–15. Available from https://doi.org/10.1080/10937401003673800.

Vet, R., Artz, R. S., Carou, S., Shaw, M., Ro, C. U., Aas, W., Baker, A., Bowersox, V. C., Dentener, F., Galy-Lacaux, C., Hou, A., Pienaar, J. J., Gillett, R., Forti, M. C., Gromov, S., Hara, H., Khodzher, T., Mahowald, N. M., Nickovic, S., Rao, P. S. P., & Reid, N. W. (2014). A global assessment of precipitation chemistry and deposition of sulfur, nitrogen, sea salt, base cations, organic acids, acidity and pH, and phosphorus. *Atmospheric Environment*, *93*, 3–100. Available from https://doi.org/10.1016/j.atmosenv.2013.10.060, http://www.elsevier.com/locate/atmosenv.

Wang, L., Slowik, J. G., Tong, Y., Duan, J., Gu, Y., Rai, P., Qi, L., Stefenelli, G., Baltensperger, U., Huang, R. J., Cao, J., & Prévôt, A. S. H. (2021). Characteristics of wintertime VOCs in urban Beijing: Composition and source apportionment. *Atmospheric Environment: X*, *9*. Available from https://doi.org/10.1016/j.aeaoa.2020.100100, https://www.journals.elsevier.com/atmospheric-environment-x.

Wang, Y., Zhuang, G., Sun, Y., & An, Z. (2006). The variation of characteristics and formation mechanisms of aerosols in dust, haze, and clear days in Beijing. *Atmospheric Environment*, *40*(34), 6579–6591. Available from https://doi.org/10.1016/j.atmosenv.2006.05.066.

Wells, K. C., Millet, D. B., Payne, V. H., Deventer, M. J., Bates, K. H., de Gouw, J. A., Graus, M., Warneke, C., Wisthaler, A., & Fuentes, J. D. (2020). Satellite isoprene retrievals constrain emissions and atmospheric oxidation. *Nature*, *585*(7824), 225–233. Available from https://doi.org/10.1038/s41586-020-2664-3, http://www.nature.com/nature/index.html.

Whitby, K. T. (1967). The physical characteristics of sulfur aerosols. *Atmospheric Environment*, 90196–90198. Available from https://doi.org/10.1016/0004-6981(78.

WHO (World Health Organization). (2018a). *Burden of disease from the joint effects of household and ambient Air pollution for 2016*. Geneva; https://www.who.int/airpollution/data/AP_joint_effect_BoD_results_May2018.pdf?ua = 1.

World Health Organization. (2018b). *Air pollution and child health: prescribing clean air: summary* (No. WHO/CED/PHE/18.01). World Health Organization.

World Health Organization. (2021). *Review of evidence on health aspects of air pollution: REVIHAAP project: technical report* (No. WHO/EURO: 2013-4101-43860-61757). World Health Organization. Regional Office for Europe.

Wu, X., Nethery, R. C., Sabath, M. B., Braun, D., & Dominici, F. (2020). Exposure to air pollution and COVID-19 mortality in the United States: A nationwide cross-sectional study. *medRxiv*. Available from https://doi.org/10.1101/2020.04.05.20054502, https://www.medrxiv.org/.

Xuan, C., Xiaoran, S., Zhaoji, S., Jiaen, Z., Zhong, Q., Huimin, X., & Hui, W. (2021). Analysis of the spatio-temporal changes in acid rain and their causes in China (1998–2018). *Journal of Resources and Ecology*, *12*(5). Available from https://doi.org/10.5814/j.issn.1674-764x.2021.05.002.

Yook, S., Thompson, D. W. J., & Solomon, S. (2022). Climate impacts and potential drivers of the unprecedented Antarctic ozone holes of 2020 and 2021. *Geophysical Research Letters*, *49*(10). Available from https://doi.org/10.1029/2022GL098064, http://agupubs.onlinelibrary.wiley.com/hub/journal/10.1002/(ISSN)1944-8007/.

Young, P. J., Naik, V., Fiore, A. M., Gaudel, A., Guo, J., Lin, M. Y., Neu, J. L., Parrish, D. D., Rieder, H. E., Schnell, J. L., Tilmes, S., Wild, O., Zhang, L., Ziemke, J., Brandt, J., Delcloo, A., Doherty, R. M., Geels, C., Hegglin, M. I., … Zeng, G. (2018). Tropospheric ozone assessment report: Assessment of global-scale model performance for global and regional ozone distributions, variability, and trends. *Elementa*, *6*. Available from https://doi.org/10.1525/elementa.265, https://www.elementascience.org/articles/.

Yu, D., Wang, L., Zhang, C., Peng, C., Yu, X., Fan, X., … Zhao, Z. (2022). Alkali metals and cerium-modified La–Co-based perovskite catalysts: facile synthesis, excellent catalytic performance, and reaction mechanisms for soot combustion. *ACS Catalysis*, *12*(24), 15056–15075. Available from https://doi.org/10.1021/acscatal.2c03418.

Zhang, R., Khalizov, A., Wang, L., Hu, M., & Xu, W. (2012). Nucleation and growth of nanoparticles in the atmosphere. *Chemical Reviews*, *112*(3), 1957–2011. Available from https://doi.org/10.1021/cr2001756.

Zhang, Y., Favez, O., Canonaco, F., Liu, D., Močnik, G., Amodeo, T., Sciare, J., Prévôt, A. S. H., Gros, V., & Albinet, A. (2018). Evidence of major secondary organic aerosol contribution to lensing effect black carbon absorption enhancement. *NPJ Climate and Atmospheric Science*, *1*(1). Available from https://doi.org/10.1038/s41612-018-0056-2, https://www.nature.com/npjclimatsci/.

Further reading

Andrade, M. D. F., Kumar, P., de Freitas, E. D., Ynoue, R. Y., Martins, J., Martins, L. D., Nogueira, T., Perez-Martinez, P., de Miranda, R. M., Albuquerque, T., Gonçalves çGonçalves, F. L. T., Oyama, B., & Zhang, Y. (2017). Air quality in the megacity of São Paulo: Evolution over the last 30 years and future perspectives. *Atmospheric Environment*, *159*, 66–82. Available from https://doi.org/10.1016/j.atmosenv.2017.03.051, http://www.elsevier.com/locate/atmosenv.

Bell, M., Davis, D. L., & Fletcher, T. (2004). A retrospective assessment of mortality from the London smog episode of 1952: The role of influenza and pollution. *Environmental Health Perspectives*, *112*(1), 6–8. Available from https://doi.org/10.1289/ehp.6539, http://ehp.niehs.nih.gov/.

Fowler, D., Brimblecombe, P., Burrows, J., Heal, M. R., Grennfelt, P., Stevenson, D. S., Jowett, A., Nemitz, E., Coyle, M., Lui, X., Chang, Y., Fuller, G. W., Sutton, M. A., Klimont, Z., Unsworth, M. H., & Vieno, M. (2020). A chronology of global air quality: The development of global air pollution. *Philosophical Transactions of the Royal Society A: Mathematical, Physical and Engineering Sciences*, *378*(2183). Available from https://doi.org/10.1098/rsta.2019.0314, http://rsta.royalsocietypublishing.org/.

Monks, P. S. (2005). Gas-phase radical chemistry in the troposphere. *Chemical Society Reviews*, *34*(5), 376. Available from https://doi.org/10.1039/b307982c.

Saldiva, P. H. N., Lichtenfels, A. J. F. C., Paiva, P. S. O., Barone, I. A., Martins, M. A., Massad, E., Pereira, J. C. R., Xavier, V. P., Singer, J. M., & Bohm, G. M. (1994). Association between air pollution and mortality due to respiratory diseases in children in São Paulo, Brazil: A preliminary report. *Environmental Research*, *65*(2), 218–225. Available from https://doi.org/10.1006/enrs.1994.1033.

Shim, C., Li, Q., Luo, M., Kulawik, S., Worden, H., Worden, J., Eldering, A., Diskin, G., Sachse, G., Weinheimer, A., Knapp, D., Montzca, D., & Campos, T. (2009). Satellite observations of Mexico City pollution outflow from the Tropospheric Emissions Spectrometer (TES). *Atmospheric Environment*, *43*(8), 1540–1547. Available from https://doi.org/10.1016/j.atmosenv.2008.11.026.

Slanina, J. (2004). *Air pollution from energy production and use* (pp. 39–54). Elsevier BV. Available from 10.1016/b0-12-176480-x/00387-9.

Trasande, L. (2021). Council on health, ambient air pollution: Health hazards to children. *Pediatrics*, *147*(6). Available from https://doi.org/10.1542/peds.2021-051484.

WHO (World Health Organization). (1958). Fifth report of the expert committee on environmental sanitation.

Zhang, F., Peng, J., Chen, L., Collins, D., Li, Y., Jiang, S., Liu, J., & Zhang, R. (2022). The effect of black carbon aging from NO2 oxidation of SO2 on its morphology, optical and hygroscopic properties. *Environmental Research*, *212*, 113238. Available from https://doi.org/10.1016/j.envres.2022.113238.

Chapter 5

Atmospheric dynamics and pollution transport

Helen F. Dacre
Department of Meteorology, University of Reading, Reading, United Kingdom

5.1 Atmospheric dynamics

5.1.1 Primitive equations

The atmospheric dynamics driving the transport processes described in Fig. 5.1 are governed by a set of physical laws, known as the primitive equations. These equations describe the conservation of momentum, the conservation of mass and the conservation of thermodynamic energy in the atmosphere. These equations will be referred to throughout the chapter to describe the dominant balances and thermodynamics acting in different transport processes so they are introduced before describing the pollution transport processes.

5.1.1.1 Conservation of momentum

The momentum equations describe the conservation of momentum in the atmosphere. They are obtained by applying Newton's second law of motion, which states that the total force acting per unit mass is equal to the acceleration produced. For meteorological purposes the forces acting on an air parcel include the pressure gradient force, the gravitational force and frictional forces. Since the motion of an air parcel is typically referred to in a coordinate system that is rotating with the earth then curvature terms and an additional apparent force, known as the Coriolis force, is also included in the equations.

The Coriolis force, $\boldsymbol{F}_c$, is the apparent force experienced by an air parcel due to the earth's rotation with respect to fixed axes on earth. There is no force acting on the air parcel, but if an air parcel moves, then viewed by an observer on the Earth's surface, an apparent force acting on the parcel is necessary to explain its motion. The Coriolis force is given by

$$\boldsymbol{F}_c = -2\boldsymbol{\Omega} \times \boldsymbol{V}$$

The Coriolis force acts at right angles to the earth's axes of rotation ($\boldsymbol{\Omega}$) and the wind velocity vector ($\boldsymbol{V}$). In meteorology, the vertical component of the Coriolis force is small compared to the gravitational force and vertical pressure gradient force. For horizontal air motion ($w = 0$), the horizontal Coriolis force is given by

$$\boldsymbol{F}_c = 2\Omega \mathbf{V} \sin\phi = f\boldsymbol{V}$$

where Ω is the angular velocity of the earth, ϕ is latitude and f is known as the Coriolis parameter. Thus, air moving in the northern hemisphere appears to be deflected to the right by the Coriolis force and in the southern hemisphere appears to be deflected to the left, that is, in a frame of reference fixed on the rotating earth, the air has horizontal acceleration is in these respective directions.

Air Quality. DOI: https://doi.org/10.1016/B978-0-12-822591-2.00005-6

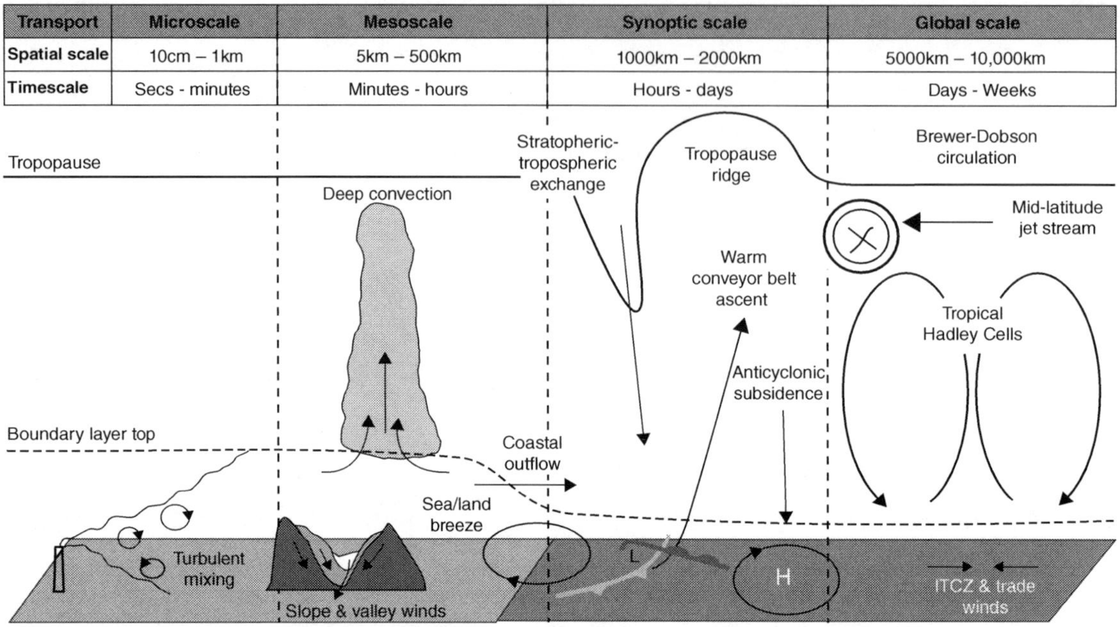

FIGURE 5.1 Atmospheric processes. Schematic of meteorological transport processes occurring on a range of spatial and temporal scales.

If we assume vertical motions are much smaller than horizontal motions ($w \ll u, v$) then the momentum equations are

$$\frac{Du}{Dt} = fv - \frac{1}{\rho}\frac{\partial p}{\partial x} + F_x$$

$$\frac{Dv}{Dt} = -fu - \frac{1}{\rho}\frac{\partial p}{\partial y} + F_y$$

$$\frac{Dw}{Dt} = -\frac{1}{\rho}\frac{\partial p}{\partial z} - g + F_z$$

where ρ is the density of air, g is the acceleration due to gravity, p is pressure, $\boldsymbol{F}$ is the frictional force acting in three-dimensions and D/Dt is the acceleration following the horizontal motion of a fluid parcel.

5.1.1.2 Conservation of mass

The continuity equation describes the conservation of mass in the atmosphere. It states that the rate at which mass enters a volume is equal to the rate at which mass leaves the volume, plus the accumulation of mass within the volume. The continuity equation following the horizontal motion of an air parcel is given by

$$\frac{1}{\rho}\frac{D\rho}{Dt} + \nabla \cdot V = 0$$

where ρ is the air density, t is time, $\boldsymbol{V}$ is the velocity vector. The first term is the rate of change of density following the motion of an air parcel and the second term is the divergence of mass from the volume.

5.1.1.3 Conservation of energy

The equation to describe the conservation of energy in a volume is obtained by applying the first law of thermodynamics for an ideal gas. The thermodynamic equation states that the rate of change of temperature following the horizontal motion of a parcel is due to the sum of the net diabatic heating by sources such as latent heat release and solar or thermal radiative heating (J) plus the adiabatic expansion or compression due to vertical motion.

$$\frac{DT}{Dt} = \frac{J}{c_p} + \frac{R_d}{c_p}\frac{Dp}{Dt}$$

where c_p is the specific heat at constant pressure, R_d is the gas constant of dry air, $\frac{DT}{Dt}$ is the rate of change of temperature following the fluid parcel and $\frac{Dp}{Dt}$ is the rate of change of pressure following the fluid parcel. Thus, the thermodynamic energy equation relates temperature, density and pressure and describes how they vary with heating.

5.2 Microscale transport

This section describes microscale circulations in the atmosphere and their impact on pollution transport. Microscale transport occurs on horizontal scales of centimetres to a kilometer and timescales of a few seconds to minutes (Fig. 5.1). Chapter 6 gives a detailed explanation boundary layer turbulence so this section will only give a brief overview of pollution transport by microscale turbulence.

5.2.1 Turbulence

Turbulent motions are generated by two mechanisms: mechanically driven turbulence and thermally driven turbulence. Mechanically driven turbulence occurs wherever there is vertical or horizontal wind shear. In the planetary boundary layer, mechanical turbulence is always present because the Earth's surface is rough. When winds blow over the Earth's surface it encounters these roughness elements (grass, trees, buildings, etc.). The roughness elements exert a force on the wind due to friction which slows it down near the surface relative to the winds aloft. This creates vertical wind shear which leads to mixing in the atmosphere. The larger the wind shear the greater the mixing.

The other mechanism that generates mixing in the atmosphere is surface heating which leads to thermally driven turbulence. Thermally driven turbulence on the microscale typically occurs in the atmospheric boundary layer during the day when the Sun causes the earth's surface to heat up. The air near the surface heats up relative to the air aloft causing the atmospheric boundary layer to become unstable (see Section 5.2.1.1 for a mathematical explanation of atmospheric stability). Positively buoyant air parcels rise through the atmospheric boundary layer since they are warmer than the surrounding air leading to vertical mixing. The opposite can occur at night. When the Sun goes down the Earth's surface starts to cool down. The air near the surface cools down relative to the air aloft causing the atmospheric boundary layer to become stable. This acts to suppress vertical mixing due to turbulence. Thus, the stability of the atmospheric boundary layer can either suppress or produce turbulence.

The balance between the mechanical production of turbulence and thermal production or suppression of turbulence influences the transport of pollutants in the atmospheric boundary layer on timescales of seconds to minutes. For example, if we consider pollution emitted from a chimney stack then the pollution plume can exhibit very different behavior depending on the environmental conditions into which the pollution is emitted. If pollution is emitted into a convectively stable atmosphere the stability will damp any vertical motions so the pollution plume will be prevented from ascending or descending and will only spread out in the horizontal (known as a fanning plume). Alternatively, if the pollution is emitted into a convectively unstable atmosphere any vertical motions will be enhanced by thermally generated turbulence causing the plume to spread out rapidly in both the horizontal and vertical directions (known as a looping plume).

5.2.1.1 Atmospheric stability

Adiabatic processes are reversible processes in which no heat is exchanged with the surroundings ($J = 0$). Thus, from the thermodynamic energy equation (Section 5.1.1.3) and using the ideal gas law ($p = \rho R_d T$),

$$0 = \frac{dT}{T} - \frac{R_d}{c_p}\frac{dp}{p}$$

Integrating this expression from a state at pressure p and temperature T to a state in which the pressure is p_s and temperature θ we obtain:

$$\int_T^\theta \frac{1}{T}dT = \frac{R_d}{c_p}\int_p^{p_s}\frac{1}{p}dp$$

$$\theta = T\left(\frac{p_s}{p}\right)^{R_d/c_p}$$

where θ is called the potential temperature and is the temperature that a parcel of dry air at pressure p and temperature T would have it were expanded or compressed adiabatically to a standard pressure p_s.

The stability of the atmosphere is primarily determined by the vertical temperature profile, known as the lapse rate, Γ. Taking the logarithm of the potential temperature, differentiating with respect to height and using the ideal gas law gives

$$\frac{T}{\theta}\frac{\partial\theta}{\partial z} = \frac{\partial T}{\partial z} + \frac{g}{c_p}$$

For adiabatic motions a dry parcel will conserve its potential temperature ($\partial\theta/\partial z = 0$), thus the adiabatic lapse rate, Γ_d, is the rate at which a rising or descending dry air parcel cools or warms due to adiabatic processes and is given by the equation:

$$\Gamma_d = -\frac{dT}{dz} = \frac{g}{c_p}$$

where dT is the change in air temperature, dz is the change in height above the ground, g is the acceleration due to gravity and c_p is the specific heat capacity of dry air at a constant pressure. On earth, c_p is approximately 1000 $\text{J kg}^{-1}\text{K}^{-1}$ and g is approximately 9.81 m s^{-2} giving a dry adiabatic lapse rate of approximately 9.8 K km^{-1}. That is, dry air parcels will cool by 9.8 K for every 1000 m they ascend.

When the temperature of an unsaturated environment decreases with height at a rate slower than the adiabatic lapse rate, the atmosphere is considered to be stable ($\Gamma < \Gamma_d$). Conversely, if the air temperature decreases with a height at a rate faster than the adiabatic lapse rate, the atmosphere is unstable ($\Gamma > \Gamma_d$). When the temperature decreases with height at the same rate as the dry adiabatic lapse rate then the atmosphere has neutral stability ($\Gamma = \Gamma_d$).

5.2.2 Diurnal evolution of boundary layer turbulence

Much of the pollution in the atmosphere originates from emissions in the atmospheric boundary layer. The boundary layer is a region of the atmosphere closest to the Earth's surface where the properties of air parcels are strongly modified by the presence of the earth's surface on time scales of an hour or less. The stability of the boundary layer is closely tied to the diurnal cycle.

The evolution of atmospheric boundary layer turbulence can be described as follows.

1. During the daytime, heating of the surface increases the air temperature close to the surface, resulting in an unstable atmosphere. The lowest few 100 m of the atmosphere therefore becomes well mixed by thermally driven mixing. Turbulent mixing works rapidly within an hour and establishes what's known as the convective mixed layer. At the top of the convective mixed layer, there's often a capping inversion (stable layer), where the temperature increases with height. This stable layer prevents air from being easily mixed from the boundary layer into the free atmosphere.
2. As the sun sets, thermally driven turbulence decreases, and the atmosphere can become stable. If winds are weak then mechanically driven turbulence is also small, and a shallow stable layer capped by a temperature inversion typically develops near to the surface overnight.
3. After sunrise, heating of the surface and thermally generated turbulence gradually erode the stable atmosphere from below until the unstable daytime profile is re-established.

5.2.2.1 *Worked example*

Question:: How will the evolution of the boundary layer influence air pollution levels near the Earth's surface?

Solution: The typical diurnal cycle of the boundary layer depth is shallower at night and deeper during the daytime. Thus, pollutants emitted at night tend to be trapped near the surface, which can lead to high concentrations. Pollutants tend to be mixed throughout a large volume during the daytime which can lead to a reduction in concentrations, assuming constant emissions. However, in winter, when solar heating is weak, breaking of the night-time inversion is difficult and the accumulation of pollutants may result in severe air pollution episodes.

5.2.3 Summary

To summarize this section on microscale transport processes. Turbulent transport of pollution occurs on timescales of seconds to minutes and spatial scales of a few centimetres to a kilometer. Within the atmospheric boundary layer, friction at the Earth's surface generates turbulence which mixes pollution vertically away from their sources. During the daytime, this mixing can be enhanced by thermally generated turbulence which mixes pollution throughout the boundary layer. At night, surface cooling can result in a stable boundary layer which suppresses turbulence and can result in the accumulation of pollution near the surface.

5.3 Mesoscale transport

This section describes mesoscale circulations in the atmosphere and their impact on pollution transport. Mesoscale meteorology ranges from a few kilometers to several hundred kilometers in scale and occurs over time scales ranging from minutes to hours. Examples of mesoscale weather systems include coastal winds, moist convection, slope and valley winds (Fig. 5.1). Mesoscale circulations often occur during high pressure and weak wind situations, allowing locally driven thermal circulations to dominate the flow and any resulting pollution transport.

5.3.1 Land and sea breezes

Land and sea breezes are examples of mesoscale thermal circulations occurring on a diurnal timescale. Thermal circulations typically occur due to horizontal differences in the diurnal heating and cooling of the earth's surface. For example, a sharp thermal contrast can occur at coastlines due to the different thermal properties of land and sea. When coastal land is warmed by the sun it quickly becomes warmer than the adjacent coastal marine air, especially if there's no large-scale advection to smooth out the temperature gradient. This is because the sea surface temperature does not change by more than a few degrees during the day and night, whereas the land temperature can vary by more than 10 degrees. The horizontal temperature gradient leads to a thermal circulation. During the daytime, this thermal circulation is known as a sea breeze, and it leads to onshore winds at the surface. At night the reverse circulation occurs, known as a land breeze, leading to offshore air flow at the surface (Fig. 5.2).

So how are land and sea breeze circulations created? During the day, heating over land causes the air above it to become warmer. As a result of the warming, it becomes less dense and rises. The air above the sea is cooler and thus denser than the air over the land. This density difference sets up a pressure gradient, with higher pressure over the sea and lower pressure over the land (Fig. 5.2A). The denser air from over the sea surface is then advected towards the lower pressure area over the land, creating a sea breeze (see Section 5.3.1.1 for a mathematical explanation of the pressure gradient force). The cooler air from the sea replaces the rising warm air over the land. The cold and moist air from over the sea warms as it is advected over the land and starts to rise, sometimes forming cumulus clouds. A weak return flow occurs aloft which cools and descends over the sea completing in the sea breeze circulation.

During the night-time, the thermal circulation is reversed. Air over the land cools relative to air over the sea resulting in higher pressure over land than over the sea near the surface (Fig. 5.2b). Thus, an offshore flow from land to the sea. Aloft the reverse occurs resulting in onshore flow. Usually, the night-time land breeze is weaker and shallower than the daytime sea breeze circulation. These thermal circulations are typically only a few hundred meters deep. The head of the circulation is known as the sea breeze or land breeze front. We can detect the passage of these fronts by

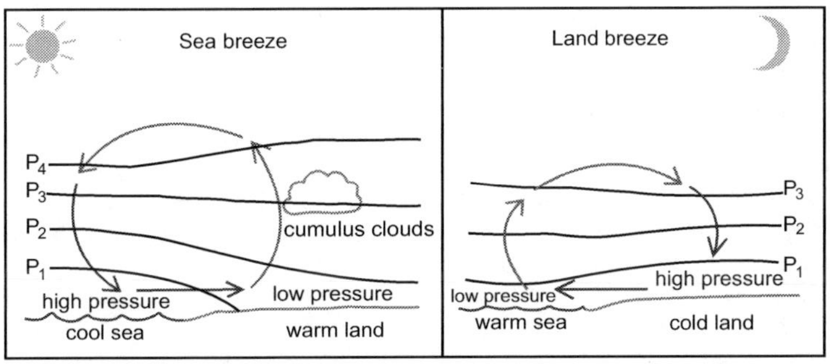

FIGURE 5.2 Sea breeze circulation (*left*) and land breeze circulation (*right*).

measuring changes in wind direction, temperature, humidity, and gustiness. Sea breezes can have a considerable effect on coastal regions, lowering temperatures and producing clouds and rain.

In places such as Israel, bordering the Mediterranean, sea breezes are a daily occurrence. Ashdod is a port town on the edge of the Mediterranean (Fig. 5.3A). The coast lies approximately south-west to north-east (Fig. 5.3B). We can examine the meteorological conditions during the passage of a sea breeze front. Fig. 5.3C–F shows the change in temperature, relative humidity, wind direction and wind speed on the 1st of April 2004. The time series start at 6:00 a.m. local time and finish at 2:00 p.m. local time. There is a gradual increase in temperature through the morning until around 10:00 a.m. when temperatures start to drop and then level out at around 22°C. The levelling off in temperature coincides with a sharp increase in relative humidity from 40% to 55% and a large change in wind from South-easterlies to North-westerlies. This is almost a 180-degree shift in the wind direction occurring over a few minutes. There is also a gradual increase in wind speed over the same period. Ashdod is on the coast of Israel where the coastline is orientated South-west to North-east. Thus, the change in wind direction indicates a change from offshore winds during the night to onshore winds during the morning. Combined, these meteorological variables indicate the passage of a sea breeze front over the meteorological station at 10:00 a.m. local time.

Worked example

Question: Examine Fig. 5.3G and determine what affect the sea-breeze has on SO_2 concentrations at Ashdod, Israel?

Solution: Fig. 5.3G shows the average diurnal evolution of SO_2 concentrations at Ashdod, Israel. It is shown that during the summer, initially the SO_2 concentrations are low, around 2 ppb. Between 7:00 and 11:00 a.m. the SO_2 concentrations increase, reaching a peak of 18 ppb at 11:00 a.m. local time. After this time the SO_2 concentrations decrease. Similarly, in the winter, SO_2 concentrations are low overnight, but start to increase slightly later than in the summer reaching a peak at around 12:00 p.m. local time. Since Ashdod is on the coast of the Mediterranean and is a port town, the sea air is heavily polluted by SO_2 emitted from ships in the port. As the wind changes from offshore to onshore this polluted air is advected onshore leading to increased SO_2 concentrations inland.

5.3.1.1 Pressure gradient force

When there is a difference in pressure along a surface, the resulting force is known as the pressure gradient force, $\boldsymbol{F}_{\text{PGF}}$. In the vertical direction, the pressure gradient force is balanced by the gravitational force (known as hydrostatic balance, see Section 5.3.3.1). In the horizontal directions, the pressure gradient force per unit mass is described by the following equation:

$$\boldsymbol{F}_{\text{PGF}} = -\frac{1}{\rho}\nabla_H p$$

where ρ is air density, p is atmospheric pressure, ∇_H is the del operator. In Cartesian coordinates $\nabla_H = \left(\frac{\partial}{\partial x}, \frac{\partial}{\partial y}\right)$. The pressure gradient forces a horizontal acceleration of the wind in the direction from high-pressure areas to low-pressure areas.

5.3.2 Coastal outflow

Coastal outflow occurs on a diurnal timescale and is caused by the thermal contrast between the land and the sea. During the night-time the land surface can cool faster than the sea surface. The boundary layer thus becomes stable over the land surface which suppresses turbulence and leads to a very shallow boundary layer depth. Over the sea, the atmosphere is often less stable, so the boundary layer over the sea is slightly deeper than that over land. As the sun rises the land typically heats up more quickly than the sea so the air above the land becomes unstable and the boundary layer starts to deepen more quickly than the boundary layer over the sea, which remains fairly constant in depth. If pollution is emitted at the surface over the land then this pollution will get mixed to a greater height than that over the marine boundary layer (Fig. 5.4). If there's an offshore wind direction then this polluted air can be advected above the marine boundary layer. Because there's typically a temperature inversion at the top of the marine boundary layer, the polluted air in this layer is prevented from being mixed down to the surface (Fig. 5.4B). Thus, pollutants are prevented from reaching the surface and, as a result, layers of polluted air at elevations between 500 m and 2 km from the surface have been observed hundreds of kilometers downwind from coastlines. Coastal outflow is a very efficient mechanism for transporting pollution long distances, and it exerts control on surface concentrations in many highly populated coastal regions, transporting pollution emitted over land offshore (Peake et al., 2014). As a result of the bounding by stable layers above and below, polluted layers have been observed to reach the Azores and even Europe and traced back to their origins over North America (e.g., Methven et al., 2006) using their chemical fingerprint combined with calculation of air mass back trajectories.

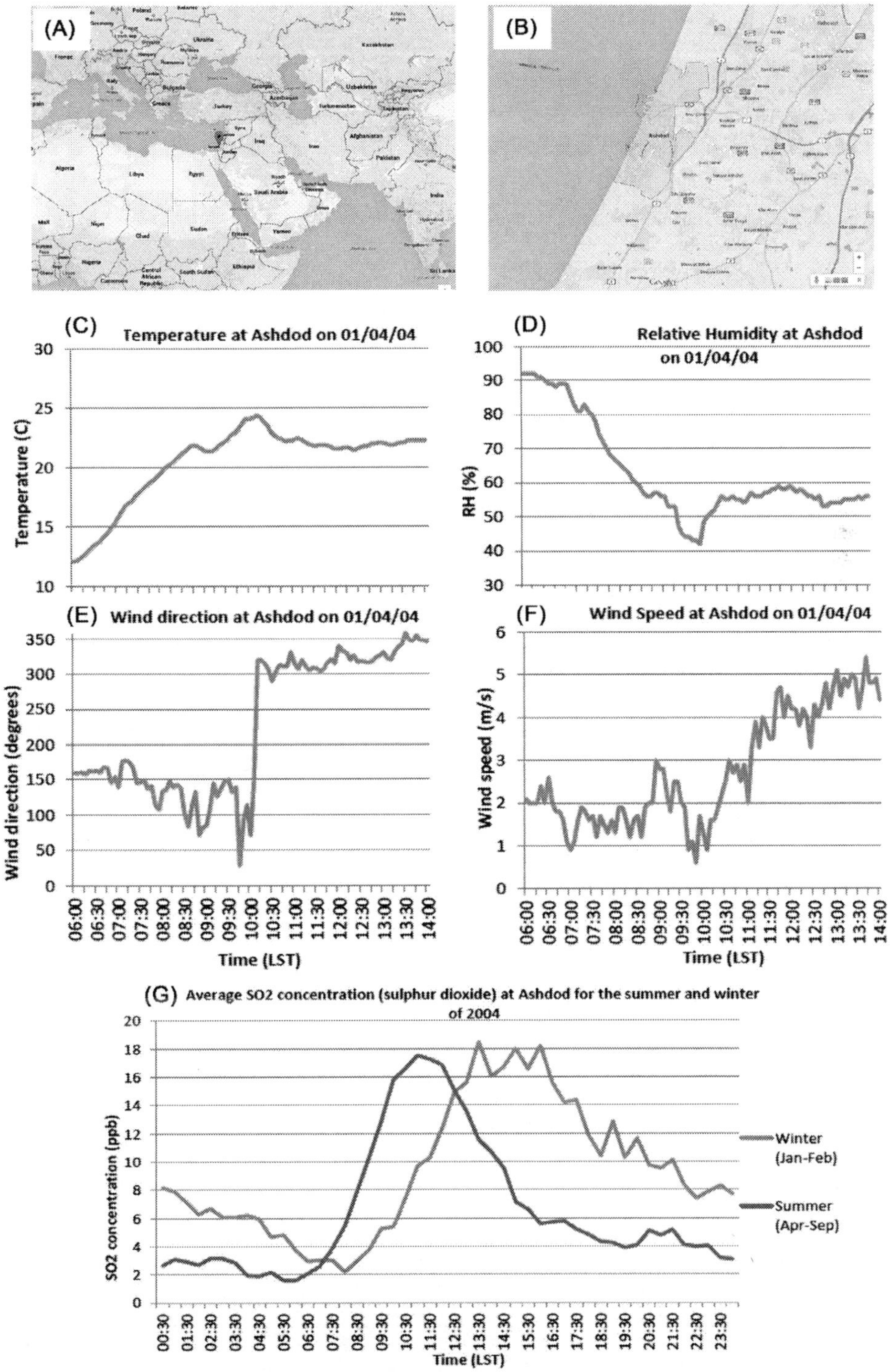

FIGURE 5.3 Sea breeze example Case study of a sea breeze passage at Ashdod, Israel. (A) Map of the Mediterranean showing location of Israel (red pin), (B) map of the coastline of Israel showing location of port of Ashdod (red shading). Timeseries of meteorological variables from 06:00 local time (LST) to 14:00 LST on April 1, 2004, (C) 2 m temperature, (D) relative humidity, (E) wind direction, (F) wind speed. (G) Seasonal average timeseries of SO_2 concentrations from 00:00 LST to 24:00 LST at Ashdod for summer 2004 (April–September) and winter 2004 (January–February). Figures produced by M. Pronger.

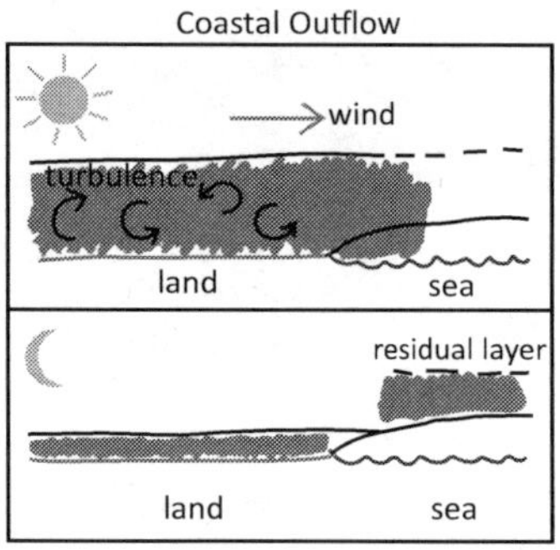

FIGURE 5.4 Schematic of coastal outflow. (*top*) During the daytime, pollution (gray) is mixed by turbulence in the land boundary layer and advected offshore. (*bottom*) During the nightime, pollution in the marine boundary layer may be deposited to the surface leaving a layer of pollution in a residual layer aloft.

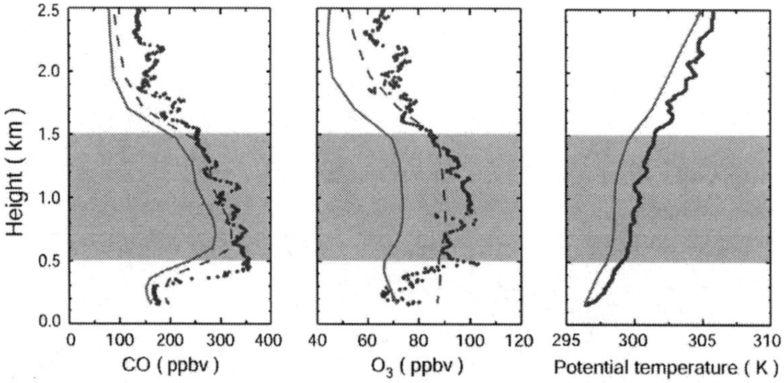

FIGURE 5.5 Coastal pollution layer. Profiles of CO, O_3 and potential temperature approximately 100 km offshore from New York City (NYC) taken at 18:45 UTC on July 20, 2004 during the Intercontinental Transport and Chemical Transformation (ITCT) field campaign. Observations (dots purple line) and simulated (solid red and dashed blue lines). Blue shading shows NYC urban plume layer between 0.5 and 1.5 km (Lee et al., 2011). *Figure from Lee, S.H., Kim, S.W., Trainer, M., Frost, G.J., McKeen, S.A., Cooper, O.R., Flocke, F., Holloway, J.S., Neuman, J.A., Ryerson, T., Senff, C.J., Swanson, A.L., & Thompson, A.M. (2011). Modeling ozone plumes observed downwind of New York City over the North Atlantic Ocean during the ICARTT field campaign. Atmospheric Chemistry and Physics,* 11*(14), 7375 –7397, https://doi.org/10.5194/acp-11-7375-2011. [permission needed].*

Worked example

Question:: Fig. 5.5 shows profiles taken off the coast of the US during the summer of 2004. Identify the layer of polluted air associated with coastal outflow.

Solution: Relatively clean air with low concentrations of CO and O_3 are seen below 500 m. Well-mixed layers of elevated ozone and carbon monoxide at heights between 500 m and 1.5 km in the neutral residual layer ($d\theta/dz = 0$) above this boundary layer associated with coastal outflow. This layer is bounded above by the stable free troposphere ($d\theta/dz > 0$).

5.3.3 Slope and valley winds

Slope winds and valley winds are also mesoscale phenomena occurring on the time scales of hours with horizontal scales of a few kilometers. Slope and valley winds are another example of a thermal circulation that's caused by the heating and cooling of the underlying surface. Other names for upslope and downslope winds are anabatic winds and katabatic winds. Upslope anabatic winds are usually quite gentle whereas downslope katabatic winds can be very strong. The strength of the winds depends on the temperature gradient driving these winds.

Slope winds develop due to the diurnal heating and cooling of mountain slopes during the daytime. When slopes are warmed by the sun, the air in contact with the surface also warms increasing its potential temperature. The heating causes the air to become less dense and a low-pressure perturbation to form near the surface. Both the potential temperature and pressure perturbations lead to a vertical acceleration of the flow upslope (see Section 5.3.3.1 for a mathematical explanation of nonhydrostatic effects). Conversely, at night, the temperature of the air near the slope cools down resulting in a decrease in its potential temperature and descent which is enhanced by the physical slope between the slope and the valley.

The thickness of slope winds usually lie between 100 and 200 m normal to the slope although it varies with the wind intensity. Slope winds can transport pollutants. Anabatic winds can ventilate pollutants out of the valley boundary layer if they are strong enough. This is known as the chimney effect since pollutants are able to escape into the free troposphere in a diurnal puffing action. On the other hand, if the upslope winds are weak or the mountains are high, then pollutants will move upslope during the day away from the towns in the valleys, only to be recirculated back down into the town on the valley floor at night resulting in very high pollution concentrations (Chemel & Burns, 2015).

Valley winds are caused by the heating differential between mountain valleys and the surrounding plains. The valley and the plains receive a similar amount of solar heating, but in the valley, the volume of air heated by the surface is less than over the plains, so the temperature response of the air above the valley is faster than above the plain. As a result, during the day, low pressure forms over the valley driving a thermal circulation from the plain to the valley and vice versa at night. This circulation can be enhanced by the physical slope between the valley in the plains.

5.3.3.1 *Hydrostatic balance and non-hydrostatic effects*

In hydrostatic balance, vertical accelerations are small compared to other forces acting in the atmosphere ($Dw/Dt = 0$). Thus, the vertical momentum equation (Section 5.1.1.3) states that for small frictional force, hydrostatic balance is an equilibrium state that occurs when the vertical pressure gradient force is balanced by the gravitational force.

$$\frac{dp}{dz} = -\rho g$$

where dp is the change in pressure, dz is the change in altitude, ρ is the air density and g is the acceleration due to gravity. Hydrostatic balance is valid for large-scale atmospheric motions where the horizontal length scale (L) of the circulation is much larger than the vertical depth (H), That is, $H/L \ll 1$.For mesoscale features, $H/L \sim 1$ so the atmosphere is not in hydrostatic balance and vertical accelerations are nonnegligible,

$$\frac{Dw}{Dt} = -\frac{1}{\rho_0}\frac{dp'}{dz} + g\frac{\rho'}{\rho_0} \approx -\frac{d}{dz}\left(\frac{p'}{\rho_0}\right) + g\frac{\theta'}{\theta_0}$$

where Dw/Dt is the vertical acceleration of the flow, p' is the pressure deviation from hydrostatic balance, ρ_0 is a reference air density, ρ' is the density perturbation from hydrostatic balance, θ_0 is a reference potential temperature and θ' is the potential temperature deviation. The first term on the right-hand side is the vertical gradient of pressure perturbations over a reference density and the second term is the buoyancy acceleration which is due to local differences in potential temperature. Thus in a nonhydrostatic atmosphere both pressure perturbations and temperature perturbations lead to vertical acceleration.

5.3.4 Convection

So far we have discussed pollution transport in dry boundary layers, what happens in a moist environment? In a dry atmosphere the stability of a dry parcel with respect to any vertical displacement depends on whether the environmental lapse rate is larger or smaller than the dry adiabatic lapse rate (see Section 5.2.1.1 for a mathematical explanation of the atmospheric stability). However, if a parcel of air is saturated, condensation and latent heat release will cause the parcel to cool at the saturated adiabatic lapse rate which is a smaller rate than the dry adiabatic lapse rate (see Section 5.3.4.1 for a mathematical explanation of the saturated adiabatic lapse rate). This is because when a parcel becomes saturated, the condensation of water vapor leads to latent heat release causing the saturated air parcel to cool more slowly than an unsaturated parcel of air as it ascends.

In order to diagnose whether a moist environment is stable or unstable we compare the environmental lapse rate (Γ) to the dry adiabatic lapse rate (Γ_d) and saturated adiabatic lapse rate (Γ_s). If the environmental lapse is larger than both the dry unsaturated lapse rates ($\Gamma > \Gamma_s, \Gamma_d$) then a lifted dry or saturated parcel will cool more slowly than the environment so the parcel will be positively buoyant and will continue to rise. Hence the atmosphere is absolutely unstable if the environmental lapse rate is larger than both the dry and saturated lapse rates. If the environmental lapse rate is smaller than both the dry and saturated lapse rate ($\Gamma < \Gamma_s, \Gamma_d$) then a lifted parcel will cool more quickly than the environment so the parcel will be negatively buoyant and will therefore return to its original position and the atmosphere is absolutely stable. Finally, if the environmental lapse rate is smaller than the dry adiabatic lapse rate but larger than the saturated lapse rate ($\Gamma_d > \Gamma > \Gamma_s$) then the environment is stable to dry ascent, but unstable to moist ascent. This is known as a conditionally unstable atmosphere.

5.3.4.1 Saturated adiabatic lapse rate

The saturated adiabatic lapse rate, Γ_s is the adiabatic cooling rate of a rising parcel of moist air. The saturated adiabatic lapse rate is smaller than the dry adiabatic lapse rate because condensation takes place as the moist air parcel rises, the energy release of the latent heat of vaporization moderates the adiabatic cooling. The saturated adiabatic lapse rate is given by the equation:

$$\Gamma_s = \Gamma_d \left[\frac{1 + \frac{L_v r_v}{R_d T}}{1 + \frac{L_v^2 r_v}{c_p R_v T^2}} \right]$$

where Γ_d is the dry adiabatic lapse rate, c_p is the specific heat of dry air at constant pressure, r_v is the mixing ratio of water vapor, L_v is the latent heat of vapourization, R_d is the gas constant of dry air, R_v is the gas constant of water vapor, T is the air temperature. Unlike the dry adiabatic lapse rate, the saturated adiabatic lapse rate is not constant but is dependent on the air temperature and pressure. At 20°C the saturated adiabatic lapse rate can be close to 4°C km^{-1}, whereas at -40°C the saturated adiabatic lapse rate is closer to 9°C km^{-1}. Conditional stability occurs when the environmental lapse rate is between the dry adiabatic lapse rate and the saturated adiabatic lapse rate for a given layer of the atmosphere.

5.3.4.2 Thermodynamic diagrams

In order to diagnose whether the environment is unstable to moist ascent we can use a thermodynamic diagram. Thermodynamic diagrams are graphs of observed pressure, temperature, and humidity, made in a vertical sounding of the atmosphere, usually taken during a radiosonde ascent. There are different types of thermodynamic diagrams, but all have reference lines to aid analysis. These include isobars (constant pressure, p), isotherms (constant temperature, T), dry adiabats (constant dry-bulb potential temperature, θ), saturated adiabats (constant wet-bulb potential temperature, θ_w) and isohumes (constant saturation mixing ratio, r_v). On a "$T - \phi$" thermodynamic diagram (tephigram) the major axes are temperature and entropy (ϕ) where $\phi = \log\theta$. Isotherms are perpendicular to the dry adiabats. On a "skew-T log p" thermodynamic diagram the major axis is temperature and log$p$. Isotherms and log$p$ axes meet at an angle of 45 degrees making the dry adiabats (which are slightly curved) and isotherms meet at an angle of about 90 degrees.

There are usually two profiles drawn on a thermodynamic diagram (Fig. 5.6). The profile to the right is the dry-bulb temperature (T) which represents the air temperature. The profile to the left is the dew-point temperature (T_d) which is the temperature to which air must be cooled in order that it will become saturated with respect to water and at its existing pressure and humidity mixing ratio. If $T = T_d$ then the air is saturated, and clouds will form. Also typically shown with thermodynamic diagrams are the wind speed and direction at different heights in the atmosphere (Fig. 5.6). These are denoted using a symbol known as a wind barb consisting of a line pointing in the direction from which the wind is flowing, and barbs attached to the line to represent the wind speed. A short barb represents 5 knots of wind speed, a long barb represents 10 knots, and a triangle pennant represents 50 knots.

We can diagnose whether the environment is unstable to moist ascent by performing Normand's construction using a thermodynamic diagram. This process is the same regardless of the kind of thermodynamic diagram used. For example, on a tephigram (Fig. 5.6) Normand's construction is started from the surface level by cooling a parcel of air by adiabatic lifting. Since the parcel is unsaturated, it cools at the dry adiabatic lapse rate. The moisture in the parcel during adiabatic lifting is conserved, That is, has a constant humidity mixing ratio. The height at which the parcel is cooled enough to become saturated is known as the lifting condensation level (LCL). This is the level at which the water vapor will start to condense into liquid water. If the parcel is lifted higher it will now cool at the saturated adiabatic lapse rate until it reaches the level of neutral buoyancy (LNB). This is the level at which the parcel temperature equals that of the surrounding air and hence the level at which the parcel stops rising. This represents the height of the cloud top. If there's a positive area between the parcel lapse rate and the environmental lapse rate then the environment contains convective available potential energy (CAPE). CAPE is the energy contained in the atmosphere that could be released if a parcel there were lifted to the level of free convection (LFC; see Section 5.3.4.2 for a mathematical explanation of CAPE). If there is a negative area between the parcel lapse rate and the environmental lapse rate then the environment

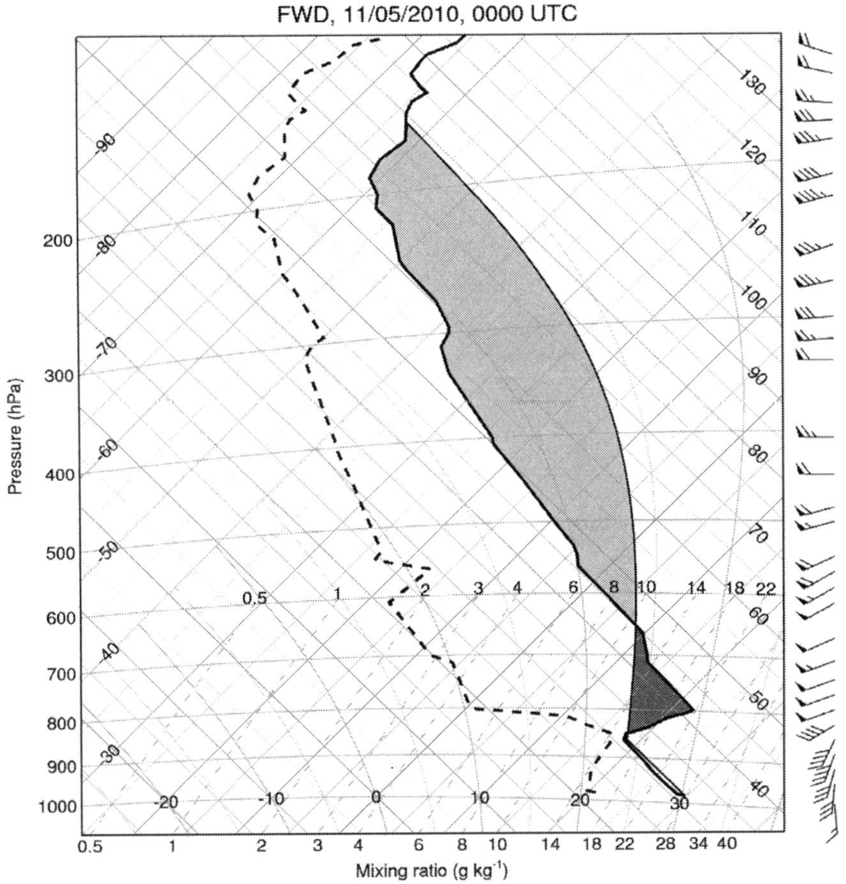

FIGURE 5.6 Tephigram showing dry-bulb temperature profile (thick solid black line), dew-point temperature profile (thick dashed black line), wind speed and direction profile (wind barbs). Normand's construction showing parcel ascent (thin black line). Lifting condensation level (LCL), level of free convection (LFC) and level of neutral buoyancy (LNB). Area of convective available potential energy (CAPE) (light gray shading) and convective inhibition (CIN) (dark gray shading).

contains convective inhibition (CIN). CIN is the energy required by a parcel to overcome negative buoyancy and reach the LFC (see Section 5.3.4.2 for a mathematical explanation of CIN).

Convective storms can take a variety of forms, they can range in scale from an isolated thunderstorm involving a single convective cloud to a mesoscale convective complex consisting of a multicell thunderstorm. The type of convection depends on the low-level wind shear. Convection is usually classed as being deep or shallow depending on the height to which it penetrates into the atmosphere, which in turn depends on the amount of CAPE in the environment. This can be limited if there are any preexisting stable layers in the atmosphere which can act as a lid and prevent convective thermals from freely ascending to the tropopause. Convection can transport pollutants away from the surface lofting them rapidly up to the upper troposphere in a matter of minutes where they are often detrained in the cloud anvil to form a layer of polluted air aloft. However, some pollutants are soluble and therefore they can be efficiently removed from the atmosphere via washout by falling precipitation.

Worked example

Question: Fig. 5.7A shows a visible image from the UK taken from onboard the MODIS Aqua satellite at 12:45 on the 9th of May 2005. Convection is occurring everywhere over the whole of the UK. Using the tephigram in Fig. 5.7C explain why there was an elevated layer of CO concentrations observed at heights between 1500 m and 3000 m above the surface by the Facility for Airborne Atmospheric Measurements (FAAM) research aircraft at 3:34 pm on this day.

Solution: The convective thermals reached a height of 850 hPa as shown by the LNB in Fig. 5.7C where they encountered a stable layer which prevented them from ascending any further. This is a case of shallow convection since

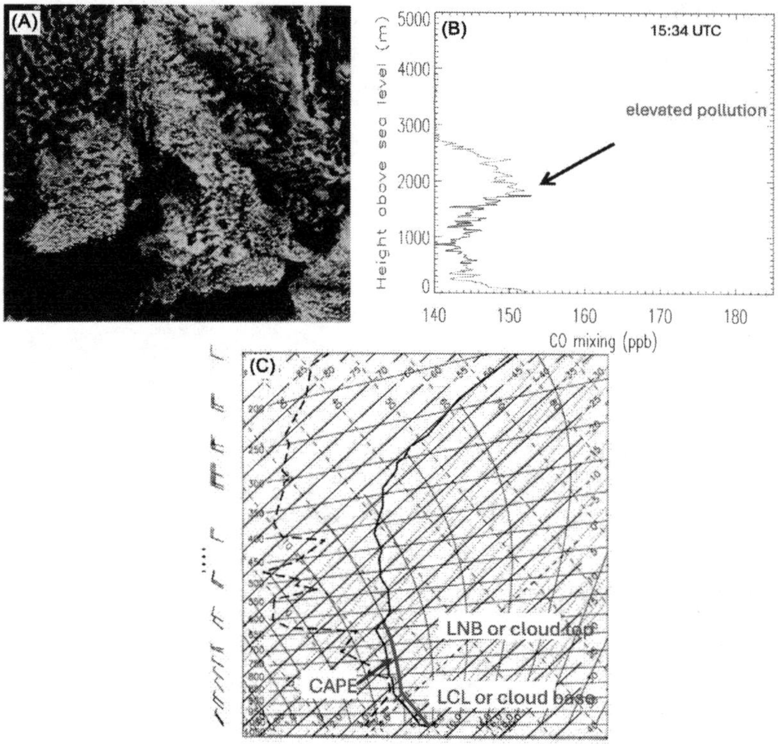

FIGURE 5.7 Shallow convection Case study of shallow convection over the United Kingdom on May 9, 2005. (A) Visible image of United Kingdom from Modis Aqua satellite at 12:45 UTC. (B) CO mixing profile at 15:34 UTC taken 50 km off south coast of England. (C) Tephigram from 12:00 UTC from Waddington RAF (53°N, 0°W) (Dacre et al., 2007). *Figure from Dacre, H.F., Gray, S.L., & Belcher, S.E. (2007). A case study of boundary layer ventilation by convection and coastal processes. Journal of Geophysical Research: Atmospheres, 112 (D17). https://doi.org/10.1029/2006jd007984.*

the convection is capped below 800 hPa. At 3:34 p.m. measurements of high CO concentrations were observed at heights above the surface by the Facility for Airborne Atmospheric Measurements (FAAM) research aircraft between 1500 and 3000 m. Typically, CO concentrations at this height are background concentrations of around 140 ppb. On this day the shallow convection has transported anthropogenic CO out of the boundary layer up to heights of around 3 km. This pollution was then advected offshore, where it was detected in a distinct layer.

5.3.4.3 Convective available potential energy and convective inhibition

CAPE is a measure of the amount of energy in the atmosphere that is available for the formation of convective clouds. It is calculated by integrating the vertical difference in temperature between a parcel of air and the surrounding environment as the parcel is lifted from its LFC to its LNB.

$$\mathrm{CAPE} \approx g \int_{\mathrm{LFC}}^{\mathrm{LNB}} \left(\frac{T - T_e}{T_e} \right) dz$$

where g is the acceleration due to gravity, T is the temperature of the lifted parcel of air, T_e is the temperature of the environment. On a tephigram, CAPE is represented by the positive area between an ascending air parcel and the environmental lapse rate.

CIN is the energy required by a parcel to overcome negative buoyancy and reach the LFC. It is calculated by integrating the vertical difference in temperature between a parcel of air and the surrounding environment as the parcel is lifted from the surface to the LFC.

$$\mathrm{CIN} = g \int_{0}^{\mathrm{LFC}} \left(\frac{T - T_e}{T_e} \right) dz$$

5.3.5 Summary

To summarize this section on mesoscale transport processes. These processes influence the transport of pollution by thermal circulations associated with horizontal temperature gradients or via convective instability. Transport occurs on time scales of a few hours and includes processes such as land and sea breezes, slope and valley winds, shallow convection, deep convection and coastal outflow. Once pollution has been ventilated from the boundary layer it can be advected downwind at a faster speed. Pollution may also be subject to less deposition to the surface and fewer chemical transformations above the boundary layer due to colder temperatures and lower humidity.

5.4 Synoptic scale transport

The synoptic scale in meteorology has a horizontal length scale of order 500 to around 3000 km. This corresponds to the scales typical of mid-latitude cyclones and high-pressure systems. The time scales of transport by synoptic scale features range from a day to a week (Fig. 5.1). Synoptic transport processes are larger than mesoscale transport processes which were discussed in the previous section. Therefore, synoptic scale processes are able to transport pollutants over greater horizontal scales, up to intercontinental scales. On synoptic time scales we assume that microscale and mesoscale processes have already transported pollutants away from their sources and mixed them throughout the depth of the boundary layer.

To identify synoptic scale features in the atmosphere we typically use synoptic charts (e.g., Fig. 5.8A). Synoptic charts summarize the weather conditions at a particular time. They include information about the pressure pattern and fronts. Solid lines are contours of constant pressure (known as isobars). Isobars can be used to infer the direction and strength of the wind and to identify regions of high and low pressure. Also on a synoptic chart are the locations of surface fronts. Fronts denote the boundaries between air masses with different temperature and moisture characteristics. Cold fronts are represented by a line with triangles. The direction in which the triangles point shows the direction in which the front is moving. Warm fronts are represented by lines with semicircles. Cold air follows a cold front and warm air follows a warm front. Sometimes the cold and warm fronts are broken by crosses. This indicates that the front is weakening (known as frontolysis). Occasionally the frontal symbols are open and not filled. These open symbols represent fronts which are not identified at the surface but are higher up in the atmosphere. Upper-level fronts can also be frontolytic. Fronts with both a triangle and a semicircle next to each other are known as occluded fronts. Occluded fronts typically occur when cold fronts catch up with warm fronts. Finally, black lines that have no triangles or semicircles are known as troughs and they denote areas where the air is very unstable.

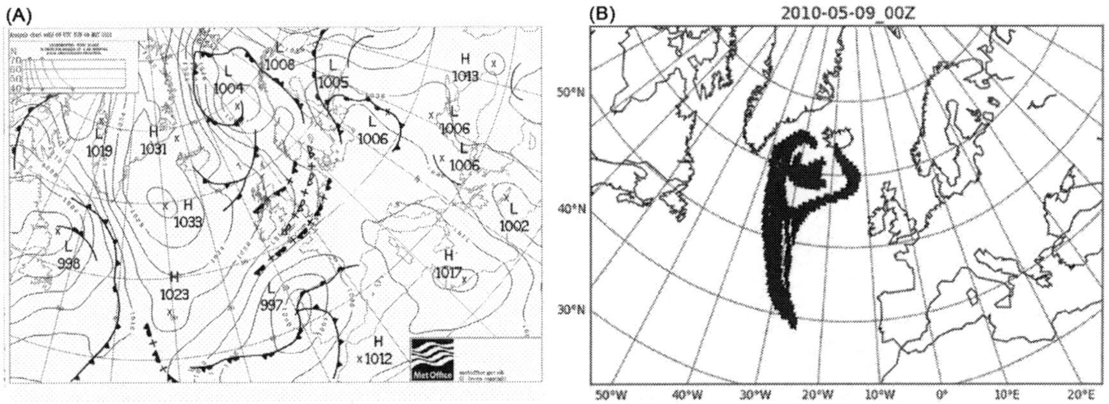

FIGURE 5.8 (A) Anticyclonic advection UK Met Office synoptic chart at 00 UTC on May 9, 2010. (B) Volcanic ash cloud emitted from the Eyjafjallajokull volcanic eruption, Iceland at 00 UTC on May 9, 2010 (Dacre et al., 2016). *Figure from Dacre, H.F., Harvey, N.J., Webley, P.W., & Morton, D. (2016). How accurate are volcanic ash simulations of the 2010 Eyjafjallajökull eruption?. Journal of Geophysical Research: Atmospheres, 121(7), 3534 –3547. https://doi.org/10.1002/2015jd024265.*

5.4.1 Anticyclonic and cyclonic advection

Anticyclonic regions are regions of relatively high pressure around which air circulates clockwise in the northern hemisphere. Extratropical cyclones are regions that are relatively low-pressure. The circulation around low-pressure systems is anticlockwise in the Northern Hemisphere. Fig. 5.8A is a synoptic chart showing two high-pressure systems on May 9, 2010. One region of high pressure is in the North Atlantic, South of Iceland, and one is located in the Mediterranean. This was during the eruption of the 2010 Eyjafjallajokull volcano. There is a clockwise advection of air around the anticyclone and anticlockwise advection of air around the low-pressure center off the coast of Portugal (see Section 5.4.1.2 for a mathematical explanation of geostrophic wind balance).

Fig. 5.8B shows volcanic ash emitted from the 2010 Eyjafjallajokull eruption. Some ash travels anticyclonically around the high-pressure center near Iceland (Fig. 5.8A), while a second branch splits off and travels cyclonically towards the Mediterranean. Broadly speaking the position of the high and the low-pressure centers explains the dispersion pattern of the ash during the initial phase of the eruption. The strongest winds associated with extratropical cyclones are typically found ahead of the cold front and to the south of the cyclone center. Strong winds can advect polluted air anticlockwise around the low center.

Worked Example

Fig. 5.9B shows measurements of pollution from the London air quality network between March 31 and April 4, 2014. High PM_{10} and $PM_{2.5}$ concentrations were detected across the entire London network. Measurements from the

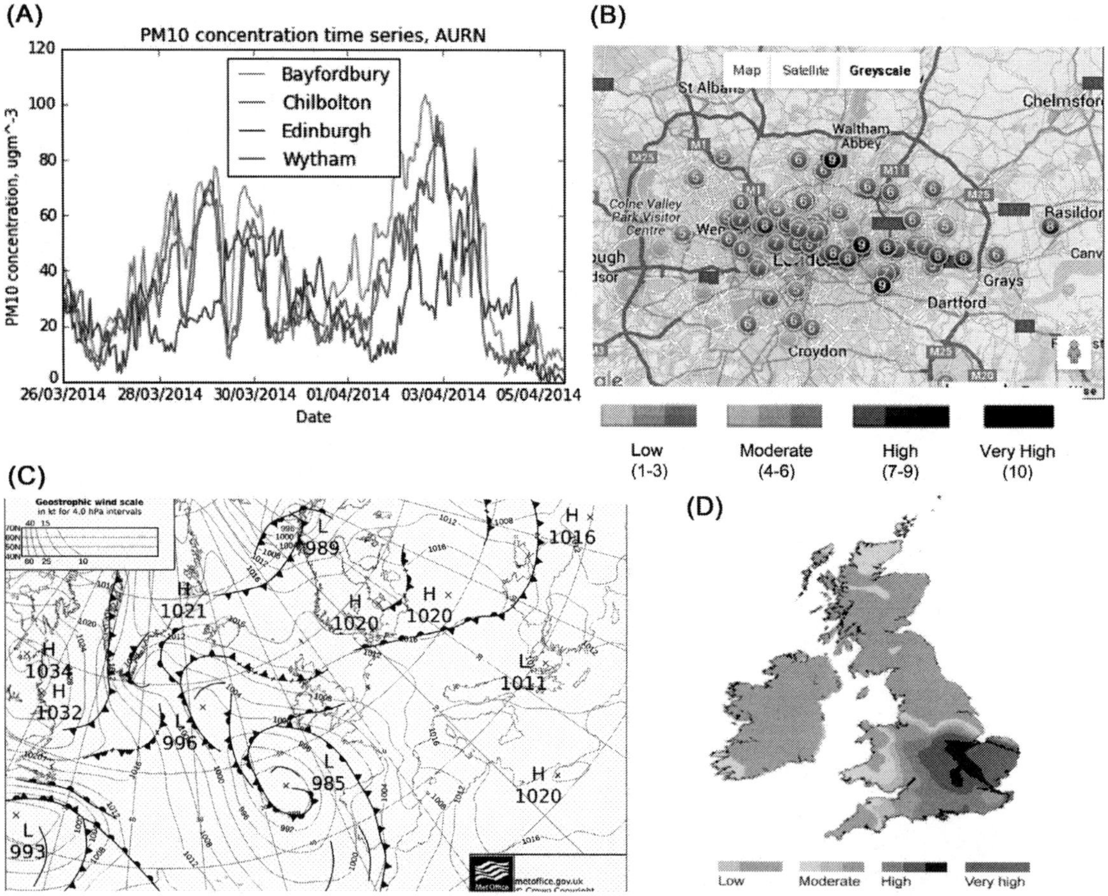

FIGURE 5.9 (A) Timeseries of automatic urban and rural network (AURN) measurements of PM10 concentrations between March 16 and April 5, 2014 taken at Bayfordbury (green), Chilbolton (red), Edinburgh (blue), and Wytham (purple). Figure produced by E. Sainsbury. (B) London Air Quality Network (LAQN) peak PM10 index between March 31 and April 4, 2014. (C) UK Met Office synoptic analysis chart for April 2, 2014. (D) Nimrod radar network at 18 UTC on April 1, 2014. *Data from London Air Quality Network (LAQN).*

nationwide automatic urban and rural network (AURN) for the same period are shown in Fig. 5.9A. PM_{10} levels were above WMO health levels in the South of England, in Hertfordshire, Oxfordshire and Hampshire, but also in Edinburgh in Scotland. Use the synoptic chart in Fig. 5.9C to suggest an explanation for the elevated concentrations of PM_{10} on this day.

Solution: The synoptic chart for April 22, 2014, Fig. 5.9C indicates that there is a low-pressure center to the West of the UK with southerly flow ahead of the cold front. The southerly flow advects air from Spain and North Africa resulting in long-range transport of PM (Saharan dust) to the UK. In addition, over northern France, and Belgium there are many industrial regions which also emit nitrates. This likely led to the high levels of PM_{10} observed in London and elsewhere.

5.4.1.1 Geostrophic wind balance

In geostrophic balance, horizontal accelerations are small compared to other forces acting in the atmosphere. Thus, the horizontal equations of motion (Section 5.1.1.1) state that for small frictional forces, geostrophic balance is an equilibrium state that occurs when the horizontal pressure gradient force balances the horizontal component of the Coriolis force.

$$u_g = -\frac{1}{\rho f}\frac{\partial p}{\partial y}$$

$$v_g = \frac{1}{\rho f}\frac{\partial p}{\partial x}$$

where ρ is the air density, p is pressure, f is the Coriolis parameter and u_g, v_g are the geostrophic wind velocities. Thus, in equilibrium the geostrophic wind blows at right angles to the horizontal pressure gradient and the strength of the geostrophic wind is proportional to the horizontal pressure gradient. In terms of wind direction in the Northern Hemisphere, geostrophic balance results in air moving around high pressure in a clockwise (anticyclonic) direction and around low pressure in an anticlockwise (cyclonic) direction.

5.4.2 Anticyclonic subsidence

Convergence (see Section 5.4.2.1 for a mathematical explanation of convergence) of mass in the upper troposphere is the primary mechanism for generating high-pressure at the surface. Mass convergence aloft is associated with descending air below due to mass conservation. Air sinking below convergence aloft is balanced by divergence at the surface. The slow subsidence of air warms adiabatically as it descends but the humidity mixing ratio is conserved so that the air becomes subsaturated. This leads to a suppression of the formation of clouds and precipitation. It also helps in the formation of a low-level inversion which can confine pollutants to the boundary layer underneath.

Fig. 5.10A shows a typical daytime radiosonde sounding during anticyclonic conditions. We typically find a shallow unstable layer at the surface ($\Gamma > \Gamma_d$) with a well-mixed neutral boundary layer above ($\Gamma \approx \Gamma_d$). During anticyclonic conditions, the subsiding air aloft warms adiabatically as it sinks. This leads to warmer air aloft and the formation of a stable inversion layer at the top of the atmospheric boundary layer ($\Gamma \ll \Gamma_d$). During the night the clear skies associated with anticyclonic conditions typically lead to rapid cooling of the surface, a strong nocturnal inversion layer near the surface and a stable layer above. Thus, air pollution emitted at the surface is trapped below the capping inversion during the day and below the nocturnal inversion at night. This can result in poor air quality, particularly when winds are light, and air stagnates.

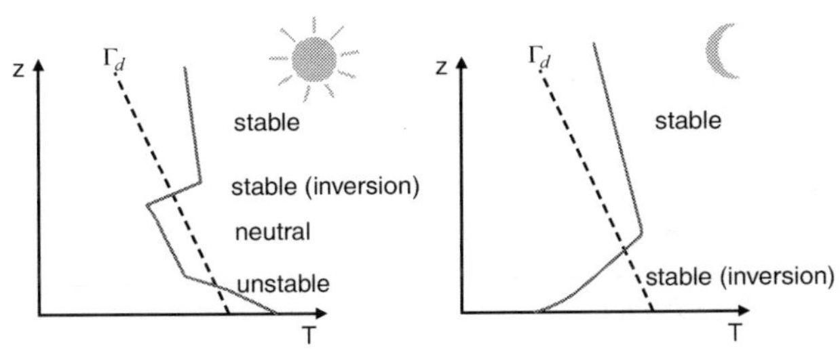

FIGURE 5.10 Schematic environmental lapse rates (red) under anticyclonic conditions during (*left*) the daytime and (*right*) nighttime. Dry adiabatic lapse rate, Γ_d (dashed line).

5.4.2.1 *Convergence and divergence*

The second term in the continuity equation (Section 5.1.1.2) is the divergence of velocity, a three-dimensional property which expresses the rate of expansion of the air per unit volume. In Cartesian coordinates the divergence is given by

$$\nabla \cdot \boldsymbol{V} = \frac{\partial u}{\partial x} + \frac{\partial v}{\partial y} + \frac{\partial w}{\partial z}$$

In the atmosphere we are mainly interested in the horizontal divergence of velocity which expresses the rate of horizontal expansion of the air per unit area. Since local density changes are small in the atmosphere, the continuity equation has the approximate form $\nabla \cdot \boldsymbol{V} = 0$

$$\frac{\partial u}{\partial x} + \frac{\partial v}{\partial y} = -\frac{\partial w}{\partial z}$$

that is, if there is horizontal expansion of an air parcel (positive divergence) then it will contract vertically. Thus, divergence in the upper levels of the atmosphere must be associated with ascending air beneath. Negative divergence is termed convergence. When convergence occurs in the upper levels of the atmosphere it must be associated with descending air beneath.

Worked example

Question: Describe five ways in which anticyclonic conditions can lead to high pollution concentrations?

Solution:

1. The light winds associated with weak horizontal pressure gradients around anticyclones typically result in weak dispersion allowing pollution to accumulate near sources.
2. Convergence aloft is associated with subsidence above anticyclones which can create a capping inversion preventing pollution emitted at the surface from being ventilated into the free atmosphere.
3. Anticyclonic subsidence can also suppress precipitation which reduces the wet deposition of soluble pollutants to the surface.
4. The anticyclone may be positioned such that anticyclonic advection can transport nonlocal emissions to the receptor adding to local emissions.
5. Surface cooling due to clear skies during anticyclonic conditions can create a stable surface layer which traps pollution near its source resulting in high pollution concentrations.

5.4.3 Extratropical cyclone warm conveyor belt ascent

Extratropical cyclones are characterized by divergence aloft which is associated with ascending motion in the mid-troposphere (see Section 5.4.2.1 for a mathematical explanation of divergence). This ascending motion has an influence on the vertical distribution of pollutants in the vicinity of the extratropical cyclone. Cyclones are typically composed of three air streams, the warm conveyor belt, the cold conveyor belt, and the dry intrusion (or dry conveyor belt). The warm conveyor belt is located on the equatorward side of the cyclone ahead of the cold front at the surface. The air flowing along the warm conveyor belt originates at low altitudes in the warm sector of the cyclone and then ascends as it travels poleward (purple arrow in Fig. 5.11). After a day or two, air from the boundary layer has ascended into the mid and upper troposphere. The warm conveyor belt is the most prominent airstream for rapid intercontinental transport

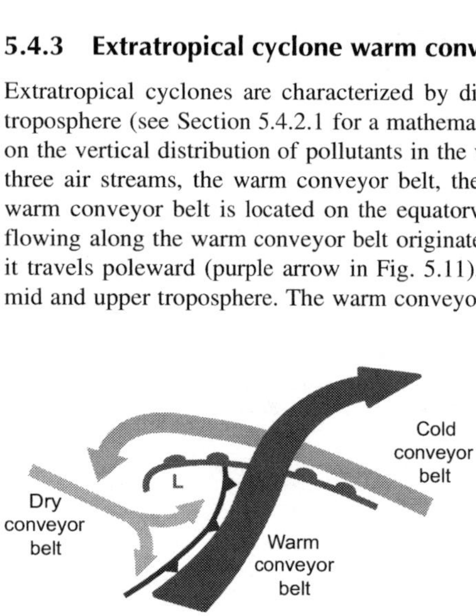

FIGURE 5.11 Schematic of extratropical cyclone airflows. Low pressure center (L), cold and warm fronts. Overlaid with descending dry conveyor belt (yellow arrows), low-level cold conveyor belt (blue arrow) and ascending warm conveyor belt (purple arrow).

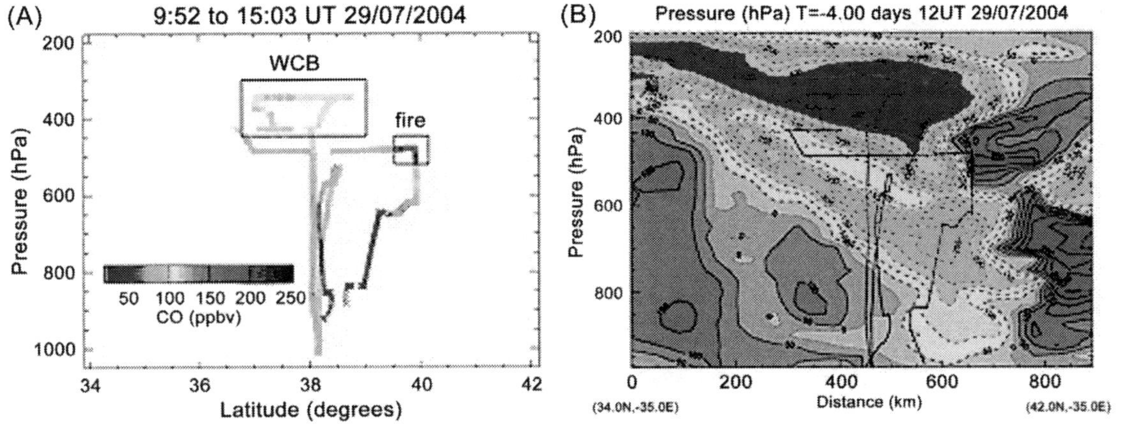

FIGURE 5.12 (A) CO concentrations (ppbv) along the flight track as a function of latitude-pressure. High concentrations of CO are shown above 500 hPa. (B) Back trajectories from each latitude-pressure position showing the ascent (green/blue) and descent (orange/red) of the trajectory over the previous 4 days (Lewis et al., 2007). *From Lewis, A.C., Evans, M.J., Methven, J., Watson, N., Lee, J.D., Hopkins, J.R., Purvis, R.M., Arnold, S.R., McQuaid, J.B., Whalley, L.K., Pilling, M.J., Heard, D.E., Monks, P.S., Parker, A.E., Reeves, C.E., Oram, D.E., Mills, G., Bandy, B.J., Stewart, D., . . . Crosier, J. (2007). Chemical composition observed over the mid-Atlantic and the detection of pollution signatures far from source regions. Journal of Geophysical Research: Atmospheres,* 112*(D10). https://doi.org/10.1029/2006jd007584.*

of pollution because of its ability to loft polluted planetary boundary layer air from the cyclone's warm sector into the upper troposphere where it then gets entrained into the jet stream. Typical transport time scales are of the order of three to four days.

Fig. 5.12A shows latitude versus height CO measurements taken on a flight over the North Atlantic during the International Consortium for Atmospheric Research on Transport and Transformation (ICARTT) field campaign in 2004. Typically, CO is emitted near the Earth's surface so concentrations of CO decrease with height. However, on this flight high values of CO (yellow colors) were observed above 400 hPa along the flight track. Fig. 5.12B shows the back trajectory ascent over the preceding 4 days. To calculate back trajectories, we use wind fields over the preceding time window to track air parcels backward from their final destination. The blue colors show back trajectories that have ascended more than 500 hPa over the preceding four days and the red color show back trajectories that have experienced net decent over the preceding 4 days. In the region of high CO, the air had ascended more than 500 hPa. This suggests that 4 days ago this air resided within the boundary layer over the continental United states. The synoptic chart at this time (not shown) showed that this gradual ascent was associated with a warm conveyor belt. There is also evidence of coastal outflow and long-range trans-Atlantic transport of pollution at low-level during this case (Methven et al., 2006).

5.4.4 Extratropical cyclone dry intrusion descent

The dry intrusion airstream (yellow arrow in Fig. 5.15) is located behind the cyclone and is associated with upper-troposphere and stratospheric air descending down behind the cold front. There's often an exchange of stratospheric and tropospheric air at mid-latitudes as a result of this descending motion and mixing along the edge of the different air mass boundaries. Stratospheric air from above the tropopause can become mixed with the turbulent tropospheric air. Because the stratospheric air is high in stratospheric ozone this process can lead to high levels of ozone near the surface which can be harmful to humans and crops. In the region behind extratropical cyclones, we often see tropopause folds which are depressions in the height of the tropopause in which stratospheric air descends down to low altitudes.

Fig. 5.13 shows measurements taken onboard a research flight aircraft through a tropopause fold. The yellow surface represents the dynamical tropopause. The tropopause fold is shown where this surface dips down from an altitude of 10 to 3 km. Low values of ozone (light blue colors) are found in the troposphere ahead of tropopause fold. High values of ozone are found in the stratosphere above the tropopause fold (yellow colors). Finally intermediate values of ozone are found in the vicinity of the tropopause fold as shown by the red and dark blue colors.

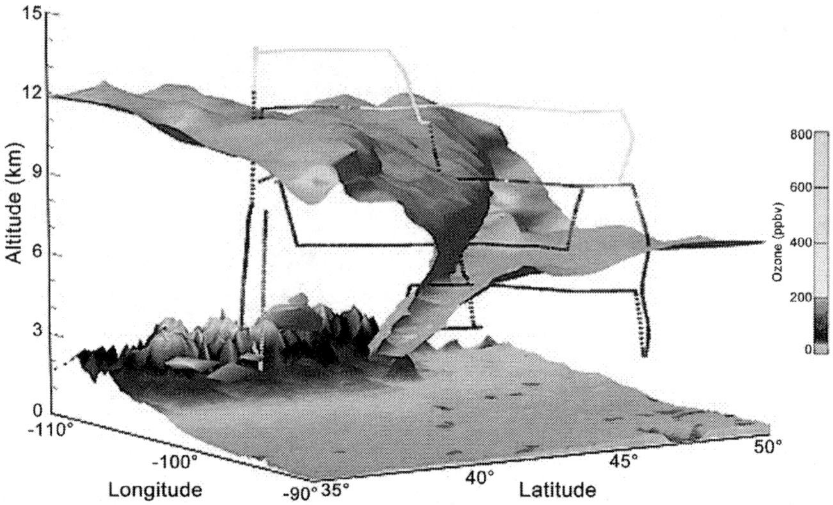

FIGURE 5.13 Tropopause fold Tropopause height denoted by the PV = 2 surface at 18UTC on 1/12/2005. Ozone concentrations are shown along the HIAPER aircraft flight. Yellow indicates high values of ozone (> 200 ppbv), red indicates intermediate values and blue indicates low values of ozone (<60 ppbv). *From Pan, L. L., Bowman, K. P., Shapiro, M., Randel, W. J., Gao, R. S., Campos, T., Davis, C., Schauffler, S., Ridley, B. A., Wei, J. C., & Barnet, C. (2007). Chemical behavior of the tropopause observed during the Stratosphere-Troposphere Analyses of Regional Transport experiment. Journal of Geophysical Research: Atmospheres, 112(D18).*

5.4.5 Summary

The synoptic scale in meteorology has a horizontal length scale of order 500 km or more. This corresponds to the scales typical of mid-latitude cyclones and high-pressure systems and the advection of pollutants on scales of 500 km or more. Subsidence associated with anticyclones can result in an inversion which can trap pollutants near the surface. In addition, clear skies associated with anticyclones can result in extreme surface cooling at night again creating an inversion which can trap pollutants. Anticyclones are associated with light winds which leads to lack of dispersion of pollutants and high pollution levels. Finally, anticyclones are associated with suppressed precipitation. Since precipitation can remove pollutants from the atmosphere a lack of precipitation can often mean that pollutant concentrations remain high.

Extratropical cyclones on the other hand are associated with high winds and precipitation. These high winds can disperse pollutants reducing concentrations and the precipitation can remove pollutants from the atmosphere via wet deposition. Both lower and upper-level tropospheric pathways for pollution are commonly observed near extratropical cyclones. Cyclones can enhance the ventilation of pollution from the boundary layer if pollution is entrained into the warm conveyor belt where they can be transported up to the upper troposphere and entrained into the jet to be transported long distances. Behind extratropical cyclones, there is a descending dry conveyor belt. This can transport stratospheric ozone down to the surface and result in ozone episodes at the Earth's surface.

5.5 Global-scale transport

5.5.1 Global-scale circulations

In this section, we consider time scales of pollution transport from weeks to months and over spatial scales of several thousands of kilometers. Global-scale circulations include transport by monsoons, the trade winds, the intertropical convergence zone and mid-latitude jet streams (Fig. 5.1). In order to understand the formation of global-scale circulations it is helpful to first consider an aqua planet. An aqua planet is one in which the Earth has no landmasses. The first stage is to consider the differential solar heating of the aqua planet that occurs between the equator in the poles (Fig. 5.18A) in a hypothetical situation where the planet has no rotation. This differential heating causes the tropics to heat up more than the poles. Thus, the atmosphere in the tropics expands relative to the atmosphere at the poles. This results in relatively low surface pressure at the tropics and high surface pressure at the poles (Fig. 5.14B). Near to the Earth's surface air flows from high to low pressure, that is, from the poles to the equator. The opposite effect occurs aloft with air flowing from the equator to the poles. This creates a global thermally driven circulation.

This global circulation becomes more interesting when we consider that the Earth is rotating. The associated Coriolis force causes air to be deflected to the right in the Northern Hemisphere and to the left in the Southern Hemisphere (see Section 5.4.1.2 for a mathematical explanation of the Coriolis force). Air at the Earth's surface is thus deflected to produce Easterlies in the tropics (Fig. 5.14C). In the tropics, the Easterlies lead to convergence which, due to mass conservation, leads to ascent and cloud formation. This region of convergence is known as the Inter-Tropical Convergence Zone (ITCZ). Aloft, poleward moving air in the Northern Hemispheres is also deflected to the right by the Coriolis force. This deflection results in westerly flow aloft in the mid-latitudes (Fig. 5.14D). Similarly, poleward moving air in the Southern Hemisphere is deflected to the left resulting in westerly flow aloft. Westerlies aloft and easterlies at the surface create a region of strong vertical wind shear which, by thermal wind balance, is associated with a horizontal temperature gradient (see Section 5.4.1.1 for a mathematical explanation of thermal wind balance). This temperature gradient enhances the existing equator-to-pole temperature gradient in Fig. 5.14A. These sharp temperature gradients can become unstable leading to waves which modify the general circulation forming two smaller circulations known as the Hadley cells in the tropics and the Ferrel cells in the mid-latitudes. Thus, at the surface, we observe easterlies in the tropics which are known as the trade winds. Aloft between the Hadley cells and the Ferrel cells, we find the westerly jets known as the subtropical jets.

Now what happens when we add in some landmasses? Fig. 5.15A shows a 40-year average of 10 m wind speed and direction from the European Center for Medium Range Weather Forecasting (ECMWF) Re-Analysis (ERA-40). Averaging over many years allows the general circulation to be identified by averaging out all of the day-to-day weather variability. In the subtropics, Fig. 5.15A shows the easterly Tradewinds in both the Pacific and the Atlantic basins. In the mid-latitudes surface westerly winds can be identified in the Southern Ocean extending in a fairly zonal band around the whole globe due to the lack of land masses to disturb the flow. In the northern hemisphere the surface flow is more disturbed and westerly winds are only evident in the North Atlantic and the North Pacific basins.

Further global averaging of the wind at each latitude smooths the effect of the landmasses. Fig. 5.15B shows the zonal mean zonal wind. Green colors indicate westerly winds and blue colors indicate easterly winds. In the mid-latitudes the westerly winds increase with height. The westerly jets are maximum at the height of the tropopause 200 hPa.

Next, the meridional component of the wind can be selected to focus on the Hadley cells and Ferrel cells. Fig. 5.16 shows the zonal mean meridional winds. The Hadley cells are identified on either side of the equator the Ferrel cells are found in the mid-latitudes (Fig. 5.16A). Fig. 5.16B shows the December, January, and February (DJF) averages and Fig. 5.16C shows the June, July, and August (JJA) averages. The location and strength of the Hadley cells depend on the season. In the northern hemisphere winter (DJF) the northern branch of the Hadley cell is strongest and the ITCZ between the Hadley cells has moved South, relative to the annual mean, into the southern hemisphere following the sun. The opposite occurs in the northern hemisphere summer (JJA). In JJA, the ITCZ is found north of the equator. The seasonal movement of the ITCZ is partially responsible for the different monsoons described in Section 5.4.3.

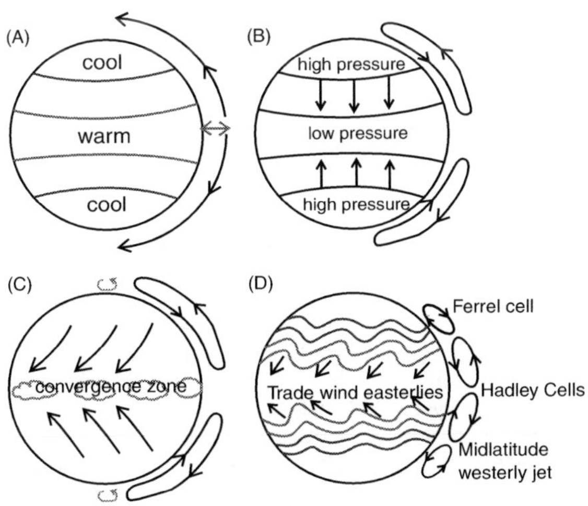

FIGURE 5.14 Global circulation Schematic of global scale circulation on an aqua planet. (A) Differential heating of tropics and poles. (B) Pressure dipoles and meridional flow. (C) Effects of rotation. (D) Effects of wind shear and formation of baroclinic waves.

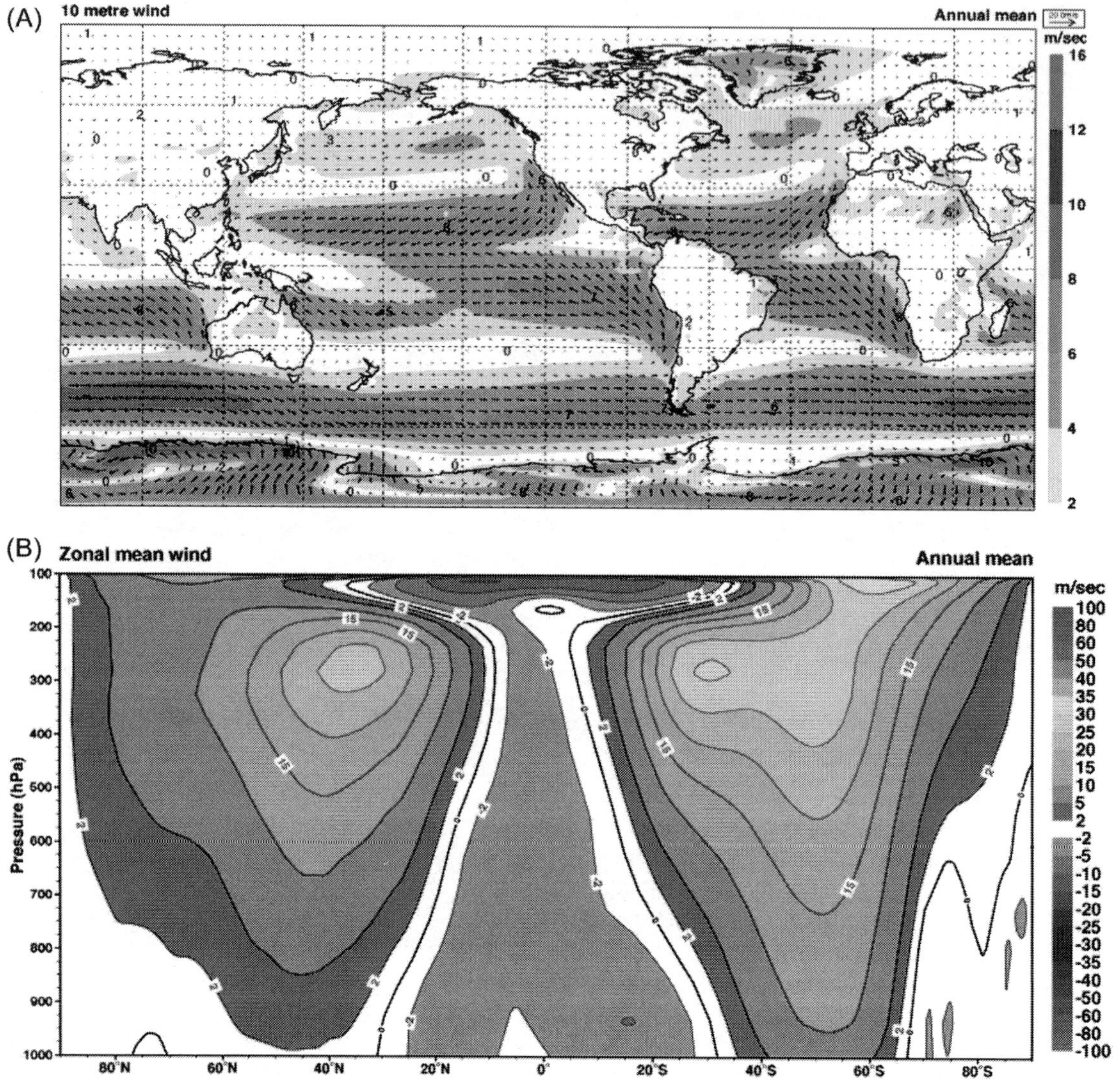

FIGURE 5.15 Zonal winds ERA-40 Atlas (A) annual mean 10 m winds and (B) annual mean Zonal mean zonal wind. *Data from https://sites.ecmwf.int/era/40-atlas CC BY 4.0.*

The circulation is dominated by thermally direct motion, with rising motion in the ITCZ and subsidence in the subtropics and extratropics of both hemispheres. In the extratropics of both hemispheres, air disperses relatively quickly throughout the region between approximately 30 degrees latitude and the pole. Air from the extratropics moves into the tropics at low levels in the trade wind circulation. Upon reaching the ITCZ, this air ascends into the middle and upper tropical troposphere. There is a fairly rapid dispersion of air within the tropics. As air spreads out in the upper branches of the Hadley cells and descends in the subtropics, some subsiding air mixes into the extratropics of both hemispheres. From a pollution transport point of view, dispersion within the northern and southern hemispheres is relatively rapid, but the exchange between the hemispheres is slower. Interhemispheric transport rates can be computed from measurements of long-lived trace species such as CFCs and suggest an interhemispheric time lag of around 2 years.

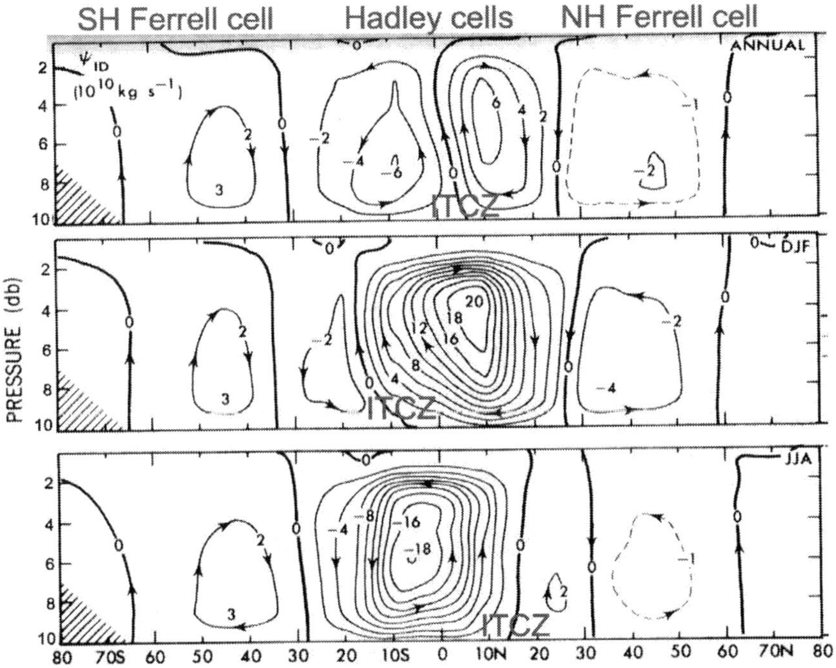

FIGURE 5.16 Meridional winds Zonal mean meridional winds (A) Annual mean, (B) December, January, February (DJF) mean, (C) June, July, August (JJA) mean (Peixoto et al., 1992). *Figure from Peixoto, J. P., Oort, A. H., Covey, C., & Taylor, K. (1992). Physics of climate. Physics Today, 45(8), 67–67. https://doi.org/10.1063/1.2809772.*

5.5.1.1 *Thermal wind balance*

Combining the geostrophic balance (Section 5.4.1.2) and hydrostatic balance (Section 5.3.3.1) with the ideal gas law, we find a relationship between the horizontal temperature gradient and the vertical gradient of the geostrophic wind in the atmosphere. This is known as thermal wind balance:

$$\frac{\partial u_g}{\partial p} = -\frac{R_d}{pf}\frac{\partial T}{\partial x} \quad \frac{\partial v_g}{\partial p} = \frac{R_d}{pf}\frac{\partial T}{\partial y}$$

where $\partial v_g/\partial p$ is the vertical wind shear, p is pressure, f is the Coriolis parameter and T is temperature. Thermal wind balance states that the change in geostrophic wind between two pressure levels is directly proportional to the horizontal layer-mean temperature gradient. Therefore, in the atmosphere, equator to pole temperature gradients ($\partial T/\partial y < 0$) are associated with westerly zonal jets aloft.

5.5.2 Intertropical convergence zone and trade winds

What has the ITCZ described in Section 5.6.1 to do with pollution transport? The ITCZ plays a role in the seasonal cycle of dust transport. Fig. 5.17 shows the seasonal cycle of dust transport in the tropics and subtropics. Aerosol optical depth (AOD) is a measure of the extinction of solar radiation by dust layers. Dust particles prevent direct sunlight from reaching the surface by absorbing and scattering light. Thus, the AOD tells us how much sunlight is prevented from reaching the ground. A clear seasonal cycle in AOD exists, varying with the strength and location of the trade winds and the position of the ITCZ. In the northern hemisphere summer, the ITCZ moves north. Convergence in West Africa leads to enhanced gustiness and dust uplift. This dust is advected across the Atlantic to Central America by the easterly trade winds. In the Northern Hemisphere winter, the ITCZ moves further south. Convergence occurs in sub-Saharan Africa where dust sources are more limited.

Worked Example

How does the movement of the ITCZ control the seasonal cycle of AOD over the North Atlantic shown in Fig. 5.17.

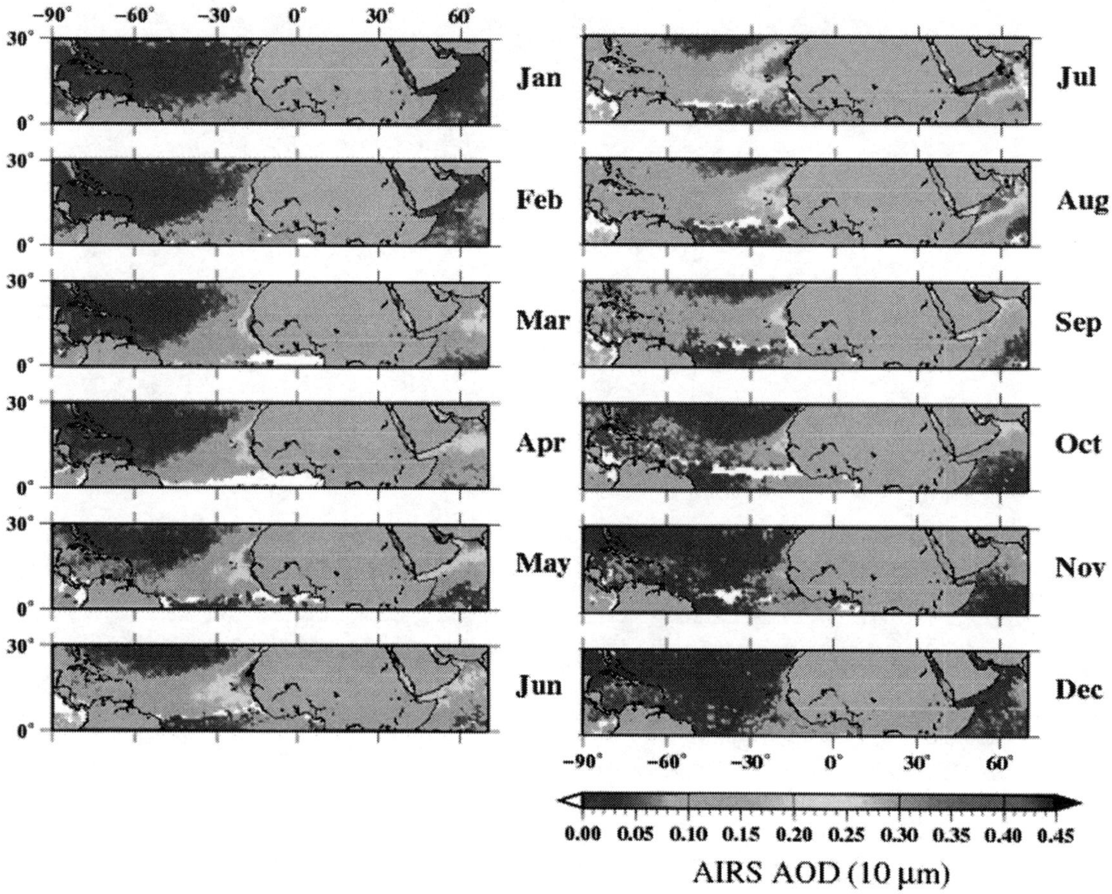

FIGURE 5.17 Trade winds Monthly climatology of the aerosol layer mean optical depth seen by AIRS from 2003 to 2008 (Peyridieu et al., 2010). *Data from Peyridieu, S., Chedin, A., Tanre, D., Capelle, V., Pierangelo, C., Lamquin, N., & Armante, R. (2010). Saharan dust infrared optical depth and altitude retrieved from AIRS: A focus over North Atlantic—comparison to MODIS and CALIPSO. Atmospheric Chemistry and Physics, 10, 1953–1967 (Fig. 4). CC BY 4.0.*

Solution: In the Northern Hemisphere summer the ITCZ moves north where low-level convergence leads to enhanced gustiness and dust uplift in the Sahara. Dust is advected across the Atlantic to Central America by the easterly trade winds. In the Northern Hemisphere winter, the ITCZ moves south and low-level convergence occurs in sub-Saharan Africa where dust sources are limited. This results in reduced transport of Saharan dust across the Atlantic.

5.5.3 Intertropical convergence zone and monsoon circulations

Monsoons occur in seasonally arid climates in the subtropics and the tropics. There are three main monsoon regions: Asia, Australia, and West Africa. Monsoons are characterized by a reversal of wind directions between seasons, typically driven by differential heating of the land and ocean. During the monsoon season the prevailing wind direction shifts by around 120 degrees between January–February (Fig. 5.18, top panel) and July–August (Fig. 5.18, bottom panel). The main driver of the monsoons is the seasonal migration of the ITCZ. Therefore, the start and end of monsoon season begins and ends at approximately the same time each year.

During the northern hemisphere summer months, the ITCZ shifts northward, causing the monsoon wind to blow from the ocean towards the land in the Asian monsoon region (Fig. 5.18A). As the moisture laden air reaches the warm land it rises due to the presence of the coastal hills and the inland mountains. This moisture condenses and the rainy

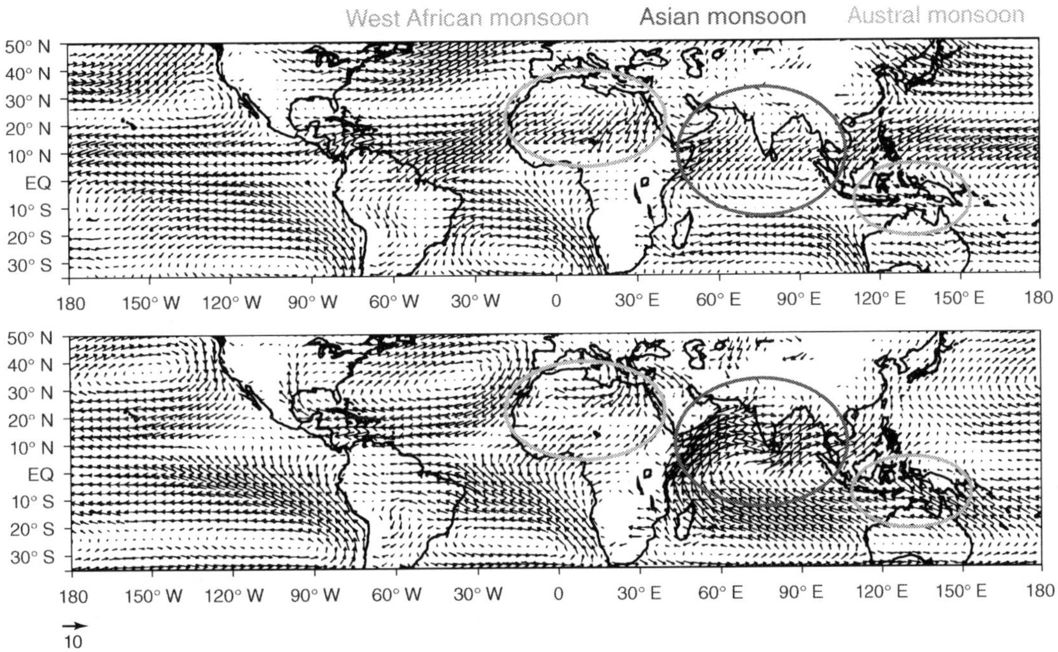

FIGURE 5.18 Monsoons ERA-40 Mean winds (m s^{-1}) at 925 hPa for northern hemisphere winter (January–February, upper panel) and northern hemisphere summer (July–August, lower panel). The circles highlight the West African monsoon region (blue), Asian monsoon region (red) and Australian monsoon region (yellow). *Data from https://sites.ecmwf.int/era/40-atlas CC BY 4.0*

season begins along the coast and at the leading edge of the monsoon flow where convergence leads to ascent and rainfall. By contrast, in the northern hemisphere winter, the ITCZ moves southward, and the Asian monsoon winds reverse their direction, blowing from the land towards the ocean (Fig. 5.18B). Obviously this flow and the rainfall it brings is extremely important for agriculture, but it also plays a role in controlling air quality.

We can examine the annual PM concentrations at Gual Pahari in Northern India (Fig. 5.19B). Fig. 5.19A shows typical onset times of the Indian Monsoon. The Indian monsoon typically starts in Northern India in late June-early July. How would we detect the monsoon onset and what will the monsoon do to PM_{10} concentrations? When the monsoon reaches Gual Pahari we expect a shift in the prevailing wind direction. We also expect a corresponding change in the air mass characteristics such as an increase in the humidity and associated precipitation. Fig. 5.19C shows three times series of measurements at Gual Pahari from January 1, 2008 to January 1, 2010. The PM_{10} concentrations timeseries show an annual cycle, with maximum surface concentrations occurring during the winter months and minimum concentrations from July to September. What causes the seasonal cycle in PM_{10} concentrations? From Section 5.2 we expect that if the planetary boundary layer is shallow, then PM_{10} concentrations will be high and as the planetary boundary layer grows the PM_{10} concentrations will decrease. However, the maximum boundary height doesn't coincide with the minimum PM_{10} concentrations. The minimum PM_{10} concentrations coincide with the peak in the rainfall suggesting loss due to wet deposition. However, the PM_{10} concentrations start to rise again in the winter as the rain abruptly ceases and the boundary layer heights are low. It's also quite likely that emissions are increasing in the winter due to increased domestic wood burning for heating.

5.5.4 Mid-latitude jet streams

In this section, the focus is on examples of intercontinental pollution transport by the mid-latitude jet streams described in Section 5.5.1. Jet streams are strong flows in the upper troposphere characterized by strong wind shear and one or more wind speed maxima. They extend thousands of kilometers from West to east and are several 100 km wide and a few kilometers deep. Wind speeds in the central core of the jet stream can exceed 50 m s^{-1} and occasionally 100 m s^{-1}. They form at the boundaries of the meridional circulation cells. However, due to large meanders in the jet stream, they

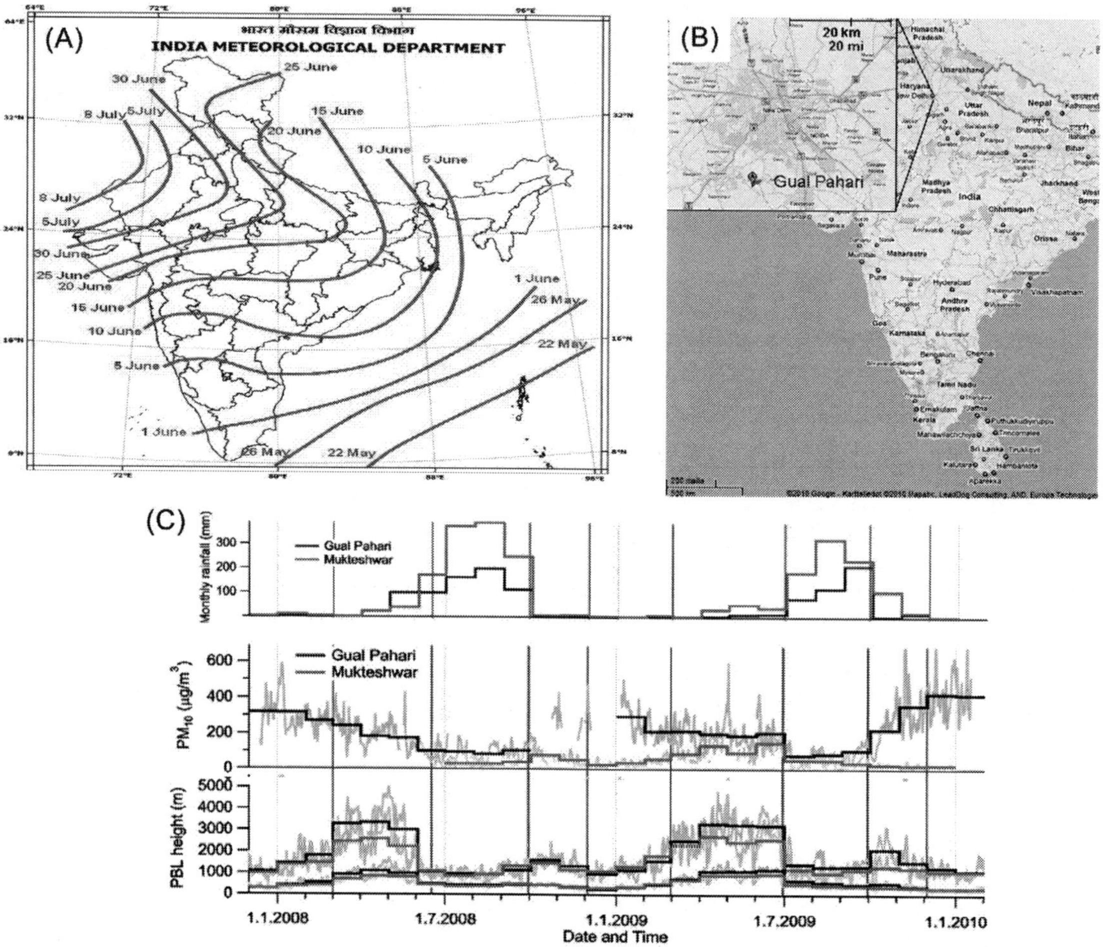

FIGURE 5.19 Indian Monsoon. (A) Typical onset dates of the Indian Monsoon, (B) location of Gual Pahari in Northern India. (C) Timeseries of monthly rainfall, PM10 concentrations and planetary boundary layer heights in Gual Pahari (Raatikainen et al., 2014). *(A and B) Komppula (2012). (C) Data from Raatikainen, T., Hyvärinen, A.P., Hatakka, J., Panwar, T.S., Hooda, R.K., Sharma, V.P., & Lihavainen, H. (2014). The effect of boundary layer dynamics on aerosol properties at the Indo-Gangetic plains and at the foothills of the Himalayas. Atmospheric Environment, 89, 548–555, http://www.elsevier.com/locate/atmosenv. https://doi.org/10.1016/j.atmosenv.2014.02.058. Modified from Figs. 4 and 5.*

can merge in regions of semipermanent troughs. In other locations the subtropical and the polar jets can be distinguished as separate wind speed maximum.

Fig. 5.20 shows annual mean wind vectors and isotachs at 200 hPa. Jet streams can be observed in the mid-latitudes as bands of high wind speeds in the upper troposphere. If pollutants reach these altitudes then they can be transported thousands of kilometers very quickly. Thus, jet streams can cause pollution to be transported on intercontinental scales. For example, pollution emissions from East Asia can travel across the North Pacific to North America and from North America across the North Atlantic towards Europe and finally from Europe to the Arctic and Asia. Average intercontinental transport timescales are on average 2–3 weeks (Liu & Mauzerall, 2005).

How can upper-level jet streams influence pollution transport? A good example of this is the June 2011 eruption of the Puheyue-Cordon Caulle in Chile (Fig. 5.21A). This volcanic eruption injected particulate matter and SO_2 into the mid-troposphere. Once volcanic ash was in the upper troposphere it was transported by the southern hemisphere jet stream west towards Australia (Fig. 5.21B), where it caused flight disruptions for several days. Fig. 5.21C shows a

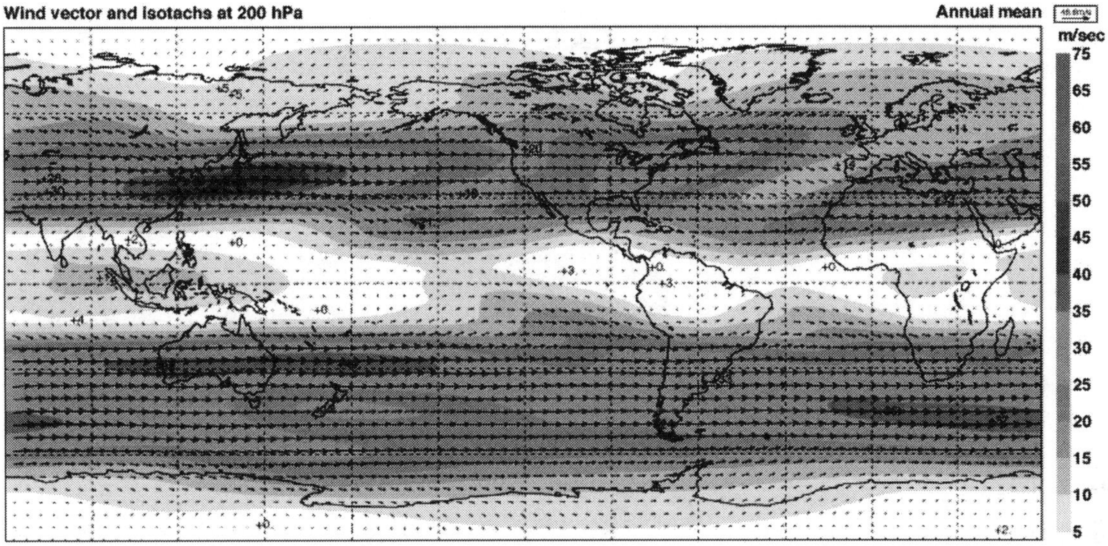

FIGURE 5.20 Upper-level winds ERA-40 annual mean wind vectors and isotachs at 200 hPa (Kållberg et al., 2005). *Data from Kållberg, P., Berrisford, P., Hoskins, B., Simmons, A., Lamy-thépaut, R., & Hine, S. (2005). ERA-40 Atlas. https://sites.ecmwf.int/era/40-atlas CC BY 4.0.*

model simulation of SO_2 emitted from the volcano. There is a relatively uninterrupted zonal southern hemisphere jet during this time allowing SO_2 to circumnavigate the globe, going all the way to Australia and then all the way background to Chile again within 10 to 15 days.

5.5.5 Summary

In this section we looked at global circulations and annual cycles in pollution concentrations and showed how they linked to the movement of the ITCZ. Examples of long-range transport of pollutants include low-level transport of Saharan dust to Central America by the tropical trade winds, upper-atmosphere transport of volcanic ash by the mid latitude jet stream and the seasonal cycle of pollution concentrations caused by monsoon circulations

5.6 Questions

5.6.1 Atmospheric stability and pollution transport

Profiles A, B, and C in Table 5.1 are taken in the UK. Plot the profiles on separate height vs θ graphs. Identify each layer (e.g., mixed layer, residual layer, stable layer, free atmosphere). Estimate what time of day and year the soundings were made. For profile C, sketch the typical plume dispersion from chimneys with heights of 50 and 200 m.

5.6.2 Convective instability and CO transport

What do the LCL and CAPE represent and how can they be found on a tephigram? If the atmosphere is absolutely unstable what will happen to a parcel of polluted air near the surface?

5.6.3 Synoptic advection and pollution transport

Fig. 5.22C shows PM_{10} measurements from 3 Reading urban roadside sites for the period of March 1–25, 2015. All three sites show high variability because they are located close to traffic road sources of PM. There is a double rush hour peak each day and also some longer time scale variability in which PM_{10} gradually builds up over a few days to a

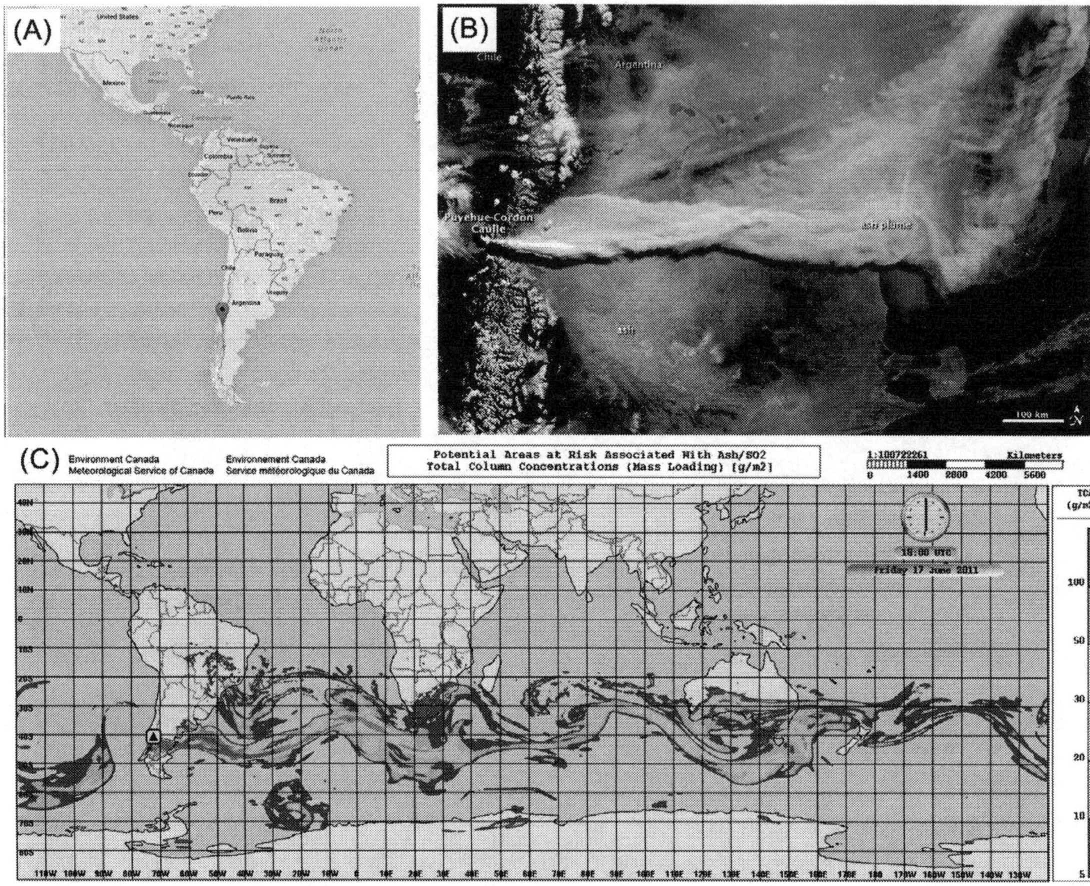

FIGURE 5.21 Jet stream (A) Map of South America showing the location of the Puyehue-Cordon Caulle volcano in Chile, (B) dispersion of volcanic ash following the 2012 eruption, (C) simulation of SO_2 column loading using the MLDPO model by the meteorological service of Canada. *From https://all-geo.org/volcan01010/2011/06/puyehue-cordon-caulle-ash-circles-the-globe.*

TABLE 5.1 Table of potential temperature profiles.

		Profile A		Profile B		Profile C	
Height (m)	Pressure (hPa)	T (°C)	θ (K)	T (°C)	θ (K)	T (°C)	θ (K)
0	1013	25	297	15	287	7	279
100	1000	20	293	18	291	4	277
500	950	15	292	18	295	0	277
1000	890	10	292	15	297	−5	277
1500	840	5	292	10	298	−10	276
1750	810	3	292	8	298	−8	282
2000	780	0	293	10	303	−10	283
2500	740	5	303	8	306	−11	287
3000	700	2	305	4	307	−12	289

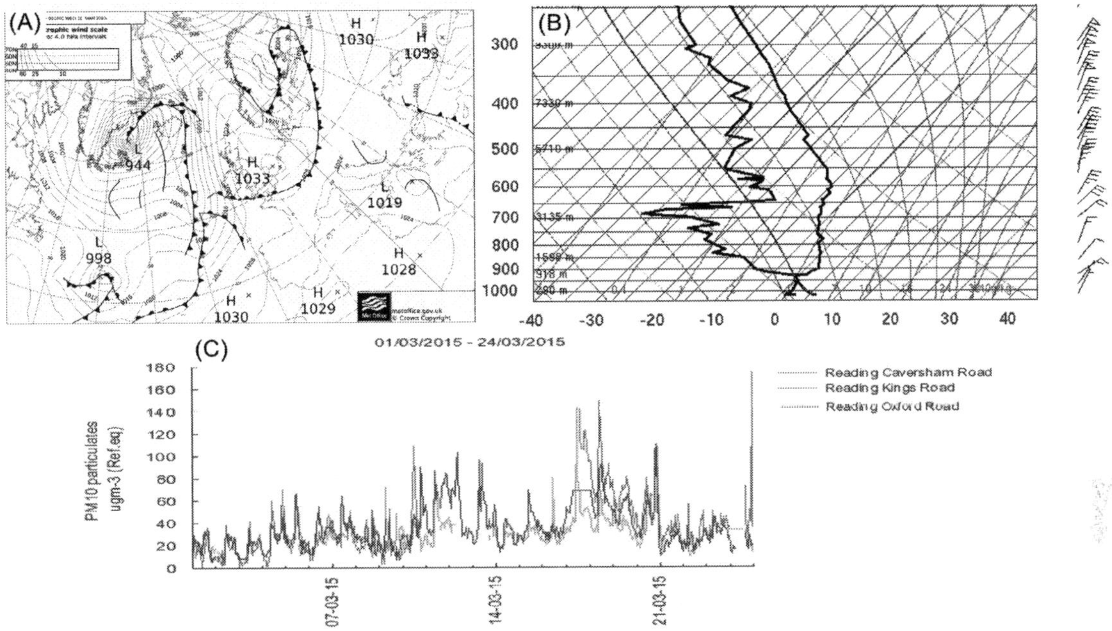

FIGURE 5.22 Anticyclonic subsidence (A) UK Met Office synoptic chart at 00 UTC on March 19, 2015. (B) Radiosonde ascent from Nottingham at 12 UTC on March 19, 2015, (C) PM10 measurements at 3 Reading urban roadside sites from March 1–25, 2015. © *Crown copyright 2021 Defra via uk-air.defra.gov.uk, licensed under the Open Government Licence.*

week. In particular between March 1–11 and from March 17–20. During the latter period PM_{10} measurements are close to the EU guidelines for health of 50 $\mu g\,m^{-3}$. Use the synoptic chart in Fig. 5.22A and skew-T-log p diagram (see Section 5.3.4 for a description of skew-T log p thermodynamic diagram) in Fig. 5.22B to explain the meteorological processes leading to high PM_{10} concentrations during this period.

5.6.4 Jet stream and volcanic ash transport

The volcanic ash in Fig. 5.21B is emitted into the mid-latitude atmosphere (40°S) at a height of approximately 10 km above the surface. How long would it take to reach the surface via sedimentation if it was falling at a terminal velocity of 0.4 $cm\,s^{-1}$? Would you expect the ash particles to circumnavigate the Earth before they are deposited on the surface? What other factors might affect whether the volcanic ash is transported the entire way around the globe?

Solutions to the end-of-chapter questions will be available at https://www.elsevier.com/books-and-journals/book-companion/9780128225912.

References

Chemel, C., & Burns, P. (2015). Pollutant dispersion in a developing valley cold-air pool. *Boundary-Layer Meteorology, 154*(3), 391–408. Available from https://doi.org/10.1007/s10546-014-9984-5, http://www.wkap.nl/journalhome.htm/0006-8314.

Dacre, H. F., Gray, S. L., & Belcher, S. E. (2007). A case study of boundary layer ventilation by convection and coastal processes. *Journal of Geophysical Research: Atmospheres, 112*(D17). Available from https://doi.org/10.1029/2006jd007984.

Dacre, H. F., Harvey, N. J., Webley, P. W., & Morton, D. (2016). How accurate are volcanic ash simulations of the 2010 Eyjafjallajökull eruption? *Journal of Geophysical Research: Atmospheres, 121*(7), 3534–3547. Available from https://doi.org/10.1002/2015jd024265.

Kållberg, P., Berrisford, P., Hoskins, B., Simmons, A., Lamy-thépaut éLamy-thépaut, S., & Hine, R. (2005). *ERA-40 Atlas.*

Lee, S. H., Kim, S. W., Trainer, M., Frost, G. J., McKeen, S. A., Cooper, O. R., Flocke, F., Holloway, J. S., Neuman, J. A., Ryerson, T., Senff, C. J., Swanson, A. L., & Thompson, A. M. (2011). Modeling ozone plumes observed downwind of New York City over the North Atlantic Ocean during the ICARTT field campaign. *Atmospheric Chemistry and Physics, 11*(14), 7375–7397. Available from https://doi.org/10.5194/acp-11-7375-2011.

Lewis, A. C., Evans, M. J., Methven, J., Watson, N., Lee, J. D., Hopkins, J. R., Purvis, R. M., Arnold, S. R., McQuaid, J. B., Whalley, L. K., Pilling, M. J., Heard, D. E., Monks, P. S., Parker, A. E., Reeves, C. E., Oram, D. E., Mills, G., Bandy, B. J., Stewart, D., ... Crosier, J. (2007). Chemical composition observed over the mid-Atlantic and the detection of pollution signatures far from source regions. *Journal of Geophysical Research: Atmospheres, 112*(D10). Available from https://doi.org/10.1029/2006jd007584.

Liu, J., & Mauzerall, D. L. (2005). Estimating the average time for inter-continental transport of air pollutants. *Geophysical Research Letters, 32*(11), 1–4. Available from https://doi.org/10.1029/2005GL022619.

Methven, J., Arnold, S. R., Stohl, A., Evans, M. J., Avery, M., Law, K., Lewis, A. C., Monks, P. S., Parrish, D. D., Reeves, C. E., Schlager, H., Atlas, E., Blake, D. R., Coe, H., Crosier, J., Flocke, F. M., Holloway, J. S., Hopkins, J. R., McQuaid, J., ... Williams, P. I. (2006). Establishing Lagrangian connections between observations within air masses crossing the Atlantic during the international consortium for atmospheric research on transport and transformation experiment. *Journal of Geophysical Research: Atmospheres, 111*(D23). Available from https://doi.org/10.1029/2006jd007540.

Peake, D. L., Dacre, H. F., Methven, J., & Coceal, O. (2014). Meteorological factors controlling low-level continental pollutant outflow across a coast. *Atmospheric Chemistry and Physics, 14*(23), 13295–13312. Available from https://doi.org/10.5194/acp-14-13295-2014, http://www.atmos-chem-phys.net/volumes_and_issues.html.

Peixoto, J. P., Oort, A. H., Covey, C., & Taylor, K. (1992). Physics of climate. *Physics Today, 45*(8), 67. Available from https://doi.org/10.1063/1.2809772, −67.

Peyridieu, S., Chedin, A., Tanre, D., Capelle, V., Pierangelo, C., Lamquin, N., & Armante, R. (2010). Saharan dust infrared optical depth and altitude retrieved from AIRS: A focus over North Atlantic—comparison to MODIS and CALIPSO. *Atmospheric Chemistry and Physics, 10*, 1953–1967.

Raatikainen, T., Hyvärinen, A. P., Hatakka, J., Panwar, T. S., Hooda, R. K., Sharma, V. P., & Lihavainen, H. (2014). The effect of boundary layer dynamics on aerosol properties at the Indo-Gangetic plains and at the foothills of the Himalayas. *Atmospheric Environment, 89*, 548–555. Available from https://doi.org/10.1016/j.atmosenv.2014.02.058, http://www.elsevier.com/locate/atmosenv.

Chapter 6

Physics of the planetary boundary layer

Alexander Baklanov

Climate and Geophysics, Niels Bohr Institute, University of Copenhagen, Copenhagen, Denmark

6.1 Planetary boundary layer structure and properties

6.1.1 Definitions

The planetary boundary layer (PBL), also called the atmospheric boundary layer (ABL), is the lower, essentially turbulent atmospheric layer that responds to spatial or temporal changes in properties of, or impacts from, Earth's surface (see Fig. 6.1). PBL creates an interface between the surface and free troposphere extended up to the tropopause. There are several criteria and definitions of the PBL upper boundary. One of the characteristics of the top of the PBL is that the turbulence is near zero. The principle of the lidar-based PBL height retrieval is that a sharp decreasing of the aerosol load from the boundary layer to the free troposphere exists at the top of the PBL. It can also be defined as the height at which wind velocity approaches 95% of its value in the free-flow. It strongly varies in space and time, ranging from hundreds of meters (or tens of meters in polar regions) to a few kilometers. Over deserts, in equatorial or tropical regions, the PBL can reach up to 4–5 km.

The European Cooperation in Science and Technology (COST) Action 710 (Harmonization of the preprocessing of meteorological data for atmospheric dispersion models) defined the daytime PBL height as "the height of the layer adjacent to the ground over which pollutants or any constituents emitted within this layer or entrained into it become vertically dispersed by convection or mechanical turbulence within a time scale of about an hour" (COST-710, 1998).

The height of the PBL or ABL is a key parameter for air quality analysis, pollutant dispersion and quantification of pollutant emissions and sources. The PBL controls the interactions of the atmosphere with the oceans and land and determines the air volume available for the dispersion of all atmospheric constituents, including anthropogenic pollution and water vapor, emitted at the Earth's surface. Hence the PBL contributes to the assessment of the pollutant concentration near the surface and the PBL height is thus a key parameter of all air pollution models.

Despite its critical importance, the PBL cannot be directly measured but has to be estimated by upper-air instruments. The PBL height can consequently be estimated by the measurement of mechanical turbulence, of the temperature enabling

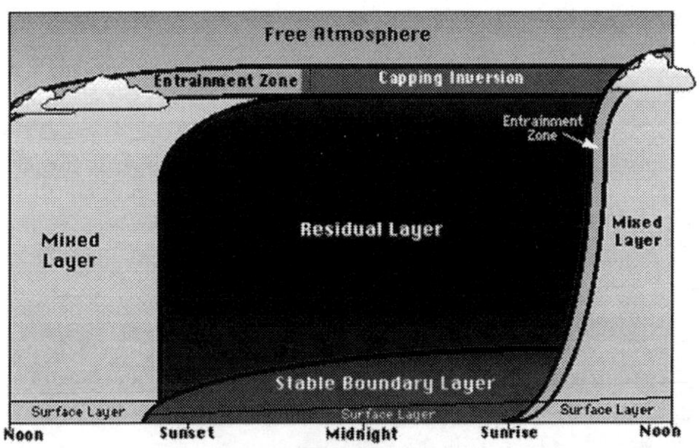

FIGURE 6.1 Structure and diurnal evolution of planetary boundary layer. *Adapted from Stull, R. B. (1988).* An Introduction to Boundary Layer Meteorology. *Kluwer Academic Publishers, Dordrecht.*

Air Quality. DOI: https://doi.org/10.1016/B978-0-12-822591-2.00006-8

convection or of the concentration of PBL constituents. These detection methods are based on various atmospheric parameters, various measuring instruments and different analysis algorithms, leading to several PBL height estimations that are not always consistent with each other.

6.1.2 Vertical structure, stability and temporal variability of planetary boundary layer

The PBL can be subdivided into three separate component layers: the surface layer, the mixed layer, and the stable layer. The surface layer is the sublayer closest to the earth where turbulent stresses are relatively constant (varying by less than 10% of their magnitudes). Therefore, the height of the surface layer is roughly estimated as 10% of the height of the PBL. Between the surface layer and the earth's surface is a very thin layer called the microlayer. In this layer (which is only a few centimeters thick), molecular viscosity dominates over turbulent motion. Above the surface layer is either the mixed layer or the stable layer, depending on the temperature structure of the PBL.

As a general rule, the surface layer is the bottom 20 to 200 m of the PBL. Sometimes this layer is known as the constant flux layer, since whatever turbulent fluxes there might be are constant, or uniform, with height within this layer. There are a number of meteorological variables that are measured in the surface layer. Certainly, temperature is an important quantity of interest in the surface layer. The temperature structure depends on heating during the day and cooling during the night.

As demonstrated in Fig. 6.1, the PBL varies between day and night. During the day, inversion layers formed during the night break up as a consequence of the turbulent rise of heated air. The boundary layer stabilizes shortly before sunset, and remains stable during the night, forming a daily cycle. During winter and cloudy days, incomplete breakup of nighttime layering can lead to persistence of atmospheric conditions established in previous days. In Polar regions during Polar nights in wintertime, the stable PBL can persist for several days or weeks.

During the daytime, surface heating leads to convective motion in the PBL. This surface heating results in the formation of rising warm air, while radiative cooling from clouds generates sinking cooler air. The convective motion induces significant turbulence, effectively mixing the air within this layer. Because of the convective motion and significant mixing of air, this sub-layer is called the convective layer or mixed layer (usually actively used for dispersion modeling). The height (or depth) of the mixed layer is typically denoted as h. Above the mixed layer lies a stable layer which prevents the continued upward motion of thermals. This stable layer also restricts turbulence, preventing frictional influences from reaching above the PBL. Known as the entrainment zone (EZ), this stable layer is where air from above the PBL becomes incorporated into the mixed layer. During the day, the mixed layer reaches heights over one kilometer and makes up the entire layer of the PBL above the surface layer. However, the mixed layer vanishes with the sun as the thermally driven convection ceases.

Until recently, atmospheric PBLs were classified as stable, neutral, or unstable, based on near-surface stratification, without consideration for processes at its upper boundary. This classification is understandable, as boundary-layer meteorology primarily focuses on mid-latitudinal PBLs over land, characterized by large diurnal variations, with typically unstable stratification during daytime and stable stratification at night. Typical *nocturnal stable* PBLs develop against the well-mixed, and thus almost neutrally stratified *residual layer* that remains from the daytime convective PBL (CPBL). This scheme is presented in all textbooks, with the result that "*nocturnal*" and "*stable*" PBL (SPBL) are often considered synonymously.

These simple categories were, however, shown as more complex by Zilitinkevich et al. (2007), who realized that diurnal variations of surface temperature are weak during long polar nights and over the open sea. These periods can last several days or weeks over oceans and can extend to more than a month over both oceans and land during the polar night or day. During such periods, residual layers do not occur, and the PBL then develops against the free troposphere, where global horizontal heat transfer *permanently* reestablishes stable stratification with a typical Brunt-Vaisala frequency (N) of approximately $10^{-2}\,s^{-1}$ (Stone, 1972). Interactions between the PBL and free troposphere thus suppress turbulence in the upper PBL and dramatically reduce its height (Zilitinkevich & Baklanov, 2002). Long-lived PBLs with no buoyancy flux at the surface were traditionally considered neutral, but they are in fact strongly affected by the background stratification in the troposphere, which makes them several times shallower than truly neutral PBLs (Zilitinkevich & Esau, 2002). Similarly, long-lived stable PBLs are also several times shallower than usual nocturnal stable PBLs (Zilitinkevich et al., 2007).

6.1.3 Monin-Obukhov similarity theory and stability functions

In general, the vertical turbulent surface fluxes of heat (H_0), moisture (E_0), and momentum (τ_0) are described with the mean flow characteristics through exchange coefficients, with a closure of first order applied for the turbulence flow. The most widely used first-order closure is the Monin–Obukhov similarity theory (MOST).

The Monin and Obukhov (1954) similarity theory (MOST) describes the nondimensionalized mean flow and mean temperature in the surface layer under nonneutral conditions as a function of the dimensionless height parameter. Similarity theory is an empirical method that describes universal relationships between nondimensionalized variables of the atmosphere. It is extensively used in boundary layer meteorology since relations in turbulent processes are not always resolvable from first principles.

An idealized vertical profile of the mean flow for a neutral PBL is the logarithmic wind profile derived from Prandtl's mixing length theory, which states that the horizontal component of mean flow is proportional to the logarithm of height. The MOST further generalizes the mixing length theory to nonneutral conditions by using so-called "universal functions" of dimensionless height to characterize vertical distributions of mean flow and temperature. The Obukhov length (L), a characteristic length scale of surface layer turbulence derived by Obukhov (1946), is used for the nondimensional scaling of the actual height. MOST represents a significant milestone in modern micrometeorology, providing a theoretical basis for micrometeorological experiments and measurement techniques.

MOST relates the vertical behaviour of nondimensionalized mean flow and turbulence properties within the atmospheric surface layer (the lowest 10% or so of the ABL) to the Monin–Obukhov key parameters, namely the height above the surface, z, the buoyancy parameter, g/θ (where g is gravitational acceleration, and θ is potential temperature), the kinematic surface stress τ_0, and the surface potential temperature flux, so-called the kinematic sensible heat flux (Q_0) is calculated as

$$Q_0 = H_0/\left(\rho C_p\right) = \overline{(w'\theta')}_0, \tag{6.1}$$

where ρ is the air density, H_0 is a dynamic sensible heat flux at the surface, C_p is the specific heat of air at constant pressure, and $\left(\overline{w'\theta'}\right)_0$ is the covariance of vertical velocity w with the potential temperature near the surface (the virtual potential temperature and corresponding fluxes can be used instead for humid atmosphere), an overbar denotes an average, and a prime denotes a deviation from the average. The key parameters can be used to define a set of four scales for the surface layer:

$u_* = \left(\tau_0/\rho\right)^{1/2}$, a velocity scale, the friction or shearing velocity;

$T_{*SL} = Q_0/u_*$, a surface-layer temperature scale;

$$L = \frac{-u_*^3\theta}{\kappa g Q_0}, \text{ a length scale called the Obukhov length;} \tag{6.2}$$

z, a length scale corresponding to the height above the ground,

where κ is the von Kármán constant. These key scales can then be used in dimensional analysis to express all surface-layer flow properties as dimensionless universal functions of z/L.

The nondimensional gradients of wind, temperature and humidity are given by

$$\phi_m = \frac{dU}{dz}\frac{\kappa z}{u_*}, \quad \phi_h = \frac{d\theta}{dz}\frac{\kappa z}{T_*}, \quad \phi_q = \frac{dq}{dz}\frac{\kappa z}{q_*}. \tag{6.3}$$

are expressed by empirical relations, with slightly different behavior from different studies. In a review article, Högström (1996) suggested:

$$\phi_m = \left(1 - 19z/L\right)^{-1/4} \tag{6.4}$$

$\varphi_h = \left(1 - xz/L\right)^{-1/2}$ for unstable conditions, and

$$\phi_m = \phi_h = 1 + xz/L. \tag{6.5}$$

for stable stratification, where x is about 5 (Businger et al., 1971).

6.1.3.1 Worked example 6.1: definitions of surface and planetary bounder layers, their structure and height/depth

Three questions for this exercise:

#1: PBL structure

The PBL can be subdivided into four separate component layers: the surface layer, the mixed layer, the stable layer, and the residual layer (see Fig. 6.1).

Question: Which of these is not a component of the PBL?

1. stable layer
2. unstable layer
3. mixed layer
4. residual layer

Answer: The residual layer, as it is a near stable layer formed at nighttime as a remanent from the daytime mixing layer. The residual layer characterized by weak sporadic turbulence and initially uniformly mixed potential temperature and pollutants remaining from the mixed layer of the previous day. Below the statically neutral residual layer is the stable boundary layer in contact with the radiatively cooled ground, and above it is the capping inversion that together with the residual layer separate the boundary layer air from free-atmosphere air (see Fig. 6.1).

#2: PBL height

PBL height is depending on temperature stratification, diurnal and seasonal cycles and longitude of the position.

Question: In what period and in what region the PBL height will be minimal?

1. Tropic region nighttime
2. Polar region daytime summer
3. Polar region nighttime winter
4. Tropic region daytime
5. Polar region daytime winter
6. Polar region nighttime summer

Answer: Polar region nighttime winter

#3: Surface layer depth

There are many meteorological and surface parameters that effect on the surface layer characteristics, including its depth/height.

Question: In what conditions the surface layer depth will be the highest?

1. Overcooling grass surface
2. Overwarming daytime urban area
3. Overcooling nighttime urban area
4. Over water surface

Answer: Overwarming daytime urban area.

6.2 Planetary boundary layer representation for numerical modeling

Ensuring adequate representation of PBL processes is crucial for numerical modeling of weather, climate, air pollution, and atmospheric dispersion. Thanks to continuously increasing computer power, the model-grid resolution is continually enhancing. This allows for better representation of PBL processes and calls for better knowledge of physical, chemical, and biological mechanisms controlling these processes. Thus, the development and requirements of advanced PBL parameterizations in numerical weather prediction (NWP) and Chemical weather-air quality forecasting (CW-AQF) models are closely linked with advances in computing capability, ensuring they are suitable for the purpose and within the capabilities of computers. A short summary of PBL parameterizations or schemes is given in the following sections. For more comprehensive discussions on these topics, refer to Stull (1988), Garratt (1994), COST-710 (1998), Baklanov and Grisogono (2007), Baklanov et al. (2011), Holton (2012), Salby (2012), Zilitinkevich (2013), Pielke (2013), and Sokhi et al. (2018).

Despite increasing resolution in mesoscale models, it remains insufficient to resolve crucially important processes controlling the heat and mass transfer. Examples where the resolution of models can be particularly limiting include deep canopies, complex terrain, stormy seas, PBLs in very stable stratification typical of wintertime at high latitudes, and long-lived PBLs typical of the polar night (stable stratification) or polar day (unstable stratification). Additionally, contemporary models are incapable of resolving the PBL upper boundary and quantifying vertical exchanges between the PBL and free troposphere.

In numerical models, PBL can be considered as a single layer which is typical in global or simplified models (see Deardorff, 1972; Mahrt, 1974; Smith & Mahrt, 1981)). However, in most regional models, the PBL is resolved into several sublayers to more accurately represent specific features over complex terrain or across land-water borders characterized by large horizontal gradients of meteorological variables (see overview in Sokhi et al., 2018). Traditionally, the PBL consists of three essentially different parts: roughness layer, surface layer, and PBL core:

- The roughness layer, immediately adjacent to the Earth's surface, is defined for shallow canopies as $0 < z < z_0$ where z_0 is roughness length for momentum (typically an order of magnitude larger than the typical height of roughness elements, h_0). Deep canopy causes a kind of stagnation zone quantified by so-called displacement height, d (essentially shallower than the canopy depth, h_0), so that the roughness layer is defined as $0 < z < (d + z_0)$ (e.g., Pielke, 2013).
- The surface layer is defined as the layer where vertical turbulent fluxes are practically independent of height. It is described by the Monin and Obukhov (1954) similarity theory (MOST). It is conventionally defined as the lower 10% of PBL.
- The PBL core is defined as the upper 90% of the PBL, extending up to the PBL upper boundary.

Notably, the free troposphere above the PBL is characterized by typically weak velocity shears and a highly stable stratification, which significantly reduces the intensity of turbulence. This in turn strongly reduces vertical exchange processes at the PBL upper boundary, so that any impacts from Earth's surface, such as warming or cooling, stronger or weaker surface drag, or emissions of pollutants, rapidly affect the entire PBL, but take longer to penetrate into the free troposphere and considerably weakened. The difference in nature between the free troposphere and unstably stratified and, hence, strongly mixed PBL is especially dramatic. In the case of an unstably stratified PBL, the PBL height is clearly seen in vertical profiles of mean potential temperature and concentration of pollutants emitted at Earth's surface (which almost completely remain within PBL).

6.2.1 Worked example 6.2: wind profile over rough surface

Let's consider the surface layer and boundary layer over a very rough surface (e.g., forest or urban canopies). For simplicity, assume that the surface across our domain is homogeneous. How can we calculate meteorological characteristics, for example, wind profile in such cases? (Fig. 6.2).

Solution:

One commonly used method is to consider the roughness length, z_{0u}, and displacement height, d_{0u}, for the logarithmic formula for the vertical profile of horizontal wind (see Section 6.2 and Chapter 2):

$$u(z) = \frac{u_*}{k}\left[\ln\frac{z - d_{0u}}{z_{0u}} + \Psi_u\left(\frac{z}{L}\right)\right], \tag{6.6}$$

where u_* is the friction velocity, L is the Monin-Obukhov length, Ψ_u is the stability function (for neutral stratification $\Psi_u = 0$).

For the roughness sublayer height, z_0 is typically between 2 and 5 h_0 for forest canopies and around 2 $h0$ for urban canopies. The maximum Reynolds stress occurs within the range of 1.5–2.5 h_0 (COST-715, 2005).

Depending on the building density the displacement height: $d_0 = 0.7h_b$ for $0.3 < \lambda_p < 0.5$; $0.1 < \lambda_f < 0$ (Grimmond & Oke, 1999) and $d_0 = h\lambda_{p0.29}$ for low building density $\lambda_p < 0.29$ (Kutzbach, 1961), where plan areal fraction $\lambda_p = A_p/A_t$; $\lambda_f = A_p/A_t$. For the displacement height, d_0, in most of the urban districts the rule of thumb $d_0 \sim 2/3h_b$ (h_b is the mean building height) applies quite well.

For more precise modeling, the key parameters for urban models of different scales have more sophisticated requirements presented in Table 6.1 (COST-715, 2005).

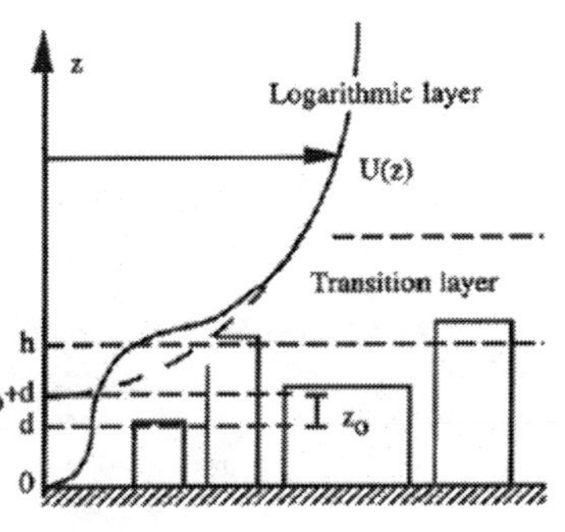

FIGURE 6.2 Schematic representations of the urban boundary layer (COST-715, 2005).

TABLE 6.1 Key parameters to be studied for models of different scales in the urban environment.

Mesoscale models	Submeso scale models	Street canyon scale models
Z_o, Z_{OT}	$z_o(x)$,d(x)	
h_{UBL}	L_o, L_a, Z_*	Detailed geometry
"Surface" fluxes (effective)	u^{IS}, H^{IS}, general: X_*^{IS}	$\bar{u}(h)$ second velocity scale for horizontal transport
Anthropogenic heat flux (nonsurface) at some representative height	Dispersive fluxes	Heat exchange at vertical and horizontal building surfaces
Profiles of turbulent fluxes	Profiles of turbulent fluxes	Characteristic velocity variance in street canyon
Higher-order moments?	Higher-order moments (skewness, . . .)	Higher order moments?
Synoptic forcing, average albedo	Mesoscale stability, albedo(x)	

Notes: z_o = roughness length (subscript T for temperature); d = zero plane displacement; (x) denotes a spatial variability; h_{UBL} = height of the UBL; L_C, L_a are characteristic length scales; z_* is the height of the roughness sublayer, and lowest height within the inertial sublayer; u_*^{IS} refers to the friction velocity in the inertial sublayer (and correspondingly for the other fluxes; cf. x_*^{IS}); and h = average building height.
Source: From COST-715 (2005). Meteorology applied to urban air pollution problems. In Fisher, B., Joffre, S., Kukkonen, J., Piringer, M., Rotach, M., & Schatzmann, M. (eds.), Final Report ofCOST Action 715 (276 pp.). Demetra Ltd Publishers.

6.3 Representation of interactions between the atmosphere and various earth surfaces

Accurate parameterizations of surface fluxes of heat, moisture, and momentum are crucial in modeling of practically all meteorological processes, by controlling the mass-energy, and momentum exchanges, between the atmosphere and Earth's surface. Turbulent fluxes and turbulent energies, both near the surface and within the PBL, to a large extent, control meteorological fields, namely, the mean flow and turbulent mixing. Hence, in practice, this includes the transport and dispersion of pollutants, thermal and moisture regimes, and practically the most important near-surface phenomena, such as fogs, frosts, storms and gustiness, all of which are essential for aviation, energy sector, transportation and agriculture. Surface fluxes are key factors in the initiation and development of hurricanes. Below, we briefly describe the parameterization of turbulent fluxes over various surfaces of Earth for CW-AQF models. Additional information can be found, for example, in Chapter 3 of Sokhi et al. (2018).

6.3.1 Homogeneous terrains

In a broad framework, vertical turbulent fluxes of momentum (τ_0), heat (H_0), and moisture (E_0) at Earth's surface are calculated through the mean flow characteristics using the drag and heat- or moisture-transfer coefficients derived from various (usually first-order) turbulence closure models. Presently, only vertical turbulent fluxes are parameterized for the grids of mesoscale models, whereas very little is known about horizontal fluxes. As emphasized by Pielke (2013), subgrid-scale fluctuations of modeled variables can be of the same order or even larger than resolvable variations. As an example, both resolved wind speed and fluctuations of wind-speed could be of the order of 5 $m\,s^{-1}$. Similarly, in a grid volume where the averaged vertical wind speed is close to zero, vertical turbulent heat flux can still be significant due to pronounced fluctuations in vertical-velocity and potential-temperature (especially in convective PBLs).

For homogeneous terrains the widely-used instrument for calculating surface fluxes is MOST, which is treated as a type of first-order closure model and, thus, expressing turbulent fluxes via the calculated vertical profiles of mean-flow parameters. Examples of the surface flux schemes commonly used in mesoscale meteorological models, for example, in the weather research and forecast (WRF) model, are discussed in Jimenez et al. (2012) and ARW-WRF (2015). Most parameterizations, such as Blackadar (1962), MRF and Eta schemes, employ MOST. However, the Burk and Thompson (1989) PBL scheme, utilized in models like HIRLAM, employs Mellor and Yamada (1982) closure level-2.5, which includes the prognostic equation for turbulent kinetic energy (TKE). Studies by Quante et al. (2009) and Gryning et al. (2011) have shown significant variations in the ability of models to calculate the surface fluxes of sensible heat and momentum. It is not surprising that modern NWP and CW-AQF models, for example, WRF-Chem, are equipped with many different options for the surface-layer scheme to be chosen by the user.

6.3.2 Impact of surface inhomogeneities

The following paragraphs briefly explore how different surfaces are treated in air quality forecasting models.

Urban air pollution and atmospheric chemical transport models have distinct primary requirements when incorporating urbanization effects. These requirements depend on the scale of the models and their functional purposes, such as whether they are intended to be used for forecasting, assessment, environmental and air quality applications or emergency preparedness. In the case of incorporating urban effects into urban- and regional-scale atmospheric pollution models, a key primary requirement is the improvement of urban meteorological fields. Urban cases are essential for air pollution forecasting applications, particularly when information is targeted at the public, and are discussed separately in Section 6.6.

As forests are typically taller than other vegetation surface types, air flow is influenced by larger aerodynamic roughness and displacement heights (Dupont & Patton, 2012; Finnigan, 2000). Both the height and density of the forest are important for shaping airflow (Gayev & Hunt, 2007). The forest canopy reduces the amount of shortwave radiation reaching the ground while simultaneously emitting and absorbing longwave radiation as a function of forest type, which is characterized by the leaf area index (LAI). The aerodynamic and thermal parameters change during the year; for example, snow cover strongly affects the surface albedo. Recent advancements in micrometeorological studies which have resulted improvements in parameterizations (e.g., Ershadi et al., 2014) include: (i) the exchange of heat and moisture between trees and air, (ii) local momentum flux over the forest, including stability effects on it, and (iii) transfer of solar radiation within the forest.

It seems that while there are improvements in the horizontal resolution of models, less attention has been paid to the treatment of surface heterogeneity within a grid square. Heterogeneity is commonly observed in various environments such as forests, fields, swamps, and lakes with scales smaller than a few kilometers. Particularly for forests and cities, many models do not take into account certain important effects, such as displacement height and roughness sublayer. These omissions present major challenge for the parameterization of turbulent surface fluxes, especially in conditions of stable stratification, where the constant-flux layer may be extremely thin or entirely absent. Other effects that need further attention include:

- effects of forest on the transfer of longwave and shortwave radiation;
- representation of vertical profiles within forests as current models often treat forest effects as part of the exchange between the surface and the lowest model level, despite forests extending to higher altitudes;
- application of the tile/mosaic method in its basic form, without paying attention to the subgrid-scale variability at the lowest atmospheric model level.

The turbulent flux parametrizations over heterogeneous surfaces are implemented in different models through different surface layer formulations and canopy models (for an overview, see Baklanov et al., 2014; Kukkonen et al., 2012; Zhang, 2008). Surface fluxes depend considerably on the land cover. While some models consider only one land cover type per grid cell, others parameterize the effects of subgrid-scale land use on turbulent fluxes (e.g., Schlünzen & Katzfey, 2003). Above the sea surface, drag is often parameterized using the Charnock formula (Charnock, 1955), particularly in areas where wave data is lacking. It works reasonably well for flat coastal regions, while recent studies suggest a different approach for deeper water (Foreman & Emeis, 2010).

6.4 Turbulence closures and boundary-layer parameterizations

The key aspect of PBL is its turbulent nature. Turbulence is several orders of magnitude more effective at transporting quantities than molecular diffusivity. It is turbulence that allows the PBL to respond to changing surface forces, such as frictional drag, evaporation and transpiration, heat transfer, pollutant emission, and terrain-induced modification. As a result, about fifty percent of the atmosphere's kinetic energy is dissipated within the PBL. Understanding the PBL is fundamental for the numerical modeling of meteorological and transport processes, from micro to global scales.

The levels of turbulence in an ABL with zero heat flux at the surface in uniform, homogeneous, steady conditions are determined by the following fundamental parameters (COST-710, 1998):

- wind velocity at the upper boundary of the layer (often identified with geostrophic wind speed G);
- Coriolis parameter f quantifying the influence of Earth's rotation;
- roughness lengths for momentum and scalars characterizing the surface drag and heat/mass transfer and dependent on the height and geometry of roughness elements from very small (sand particles in a desert) to very large (buildings);
- the background thermal stratification of the atmosphere characterized by the Brunt-Väisälä frequency in the atmosphere above the PBL layer.

When the surface heating is nonzero, the surface heat flux H becomes the other driving force setting up the structure of the boundary layer. During the day, when the flux of heat carried from the surface into the atmosphere by convection is usually positive, the heat flux acts as an additional source of turbulence over and above that caused by the wind. At night, the heat flux is typically negative, leading to the dissipation of energy from wind-induced turbulence, resulting in considerably lower turbulence levels for a given wind speed. Since the interests of boundary layer meteorology and dispersion modeling are in the main velocity and length scales, it is usual to introduce a new length scale L_* into the equations describing wind, temperature, and turbulence profiles. L_* is the Monin-Obukhov length, equal to u_*^3/H apart from some constant of proportionality. In convective boundary layers it is usual to introduce the convective velocity scale w_*, which is proportional to $(hH)^{1/3}$.

Most of the operationally-used closures employ the concept of down-gradient turbulent transport, implying that turbulent fluxes are proportional to and oriented along mean gradients of the transporting properties. The proportionality coefficients in such relations, called eddy viscosity K_M, eddy conductivity K_H and eddy diffusivity K_D, are considered the major (or only) unknowns to be determined from turbulence closure theory. The modern framework of this theory is based on the paradigm initiated by Kolmogorov (1941/1942). His turbulence closure theory was based on (i) budget equation for the TKE, E_K, (ii) the definition of the turbulent velocity scale, u_T, as the square root of TKE: $u_T = E_K^{1/2}$, (iii) Prandtl's concept of turbulent exchange coefficients, $K_M \sim K_H \sim K_D$, as proportional to the product $u_T l_T$, where l_T is turbulent length scale identified in neutrally stratified boundary-layer flow with the height above the surface, z. This revolutionary idea has resolved the problem of the dissipation rate of TKE: $\varepsilon_K \sim E_K^{3/2}/l_T$ and yielded constructive turbulence-closure theory for neutrally stratified sheared flows.

In due time, this vision of turbulence has revolutionized the understanding and modeling of turbulence in neutrally stratified flows. Later on, followers of Kolmogorov extended it without proof to stably and unstably stratified flows. This extended version gave rise to the streamlining of turbulence closure theory, in particular, to the models in operational use. In moderately stable or unstable stratifications, this approach shows quite good results. However, it fails in strongly stable stratification. In particular, the conventional approach erroneously prescribes degeneration of turbulence in supercritical stratification typical of the free troposphere and ocean thermocline (yet corrected only approximately and without physical explanation). Moreover, conventional theory and closure models become fully erroneous when applied to horizontal diffusion in unstable stratification. The problem of stratified turbulence, typical of the atmosphere problem, is now being resolved. Stable stratification is done via the energy- and flux-budget turbulence closure theory (Zilitinkevich et al., 2007; Zilitinkevich et al., 2009; Zilitinkevich et al., 2013), and unstable stratification, via the concept of self-organization of convective turbulence (Zilitinkevich, 2013).

Parameterization of processes in the stable PBL constitutes a considerable source of uncertainty in modeling air pollution. Despite their sophistication, schemes used in NWP models have limitations that are critical in CW-AQF applications. For example, CW-AQF models require higher vertical resolution within PBLs and improved surface-layer parameterizations, especially in urban-scale air pollution modeling where low-level emissions (e.g., from traffic) within the surface layer are prevalent.

6.4.1 Worked example 6.3: planetary boundary layer type and formation mechanism

As described in Sections 6.1 and 6.5, there are several types of stratified PBLs depending on the mechanisms of their formation. Please recognize, based on the vertical profiles of potential temperature (on graphs *a* and *b*), what are the types of PBL on Fig. 6.3 and suggest what kind of formation mechanisms are important for this case and what theory is the most relevant for modeling such PBL developments?

Solution:

First of all, as we can see, the potential temperature increases with the height in both cases ($Bs < 0$), so in both cases, we have a stably stratified atmosphere. Proceeding from the classification of stratified PBLs:

Simple classification by the sign of the surface buoyancy flux, Bs:

Stable $Bs < 0$, Neutral $Bs = 0$, and Unstable (convective) $Bs > 0$.

This classification is based on surface-flux dominating stratification and disregards the free-flow impact. If we account for mechanisms essentially controlled by persistent stable stratification inherent to the free troposphere, characterized by the Brunt-Väisälä frequency, N, the PBL classification (as suggested by Zilitinkevich et al., 2007) distinguishes:

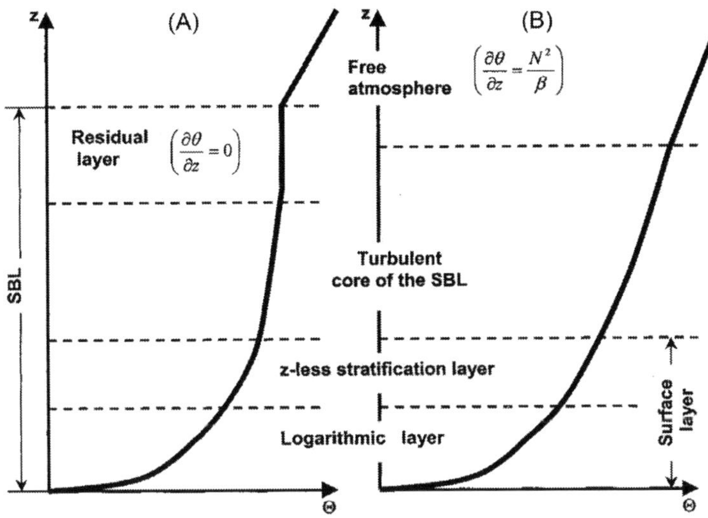

FIGURE 6.3 Thermal stratification for (A) nocturnal and (B) long-living SBL (typical for Arctic winter). *Source: Adapted from Zilitinkevich, S., Esau, I. and Baklanov, A. (2007). Further comments on the equilibrium height of neutral and stable planetary boundary layers. Quarterly Journal of Royal Meteorological Society, 133, 265–271.*

Stable: nocturnal stable (NS) $N=0$
long-lived stable (LS) $N>0$
Neutral: truly neutral (TN) $N=0$
conventionally neutral (CN) $N>0$
Unstable shear-free (convective cells): in two-layer fluid $N=0$
in stratified fluid $N>0$
Unstable sheared (convective rolls): in two-layer fluid $N=0$
in stratified fluid $N>0$

On the left profile we see a well-formed residual layer, preventing impacts of free-flow stability ($N=0$), so it is a classical case of nocturnal stable (NS) SBL. On the right profile the residual layer is almost disappearing ($N>0$), and the potential temperature is still increasing. So, here we have typical case for Arctic winter long-lived stable (LS) PBL. In the latter case, in our models, the mechanism of free-flow stability is very important to consider.

6.5 Planetary boundary layer height and turbulent exchanges through the planetary boundary layer upper boundary

The PBL upper boundary acts as a barrier preventing (or strongly reducing) vertical dispersion in most air quality and dispersion models and essentially controlling extreme weather events and microclimates in NWP and climate models. Methods of calculation of PBL height, h_{PBL}, have been the subject of numerous theoretical and experimental studies since Ekman (1905), who identified the PBL as the steady-state boundary layer in a rotating fluid and derived the relation: $h_{PBL}\sim\sqrt{K_M/f}$, where K_M is a reference value of eddy viscosity and f is Coriolis parameter quantifying the effect of Earth's rotation; and Rossby and Montgomery (1935) whose formulation; $h_{PBL}\sim u_*/f$, where u_* is friction velocity, is still used in meteorological practice as a rough approximation. In reality, the PBL height (as well as other features of PBL) strongly depends on static stability. Its effect is traditionally characterized by just one parameter: vertical turbulent flux of buoyancy at Earth's surface, $F_B=(g/T)F_\theta+0.61gF_q$, where $g=9.81\ \mathrm{ms}^{-1}$ is the acceleration due to gravity, T is absolute temperature, F_θ is the vertical turbulent flux of potential temperature and F_q is the vertical turbulent flux of specific humidity. Accordingly, the three basic types of PBL are distinguished, namely, stable: $F_B<0$ reducing turbulence, neutral: $F_B=0$, and unstable (or convective): $F_B>0$ enhancing turbulence.

For the convective PBL (CPBL), Zubov (1945) has derived a prognostic equation for the CPBL height, h_{CPBL}, indicating its growth as long as F_B remains positive: $dh^2_{CPBL}/dt=2F_B/N^2=0$, where N is Brunt-Vaisala frequency in the free flow beyond the ever-developing CPBL. A more general formulation of the CPBL height equation accounting for turbulent entrainment and the CPBL's upper boundary can be found in Zilitinkevich (1991; 2012).

For the stable PBL (SPBL) that forms in the atmosphere over land after sunset due to the change from $F_B > 0$ to $F_B < 0$, Zilitinkevich (1972) has derived a diagnostic formulation: $h_{SPBL} = C_{SPBL}u_*^2(F_Bf)^{-1/2}$, where $C_{SPBL} = 0.5$ is a dimensionless empirical constant.

In later publications, it has been revealed that the major properties of PBLs, particularly their heights, are essentially different for the two principally distinct PBL types (Zilitinkevich & Esau, 2002; Zilitinkevich et al., 2007):

(i) Short-lived mid- and low-latitude continental PBLs are subjected to pronounced diurnal variations and are, therefore, almost fully controlled by the surface buoyancy flux, F_B.
(ii) Long-lived high-latitude PBLs and those over open ocean keeping their type of stratification (stable or unstable) over long periods from several days or weeks (over ocean) to more than a month (over both ocean and land during polar night or polar day). We recall that long-lived PBLs are essentially controlled by persistent stable stratification inherent to the free troposphere and characterized by the Brunt-Vaisala frequency, N (typically of the order of $10^{-2}\,s^{-1}$) which makes them very shallow and, hence, very sensitive to any external impacts.

In practical CW-AQF applications, the height of the PBL can be derived from meteorological measurements as well as model data. Several methods have also been suggested based on the analysis of turbulence processes. For a comprehensive treatment see the overview in COST-710 (1998); Zilitinkevich and Baklanov (2002). These methods include:

- Bulk Richardson number method;
- Gradient Richardson number method;
- Profiles of TKE, momentum and heat fluxes;
- Theoretical approaches.

In CW-AQF models, turbulent mixing takes place within the PBL, and therefore, the PBL height is estimated at each time step to define up to which model level turbulence mixing exists. When first order closure is used, the height of the PBL is an essential parameter (implicit or explicit) in the description of mixing of meteorological properties (see overview in Seibert et al., 2000).

6.6 Methods for estimating the planetary boundary layer height

In this section, a brief overview of methods to estimate PBL height [or mixing height (MH) which is important for AQ modeling] will be given, as they are mostly well-known (see an extended discussion in Baklanov et al., 2006; Piringer & Joffre, 2005). The MH can be estimated or parameterized based on:

(i) vertical profiles using various probing methodologies,
(ii) surface meteorological measurements (e.g., friction velocity and stability), or
(iii) NWP or other numerical models.

6.6.1 Experimental methods

In general terms, the MH can be estimated experimentally from measured vertical profiles by several means or criteria, for example,

- The level where turbulence or the heat flux diminishes to, say, approximately 5%, of its surface value.
- A level of discontinuity in the wind/temperature/dew point profiles (from radiosonde data).
- The level of strong back-returns from thermal discontinuities (inversions) or of strong decay of back-returns from thermal fluctuations (top of turbulent layer), obtained from sodars and wind-profilers/radio acoustic sounding system (RASS).
- The level of decay in aerosol backscatter signal from lidar probing or ceilometer.
- The level of decay in turbulent motions as measured by pulsed Doppler lidars.

The first method would be the ideal, but it requires turbulence profiles, which are rarely available.

6.6.1.1 Diagnosing the mixing height from radiosonde data

Radiosoundings could, in principle, be used for estimating the MH (including in operational modeling), but sounding stations are usually located outside urban areas. However, in many European countries, they are getting ever closer to or inside of growing urban agglomerations and therefore, for many cities, it is becoming reasonable to analyze radiosonde data for urban MH estimation by selecting wind direction sectors downwind of the city (see Section 6.4.2).

Under unstable conditions, the MH is often identified as the base of an elevated inversion or stable layer using the parcel method (e.g., Holzworth, 1967) or advanced parcel method (e.g., Stull, 1991). The MH is sometimes identified as the height of a significant reduction of air moisture (the "humidity-jump" method, e.g., Lyra et al., 1992).

A standard generic method for estimating the MH is based on the Richardson number approach. This approach includes several variants differing in the formulation, choice of the levels over which the gradients are determined and the value of the critical Richardson number, Ri_c: (i) the gradient Richardson number *Ri*, (ii) the finite-difference *Ri*-number, (iii) the bulk Richardson number, and (iv) the modified Richardson number method. However, the value of Ri_c is not a constant but apparently increases with increasing free flow stability *N* (the Brunt-Väisälä frequency) and most probably depends on factors such as the surface roughness length, the Coriolis parameter and the geostrophic wind shear in baroclinic flows (Zilitinkevich & Baklanov, 2002).

6.6.1.2 Mixing height interpreted from sodar/lidar/radar/ceilometer

The various types of vertical profilers (such as sodars, lidars, radars, ceilometers, etc.) have the potential to provide better and more continuous information on the vertical structure of the ABL and to estimate the MH than radiosondes. Interpretation can be difficult when the lower atmosphere exhibits multiple layers. Their main weaknesses arise from the limitations on their use in urban areas (e.g., due to noise) and the need of expert personnel so they are not usually in continuous operational use. However, they are sometimes used for other purposes (e.g., for nuclear emergency preparedness systems or other monitoring purposes) and thus could be used for MH estimation in urban or semiurban areas.

For **sodars**, the MH can be deduced from the vertical profile of the echo intensity or from a spectral analysis of the vertical velocity. Methods based on the former have been summarized by Beyrich (1997).

The **lidar** measures both Doppler wind velocity and back-scatter intensity. The back-scatter intensity is a function of the system properties, atmospheric attenuation (due to the atmospheric water vapor content), the returned power (and thus the measurement range), and the aerosol loading of the atmosphere. In urban regions, where atmospheric aerosol concentrations are high, there is a large back-scatter signal in most meteorological conditions.

Under well-established convection, **wind profiler radars** (WPRs) can find the mixing height from the reflectivity (roughly the product of humidity gradient and turbulence intensity) profiles. Several examples of the MH estimation by wind profilers in different meteorological conditions are presented by Angevine (2003). He concluded that WPRs are very useful in determining the MH, but other instruments (e.g., ceilometers etc.) greatly help in the interpretation, since simple automatic procedures are expected to work in only very simple situations.

Electromagnetic backscatter is generally the best option for determining the (high) afternoon CBL heights. Within this category, the **ceilometer** measures backscatter intensity. It is found to be a very useful instrument in clear sky conditions as it is the only instrument (compared to RASS/sodar) that directly provide the vertical aerosol profile. Ceilometers, originally designed just for cloud base height measurements, have become available also for atmospheric profiling (Münkel et al., 2002; Räsänen et al., 2000; Rogers et al., 1997). They have an improved vertical resolution (now about 15 m) and a much smaller lower detection range limit (now about 30 m above ground; Emeis et al., 2004).

The **RASS** is an extension of a coherent radar that allows for the measurement of the virtual temperature profile above the radar. RASS temperature retrievals are sensitive to the wind component parallel to the sound beam and the humidity. The influence of the wind on these measurements can be used to retrieve profiles of the horizontal wind components.

6.6.1.3 Commercial aircraft measurements/aircraft meteorological data reporting data processing

New opportunities for estimating the MH and profiles of potential interest to meteorologists and air pollution scientists have recently arisen from the aircraft meteorological data reporting system (AMDAR, 2003). A joint European EUMETNET-AMDAR project (E-AMDAR, 2002) is looking at the use of data from civil aircrafts to complement existing radiosoundings. This E-AMDAR system uses aircraft sensors for measuring wind speed and direction, air temperature, altitude, a measure of turbulence and the aircraft position. Errors in reported wind and temperatures are comparable with those of radiosonde systems.

However, some issues need further consideration, including improvements in processing and the evaluation of AMDAR data relative to other observations/model products. In the absence of radiosonde ascents over cities, AMDAR data offer useful profiles in the lower atmosphere. Standard objective methods for the determination of the MH from profiles (e.g., parcel method, Ri-number) may then be applicable to the AMDAR data (Piringer & Joffre, 2005).

6.6.2 Mixing height parameterizations

The most common approach used in dispersion models to obtain the MH values is its calculation from specific parameterizations and preprocessors. This approach can use *in situ* measurements or NWP-derived profiles. So far, parameterizations of the MH have been developed and validated mostly using rural homogeneous conditions, so their applicability to urban conditions should be verified. Some authors have suggested specific methods for MH determination in urban areas which can be classified into two main categories: (i) using a local correction to the heat fluxes and roughness to incorporate urban effects and, (ii) estimating the Internal Boundary Layer (IBL) height growth as the flow moves over a city. In this section, we address the first category, usually based on classical methods for homogeneous terrain, but using urban values for heat flux and roughness. The second category is based on the general methods for the growth of the IBL height following an abrupt change of surface roughness and heat, and will be addressed in Section 6.6.3.

6.6.2.1 ABL formulae

For the estimation of the MH in urban areas, most authors use a standard "rural" diagnostic method for stable conditions and a prognostic equation for unstable conditions (e.g., Seibert et al., 2000), but without any corrections for urban features. The most commonly used formulae are listed in Appendix A. The properties of the ABL are primarily described within the theoretical framework of similarity relations, by which the variability of the structure of the ABL can be explained by the variation in the scales of the phenomena.

The basic internal scaling quantities are the friction velocity u_* (or the convective velocity scale w_*), the temperature scale T_*, and three length scales: the Monin-Obukhov (MO) length L, the vertical height z, and the ABL height h. Additionally, the ABL structure is determined by the following external parameters: (i) the wind velocity at the top of the layer (or the geostrophic wind speed G), (ii) the Coriolis parameter f arising from Earth's rotation, (iii) the roughness of the surface described by the roughness length z_o, a measure of the height of typical surface irregularities, and (iv) the Brunt-Väisälä frequency N ($N^2 = \beta\ \gamma$, with γ the potential temperature gradient above the ABL), representative of the background stratification into which the ABL evolves.

In homogeneous steady conditions, the ABL structure and variability are expressed as a combination of dimensionless ratios of these scales. The most recent theories have used the length scale $L_N = u_*/N$ alongside the classical L to represent nonlocal effects through background stratification at the top of the ABL (Kitaigorodskii & Joffre, 1988; Zilitinkevich, 2002). Such expressions have been shown to perform better than classical ones for rural data (Joffre et al., 2001; Zilitinkevich & Baklanov, 2002; Zilitinkevich, Baklanov, Rost, et al., 2002). If these parameters are to be useful for describing the ABL structure and pollutant dispersion in a practical way, they must be available at any site.

6.6.2.2 Internal boundary layer development over an urban area

The second category of MH schemes is based on a general method describing the growth of the IBL height following an abrupt change (smooth-to-rough) of surface roughness (Fig. 6.4). Just a few authors have suggested such a method specifically for urban MH.

Formulae for the IBL height developed principally for coastal sites (e.g., Garratt, 1990; Panovsky & Dutton, 1984; Walmsley, 1989; Wright et al., 1998) can be used, in principle for the height of the urban IBL. For instance, Henderson-Sellers (1980) developed a simple model for the urban MH as a function of distance downwind into the city. Nkemdirim (1986) tested and further improved the Summers (1965) formulation for the urban MH:

$$h_\alpha = \left[\frac{2Q_{Hc}x_k}{\rho c_p \alpha U_a}\right]^{1/2} \tag{6.7}$$

where Q_{Hc} is the cumulative heat flux over the urban fetch between x_O and x_k, α is the difference between the dry adiabatic rate and the prevailing lapse rate at x_O, ρ air density, c_p is specific heat and U_a is the reference wind speed upwind of the city (x_O). Formula (6.7) was verified for the cities of New York and Calgary. It was shown that it could be used for a rough MH estimation, but only for wind velocities $U_a < 4\ \mathrm{m\,s^{-1}}$.

Gryning and Batchvarova (1996) developed a zero-order scheme slab model for the IBL over terrain with abrupt changes of surface for near neutral and unstable atmospheric conditions [Eq. (12) in Appendix A], which uses the mean wind components μ and ν in the IBL, and the mean vertical air motion w_s above the boundary layer that can be estimated from NWP data.

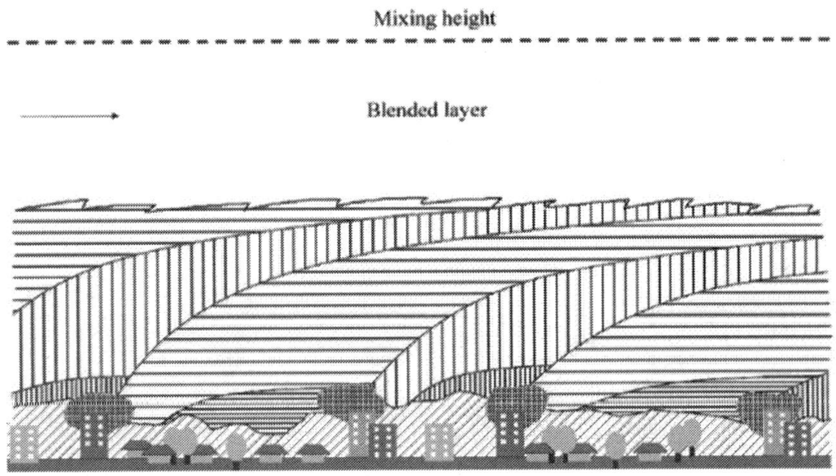

FIGURE 6.4 Schematic of the boundary layer structure over an urban area. The vertical and horizontal patterns represent the underlying surface of the neighborhoods of tall and low buildings, respectively. Broad-spaced patterns represent the urban internal boundary layers where advection processes are important. Fine-spaced patterns show the inertial sublayers that are in equilibrium with the underlying surface and where Monin-Obukhov scaling applies. The forward slash pattern represents the roughness sublayer that is highly inhomogeneous both in its vertical and horizontal structure. The dotted pattern represents adjustment zones between neighborhoods with large accelerations and shears in the flow near the top of the canopy. Above the height where the internal boundary layers are intermixed the effects of the individual neighborhoods cannot be distinguished any more—the so-called blended layer. *Adapted from S.-E. Gryning, in: COST-715 (2005). Meteorology applied to urban air pollution problems. In Fisher, B., Joffre, S., Kukkonen, J., Piringer, M., Rotach, M., & Schatzmann. M. (eds.),* Final Report of COST Action 715 *(276 pp.). Demetra Ltd Publishers.*

For example, Gryning and Batchvarova (1990) have suggested an often-used model analytical formulation for the CPBL height, z_i, accounting for mechanical and convective turbulence mixing and including a diagnostic parameterization for $\Delta\theta_i$ the jump model:

$$\frac{dz_i}{dt} = \frac{(1+2A)Q_0}{\gamma h} + \frac{2B\,u_*^3}{\gamma\beta z_i^2} \tag{6.8}$$

where A and B are model constants, β is the buoyancy parameter and γ is the potential temperature gradient. Batchvarova and Gryning (1994) also introduced an additional parameterization of Δz_i, the depth of the entrainment zone.

The constants most suitable for this model are discussed in Källstrand and Smedman (1997). The model was solved numerically for the IBL height of an urban area with a coastline and showed good applicability (Batchvarova & Gryning, 1998; Batchvarova et al., 1999; Gryning & Batchvarova, 2001).

6.6.3 Applications of NWP model outputs

There is increasing interest in using numerical outputs from NWP models to generate formatted data that can be then easily input into air quality forecasting or environmental impact assessment models, where the MH is an important requirement, alongside the stability. Although modern NWP models can approach the necessary resolution for the urban scale, the parameterizations of urban effects in most of the existing operational models are absent or greatly simplified (Baklanov, 2002). Therefore, it becomes important to "urbanize" NWP models to enhance their suitability for urban MH calculation (Baklanov et al., 2002).

Nevertheless, one potentially promising method to estimate the MH or turbulence profiles for dispersion models is using output from mesoscale NWP models. This can be done using state-of-the-art turbulence closures in the meteorological models (with similarity formulations of the turbulent viscosity and diffusion) or directly in 3D atmospheric pollution models. Many advanced atmospheric dynamics and pollution models already follow this approach (e.g., Baklanov, 2001; Kurbatskii, 2001; Zhang et al., 2001). However, the complexity of prognostic turbulence closure schemes, the requirements for high grid spatial resolution and problems due to numerical diffusion in Eulerian models make this method too expensive for practical application for the time being. So, many dispersion models, especially regulatory models, are still not coupled to meteorological models. They are based on in situ measurements of meteorological characteristics and need the MH as an input parameter.

TABLE 6.2 Mixing heights for the Copenhagen area: mean value (in m), standard deviation (Std Dev) and number of profiles (*N*) for urban, and semi-urban/rural sectors, assessed separately for the SBL (mostly night-time cases) and the CBL (mostly daytime cases).

		SBL					CBL				
Sector	Statistics	*Ri*	*AR81*	*BeSc79*	*Vmax*	*Hu*	*Ri*	*BeSc79*	*Pa*	*APa*	*Hu*
Urban	Mean	261	783	764	593	456	799	918	718	900	532
	Std dev	197	444	474	192	260	415	440	445	494	238
	N	114	114	114	114	44	61	61	61	61	29
Rural	Mean	189	451	595	552	462	965	920	898	1130	634
	Std. dev	175	255	410	199	260	368	387	403	403	255
	N	122	122	121	122	59	80	80	80	80	21

6.6.3.1 *Worked example 6.4 Estimation of mixing height for Copenhagen using radiosounding data*

Let us use the radiosounding data from the year 2001 at the Jægersborg station in Copenhagen (Baklanov & Kuchin, 2004) to estimate the MH for SBL and CBL cases and to test different diagnostic MH methods discussed in Section 6.6 and presented in Appendix A: the bulk Richardson number (*Ri*), the method of Arya (1981: AR81), the empirical method of Benkley and Schulman (1979, BeSc79), the maximum low-level jet velocity method (V_{max}), humidity-jump methods (Hu), the parcel (Pa) and advanced parcel (APa). The bulk Richardson number method (*Ri*) using $Ri_c = 0.25$.

Results are presented in Table 6.2 for Worked Example 6.4.

Table 6.2 suggests that the effect of the urban sector on the average SBL MH values is significant for all the tested methods except the Hu method. The average MH values of the urban sector are *approximately* 100 m higher than the average rural MH. The Ri-method gives the lowest MH values in both sectors (261/189 m). Of course, without direct measurements of the MH, no conclusion about the correct value of the MH can be made. However, a previous analysis of the bulk Ri-number method and a comparison with MH measurements by sodar (Zilitinkevich & Baklanov, 2002) showed that this method underestimates the MH for the SBL. This should require a revision of the critical value of the Ri-number. Two other methods (AR81 and BeSc79), based on mere surface-level measurements, probably strongly overestimate the urban MH (783 and 764 m for the urban sector). The Vmax method yields intermediate MH values. The Hu method shows unexpectedly similar results. However, it is not recommended to use this method, because it is usually used only for neutral and convective conditions.

In contrast to the SBL cases that displayed significant urban effects, the CBL cases showed the opposite effect with higher MHs observed in the rural/semiurban cases. This, at first sight, unexpected result, can be partly explained for the Copenhagen area by its coastal position. For convective conditions, the anthropogenic heat flux does not play a dominant role compared to the sensible heat flux, so sea breeze effects, which are important for the rural sector, are stronger than the urban effects.

To minimize uncertainties, it is thus recommended to combine the Ri-number and parcel methods for diagnosing the MH from radiosounding profiles for both urban and nonurban CBL conditions.

6.7 Urban boundary layer structures and modeling

6.7.1 Urban boundary layer structure

A wide range of micro- and mesoscale urban features can influence the atmospheric flow, including its turbulence regime, the microclimate, and, accordingly, modify the transport, dispersion, and deposition of atmospheric pollutants within urban areas and the vicinity downstream (Barlow et al., 2014). For example:

- the heterogeneity of buildings and other rough elements of the surface, affects the turbulence regime of the flow;
- the large amount of impervious materials and reduced pervious fractions affect the hydro-meteorological regime and pollutants deposition;
- the release of anthropogenic heat fluxes by human activities affect the thermal regime;

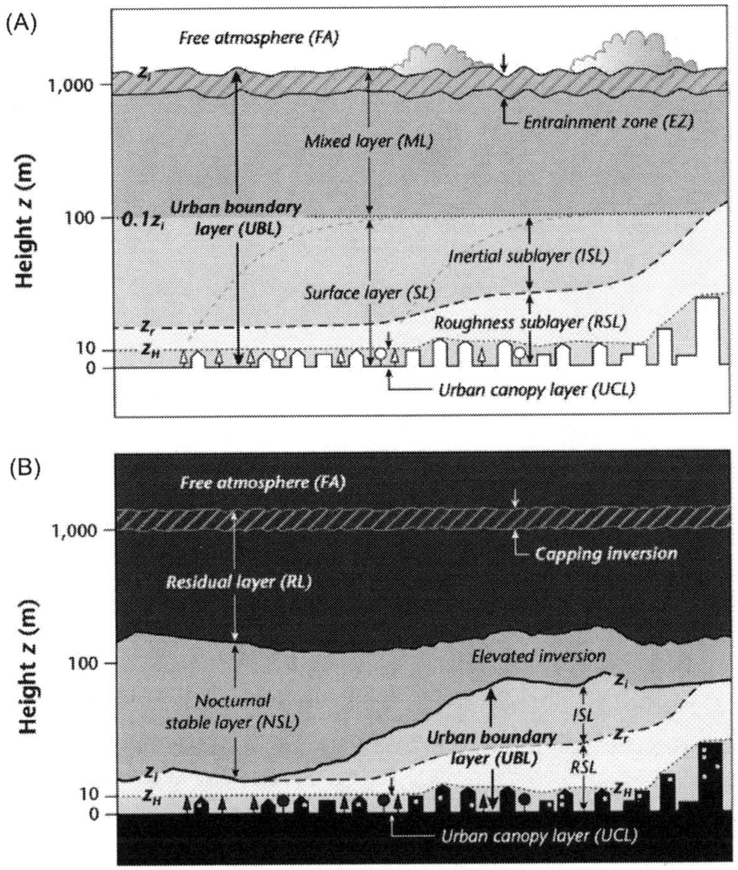

FIGURE 6.5 Schematic representation of urban planetary boundary layer structure during (A) day and (B) night. *Adapted from Oke, T.R., Mills, G., Christen, A., & Voogt, J.A. (2017).* Urban Climates. *Cambridge University Press, Cambridge. https://doi.org/10.1017/9781139016476.*

- the release of pollutants such as aerosols impact transfer of radiation, clouds and precipitation formation;
- the streets geometry ("street canyons") impacts the flow regime and the exchanges of heat between the different surfaces (road, walls).

So, urban PBLs (UPBLs) are more complex than rural ones Oke et al. (2017). During daytime (Fig. 6.5A), the warm /turbulent/polluted mesoscale UBL consists of a surface layer (SL) and mixed (sometimes "mixing") layer. Acronyms and symbols from the figure are used here for convenience, even if they do not coincide with those in other parts of the chapter. With a mean wind, the UBL forms an advected plume over a downwind rural boundary layer, while without a mean wind, an urban dome-shaped PBL forms instead (Bornstein, 1968). On the local scale, the SL is comprised of the urban canopy layer (UCL, with a top at Z_H) between its buildings, which forms the lower half of the roughness sublayer (RSL, with a top at Z_r). The upper most sublayer of the SL is the inertial sublayer (ISL, with a top at Z_L). As for the PBL (Fig. 6.1), the UBL is frequently capped by an Entrainment Zone (EZ), above which is the free atmosphere (FA). The EZ is frequently centered on the base of an elevated capping inversion (CI, not shown in Fig. 6.5A). At night (Fig. 6.5B), the upwind rural nocturnal surface inversion layer (NSL) is advected over the city, producing a weak shallow elevated inversion (first observed by Bornstein, 1968) and sequentially capped by the nocturnal RL, sometimes a CI, and then the FA.

6.7.2 Parameterization of urban canopy layer for numerical NWP and AQ models

While the urban UCL can be simulated via Computer Fluid Dynamics (CFD), Large Eddi Simulations (LES), or Direct Numerical Simulation (DNS) simulations for a single canyon up to a full neighborhood, the small grids required (1 m for the first two and 1 mm for the latter) negate these approaches for NWP models. As urban UCLs and RSLs are not in TKE equilibrium, MOST theories are not valid (Oke et al., 2017). Their analytical parameterizations in NWP models are thus modified rural SBL formulations developed from observations in urban areas (see, e.g., Baklanov, 2002; Masson et al., 2024).

Urban NWP and AQ modeling utilized different PBL schemes and improvements and concluded that improved forecast accuracy requires parameterizations of: (i) assimilated surface characteristics from satellite data, (ii) high-resolution land-use classifications and algorithms for roughness parameters based on morphologic methods, (iii) model downscaling, including increased vertical and horizontal resolutions and (one- or two-way) nesting, (iv) UCL meteorological fields, and (v) urban turbulent fluxes (Baklanov et al., 2008).

The urbanization options in WRF are summarized in Skamarock et al. (2019), they include the: (i) default Noah land use scheme (Liu et al., 2006), which represents urban areas as only changes in surface parameters, such as z_o, specific heat, and albedo, as first done in Bornstein (1975). This provides reasonable Urban Heat Island (UHI) estimates but cannot reproduce building drag effects on PBL winds, (ii) single layer urban canopy model (SLUCM, Kusaka & Kimura, 2004; Kusaka et al., 2001), which solves complex surface energy balance equations within a single urban canyon, and (iii) multilevel PBL building environment parameterization (BEP) of Martilli et al. (2002).

As buildings in mega-metropolises can reach hundreds of meters, NWP models need to address the building interaction at multiple model levels. The BEP "porous building" interactions use known building fractions in each numerical cell. It can be combined with the building energy model (BEM) option (Salamanca & Martilli, 2010), which parameterizes building momentum and energy balances. For each NWP model partial differential equation (PDE) (i.e., momentum, heat, and TKE), BEP adds three new terms to parameterize impacts from streets, roofs, and walls. The first two terms are functions of u_*, while the last is a function of building drag coefficient (C_d). As BEP + BEM can overestimate both 2-m UHIs and wind speeds, Gutierrez, Martilli et al. (2015); Gutiérrez, González et al. (2015) added new variable C_d and cooling tower (CT) parameterizations, the latter of which divides anthropogenic heat (Q_A) from commercial buildings into its latent and sensible heat components.

The urban morphological data for each of the three urbanization schemes (SLUCM, BEP, and BEP + BEM) can be applied to the one, three, or 10 urban classes in the Noah, BEP, and local climate zone (LCZ, Stewart & Oke, 2012) schemes, respectively. LCZ data can be obtained from: (i) WRF lookup tables, (ii) the World Urban Database and Access Portal Tools (WUDAPT) methodology (Bechtel et al., 2019; Ching et al., 2019) that uses remotely sensed surface data to produce grid specific distributions of some required parameters, or (iii) building specific data for use in WUDAPT, for example, the NYC Primary Land Use Tax Output (PLUTO) data in Ortiz et al. (2018). Global LCZs maps by Demuzere et al. (2022) can support earth system modeling and urban-scale environmental science.

While the use of BEP + BEM can double WRF CPU times, the results of Martilli et al. (2002) show its value. They used the BEP and Noah urbanization schemes and compared both with urban canyon TKE observations. The data were a synthesis by Rotach (1995) of observations over Zurich, Basel, and Sapporo, and of wind tunnel results. The Noah results showed expected TKE decreases with height during both day and night hours, with the latter values reduced over the former. The new results showed TKE: (i) reaching a maximum just above the grid level with the maximum number of buildings, (ii) reduced within the UCL and increased above rooftops, as compared to Noah values, (iii) with identical profile day and night, showing dominance of mechanical over thermal TKE production, and (iv) with a good match with the observations.

The field of urban modeling has rapidly advanced, now able to correctly simulate two extreme urban situations: its impacts on thunderstorms and conditions in narrow urbanized valleys. The urbanized WRF modeling study by Tomasi et al. (2019) focused on Bolzano Italy, located at an altitude of 2562 m in a 2–3-km wide, deep valley. The PBL TKE scheme of Nakanishi and Niino (2009) was used with a 300-m horizontal grid spacing. Outputs were used to drive the CALPUFF dispersion model and results were compared to observations from the Bolzano Tracer Experiment (BTEX). A challenging urban weather feature to predict is when urban areas initiate or intensity a convective thunderstorm. The climatological observations of Dou et al. (2015) found that an UHI of 1.5 K was necessary for this in Beijing; with lesser values existing storms bifurcated around the city. Dou et al. (2020) reproduced bifurcation for a Beijing case that used the BEP urbanization in WRF.

A summary report on the state of the art of urban climate research and the resiliency of cities under climate change stresses also included required future research directions (González et al., 2021). It included the following grand challenges for improved urban modeling: (i) a capability for simulation of 3-D urban PBL turbulence, (ii) representation of the nexus between urbanization, water, and energy, (iii) incorporation of human processes, (iv) prediction of urban weather and climate extremes, (v) focus on cities in coastal and complex terrain areas, and (vi) harnessing of big data.

6.7.2.1 Worked example 6.5: choice of suitable parameterizations of urban canopy layer for numerical NWP and AQ models

You adjust and apply an NWP and AQ model for your specific study which includes urban areas. You need to choose and implement (if not included in the model yet) the most suitable and relevant parameterization of UCL for your study. What criteria should you use and how should you do it?

Solution:

As described in Section 6.7 there are many possibilities for incorporating urban effects into numerical models.

Firstly, let's analyze the requirements for your task. The chosen parameterization should align with your purposes (see Baklanov et al., 2009). Depending on temporal and spatial scales, as well as the key processes and their interaction types, the sophistication of UCL description may vary. This can include single-layer and slab/bulk-type UC schemes, multilayer UCL schemes and obstacle-resolved microscale models:

- For micro-scale (up to 1 km) the obstacle-resolved approach is recommended, with a focus on pollutant gas density feedbacks.
- For the city scale (1–100 km) a statistical description of urban characteristics is essential, with semi-direct and second indirect aerosol feedbacks being dominated.
- For regional scales (more than 100 km) all the aforementioned gas and aerosol feedbacks represent the highest interest, and the urban effects could be simply parameterized.

Fig. 6.6 below presents a suggested hierarchy of urban canopy schemes for different types and scales of models, which users can choose to suit their purposes:

where

- Simple modification of land surface schemes (AHF + R + A);
- Medium-Range Forecast Urban Scheme (MRF-Urban);
- Building Effect Parameterization (BEP);
- Town Energy Budget (TEB) scheme;
- Soil Model for Submeso scales Urbanized version (SM2-U);
- UM Surface Exchange Scheme (MOSES);
- Urbanized Large-Eddy Simulation Model (PALM);
- CFD type Micro-scale model for urban environment (M2UE).

See more detailed description of the parameterizations in MEGAPOLI reports, for example, Baklanov and Mahura (2011) and recent WMO report: Masson et al. (2024).

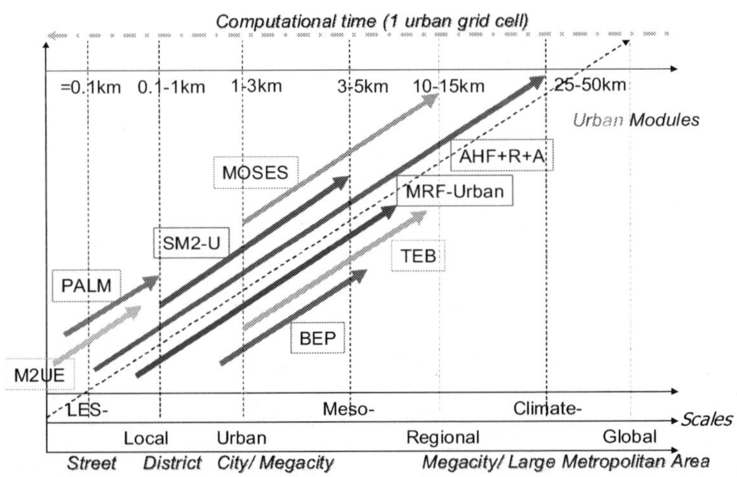

FIGURE 6.6 MEGAPOLI suggested hierarchy of urban canopy schemes for different types and scales of models.

6.8 Parameterizations of radiation

The presence of longwave and shortwave radiation in the atmosphere profoundly influences weather conditions and correspondingly atmospheric pollution processes at all spatial and temporal scales. Differential heating of Earth's surface by the sun creates horizontal temperature gradients, and thus pressure differences, driving the circulation of the atmosphere.

NWP models usually utilize two separate radiation schemes: for the shortwave (SW) and for the longwave (LW) radiation. Both shortwave and longwave radiation schemes must account for the absorption of radiation by water vapor, ozone, carbon dioxide (and other greenhouse gases), clouds, and Earth's surface. Additionally, models must demonstrate other phenomena such as reflection of radiation by clouds and Earth's surface, scattering by clouds and aerosols, and re-emission of longwave radiation by greenhouse gases. While radiation schemes may vary across different models (see an overview and tables for different AQF models in Baklanov et al., 2014; Kukkonen et al., 2012), they perform similar functions. Dealing appropriately with the energy input and the energy budget of the atmosphere is essential for the accuracy of model results. Most radiation schemes currently in use are column (one-dimensional) schemes, treating each column independently, with fluxes correspond to those in infinitely horizontally uniform planes. This approach is a good approximation when the vertical thickness of model layers is much smaller than the horizontal grid length. However, this assumption may become less accurate at higher horizontal resolutions.

In the case of meteorological models used as the drivers (i.e., offline models) or as host models (i.e., online models) for CW-AQF, the formulation of the radiation parameterization is of lesser importance in terms of atmospheric dispersion. However, the methodology used for calculating the coupling constant value (J) in the chemical solver might have significant impacts on modeled chemical processes. The concentrations and distribution of the trace species that affect the radiative transfer in the atmosphere are derived from the climatological datasets.

As our understanding of the radiative properties of chemical compounds (especially aerosols) increases as well as the continuous increase in available computation resources, further enhancements and improvements can be introduced to radiation schemes. Ongoing development of the fully-coupled online atmospheric modeling systems focuses mainly on feedback, namely, the effects of the scattering on aerosols as well as absorption of atmospheric gases are treated interactively. Studies conducted with the advanced coupled CW-AQF models show the mechanism and the magnitude of meteorological and chemical responses to aerosol radiative forcings. During short-term air quality episodes, the interactive treatment of the direct aerosol effect leads to a decrease in solar radiation flux, a reduction in the daytime diurnal temperature amplitude, an increase in night-time temperature, and a decrease in the PBL height.

6.9 Parameterization of convection and clouds

A precise representation of convection in NWP, air quality, and climate models has long been acknowledged as one of the most significant challenges in meteorological research. Cloud processes play a crucial role in both air quality and weather prediction. Clouds couple dynamical and hydrological processes in the atmosphere and on the ground through the heat of condensation and evaporation, through redistributions of sensible and latent heat and momentum, and through precipitation. Clouds serve as an important link between weather/air quality and climate by coupling radiative, microphysical, and aerosol processes in the atmosphere through the reflection, absorption, and emission of radiation as well as the interaction of cloud condensation nuclei (CCN) with the microphysical properties. Additionally, photolysis rates directly influence chemical transformations, and wet deposition by rainout removes pollutants. Even though most of these processes are highly coupled, most modeling systems artificially separate them, or omit some entirely.

Clouds encompass a wide range of motion scales, including shallow convection, deep convection, and larger-scale stable stratiform clouds. Almost all types of clouds require some form of parameterizations, even in the most high-resolution NWP models. The increase in resolution of the models, however, has introduced additional problems for cloud parameterizations. As the grid resolution increases, more and more of the convective clouds can be resolved by the model. The resulting heating and moistening profiles from the microphysics in a cloud resolving model are very different than the tendencies from a convective parameterization. A large part of the parameterized heating and drying tendencies are caused by compensating subsidence. In cloud resolving models the subsidence is caused by explicit vertical motion away from the heat and moisture source from the microphysical parameterization. This scale separation problem is one of the biggest challenges with current NWP and online integrated CW-AQF models. Previously, clouds were commonly classified as (A) large scale stable precipitating clouds, (B) deep convective clouds, and (C) shallow

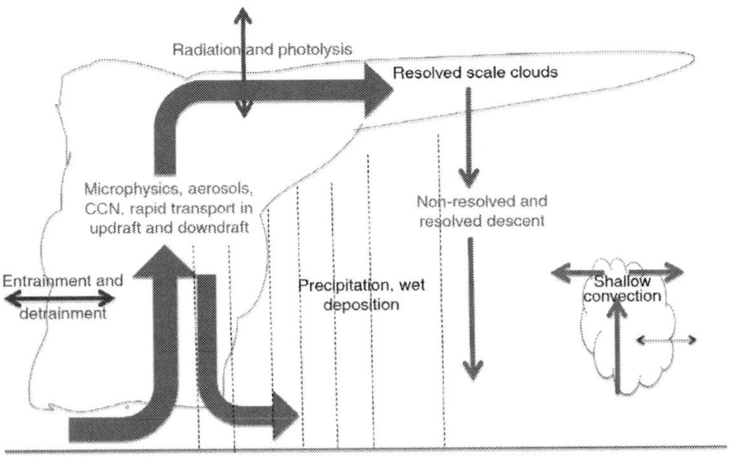

FIGURE 6.7 Cloud processes important for air quality and weather forecasting. *Adapted from Sokhi, R., Baklanov, A., Schluenzen, H. (Eds) (2018).* Mesoscale modeling for meteorological and air pollution applications *(76 pp.). Anthem Press. ISBN-10:1-78308-826-5.*

nonprecipitating convection. With the increase in resolution (B) are now partially resolved in NWP models. Hence a more refined classification, as used in models (Sokhi et al., 2018) would therefore be:

1. Unresolved precipitating convective clouds;
2. Resolved precipitating and nonprecipitating clouds (microphysical parameterizations);
3. Nonprecipitating shallow convection.

Classes (1) and (3) require convective parameterizations, (2) and possibly (1) require microphysical parameterizations. Fig. 6.7 shows a schematic view of processes that need to be parameterized or resolved in the models.

There are many different types of parameterizations of nonresolved deep convection (see relevant reviews in Grell (1993), Yano et al. (2013), and Sokhi et al. (2018). For example, the "convective adjustment" approach simply adjusts the thermo-dynamic sounding back to some neutral condition. This approach does not require any understanding of explicit mass flux processes as displayed in Fig. 6.7.

The most difficult challenge is usually the parameterization of precipitating convective clouds (1). Additional attention is required to parameterization of cloud cover, which is a very important parameter for air quality applications, since it links the clouds with the radiation and deposition schemes (both atmospheric radiation and photolysis schemes). In Sokhi et al. (2018), different approaches for (1) through (3) are described. While a number of studies focusing on specific aspects of convection are cited there, the reader is directed to works that provide an overview of convection parameterization for online CW-AQF, meteorological and climate models (e.g., Plant & Yano, 2015; Yano et al., 2015 and references therein).

The microphysical cloud parameterizations and mechanisms of interaction of clouds with aerosols and gases are more sophisticated and computationally demanding, but also one of the key components of modern CW-AQF and NWP models. These issues are considered by Zhang and Baklanov (2019).

6.10 Summary

(i) Adequate representation of atmospheric PBL processes is crucial for numerical modeling of weather, climate, air pollution, and atmospheric dispersion.

(ii) The PBL upper boundary acts as a barrier preventing (or strongly reducing) vertical dispersion in most air quality and dispersion models, making the PBL height extremely important for CW-AQF modeling.

(iii) The development and requirements of advanced PBL parameterizations in NWP and CW-AQF models are closely connected with advances in computing capability, so should be tailored to both the purpose and the computational resources available.

(iv) This section examines key PBL processes described or parameterized in meteorological and atmospheric pollution models for CW-AQF.

(v) Representation of interactions between the atmosphere and various Earth surfaces in CW-AQF models necessitates different schemes of turbulence closure and parameterizations of surface and boundary-layers.
(vi) Parameterizations of radiation processes, convection and cloud microphysics are also important for CW-AQF.

6.11 Questions

6.1. Explain the differences between the surface layer, the boundary layer, the mixed layer, the residual layer, and free flow in the atmosphere?
6.2. Discuss why accurate representation of atmospheric PBL processes is crucial not only for numerical modeling of weather and climate but also for air pollution and atmospheric dispersion?
6.3. How are surface fluxes calculated for different types of underlying surfaces in numerical models? Provide examples of such parameterizations.
6.4. Explain why turbulence closures and boundary-layer parameterizations are important for atmospheric dynamics and AQ models. What are the differences between turbulent characteristics for atmospheric pollutants and meteorological variables?
6.5. Describe the processes responsible for PBL height and explain why the PBL height and mixing height are important for CW-AQF. Provide examples of PBL heights controlled by different mechanisms.
6.6. What are the methods for estimating the PBL height in practical applications, AQ modeling and CW-AQF? Explain why it is difficult to measure the PBL height directly and outline the criteria for estimating it. Provide a few comparative calculations of SBL and CPB heights by different methods and analyze the differences between them (see the methods in Appendix A).
6.7. What are the key processes influencing the Urban Boundary Layer (UBL) structure? Why different levels of sophistication for UCL parameterizations are used for urban meteorology and AQ modeling? What are the main criteria for choosing a suitable parameterization to fit for purpose?
6.8. Explain why parameterizations of radiation are important not only for meteorological models but also for chemical transport and coupled meteorology-chemistry models? Provide several examples of processes and cases where is critically important.
6.9. What are the main convection and cloud processes contributing to mechanisms of atmospheric chemistry, aerosol formation and removal from the atmosphere? Explain why parameterization of convection and clouds are important in CW-AQF and especially in online integrated CCMMs?

Solutions to the end-of-chapter questions will be available at https://www.elsevier.com/books-and-journals/book-companion/9780128225912.

Appendix A

Some formulations for estimating the MH (Table A1).

TABLE A1 Stable boundary layer height formulations.

Reference [code]	SBL height equations
1. Zilitinkevich (1972)	$h = c_2\left(\frac{u_* L}{f}\right)^{1/2}, c_2 \approx 0.4$ (varies as 0.13÷0.72 according to different authors)
2. Venkatram (1980) [Ven80a]	$h = 2300\, u_*^{1.5}$
3. Venkatram (1980) [Ven80b]	$h = u_* \sqrt{\frac{2}{fN}}$
4. Arya (1981) (after Zilitinkevich, 1972)	$h = a\left(\frac{u_* L}{f}\right)^{1/2} + b;\ a = 0.43, b = 29.3$
5. Nieuwstadt (1981) [Nieu81]	$h = L \frac{0.3 u_*}{\lvert f \rvert L} \frac{1.0}{1.0 + 1.9\, h/L}$
6. Zilitinkevich and Mironov (1996) [ZiMi96]	$\left(\frac{fh}{0.5 u_*}\right)^2 + \frac{h}{10L} + \frac{Nh}{20 u_*} + \frac{h\lvert f \rvert^{1/2}}{(u_* L)^{1/2}} + \frac{h\lvert Nf \rvert^{1/2}}{1.7 u_*} = 1$
7. Zilitinkevich, Baklanov, Rost, et al. (2002) [Zili02]	$h = \frac{C_R u_*}{\lvert f \rvert}\left[\left(1 + C_h \frac{w_h}{u_*}\right) / \left(1 + \frac{C_R^2 u_* (1 + C_{uN} NL/u_*)}{C_S^2 L \lvert f \rvert}\right)\right]^{1/2}$ with: $C_R = 0.4$, $C_S = 0.74$, $C_{uN} = 0.25$ and $C_h = 0.3$.
8. Zilitinkevich and Baklanov (2002)	$\frac{\partial h}{\partial t} + \mathbf{V} \cdot \nabla h = -C_E \lvert f \rvert (h - h_{CQE}) + K_h \nabla^2 h$ with $C_E \approx 1$.
9. Joffre et al. (2001) [staJof]	$h = \frac{b'}{2a'} \mu_N \left\{-1 + \left[1 + \frac{4 a' m'}{b'^2} \mu_N^{-2}\right]^{1/2}\right\} L_N$ with: $a' = 0.12$; $b' = 2.85$; $m' = 24$ (very rough surface)

Where u_{10} is the wind speed at 10 m height; $\mathbf{V} = (u,v)$ is the horizontal velocity vector, K_h is the horizontal diffusivity, h_{CQE} is the equilibrium MH calculated from a diagnostic formulation (e.g., Zilitinkevich et al., 2002a), $\mu_N = L_N/L$, $L_N = u_*/N$ (N is the Brunt-Väisälä frequency), f is the Coriolis parameter, and w_h is the large-scale vertical velocity at the SBL upper boundary (Table A1).

TABLE A2 Convective boundary layer height formulations.

Reference [code]	CBL height equations
10. Tennekes (1973); Tennekes and Driedonks (1981) [TeDr82]	$\frac{dh}{dt} = \frac{S}{\Delta\theta}$; where $S = A\,\overline{(w'\theta')}_o + \frac{B\,u_*^3}{\beta h}$ $\frac{d\Delta\theta}{dt} = \frac{\gamma S}{\Delta\theta} - \frac{\overline{(w'\theta')}_o}{h} - \frac{S}{h}$; where: $A = 0.2$; $B = 5$.
11a. Batchvarova and Gryning (1991) [BaG91a]	$\frac{\partial h}{\partial t} = (1+2A)\,\frac{\overline{(w'\theta')}_o}{\gamma h} + 2B\frac{u_*^3}{\gamma \beta h^2}$, where: $A = 0.2$, $B = 2.5$
11b. Batchvarova and Gryning (1991) [BaG91b]	$\frac{\partial h}{\partial t} = \overline{(w'\theta')}_0\left[\frac{\gamma h^2}{(1+2A)\,h - 2kBL_*} + \frac{Cu_*^2}{\beta(1+A)\,h - kBL_*}\right]^{-1}$ *where:* $A = 0.2$; $B = 2.5$; $C = 8$.
12. Gryning and Batchvarova (1996)	$\left\{\frac{h^2}{(1+2A)h - 2B\kappa L} + \frac{C\,u_*^2}{\gamma\beta[(1+A)h - B\kappa L]}\right\}\left(\frac{\partial h}{\partial t} + u\frac{\partial h}{\partial x} + v\frac{\partial h}{\partial y} - w_s\right) = \frac{\overline{(w'\theta')}_o}{\gamma}$, with $A = 0.2$, $B = 5$, $C = 1.3$
13. Joffre et al. (2001) [unsJof]	$h = \frac{b''}{2a''}\mu_N\left\{1 + \left[1 + \frac{4\,a''m''}{b''^2}\mu_N^{-2}\right]^{1/2}\right\}L_N$ with a″ = 0.1; b″ = - 0.85; m″ = 12

Where γ is the potential temperature gradient above the ABL ($N^2 = \beta\,\gamma$), $\beta = g/T_0$ the buoyancy parameter, $\Delta\theta$ the temperature jump in the entrainment layer at the ABL top, $\overline{(w'\theta')}_0$-the surface sensible heat flux ($=Q_H/\rho c_p$), w_s the vertical synoptic velocity above the ABL. Equation (13) was derived for moderate unstable conditions (both buoyant and mechanical production of turbulence) but not for convective conditions (Table A2).

TABLE A3 Neutrally stratified boundary layer height formulations.

Reference [code]	Neutral ABL height equations
14. Rossby and Montgomery (1935) [RoMo35]	$h = c_N \frac{u_*}{\lvert f \rvert}$, where: $c_N = 0.3$ ($c_N = 0.04$–0.3 according to Seibert et al., 2000)
15. Arya (1981) [Arya81] (1)	$h = 0.089\frac{u_*}{f} + 85.1$
16. Mahrt (1981) [Mahr82]	$h = 0.06\frac{u_*}{f}$
17. Benkley and Schulman (1979) [BeSc79]	$h = 125u_{10}$
18. Nieuwstadt (1984a,b) [Nieu84]	$h = 28u_{10}^{3/2}$

Developed for neutral conditions, Eqs. (14–18) have been used also for stable conditions by some authors.
1. In some studies, formulae (14–18) have been applied also to the SBL.

References

AMDAR (2003). International AMDAR program: Aircraft meteorological data relay, WMO, AMDAR Panel, 7 pp.

Angevine, W. (2003). Wind profiler, applications. Lecture material for: short course on the fundamentals of boundary layer wind and temperature profiling using radar and acoustic techniques. *AMS Annual Meeting*, Feburary 8-9, 2003 Long Beach, CA. Available from http://www.etl.noaa.gov/ams_measurement/short_course_2003.htm.

ARW-WRF (2015). ARW-WRF weather forecasting and research modeling system. Version3 User's Guide.

Arya, S. P. S. (1981). Parameterizing the height of the stable atmospheric boundary layer. *Journal of Applied Meteorology.*, *20*, 1192–1202.

Baklanov, A. (2001). Parameterisation of SBL height in atmospheric pollution models. In X. V. C. Borrego, & G. Schayes (Eds.), *Air Pollution Modelling and its Application* (pp. 415–424). New York: Kluwer Academic / Plenum Publishers.

Baklanov, A. (2002). The mixing height in urban areas – A review. In Piringer M. & Kukkonen J. (Eds.), *Proceedings of the Workshop "Mixing height and inversions in urban areas* (pp. 9–28). Toulouse, France, 3–4 October 2001, COST Action 715. Office for Official Publications of the European Communities, EUR 20451.

Baklanov, A., Rasmussen, A., Fay, B., Berge, E., & Finardi, S. (2002). Potential and shortcomings of numerical weather prediction models in providing meteorological data for urban air pollution forecasting. *Water, Air and Soil Pollution: Focus*, *2*(5-6), 43–60.

Baklanov, A., Mestayer, P., Clappier, A., Zilitinkevich, S., Joffre, S., Mahura, A., & Nielsen, N. W. (2008). Towards improving the simulation of meteorological fields in urban areas through updated/advanced surface fluxes description. *Atmospheric Chemistry and Physics*, *8*, 523–543.

Baklanov, A., Grimmond, S., Mahura, A., Athanassiadou, M. (2009). Meteorological and Air Quality Models for Urban Areas. Springer, 169p., ISBN 978-3-642-00297-7.

Baklanov, A., Mahura, A., & Sokhi, R. (Eds.), (2011). *Integrated systems of meso-meteorological and chemical transport models*. Springer. Available from 10.1007/978-3-642-13980-2.

Atmospheric boundary layers: Nature, theory and applications to environmental modeling and security. In A. Baklanov, & B. Grisogono (Eds.), *Boundary- Layer Meteorology* (125, pp. 157–397). .

Baklanov, A., Schlünzen, K., Suppan, P., Baldasano, J., Brunner, D., Aksoyoglu, S., Carmichael, G., Douros, J., Flemming, J., Forkel, R., Galmarini, S., Gauss, M., Grell, G., Hirtl, M., Joffre, S., Jorba, O., Kaas, E., Kaasik, M., Kallos, G., . . . Zhang, Y. (2014). Online coupled regional meteorology chemistry models in Europe: Current status and prospects. *Atmospheric Chemistry and Physics, 14*, 317–398.

Baklanov, A., & Kuchin, A. (2004). The mixing height in urban areas: Comparative study for Copenhagen. *Atmosphere Chemical Physic. Discus., 4*, 2839–2866.

Baklanov A., Mahura A. (Eds) 2011. Third Year MEGAPOLI Dissemination Report. Del. D9.4. 3, FP7 MegapolI Sci.Report 11-24, MEGAPOLI-50-REP-2011-10, 104p.

Baklanov, A., Joffre, S. M., Piringer, M., Deserti, M., Middleton, D. R., Tombrou, M., Karppinen, A., Emeis, S., Prior, V., Rotach, M. W., Bonafè, G., Baumann-Stanzer, K., & Kuchin, A. (2006). Towards estimating the mixing height in urban areas. Recent experimental and modelling results from the COST-715 Action and FUMAPEX project. *DMI Scientific Report 06-06*, 41, ISBN: 87-7478-540-0.

Barlow, J. F. (2014). Progress in observing and modelling the urban boundary layer. *Urban Climate, 10*, 216–240.

Batchvarova, E., Cai, X., Gryning, S.-E., & Steyn, D. (1999). Modelling internal boundary-layer development in a region with a complex coastline. *Boundary-Layer Meteorol., 90*, 1–20.

Batchvarova, E., & Gryning, S.-E. (1998). Wind climatology, atmospheric turbulence and internal boundary-layer development in athens during the MEDCAPHOT-trace experiment. *Atmospheric Environment, 32*, 2055–2069.

Batchvarova, E., & Gryning, S. E. (1994). An applied model for the height of the daytime mixed layer and the entrainment zone'. *Boundary- Layer Meteorology, 71*, 311–323.

Bechtel, B., Alexander, P. J., Beck, C., Böhner, J., Brousse, O., Ching, J., & Xu, Y. (2019). Generating WUDAPT Level 0 data: Current status of production and evaluation. *Urban Climate, 27*, 24–45.

Benkley, C. W., & Schulman, L. L. (1979). Estimating mixing depths from historical meteorological data. *Journal of Applied Meteorology, 18*, 772–780.

Beyrich, F. (1997). Mixing height estimation from sodar data—A critical discussion. *Atmospheric Environment, 31*(23), 3941–3953. Available from https://doi.org/10.1016/S1352-2310(97)00231-8.

Blackadar, A. K. (1962). The vertical distribution of wind and turbulent exchange in a neutral atmosphere. *Journal of Geophysical Research, 67*, 3095–3102.

Bornstein, R. D. (1968). Observations of the urban heat island effect in New York City. *Journal of Applied Meteorogoy, 7*, 575–582, 10.1175/1520-0450(1968)007,0575:OOTUHI.2.0. CO;2.

Bornstein, R. D. (1975). The two-dimensional URBMET urban boundary layer model. *Journal of Applied Meteorology, 14*(8), 1459–1477, 10.1175/1520-0450(1975)014\textless1459:TTDUUB\textgreater2.0.CO;2.

Burk, S. D., & Thompson, W. T. (1989). A vertically nested regional numerical prediction model with second- order closure physics. *Monthly Weather Review, 117*, 2305–2324.

Businger, J. A., Wyngaard, J. C., Izumi, Y., & Bradley, E. F. (1971). Flux-profile relationships in the atmospheric. *Surface Layer. Journal of Atomic Science*, 181–189. Available from https://doi.org/10.1175/1520-0469.

Charnock, H. (1955). Wind Stress on a Water Surface. *Quarterly Journal of Royal Meteorology Society., 81*, 639–640.

Ching, J., Aliaga, D., Mills, G., Masson, V., See, L., Neophytou, M., Middel, A., Baklanov, A., Ren, C., Ng, E., Fung, J., Wong, M., Huang, Y., Martilli, A., Brousse, O., Stewart, I., Zhang, X., Shehata, A., Miao, S., & Wang, X. (2019). Pathway using WUDAPT's digital synthetic city tool towards generating urban canopy parameters for multi-scale urban atmospheric modeling. *Urban Climate, 28*, 100459. Available from https://doi.org/10.1016/j.uclim.2019.100459.

COST-710 (1998). Harmonisation of the pre-processing of meteorological data for atmospheric dispersion models. Final report of COST Action 710. Directorate-General for Research and Innovation, European Commission, ISBN: 92-828-3302-X, 456 p.

COST-715 (2005). Meteorology applied to urban air pollution problems. In Fisher, B., Joffre, S., Kukkonen, J., Piringer, M., Rotach, M., & Schatzmann, M. (eds.), Final Report ofCOST Action 715 (276 pp.). Demetra Ltd Publishers.

Deardorff, J. W. (1972). Parameterization of the planetary boundary layer for use in general circulation models. *Monthly Weather Review, 100*, 93–106.

Demuzere, M., Kittner, J., Martilli, A., Mills, G., Moede, C., Stewart, I. D., & Bechtel, B. (2022). A global map of local climate zones to support earth system modelling and urban scale environmental science. *Earth System Science Data Discussions*, 1–57.

Dou, J. J., Bornstein, R., Miao, S., Sun, J., & Zhang, Y. (2020). Observation and simulation of a bifurcating thunderstorm over Beijing. *Journal of Applied Meteorology and Climatology*, 2129–2148. Available from https://doi.org/10.1175/JAMC-D-20-0056.1.

Dou, J. J., Wang, Y., Bornstein, R., & Miao, S. (2015). Observed spatial characteristics of Beijing urban-climate impacts on summer thunderstorms. *Journal of Applied Meteorogy Climatology, 56*, 95–105. Available from https://doi.org/10.1175/JAMC-D-13-0355.1.

Dupont, S., & Patton, E. G. (2012). Influence of stability and seasonal canopy changes on micrometeorology within and above an orchard canopy: The CHATS experiment. *Agricultural and Forest Meteorology, 157*, 11–29.

E-AMDAR (2002). EUMETNET-AMDAR Project: Aircraft Meteorological Data Reporting. http://www.metoffice.com/research/interproj/amdar/index.html.

Ekman V.W. (1905). On the influence of the Earth's rotation on ocean currents, Arkiv Mat. Astron. Fys., Uppsala – Stockholm, 2, No.11, 1-52.

Emeis, S., Ch., Münkel, S., Vogt, W. J., Müller., & Schäfer, K. (2004). Atmospheric boundary-layer structure from simultaneous SODAR, RASS, and ceilometer measurements. *Atmospheric Environment*, *38*, 273–286.

Ershadi, A., McCabe, M. F., Evans, J. P., Chaney, N. W., & Wood, E. F. (2014). Multi- site evaluation of terrestrial evaporation models using fluxnet data. *Agricultural and Forest Meteorology*, *187*, 46–61.

Finnigan, J. (2000). Turbulence in plant canopies. *Annual Review of Fluid Mechanics*, *32*, 519–571.

Foreman, R. J., & Emeis, S. (2010). Revisiting the definition of the drag coefficient in the marine atmospheric boundary layer. *Journal of Physical Oceanography 40, no. 10*, 2325–2332.

Garratt, J. R. (1990). The internal boundary layer - A review. *Boundary-Layer Meteorology.*, *50*, 171–203.

Garratt, J. R. (1994). *The atmospheric boundary layer*. Cambridge: Cambridge University Press.

Gayev, Y. A., & Hunt, J. C. R. (2007). *Flow and transport processes with complex obstructions. NATO science series book* (p. 414) Springer.

González, J. E., Ramamurthy, R., Bornstein, R. D., Chen, F., Bou-Zeid, E. R., Ghandehari, M., Luvall, J., Mitra, C., & Niyogi, D. (2021). Urban climate and resiliency: A synthesis report of state of the art and future research directions. *Urban Climate*, *38*, 100858. Available from https://doi.org/10.1016/j.uclim.2021.100858.

Grell, G. A. (1993). Prognostic evaluation of assumptions used by cumulus parameterizations within a generalized framework'. *Monthly Weather Review*, *121*, 764–787.

Gryning, S.- E., & Batchvarova, E. (1990). Analytical model for growth of the coastal internal boundary layer during onshore flow. *Quarterly Journal of Royal Meteorological Society*, *116*, 187–203.

Gryning, S.- E., Batchvarova, E., Quante, M., & Matthias, V. (2011). Evaluation of vertical profiles in mesoscale meteorological models based on observations for the COST728 study of winter 2003 PM episodes in Europe. In *Air pollution modeling and its application* XXIIn D. G. Steyn, & S. Trini- Castelli (Eds.), (pp. 499–503). Dordrecht: Springer.

Gryning, S.-E., & Batchvarova, E. (1996). A model for the height of the internal boundary layer over an area with an irregular coastline. *Boundary-Layer Meteorology*, *78*, 405–413.

Gryning, S.-E., & Batchvarova, E. (2001). Mixing height in urban areas: Will 'rural' parameterisations work? In M. Rotach, B. Fisher, & M. Piringer (Eds.), *COST Action 715 workshop on urban boundary layer parameterisations (Zurich, 24-25 May 2001). Office for Official Publications of the European Communities, EUR 20355* (pp. 99–109). .

Gutiérrez, E., González, J. E., Martilli, A., & Bornstein, R. (2015b). On the anthropogenic heat fluxes using an air conditioning evaporative cooling parameterization for mesoscale urban canopy models. *Journal of Solar Energy Engineering*, *137*(5), 051005. Available from https://doi.org/10.1115/1.4030854.

Gutiérrez, E., Martilli, A., Santiago, J. L., & González, J. E. (2015a). A mechanical drag coefficient formulation and urban canopy parameter assimilation technique for complex urban environments. *Boundary-Layer Meteorology*. Available from https://doi.org/10.1007/s10546-015-0051-7.

Henderson-Sellers, A. (1980). A simple numerical simulation of urban mixing depths. *Journal of Applied Meteorology.*, *19*, 215–218.

Högström, U. (1996). Review of some basic characteristics of the atmospheric surface layer. *Boundary-Layer Meteorology*, *78*, 215–246.

Holton, J.R. (2012). Introduction to dynamic meteorology. 5th ed. Amsterdam, Boston, Heidelberg, London, New York, Oxford, Paris, San Diego, San Francisco, Singapore, Sydney, Tokyo: Elsevier.

Holzworth, C. C. (1967). Mixing depths, wind speeds and air pollution potential for selected locations in the United States. *Journal of Applied Meteorology*, *6*, 1039–1044.

Jimenez, P. A., Dudhia, J., González-Rouco, J. F., Navarro, J., Montavez, J. P., & Garcia–Bustamante, E. (2012). A revised scheme for the WRF surface layer formulation. *Monthly Weather Review*, *140*, 898–918.

Joffre, S. M., Kangas, M., Heikinheimo, M., & Kitaigorodskii, S. A. (2001). Variability of the stable and unstable atmospheric boundary layer height and its scales over a boreal forest. *Boundary-Layer Meteorology*, *99*(3), 429–450.

Källstrand, B., & Smedman, A.-S. (1997). A case study of the near-neutral coastal boundary layer growth: Aircraft measurements compared with different model estimates. *Boundary-Layer Meteorology*, *85*, 1–33.

Kitaigorodskii, S. A., & Joffre, S. M. (1988). In search of a simple scaling for the height of the stratified atmospheric boundary layer. *Tellus*, *40A*(5), 419–433.

Kolmogorov, A. N. (1941). The local structure of turbulence in incompressible viscous fluid for very large Reynolds numbers. *C. R. Academic Science URSS*, *30*(301), 1941.

Kolmogorov, A. N. (1942). Equations of turbulent motion in an incompressible fluid. Izv. *AN SSSR, Ser. Fiz*, *6*(1-2), 56–58.

Kukkonen, J., Balk, T., Schultz, A. D. M., Baklanov, A., Klein, T., Miranda, A. I., Monteiro, A., Hirtl, M., Tarvainen., V., Boy, M., Sokhi, R., Lehtinen, K., Karatzas, K., San José, R., Astitha, M., Kallos, G., Schaap, M., Reimer, E., Jakobs, H., & Eben, K. (2012). A review of operational, regional-scale, chemical weather forecasting models in Europe. *Atmospheric Chemistry and Physics*, *12*(12), 1–87, 01/ 2012. Available from http://www.atmos-chem-phys.net/12/1/2012/.

Kurbatskii, A. F. (2001). Computational modelling of the turbulent penetrative convection above the urban heat island in a stably stratified environment. *Journal of Applied Meteorology.*, *40*, 1848-1761.

Kusaka, H., & Kimura, F. (2004). Coupling a single-layer urban canopy model with a simple atmospheric model: Impact on urban heat island simulation for an idealized case. *Journal of the Meteorological Society of Japan*, *82*(1), 67–80. Available from https://doi.org/10.2151/jmsj.82.67.

Kusaka, H., Kondo, H., Kikegawa, Y., & Kimura, F. (2001). A simple single-layer urban canopy model for atmospheric models: Comparison with multi-layer and slab models. *Boundary-Layer Meteorology*, *101*(3), 329–358. Available from https://doi.org/10.1023/A:1019207923078.

Liu, Y., Chen, F., Warner, T., & Basara, J. (2006). Verification of a mesoscale data-assimilation and forecasting system for the Oklahoma City area during the joint urban 2003 field project. *Journal of Applied Meteorology and Climatology*, *45*(7), 912–929. Available from https://doi.org/10.1175/JAM2383.1.

Lyra, R., Druilhet, A., Benech, B., & Bouka Biona, C. (1992). Dynamics above a dense equatorial rain forest from the surface boundary layer to the free atmosphere. *Journal of Geophysical Research*, *97*(12), 953-12, 965.

Mahrt, L. J. (1974). Time-dependent integrated planetary boundary layer flow. *Journal of the Atmospheric Sciences*, *31*, 457–464.

Mahrt, L. (1981). Modelling the depth of the stable boundary layer. *Boundary-Layer Meteorology*, *21*, 3–19.

Martilli, A., Clappier, A., & Rotach, M. W. (2002). An urban surface exchange parameterization for mesoscale models. *Boundary-Layer Meteorology*, *104*(2), 261–304.

Masson, V., Re, M., Chen, D., Baklanov, A., Liang, F., et al. (2024). *Good practices in high-resolution modelling for integrated urban services.* World Meteorological Origanization (p. 47) Geneva, Switzerland: WMO publication #1313.

Mellor, G. L., & Yamada, T. (1982). Development of a turbulence closure model for geophysical fluid problems. *Reviews of Geophysics*, *20*(4), 851–875.

Monin, A. S., & Obukhov, A. M. (1954). Main characteristics of the turbulent mixing in the atmospheric surface layer. *Trudy Geophys. Inst. AN. SSSR*, *24*(151), 153–187.

Münkel, C., Emeis, S., Müller, W.J., & Schäfer, K. (2002, September 24–27). Observation of aerosol in the mixing layer by a ground-based LIDAR ceilometer. In K. Schäfer, O. Lado-Bordowsky, A. Comeron, R.H. Picard (eds.), *9th International Symposium on Remote Sensing of Clouds and the Atmosphere VII, Proceedings of SPIE* (Vol. 4882, pp. 344–352). Agia Pelagia, Crete, Greece.

Nakanishi, M., & Niino, H. (2009). Development of an improved turbulence closure model for the atmospheric boundary layer. *Journal of the Meteorological Society of Japan*, *87*(5), 895–912. Available from https://doi.org/10.2151/jmsj.87.895.

Nkemdirim, L. C. (1986). Mixing depths and wind speed in an urban airshed: A note on the limits to a relationship. *Atmospheric Environment*, *20*(9), 1829–1830.

Obukhov, A. M. (1946). Turbulence in an atmosphere with a non-uniform temperature. *Boundary-Layer Meteorology*, *2*, 7–29. Available from https://doi.org/10.1007/BF00718085.

Oke, T. R., Mills, G., Christen, A., & Voogt, J. A. (2017). *Urban climates.* Cambridge: Cambridge University Press. Available from https://doi.org/10.1017/9781139016476.

Ortiz, L. E., Gonzalez, J. E., Wu, W., Schoonen, M., Tongue, J., & Bornstein, R. (2018). New York City impacts on a regional heat wave. *Journal of Applied Meteorology and Climatology*, *57*(4), 837–851. Available from https://doi.org/10.1175/JAMC-D-17-0125.1.

Panovsky, H. A., & Dutton, J. A. (1984). *Atmospheric turbulence* (p. 397) NY: Wiley.

Pielke, R. A. (2013). *Mesoscale meteorological modeling.* (2nd ed.). San Diego: Academic Press.

Piringer, M., & Joffre, S. (Eds.) (2005). The urban surface energy budget and mixing height in European cities: Data, models and challenges for urban meteorology and air quality. In Baklanov, A., J. Burzynski, A. Christen, M. Deserti, K. De Ridder, S. Emeis, S. Joffre, A. Karppinen, P. Mestayer, D. Middleton, M. Piringer and M. Tombrou (eds.), *Final Report of WG2 COST Action 715* (194 pp.).

Parameterization of atmospheric convection. In R. S. Plant, & J.-I. Yano (Eds.), *Theoretical background and formulation* (Volume 1). World Scientific, Imperial College Press.

Quante, M., Matthias, V., Gryning, S.E., Batchvarova, E., Aulinger, A., Chemel, C., Geertsema, G., Geyer, B., Jakobs, H., Kerschbaumer, A., Frank, M., San José, R., Schlünzen, H., Struzewska, J., Szintai, B., & Wolke, R. (2009). Using wind profiler data in time and frequency domain for the evaluation of meteorological drivers employed in chemistry transport modeling. In *Proceedings: 8th international symposium on tropospheric profiling*. RIVM, Delft (NL).

Räsänen, J., Lönnqvist, J., & Piironen, A.K. (2000, August 15–18). Urban boundary layer measurements with a commercial ceilometer. In *Proceedings of the 3rd symposium on urban environment* (pp. 34–35). American Meteorological Society, Boston, Davis CA.

Rogers, R. R., Lamoureux, M.-F., Bissonnette, L., & Peters, R. M. (1997). Quantitative interpretation of laser ceilometer intensity profiles. *Journal of Atmospheric Ocean Tech.*, *14*, 396–411.

Rossby, C. G., & Montgomery, R. B. (1935). The layer of frictional influence in wind and ocean currents. *Pap. Physics Oceanography Meteorological.*, *3*(3), 1–101.

Rotach, M. W. (1995). Profiles of turbulence statistics in and above an urban street canyon. *Atmospheric Environment.*, *29*, 1473–1486.

Salamanca, F., & Martilli, A. (2010). A new building energy model coupled with an urban canopy parameterization for urban climate simulations. part II. Validation with one-dimension off-line simulations. *Theoretical and Applied Climatology*, *99*(3–4), 345.

Salby, M. L. (2012). *Physics of the atmosphere and climate*. Cambridge: Cambridge University Press.

Schlünzen, K. H., & Katzfey, J. J. (2003). Relevance of sub-grid-scale land-use effects for mesoscale models. *Tellus*, *55A*, 232–246. Available from https://doi.org/10.1034/j.1600-0870.2003.00017.x.

Seibert, P., Beyrich, F., Gryning, S.-E., Joffre, S., Rasmussen, A., & Tercier, P. H. (2000). Review and intercomparison of operational methods for the determination of the mixing height. *Atmospheric Environment*, *34*, 1001–1027.

Skamarock, W., Klemp, J., Dudhia, J., Gill, D., Liu, Z., Berner, J., Wang, W., Powers, J., Duda, M., Barker, D., & Huang, X. Y. (2019). A Description of the advanced research WRF version 4. UCAR/NCAR. https://doi.org/10.5065/1dfh-6p97.

Smith, B., & Mahrt, L. (1981). A study of boundary-layer pressure adjustments. *Journal of the Atmospheric Sciences*, *38*, 334–346.

Sokhi, R., Baklanov, A., & Schluenzen, H. (Eds.), (2018). *Mesoscale modeling for meteorological and air pollution applications.* Anthem Press, ISBN-10:1-78308-826-5.

Stewart, I. D., & Oke, T. R. (2012). Local climate zones for urban temperature studies. *Bulletin of the American Meteorological Society, 93*(12), 1879–1900. Available from https://doi.org/10.1175/BAMS-D-11-00019.1.

Stone, P. H. (1972). A simplified radiative-dynamical model for the static stability of rotating atmospheres. *Journal of Atmospheric Science, 29*, 405–418.

Stull, R. B. (1988). *An introduction to boundary layer meteorology*. Dordrecht: Kluwer Academic Publishers.

Stull, R. B. (1991). Static stability—An update. *Bulletin American Meteorological Society., 72*, 1521–1529.

Summers, P.W. (1965). An urban heat island model: Its role in air pollution problems with applications to Montreal. In *First Canadian conference on micrometeorology* (22 pp.). Toronto, CMOS.

Tomasi, E., Giovannini, L., Falocchi, M., Antonacci, G., Jiménez, P. A., Kosovic, B., Alessandrini, S., Zardi, D., Delle Monache, L., & Ferrero, E. (2019). Turbulence parameterizations in sub-kilometer horizontal non-homogeneous flows. *Atmospheric Research, 228*, 122–136.

Walmsley, J. L. (1989). Internal boundary-layer height formulae – A comparison with atmospheric data. *Boundary-Layer Meterology, 47*, 251–262.

Wright, S. D., Elliot, L., Ingham, D. B., & Hewson, M. J. C. (1998). The adaptation of the atmospheric boundary layer to a change in surface roughness. *Boundary-Layer Meteorology, 89*, 175–195.

Yano, J.- I., Bister, M., Fuchs, Z., Gerard, L., Phillips, V., Barkidija, S., & Piriou, J.- M. (2013). Phenomenology of convection-Parameterization closure. *Atmospheric Chemistry and Physics, 13*, 4111–4131.

Yano, J.-I., Geleyn, J. F., Köller, M., Mironov, D., Quaas, J., Soares, P. M. M., Phillips, V. T. J., et al. (2015). Basic concepts for convection parameterization in weather forecast and climate models: COST action ES0905 final report. *Atmosphere, 6*, 88–147. Available from https://doi.org/10.3390/atmos6010088.

Zhang, K., Mao, H., Civerolo, K., Berman, S., Ku, Y.-Y., Rao, S. T., Doddridge, B., Philbrick, C. R., & Clark, R. (2001). Numerical investigation of boundary layer evolution and nocturnal low-level jets: Local versus non-local PBL schemes. *Environmental Fluid Mechanics, 1*, 171–208.

Zhang, Y. (2008). Online: Coupled meteorology and chemistry models: History, current status, and outlook. *Atmospheric Chemistry and Physics, 8*, 2895–2932.

Zhang, Y., & Baklanov, A. (Sci. eds.) (2019). *Training materials and best practices for chemical weather/air quality forecasting (CW-AQF)* (ETR-26, p. 565). World Meteorological Organization. <https://library.wmo.int/doc_num.php?explnum_id = 10439>.

Zilitinkevich, S. (2002). Third-order transport due to internal waves and non-local turbulence in the stably stratified surface layer. *Quarterly Journal of Royal Meteorological. Society, 128*, 913–925.

Zilitinkevich, S. (2013). *Atmospheric turbulence and planetary boundary layers* (p. 252) Moscow: FIZMATLIT, ISBN 978-5-9221-1519-3.

Zilitinkevich, S., & Baklanov, A. (2002). Calculation of the height of stable boundary layers in practical applications. *Boundary-Layer Meteorology, 105*(3), 389–409.

Zilitinkevich, S., Baklanov, A., Rost, J., Smedman, A.-S., Lykosov, V., & Calanca, P. (2002). Diagnostic and prognostic equations for the depth of the stably stratified Ekman boundary layer. *Quartely Journal of Royal Meteorological Society., 128*, 25–46.

Zilitinkevich, S., Esau, I., & Baklanov, A. (2007). Further comments on the equilibrium height of neutral and stable planetary boundary layers. *Quartely Journal of Royal Meteorological Society., 133*, 265–271.

Zilitinkevich, S. S. (1972). On the determination of the height of the Ekman boundary layer. *Boundary-Layer Meteorology, 3*, 141–145.

Zilitinkevich, S. S. (1991). Turbulent penetrative convection. *Avebury Technical, Aldershot*, 180.

Zilitinkevich, S. S., Elperin, T., Kleeorin, N., et al. (2013). A hierarchy of energy- and flux-budget (EFB) turbulence closure models for stably stratified geophysical flows. *Boundary-Layer Meteorology., 146*, 341. Available from https://doi.org/10.1007/s10546-012-9768-8.

Zilitinkevich, S. S., Elperin, T., Kleeorin, N., L'vov, V., & Rogachevskii, I. (2009). Energy- and flux-budget (EFB) turbulence closure model for stably stratified flows. Part II: The role of internal gravity waves. *Boundary-layer meteorology., 133*, 139–164.

Zilitinkevich, S. S., Elperin, T., Kleeorin, N., & Rogachevskii, I. (2007). Energy- and flux-budget (EFB) turbulence closure model for the stably stratified flows. Part I: Steady-state, homogeneous regimes. *Boundary-layer meteorology., 125*, 167–192.

Zilitinkevich, S. S., & Esau, I. N. (2002). On integral measures of the neutral, barotropic planetary boundary layers. *Boundary-layer meteorology., 104*, 371–379.

Zilitinkevich, S. S. (2012). The height of the atmospheric planetary boundary layer: State of the art and new development – Chapter 13 in "National Security and Human Health Implications of Climate Change. In H. J. S. Fernando, Z. Klaić, & J. L. McKulley (Eds.), *NATO Science for Peace and Security Series – C: Environmental Security* (pp. 147–161). Springer, ISBN 978-94-007-2429-7.

Zubov N.N., 1945. Arctic ice, Glavsevmorput, Moscow, 360 pp.

Chapter 7

Theory of atmospheric pollution dispersion

Alexander Mahura[1] and Roman Nuterman[2]

[1]*Faculty of Science, Institute for Atmospheric and Earth System Research, University of Helsinki, Helsinki, Finland,* [2]*Niels Bohr Institute, University of Copenhagen, Copenhagen, Denmark*

7.1 Basic processes

Basic processes contributing to dispersion of pollutants in the atmosphere include atmospheric transport and advection, buoyancy, turbulent diffusion, dry deposition, sedimentation, and wet deposition (see Fig. 7.1).

7.1.1 Atmospheric transport and advection

Atmospheric transport is a mechanism that moves air pollutants from a source to a receptor. The simplest source-receptor connection could be an isolated point source and the receptor. An example of a point source might be a stack emitting pollutants into the atmosphere. An example of a receptor might be a rural settlement without local sources of pollution. The pollution plume will be transported from the source to the receptor following wind direction. During its transport, the shape of the plume will not remain the same as it was initially formed at emitting from the stack. During its traveling, the atmospheric turbulent eddies will move air parcels from the plume's edges into surrounding air as well as from surrounding air into the plume. In case of wind speed greater than the speed of emitting pollutants from the stack, the wind will stretch out the plume in all directions until the plume speed becomes equal to the wind speed. There are three processes influencing the plume evolution. The first process is the turbulent mixing of the plume. The second process is the stretching out of the plume at the emitting stage. The third process is the fluctuating (during transport between the source and the receptor) of the plume from its mean value due to the wind direction changes. These lead to changes (decreasing) in the concentration of pollutants at the source site vs. the receptor area. The diffusion is a sum of these three above-mentioned processes. Note, in chemistry, the diffusion term has a different meaning, where chemical species are diffused according to Fick's law of diffusion, wherein the concentrations decrease with a distance from the source. Therefore, dispersion is a preferred term.

The pollution plume can be transported through the atmosphere (due to the advective, convective and turbulent processes) for hundreds of kilometers before being brought to a remote region. Such transport is called the long-range transport (LRT).

7.1.2 Buoyancy

Gaseous-phase pollutants, which leave the stack top, rise higher than the stack top. This might happen when such pollutants have a lower density compared with the surrounding air or they are emitted at a velocity higher when in the surrounding air. It gives enough upward kinetic energy for gases to exit the stack. The first one is called the buoyancy (or thermal) rise, and the second—the momentum rise. For the buoyancy thermal rise, a higher temperature leads to a lower density. Exceptions are the emissions of gases having higher density than the surrounding air, and the lowering parts of the plume below the stack top, which are shed by the vortices downwind of the stack.

7.1.3 Turbulent diffusion

There are highly irregular sporadic changes/fluctuations of the wind in the atmosphere. These are associated with turbulence. Such averaged fluctuations represent a mean flow with swirls or eddies of various sizes. Basically, there are two different causes of turbulent eddies. First, the mechanical turbulence, that is, eddies set in a motion by air parcels

Air Quality. DOI: https://doi.org/10.1016/B978-0-12-822591-2.00007-X

FIGURE 7.1 Illustration of basic processes contributing to the dispersion of pollutants in the atmosphere.

moving past obstacles. Second, the thermal turbulence, that is, superheated air parcels rising from the heated underlying surface, and the slower descent of cooler air parcels. Note that the sizes of the eddies caused by the thermal turbulence are larger compared with the mechanical turbulence. The manifestation of the turbulent eddies is gustiness. It is observed in the fluctuations, for example, on continuous records of the wind speed and direction or the air temperature. The fluctuations due to the mechanical turbulence tend to be quite regular and generated eddies are of nearly a similar size or scale. The eddies generated due to the thermal turbulence are both larger and more variable in size than those generated due to the mechanical turbulence. The most important mixing process in the atmosphere causing the dispersion of air pollutants is called eddy diffusion, or turbulent diffusion.

The plume is diluted in the most efficient way when the eddy scale is similar to the plume scale. Smaller eddies are effective only at tearing at the edges of the pollution plume. Larger eddies are usually only transport the entire plume.

The size and influence of the eddies on the vertical expansion of the continuous pollution plume are related to the vertical profile of the air temperature. There are situations for plumes with different combinations of the lapse rate (from strong to weak), that is, the air temperature's increase with a height (inversion), no change (isothermal), or decrease (typical or regularly observed).

7.1.4 Dry deposition

Dry deposition is a direct transfer (without the aid of precipitation) of both gaseous and particulate chemical species to the Earth's surface. It is one of the key processes for the removal of pollutants from the atmosphere. It plays an important role in controlling their lifetime in the atmosphere, and thus, has an impact on climate and air quality. The concentration gradient between the atmosphere and the surface determines the deposition of chemical species to the surface. There are three contributing processes: (1) turbulent diffusion bringing chemical species close to the surface (within a few mm); (2) molecular diffusion adjacent to the underlying surface; and (3) chemical reactions with the surface. Such deposition depends on deposition velocity, type of underlying surface (e.g., vegetative, water, anthropogenic, etc.), particles' size. The dry deposition takes place continuously compared with the wet deposition which occurs episodically (e.g., the passage of cold or warm fronts with precipitation, fog).

7.1.5 Sedimentation

Sedimentation is a process occurred under the action of gravity. In particular, gravitational sedimentation is the settling of the particles onto the surface. Such a process depends on the size of particles, their density, and settling velocity. Note that large-size particles have sufficiently large settling velocities, and hence, these settle out from the atmosphere in a reasonably short time.

7.1.6 Wet deposition

Wet deposition is a process whereby particles and gases are mixed with suspended water in the atmosphere, and then, are washed out by rain, snow or fog. Scavenging of both particles and gases takes place in clouds (rainout) by cloud droplets or below clouds (washout) by precipitation (liquid/ rain, fog and solid/ snow forms). Such deposition depends on the size of particles (e.g., diameter) and intensity of precipitation (e.g., rate). Due to the wet deposition, concentrations of air pollutants can be reduced rapidly compared with the dry deposition which is a relatively slow process.

7.2 Basic dispersion equation

The turbulent dispersion in the atmosphere can be described in the Eulerian and Lagrangian representations in the Cartesian coordinate system (x, y, z). The equations used in these representations can be in diagnostic or prognostic forms. In the context of physical modeling, the prognostic equation predicts the value of a variable in the future based on values of this variable at the current or previous times. In opposite, the diagnostic equation is time-independent and predicts all variables simultaneously at the same moment of the time.

7.2.1 Eulerian representation

For simplicity, the continuity equation for an air pollutant (gas or particulate) can be written similarly to that written for any meteorological parameter (see Chapter 5):

$$\underbrace{\frac{\partial C}{\partial t}}_{\text{non-stationary term}} + \underbrace{\frac{\partial(uC)}{\partial x} + \frac{\partial(vC)}{\partial y} + \frac{\partial(wC)}{\partial z}}_{\text{advection terms}} = 0 \tag{7.1}$$

Eq. (7.1) is sometimes called the atmospheric dispersion equation. This equation includes non-stationary and advection terms. Because molecular diffusion is negligible (except near surfaces) in the atmosphere compared to the turbulent diffusion, it is not taken into account. Turbulent flow affects momentum, heat, etc. as well as pollutant concentration. Following the basic approach for the turbulent dispersion (i.e., the Reynolds averaging) the concentration can be assumed as a sum of the averaged and turbulent fluctuating components: $C = \overline{C} + C'$, and similarly for the wind components (u,v,w): $u = \overline{u} + u'$, $v = \overline{v} + v'$, $w = \overline{w} + w'$. Given that the averaged terms for the fluctuations are equal to 0, that is, $\overline{\overline{u}C'} = \overline{u'\overline{C}} = \overline{\overline{v}C'} = \overline{v'\overline{C}} = \overline{\overline{w}C'} = \overline{w'\overline{C}} = 0$, Eq. (7.1) will become:

$$\frac{\partial \overline{C}}{\partial t} + \frac{\partial\left(\overline{u}\overline{C} + \overline{u'C'}\right)}{\partial x} + \frac{\partial\left(\overline{v}\overline{C} + \overline{v'C'}\right)}{\partial y} + \frac{\partial\left(\overline{w}\overline{C} + \overline{w'C'}\right)}{\partial z} = 0 \tag{7.2}$$

Based on the continuity equation (using the Boussinesq hypothesis) for the momentum $\frac{\partial u}{\partial x} + \frac{\partial v}{\partial y} + \frac{\partial w}{\partial z} = 0$, the simplified version of Eq. (7.2) is the following:

$$\frac{\partial \overline{C}}{\partial t} + \overline{u}\frac{\partial \overline{C}}{\partial x} + \frac{\partial\left(\overline{u'C'}\right)}{\partial y} + \overline{v}\frac{\partial \overline{C}}{\partial y} + \frac{\partial\left(\overline{v'C'}\right)}{\partial y} + \overline{w}\frac{\partial \overline{C}}{\partial z} + \frac{\partial\left(\overline{w'C'}\right)}{\partial z} = 0 \tag{7.3}$$

Applying the first-order closure by the analogy with Fick's law for molecular diffusion (Bird et al., 2006), the averaged fluctuations terms such as $\overline{u'C'}$, $\overline{v'C'}$, $\overline{w'C'}$ become the following:

$$\overline{u'C'} = -K_{xx}\frac{\partial\overline{C}}{\partial x} - K_{xy}\frac{\partial\overline{C}}{\partial y} - K_{xz}\frac{\partial\overline{C}}{\partial z} \tag{7.4a}$$

$$\overline{v'C'} = -K_{yx}\frac{\partial\overline{C}}{\partial x} - K_{yy}\frac{\partial\overline{C}}{\partial y} - K_{yz}\frac{\partial\overline{C}}{\partial z} \tag{7.4b}$$

$$\overline{w'C'} = -K_{zx}\frac{\partial\overline{C}}{\partial x} - K_{zy}\frac{\partial\overline{C}}{\partial y} - K_{zz}\frac{\partial\overline{C}}{\partial z} \tag{7.4c}$$

where $K_{xx}, K_{xy}, K_{xz}, K_{yx}, K_{yy}, K_{yz}, K_{zx}, K_{zy}, K_{zz}$ are the turbulent diffusion coefficients. In general, it is assumed that the nondiagonal terms $(K_{xy}, K_{xz}, K_{yx}, K_{yz}, K_{zx}, K_{zy})$ of the matrix of the turbulent diffusion coefficients are equal to zero, and hence, Eqs. (7.4a–c) become as the following:

$$\overline{u'C'} = -K_{xx}\frac{\partial\overline{C}}{\partial x} \tag{7.5a}$$

$$\overline{v'C'} = -K_{yy}\frac{\partial\overline{C}}{\partial y} \tag{7.5b}$$

$$\overline{w'C'} = -K_{zz}\frac{\partial\overline{C}}{\partial z} \tag{7.5c}$$

Because the molecular diffusion coefficient is negligible, the contribution of the molecular diffusion can be omitted. As seen above (Eq. 7.5), the turbulent diffusion is proportional to the concentration gradients along each spatial direction.

Let's use the following notation $C = \overline{C}$, $u = \overline{u}$, $v = \overline{v}$, $w = \overline{w}$, and thereafter, the atmospheric diffusion equation can be re-written as the following:

$$\frac{\partial C}{\partial t} + u\frac{\partial C}{\partial x} - \frac{\partial}{\partial x}K_{xx}\frac{\partial C}{\partial x} + v\frac{\partial C}{\partial y} - \frac{\partial}{\partial y}K_{yy}\frac{\partial C}{\partial y} + w\frac{\partial C}{\partial z} - \frac{\partial}{\partial z}K_{zz}\frac{\partial C}{\partial z} = 0 \tag{7.6}$$

By regrouping the advection and the turbulent diffusion terms in Eq. (7.6), the resulting equation became as the following:

$$\underbrace{\frac{\partial C}{\partial t}}_{\text{non-stationary term}} + \underbrace{u\frac{\partial C}{\partial x} + v\frac{\partial C}{\partial y} + w\frac{\partial C}{\partial z}}_{\text{advection terms}} - \underbrace{\frac{\partial}{\partial x}K_{xx}\frac{\partial C}{\partial x} - \frac{\partial}{\partial y}K_{yy}\frac{\partial C}{\partial y} - \frac{\partial}{\partial z}K_{zz}\frac{\partial C}{\partial z}}_{\text{turbulent diffusion terms}} = 0 \tag{7.7}$$

7.2.2 Lagrangian representation

In the Lagrangian representation, the evolution of the turbulent dispersion characteristics in space and time can be also considered. A changing distance between two particles (or molecules) originating from the same source, for example, shows such evolution. Let's assume that a particle is at location $\boldsymbol{x}'$ in a three-dimensional space at time $t = 0$ and at another location $\boldsymbol{x}$ at time $t > 0$. Particle displacement $\boldsymbol{x} - \boldsymbol{x}'$ can be represented as a result of a random motion by the spatial probability density function (*pdf*) as follows $P(\boldsymbol{x} - \boldsymbol{x}', t)$. Therefore, there is a small probability $P(\boldsymbol{x} - \boldsymbol{x}', t)\, d\boldsymbol{x}$ that the particle's motion will lead to the particle's arrival at some point surrounding the location $\boldsymbol{x}$ during time $t > 0$. Here, $d\boldsymbol{x}$ is a volume surrounding the location $\boldsymbol{x}$. Then, the concentration of particles, C, observed at time t at the location $\boldsymbol{x}$ (i.e., the mass of particles within a volume $d\boldsymbol{x}$) will be the following:

$$C(x, t|x') = S_0 P(x - x', t) \tag{7.8}$$

where S_0 is the mass of particles emitted at $t = 0$ at the location $\boldsymbol{x}'$. Process of the turbulent diffusion is random, and it leads to a specific distribution of the particle concentrations in the space. The width of such distribution can be described by a characteristic scale, known as the standard deviation, σ (where σ is maximum at the location $\boldsymbol{x}'$). Assuming that the source is at location $\boldsymbol{x}' = (x', y', z') = (0, 0, 0)$, and σ in a three-dimensional space undergoes the turbulent diffusion for the entire population of particles, in one direction (here, x) σ can be defined through the concentration C as follows:

$$\sigma_x^2 = \frac{1}{S_0}\iiint x^2 C(x, y, z, t)\,dx\,dy\,dz \tag{7.9}$$

at the same time σ can be also defined through *pdf* (*P*) as follows:

$$\sigma_x^2 = \iiint x^2 P(x, y, z, t) dx\, dy\, dz = \overline{x^2} \tag{7.10}$$

Therefore, the standard deviation, σ_x, of the particle distribution in the *x* horizontal direction follows the mean wind direction, and it is equivalent to the root mean square of particles' motions in that direction, that is, $\sigma_x = \sqrt{\overline{x^2}}$. Similarly, $\sigma_y = \sqrt{\overline{y^2}}$ in the horizontal crosswind direction, $\sigma_z = \sqrt{\overline{z^2}}$ in the vertical direction. In case of the uniform and isotropic turbulence (i.e., $\sigma_x = \sigma_y = \sigma_z = \sigma$), a travel distance (or the displacement) of an individual dispersing particle at velocity *u* during time *t* is

$$x(t) = \int_0^t u(t')dt' \tag{7.11}$$

As observations show, with a time passing by, more probably such displacements, $x(t)$, become larger, and this process is non-stationary. The properties of such displacement can be related to changes in the velocity with time (i.e., evolution of the velocity). Hence, the rate of change, $x^2(t)$, can be written as follows:

$$\frac{dx^2(t)}{dt} = 2x\frac{dx}{dt} = 2\int_0^t u(t)u(t')dt' \tag{7.12}$$

After applying the arithmetic averaging (the ensemble averaging) to the both sides of Eq. (7.12), the following velocity covariance is obtained:

$$\frac{\overline{dx^2(t)}}{dt} = 2\int_0^t \overline{u(t)u(t')}dt' \tag{7.13}$$

An important function characterizing stationary stochastic processes is their autocorrelation function. Once a particle possesses a certain velocity, it is still likely that for a short time, τ, it will later have a velocity of a similar magnitude and sign. For a given time interval τ a mean product may be formed $\overline{u(t)u(t+\tau)}$ known as the velocity covariance, which evidently at $\tau \to 0$ tends to $\overline{u^2}$.

Given that the process is stationary, the ensemble-average product $\overline{u(t)u(t+\tau)}$ is independent of *t* for computing this covariance, and hence,

$$\overline{u(t)u(t+\tau)} = \overline{u(0)u(\tau)} = \overline{u(-\tau)u(0)}$$

which shows also that the velocity covariance is an even function of τ. A convenient nondimensional representation of the covariance between the velocities of two particles at times *t* and $t' = t + \tau$ is the autocorrelation function as follows:

$$R(\tau) = \frac{\overline{u(t)u(t+\tau)}}{\overline{u^2}} \tag{7.14}$$

Therefore, the change of the standard deviation as a function of time can be written as follows (dt' from Eq. 7.13 becomes $d\tau$):

$$\frac{d\sigma^2(t)}{dt} = 2\overline{u^2}\int_0^t R(\tau)d\tau \tag{7.15}$$

This expression is known as the Taylor's theorem (Taylor, 1922). After integration of Eq. (7.15) it becomes as follows:

$$\sigma^2(t) = 2\overline{u^2}\int_0^t \int_0^{t'} R(\tau)d\tau dt' \tag{7.16}$$

Integrating by parts provides the solution for the square of the standard deviation of the distribution of particles for a steady-state dispersion process:

$$\sigma^2(t) = 2\overline{u^2}\left[t\int_0^t R(\tau)d\tau - \int_o^t \tau R(\tau)d\tau\right] = 2\overline{u^2}\int_0^t (t-\tau)R(\tau)d\tau \tag{7.17}$$

In the absence of any organized drift and neglecting molecular velocities, the autocorrelation function of velocity (see Eq. 7.14), may be approximated by the Lagrangian velocity autocorrelation of moving air parcel in which particles are suspended. For such an idealized situation (constant in time and space), let's assume that values of the velocity can be represented by the Markov process. In this case, the prediction is based on information only about the present state. Then, the autocorrelation function can be represented by the exponential decay function as follows:

$$R(\tau) = e^{-\frac{\tau}{t_L}} \tag{7.18}$$

where t_L is the Lagrangian time scale (on an order of 60–100 seconds in the atmospheric boundary layer), as shown in Chapter 6. At a short distance from the source location (i.e., as $\tau \to 0$, $R(\tau) \to 1$) the standard deviation for $t \to 0$ is

$$\sigma^2(t) = 2\overline{u^2} \int_0^t (t-\tau)d\tau = 2\overline{u^2}\left(t^2 - \frac{t^2}{2}\right) = \overline{u^2}t^2 \tag{7.19}$$

The standard deviation σ is proportional to the standard deviation of the instantaneous velocities as well as to the distance ($x = ut$) traveled by the particle from the source with constant velocity. At a large distance from the source (i.e., as $\tau \to \infty$, $R(\tau) \to 0$) the standard deviation for $t \to \infty$ is

$$\sigma^2(t) = 2\overline{u^2} \int_0^\infty (t-\tau)R(\tau)d\tau = 2\overline{u^2}t \int_0^\infty R(\tau)d\tau - 2\overline{u^2} \int_0^\infty \tau R(\tau)d\tau \tag{7.20}$$

By integral definition let's define that t_0 is the area under the $R(\tau)$ curve provided that $t_0 \neq 0$, and t_1 is its center of gravity:

$$t_0 = \int_0^\infty R(\tau)d\tau \tag{7.21a}$$

$$t_1 = \frac{1}{t_0}\int_0^\infty \tau R(\tau)d\tau \tag{7.21b}$$

Hence, Eq. (7.20) can be re-written as a function of t, t_0, t_1 as the follows:

$$\sigma^2(t) = 2\overline{u^2}tt_0 - 2\overline{u^2}t_0t_1 = 2\overline{u^2}t_0(t-t_1) \tag{7.22}$$

The autocorrelation function (Eq. 7.18) has an asymptotic behavior, with $t_0 = t_L$, and hence,

$$t_0 = \int_0^\infty R(\tau)d\tau = \int_0^\infty e^{-\frac{\tau}{t_L}}d\tau = -t_L e^{-\frac{\tau}{t_L}}\Big|_0^\infty = -0-(-t_L)=t_L \tag{7.23a}$$

and integrating by parts Eq. (7.21b), one obtains:

$$t_1 = \frac{1}{t_0}\int_0^\infty \tau R(\tau)d\tau = t_L. \tag{7.23b}$$

Example 7.1: Lagrangian Representation for Turbulent Dispersion—Autocorrelation Function

In the Lagrangian representation for the evolution of the turbulent dispersion characteristics in space and time, the autocorrelation function, $R(\tau)$ (see Eq. 7.18) has an asymptotic behavior. Since t_0 is defined as the area under the $R(\tau)$ curve, given that $t_0 \neq 0$, and t_1 is its center of gravity (see Eqs. 7.21a and 7.21b). Please, integrate step-by-step the equation for t_1 by parts to obtain the result shown in Eq. (7.23b).

Solution:

$$t_1 = \frac{1}{t_0}\int_0^\infty \tau R(\tau)d\tau = \frac{1}{t_0}\int_0^\infty \tau e^{-\frac{\tau}{t_L}}d\tau = \frac{1}{t_0}\left(-t_L\tau e^{-\frac{\tau}{t_L}}\Big|_0^\infty + t_L \underbrace{\int_0^\infty e^{-\frac{\tau}{t_L}}d\tau}_{t_0}\right) = \frac{1}{t_0}\left(-t_L\tau e^{-\frac{\tau}{t_L}}\Big|_0^\infty + t_L\left(-t_L e^{-\frac{\tau}{t_L}}\Big|_0^\infty\right)\right)$$

$$= \frac{1}{t_0}(-0 + t_L \cdot 0 \cdot 1 - t_L \cdot 0 + t_L t_L) = \frac{t_L^2}{t_0} = \frac{t_L^2}{t_L} = t_L$$

As seen, $t_1 = t_L$, and hence, $\sigma^2(t) = 2\overline{u^2}\left(t_L t - t_L^2\right)$. Since $t \gg t_L$, one can neglect the second term in the expression for $\sigma^2(t)$ and it becomes as follows:

$$\sigma^2(t) = 2\overline{u^2}t_L t \tag{7.24}$$

Note that the standard deviations σ_y^2 and σ_z^2 in y and z, respectively, can be estimated similarly to the standard deviation σ_x^2 in x-direction described above. The Lagrangian representation leads to a dispersion process in each direction that increases faster near the source than farther downstream. At shorter distances from the source the plume has a conic shape (i.e., σ proportional to x), whereas at larger distances (downstream from the source) the plume has an elliptical paraboloidal shape (i.e., σ proportional to $x^{1/2}$). The described above Lagrangian representation follows an assumption of the Gaussian distribution for the instantaneous velocities about the mean velocity. The latter is generally valid for both horizontal directions, but it is rarely accurate assumption in the vertical direction because of upward/downward (having higher/lower velocities) motions above the Earth's surface induced by, for example, the orography and atmospheric stability effects.

7.3 Turbulence and closure schemes

The atmospheric dispersion depends on the atmospheric conditions, and the dispersion process includes the full spectrum of the turbulent eddies. The turbulent diffusion coefficients are defined as the functions of the meteorological conditions in the horizontal (K_{xx}, K_{yy}) and vertical (K_{zz}) directions.

7.3.1 Vertical dispersion

The vertical dispersion process leads to the mixing of atmospheric pollutants in the planetary boundary layer (PBL) [or atmospheric boundary layer (ABL)]. There are two main categories of formulations (or parameterizations)—local and nonlocal. The local parameterizations are the first-order closure of the turbulent diffusion (so-called K-theory or the gradient theory formulations), which relates the turbulent fluxes to the gradients of averaged variables. In this case, a mass transfer between adjacent atmospheric layers is occurred due to the vertical dispersion. The nonlocal parameterizations, on the other hand, are usually applied to convective conditions with a strong vertical mass transfer and the mixing between multiple atmospheric layers. Here, the transfer of pollutants may occur upward (or downward) among nonadjacent layers as well. Note, in numerical modeling, the atmospheric layers became the model levels.

Following the similarity theory approach and by analogy with the turbulent kinematic viscosity formulation (i.e., the coefficient of the turbulent transfer of the concentration $\overline{w'C'} = -K_{zz}\frac{\partial \overline{C}}{\partial z}$), the vertical turbulent diffusion coefficient, K_z, can be written as follows:

$$K_z(z) = \kappa \frac{u_*}{\Phi\left(\frac{z'}{L_{MO}}\right)} z\left(1 - \frac{z}{z_{pbl}}\right)^2 \tag{7.25}$$

where κ is the dimensionless von Karman constant ($=0.4$), u_* is the friction velocity (representing an amplitude of the vertical turbulent flux of the horizontal momentum within the surface layer), L_{MO} is the Monin–Obukhov length (characterizing the atmospheric stability), z is the altitude (or height), z_{pbl} is the PBL height (obtained from simulations or parameterized), and z' is equal to z within PBL for the stable, neutral, and unstable conditions and equal to $0.1z_{pbl}$ above PBL for the unstable conditions. Φ is the empirically determined function. The Φ function varies depending on atmospheric stability: $\Phi = 1$ (for the neutral conditions), $\Phi = 1 + 6\zeta$ (for the stable conditions), and $\Phi = (1 - 19.3\zeta)^{-1/4}$ (for the unstable conditions) with $\zeta = z'/L_{MO}$ according to Högström (1988).

Example 7.2: Vertical Turbulent Diffusion for Various Atmospheric Stability Conditions

The vertical turbulent diffusion coefficient, K_z, is given by Eq. (7.25). Please, calculate and plot K_z as a function of the altitude, z, for various atmospheric stability conditions. Assume that (1) the boundary layer height, z_{pbl} is 500 m; (2) the Monin–Obukhov length, L_{MO} is equal to 50, -12, and 150 m for the stable, unstable and neutral conditions, respectively (on examples of values from Table 7.2); (3) the friction velocity, u_* is equal to 0.3, 0.6, and 0.4 m s^{-1} for the stable, unstable and neutral conditions, respectively.

Solution:

(1) Using Eq. (7.25) and empirical Φ function (see above), the vertical turbulent diffusion coefficients are calculated at different altitudes for different values of the Monin–Obukhov length and the friction velocity for the stable, unstable and neutral atmospheric conditions.

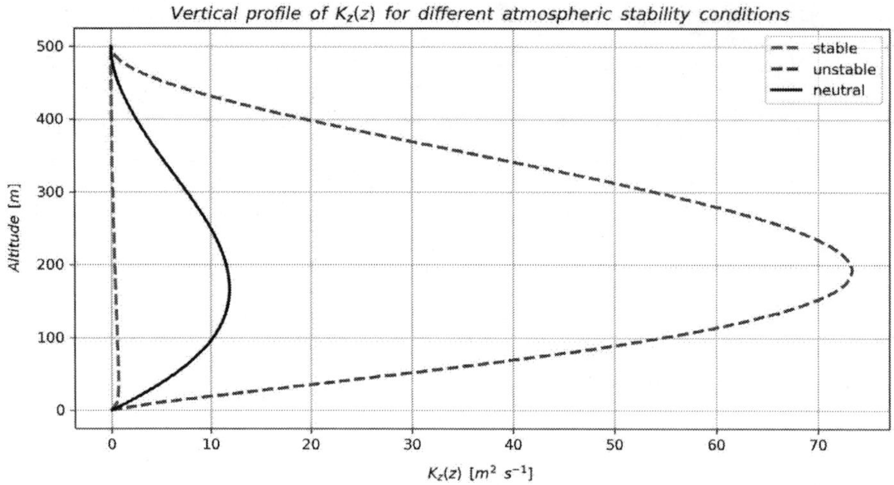

FIGURE 7.2 Profiles of vertical turbulent diffusion coefficients for stable, unstable, and neutral atmospheric stability conditions.

We are calculating the vertical turbulent diffusion coefficient at selected altitudes for the neutral conditions (within the boundary layer). For example:

$$\kappa = 0.4, u_* = 0.4\ \mathrm{m\,s^{-1}}, L_{MO} = 150\ \mathrm{m}, z_{pbl} = 500\ \mathrm{m}, z = z' = 150\ \mathrm{m}, \Phi\left(\frac{z'}{L_{MO}}\right) = 1$$

$$K_z(z = 150) = 0.4 \cdot \frac{0.4}{1} \cdot 150 \cdot \left(1 - \frac{150}{500}\right)^2 = 11.76\ \mathrm{m^2\,s^{-1}}$$

(2) By calculating and plotting the vertical turbulent diffusion coefficients as a function of altitude you will obtain the following (Fig. 7.2).

7.3.2 Horizontal dispersion

The horizontal dispersion process also contributes to the atmospheric pollution mixing within PBL. Let's apply the turbulent kinematic viscosity formulation (i.e., the coefficients of the turbulent transfer of the concentration $\overline{u'C'} = -K_{xx}\frac{\partial \overline{C}}{\partial x}$ and $\overline{v'C'} = -K_{yy}\frac{\partial \overline{C}}{\partial y}$), and assume that the pollution plume changes due to airflow (wind) stretching and shearing in the horizontal directions. Hence, assuming homogeneous diffusion in both horizontal directions (i.e., $K_{xx} = K_{yy}$), the nonlinear horizontal diffusion coefficient, K_h, following Smagorinsky (1963) can be written as follows:

$$K_h = l_s^2\sqrt{T^2 + S^2} \tag{7.26a}$$

$$T = \frac{\partial u}{\partial x} - \frac{\partial v}{\partial y} \tag{7.26b}$$

$$S = \frac{\partial u}{\partial y} - \frac{\partial v}{\partial x} \tag{7.26c}$$

where T and S are the horizontal tension and horizontal shearing strains, respectively; l_s is the characteristic length scale (a sort of a mixing length), which in numerical modeling is proportional to the horizontal resolution (i.e., dimensions of a model grid cell). Note, the defined above K_h can be considered relatively accurate at resolutions of less than 10 km, and at the cruder resolutions a correction is required. Byun & Schere (2006) proposed to consider K_h as inversely proportional to the area of the horizontal grid cell. Finally, the combination of both—the proportional and inversely proportional parts—allows to cover a full spectrum of resolutions from fine to coarse.

7.4 Atmospheric stability regimes and dispersion

The stability regimes (or the atmospheric stability categories) can be obtained based on routine simple meteorological data from ground-based observations (Fig. 7.3). Among these categories are the following: A (very unstable), B (unstable), C (moderately unstable), D (neutral), E (stable), and F (very stable). These categories are defined by wind speed, cloudiness (nighttime) and solar radiation (daytime) on a diurnal cycle (see Table 7.1), and are associated with various natural phenomena.

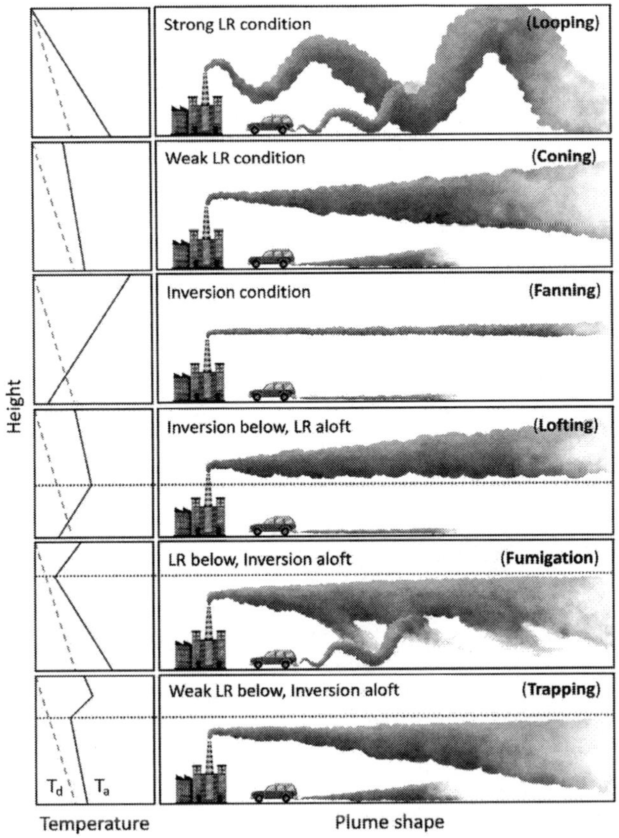

FIGURE 7.3 Spatial dispersion of pollution plume (plume shape) related to the temperature vertical structure. LR—lapse rate, T_a—air temperature, T_d—dew point temperature.

TABLE 7.1 Atmospheric stability categories (A, B, C, D, E, F) as a function of the meteorological data (Turner, 1970).

Wind speed	Solar radiation (daytime)			Cloudiness (nighttime)	
	Strong	Moderate	Weak	0–3	4–8
<2	A	A/B	B	F	F
2–3	A/B	B	C	F	E
3–5	B	B/C	C	E	D
5–6	C	C/D	D	D	D
>6	C	D	D	D	D

Notes: The wind speed at 10 m (in $m\,s^{-1}$), solar radiation (in $W\,m^{-2}$ strong; >700, moderate 350–700, weak <350), and cloudiness (in oktas; 0—clear sky or complete absence of clouds, 8—overcast or full cloud cover with no breaks).
Source: From Turner, J. S. (1970). *Workbook of atmospheric dispersion estimates*. U.S. Environmental Protection Agency, Research Triangle Park, NC, USA.

For example, the A category is observed at strong thermal instability, which might occur from the late morning to middle afternoon hours. The B–C categories are associated with moderate-slight mixing and transitional periods. The D category is related to strong winds, and day-night transitions of cloudiness. The E category is linked to moderate winds, night hours, and transitional periods. The F category is connected to very limited vertical mixing, light winds, clear sky conditions.

Moreover, the atmospheric stability categories can be represented as functions of the bulk Richardson number (Ri_b), the vertical air temperature gradient ($\Delta T/\Delta z$) measured, at least, for a 100 m layer depth, and the Monin–Obukhov length (L_{MO}) over open terrain (i.e., a roughness length of less than 0.1 m, e.g., grass, low crops, etc.). As seen in Table 7.2, Ri_b is negative for the A-B-C categories, positive for E–F, and can be both negative and positive for D. The $\Delta T/\Delta z$ is positive only for the F category, both signs for the E category, and negative for all other categories. L_{MO} is negative for the A–B–C categories, positive for the E–F categories, and it can be both negative and positive for the D category (i.e., $L_{MO} < -100$ and $L_{MO} > +100$). In case of larger values of the roughness length (e.g., agricultural field, water surface compared with bushes, forest, urban area) the range of values will increase because there is more influence of the mechanical turbulence.

Weather-related information such as the wind speed and direction, precipitation as well as atmospheric stability is important for the treatment of the dispersion. The effects of the wind include the wind direction affecting a pathway of the atmospheric transport and dispersion of the pollution plume; the wind speed (e.g., calm, low, moderate, high, or extreme) affecting the time to reach a certain area/region. The precipitation occurs on a pathway of the pollution plume causing the removal of pollutants from the plume and their deposition on the underlying surface. The extent and rate of the plume dispersion during the atmospheric transport are related to the turbulence, which depends on the atmospheric stability. The turbulence is enhanced when the atmosphere is unstable. The turbulence is suppressed when the atmosphere is stable. The turbulence is developed when the atmosphere is neutral.

The A and D categories are mostly related to a strong dispersion in the vertical and horizontal directions, respectively, with a relatively quick decrease of pollutants concentrations in the plume. The B and C categories are mostly associated with a moderate dispersion in the vertical leading to a moderate decrease of concentrations. The E and F categories are mostly connected with a low dispersion in both the vertical and horizontal leading to a weak decrease of concentrations.

The various chemical species both gases and particulates have different lifetimes in the atmosphere due to their reactivity (the rate of chemical reactions). For example, CO is the nonreactive chemical species compared with O_3 and NO_2 being the reactive species. In addition to the chemical reactions, the dispersion influences continuously the spatial-temporal distribution leading to further changes in concentrations. Moreover, the chemical species can be short-lived (i.e., up to a few days-weeks) and long-lived (i.e., up to tens or hundreds of years) in the atmosphere. For example, the black carbon (soot) and O_3 are short-lived up to days and weeks, respectively; CH_4 and CO_2 are long-lived up to 12 and more than 300 years, respectively. Due to the LRT and continuous dispersion, the long-lived pollutants will be well mixed within the atmosphere on a hemispheric scale. On the same scale, the short-lived pollutants will be also dispersed during the atmospheric transport, although such transport might occur on shorter distances.

TABLE 7.2 Atmospheric stability categories as a function of the bulk Richardson number, the vertical temperature gradient, and the Monin–Obukhov length.

Atmospheric stability category	Bulk Richardson number, Ri_b	Vertical temperature gradient, $\Delta T/\Delta z$	Monin–Obukhov length, L_{MO}
A	$Ri_b < -0.86$	$\Delta T/\Delta z < -1.9$	$-10 \le L_{MO} < 0$
B	$-0.86 \le Ri_b < -0.37$	$-1.9 \le \Delta T/\Delta z < -1.7$	$-15 \le L_{MO} < -10$
C	$-0.37 \le Ri_b < -0.10$	$-1.7 \le \Delta T/\Delta z < -1.5$	$-100 \le L_{MO} < -15$
D	$-0.10 \le Ri_b < 0.053$	$-1.5 \le \Delta T/\Delta z < -0.5$	$L_{MO} < -100$; $L_{MO} \ge 100$
E	$0.053 \le Ri_b < 0.134$	$-0.5 \le \Delta T/\Delta z < 1.5$	$30 \le L_{MO} < 100$
F	$0.134 \le Ri_b$	$1.5 \le \Delta T/\Delta z$	$0 \le L_{MO} < 30$

Source: From Seigneur, C. (2019). *Air pollution: Concepts, theory, and applications* (pp. 1–354). Cambridge University Press..

7.5 Gaussian plume concept

7.5.1 General approach

A concept of a reference point is used in both the Eulerian and Lagrangian methods. For the first, this reference point is fixed in time and space, and the air flows relatively to this point. For the latter, the point moves with the airflow. As a simplification, the Gaussian method considers an instantaneous release of pollutant from a point source. In other words, an air parcel with pollutant released from such a source is called a puff. Once released, such puff moves downwind along the path, which corresponds to the average wind direction. During movement, the puff expands in volume. Air parcels move from the puffs' edges into surrounding air as well as from surrounding air into the puff, hence, reducing its concentration. During atmospheric transport such puff experiences small random movements due to the turbulence. An infinitely rapid series of small individual puffs can be considered as a continuous emission. The Gaussian dispersion theory enables to calculate the concentration of pollutant along the path (or trajectory), for example, at any point downwind. To predict the concentration of a pollutant at any point in the atmosphere, a system of equations describing physical processes is required, and it will be derived below from the first principles.

At first, a position of the point, where the concentration is estimated, should be described relatively to the source and the Earth's surface. Here, a Cartesian coordinate system (x, y, z) can be used. In such system, the physical source (e.g., base of a stack) is placed at the origin (0, 0, 0), where x-axis is oriented along the mean wind direction (see Fig. 7.4A). The x describes the distance downwind from the source; y is the lateral distance from the mean wind direction; z is the height above the surface; h_s is the physical height of the stack; $\overline{u}$ is the mean wind speed at the plume height. The effective height of the release, H, can be calculated as $H = h_s + \Delta h$, where Δh is the additional height of the plume rise due to its buoyancy and/or momentum.

The starting point for the dispersion theory is to consider the diffusion of a plume of pollutant released into the atmosphere instantaneously from a point source. For a sake of simplicity, all three advection terms and the turbulent diffusion terms in the y- and z-directions are considered as negligible in Eq. (7.7) in the Eulerian representation. Assuming that the dominant flow direction of the plume is along the x-direction, the turbulent transfer in the atmosphere can be described by the Fickian equation as

$$\frac{\partial \overline{C}}{\partial t} = K\frac{\partial^2 \overline{C}}{\partial x^2} \tag{7.27}$$

where $\overline{C}$ is the averaged concentration of pollutant, and K is the constant turbulent diffusivity ($K = K_{xx} = K_{yy} = K_{zz}$). The analytical solution of Eq. (7.27) in the x-dimension is

$$\overline{C} = \frac{S_0}{(4Kt)^{1/2}}\exp\left(\frac{-x^2}{4Kt}\right) \tag{7.28}$$

where S_0 is the mass of particles emitted from the source (same in both the Eulerian and Lagrangian representations), t is the time. The concentration follows the so-called Gaussian bell-shaped distribution (Fig. 7.4B). The concentration has its maximum at the source (or the point of the release), and then, it decreases in all directions with increasing distance from the source.

7.5.2 Instantaneous point source

Let's consider an instantaneous point source located at (x', y', z') for both the Eulerian and Lagrangian representations. In this case, the pollution is distributed isotropically in all 3D, and it is simultaneously transported away from the source. Under the initial and boundary conditions such as:

$$C(x,y,z,0) = S_0\delta(x-x')\delta(y-y')\delta(z-z')$$

$$C(x,y,z,t) = 0 \text{ for } x,y,z \rightarrow \infty$$

where δ is the Dirac function. For example, for $\delta(x - x')$:

$$\delta(x-x') = \begin{cases} 0 \text{ for } x \neq x' \\ +\infty \text{ for } x = x' \end{cases}; \int_{-\infty}^{+\infty} \delta(x-x')dx = 1$$

and similarly, for two other dimensions.

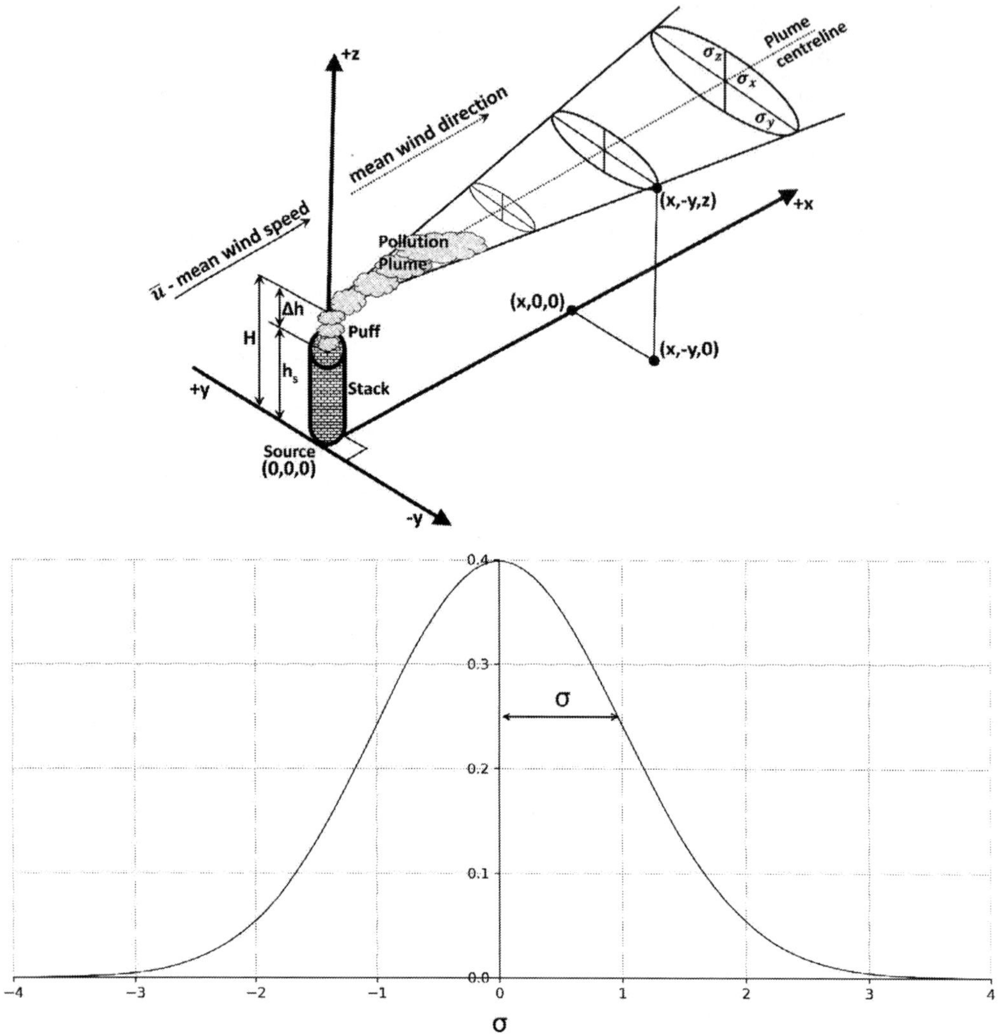

FIGURE 7.4 (A) Conceptual schematics (in the Cartesian coordinate system) of the pollution plume release, movement, and dispersion in the air and (B) pollutant concentration profiles following the Gaussian (normal) distribution showing standard deviations, σ.

Following the continuity equation $\frac{\partial u}{\partial x} + \frac{\partial v}{\partial y} + \frac{\partial w}{\partial z} = 0$ in the Eulerian representation, the solution for the pollutant concentration is as follows:

$$C(x, y, z, t) = \frac{S_0}{(2\pi)^{3/2}\sigma_x\sigma_y\sigma_z} \cdot \exp\left(-\frac{(x-x')^2}{2\sigma_x^2} - \frac{(y-y')^2}{2\sigma_y^2} - \frac{(z-z')^2}{2\sigma_z^2} \right) \tag{7.29}$$

where the standard deviations correspond to the Eulerian turbulent dispersion coefficients $\sigma_x = \sqrt{2K_{xx}t}, \sigma_y = \sqrt{2K_{yy}t}, \sigma_z = \sqrt{2K_{zz}t}$.

7.5.3 Continuous point source

Let's consider a continuous point source located at (x', y', z') under the stationary and uniform meteorological conditions (i.e., constant wind speed and direction). In this case, the concentration of pollutant is calculated with the following mass conservation equation:

$$u\frac{\partial C}{\partial x} - \frac{\partial}{\partial y}K_{yy}\frac{\partial C}{\partial y} - \frac{\partial}{\partial z}K_{zz}\frac{\partial C}{\partial z} = Q\delta(x-x')\delta(y-y')\delta(z-z') \tag{7.30}$$

Here, the turbulence in x-direction is neglected assuming that the atmospheric transport is dominated by the advection only in this direction. The left-hand side of Eq. (7.30) is the atmospheric transport of the concentration along x-direction with 1D advection and 2D diffusion of the pollutant. The right-hand side is the source term with Q being the source strength (or the emission rate, e.g., in $kg\,s^{-1}$). The solution of Eq. (7.30) according to the Gaussian form is the following:

$$C(x,y,z) = \frac{Q}{2\pi u\sigma_y\sigma_z}\cdot\exp\left(-\frac{(y-y')^2}{2\sigma_y^2} - \frac{(z-z')^2}{2\sigma_z^2}\right) \tag{7.31}$$

As seen from Eq. (7.31), the concentration of pollutant emitted from a continuous source is inversely proportional to the wind speed, u, following the Gaussian distribution. The values of σ_y and σ_z at any distance from the source represent an extent of the plume's half width in the vertical or horizontal directions from the plume centerline (see also in Fig. 7.4A).

The proposed Gaussian formulation for the plume dispersion is a good representation of the atmosphere when conditions are relatively stationary. The above-given derivations are for the point source. The same approach can be applied for nonpoint sources such as the line (e.g., road traffic), area (e.g., forest fires), and volume (e.g., chemical refineries) sources of emissions. However, the integration over the line, area or volume rarely leads to an analytical solution, and in these cases numerical solutions are required.

7.6 Plume rise

Most pollutants in industrial areas emitted from the stacks form plumes. These plumes are transported in the vertical direction due to exit velocity (e.g., buoyancy dependent on the temperature or density difference with the ambient air) and in the horizontal due to the speed of dominating airflows. The thermal buoyancy and the vertical momentum contribute to increasing the plume height above the stacks. The value of the initial plume height depends on the stack height and on the plume rise above the stack, and the effective plume height is higher than the stack height. It is quantified by a single parameter called the plume rise, Δh, and defined as a vertical displacement of the plume at the initial phase of emissions. This value is derived from semiempirical studies, and in a generalized form it can be written as follows:

$$\Delta h(x) = AQ_h^a x^b u^c \tag{7.32}$$

where A, a, b, c are constants, x is the distance downwind from the source, u is the wind speed at the source height (z_s), and Q_h is the heat emission rate of the source. It is defined as follows: $Q_h = Q_m c_p(T_s - T_a)$, where c_p is the specific heat at constant pressure, T_s and T_a are the emitted gases exit temperature and the ambient air temperature (both in degree K) at the source height z_s, respectively. Q_m is the total mass emission rate at the source height z_s given as: $Q_m = \rho_s \pi r_s^2 v_s$, where ρ_s is the density of total emissions, r_s is the inside radius of the stack, and v_s is the vertical speed of emitted pollutants. The plumes are considered the buoyant plumes when $T_s > T_a$, and the nonbuoyant plumes when $T_s \approx T_a$.

Among various schemes, the Briggs (1969) method is one of the well-known. It defines the buoyancy flux parameter, F_b, as: $F_b = g v_s r_s^2 \frac{(T_s - T_a)}{T_s}$ (with SI units of $m^4\,s^{-3}$) and the critical downwind distance, x_c, as

$$x_c = \begin{cases} 2.16F_b^{2/5} z_s^{3/5} & \text{for } z_s < 305\text{ m} \\ 67F_b^{2/5} & \text{for } z_s \geq 305\text{ m} \end{cases}$$

which separates two stages of the plume rise.

For distances less than or equal to the critical ($x \leq x_c$) and all atmospheric stability regimes the plume disperses according to the generalized form (Eq. 7.32) as follows:

$$\Delta h(x) = 1.6F_b^{1/3} x^{2/3} u^{-1} \tag{7.33}$$

where constant A is equal to 1.6 (Briggs, 1972).

Similarly, for all atmospheric stability regimes, for distances more than the critical ($x > x_c$), where turbulence contributes more to the dispersion of the plume, Eq. (7.32) becomes as follows:

$$\Delta h(x) = 1.6 F_b^{1/3} x^{2/3} u^{-1} \cdot f(x, x_c) \tag{7.34}$$

where $f(x, x_c)$ is a polynomial function constructed based on empirical data.

There are also other classical semiempirical formulations, e.g., proposed by Bringfelt (1969), Brummage (1968), Carpenter et al. (1971), Fay et al. (1970), Holland (1953), and others.

Example 7.3: Plume rise

Following Eq. (7.33) and assuming that the pollution plume is the buoyant plume, please, calculate and plot the plume rise as a function of the distance and the ambient air temperature and the wind speed at the stack height for the given input conditions: z_s is the source (stack) height is 100 m, r_s is the inside radius at the stack height is 4 m, T_s is the emitted gases exit temperature at the stack height is 120°C, v_s is the vertical speed of emitted pollutants at the stack height is 20 m s^{-1}, T_a is the ambient air temperature at the stack height changes: 5°C, 15°C, 25°C, and u is the wind speed at the stack height changes: 3, 5, and 10 m s^{-1}.

Solution:

For example:

(1) We are calculating the buoyancy flux parameter, F_b. Assuming $g = 9.8067$ m s^{-2}, $v_s = 20$ m s^{-1}, $r_s = 4$ m and after converting temperatures from degrees C to K: $T_s = 120$°C or $T_s = 273.15 + 120 = 393.15$ K; $T_a = 5$°C or $T_a = 273.15 + 5 = 278.15$ K, we have:

$$F_b = g v_s r_s^2 \frac{(T_s - T_a)}{T_s} = 9.8067 \cdot 20 \cdot 4^2 \cdot \frac{(393.15 - 278.15)}{393.15} = 917.93605 \text{ m}^4 \text{ s}^{-3}$$

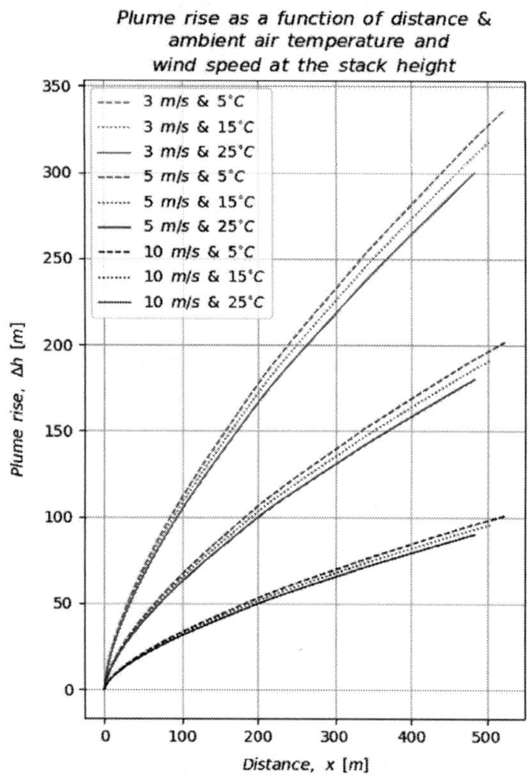

FIGURE 7.5 Profiles of plume rise as a function of distance from a stack (source) depending on ambient air temperature and wind speed at the stack height.

(2) Since, z_s, source or stack height is 100 m (or $z_s < 305$ m), we are calculating the critical downwind distance as follows:

$$x_c = 2.16F_b^{2/5}z_s^{3/5} = 2.16\cdot(917.93605)^{\frac{2}{5}}\cdot(100)^{\frac{3}{5}} = 524.29862 \text{ m}$$

(3) We are calculating the plume rise for conditions of a distance $x = 400$ m (which is less than the above calculated critical distance, x_c, and hence, corresponds to the first stage of the plume rise) and wind speed at the stack height $u = 10 \text{ m s}^{-1}$ as follows:

$$\Delta h(x) = 1.6F_b^{1/3}x^{2/3}u^{-1} = 1.6\cdot(917.93605)^{\frac{1}{3}}\cdot(400)^{\frac{2}{3}}\cdot 10^{-1} = 84.41717 \text{ m}$$

(4) By calculating and plotting the plume rise as a function of distance from a stack (source) depending on ambient air temperature and wind speed at the stack height you will obtain the following (Fig. 7.5).

7.7 Removal processes

Removal of pollutants from the atmosphere takes place due to the dry and wet depositions and the sedimentation processes. The dry deposition is uptake by soil, water or vegetation at the Earth's surface. The sedimentation is the settling of pollutants (in the form of particles) on the underlying surface due to the gravitational effects. The wet deposition is absorption by liquid/solid droplets during rain/snowfall precipitation or fog (within the surface layer of the atmosphere).

These processes are observed through all temporal and spatial scales. However the wet deposition is observed physically only where the precipitation geographically takes place contrary to the dry deposition and the sedimentation. These processes for gases and aerosols are parameterized in numerical modeling, but still have large uncertainties.

7.7.1 Dry deposition

Assuming that the vertical velocity component of wind, w, is equal to the deposition velocity of pollutant, V_d, and following $\overline{wC} = -K_{zz}\frac{\partial \overline{C}}{\partial z}$, the dry deposition at the ground level ($z = 0$) is defined by the ratio between the pollutant deposition flux F ($\text{kg m}^{-2}\text{s}^{-1}$) and the pollutant averaged concentration $\overline{C}$ (kg m^{-3}):

$$V_d = \frac{F}{\overline{C}} \tag{7.35}$$

Note that V_d is a function of the meteorological conditions, for example, the atmospheric stability, the pollutant type, and the type of the underlying surface.

For gases, the dry deposition is strongly affected by their chemical interactions with the underlying surface. For particles of smaller sizes (i.e., with a diameter of less than 0.1 μm) the Brownian motion allows rapid movements across the viscous air layers just above the surface. For particles of large sizes (greater than 1 μm), the dry deposition is dominated by the gravitational effects. For particles of intermediate sizes, the dry deposition is strongly affected by impaction and interception. The theoretically predicted deposition velocities are low due to relative weakness of the Brownian motion and the gravitational settling effects, but field measurements show high depositions.

The theory of gaseous and particle dry deposition is governed both by the transfer in the gas phase and by the absorption at the underlying surface, which is generally assumed to be an irreversible process. The deposition velocity can be expressed as the inverse of a sum of so-called "resistances" in three sequential layers (as shown in Fig. 7.6). The surface layer (up to 20 m above the ground) is a layer, where the atmospheric fluxes are assumed to be constant, and the pollution transport is characterized by the turbulence properties. The deposition layer is a thin viscous boundary layer adjacent to the surface of an obstacle on which deposition is occurring, and it is characterized by intermittent turbulence. In this layer the gases are influenced by molecular diffusion; and the particles are affected by the Brownian diffusion and inertial impaction. The vegetation layer is a layer where pollutants are uptaken (reacted) with vegetation/by canopy, and deposited directly onto the ground or water surface.

For gases, the deposition velocity, V_d, can be represented by resistances from these three above-mentioned layers:

$$V_d = \frac{1}{r_a + r_d + r_c} \tag{7.36}$$

where r_a is the aerodynamic resistance, r_d is the deposition resistance, and r_c is the canopy resistance. All these resistances have units of (s m^{-1}).

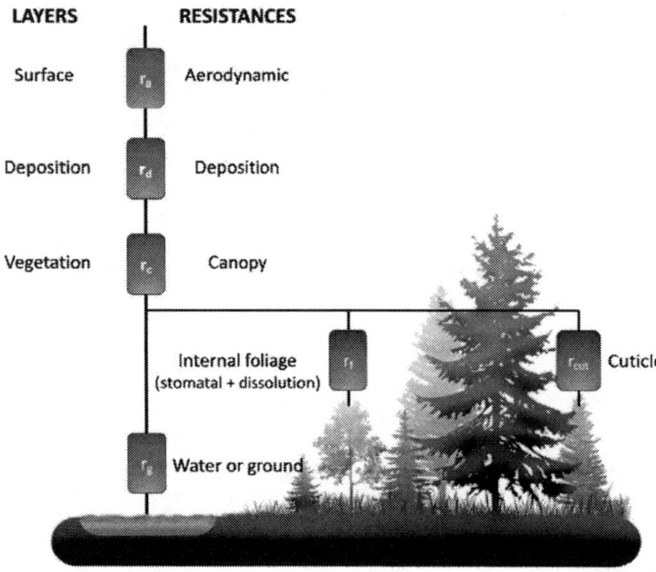

FIGURE 7.6 Dry deposition of pollutants through the sequential layers with their corresponding resistances (see details and explanation in the text below).

According to Wesely and Hicks (1977, 2000), the aerodynamic resistance r_a is

$$r_a = \frac{1}{\kappa u_*}\left(\ln\left(\frac{z_{ref}}{z_0}\right) - \Psi_c\right) \quad (7.37)$$

where z_{ref} is the reference height (m), z_0 is the roughness length (m) associated with the momentum transfer, κ is the von Karman constant, u_* is the friction velocity (m s^{-1}), and Ψ_c is a stability correction term (accounting for the effects of buoyancy-induced changes in the flux-gradient relationships) for different atmospheric conditions

$$\Psi_c = \begin{cases} 0 & \text{for neutral} \\ -5\dfrac{z_{ref}}{L_{MO}} & \text{for stable} \\ ln[f(x)] - 2\arctan(x) + \dfrac{\pi}{2} & \text{for unstable} \end{cases}$$

where $f(x)$ is a polynomial function of x and $x = \left(1 - 16\frac{z_{ref}}{L_{MO}}\right)^{1/4}$.

According to Hicks et al. (1987), the deposition resistance r_d near the surface is

$$r_d = b_1 \frac{Sc^{b_2}}{\kappa u_*} \quad (7.38)$$

where Sc is the Schmidt number ($Sc = \frac{\nu}{D}$), ν is the kinematic viscosity of the air (m^2 s^{-1}), D is the molecular diffusivity of a gaseous pollutant (m^2 s^{-1}), b_1 and b_2 are the empirical parameters ($b_1 \approx 1.6 \div 16.7$, $b_2 \approx 0.4 \div 0.8$, with the suggested choice of $b_1 = 5$ and $b_2 = 2/3$). Such resistance is associated with the transfer through the quasilaminar layer in contact with the surface.

Example 7.4: Dry deposition—deposition resistance

Following Eq. (7.38), please, calculate the deposition resistance r_d near the surface for different atmospheric stabilities and chemical species for the given input conditions: u_* is the friction velocity which is equal to 0.3, 0.6, and 0.4 (in m s^{-1}) for the stable, unstable, and neutral atmospheric conditions; ν is the kinematic viscosity of the air which is 1.48×10^{-5} (in m^2 s^{-1}); D is the coefficients of diffusivity (or the molecular diffusivity) for chemical species like

CO_2, CH_4, CO, SO_2, and N_2O which are 0.1381, 0.1952, 0.1807, 0.1089 and 0.1436 (in $cm^2 s^{-1}$), respectively; b_1 and b_2 are the empirical parameters (use suggested values from the Section 7.7).

Solution:

For example: (1) We are calculating the Schmidt number, Sc, for CO_2 for given conditions $\nu = 1.48 \times 10^{-5}\ m^2 s^{-1}$ and $D = 0.1381\ cm^2 s^{-1}$ with converting D units from $cm^2 s^{-1}$ to $m^2 s^{-1}$ as follows: $Sc = \frac{\nu}{D} = \frac{1.48 \cdot 10^{-5}}{0.1381/1000} = 0.1072$

(2) For CO_2, we are calculating the deposition resistance, r_d, for given conditions $\kappa = 0.4, u_* = 0.4\ m\,s^{-1}$, $b_1 = 5, b_2 = 2/3$, and assuming neutral atmosphere, as follows:

$$r_d = b_1 \frac{Sc^{b_2}}{\kappa u_*} = 5 \cdot \frac{0.1072^{\frac{2}{3}}}{0.4 \cdot 0.4} = 7.0506\ s\,m^{-1}$$

(3) By calculating the Schmidt number and the deposition resistance for different chemical species and atmospheric conditions you will obtain the following summary (Table 7.3).

According to Hicks et al. (1987), the canopy resistance r_c is consist of three resistances. The r_f is the internal foliage (i.e., leaves inside trees/plants, rather than the leaves that are visible outside) resistance, r_{cut} is the cuticle (i.e., a thin waxy layer covering the leaves' surface) resistance, and r_g is the ground or water surface resistance. Moreover, the internal foliage resistance, r_f, also includes two other resistances. The r_s is the resistance for the transport through the stomatal pore (i.e., small openings on the plants' surfaces allowing exchange of gases between the plants and the environment), and r_m is the resistance to dissolution or reaction of a pollutant in the mesophyll cells (i.e., cells responsible for photosynthesis when plants convert the light energy into chemical energy). Hence, the summation of all these resistances is the canopy resistance r_c:

$$r_c = \left(\frac{LAI}{r_s + r_m} + \frac{LAI}{r_{cut}} + \frac{1}{r_g} \right)^{-1} \tag{7.39}$$

where r_f is the sum of $r_s + r_m$, LAI is the Leaf Area Index (i.e., the ratio of the leaf surface area divided by the ground surface area). In particular, the parameterizations and the range of values for r_s and r_m are discussed by O'Dell et al. (1977), r_{cut} and r_g by Pleim et al. (1984).

For particles having different weights and diameters, the deposition velocity in the atmosphere will differ. For very small particles (i.e., with a diameter less than 0.1 μm), the Brownian motion allows rapid movement of particles across the viscous layers of the air just above the surface. For large particles (diameter greater than 1 μm) this motion is

TABLE 7.3 Summary table of calculated deposition resistances (based on coefficients of diffusivity and the Schmidt number) for selected chemical species (pollutants) and under different atmospheric stability conditions.

Pollutant	D	Sc	rd	AtmStab
CO_2	0.1381	0.1072	9.4009	Stable
CO_2	0.1381	0.1072	4.7004	Unstable
CO_2	0.1381	0.1072	7.0506	Neutral
CH_4	0.1952	0.0758	7.4641	Stable
CH_4	0.1952	0.0758	3.7320	Unstable
CH_4	0.1952	0.0758	5.5981	Neutral
CO	0.1807	0.0819	7.8582	Stable
CO	0.1807	0.0819	3.9291	Unstable
CO	0.1807	0.0819	5.8937	Neutral
SO_2	0.1089	0.1359	11.0140	Stable
SO_2	0.1089	0.1359	5.5070	Unstable
SO_2	0.1089	0.1359	8.2605	Neutral
N_2O	0.1436	0.1031	9.1593	Stable
N_2O	0.1436	0.1031	4.5796	Unstable
N_2O	0.1436	0.1031	6.8694	Neutral

dominated by the sedimentation (gravitational) effects. Hence, the deposition velocity, V_d, can be represented by the resistances from the surface and the deposition layers (except the vegetation layer, where r_c is practically zero, because particles penetrated the deposition layer stuck to the surface) and the sedimentation or the gravitational settling, V_g (i.e., alternative pathway to the turbulent exchange for reaching the ground):

$$V_d = \frac{1}{r_a + r_d + r_a r_d V_g} + V_g \tag{7.40}$$

where r_a is calculated following Eq. (7.37), and r_d has the following form:

$$r_d = \frac{\left(Sc^{-2/3} + 10^{-3/St}\right)}{u_*} \tag{7.41}$$

In Eq. (7.41) the Sc is the Schmidt number ($Sc = \frac{\nu}{D}$), ν is the kinematic viscosity of the air ($m^2 s^{-1}$), D is the Brownian diffusivity of a particulate pollutant in the air ($m^2 s^{-1}$) instead of the molecular diffusivity for gases, St is the Stokes number:

$$St = \frac{V_g u_*^2}{g\nu} \tag{7.42}$$

where g is the gravity acceleration ($m\,s^{-2}$).

It should be noted that the gravitational settling depends on the size, shape and density of a particle. Assuming that particles can be approximated by spheres, the Stokes equation is the following:

$$V_g = \frac{d_p^2 g\left(\rho_p - \rho_{air}\right)}{18\mu} \tag{7.43}$$

where d_p is the particle diameter (m), μ is the viscosity ($N\,s\,m^{-2}$), ρ_p and ρ_{air} are the particle and the air densities ($kg\,m^{-3}$), respectively. In cases of small particles, the calculated V_g should be multiplied the Cunningham correction factor, see Cunningham (1910) and Seigneur (2019).

7.7.2 Wet deposition

Several terms are considered as synonyms with the wet deposition. These include precipitation scavenging, wet removal, washout, and rainout. The precipitation scavenging is the removal of pollutants by a raining cloud. Removal of pollutants by settling fog droplets (e.g., radiation fog) is called fog deposition. Removal of pollutants by ice crystals (e.g., snowstorm) is called the snow deposition. Removal of pollutants by cloud droplets on the terrain (for example, observed on top of high mountains) is called the cloud interception. Removal of pollutants due to in-cloud scavenging is called the rainout. Removal of pollutants due to below-cloud scavenging (by falling rain, snow) is called the washout. Note that the scavenging of pollutants by snow is less efficient than the scavenging by rain.

All these above-mentioned deposition processes have three conditions or steps, which are necessary for the wet deposition of pollutants (gases or aerosols). At first, condensed water must be presented in the air. Second, pollutants must be scavenged in the air by water or ice particles and transported to the Earth's surface. Third, pollutants may undergo chemical transformations during previous steps. A conceptual detalization of the processes—such as the scavenging, evaporation, nucleation, collision, dissolution, interception, and chemical reactions—leading to the wet deposition is shown in Fig. 7.7. For example: (1) rain may scavenge particles below cloud, but raindrops that evaporate produce new aerosols; (2) gaseous pollutants can be dissolved by cloud droplets, but cloud droplets that evaporate produce gas-phase pollutants; (3) both aerosols and gases can participate in chemical reactions in both in- and below-the-cloud environments resulting in new chemical species with increasing or decreasing concentrations; (4) aerosols in-the-cloud enlarge cloud droplets, and the falling of these lager size droplets accelerate, leading to more effective collisions with smaller droplets; and others.

Pathways of the wet deposition depend on multiple and complex processes, involve various physical phases, and are influenced by phenomena on a variety of physical scales (from micrometers to 1000s km). There might be considered three usual phases such as gaseous, aerosol, and aqueous. In addition, the later one can be presented in several forms such as rain, snow, cloud water, ice crystals, etc., having different characteristic sizes. Some processes take place inside or below a cloud. Moreover, there are differences in a size distribution of the cloud droplets or the rain droplets.

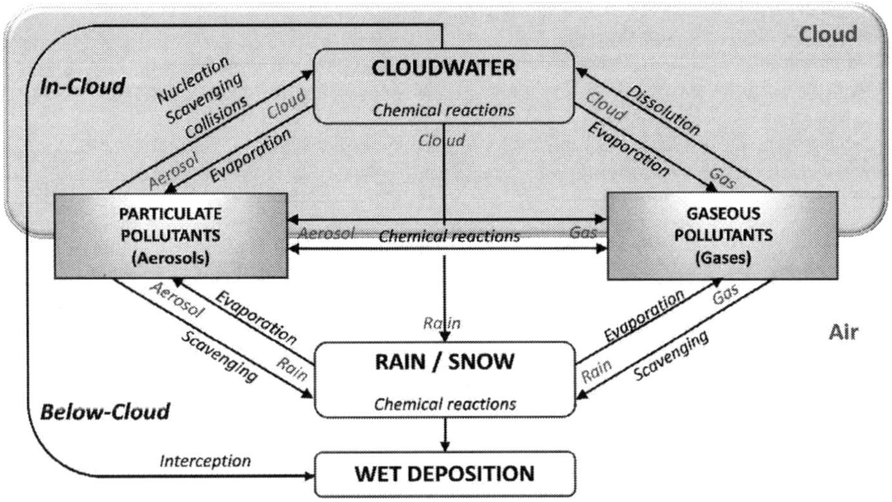

FIGURE 7.7 The wet deposition processes of pollutants through the transformation pathways between the pollutants (aerosols and gases) and rain/snow/cloud water occurred within the below-/in-the-cloud environments. *Modified from Seinfeld, J. H., & Pandis, S.N. (2006).* Atmospheric chemistry and physics: From air pollution to climate change. *John Wiley & Sons.*

This provides additional complexity. All these mentioned create a rather complicated picture. Therefore, some simplifications might be applied/ used to better understand the wet deposition.

Let's consider a "warm" raining cloud (i.e., without ice and snow crystals). For such a cloud, there are three basic phases—gaseous (air), liquid (cloud droplets and rain droplets) and solid (particles). These phases are continuum and can be represented by the following media (*m*)—air (*a*), cloudwater (*cw*), rainwater (*rw*), and aerosol particles (*ap*). Pollutants may exist in each of these media. Therefore, concentrations of pollutants at a location in space and time in these four media can be described by the atmospheric dispersion equation:

$$\frac{\partial C_{i,m}}{\partial t} = \underbrace{V_m \cdot \nabla C_{i,m}}_{\substack{\text{advection}\\ \text{(I)}}} + \underbrace{\nabla \cdot (K \nabla C_{i,m})}_{\substack{\text{diffusion}\\ \text{(II)}}} + \underbrace{\sum\nolimits_n W^i_{n\to m}}_{\substack{\text{transport}\\ \text{between}\\ \text{phases}\\ \text{(III)}}} + \underbrace{R_{i,m}}_{\substack{\text{chemical}\\ \text{reactions}\\ \text{(IV)}}} + \underbrace{E_{i,m}}_{\substack{\text{sources}\\ \text{(V)}}} \tag{7.44}$$

where $C_{i,m}$ is the concentration of pollutant *i* in the *m*-medium (*a, cw, rw, ap*), V_m is the velocity of the medium, *K* is the turbulent diffusivity, $W^i_{n\to m}$ is the transport flux between different *m*-media, $R_{i,m}$ and $E_{i,m}$ are the production (chemical reactions) and emission (sources) rates of pollutant *i* in each medium.

The velocities of the media can be different. For example, the rainwater velocity, V_{rw}, is a sum of the raindrops' velocity during rain precipitation event and the horizontal wind velocity. The wet deposition is more complex than the dry deposition in terms of many physical processes considered. It is because of multiple combinations of the phases and the transport between the media as well as of chemical reactions. The advection and the turbulent diffusion terms (I, II) in Eq. (7.44) were described in Section 7.2 in Eq. (7.7). The chemical reactions term (IV) includes a contribution from various types of reactions in the aqueous phase between gaseous and particulate pollutants occurred in- and below-the-cloud. The source term (V) includes a contribution to pollutant concentration from local and remote sources of natural and anthropogenic origins. For the term (III) with transport flux between the phases, the terms $W^i_{n\to p}$ can include in total of 11 possible transformation pathways (as seen in Fig. 7.7) occurred below- and in-the-cloud.

Below-the-cloud, the rate of transfer of a soluble gas or a particle into the rain can be approximated by the first-order relationships:

$$W^i_{a\to rw} = -\Lambda_{i,a} C_{i,a} \tag{7.45a}$$

$$W^i_{ap\to rw} = -\Lambda_{i,ap} C_{i,ap} \tag{7.45b}$$

where $\Lambda_{i,a}$ and $\Lambda_{i,ap}$ are the scavenging coefficients for pollutant i in the a- and ap- media (of the gas and solid phases), respectively. The minus sign here indicates that these terms are sinks of soluble gases and particles. Each scavenging coefficient is a function of space, time, characteristics of cloud/rain, and size distribution (in case of aerosol particles). Note Eq. (7.45a, b) are valid in a case of the irreversibility and independence of the scavenging in the layers above or in the atmospheric column.

Assuming the horizontally homogeneous atmosphere, the integrated washout of pollutant i by rain, can be represented by the below-the-cloud (bc) scavenging rate, $F_{a,i,bc}$ (for gaseous pollutants) and $F_{ap,i,bc}$ (for particulate pollutants) as follows:

$$F_{a,i,bc}(t) = \int_0^{H_{cb}} \Lambda_{i,a}(z,t)C_{i,a}(z,t)dz \tag{7.46a}$$

$$F_{ap,i,bc}(t) = \int_0^{H_{cb}} \Lambda_{i,ap}\left(d_{ap},z,t\right)C_{i,ap}\left(d_{ap},z,t\right)dz \tag{7.46b}$$

where H_{cb} is a height of the cloud base, the scavenging coefficients—$\Lambda_{i,a}$ (height and time-dependent) and $\Lambda_{i,ap}$ (height, time, and aerosol particle diameter dependent) for pollutant i, and d_{ap} is the diameter of aerosol particles. For the gaseous pollutant, the wet deposition flux will be equal to the below-the-cloud scavenging rate only if the rainout contribution is negligible.

Example 7.5: Wet deposition—below-the-cloud scavenging

Using Eq. (7.44), please, derive an equation for a concentration by neglecting the advection, diffusion, chemical reactions, and emission terms. Assuming that the scavenging coefficient is constant with time, integrate the resulting ordinary differential equation (ODE) with the remaining term. Based on the obtained solution, calculate the amount of a gaseous pollutant removed per volume of the air and plot its concentration in the atmosphere and the wet deposition flux as a function of time. Note, the parameters—the scavenging coefficient and the time—in the ODE solution should have the same time units (e.g., seconds, or hours).

Assume that (1) initial concentration of a gaseous pollutant, C_0, is 20 μg m^{-3} below a raining cloud; (2) the cloud base height, H_{cb}, is at 2000 m; and (3) the below-the-cloud scavenging coefficients, Λ, are constants ranging 1, 2, 3, 4, and 5 h^{-1}.

Assuming that initial rainwater concentration of the pollutant is negligible (zero), please, calculate and plot the concentration and the overall wet deposition flux of the pollutant as a function of time for the given cloud base height and the below-the-cloud scavenging coefficients.

Solution:

(1) After neglecting the advection, diffusion, chemical reactions, and emission terms, Eq. (7.44) became as

$$\frac{\partial C_{i,m}}{\partial t} = \sum_n W^i_{n\to m}$$

and in a raining cloud for a soluble gas (gaseous pollutant) it became as

$$W^i_{a\to rw} = -\Lambda_{i,a}C_{i,a}$$

$$\frac{\partial C_{i,rw}}{\partial t} = -\Lambda_{i,a}C_{i,a}$$

and after integration, it became as

$$C(t) = C_0 e^{-\Lambda t}$$

Using the last equation, we are calculating the concentrations as a function of time and the scavenging coefficients. For example:

$$C_0 = 20\ \mu g\ m^{-3},\ \Lambda = 1h^{-1},\ t = 1h,\ C(t=1) = C_0 e^{-\Lambda t} = 20\cdot e^{-1\cdot 1} = 7.35759\ \mu g\ m^{-3}$$

$$C_0 = 20\ \mu g\ m^{-3},\ \Lambda = 5h^{-1},\ t = 1h,\ C(t=1) = C_0 e^{-\Lambda t} = 20\cdot e^{-5\cdot 1} = 0.13476\ \mu g\ m^{-3}$$

(2) Following Eq. (7.46a), the wet deposition flux in a discrete form became as

$$F(t) = H_{cb}(C_0 - C(t))$$

We are calculating the wet deposition flux as a function of time and the scavenging coefficients. For example:

$C_0 = 20\ \mu g\ m^{-3},\ C(t=1) = 7.35759\ \mu g\ m^{-3},\ H_{cb} = 2000\ m,\ F(t=1) = 2000 \cdot (20 - 7.35759) = 25284.82\ \mu g\ m^{-2} = 25.28482\ mg\ m^{-2}$

$C_0 = 20\ \mu g\ m^{-3},\ C(t=1) = 0.13476\ \mu g\ m^{-3},\ H_{cb} = 2000\ m,\ F(t=1) = 2000 \cdot (20 - 0.13476) = 39730.48\ \mu g\ m^{-2} = 39.73048\ mg\ m^{-2}$

(3) By calculating and plotting the concentrations and the wet deposition fluxes of a gaseous pollutant as a function of time you will obtain the following (Fig. 7.8).

Concerning pollutants as soluble gases, it should be noted that during rain such pollutants that exist below clouds will be dissolved into falling rain droplets and hence, removed from the atmosphere. The rate of removal of these pollutants depends on the rain event characteristics (intensity, size of droplets) and physical-chemical properties of pollutants. In particular, the rate of transfer of pollutant to the surface of falling rain droplets is defined by the mass transfer coefficient, concentration in a gas phase, and concentration of the dissolved pollutant in an equilibrium with the

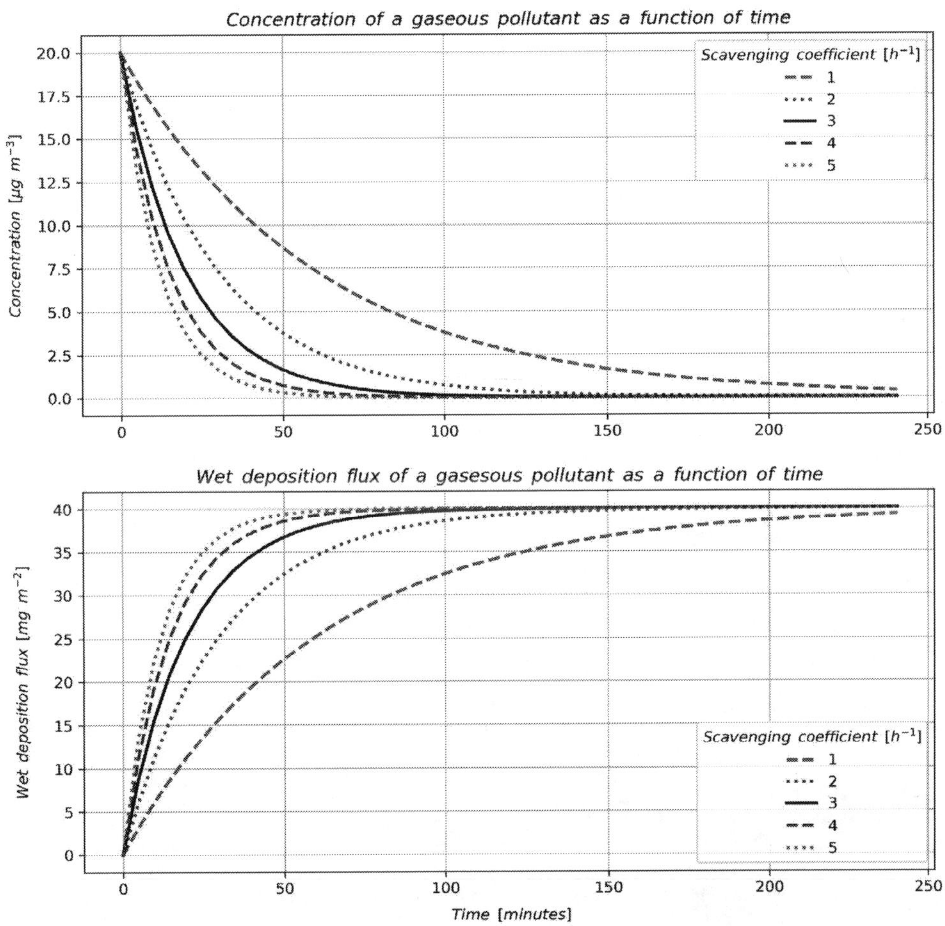

FIGURE 7.8 Profiles of concentrations and the wet deposition fluxes of a gaseous pollutant as a function of time.

aqueous phase. One may find more details in Seinfeld and Pandis (2006) for reversible and irreversible scavenging processes below-the-cloud.

Concerning pollutants as soluble particles, it should be noted that when a rain droplet falls through a cylindrical volume of the air, it collides with particles and collects them passing through the air. As a first approximation, due to collision the rain droplet would collect all the particles in such volume. In the reality, the process is more complex because the falling droplet perturbs the surrounding air, and hence, an airflow will diverge around the droplet. The occurrence of the collision will depend on the droplet and particle sizes, their relative locations, and collision efficiency. The latter is a correction factor describing interaction between the falling rain droplet and particle. The aerosol scavenging rate will depend on the aerosol particle diameter, the droplet size distribution, and the collision efficiency (or the scavenging efficiency). One may find more details on the rain droplets size distribution, the collision efficiency between rain droplets and aerosol particles, and the scavenging rates for the processes below-the-cloud in Seinfeld and Pandis (2006).

In-the-cloud, pollutants can be incorporated into the raining cloud and rain droplets. These determine the initial concentration of pollutant in the rain droplets before droplets start falling below the cloud base. The soluble gaseous pollutants can be removed from the cloud by the dissolution into cloud droplets.

For highly soluble gases (like HNO_3 and HCl), there is a rapid uptake process in a cloud compared with the air movement through the cloud and the change of cloud liquid water content due to condensation and precipitation.

For less soluble gases (like SO_2), uptake is a more complex process which depends on the pollutant properties, cloud droplets distribution, involvement of other pollutants in the aqueous-phase reactions (e.g., H_2O_2), and control the cloud droplets acidity and/or alkalinity (e.g., NH_3).

For aerosol particles, in-the-cloud scavenging is a result of two processes. The first is the nucleation scavenging (growth of the cloud condensation nuclei, CCN to the cloud droplets). The nucleation scavenging is an efficient process, which often incorporates most of the aerosol mass in the cloud droplets. The second is the collection of a fraction of the remaining aerosols by cloud or rain droplets. Usually, it is a rather slow process, and hence, a negligible fraction of the aerosol mass is scavenged by the cloud droplets.

The existing deposition modules (or parameterizations) in the air quality and atmospheric chemical transport models have large uncertainties. To better predict the complex deposition phenomena, a better understanding of the basic and key underlying processes influencing the removal (dry, wet, and sedimentation) of pollution and the evaluation of various parameterizations are required. Different modeling aspects including parameterized removal processes are given in Chapter 8.

7.8 Outlook for dispersion modeling approaches

There are several approaches applied in dispersion modeling for air quality applications. First of all, based on existing knowledge, there are the deterministic, physical, and statistical (see Fig. 7.9). The physical one is based on the wind tunnel experiments (studies on facilities to investigate various aerodynamic effects of artificially constructed terrain or canopy on turbulent airflow characteristics). The statistical one is based on the empirical evidence (information acquired

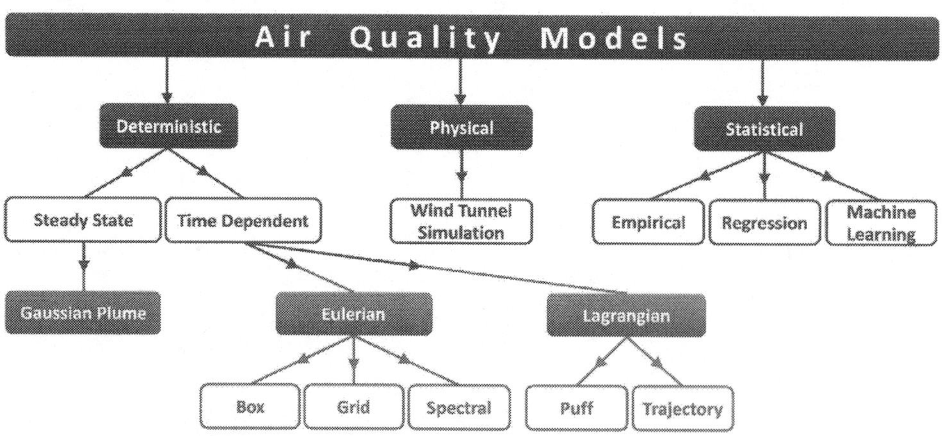

FIGURE 7.9 Basic approaches in the dispersion modeling for air quality.

through observation and experimentation), the regression analysis (set of statistical methods used for estimation of relationships between various variables), and the machine learning (based on methods and algorithms with automated statistical data analysis leading to analytical solutions and uncover key insights in data mining). The deterministic one includes the steady state and the unsteady (time-dependent) techniques. For the steady state, the Gaussian plume is the simplest example. For the time-dependent technique, the Eulerian and the Lagrangian methods are commonly used. For the Eulerian method, the box, grid, and spectral models are developed. For the Lagrangian method, the puff and trajectory models are elaborated. See more details in Chapter 8.

7.9 Questions

1. Describe and explain key processes involved into the atmospheric pollution dispersion.
2. What are the two main representations of the turbulent dispersion in the atmosphere?
3. What are the categories of the turbulent closure parameterizations describing the vertical dispersion processes?
4. Describe and explain the main atmospheric stability regimes and associated pollution plume shape related to the temperature vertical structure.
5. Explain the basic concept of the pollution plume release from the source (stack), its movement, and dispersion in the air.
6. What are the main processes contributing to increase of the height of the pollution plume above the point of release (stack)?

7a. Describe and explain the overall dry deposition of gaseous and particulate (aerosols) pollutants through the sequential layers with corresponding resistances.

7b. Describe and explain the wet deposition processes of pollutants (aerosols and gases) through the transformation pathways between pollutants and rain/ snow/ cloud water occurred within the below-/in-the-cloud environments.

8. Describe and explain the basic approaches in dispersion modeling for air quality.

Solutions to the end-of-chapter questions will be available at: https://www.elsevier.com/books-and-journals/book-companion/9780128225912.

References

Bird, R. B., Stewart, W. E., & Lightfoot, E. N. (2006). *Transport phenomena*. USA: John Wiley & Sons, Inc.

Briggs, G. A. (1969). Plume rise. *AEC Critical Review Series*, TID-25075. *US atmospheric energy comminity*, Division of Technical Information Extension, Oak Ridge, Tennessee, USA.

Briggs, G. A. (1972). Chimney plumes in neutral and stable surroundings. *Atmospheric Environment*, *6*(7), 507–510. Available from https://doi.org/10.1016/0004-6981(72)90120-5.

Bringfelt, B. (1969). A study of buoyant chimney plumes in neutral and stable atmospheres. *Atmospheric Environment*, *3*(6), 609–623. Available from https://doi.org/10.1016/0004-6981(69)90018-3.

Brummage, K. G. (1968). The calculation of atmospheric dispersion from a stack. *Atmospheric Environment*, *2*(3), 197–224. Available from https://doi.org/10.1016/0004-6981(68)90049-8.

Byun, D., & Schere, K. L. (2006). Review of the governing equations, computational algorithms, and other components of the Models-3 Community Multiscale Air Quality (CMAQ) modeling system. *Applied Mechanics Reviews*, *59*(2), 51–76. Available from https://doi.org/10.1115/1.2128636.

Carpenter, S. B., Montgomery, T. L., Leavitt, J. M., Colbaugh, W. C., & Thomas, F. W. (1971). Principal plume dispersion models: TVA power plants. *Journal of the Air Pollution Control Association*, *21*(8), 491–495. Available from https://doi.org/10.1080/00022470.1971.10469560.

Cunningham, E. (1910). On the velocity of steady fall of spherical particles through fluid medium. *Proceedings of the Royal Society of London. Series A, Containing Papers of a Mathematical and Physical Character*, *83*(563), 357–365. Available from https://doi.org/10.1098/rspa.1910.0024.

Fay, J. A., Escudier, M., & Hoult, D. P. (1970). A correlation of field observations of plume rise. *Journal of the Air Pollution Control Association*, *20*(6), 391–397. Available from https://doi.org/10.1080/00022470.1970.10469418.

Hicks, B. B., Baldocchi, D. D., Meyers, T. P., Hosker, R. P., & Matt, D. R. (1987). A preliminary multiple resistance routine for deriving dry deposition velocities from measured quantities. *Water, Air, and Soil Pollution*, *36*, 311–330. Available from https://doi.org/10.1007/bf00229675.

Holland, J. Z. (1953). *Meteorology survey of the Oak Ridge area*. ORO-99. U.S. At. Energy Comm., Oak Ridge, Tennessee.

Högström, U. (1988). Non-dimensional wind and temperature profiles in the atmospheric surface layer: A re-evaluation. *Boundary-Layer Meteorology*, *42*, 55–78. Available from https://doi.org/10.1007/BF00119875.

O'Dell, R. A., Taheri, M., & Kabel, R. L. (1977). A model for uptake of pollutants by vegetation. *Journal of the Air Pollution Control Association*, *27*(11), 1104–1109. Available from https://doi.org/10.1080/00022470.1977.10470533.

Pleim, J. E., Venkatram, A., & Yamartino, R. J. (1984). *The dry deposition model, V4. ADOMITADAP model development program*. Rexdale, Ontario, Canada: Ontario Ministry of the Environment.

Seigneur, C. (2019). *Air pollution: Concepts, theory, and applications* (pp. 1–354). Cambridge University Press. Available from http://doi.org/10.1017/9781108674614.

Seinfeld, J. H., & Pandis, S. N. (2006). *Atmospheric chemistry and physics: From air pollution to climate change*. New York, USA: John Wiley & Sons.

Smagorinsky, J. (1963). General circulation experiments with the primitive equations. *Monthly Weather Review*, *91*(3), 99–164. Available from https://doi.org/10.1175/1520-0493(1963)091 < 0099:gcewtp > 2.3.co;2.

Taylor, G. I. (1922). Diffusion by continuous movements. *Proceedings of the London mathematical society*, *2–20*(1), 196–212. Available from https://doi.org/10.1112/plms/s2-20.1.196.

Turner, J. S. (1970). *Workbook of atmospheric dispersion estimates*. NC, USA: U.S. Environmental Protection Agency, Research Triangle Park.

Wesely, M. L., & Hicks, B. B. (1977). Some factors that affect the deposition rates of sulfur dioxide and similar gases on vegetation. *Journal of the air pollution control association*, *27*(11), 1110–1116. Available from https://doi.org/10.1080/00022470.1977.10470534.

Wesely, M. L., & Hicks, B. B. (2000). A review of the current status of knowledge on dry deposition. *Atmospheric Environment*, *34*(12-14), 2261–2282. Available from https://doi.org/10.1016/S1352-2310(99)00467-7.

Chapter 8

Advanced modeling approaches to predict air quality and applications

George Tsegas[1], Evangelia Fragkou[1] and Nicolas Moussiopoulos[2]
[1]Department of Mechanical Engineering, Aristotle University of Thessaloniki, Thessaloniki, Greece, [2]Aristotle University of Thessaloniki, Thessaloniki, Greece

8.1 Introduction to modeling of meteorology and air quality

This chapter provides an introduction to atmospheric dispersion modeling, the mathematical simulation of how air pollutants disperse in the ambient atmosphere. It is performed with computer programs that include algorithms to solve the mathematical equations that govern pollutant dispersion. The models are used to estimate the ambient concentration of air pollutants or toxins emitted from sources such as industrial plants, vehicular traffic, or accidental chemical releases. They can also be used to predict future concentrations under specific scenarios, such as changes in emission sources.

In atmospheric science, the dispersion of pollutants is a multifaceted process governed by an array of physical effects and interactions (Zannetti, 2013). Key processes that influence the dispersion of pollutants include:

Emission Sources: The process of pollutant dispersion initiates at the emission source, which could range from mobile sources like vehicles to stationary ones such as factories or heating systems, or area sources like a cluster of idling vehicles. The volume and composition of pollutants discharged per unit time can vary extensively based on the source.

Meteorological Conditions: Meteorological parameters play a pivotal role in the dispersion of pollutants. Wind velocity and direction can transport pollutants away from their origin and spread them over a broader area. Temperature, pressure, and humidity can also modulate how pollutants behave in the atmosphere. For instance, temperature inversions can encapsulate pollutants near the ground, resulting in higher concentrations in those zones.

Atmospheric Turbulence: Atmospheric turbulence can enhance the integration of unpolluted air into a plume of pollutants, thus decreasing the pollutant concentration. The turbulence intensity can fluctuate based on the prevailing weather conditions and the diurnal cycle.

Physical and Chemical Transformations: As pollutants journey through the atmosphere, they may undergo a series of physical and chemical transformations. They can interact with other substances in the air to generate secondary pollutants. Furthermore, they can be eliminated from the atmosphere via deposition or precipitation processes.

Terrain and Land Use Characteristics can also impact pollutant dispersion. For instance, valleys may trap pollutants, resulting in elevated concentrations in these regions. Conversely, open terrains with minimal obstructions can facilitate efficient dispersion. Land surface characteristics can also directly or indirectly influence thermal and turbulent transport that strongly affects pollutant dispersion.

Long-Range Transport: Atmospheric currents can transport pollutants over significant distances, leading to the dispersion of pollution over extensive areas. This transport mechanism can also lead to the contamination of regions remote from the original pollutant source.

Atmospheric Stability: The stability of the atmosphere can influence the vertical dispersion of pollutants. During unstable conditions, warmer air near the surface can ascend and mix with the cooler air above, dispersing the pollutants. Conversely, during stable conditions, where warm air overlies cooler air, vertical mixing is curtailed, potentially leading to elevated concentrations of pollutants near the ground.

Air Quality. DOI: https://doi.org/10.1016/B978-0-12-822591-2.00008-1

These processes collectively and intricately dictate how pollutants are dispersed in the atmosphere. Comprehension of these processes and their interplay is vital not only for formulating new pollutant dispersion models but also for effectively applying them in practice understanding the limitations and being able to interpret the results (Sokhi et al., 2022).

8.1.1 Modeling approaches and science

There are several types of models used for atmospheric dispersion, each with its own strengths and limitations. Traditionally, atmospheric modeling has developed along approaches that aim to simulate the various atmospheric processes that govern pollutant dispersion and chemical transformation. The basic concept underlying most parameter-based models is that of a quantity of pollutant being transported, dispersed and potentially chemically transformed across space under the influence of a wind field. Depending on the underlying mathematical formulation, dispersion models can be categorized as Gaussian, Lagrangian and Eulerian approaches. The following sections provide an introduction to the main characteristics of each type of model, together with a description of typical use cases, required parameters, limitations, and practical considerations.

8.1.1.1 Gaussian models

These models are based on approximating a pollutant plume downwind of the source using a Gaussian distribution. They are often used for simple, steady-state situations where the wind speed and direction are relatively constant. Examples of Gaussian models include AEOLIUSF, AERMOD, AUSPLUME, CALINE3, CAL3QHC, and CAL3QHCR, CTDMPLUS, ISC, and OCD. Table 8.1 provides a list of the most commonly used Gaussian models for air pollution dispersion, along with some further information about each model.

It is important to mention that Gaussian models are used to predict the concentration of atmospheric emissions downwind of a point source by approximating the pollutant plume using a Gaussian distribution. These models are often used for simple, steady-state situations where wind speed and direction are relatively constant. The basic underlying assumption of the Gaussian approximation is that the concentration of pollutants decreases symmetrically in all directions from the centerline of dispersion, forming a bell-shaped curve when plotted against any of the spatial coordinates (Fig. 8.1).

8.1.1.1.1 Gaussian model parameters

The effective stack height is computed by summing the physical stack height and the plume rise, which is influenced by the buoyancy and momentum of the exhaust gases, as well as atmospheric stability. The dispersion coefficients, which describe the horizontal and vertical spread of the plume, are dependent on atmospheric stability conditions classified from A (very unstable) to F (very stable). Wind speed at varying heights can be estimated using a power law expression, which accounts for the change in speed with elevation. The model can also calculate the maximum concentration of pollutants, which is directly proportional to the emission rate and inversely proportional to wind velocity and effective stack height.

TABLE 8.1 A list of the most widely-used Gaussian air quality models.

Acronym	Full name	Description
AERMOD	AMS/EPA Regulatory Model	A widely used Gaussian dispersion model for air quality analysis. It is used to estimate pollutant concentrations from industrial sources and is recommended by the US Environmental Protection Agency (EPA) for regulatory purposes.
CALINE3	California Line Source Model	A steady-state Gaussian dispersion model designed to determine air pollution concentrations at receptor locations downwind of highways located in relatively uncomplicated terrain It is incorporated into more refined models such as CAL3QHC and CAL3QHCR.
CAL3QHC	CALINE3 with Queuing and Hot Spot Calculations	A CALINE3-based CO model that includes queuing and hot spot calculations. It also incorporates a traffic model to calculate delays and queues that occur at signalized intersections.
CAL3QHCR	Refined CAL3QHC Model	A more refined version of CAL3QHC that requires local meteorological data.
ISC	Industrial Source Complex	A Gaussian dispersion model developed by the U.S. EPA for estimating air pollutant concentrations from industrial sources

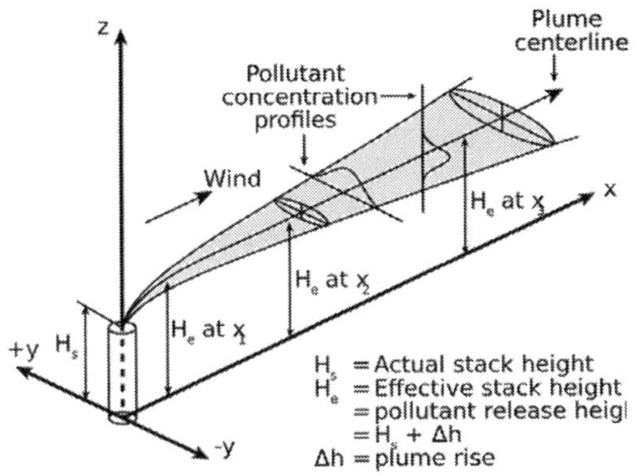

FIGURE 8.1 Gaussian modeling of a plume emitted from an industrial smokestack.

8.1.1.1.2 Applications of Gaussian models

Gaussian models find extensive application in representing the dispersion of smoke from industrial exhaust stacks and air quality dispersion modeling. Several alternative models that use Gaussian or modified Gaussian approaches are listed by the US Environmental Protection Agency (EPA), such as the Air force dispersion assessment model, atmospheric dispersion modeling system (ADMS), and Industrial Source Complex version 3 (ISC3).

Models like AERMOD, which includes treatment of multiple ground-level and elevated sources and can handle complex terrain, use Gaussian dispersion for stable conditions and non-Gaussian dispersion for unstable conditions. CALPUFF, a non-steady-state puff dispersion model, simulates the effects of time- and space-varying meteorological conditions on pollution transport, transformation, and removal.

8.1.1.1.3 Example of Gaussian model implementation

Consider a practical example of implementing a Gaussian atmospheric dispersion model to estimate the concentration of a pollutant emitted from an industrial smokestack (Fig. 8.1). Suppose an industrial facility emits sulfur dioxide (SO_2) into the atmosphere. The facility is located in a relatively flat and open area, which is ideal for applying a Gaussian model. The local environmental agency requires the facility to estimate the concentration of SO_2 at various distances from the smokestack to ensure compliance with air quality standards. Collecting the basic input data would require quantifying the following parameters:

- **Emission Rate (Q):** The facility reports an emission rate of 100 g second^{-1} of SO_2.
- **Stack Height (*H*):** The physical height of the smokestack is 50 m.
- **Plume Rise:** Due to the buoyancy and exit velocity of the emissions, the plume rise is estimated to be an additional 10 m, making the effective stack height 60 m.
- **Wind Speed (*u*):** Meteorological data indicates an average wind speed of 5 m second^{-1} at the effective stack height.
- **Dispersion Coefficients (σy and σz):** Based on the prevailing atmospheric stability class, which is determined to be class D (neutral), the dispersion coefficients are selected from standard tables.

The Gaussian plume equation for a continuous point source is given by

$$C(x, y, z) = \frac{Q}{(2\pi u \sigma_y \sigma_z)} e^{\left(-\frac{(z-H)^2}{2\sigma_z^2}\right)} e^{-\frac{y^2}{2\sigma_y^2}}$$

where

- $C(x, y, z)$ is the concentration of SO_2 at a point downwind (x) and crosswind (y) from the stack, at height (z).
- Q is the emission rate of SO_2 (100 g second^{-1}).
- u is the wind speed at the effective stack height (5 m second^{-1}).
- σ_y and σ_z are the horizontal and vertical dispersion coefficients, respectively, obtained from standard tables for stability class D.
- H is the effective stack height (60 m).

Implementation

A typical application of Gaussian models consists of several steps:

1. **Data Collection:** The first step necessitates the collection of all pertinent data, which includes emission rates, source characteristics (e.g., stack dimensions, outflow velocity), meteorological data, and information regarding atmospheric stability.
2. **Parameter Estimation:** This step involves the calculation of the effective emission height, achieved by adding the plume rise to the physical stack height. Dispersion coefficients for the given stability class are also determined during this stage.
3. **Model Application:** The Gaussian plume equation is then applied to estimate the concentration of SO_2 at various distances downwind from the stack.
4. **Concentration Estimation:** For instance, to calculate the ground-level concentration ($z = 0$) at a downwind distance of 500 m ($x = 500$, $y = 0$), the relevant values are plugged into the Gaussian plume equation.
5. **Model Validation:** The final step involves the comparison of estimated concentrations with actual measurements, if accessible, to validate the accuracy of the model.

8.1.1.1.4 Limitations and special considerations

The Gaussian plume model assumes steady-state conditions and does not consider temporal variations in emissions or meteorological conditions. It is most accurate for short-range distances (typically up to a few kilometers) and may not be suitable for complex terrains or varying wind directions. It is most appropriate for surface and elevated point sources but can also accommodate line sources in an approximate way. The model's simplicity allows for quick estimates but may necessitate adjustments or the utilization of more sophisticated models for comprehensive assessments.

8.1.1.1.5 Insights for researchers

1. **Understand Atmospheric Stability:** The model users should familiarize themselves with the concept of atmospheric stability classification and their influence on dispersion coefficients.
2. **Use Turner's Charts:** Utilize Turner's charts (Turner, 1969) to determine the dispersion coefficients based on stability class and downwind distance.
3. **Validate Model Predictions:** Compare model predictions with actual measurements to validate the model and discern its limitations.
4. **Consider Terrain:** Understand that the Gaussian plume model presumes flat terrain and may not be valid for complex terrains.
5. **Account for Wind Direction:** Recognize that the model cannot account for wind direction curvature and is generally valid only up to 10 km downwind.
6. **Use Computational Tools:** Employ computational tools for more complex scenarios, particularly when dealing with multiple sources or complex terrains.
7. **Explore Advanced Models:** Consider advanced models like ADMS or CALPUFF for situations involving complex terrains, chemical reactions, or long-range transport.
8. **Model Limitations:** Be aware that Gaussian models tend to overpredict concentration and underpredict plume spread, limiting their accuracy.

By understanding and applying these principles, researchers can effectively employ Gaussian atmospheric models to predict pollutant dispersion and assess their impact on air quality.

8.1.1.2 Lagrangian models

In principle, Lagrangian models track individual air parcels as they move through the atmosphere, providing a comprehensive understanding of their behavior and interactions. These models are particularly useful in handling complex situations with varying wind speeds and directions, as well as simulating chemical reactions that occur in the atmosphere. Let's explore some examples of Lagrangian models and their specific features:

1. **Advanced Finite-Volume Transport and Transformation** (AFTOX) is a Lagrangian model that focuses on the transport and transformation of air pollutants. It incorporates advanced numerical methods and algorithms to accurately simulate the dispersion and chemical reactions of pollutants in the atmosphere. AFTOX is widely used for studying pollutant transport in various settings, including urban areas and industrial regions.

2. **CALPUFF Atmospheric Transport Modeling System** (CALPUFF) is a Lagrangian model designed for simulating long-range and complex air pollutant transport. It considers the effects of topography, meteorological conditions, and chemical reactions on the dispersion of pollutants. CALPUFF is widely used for assessing the impacts of emissions from industrial facilities, power plants, and other sources on regional air quality.
3. **Grid Adaptative Model-GRAL** (GRAM-GRAL) is a Lagrangian model that combines a grid adaptation technique with Lagrangian particle tracking. It allows for accurate simulation of pollutant dispersion in complex terrains. GRAM-GRAL is often used for studying the dispersion of pollutants in mountainous regions, where traditional Eulerian models may not be as effective due to the irregular terrain.
4. **AUSTAL for 2000 (AUSTAL2000)** is a Lagrangian model developed in Germany under contract to the Federal Ministry for Environment, Nature Conservation and Nuclear Safety and in compliance with the requirements of Annex 3 of the federal guidelines for ambient air quality (TA Luft). It incorporates various physical and chemical processes, including turbulent diffusion and chemical reactions, to capture the complex behavior of pollutants. AUSTAL2000 is commonly used for assessing the impacts of industrial emissions on air quality and developing effective mitigation strategies.
5. **Hybrid Eulerian–Lagrangian Dispersion Models (HDMs):** Hybrid Eulerian–Lagrangian dispersion models combine the advantages of both Eulerian and Lagrangian approaches. These models use Eulerian methods for large-scale simulations, capturing the overall behavior of air pollutants, and switch to Lagrangian methods for detailed simulations in specific regions of interest. HDMs are useful for studying air pollutant dispersion in complex scenarios that require a balance between computational efficiency and accuracy.

8.1.1.2.1 Example of a Lagrangian atmospheric model implementation

Consider the problem of simulating the dispersion of pollutants (like smoke particles) from a factory chimney into the atmosphere. This is a typical problem in environmental engineering that can be solved using a Lagrangian approach.

8.1.1.2.2 Model parameters and setup

In the study of atmospheric dispersion models, particularly the Lagrangian type, we start with a set of assumptions designed to simplify the complex reality of particle dispersion. In our model, we treat the particles as point masses, which allows us to focus more directly on their movement and the forces acting upon them. The journey of these particles begins at the chimney outlet, where they are released with an initial upward velocity.

As the particles make their way through the atmosphere, they are subject to a variety of forces, with gravity and air drag being the most significant. These forces, in combination with the initial conditions of the particles, determine their paths and eventual dispersion.

Lagrangian models track the movement of each particle individually, focusing on the forces acting on the particle and its resulting motion. This approach is encapsulated in the equations of motion, which are derived from Newton's second law of motion.

The equation of motion is expressed as

$$m\frac{d\vec{v}}{dt} = \vec{F}$$

Here, m represents the mass of the particle, $d\vec{v}/dt$ is the derivative of velocity with respect to time (acceleration), and $\vec{F}$ is the net force acting on the particle. This equation describes the balance of forces acting on the particle and how they influence its acceleration.

The net force $\vec{F}$ is a combination of the force due to gravity and the force due to air drag. The gravitational force is calculated using the formula:

$$\vec{F_g} = m\vec{g}$$

where $\vec{F_g}$ is the force due to gravity, m is the mass of the particle, and $\vec{g}$ is the acceleration due to gravity. This equation shows that the gravitational force acting on a particle is directly proportional to its mass and the acceleration due to gravity.

In addition to gravity, particles experience a drag force due to air, which is given by

$$\vec{F_d} = -\frac{1}{2}C_d\rho A|\vec{v}|\vec{v}$$

In this equation, $\vec{F_d}$ is the drag force, C_d is the drag coefficient, ρ is the air density, A is the cross-sectional area of the particle (assumed to be spherical), $\vec{v}$ is the velocity of the particle. The negative sign signifies that the drag force opposes the direction of the particle's movement.

This Lagrangian approach provides a detailed, particle-by-particle representation of the dispersion process, taking into account the individual influences of gravity and air drag on each particle's path.

Implementation

The implementation of a Lagrangian model typically involves several critical steps:

1. **Initialization:** The simulation begins with the establishment of known parameters, which include the initial velocity, mass, drag coefficient, and gravitational acceleration of the particles.
2. **Time-Stepping:** The model progresses through the simulation timeline using small, discrete time steps to capture the dynamic evolution of the system accurately.
3. **Solving Equations:** At each incremental time step, the equations of motion are solved to update the velocity and position of each particle, thereby tracking their individual trajectories.
4. **Monitoring Results:** Throughout the simulation, the positions of the particles are monitored to visualize and analyze the dispersion pattern of the pollutants.

8.1.1.2.3 Postprocessing and analysis of simulation data

Upon completion of the simulation, the data undergoes postprocessing and analysis. This step is crucial to extract meaningful insights from the simulation results and assess the dispersion of pollutants.

- **Visualization techniques** are employed to depict the temporal dispersion of the particles, providing a clear representation of how the pollutants spread over time. By visualizing the dispersion patterns, researchers can gain valuable insights into the behavior and extent of pollutant transport. This visual representation aids in identifying areas that are most affected by the pollutants and understanding the temporal evolution of the dispersion process.
- **Analytical methods** are applied to the simulation data to identify affected areas and quantify the concentration of pollutants within those regions. By analyzing the simulation data, researchers can determine the areas where pollutant concentrations exceed certain thresholds or regulatory limits. This information is crucial for assessing the potential impact on air quality and identifying regions that require mitigation strategies or further investigation.
- The model is also capable of simulating various **scenarios**, such as alterations in wind speed and direction, to evaluate their effects on the dispersion of pollutants. This capability allows researchers to assess the **sensitivity** of the dispersion patterns to changes in meteorological conditions. By performing such **sensitivity analyses**, researchers can gain insights into the influence of wind speed and direction on pollutant transport, aiding in the understanding of potential risks and the development of effective mitigation strategies.

8.1.1.2.4 Considerations when utilizing Lagrangian models

Several considerations are paramount when utilizing Lagrangian models:

Computational Intensity: The Lagrangian approach can be computationally intensive, especially when simulating a large number of particles. This is because the equation of motion must be resolved for each individual particle. Researchers must carefully consider computational resources and optimize simulation parameters to achieve reasonable computational efficiency.

Accuracy-Computational Effort Trade-off: The accuracy of the produced concentration fields is proportional to the number of simulated particles. However, increasing the number of particles also increases the computational effort required. Thus researchers face a trade-off between achieving higher accuracy and managing computational resources effectively.

Quality of Input Data and Numerical Methods: The precision of the model heavily relies on the quality of the input data and the robustness of the numerical methods employed to solve the equations of motion. Accurate meteorological data and reliable numerical techniques are essential for obtaining reliable simulation results.

Interactions between Particles: In certain scenarios, interactions between particles may be relevant. These interactions can add layers of complexity to the model, necessitating more sophisticated computational techniques to accurately capture these effects. Researchers must carefully consider the inclusion of particle-particle interactions based on the specific characteristics of the pollutant being modeled.

8.1.1.3 Eulerian models

Eulerian models are a fundamental class of models in atmospheric science, dividing the atmosphere into a grid of cells and solving the governing equations of atmospheric dynamics for each cell. These models are adept at handling the complexities of atmospheric processes, including variable wind speeds and directions, as well as simulating the myriad of chemical reactions that occur within the atmosphere.

Notable examples of Eulerian models include:

1. **Transport and Dispersion Model (TAPM)** is a model commonly used for atmospheric dispersion and air quality modeling. It simulates the transport and dispersion of pollutants based on atmospheric dynamics and emissions data. TAPM is widely used for assessing the impact of pollutants on air quality in various regions.
2. **Community Multiscale Air Quality (CMAQ) Modeling System** is a model that integrates meteorological and chemical processes to simulate air quality. It is widely used by researchers and policymakers to assess the impacts of emissions from various sources, such as industrial facilities and transportation, on regional and urban air quality.
3. **CHIMERE** is a chemistry-transport model that focuses on simulating atmospheric chemistry and transport of pollutants. It incorporates detailed chemical mechanisms to accurately represent the complex chemical reactions occurring in the atmosphere. CHIMERE is widely used for regional and urban air quality studies.
4. **Comprehensive Air Quality Model with Extensions (CAMx)** is a model that combines atmospheric dynamics, chemical reactions, and emissions to simulate air quality. It is capable of modeling complex chemical processes and their interactions with meteorological conditions. CAMx is utilized for a range of applications, including regulatory assessments and research studies.
5. **LOTOS-EUROS** is an Eulerian model designed to simulate the transport and dispersion of atmospheric pollutants, including both primary emissions and secondary pollutants formed through chemical reactions. It is widely used for regional and continental-scale air quality studies in Europe.
6. **Weather Research and Forecasting model with Chemistry (WRF-Chem)** is an Eulerian model that integrates the WRF model with atmospheric chemistry. It simulates the interactions between meteorology, emissions, and chemical processes to study air quality and atmospheric composition.

8.1.1.3.1 Context and setup

Imagine a scenario where a research team is tasked with assessing the impact of emissions from an industrial complex on regional air quality. The team decides to use an Eulerian atmospheric model to simulate the dispersion of pollutants in the atmosphere over time.

The atmosphere is discretized into a 3D mesh that covers the region of interest. The mesh is refined in areas where greater accuracy is required, such as around the emission sources or in regions with complex terrain. The initial concentration of pollutants is set to background levels, except at the emission source locations, where it matches the known emission rates. Boundary conditions are defined at the limits of the domain, which may include inflow and outflow conditions, as well as conditions at the land surface and at the top of the atmosphere.

In the Eulerian framework, the atmosphere is described by fields such as velocity, pressure, temperature, and density, which are functions of both position and time. This framework provides a fixed spatial reference for observing and analyzing atmospheric properties and their changes. The Eulerian atmospheric models are governed by a set of partial differential equations (PDEs) that express the conservation of mass, momentum, and energy, as well as the transport of chemical species within a control volume fixed in space.

8.1.1.3.2 Conservation of mass (continuity equation)

The continuity equation for a compressible atmosphere ensures that mass is conserved, taking into account the density changes due to compression and expansion of air. It is expressed as

$$\frac{\partial \rho}{\partial t} + \nabla \times (\rho u) = 0$$

Here, ρ represents the density of air, and **u** is the velocity vector of the air. This equation ensures that the total mass within a given volume of air remains constant over time, accounting for the inflows and outflows of air within the fixed spatial domain.

8.1.1.3.3 Conservation of momentum (momentum equations)

The momentum equations for the atmosphere, analogous to the Navier-Stokes equations, account for the forces acting on the air. These forces include pressure gradients, Coriolis forces due to the Earth's rotation, gravitational forces, and frictional forces. The momentum equations are given by

$$\frac{\partial(\rho u)}{\partial t} + \nabla \times (\rho uu) = -\nabla p + \rho g + F$$

In this equation, p represents the atmospheric pressure, g is the gravitational acceleration, and F represents frictional and other forces. The momentum equations describe the motion and acceleration of air parcels in response to these forces, providing insights into the atmospheric circulation patterns and the behavior of winds within the fixed spatial domain.

8.1.1.3.4 Conservation of energy (thermodynamic energy equation)

The energy equation in atmospheric modeling accounts for the conservation of thermal energy, including contributions from the radiative transfer, sensible and latent heat fluxes, and adiabatic processes. The thermodynamic energy equation is expressed as

$$\frac{\partial(\rho c_p T)}{\partial t} + \nabla \times (\rho c_p T u) = \nabla \cdot (k \nabla T) + Q$$

Here, T represents the temperature, is the specific heat at constant pressure, k is the thermal conductivity, and Q represents heat sources or sinks. This equation describes how thermal energy is transported and transformed within the atmosphere, providing insights into temperature variations, heat exchange processes, and the overall energy balance within the fixed spatial domain.

8.1.1.3.5 Advection-diffusion equation (for scalar transport)

To model the transport of scalar quantities, such as pollutant concentrations (C) within the atmosphere, the advection-diffusion equation is employed. It can be written as

$$\frac{\partial(\rho C)}{\partial t} + \nabla \times (\rho C u) = \nabla \cdot (D \nabla C) + S$$

In this equation, D represents the diffusion coefficient, and S represents sources or sinks of the scalar quantity. The advection-diffusion equation accounts for the advection of pollutants by the air currents, as well as the diffusion and dispersion of the pollutants due to molecular motion within the fixed spatial domain.

The Eulerian framework provides a valuable perspective for understanding the behavior and dynamics of the atmosphere within a fixed spatial reference. By solving and analyzing these fundamental equations, atmospheric models can simulate the complex interactions and processes occurring within the atmosphere, contributing to various fields such as weather forecasting, climate studies, and air pollution management.

8.1.1.3.6 Eulerian models: implementation and application

The implementation of an Eulerian atmospheric model typically involves the following steps:

1. **Discretization:** The atmospheric domain is partitioned into a grid or mesh, and the governing equations are discretized using numerical methods such as finite difference, finite volume, or finite element methods.
2. **Initial and Boundary Conditions:** Appropriate initial conditions for the atmospheric fields are specified, and boundary conditions are applied at the domain's boundaries.
3. **Numerical Solution:** The discretized equations are solved iteratively using numerical methods to advance the solution in time.
4. **Postprocessing:** The numerical results are analyzed and visualized to interpret the behavior of the atmosphere and the transported substances.

8.1.1.3.7 Postprocessing and analysis

The simulation results are visualized to show the distribution of pollutants at different times and locations. The results are analyzed to identify areas of concern, such as regions where pollutant concentrations exceed health standards. The model can also be used to test different scenarios, such as the impact of emission control strategies or changes in meteorological conditions.

8.1.1.3.8 Considerations

The fidelity of an Eulerian atmospheric model depends on the resolution of the mesh and the quality of the input data. The model must be calibrated and validated with observational data to ensure the reliability of its predictions. Computational resources can be a constraint, as high-resolution models require significant processing power and memory. By employing an Eulerian atmospheric model, researchers and policymakers can comprehensively assess the impact of emissions on air quality in areas of intense chemical activity, aiding in mitigation efforts and informing regulatory decisions in urban areas.

8.1.2 Process representation and parameterizations

All types of process-based models incorporate an underlying foundation of methods and numerical approaches for simulating the effects of atmospheric transport, turbulence and chemical transformations. Some key aspects of these common approaches are highlighted in the following sections.

8.1.2.1 Common concepts in computational fluid dynamics-based models

Computational Fluid Dynamics (CFD) models are an essential tool in the field of fluid mechanics, employing numerical methods to solve the equations that govern fluid flow. These models are capable of simulating the flow of air around structures and are particularly useful for studying pollutant dispersion in urban environments.

The equations governing fluid dynamics are continuous PDEs. To solve these equations numerically, they must be discretized into algebraic equations that can be computed. This is achieved using numerical methods such as

1. **The Finite Volume Method (FVM)**, integrates the PDEs over discrete volume elements, ensuring conservation of fluxes across control volume boundaries.
2. **The Finite Difference Method (FDM)**, approximates the derivatives in the PDEs using differences between nodal values.
3. **The Finite Element Method (FEM)**, which constructs an approximate solution to the PDEs by minimizing an associated error function over discretized elements.

8.1.2.1.1 Time integration schemes

Time integration is a critical aspect of CFD simulations, allowing the model to evolve over time. This can be performed using either explicit methods, which are generally simpler and computationally faster but may require smaller time steps for stability, or implicit methods which are more stable for larger time steps but computationally more intensive —as they often require solving a system of equations at each step.

8.1.2.1.2 Turbulence modeling

Turbulence is a complex phenomenon in fluid flow with a potential strong contribution to the transport and dispersion of pollutants and is modeled using various approaches:

1. **Reynolds-Averaged Navier-Stokes (RANS)** equations, which model the average effect of turbulence using closure models such as k-epsilon or k-omega. This is the approach employed by most of the Eulerian transport models in use, covering scales up to thousands of kilometers.
2. **Direct Numerical Simulation (DNS)**, resolves all scales of turbulence but is computationally very expensive. Currently, this approach is only employed in studying microscopic fluid flows or inert pollutant transport in limited areas.
3. **Large Eddy Simulation (LES)**, which resolves large-scale turbulent structures while modeling smaller scales. Models that support LES to simulate turbulence can typically cover spatial scales from a few meters to tens of kilometers, but currently very few of them implement any kind of chemistry.

8.1.2.1.3 Simulation process

The CFD simulation process involves several stages:

1. **Preprocessing:** This stage involves defining the computational domain, generating the mesh, setting initial and boundary conditions, and specifying the physical and chemical properties of the atmosphere.
2. **Solving:** The simulation is run by iteratively solving the discretized equations until the solution converges or a specified simulation end time is reached.
3. **Postprocessing:** The results are analyzed to visualize flow fields and pollutant dispersion, and to interpret the data for environmental impact assessment.

8.1.2.2 *Distance-dependent parameterized modeling*

In the field of atmospheric modeling, distance-dependent parameterized models are predicated on the principle that the influence of a pollutant source diminishes with increasing distance from it. These models are essential for simulating dispersion phenomena that exhibit spatial heterogeneity in their impacts, such as the spread of pollutants from a specific source or the distribution of temperature variations across a landscape.

8.1.2.2.1 Inverse distance weighting and its applications

Inverse distance weighting (IDW) is a method where the influence of a source decreases inversely with distance. It's a common approach in atmospheric sciences for interpolating environmental data, such as pollutant concentrations from air quality monitoring stations. By interpolating data from these stations, IDW can estimate pollution levels across urban areas, with each station's influence diminishing with distance. This method is particularly useful in urban air quality management, where it helps in assessing the exposure of populations to various air pollutants and in designing strategies for air quality improvement.

8.1.2.2.2 Kernel functions in atmospheric data smoothing

Kernel functions, such as Gaussian and exponential kernels, assign weights to points based on their distance from a target location. These functions are widely used for data smoothing or kernel density estimation in atmospheric sciences. For instance, they can be applied to smooth temperature data across a region, which is helpful for meteorologists when creating weather maps that depict temperature gradients.

8.1.2.2.3 Gravity models and disease spread

Gravity models posit that the interaction between two points diminishes with distance and increases with certain attributes of the points, such as population size. In the context of atmospheric modeling, these models can simulate the spread of airborne diseases or pollutants between cities. They can be particularly relevant in studying the dispersion of pollutants between urban and rural areas or in assessing the regional transport of airborne pathogens.

8.1.2.2.4 Dispersal models in ecology and atmospheric studies

Dispersal models are used to simulate the spread of biological entities like seeds or animals, but they also have applications in atmospheric sciences. For example, these models can simulate the dispersal of particulate matter or gaseous pollutants from their origin. The probability of dispersal typically decreases with distance, and these models can help in understanding the spread of pollutants from a point source, such as a factory or a power plant.

8.1.2.2.5 Geostatistical models and their use in atmospheric prediction

Techniques such as kriging employ semivariograms to model spatial autocorrelation, which is the correlation between sample points as a function of distance (Curran, 1988). In atmospheric modeling, this approach is utilized to predict values at unsampled locations, such as estimating the concentration of pollutants in areas where direct measurements are not available. This method is particularly useful in creating detailed maps of air quality or in assessing the impact of environmental policies on air pollution.

8.1.2.2.6 Special considerations in atmospheric modeling

When applying distance-dependent parameterized models in atmospheric sciences, it's essential to understand the underlying process and its expected variation with distance to select an appropriate distance-decay function. The choice of distance metric should be context-specific, and alternatives to Euclidean distance, such as Manhattan or Great-circle distance, may be more appropriate in certain scenarios.

The scale sensitivity of the model is also crucial, as the choice of scale and units can influence the model's outcomes. It's important to ensure consistency of distance units with the scale at which the atmospheric process operates.

Validation with cross-validation techniques is necessary to ensure that the model is validated with data that are spatially independent from the training data. If the process exhibits directional dependence, this anisotropy should be integrated into the model to reflect phenomena such as preferential pollutant dispersion downwind.

Physical or environmental barriers that may affect the distance-dependent process should be considered, and adjustments made to the model accordingly. Edge effects should be cautiously addressed, particularly in ecological studies where boundaries may limit the perceived range of dispersal.

Nonlinear relationships should be explored if the process does not change linearly with distance. Spatial autocorrelation must be considered to avoid incorrect inferences about the significance of model parameters.

Finally, geographic information systems (GIS) and statistical software designed for spatial data and distance-dependent modeling should be leveraged. Tools such as the "sp," "gstat," or "spatstat" packages in R are particularly useful for these types of analyses in atmospheric modeling (Baddeley et al., 2015).

8.1.2.3 Chemistry modeling

Chemistry modeling is a multifaceted field that encompasses the representation and parameterization of chemical processes to predict and understand the behavior of chemical processes in the atmosphere. These models can vary from simple reaction kinetics to intricate networks involving a multitude of reactions and species. The development of these models is an interdisciplinary endeavor that integrates theoretical principles, computational techniques, and empirical data.

Photochemical models in particular, play a pivotal role in accurately simulating atmospheric chemistry, including the chemical composition of the atmosphere, the genesis of pollutants such as ozone, and the photodegradation of various chemical species. These models are inherently complex, encapsulating a multitude of chemical reactions and species.

8.1.2.3.1 Chemical mechanisms for atmospheric dispersion models

Most chemical mechanisms implemented in atmospheric dispersion problems introduce a fixed number of chemical species and reactions which are used to simulate the chemical processes in each computational timestep.

A usual distinction is used between methods that treat species reacting in the gas-phase chemistry and species that either appear in the solid or liquid aerosol phase or contribute to the formation of aerosols:

Gas-phase chemistry is crucial in atmospheric dispersion models, as it outlines the transformations that atmospheric gases undergo due to various chemical reactions. These mechanisms are essential for understanding the formation, transformation, and removal of key air pollutants (Seinfeld & Pandis, 2016).

The **EMEP MSC-W** model incorporates the EMEP chemical mechanism (including later versions like EmChem19a) which was adopted in other chemical transport models (CTMs). Specifically designed by the European Monitoring and Evaluation Programme's Meteorological Synthesizing Centre-West, this model simulates the transport, transformation, and deposition of several air pollutants. These include sulfur and nitrogen compounds, ammonia, volatile organic compounds (VOCs), and ground-level ozone (Simpson et al., 2012). For instance, the model has been used extensively to track the long-range transport of air pollutants across Europe, providing valuable data for policymakers to implement effective air quality strategies.

The **Regional Atmospheric Chemistry Mechanism (RACM)** and **Carbon Bond Mechanism (CBM-IV)** are other examples of gas-phase chemical mechanisms. RACM provides a detailed mechanism for the oxidation of VOCs, including alkanes, alkenes, and aromatic compounds, which are precursors to ozone and secondary organic aerosols (SOAs). It was developed by the University of California, Riverside, and has been used in several regional and global air quality studies.

The CBM-IV chemical mechanism has been widely used in air quality models for urban and regional-scale simulations. It was developed by the US Environmental Protection Agency and includes reactions for ozone precursors such as nitrogen oxides and VOCs. It has been implemented in numerous models, including the CMAQ model, which is widely used for regulatory and research applications.

Aerosol chemistry involves the interactions between gaseous components and aerosols, the microscopic particles suspended in the atmosphere that play a vital role in atmospheric chemistry and climate (Kanakidou et al., 2005). They influence the Earth's radiative balance, cloud formation, and air quality.

Several models have been developed specifically to understand the formation, transformation, and removal of aerosols in the atmosphere (Seinfeld & Pandis, 2016).

ISORROPIA is one such model, designed to simulate the thermodynamic equilibrium among gas-phase precursors and aerosol-phase species in the inorganic aerosol system (Nenes et al., 1998). Developed by the Georgia Institute of Technology, the model can predict aerosol phase state, water uptake, and the partitioning of inorganic species between the gas and aerosol phases. It has been implemented in several regional and global models, including CMAQ and the WRF-Chem, and has been used in numerous studies to understand the formation and impacts of fine particulate matter (PM2.5).

Secondary ORGanic Aerosol Model (SORGAM) is another aerosol model, specifically designed to simulate the formation and growth of SOAs in the atmosphere. It includes a detailed chemical mechanism for the oxidation of VOCs and the subsequent formation of SOA. Developed by the National Center for Atmospheric Research, SORGAM has been used in several studies to understand the sources and processes of SOA formation, which is a major component of PM2.5 and has significant impacts on air quality and climate.

Modal and Sectional Aerosol models are two complementary approaches used to represent the size distribution of aerosols (Baklanov et al.,2014). In modal models, the particle size distribution is represented by a sum of log-normal modes, whereas sectional models divide the particle size distribution into discrete size bins. Each approach has its own advantages and is chosen based on the specific requirements of the study.

The CMAQ model, for instance, uses a modal approach to represent aerosol size distribution. Developed by the US EPA, CMAQ is a comprehensive model used for air quality management and research. The modal approach allows for efficient computational representation of aerosol processes, including formation, growth, and removal, making it suitable for large-scale, long-term simulations.

On the other hand, the sectional approach, implemented in models such as the **Aerosol Inorganics Model** (AIM) (Wexler & Seinfeld, 1991), provides a detailed representation of aerosol size distribution. While more computationally intensive, the sectional approach allows for the explicit tracking of aerosol growth and is often used in studies focused on specific aerosol processes or events.

Simple Oxidative Chemistry can be used in atmospheric modeling to represent the basic reactions involving the oxidation of pollutants. This could include the oxidation of nitrogen oxides (NO_x) and VOCs in the presence of sunlight, leading to the formation of ozone. This simplified approach is often used in simpler box models or when computational resources are limited (Kanakidou et al., 2005).

For instance, the **Regional Acid Deposition Model (RADM)** uses a simple oxidative scheme to simulate the formation and deposition of acid rain (Chang et al., 1987). It includes simplified reactions for the oxidation of sulfur dioxide and nitrogen oxides, which are major precursors for acid rain. Despite its simplicity, the model has been successfully used in numerous studies to understand the sources, processes, and impacts of acid deposition.

An example of a simplified photochemistry model is the **Photolysis Rates (PHOTRATES)** model by (Madronich & Flocke, 1999). It provides photolysis rates for key atmospheric species under a range of atmospheric conditions, including different solar zenith angles and altitudes. Despite its simplicity, the model has been widely used in atmospheric research, providing valuable data for understanding the role of photolysis in atmospheric chemistry.

Another example is the **Simple Photochemistry for All Resolving Models (SPARC)** model, a simplified photochemical scheme designed for high-resolution air quality modeling (Gross & Stockwell, 2003). It includes a reduced set of photochemical reactions for the formation and removal of ozone, making it computationally efficient for high-resolution, urban-scale simulations. Despite its simplicity, the SPARC model has been shown to accurately reproduce ozone concentrations observed in several urban areas, highlighting the utility of simplified photochemistry models in atmospheric research.

8.1.2.3.2 Guidelines for researchers

Researchers engaged in photochemical modeling are advised to select a chemical mechanism that accurately reflects the conditions of the study area. For urban air quality studies, mechanisms with detailed VOC chemistry may be required, while remote areas might need mechanisms that include halogen chemistry. It is also important to validate photolysis rates against field measurements or satellite data to ensure they accurately represent the variations in sunlight.

8.1.2.3.3 Impact of photolysis rates and rate constants

Another critical component is the rate of photolytic reactions, which are influenced by actinic flux and quantum yields. Actinic flux measures the number of photons that initiate reactions, while quantum yields represent the probability of a photon causing a specific reaction. The importance of these rates is exemplified by the photodissociation of ozone, a key process in the formation of photochemical smog. Rate constants are also integral to these models. They are temperature-dependent and crucial for determining the speed of chemical processes. These constants are often parameterized using the Arrhenius equation, which relates the rate of a reaction to temperature.

Tropospheric ozone production is a vital process influencing atmospheric composition in the lower troposphere with a dominating effect over urban areas. This process involves the photochemical interaction between VOCs and nitrogen oxides (NO_x), leading to the formation of ozone under the influence of sunlight. Additionally, the cycles of NO_x and HO_x radicals, which are central to atmospheric chemistry, are largely governed by photolysis rates and are critical for the oxidative capacity of the atmosphere.

8.1.2.4 *Steady-state modeling*

Steady-state models are frequently employed across various scientific and engineering disciplines to analyze systems where certain variables, such as concentrations of chemical species, temperature, and pressure, remain constant over time.

Steady-state models hold a prominent place in the scientific modeling toolbox, especially within the domain of atmospheric sciences. These models, underpinned by the equilibrium assumption, propose that system variables remain unchanged over time, thereby creating a balance of inputs and outputs. Under steady-state conditions, the rate of input and the rate of are balanced by the rate of output.

8.1.2.4.1 Balancing mass and energy

In atmospheric chemistry, one can consider a column of the atmosphere where chemical reactions are taking place. Under steady-state conditions, the mass of a particular pollutant or chemical species entering the column from above and below equals the mass leaving it. This balance arises because the generation of new mass through chemical reactions equals the consumption of mass, and there is no net accumulation of mass within the system. This principle is fundamental in the understanding of stratospheric ozone chemistry, where the steady-state assumption has been used to estimate the global ozone depletion potential of various substances.

Similarly, in atmospheric energy balance models, the energy entering the earth-atmosphere system from the sun equals the energy leaving it through reflected sunlight and thermal radiation emitted to space. This balance results in no net accumulation of energy within the system, keeping the earth's temperature relatively stable over long periods. This concept is integral to climate science, particularly in understanding and predicting the impacts of increasing greenhouse gases on global temperatures.

8.1.2.4.2 Rate equations and boundary layer theory

Under steady-state conditions, the rates of chemical reactions equate to the flow rates of reactants and products. For instance, in a column of the atmosphere, the rate of conversion of ozone to oxygen through photodissociation balances the rate of ozone formation from oxygen. This concept is often used in atmospheric chemistry to understand the dynamics of various chemical species.

Steady-state models also play a crucial role in atmospheric boundary layer studies, where they describe the layer of the atmosphere closest to the earth's surface. In this layer, properties such as wind speed, temperature, and humidity remain relatively constant with time at any fixed position. Understanding the dynamics of the boundary layer is crucial for weather prediction, pollution dispersion modeling, and wind energy studies.

8.1.2.4.3 Special considerations for steady-state models

When implementing steady-state models in atmospheric sciences, several special considerations should be taken into account to ensure accurate and meaningful results.

8.1.2.4.4 Justifying the assumption

The first consideration is to confirm that the steady-state assumption is valid for the system under study. This often entails checking that the system has had ample time to reach equilibrium or that dynamic effects are negligible. For example, in the stratospheric ozone layer, the assumption of a steady state allows scientists to estimate the ozone depletion potential of various substances.

8.1.2.4.5 Checking time scales

Another crucial step is to compare the time scales of process dynamics to the time scale of interest. If the dynamics of the process occur significantly faster than the time scale of interest, the steady-state assumption may be suitable. This is often the case in systems where rapid reactions occur, such as in the photochemical reactions in the atmosphere.

8.1.2.4.6 Nondimensionalization and dimensional analysis

Nondimensionalizing the governing equations can simplify the analysis and highlight key dimensionless groups that dictate system behavior. For example, in atmospheric dynamics, the Rossby number, a dimensionless quantity, can indicate the relative importance of rotation versus advection. Similarly, dimensional analysis can reduce the number of variables and identify scaling laws, simplifying the problem.

8.1.2.4.7 Validation and sensitivity analysis

Where possible, the results from the steady-state model should be compared with those from a dynamic model. This comparison can ensure that the steady-state assumption does not overlook important transient behaviors. For example, in climate modeling, a steady-state model might overlook transient effects such as El Niño events.

8.1.2.4.8 Defining boundary conditions and considering multiple steady states

Clearly defining and justifying the boundary conditions used in the model is important, as these can significantly impact results. For example, in atmospheric dispersion models, the choice of boundary conditions can greatly affect the predicted pollutant concentrations.

8.1.2.4.9 Iterative methods and computational tools

For nonlinear systems, iterative numerical methods may be employed to solve steady-state equations, as analytical solutions may not be feasible. Techniques like Newton's method or the bisection method can often be applied.

Computational tools and software can also be utilized to solve steady-state equations, especially for complex systems with many variables and equations. Software packages like WRF model or CMAQ can be used to solve complex steady-state problems in atmospheric sciences.

8.1.2.5 Statistical/empirical chemical models

Empirical modeling in atmospheric science harnesses observed data to establish relationships between atmospheric variables and the concentrations of chemical species. These models, which stand apart from deterministic models that are based on the fundamental laws of physics, focus on identifying correlations and patterns within historical datasets to forecast chemical concentrations under varying conditions.

8.1.2.5.1 Development of statistical and empirical models

The development of empirical chemical models in atmospheric science involves a variety of statistical techniques. Regression analysis, including both simple and multiple regression, is used to predict pollutant concentrations based on predictor variables such as temperature, solar radiation, and traffic volume. For example, regression models have been used to estimate the concentration of nitrogen dioxide (NO_2) in urban areas based on traffic flow data and meteorological conditions.

Advanced statistical models leverage machine learning algorithms to discern complex, nonlinear relationships between input variables and chemical concentrations. These algorithms include decision trees, random forests, support vector machines, and neural networks. Machine learning techniques have been applied to predict the formation of ground-level ozone, taking into account a wide range of inputs including industrial emissions, vehicle traffic, and weather patterns.

Time series forecasting models, such as **ARIMA**, are utilized to project future levels of pollutants based on historical trends and cyclical patterns. These models have been instrumental in predicting the seasonal variations of pollutants like ozone and particulate matter, which are influenced by both anthropogenic activities and natural phenomena.

Land Use Regression (LUR) models estimate the spatial distribution of pollutants using geographic information system (GIS) data on land use, traffic, population density, and other factors. LUR models have been particularly useful in urban air quality studies, where they help to map the distribution of pollutants like particulate matter and nitrogen oxides across different urban landscapes.

Chemical Mass Balance (CMB) models determine source contributions to observed concentrations at a receptor site by comparing the known chemical composition of potential source emissions with the observed chemical composition. These models have been used to identify the contributions of various sources, such as vehicles, biomass burning, and industrial emissions, to the concentrations of particulate matter and other pollutants in the atmosphere.

8.1.2.5.2 Applications of empirical models

Empirical models have a wide range of applications in atmospheric chemistry. They can forecast daily maximum ozone concentrations using meteorological variables, with neural network models trained on historical ozone data and meteorological conditions being one example. These models are crucial for air quality management and public health advisories.

Particulate matter (PM) estimation is another application where LUR models can estimate concentrations of PM2.5 or PM10 in urban environments based on variables such as traffic density and industrial activities. These models are vital for understanding the health impacts of air pollution and for guiding policy decisions.

Source Apportionment is a critical application of CMB models, which are applied to ascertain the contributions of various sources to pollutant concentrations at specific locations. This information is essential for developing targeted emission control strategies and for assessing the effectiveness of air quality regulations.

8.1.2.5.3 Special considerations

When developing empirical models in atmospheric chemistry, several considerations are paramount. The quality, representativeness, and temporal span of data used for model training and validation are critical to capturing the variability in chemical concentrations. Feature selection, guided by domain expertise, enhances model performance and interpretability by focusing on relevant input features that influence the chemical species of interest.

Independent datasets are essential for model validation, assessing the model's predictive capabilities, and its applicability to different conditions or locales. Uncertainty quantification is necessary to gauge confidence levels in model predictions and to pinpoint influential parameters.

The temporal and spatial resolution of the model should match the resolution of the available data and the intended scale of application. Nonstationarity considerations are also crucial, as variable-concentration relationships may evolve due to changes in emission sources, regulatory measures, or climatic conditions.

For complex machine learning models, interpretative techniques such as feature importance scores or SHAP values are instrumental in understanding the influence of predictors. Cross-validation techniques should be rigorously applied to prevent overfitting, especially in high-dimensional data or complex models.

Integration with deterministic models can harness the strengths of both methodologies, combining the detailed process understanding from deterministic models with the pattern recognition capabilities of statistical models.

Lastly, model maintenance, including regular updates with new data, is imperative to sustain model accuracy, particularly for models used in regulatory or health advisory contexts.

8.1.3 Model coupling

Model coupling is a general approach employed in atmospheric modeling where models covering different spatial scales or different aspects of atmospheric phenomena are set up to exchange information, so as to improve the accuracy, spatial coverage and reliability of calculated physical parameters (Warner, 2011). In a coupling scheme, constituent models may be using different computational schemes (e.g., Eulerian vs Lagrangian), cover differing spatial scales (e.g., mesoscale vs regional) or even provide different aspects of the required physical system (e.g., meteorology vs atmospheric chemistry).

Coupling can be performed either online, which implies that information is exchanged between the models while they are executed concurrently, or offline, in which case the output of one model is used as input for the subsequent application of the other. In the following sections, the most typical coupling configurations are described in more detail.

8.1.3.1 Coupling between models of different scales

The coupling of atmospheric dispersion models across different scales is a crucial aspect of atmospheric science. It allows for a more comprehensive understanding of atmospheric processes, from large-scale weather patterns to local-scale air pollution dispersion. It has also been proposed as a way to computationally address multiscale phenomena in the atmosphere.

8.1.3.1.1 Coupling between regional-scale and mesoscale atmospheric models

The coupling between regional-scale and mesoscale models is often used to simulate the transport and dispersion of air pollutants over large geographical areas. The regional-scale model provides the large-scale meteorological conditions and pollutant concentrations, which serve as boundary conditions for the mesoscale mode (Skamarock et al., 2008).

For instance, the WRF model, a state-of-the-art mesoscale numerical weather prediction system, is often used in conjunction with regional-scale models. The WRF model can simulate a wide range of meteorological phenomena, from large-scale weather patterns to local-scale atmospheric dispersion, making it a versatile tool for atmospheric research and operational forecasting.

In a real-world application, the WRF model was used to simulate the transport and dispersion of wildfire smoke over the western United States. The model was able to accurately capture the large-scale transport of smoke by prevailing winds, as well as the local-scale dispersion of smoke influenced by topography and local wind patterns (Coen et al., 2013).

8.1.3.1.2 Coupling between mesoscale and microscale (CFD) models

Microscale models, often based on CFD, are used to simulate the dispersion of air pollutants at very fine spatial scales, such as within urban areas or around industrial facilities. These models can capture the complex flow patterns induced by buildings and other obstacles, which would not be resolved by the mesoscale model itself, thus providing detailed information on pollutant concentrations at the street level (Lateb et al., 2016).

In one study, a CFD model was used to simulate the dispersion of traffic-related air pollutants in a busy urban area. The model was able to capture the complex flow patterns around buildings and the resulting high pollutant concentrations along busy streets. This information was crucial for urban planners and public health officials, helping them identify areas of high air pollution exposure and develop strategies to mitigate these exposures (Tsegas et al., 2011).

8.1.3.1.3 Surrogate model approaches for mesoscale-microscale coupling

The surrogate model approach is a method for coupling mesoscale and microscale models in an almost online fashion while economizing in computational load compared to a traditional online coupling scheme (Tsegas et al., 2011). In this approach, the mesoscale model provides the large-scale meteorological conditions and pollutant concentrations, which are then used as boundary conditions for the microscale CFD model. The microscale model, in return, provides an output that is used to "train" a surrogate model, or so-called metamodel, that can approximate the response of the CFD model but using orders of magnitude less computational resources. During the coupled execution, the mesoscale model is actually coupled to the simplified surrogate model, rather than the full CFD model, with which it interacts in a bidirectional (two-way) fashion (Fig. 8.2).

In a real-world application, the surrogate approach was used to simulate the dispersion of pollutants from a large industrial facility. The mesoscale model was used to simulate the regional weather patterns and the large-scale transport of pollutants, while the microscale model was used to simulate the dispersion of pollutants around the facility and the resulting local-scale pollutant concentrations. This approach allowed for a comprehensive simulation of pollutant dispersion, from the regional scale to the local scale, providing valuable information for air quality management and regulatory compliance.

8.1.3.2 Coupling of meteorological and chemistry transport models (chemical weather models)

The accuracy of atmospheric chemistry and air quality simulations can be improved by integrating meteorological and chemistry transport models. These coupled models, often referred to as chemical weather models, can account for the complex interactions between atmospheric chemistry and meteorology (Grell & Baklanov, 2011). They find extensive use in understanding and predicting a range of atmospheric phenomena, including air pollution episodes, long-range pollutant transport, and climate change (Fast et al., 2006).

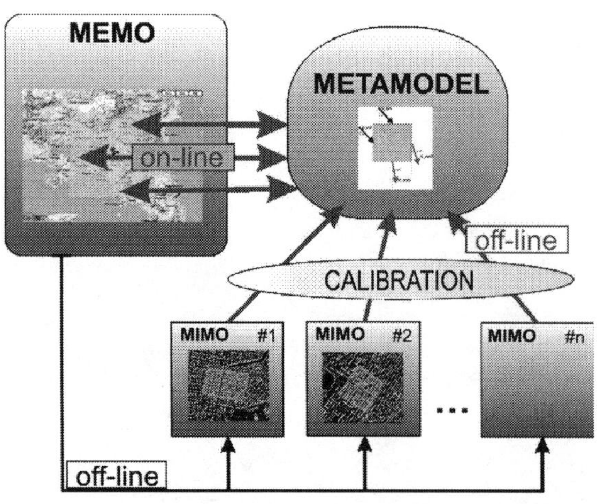

FIGURE 8.2 Example of the use of a surrogate model (metamodel) for coupling a mesoscale model (memo) and a microscale/CFD model (MIMO). By simulating a sample configuration set of flows, MIMO is used to calibrate the metamodel, which is then coupled online with MEMO. *CFD*, Computational fluid dynamics.

8.1.3.2.1 Online coupled chemical weather models

Online coupled models seamlessly integrate meteorological and chemical processes into a single modeling system. In these models, the meteorological and chemical components interact with each other at every time step, allowing for explicit feedbacks between the two. This ensures that changes in atmospheric chemistry, such as the formation of aerosols, can influence meteorological variables, such as radiation and cloud formation, and vice versa.

One of the most widely used online coupled models is the WRF-Chem. Developed by the NCAR, WRF-Chem simulates the emission, transport, mixing, and chemical transformation of trace gases and aerosols simultaneously with meteorology. The model has been used in numerous studies to understand and predict a range of atmospheric phenomena, from urban air pollution to dust storms and wildfire smoke dispersion.

Another example is the CMAQ model, developed by the US EPA. CMAQ is a comprehensive model that simulates the interactions of meteorology and atmospheric chemistry across scales ranging from urban to continental. It has been widely used for air quality management, policy decision support, and atmospheric science research.

A review of online chemical weather models and their application in Europe was undertaken by (Baklanov et al., 2014). The review also contains useful suggestions for implementing and applying online coupled approaches in real-world applications. An earlier review covering a wider scope of such models applied worldwide can be found in (Zhang, 2008).

8.1.3.2.2 Offline coupled chemical weather models

In offline coupled models, the meteorological and chemistry transport models are run separately, with the meteorological model typically run first to provide inputs to the chemistry transport model. This approach allows for the use of detailed, computationally intensive chemical mechanisms without the need for high-performance computing resources. However, one limitation is that the feedback of atmospheric chemistry on meteorology is not captured.

A prime example of an offline coupled model is the combination of the WRF model with the CMAQ model. In this setup, WRF is run first to generate meteorological fields, which are then used as inputs to CMAQ for simulating atmospheric chemistry and air quality. This approach has been widely used in air quality studies and allows for the use of detailed chemical mechanisms and high-resolution meteorological inputs.

Another example of an offline coupled model is the combination of the European Centre for Medium-Range Weather Forecasts (ECMWF) Integrated Forecast System (IFS) with the Tracer Model 5 (TM5) chemistry transport model. In this setup, IFS provides meteorological inputs to TM5, which simulates the global distribution of trace gases and aerosols. This model setup has been used in numerous studies to understand the global distribution and long-range transport of air pollutants.

In summary, both online and offline coupled models play crucial roles in atmospheric science and air quality management. While online coupled models provide a more detailed representation of the complex interactions between meteorology and atmospheric chemistry, offline coupled models provide a computationally efficient approach that can still capture the major processes influencing air quality. The choice between online and offline coupling depends on the specific goals of the study and the available computational resources.

Worked Example 8.1.1

Evaluate the dispersion of pollutants from an industrial smokestack of given dimensions and emissions rate

8.1.3.2.3 Background

An industrial facility emits a specific pollutant into the atmosphere from a stack. To evaluate the impact on air quality, a Gaussian dispersion model is used to estimate the concentration of the pollutant at various distances downwind from the stack.

8.1.3.2.4 Data collection

The facility provides the emission rate (Q) of the pollutant, the stack height (H), and the average wind speed (U) at the stack height. Meteorological data, including atmospheric stability classes, are obtained from a nearby weather station.

8.1.3.2.5 Model implementation

The Gaussian dispersion model is implemented in Python to calculate the concentration of the pollutant at different distances and heights. The model assumes a steady-state plume and uses Pasquill-Gifford stability classes to determine the dispersion coefficients.

Solution

See Box 8.1 for an implementation in Python.

BOX 8.1 Simple implementation of a Gaussian plume model.

```
import numpy as np
    # Gaussian dispersion model function
    def gaussian_plume(Q, U, H, x, y, z, sigma_y, sigma_z):
    """
    Calculate the concentration of a pollutant at a given point.
    Parameters:
    Q: float
    Emission rate (g s^-1)
    U: float
    Wind speed (m s^-1)
    H: float
    Effective stack height (m)
    x: float
    Downwind distance from the source (m)
    y: float
    Crosswind distance from the source (m)
    z: float
    Height above ground (m)
    sigma_y: float
    Horizontal dispersion coefficient (m)
    sigma_z: float
    Vertical dispersion coefficient (m)
    """
    C = (Q/(2 * np.pi * U * sigma_y * sigma_z)) * \
    np.exp(-y**2/(2 * sigma_y**2)) * \
    (np.exp(-(z-H)**2/(2 * sigma_z**2)) + np.exp(-(z + H)**2/(2 * sigma_z**2)))
    return C
    # Example parameters
    Q = 100 # Emission rate (g s^-1)
    U = 5 # Wind speed (m s^-1)
    H = 50 # Stack height (m)
    x = 1000 # Downwind distance (m)
    y = 0 # Crosswind distance (m)
    z = 2 # Height above ground (m)
    sigma_y = 50 # Horizontal dispersion coefficient (m)
    sigma_z = 20 # Vertical dispersion coefficient (m)
    # Calculate concentration
    concentration = gaussian_plume(Q, U, H, x, y, z, sigma_y, sigma_z)
    print(f"Concentration at ({x}, {y}, {z}) m: {concentration} g m^-3")
```

8.1.3.2.6 Model evaluation

The model's predictions are compared with air quality measurements taken at various locations around the facility. Statistical methods such as root mean square error (RMSE) or R^2 are used to quantify the model's accuracy.

8.1.3.2.7 Sensitivity analysis

A sensitivity analysis is conducted to determine how changes in model parameters like wind speed and emission rate affect the concentration predictions.

Worked Example 8.1.2

Assess pollutant dispersion emanating from road traffic in an urban area.

8.1.3.2.8 Background

A city is concerned about the dispersion of NO_x emissions from a dense network of roads. A Gaussian dispersion model is used to estimate the ground-level concentration of NO_x throughout the urban area.

8.1.3.2.9 Data collection

Traffic data is used to estimate the emission rate of NO_x from vehicles. Meteorological data is collected from an urban weather station to determine wind speed, wind direction, and atmospheric stability. Optionally, a mesoscale meteorological model can be used to provide gridded wind data for the area of application.

8.1.3.2.10 Model implementation

The Gaussian dispersion model is implemented to calculate NO_x concentrations at various points within the urban area. The model accounts for multiple line sources representing the roads.

Solution

See Box 8.2 for an implementation in Python.

BOX 8.2 Implementation of a Gaussian plume model for line sources.

```
import numpy as np
    import matplotlib.pyplot as plt
    # Gaussian dispersion model for line sources
    def gaussian_line_source(Q, U, H, x, y, sigma_y, sigma_z):
    """
    Calculate the concentration of a pollutant from a line source at a given point.
    Parameters are similar to the previous function, with Q now representing
    the emission rate per unit length of the line source (g s m^-1).
    """
    # Integrate along the line source to calculate concentration
    # For simplicity, we assume the line source is perpendicular to the wind direction
    # and passes through the origin (0, 0).
    C = 0
    for point in np.linspace(-500, 500, 100): # 1 km line source centered at the origin
    C + = gaussian_plume(Q, U, H, x, y - point, z, sigma_y, sigma_z) * (1000/100)
    return C
    # Example parameters for urban scenario
    Q = 0.1 # Emission rate per unit length (g s m^-1)
    U = 3 # Wind speed (m s^-1)
    H = 0 # Effective height of emissions (m)
    x = 100 # Downwind distance (m)
    y = 0 # Crosswind distance (m)
    z = 2 # Height above ground (m)
    sigma_y = 40 # Horizontal dispersion coefficient (m)
    sigma_z = 10 # Vertical dispersion coefficient (m)
    # Calculate concentration for a grid of points
    X, Y = np.meshgrid(np.arange(0, 1000, 50), np.arange(-500, 500, 50))
    Concentration = np.zeros_like(X)
    for i in range(X.shape[0]):
    for j in range(X.shape[1]):
    Concentration[i, j] = gaussian_line_source(Q, U, H, X[i, j], Y[i, j], sigma_y, sigma_z)
    # Plot the concentration field
    plt.contourf(X, Y, Concentration, levels = 20, cmap = "viridis")
    plt.colorbar(label = "NO_x Concentration (g m^-3)")
    plt.xlabel("Downwind Distance (m)")
    plt.ylabel("Crosswind Distance (m)")
    plt.title("NO_x Concentration Field from Urban Roads")
    plt.show()
```

8.1.3.2.11 Model evaluation

The model's predictions are evaluated against NO_x concentration measurements from air quality monitoring stations throughout the city. Discrepancies are analyzed, and the model is refined accordingly.

8.1.3.2.12 Sensitivity analysis

The model's sensitivity to traffic volume and meteorological conditions is assessed to understand the potential range of NO_x concentrations under different scenarios.

8.2 Input data for air quality models

Historically, ground-based monitoring stations have been the key sources of air quality data. Positioned in various locales, these monitors provide both instantaneous and historical data on air quality, facilitating the analysis of long-term trends and daily variations. They track a range of pollutants, including particulate matter (PM10 and PM2.5), nitrogen dioxide (NO_2), sulfur dioxide (SO_2), and other harmful substances. The collected data is often accessible through online platforms, enabling users to visualize trends, compare current air quality index (AQI) readings with past data, and download information for detailed analysis.

Satellite observations complement ground-based monitoring by providing data on aerosol optical depth, nitrogen dioxide, sulfur dioxide, and other atmospheric constituents. Accessible through web services, satellite data allows for the creation of custom visualizations and maps, essential for analyzing and researching air quality on a global scale.

In a typical application of atmospheric dispersion models, three broad categories of input information is required: Emissions data, which is typically estimated through various models and categorized according to source types such as point sources (e.g., factories), mobile sources (e.g., vehicles), area sources (e.g., residential heating), and biogenic sources (e.g., vegetation); meteorological data, which can be provided either from surface measurements or vertical profiles from sondes; and topographic and land use data for the area of study. These classes of input data are described in detail in the following sections.

8.2.1 Emissions data

8.2.1.1 Foundations of emissions data for atmospheric modeling

Emissions data serves as a crucial element in atmospheric modeling, providing quantitative insights into the types and amounts of pollutants dispersed into the atmosphere from various sources. These data are systematically collected and compiled into emission inventories, which offer comprehensive records of emissions from all sources within a defined region and time period.

8.2.1.2 Methodologies for constructing emission inventories

Two primary methodologies are employed in creating emission inventories: the bottom-up and top-down approaches. The bottom-up method calculates emissions from individual sources and then aggregates them to obtain total emissions. This is typically done using activity data (e.g., fuel consumption, industrial production rates) and emission factors, which represent the amount of a pollutant emitted per unit of activity. This method is often used for point sources, such as factories or power plants, and for area sources, such as residential heating or agriculture.

On the other hand, the top-down method uses atmospheric measurements and inverse modeling techniques to estimate total emissions. This approach is particularly advantageous for pollutants that are challenging to measure directly or for sources that are difficult to quantify using bottom-up methods, such as fugitive or biogenic emissions.

In many instances, a combined approach is adopted, where bottom-up inventories are adjusted or validated using top-down measurements. This can help to reduce uncertainties and enhance the accuracy of the emission estimates.

8.2.1.3 Data sources and calculation of annual emissions

Emission inventories can be built using a variety of data sources. For example, some organizations provide data on annual CO_2 emissions, which can be used as an initial metric in a bottom-up approach to build a comprehensive dataset of CO_2 emissions. Other sources, such as the US EPA, provide data on greenhouse gas emissions from various sectors, which can be used to calculate emissions from individual sources.

Annual emissions are typically calculated using activity data and emission factors. The activity data represents the amount of a certain activity (e.g., fuel combustion, industrial production) that occurred during the year, and the emission factor represents the amount of a pollutant emitted per unit of activity. The product of the activity data and the emission factor gives the total emissions for that activity.

8.2.1.4 *Temporal and spatial disaggregation of emissions*

Temporal and spatial disaggregation involves breaking down the total annual emissions into smaller temporal and spatial units. This is necessary for atmospheric modeling, which requires information on the distribution of emissions in space and time. Spatial disaggregation involves distributing the total emissions over the geographical area of interest, which can be done based on the location of emission sources, population density, land use data, or other relevant factors. Temporal disaggregation involves distributing the total emissions over time, which can be done based on the temporal patterns of the emission activities, such as diurnal and seasonal variations in traffic volume or energy consumption.

8.2.1.5 *Uncertainties in emission inventories*

There are various sources of uncertainty in emission inventories, including uncertainties in the activity data, emission factors, and the methodologies used for spatial and temporal disaggregation. These uncertainties can be quantified using statistical methods and should be taken into account in the interpretation of the emission estimates.

8.2.1.6 *Additional data sources for calculating emissions*

Besides the traditional sources, company-level environmental data can also be used to calculate emissions. For example, environmental-related metrics for thousands of companies worldwide, including data on CO_2 equivalent emissions, are available. Finally, emissions data can also be obtained from digital activities. For example, online website carbon calculators provide a quick way to start estimating the emissions associated with operating a website.

8.2.2 Meteorology data

Meteorological data are a critical component of air pollution modeling, providing the atmospheric state and dynamics that drive the dispersion and transformation of pollutants. In a typical modeling configuration, a meteorological model of the same scale and resolution or a reanalysis dataset will be used to provide the dispersion model with the required static or time-dependent meteorological fields. In the case of a separate meteorological model, additional larger-scale information will be needed to initialize it, in the form of initial and boundary conditions.

Various sources can be utilized to provide the air pollution modeling system with the required meteorological fields:

Reanalysis Data: Reanalysis datasets, such as the ERA5 from ECMWF, combine model data with observations to provide a consistent, long-term record of the global atmosphere at various spatial and temporal resolutions. ERA5, for instance, offers hourly data on many atmospheric parameters at a 31 km grid spacing (Hersbach et al., 2020). These datasets are invaluable for air pollution modeling, especially for historical analysis, as they provide a comprehensive atmospheric picture that includes data-sparse regions. Reanalysis data are used to understand past weather and climate conditions and to initialize air quality models.

Fields from Numerical Weather Prediction (NWP): Larger-scale meteorological models, such as those used in NWP, are designed to simulate the atmosphere's behavior over large spatial scales and extended time periods. These models incorporate complex physics and dynamics to forecast weather and are driven by equations that govern atmospheric motions, thermodynamics, radiation, and surface exchanges (Skamarock et al., 2008). NWP models like the Global Forecast System (GFS) or the ECMWF model provide gridded meteorological fields that can be downscaled for regional air quality modeling. They offer variables such as wind speed and direction, temperature, pressure, humidity, and precipitation, which are essential for predicting the transport and chemical transformation of air pollutants.

Surface Measurements: Surface meteorological measurements are taken at weather stations and include observations of temperature, wind speed and direction, relative humidity, atmospheric pressure, and precipitation. These point measurements can be used both to provide an initial consistent field for initializing the meteorological part of the modeling system or be ingested during the run to drive the calculation. They can also provide ground truth data that can be used to adjust model outputs and improve their accuracy. For air pollution modeling, surface measurements help in assessing the dispersion of pollutants under various atmospheric conditions (WMO, 2018).

Upper-air Measurements: Upper-air measurements are crucial for capturing the vertical structure of the atmosphere, which influences the transport and diffusion of air pollutants. Radiosondes, which are weather balloons equipped with sensors, are launched to measure profiles of temperature, humidity, and pressure up to the stratosphere. Wind speed and direction are also derived from the balloon's trajectory. These data are typically available on major airfield locations from two up to four times a day and can be used by the meteorological submodel to synthesize a set of self-consistent initial conditions or used as correction input throughout the computation. The University of Wyoming provides a publicly accessible repository of radiosonde data, which is widely used by the atmospheric science community (University of Wyoming, 2023).

8.2.3 Topography/land use data

8.2.3.1 The role of topography and land use data in atmospheric modeling

Topography and land use data constitute vital variables in atmospheric modeling, furnishing comprehensive information on the Earth's surface's physical characteristics and the human-induced modifications of this feature (Fig. 8.3).

Topographical data encapsulates the 3D attributes of the Earth's surface. This includes details on elevation, the shape of the landscape, and other physical features such as bodies of water and mountainous regions. There exist several forms of topography data. Digital Elevation Models (DEMs) are 3D representations of the Earth's surface, derived from elevation data collected through a variety of methods including terrestrial surveys, airborne light detection and ranging (LiDAR), and satellite radar data. Topographic maps, on the other hand, are 2D depictions of the Earth's surface that display the land's shape and elevation via contour lines, often featuring additional elements such as transportation networks, structures, and bodies of water. Lastly, topographic surveys, which can be conducted using traditional surveying techniques or modern technologies such as GPS and LiDAR, provide precise measurements of the elevation and location of specific points on the land surface.

Land use data, in contrast, provides insights into the ways in which land is utilized or altered by human activities. This includes information on agricultural practices, urban development, and forestry operations, among others. There are various sources from which land use data can be obtained. Satellite imagery, for instance, allows for land use classification based on the spectral characteristics of different types of land cover. Aerial photographs offer high-resolution, localized land use information, which can be manually interpreted or classified using automated algorithms. Ground surveys, though labor-intensive and costly, yield highly accurate and detailed land use data.

FIGURE 8.3 Artistic illustration of different land use types and their distribution, which is crucial in atmospheric modeling as it influences the exchange of energy, moisture, and momentum between the land surface and the atmosphere.

In the context of atmospheric modeling, topography and land use data are instrumental in understanding and simulating the interactions between the Earth's surface and the atmosphere. This encompasses the influence of topography on weather patterns, the role of land cover in energy balance and hydrological cycles, and the impact of human activities on atmospheric conditions. The integration of topography and land use data into atmospheric models, however, necessitates careful consideration of data resolution, accuracy, and the representation of temporal changes. As such, the selection and processing of these data types are critical to the fidelity and applicability of atmospheric models.

8.2.3.2 Topography and land use data in atmospheric modeling: sources, types, and uncertainties

Topography and land use data form the bedrock of atmospheric modeling, elucidating the Earth's surface's physical features and the anthropogenic modifications of these features. This data provides crucial information influencing weather patterns, climatic processes, and air pollutant distribution and movement.

8.2.3.3 Topographic data

Topographic data delineates the Earth's surface shape and features, including elevation and the location of natural and man-made structures. Its importance spans applications from flood modeling and environmental management to urban planning.

Digital Elevation Models (DEMs): DEMs are a common source of topographic data. They offer elevation data for the Earth's surface, constructed using techniques such as light detection and ranging (LiDAR) (Wehr & Lohr, 1999) or interferometric synthetic aperture radar (Bamler & Hartl, 1998). The Shuttle Radar Topography Mission, for instance, has provided a comprehensive dataset offering global topographic coverage extensively used across applications (Farr et al., 2007).

National Elevation Datasets: Certain countries maintain national elevation datasets derived from various sources, including the digitization of topographic maps. The United States Geological Survey (USGS), for instance, offers a National Elevation Dataset employed for a wide range of applications within the United States (Gesch, 2007).

Synthetic Aperture Radar (SAR): SAR, a form of radar used to create 2D images or 3D reconstructions of landscapes, can penetrate clouds, making it useful for all-weather topography mapping (Moreira et al., 2013).

8.2.3.4 Land use data

Land use data describes human utilization of land, which can include residential areas, commercial buildings, agricultural fields, and more. This data is essential for understanding human environmental impacts and modeling processes such as urban heat islands and surface runoff. They can also be used to simulate biogenic emissions, which are emissions from natural sources such as vegetation, and in the calculation of surface albedo, which is the fraction of solar energy reflected from the Earth back into the atmosphere. Accurate albedo values are essential for climate models as they affect the Earth's energy balance. Additionally, land use data are essential in estimating heat fluxes—the transfer of heat energy between the Earth's surface and the atmosphere. These fluxes are crucial for understanding local and regional dynamics but also the near-surface turbulence which in turn influences turbulence transport of pollutants. Processing of land use data in general uses the following sources and tools:

Ground Surveys: Ground surveys, traditional methods of collecting land use data, involve on-the-ground observations and can offer highly detailed and accurate information about land use practices.

Remote Sensing: Remote sensing technologies, including satellite and aerial imagery, are widely used to collect land use data over large areas. These methods can quickly offer data on land cover changes, urban sprawl, and deforestation.

Geographic Information Systems (GIS): GIS is a framework for gathering, managing, and analyzing spatial data. Land use data is often analyzed and visualized using GIS, which can combine various data sources to provide a comprehensive view of land use patterns.

In the last couple of decades, consolidated and standardized land cover datasets have become available, covering most of Europe and the Americas. These are usually provided in GIS-compatible formats and can be readily converted to the various input formats required by the air pollution models:

The CORINE Land Cover (CLC): The CLC dataset is provided through the Copernicus EU service and undergoes regular updates. It provides land cover for the entire EU region using a standard classification scheme at a horizontal resolution of 250 m (Büttner et al., 2017).

The National Land Cover Database (NLCD): In the United States, the NLCD provides detailed land cover information that is used to simulate meteorological variables such as temperature, humidity, and wind patterns (Homer et al., 2015).

National Cadasters: In most European countries, national authorities maintain surface maps generated based on cadaster information which at the minimum provide land use classifications for urban areas and other conurbations as well as cultivated land. In many cases, these publicly available datasets are supplanted by remote sensing information in order to incorporate nonurban land covers, including forestation, vegetation, and arid areas. Cadaster-based land use maps are usually available upon request from the responsible authority at various GIS formats in high spatial resolutions, though in many cases a nonstandard land classification is used.

8.2.3.5 Uncertainties in topography and land use data

Topography and land use data come with inherent uncertainties, which can arise from measurement errors, data processing, and the resolution of the datasets.

Measurement Errors: Errors can occur during the data collection process, whether it's from satellite misalignment, atmospheric interference, or inaccuracies in ground-based measurements.
Data Artifacts: Gridded data derived from digitization techniques or rasterization of vector datasets can be prone to artifacts, which may require careful processing and validation to ensure accuracy.
Resolution and Coverage: The spatial and temporal resolution of the data can introduce uncertainties, especially if the data is not fine-grained enough to capture small-scale features or if there are gaps in data coverage.

To mitigate these uncertainties, researchers often use a combination of datasets and validation techniques. For example, comparing DEMs from different sources can help identify and correct inconsistencies. Integrating ground-based measurements with remote sensing data can also enhance the overall quality and reliability of topography and land use datasets.

Worked example 8.2.1

Briefly describe the three categories of input data required for atmospheric modeling simulations in regard to the main information that needs to be included in each dataset.

Solution

The following three categories of input information are required for air quality simulations:

1. Emissions data, is typically estimated through various models using either a top-down or a bottom-up approach or a combination of the two. These data are systematically collected and compiled into emission inventories, which offer comprehensive records of emissions from all sources, such as point sources (e.g., factories), mobile sources (e.g., vehicles), area sources (e.g., residential heating), and biogenic sources (e.g., vegetation). Atmospheric modeling requires information on the distribution of emissions in space and time and therefore the temporal and spatial disaggregation of emissions is also necessary.
2. Meteorological data, can be provided either from larger-scale models, surface measurements or vertical profiles from sondes. These typically include pointwise or spatially resolved time series of wind speed, wind direction, pressure, temperature, and precipitation; and
3. Topographic and land use data for the area of study. Topographical data encapsulates the 3D attributes of the Earth's surface. This includes details on elevation, the shape of the landscape, and other physical features such as bodies of water and mountainous regions. Land use data, in contrast, provides insights into the ways in which land is utilized or altered by human activities. This includes information on agricultural practices, urban development, and forestry operations, among others.

8.3 Air quality model evaluation

Air quality models are increasingly being recognized as valuable tools for management and regulatory purposes. In this frame, the reliability of the model to generate realistic and representative results for the selected region is of imperative importance for relevant bodies to implement effective management strategies and make informative choices in relation to new developments so that air quality is protected and improved. For this purpose, model performance has to be evaluated on the basis of predefined criteria and goals. In general, model evaluation approaches are distinguished into four main categories (as shown in Fig. 8.4), according to the framework suggested by Dennis et al. (2010):

1. Validation of model results through the comparison with surface measurements (i.e., **operational model evaluation**), where the uncertainty for objective model estimation is calculated on the basis of the deviation between measured and modeled concentration levels.

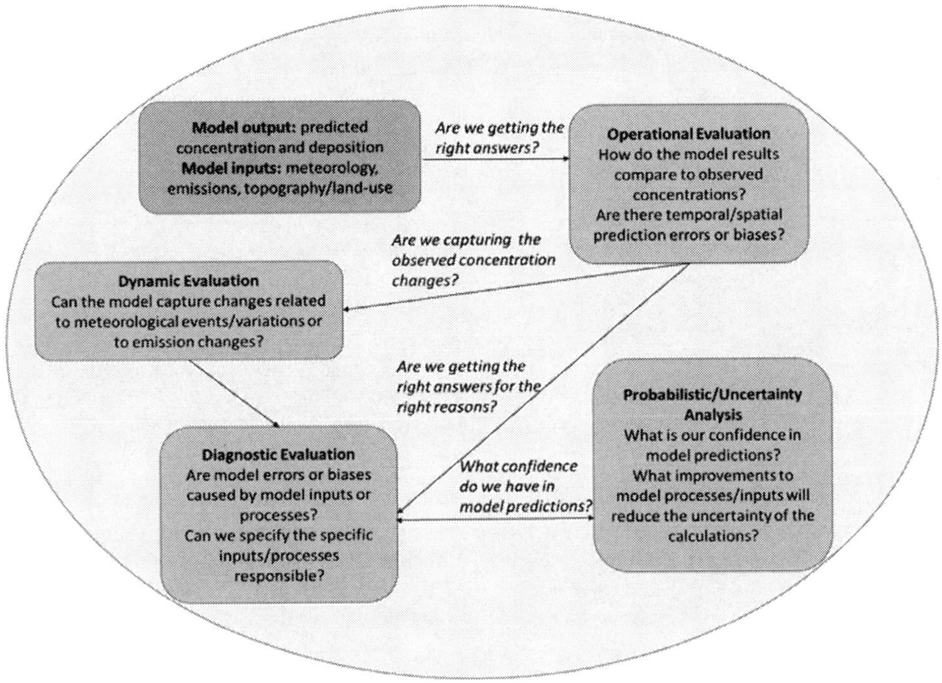

FIGURE 8.4 The components of an air quality model evaluation framework. *Adapted from Dennis, R., Fox, T., Fuentes, M., Gilliland, A., Hanna, S., Hogrefe, C., Irwin, J., Rao, S.T., Scheffe, R., Schere, K., Steyn, D., & Venkatram, A. (2010). A framework for evaluating regional-scale numerical photochemical modeling systems.* Environmental Fluid Mechanics, 10*(4), 471–489. https://doi.org/10.1007/s10652-009-9163-2.*

2. Comparison of results between different models, the so-called model intercomparison exercises, which comprises the main method of **probabilistic model evaluation**, along with ensemble modeling. Model intercomparison applications are particularly useful to assess the reliability of model results if monitoring data is not available for comparison, such as in areas where monitoring stations have not been installed or for future predictions.
3. Sensitivity analysis to changes in model software and hardware features or input data as part of **dynamic model evaluation**, in addition to the function of this methodology to study specific research-related air quality issues, as described previously.
4. Focusing on research and models advancing questions, **diagnostic model evaluation** applications aim to shed light on the reasons for model behavior by attempting to explain model results and responses on the basis of their physical and chemical parameterizations. Dispersion models are the main model type evaluated using this approach under different application conditions, however, it is appropriate for all model types in the stage of their development or updating process.

For air quality dispersion models which are based on simple parameterizations, such as Box models or Gaussian puff models, the operational evaluation is sufficient to assess the reliability of their modeled values against measurements, which are assumed to represent the "real state." More complex models, such as CTMs resolving many complex dynamical, physical and chemical atmospheric processes require a more in-depth evaluation procedure, so that error and bias causes are correctly identified, and models can be accordingly updated and improved. In this case, all four model evaluation approaches may be undertaken for a holistic assessment of the model capabilities and shortcomings. The relevant evaluation framework is shown in Fig. 8.4.

As shown, uncertainty analysis and diagnostic evaluation depend on each other because of a two-way feedback mechanism, as identifying the uncertainty sources is a prerequisite for their reduction. If the purpose of the evaluation is to estimate the quality of the model itself then data uncertainties are not relevant, but if the modeling results will be used for management support, the total model uncertainty is essential. For the estimation of uncertainty related to input data, a sensitivity analysis of individual input parameters (initial and boundary conditions, meteorological parameters, emissions, land use and topography) is an appropriate approach. For a complete knowledge of total model uncertainty, a sensitivity analysis and/or model intercomparison are employed to examine the different model modules (chemical mechanisms, physical parameterizations, and numerical algorithms).

As a minimum, an operational evaluation of air quality models intended to be used for policy applications is required by comparison with measurements that are representative of the specific model scale. Several software tools to facilitate operational model evaluation are available for use, such as the BOOT software (Chang & Hanna, 2005) and AMET (Appel et al., 2011) for dispersion models. Useful guidance on the application and evaluation of AQ models for policy purposes according to the needs of the EU Directive 2008/50/EC is provided in the EEA Technical Report No. 10/2011 (Denby et al., 2011). The topic of operational evaluation is also addressed by the Forum for Air Quality Modeling in Europe (FAIRMODE) network through the development of an evaluation and benchmarking procedure and the respective DELTA tool (Thunis et al., 2011; Thunis et al., 2012), explained in detail in the operational evaluation paragraph.

8.3.1 Model operational evaluation

Operational model evaluation is a procedure used to assess the performance of air quality models using consistency metrics and visualization techniques. This process is crucial for understanding the strengths and weaknesses of a particular model, and assessing its performance in a specific setting, given specific limitations of input data and the physical characteristics of the domain of simulation.

The core of operational model evaluation involves comparing the results generated by the model (modeled concentration levels of pollutants) with actual, ground-level measurements of these pollutants (measured concentration levels). The difference between the measured and modeled values gives an indication of the model's performance—a smaller difference suggests that the model is performing well, while a larger difference could indicate problems with the model's calculations or assumptions.

Several statistical measures can be used to quantify this difference, including:

- **Bias:** This measures the average difference between the modeled and measured values. A positive bias means the model tends to overestimate concentrations, while a negative bias means it tends to underestimate them.
- **Root Mean Square Error (RMSE):** This provides a measure of the overall difference between the modeled and measured values, taking into account both the magnitude and direction of the errors. Smaller values indicate better agreement with observations.
- **Correlation Coefficient:** This measures the degree to which the modeled and measured values change together. A high correlation coefficient suggests that the model is good at capturing the variations in pollutant concentrations.
- **Index of Agreement:** This compound indicator evaluates the agreement between modeled and observed values in terms of both systematic bias and stochastic errors, providing a normalized index of the overall performance.

8.3.1.1 The DELTA operational evaluation tool

The DELTA tool (Monteiro et al., 2018) is software developed within the framework of FAIRMODE for the evaluation and benchmarking of air quality models for regulatory purposes. It is designed to facilitate the operational model evaluation process. It provides a set of functionalities that allow users to perform statistical analyses and generate graphical outputs that can help visualize the performance of air quality models. It is aimed at atmospheric modelers, but also in regulatory authorities who can gain valuable insights into the performance of air quality models and make informed decisions regarding air quality management. Its main features include:

- **Data Import and Export:** The tool can handle various data formats and allows for easy import and export of data.
- **Data Analysis:** The tool provides a range of statistical methods for comparing modeled and observed data. These include the calculation of bias, RMSE, correlation coefficients, and other statistical measures.
- **Data Visualization:** The tool generates graphical outputs such as scatter plots, time series plots, and histogram plots, which can help in visualizing and understanding the model's performance.

The DELTA tool provides a multitude of evaluation metrics and visualization, including:

1. **Summary Statistics:** The DELTA Tool generates summary statistics that provide an overview of the model's performance compared to observed data. These statistics include metrics such as mean bias, root mean square error, correlation coefficient, and index of agreement.
2. **Scatter Plots:** Scatter plots are used to visualize the relationship between model predictions and observed data, which help identify any systematic biases or discrepancies between the model and observations.

3. **Time Series Plots:** Time series plots display the temporal variation of model predictions and observations, which allow for the examination of model performance over different time periods, such as daily, monthly, or yearly.
4. **Taylor Diagrams:** Taylor diagrams are used to assess the overall performance of a model by comparing multiple performance metrics simultaneously, providing a comprehensive evaluation of the model's accuracy, precision, and correlation with observed data.
5. **Target Diagrams:** Target diagrams are similar to Taylor diagrams but focus on specific target values or thresholds, assessing the model's ability to meet specific air quality standards or regulatory targets.

In addition to the above components, the DELTA Tool may produce other diagrams and visualizations that aid in the evaluation of air quality models. These include box plots, frequency histograms, and spatial maps.

In addition to the key components mentioned earlier, the DELTA Tool also includes the following functionality for quantitative model evaluation:

1. **Statistical Analysis:** various statistical analysis methods, including statistical metrics such as bias, error, and variability, as well as conducting hypothesis tests to determine the significance of model performance.
2. **Spatial Analysis:** spatial analysis of model results to facilitate comparison of model results to observed data at different monitoring locations. This helps identify spatial patterns and discrepancies in the model's performance, which can be crucial for understanding the accuracy of the model on a regional or local scale.
3. **Model Intercomparison:** this module facilitates the comparison of multiple air quality models by providing tools for intercomparison. This allows researchers and regulatory authorities to assess the relative strengths and weaknesses of different models and select the most appropriate one for a specific application.

Some examples of operational evaluation applications using the DELTA Tool include:

Example 1: Evaluation of Air Quality Models in Bulgaria

In a study conducted in Bulgaria, the DELTA Tool was used to evaluate the performance of an air quality model by comparing simulated and observed air pollution concentrations (Monteiro et al., 2018). The study focused on maximum daily 8-hour mean ozone concentrations, hourly nitrogen dioxide (NO_2) concentrations, and daily particulate matter (PM_{10}) concentrations The DELTA Tool provided summary statistics and performance indicators related to the model application, allowing for an objective assessment of the model's quality.

Example 2: Evaluation of EU Air Quality Standards in the East Mediterranean

Another study evaluated air quality modeling over the East Mediterranean and the island of Cyprus using the benchmarking methodology developed within FAIRMODE and implemented in the DELTA Tool (Kushta et al., 2019). The study analyzed model results for maximum daily 8-hour mean ozone concentrations, hourly nitrogen dioxide (NO_2) concentrations, and daily particulate matter (PM_{10}) concentrations. The DELTA Tool was used to assess the model performance and compare it to EU air quality standards. The study found that ozone was overestimated by the model, while NO_2 and PM_{10} were underestimated.

Example 3: Evaluation of Air Quality Models in the United States

In the United States, the Atmospheric Model Evaluation Tool (AMET) is used to evaluate air quality model output against observed data (https://www.epa.gov/measurements-modeling/learn-about-models-epa). While not specifically related to the DELTA Tool, AMET serves a similar purpose in assessing the performance of air quality models It relies on a combination of observation data, Fortran scripts, MySQL database, and R scripts to generate evaluation plots such as scatter plots, time series plots, spatial plots, and stacked bar plots.

8.3.1.2 Online operational evaluation tools

In the last couple of decades, Air Quality Management Systems (AQMS) have been deployed throughout Europe as tools to support air quality assessment and decision management. These systems typically cover urban or regional scales providing updated, near-real-time air quality information to authorities, decision-makers and citizens. AQMSs are further discussed in Section 8.4.5.

Typically, such a system comprises a static or dynamically updated emissions inventory, a meteorological model, a chemical dispersion model and postprocessing and visualization tools. In view of their online nature and given the availability of contemporaneous air quality observations in the areas where AQMSs operate, they provide an excellent opportunity to test a model's performance through the tools of operational evaluation. Many AQMSs therefore provide functionality to evaluate the accuracy of air quality forecasts and nowcasts by means of comparison to air quality and meteorological observations available in specific locations of interest. These comparisons can either be performed in near-real time or retroactively by comparing performance statistics over a past time period of a day or a week. Some

AQMSs include so-called "dashboard" interfaces which provide the end user access to visualization tools and performance metrics enabling a day-by-day overview of the model performance.

In Fig. 8.5, an example of such an operational evaluation interface is given. In this particular case, the mesoscale model MARS-aero (Moussiopoulos et al., 2012) is being applied as part of a real-time operational system to provide hourly nowcasting of air pollution as well as next-day forecasts of pollutant concentrations for five cities in Cyprus. The system implements an operational evaluation dashboard which automatically visualizes the model's performance in comparison to observed data of the previous day, at the same time providing performance metrics in terms of the absolute, normalized and fractional bias, absolute and normalized mean square error, correlation coefficient and index of agreement. For each of the statistical indicators, the system also highlights the best-performing module—either nowcasting of the previous day or forecasting from two days before.

8.3.2 Model intercomparison evaluation

Model intercomparison exercises collecting information on different models' performance under different conditions are a useful method of selecting the appropriate model that will provide more reliable results for a specific application. Model intercomparison is included in the model benchmarking guidance of FAIRMODE, recommending the ENSEMBLE system for this specific evaluation approach. FAIRMODE is an initiative that aims to ensure the quality and consistency of air quality models used across Europe.

Besides providing guidance and tools such as the DELTA tool, FAIRMODE also works towards:

- **Harmonization of Models:** By promoting the use of common methodologies and evaluation criteria, FAIRMODE aims to ensure that different models used across Europe are comparable and consistent.
- **Collaboration and Knowledge Sharing:** FAIRMODE organizes workshops, webinars, and other events that provide a platform for researchers, practitioners, and policymakers to exchange information, share experiences, and learn from each other.
- **Development of New Methodologies and Tools:** FAIRMODE also leads and participates in research projects that aim to develop new and improved methodologies and tools for air quality modeling.

Through these efforts, FAIRMODE contributes to the advancement of air quality modeling in Europe, making it a more effective tool for environmental management and policymaking. The ENSEMBLE system is a web-based platform allowing online model intercomparison and evaluation in the three spatial dimensions and time. ENSEMBLE hosts all the model results and available measurements relating to campaigns and case studies. As in operational

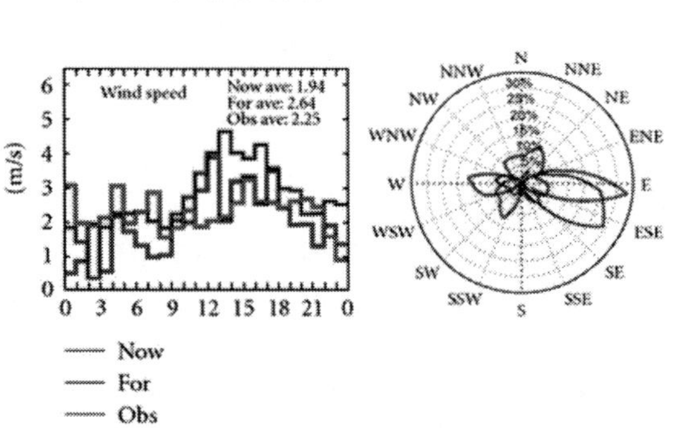

FIGURE 8.5 Operational evaluation dashboard of the MARS-aero model. *Adapted from Moussiopoulos, Nicolas, Douros, Ioannis, Tsegas, George, Kleanthous, Savvas & Chourdakis, Eleftherios. (2012). An air quality management system for policy support in Cyprus.* Advances in Meteorology, 2012, *1–8. https://doi.org/10.1155/2012/959280.*

evaluation, model-to-model comparison is based on several indicators that are predefined for the specific application and include statistics and graphs. As this evaluation method does not rely on the availability of measured data, model results are also compared to each other in application over areas without observations, allowing for a more complete analysis of model performance at a selected region. Useful indicators in model intercomparison exercises include species ratio (e.g., O_3/NO_y) which can provide useful information on chemical regimes. In addition, model-to-model comparisons of important meteorological parameters or boundary conditions for which no observations exist can be undertaken to analyze their influence on model output. In general, criteria and goals are fixed in relation to percentiles of model results, to avoid outlying model results from obscuring the analysis. An example suggested by Thunis and Clappier (2014) are indicators based on the relative potency "p" approach (i.e., a relative concentration change resulting from a relative emission change), defined for each of the simulated scenarios. Based on these relative potencies two indicators are defined to construct a comparative diagram for model performance, called the potency diagram. The first indicator, plotted on the x-axis of the diagram, represents the maximum potency which provides the maximum concentration change resulting from an emission reduction. The second indicator, plotted on the y-axis, provides information on the importance of a single precursor impact in terms of concentration change in relation to the others.

Indeed, model intercomparison evaluation stands out as the most appropriate method to better understand/flag out "strange" (not expected) model behaviors and exclude them from policy-relevant applications. In this frame, several continental and intercontinental scale model intercomparison exercises have been performed, coordinated by higher-order legislative bodies with the participation of different state-of-the-art CTMs for revealing strengths and weaknesses of the models and recommending the most appropriate modeling tools for specific regulatory needs. Examples include the Air Quality Model Evaluation International Initiative (AQMEII) jointly launched by US EPA and EU JRC, the Model Inter-Comparison Study for Asia (MICS-Asia), related COST Actions and the EU Commission supervised CityDelta and EuroDelta model intercomparison exercises, described in the "Air Quality Model Applications" section of the current chapter.

8.3.3 Model sensitivity analysis

Sensitivity analysis is useful to address a variety of factors influencing the reliability of model results, by analyzing the response of the model to the change of individual parameters, from variations in input data related to different application conditions (e.g., relating to meteorological and atmospheric conditions, emission sources, pollution levels, different application time periods or areas of different topography features) to different model parameterizations or set up conditions (hardware components or nesting and grid size variations). Sensitivity analysis can help in uncovering technical errors in the model and identify critical issues regarding input data by comparing the different sensitivity simulation results against measurements. In general, sensitivity analysis relies on describing the change in model output due to variations in model input and setup parameters. A large number of numerical approaches are used in air quality model sensitivity studies, but in general, they belong to the four following categories. A detailed technical analysis of these approaches is beyond the scope of this chapter.

8.3.3.1 Partial derivatives methods

Partial derivatives are sensitivity coefficients of the model output with respect to the input parameters, useful in assessing the reliability of the output from a complex model with many uncertainty parameters. The frequently used methods for applying sensitivity analysis to CTMs, the Brute-Force Method (BFM) and the Direct Decoupling (DDM) method are examples of this category. The BFM represents one of the simplest and most widely used methods to evaluate the sensitivity of an atmospheric dispersion model to uncertainty sources. As a first step in the procedure, a reference simulation is performed including all sources, while as a second step, simulations excluding some of these sources (or a typology of sources) are undertaken. The difference between the results of the two simulations estimates the contribution of the sources that have been removed. As the BFM can be computationally expensive, a suitable approach suggested to reduce the computational costs is to run a single simulation with tagged species. Tagging the emitted species allows them to be tracked and their origin to be identified. The DDM belongs to direct sensitivity techniques, which are based on obtaining a system of sensitivity equations from the model equations and on solving both alongside each other. The decoupled method incorporates the sensitivity equations decoupled from the model equations, enhancing the stability and efficiency of the method as compared to the coupled method. First-order coefficients, or slopes of the response curve, provide valuable information in linear ranges of response. In cases of nonlinearity, which is particularly relevant for photochemical dispersion models (e.g., in transition from NO_x-inhibited to NO_x-limited photochemical scenario) or for aerosol formation modules, the high-order direct decoupled method is used for calculation of higher-order sensitivity coefficients, to include cross sensitivities (partial derivatives regarding more than one sensitivity parameter) in the analysis.

8.3.3.2 Regression analysis

Within the frame of sensitivity studies, regression analysis is based on fitting a linear regression to the model response and using standardized regression coefficients as direct measures of sensitivity. As a result, this approach is appropriate in the case of a linear model response to the selected uncertainty parameter; the linearity hypothesis can be verified by a high value of the determination coefficient. The main advantages of regression analysis methods are their simplicity and the low computational cost, but the fitting regression can be far from the initial model if the relationship is highly nonlinear.

8.3.3.3 Variance decomposition methods

Several sensitivity analysis methods relying on the investigation of the contribution of a single (or group of) input variance to the output variance are applied for atmospheric model evaluation purposes, most of which are computationally expensive due to the variance estimation process. The most detailed methods are based on the estimation of sensitivity indexes of first and second (or higher) order and of the total effect. First-order sensitivity indexes represent the pure effect of each input factor on the variance of the output, while second (or higher) order indexes correspond to the effect of the interaction between two (or more) factor (s) on the variance of the model output. As expected, the total effect index describes the overall effect of the input change (considering also all the interactions) on the variance of the output. In most cases, first-order effect and total effect are initially computed, so that the individual (the "pure") and the total impact of each factor can be estimated. In evaluating modeling of secondary pollutants (e.g., ozone or PM_{10} formation), higher-order indexes can be computed. An example is the evaluation of modeling the combined effect of NO_x-VOC in ozone formation/accumulation or the link between inorganic gas precursors for PM_{10}.

8.3.3.4 Elementary effects

The elementary effect sensitivity approach is a screening method aiming to highlight a limited number of the most influential input factors among the many affecting the model output. This is achieved by defining two sensitivity indexes to identify which input can have negligible, linear and additive or nonlinear effects on the output through the computation of the so-called elementary effects. For this purpose, the elementary effect is estimated for a subset of input values obtained by randomly sampling the input hypercube, so that a distribution function for each of the input factors is obtained. The two sensitivity indexes are then respectively the estimation of the mean and the standard deviation for each distribution. The mean values the overall influence of the factor on the output, while the standard deviation represents nonlinearities and interactions with the other input.

Irrespective of the sensitivity method applied, the stages shown in Fig. 8.6 are followed in designing a sensitivity study for evaluating air quality models.

In some cases, a two-stage approach can be implemented, where the first sensitivity iteration identifies the most important uncertainty sources to be better investigated in the following iteration, which may also apply a different sensitivity analysis method from the first stage.

8.3.4 Diagnostic model evaluation

Diagnostic evaluation is considered an in-depth evaluation focusing on identifying the cause of the model errors or biases. In this frame, it is useful to design the diagnostic evaluation procedure by considering specific questions that are more relevant to specific pollutants, as suggested by the UK Environment Department (DEFRA, 2010). In the case of photochemical models' ozone predictions, diagnostic evaluation is based on "indicator species ratios," employed to characterize whether ozone formation is VOC- or NO_x -limited (or in between), thus responding more strongly to reductions in VOC or NO_x emissions, respectively. In this way, comparing model sensitivities to VOC or NO_x emission reductions with our theoretical expectations allows for evaluating model performance. A map can then be produced demonstrating the locations in the studied domain where the model response to the specific VOC reduction is greater than that to the same NO_x reduction (VOC-limited) and vice versa (NO_x -limited), providing valuable information to policymakers.

Diagnostic evaluation of the ability of dispersion models to describe urban NO_2 concentration patterns focuses on their capability to represent the effect of pollutant emissions on the partitioning between the oxidant species. For this purpose, appropriate graphs (e.g., scatter plots) of modeled mean NO, NO_2, O_3, and O_x concentration levels plotted against mean NO_x concentrations for different locations (including rural, suburban, and urban background, traffic) of the studied domain and years, and in comparison to available measurements, can be produced and studied.

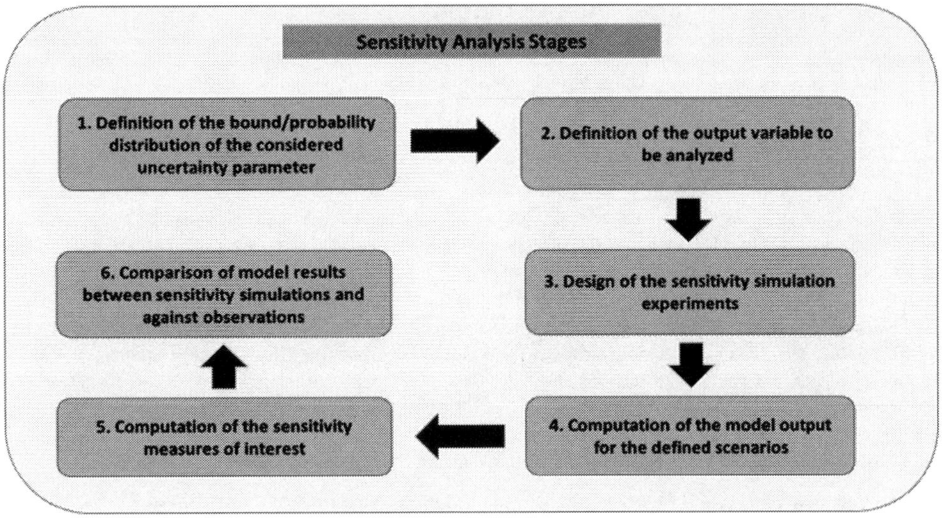

FIGURE 8.6 Main stages of conducting sensitivity analysis in air quality modeling applications.

Diagnostic evaluation of urban PM modeling addresses the models' ability to adequately represent the effect of emission changes on $PM_{2.5}$ and PM_{10} concentrations. The model is considered of acceptable predictive quality if it can indicate whether PM concentrations respond more strongly to reductions in NH_3 emissions (ammonia-limited) or to SO_2 and NO_x (strong acid-limited) emissions reductions.

In addition to the above model evaluation methods, a *retrospective analysis* is often used in assessment applications to examine the ability of an air quality model to respond to emission changes and describe dispersion patterns and concentration trends by comparing recent trends in observed pollutant concentrations to the model-predicted trends over the same period. In this frame, the model simulations may be performed for current episodes and episodes in one or more historical periods using the relevant emissions and meteorological data.

Worked example 8.3.1

Briefly describe the four main air quality model evaluation types. Which is the most suitable approach for (a) a consultant who has to support the use of a model tool for management purposes to local regulators; (b) a scientific team examining the behavior and reliability of an atmospheric model.

Solution

The four different model evaluation methods are the following: (1) **operational model evaluation**, which is the validation of model results through the comparison with available measurements. In this approach, the uncertainty for objective model estimation is calculated on the basis of the deviation between measured and modeled concentration levels; (2) **probabilistic model evaluation**, which mainly includes a comparison of results between different models, the so-called model intercomparison exercises, along with ensemble modeling. Model intercomparison applications are particularly useful to assess the reliability of model results if monitoring data is not available for comparison, such as in areas where monitoring stations have not been installed or for future predictions; (3) **dynamic model evaluation**, mainly including sensitivity analysis to changes in model software and hardware features or input data, in addition to the function of this methodology to research specific research-related air quality issues; (4) **diagnostic model evaluation**, focusing on research and models advancing questions, aiming to shed light on the reasons for model behavior by attempting to explain model results and responses on the basis of their physical and chemical parameterizations. Dispersion models are the main model type evaluated using this approach under different application conditions, however it is appropriate for all model types in the stage of their development or updating process.

The consultant will use operational model evaluation to demonstrate the model's ability to realistically simulate air quality, and model intercomparison with widely used and validated models may be additionally applied for areas/periods for which measurement data is not available. The scientific team will assess the model's potential strengths and weaknesses by applying ideally all four evaluation methods, in order to interpret the results of the dynamic and diagnostic model evaluation.

8.4 Areas of application of air quality models

Atmospheric model applications are performed in many different cases, but they serve three main purposes: **(1) air quality research, (2) management and (3) model evaluation** (Fig. 8.7). Research and model evaluation exercises are undertaken by atmospheric scientists, while air quality model simulations for management needs may be applied by scientists but also by policymakers.

In Europe, the current Air Quality Directive 2008/50/EC on ambient air quality and cleaner air for Europe (http://data.europa.eu/eli/dir/2008/50/2015-09-18) recognizes that models are scientifically appropriate tools to support all management and planning decisions facilitating the improvement of the atmospheric environment in both urban and rural areas. The US Environmental Protection Agency (US EPA) also recommends the use of models for decision-making and encourages their use in science policy through the development of technical support facilities and guidance documents (https://www.epa.gov/cmaq/atmospheric-model-evaluation-tool). The National Academy of Science report on regulatory models describes the different categories of regulatory air quality models used by the EPA (National Academy of Sciences, 2007), including emissions models, fate and transport models and exposure estimation models. Similarly to Europe, the US EPA recommended models are applied for current assessment and future prediction needs in relation to the establishment and attainment of National Ambient Air Quality Standards (NAAQS) and to the development of mitigation planning (State Implementation Plans). In this frame, the use of models is recommended for all the following management needs, in combination with measurements, where available:

1. **Air Quality Assessment:** Assessment of the current pollution state for regulatory purposes focuses on assessing compliance with the limit values and is specifically foreseen in the case of exceedance, in combination with measurements, to provide adequate information on the spatial distribution of the ambient air quality. Modeling may be used as a stand-alone assessment method in areas where anthropogenic emissions are limited, and pollutants are below the lower assessment threshold.
2. **Air Quality Prediction:** Air quality measurements can only provide information on the current pollutant levels, while models can simulate future dispersion, transport and fate of pollutants to produce future pollutant estimations. Future predictions are useful in air quality management for issuing alerts and informing the public and policymakers on impaired or accidental air quality situations so that both individual behavior patterns and strategic abatement measures can be timely adopted to reduce pollutant risks and costs. Air quality model forecasts usually disseminate predictions in the form of AQI, through dedicated websites which are easily and freely accessible to the public.

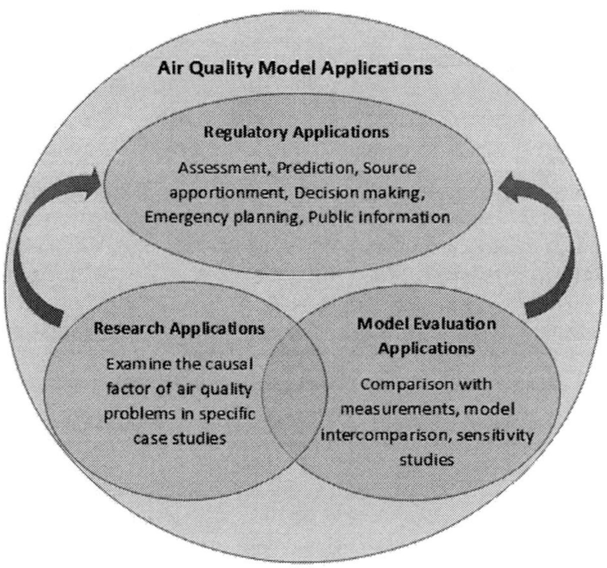

FIGURE 8.7 Areas of application of air quality models, including regulatory applications, research applications, and model evaluation applications.

3. **Source Apportionment:** The ability of models to follow the trajectories of emitted pollutants allows for the identification and quantification of the relative contributions of pollution sources. This information enables local authorities to define appropriate emission limits for new and existing sources in order to ensure that legislative limit values are attained.
4. **Strategic Decision-Making:** Model use for assessment, prediction and source apportionment provides the scientific background to relevant authorities to suggest appropriate air pollution abatement measures. Policymakers consult model results to investigate the efficiency of alternative pollution control strategies and air quality plans.
5. **Episodes and emergency planning:** Timely and accurate prediction of air pollution episodes and emergency situations (e.g., dispersion of pollutants from a wildfire or accidental releases from industry facilities) by appropriate models allows authorities to apply effective mitigation measures for minimizing health and socio-economic impacts.
6. **Dissemination of information to the public:** The development of AQMSs and their application by relevant authorities is an essential tool for air quality management as it incorporates both assessment of current status in selected locations, air quality alerts, as well as scenario analysis for forecasting purposes.

Models are indispensable tools in atmospheric research, as they combine current scientific knowledge on individual physical and chemical phenomena to describe dynamic processes and complex chemical reactions influencing the dispersion and fate of air pollutants. In this way, model results can provide a complete picture of air quality through the characterization of temporal and spatial dispersion patterns for the entire study domain and period, in contrast to measurements which are locally and temporally confined. Also, numerical modeling draws information regarding meteorology, pollutant emissions, background pollution levels and terrain/land use conditions in the form of input data, which allows simulations to reveal the causal relationships between all the above factors and air quality. The most characteristic example of research applications to examine the relative importance of different processes in an air quality problem is sensitivity analysis showing how the model responds to changes in input data and parameterizations. Sensitivity analysis is also an integral part of dynamic model evaluation, assessing the model's ability to capture observed changes in air quality.

Regarding model evaluation, relevant model applications aim to examine the model performance under different conditions (e.g., spatial/temporal scales, different terrain/land use, different meteorological conditions, different pollutants) and, as a next step, identify and explain any problems or inconsistencies observed in model results. In the relevant chapters of the present manual, the sources of model uncertainties (theoretical background, parameterizations, input data) are thoroughly described. Considering these uncertainties, the results of experimental model simulations can highlight the underlying reasons for poor model performance and facilitate model improvements. Model evaluation applications target all four model evaluation types described in Section 8.3.

In all the above cases, model simulations are ideally performed for different seasons and appropriate time periods, so that hourly, daily, seasonal and annual patterns can be detected. Model evaluation also focuses on the spatial distribution of concentrations and, particularly, model results for hot spot areas, to identify local-scale parameters influencing model performance. It is important to remember the connection between model evaluation applications and management applications, as model evaluation reveals the advantages/disadvantages and thus the suitability of individual models for management purposes under different conditions.

8.4.1 Air quality model application initiatives and support sources

As the importance of model applications for air quality management is increasingly acknowledged by policymakers and the scientific community, several coordinated efforts worldwide (mainly in Europe and the USA) have emerged to harmonize the use of models and provide recommendations for best practices. Relevant legislation, i.e. the Air Quality Directive in Europe and the US Clean Air Act, supports the use of air quality models for regulatory needs on the basis of common methods and evaluation criteria.

The most well-known and established networks and initiatives are presented in Table 8.2.

Other air quality model initiatives were established focusing on the coordination of model intercomparison exercises. The AQMEII was launched in 2008 by the US EPA, Environment and Climate Change Canada, and the European Commission Joint Research Center (JRC), to promote a common regional-scale air quality model evaluation framework through the collaboration of the European and North American modeling communities (Rao et al., 2011). The three first phases of AQMEII have been successfully completed and the fourth phase which started in 2019 is in

TABLE 8.2 Air quality model application initiatives and support sources.

Name	Description
FAIRMODE	The Forum for Air quality Modeling (FAIRMODE) was established in 2007 as a joint initiative of the European Environment Agency (EEA) and the European Commission Joint Research Center (JRC) to support the harmonized model application by EU Member States for the purposes of the Air Quality Directive. The forum is still active (https://fairmode.jrc.ec.europa.eu/) and provides guidance through the organization of exercises, workshops and conferences on modeling issues such as source apportionment, QA/QA for assessment and forecasting and high-resolution input data inventories. Valuable tools for modeling within FAIRMODE are freely available through its website, such as the DELTA model evaluation software and the SHERPA tool for scenario analysis to support air quality planning.
Initiative on Model Harmonization	In 1991 a European initiative was established by renowned research and higher education bodies for increased cooperation and standardization of atmospheric dispersion models for regulatory purposes. Its main activity is to organize regular international conferences which have become recognized meeting place for model developers, model users and stakeholders (https://www.harmo.org/).
COST Actions	The European Cooperation in Science and Technology (COST) is an EU funded intergovernmental framework promoting the creation of research networks, called COST Actions. COST Actions bring researchers and innovators together to investigate a science topic of their choice for 4 years, thus helping to advance knowledge and strengthen the research and development sector. A number of COST Actions focused on air quality modeling including Actions 710,715, 728, 732, and ES 1004 and ES1006.
CREM	Officially established by the EPA in 2000, the Council for Regulatory Environmental Modeling (CREM) was the first agency program to comprehensively address and provide guidance on environmental regulatory modeling issues. Its main accomplishment was the guidance document on "Development, Evaluation and Application of Environmental Models" (CREM, 2008). https://digitalcommons.mtu.edu/cgi/viewcontent.cgi?referer = &httpsredir = 1&article = 1290&context = etds. The organization is no longer active, and its role has been taken over by SCRAM.
SCRAM	The EPA's Support Center for Regulatory Atmospheric Modeling (SCRAM) is a resource for atmospheric model users, providing model information and guidance documentation, as well as model codes, input data, and model processors.
APSI	The Air Pollution Scientific Initiative (APSI) was launched in May 2020 as part of the activities of the nonprofit EnviroComp Institute. The main goal of APSI is the creation of an evolving, multilingual repository of learning guides in air pollution sciences. The guides are organized into a MATRIX containing state-of-the-science information on various air pollution related topics, targeting the needs of users with different knowledge levels. The MATRIX information is freely available at https://www.apsi.tech/.

progress (Table 8.3). The unprecedented scale of this exercise (1 year, two continents) allows for revealing the general characteristics of models' skill and uncertainty. The exercises placed emphasis on photochemical modeling for simulating ozone distribution and transboundary pollution and on PM simulations. To facilitate the comparison of model results, common inventories were developed and used by all participating groups for baseline emissions and lateral air quality boundary conditions and common analysis grids have been defined.

The intercomparison and operational model evaluation of Phase 1 revealed that the main challenges for meteorological models providing input data for air quality simulations are nighttime meteorology influencing calculations of primary pollutants and aerosols, and shortwave radiation effect on cloud processes as the most critical process controlling model accuracy of ozone build-up (Vautard et al., 2012).

CityDelta (February 2001 to October 2004) and Eurodelta model intercomparison exercises were organized by the Joint Research Center (European Commission) in response to the needs of the Clean Air for Europe Program of the DG Environment of the European Commission.

A large number of modeling groups (over 20 teams) from research centers all over Europe have participated in the two projects, using different model configurations to explore the efficiency of different emission scenarios in 2010 regarding ozone and PM air quality improvement. The CityDelta exercise focused on the influence of reduction scenarios in urban air pollutant emissions on ozone and PM concentrations (called the "Deltas") in eight selected European

TABLE 8.3 AQMEII phases, focus, and models used in each phase.

Phase	Focus	Met models—air quality models
1 2010–12	Chemical transport models were applied for the entire 2006 for simulating ozone and PM dispersion patterns and evaluated based on the comprehensive model evaluation framework presented by Dennis et al. (2010), but most of the contributions focused on operational and probabilistic evaluation. A special Issue in Atmospheric Environment (2012) on publications from this phase entitled "AQMEII: An International Initiative for the Evaluation of Regional-Scale Air Quality Models - Phase 1" can be found at (https://www.sciencedirect.com/journal/atmospheric-environment/vol/53).	WRF, MM5– Chimere MM5-Polyphemus MM5-CAMx WRF, MM5-Muscat PARLAM_PS-EMEP COSMO-Muscat ECMWF-SILAM MM5-DEHM WRF-CMAQ ECMWF-LOTOS_EUROS GEM-AURAMS COSMO_CLM-CMAQ WRF-WRF/Chem LOTOS-EUROS
2 2012–14	On-line coupled or integrated chemistry transport models and meteorological models were applied for 2006 and 2010 and evaluated. Online coupled meteorology-chemistry modeling systems attempt to close feedback loops between atmospheric dynamics and composition. A special Issue in Atmospheric Environment (2015) on publications from this phase entitled "Evaluating Coupled Models (AQMEII P2)" can be found at (https://www.sciencedirect.com/journal/atmospheric-environment/vol/115/).	GEM-MACH WRF-Chem NCSU WRF-Chem WRF-CMAQ MetUM-UKCA RAQ RACMO LOTOS-EUROS NMMB-BSC-CTM COSMO-MUSCAT COSMO-ART
3 2014–18	Linked global-regional scale modeling systems were applied to study long-range transport by simulating relevant processes at hemispheric-to-regional scales. AQMEII network performed the intercomparison applications in partnership with the Task Force on Hemispheric Transport of Air Pollution (HTAP) and the Model Intercomparison Study for Asia (MICS-Asia). Results were published in the journal Atmospheric Chemistry and Physics in a special issue entitled "Global and regional assessment of intercontinental transport of air pollution: results from HTAP, AQMEII and MICS."	ECMWF-SILAM ECMWF-LOTOS EUROS ECMWF-CHIMERE WRF-WRF/Chem WRF-CAMx WRF-DEHM WRF-CMAQ CCLM-CMAQ ECMWF and NCEP were used as global meteorological data providers.
4 2019–	The model intercomparison exercise of phase 4 focuses on the description of pollutant wet and dry deposition processes. (Galmarini et al., 2021).	COSMO-Muscat GEM-MACH WRF-Chem LOTOS-EUROS WRF-CMAQ

cities (Berlin, Katowice, London, Milan, Paris, Prague, Copenhagen and Marseille, shown in Fig. 8.8). For this purpose, 22 different configurations of 3D Eulerian chemistry-transport models (CTMs) were applied, using the same emissions inventory for the baseline year 2000 and meteorological data from different sources for the year 1999. The first stage of the model intercomparison exercise was the validation of model results through the comparison of observations from the monitoring networks operated by local air quality organizations of the participating cities.

In CityDelta model evaluation process, particular emphasis was placed on the use of Taylor plots, which allow for the visualization of three statistical indicators in a single plot. The main scope of this stage was to detect models with insufficient skills for predicting future air quality under the different emission scenario simulations. The sensitivity of model results to the spatial resolution (regional-scale simulations with resolution of 50km, as well as city-scale simulations, with a typical resolution of 5–10 km) was also examined. The second stage of the CityDelta methodology was to study the different predicted changes of pollutant concentrations (the "Deltas") in response to the hypothetical adoption of the different emission reduction measures (scenarios). The ensemble mean approach for the individual Deltas was followed to address the uncertainty of the different model results and was proposed as the most representative response to the different emission scenarios

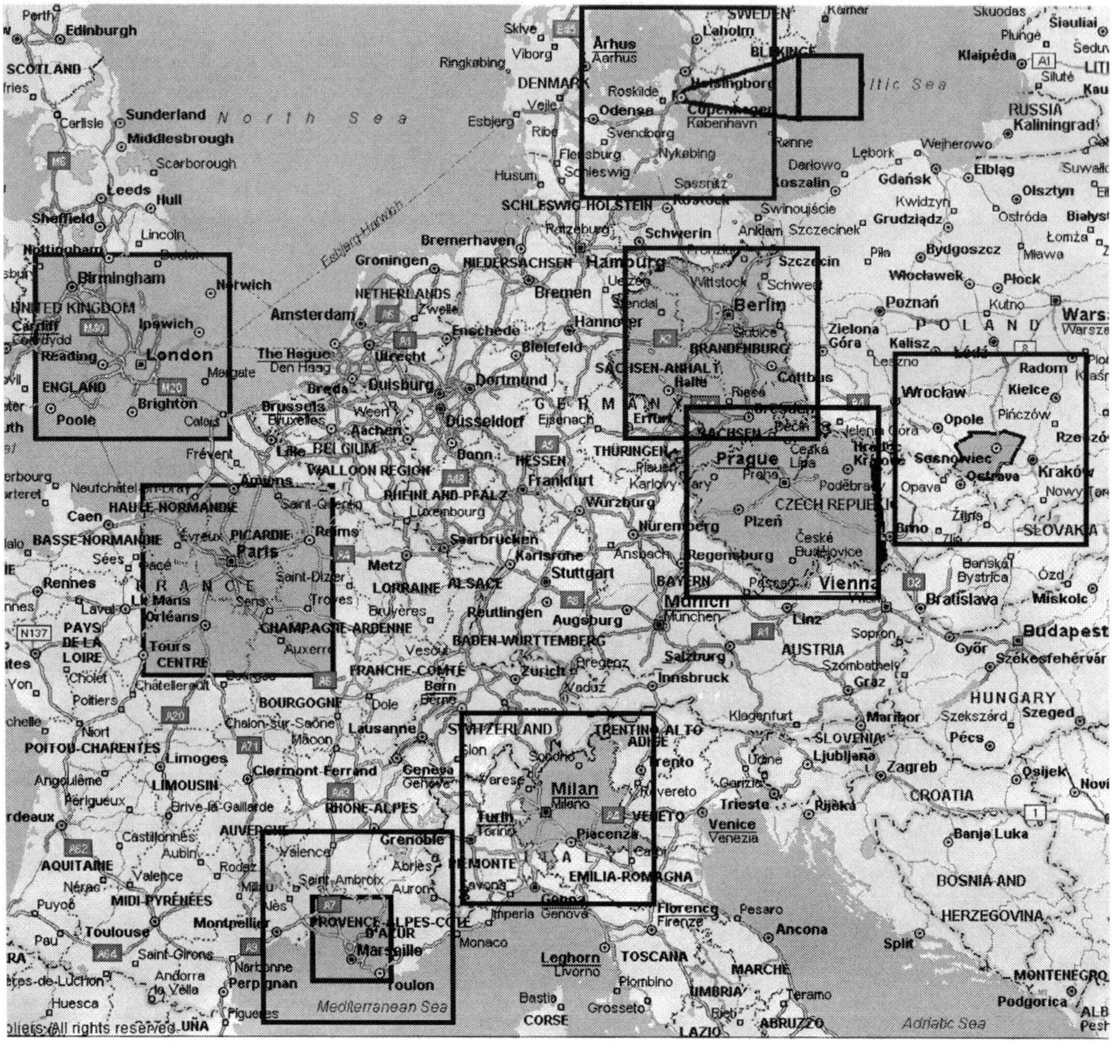

FIGURE 8.8 Six selected CityDelta exercise cities. Spatial coverage of the high-resolution emission inventory (green shaded area) of six selected CityDelta exercise cities, within the computational domain of $300 \times 300\ km^2$ indicated by the solid squares around each city. *From Cuvelier, C., Thunis, P., Vautard, R., Amann, M., Bessagnet, B., Bedogni, M., Berkowicz, R., Brandt, J., Brocheton, F., Builtjes, P., Carnavale, C., Coppalle, A., Denby, B., Douros, J., Graf, A., Hellmuth, O., Hodzic, A., Honoré, C., Jonson, J., … Zuber, A. (2007). CityDelta: A model intercomparison study to explore the impact of emission reductions in European cities in 2010.* Atmospheric Environment, 41*(1), 189–207. https://doi.org/10.1016/j.atmosenv.2006.07.036.*

(Cuvelier et al., 2007). A common visualization framework of the model results was used, comprised of different plot types to indicate inter-model variability in results and differences with measured values (Fig. 8.9).

The scientific interest raised in response to the CityDelta experiment led to the follow-up EuroDelta model intercomparison focusing on the evaluation of air pollution mitigation strategies at the regional scale. EuroDelta was initiated by the Task Force on Measurements and Monitoring (UNECE_TFMM) and the Task Force on Integrated Assessment Modeling (UNECE_TFIAM). The first EuroDelta model intercomparison exercise (EuroDelta I, 2004) focused on the performance of regional dispersion models in predicting 2000 and 2020 air quality in Europe, using the concept of a model ensemble as a measure of uncertainty for individual models and strengthening the robustness of predictions. EURODELTA II (2006) examined the capability of the different models to simulate the changes in pollutant levels (ozone, OM, and NO_x) on a European scale resulting from emission reductions to individual emission sectors.

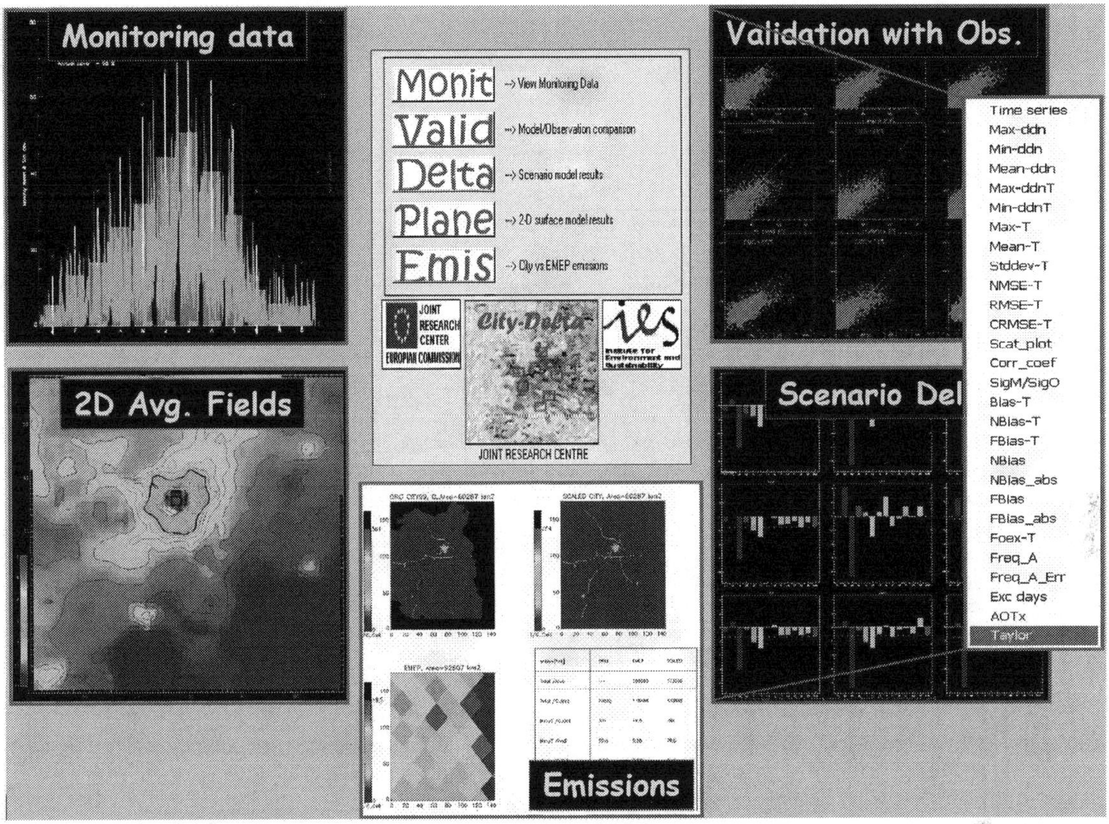

FIGURE 8.9 Visualization tool in CityDelta model intercomparison exercise. From Cuvelier, C., & Thunis, P. (n.d.). CityDelta: Objectives, methodology, Results. Retrieved March 23, 2024, from https://www.pecj.or.jp/japanese/jcap/jcap2/pdf/4th/1_5.pdf.

EuroDelta III exercise contributed to the assessment of the Gothenburg Protocol to Abate Acidification, Eutrophication and Ground-level Ozone by examining the modeled changes in pollutant concentrations in response to the different emission reduction policies adopted in the previous two decades. In EuroDelta III particular emphasis was given to regional modeling of carbonaceous aerosol. For model evaluation in EuroDelta, observational data were derived from intensive measurement campaigns undertaken within the European Monitoring and Evaluation Programme (EMEP) and the European Integrated Project on Aerosol Cloud Climate and Air Quality Interactions (EUCAARI).

Both in the CityDelta and the EuroDelta model intercomparison applications, the purpose was threefold, namely (1) model performance evaluation on the basis of predefined indicators and common evaluation tools, (2) support on the decision for adopting future emission reduction policies through the prediction of related changes in air quality levels, and (3) gaining scientific information on the capability of regional atmospheric dispersion models to reproduce chemical processes in response to emission changes.

8.4.2 Assessment applications and challenges

Air quality assessment at the regional scale but, most importantly in urban agglomerations, due to the higher number of people exposed to air pollution, is a necessary prerequirement for management purposes, aiming to improve air quality and reduce induced health problems in most affected areas. Therefore, model assessment applications are mainly directed towards producing maps for describing pollutant dispersion patterns in space and time to provide a comprehensive evaluation of pollution levels in comparison to legislation values for the examined domain. The model output provides indispensable information on air quality in areas not covered by monitoring stations but also reveals areas of particular interest, the so-called hot spots, or conditions favoring air pollution episodes. The realistic implementation of **scale interaction processes** is of particular importance in air quality assessment simulations, particularly for urban areas with complex topography features (in the vicinity of sea or mountains), where mesoscale circulations strongly influence dispersion at smaller scales. In view of this requirement, mesoscale Eulerian dispersion models are most frequently used in urban assessment studies, using appropriate methodologies to address scale interaction processes. Such methods include mainly (1) coupling with a smaller scale or CFD model for simulations zooming into hot spot areas or in street-canyons and (2) applying model nesting over the examined region, using different grid resolutions for each nest. In the case of nesting, larger-scale processes in the model are numerically described by introducing an additional simulation for a larger domain, and spatial resolution is increased only in the domains where small-scale phenomena might occur that affect the forcing mechanisms within the study area. Also, a non-uniform model grid structure can be applied, by including fine-resolution scales within a coarse grid mesh to effectively capture atmospheric processes occurring in dense source regions of the urban area. The general idea is to place the finest grids over the target area of interest and over the adjacent areas expected to influence the air quality in that region, while coarser grid cells cover areas not considered to significantly contribute to the air quality in the target area with the coarsest cells typically near the boundaries of the domain. The variable grid modeling approach is particularly recommended in air quality studies, where the complex nature of pollutant species' interaction with the prevailing meteorological conditions can result in high concentration levels at a significant distance from the emission source. From a policy perspective, risk assessment due to air pollution exposure can be obtained at the regional or local scale using different grid resolution levels. As a final thought, it has been concluded by several studies examining the influence of grid resolution on model performance that increasing the resolution level may increase the uncertainty level, however, a moderate resolution improvement can improve the reliability of model results.

As an example, the EPISODE CTM has been developed for air quality assessment and urban air quality management to simulate the emission, photochemistry, and transport of pollutants in urban areas. The structure of the EPISODE model was designed to address the weaknesses of regional, mesoscale and microscale models in simulating pollution dispersion in urban areas at scales relevant to exposure mapping and assessment. EPISODE is a 3D Eulerian grid CTM incorporating several subgrid-scale processes (i.e., emissions represented as line sources and point sources) addressed by the means of Gaussian dispersion, thus allowing for the estimation of concentrations at the subgrid scale in locations specified by the user or policy maker. EPISODE is typically run at 1 km $\times$ 1 km resolution over an entire city with domains covering up to approximately 1000 km^2 in size. Because of these features, EPISODE combines regional and microscale abilities, allowing for simulating pollutant dispersion both at the urban scale and microscale at the same time. EPISODE's typical model resolution, scale of representation (i.e., down to tens of meters), size of domain (i.e., city scale), level of detail of its subgrid-scale transport processes (i.e., Gaussian dispersion), and receptor point sampling, facilitate the realistic representation of regional-scale physical and chemical atmospheric processes and explicit simulation of mean flow and turbulent dispersion due to microscale urban surface characteristics. The model consists of an Eulerian grid model and two subgrid Gaussian dispersion models providing receptor point calculations of pollutant levels (Fig. 8.10).

The model and its updated CityChem version have been applied and evaluated for episodic and nonepisodic (seasonal) periods for air quality assessment over the city of Hamburg and several cities in Norway, demonstrating reliable results, particularly during autumn and winter periods, as a result of its photochemical mechanism. The model was therefore found to be suitable for episode analysis, revealing air quality trends and air quality policy and management purposes in areas of less intense solar radiation.

Another approach to resolving microscale urban features in urban dispersion modeling is by applying data assimilation to "tune" the model results to the measured "real" conditions. A relevant example is the operational multiscale air quality modeling system called ENFUSER (Environmental information fusion service), used in the Helsinki Metropolitan area in Finland, as described by Johansson et al. (2022). The modeling system combines Gaussian plume- and puff modeling techniques with data assimilation which results in a set of hourly adjustments helping the model to achieve a higher degree of agreement at the measurement locations. This combination allows for performing high-resolution simulations and providing hot spot information on air pollutant levels, at a moderate computational cost.

8.4.3 Source apportionment applications and challenges

Several air quality management objectives may require the assessment of source contribution at regional or urban scales, such as determining compliance with current or future legislation limits in the vicinity of a major source, impact assessment for new industrial developments and emergency planning, for which the realistic prediction of the impact caused by a localized emission source is of particular importance. Also, in epidemiological studies, it is important to

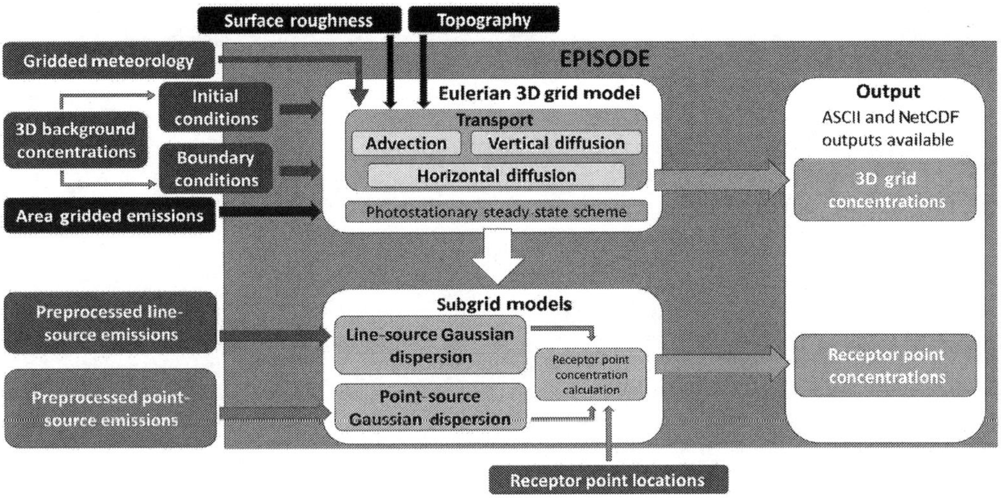

FIGURE 8.10 The structure of the EPISODE model (Hamer et al., 2020). *From Hamer, P.D., Walker, S.E., Sousa-Santos, G., Vogt, M., Vo-Thanh, D., Lopez-Aparicio, S., Schneider, P., Ramacher, M.O.P., & Karl, M. (2020). The urban dispersion model EPISODE v10.0—Part 1: An Eulerian and sub-grid-scale air quality model and its application in Nordic winter conditions.* Geoscientific Model Development, 13*(9), 4323–4353. https://doi.org/10.5194/gmd-13-4323-2020.*

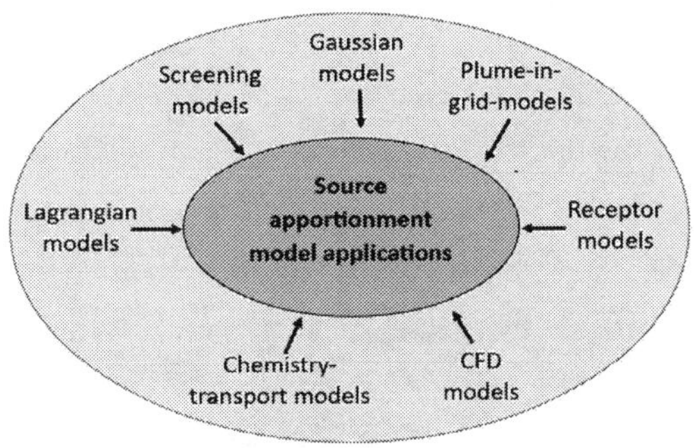

FIGURE 8.11 Air quality model types used in source apportionment applications.

calculate population exposure to air pollutants emitted from a particular source and the subsequent damage to health. In urban areas, both compliance and exposure calculations require spatial resolutions ranging from meters, to assess hot-spots, to kilometers, for an overall spatial assessment. Large spatial resolution simulations are used for biogenic sources or for investigating the relative source contribution of long-range transported ozone. Scale interaction modeling methods are particularly relevant in source apportionment modeling applications and correspond to the scale impact of the source. For example, local scale interactions and high grid resolution are not of great importance for wood-burning and soil-dust contributions which are spatially representative over larger domains. On the contrary, power plant sources exhibit local impacts, in which case interactions between the mesoscale and local scales should also be considered in the simulation.

Different model types are employed for source contribution applications (Fig. 8.11), depending on the aim of the study as well as on the source and pollutant characteristics. The theoretical approach and characteristics of the different model types are described in detail in Section 8.1 of the present chapter. Receptor models are preferred when measured chemical composition data are available and for smaller scales, while Eulerian CTMs are appropriate for source apportionment of transboundary pollutants and for secondary pollutants which require adequate representation of the relevant chemical reactions in the atmosphere (Fig. 8.11).

However, all model types included fall under **two principally different source apportionment models approaches: (1) the receptor models** correlating the measured mass of an atmospheric pollutant at a given site to its emission sources and **(2) the source-oriented models** based on emissions sensitivity analyses performed with different types of air quality models (Gaussian, Lagrangian, or Eulerian chemistry–transport models). Source apportionment at smaller scales (urban or local scales) is frequently undertaken using receptor models, such as the US EPA recommended "Positive Matrix Factorization (PMF) Model for Environmental Data Analyses" receptor model (https://www.epa.gov/air-research/positive-matrix-factorization-model-environmental-data-analyses). In many cases, these methods are complementary to each other, and the implementation of a combination of different methods is used to decrease the uncertainties inherent to individual model types. In the last decade, scientific and management interest in source apportionment has been directed to particulate matter pollution, due to the increasing evidence on health impacts and the complexities characterizing their transport and transformation processes. For example, the combined application of the CAPITA Monte Carlo Lagrangian CTM and a PMF receptor model has been used to quantify the contribution of wildfires to carbonaceous aerosol levels at Western US rural protected areas. The hybrid system was able to separately identify major contributions of total carbon from wildfires and minor contributions from biogenic sources. The predictions demonstrated lower statistical error compared to measurements than those from either PMF or CTM alone. CTM simulations are particularly valuable for source apportionment at the regional scale, focusing on the urban areas of the examined domain, as scale interaction processes are intrinsically considered using different grid resolutions in nested application mode. Coupled mesoscale WRF-Chem simulations were conducted in the Yangtze River Delta (YRD) region of eastern China. A series of emission sensitivity simulations were performed to quantify the contributions from individual source sectors and from different regions to $PM_{2.5}$ levels in the studied region. Model results revealed the high local source contribution (of 64%) to the regional $PM_{2.5}$ concentrations. Among the local sources, industry activity was the most significant sector in spring (25%), summer (36%) and fall (33%), while residential sources were more important in winter (38%) (Fig. 8.12).

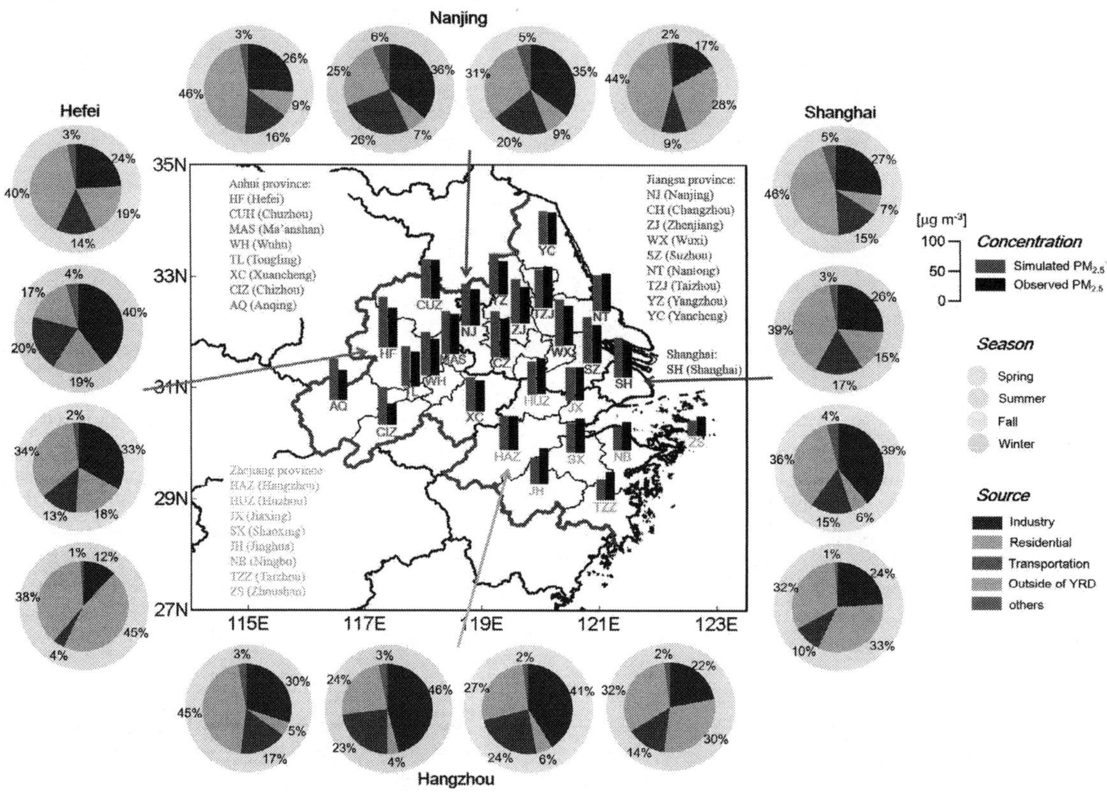

FIGURE 8.12 *$PM_{2.5}$* source apportionment modeled results from WRF-Chem for the Yangtze River Delta (YRD) region of eastern China (Li et al., 2018). *From Li, Nan, Lu, Yilei, Liao, Hong, He, Qingyang, Li, Jingyi, Long, Xin & Ko, Fung-Chi. (2018). WRF-Chem modeling of particulate matter in the Yangtze River Delta region: Source apportionment and its sensitivity to emission changes.* PLOS ONE, 13*(12). https://doi.org/10.1371/journal.pone.0208944.*

The sensitivity of predicted NO_2 concentrations of the SIRANE Gaussian plume dispersion model to emission changes has been addressed for source attribution purposes in a study focusing on the Lyon urban agglomeration (approximately 1.4 million people), for the year 2008. The BFM using tagging of the emitted species to reduce the computational cost was employed. Statistical evaluation of model performance after applying a data assimilation method showed improved agreement with measurements, supporting the usefulness of data assimilation and fusion methodologies for routine urban air quality model assessments. Source apportionment analysis revealed the low contribution of industrial sources to NO_2 pollution over the Lyon agglomeration. The contribution of miscellaneous distributed sources was found to be larger in the city center, while traffic contribution was prominent close to the roads and in the center, in contrast to the lowest levels of the background sources contribution in the center and close to the main roads. The model results highlight the reduction of traffic emission sources as the most appropriate mitigation strategy for the attainment of the NO_2 limit values in the urban agglomeration of Lyon (Fig. 8.13).

8.4.4 Forecasting applications and challenges

Air quality model prediction applications extend from short-term predictions at the local scale as a response to source-related emission changes (including accidental releases for emergency planning) to national-scale operational

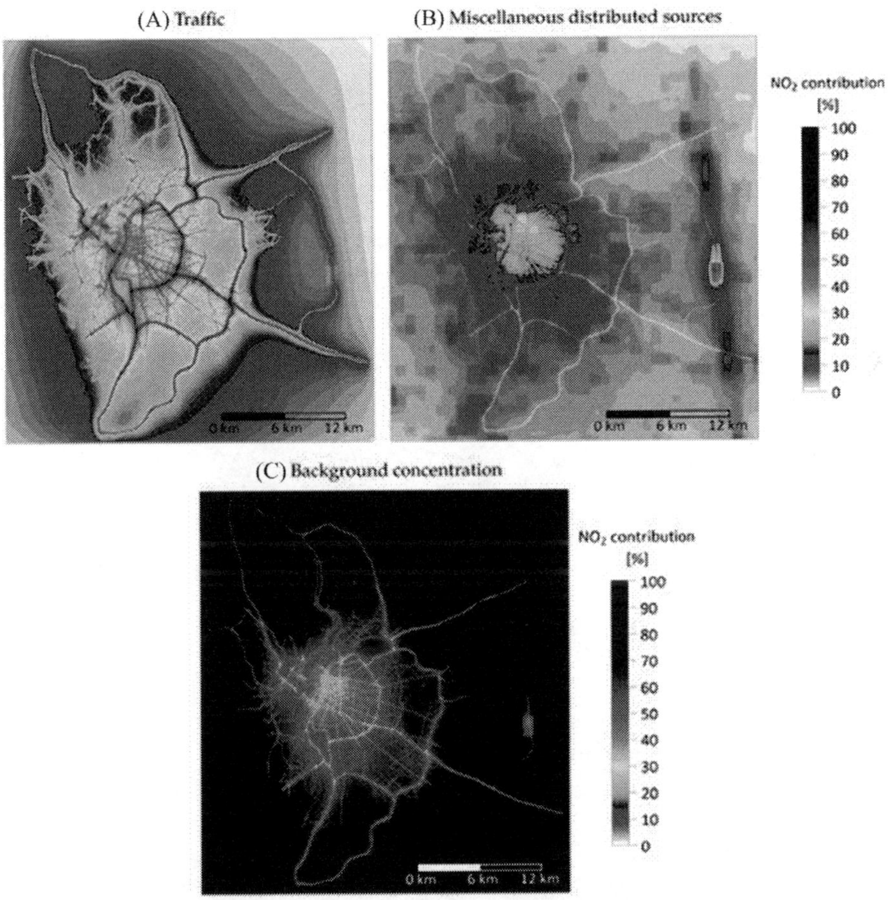

FIGURE 8.13 (A) Traffic, (B) miscellaneous distributed sources, and (C) background relative contribution [%] to the NO_2 annual mean concentration on the Lyon agglomeration in 2008 estimated with the SIRANE Gaussian plume model (Nguyen et al., 2018). *From Nguyen, C.V., Soulhac, L., & Salizzoni, P. (2018). Source apportionment and data assimilation in urban air quality modeling for NO_2: The Lyon case study.* Atmosphere, *9(1). https://doi.org/10.3390/atmos9010008.*

Daily AQI Color	Levels of Concern	Values of Index	Description of Air Quality
Green	Good	0–50	Air quality is satisfactory, and air pollution poses little or no risk.
Yellow	Moderate	51–100	Air quality is acceptable. However, there may be a risk for some people, particularly those who are unusually sensitive to air pollution.
Orange	Unhealthy for Sensitive Groups	101–150	Members of sensitive groups may experience health effects. The general public is less likely to be affected.
Red	Unhealthy	151–200	Some members of the general public may experience health effects; members of sensitive groups may experience more serious health effects.
Purple	Very Unhealthy	201–300	Health alert: The risk of health effects is increased for everyone.
Maroon	Hazardous	301 and higher	Health warning of emergency conditions: everyone is more likely to be affected.

FIGURE 8.14 Air quality categories of the US EPA AQI tool. *AQI*, Air quality index; *EPA*, Environmental Protection Agency. *From Home of the U.S. Air Quality Index. (n.d.). Retrieved March 23, 2024, from https://www.airnow.gov/.*

daily forecasts for public information on possible health risks. The US EPA AIRNOW application involves a coordinated effort to combine air quality forecasts submitted by state or local air quality agencies, using different forecasting techniques, into one forecast file covering the most important legislated pollutants and up to 4 forecasting days. Through the relevant website (https://www.airnow.gov/), air quality information and related health risks are shown for the area selected by the user. Similarly, hourly air quality forecasts for up to four days at a European scale are generated by the EU Copernicus Atmosphere Monitoring Service (CAMS), using an Ensemble model (so-called CAMS regional Ensemble) approach, based on the median value of the individual outputs of nine state-of-the-art numerical air quality models (https://regional.atmosphere.copernicus.eu/). Forecasting modeling systems aiming to communicate air quality information and health-related warnings to the general public demonstrate results using the AQI approach. AQI is a classification scale for different states of air quality (from good to hazardous), using predefined color coding to generate general statements describing the predicted health risk associated with predicted pollutant levels. The US EPA national AQI shown in Fig. 8.14 is among the most commonly used indexes in air quality model forecasting systems and assigns values from 0 to 500 to concentration ranges characterizing the air quality categories.

A common challenge for all model applications is **improving reliability and reducing uncertainty** of model results. Assimilation from ground station measurements and satellite data is widely recognized in the scientific community as a technique to improve the nowcasting and forecasting performance of mesoscale air quality models, primarily by modifying the initial conditions of the simulations. Several statistical methodologies for introducing monitoring data into air quality simulations are deployed, such as Bias Adjustment Techniques (BAT), the Best Linear Unbiased Estimator (BLUE), and the Source Apportionment Least Square (SALS) method. Data assimilation experiments on urban-scale air quality models, such as Gaussian puff dispersion models, have recently also gained scientific interest (Nguyen & Soulhac, 2021). Another promising methodology for injecting air quality observations in forecasting simulations entails the use of artificial intelligence, neural networks, and machine learning methods, either as stand-alone approaches or as hybrid methods in combination with statistical data assimilation (Sokhi et al., 2022).

8.4.5 Policy and management support operational systems

The use of air quality dispersion models for management purposes should ideally serve: **(1) nowcasting** of the current pollutant levels at background and hot spot locations, **(2) forecasting** of pollutant changes in response to mitigation measures or to accidental releases/episodes, and **(3) communication** of daily ambient air information to the general public. In this direction, the AQMS for policy support in Cyprus (Moussiopoulos et al., 2012) is operated by the Department of Labor Inspection of the Ministry of Labor and Social Insurance, which is the management body for air quality in the Republic of Cyprus. The system is based on the operational application of the Eulerian chemistry-transport model MARS-aero which runs back-to-back with the mesoscale prognostic meteorological model MEMO, featuring real-time integration of ambient pollutant concentration data from measurements and larger-scale models (Fig. 8.15).

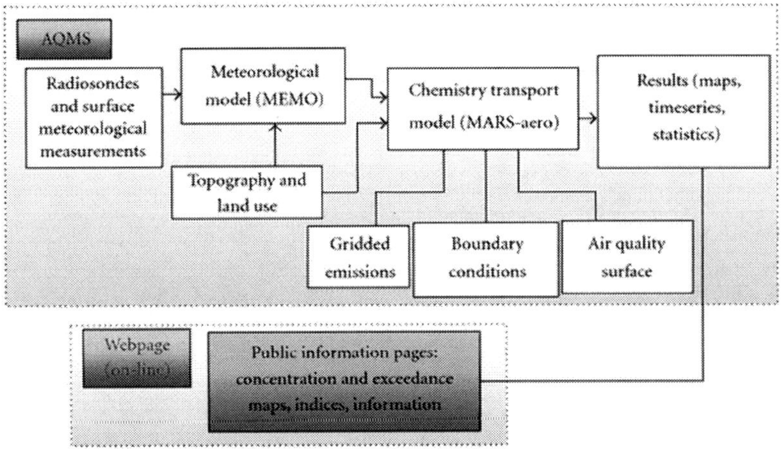

FIGURE 8.15 Structure of the AQMS used for AQ management in Cyprus. *AQMS*, Air quality management systems.

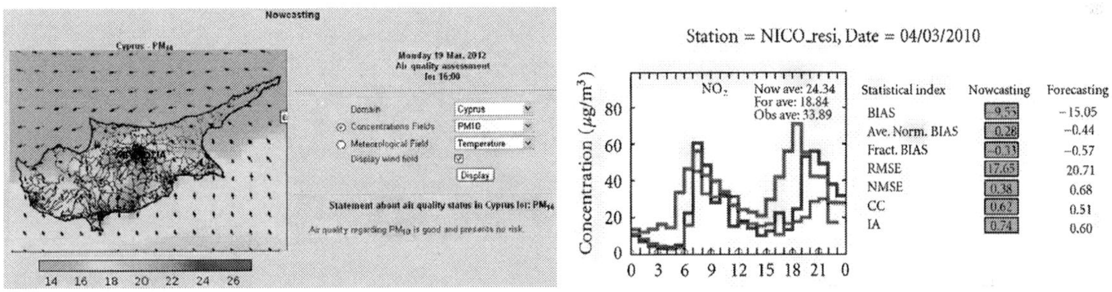

FIGURE 8.16 AQMS nowcasting plots (*left*) and validation report (*right*). *AQMS*, Air quality management systems.

Through the public web site (http://212.31.101.42:7080/cyprus/), hourly concentration fields of PM_{10}, $PM_{2.5}$, NO_2, O_3, and benzene for the entire Cyprus and, at higher resolution, for the five largest cities, are provided, along with color coding information on the expected impact on human health. Daily forecasting maps of pollutant levels are accessible via the same web page. The AQMS supports decision-making by allowing interactive configuration of custom emission scenarios and corresponding model runs to authorized regulatory users and by predicting episodic conditions, such as Saharan dust episodes which are particularly relevant for Cyprus. Confidence in the reliability of AQMS results is assessed on the basis of daily validation against observations from the national air quality monitoring network. A number of relevant statistical indices for the evaluation of model performance are reported in the form of graphs and validation charts (Fig. 8.16).

Worked example 8.4.1

Briefly describe the three main purposes underlying atmospheric model applications. What are the main advantages of using modeling tools for air quality management purposes?

Solution

Air quality models are used in research, regulatory and model evaluation applications. In research applications, the causal factor of air quality problems in specific case studies is most usually studied. Regulatory applications may be used for different management needs including air quality assessment, prediction, source apportionment, decision-making, emergency planning and public information. Model evaluation applications most commonly entail comparison of model results with measurements, model intercomparison and sensitivity studies.

The use of models allows for air quality assessment over areas and time periods for which measurement data are not available. Models are a cost-efficient tool to assess air quality in different seasons and appropriate time periods, so that hourly, daily, seasonal and annual patterns can be detected. Model applications provide visualized information on the spatial distribution of concentrations and, particularly, model results for hot spot areas, facilitate the identification of local-scale parameters influencing local air quality. Source identification and prediction of air pollution levels supported by the application of modeling tools is of indispensable value for policymaking.

Worked example 8.4.2

Perform an operation evaluation of two different atmospheric dispersion models in predicting ozone concentrations in an urban area. The observed ozone concentrations (Table 8.4, last column) come from a suburban measuring station and are representative of a typical diurnal concentration profile, while the corresponding concentrations calculated for the same location by the two models are given in the second and third columns, respectively:

To assess the performance of the models, calculate the following agreement metrics for each model:

- **Bias:** The average difference between the model predictions and the observed concentrations
- **Normalized Mean Square Error (NMSE):** The mean square error normalized by the product of the mean observed and predicted concentrations.
- **Index of Agreement (d):** Proposed by Willmott (1981), it is a standardized measure of the degree of model prediction error and ranges from 0 (no agreement) to 1 (perfect agreement).

Solution

The calculation formulas for the agreement metrics are as follows:

$$\text{Bias} = \frac{\sum(\text{Model} - \text{Observed})}{N}$$

$$\text{NMSE} = \frac{\sum(\text{Model} - \text{Observed})^2}{\sum(\text{Model}) \times \sum(\text{Observed})}$$

$$d = 1 - \frac{\sum(\text{Model} - \text{Observed})^2}{\sum\left(\left(|\text{Model} - \overline{\text{Observed}}| + |\text{Observed} - \overline{\text{Observed}}|\right)^2\right)}$$

where Σ denotes the sum over all time points, N is the number of time points and subsequently denoted as Mean Observed is the average of the observed concentrations.

Based on the concentration data in the table, we first calculate the necessary sums and means for the observed concentrations:

- Sum of Observed Concentrations (ΣObserved): $48 + 59 + 78 + 88 + 98 + 92 + 82 + 68 + 53 + 38 = 704\ \mu g\, m^{-3}$ ΣSum of Observed Concentrations (ΣObserved): $48 + 59 + 78 + 88 + 98 + 92 + 82 + 68 + 53 + 38 = 704\ \mu g\, m^{-3}$
- Mean of Observed Concentrations (MeanObserved): ΣObserved/N $= 704\ \mu g\, m^{-3}/10 = 70.4\ \mu g\, m^{-3}$ ΣMean of Observed Concentrations (MeanObserved): ΣObserved/N $= 704\ \mu g\, m^{-3}\ 10 = 70.4\ \mu g\, m^{-3}$

TABLE 8.4 Ozone observed concentrations and model predictions for different times of a single day.

Time of day	Model 1 ($\mu g\, m^{-3}$)	Model 2 ($\mu g\, m^{-3}$)	Observed ($\mu g\, m^{-3}$)
06:00	45	50	48
08:00	60	55	59
10:00	75	80	78
12:00	90	85	88
14:00	105	100	98
16:00	95	90	92
18:00	80	85	82
20:00	65	70	68
22:00	50	55	53
00:00	35	40	38

Now, let's calculate the metrics for each model:

Model 1

Bias

$$\Sigma\Sigma(\text{Model1} - \text{Observed}) = (45-48) + (60-59) + (75-78) + (90-88) + (105-98) + (95-92) + (80-82) + (65-68) + (50-53) + (35-38) = -4\,\mu\text{g}\,\text{m}^{-3}$$

$$\text{Bias} = \frac{\Sigma(\text{Model1} - \text{Observed})}{N} = -\frac{\frac{4\,\mu\text{g}}{\text{m}^3}}{10} = -\frac{0.4\,\mu\text{g}}{\text{m}^3}$$

$$\Sigma\text{Bias} = \frac{\Sigma(\text{Model1} - \text{Observed})}{N} = -\frac{\frac{4\,\mu\text{g}}{\text{m}^3}}{10} = -\frac{0.4\,\mu\text{g}}{\text{m}^3}$$

NMSE

$$\Sigma\Sigma\left((\text{Model1} - \text{Observed})^2\right) = (45-48)^2 + (60-59)^2 + . + (35-38)^2 = 74\,\mu\text{g}\,\text{m}^{-2}$$

$$\begin{aligned}
\text{NMSE} &= \Sigma\left((\text{Model1}-\text{Observed})^2\right)/(\Sigma(\text{Model1}) \times \Sigma(\text{Observed}))\\
&= 74\,\mu\text{g}\,\text{m}^{-2}/\left(704\,\mu\text{g}\,\text{m}^{-3} \times 704\,\mu\text{g}\,\text{m}^{-3}\right) = 74\,\mu\text{g}\,\text{m}^{-2}/495616\,\mu\text{g}^2\,\text{m}^{-6} \approx 0.000149\Sigma\text{NMSE}\\
&= \Sigma\left((\text{Model1} - \text{Observed})^2\right)/(\Sigma(\text{Model1}) \times \Sigma(\text{Observed}))\\
&= 74\,\mu\text{g}\,\text{m}^{-2}/\left(704\,\mu\text{g}\,\text{m}^{-3} \times 704\,\mu\text{g}\,\text{m}^{-3}\right)\\
&= 74\,\mu\text{g}\,\text{m}^{-2}/495616\,\mu\text{g}^2\,\text{m}^{-6} \approx 0.000149\Sigma\text{NMSE}\\
&= \Sigma\left((\text{Model1}-\text{Observed})^2\right)/(\Sigma(\text{Model1}) \times \Sigma(\text{Observed}))\\
&= 74\,\mu\text{g}\,\text{m}^{-2}/\left(704\,\mu\text{g}\,\text{m}^{-3} \times 704\,\mu\text{g}\,\text{m}^{-3}\right)\\
&= 74\,\mu\text{g}\,\text{m}^{-2}/495616\,\text{g}^2\,\text{m}^{-6} \approx 0.000149\Sigma\text{NMSE}\\
&= \Sigma\left((\text{Model1} - \text{Observed})^2\right)/(\Sigma(\text{Model1}) \times \Sigma(\text{Observed}))\\
&= 74\,\mu\text{g}\,\text{m}^{-2}/\left(704\,\mu\text{g}\,\text{m}^{-3} \times 704\,\mu\text{g}\,\text{m}^{-3}\right) = 74\,\mu\text{g}\,\text{m}^{-2}/495616\,\text{g}^2\,\text{m}^{-6} \approx 0.000149 \approx \text{NMSE}\\
&= \Sigma\left((\text{Model1}-\text{Observed})^2\right)/(\Sigma(\text{Model1}) \times \Sigma(\text{Observed}))\\
&= 74\,\mu\text{g}\,\text{m}^{-2}/\left(704\,\mu\text{g}\,\text{m}^{-3} \times 704\,\mu\text{g}\,\text{m}^{-3}\right) = 74\,\mu\text{g}\,\text{m}^{-2}/495616\,\text{g}^2\,\text{m}^{-6} \approx 0.000149
\end{aligned}$$

Index of Agreement (d):

$$\Sigma\Sigma\left((|\text{Model1} - \text{MeanObserved}| + |\text{Observed} - \text{MeanObserved}|)^2\right) = (|45-70.4| + |48-70.4|)^2 + . + (|35-70.4| + |38-70.4|)^2 = 8620.32\,\mu\text{g}\,\text{m}^{-2}$$

$$\begin{aligned}
d &= 1 - \left(\Sigma\left((\text{Model1} - \text{Observed})^2\right)/\Sigma\left((|\text{Model1} - \text{MeanObserved}| + |\text{Observed} - \text{MeanObserved}|)^2\right)\right)\\
&= 1 - \left(74\,\mu\text{g}\,\text{m}^{-2}/8620.32\,\mu\text{g}\,\text{m}^{-2}\right) \approx 0.9914\\
\Sigma d &= 1 - \left(\Sigma\left((\text{Model1} - \text{Observed})^2\right)/\Sigma\left((|\text{Model1} - \text{MeanObserved}| + |\text{Observed}-\text{MeanObserved}|)^2\right)\right)\\
&= 1 - \left(74\,\mu\text{g}\,\text{m}^{-2}/8620.32\,\mu\text{g}\,\text{m}^{-2}\right) \approx 0.9914\\
\Sigma d &= 1 - \left(\Sigma\left((\text{Model1} - \text{Observed})^2\right)/\Sigma\left((|\text{Model1} - \text{MeanObserved}| + |\text{Observed}-\text{MeanObserved}|)^2\right)\right)\\
&= 1 - \left(74\,\mu\text{g}\,\text{m}^{-2}/8620.32\,\mu\text{g}\,\text{m}^{-2}\right) \approx 0.9914 \approx d = 1 - \left(\Sigma\left((\text{Model1} - \text{Observed})^2\right)/\Sigma((|\text{Model1} - \text{MeanObserved}|\right.\\
&\left. + |\text{Observed} - \text{MeanObserved}|)^2)\right) = 1 - \left(74\,\mu\text{g}\,\text{m}^{-2}/8620.32\,\mu\text{g}\,\text{m}^{-2}\right) \approx 0.9914
\end{aligned}$$

Model 2

Bias

$$\Sigma\Sigma(\text{Model2} - \text{Observed}) = (50-48) + (55-59) + . + (40-38) = 2\,\mu\text{g}\,\text{m}^{-3}$$

$$\text{Bias} = \Sigma(\text{Model2} - \text{Observed})/N = 2\,\mu\text{g}\,\text{m}^{-3}/10 = 0.2\,\mu\text{g}\,\text{m}^{-3}\Sigma\text{Bias} = \Sigma(\text{Model2} - \text{Observed})/N = 2\,\mu\text{g}\,\text{m}^{-3}/10 = 0.2\,\mu\text{g}\,\text{m}^{-3}$$

NMSE

$$\Sigma\Sigma\left((\text{Model2} - \text{Observed})^2\right) = (50-48)^2 + (55-59)^2 + . + (40-38)^2 = 62\ \mu\text{g m}^{-2}$$

$$\text{NMSE} = \Sigma\left((\text{Model2} - \text{Observed})^2\right) / (\Sigma(\text{Model2}) \times \Sigma(\text{Observed}))$$
$$= 62\ \mu\text{g m}^{-2} / \left(495616\ \mu\text{g}^2\ \text{m}^{-6}\right) \approx 0.000125$$
$$\text{NMSE} = \Sigma\left((\text{Model2} - \text{Observed})^2\right) / (\Sigma(\text{Model2}) \times \Sigma(\text{Observed}))$$
$$= 62\ \mu\text{g m}^{-2} / \left(495616\ \mu\text{g}^2\ \text{m}^{-6}\right) \approx 0.00012$$
$$\Sigma\text{NMSE} = \Sigma\left((\text{Model2} - \text{Observed})^2\right) / (\Sigma(\text{Model2}) \times \Sigma(\text{Observed}))$$
$$= 62\ \mu\text{g m}^{-2} / \left(495616\ \mu\text{g}^2\ \text{m}^{-6}\right) \approx 0.000125$$
$$\Sigma\text{NMSE} = \Sigma\left((\text{Model2} - \text{Observed})^2\right) / (\Sigma(\text{Model2}) \times \Sigma(\text{Observed}))$$
$$= 62\ \mu\text{g m}^{-2} / \left(495616\ \mu\text{g}^2\ \text{m}^{-6}\right) \approx 0.000125$$
$$\approx \text{NMSE} = \Sigma\left((\text{Model2} - \text{Observed})^2\right) / (\Sigma(\text{Model2}) \times \Sigma(\text{Observed}))$$
$$= 62\ \mu\text{g m}^{-2} / \left(495616\ \mu\text{g}^2\ \text{m}^{-6}\right) \approx 0.000125$$

Index of Agreement (d):

$$\Sigma\Sigma\left((|\text{Model2} - \text{MeanObserved}| + |\text{Observed} - \text{MeanObserved}|)^2\right) = (|50-70.4| + |48-70.4|)^2 + .$$
$$+ (|40-70.4| + |38-70.4|)^2 = 8464.32\ \mu\text{g m}^{-2}$$
$$d = 1 - \left(\Sigma\left((\text{Model2} - \text{Observed})^2\right) / \Sigma\left((|\text{Model2} - \text{MeanObserved}| + |\text{Observed} - \text{MeanObserved}|)^2\right)\right)$$
$$= 1 - \left(62\ \mu\text{g m}^{-2} / 8464.32\ \mu\text{g m}^{-2}\right) \approx 0.9927$$
$$\Sigma d = 1 - \left(\Sigma\left((\text{Model2} - \text{Observed})^2\right) / \Sigma\left((|\text{Model2} - \text{MeanObserved}| + |\text{Observed} - \text{MeanObserved}|)^2\right)\right)$$
$$= 1 - \left(62\ \mu\text{g m}^{-2} / 8464.32\ \mu\text{g m}^{-2}\right) \approx 0.9927$$
$$\Sigma d = 1 - \left(\Sigma\left((\text{Model2} - \text{Observed})^2\right) / \Sigma\left((|\text{Model2} - \text{MeanObserved}| + |\text{Observed} - \text{MeanObserved}|)^2\right)\right)$$
$$= 1 - \left(62\ \mu\text{g m}^{-2} / 8464.32\ \mu\text{g m}^{-2}\right) \approx 0.9927 \approx$$
$$d = 1 - \left(\Sigma\left((\text{Model2} - \text{Observed})^2\right) / \Sigma\left((|\text{Model2} - \text{MeanObserved}| + |\text{Observed} - \text{MeanObserved}|)^2\right)\right)$$
$$= 1 - \left(62\ \mu\text{g m}^{-2} / 8464.32\ \mu\text{g m}^{-2}\right) \approx 0.9927$$

By comparing the metrics:

- **Bias:** Model 1 has a bias of $-0.4\ \mu\text{g m}^{-3}$, and Model 2 has a bias of $0.2\ \mu\text{g m}^{-3}$. A smaller absolute bias value indicates better performance, so Model 2 is slightly better in terms of bias.
- **NMSE:** Model 1 has an NMSE of approximately 0.000149, and Model 2 has an NMSE of approximately 0.000125. A lower NMSE indicates better performance, so Model 2 is better in terms of NMSE.
- **Index of Agreement (d):** Model 1 has a d of approximately 0.9914, and Model 2 has a d of approximately 0.9927. A higher d value indicates better performance, so Model 2 is better in terms of the index of agreement.

It is evident that, based on the selected metrics, Model 2 is more accurate in predicting the actual ozone concentrations than Model 1.

8.5 Questions

Q 8.1—Briefly describe the main types of air pollution dispersion models, on the basis of their underlying mathematical formulation. Please indicate and briefly explain which modeling approach would be more suitable for: (1) quickly deriving information regarding the impact of a pollutant plume originating from an industrial unit on the air quality of a nearby village, assuming low-intensity winds in the northeastern direction, (2) identifying the regions downwind of a factory unit that will be mostly affected by its annual operation. The factory unit operates in a mountainous area characterized by local meteorological conditions with high temporal variations during a calendar year, and (3) assessing air quality over an urban area, highlighting spatial and temporal variations.

Q 8.2—Describe the kind of data sources that you would choose for obtaining meteorological data for the purposes of providing input to (1) a model used for assessing air quality of an urban area of a size of about 10km, over the last 10 years, (2) a model used operationally, that is, to provide hourly nowcasts of air quality in near-real time, and (3) a modeling application for assessing future air quality for the same area, under different emission reduction scenarios.

Q 8.3—Which performance metrics would be more relevant for evaluating a real-time operational assessment system which is used by local authorities to support reporting of compliance with EU air quality legislation? Take into account the fact that legislated limit values are often defined in terms of average concentrations over different intervals but also as a maximum number of **exceedances** of concentration limit values.

Q 8.4a—Briefly explain which modeling tool (s) application you would suggest as an air quality consultant of a national regulatory body for day-to-day air quality management at a national level.

Q 8.4b—Describe the modeling methodology you would follow as an air quality consultant of an urban authoritative body and suggest an appropriate strategy for reducing NO_2 pollution levels, which are the main concern of the area.

Q 8.4c—What modeling approach would you employ to explain frequent ozone episodes over an urban area and which considerations would guide your model evaluation methodology?

Solutions to the end-of-chapter questions will be available at: https://www.elsevier.com/books-and-journals/book-companion/9780128225912.

References

Appel, K. W., Gilliam, R. C., Davis, N., Zubrow, A., & Howard, S. C. (2011). Overview of the atmospheric model evaluation tool (AMET) v1.1 for evaluating meteorological and air quality models. *Environmental Modelling & Software 26.4*, 434–443.

Baddeley, A., Rubak, E., & Turner, R. (2015). *Spatial point patterns: Methodology and applications with R*. Chapman and Hall/CRC Press.

Baklanov, A., Schlünzen, K., Suppan, P., Baldasano, J., Brunner, D., Aksoyoglu, S., Carmichael, G., Douros, J., Flemming, J., Forkel, R., Galmarini, S., Gauss, M., Grell, G., Hirtl, M., Joffre, S., Jorba, O., Kaas, E., Kaasik, M., Kallos, G., … Zhang, Y. (2014). Online coupled regional meteorology chemistry models in Europe: Current status and prospects. *Atmospheric Chemistry and Physics*, *14*(1), 317–398. Available from https://doi.org/10.5194/acp-14-317-2014.

Bamler, R., & Hartl, P. (1998). Synthetic aperture radar interferometry. *Inverse Problems*, *14*(4), R1–R54. Available from https://doi.org/10.1088/0266-5611/14/4/001.

Büttner, G., Kosztra, B., Maucha, G., & Pataki, R. (2017). Implementation and achievements of CLC2006. *International Journal of Applied Earth Observation and Geoinformation*, *23*, 1–9.

Chang, J. S., Brost, R. A., Isaksen, I. S. A., Madronich, S., Middleton, P., Stockwell, W. R., & Walcek, C. J. (1987). A three-dimensional Eulerian acid deposition model: Physical concepts and formulation. *Journal of Geophysical Research: Atmospheres*, *92*(D12), 14681–14700. Available from https://doi.org/10.1029/jd092id12p14681.

Chang, J. C., & Hanna, S. R. (2005). Technical Descriptions and User's Guide for the Boot Statistical Model Evaluation Software Package, Version 2.0.

Coen, J. L., Cameron, M., Michalakes, J., Patton, E. G., Riggan, P. J., & Yedinak, K. M. (2013). Wrf-fire: Coupled weather-wildland fire modeling with the weather research and forecasting model. *Journal of Applied Meteorology and Climatology*, *52*(1), 16–38. Available from https://doi.org/10.1175/JAMC-D-12-023.1, http://journals.ametsoc.org/doi/pdf/10.1175/JAMC-D-12-023.1.

Curran, P. J. (1988). The semivariogram in remote sensing: An introduction. *Remote Sensing of Environment*, *24*(3), 493–507. Available from doi:10.1016/0034-4257(88)90021-1.

Cuvelier, C., Thunis, P., Vautard, R., Amann, M., Bessagnet, B., Bedogni, M., Berkowicz, R., Brandt, J., Brocheton, F., Builtjes, P., Carnavale, C., Coppalle, A., Denby, B., Douros, J., Graf, A., Hellmuth, O., Hodzic, A., Honoré, C., Jonson, J., … Zuber, A. (2007). CityDelta: A model intercomparison study to explore the impact of emission reductions in European cities in 2010. *Atmospheric Environment*, *41*(1), 189–207. Available from https://doi.org/10.1016/j.atmosenv.2006.07.036.

DEFRA (2010). *Evaluating the performance of air quality models*. <https://uk-air.defra.gov.uk/assets/documents/reports/cat05/1006241607_100608_MIP_Final_Version.pdf>.

Denby, B., Georgieva, E., & Lükewille, A. (2011). EEA Technical report EEA Technical report, No 10/2011, The application of models under the European Union's Air Quality Directive: A technical reference guide, ISSN 1725-2237.

Dennis, R., Fox, T., Fuentes, M., Gilliland, A., Hanna, S., Hogrefe, C., Irwin, J., Rao, S. T., Scheffe, R., Schere, K., Steyn, D., & Venkatram, A. (2010). A framework for evaluating regional-scale numerical photochemical modeling systems. *Environmental Fluid Mechanics*, *10*(4), 471–489. Available from https://doi.org/10.1007/s10652-009-9163-2.

Farr, T. G., Rosen, P. A., Caro, E., Crippen, R., Duren, R., Hensley, S., Kobrick, M., Paller, M., Rodriguez, E., Roth, L., Seal, D., Shaffer, S., Shimada, J., Umland, J., Werner, M., Oskin, M., Burbank, D., & Alsdorf, D. E. (2007). The shuttle radar topography mission. *Reviews of Geophysics*, *45*(2). Available from https://doi.org/10.1029/2005RG000183.

Fast, J. D., Gustafson, W. I., Easter, R. C., Zaveri, R. A., Barnard, J. C., Chapman, E. G., Grell, G. A., & Peckham, S. E. (2006). Evolution of ozone, particulates, and aerosol direct radiative forcing in the vicinity of Houston using a fully coupled meteorology-chemistry-aerosol model. *Journal of Geophysical Research Atmospheres*, *111*(21). Available from https://doi.org/10.1029/2005JD006721, http://onlinelibrary.wiley.com/journal/10.1002/(ISSN)2169-8996.

Galmarini, S., Makar, P., Clifton, O. E., Hogrefe, C., Bash, J. O., Bellasio, R., Bianconi, R., Bieser, J., Butler, T., Ducker, J., Flemming, J., Hodzic, A., Holmes, C. D., Kioutsioukis, I., Kranenburg, R., Lupascu, A., Perez-Camanyo, J. L., Pleim, J., Ryu, Y. H., ... Wolke, R. (2021). Technical note: AQMEII4 Activity 1: Evaluation of wet and dry deposition schemes as an integral part of regional-scale air quality models. *Atmospheric Chemistry and Physics*, *21*(20), 15663–15697. Available from https://doi.org/10.5194/acp-21-15663-2021, http://www.atmos-chem-phys.net/volumes_and_issues.html.

Gesch, D. B. (2007). The national elevation dataset. *Geocarto International*, *22*(1), 3–11.

Grell, G., & Baklanov, A. (2011). Integrated modeling for forecasting weather and air quality: A call for fully coupled approaches. *Atmospheric Environment*, *45*(38), 6845–6851. Available from https://doi.org/10.1016/j.atmosenv.2011.01.017.

Gross, Stockwell W. R. (2003). Preprints, 5th *Conference on atmospheric chemistry: Gases, aerosols, and clouds American meteorological society the simple photochemistry for all resolving models (SPARC) mechanism.*

Hamer, P. D., Walker, S. E., Sousa-Santos, G., Vogt, M., Vo-Thanh, D., Lopez-Aparicio, S., Schneider, P., Ramacher, M. O. P., & Karl, M. (2020). The urban dispersion model EPISODEv10.0 - Part 1: An Eulerian and sub-grid-scale air quality model and its application in Nordic winter conditions. *Geoscientific Model Development*, *13*(9), 4323–4353. Available from https://doi.org/10.5194/gmd-13-4323-2020, http://www.geosci-model-dev.net/volumes_and_issues.html.

Hersbach, H., Bell, B., Berrisford, P., Hirahara, S., Horányi, A., Muñoz-Sabater, J., Nicolas, J., Peubey, C., Radu, R., Schepers, D., Simmons, A., Soci, C., Abdalla, S., Abellan, X., Balsamo, G., Bechtold, P., Biavati, G., Bidlot, J., Bonavita, M., ... Thépaut, J. -N. (2020). The ERA5 global reanalysis. *Quarterly Journal of the Royal Meteorological Society*, *146*(730), 1999–2049. Available from https://doi.org/10.1002/qj.3803.

Homer, C., Dewitz, J., Yang, L., Jin, S., Danielson, P., Xian, G., Coulston, J., Herold, N., Wickham, J., & Megown, K. (2015). Completion of the 2011 national land cover database for the conterminous United States– Representing a decade of land cover change information. *Photogrammetric Engineering and Remote Sensing*, *81*(5), 345–354. Available from http://www.asprs.org/a/publications/pers/2015journals/PERS_May_2015/HTML/files/assets/common/downloads/PE&RS%20May%202015.pdf.

Johansson, L., Karppinen, A., Kurppa, M., Kousa, A., Niemi, J. V., & Kukkonen, J. (2022). An operational urban air quality model ENFUSER, based on dispersion modelling and data assimilation. *Environmental Modelling & Software*, *156*. Available from https://doi.org/10.1016/j.envsoft.2022.105460.

Kanakidou, M., Seinfeld, J. H., Pandis, S. N., Barnes, I., Dentener, F. J., Facchini, M. C., Van Dingenen, R., Ervens, B., Nenes, A., Nielsen, C. J., Swietlicki, E., Putaud, J. P., Balkanski, Y., Fuzzi, S., Horth, J., Moortgat, G. K., Winterhalter, R., Myhre, C. E. L., Tsigaridis, K., ... Wilson, J. (2005). Organic aerosol and global climate modelling: A review. *Atmospheric Chemistry and Physics*, *5*(4), 1053–1123. Available from https://doi.org/10.5194/acp-5-1053-2005.

Kushta, J., Georgiou, G. K., Proestos, Y., Christoudias, T., Thunis, P., Savvides, C., Papadopoulos, C., & Lelieveld, J. (2019). Evaluation of EU air quality standards through modeling and the FAIRMODE benchmarking methodology. *Cyprus Air Quality, Atmosphere and Health*, *12*(1), 73–86. Available from https://doi.org/10.1007/s11869-018-0631-z, http://www.springer.com/environment/environmental + health + - + public + health/journal/11869.

Lateb, M., Meroney, R. N., Yataghene, M., Fellouah, H., Saleh, F., & Boufadel, M. C. (2016). On the use of numerical modeling for near-field pollutant dispersion in urban environments – A review. *Environmental Pollution*, *208*, 271–283. Available from https://doi.org/10.1016/j.envpol.2015.07.039.

Li, N., Lu, Y., Liao, H., He, Q., Li, J., & Long, X. (2018). WRF-Chem modeling of particulate matter in the Yangtze River Delta region: Source apportionment and its sensitivity to emission changes. *PLoS One*, *13*(12), e0208944. Available from https://doi.org/10.1371/journal.pone.0208944.

Madronich, S., & Flocke, S. (1999). *The role of solar radiation in atmospheric chemistry* (pp. 1–26). Springer Science and Business Media LLC. Available from https://doi.org/10.1007/978-3-540-69044-3_1.

Monteiro, A., Durka, P., Flandorfer, C., Georgieva, E., Guerreiro, C., Kushta, J., Malherbe, L., Maiheu, B., Miranda, A. I., Santos, G., Stocker, J., Trimpeneers, E., Tognet, F., Stortini, M., Wesseling, J., Janssen, S., & Thunis, P. (2018). Strengths and weaknesses of the FAIRMODE benchmarking methodology for the evaluation of air quality models. *Air Quality, Atmosphere & Health*, *11*(4), 373–383. Available from https://doi.org/10.1007/s11869-018-0554-8.

Moreira, A., Prats-Iraola, P., Younis, M., Krieger, G., Hajnsek, I., & Papathanassiou, K. P. (2013). A tutorial on synthetic aperture radar. *IEEE Geoscience and Remote Sensing Magazine*, *1*(1), 6–43. Available from https://doi.org/10.1109/MGRS.2013.2248301, http://ieeexplore.ieee.org/xpl/RecentIssue.jsp?punumber = 6245518.

Moussiopoulos, N., Douros, I., Tsegas, G., Kleanthous, S., & Chourdakis, E. (2012). An air quality management system for policy support in Cyprus. *Advances in Meteorology*, *2012*, 1–8. Available from https://doi.org/10.1155/2012/959280.

National Academy of Sciences. (2007). *Models in environmental regulatory decision making*. National Academy Press.

Nenes, A., Pandis, S. N., & Pilinis, C. (1998). ISORROPIA: A new thermodynamic equilibrium model for multiphase multicomponent inorganic aerosols. *Aquatic geochemistry*, *4*, 123–152.

Nguyen, C. V., Soulhac, L., & Salizzoni, P. (2018). Source apportionment and data assimilation in urban air quality Modelling for NO_2: The Lyon case study. *Atmosphere*, *9*(1), 8. Available from https://doi.org/10.3390/atmos9010008.

Nguyen, C. V., & Soulhac, L. (2021). Data assimilation methods for urban air quality at the local scale. *Atmospheric Environment*, *253*. Available from https://doi.org/10.1016/j.atmosenv.2021.118366, http://www.elsevier.com/locate/atmosenv.

Rao, S. T., Galmarini, S., & Puckett, K. (2011). Air quality model evaluation international initiative (AQMEII): Advancing the state of the science in regional photochemical modeling and its applications. *Bulletin of the American Meteorological Society*, *92*, 23–30. Available from https://doi.org/10.1175/2010BAMS3069.1.

Seinfeld, J. H., & Pandis, S. N. (2016). *Atmospheric chemistry and physics: From air pollution to climate change*. John Wiley & Sons.

Simpson, D., Benedictow, A., Berge, H., Bergström, R., Emberson, L. D., Fagerli, H., Flechard, C. R., Hayman, G. D., Gauss, M., Jonson, J. E., Jenkin, M. E., Nyíri íNyíri, A., Richter, C., Semeena, V. S., Tsyro, S., Tuovinen, J.-P., Valdebenito, Á., & Wind, P. (2012). The EMEP MSC-W chemical transport model—Technical description. *Atmospheric Chemistry and Physics*, *12*(16), 7825—7865. Available from https://doi.org/10.5194/acp-12-7825-2012.

Skamarock, W. C., Klemp, J. B., Dudhia, J., Gill, D. O., Barker, D. M., Duda, M. G., & Powers, J. G. (2008). A description of the Advanced Research WRF version 3.

Sokhi, R. S., Moussiopoulos, N., Baklanov, A., Bartzis, J., Coll, I., Finardi, S., Friedrich, R., Geels, C., Grönholm, T., Halenka, T., Ketzel, M., Maragkidou, A., Matthias, V., Moldanova, J., Ntziachristos, L., Schäfer, K., Suppan, P., Tsegas, G., Carmichael, G., ... Kukkonen, J. (2022). Advances in air quality research - Current and emerging challenges. *Atmospheric Chemistry and Physics*, *22*(7), 4615—4703. Available from https://doi.org/10.5194/acp-22-4615-2022, http://www.atmos-chem-phys.net/volumes_and_issues.html.

Thunis, P., & Clappier, A. (2014). Indicators to support the dynamic evaluation of air quality models. *Atmospheric Environment*, *98*, 402—409. Available from https://doi.org/10.1016/j.atmosenv.2014.09.016.

Thunis, P., Georgieva, E., & Pederzoli, A. (2011). The DELTA tool and Benchmarking Report template, Concepts and User guide, Version 2.

Thunis, P., Georgieva, E., & Pederzoli, A. (2012). A tool to evaluate air quality model performances in regulatory applications. *Environmental Modelling & Software*, *38*, 220—230. Available from https://doi.org/10.1016/j.envsoft.2012.06.005.

Turner, D. B. (1969). Workbook of Atmospheric Dispersion Estimates, 999-AP-26, U.S. Public Health Service, Cincinnati, Ohio, 8U pp.

Tsegas, G., Barmpas, P., Douros, I., & Moussiopoulos, N. (2011). A metamodelling implementation of a two-way coupled mesoscale-microscale flow model for urban area simulations. *International Journal of Environment and Pollution*, *47*(1-4), 278—289. Available from https://doi.org/10.1504/IJEP.2011.047342, http://www.inderscience.com/ijep.

U.S. Environmental Protection Agency. (n.d.). *Community multiscale air quality (CMAQ) modeling system*. Retrieved from [EPA website].

University of Wyoming (2023), *Upper air data*. <http://weather.uwyo.edu/upperair/sounding.html> (accessed 18.12.24).

Vautard, R., Moran, M. D., Solazzo, E., Gilliam, R. C., Matthias, V., Bianconi, R., Chemel, C., Ferreira, J., Geyer, B., Hansen, A. B., Jericevic, A., Prank, M., Segers, A., Silver, J. D., Werhahn, J., Wolke, R., Rao, S. T., & Galmarini, S. (2012). Evaluation of the meteorological forcing used for the air quality model evaluation international initiative (AQMEII) air quality simulations. *Atmospheric Environment*, *53*, 15—37. Available from https://doi.org/10.1016/j.atmosenv.2011.10.065.

Warner, T. T. (2011). Quality assurance in atmospheric modeling. *Bulletin of the American Meteorological Society*, *92*(12), 1601—1610. Available from https://doi.org/10.1175/BAMS-D-11-00054.1, http://journals.ametsoc.org/doi/pdf/10.1175/BAMS-D-11-00054.1.

Wehr, A., & Lohr, U. (1999). Airborne laser scanning—An introduction and overview. *ISPRS Journal of Photogrammetry and Remote Sensing*, *54*(2-3), 68—82. Available from https://doi.org/10.1016/s0924-2716(99)00011-8.

Wexler, A. S., & Seinfeld, J. H. (1991). Second-generation inorganic aerosol model. *Atmospheric Environment Part A, General Topics*, *25*(12), 2731—2748. Available from https://doi.org/10.1016/0960-1686(91)90203-J.

Willmott, C. J. (1981). On the validation of models. *Physical Geography*, *2*, 184—194.

WMO. (2018). *Guide to meteorological instruments and methods of observation*. World Meteorological Organization.

Zannetti, P. (Ed.), (2013). *Air pollution modeling: Theories, computational methods and available software*. Springer Science & Business Media.

Zhang, Y. (2008). Online-coupled meteorology and chemistry models: History, current status, and outlook. *Atmospheric Chemistry and Physics*, *8*(11), 2895—2932. Available from https://doi.org/10.5194/acp-8-2895-2008.

Chapter 9

Methods for measuring air quality

Ranjeet S. Sokhi[1], Ummugulsum Alyuz[1], Somayyeh Arghavani[1], Kostas Karatzas[2], Evangelos Bagkis[2], Theodosios Kassandros[2], Bhishma Tyagi[3] and Vikas Singh[4]

[1]*Centre for Climate Change Research (C3R), School of Physics, Engineering and Computer Science, University of Hertfordshire, Hatfield, United Kingdom,* [2]*Environmental Informatics Research Group, School of Mechanical Engineering, Aristotle University of Thessaloniki, Thessaloniki, Greece,* [3]*Department of Earth and Atmospheric Sciences, National Institute of Technology Rourkela, Rourkela, India,* [4]*National Atmospheric Research Laboratory (NARL), Gadanki, Andhra Pradesh, India*

9.1 Introduction

As introduced in Chapter 1, air pollution has been designated by the World Health Organization (WHO) as the largest environmental risk to people across the planet with approximately 7 million premature deaths attributed to air pollution (WHO, 2014). Air quality observations are a critical tool to assess the impacts of air pollution and to monitor its trends to gauge the effectiveness of policy control measures. Because of the economic cost, it would be unrealistic to expect monitoring to be taking place in all locations across the world. However, any monitoring network is only as efficient as its representation of the main ambient environments in order to accurately capture the variation in concentrations, exposures and source emissions. The distribution of air pollution monitoring, unfortunately, is highly variable across the main regions of the world (WHO, 2023).

The terms **monitoring** and **measurements** are often used interchangeably. Although they are linked, they are not the same for our purposes. In our context, **measurement** refers to the direct or indirect quantification of one or more air quality metrics, usually concentrations of air pollutant species in the air, with an instrument over a given time resolution. **Monitoring**, on the other hand, refers to regular, longer-term measurements of the same air quality metric or metrics, usually concentrations of air pollutant species in the air, with the same method or methodology. Monitoring is usually used in the context of air quality measurement networks operated, for example by local and national governments. These networks usually consist of a number of instruments placed in the same locations for a longer period of time to assess trends of concentrations (e.g., to check if they meet or exceed air quality standards or to check the efficacy of control measures). In this chapter, we will review some of the main methods for measuring air pollutant concentrations in ambient air and do not discuss actual monitoring networks. Examples of air pollution monitoring networks, for the interested reader, can be found in iQAir (2022) and EAQP (2024).

Another term that is often treated synonymously with air pollutant concentrations is **exposure**. However, even in this case, the two terms are not strictly equivalent. The definition of **exposure** from breathing air pollution is the product of the air pollutant concentration in a given environment and the time duration that the person spends in that environment (WHO, 2021). As air pollutant concentrations at any given time can be different for the same air pollutant species, (e.g., near a busy road, in a shopping mall, in a park or indoors), we usually quantify exposure for specific environments (e.g., Singh, Sokhi, et al., 2020). For these reasons, the reader should be cautiously aware of the difference between **concentration** and **exposure**, as in many cases the two terms are used interchangeably in the literature. An overview of methods to measure air pollution and their applications can be found in WHO (2023).

There are many reasons why air pollution needs to be measured and monitored. Some of the important ones include:

(i) To assess compliance with air quality standards and legislation;
(ii) To undertake analysis of long-term trends to assess the effectiveness of policy and control measures;
(iii) To support source apportionment and identification;
(iv) To support air quality assessment for managing local and urban air quality including urban planning;
(v) To quantify environmental, exposure and health impacts;

Air Quality. DOI: https://doi.org/10.1016/B978-0-12-822591-2.00009-3

(vi) To identify hotspots (that is, locations of particularly high concentrations);
(vii) To support air quality model development and evaluation; and
(viii) To support scientific research (e.g., chemical and physical transformations of air pollutants, the influence of meteorology, interactions with climate change, process analysis, and parametrization).
(ix) To raise public awareness on air pollution exposure, especially for sensitive and vulnerable groups (e.g., children, elderly and those with respiratory and cardiovascular diseases).

9.2 Overview of methods for measuring air quality

While there are many reasons for measuring air pollution, there are equally many different methods of air quality measurement. These methods can be broadly classified as follows:

(i) In-situ methods and
(ii) Remote sensing methods.

In-situ methods refer to those that are taken on-site or at the place of the air pollution and usually relate to ground-based measurements. These are subdivided into **active** and **passive** methods. **Active** methods refer to those that require a pump to collect a sample of air that is analyzed to estimate the concentration of the air pollutant species on site. Examples of active methods include particulate matter (PM) samplers, nitrogen oxide analyzer and ozone analyzer which are discussed below. **Passive** methods on the other hand require subsequent chemical analysis, usually in a certified laboratory. The most common example is that of a diffusion (Palmes) tube used for measuring nitrogen dioxide concentrations over typically a week or a 2-week period. These are relatively cheap and can be used for determining longer-term averages of air pollutant concentrations. We do not discuss passive methods in this chapter and the reader is referred to Butterfield et al. (2021) and Rowell et al. (2021) for further information.

In-situ methods include reference standard methods, instruments that are equivalent to the reference methods, research-orientated instruments and low-cost sensors. This chapter discusses some of the key reference methods and low-cost sensors for measuring air pollutants. Information on some of the equivalent methods can be found in Gilliam and Hall (2016).

Remote sensing methods refer to those where the approach does not involve taking an actual sample and the measurement is conducted remotely at a distance from the site of interest. These include measurements with satellites, aircrafts, unmanned aerial vehicles (UAVs) (e.g., drones) and even some which are placed at ground level (e.g., Zhou et al., 2023)

When conducting ground-based measurements, the choice of the station's location and surrounding environment is crucial. Usually, the overriding criterion is to measure air pollution in environments that will represent the general exposure of the population in that area, and/or ecosystems in the surrounding environment. It should be noted that in the wider literature and in different regions of the world, the terminology for station types can vary significantly. Generally, however, station types are classified according to their vicinity to air pollution sources, such as near road traffic or industrial sources, those that reflect the general background in an urban area or those that represent rural air pollution outside a city or town (DEFRA, 2024; European Parliament, 2019).

9.3 Scope of the chapter

This chapter focuses on ambient (outdoor) air quality measurement methods. It does not include methods for measuring source emissions (e.g., stack emissions) or measurement if indoor air quality. The reader is referred to other references such as Environment Agency (2022a) and Pipal and Taneja (2023). Usually, at most sites, air quality measurements are conducted alongside meteorological measurements, as already explained in Chapter 2, weather has a major influence on air pollutant concentrations. However, given that the focus of this book is on air quality, we do not examine meteorological measurements and the reader is directed to other pieces of works such as Foken (2022).

We begin with an introduction to standardized reference methods, followed by examining low-cost sensors and satellite methods and finally ending with a summary of some specialized measurement approaches. An understanding of reference methods is important to appreciate their role in accurately measuring air pollutant concentrations according to agreed protocols. Low-cost sensors are increasingly being used in many parts of the world especially where reference-grade instruments are not available and to increase the number of measurement sites without incurring excessive costs (Kang et al., 2022). Satellite observations are now routinely used to provide greater spatial coverage and to supplement ground-based measurements (Holloway et al., 2021). Selected specialized approaches are mentioned as they are often used for research-based studies. It is not possible to cover all possible instruments and approaches, but the selection covered below should provide the reader with a broad introduction into the area of air pollution measurements.

9.4 Standard reference instruments for measuring air quality

9.4.1 Protocols for reference methods

Some regions across the world have set specific protocols or standardization procedures on how measurements of regulated air pollutants should be undertaken to ensure that the data is accurate, reliable and verifiable. We will not discuss these in detail as it will be out of the scope of this book, but examples are given here to enable the reader to appreciate the wider context when conducting and using air quality measurements.

Standard procedures ensure that the monitoring equipment and related processes comply with appropriate air quality and environmental legislation. For example, the European Committee for Standardization (CEN), acting on behalf of the European Commission, have produced standard methods for monitoring air pollutants. These include CEN standard EN14211:2012 for measuring nitrogen oxides, EN14625:2012 for measuring ozone and EN12341:2014 for measuring PM. The British Standards Institution (BSI) have adopted these standards and produced detailed documentation outlining minimum performance requirements for analyzers. For PM (PM_{10} and $PM_{2.5}$), these include BS EN12341:2014 Ambient Air (for reference gravimetric method) and BS EN16450:2017 Ambient Air (for automated measuring systems) (CEN - European Committee for Standardization, 2023).

These standards are routinely reviewed, and future updates are expected. For example, there are recently updated standards on measuring air quality pollutants such as EN15267-1:2023 which documents procedures related to air quality monitoring equipment and EN15267-2:2023 which focuses on manufacturer's quality management systems. Importantly, these standards serve to ensure that instruments measure specific air pollutants and exhibit minimum performance characteristics through laboratory and field testing. Countries have their own procedures to ensure that these international standards are met. In the UK, the Environment Agency operates the Monitoring Certification Scheme (MCERTS) which provides a framework for manufacturers to meet standards when monitoring air quality for example. This framework includes standards of performance of the equipment as well as the competence of staff to ensure that laboratories operate according to European and international standards (see, e.g., Environment Agency, 2017).

As another example, within the USA, there are specific standard or reference methods for ambient air pollutant measurements which are regulated through the Code of Federal Regulation (CFR). Reference methods have been developed by the US Environmental Protection Agency (US EPA) for measuring ambient air pollutants such as PM (e.g., US EPA, 2016a) and a list of these along with equivalent methods is given in US EPA (2023). These methods when implemented according to their user manuals and quality assurance procedures allow them to be used as part of local or state-wide monitoring networks and cover all the main operating conditions that determine their performance characteristics. Access to the data and summaries of the air pollutant trends can be found online (e.g., https://gispub.epa.gov/air/trendsreport/2023/#home) and US EPA (2022).

9.4.2 Measurement of nitrogen oxides by chemiluminescence

The reference method for measuring nitrogen oxides (NO_x) is based on the chemiluminescence reaction between nitric oxide (NO) and ozone (O_3) (Demerjian, 2000; described in detail in Fontjin et al., 1970). A molecule can pass from the electronic ground state to an excited electronic state through energy absorption. The absorbed energy is emitted as the molecule reverts back to the ground state again. This happens in the form of heat (nonradiative deactivation) and the emission of light (luminescence). NO molecules react with O_3 and produce excited nitrogen dioxide (NO_2*). When these exited molecules decay to the ground state, photons are emitted and measured as their intensity is proportional to the NO_x concentrations. These processes, first described by Clyne et al. (1964) are as follows:

$$NO + O_3 \rightarrow NO_2^*(\text{excited state}) + O_2 \tag{9.1}$$

$$NO_2^*(\text{excited state}) \rightarrow NO_2(\text{ground state}) + h\nu(\lambda = 0.6 - 3.0\,\mu\text{m}) \tag{9.2}$$

$$NO_2^*(\text{excited state}) + \text{M(other gas molecules)} \rightarrow NO_2(\text{ground state}) + M + \text{heat} \tag{9.3}$$

As schematic of the chemiluminescence analyzer is shown in Fig. 9.1. It consists of five main sections including a convertor (to convert NO_x molecules to NO), ozone generator (to produce O_3 for the reaction in high-voltage electric discharge), three-way solenoid valve, reaction chamber (where NO and O_3 react, and photons are emitted) and a photomultiplier tube (as a light-detecting sensor).

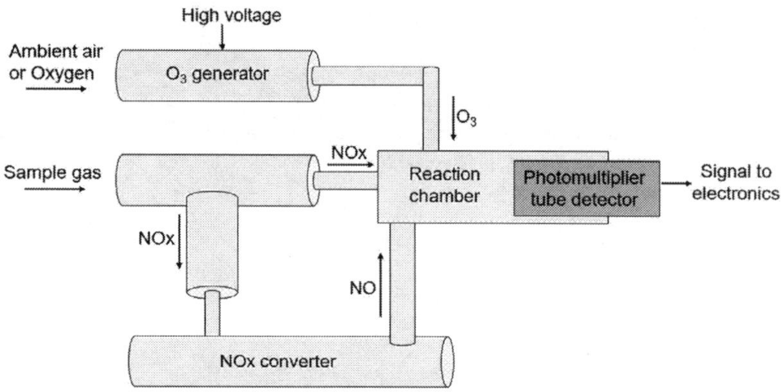

FIGURE 9.1 Schematic of chemiluminescent method for NO_x and NO measurements. *Adapted from Fontjin, A., Sabadell, A. J., & Ronco, R. J. (1970). Homogeneous chemiluminescence measurement of nitric oxide with ozone.* Analytical Chemistry, 42 *(5), 575–579 and Air Quality Expert Group (AQEG) (2004). Nitrogen Dioxide in the United Kingdom. Department for Environment, Food and Rural Affairs. <http://www.defra.gov.uk/environment/air-quality/aqeg>.*

To measure nitrogen oxides, all nitrogen oxides in the air sample are converted to NO before it enters the reaction chamber. This process is conducted in the converter which contains a heated metallic catalyst, such as molybdenum (Mo) to remove NO_2 molecules as follows (AQEG, 2004):

$$3NO_2 + Mo \rightarrow MoO_3 + 3NO \tag{9.4}$$

The use of a catalyst avoids other nitrogen-containing species, such as ammonia, also being converted into nitric oxide at high temperature. The NO from the converter is made to react with O_3 in the reaction chamber forming NO_2 in its excited state. As all of NO has now been converted to NO_2, the emitted photons are proportional to the NO_x concentration. To estimate the NO concentration, sample air directly enters the reaction chamber, and only NO molecules react with O_3 and get measured. The difference between NO_x and NO concentrations gives the NO_2 concentration in the sample air.

Despite the high operation costs, this chemiluminescence method has a low detection limit of around 1 $\mu g\ m^{-3}$ and provides real-time data with a time resolution of 1 hour or shorter (AQEG; 2004; BS EN 14211, 2012).

9.4.3 Ozone measurement method by ultraviolet absorption

Ultraviolet (UV) absorption is the standard method for measuring ozone (O_3) concentrations in ambient air (Long et al., 2014; Williams et al., 2006). This method is based on the UV light absorption by ozone in the wavelength range of 200–310 nm (Hartley band) with a peak at 253.7 nm (Hearn, 1961).

Fig. 9.2 shows the ozone analyzer configuration based on this UV absorption method. A sample of air enters a detection cell (or optical absorption cell) where it is irradiated by UV light at 253.7 nm from a low-pressure mercury lamp. The absorption of this radiation by the air sample is a measure of the ambient air ozone concentration. To avoid interference from other gases absorbing light at the same wavelength, an ozone catalytic converter is applied to remove ozone from the sample by alternately fill the sample cell with unscrubbed (sample gas with ozone molecules) and scrubbed sample air (sample gas without ozone molecules) (Jodpimai et al., 2016; Turnipseed et al., 2017).

The governing Beer-Lambert law equation is used to calculate the concentration of ozone from the ratio of two measured light intensities (DEC, 2016a):

$$O_3 \text{ concentration (molecules cm}^{-3}) = \frac{1}{\sigma l} \ln\left(\frac{I_0}{I}\right) \tag{9.5}$$

where I_0 is the light intensity measured with no ozone in the gas sample, I is the light intensity measured with ozone in the gas sample (it shows how much UV light is absorbed or blocked by ozone molecules in the detection cell), σ is the absorption cross-section of O_3 molecules or effective cross-sectional area blocked by a single molecule of O_3 where $\sigma_{253.7nm} = 1.147 \times 10^{-17}$ cm^2 $molecule^{-1}$ (first reported by Hearn (1961) and l is the length of the detection cell in cm).

To convert the concentration of ozone (C_{O3}) from molecules cm^{-3} unit to ppb we use:

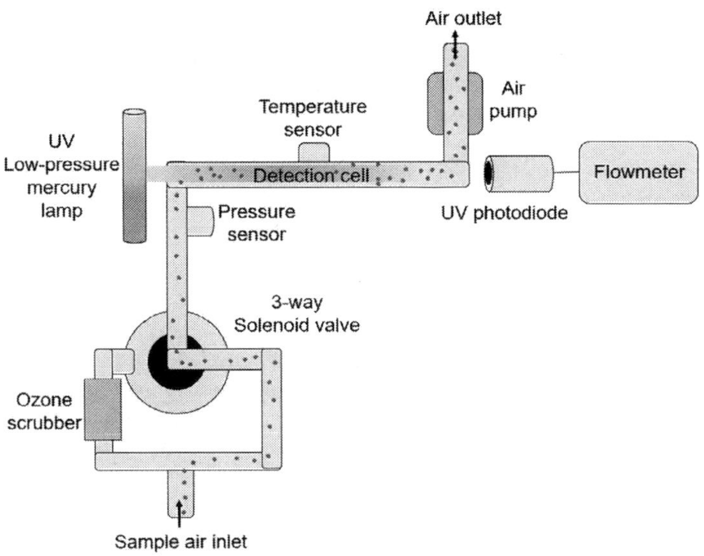

FIGURE 9.2 Schematic of UV absorption method for O_3 measurements. *Adapted from Turnipseed, A. A., Andersen, P. C., Williford, C. J., Ennis, C. A., & Birks, J. W. (2017). Use of a heated graphite scrubber as a means of reducing interferences in UV-absorbance measurements of atmospheric ozone.* Atmospheric Measurement Techniques, 10*(6), 2253–2269 and DEC (2016a). Standard operating procedures ozone (O_3) monitoring in ambient air by ultraviolet absorption spectrophotometry.* Air Quality Division, Air Monitoring & Quality Assurance Program, *The State of Alaska Department of Environmental Conservation, Revision 1.*

$$C_{O_3}(\text{ppb}) = \frac{C_{O_3}}{C_{\text{air}}} \times 10^9 \tag{9.6}$$

where C_{air} is the concentrations of air molecules cm^{-3}. Using the ideal gas law given in Equation 2.14, we have:

$$C_{\text{air}}\left(\frac{\text{molecules}}{\text{cm}^3}\right) = \frac{pN_A}{RT} \tag{9.7}$$

where Avogadro's number $= 6.023 \times 10^{23}$, T (temperature in Kelvin), p (air pressure in atmospheres), R (gas constant) $= 82.06\ \text{cm}^3\ \text{atm}\ \text{K}^{-1}\ \text{mole}^{-1}$. To estimate C_{air}, temperature and pressure sensors are applied as part of the instrument assembly.

By combining Beer–Lambert Law [Eq. (9.5)] with Eq. (9.7), we arrive at Eq. (9.8) which is used in all modern O_3 monitors to calculate O_3 concentration:

$$C_{O_3}(\text{ppb}) = \frac{RT10^9}{pN_A\sigma l}\ln\left(\frac{I_0}{I}\right) \tag{9.8}$$

The UV absorption method has high accuracy and reliability, and is routinely used for continuous measurements of ozone. However, it is an expensive method, and the lifetime of a mercury lamp is rather short (Jodpimai et al., 2016). Further details can be found in references such as (Long et al., 2021; Parrish & Fehsenfeld, 2000; Petruci et al., 2022).

Worked Example 9.1

A modern O_3 monitor measures the incident intensity of light, I_0, as 100 mW while the transmitted intensity of light, I, 50 mW where the path length is 40 cm and absorption cross-section of ozone is $2.5 \times 10^{-19}\ \text{cm}^2\,\text{molecule}^{-1}$. Calculate the concentration of O_3 by using gas constant, R, 82.06 in the unit of $\text{cm}^3\,\text{atm}\,\text{K}^{-1}\,\text{mole}^{-1}$; temperature, T, 298 K; pressure, p, 1 atm and Avogadro's number as 6.022×10^{23} molecules mol^{-1}.

Solution:

From Eq. (9.8),

$$C_{O_3}(\text{ppb}) = \frac{R \times T \times 10^9}{p \times N_A \times \sigma \times l} \times \ln\left(\frac{I_0}{I}\right) = \frac{82.06\frac{\text{cm}^3\text{atm}}{\text{Kmole}} \times 298\ \text{K} \times 10^9}{1\text{atm} \times \left(6.022 \times 10^{23}\frac{\text{molecule}}{\text{mole}}\right) \times \left(2.5 \times 10^{19}\frac{\text{cm}^2}{\text{molecule}}\right) \times 40\ \text{cm}} \times \ln\left(\frac{100}{50}\right) = 2.8 \times 10^6\ \text{ppb}$$

9.4.4 Sulfur dioxide measurement methods by ultraviolet fluorescence

UV fluorescence is the standard method for measuring sulfur dioxide in air (Matsumi et al., 2005; Pellikka et al., 2019). This widely used method is based on the emission of fluorescence radiation when SO_2 molecules are irradiated by a pulsed UV lamp (e.g., xenon). Fig. 9.3 shows the simplified schematic for this method. When the air sample is exposed to UV radiation in the reaction chamber, the SO_2 molecules get excited by the absorption of the light in the region of 190–230 nm (maximum wavelength of 214 nm). These excited molecules emit fluorescence radiation of 320–380 nm wavelength (mostly 330 nm) when they decay to lower energy states as below:

$$SO_2(\text{ground state}) + h\nu_1 \rightarrow SO_2^*(\text{excited state}) \rightarrow SO_2(\text{ground state}) + h\nu_2 \tag{9.9}$$

Where $h\nu_1$ is incident UV light at 214 nm and $h\nu_2$ is fluoresced UV light emitted at 330 nm. The emitted fluorescence radiation is proportional to the SO_2 concentration in the sample gas. A detector (photomultiplier tube) records the light filtered at these wavelengths.

The primary interfering gases in the UV fluorescence method for SO_2 measurement are polynuclear aromatics such as xylene and naphthalene. A "hydrocarbon kicker" is a device applied to remove the hydrocarbons from the sample air. Nitric oxide (NO) also fluoresces in a similar spectral range as SO_2. This interference is rejected by using an optical filter that only allows the UV decay radiation for SO_2 to pass into the photomultiplier tube. Decay of the excited SO_2 molecule to a lower energy state through kinetic energy losses by collision with a gaseous molecule is also an interfering in this method that results in heat loss rather than an emission of a photon which is not significant (Oil Sands Monitoring Program Report, 2020). Another source of interference is water vapors (H_2O) but in many situations, this is generally not significant. In a high H_2O concentration case, a dryer system, which removes moisture from the sample gas, is applied (US EPA, 2004).

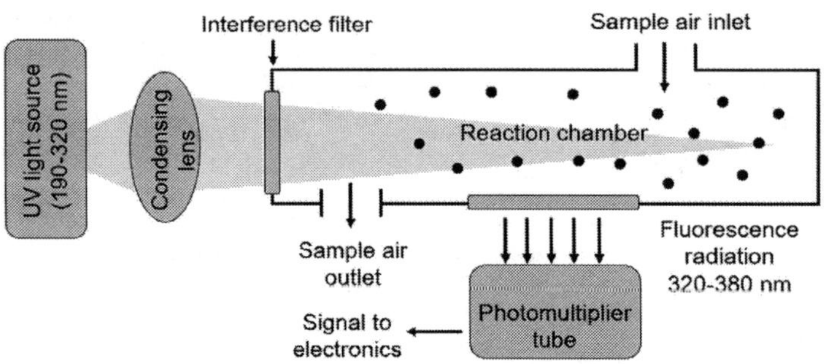

FIGURE 9.3 Schematic of ultraviolet fluorescence method for SO_2 measurement. *Adapted from Oil Sands Monitoring Program. (2020). Standard operating procedure for measurement of SO_2 by ultraviolet fluorescence SO_2 analyzer. https://open.alberta.ca/publications/sop-measurement-so2-by-ultraviolet-fluorescence-so2-analyzer and DEC (2016b). Standard operating procedures sulfur dioxide (SO_2) monitoring in ambient air by ultraviolet (UV) fluorescence spectroscopy. (2016). Air Quality Division, Air Monitoring & Quality Assurance Program, The State of Alaska Department of Environmental Conservation, Revision 3.*

9.4.5 Carbon monoxide measurement methods by infrared absorption

The usual standard method for measuring ambient carbon monoxide is infrared (IR) absorption (Liang et al., 2023). The IR absorption method, schematically illustrated in Fig. 9.4, is based on the gas molecular absorption spectroscopy according to Planck's law as below:

$$\Delta E = E_1 - E_2 = h\nu = h \times \frac{c}{\lambda} \tag{9.10}$$

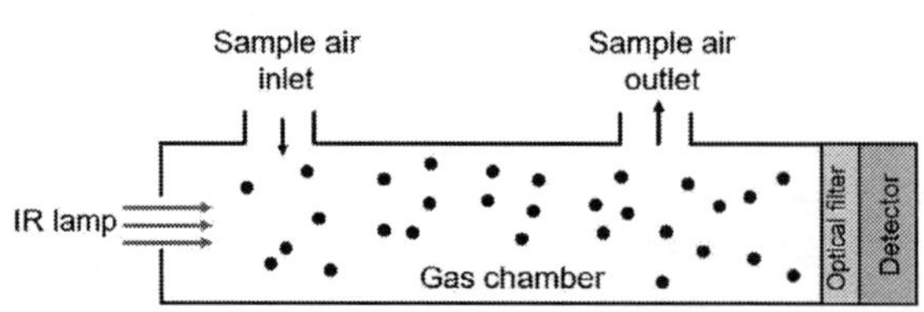

FIGURE 9.4 Schematic of infrared absorption method for CO measurement. *Adapted from Liang, J. G., Jiang, Y., Wu, J. K., Wang, C., Gratowski, S., Gu, X., & Pan, L. (2023). Multiplex-gas detection based on non-dispersive infrared technique: A review.* Sensors and Actuators A: Physical, 356, *114318 and Dinh, T. V., Choi, I. Y., Son, Y. S., & Jo-Chun Kim, J. C. (2016). A review on non-dispersive infrared gas sensors: Improvement of sensor detection limit and interference correction.* Sensors and Actuators B: Chemical, 231, *529-538.*

where E_1 and E_2 show the energy level before and after the transition, respectively. ΔE represents the energy released or absorbed during the transition. h (Plank constant) equals to 6.624×10^{-34}, ν is the frequency of the photon, c is the speed of light (3×10^8 m s^{-1}), and λ is the wavelength of the photon.

Carbon monoxide absorbs infrared radiation at specific wavelengths (4.5–5.0 μm; maximum at 4.67 μm). To avoid interference from other IR radiation, a filter is applied which only allows specific IR wavelengths to pass. When analyzing the sample of air, the infrared radiation absorbed by the CO molecules in the air sample is measured by the detector and t Beer–Lambert law equation (see Section 9.4.3) is used to estimate the CO concentration in the ambient air. A reflective gas chamber is used to increase the optical path length, which is a key parameter affecting the CO detection quality and allows the sample gas to absorb higher IR radiation (Sklorz et al., 2012).

There can be potential interferences from other gases and water vapor that absorb infrared radiation in similar frequencies and can affect CO detection accuracy. Optical filters, for example, to eliminate the absorption wavelength of water vapor 5–7 μm, are used to reduce this interference (Dinh et al., 2016). This method can also be used to measure the concentration of CO_2, SO_2, NO_x, nitrous oxide (N_2O), ammonia (NH_3) and methane (CH_4) in multiple gas detectors (Dinh et al., 2016; Liang et al., 2023).

9.4.6 Particulate matter measurement methods

This section provides an overview of commonly used ground-based methods for measuring PM. The methods covered include the standard gravimetric method, tapered element oscillating microbalance (TEOM), filter dynamics measurement system (FDMS), beta attenuation monitor (BAM), and fine dust analysis system (FIDAS).

9.4.6.1 *Gravimetric method*

The gravimetric method is considered the standard reference method by the European Committee for Standardization (2023) for measuring PM concentrations because it provides a direct measurement of mass with the use of a conditions filter and controlled operation. It is widely used for calibration and validation purposes (Aggarwal et al., 2013) to ensure that other methods can be traced back to this reference method.

A schematic illustration of the gravimetric method is shown in Fig. 9.5. The method involves drawing a specific volume of air through a size-selective inlet (that is, for sampling PM_{10} or $PM_{2.5}$), where airborne particles are separated based on their aerodynamic behavior when introduced into a moving airstream (Environment Agency, 2022b). The air passes through a pre-conditioned filter (e.g., for humidity variations) which is prepared and weighed both before and after sampling to determine the mass of PM present in that volume of air. This process is conducted for a period of a few hours to 24 hours or even days (US EPA, 2016b), yielding time-averaged concentration data.

An accurate and consistent volume flow rate, set by the user, is ensured with an active volumetric flow control system. A combination of a mass flow controller and sensors for measuring ambient temperature and pressure are employed. As the filter material can absorb water vapor in the air, hence changing its weight, the filters require conditioning in an environment with controlled temperature and humidity for 48 hours before and after exposure (Bureau Veritas, 2015; Environment Agency, 2022b).

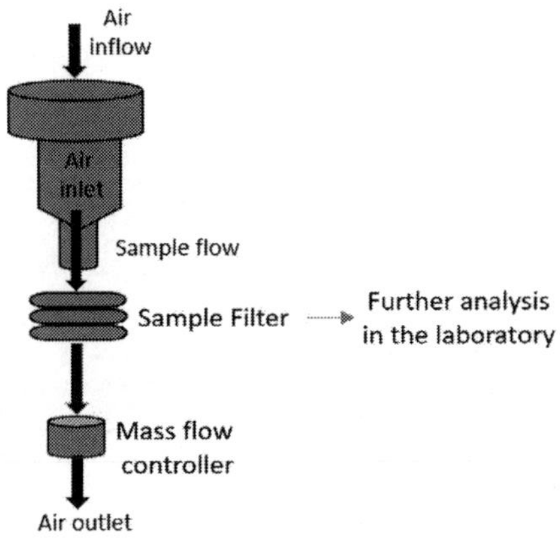

FIGURE 9.5 Simplified schematic of a gravimetric PM analyzer.

The gravimetric method, while considered a standard approach, is not suitable for routine automated measurements (Environment Agency, 2022b). It typically takes around 24 hours to complete a measurement, with results taking several days to report (AQEG, 2006; AQEG, 2012) as the filter has to undergo weighing in controlled laboratory environments. Conditioning before and after sampling is critical to avoid compromising accuracy due to the potential of moisture binding on the sampled particles and the loss of semivolatile compounds during filtration (Soysal et al., 2017). Without standardized procedures, there is a real risk that the measured mass may either over or underestimate the actual particle concentration in the air.

For the reasons discussed above, the gravimetric method is unsuitable for tracking short-lived peak concentration periods or observing short-term changes in concentration levels. However, gravimetric measurements are used effectively for validating the performance of other particle instruments including the automatic analyzers as required by the European Committee for Standardization (2023). As the gravimetric method requires more manual intervention, it is common to find more automatic measurements than gravimetric sites (Environment Agency, 2022c).

Worked Example 9.2

Suppose we want to measure PM_{10} concentration by gravimetric method and collect PM_{10} on a filter for 24 hours at a constant flow rate of 10 Liters per minute ($L\,minute^{-1}$). The initial weight of the clean filter was 5.1579 g, and after sampling, the final weight of the filter with collected PM was 5.1571 g. Calculate the concentration of PM_{10} in the air in the unit of μg mμ in the air in the unit of $\mu g\,m^{-3}$.

Solution:

Mass of PM collected = Final weight of the filter−Initial weight of the filter = 5.1579 g−5.1571 g = 0.0008 g = 800 μg.

$$\text{Volume of sampled air}\ (m^3) = \text{Flow rate}\left(\frac{m^3}{min}\right) \times \text{Sampling period (min)}$$

$$= \left[10\frac{L}{min} \times \frac{m^3}{1000L}\right] \times \left[24\,h \times \frac{60\ min}{1\ h}\right] = 14.4\ m^3$$

PM concentration is calculated below by using Equation 2.101;

$$PM_{10}\ \text{concentration}\left(\frac{\mu g}{m^3}\right) = \frac{\text{Mass of } PM_{10} \text{ collected}(\mu g)}{\text{Volume of sampled air}(m^3)} = \frac{800\ \mu g}{14.4\ m^3} = 55.5\ \mu g\ m^{-3}$$

9.4.6.2 *Tapered element oscillating microbalance*

The tapered element oscillating microbalance (TEOM) was originally developed by Ruppecht et al. (1992). Although the TEOM does not conform to the standard gravimetric method for reasons explained below, it is discussed here as it has been a commonly used PM monitor throughout the world. This device is utilized to continuously measure particulate concentrations by detecting fluctuations in oscillation frequency as aerosol particles accumulate on a small vibrating glass tube.

The working principle of TEOM is schematically illustrated in Fig. 9.6. The TEOM quantifies the particulate mass on a replaceable filter cartridge sitting on a tapered hollow tube by monitoring the variations in the frequency of the vibrating tapered hollow element. A sample of air flows through the filter and collected PM causes changes in the frequency in the tapered tube. The TEOM system also comprises a sample inlet head, control unit, and carbon vane pump. Flow is regulated by thermal mass flow controllers. It calculates the particulate concentrations by continually measuring particle mass (m) on the filter, utilizing the formula:

$$m = \frac{k_0}{f^2} \tag{9.11}$$

where k_0 denotes a constant established during TEOM analyzer calibration, f represents the frequency of the vibrating tapered element. A specific size-selective sampling head is used to ensure that the appropriate concentrations (that is $PM_{2.5}$ or PM_{10}) are measured. Further details are outlined in the report issued by the Environment Agency (2022b).

A primary concern regarding the TEOM instrument is its utilization of heated filters at an elevated temperature (50°C) to prevent the accumulation of moisture on particles (Long et al., 2003; Mignacca & Stubbs 1999). These heated filters can result in the loss of volatile species associated with fine particles, as highlighted by AQEG (2006), Long et al. (2003), and Mignacca and Stubbs (1999). This issue became apparent during comparisons between measurements obtained from co-located TEOM and gravimetric methods. Historically, a conversion factor of 1.3 was applied to all

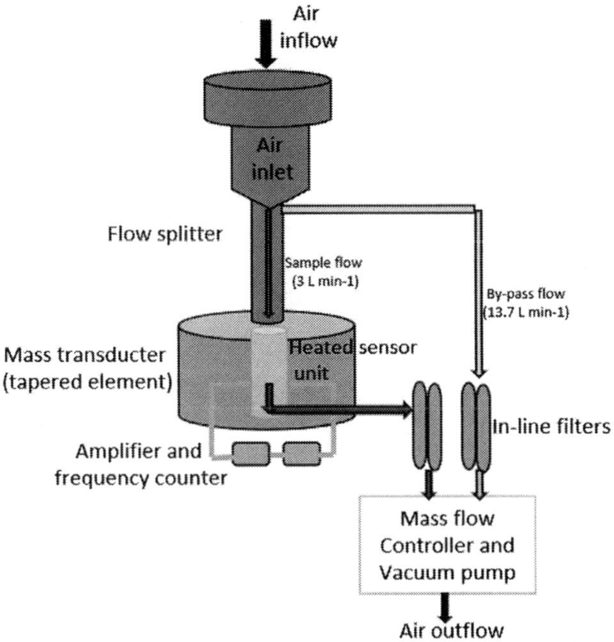

FIGURE 9.6 Simple schematic of TEOM particulate matter analyser. *Adapted from Clements, N., Milford, J. B., Miller, S. L., Navidi, W., Peel, J. L., & Hannigan M. P. (2013). Errors in coarse particulate matter mass concentrations and spatiotemporal characteristics when using subtraction estimation methods, Journal of the Air & Waste Management Association, 63(12), 1386–1398. https://doi.org/10.1080/10962247.2013.816643.*

TEOM-derived concentrations to estimate the gravimetric equivalent, aligning with the EU's gravimetric reference method and its air quality limit values (LVs) (AQEG, 2006). Studies have shown that there is variability in this correction across different locations (e.g., Environment Agency, 2022b; Green & Fuller 2006) making it incompatible with the standard gravimetric method. Consequently, TEOM analyzers are being replaced within monitoring networks in Europe and the UK. Previously, the TEOM analyzer was extensively utilized in the UK, Europe, and the USA due to its capability to provide immediate results to the public.

9.4.6.3 Filter dynamics measurement system

The filter dynamics measurement system (FDMS) was developed with the objective of addressing the loss of semivolatile substances from the TEOM filter, as described by Patashnick et al. (2001) and Meyer et al. (2002). The FDMS is a self-referencing airborne particulate monitor based on TEOM technology, capable of measuring both core and volatile components of particulates. As a retrofit option, it can be implemented on existing TEOM analyzers.

FDMS provides PM measurements that are equivalent to the EU reference method, utilizing the TEOM. This instrument determines PM mass concentrations by utilizing separate measurements of "nonvolatile" and "volatile" mass concentrations, as detailed by Ray and Vaughn (2009) and Stevenson et al. (2013).

A simplified schematic of the FDMS instrument is given in Fig. 9.7. Operation of the FDMS analyzer involves continual sampling of ambient air by utilizing a switch valve to redirect the main airflow every six minutes. To assess both volatile and nonvolatile components of PM, the FDMS employs a switching valve, which toggles between "base" and "reference" measurements every six minutes. In the "base" measurement, the FDMS samples as a regular TEOM, weighing the PM on the sensor unit filter. In the "reference" measurement, the FDMS redirects the airflow through a purge filter to eliminate all PM, followed by re-weighing the filter. The total PM measured over the 12-minute cycle is determined from the difference between the base and reference concentrations.

In the "reference" measurement, if any volatile substances trapped on the sensor unit filter evaporate, it can lead to a negative mass concentration. However, this negative reference concentration is subtracted from the "base" measurement concentration to ascertain the overall PM content. The instrument provides precise readings of PM mass concentration suitable for both short-term (hourly) and 24-hour averages and providing corresponding concentrations of "nonvolatile" ("base") and "volatile" ("reference") components (Bureau Veritas 2006; Environment Agency 2022c).

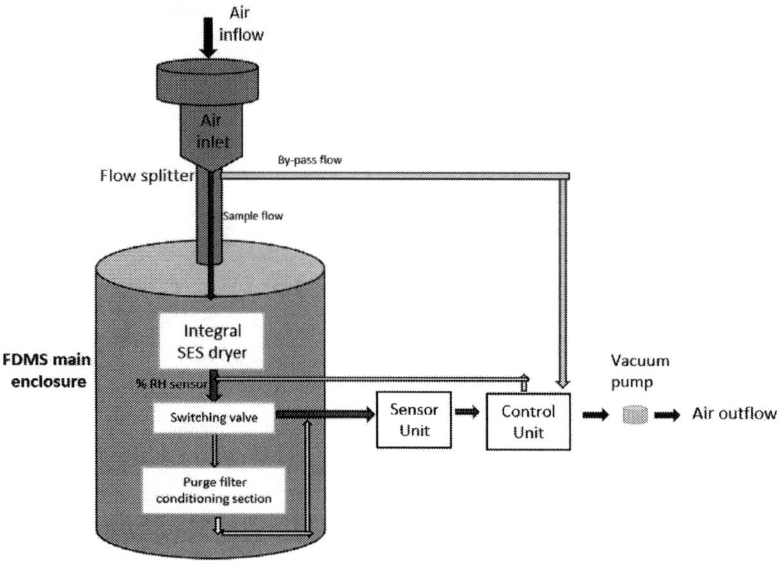

FIGURE 9.7 Simple schematic of the FDMS analyzer process. *Adapted from Clements, N., Milford, J. B., Miller, S. L., Navidi, W., Peel, J. L., & Hannigan M. P. (2013). Errors in coarse particulate matter mass concentrations and spatiotemporal characteristics when using subtraction estimation methods, Journal of the Air & Waste Management Association, 63(12), 1386–1398. https://doi.org/10.1080/10962247.2013.816643.*

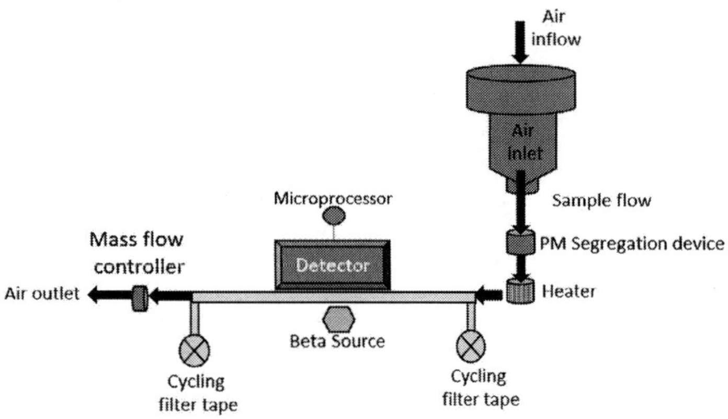

FIGURE 9.8 A schematic illustration of the Beta Attenuation Monitor. *Adapted from Shukla, K., & Aggarwal, S.G. (2022). A technical overview on beta-attenuation method for the monitoring of particulate matter in ambient air.* Aerosol Air Quality Research, *22, 220195. https://doi.org/10.4209/aaqr.220195.*

All long-term monitoring networks have to maintain instrument calibration, and performance integrity and importantly keep abreast of updates in technology. As an example, in the UK most FDMS instruments have reached the end of their operational lifespan and most of them were replaced with other methods as of 2021 (Ricardo Energy and Environment, 2022).

9.4.6.4 Beta attenuation monitor

The BAM is one of the US EPA Equivalent methods for automatic PM monitoring (see US EPA, 2023). It is extensively employed as a real-time approach for measuring ambient PM due to its ability to provide continuous measurements with limited need for operator supervision. Beta particles, which are high-energy electrons or positrons emitted during radioactive decay, can be absorbed, reflected, or passed through materials. The intensity of beta rays is attenuated depending on the amount of material present in the measurement path (Shukla & Aggarwal, 2022).

A schematic of the BAM instrument is shown in Fig. 9.8. The monitor comprises a carbon-14 beta radiation source impinging on a movable glass fiber filter tape with a scintillation detector to count the beta particles (Shukla & Aggarwal, 2022; US EPA, 2016a, 2016b). Measurements are taken pre- (without PM loading) and post- (with PM loading) and the change in the beta particle attenuation is used to calculate the mass concentration of the airborne PM (US EPA, 2016a, 2016b) using the following expression:

$$I = I_0 e^{-\mu x} \tag{9.12}$$

where I denotes the beta particle counts attenuated by the tape laden with PM a substrate of mass density x ($mg\,cm^{-2}$), I_0 is the incident unattenuated (clean tape) beta counts, μ signifies the mass absorption or attenuation coefficient ($cm^2\,mg^{-1}$) which is constant specific to the absorbed material on filter tape.

Eq. (9.12) can be rearranged to express the mass density (x) of the substrate as

$$x = \frac{1}{\mu}\ln\frac{I_0}{I} \tag{9.13}$$

Specific size-selective sampling heads through which air is drawn at a defined volume rate, are used to ensure that the concentrations of specific size fractions, $PM_{2.5}$ and PM_{10}, are measured. As with other ambient PM samplers, concentrations of PM are influenced by factors such as relative humidity, precipitation (e.g., rain or snow), and temperature which may affect the measured values (Shin et al., 2012). As with TEOM and FDMS, BAM provides nondestructive measurement and rapid automation necessary for long term monitoring needs.

Worked Example 9.3

Deduce the mass concentration in $\mu g\,m^{-3}$ from Eq. (9.13) for a surface area, A, when sampling the ambient air with a flow of Q, over a period, t.

Solution:

To convert the mass concentration to micrograms per cubic meter ($\mu g\,m^{-3}$), we need to consider the volume of air sampled over time ($Q \times t$).

The concentration C in micrograms per cubic meter can be obtained by dividing the mass x by the sampled volume of air V over time t, represented by $Q \times t$. Mathematically, it can be expressed as $C = x/V$.

For a specific sampling time, t (minutes), air is drawn at a constant flow rate Q ($L\,minute^{-1}$) and directed through a filter with a surface area A (cm^2). The mass of sampled PM on the area, A, is given by xA and hence the concentration (C) is then expressed xA divided by the sampled volume of air over time t (Qt); $C = xA/Qt$.

Hence the ambient concentration C ($mg\,m^{-3}$) of PM can be calculated by substituting the value of x from Eq. (9.13), we get:

$$C\left(\frac{mg}{m^3}\right) = \frac{\frac{1}{\mu\left(\frac{cm^2}{mg}\right)} \times \ln\frac{I_0}{I} \times A\left(cm^2\right)}{Q\left(\frac{m^3}{min}\right) \times t(min)} \tag{9.14}$$

where A is measured in cm^2, μ is expressed in $cm^{-2}\,mg$, Q is indicated in $m^3\,minute^{-1}$, and time is measured in minutes. In most circumstances, concentrations of PM are expressed in the units of $\mu g\,m^{-3}$. Hence to obtain the concentration in $\mu g\,m^{-3}$, the above expression should be multiplied by 1000.

Worked Example 9.4

Calculate the ambient PM concentration in $mg\,m^{-3}$ using measurement results with the beta-gauge method. Consider filter tape's surface area of $9\,cm^2$, a sampling flow rate of $16.67\,L\,minute^{-1}$, and a sampling duration of 42 minutes while the incident unattenuated beta count of 634,000 and an attenuated count of 627,234, along with a mass absorption coefficient of $0.8\,cm^{-2}\,mg^{-1}$.

Solution:

Given values:

Incident unattenuated beta count (I_o) = 634,000 counts;
Attenuated beta count (I) = 627,234 counts;
Mass absorption coefficient (μ) = $0.8\,cm^2\,mg^{-1}$;
Surface area of the filter tape (A) = $9\,cm^2$;
Sampling flow rate (Q) = $16.67\,L\,minute^{-1} = 0.01667\,m^3\,min^{-1}$;
Sampling duration (t) = 10 minutes.

Calculate the ambient PM concentration using Eq. (9.14).

$$C\left(\frac{mg}{m^3}\right) = \frac{\ln\frac{I_0}{I}\left(A\left(cm^2\right)\right)}{\mu\left(\frac{cm^2}{mg}\right) * Q\left(\frac{m^3}{min}\right) * t(min)} = \frac{\ln\frac{634{,}000}{627{,}234} * 9\,cm^2}{0.8\,\frac{cm^2}{mg} * 0.01667\,\frac{m^3}{min} * 47\,min} = 0.154\frac{mg}{m^3}$$

Answer is given to 3 decimal places. As $\mu g\,m^{-3}$ is commonly used as units for PM concentrations, $0.154\,mg\,m^{-3}$ can be written as $154\,\mu g\,m^{-3}$.

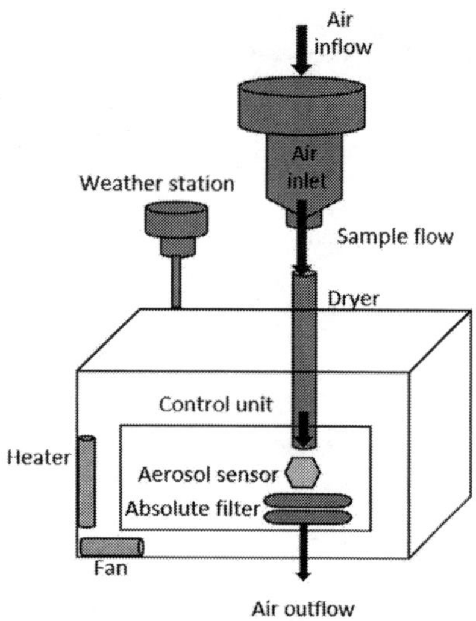

FIGURE 9.9 Simple schematic of FIDAS. *Adapted from Bureau Veritas. (2016). UK Report on the equivalence of the Palas Fidas 200 method 11 for PM10 and PM2.5. Bureau Veritas, UK.*

9.4.6.5 Fine dust analysis system

The fine dust analysis system (FIDAS) is an equivalent method based on optical particle counting technology and has the UK's MCERT accreditation (Bureau Veritas, 2016) and European standards approval. It is a continuous monitoring system for ambient air quality, providing data on all regulated PM fractions as well as particle size and number concentrations (Spielvogel & Weiss 2013). As an optical aerosol spectrometer, the FIDAS system assesses particle size and number through scattered light analysis. It then employs a proprietary algorithm devised by the manufacturer to compute the mass concentration ($\mu g\, m^{-3}$) of PM_{10} and $PM_{2.5}$ (Environment Agency, 2022b).

A schematic of the FIDAS system is given in Fig. 9.9. Ambient air is drawn through a sampling head, where it undergoes a drying process to remove moisture and prevent condensation within the instrument. During this stage, data regarding ambient temperature and relative humidity from a weather station are incorporated. The air advances to the aerosol sensor, where particles pass through an illuminated measurement chamber. The light scattered by the particles is detected at an angle ranging from 85° to 90°. Particle counts per unit volume of air are deduced from the number of scattered light pulses. Along with particle number, the instrument determines the diameter and volume of each particle by analyzing the intensity of the scattered light. Mass concentration is calculated using equations derived from "Lorenz-Mie scattered light analysis"; a method utilized for particle sizing and characterization (Mie, 1908).

9.5 Low-cost air quality sensors, their computational calibration and indicative applications

9.5.1 Introduction

Low-cost air quality sensors (LCAQS) are small, lightweight, portable devices that make use of physics and/or chemistry-related principles for measuring the concentration levels of different air pollutants in the air (indoors or outdoors). These sensors are often used in areas where air quality monitoring equipment is not available or too expensive, or in cases where additional spatial and temporal coverage is required in parallel with other existing air quality (AQ) monitoring devices. The low-cost characteristic of these sensors makes them affordable to communities and individuals who might not have had access to air quality monitoring capabilities and relevant air pollution information in the past. In addition, the fact that such devices are straightforward to use and easy in their setup and deployment, render them as appropriate for various citizen science[1] projects and applications.

The basic aspects, technology background, ways of use and limitations of the LCAQS have been discussed in numerous publications (like in Clements et al., 2022; deSouza, 2022; Gurudath & Srinivasa, 2021; Karagulian et al.,

1. Among the many definitions available, citizen science may be considered as the participation of members of the general public that have no professional relationship to scientific work, to any one of the phases of a scientific project (i.e., from its conception to the analysis and interpretation of results), in order to increase scientific knowledge and to also make science more understandable to citizens and more relevant to their everyday life.

2019, Karagulian, 2023, Oyola et al., 2022; WMO, 2020). In the following sections, their basic characteristics are discussed (section 9.5.2), followed by the operating principles of the most commonly used LCAQS (section 9.5.3). Section 9.5.4 discusses new developments in their calibration focusing on machine learning methods, while section 9.5.5 presents with a characteristic application of such LCAQS for urban air pollution assessment and for fueling urban air quality data fusion and modeling.

9.5.2 The basics of the sensor system

In the context of low-cost air quality sensors, there are several terms that are commonly used: sensor element, sensor, and sensor system. For reasons of clarity, these terms can be defined as follows (National Research Council, 1995):

- The sensor's component that directly interacts with the pollutant is known as the **sensor element**. This element (sometimes also referred to as **a sensor module**, Kang et al. (2022)), is the receiver of the physical or chemical stimulation that the pollutant causes to the element's material(s). This stimulus commonly results in a form of energy, and the sensor element converts this energy into another form, typically electrical.
- A **sensor** refers to a device that houses a sensor element, along with its physical packaging, electronics for processing electrical signals, and a means of communicating data through external connections.
- A **sensor system** (SS), **sensor device**, or **node**, is a complete (air quality) monitoring solution that comprises several sensors, associated electronics, and software-signal processing capabilities working together. The sensor system includes a control unit, power supply, telecommunications and/or storage module, and an enclosure, while in some cases it also includes a data storage module. It is commonly referred to as a device.

On the basis of the above, **the sensor system** includes the following components/functionalities (WMO, 2020):

- **Sensors** contain sensor elements that are responsible for providing electric signals that correspond to the concentration levels of the air pollutants of interest. Additionally, some sensors may also include sensor elements that provide signals for other atmospheric parameters, such as air temperature, relative humidity, air pressure, wind speed and direction.
- **Sampling capability** refers to the mechanism used to expose the sensor to the environment being monitored. Some sensors require direct exposure through passive methods, such as inlets commonly used for gaseous pollutants, or active methods, such as pumps or fans which are common for PM sensors.
- **Operational hardware and software** include microcontrollers or single-board computers, such as Arduino or Raspberry Pi, for controlling and managing sensor operations. They also include analog-to-digital conversion when required, on-board processing, local data management, and remote device management capabilities.
- **Power systems** provide the necessary power and may include one or a combination of batteries, direct electricity supply, solar panels, etc.
- **Data management and communication** refer to storage and data transmission. WiFi, cellular mobile data networks, and low-power wide-area networks, such as LoRa, are common examples. A sensor system may be served by one or more data management and communication modules.
- **Spatial and temporal reference module** may include a GPS for accurate spatial knowledge and a module for the provision of accurate time.
- **Data analytics module** converts raw data into valuable information and may provide additional statistical analysis and visualization capabilities in combination with cloud infrastructures.
- **Calibration** is another module included in many SS, which can be done as part of the operational software or via the data analytics module. It refers to the comparison of the SS with a high-quality instrument reading, and to the procedure that would allow the former to reflect the latter. Local calibration with the aid of ground truth measurements is a common procedure for SS.

Apart from the basic components of a sensor system, it is important to take into account that every sensor or generally every element that is used for measuring a quantity of interest, possesses some measurement characteristics that are important for identifying its capabilities as well as its limitations during operation. Such characteristics include (De Silva, 2017):

- Its *sensitivity* characterizes the minimum amount of input quantity (here air pollutant concentration) that is necessary to produce a detectable difference in the output signal. This might be related to the minimum concentration levels, for example, of a given air pollutant, in order for the SS to produce a signal that is different to the signal that it produced when exposed to a zero-concentration atmosphere.

- Its *selectivity* refers to the ability of the sensor system to distinguish the target pollutant from other pollutants and to respond only to its concentration changes. If the SS responds also to the existence of other pollutants, then it is said to have a high cross-sensitivity, this behavior is observed commonly in the case of gaseous pollutants (Isaac et al., 2022).
- Its *resolution* refers to the minimum incremental change to the input that is necessary to produce an identifiable difference in the output signal. This corresponds to an increase (or decrease) in the concentration level of a given air pollutant, in order for the SS to generate a signal that can be identified as different to the one produced when exposed to the initial concentration levels.
- Its *precision* refers to the capability of the SS to reproduce the same output as a response to the same input. This should not be confused with accuracy. When a SS has a big offset but produces the same output under similar input conditions, then it has a high precision.
- Its *accuracy* corresponds to the overall difference between the SS output and the output of a high-quality (reference) instrument, when repeatedly exposed to the same input stimuli (air pollutant concentration levels) under similar operational conditions. High accuracy means that the SS produces an output that is very close to the output of a reference instrument. Ideally, an SS should possess both a high accuracy and a high precision.
- Its *offset*, that is, the difference between zero and the SS output, when exposed to zero input signal (stimuli).
- Its *linearity* refers to the degree of difference or deviation between an ideal linear response of the SS to an input signal, and to the actual response of the SS to this signal.

Due to the different measuring characteristics of different sensor elements (attributed to their operating principles) and of different sensor systems (resulting from their design and operational characteristics), it is important to note that if air quality standards-AQS (target values, limit values, etc.) have been defined, then the LCAQS should be able to address the following minimum requirements under operational conditions:

(i) In terms of sensitivity and resolution, the lowest concentration that can be identified should be below the relevant AQS.
(ii) In terms of span, that is, the value range of operation, this range should be larger than the AQS under consideration.
(iii) Both the response time (i.e., the time required to provide an output signal when exposed to a certain concentration level) and the sampling interval (the minimum time interval between two consecutive measurements) should be smaller than the averaging interval of the AQS.

9.5.3 Operational principles of low-cost air quality sensors

The operational principles of low-cost air quality sensors vary depending on the type of pollutant being measured (Mendez & Sanchez, 2023; WMO, 2020; Yatkin et al., 2022). In the following, we focus on low-cost sensors (LCS) that are used for estimating concentrations of some of the gaseous pollutants in ambient air, as well as of PM.

9.5.3.1 Reactive gaseous air pollutants

The reactive gaseous air pollutants mostly monitored via LCAQS include nitrogen monoxide (NO), nitrogen dioxide (NO_2), ozone (O_3), sulfur dioxide (SO_2), carbon monoxide (CO), total volatile organic compounds (VOCs), and non-methane hydrocarbons (NMHC) (Lewis et al., 2018).

Electrochemical sensors (ECS) or electrochemical cell sensors depicted in Fig. 9.10 involve a gaseous pollutant undergoing an electrochemical reaction that produces a signal, typically a current, proportional to the target gas's concentration in the air. This is usually done in an electrolyte solution confined in a chamber, and being accessible via a porous membrane that allows gas diffusion. Electrodes are being dipped into the solution. When the gas reaches the working electrode, an electrochemical reaction takes place, resulting in an electron flow towards the counter electrode. This flow constitutes an electrical current, analogous to the amount (i.e., concentration) of the gas, which is generally used as the output signal of the sensor. ECS vary in their accuracy and reliability depending on the measured species, and additional measurements of relative humidity and temperature are required to obtain reliable results due to interferences (Baron & Saffell, 2017). Pollutants usually monitored via ECS include CO, NO, NO_2, O_3 and SO_2. The unit cost of the sensor element is in the order of magnitude of 50–500 Euros per sensor. They have good sensitivity and a very short response time (commonly between 30 and 300 seconds), while their power consumption is negligible. On the other hand, they are quite sensitive to temperature and humidity variations (as already stated), and they demonstrate a relatively short lifespan (usually approximately 1 year), depending on the exposed gas levels and on operational conditions like the temperature and humidity range of their operational environment (therefore replacement might be needed two times per year, e.g.). Moreover, they have quite poor selectivity, responding to similar molecule types.

FIGURE 9.10 Two electrochemical sensors compared in size with the 10 Euro cent coin. The light color surface on the top identifies the entry area (porous membrane) of gases to the chamber hosting the electrolyte solution and the electrodes.

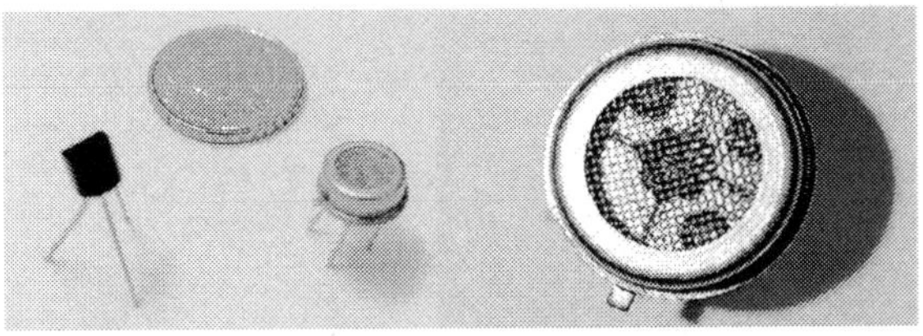

FIGURE 9.11 Left: A solid state temperature sensor (left part) and a metal oxide sensor (right part) compared in size with the 10 Euro cent coin. Right: the metal oxide sensor. The actual sensor element covers the dark area of the center of the circular sensor geometry.

Metal oxide sensors (MOS) have an exposed surface (resistive or semiconductor MO_x) film that absorbs the target gas (Fig. 9.11), resulting in a change in conductivity or resistance of the film. The reaction is supported by heating the sensor material. This adsorption is proportional to the concentration of the gas in the surrounding environment. The sensing mechanism of MOS gas sensors can be divided into two steps: the gas–solid interfacial interaction, also known as the receptor function, and the transduction of this interaction into an electrical signal through the resistance change of the sensing material, also known as the transducer function (Isaac et al., 2022). Pollutants commonly monitored via MOS include oxidative gas(es) in the air like NMHC, CO, CO_2, NO, NO_x, and O_3. The unit cost of the sensor element is in the order of magnitude of 10–30 Euros per sensor. They have good sensitivity and a long lifespan (some years) with quite satisfactory resilience in external (like weather) conditions. On the other hand, they have a long response time (in the order of magnitude of 5 to 50 minutes), and they can be influenced in case of significant changes in the temperature and humidity of their working environment (yet not as much as ECS), and may demonstrate drift in terms of their sensitivity, therefore, requiring quite a frequent recalibration. Their relatively large power consumption due to heating should also be considered in accordance with their application domain.

Other types of AQ sensors are photoionization detectors (PIDs), which use ultraviolet light to break organic molecules apart, releasing electrons and forming a positive ion, therefore inducing a small current that the sensor measures. They are typically used for detecting VOCs, while compounds with similar or lower ionization energies as the PID lamps' specific photon energy levels can also be ionized and detected. However, PID-based sensors have limitations, as they cannot ionize VOCs with equal efficiency across different compounds, resulting in total ambient VOC values that are influenced by the actual VOC mixture (Bílek et al., 2022). Their unit cost is in the order of magnitude of 500 Euros per sensor; they have a fast response time and good sensitivity, while they demonstrate small dependence on temperature and sensitivity to humidity. On the other hand, they have a high power consumption, while being characterized by their inability to identify individual VOCs and by their considerable short-term drift requiring calibration and maintenance.

Additional gaseous pollutant monitoring methods employ nondispersive infrared (NDIR) sensors that make use of the property of certain gases to absorb infrared light at certain wavelengths and get attenuated. Such sensors are commonly used for ambient as well as for indoor CO_2 concentration level estimation. The unit cost of the sensor element is in the order of magnitude of 100–400 Euros per sensor. They have a good sensitivity and selectivity for CO_2 (between 350 and 2000 ppm) and a very short response time (in the order of magnitude of 20–120 seconds). They also have a long lifespan and negligible power (and maintenance) requirements. Nevertheless, they are influenced by environmental conditions like temperature, humidity and even pressure, and if used for other gases than CO_2, they are characterized by high detection limits.

One more sensor category used is catalytic or pellistor sensors, employed for combustible gases. Methane detection is commonly based on such sensors.

The selection of sensor technologies generally depends on the deployment characteristics. MOS and ECS sensors are the least expensive and therefore most suitable for large-scale deployments, while PID and NDIR sensors are better suited for less dense deployments. When it comes to accuracy, MOS and EC sensors perform similarly, but ECS sensors are more energy-efficient, whereas MOS sensors are more durable. However, ECS sensors can demonstrate problematic behaviors in high temperatures and low humidity conditions, whereas MOS sensors are more resistant. Both sensors are affected by weather conditions and the presence of other gases, which suggests that these variables should be taken into account when calibrating them (Concas et al., 2021). In terms of integration into commercial sensor nodes for the ambient environment, the vast majority of sensors used are ECS.

9.5.3.2 Particulate matter

PM sensors use optical technology to measure the amount of PM in the air. They work by shining a light through the air and measuring how much of the light is scattered by the PM. This is done via a light beam (commonly being an infrared laser) that illuminates an air sample entering the narrow chamber of the sensor element, resulting in light scattering, which is detected by a photodiode. For this reason, these types of sensors are called optical particle sensors and are divided into two categories: Optical particle counters (OPCs) and nephelometers. The apparatus of a broken PM LCS can be seen in Fig. 9.12.

According to the sensor element-specific design characteristics, the electronic signal produced is interpreted as follows (Li et al., 2020):

1. Pulse characteristics, with its frequency corresponding to particle number and its height corresponding to particle size. Therefore, different classes of particle sizes can be identified (bins), and their overall concentration is calculated on the basis of their number, mean size per bin and density estimations, with the aid of proprietary algorithms. This is the operating principle of OPCs, while the number of bins can range from three to twenty-four, according to the manufacturer.
2. Total scattered light intensity, which is correlated to the total concentration of the particles. This is the operating principle of nephelometers, which measure particles as an ensemble, where the total scattering amplitude is correlated to a mass concentration of the particles, according to algorithms developed by the sensor manufacturer.

The unit cost of the PM sensor element is in the order of magnitude of 20–500 Euros per sensor. They have good sensitivity and a long lifespan (several years). They also have a very short response time (in the order of magnitude of 1 second). The OPC method has its limitations as detection is greatly influenced by the optical properties of the

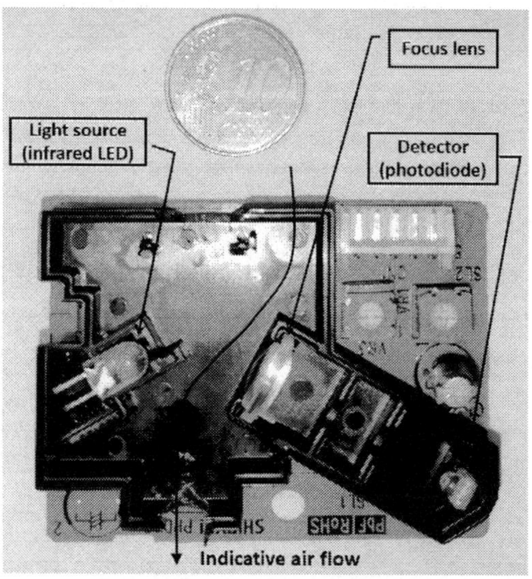

FIGURE 9.12 A broken optical particle counter reveals its interior and is compared in size with the 10 Euro cent coin.

particles (and therefore by their chemical composition among others), as well as by humidity (leading to their hygroscopic growth), and by their size. Moreover, it has been made evident that relevant low-cost sensors do not adhere to the detection ranges declared by their manufacturers (Kuula et al., 2020). Nevertheless, they have been successfully used in various AQ monitoring projects.

9.5.4 Calibration methods of LC sensor systems and networks

Despite the potential added value in air quality monitoring networks, low-cost sensors suffer from a number of operational limitations such as low data quality, cross-sensitivity, data loss, temporal and/or sensor drifts and other. It is widely accepted in the literature (Kang et al., 2022) that these systems should be calibrated and validated before deployment to obtain reliable readings.

The basic calibration is usually done in a laboratory environment (Northcross et al., 2013, Sousan et al., 2016, Wang et al., 2015): initially, the sensor element or device is exposed to a controlled environment with known concentrations of the pollutant(s) of interest. This exposure is repeated in order to estimate accuracy and precision. In addition, such exposure to different levels of the pollutant of interest allows for the estimation of the linearity of the sensor behavior, and especially for the verification of the concentration levels within which the sensor readings demonstrate a linear dependency from the actual concentrations. In this way, a simple linear model, which is presupposed for the sensor behavior, is specified, in terms of its slope and intercept.

An additional calibration step is frequently repeated on the field (often referred to as on-site or in-situ), via collocating the sensor device with a reference instrument (Cavaliere et al., 2018, Nowack et al., 2021). The approach is illustrated in Fig. 9.13, left panel. This allows for acquiring real-world feedback of the sensor behavior, and also provides further data for updating the linear behavior function of the sensor in order to better reflect the actual conditions of the specific environment within which it is expected to operate. Thus, for example, differences in the urban microclimate (temperature, relative humidity), cross-sensitivity with other (gaseous) pollutants like in the case of NO_x and O_3 sensors, and other parameters, may result in different behavior of the sensor in comparison to its performance in the lab, and therefore field-based collocation is an important (even necessary) step in order to establish a more robust and stable sensor behavior.

After collocation, the sensor node is placed in its actual operational location. This may still be different from the environment of collocation (e.g., in terms of urban photochemistry or particle formation mechanisms), and for this

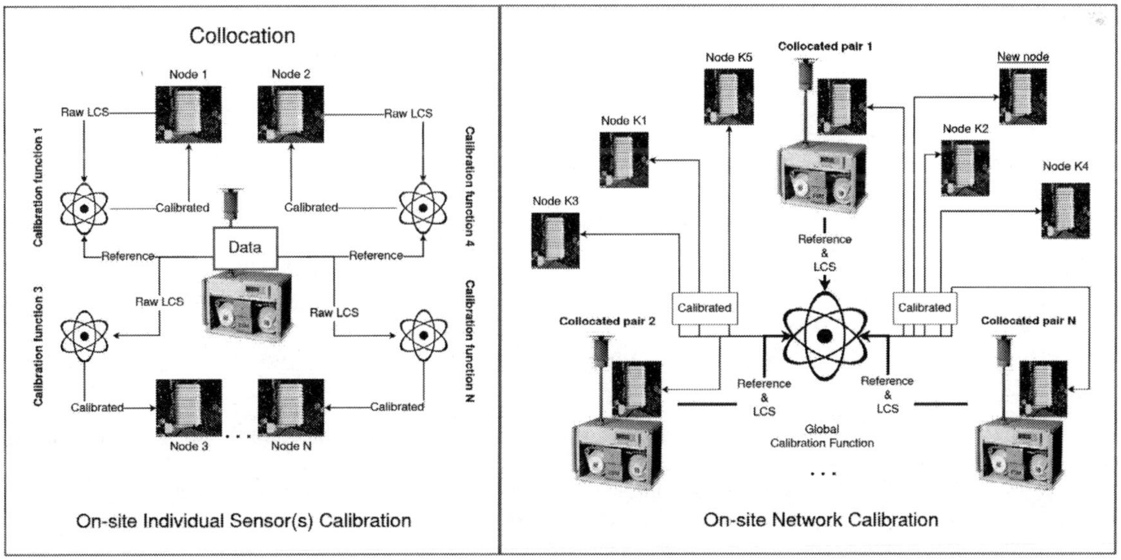

FIGURE 9.13 Schematic of the on-site calibration for (A) multiple nodes collocated with the same reference instrument where individual calibration functions are learned (*left panel*) and (B) partially overlapping nodes and reference networks where a single global calibration function is learned.

reason, the sensor will demonstrate a deviation from its performance and behavior (De Vito et al., 2020b) in comparison to the previous operational phase (of collocation to reference instruments). In this context, it is necessary to further apply calibration procedures, in order to ensure the performance of a low-cost sensor system under real/local conditions where the composition and physical properties of the atmosphere will vary greatly from the laboratory or from the initial field calibration conditions. Hence, the use of LCS requires extra steps to ensure that the readings are reflecting reality.

Taking into account all the above, on-site calibration is a postprocessing solution with the potential to address the aforementioned problem. Generally, on-site calibration can be divided into three categories (a) reference-based where at least one reference grade instrument operates inside the area of interest, (b) blind, where the network comprises only low-cost sensors, and (c) partially blind where the reference and low-cost sensor locations are far apart. Regarding the reference-based on-site calibration, the simplest conceived way of performing this type of calibration involves the collocation of a low-cost sensor system with a reference grade instrument on the field for a period of some days to a few weeks. The two systems measure independently (different measuring principles) the same pollutants and the data are collected via the data management and communication module. Considering the collected data (collocated measurements), a calibration function can be obtained to correct the low-cost sensor response to match the reference grade measurement as closely as possible. The calibration functions are usually modeled with parametric (e.g., linear regression, ARIMA), nonparametric (e.g., random forest, K-nearest neighbors) and deep learning (e.g., recurrent, convolutional, graph neural networks) algorithms, with a varying degree of success. These models are all learned via the machine learning paradigm from data. Once the calibration function is obtained, the low-cost system can be relocated to a designated location and apply the learned function (Bagkis et al., 2021).

The collocation approach improves the data quality of the calibrated sensor but gives rise to scaling issues. For example, even a medium-sized network of about 100 devices requires all the sensors to be deployed near the reference station location to obtain collocated measurements and derive the calibration function for each sensor separately. Therefore, special care should be paid when scheduling the collocation period for each sensor, to account for seasonal and other local effects. Another issue is the relocation and recalibration. Relocation requires personnel with adequate knowledge of operating, installing and performing tests on these systems. Furthermore, there aren't any guarantees that the collocation period or the range of the chemical (air composition) and physical conditions (meteorology) of the atmosphere will be sufficient to obtain a good calibration function (De Vito et al., 2020a). Thus, extending the calibration periods can make the calibration schedule difficult to follow and the scaling issues even worse. Finally, the seasonality of the atmospheric conditions will eventually cause the calibration to lose accuracy and lower the data quality and the sensor will need recalibration (relocation to the reference location, measurement period, relocation to designated location).

It becomes evident that the pairwise collocation approach as described above addresses the low data quality problem but the solution is rather time and resource consuming for larger networks. Another approach is to allow the nodes to remain in their location thus avoiding any relocation issues. For instance, on a fixed network, the node located the closest to a reference station location can be calibrated as if they were collocated. Another node can be calibrated on the calibrated data from the first node and so on. This approach is usually referred to as chain calibration and it predominantly depends on the chain length and the distance between the nodes. Remote or transfer calibration has been proven to be a better alternative than a chain of blind calibration. In this approach a calibration function is obtained on a collocated node and the same function is then applied to other nodes. It requires that the sensor technology that is used inside the systems is the same for all the nodes and the technology is reliable enough that when two nodes are operating under the same conditions the readings should be almost identical. For instance, the OPCs rely on the light scattering principle and show small intra-node variability. On the other hand, electrochemical gas sensors exhibit high intra-node variability and special care should be given to select sensors that behave as similarly as possible.

Finally, the network can be seen as a unit and a generalized calibration function can be learned (De Vito et al., 2022). In Fig. 9.13, right panel, the conceptual framework for this approach is depicted. For example, a reference network operates a small number of instruments and another LCAQS network supports it (Bagkis et al., 2022). Nodes are continuously collocated on every reference location while the rest are located on their specific locations. A single calibration function is learned on all the data provided by all the collocated pairs and is applied to all the single nodes in the network. This approach offers a number of advantages. Firstly, the need for relocation is ameliorated. Secondly, the model acting as the function is offered an increased number of data points for a single timestamp thus, extending the validity domain of the model and enabling faster learning. Consequently, the function does not rely only on a single collocation pair but on a network of pairs that measure continuously. Additionally, the collocated nodes can act as if they could provide high-quality readings when the reference instruments are maintained or are offline due to malfunctions. Recalibration can either be scheduled or triggered automatically when drifts from the expected performance are detected.

Building learned models for on-site calibration also calls for good validation strategies. Because many of the problems arise during the operational phase, choosing a random cross-validation strategy on historical data for example will lead to overoptimistic performance metrics. In operational mode, calibration is inherently a time series problem and the practitioner should consider the temporal aspect of obtaining readings and take actions to avoid data leakage. Furthermore, the spatial dimension is also important because the model will eventually be applied to an unknown location (never before seen in the training data). However, a spatial cross-validation strategy inevitably introduces temporal data leaks when it is performed on historical data in a batch-like manner. Therefore, when validating a spatiotemporal system (e.g., environmental monitoring networks) a combination of spatial and forward (temporal) cross-validation strategy should be devised to obtain reliable metrics and quantify the data quality of the network.

The validation procedures serve the model selection. The no-free lunch theorem (Wolpert & Macready, 1997) states that all optimization algorithms and subsequently machine learning (ML) models are equivalent across all datasets and problems; thus, model selection is a necessary step for all deployments. Factors such as the dataset size, the number of variables, the type of variables (continuous, nominal, categorical, binary, etc.), the preprocessing steps (normalization, feature engineering, feature selection, etc.), seasonality and more can drastically change the performance of the models. Therefore, a set of different models should be considered (Wang et al., 2023). Additionally, combinations of models (ensembles) have been proven to outperform the use of single models. Bagging (e.g., random forest) and boosting (e.g., gradient boosting) are two common approaches when combining the same algorithm (Bagkis et al., 2022). Stacking ensembles can combine different algorithms and lead to increased performance metrics (Kassandros et al., 2023). Overall, these approaches reduce the variance compared to single models but are prone to overfitting and increase the computational requirements. They also open up the potential to devise adaptability strategies to concept drifts (i.e. changes on the distribution characteristics of the measurements).

To establish a valid performance classification of the LCS systems, apart from standard metrics such as the root mean squared error and the Pearson correlation coefficient, relative expanded uncertainty (REU) should be considered. REU defines the acceptable uncertainty of an LCS system measurement when compared with a reference-grade or equivalent instrument. Stems from the European directive 2008/50/EC[2] and the associated ISO for gaseous pollutants (CEN/TS 17660-1:2021) aim to provide an uncertainty estimation in a formalized and unified way. According to the aforementioned documents a LCS system can be classified as fixed measurement, indicative measurement or objective estimation (Table 9.1).

The basic calculation of the expanded uncertainty at the limit value (LV) is defined in the ISO for gaseous pollutants. Moreover, the expanded uncertainty can be calculated at the lower assessment threshold or the upper assessment threshold. This calculation provides a single value for a specified evaluation period i.

$$U_r = 2\left(\frac{RSS}{n-2} - u^2(\text{bs}, \text{RM}) + [b_0 + (b_1 - 1)\text{LV}]^2\right)^{\left(\frac{1}{2}\right)} \tag{9.15}$$

TABLE 9.1 The relative expanded uncertainty (REU) limits for classifying LCS system measurements.

	SO_2, NO/NO_2, CO (%)	$PM_{2.5}$, PM_{10} (%)	Ozone (%)
Fixed measurements	15	25	15
Indicative measurements	25	50	30
Objective estimations	75	100	75

$$\text{RSS} = \sum_{i}^{n} (y_i - b_0 - b_1 x_i)^2 \tag{9.16}$$

Here U_r, is the expanded uncertainty, $u(bs, \text{RM})$, is the between reference method standard uncertainty, RSS represents the sum of (relative) residuals, x_i is the average result of the reference method over period i, y_i is the average result of the model over period i, while b_0 and b_1, are the coefficients of the regression $y = b_1\, x + b_0$. Note that the

2. European Union Directive (EUD). Directive 2008/50/EC of the European Parliament and of the Council of 21 May 2008 on ambient air quality and cleaner air for Europe. Off. J. Eur. Union 2008, L152. Directive 2008/50/EC has been updated by Directive 2024/2881/EU on ambient air quality and cleaner air for Europe

calculations are based on the defined LVs and differ for different pollutants. For example, to calculate REU for particles one should use daily averages while for Ozone 8-hour averages apply.

In addition, the research community has adopted a more analytical approach to calculating the REU for the full range of the reference concentration values. Based on the methodology presented by Spinelle et al. (2015) using Eqs. (9.15) and (9.16) the REU can be calculated relative to the full range of observed values.

$$U_r(y_i) = \frac{2\left(\frac{RSS}{n-2} - u^2(\text{bs, RM}) + [b_0 + (b_1 - 1)x_i]^2\right)^{\left(\frac{1}{2}\right)}}{y_i} \times 100\% \tag{9.17}$$

Worked Example 9.5

In a collocation experiment, you collected O_3 measurements from a reference analyzer (manufacturer: HORIBA) and an electrochemical low-cost sensor and aggregated them to 8-hour averages. The data are collected into three vectors *R*, *L*, and *C*. You also built a machine learning calibration function and calibrated the raw LCS data in situ. Calculate the REU for the full dataset at the LV.

Reference values

R = [92.0, 51.0, 118.0, 42.0, 97.0, 125.0, 58.0, 91.0, 111.0, 55.0, 79.0, 107.0, 52.0, 92.0, 94.0, 66.0, 59.0, 35.0, 86.0, 81.0]

Raw LCS data

L = [58.0, 64.0, 96.0, 27.0, 121.0, 84.0, 29.0, 113.0, 78.0, 16.0, 104.0, 76.0, 63.0, 77.0, 86.0, 60.0, 34.0, 32.0, 98.0, 39.0]

Calibrated LCS data

C = [83.0, 63.0, 100.0, 38.0, 90.0, 95.0, 47.0, 86.0, 99.0, 45.0, 73.0, 94.0, 62.0, 88.0, 97.0, 80.0, 57.0, 47.0, 90.0, 74.0]

Solution

The LV for O_3 is LV = 120 $\mu g\, m^{-3}$. Since the experiment is in-situ and you couldn't get access to the data from two reference analyzers, the random uncertainty of the standard method can't be computed directly therefore you will use the predefined value $u(bs, \text{RM}) = 0.41$ (for the HORIBA analyzer) based on Annex j of the ISO standard. Due to the "lack of comparability" between the candidate instrument (electrochemical LCS) and the reference instrument (HORIBA) it is assumed that the relationship between the two can be described by a linear relation. The number of datapoints is $n = 20$.

Raw LCS REU calculation: First we compute the regression coefficients b_0 and b_1 by applying the orthogonal regression.

$$Sxx = \sum_i^n (R_i - \overline{R})^2 = 12838$$

$$Sxy = \sum_i^n (R_i - \overline{R}) \times (L_i - \overline{L}) = 10428$$

$$b_1 = \frac{S_{xy}}{S_{xx}} = 0.81$$

$$b_0 = \overline{y} - b1 \times \overline{x} = 24.51$$

$$RSS = 7810$$

The period *i* here is the whole period containing all the 20 values and the REU will be calculated at the LV.

$$U_r = 2\left(\frac{7810}{20-2} - 0.41^2 + [24.51 + (0.81 - 1) \times 120]^2\right)^{\left(\frac{1}{2}\right)} = 51.47\%$$

Calibrated LCS REU calculation: We repeat the same process for the calibrated data.

$$Sxx = \sum_i^{20} (R_i - \overline{R})^2 = 12838$$

$$Sxy = \sum_i^{20} (R_i - \overline{R}) \times (C_i - \overline{C}) = 9148$$

$$b1 = \frac{S_{xy}}{S_{xx}} = 0.71$$

$$b_0 = \bar{y} - b_1 \times \bar{x} = 18.3$$

$$RSS = 3111$$

$$U_r = 2\left(\frac{3111}{20-2} - 0.41^2 + [18.3 + (0.71-1) \times 120]^2\right)^{\left(\frac{1}{2}\right)} = 27.41\%$$

Worked Example 9.6

Plot the REU for the full range of 8-hour averages included in the above dataset. Comment on the plot.

Solution:

In this example, you should repeat the above calculations 20 times, one for each data point using Eqs. (9.16) and (9.17). Note that the LV will not be used here but the 8-hour averages of the reference measurements will take its place. To reproduce the REU plot, sort both the REU and the reference values according to the reference data. It is evident that the REU can significantly vary for different concentrations. Even though the expanded uncertainty from Worked Example 9.5 showed that the raw and the calibrated LCS readings would be classified as objective estimations, a closer look at the full range reveals that the calibrated data are much closer to indicative measurements and the raw data are far from objective. Especially for the lower measurement range, the raw data are unreliable and should not be trusted.

As it is evident from Fig. 9.14, the use of the REU as a criterion for sensor calibration is an essential improvement of the calibration procedure per se. Not only does it address the issue of improved sensor performance statistics but may actually be used for upgrading a sensor system under operational conditions. This means that the use of the REU as a key component for accessing the success, robustness, applicability and advantage (in operational terms) of a sensor calibration procedure, should be a key requirement from any in-situ sensor calibration procedure. This can actually lead to product improvement and upgrade, taking into account the complexities and limitations of LCS-oriented air quality monitoring.

9.5.5 The use of low-cost air quality sensor systems

Low-cost air quality sensor systems (LCAQSS) have been used in various ways so far, while sensor elements have been integrated into even more devices and have appeared in even more applications. Indicative uses of LCAQSS include the following:

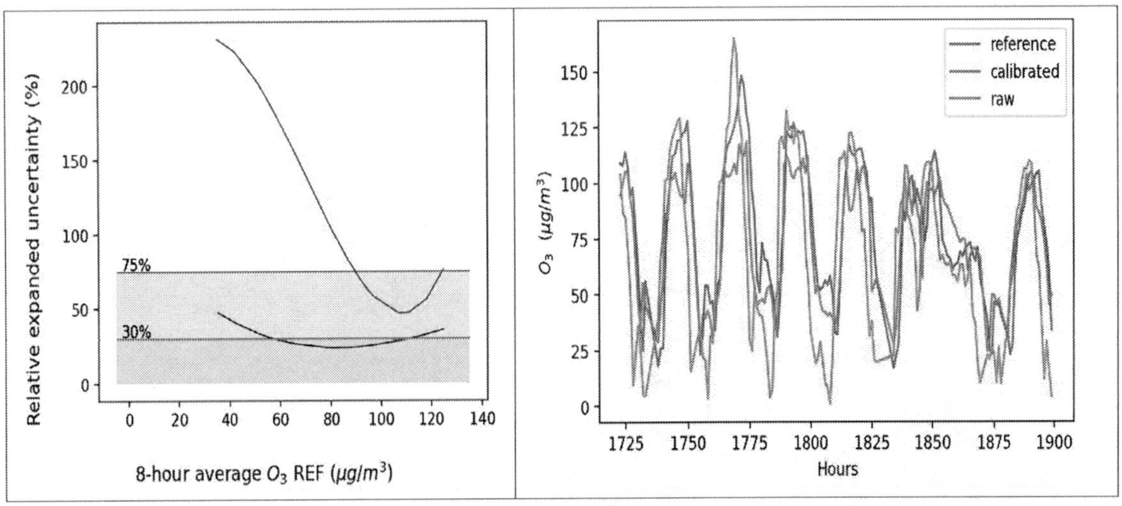

FIGURE 9.14 Left panel: Plot comparing the REU of the raw (blue line) and the calibrated (black line) LCS O_3 data. The lines are third degree polynomials on the obtained REU values fitted for clearance. Right panel: Ozone data before averaging.

1. Individuals, for monitoring outdoor (and/or indoor) air pollution levels. These initiatives are usually based on personal interest, which can be subdivided as follows:
   a. ephemerous or continuous, in accordance to citizen needs or as
   b. an organized activity, due to the participation in a citizen science initiative or project.

In both cases, a number of best practices that have resulted from relevant projects and activities, have taken the form of guidelines and suggestions from organizations like the Joined Research Center of the European Commission (Yatkin et al., 2022) and the Environmental Protection Agency (Clements et al., 2022; Environmental Protection Agency EPA, 2022), providing with practical advice concerning the installation, use, data analysis and interpretation measurements of such devices.

2. Scientific organizations (universities, research institutes, etc.), that employ them as part of an air pollution monitoring research activity.
3. Companies that have a business relationship to, or are interested in, air pollution, due to their activities (construction companies, mining companies, petroleum industries, etc.).
4. Public organizations like the environment department of local or central governments, as part of or even as a replacement of an "official" air quality monitoring network.

In addition to their wide applicability, LCAQSS have been utilized in various settings. For instance, they can be made mobile and mounted on vehicles or carried by individuals for personal exposure measurements or mapping of traffic emissions. Alternatively, they can be incorporated into drones for the purpose of measuring vertical profiles or for remote sensing purposes (like industrial, volcanic and shipping emissions). Another effective approach for LCAQSS is the use of dense networks to estimate air quality levels in urban environments. Of these, the most prevalent and useful is to set up LCAQSS as a fixed topology network. To make the best use of LCAQSS and minimize their limitations, the devices should be calibrated in the field and their calibrated values used as inputs for data fusion models in order to map air quality with high spatial resolution. A number of projects have used this approach, one of which (the KASTOM project, http://app.air4me.eu/) is presented in the next section as a characteristic example of the use of a state-of-the-art LCAQSS network.

9.5.5.1 A machine learning powered state-of-the-art LCAQSS network: The KASTOM project

The KASTOM project, a research initiative led by the Aristotle University of Thessaloniki, Greece, between 2019 and 2022, aimed to develop a cutting-edge air quality monitoring and modeling system. To this end, KASTOM deployed low-cost air quality sensor systems (LCAQSS) in the Greater Thessaloniki Area (GTA) to provide hourly data on the concentration of CO, NO_2, O_3, PM_{10}, $PM_{2.5}$, as well as temperature, relative humidity, and atmospheric pressure readings. The KASTOM node, as shown in Fig. 9.15, is a typical outdoor LCAQSS equipped with ECSs for gaseous pollutants (Alphasense, Ltd), an OPC for $PM_{2.5}$ and PM_{10} (Plantower Co., Ltd), and a BME280 BOSCH Sensortec solid-state sensor for meteorological parameters. Data transmission was accomplished via LoRa, and The Things Network's global collaborative IoT ecosystem supported data transfer through relevant LoRa gateways that were also installed in the GTA and Heraklion on the island of Crete, which was the second application domain.

As part of the project, the KASTOM nodes underwent a three-month collocation with reference instruments in the GTA before being deployed to 33 locations in the area. During this time, a device-level, machine learning-based computational

FIGURE 9.15 The KASTOM project LCAQSS. Left: inside of the node (1: electrochemical sensors; 2: OPC sensor; 3: meteorological parameters sensor; 4: battery). Right: the node in one of its installation locations in the city of Thessaloniki (OPC inlet visible on the bottom; LoRa antenna visible on the top).

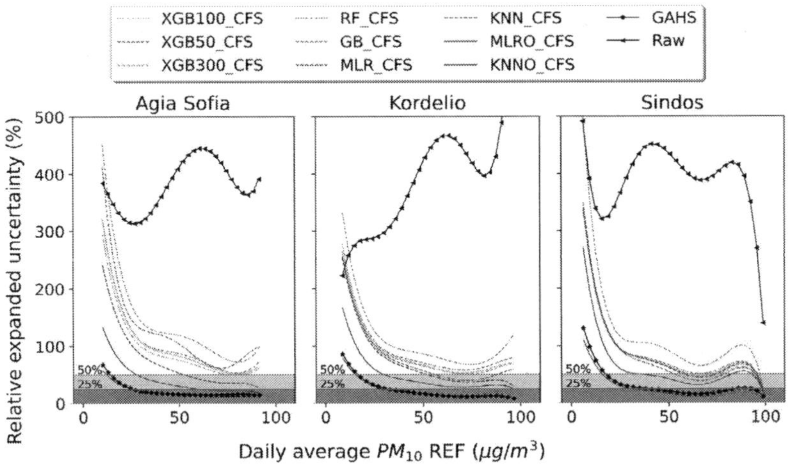

FIGURE 9.16 Relative expanded uncertainty plots comparing raw LCS data with 10 different machine learning algorithms with 3 reference instruments. The levels for indicative (50%) and fixed (25%) measurements regarding PM_{10} are colored (Bagkis et al., 2022).

calibration procedure was developed to improve the expanded uncertainty of the sensor systems in accordance with the standard CEN/TS 17660–1:2021 (Bagkis et al., 2021). In addition, a state-of-the-art meteorological and photochemical dispersion modeling system (WRF-CAMEX) was developed for the same application domain, supported by a cutting-edge emissions modeling system (NEMO), providing AQ mode results for all pollutants (NO_2, O_3, CO, $PM_{2.5}$, PM_{10}) on an hourly basis, with forecasts up to 72 hours in advance and a spatial resolution of 2 × 2 km (Liora et al., 2015, 2022)

To spatially bring AQ information to a neighborhood level, a data fusion approach was developed that combined traffic, local climate zone, land use, and urban morphology data, resulting in a fine-scale AQ estimation with a resolution of 100 × 100 m initially that was further improved to 30 × 30 m in later development stages for both nowcasting purposes. The public can access this information through a mobile phone application and a website (http://app.air4me.eu/).

The KASTOM LCS network operates along with the official national monitoring network. Three LCS are continuously collocated with three reference instruments in the areas of Agia Sofia (traffic), Kordelio (rural close to industrial) and Sindos (industrial). Incorporating the REU demonstrated that most ML-based calibration functions failed to satisfy the criterion in operation and further research was needed which led to the development of the genetic algorithm hybrid stacking (GAHS) framework (Fig. 9.16) that is able to provide highly accurate data (Bagkis et al., 2022; Kassandros et al., 2023). An advantage of building a generalized calibration function is that new sensor nodes can be seamlessly added to the calibration procedure as colocation is not a hard requirement. Moreover, if a collocated LCS fails the calibration will continue to be accurate based on the remaining collocated pairs. One caveat of the approach is that it requires a continuous flow of reference data in order to be highly accurate and slowly degrades if the flow is cut off. To demonstrate the usefulness of the calibrated LCS network, Fig. 9.17 depicts the PM2.5 contribution in the area, in an example where evidently traffic plays a major role in the particle generation and formation inside the city web.

9.5.6 Conclusions

The emergence of LCAQSS technologies and their increased availability and use by experts, governmental organizations, citizen groups and individuals, is proof of their capabilities (US GAO, 2020). The provision of direct access to AQ measurements is the key element to their growing use. Nevertheless, their popularity (both within the scientific community as well as in the "general public") should not be interpreted as an equivalence of validity and ground truth. Such devices are error-prone and their measurements may vary in terms of quality and representativity of the actual air pollution levels, while their improvement is one of the key future challenges in AQ monitoring (Sokhi et al., 2022). Moreover, there is still no unanimous acceptance of harmonized performance criteria for such devices, while, importantly, there are no guidelines on their use for official AQ monitoring. Nevertheless, the emergence of the first ISO standard on gaseous air pollution LCS, which used the REU as a key criterion for sensor categorization, paves the way towards the future (Illueca Fernández et al., 2023). On this basis, the scientific community has invested in the improvement of all components participating in their development, has suggested best practices and guidelines for their use, as

FIGURE 9.17 Data fusion instances generated by combining the LCS calibrated data with auxiliary data sources (traffic congestion, road network, local climate zones) via universal kriging. Areas of Thessaloniki, Greece.

well as advanced ways for their calibration and continuous reading correction (Brown & Martin, 2023). Such initiatives aim to serve a new paradigm of air pollution monitoring that has emerged, which makes use of low-cost instruments, in order to increase the population of monitoring instruments, therefore generating a vivid, heterogeneous "crowd" of measuring devices and leading to unprecedented insight in terms of the spatial (and temporal) resolution of provided measurements.

These developments have made it possible to improve and enhance data modeling and fusion approaches adding more air quality monitoring and forecasting capabilities to existing air pollution measurement and management systems and allowing such capabilities to appear in cases where no alternative is available.

It is expected that LCAQSS will play a cardinal role in the future of air quality monitoring (Narayana et al., 2022) in combination with reference instruments and/or in line with a hybrid (reference and nonreference) AQ monitoring network setup. In addition, their improvement is expected to fuel new modeling, data fusion and service provision projects supporting quality of life for citizens as well as decision-making for environmental managers and decision-makers.

9.6 Remote sensing of air quality

9.6.1 Introduction

Air pollution monitoring involves both in-situ and remote sensing methods. The in-situ observations are mainly based on fixed observations (e.g. tower observations), mobile measurements (using ground transportation and aircraft) and wind-flowing observations (e.g., ozone sondes) (Oke et al., 2017). Though the in-situ observations are suitable for providing the surface concentrations, due to assistance requirements and the cost of setting up the observatories, the in-situ observations are limited globally (Carpenter et al., 2022). The spatial coverage of in-situ observations is thus limited to certain latitudes and longitudes globally. The spatial coverage limitation of in-situ observations has been overcome using remote sensing observations of air quality. Remote sensing observations of air quality utilize both active (using their own source of emission/light) and passive (utilizing natural energy/sunrays) remote sensing methods (Devara, 1998). They can collect information for a wider path, either vertically, horizontally or both simultaneously (Martin, 2008). Using remote sensing observations, one can assess the state of air quality across the globe (Gupta et al., 2006) and compare the variability during the episodic changes in air quality and its impacts during the same time (Biswal et al., 2020; Singh, et al., 2020; Sokhi et al., 2021). The present section explores the remote sensing observations of air quality with a focus on different data sets and their availability for various air pollutants. The work also discusses satellite retrieval algorithms and uncertainties in the remote sensing of air quality to understand the air quality assessment better using the remotely sensed data.

9.6.2 Basic principles of remote sensing of air quality

The study of identifying earth surface features and estimating their geo-biophysical attributes using electromagnetic radiation as a medium of interaction is known as remote sensing (Navalgund et al., 2007). In common language, remote sensing usually refers to satellite, aircraft and drone platforms for measuring atmospheric species.

In all remote sensing procedures, incident radiation falling on an object of interest interacts with the surrounding environment so that the target's physical characteristics cause the incident radiation to change and reflect the incident radiation that the sensor records. An example is the application of imaging systems, also known as optical remote sensing. Using nonimaging sensors and detecting released energy are other aspects of remote sensing (thermal remote sensing). One can understand the general principle of remote sensing starting with the illumination source, a fundamental necessity for any remote sensing procedure. This energy interacts with the atmosphere as it travels from source to destination. This interaction might happen twice as the energy moves from the target back to the sensor. Using a sensor, one can record the scattered or emitted radiation from the target, which must be sent, frequently electronically, to a station for receiving and processing. Once the information is recorded, the information about the lit target is extracted from the processed image using visual, digital, or electronic interpretation involving various analysis techniques. The extracted information solves problems or discloses new information about the objective (Campbell & Wynne, 2011).

Because the technique can acquire information from a distance without interacting with the physical space, its applications are varied, including defense, agriculture monitoring, weather and climate, natural hazards monitoring and assessment, air quality, and acquiring information for remote places (Liang, 2017). The sensors aboard the platform (e.g., aircraft or satellite) can be categorized as active or passive. The Sun's energy acts as a source of radiation in the atmosphere, and the sensors using the Sun's energy (either reflected in the visible ranges or re-emitted in the thermal infrared wavelengths) are called passive sensors. However, if the sensor provides its energy source for illumination and records the information without depending on the Sun's energy, it is termed an active remote sensor (Tedesco, 2015). Thus, for passive remote sensing, the preferred ranges to operate are visible, infrared, and thermal infrared, with some microwave portions too. In contrast, for the active remote sensors, the microwave band of the electromagnetic spectrum is used primarily (NASA Earth Data, 2024).

For applications relating to air quality, remote sensing of air pollutants has a long history from the Geostationary Operational Environmental Satellite (GOES) satellite for nearly five decades (Lyons & Husar, 1976). However, remote sensing of air pollution involves instruments based on the ground, aircraft and satellites. Remote sensing of air quality has the advantage of noninterfering for source effluent monitoring and integrated-path measurements over large geographical areas (Melfi, 2005). The air quality remote sensing based on sensor location is discussed hereunder.

9.6.2.1 Ground-based observations

Ground-based remote sensing employs instruments at Earth's surface for air quality observations. The instruments can be either fixed at a particular location or can be mounted in a movable platform, hence the area of extent covered by these instruments varies. These ground-based observations can be further discussed specifically by the instrument types as follows:

9.6.2.1.1 Light detection and ranging

One of the popular instruments for ground-based remote sensing is light detection and ranging (LIDAR), a single or double-ended active remote sensing instrument based on laser light scattering (Melfi, 2005). LIDAR uses a monochromatic laser facing upwards primarily to locate aerosols and cloud droplets. LIDAR can also infer wind and turbulence by tracking the movements of aerosols in some systems and digital surface models of an urban environment (e.g., buildings, trees) while facing downwards (Oke et al., 2017). The use of LIDAR goes back more than five decades in atmospheric science, which started with aerosol measurements initially (Barrett & Ben-Dov, 1967). The advantage of using LIDAR for ground-based remote sensing includes its broad coverage of the area and its possibility of monitoring changes from the surface to a few kilometers above the ground vertically, with the possibility of mounting it on a vehicle for movable measurements (Xiang et al., 2020). With its advantages of monitoring, LIDAR has been modified for observing ozone (Papayannis et al., 1999), CO_2 flux measurements (Aiuppa et al., 2015; Wang et al., 2022), and with many other trace gases, e.g., NO, H_2O, O_3, SO_2, and CH_4 (Argall & Sica, 2015).

9.6.2.1.2 Sun photometer

Another popular ground-based remote sensing instrument is the Sun-photometer, which is based on a passive remote sensing technique (Shaw, 1983). Sun photometers are either fixed at a particular location or can be mounted on a moving platform like a ship or with handheld instruments (microtops) for field observations (Porter et al., 2001). Sun photometers are primarily used to measure either column aerosol optical depths (AODs) or total column ozone in the atmosphere above the instrument. The sensor is pointed at the sun with a narrow field of view and calculates the total column measurement based on Beer's Law.

$$\mathrm{I} = \mathrm{Io}\ \exp\left(\tau/\text{air mass}\right) \tag{9.18}$$

where τ = sum of the aerosol and molecular optical depths, air mass = the one over the cosine of the solar zenith angle and varies from 1 when the Sun is directly overhead to 20 or more when the Sun is on the horizon, and I and Io = solar direct beam at the top and bottom of the atmosphere.

As the aerosol loading increases, more light is scattered out of the direct solar beam and the sun photometer measurement decreases.

Aerosol Robotic Network (AERONET), maintained by the National Aeronautics and Space Administration (NASA), is one example of using Sun-photometers for monitoring aerosols globally (https://aeronet.gsfc.nasa.gov/). AERONET has more than 25 years of data collection, available publicly to users for research purposes through its site. The ocean observations of aerosols are available through Maritime Aerosol Network (https://aeronet.gsfc.nasa.gov/new_web/maritime_aerosol_network.html). WMO Precision Filter Radiometer (PFR) network is another example of Sun photometer data collection (https://gml.noaa.gov/obop/mlo/programs/coop/pfr/pfr.html), whereas for the column ozone and SO_2 measurements, Brewer spectrophotometer networks collected the data (Zhao et al., 2021).

9.6.2.1.3 Scintillometers

There are also indirect measurements of air pollutants using ground remote sensing of meteorological parameters, for example, Scintillometers. Scintillometers, which monitor turbulent fluxes of heat and moisture in the atmospheric boundary layer, are another popular ground-based remote sensing instrument. Scintillometers are based on transmitter-receiver sets, which are based on electromagnetic signal scattering by the turbulent eddies in the atmosphere. Scattering in signal by eddies is detected as fluctuations recorded by the scintillometer's receiver (i.e., scintillations). The turbulent eddies are driven by surface forcing (e.g., frictional drag, evaporation, wind shear, radiation and turbulent wakes) (Stull, 1988). To derive the surface heat fluxes, atmospheric turbulence and electromagnetic wave propagation theoretical principles are combined (Van Kesteren, 2012). The large-aperture scintillometer can provide the area-averaged fluxes up to 10 km (Hartogensis et al., 2003). The knowledge of these fluxes is important along with boundary layer height to the pollutants, for calculating the fluxes of trace gases in the atmosphere (DeBruin, 2009).

9.6.2.2 Aircraft based observations

Aircraft observations of air quality are another critical remote sensing approach and have existed for more than three decades (Joranger & Ottar, 1984). Aircraft observations typically mount the instruments over an aircraft, and depending on the size and capability, the plane flies to a certain height in the atmosphere for collecting various observations (Johnson et al., 2012). Many research agencies around the world have their dedicated research aircraft, e.g., the BAe-146 four-engined jet at Facility for Airborne Atmospheric Measurements at National Center for Atmospheric Science, United Kingdom (https://www.faam.ac.uk); C-20A, DC-8, P-3 and B-200 aircraft at NASA (https://www.nasa.gov/centers/armstrong/home/index.html); Environmental Research Aircraft- SkyArrow at National Research Council of Italy (https://www.eufar.net/aircrafts/74; Gioli et al., 2006); and B-200 aircrafts at National Remote Sensing Center, Indian Space research Organization (ISRO) (https://www.nrsc.gov.in/EOP_Aerial_AirbornePlatform?language_content_entity = en). These aircraft mount various instruments on board, including LIDAR, which allow them to record the state of meteorology and pollution from the sky and provide better spatial coverage at a particular time. The aircraft observations are also crucial for cloud seeding globally (Andronache, 2018). There are several field campaigns for air quality measurements, where such aircraft observations provide vital information about dispersion and concentration variability over a region, for example, Pacific Dust Experiment Campaign (Stith et al., 2009); CAIPEEX-2009 (Padmakumari et al., 2013); Fennec (Ryder et al., 2015); GAUGE (Palmer et al., 2018); SCOUT-AMMA (Cairo et al., 2010); CLARIFY-2017 (Haywood et al., 2021). However, aircraft observations are not cost-effective, so their use in field campaigns is limited globally.

9.6.2.3 Unmanned aerial vehicles

An alternative approach for remote sensing observations from the sky, which has been gaining popularity in recent years, is the use of UAVs, where small size aircraft or drones are equipped with instruments and controlled from the ground (Kezoudi et al., 2021; Ramanathan & Carmichael, 2008). UAVs are particularly useful for recording observations over a smaller region compared to satellites and aircraft and offer accurate information on emissions and composition in a particular layer of the atmosphere (Levy et al., 2004; Lambey & Prasad, 2021). The UAVs also have the advantage of approaching the sources closer than aircraft or satellites, particularly in case of emissions, mainly from hazardous effluents (Corrigan et al., 2008). The UAVs are more versatile and affordable with the advancements of drones and are widely being used in support of satellites and aircraft observations for release control, global warming, and urban and industrial air safety (Villa et al., 2016). The importance of UAV observations is unmatched when the interest is near surface observations (Casazza et al., 2019).

9.6.2.4 *Satellite-based observations*

Remote sensing of air quality largely depends on the use of satellite observations. Various satellites with geostationary or polar orbiting have different onboard sensors, which can measure air pollutants for the tropospheric and total column air (Martin, 2008; Zhang et al., 2020a; Zhang et al., 2020b). In the 1970s, first-time aerosol measurements using satellites started using instruments such as AVHRR, Landsat and GOES for transported aerosols from the desert over oceans (Fraser, 1976; Mekler et al., 1977), intended to monitor aerosols over oceans. With the operation of the TOMS instrument onboard the Nimbus 7 satellite, the stratospheric O_3 measurement started in 1978 (Martin, 2008). With more than fifty years of satellite measurements, satellite-based remote sensing has developed as a crucial platform for analyzing air quality information around the globe with many pollutant concentrations and other indices observed at the surface, tropospheric, stratospheric, and total column of the atmosphere (De Sherbinin et al., 2014). The advantages of satellite-based observations of pollutants include broad spatial coverage, freely accessible data for research, a relatively long period of dataset availability, and higher resolution observations with the recent advancement of technology (Bodah et al., 2022; Douros et al., 2023). The overview of various satellite-based instruments for monitoring air quality is given in Table 9.2, with their platform, monitored parameters, and observational frequency. It is to be noted that the most advanced ones in the list are Sentinel satellites, where the monitoring is a finer resolution (i.e., 300×300 m in Sentinel 3 and 7×3.5 km in Sentinel 5p), with approximately 14 orbits of earth each day. The advancement in spatial and temporal resolutions of remote sensing air quality observations can solve many issues related to earlier coarser resolution patterns and temporal limitations of daily average values.

9.6.2.5 *Conversions between ground-based and satellite-based observations*

The ground-based observations of various pollutants often account for their mass per unit volume (e.g., $\mu g\,m^{-3}$), whereas the satellite observations are accounting volume mixing ratio (e.g., parts per million, parts per billion volume—ppmv, ppbv) for gaseous pollutants. The aerosol measurement from satellite gives a columnar measurement known as AOD, which is a quantitative estimate of the amount of aerosol present in the atmosphere, based on their extension properties (Li et al., 2021). The surface observations for aerosols are based on concentrations measured for specific classes of aerosols based on their size, i.e., $PM_{2.5}$ or PM_{10}, which defines the PM concentrations with diameters up to 2.5 μm or 10 μm, respectively.

Often, comparison of surface-based observations to satellite observation is not direct, and one need to employ the mathematical formulae to convert the units. To convert the units between the mass per unit volume and volume mixing ratio, one must calculate the conversion as explained in Chapter 2, Section 2.6.3.

Table 9.3 lists commonly used factors that allow the conversion of gaseous air pollutants between units of ppb or ppm and $\mu g\,m^{-3}$. The conversion factors listed in Table 9.3 are calculated for an air temperature 20°C and pressure of 1013 hPa (WHO, 2021). The reader should note that if different atmospheric conditions are used, then appropriate conversion factors should be used (see Section 2.6.3). The reader should note that if different atmospheric conditions are used, then appropriate conversion factors should be used (see Section 2.6.3).

Here, it is to be noted that NO_x in $\mu g\,m^{-3}$ is usually expressed as NO_2, that is, $[NO\ (ppb) + NO_2\ (ppb)] \times 1.914 = NO_x\ \mu g\,m^{-3}$.

Worked Example 9.7

If we have 40 ppb of O_3 and 60 ppm of CO derived from satellite observations, what are the corresponding values in mass per unit volume?

Solution:

Using the data in Table 9.3, we can find the mass per unit volume ($\mu g\,m^{-3}$) values.

For O_3, the mass per unit volume ($\mu g\,m^{-3}$) value will be:

TABLE 9.2 Main satellite remote sensing-based instruments and their platform, parameter observed, and frequency.

Instrument	Platform	Start year	Nadir resolution (km)	Spectral range (μm)	Parameters observed
GOME	ERS-2	1995	320 × 40	0.23–0.79	NO_2, HCHO, SO_2, O_3
MOPITT	Terra	2000	22 × 22	4.7	CO
MISR	Terra	2000	18 × 18	4[a] λ, 0.45–0.87	Aerosol Optical Thickness (AOT)
MODIS	Terra Aqua	2000 2002	1 × 1; 10 × 10	36[a] λ 0.41–14.2	AOT
AIRS	Aqua	2002	14 × 14	3.6–16	SO_2, CO
SCIAMACHY	ENVISAT	2002	60 × 30	0.23–2.3	NO_2, HCHO, SO_2, CO, O_3
OMI	Aura	2004	24 × 13	0.27–0.50	NO_2, HCHO, SO_2, O_3
TES	Aura	2004	8 × 5	3.3–15.4	CO, O_3
PARASOL	PARASOL	2004	18 × 16	9[a] λ, 0.44–1.0	AOT
CALIOP	CALIPSO	2006	40 × 40	0.53, 1.06	AOT
GOME-2	MetOp	2006	80 × 40	0.24–0.79	NO_2, HCHO, SO_2, O_3
IASI	MetOp	2006	12 × 12	3.6–15.5	CO, O_3
ETM +	Landsat 7	1999	30 m	0.45–12.50 in 8 bands	Particulate Matter (PM)
VIIRS	Suomi	2011	Polar	412 nm–12 μm in 22 bands	AOT
OMPS	Suomi-NPP	2011	50	250–310 nm	Aerosols, SO_2, O_3
ABI L2 +	GOES-16	2016	2	0.47–2.26 in 6 bands	Aerosols, dust storms, volcanic eruptions, and forest fires
OCO	OCO-2	2014	Polar, 1.29 × 2.25	1594–1619 nm	CO_2
AHI	Himawari9	2016	0.5 × 2	470 nm–13.30 μm in 16 bands	Aerosols, SO_2, CO_2 & O_3
TROPOMI	Sentinel 5p	2017	7 × 3.5	270–2385 nm	CO, NO_2, SO_2, O_3, CH_4 & Aerosols
Orbiting Carbon Observatory (OCO)	OCO-3	2019	2.25 × 1.29	0.765–2.06 in 3 bands	CO_2
IMAGER and SOUNDER	INSAT-3D	2013	1, 4, and 8	0.55–12.5 in 6 bands	Aerosol, Total O_3
OLI	Landsat-8	2013	30 m	0.43–12.51 in 11 bands	PM_{10}
MIPAS	ENVISAT	2002	Vertical, 3	4.15–14.6	N_2O_5, CFC-12, COF_2, CCl_4, HCN, CFC-14, HCFC-22
OLCI and SLSTR	Sentinel 3	2022	300 × 300 m; 500 × 500 m;	400 nm–1020 nm in 21 bands	Aerosols

Abbreviations are listed in Appendix 9.1.
[a]*Number of discrete wavelengths.*
Source: Modified after Singh et al. (2021).

TABLE 9.3 Conversion factors for gaseous air pollutants concentrations for air temperature of 20°C and pressure 1013 hPa (WHO, 2021).

Air pollutant	Conversion factor
O_3	1 ppb = 1.9957 $\mu g\,m^{-3}$
NO_2	1 ppb = 1.914 $\mu g\,m^{-3}$
CO	1 ppm = 1.165 $mg\,m^{-3}$
SO_2	1 ppb = 2.660 $\mu g\,m^{-3}$

$$40 \times 1.9957\ \mu g\,m^{-3} = 79.828\ \mu g\,m^{-3}$$

For CO, mass per unit volume ($mg\,m^{-3}$) value will be:

$$60 \times 1.165\ mg\,m^{-3} = 69.900\ mg\,m^{-3}$$

However, when it comes for the conversion of aerosol concentration, the process is not straightforward. The conversion formulae are generally found based on regression methods, which involve several other quantities, including boundary layer height (BLH), surface temperature (*T*), relative humidity (RH), wind speed (WS) over the area of interest. The simplest algorithm for deriving surface $PM_{2.5}$ concentration from satellite-derived AOD can be understood with the van Donkelaar's algorithm (Van Donkelaar et al., 2012; Zhang & Kondragunta, 2021):

$$PM_{2.5} = \text{offset} + \text{slope} \times AOD \tag{9.19}$$

Here the calculation of slope and offsets are obtained from a look up table, which records for daily offsets and slopes based on linear regression relationships between $PM_{2.5}$ and AOD for a particular region. The relationship is also explored by Zeydan and Wang (2019) who stated this simplest relation between PM2.5 and AOD based on regression coefficients as:

$$PM_{2.5} = \alpha + \beta_1 \times AOD \tag{9.20}$$

where α is intercept and β_1 is regression coefficient. However, these simplest models based solely on intercept and regression coefficient to convert $PM_{2.5}$ to AOD or vice versa are not getting good accuracy due to nonconsideration of meteorological parameters. The inclusion of various parameters leads to different relations for the conversion. Some of these formulations are provided here, as adapted from Zeydan and Wang (2019):

$$PM_{2.5} = \alpha + \beta_1 \times \left(\frac{AOD}{BLH}\right) + \beta_2 \times \left(\frac{AOD}{WS}\right) \tag{9.21}$$

$$\ln(PM_{2.5}) = \alpha + \beta_1 \times \ln(BLH) + \beta_2 \times (RH) + \beta_3 \times \ln(AOD) \tag{9.22}$$

$$\ln(PM_{2.5}) = \alpha + \beta_1 \times \ln(BLH) + \beta_2 \times \ln(T) + \beta_3 \times \ln(WS) + \beta_4 \times \ln(RH) + \beta_5 \times \ln(AOD) \tag{9.23}$$

where $PM_{2.5}$ is hourly measured ground-based mass concentration ($\mu g\,m^{-3}$), AOD is Aerosol Optical Depth (unitless) retrieved from satellite data, α is intercept and α), AOD (unitless) retrieved from satellite data, α is intercept and β_1-β_5 are regression coefficients. Now if one wants to derive continuous information about the surface $PM_{2.5}$, the regression should not be solely based on satellite-derived AOD, but have to account for model simulations, and relative uncertainties calculated from ground-based sun photometers. One such effort made by Van Donkelaar et al. (2016) resulted in global maps of monthly $PM_{2.5}$, where the authors accounted multisatellites observation along with model simulations and uncertainty corrections. Some of the regression models also used normalized difference vegetation index (NDVI), which quantifies vegetation by measuring the difference between near-infrared (which vegetation strongly reflects) and red light (which vegetation absorbs), and ranges between -1 to $+1$ (Drisya & Roshni, 2018).

Worked Example 9.8

If one assumes that monthly averaged value of AOD over Rome, Italy is 0.4; BLH is 1500 meters, RH is 76%, WS is $0.2\ m\,s^{-1}$, intercept (offset) value is 9.21, and the regression coefficients β_1, β_2, and β_3 are 18, 12, and 57, respectively, what will be the surface $PM_{2.5}$ concentration if one takes surface $PM_{2.5}$ as a function of (i) only AOD, (ii) AOD, BLH, and RH?

Solution:

Using Eqs. (9.20) and (9.22), one can calculate the monthly surface $PM_{2.5}$ values for the situations. In case (i):

surface $PM_{2.5} = 9.21 + 18 \times 0.4 = 16.41\ \mu g\, m^{-3}$
in case (ii), surface $PM_{2.5} = 9.21 + 18 \times \ln(500) + 12 \times \ln(76) + 57 \times \ln(0.4)$
$= 9.21 + 18 \times 6.21 + 12 \times 4.33 + 57 \times (-0.91) = 121.08\ \mu g\, m^{-3}$

9.6.3 Retrieval algorithms

9.6.3.1 Satellite retrieval algorithms

For the remote sensing measurements (either active or passive), based on the measured radiation, one may account for refraction, doppler shift or spectral radiance, whereas the requirement may be temperature distribution, wind, pressure, air pollutant or any other quantity. This situation refers to the common problem of remote sensing measurement, where the direct measurement of the parameters of interest is not observed, formulating the "retrieval" problem. The retrieval problem is based on measurements and the parameters' relation, based on finding the inverse of a forward function. Often these inverse problems are not trivial with either a nonexclusive or no mathematical solution (Rodgers, 1997). This situation leads to finding the best possible solution, consistent with the measurements, supported by a range of optimization criteria. The process is known as the retrieval algorithm for satellites, which allows you to finally find the unknown atmospheric quantities based on the available measurements. Regarding air pollutants, the viewing condition of satellites is known. However, for the atmospheric and surface properties, assumptions must be made. Meteorological inputs required (e.g., pressure or temperature) can be obtained from global model outputs, and the spectral range is chosen with the dominant absorption by the quantity of interest (Veefkind et al., 2007). Because of high temporal and spatial variability, aerosols are always challenging to retrieve. Retrieval methods often involve various sophisticated mathematical techniques, including the Gauss–Newton, Jacobian, Levenberg–Marquardt, and optimal estimation (Torres et al., 1998). These retrieval algorithms evolved with technological advancements, and the different retrieval algorithms have strengths and limitations (Kahn & Gaitley, 2015; Mishchenko et al., 2010; Wei et al., 2020).

9.6.3.2 Retrieval algorithm for air pollutants

The information about the products needed for air quality analysis from satellite data is obtained through the retrieval algorithms, which generally use some formulation based on the information collected and deliver a parameter at a particular resolution. For different air quality variables, various retrieval algorithms are available with multiple options for a particular instrument (Wei et al., 2020). For AOD, one has more than fifteen well-established retrieval algorithms (Zhang et al., 2020a, 2020b) along with many research publications proposing an improvement in all these algorithms (e.g., Garcia et al., 2016; Guang et al., 2012; Han et al., 2017; Huang et al., 2018). The further distinctions in retrieval algorithms are due to instrument design, e.g., single view methods: dark target (DT), deep blue (DB), multi-angle implementation of atmospheric correction (MAIAC), the simplified aerosol retrieval algorithm, contrast reduction method; Multiple view angle methods; active LIDAR sensor algorithms; multisensor synergy retrieval methods; algorithms for polarization sensor; algorithms for geostationary satellites etc. (Wei et al., 2020).

These algorithms have clear methodologies, and the algorithm's processing starts with data collection by the satellite sensors. One example is the MAIAC retrieval algorithm for AOD, which provides the AOD at a spatial resolution of 1 km (Lyapustin et al., 2018). Fig. 9.18 describes the systematic flow chart of operations. The complete equations utilized in all the steps mentioned in Fig. 9.18 are detailed in Lyapustin et al. (2018). The MAIAC algorithm has its own advantages over standard algorithms but needs enormous computational resources due to the requirement of multiday observations for retrieval and high resolution (Wei et al., 2020). Therefore, MAIAC is more susceptible to error from the land surface than the other popular retrieval algorithms, such as DT and DB (Lyapustin et al., 2018)

9.6.4 Data accessibility

The satellite data for various remote sensing satellites are available via their homepage and operating agency links. Table 9.4 lists data links for accessing these data sets from various instruments/onboard satellites, as mentioned in Table 9.2. Some data-providing platforms offer programming interfaces and visual interfaces for accessing stored data.

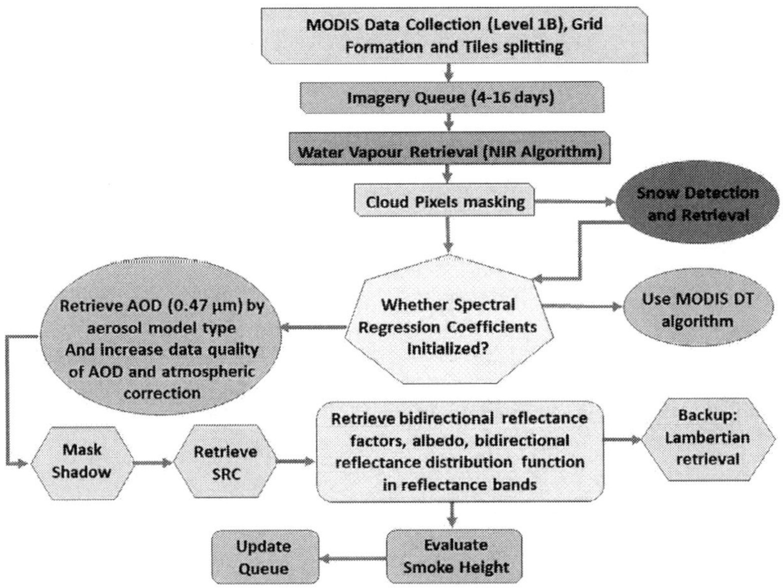

FIGURE 9.18 Flow chart of the algorithm MAIAC for AOD retrieval from MODIS. *Modified after Lyapustin, A., Wang, Y., Korkin, S., & Huang, D. (2018). MODIS collection 6 MAIAC algorithm. Atmospheric Measurement Techniques, 11 (10), 5741-5765 and Wei, X., Chang, N. B., Bai, K., & Gao, W. (2020). Satellite remote sensing of aerosol optical depth: advances, challenges, and perspectives.* Critical Reviews in Environmental Science and Technology, *50(16), 1640–1725.*

These user interfaces give users a simple method for choosing and obtaining the data. The offered data type and quantity of satellite products are where there are the most differences.

Apart from the above-listed sources, there are some platforms where multiple datasets of a particular agency (e.g., NASA, ESA) are available and some platforms where one can access all the datasets from different agencies. One such platform is Google Earth Engine (https://developers.google.com/earth-engine/datasets), where users can select the data from various satellites and model outputs and create a desired plot of the variable over a selected region. Asia Pacific Data Research Center (APDRC) is another platform where different datasets are available for users (http://apdrc.soest.hawaii.edu/data/data.php). Here the user can select any platform to access the data from the options provided next to the data and visualize the variable before downloading it. Two popular platforms for accessing the NASA datasets are Earth data (https://www.earthdata.nasa.gov/) and Giovanni (https://giovanni.gsfc.nasa.gov/giovanni/), respectively.

All these platforms usually ask for a user login to be created, not involving any payment from the user. With the user login, one can access and create a visualization of various data sets as required. One can understand the simplicity of the process by the example of using the Giovanni portal. Let us say one wants to check the AOD values observed using MODIS (Aqua) over the African region for December 1–10, 2022 in these three steps:

Step 1: Select the variable/parameters for a period and over an area (Fig. 9.19).
Step 2: Plot the results in the portal (Fig. 9.20).
Step 3: Download the final image or data for the results for further analysis or plotting by a different software (Fig. 9.21).

9.6.5 New developments, uncertainties and limitations

Remote sensing of air quality had developed a long way from its initial phases of the 1970s when space-based instruments were installed in polar-orbiting satellites to assess air pollution only daily. Ground-based remote sensing, especially with the advancement of drones in current years, has improved unprecedentedly by reducing the cost of UAVs and improving flight and battery timings (Lambey & Prasad, 2021). With new algorithms like Multi-Angle Implementation of Atmospheric Correction (MAIAC), aerosol products are available from MODIS at a 1 km spatial resolution, which greatly supports air quality scientists (Lyapustin et al., 2018). The new launches of geostationary satellites, including Geostationary Environment Monitoring Spectrometer (GEMS), Tropospheric Emissions: Monitoring Pollution (TEMPO), and Sentinel 4, make it possible for hourly air quality measurements from space (http://tempo.si.edu/index.html). With its launch in 2020, GEMS captures hourly air pollution levels for almost 20 countries in Asia, whereas North American and European measurements will be covered by TEMPO and Sentinel-4, with their operations

TABLE 9.4 Main satellite remote sensing-based instruments, and their data links.

Instrument	Operating agency[a]	Data link
GOME	ESA	https://earth.esa.int/eogateway/
MOPITT	NASA	https://www2.acom.ucar.edu/mopitt; https://opendap.larc.nasa.gov/opendap/MOPITT/contents.html
MISR	NASA	https://eosweb.larc.nasa.gov/project/misr/mil3maen_table.
MODIS	NASA	https://modis.gsfc.nasa.gov/data/dataprod/
AIRS	NASA	http://www.airs.jpl.nasa.gov/.
SCIAMACHY	ESA	https://earth.esa.int/eogateway/
OMI	NASA	http://eos.gsfc.nasa.gov/projects.html.
TES	NASA	https://tes.jpl.nasa.gov/tes/data/
PARASOL	AERIS, CNES	https://www.icare.univ-lille.fr/parasol/
CALIOP	NASA	http://www.calipso.larc.nasa.gov/about/.
GOME-2	ESA	https://earth.esa.int/eogateway/
IASI	AERIS, CNES	https://iasi.aeris-data.fr/
ETM +	NASA	http://earthexplorer.usgs.gov
VIIRS	NOAA	https://ladsweb.modaps.eosdis.nasa.gov/search/order/1/VIIRS:Suomi-NPP.
OMPS	NASA	https://earthdata.nasa.gov/earth-observation-data/near-real-time/download-nrt-data/omps-nrt.
ABI L2 +	NASA	https://www.ncdc.noaa.gov/airs-web/search.
OCO	NASA	https://www.esrl.noaa.gov/gmd/ccgg/OCO2/fluxmaps.php; https://goldsmr5.gesdisc.eosdis.nasa.gov/data/MERRA2/M2I3NPASM.5.12.4/.
AHI	JMA	https://www.data.jma.go.jp/mscweb/data/himawari/sat_img.php.
TROPOMI	ESA	https://earth.esa.int/eogateway/. https://scihub.copernicus.eu/dhus/#/home
Orbiting Carbon Observatory (OCO-3)	NASA	https://disc.gsfc.nasa.gov/datasets?keywords = oco3&sort = -dataset.
Meteorological (MET) - IMAGER and SOUNDER	ISRO	https://www.mosdac.gov.in/catalog/satellite.php?mission = INSAT-3D
OLI	NASA	http://earthexplorer.usgs.gov.
MIPAS	ESA	https://earth.esa.int/eogateway/instruments/mipas
OLCI and SLSTR	ESA	https://scihub.copernicus.eu/

[a]*ESA: European Earth Agency; NASA: National Aeronautics and Space Administration; AERIS, CNES: French Data Infrastructure for the Atmosphere, French National Space Agency; JMA: Japan Meteorological Agency; ISRO: Indian Space Research Organization.*

from 2022 and 2023, respectively. The hourly satellite measurements of aerosols, O_3, NO_X, volatile organic compounds, SO_2 and other pollutants will improve the accuracy of air quality analysis and forecasts over these regions.

The satellite measurements need to be validated using ground data. Many studies have confirmed the biases in the remote sensing measurements over a region in a particular time frame. When we discuss uncertainty in remotely sensed air quality parameters, we refer to the differences between the observed values at ground and difference by satellite observations. The uncertainty analysis is less popular than the term "error estimation," or "error analysis," which is popular terminology among researchers. Though the terms "error analysis" and "uncertainty analysis" are used as synonyms, von Clarmann et al. (2020) distinguish these two terms by defining "error" as the difference between an unknown truth and a value inferred from measurements, whereas distribution of an error is defined as "Uncertainty."

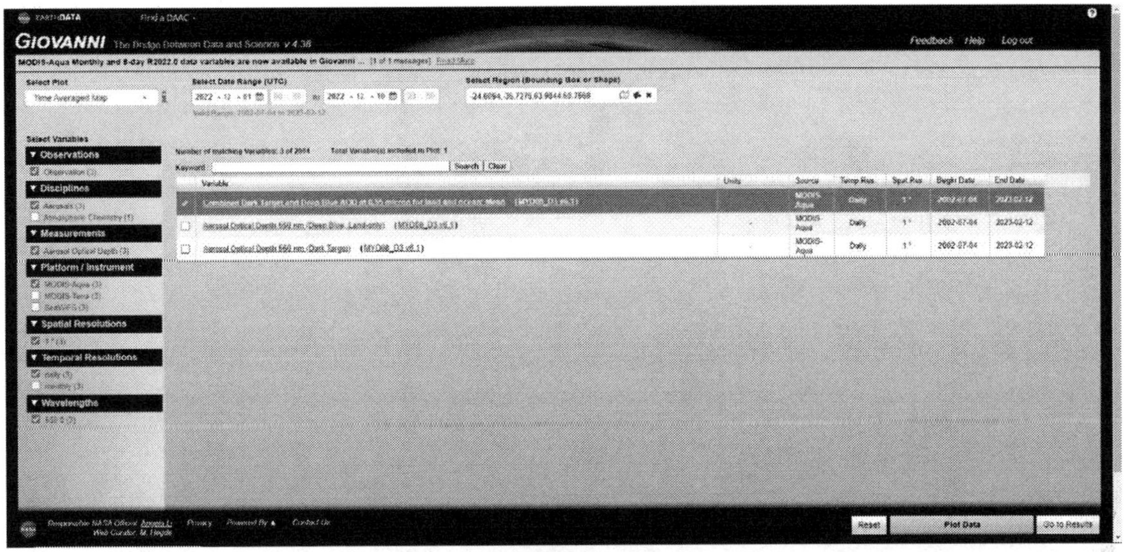

FIGURE 9.19 Selection of dataset, variable, platform, area (latitude, longitude) and other details in Giovanni portal, as mentioned in Step 1.

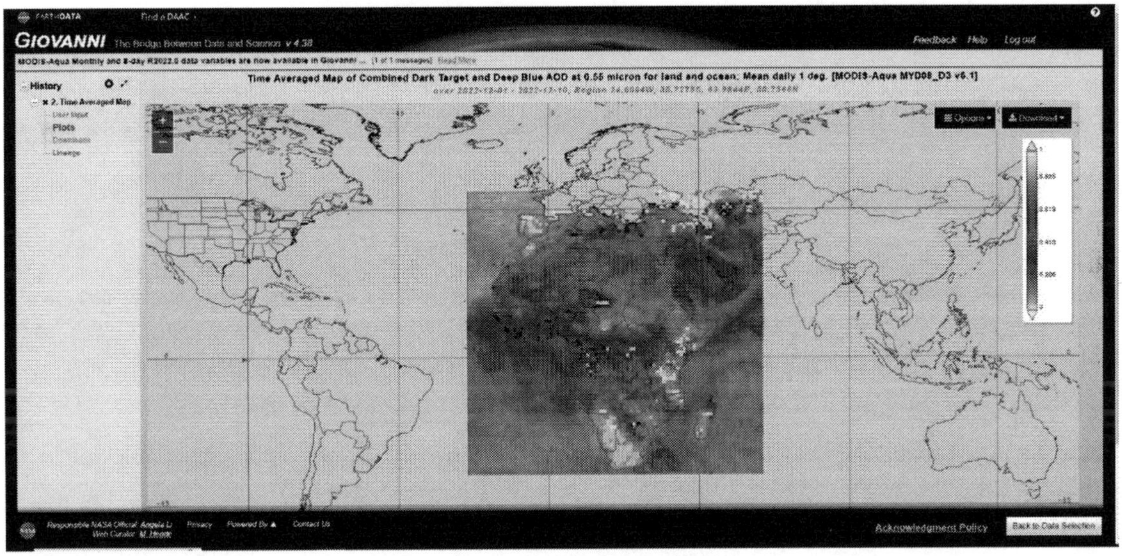

FIGURE 9.20 Plotting the result in Giovanni Portal, discussed in Step 2. The upper right corner has options to download the result in GeoTIFF, KMZ, PNG and NetCDF formats.

Detailed validations of currently available high-resolution data sets need to be attempted. Though satellite remote sensing is emerging and developing at a fast pace, there are still some issues that still need to be resolved in the dispersal of future progress. One possible direction is to create a data fusion of various sensors/satellites for variable data to improve the spatiotemporal resolution of the variable, which is currently unavailable (Chang et al., 2015). Another essential step will be to further improve the image reconstruction techniques at level 1 or 2 data, utilizing previous data sets. Data mining techniques and the use of artificial intelligence will be crucial for attaining these goals (Wei et al., 2020).

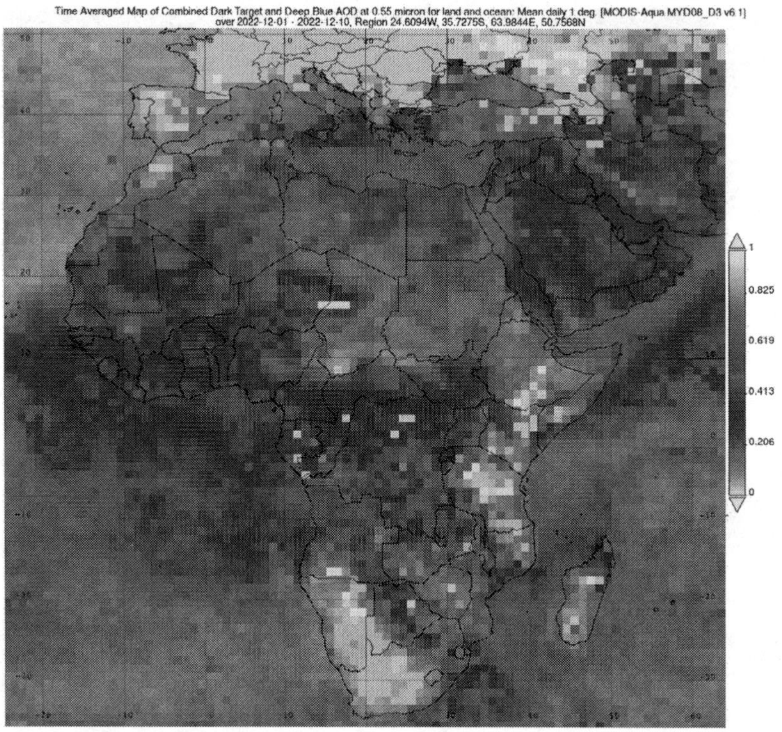

FIGURE 9.21 Downloaded PNG image of the result (Step 3).

9.7 Advanced research methods for measuring atmospheric species

In addition to the approaches examined in the previous sections, there are many specialist instruments used for research applications such as improving process understanding that underpins air quality and atmospheric composition science. Here we provide a brief survey of some of these advanced measurement methods and analytical techniques that are relevant for air quality science. A summary of the advantages and disadvantages of these methods and further reading sources are given in Table 9.5.

9.7.1 Gas chromatography-mass spectrometry

Gas chromatography-mass spectrometry (GC-Ms) was initially introduced in 1952 (James & Martin, 1952) and widely used for the detection and quantification of trace gases and volatile organic compounds (VOCs). This method combines gas chromatography with mass spectrometry to isolate and identify individual components existing in complex gas mixtures. GC-Ms is known for its precision and selectivity, making them valuable for measuring concentrations of trace gases, volatile organic compounds (VOCs), and semivolatile organic compounds (SVOCs). Components of a gas mixture sample are separated and analyzed for identification of each component based on its unique mass spectrum. An example of the application of this technique for atmospheric composition analysis is given by Johnston and Kerecman (2019).

9.7.2 Fourier transform infrared spectroscopy

Fourier transform infrared spectroscopy (FTIR) initially introduced by Peter B. Fellgett in the late 1940s (Fellgett, 1949), is a technique used to analyze chemical compounds by measuring their absorption of infrared radiation. In this method, an infrared radiation source emits a broad spectrum of infrared light, which is then split into two beams as it passes through an interferometer. One beam interacts with the sample while the other serves as a reference.

TABLE 9.5 Advantages and limitations of advanced specialized atmospheric measurement methods.

No	Method	Parameter measured	Advantages	Limitations	References
1	GC-MS	A broad range of volatile and semivolatile compounds, including VOCs, polycyclic aromatic hydrocarbons (PAHs), halogenated compounds, and more.	High sensitivity and selectivity Wide analyte coverage Robust and reliable Research flexibility through compatibility with various sampling techniques	Limited real-time monitoring due to sample preparation and analysis time Potential sample loss or contamination during handling High complexity and cost Requires frequent calibration and quality control. Demands skilled personnel. Limited capability to analyze certain types of atmospheric compounds. Restricted spatial coverage at specific sampling sites Less suitable for inorganic species analysis Not suitable for ultra-trace analysis (e.g., ppt detection limits)	Sparkman et al. (2011), McMaster (2008), Skoog et al. (2014), Johnston and Kerecman (2019).
2	FTIR	Gases	High sensitivity, reaching as low as parts per trillion (ppt). Covers a wide spectral range, typically from the far infrared to the near ultraviolet. Enables quantitative analysis of gas concentrations. Capable of measuring multiple gases simultaneously. Nondestructive sampling. Suitable for remote sensing applications. Provides high spatial coverage.	Sensitive to environmental factors such as temperature, humidity, and CO_2 levels due to strong absorption bands in the infrared region. Complex to operate. Limited ability to distinguish compounds with similar functional groups or overlapping absorption spectra.	Dutta (2017), Griffiths and de Hasseth (2007), Flood et al. (2024)
3	PTRMS	VOCs and other trace gases	High sensitivity. Real-time monitoring. Compact and portable. Suitable for long-term monitoring.	Requires regular calibration using standard compounds or reference materials. Detection range may be limited for compounds with very low or very high volatility. Expensive to purchase, operate, and maintain. Requires specialized expertise in mass spectrometry and atmospheric chemistry. Provides limited spatial coverage when operated in-situ.	Hansel et al. (1995), Pleil et al. (2019)

(Continued)

TABLE 9.5 (Continued)

No	Method	Parameter measured	Advantages	Limitations	References
4	CRDS	Greenhouse gases, ozone-depleting substances, volatile organic compounds (VOCs), and other trace gases	High sensitivity and selectivity. Real-time monitoring. High precision. Wide analyte range. Remote sensing capability. Nondestructive sampling.	Expensive. Limited portability. Sophisticated and require specialized expertise to operate and maintain. Often performed in controlled laboratory settings.	Romanini et al. (1997), Berden et al. (2000), Seidl et al. (2023)
5	DOAS	Gases	Noninvasive measurement (no need for direct contact or sampling). High sensitivity (ppb and ppt level). Real-time monitoring. Versatile as it is deployable even in mobile platforms such as aircraft and satellites. Enables quantitative analysis.	Dependence on sunlight. Influence of aerosols on gas concentration retrievals. Regular calibration requirements. Limited detection range. Complex instrument that requires specialized expertise to operate, calibrate, and maintain.	Platt and Stutz (2024), Platt et al. (1979), Ryan et al. (2023)
6	IC	A wide range of ions, including both inorganic and organic species	High sensitivity (ppb and ppt level). Highly accurate and reproducible. Automated sample injection and data analysis in new models. Requires minimal sample preparation compared to other analytical techniques. Provides fast analysis. Sample dilution is not needed for higher concentrations.	Primarily suited for the analysis of ions and ionizable compounds. Expensive to purchase, operate, and maintain. Limited compatibility with nonaqueous samples. Limited sensitivity to certain ions with low conductivity or weak ion-exchange properties. Interference from contaminants such as trace metals.	Fritz (2004), Liu et al. (2021), Haddad et al. (2008)
7	PID	VOCs and other compounds that can be ionized by UV radiation	Rapid analysis. Sensitivity in ppm and ppb levels. Portable and field deployable. Simple operation. Real-time monitoring. Cost-effective. Nondestructive and noninvasive, enabling multiple measurements on a sample without changing its composition.	Nonselective measurement (without identifying individual species). Interference from background contaminants, leading to false positives or inaccurate readings. Limited detection range. Sensitivity to humidity and temperature. Requires regular calibration. PID instruments and consumables can be expensive to purchase, operate, and maintain. Can pose a safety hazard due to emitting UV radiation.	Poole (2015), Spinelle et al. (2017)

8	SEM	Particulate matter and aerosols, best for > 100 nm particle	High-resolution imaging provides detailed qualitative information about particle morphology and composition. Elemental analysis is possible when combined with other methods. Advanced SEM techniques can also provide 3D images. Can analyze a wide range of sample types. Well-suited for analyzing particles ranging from submicron to tens of microns in size, making it ideal for studying a wide range of atmospheric aerosols and dust particles.	Quantitative analysis may be challenging. High instrumentation and operational costs. Sample preparation requirements. Only the surface features of the sample are in focus. Distortions or artifacts may occur in the images, especially for organic species exposed to the electron beam.	Vernon-Parry (2000), Goldstein et al. (2017), Li et al. (2021), Panta et al. (2023)
9	TEM	Particulate matter and aerosols, for particle of a few nm size	Suitable for particles smaller than a few nanometers. High resolution. Internal imaging capability. Elemental analysis.	Complex sample preparation. High instrumentation and operational costs. Limited depth of field, as only a thin slice of the sample is in focus at any given time. Sample damage from electron beam radiation, leading to radiation damage from high-energy electron beams.	Williams and Carter (2009), Shao et al. (2022), Ott et al. (2021)
10	LD	Particulate matter	Cost-effective. Minimal sample preparation required. Rapid analysis. Nondestructive to the sample. High sensitivity. Automated and user-friendly in new instruments. New models can analyze multiple samples sequentially or simultaneously. Accurate and reproducible results.	Assumption of spherical particles. Results may be affected by sample homogeneity. Can be influenced by variations in the refractive index of particles and dispersing medium, particularly for materials with high refractive index differences. High-quality instruments can be expensive to purchase and maintain.	Merkus (2009), Black et al. (1996), Yu et al. (2017)
11	AMS	Particulate matter	Real-time measurement capability. High sensitivity. High temporal resolution. Ability to detect a wide range of chemical species within aerosol particles. No need for extensive sample preparation. Field deployable.	May require relatively large sample volumes. Doesn't provide comprehensive coverage of all aerosol components. Data analysis can be complex and require specialized expertise. Expensive instrument costs and maintenance.	Wexler and Johnston (2011), Fuzzi et al. (2015), Cash et al. (2023)

A detector measures the combined beam's intensity across different wavelengths. By moving a mirror in the interferometer, the path length of the beam changes, creating an interferogram. This interferogram undergoes Fourier transformation, converting it into a spectrum that reveals the sample's absorption at different infrared wavelengths. Peaks in this spectrum correspond to specific chemical bonds in the sample. By comparing the spectrum with reference data, analysts can identify the chemical functional groups and compounds present in the sample. FTIR is widely used in atmospheric science, for example, it is employed for measuring solar infrared spectra at the Mauna Loa Observatory (Global Monitoring Laboratory, 2024) which hosts one of the longest-running continuous measurements of atmospheric concentrations, particularly CO_2 levels. A recent application of this technique for atmospheric science can be found in Flood et al. (2024).

9.7.3 Proton-transfer-reaction mass spectrometry

Proton-transfer-reaction mass spectrometry (PTRMS) was first developed in 1995 (Hansel et al., 1995) as a sensitive analytic technique for real-time monitoring of VOCs and other trace gases in ambient air. It operates by ionizing gas-phase molecules through proton transfer reactions with reagent ions (e.g., H_3O^+ or NO^+). The resulting protonated analyte ions are then analyzed and quantified by a mass spectrometer. This instrument has proven to be invaluable for atmospheric researchers studying the composition and sources of volatile organic compounds (VOCs) in urban air pollution across various locations (e.g., Pleil et al., 2019).

9.7.4 Cavity ring-down spectroscopy

Cavity ring-down spectroscopy (CRDS) involves laser light directed into an optical cavity containing the sample gas. The laser light interacts with the gas molecules, and the absorption of light by the molecules causes the light intensity inside the cavity, leading to a gradual decrease in light intensity over time, known as the ring-down time. By measuring this decay or "ring-down" time of light intensity, CRDS enables the determination of the absorption coefficient of the sample gas, which is directly related to the concentration of the gas molecules. CRDS method is widely used by researchers and institutions around the world to study atmospheric composition, trace gases, air quality, and climate change (e.g., Seidl et al., 2023).

9.7.5 Differential optical absorption spectroscopy

Differential optical absorption spectroscopy (DOAS) measures trace gases and pollutants in the atmosphere by observing the absorption of light by these species. It utilizes sunlight as the light source, which interacts with gases and aerosols in the atmosphere. In a DOAS setup, a spectrometer disperses sunlight into its constituent wavelengths. Then the light's intensity at various wavelengths is measured along two optical paths: the reference and measurement paths. The reference path typically represents a direct sunlight beam passing through a clean atmospheric segment, while the measurement path includes the portion of the atmosphere containing the target gas or pollutant. By subtracting the spectrum observed along the reference path from that along the measurement path, the absorption characteristics specific to the target gas or pollutant are then analyzed (Platt & Stutz, 2024) to quantify the gas concentration in the atmosphere. DOAS is widely used for remote sensing of atmospheric pollutants, providing information on pollutant concentrations, vertical profiles, and spatial distribution. An example of the use of this technique to characterize an ozone episode can be found in Ryan et al. (2023).

9.7.6 Ion chromatography

Ion chromatography (IC) technique is a widely used method for separating and analyzing ions in a sample (such as sulfate, chloride and nitrate ions in ambient air samples) (see e.g., Fritz, 2004). The process starts with sample injection, usually dissolved in a liquid. As the sample travels through a column, ions interact uniquely, resulting in their separation based on distinctive properties. Certain ions exhibit longer retention within the column compared to others. Following separation, each ion moves through the column at different rates, depending on their interaction with the column material, and eventually exits the column separately, through a detector. The detector measures various properties of the ions, such as their electrical conductivity or absorbance of light, facilitating their identification and quantification. Through comparison of detector signals with prepared chemical standards, the type and concentration of ions within the sample are determined. Finally, detector data is analyzed to understand the composition of the sample, aiding in the identification and quantification of specific ions present (Haddad et al., 2008, Liu et al., 2021).

9.7.7 Photoionization detection

Photoionization detection (PID) is a rapid and sensitive method for measuring VOCs and other ionizable compounds in air and gas samples. In the PID method, samples are collected using an appropriate sampling device, such as a passive sampling badge or a pump, and then introduced into the instrument. The sample passes through a chamber or cell within the PID instrument, where it is exposed to UV radiation emitted by an internal lamp. This UV radiation possesses sufficient energy to ionize the compounds in the sample. The ionized compounds are then drawn towards an electrode, creating an electric current proportional to the concentration of the ionized compounds. This electric current is converted into a signal, which corresponds to the concentration of the compounds in the sample (e.g., Spinelle et al., 2017).

9.7.8 Scanning electron microscopy

Scanning electron microscopy (SEM) is a common electron microscope technique used to provide high-resolution images of PM to understand its structure. In this method, an electron source, typically a heated tungsten filament or a field emission gun, emits electrons towards the sample, creating a focused electron beam that scan across the surface of the sample while interacting with its atoms. During this interaction process, electrons are scattered and collected by detectors to produce imagery for analysis. Different detectors can be employed to capture the signals produced by the atom-electron interactions and collected for producing the images (Vernon-Parry, 2000). An example of the use of SEM to characterize mineral dust can be found in Panta et al. (2023).

9.7.9 Transmission electron microscopy

Transmission electron microscopy (TEM) is used to examine the internal structure and composition of materials at the atomic and nanoscale levels, achieving even higher resolution than SEM, which primarily provides surface imaging.

TEM, unlike SEM, focuses and shapes the electron beam using a series of electromagnetic lenses to penetrate through the sample. It requires specific specimen preparation for TEM analysis. Electrons pass through the specimen, interacting with its atoms, forming an image that provides detailed information about its internal structure and composition. Additionally, TEM offers various analytical techniques that provide valuable information, such as elemental distribution and chemical bonding within the specimen (Ott et al., 2021; Shao et al., 2022).

9.7.10 Laser diffraction

Laser diffraction (LD) provides rapid and nondestructive laboratory-based analysis of particle size distributions in solution. LD operates by illuminating a laser beam onto a sample containing particles. The laser light interacts with the particles in the sample, causing scattering of light in diverse directions. Subsequently, a detector collects the scattered light and measures its intensity at different angles, which is then employed in determining the particle size distribution using mathematical models, such as Mie theory (Black et al., 1996; Merkus, 2009). Dust analysis example with this approach can be found in Yu et al. (2017).

9.7.11 Aerosol mass spectrometry

Aerosol mass spectrometry (AMS) is a commonly used instrument that employs aerodynamic particle sizing coupled with time-of-flight mass spectrometry technique to measure the chemical composition of aerosols. In this method, aerosol particles are initially treated with a narrow beam and then undergo vaporization by either a laser beam or a heated surface. Subsequently, the resulting ions are subjected to analysis via a mass spectrometer. It is a widely used technique with real-time measuring capability, offering valuable insights into aerosol sources, transformations, and their impacts on air quality and climate. The single particle mass spectrometer is a specialized type of AMS that analyses individual aerosol particles sequentially through a mass spectrometer (Wexler & Johnston, 2011). An example can be found in Cash et al. (2023) reporting on a study of air pollution in Delhi using AMS.

There are many other techniques that could not be included here but we hope that this selection of advanced measurement techniques demonstrates their role in advancing air quality and pollution science.

9.8 End of chapter questions

1. An atmospheric scientist measures the concentration of ozone in a sample using ultraviolet absorption and finds it to be 2×10^8 molecules cm^{-3}. Knowing that Avogadro's number is 6.022×10^{23} molecules mol^{-1}, calculate the concentration of ozone in parts per billion (ppb). Assume atmospheric pressure as 1 atm and temperature as 25 °C.

Number	Characteristic
1	Low accuracy high precision
2	High accuracy, high precision
3	High accuracy, low precision
4	Low accuracy, Low precision

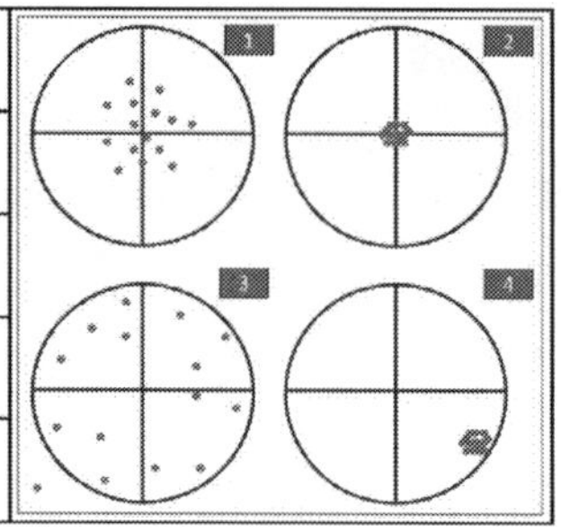

FIGURE 9.22 Illustration of the concepts of accuracy and precision.

2. A researcher is using a TEOM (Tapered Element Oscillating Microbalance) system to measure PM concentrations in the air. TEOM system is measuring PM_{10} concentrations, and during calibration, the constant k_0 was determined to be 0.002 g Hz^{-2}. If the frequency of the vibrating tapered element (f) is measured to be 100 Hz, calculate the PM (PM_{10}) concentration.
3. If the TEOM system collected 2×10^{-7} g of PM_{10} on the filter over a period of 24 hours and the volume of air sampled during this time period was 1000 L, what is the concentration of PM_{10} in the air sample in the unit of $\mu g\, m^{-3}$?
4. What are the differences between sensors and sensor systems?
5. In the following Fig. 9.22, representing with the aid of a target and relevant points the concepts of accuracy and precision, draw a line connection the correct graph with the correct explanation.
6. Sometimes fast photochemistry occurs in urban areas resulting in rapid growth of relevant pollution levels. Which LCS technology would you favor for monitoring corresponding concentrations?
7. Anemophilous pollen grain sizes are generally between 10 and 100 μm, while mold spore size can be between 2 and 10 μm (whole and not fractured grain and spore sizes respectively). Which type of LCS would you suggest as a way to estimate their concentration?
8. If a low-cost air quality sensor has a response time of 12 hours, is it capable of identifying hourly NO_2 values? Explain.
9. In an air quality monitoring project, you are asked to select devices to be used for ambient air quality monitoring, that make use of either electrochemical or metal-oxide sensors for O_3 and NO_2 concentration estimations, under harsh environmental conditions (high temperature and humidity range). Which sensor technology would you choose and why?
10. You are asked to deploy five low-cost sensor devices in a small urban area. Given that you can access only 1 reference station and only for a specific amount of time, which calibration method is the fittest for this application and why? What if you have 100 devices to deploy?
11. Why is remote sensing of air quality essential?
12. What are the three main categories of remote sensing approaches for acquiring air quality datasets? Explain your choice of categories.
13. What is the highest resolution of available satellite remote sensing product, and for which variable? Support your answer with examples of air quality variables.
14. What do you understand by the retrieval algorithm? Provide examples in your answer.
15. What are the comparative advantages of using GEMS, TEMPO and Sentinel 4?
16. What will be the surface concentration of NO_x ($\mu g\, m^{-3}$) if one measures the satellite derived values of NO as 30 ppm and NO_2 as 38 ppm over a place?
17. If one assumes that for Delhi, India, monthly mean values of boundary layer height (BLH) is 1500 m, relative humidity (RH) is 52%, wind speed (WS) is 0.9 m s^{-1}, air temperature (T) is 313 K, AOD is 0.8, intercept (offset) value is 21.12, and the regression coefficients β_1–β_5 are 18, 12, 57, 21 and 30, respectively, what will be the surface $PM_{2.5}$ concentrations, considering that they are dependent on BLH, WS, RH, T and AOD?

Solutions to the end-of-chapter questions will be available at: https://www.elsevier.com/books-and-journals/book-companion/9780128225912

Appendix 9.1 List of abbreviations used in Table 9.2

Acronym	Full Name
ABI L2 +	Advanced Baseline Imager Level 2 Plus
AERONET	Aerosol Robotic Network
AHI	Advanced Himawari Imager
AIRS	Atmospheric infrared sounder
AMMA	African Monsoon Multidisciplinary Analysis
AMS	Aerosol Mass Spectrometry
AVHRR	Advanced Very High-Resolution Radiometer
CAIPEEX	Cloud Aerosol Interaction and Precipitation Enhancement Experiment
CALIOP	Cloud-Aerosol Lidar with Orthogonal Polarization
CLARIFY-2017	CLoud–Aerosol–Radiation interaction and forcing: Year 2017
CRDS	Cavity Ring-Down Spectroscopy
DOAS	Differential Optical Absorption Spectroscopy
ETM +	Enhanced Thematic Mapper Plus
FTIR	Fourier Transform Infrared Spectroscopy
GAUGE	Greenhouse gAs Uk and Global Emissions
GC-MS	Gas Chromatography-Mass Spectrometry
GOES	Geostationary Operational Environmental Satellite
GOME	Global Ozone Monitoring Experiment
GOME-2	Global Ozone Monitoring Experiment–2
IASI	Infrared Atmospheric Sounding Interferometer
IC	Ion Chromatography
LD	Laser Diffraction
MIPAS	Michelson Interferometer for Passive Atmospheric Sounding
MISR	Multi-angle Imaging SpectroRadiometer
MODIS	Moderate Resolution Imaging Spectroradiometer
MOPITT	Measurements Of Pollution in The Troposphere
NASA	National Aeronautics and Space Administration
OCO	Orbiting Carbon Observatory
OLCI	Ocean and Land Color Instrument
OLI	Operational Land Imager
OMI	Ozone Monitoring Instrument
OMPS	Ozone Mapping and Profiler Suite
PARASOL	Polarization and Anisotropy of Reflectances for Atmospheric Science coupled with Observations from a Lidar
PID	Photoionization Detection
PTRMS	Proton-Transfer-Reaction Mass Spectrometry
SCIAMACHY	SCanning Imaging Absorption spectroMeter for Atmospheric CartograpHY
SCOUT	Stratosphere-Climate links with emphasis On the Upper Troposphere and Lower stratosphere-O_3

SEM	Scanning Electron Microscopy
SLSTR	Sea and Land Surface Temperature Radiometer
TEM	Transmission Electron Microscopy
TES	Technology Experiment Satellite
TROPOMI	TROPOspheric Monitoring Instrument
VIIRS	Visible Infrared Imaging Radiometer Suite
WHO	World Health Organization

References

Aggarwal, S. G., Kumar, S., Mandal, P., Sarangi, B., Singh, K., Pokhariyal, J., Mishra, S. K., Agarwal, S., Sinha, D., Singh, S., Sharma, C., & Gupta, P. K. (2013). Traceability issue in $PM_{2.5}$ and PM_{10} measurements. *MAPAN, 28*, 153–166. Available from https://doi.org/10.1007/s12647-013-0073-x.

Aiuppa, A., Fiorani, L., Santoro, S., Parracino, S., Nuvoli, M., Chiodini, G., Minopoli, C., & Tamburello, G. (2015). New ground-based lidar enables volcanic CO_2 flux measurements. *Scientific Reports, 5*(1), 1–12.

Andronache, C. (2018). Characterization of mixed-phase clouds: Contributions from the field campaigns and ground-based networks. *In* Mixed-phase clouds (pp. 97–120). Elsevier.

(AQEG) (2004). Nitrogen Dioxide in the United Kingdom. Department for Environment, Food and Rural Affairs. <http://www.defra.gov.uk/environment/airquality/aqeg>.

(AQEG) (2006). Air quality expert group. Particulate Matter in the United Kingdom. London: Environment Agency. <https://webarchive.nationalarchives.gov.uk/ukgwa/20060811120000/http://www.defra.gov.uk/environment/airquality/aqeg/particulate-matter/index.htm>.

(AQEG) (2012). Fine particulate matter ($PM_{2.5}$) in the United Kingdom. Report prepared for the Department for Environment, Food and Rural Affairs; Scottish Executive; Welsh Government; and Department of the Environment in Northern Ireland.

Argall, P. S., & Sica, R. (2015). *LIDAR. Atmospheric sounding introduction. Encyclopedia of Atmospheric Sciences* (Second Edition, pp. 262–269). Academic Press. Available from https://doi.org/10.1016/B978-0-12-382225-3.00203-6.

Bagkis, E., Kassandros, T., & Karatzas, K. (2022). Learning calibration functions on the fly: Hybrid batch online stacking ensembles for the calibration of low-cost air quality sensor networks in the presence of concept drift. *Atmosphere, 13*(3), 416. Available from https://doi.org/10.3390/atmos13030416.

Bagkis, E., Kassandros, T., Karteris, M., Karteris, A., & Karatzas, K. (2021). Analyzing and improving the performance of a particulate matter low-cost air quality monitoring device. *Atmosphere, 12*(2), 251. Available from https://doi.org/10.3390/atmos12020251.

Baron, R., & Saffell, J. (2017). Amperometric gas sensors as a low-cost emerging technology platform for air quality monitoring applications: A review. *ACS Sensors, 2*, 1553–1566.

Barrett, E. W., & Ben-Dov, O. (1967). Application of the lidar to air pollution measurements. *Journal of Applied Meteorology and Climatology, 6*(3), 500–515.

Berden, G., Peeters, R., & Meijer, G. (2000). Cavity ring-down spectroscopy: Experimental schemes and applications. *International Reviews in Physical Chemistry, 19*(4), 565–607. Available from https://doi.org/10.1080/014423500750040627.

Bílek, J., Maršolek, P., Bílek, O., & Buček, P. (2022). Field test of mini photoionization detector-based sensors—Monitoring of volatile organic pollutants in ambient air. *Environments, 9*(4), 49. Available from https://doi.org/10.3390/environments9040049.

Biswal, A., Singh, T., Singh, V., Ravindra, K., & Mor, S. (2020). COVID-19 lockdown and its impact on tropospheric NO_2 concentrations over India using satellite-based data. *Heliyon, 6*(9), e04764.

Black, D. L., McQuay, M. Q., & Bonin, M. P. (1996). Laser-based techniques for particle-size measurement: A review of sizing methods and their industrial applications. *Progress in Energy and Combustion Science, 22*(3), 267–306.

Bodah, B. W., Neckel, A., Maculan, L. S., Milanes, C. B., Korcelski, C., Ramírez, O., Mendez-Espinosa, J. F., Bodah, E. T., & Oliveira, M. L. (2022). Sentinel-5P TROPOMI satellite application for NO2 and CO studies aiming at environmental valuation. *Journal of Cleaner Production, 357*, 131960.

Brown, R. J. C., & Martin, N. A. (2023). How standardizing 'low-cost' air quality monitors will help measure pollution. *Nature Review Physics, 5*, 139–140. Available from https://doi.org/10.1038/s42254-023-00561-8.

BS EN 14211 (2012). British Standard on Ambient Air. Standard method for the measurement of the concentration of nitrogen dioxide and nitrogen monoxide by chemiluminescence.

Bureau Veritas (2006). *UK equivalence programme for monitoring of particulate matter: Final report.* Report prepared for the Department for the Environment, Food and Rural Affairs; Welsh Assembly Government; Scottish Executive; Department of Environment for Northern Ireland. Ref: BV/AQ/AD202209/DH/2396.

Bureau Veritas (2015). *Thermo Fisher scientific UK report on the equivalence of the $PM_{2.5}$* Partisol 2025.

Bureau Veritas. (2016). *UK Report on the equivalence of the Palas Fidas 200 method 11 for PM10 and $PM_{2.5}$*. Bureau Veritas, UK.

Butterfield, D., Martin, N. A., Coppin, G., & Fryer, D. E. (2021). Equivalence of UK nitrogen dioxide diffusion tube data to the EU reference method. *Atmospheric Environment, 262*, 118614. Available from https://doi.org/10.1016/j.atmosenv.2021.118614.

Cairo, F., Pommereau, J. P., Law, K. S., Schlager, H., Garnier, A., Fierli, F., Ern, M., Streibel, M., Arabas, S., Borrmann, S., Berthelier, J. J., Blom, C., Christensen, T., D'Amato, F., Di Donfrancesco, G., Deshler, T., Diedhiou, A., Durry, G., Engelsen, O., Goutail, F., Harris, N. R. P., Kerstel, E. R. T., Khaykin, S., Konopka, P., Kylling, A., Larsen, N., Lebel, T., Liu, X., MacKenzie, A. R., Nielsen, J., Oulanowski, A., Parker, D. J., Pelon, J., Polcher, J., Pyle, J. A., Ravegnani, F., Rivière, E. D., Robinson, A. D., Röckmann, T., Schiller, C., Simões, F., Stefanutti, L., Stroh, F., Some, L., Siegmund, P., Sitnikov, N., Vernier, J. P., Volk, C. M., Voigt, C., von Hobe, M., Viciani, S., & Yushkov, V. (2010). An introduction to the SCOUT-AMMA stratospheric aircraft, balloons and sondes campaign in West Africa, August 2006: Rationale and roadmap. *Atmospheric Chemistry and Physics, 10*(5), 2237–2256. Available from https://doi.org/10.5194/acp-10-2237-2010.

Campbell, J. B., & Wynne, R. H. (2011). *Introduction to remote sensing*. Guilford press.

Carpenter, L. J., Simpson, I. J., & Cooper, O. R. (2022). Ground-based reactive gas observations within the global atmosphere watch (GAW) network. *In* Handbook of air quality and climate change (pp. 1–21). Singapore: Springer Nature.

Casazza, M., Lega, M., Jannelli, E., Minutillo, M., Jaffe, D., Severino, V., & Ulgiati, S. (2019). 3D monitoring and modelling of air quality for sustainable urban port planning: Review and perspectives. *Journal of Cleaner Production*. Available from https://doi.org/10.1016/j.jclepro.2019.05.257.

Cash, J. M., Di Marco, C., Langford, B., Heal, M. R., Mandal, T. K., Sharma, S. K., Gurjar, B. R., & Nemitz, E. (2023). Response of organic aerosol to Delhi's pollution control measures over the period 2011–2018. *Atmospheric Environment, 315*, 120123. Available from https://doi.org/10.1016/j.atmosenv.2023.120123.

Cavaliere, A., Carotenuto, F., Di Gennaro, F., Gioli, B., Gualtieri, G., Martelli, F., & Zaldei, A. (2018). Development of low-cost air quality stations for next generation monitoring networks: Calibration and validation of $PM_{2.5}$ and PM_{10} sensors. *Sensors, 18*(9), 2843. Available from https://doi.org/10.3390/s18092843.

CEN - European Committee for Standardization. (2023). *BS EN 12341:2023. Ambient air. Standard gravimetric measurement method for the determination of the PM_{10} or $PM_{2.5}$ mass concentration of suspended particulate matter*.

Chang, N. B., Bai, K., & Chen, C. F. (2015). Smart information reconstruction via time-space-spectrum continuum for cloud removal in satellite images. *IEEE Journal of Selected Topics in Applied Earth Observations and Remote Sensing, 8*(5), 1898–1912.

Clements, A., Duvall, R., Greene, D., & Dye, T. (2022). *The enhanced air sensor guidebook*. U.S. Environmental Protection Agency, Washington, DC.

Clements, N., Milford, J. B., Miller, S. L., Navidi, W., Peel, J. L., & Hannigan M. P. (2013). Errors in coarse particulate matter mass concentrations and spatiotemporal characteristics when using subtraction estimation methods. *Journal of the Air & Waste Management Association, 63*(12), 1386–1398. Available from https://doi.org/10.1080/10962247.2013.816643.

Clyne, M. A. A., Thrush, B. A., & Wayne, R. P. (1964). Kinetics of the chemiluminescent reaction between nitric oxide and ozone. *Transactions of the Faraday Society, 60*, 359–370.

Concas, F., Mineraud, J., Lagerspetz, E., Varjonen, S., Liu, X., Puolamäki, K., Nurmi, P., & Tarkoma, S. (2021). Low-cost outdoor air quality monitoring and sensor calibration: A survey and critical analysis. *ACM Transactions on Sensor Networks, 17*(2), Article 20. Available from https://doi.org/10.1145/3446005.

Corrigan, C. E., Roberts, G. C., Ramana, M. V., Kim, D., & Ramanathan, V. (2008). Capturing vertical profiles of aerosols and black carbon over the Indian ocean using autonomous unmanned aerial vehicles. *Atmospheric Chemistry and Physics, 8*, 737–747.

De Sherbinin, A., Levy, M. A., Zell, E., Weber, S., & Jaiteh, M. (2014). Using satellite data to develop environmental indicators. *Environmental Research Letters, 9*(8), 084013.

De Silva, C. (2017). *Sensor systems: Fundamentals and applications*. Taylor & Francis: CRC Press.

De Vito, S., D'Elia, G., & Di Francia, G. (2022). Global calibration models match ad-hoc calibrations field performances in low-cost particulate matter sensors. In *2022 IEEE international symposium on olfaction and electronic nose (ISOEN)*. https://doi.org/10.1109/ISOEN54820.2022.9789669.

De Vito, S., Di Francia, G., Esposito, E., Ferlito, S., Formisano, F., & Massera, E. (2020a). Adaptive machine learning strategies for network calibration of IOT smart air quality monitoring devices. *Pattern Recognition Letters, 136*, 264–271. Available from https://doi.org/10.1016/j.patrec.2020.04.032.

De Vito, S., Esposito, E., Castell, N., Schneider, P., & Bartonova, A. (2020b). On the robustness of field calibration for smart air quality monitors. *Sensors and Actuators B: Chemical, 310*, 127869. Available from https://doi.org/10.1016/j.snb.2020.127869.

DeBruin, H. A. (2009). Supplement: Time to think reflections of a pre-pensioned scintillometer researcher. *Bulletin of the American Meteorological Society, 90*(5), ES17–ES26.

DEC (2016a). Standard operating procedures ozone (O_3) monitoring in ambient air by ultraviolet absorption spectrophotometry. (2016). Air Quality Division, Air Monitoring & Quality Assurance Program, The State of Alaska Department of Environmental Conservation, Revision 1.

DEC (2016b). Standard operating procedures sulfur dioxide (SO_2) monitoring in ambient air by ultraviolet (UV) fluorescence spectroscopy. (2016). Air Quality Division, Air Monitoring & Quality Assurance Program, The State of Alaska Department of Environmental Conservation, Revision 3.

DEFRA. (2024). Department for Environment, Food & Rural Affairs. Site Types - UK Air. <https://uk-air.defra.gov.uk/networks/site-types>.

Demerjian, K. L. (2000). A review of national monitoring networks in North America. *Atmospheric Environment, 34*(11), 1861–1884.

deSouza, P. N. (2022). Key concerns and drivers of low-cost air quality sensor use. *Sustainability, 14*, 584. Available from https://doi.org/10.3390/su14010584.

Devara, P. S. (1998). Review article remote sensing of atmospheric aerosols from active and passive optical techniques. *International Journal of Remote Sensing, 19*(17), 3271–3288.

Dinh, T. V., Choi, I. Y., Son, Y. S., & Jo-Chun Kim, J. C. (2016). A review on non-dispersive infrared gas sensors: Improvement of sensor detection limit and interference correction. *Sensors and Actuators B: Chemical, 231*, 529–538.

Douros, J., Eskes, H., van Geffen, J., Boersma, K. F., Compernolle, S., Pinardi, G., Blechschmidt, A. M., Peuch, V. H., Colette, A., & Veefkind, P. (2023). Comparing Sentinel-5P TROPOMI NO_2 column observations with the CAMS regional air quality ensemble. *Geoscientific Model Development, 16*(2), 509–534.

Drisya, J., & Roshni, T. (2018). Spatiotemporal variability of soil moisture and drought estimation using a distributed hydrological model. *In* Integrating disaster science and management (pp. 451–460). Elsevier.

Dutta, A. (2017). Chapter 4 - Fourier transform infrared spectroscopy. In S. Thomas, R. Thomas, A. K. Zachariah, & R. K. Mishra (Eds.), *Spectroscopic Methods for Nanomaterials Characterization* (pp. 73–93). Elsevier. Available from https://doi.org/10.1016/B978-0-323-46140-5.00004-2.

EAQP. (2024). *The European air quality (AQ) portal digital platform centered around the air quality eReporting system for the European Commission (EC).* <https://aqportal.discomap.eea.europa.eu/> (accessed 19.02.24).

Environment Agency (2017*). Performance standards for indicative ambient particulate monitors (version 4).* UK.

Environment Agency. (2022a). *Monitoring stack emissions: Techniques and standards for periodic monitoring.*

Environment Agency (2022b). *UK automatic urban and rural network (AURN) LSO manual - essential background reading (version 1.2).*

Environment Agency (2022c). *UK AURN QAQC annual technical report 2022*. Report for: The Environment Agency. Ref. 25952 Lot 2&5 ATR 2022. Ricardo ref. ED12695201_ATR_2022 Issue 1 26/07/2023.

Environmental Protection Agency (EPA). (2022). A guide to siting and installing air sensors. <https://www.epa.gov/air-sensor-toolbox/guide-siting-and-installing-air-sensors> (accessed 04.05.23).

European Parliament. (2019, March). Sampling points for air quality: Representativeness and comparability of measurement in accordance with Directive 2008/50/EC on ambient air quality and cleaner air for Europe (Policy Department for Economic, Scientific and Quality of Life Policies, Directorate-General for Internal Policies, Authors: Christian Nagl, Wolfgang Spangl, Iris Buxbaum, PE 631.055).

Fellgett, P. B. (1949). On the ultimate sensitivity and practical performance of radiation detectors. *Journal of the Optical Society of America, 39*(12), 970–976. Available from https://doi.org/10.1364/JOSA.39.000970.

Flood, V. A., Strong, K., Whaley, C. H., Walker, K. A., Blumenstock, T., Hannigan, J. W., Mellqvist, J., Notholt, J., Palm, M., Röhling, A. N., Arnold, S., Beagley, S., Chien, R.-Y., Christensen, J., Deushi, M., Dobricic, S., Dong, X., Fu, J. S., Gauss, M., ... Turnock, S. (2024). Evaluating modelled tropospheric columns of CH_4, CO, and O_3 in the Arctic using ground-based Fourier transform infrared (FTIR) measurements. *Atmospheric Chemistry and Physics, 24*, 1079–1118. Available from https://doi.org/10.5194/acp-24-1079-2024.

Foken, T. (Ed.), (2022). *Springer handbook of atmospheric measurements* (1st ed.). Cham: Springer. Available from https://doi.org/10.1007/978-3-030-52171-4.

Fontjin, A., Sabadell, A. J., & Ronco, R. J. (1970). Homogeneous chemiluminescence measurement of nitric oxide with ozone. *Analytical Chemistry, 42*(5), 575–579.

Fraser, R. S. (1976). Satellite measurement of mass of Sahara dust in the atmosphere. *Applied Optics, 15*, 2471–2479.

Fritz, J. S. (2004). Early milestones in the development of ion-exchange chromatography: A personal account. *Journal of Chromatography. A, 1039* (1–2), 3–12. Available from https://doi.org/10.1016/j.chroma.2004.02.001.

Fuzzi, S., Baltensperger, U., Carslaw, K., Decesari, S., Denier van der Gon, H., Facchini, M. C., Fowler, D., Koren, I., Langford, B., Lohmann, U., Nemitz, E., Pandis, S., Riipinen, I., Rudich, Y., Schaap, M., Slowik, J. G., Spracklen, D. V., Vignati, E., Wild, M., ... Gilardoni, S. (2015). Particulate matter, air quality and climate: Lessons learned and future needs. *Atmospheric Chemistry and Physics, 15*(15), 8217–8299. Available from https://doi.org/10.5194/acp-15-8217-2015.

Garcia, R. D., Garcia, O. E., Cuevas, E., Cachorro, V. E., Barreto, A., Guirado-Fuentes, C., & Kouremeti, N. (2016). Aerosol optical depth retrievals at the Izana Atmospheric Observatory from 1941 to 2013 by using artificial neural networks. *Atmospheric Measurement Techniques, 9*(1), 53–62. Available from https://doi.org/10.5194/amt-9-53-2016.

Gilliam, J., & E. Hall. (2016) *Reference and equivalent methods used to measure national ambient air quality standards (NAAQS) criteria air pollutants - Volume I.* U.S. Environmental Protection Agency, Washington, DC, EPA/600/R-16/139, 2016.

Gioli, B., Miglietta, F., Vaccari, F. P., Zaldei, A., & De Martino, B. (2006). The Sky Arrow ERA, an innovative airborne platform to monitor mass, momentum and energy exchange of ecosystems. *Annals of Geophysics, 49*(1), 109–116.

Global Monitoring Laboratory. (2024). *Fourier transform infrared (FTIR) spectrometer program at Mauna Loa Observatory (MLO).* Retrieved February 12, 2024, from https://gml.noaa.gov/obop/mlo/programs/ndacc/ftir/ftir.html.

Goldstein, J. I., Newbury, D. E., Michael, J. R., Ritchie, N. W. M., Scott, J. H. J., & Joy, D. C. (2017). *Scanning electron microscope (SEM) instrumentation. In* Scanning electron microscopy and X-ray microanalysis (pp. 65–91). Springer. Available from https://doi.org/10.1007/978-1-4939-6676-0_3.

Green, D., & Fuller, G. W. (2006). The implications of tapered element oscillating microbalance (TEOM) software configuration on particulate matter measurements in the UK and Europe. *Atmospheric Environment, 40*(29), 5608–5616. Available from https://doi.org/10.1016/j.atmosenv.2006.04.052.

Griffiths, P., & de Hasseth, J. A. (2007). *Fourier transform infrared spectrometry* (2nd ed.). Wiley-Blackwell, ISBN 978-0-471-19404-0.

Guang, J., Xue, Y., Li, Y., Liang, S., Mei, L., & Xu, H. (2012). Retrieval of aerosol optical depth over bright land surfaces by coupling bidirectional reflectance distribution function model and aerosol retrieval model. *Remote Sensing Letters, 3*(7), 577–584. Available from https://doi.org/10.1080/01431161.2011.642322.

Gupta, P., Christopher, S. A., Wang, J., Gehrig, R., Lee, Y. C., & Kumar, N. (2006). Satellite remote sensing of particulate matter and air quality assessment over global cities. *Atmospheric Environment, 40*(30), 5880–5892.

Gurudath, S., & Srinivasa, K. G. (2021). Air quality measurement using low-cost sensors—A review. In A. Abraham, O. Castillo, & D. Virmani (Eds.), *Proceedings of 3rd international conference on computing informatics and networks* (pp. 687–697). Springer. Available from https://doi.org/10.1007/978-981-15-9712-1_43.

Haddad, P. R., Nesterenko, P. N., & Buchberger, W. (2008). Recent developments and emerging directions in ion chromatography. *Journal of Chromatography. A*, *1184*(1–2), 456–473. Available from https://doi.org/10.1016/j.chroma.2007.10.022.

Han, B., Ding, H., Ma, Y., & Gong, W. (2017). Improving retrieval accuracy for aerosol optical depth by fusion of MODIS and CALIOP data. *Tehnicki vjesnik–Technical Gazette*, *24*(3), 791–800. Available from https://doi.org/10.17559/TV-20160429044233.

Hansel, A., Jordan, A., Holzinger, R., Prazeller, P., Vogel, W., & Lindinger, W. (1995). Proton transfer reaction mass spectrometry: On-line trace gas analysis at the ppb level. *International Journal of Mass Spectrometry and Ion Processes*, *149–150*, 609–619. Available from https://doi.org/10.1016/0168-1176(95)04294-U.

Hartogensis, O. K., Watts, C. J., Rodriguez, J. C., & De Bruin, H. A. R. (2003). Derivation of an effective height for scintillometers: La Poza experiment in Northwest Mexico. *Journal of Hydrometeorology*, *4*(5), 915–928.

Haywood, J. M., Abel, S. J., Barrett, P. A., Bellouin, N., Blyth, A., Bower, K. N., & Cotterell, M. I. (2021). The CLoud–Aerosol–Radiation interaction and forcing: Year 2017 (CLARIFY-2017) measurement campaign. *Atmospheric Chemistry and Physics*, *21*(2), 1049–1084.

Hearn, A. G. (1961). The absorption of ozone in the ultra-violet and visible regions of the spectrum. *Proceedings of the Physical Society*, *78*(5), 932–940.

Holloway, T., Miller, D., Anenberg, S., Diao, M., Duncan, B., Fiore, A. M., Henze, D. K., Hess, J., Kinney, P. L., Liu, Y., Neu, J. L., O'Neill, S. M., Odman, M. T., Pierce, R. B., Russell, A. G., Tong, D., West, J. J., & Zondlo, M. A. (2021). Satellite monitoring for air quality and health. *Annual Review of Biomedical Data Science*, *4*(1), 417–447. Available from https://doi.org/10.1146/annurev-biodatasci-110920-093120.

Huang, C., Ho, H., & Lin, T. (2018). Improving the image fusion procedure for high spatiotemporal aerosol optical depth retrieval - A case study of urban area in Taiwan. *Journal of Applied Remote Sensing*, *12*(4), 1. Available from https://doi.org/10.1117/1.JRS.12.042605.

Illueca Fernández, E., Bernabé Mulero, N., Pujante Pérez, A., Merino García, J. M., Cuevas Martínez, I., & Jara Valera, A. J. (2023). CEN/TS 17660 in Air Quality Systems for Data Quality Validation and Certification over Smart Spot Air Quality Systems. In J. Bravo, S. Ochoa, & J. Favela (Eds.), *Proceedings of the international conference on ubiquitous computing & ambient intelligence (UCAmI 2022)* (pp. 718–729). Springer. Available from https://doi.org/10.1007/978-3-031-21333-5_65.

iQAir (2022). *2022 World air quality report, region and city PM2.5 ranking (v1.1)*. Published by iQAir.

Isaac, N. A., Pikaar, I., & Biskos, G. (2022). Metal oxide semiconducting nanomaterials for air quality gas sensors: operating principles, performance, and synthesis techniques. *Microchimica Acta*, *189*(5), 196. Available from https://doi.org/10.1007/s00604-022-05254-0.

James, A. T., & Martin, A. J. P. (1952). Gas-liquid partition chromatography: the separation and micro-estimation of volatile fatty acids from formic acid to dodecanoic acid. *Biochemical Journal*, *50*(5), 679–690. Available from https://doi.org/10.1042/bj0500679.

Jodpimai, S., Boonduang, S., & Limsuwan, P. (2016). Inline ozone concentration measurement by a visible absorption method at wavelength 605 nm. *Sensors and Actuators B*, *222*, 8–14.

Johnson, B., Turnbull, K., Brown, P., Burgess, R., Dorsey, J., Baran, A. J., Webster, H., Haywood, J., Cotton, R., Ulanowski, Z., & Hesse, E. (2012). In situ observations of volcanic ash clouds from the FAAM aircraft during the eruption of Eyjafjallajökull in 2010. *Journal of Geophysical Research: Atmospheres*, *117*(D20).

Johnston, M. V., & Kerecman, D. E. (2019). Molecular characterization of atmospheric organic aerosol by mass spectrometry. *Annual Review of Analytical Chemistry*, *12*(1), 247–274. Available from https://doi.org/10.1146/annurev-anchem-061516-045135.

Joranger, E., & Ottar, B. (1984). Air pollution studies in the Norwegian Arctic. *Geophysical Research Letters*, *11*(5), 365–368.

Kahn, R. A., & Gaitley, B. J. (2015). An analysis of global aerosol type as retrieved by MISR. *Journal of Geophysical Research: Atmospheres*, *120*, 4248–4281. Available from https://doi.org/10.1002/2015JD023322.

Kang, Y., Aye, L., Ngo, T. D., & Zhou, J. (2022). Performance evaluation of low-cost air quality sensors: A review. *Science of The Total Environment*, *818*, 151769. Available from https://doi.org/10.1016/j.scitotenv.2021.151769.

Karagulian, F. (2023). New challenges in air quality measurements. In S. De Vito, K. Karatzas, A. Bartonova, & G. Fattoruso (Eds.), *Air quality networks* (pp. 3–20). Springer. Available from https://doi.org/10.1007/978-3-031-08476-8_1.

Karagulian, F., Barbiere, M., Kotsev, A., Spinelle, L., Gerboles, M., Lagler, F., & Borowiak, A. (2019). Review of the performance of low-cost sensors for air quality monitoring. *Atmosphere*, *10*(9), 506. Available from https://doi.org/10.3390/atmos10090506.

Kassandros, T., Bagkis, E., Johansson, L., Kontos, Y., Katsifarakis, K.L., Karppinen, A., & Karatzas, K. (2023). Machine learning-assisted dispersion modelling based on genetic algorithm-driven ensembles: An application for road dust in Helsinki. *Atmospheric Environment*, *307*, 119818. Available from https://doi.org/10.1016/j.atmosenv.2023.119818.

Kezoudi, M., Keleshis, C., Antoniou, P., Biskos, G., Bronz, M., Constantinides, C., Desservettaz, M., Gao, R. S., Girdwood, J., Harnetiaux, J., & Kandler, K. (2021). The unmanned systems research laboratory (USRL): A new facility for UAV-based atmospheric observations. *Atmosphere*, *12*(8), 1042.

Kuula, J., Mäkelä, T., Aurela, M., Teinilä, K., Varjonen, S., González, Ó., & Timonen, H. (2020). Laboratory evaluation of particle-size selectivity of optical low-cost particulate matter sensors. *Atmospheric Measurement Techniques*, *13*(5), 2413–2423. Available from https://doi.org/10.5194/amt-13-2413-2020.

Lambey, V., & Prasad, A. D. (2021). A review on air quality measurement using an unmanned aerial vehicle. *Water, Air, & Soil Pollution*, *232*, 1–32.

Levy, R. C., Remer, L. A., & Kaufman, Y. J. (2004). Effects of neglecting polarization on the MODIS aerosol retrieval over land. *IEEE Transactions on Geoscience and Remote Sensing, 42*(11), 2576–2583. Available from https://doi.org/10.1109/TGRS.2004.837336.

Lewis, A. C., von Schneidemesser, E., & Peltier, R. (2018). Low-cost sensors for the measurement of atmospheric composition: Overview of topic and future applications. World Meteorological Organization (WMO-No. 1215).

Li, J., Ge, X., He, Q., & Abbas, A. (2021). Aerosol optical depth (AOD): Spatial and temporal variations and association with meteorological covariates in Taklimakan desert, China. *PeerJ, 9*, e10542.

Li, J., Mattewal, S. K., Patel, S., & Biswas, P. (2020). Evaluation of nine low-cost-sensor-based particulate matter monitors. *Aerosol and Air Quality Research, 20*(2), 254–270. Available from https://doi.org/10.4209/aaqr.2018.12.0485.

Liang, J. G., Jiang, Y., Wu, J. K., Wang, C., Gratowski, S., Gu, X., & Pan, L. (2023). Multiplex-gas detection based on non-dispersive infrared technique: A review. *Sensors and Actuators A: Physical, 356*, 114318.

Liang, S. (Ed.), (2017). *Comprehensive remote sensing*. Elsevier.

Liora, N., Kontos, S., Parliari, D., Akritidis, D., Poupkou, A., Papanastasiou, D. K., & Melas, D. (2022). "On-line" heating emissions based on WRF meteorology—Application and evaluation of a modeling system over Greece. *Atmosphere, 13*(4), 568. Available from https://doi.org/10.3390/atmos13040568.

Liora, N., Markakis, K., Poupkou, A., Giannaros, T. M., & Melas, D. (2015). The natural emissions model (NEMO): Description, application and model evaluation. *Atmospheric Environment, 122*, 493–504.

Liu, X., Wang, Y., Cong, H., Shen, Y., & Yu, B. (2021). A review of the design of packing materials for ion chromatography. *Journal of Chromatography A. Advance online publication.* Available from https://doi.org/10.1016/j.chroma.2021.462313.

Long, R. W., Hall, E., Beaver, M., Duvall, R., Kaushik, S., Kronmiller, K., Wheeler, M., Garvey, S., Drake, Z., & McElroy, F. (2014). Performance of the proposed new federal reference methods for measuring ozone concentrations in ambient air (EPA/600/R-14/432).

Long, R. W., Whitehill, A., Habel, A., Urbanski, S., Halliday, H., Colón, M., Kaushik, S., & Landis, M. S. (2021). Comparison of ozone measurement methods in biomass burning smoke: An evaluation under field and laboratory conditions. *Atmospheric Measurement Techniques, 14*(3), 1783–1800.

Long, R. W., Eatough, N. L., Mangelson, N. F., Thompson, W., Fiet, K., Smith, S., Smith, R., Eatough, D. J., Pope, C. A., & Wilson, W. E. (2003). The measurement of $PM_{2.5}$, including semi-volatile components, in the EMPACT program: Results from the Salt Lake City study and implications for public awareness, health effects, and control strategies. *Atmospheric Environment, 37*, 4407–4417.

Lyapustin, A., Wang, Y., Korkin, S., & Huang, D. (2018). MODIS collection 6 MAIAC algorithm. *Atmospheric Measurement Techniques, 11*(10), 5741–5765.

Lyons, W. A., & Husar, R. B. (1976). SMS/GOES visible images detect a synoptic-scale air pollution episode. *Monthly Weather Review, 104*, 1623.

Martin, R. V. (2008). Satellite remote sensing of surface air quality. *Atmospheric Environment, 42*(34), 7823–7843.

Matsumi, Y., Shigemori, H., & Takahashi, K. (2005). Laser-induced fluorescence instrument for measuring atmospheric SO_2. *Atmospheric Environment, 39*, 3177–3185.

McMaster, M. C. (2008). *GC/MS: A practical user's guide* (2nd ed.). New York: Wiley Measurement methods and UK monitoring networks for NO2. (n.d.). Retrieved from. Available from https://uk-air.defra.gov.uk/library/assets/documents/reports/aqeg/chapter4.pdf.

Mekler, Y., Quenzel, H., Ohring, G., & Marcus, I. (1977). Relative atmospheric aerosol content from ERTS observations. *Journal of Geophysical Research, 82*, 967–970.

Melfi, S. H. (2005). Remote sensing for air quality management. *Laser Monitoring of the Atmosphere*, 9–28.

Mendez, D., & Sanchez, M. (2023). AQ mapping through low-cost sensor networks. In S. De Vito, K. Karatzas, A. Bartonova, & G. Fattoruso (Eds.), *Air quality networks. Environmental informatics and modeling*. Springer. Available from https://doi.org/10.1007/978-3-031-08476-8_4.

Merkus, H. G. (2009). *Particle size measurements: Fundamentals, practice, quality*. Berlin: Springer Netherlands.

Meyer, M. B., Patashnick, H., & Ambs, J. L. (2002). Ongoing development of a continuous reference standard particulate matter mass monitor for ambient air. In *Paper presented at the symposium on air quality measurement methods and technology*. Air and Waste Management Association, San Francisco, California.

Mie, G. (1908). Beiträge zur Optik trüber Medien, speziell kolloidaler metallösungen. *Annalen der Physik, 330*(3), 377–445.

Mignacca, D., & Stubbs, K. (1999). Effects of equilibration temperature on PM_{10} concentrations from the TEOM method in the Lower Fraser Valley. *Journal of the Air & Waste Management Association, 49*, 1250–1254.

Mishchenko, M. I., Liu, L., Geogdzhayev, I. V., Travis, L. D., Cairns, B., & Lacis, A. A. (2010). Toward unified satellite climatology of aerosol properties. 3. MODIS versus MISR versus AERONET. *Journal of Quantitative Spectroscopy and Radiative Transfer, 111*(4), 540–552. Available from https://doi.org/10.1016/j.jqsrt.2009.11.003.

Narayana, M. V., Jalihal, D., & Nagendra, S. M. S. (2022). Establishing a sustainable low-cost air quality monitoring setup: A survey of the state-of-the-art. *Sensors, 22*(1), 394. Available from https://doi.org/10.3390/s22010394.

NASA Earth Data. (2024). Open Access for Open Science: What is remote sensing? Tutorial on remotely-sensed data, from sensor characteristics, to different types of resolution, to data processing and analysis. <https://www.earthdata.nasa.gov/learn/backgrounders/remote-sensing>.

National Research Council. (1995). *Expanding the Vision of Sensor Materials*. Washington, DC: The National Academies Press. Available from https://doi.org/10.17226/4782.

Navalgund, R. R., Jayaraman, V., & Roy, P. S. (2007). Remote sensing applications: An overview. *Current Science*, 1747–1766.

Northcross, A. L., Edwards, R. J., Johnson, M. A., Wang, Z.-M., Zhu, K., Allen, T., & Smith, K. R. (2013). A low-cost particle counter as a real-time fine-particle mass monitor. *Environmental Science: Processes & Impacts, 15*(2), 433–439. Available from https://doi.org/10.1039/C2EM30568B.

Nowack, P., Konstantinovskiy, L., Gardiner, H., & Cant, J. (2021). Machine learning calibration of low-cost NO_2 and PM_{10} sensors: Non-linear algorithms and their impact on site transferability. *Atmospheric Measurement Techniques*, *14*(8), 5637–5655. Available from https://doi.org/10.5194/amt-14-5637-2021.

Oil Sands Monitoring Program. (2020). *Standard operating procedure for measurement of SO_2 by ultraviolet fluorescence SO_2 analyzer*. <https://open.alberta.ca/publications/sop-measurement-so2-by-ultraviolet-fluorescence-so2-analyzer>.

Oke, T. R., Mills, G., Christen, A., & Voogt, J. A. (2017). *Urban climates*. Cambridge University Press.

Ott, E.-J. E., Kucinski, T. M., Dawson, J. N., & Freedman, M. A. (2021). Use of transmission electron microscopy for analysis of aerosol particles and strategies for imaging fragile particles. *Analytical Chemistry*, *93*(33), 11347–11356. Available from https://doi.org/10.1021/acs.analchem.0c05225.

Oyola, P., Carbone, S., Timonen, H., Torkmahalleh, M., & Lindén, J. (2022). Editorial: Rise of low-cost sensors and citizen science in air quality studies. *Frontiers in Environmental Science*, *10*. Available from https://doi.org/10.3389/fenvs.2022.868543, Article 868543.

Padmakumari, B., Maheskumar, R. S., Morwal, S. B., Harikishan, G., Konwar, M., Kulkarni, J. R., & Goswami, B. N. (2013). Aircraft observations of elevated pollution layers near the foothills of the Himalayas during CAIPEEX-2009. *Quarterly Journal of the Royal Meteorological Society*, *139*(672), 625–638.

Palmer, P. I., O'Doherty, S., Allen, G., Bower, K., Bösch, H., Chipperfield, M. P., & Gallagher, M. W. (2018). A measurement-based verification framework for UK greenhouse gas emissions: an overview of the greenhouse gas UK and global emissions (GAUGE) project. *Atmospheric Chemistry and Physics*, *18*(16), 11753–11777.

Panta, A., Kandler, K., Alastuey, A., González-Flórez, C., González-Romero, A., Klose, M., Querol, X., Reche, C., Yus-Díez, J., & Pérez García-Pando, C. (2023). Insights into the single-particle composition, size, mixing state, and aspect ratio of freshly emitted mineral dust from field measurements in the Moroccan Sahara using electron microscopy. *Atmospheric Chemistry and Physics*, *23*, 3861–3885. Available from https://doi.org/10.5194/acp-23-3861-2023.

Papayannis, A. D., Porteneuve, J., Balis, D., Zerefos, C., & Galani, E. (1999). Design of a new DIAL system for tropospheric and lower stratospheric ozone monitoring in Northern Greece. *Physics and Chemistry of the Earth, Part C: Solar, Terrestrial & Planetary Science*, *24*(5), 439–442.

Parrish, D. D., & Fehsenfeld, F. C. (2000). Methods for gas-phase measurements of ozone, ozone precursors and aerosol precursors. *Atmospheric Environment*, *34*(11), 1921–1957.

Patashnick, H., Rupprecht, G., Ambs, J. L., & Meyer, M. B. (2001). Development of a reference standard for particulate matter mass in ambient air. *Aerosol Science and Technology: the Journal of the American Association for Aerosol Research*, *34*, 42–45.

Pellikka, T., Kajolinna, T., & Perälä, M. (2019). SO_2 emission measurement with the European standard reference method, EN 14791, and alternative methods—observations from laboratory and field studies. *Journal of the Air & Waste Management Association*, *69*(9), 1122–1131. Available from https://doi.org/10.1080/10962247.2019.1640809.

Petruci, J. F. S., Barreto, D. N., Dias, M. A., & Felix, E. P. (2022). Analytical methods applied for ozone gas detection: A review. *TrAC Trends in Analytical Chemistry*, *149*, 116552.

Pipal, A. S., & Taneja, A. (2023). Measurements of indoor air quality. In D. K. Aswal, S. Yadav, T. Takatsuji, P. Rachakonda, & H. Kumar (Eds.), *Handbook of metrology and applications*. Singapore: Springer. Available from https://doi.org/10.1007/978-981-19-1550-5_90-1.

Platt, U., & Stutz, J. (2024). *Differential optical absorption spectroscopy: Principles and applications* (2nd ed.). Berlin, Heidelberg: Springer.

Platt, U., Perner, D., & Paetz, H. W. (1979). Simultaneous measurement of atmospheric CH_2O, O_3, and NO_2 by differential optical absorption. *Journal of Geophysical Research*, *84*(C10), 6329–6335. Available from https://doi.org/10.1029/JC084iC10p06329.

Pleil, J. D., Hansel, A., & Beauchamp, J. (2019). Advances in proton transfer reaction mass spectrometry (PTR-MS): Applications in exhaled breath analysis, food science, and atmospheric chemistry. *Journal of Breath Research*, *13*(3), 039002. Available from https://doi.org/10.1088/1752-7163/ab21a7.

Poole, C. F. (2015). Ionization-based detectors for gas chromatography. *Journal of Chromatography. A*, *1382*, 24–36. Available from https://doi.org/10.1016/j.chroma.2015.02.035.

Porter, J. N., Miller, M., Pietras, C., & Motell, C. (2001). Ship-based sun photometer measurements using microtops sun photometers. *Journal of Atmospheric and Oceanic Technology*, *18*(5), 765–774.

Ramanathan, V., & Carmichael, G. (2008). Global and regional climate changes due to black carbon. *Nature Geoscience*, *1*(4), 221–227.

Ray, A. E., & Vaughn, D. L. (2009, September 1). Standard operating procedure for the continuous measurement of particulate matter (PDF). U.S. Environmental Protection Agency, pp. 1–1, 3-1–3-2, 6-1–6-2. Retrieved June 28, 2017.

Ricardo Energy and Environment (2022). AURN QAQC Annual Technical Report 2021: Report for The Environment Agency, Ref. 25952 Lot 2&5 ATR 2021, Ricardo ref. ED12695100_ATR_2021 Issue1 24/08/2022.

Rodgers, C. (1997). *Satellite remote sensing: The retrieval problem. In* The stratosphere and its role in the climate system (pp. 133–171). Berlin, Heidelberg: Springer.

Romanini, D., Kachanov, A. A., Sadeghi, N., & Stoeckel, E. (1997). CW cavity ring down spectroscopy. *Chemical Physics Letters*, *264*, 316–322. Available from https://doi.org/10.1016/S0009-2614(96)01335-8.

Rowell, A., Terry, M. E., & Deary, M. E. (2021). Comparison of diffusion tube—measured nitrogen dioxide concentrations at child and adult breathing heights: who are we monitoring for? *Air Quality, Atmosphere & Health*, *14*, 27–36. Available from https://doi.org/10.1007/s11869-020-00909-4.

Ruppecht, E., Meyer, M., & Patashnick, H. (1992). The tapered element oscillating microbalance as a tool for measuring ambient particulate concentrations in real time. *Journal of Aerosol Science*, *23*(Supplement 1), 635–638. Available from https://doi.org/10.1016/0021-8502(92)90492-E.

Ryan, R. G., Marais, E. A., Gershenson-Smith, E., Ramsay, R., Muller, J.-P., Tirpitz, J.-L., & Frieß, U. (2023). Measurement report: MAX-DOAS measurements characterise Central London ozone pollution episodes during 2022 heatwaves. *Atmospheric Chemistry and Physics, 23*, 7121–7139. Available from https://doi.org/10.5194/acp-23-7121-2023.

Ryder, C. L., McQuaid, J. B., Flamant, C., Rosenberg, P. D., Washington, R., Brindley, H. E., & Banks, J. R. (2015). Advances in understanding mineral dust and boundary layer processes over the Sahara from Fennec aircraft observations. *Atmospheric Chemistry and Physics, 15*(14), 8479–8520.

Seidl, A. W., Sodemann, H., & Steen-Larsen, H. C. (2023). A modular field system for near-surface, vertical profiling of the atmospheric composition in harsh environments using cavity ring-down spectroscopy. *Atmospheric Measure Technique, 16*, 769–790. Available from https://doi.org/10.5194/amt-16-769-2023.

Shao, L., Liu, P., Jones, T., Yang, S., Wang, W., Zhang, D., Li, Y., Yang, C.-X., Xing, J., Hou, C., Zhang, M., Feng, X., Li, W., & BéruBé, K. (2022). A review of atmospheric individual particle analyses: Methodologies and applications in environmental research. *Gondwana Research, 110*, 347–369.

Shaw, G. E. (1983). Sun photometry. *Bulletin of the American Meteorological Society, 64*(1), 4–10.

Shin, S. E., Jung, C. H., & Kim, Y. P. (2012). Estimation of the optimal heated inlet air temperature for the beta-ray absorption method: analysis of the PM_{10} concentration difference by different methods in coastal areas. *Advanced Environmental Research, 1*, 69–82. Available from https://doi.org/10.12989/aer.2012.1.1.069.

Shukla, K., & Aggarwal, S. G. (2022). A technical overview on beta-attenuation method for the monitoring of particulate matter in ambient air. *Aerosol Air Quality Research, 22*, 220195. Available from https://doi.org/10.4209/aaqr.220195.

Singh, D., Dahiya, M., Kumar, R., & Nanda, C. (2021). Sensors and systems for air quality assessment, monitoring, and management: A review. *Journal of Environmental Management, 289*, 112510. Available from https://doi.org/10.1016/j.jenvman.2021.112510.

Singh, V., Singh, S., Biswal, A., Kesarkar, A. P., Mor, S., & Ravindra, K. (2020). Diurnal and temporal changes in air pollution during COVID-19 strict lockdown over different regions of India. *Environmental Pollution, 266*, 115368.

Singh, V., Sokhi, R. S., & Kukkonen, J. (2020). An approach to predict population exposure to ambient air $PM_{2.5}$ concentrations and its dependence on population activity for the megacity London. *Environmental Pollution, 257*, 113623. Available from https://doi.org/10.1016/j.envpol.2019.113623, 2020.

Sklorz, A., Janßen, S., & Lang, W. (2012). Detection limit improvement for NDIR ethylene gas detectors using passive approaches. *Sensors and Actuators B: Chemical, 175*, 246–254.

Skoog, D. A., West, D. M., Holler, F. J., & Crouch, S. R. (2014). *Fundamentals of analytical chemistry* (9th ed.). Cengage Learning.

Sokhi, R. S., Moussiopoulos, N., Baklanov, A., Bartzis, J., Coll, I., Finardi, S., & Kukkonen, J. (2022). Advances in air quality research–current and emerging challenges. *Atmospheric Chemistry and Physics, 22*, 4615–4703. Available from https://doi.org/10.5194/acp-22-4615-2022.

Sokhi, R. S., Singh, V., Querol, X., Finardi, S., Targino, A. C., de Fatima Andrade, M., & Baklanov, A. (2021). A global observational analysis to understand changes in air quality during exceptionally low anthropogenic emission conditions. *Environment International, 157*, 106818.

Sousan, S., Koehler, K., Thomas, G., Park, J. H., Hillman, M., Halterman, A., & Peters, T. M. (2016). Inter-comparison of low-cost sensors for measuring the mass concentration of occupational aerosols. *Aerosol Science and Technology, 50*(5), 462–473. Available from https://doi.org/10.1080/02786826.2016.1162901.

Soysal, U., Géhin, E., Algré, E., Berthelot, B., Da, G., & Robine, E. (2017). Aerosol mass concentration measurements: Recent advancements of real-time nano/micro systems. *Journal of Aerosol Science, 114*, 42–54. Available from https://doi.org/10.1016/j.jaerosci.2017.09.008.

Sparkman, O. D., Penton, Z. E., & Kitson, F. G. (2011). *Gas chromatography and mass spectrometry* (2nd ed.). Amsterdam: Elsevier.

Spielvogel, J., & Weiss, M. (2013). The Fidas®—A new continuous ambient air quality monitoring system that additionally reports particle size and number concentration. In *Urban environment* (pp. 243–251).

Spinelle, L., Gerboles, M., Kok, G., Persijn, S., & Sauerwald, T. (2017). Review of portable and low-cost sensors for the ambient air monitoring of benzene and other volatile organic compounds. *Sensors (Basel), 17*(7), 1520. Available from https://doi.org/10.3390/s17071520.

Spinelle, L., Gerboles, M., Villani, M., Aleixandre, M., & Bonavitacola, F. (2015). Field calibration of a cluster of low-cost available sensors for air quality monitoring. Part A: Ozone and nitrogen dioxide. *Sensors and Actuators B: Chemical, 215*, 249–257. Available from https://doi.org/10.1016/j.snb.2015.03.031.

Stevenson, J. A., Loughlin, S. C., Font, A., Fuller, G. W., MacLeod, A., Oliver, I. W., & Thordarson, T. (2013). UK monitoring and deposition of tephra from the May 2011 eruption of Grímsvötn, Iceland. *Journal of Applied Volcanology, 2*(1). Available from https://doi.org/10.1186/2191-5040-2-3, Additional file 3: Detailed air quality monitoring methodology.

Stith, J. L., Ramanathan, V., Cooper, W. A., Roberts, G. C., DeMott, P. J., Carmichael, G., & Baumgardner, D. (2009). An overview of aircraft observations from the Pacific dust experiment campaign. *Journal of Geophysical Research: Atmospheres, 114*(D5).

Stull, R. B. (1988). *An introduction to boundary layer meteorology* (Vol. 13). Springer. Available from http://doi.org/10.1007/978-94-009-3027-8.

Tedesco, M. (2015). Remote sensing and the cryosphere. *Remote Sensing of the Cryosphere*, 1-16.

Torres, O., Bhartia, P. K., Herman, J. R., Ahmad, Z., & Gleason, J. (1998). Derivation of aerosol properties from satellite measurements of backscattered ultraviolet radiation: Theoretical basis. *Journal of Geophysical Research: Atmospheres, 103*(D14), 17099–17110.

Turnipseed, A. A., Andersen, P. C., Williford, C. J., Ennis, C. A., & Birks, J. W. (2017). Use of a heated graphite scrubber as a means of reducing interferences in UV-absorbance measurements of atmospheric ozone. *Atmospheric Measurement Techniques, 10*(6), 2253–2269.

US EPA. (2004). Fact sheet for trace level SO2 monitoring method. (2004). United States Environmental Protection Agency. <https://www3.epa.gov/ttnamti1/files/ambient/pm25/spec/so2sum2.pdf>.

US EPA. (2016a). U.S. Environmental Protection Agency. Quality Assurance Guidance Document 2.12: Monitoring $PM_{2.5}$ in Ambient Air Using Designated Reference or Class I Equivalent Methods (EPA-454/B-16-001).

US EPA. (2016b). U.S. Environmental Protection Agency. Monitoring $PM_{2.5}$ in ambient air using designated reference or class I equivalent methods. In: *Quality Assurance Handbook for Air Pollution Measurement Systems, vol. II*, pp. 2–12. Part II, Section.

US EPA. (2022). U.S. Environmental Protection Agency. Our Nation's Air: Status and Trends Through 2022. <https://gispub.epa.gov/air/trendsreport/2023/#home>.

US EPA. (2023). U.S. Environmental Protection Agency. List of Designated Reference and Equivalent Methods. Center for Environmental Measurements & Modeling, Air Methods & Characterization Division (MD-D205-03), Office of Research and Development. <https://www.epa.gov/ttn/amtic/criteria.html>.

US GAO. (2020). U.S. Government Accountability Office. Air Pollution-Opportunities to Better Sustain and Modernize the National Air Quality Monitoring System. Report to Congressional Requesters. <https://www.gao.gov/assets/gao-21-38.pdf>.

Van Donkelaar, A., Martin, R. V., Brauer, M., Hsu, N. C., Kahn, R. A., Levy, R. C., & Winker, D. M. (2016). Global estimates of fine particulate matter using a combined geophysical-statistical method with information from satellites, models, and monitors. *Environmental Science & Technology*, *50*(7), 3762–3772.

Van Donkelaar, A., Martin, R. V., Pasch, A. N., Szykman, J. J., Zhang, L., Wang, Y. X., & Chen, D. (2012). Improving the accuracy of daily satellite-derived ground-level fine aerosol concentration estimates for North America. *Environmental Science & Technology*, *46*(21), 11971–11978.

Van Kesteren, B. (2012). Measuring Water-Vapour and Carbon-Dioxide Fluxes at Field Scales with Scintillometry (Sl: sn). http://edepot.wur.nl/239949.

Veefkind, P., Van Oss, R.F., Eskes, H., Borowiak, A., Dentner, F., & Wilson, J. (2007). *The applicability of remote sensing in the field of air pollution* (p. 59). Institute for Environment and Sustainability, Italy.

Vernon-Parry, K. D. (2000). Scanning electron microscopy: An introduction. *III-Vs Review*, *13*(4), 40–44. Available from https://doi.org/10.1016/S0961-1290(00)80006-X.

Villa, T. F., Gonzalez, F., Miljievic, B., Ristovski, Z. D., & Morawska, L. (2016). An overview of small unmanned aerial vehicles for air quality measurements: Present applications and future prospectives. *Sensors*, *16*(7), 1072.

von Clarmann, T., Degenstein, D. A., Livesey, N. J., Bender, S., Braverman, A., Butz, A., & Funke, B. (2020). Overview: Estimating and reporting uncertainties in remotely sensed atmospheric composition and temperature. *Atmospheric Measurement Techniques*, *13*(8), 4393–4436.

Wang, A., Machida, Y., deSouza, P., Mora, S., Duhl, T., Hudda, N., Durant, J. L., Duarte, F., & Ratti, C. (2023). Leveraging machine learning algorithms to advance low-cost air sensor calibration in stationary and mobile settings. *Atmospheric Environment*, 119692. Available from https://doi.org/10.1016/j.atmosenv.2023.119692.

Wang, Q., Mustafa, F., Bu, L., Yang, J., Fan, C., Liu, J., & Chen, W. (2022). Monitoring of atmospheric carbon dioxide over a desert site using airborne and ground measurements. *Remote Sensing*, *14*(20), 5224.

Wang, Y., Li, J., Jing, H., Zhang, Q., Jiang, J., & Biswas, P. (2015). Laboratory evaluation and calibration of three low-cost particle sensors for particulate matter measurement. *Aerosol Science and Technology*, *49*(11), 1063–1077. Available from https://doi.org/10.1080/02786826.2015.1100710.

Wei, X., Chang, N. B., Bai, K., & Gao, W. (2020). Satellite remote sensing of aerosol optical depth: advances, challenges, and perspectives. *Critical Reviews in Environmental Science and Technology*, *50*(16), 1640–1725.

Wexler, A. S., & Johnston, M. V. (2011). Real-Time Particle Analysis by Mass Spectrometry. In P. Kulkarni, P. A. Baron, & K. Willeke (Eds.), *Aerosol measurement: Principles, techniques, and applications, third edition* (Chapter 11). <https://doi.org/10.1002/9781118001684.ch11>.

(WHO) (2014, March 25). 7 million premature deaths annually linked to air pollution. [News release]. Geneva. <https://www.who.int/news/item/25-03-2014-7-million-premature-deaths-annually-linked-to-air-pollution>.

WHO (2021). WHO global air quality guidelines. Particulate matter (PM2.5 and PM10), ozone, nitrogen dioxide, sulfur dioxide and carbon monoxide. Geneva: World Health Organization; 2021. Licence: CC BY-NC-SA 3.0 IGO.

(WHO) (2023). WHO ambient air quality database, 2022 update: Status report. Geneva: World Health Organization. License: CC BY-NC-SA 3.0 IGO.

Williams, D. B., & Carter, C. B. (2009). *Transmission electron microscopy: A textbook for materials science*. Springer.

Williams, E. J., Fehsenfeld, F. C., Jobson, B. T., Kuster, W. C., Goldan, P. D., Stutz, J., & McClenny, W. A. (2006). Comparison of ultraviolet absorbance, chemiluminescence, and DOAS instruments for ambient ozone monitoring. *Environmental Science & Technology*, *40*(19), 5755–5762.

(WMO) (2020). An update on low-cost sensors for the measurement of atmospheric composition. World Meteorological Organization, Geneva. <https://library.wmo.int/doc_num.php?explnum_id = 10620>.

(WHO) (2023). Overview of methods to assess population exposure to ambient air pollution. Geneva: World Health Organization. ISBN 978-92-4-007349-4 (electronic version), ISBN 978-92-4-007350-0(print version).

Wolpert, D., & Macready, W. (1997). No free lunch theorems for optimization. *IEEE Transactions on Evolutionary Computation*, *1*(1), 67–82. Available from https://doi.org/10.1109/4235.585893.

Xiang, Y., Lv, L., Chai, W., Zhang, T., Liu, J., & Liu, W. (2020). Using LIDAR technology to assess regional air pollution and improve estimates of $PM_{2.5}$ transport in the North China Plain. *Environmental Research Letters*, *15*(9), 094071.

Yatkin, S., Gerboles, M., Borowiak, A., & Signorini, M. (2022). Guidance on low-cost air quality sensor deployment for nonexperts based on the AirSensEUR experience. Publications Office of the European Union, Luxembourg. doi:10.2760/180094, JRC130628.

Yu, X., Shi, Y., Wang, T., & Sun, X. (2017). Dust-concentration measurement based on Mie scattering of a laser beam. *PLoS One*, *12*(8), e0181575. Available from https://doi.org/10.1371/journal.pone.0181575.

Zeydan, Ö., & Wang, Y. (2019). Using MODIS-derived aerosol optical depth to estimate ground-level $PM_{2.5}$ concentrations over Turkey. *Atmospheric Pollution Research, 10*(5), 1565–1576. Available from https://doi.org/10.1016/j.apr.2019.05.005.

Zhang, H., & Kondragunta, S. (2021). Daily and hourly surface $PM_{2.5}$ estimation from satellite AOD. *Earth and Space Science*, *8*(3), e2020EA001599.

Zhang, S., Wu, J., Fan, W., Yang, Q., & Zhao, D. (2020a). Review of aerosol optical depth retrieval using visibility data. *Earth-Science Reviews*, *200*, 102986.

Zhang, X., Wang, F., Wang, W., Huang, F., Chen, B., Gao, L., Wang, S., Yan, H., Ye, H., Si, F., & Hong, J. (2020b). The development and application of satellite remote sensing for atmospheric compositions in China. *Atmospheric Research*, *245*, 105056.

Zhao, X., Fioletov, V., Brohart, M., Savastiouk, V., Abboud, I., Ogyu, A., Davies, J., Sit, R., Lee, S. C., Cede, A., & Tiefengraber, M. (2021). The world Brewer reference triad—updated performance assessment and new double triad. *Atmospheric Measurement Techniques*, *14*(3), 2261—2283.

Zhou, B., Zhang, S., Xue, R., Li, J., & Wang, S. (2023). A review of space-air-ground integrated remote sensing techniques for atmospheric monitoring. *Journal of Environmental Sciences*, *123*, 3—14.

Chapter 10

Regional and global air quality

Oliver Wild
Lancaster Environment Centre, Lancaster University, Lancaster, United Kingdom

10.1 Introduction

Poor air quality is usually most apparent at local and urban scales, where local sources contribute to sulfurous or photochemical smog, impairing visibility and creating conditions hazardous to health. Stagnant meteorological conditions may support the development of major pollution episodes lasting several days, and these may lead to severe episodes such as the London winter smogs or the summertime oxidant episodes of Los Angeles in the first half of the 20th century. However, local pollution sources typically build on a background of pollutants from a much wider region, incorporating contributions from both natural and anthropogenic sources over regional and even continental scales. In addition, longer-lived pollutants may be substantially influenced by background pollutant concentrations that reflect atmospheric composition at a hemispheric or global scale.

This chapter introduces the importance of air pollution on regional and global scales and explores how changes on these larger scales influence local air quality. It starts by considering the concept of global air quality, which reflects the relatively well-mixed contributions of long-lived pollutants present in the global atmosphere, and then explores the differing characteristics of regional scale pollution in different parts of the world. Long-range transport of pollutants in the atmosphere provides a mechanism that links air quality in different regions, such that they are not independent of each other. The chapter then explores the historical and likely future changes in air quality at regional and global scales and highlights the implications that this may have for local air quality.

10.2 Global air quality

The large-scale composition of the lower atmosphere is relatively uniform, reflecting the long lifetimes of the principal constituent gases and the timescales for transport and turbulent mixing (see Chapter 2). The main exception to this is water vapor, where the large surface source and strong dependence of its vapor pressure on temperature leads to large differences in abundance between the lower and upper troposphere. However, atmospheric pollutants with lifetimes of weeks to months, exceeding the timescales for transport from one continental region to another, are effectively global in nature and lead to degraded air quality at a hemispheric or global scale.

Table 10.1 outlines some of the major air pollutants that are important on a regional or global scale, summarising their approximate lifetimes in the troposphere and their surface abundance away from major emission regions. The atmospheric lifetime of a pollutant is typically governed by its physical and chemical properties. Pollutants such as NO_2, SO_2, and larger VOC are photochemically reactive, and many of their oxidation products are susceptible to dry or wet scavenging, and thus their lifetimes in the atmosphere are relatively short, ranging from hours to days. These pollutants are therefore important on local and regional scales but are of relatively minor importance on a global scale. However, these species may still influence air quality on a global scale through their oxidation products, many of which are longer lived. An important example of this is NO_2, which acts as a precursor to O_3 on a regional scale, but may also react with VOC to form peroxyacyl nitrate (PAN) compounds, long-lived nitrogen-carrying species that are not only pollutants in their own right but which serve to transport NO_2 to cleaner parts of the troposphere where subsequent O_3 production can be substantially more effective than it is in polluted regions.

In contrast, pollutants that are less reactive away from source regions such as CO, O_3 and simple VOC have lifetimes of several months and may be transported around the globe, influencing air quality at a hemispheric or global

Air Quality. DOI: https://doi.org/10.1016/B978-0-12-822591-2.00010-X

TABLE 10.1 Major global air pollutants, their sources and approximate surface and tropospheric lifetimes and their surface abundances away from major emission regions.

Pollutant	Source type	Major global sources	Surface lifetime	Tropospheric lifetime	Surface abundance
NO_2	Primary	Combustion, lightning, soils	Hours	Hours to days	0.1–10 ppb
CO	Primary/ secondary	Combustion, photochemistry	Days to weeks	1–4 months	100–400 ppb
VOC	Primary/ secondary	Biogenic, industry, solvents	Hours to weeks	Hours to months	1–20 ppb
CH_4	Primary	Wetlands, agriculture, energy	~10 yr	~10 yr	~1920 ppb
H_2	Primary/ secondary	Combustion, biogenic	~2 yr	~2 yr	~550 ppb
O_3	Secondary	Stratosphere, photochemistry	Hours to days	Days to weeks	10–60 ppb
SO_2	Primary	Coal combustion, natural	Days	Days	1–20 ppb
NH_3	Primary	Agriculture, fertilizer use	Hours to days	Hours to days	0.1–5 ppb
PM_{10}	Primary	Combustion, dust, natural	Hours to days	Hours to days	5–100 $\mu g\,m^{-3}$
PM_1	Primary/ secondary	Secondary formation	Days	Days to weeks	1–20 $\mu g\,m^{-3}$
Hg	Primary	Combustion, refining, natural	0.5–2.0 yr	0.5–2 yr	1–4 $ng\,m^{-3}$
POPs	Primary	Industry, agriculture	Days to weeks	Days to months	<1 ppb
HFCs	Primary	Industry, consumer products	Days to years	Days to years	10–100 ppt

scale. The longest-lived species, such as CH_4 and HFCs, are sufficiently unreactive not to be classified as air pollutants at all, despite having important roles as greenhouse gases, but they may still play a key role in global air quality, particularly for CH_4 which serves as a major source of O_3 in the global troposphere. The lifetime of particulate matter in the atmosphere is principally determined by its size and hygroscopicity, and while coarse particulates (e.g., PM_{10}) have a relatively short lifetime, fine particulate matter (e.g., $PM_{2.5}$ and PM_1) may remain suspended sufficiently long to contribute to poor air quality over regional or hemispheric scales.

The importance of a pollutant in influencing regional and global air quality depends strongly on the nature of its emissions or formation as well as on its atmospheric lifetime. Emissions of a pollutant or its precursors may be highly localized and episodic, such as those from fires or volcanic activity, or may be widely distributed and varying but essentially continuous, such as natural emissions from vegetation and soils. Global impacts on air quality may thus take the form of occasional events that affect particular regions at specific times of year or may be more pervasive leading to elevated concentrations year-round on a hemispheric scale. In the following sections, we consider two examples of this: tropospheric ozone, a key air pollutant that has increased substantially throughout the troposphere, and fine particulate matter, a more episodic pollutant that has regional and sometimes global impacts.

10.2.1 Global ozone

Ozone is a strong oxidant which has damaging impacts on vegetation and human health when present at high concentrations at the Earth's surface. It is a major constituent of photochemical smog formed in urban regions in summertime, but it is also naturally present throughout the atmosphere. While about 90% of atmospheric ozone resides in the stratosphere, where it is formed through the photolysis of oxygen at short wavelengths of ultraviolet light, about 10% is found in the troposphere, where the average mixing ratio is 50–60 ppbv. Most of this ozone is formed from the oxidation of CH_4, CO, and VOC in the troposphere in the presence of NO_x (approximately 5000 Tg year^{-1}), but a proportion is transported from the stratosphere, where it is more abundant (approximately 500 Tg year^{-1}). The principal sinks of ozone are through chemical removal (approximately 4500 Tg year^{-1}) and deposition to the Earth's surface (approximately 1000 Tg year^{-1}) (Young et al., 2018). Anthropogenic sources of ozone precursors make a large contribution to chemical production, but there are substantial natural sources, in particular NO_x from lightning, soils and wildfires, VOC from vegetation, and natural CH_4. These natural sources may be responsible for as much as 20–30 ppbv of ozone in northern midlatitude continental regions, making a substantial contribution to its current surface abundance (typically 20–60 ppbv), and therefore surface pollutant levels are much more sensitive to global background concentrations for ozone than they are for other major pollutants. This is particularly a problem given that there is no recognized safe level of surface ozone below which damage to crops or human health is avoided.

The abundance of ozone at the surface shows a natural seasonality that differs around the globe. Ozone concentrations are typically lower over the oceans than over continental regions due to the higher humidity, which enhances chemical destruction. The chemical formation of ozone is greatest during summertime when temperatures are higher and insolation is greatest, and this is enhanced by higher emissions of VOC from vegetation. However, the removal of ozone through chemical destruction and dry deposition to the surface is also faster in summer. In contrast, the influx of ozone from the stratosphere is typically greatest in mid-latitude regions in springtime, when the jet streams in the upper troposphere are strongest. Consequently, the annual cycle of ozone at the surface typically shows a springtime peak in remote or oceanic regions, but a summer peak in continental regions where there are fresh precursor emissions. The contribution of global background ozone to local air quality is thus typically greatest in late springtime, somewhat earlier than the peak in ozone produced from local precursor emissions, which tends to be maximum in midsummer in many locations (Monks et al., 2015).

The large spatial and temporal variability in ozone is revealed in a global model simulation of the instantaneous surface ozone distribution shown in Fig. 10.1. There is a strong latitudinal gradient, with ozone higher in the northern hemisphere, and the highest concentrations are found in high-altitude regions such as Tibet and Greenland and

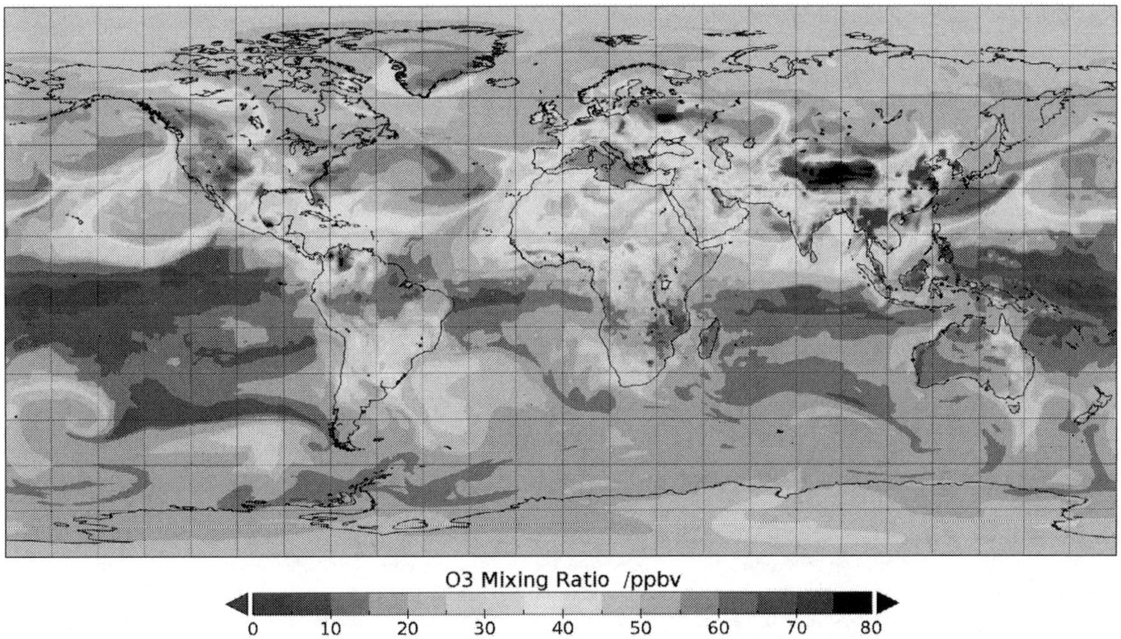

FIGURE 10.1 The surface ozone distribution (in ppbv) at 00:00 GMT on March 2, 2001 from a model simulation with the FRSGC/UCI chemistry transport model (Wild et al., 2003). The abundance of ozone is high in daytime over heavily populated industrial regions and is low over remote marine regions, but the global nature of ozone pollution is evident in the transport of ozone over the Atlantic and Pacific Oceans between the continents at Northern mid-latitudes.

downwind of the continents. Photochemical production is highest in continental regions during the daytime, notable over North America in this example, and destruction is greatest over polluted regions in the night time and early morning, as seen over northern Europe and East Asia. The outflow of polluted air from the continents is clearly seen over the Atlantic and Pacific oceans, and there is clear evidence of synoptic patterns associated with the passage of frontal systems in the mid-latitude westerly winds in each hemisphere. Surface ozone is lowest over the equatorial oceans, where slower transport and photochemical destruction in the humid marine boundary layer lead to concentrations below 10 ppbv. The relatively long lifetime of ozone away from polluted regions, typically about 1–3 weeks, exceeds the timescale for transport between continents, and thus ozone is a pollutant on a global scale (Akimoto, 2003).

10.2.2 Particulate matter

Particulate matter is present throughout the troposphere, although the origin and composition of particles differ greatly. The largest sources are typically regional in nature, and include forest fires, desert dust storms and volcanic activity. However, smaller sources are more ubiquitous and include wind-blown sea salt from spray over ocean regions, soil lofted from continental surfaces, and biogenic particles originating from vegetation. Much of the fine particulate matter is secondary in nature, formed from the condensation of involatile oxidation products in the atmosphere, including inorganic nitrate and sulfates and organics from oxidation of biogenic VOCs. These finer particles are typically longer-lived as they are less readily removed through sedimentation processes, and while they constitute a relatively small proportion of the total mass of particulates, they are an important component of particulate air quality on a global scale.

The global background concentration of particulate matter is far smaller than that present in polluted urban regions, and background levels therefore contribute little to local air quality under most conditions. However, substantially elevated levels of particulate matter may occur on a regional scale in discrete events or episodes, such as those associated with large forest fires or desert dust storms. Major volcanic eruptions may generate and inject a large quantity of ash and other particulate matter into the upper troposphere and lower stratosphere. In these regions, the lifetime of the particulate matter is relatively long, due to reduced scavenging, and transport can be rapid, permitting the material to be transported around the globe. This material gradually settles out of the atmosphere on a long timescale, but the impacts of this on global surface air quality are thought to be relatively minor.

10.2.3 Other global pollutants

Any air pollutant with an atmospheric lifetime of 1–3 months or more, matching or exceeding the meteorological timescales for transport around a hemisphere, can be classified as a global pollutant, capable of influencing air quality worldwide. Examples include major oxidant precursors such as methane, oxidation products such as carbon monoxide, toxic compounds containing heavy metals such as mercury, persistent organic compounds from agricultural or industrial sources, and long-lived greenhouse gases, such as HFCs and N_2O. The sources, abundance, distribution, and lifetime of these pollutants differ greatly, as do their toxicity and importance from a global health perspective. Global annual mean distributions of some major pollutants derived from satellite observations are shown in Fig. 10.2. These images show tropospheric columns, the total abundance of a pollutant through the depth of the troposphere rather than its concentration at the surface, but they clearly reveal how the global distribution relates to pollutant lifetime. Short-lived pollutants such as NO_2 and aerosol PM are strongly localized in emission regions, while longer-lived pollutants such as CO, although still greatest close to source regions, are much more widely distributed. NO_2, in particular, has a sufficiently short lifetime that it clearly reveals hotspots associated with large cities, and its regional distribution tracks population density reasonably well, although there is also a strong signal from biomass burning in Africa and other tropical regions. In contrast, CH_4, as a much longer-lived pollutant, is relatively well-mixed around each hemisphere, and the major feature of its global distribution is the gradient between hemispheres. The abundance of CH_4 in the northern hemisphere is typically 5%–10% higher than that in the southern hemisphere, reflecting the greater sources associated with continental regions.

Methane is a major atmospheric constituent, and while too inert to be considered harmful to health at surface concentrations, typically averaging 1900 ppbv, it is an important source of O_3 and a sink of the hydroxyl radical, OH. Oxidation of CH_4 provides a source of O_3 throughout the troposphere, and this oxidation is fastest in the tropical lower troposphere where strong sunlight, warm temperatures and high humidity support high concentrations of OH. Net production of O_3 then occurs wherever NO_x concentrations exceed 30–50 pptv, typically in the lower troposphere over and downwind of continents and in the upper troposphere, where its lifetime is longer. This relatively well-mixed tropospheric source of O_3 ensures that there is a persistent background of ozone even in regions of the atmosphere remote

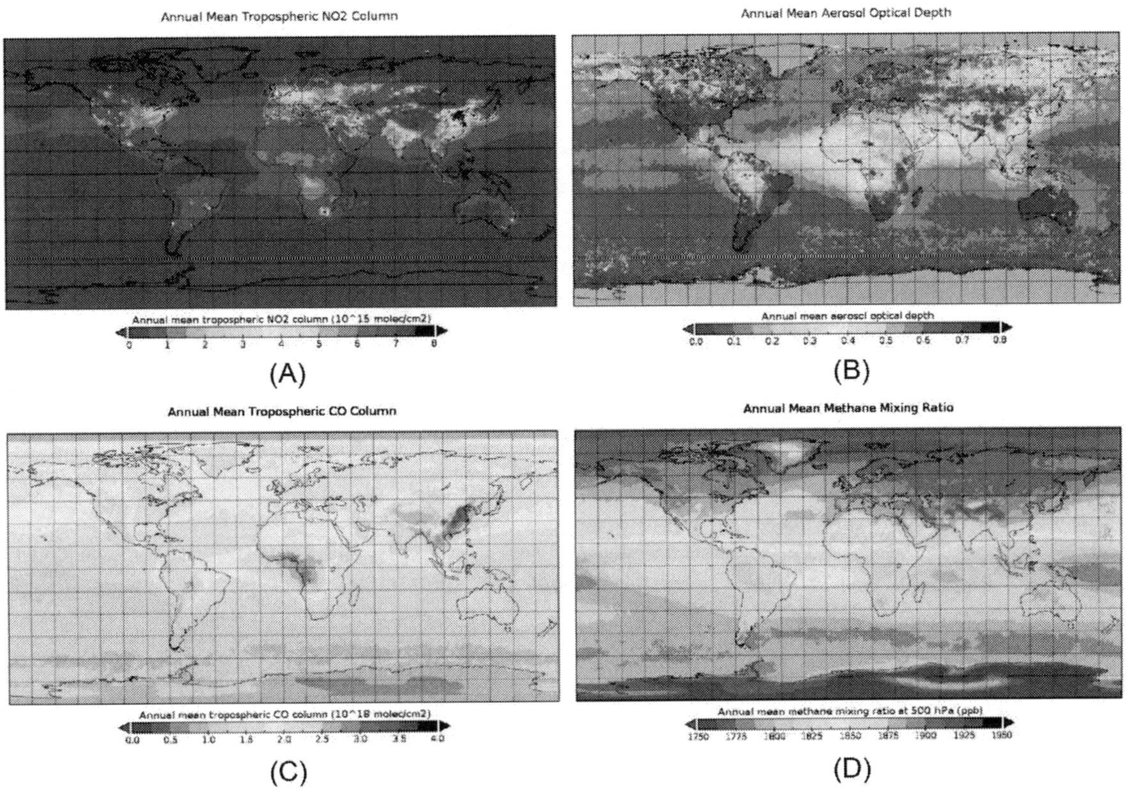

FIGURE 10.2 Annual mean tropospheric abundance of key regional and global pollutants in 2019 derived from satellite measurements, showing (A) NO_2 columns from the TROPOMI instrument on the Sentinel-5P satellite, (B) aerosol optical depth (AOD) from the MODIS instrument on the Aqua satellite, (C) CO columns from the MOPITT instrument on the Terra satellite, and (D) CH_4 mixing ratios at 500 hPa (about 5 km) from the AIRS instrument on the Aqua satellite. Past and near real time satellite retrievals of these and other pollutants are available from NASA Worldview (NASA, 2022). *We acknowledge the free use of tropospheric NO_2 column data from the TROPOMI sensor from http://www.temis.nl (van Geffen et al., 2022). MODIS data are available from the MODIS Atmosphere Science Team, NASA Goddard Space Flight Center (NASA Earth Observations, 2022). MOPITT data were obtained from the NASA Langley Research Center Atmospheric Science Data Center (NASA/LEARC/SD/ASDC, 2000). AIRS data are available from the Goddard Earth Sciences Data and Information Services Center (AIRS Science Team/Joao Teixeira, 2013).*

from surface precursor emissions and transport from the stratosphere. The relatively long lifetime of CH_4 ensures that it is uniformly mixed around each hemisphere, although concentrations are slightly higher in the Northern Hemisphere reflecting the greater source over continental regions. There is a small seasonal variation of about $\pm 1\%$ caused by seasonality in sources and by the faster oxidation in the summertime hemisphere. A more detailed assessment of the global methane budget has been published recently (Saunois et al., 2020).

Carbon monoxide is a product of incomplete combustion, and is generated through anthropogenic activities associated with industry, energy generation, domestic sources and transportation. However, it also has major natural sources such as forest fires, and it is generated in situ in the atmosphere from oxidation of CH_4 and VOC, which accounts for about half of its global source. With an average atmospheric lifetime of 2–3 months, it is readily transported on a global scale. While global surface background concentrations are relatively low, 100–400 ppb, below the levels where it is considered toxic and much less than levels typically found in cities, where it may reach 10 ppm, this background contributes to oxidant formation, as in the presence of NO_x it is responsible for regional and global ozone formation. It therefore makes a small but notable contribution to regional air quality as well as influencing global ozone concentrations. The global distribution of CO shown in Fig. 10.2 highlights its widespread presence throughout the troposphere, and while the highest concentrations are found over industrial regions and in equatorial regions affected by forest fires, appreciable concentrations are also transported over remote marine regions.

Worked Example 10.1: What factors control whether an air pollutant is important on a global scale?

For an air pollutant to be important on a global scale it requires a sufficiently long lifetime to be transported over regional or intercontinental scales, typically 5 days or more. The lifetime depends on the physical and chemical properties of the pollutant and is typically determined by chemical oxidation or physical scavenging processes. Highly reactive pollutants are usually the most harmful but are removed close to their sources by chemical processes and are not transported beyond local scales. Long-lived pollutants with lifetimes of years are transported globally but are typically relatively unreactive and not very harmful. The pollutants of greatest concern are those with intermediate lifetimes of 2–3 weeks, like ozone; these are sufficiently long-lived to be transported very effectively, but reactive enough to cause harm or damage to the environment.

10.3 Regional air quality on continental scales

The location and climate of different continental regions, along with their differing stages of economic and cultural development, lead to substantial differences in regional background air quality. This is associated with both differing natural emissions, driven by climate conditions, vegetation and agricultural practices, and by differing anthropogenic emissions governed by the patterns and intensity of regional development. The main criteria pollutants on a regional basis include NO_2, O_3, and $PM_{2.5}$, and these form the principal focus of this section. The current distribution of air quality problems globally is revealed in Fig. 10.3, which shows the national annual mean population-weighted concentration of $PM_{2.5}$ in 2019 based on surface monitoring data. Remote and more sparsely populated parts of the world experience relatively good air quality, with $PM_{2.5}$ concentrations less than $10\ \mu g\ m^{-3}$, within a factor of two of the WHO guidelines of $5\ \mu g\ m^{-3}$. However, rapidly-developing and densely populated countries in tropical regions currently experience very poor air quality, with annual mean concentrations exceeding $50\ \mu g\ m^{-3}$, associated with rapid economic development and relatively weak emission control policies. The lack of reliable surface air quality measurements from many countries, along with sparsity of measurements from others, means that there is substantial uncertainty in this national-level assessment, but nevertheless it provides a valuable overview of current conditions. This section addresses the

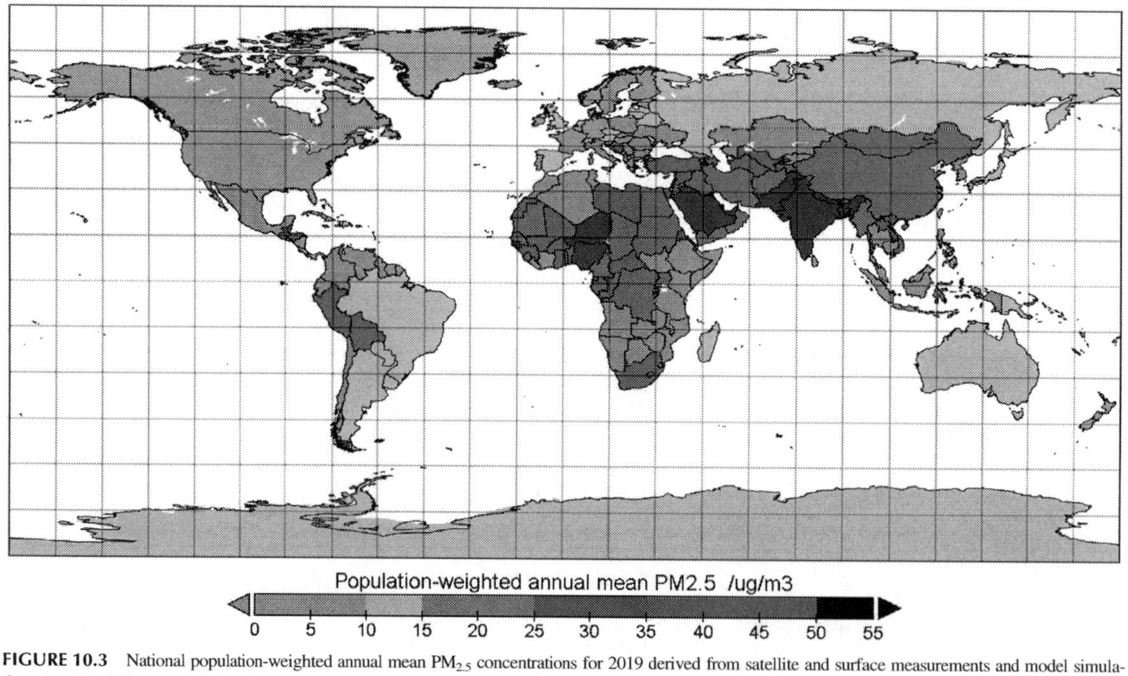

FIGURE 10.3 National population-weighted annual mean $PM_{2.5}$ concentrations for 2019 derived from satellite and surface measurements and model simulations (van Donkelaar et al., 2021). The color scale reflects the WHO annual mean $PM_{2.5}$ guideline ($5\ \mu g\ m^{-3}$) and interim targets (10, 15, 25, and $35\ \mu g\ m^{-3}$) along with concentrations substantially hazardous to health ($>50\ \mu g\ m^{-3}$). *Data from van Donkelaar et al. (2021).*

broad characteristics of regional air quality in each major continental region, noting the common aspects, distinctive features and particular challenges that distinguish air quality in different regions.

10.3.1 North America

The distribution of shorter-lived anthropogenic pollutants over North America principally reflects population density and the location of major industrial activities and energy generation, as seen in the tropospheric NO_2 column in Fig. 10.4. Regional concentrations of NO_2 (and similarly of $PM_{2.5}$, not shown here) are typically higher over the eastern side of the continent than over the west, with the highest concentrations along the northeast coast of the US. The western states of the US show lower levels of these pollutants, reflecting both low population density over mountainous and desert regions and the prevailing westerly flow of cleaner air from the Pacific. The relatively short lifetime of NO_2 means that major urban regions stand out clearly as hotspots of anthropogenic pollution, and this is particularly evident for Los Angeles, Chicago, New York and Mexico City. Concentrations in rural regions are much lower and are lowest in Canada away from major populated regions along the US border. In contrast, mean daytime surface concentrations of ozone in summer show a rather different pattern, with a more uniform distribution over the continent but generally lower concentrations in coastal regions than inland, reflecting continental sources. Efficient ozone production in dry, sunny conditions leads to high concentrations in parts of California away from the coast, as well as at elevated sites inland around the Rocky Mountains. Major cities such as Los Angeles and Mexico City clearly stand out, with summertime mean ozone mixing ratios exceeding 50 ppbv. There is a notably greater density of surface observation stations on the eastern side of the continent, reflecting the greater population in these areas.

Populated urban areas are the largest sources of pollutants and their precursors, particularly from traffic, residential and industrial sources. Power generation in surrounding areas was once a major additional source of NO_2, SO_2, CO, BC, and other pollutants, but effective emission controls and a gradual shift from coal to gas and renewable sources since 2007 have greatly improved air quality. This has been particularly evident in the Ohio River Valley where controls on coal-fired power stations have greatly reduced regional NO_2 and SO_2 concentrations. Other major sources of pollutants outside populated regions include those from oil extraction and refining, notably in the southern US and Gulf of Mexico region, and more recently from hydraulic fracturing activities, particularly in the western US. These release fugitive CH_4 and light VOC emissions, and combined with NO_x from the associated anthropogenic

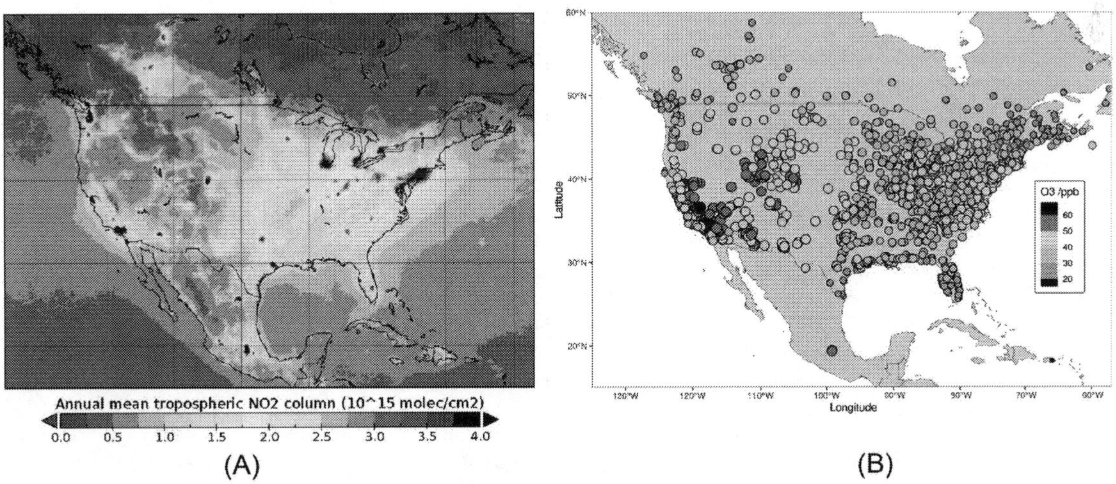

FIGURE 10.4 Regional pollution levels over North America showing (A) the annual mean NO_2 column and (B) mean daytime surface O_3 mixing ratios in summer (in ppbv) from monitoring stations across the continent. NO_2 columns are derived from 2019 measurements from the TROPOMI instrument; surface O_3 measurements for 2015 are from the TOAR ozone database (Schröder et al., 2021). *We acknowledge the free use of tropospheric NO_2 column data from the TROPOMI sensor from http://www.temis.nl, van Geffen et al. (2022).*

activity this has been shown to lead to substantial regional ozone formation that may exceed 3 ppbv in summertime (Pozzer et al., 2020).

Wildfires are a major natural source of NO_x, CO, VOC, and PM, mainly in summertime, and these are particularly prevalent over the forested areas of the western US. Fires may be ignited by lightning, but are also heavily influenced by anthropogenic activity, which affects both fire ignition and suppression. Large wildfires enhance O_3, PM and other pollutants over an extended area, and these pollutants may be transported east over populated regions and out over the Atlantic Ocean. Wildfires are thought to contribute a 3 ppbv enhancement in O_3 over the western US during the fire season, although substantially more during major fire episodes (Xu et al., 2021). There is a particular concern that changes in climate will lead to drier summertime conditions over the western US and thus more frequent and severe fire activity, with detrimental impacts on air quality over much of the continent in summer.

Biogenic emissions also make a major contribution to poor air quality in some regions, most notably in the southeastern US where there is a natural abundance of high isoprene emitting vegetation. There is a notable hotspot over the Ozarks in Missouri and Arkansas. Summertime isoprene emissions combine with anthropogenic NO_x to generate high levels of ozone over large parts of the southeastern US. Forested regions in other parts of north America also release substantial biogenic VOC, but much of this is in the form of monoterpenes. These contribute little to ozone formation, but are responsible for the formation of secondary particulate matter, and may therefore contribute to the natural background of $PM_{2.5}$ over the region. This haze of natural particles is responsible for the "smoky" characteristics of the air in forested mountain regions such as the Appalachians.

10.3.2 Europe

With a high population density and heavy cultivation of land over much of the continent, Europe is strongly influenced by anthropogenic emissions, particularly in Western Europe. A high density of small cities located close together and separated by intensively farmed agricultural land has led to strong anthropogenic emissions over extended areas. The greatest regional air pollution is found in the heavily populated urban and industrial regions of the Netherlands, Belgium and Northern Germany, centered on the Ruhr valley, and in the industrial regions of Northern Italy centered on the Po valley, where locally emitted pollutants are trapped by the Alps. These regions stand out clearly in tropospheric NO_2 columns (see Fig. 10.5) and both have substantial air quality problems. The industrialized regions of northern Germany have high levels of NO_x and $PM_{2.5}$ year-round, and under warm sunny conditions, this may lead to ozone formation, particularly under anticyclonic conditions in summertime. However, the warmer conditions and greater photochemical activity in northern Italy lead to high levels of ozone for much of the year, and this is the principal hotspot for ozone in Europe, with summertime mean ozone in major cities exceeding 60 ppbv on a seasonal average basis. There is a notable gradient in surface ozone between northwestern Europe, where ozone concentrations in cool cloudy conditions remain relatively low, and southeastern Europe, where warm dry stable conditions support much greater production and accumulation.

Meteorological conditions have a particularly strong influence on air quality over Europe. Along the west coast of the continent, the weather is dominated by westerly flow from across the Atlantic and by the passage of midlatitude depressions that bring cleaner air from over the ocean. The unsettled weather that this typically brings, characterized by wind and rainfall, prevent substantial local build-up of pollutants, and this benefits regional air quality over the UK, northwestern France and Scandinavia. However, this prevailing flow leads to substantial transboundary transport of pollutants across the region. The most well-known example of this in the mid-20th Century was the transport of NO_2 and SO_2 derived from coal combustion in the UK for power generation and industry towards Scandinavia, where acidification of rainfall led to severe impacts on natural ecosystems. This was recognized in the 1970s and led to the Convention on Long-range Transboundary Air Pollution, stimulating regional cooperation on air quality and ultimately leading to regional emission controls.

Over continental Europe, anticyclonic conditions are more prevalent, particularly in wintertime, and these more stable conditions lead to the trapping of local pollutants and to poor regional air quality. These systems may persist for 1–2 weeks, leading to extended periods of poor air quality. Stable wintertime conditions of this type were responsible for the London smogs of the 19th and early 20th centuries, when coal burning for industry and domestic heating led to the accumulation of high levels of SO_2, NO_2 and $PM_{2.5}$. These combined with fog formed under the cold conditions to generate dense and acrid smog which reduced visibility to as little as 10 m and caused breathing difficulties for many urban residents. The particularly bad smog episode of December 5–9, 1952 is now believed to have been responsible for as many as 12000 premature deaths, and spurred the development of urban air quality legislation to limit urban emissions from coal combustion. While these intense pollution episodes have now largely been resolved through

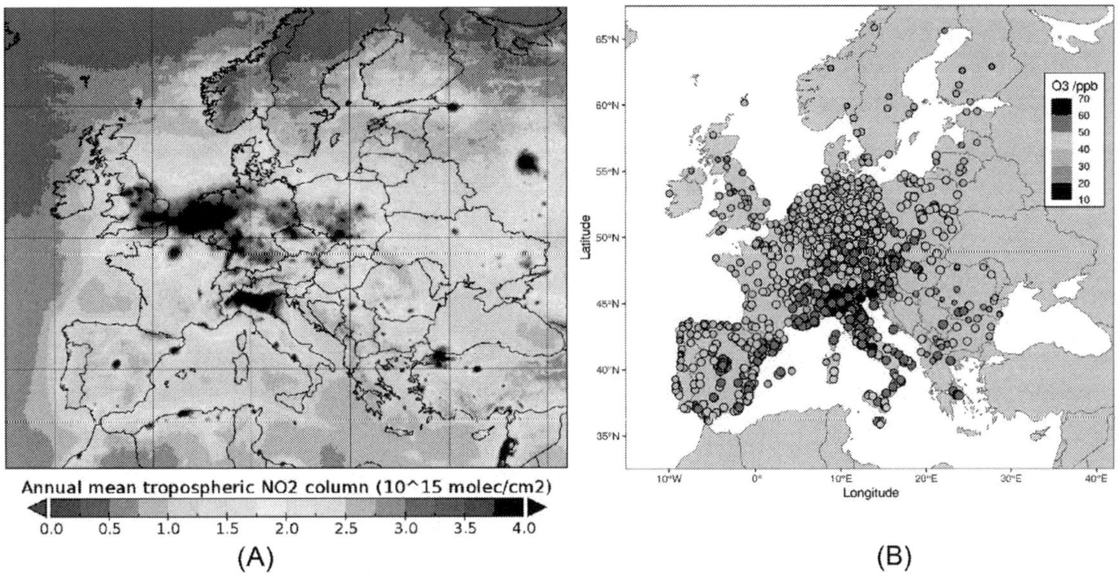

FIGURE 10.5 Regional pollution levels over Europe showing (A) the annual mean NO_2 column and (B) mean daytime surface O_3 mixing ratios in summer (in ppbv) from monitoring stations across the continent. NO_2 columns are derived from 2019 measurements from the TROPOMI instrument; surface O_3 measurements for 2015 are from the TOAR ozone database. *We acknowledge the free use of tropospheric NO_2 column data from the TROPOMI sensor from http://www.temis.nl, van Geffen et al. (2022).*

changes in fuel usage, stable wintertime high-pressure systems continue to bring poor air quality to Northern Europe through trapping and recirculation of freshly-emitted NO_2 and $PM_{2.5}$ over the region.

Anticyclonic conditions are also common over the continent in summertime, when clear skies and high temperatures lead to strong oxidant formation. Calm, settled conditions support the development of pollution episodes with levels of ozone greatly exceeding air quality standards. Examples of this include the heatwaves of July–August 2003, July 2006 and June–July 2018. Under these conditions, high temperatures increase the emissions of volatile compounds from both anthropogenic and natural sources, most notably isoprene from vegetation, and strong sunlight supports efficient O_3 production. In addition, the drought conditions that often accompany heatwaves support the outbreak of wildfires, increasing emissions of NO_x and PM, and reducing the deposition of oxidants to the surface. Daily maximum ozone mixing ratios over extended regions of continental Europe exceeded 90 ppb during the August 2003 heatwave, approaching levels expected in the most polluted cities in summertime (Solberg et al., 2008).

Southern Europe and the Mediterranean region are less heavily populated than northern Europe, but warm sunny conditions in summer favor oxidant formation, and the region often has extended periods of high ozone exceeding 60 ppb. In common with the coastal regions of Western Europe, the region is also affected by NO_x, SO_2, and PM emissions from shipping, which enhance concentrations of O_3 and $PM_{2.5}$ throughout the year. The importance of shipping in influencing air quality in Europe is evident from the NO_2 columns shown in Fig. 10.5 which reveal the main routes for marine transport, and shipping is responsible for about 5% of ozone over Europe and as much as 30% over the Mediterranean (Tagaris et al., 2017). Southern Europe is also influenced by transport of desert dust from the Sahara between spring and autumn, leading to high PM levels and substantial dust deposition. Under strong southerly air flows dust from the Sahara may reach further north and affect extended parts of Europe. A particularly strong episode occurred in March 2022, when dust was carried north over Spain and France, and surface concentrations of PM_{10} over Spain exceeded 1000 $\mu g\, m^{-3}$ over much of the country.

There are a number of remaining challenges for regional air quality in Europe. With anthropogenic emissions of many pollutants dropping under the effects of emission controls, there is particular concern over changes in natural emissions under the influence of climate change. Warmer, drier summer conditions are expected to lead to increased biogenic emissions and more frequent and severe wildfires, particularly in Portugal, Spain and southern France. An additional concern is that warmer conditions may lead to higher emissions of volatile ammonia (NH_3) from agricultural sources. In the presence of NO_x from urban or transportation sources, this may exacerbate formation of secondary inorganic aerosol particles, and hinder the attainment of $PM_{2.5}$ standards over the region.

10.3.3 Central, South and East Asia

The Asian continent is characterized by a wide variety of conditions ranging from desert and mountain environments in the west and central regions to heavily populated and increasingly industrialized regions along the south, east and southeast. The heavily industrialized and rapidly developing regions of north China stand out strongly in the annual mean NO_2 column (see Fig. 10.6), although heavily populated regions of Northern India, South Korea and Japan also experience high NO_2 concentrations. In contrast, surface measurements show the highest levels of $PM_{2.5}$ exceeding 80 $\mu g\,m^{-3}$ over northern India, with lower but still elevated concentrations averaging about 60 $\mu g\,m^{-3}$ over northern China, and much lower concentrations in most other regions.

Arid Middle Eastern regions are strongly influenced by desert dust throughout the year, and this dust dominates particulate matter concentrations over much of the region. Major anthropogenic emission sources in this region include those from oil extraction and refining, industrial activity, construction and transportation. The highest pollution levels are typically found around the Persian Gulf, where annual mean $PM_{2.5}$ concentrations are typically 40–50 $\mu g\,m^{-3}$. Hot, sunny conditions over the region support very active photochemistry, and thus the combination of fugitive VOC emissions from refining and industry with NO_x from associated anthropogenic activity leads to rapid and efficient O_3 production, particularly in summertime. Model studies of global air quality suggest that some of the highest concentrations of surface ozone globally are found in this region, although there are few measurement stations outside populated regions to provide strong support for this yet.

Continental and mountainous inland regions of central Asia are typically characterized by low population densities. Urban regions are influenced by traffic, residential and industrial sources that lead to elevated NO_2 and $PM_{2.5}$ concentrations. In rural regions, air quality is typically defined by combustion of biomass and agricultural waste, leading to high levels of $PM_{2.5}$ that in more mountainous regions may be trapped in valleys leading to elevated surface pollution levels.

Over the heavily populated Indian subcontinent, the seasonality of regional air pollution is strongly influenced by the Asian Monsoon. In the wintertime, dry, northeasterly winds bring continental air from around the southeasterly flank of Tibet. High temperatures, light winds and low humidity over the subcontinent permit the build-up of pollutants from local and regional sources, and photochemistry remains very active throughout the period. During the Monsoon season, between June and September, southwesterly winds from the Arabian Sea or the Bay of Bengal bring clean, marine air over the subcontinent, and the high rainfall has a cleansing effect over much of the region, suppressing the

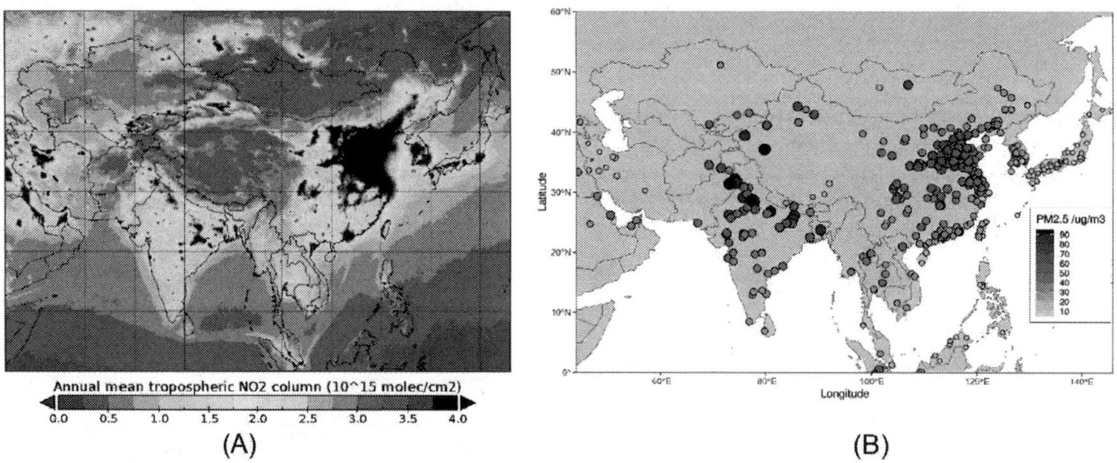

FIGURE 10.6 Regional pollution levels over Asia showing (A) the annual mean NO_2 column, and (B) annual mean surface $PM_{2.5}$ concentrations (in $\mu g\,m^{-3}$) from monitoring stations across the continent. NO_2 columns are derived from 2019 measurements from the TROPOMI instrument; surface $PM_{2.5}$ concentrations are also for 2019. *We acknowledge the free use of tropospheric NO_2 column data from the TROPOMI sensor from http://www.temis.nl, van Geffen et al. (2022).*

lifting of dust, scavenging PM and reducing photochemical activity. Air quality is generally at its best in this season, but deteriorates as the Monsoon recedes, and air quality is typically worst in mid-Autumn when hot, dry conditions return and post-harvest agricultural burning provides a major local source of $PM_{2.5}$.

The high population density across the Indo-Gangetic Plain, spanning Pakistan, Northern India and Bangladesh, along with the trapping effect of the Himalaya and Karakorum mountain ranges to the North, make this region a hotspot for pollution. Many of the world's most polluted cities lie across the Indo-Gangetic Plain, and annual mean $PM_{2.5}$ concentrations exceed 90 $\mu g\,m^{-3}$ over an extended region. Intensive industrial and agricultural activities provide a widespread source of many pollutants that accumulate over the region. In cities, the high volume of light vehicles generates high concentrations of NO_2, VOC, and $PM_{2.5}$, leading to the formation of both secondary aerosol and, increasingly, ozone. High temperatures encourage evaporation of volatile pollutants and lead to large emissions of VOCs from industrial activities, and the very dry conditions for much of the year support a substantial source of primary PM_{10} from road dust. Open waste burning in and around major cities provides a large source of soot, NO_x and a range of toxic pollutants including metals and chlorinated compounds. Many of these sources are most active in the pre-monsoon season when conditions are hottest. Annual mean $PM_{2.5}$ concentrations in Delhi have fallen recently, but still lie close to 100 $\mu g\,m^{-3}$, and hourly concentrations exceed 50 $\mu g\,m^{-3}$ for more than half of the year.

Outside cities, rural industrial activity provides a major and poorly regulated source of regional air pollution, and in particular traditional brick kilns fueled with low-grade coal generate vast quantities of NO_2, SO_2, CO, soot and other particulate matter. $PM_{2.5}$ concentrations in the vicinity of these kilns typically exceed 1000 $\mu g\,m^{-3}$ and have been demonstrated to have substantial health impacts on workers and nearby residents. Agricultural practices present another major source, and seasonal burning of crop residues following harvesting leads to concentrations of $PM_{2.5}$ substantially exceeding 100 $\mu g\,m^{-3}$ across the whole of Northern India for a several-week period in October and November. Regional pollution is typically at its worst in this postmonsoon period, and transport of smoke from agricultural fires across northeastern India makes a large contribution to urban air quality in Delhi and other cities over the Indo-Gangetic Plain. There is a particular problem when this burning coincides with the annual festival of Diwali, a 5-day period in late October or early November when fireworks are traditionally set off during celebrations. The combination of regional smoke from crop burning and intense local particulate emissions from fireworks can lead to very high levels of $PM_{2.5}$. Concentrations exceeding 1000 $\mu g\,m^{-3}$ may occur in Delhi over this period, leading to severe pollution episodes such as the major smog events of November 2016 and 2017 when high $PM_{2.5}$ levels persisted for more than a week. While emission controls on vehicles and industrial sources are starting to tackle these high pollution levels in urban regions, the large contribution from rural sources presents a particular challenge for regulation.

East Asia, comprising midlatitude regions of China, the Korean Peninsula and Japan, has its own distinct air quality challenges. Meteorologically, the region is characterized by the Siberian High, a large, stable region of high pressure that is firmly established over Central Asia in wintertime, and the East Asian monsoon, which dominates south-eastern coastal regions in summertime. Northern parts of China experience a severe continental climate, with cold winters and hot summers, while coastal regions are milder, influenced by marine flow from the southwestern Pacific during the summer season. Air pollution over the region is poorest over the heavily populated regions of the North China Plain, where there is heavy industry and intensive agriculture. Heavy industry, power generation, residential and traffic sources lead to high atmospheric concentrations of NO_2 and $PM_{2.5}$, and under stable anticyclonic conditions in wintertime much of this pollution accumulates, trapped by the Yanshan mountains north of Beijing. Recirculation of this polluted air leads to very high levels of $PM_{2.5}$ which can often reach 300–500 $\mu g\,m^{-3}$ over an extended region during pollution episodes. These high pollutant concentrations are typically built up over 3–4 days and are then swept away by cold, dry air from the northwest, following the development and movement of cold fronts over the region. This air is swept eastwards in low-level flow towards the Korean peninsular and Japan. Consequently, air pollution over the North China Plain in wintertime is characterized by a sawtooth pattern of regional pollutant build-up followed by rapid cleansing (Guo et al., 2014).

In springtime, the weakening of the Siberian High allows the greater development of low pressure systems, and Northern China becomes increasingly influenced by cleaner continental flow. However, these conditions also support the lofting of dust from the Taklamakan and Gobi regions of western China and Mongolia, and the strongest of these systems generate large dust storms which may sweep across northern China, bringing PM_{10} levels greater than 1000 $\mu g\,m^{-3}$ for a number of hours over extended regions, and severely degrading air quality across many cities at the same time. These "yellow dust" conditions typically migrate across the region in 1–2 days, bringing very poor visibility and presenting a major hazard to health. While coarser particles are deposited over the region, leaving a coating of dust on exposed surfaces, finer particles may remain suspended and be carried great distances, notably to Korea and Japan, but occasionally as far as the western US. A particularly large and well-documented event in April 2001 (Liu et al., 2003) brought a plume of dust across the Pacific that was more than 2000 km long by the time it reached North America, where it could be readily detected in US measurement networks as a low-level haze. While these dust storms are essentially natural and have occurred for thousands of years, they have been exacerbated by anthropogenically-driven desertification in the 20th Century. Recent efforts to tackle desertification in parts of western China are paying off, and the severity of dust storms is now starting to decline. While these dust events present a regional challenge for meeting urban air quality targets, they also provide a particular hazard for human health. Although principally of mineral origin, dust particles may interact with air pollutants from anthropogenic sources, which coat the surface of the particles, altering their properties and potentially increasing their toxicity. Dust thus provides a medium for the transport of anthropogenic pollutants on regional scales.

In summertime, East Asian monsoon flow from the southeast reaches as far as the North China Plain, bringing warm humid air from eastern China along with rainfall. However, stable anticyclonic conditions at this time of year accompanied by strong sunshine and high humidity may lead to rapid photochemistry, and extended pollution episodes with high ozone may occur in summer. As strong emission controls have started to take effect in China, bringing down concentrations of NO_2 and $PM_{2.5}$ throughout the year, summertime O_3 concentrations have increased notably, as little change in VOC emissions and reduced photochemical titration by NO have led to greater local chemical production. Ozone now presents a much greater problem for air quality in summer than $PM_{2.5}$.

Although urban air pollution over Beijing has attracted widespread attention, annual mean concentrations of $PM_{2.5}$ have fallen substantially under the effect of emission controls, from almost 100 $\mu g\,m^{-3}$ in 2013 to less than 40 $\mu g\,m^{-3}$ in 2020. Particulate concentrations are now substantially lower than those seen over parts of Northern India. However, the high population density and widespread emissions over the North China Plain make the whole region a major pollution source, and air quality episodes are regional in scale. This has become particularly apparent following the implementation of emission controls for major international events such as the 2008 Olympics and the 2014 APEC Summit. These strict local and urban emission controls have been effective in reducing high levels of $PM_{2.5}$, but they reveal that there is a large background from more distant sources that makes control of air quality over the wider North China Plain a major challenge (Sun et al., 2016).

In South Korea and Japan, the implementation of effective emission controls has led to much lower concentrations of $PM_{2.5}$, which average about 20 $\mu g\,m^{-3}$ in South Korea and 10 $\mu g\,m^{-3}$ in Japan, as shown in Fig. 10.6. However, summertime O_3 remains a challenge in both countries, and is increasing in Korea at about 0.6 ppb $year^{-1}$. While local NO_2 and PM emissions from industrial, domestic and traffic sources in major urban regions dominate the distribution of key pollutants across each country, there is also substantial concern about transport of O_3 and fine particulates over regional scales.

In southern China and along the western shores of the Pacific the air quality is strongly influenced by east Asian monsoon flow, which replaces polluted continental outflow from China with cleaner, humid air from the Western Pacific. The northward extent of this air mass is marked by the movement of the Meiyu (or Baiu, "plum rain") front which brings strong rainfall that migrates northwards from spring into summer, bringing cleaner conditions over southern and eastern China in summertime. The front migrates as far as Japan, bringing a strong seasonality to air pollutants such as O_3 and $PM_{2.5}$ which show a rapid decline in July and August each year.

Over the tropical regions of southeast Asia air quality is principally a problem in major cities such as Jakarta and Hanoi where increasing vehicle usage and expanding industrial activity have led to $PM_{2.5}$ concentrations averaging 20–40 $\mu g\,m^{-3}$. Rural areas are typically cleaner, particularly in coastal regions, except where open agricultural burning is practised. The seasonality of air pollution in this region is largely driven by the movement of the intertropical convergence zone, which governs the location and timing of the wet season. Rainfall typically occurs in the summer months, coinciding with July to October in the northern parts of southeast Asia, bringing the cleanest conditions. In this region there is widespread crop burning in the dry season, typically between February and April, and this produces large quantities of soot, organic matter and ozone precursors which are transported across the region. This contributes substantially to urban air pollution, but pollutants are frequently lifted into the atmosphere and transported eastwards from Myanmar and Thailand towards Vietnam, southern China and the Western Pacific. Subsequent descent of aged biomass-burning plumes has been shown to influence surface air quality as far away as the Philippines and Taiwan during this season, with enhancements in O_3 of as much as 8 ppb in Taiwan (Lin et al., 2013).

In the southern part of southeast Asia, the dry winter season occurs later in the year, and fires are most common between August and October. Land clearance in parts of Indonesia and Malaysia, coupled with increasing drainage of peatland for agriculture, leads to major forest fires that may also ignite underground peat, leading to extensive smoldering fires that emit high quantities of BC, OC, and CO for long periods. These conditions are strongly influenced by El Nino conditions over the Pacific, when southeast Asia is drier than usual, and the strong El Nino years of 1997 and 2015 led to major fires that generated extended areas of haze over much of South East Asia. Major fires in September-October 2015 generated a thick haze that affected much of southeast Asia, and $PM_{2.5}$ concentrations in Singapore, usually averaging about 15 μg m^{-3}, exceeded 250 μg m^{-3}. The fires and accompanying haze persisted for about two months and galvanized efforts for regional cooperation to address transboundary haze.

10.3.4 Central and South America

Central and southern America are characterized by increasing populations and a high level of urbanization that presents challenges for local and regional air quality. While air quality monitoring networks are now being established across a number of countries, reliable long-term measurements of pollutant levels remain sparse. The cleanest air typically occurs in parts of the Caribbean and in coastal regions of Central America, where $PM_{2.5}$ concentrations are typically less than 20 μg m^{-3} and where sea breezes bring cleaner marine air, hindering the build-up of local pollutants. The poorest air quality over the region lies in a narrow strip along the west coast of South America associated with major cities and industrial activity in Peru and Chile. Here energy generation from coal combustion and high use of solid fuels or firewood for heating and cooking present major local emission sources and lead to high levels of $PM_{2.5}$ in urban regions, averaging 30–40 μg m^{-3} on an annual mean basis. $PM_{2.5}$ concentrations can be substantially higher than this in major cities, and hourly concentrations during pollution episodes can reach 300 μg m^{-3} in Santiago. Pollution levels are typically highest in the winter months of July and August associated with the additional heating requirements, and with more stable meteorological conditions which allow urban pollutants to build up. The topography of the region also aggravates pollution levels, as many cities are located in the valleys between the Andes and the coast, and these trap pollutants from local sources, limiting their transport downwind. High levels of O_3 are common in major cities during summertime, and the Santiago region exceeds the Chilean 8-hour average ozone standard of 61 ppb around 43 days per year, with traffic sources and increasing private vehicle usage major contributors (Seguel et al., 2020).

On the eastern side of the continent, rapid urbanization is also a problem, with very high urban populations and heavy industrial activities leading to localized pollution issues. In Brazil, coastal cities such as Sao Paulo and Rio de Janeiro are influenced by marine conditions and show relatively low annual mean $PM_{2.5}$ concentrations, typically 15–20 μg m^{-3}. Traffic sources dominate in urban regions, with power generation and industrial sources also playing a role. A study of urban air pollution across the continent showed that large developed cities with high vehicle usage and major congestion problems had the highest $PM_{2.5}$, while poorer cities with higher population densities had lower concentrations (Gouveia et al., 2021). While $PM_{2.5}$ concentrations remain notably lower than those found in urban areas in south and east Asia, the majority of urban residents across the continent experience annual mean $PM_{2.5}$ concentrations greater than 10 μg m^{-3}.

Fires associated with forest clearance around the Amazon present a major regional source of PM and NO_2 in the dry season of July–November. Fires may be associated with deforestation, controlled burning or agricultural clearance, and although many fires are short-term, major uncontrolled forest fires are also common. While these regions are situated a long way from the largest population centers, smoke plumes are often transported southwards towards move heavily populated regions of Brazil. PM levels may be elevated over an extended area, and fires are estimated to be responsible for 10% of the $PM_{2.5}$-related premature mortality in the region (Nawaz & Henze, 2020). The Amazon rainforest also provides the world's largest source of natural VOC emissions, much of which are produced as isoprene. While this is an important precursor for ozone, the relatively low concentrations of NO_x in this region limit total oxidant production. However, encroaching urbanization and forest clearance through burning provide increasing local sources of NO_x, and this may lead to increases in ozone in this region in future.

10.3.5 Africa

Relatively sparse monitoring of air quality over many parts of Africa makes a detailed assessment of continental air quality difficult. However, the region contains some of the world's largest and fastest growing cities, and relatively poorly developed infrastructure leads to large emissions from industrial, traffic and residential sources. West African countries typically experience the highest levels of $PM_{2.5}$, with lower concentrations on the eastern side of the continent, and lowest levels further south in Namibia, Botswana and Mozambique. Many of the most polluted cities lie close to equatorial regions, including along the coast of West Africa, and in parts of the Sudan and Egypt in the northeast, and in these regions annual mean $PM_{2.5}$ concentrations typically exceed 35 μg m^{-3}, with concentrations in the largest cities such as Lagos and Accra exceeding 50 μg m^{-3}. Urban emissions from poorly-regulated traffic sources are aggravated by biofuel and waste burning, and these lead to widespread pollution. Major industrial activities associated with oil and gas extraction and the mining of coal and minerals provide strong regional pollution sources in parts of the continent. In west Africa, in particular, gas flaring associated with oil extraction is a major local source of soot and combustion products. Poorly regulated industrial activity and artisanal mining in more rural areas also constitute major pollution sources. The widespread burning of biofuels for heating and cooking, along with agricultural waste burning, contribute to air pollution in rural regions. However, the greatest hazard to health in these regions often comes from indoor air pollution, and domestic combustion may make a greater contribution to human exposure than sources of ambient air pollution from outside the home.

Natural sources of pollutants also play an important role. Sand and dust from desert regions along the tropics, particularly from the Sahara but also the Kalahari, present a major regional source of PM_{10} and $PM_{2.5}$ across the continent. While this contributes to high concentrations of PM across North Africa and parts of the Middle East, it also brings substantial $PM_{2.5}$ to more heavily populated parts of sub-Saharan Africa. Wind-blown dust is thus a major contributor to $PM_{2.5}$ in many cities in central and equatorial Africa. Wildfires associated with forest clearance and savanna burning are the largest source of soot, NO_x and CO on a regional scale, and there is a distinct seasonality in this source, which is greatest in northern savanna regions in January-February and southern Congo, Angola and Zambia in July-August, following the drier, winter season governed by the movement of the ITCZ. Plumes from these fires affect air quality over an extended region and can be lifted by convection high into the free troposphere where they affect atmospheric composition on a global scale. In addition to enhancing regional surface $PM_{2.5}$ concentrations, wildfires contribute to the formation of ozone and other oxidants. Ozone concentrations are typically highest in the dry, sunny regions of North Africa, but rapid industrialization and urbanization are leading to increasing concentrations across the continent.

Air pollution in South Africa and the more economically developed parts of the continent is better characterized. Poor air quality affects many of the major cities, as elsewhere in Africa, although the cities along the western and southern coast, such as Cape Town, are relatively clean under the influence of marine air masses. Pollutant concentrations are higher inland, and particularly in the heavily industrialized upland region of the Highveld around Pretoria and Johannesburg. Major mining activities and coal-fired power stations emit high levels of soot, NO_x and SO_2, making this one of the most polluted regions in Africa. This region stands out clearly in the tropospheric NO_2 column (see Fig. 10.2), emphasizing the magnitude of the local sources. However, although pollutant concentrations in this area are relatively high, with annual mean $PM_{2.5}$ averaging $25-30\,\mu g\,m^{-3}$, they remain much less than those in the more heavily populated cities of southern and eastern Asia.

10.3.6 Australasia

Due to its location and low population, air over most of Australasia is relatively clean. Anthropogenic emissions from industrial, residential and transportation sources impact air quality in larger cities, although many of these are situated along the east coast of Australia and benefit from sea breezes that prevent substantial build-up of pollutants from local sources. Annual mean urban $PM_{2.5}$ concentrations are typically less than $10\,\mu g\,m^{-3}$ across both Australia and New Zealand. Current challenges for air quality in Australia include a greater reliance on coal combustion for energy generation than in most industrialized countries, heavy vehicle usage, and widespread use of wood burners for heating during the wintertime. These sources lead to substantial emissions of NO_x, SO_2 and $PM_{2.5}$, impacting air quality, particularly in the populated southeast of the country. Dry, sunny conditions are conducive to the formation of photochemical pollutants, and ozone concentrations can exceed the Australian 4-hour standard of 80 ppb in summertime.

However, Australia also experiences major pollution episodes associated with natural sources, in particular from dust and wildfires. Strong winds associated with the passage of cold fronts in springtime can lift dust from the deserts and cause large dust plumes that raise PM_{10} to high levels in coastal regions and these may be transported eastwards as far as New Zealand. This effect is exacerbated by drought conditions that leave the land surface dry and loose. When hot, dry conditions persist over the summer months there is a high likelihood of wildfires across Australia's forested regions, and the smoke from major fires may lead to extended periods of poor air quality associated with high concentrations of $PM_{2.5}$, CO, NO_x, and other pollutants. A particularly dry summer in 2019–20 led to a severe drought and to major wildfires in December and January that burned an area of more than 5.5 million hectares and led directly to 33 deaths and a further 400 premature deaths associated with smoke inhalation. New South Wales and Victoria were particularly badly affected by the smoke and pollutants from these fires, and 24-hour average $PM_{2.5}$ concentrations reached nearly $1000\,\mu g\,m^{-3}$ in Canberra on January 1, with concentrations exceeding $50\,\mu g\,m^{-3}$ on more than half of the days in the December–January period. The smoke plume was transported many thousands of kilometers downwind in the westerly flow, affecting New Zealand and reaching as far as South America, and trace amounts of smoke were detected by satellite instruments to have circumnavigated the globe. While these episodes were unusually bad, wildfires remain an annual occurrence and present a substantial challenge for regional air quality, particularly given the likely increase in drought conditions under climate change.

10.3.7 Polar and marine regions

The polar regions are the most remote from anthropogenic activity and thus constitute some of the cleanest regions on the Earth's surface. However, air pollution remains a concern at some times of year, particularly over the Arctic, under the influence of both transport processes and local sources. The strongest feature over this region occurs in winter and springtime and is referred to as Arctic haze, comprising pollutants that are predominantly transported from mid-latitude source regions. The origin of this haze is governed by regional meteorology, which is characterized by anticyclonic conditions, with air cooled by the cold surface, particularly in wintertime, forming a shallow, stable boundary layer. This layer isolates the region from air aloft, as air transported into the Arctic circle from lower latitudes arrives above this layer and remains away from the surface. This barrier to transport is commonly referred to as the polar dome. In wintertime, the geographical extent of this dome increases, and it extends towards northern midlatitudes, incorporating parts of Canada, Russia and Northern Europe. Under these conditions, pollutants from anthropogenic sources can be released inside the stable polar dome and transported directly into the Arctic.

While local emission sources inside the Arctic are not extensive, parts of the region are strongly influenced by mining, refining, industrial activity, urban settlements, and shipping (Schmale et al., 2018). Freshly-emitted pollutants may be trapped in the stable layer near the surface by a low-level temperature inversion, leading to high pollutant concentrations close to source regions. There is also an episodic source from boreal forest fires over Siberia, Alaska and Canada during the summer months, which release NO_x, CO and $PM_{2.5}$. Transport into the Arctic is less effective in this season, and more active photochemistry in the summertime sunlight enhances the removal of key pollutants. However, NO_x from fires and other sources may be preserved in the form of PAN under cold Arctic conditions, and subsequent release of NO_2 may enhance regional ozone production. The lifetime of ozone is relatively long in the cold dry conditions of the Arctic, and at the surface it typically shows a maximum in springtime and a minimum in late summer, with mixing ratios typically 20–30 ppb. However, in coastal regions there may be a distinct springtime minimum associated with the release of halogens, particularly bromine and iodine, from the breakup of sea-ice. While the abundance of halogens may not exceed 10–20 ppt, this is sufficient to lead to very rapid surface O_3 depletion events that bring O_3 concentrations close to zero for a period of a few days.

Air pollution may also present a problem over marine regions. While remote parts of the oceans may be far removed from the major continental source regions, the prevailing westerly transport at midlatitudes carries pollutants from one region to another. Transport is most effective in the free troposphere, where pollutants are isolated from the surface and winds are stronger, but low-level trans-oceanic transport can still occur in the boundary layer, particularly for longer-lived pollutants such as ozone, as shown in Fig. 10.1. Coastal oceans are particularly susceptible to air pollution, both from polluted outflow from the continents and from national and international shipping, which releases large amounts of NO_x, SO_2 and PM into the marine boundary layer. The main international shipping lanes are clearly visible from satellite, both through their effects on PM and its contribution to cloud formation, which generate ship tracks along the major trading routes, and through their effects on NO_2 (see Fig. 10.2). Shipping provides the principal local source of O_3 in marine regions, and may contribute 10%–20% of surface O_3 over the oceans, and to substantially more in heavily trafficked regions such as the Mediterranean and the Pacific coast of East Asia. Coastal oceans are also influenced by natural sources, particularly the release of halogens from seaweed which contributes to particle formation and to depletion of surface ozone. They are also influenced by the emission of reduced sulfur species, principally as dimethylsulfide (DMS), from phytoplankton which enhance particle formation in marine environments. However, while these processes are important components of natural biogeochemical cycling in the environment, they do not have substantial detrimental effects on local air quality.

Worked Example 10.2: What factors govern regional differences in air quality?

While emissions from combustion processes lead to similar anthropogenic contributions to poor air quality in many parts if the world, this is moderated by social, economic and industrial development, by population density and by the extent of emissions regulation in different regions. These effects are compounded by different natural emissions, particularly those from fires and desert dust, but also from soils and vegetation, that may be characteristic of a region. Differences in regional meteorology are also important, as these may degrade air quality through accumulation of pollution under stable continental anticyclonic conditions, or improve it through rapid dispersion or inflow of clean air.

10.4 Intercontinental transport of pollutants

Long-range atmospheric transport of pollutants connects the air quality over different regions, particularly at midlatitudes, and thus neighboring regions can influence each other's air quality, even over continental scales. Pollutants with lifetimes of a few days can be transported over regional scales, while those with a lifetime of a week or more can undergo more extensive transport. The atmospheric lifetime of many pollutants is greater in the free troposphere, where they are isolated from deposition processes occurring at the surface of the Earth and from high concentrations of oxidizing species, and this permits more widespread transport. In addition, some secondary pollutants such as ozone can be formed more efficiently from their precursors in cleaner regions above the boundary layer, and thus precursors that are lofted into the atmosphere can have a much larger impact on global pollutant concentrations. Horizontal transport of airmasses is typically faster in the free troposphere than in the boundary layer, and thus lifting of pollutants greatly extends the region of influence of major pollutant sources. A detailed overview of the role of meteorological processes in governing atmospheric transport is provided in Chapter 5.

10.4.1 Transport processes

The principal mechanisms for intercontinental transport of pollutants are illustrated in Fig. 10.7. Rapid lifting of air out of the polluted boundary layer through deep convection provides the fastest pathway for pollutants and their precursors to reach the free troposphere, and lifting in large cumulonimbus may carry pollutants into the upper troposphere in a few hours. However, cloud droplets and precipitation in these rising pockets of air may remove particles and soluble gases through wet deposition, reducing the efficiency of this export process. Pollutants and their precursors lifted to the middle and upper troposphere typically have longer lifetimes and may be transported over continental or hemispheric scales, particularly if they encounter the westerly jet streams. Ozone production from cycling of NO_x is about ten times more efficient at these altitudes, and along with a longer lifetime associated with much drier conditions (1–2 months compared with a few days in the polluted boundary layer), this lifting mechanism has a disproportionately large impact on the global tropospheric ozone burden.

A slower but more widespread transport mechanism involves frontal lifting associated with cyclonic systems. Polluted air trapped in the warm sector of a mid-latitude cyclone may be lifted into the free troposphere by the movement of the frontal systems, typically in a tight band of air ahead of the cold front termed the warm conveyor belt. Much of this lifting is associated with clouds along the cold front, limiting the efficiency of transport of soluble species, and lifting typically occurs at a slower rate, rarely reaching above the mid troposphere. However, this mechanism can carry large quantities of polluted air into the free troposphere at mid latitudes, particularly over the western Pacific and western Atlantic, leading to cross-oceanic transport with a timescale of about a week. This air may be brought down towards the surface in subsiding air associated with anticyclones, providing a direct mechanism for influence on surface air quality. For ozone, this mechanism is particularly effective as some of the precursor NO_x may be converted to thermally-stable PANs which are long-lived in the colder conditions of the mid to upper troposphere, and these may subsequently release NO_x and VOC on descent towards the surface, providing an additional burst of ozone production.

Polluted airmasses that do not encounter convective or frontal lifting may be directly transported around the globe in the boundary layer and lower troposphere. Slower transport at low altitudes leads to transit times across the major oceans of 1–3 weeks, and greater chemical removal and surface deposition to the Earth's surface deplete the

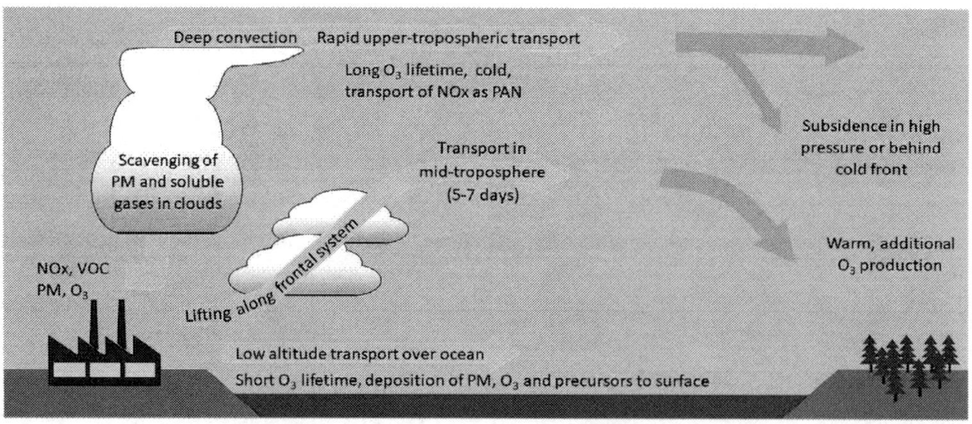

FIGURE 10.7 Principal meteorological mechanisms governing the intercontinental transport of pollutants.

concentrations of most pollutants during transit. Nevertheless, more polluted air is transported through this process than through convective or frontal lifting, and the mixing and dispersion of pollutants in transit can lead to an enhanced background of pollutants reaching distant continents. While low altitude transport of pollutants always occurs, even in the absence of convective or frontal weather systems, all three mechanisms may occur simultaneously, and their relative importance varies with region and season. Typically, convective lifting is responsible for rapid transport associated with long-range transport events, while low-level transport is pervasive and leads to enhanced background concentrations.

10.4.2 Contributions to regional air quality

The importance of intercontinental transport from major source regions is illustrated for ozone in Fig. 10.8. The figure shows the contribution of anthropogenic precursor emission sources to surface ozone in summer and is based on model simulations of the effect of regional 20% emission changes conducted by the United Nations Economic Commission for Europe's Task Force on Hemispheric Transport of Air Pollution, (HTAP, 2010). These studies were representative of 2001 conditions, so the 20% emission changes roughly reflect the actual decrease in NO_x emissions between 2001 and 2008 over North America and Europe, and the increase in emissions between 2001 and 2005 over East Asia driven by rapid industrial development in China.

While the effect of precursor emission changes is largest over the respective source region, there is a substantial footprint of as much as 1 ppbv O_3 at the surface extending downwind of each region associated with the midlatitude westerlies. This typically leads to elevated O_3 concentrations in the immediate downwind continent, and this contribution varies in strength and location by season, with a greater poleward extent in the summer months, and with transport greatest in the springtime when winds are stronger and the lifetime of O_3 is greater. Given the lifetime of O_3 there is relatively little direct transport between hemispheres, although the general increase in longer-lived precursors ultimately leads to greater O_3 worldwide. The contributions to O_3 in the mid-troposphere are larger than those at the surface, reflecting faster transport and longer lifetimes, and thus the footprints seen here reflect the effect of subsidence from higher altitudes as well as direct low-altitude transport. This effect is responsible for the contributions from north American sources over central Asia and the notable contribution of East Asian sources over the west coast of North America. The footprint of southern Asian sources is substantially smaller than that from the three midlatitude regions. This reflects stronger regional formation and destruction associated with more active photochemistry, weaker horizontal transport, and stronger lifting by deep convection into the mid and upper troposphere. The surface contributions from southern Asian sources are strongest in winter and spring, and are smallest in summer (shown here) due to the regional monsoon.

For comparison, the contributions to surface O_3 from a 20% increase in atmospheric CH_4 abundance, roughly corresponding to the increase observed between 1980 and 2020, is shown in Fig. 10.9. The long lifetime of CH_4 of about 10 years means that ozone formation can occur throughout the troposphere wherever sufficient NO_x is present, and

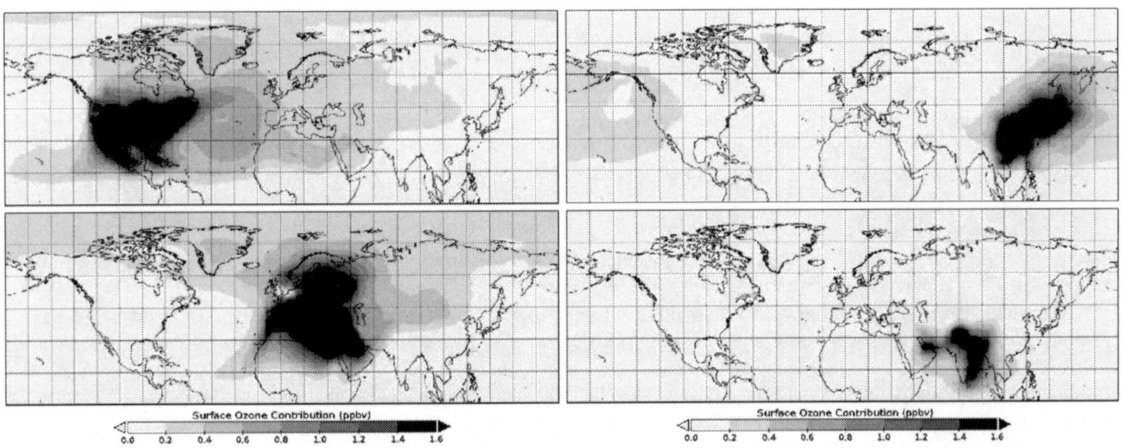

FIGURE 10.8 Surface ozone contributions in July (in ppbv) associated with a 20% change in anthropogenic precursor emissions over four major northern hemisphere source regions. These ozone contributions are the ensemble mean of 14 models contributing to the HTAP model intercomparison (Fiore et al., 2009).

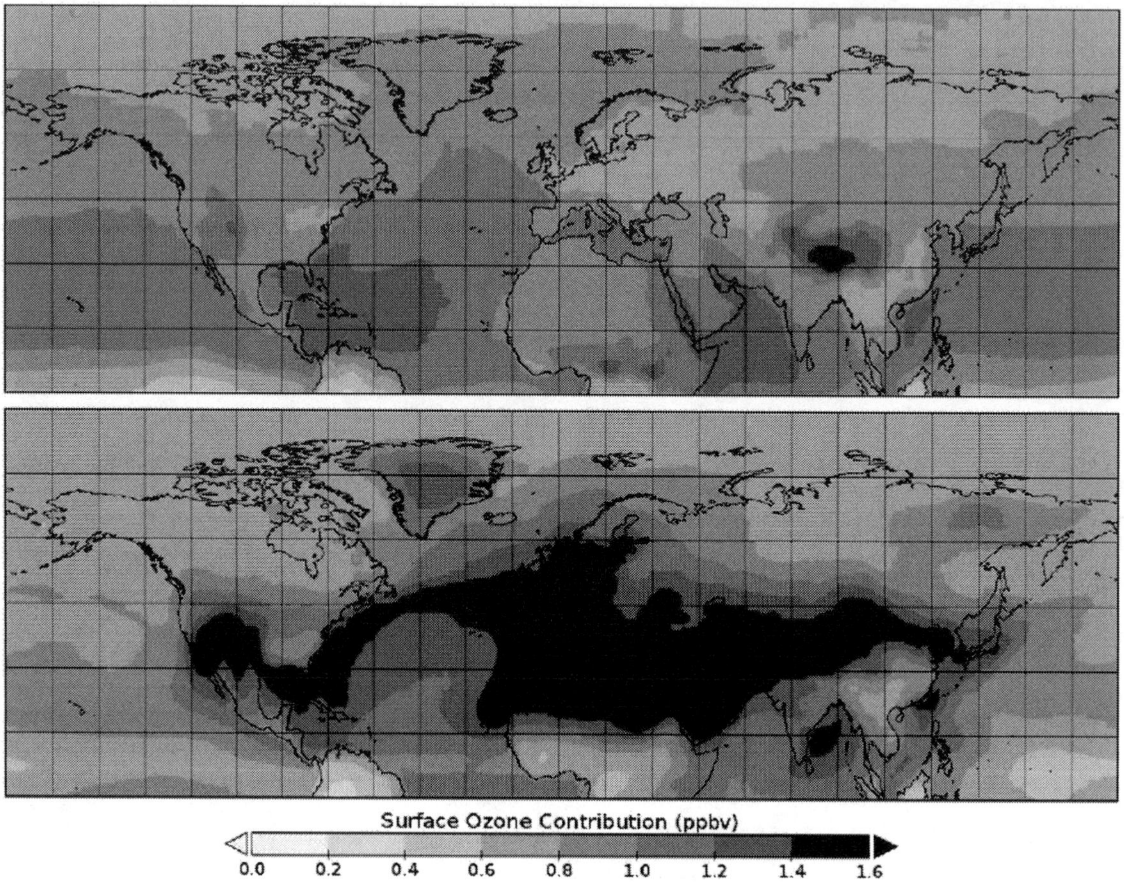

FIGURE 10.9 Change in surface ozone (in ppbv) in January (top) and July (bottom) caused by the 20% increase in atmospheric CH_4 concentrations that has occurred between 1980 and 2020. These distributions are from 17 models contributing to the HTAP model intercomparison (Fiore et al., 2009).

consequently the ozone response is not very sensitive to the location of the CH_4 emissions (Fiore et al., 2008). There is a hemispheric-scale increase in surface O_3 that is greatest in tropical regions, and the contributions are larger at high altitudes, as evidenced by the higher surface responses over Tibet and Greenland. Contributions are smallest over regions with high anthropogenic precursor emissions, where local chemical destruction of ozone is faster. The contributions over continental regions are highest in summertime, reflecting active photochemistry and convective mixing. This seasonality is reversed over the oceans in the Tropics, where higher humidity and more rapid destruction outweigh the enhanced formation occurring in continental outflow regions. The contribution of changes in CH_4 to surface ozone, reaching 1.5 ppbv over the continents, is comparable to that from anthropogenic emission changes, highlighting the importance of considering this global pollutant when assessing changes in regional air quality.

Other long-lived anthropogenic pollutants, such as carbon monoxide, also undergo intercontinental transport, although as CO has a longer lifetime and less sensitivity in chemical formation and destruction to lifting processes, the distributions are more uniform. However, in contrast to ozone, the contribution of global background concentrations to regional or urban air pollution is relatively small, given the dominant effect of primary local emission sources, and therefore intercontinental transport is a smaller concern for regional air quality.

In contrast to ozone, the shorter lifetime of most particulate matter means that much of it is removed over the timescales required for intercontinental transport. However, when emission sources are sufficiently intense, such as in major forest fires, large dust storms, or volcanic eruptions, particulate matter can be lifted into the mid and upper troposphere and can be transported over a hemispheric scale. Important examples of this include springtime dust storms in the Taklamakan and Gobi deserts which can lift dust into the troposphere where it may be carried over the Pacific Ocean on the prevailing westerlies as far the west coast of the United States. The major dust storm in April 2001 described in Section 10.3.3 produced a noticeable haze and substantial enhancement in $PM_{2.5}$ concentrations across the western US. Typical 24-hour average contributions to $PM_{2.5}$ concentrations over this region are 0.2–1.0 $\mu g\,m^{-3}$ but can reach 5 $\mu g\,m^{-3}$. About 50% of the variability in springtime $PM_{2.5}$ at background sites in the northwestern US has been attributed to Asian dust events (Fischer et al., 2009). Similarly, fine PM from major forest fires may produce an extensive haze that can influence air quality on regional scales and sometimes hemispherically. Examples of this include haze over Southeast Asia associated with forest clearance, which may be particularly severe under dry conditions associated with El Nino (see Section 10.3.3) and smoke plumes from high latitude boreal forest fires in Alaska, Canada and Siberia, which peak in the summer season and may spread soot and organic aerosol over thousands of kilometers.

Volcanic activity can be a major source of ash and sulfate aerosols, and strong eruptions can inject these into the stratosphere where they may circulate around the globe for 1–2 years before they settle out. The eruption of Mt. Pinatubo in the Philippines in June 1991 led to a large injection of reflective sulfate aerosol into the stratosphere that is estimated to be responsible for a 0.4°C drop in mean global surface temperatures the following year. While these particles also alter photochemistry in the troposphere, the effect on surface air quality on a global scale is thought to be small. Smaller eruptions, such as that of Eyjafjallajökull in Iceland in April 2010, inject ash into the upper troposphere where it has a shorter lifetime and the impacts are more regional in scale. In this event the plume was transported over the North Atlantic and Europe, severely interrupting air travel. While suspected dust fall was detected over parts of Western Europe, the impact on regional air quality was relatively minor. However, modeling studies suggest that larger eruptions, such as that of Laki in Iceland in 1783–84, might be expected to double $PM_{2.5}$ concentrations over much of Europe for an extended period (Schmidt et al., 2011).

Worked Example 10.3: How do convective lifting processes enhance intercontinental transport of air pollutants?

Convective lifting processes carry pollutants and their precursors out of the polluted boundary layer and into the free troposphere. While cloud formation and precipitation associated with convection may lead to removal of soluble pollutants and particulate matter, pollutants which survive this typically have a longer lifetime in the cleaner free troposphere, isolated from rapid chemical removal and surface deposition processes in the boundary layer. Horizontal transport is more rapid as winds in the free troposphere are typically stronger than in the boundary layer. Colder temperatures extend the lifetime of nitrogen-carrying pollutants such as PAN, allowing them to be transported greater distances. The chemical production of oxidants such as O_3 from their precursors is also more effective in the cleaner, colder conditions of the free troposphere, and the longer lifetimes at this altitude mean that convectively lifted precursors can have a disproportionately large effect on oxidant concentrations over intercontinental distances.

10.5 Trends in regional and global air quality

The surface concentration of many atmospheric pollutants has changed markedly since preindustrial times under the influence of both natural emission changes and rapid socioeconomic development around the globe. Further substantial changes are expected over the coming century under the effects of changing climate and human activity, although for many air pollutants these changes depend most strongly on how well current and future policies targeting air quality and climate change are implemented.

10.5.1 Global trends

The strongest drivers of changes in global composition are associated with anthropogenic emissions. Increased human population and the accompanying industrial and agricultural activity has led to increased global emissions of CO_2, CH_4, and O_3 precursors, and to changes in atmospheric composition that influence background air quality. The global mean tropospheric abundance of CH_4 derived from surface observations increased from about 1645 ppb in 1984 to about 1880 ppb in 2020, a rise of almost 15% (see Fig. 10.10), and longer-term records from ice cores show an abundance in the mid-18th century of about 720 ppb, indicating that it has increased by more than two and half times since the preindustrial period. Current observations show that it continues to rise by 5–10 ppb year^{-1} (approximately 0.3%–0.5% year^{-1}), although this growth rate has varied with time, and was close to zero for a short period in 1999–2006 for reasons that are not yet fully understood. Over 2020, the increase reached 15 ppb year^{-1}, reflecting both increased CH_4 sources and slower photochemical destruction associated with emissions changes under COVID-19 lockdowns. As an important source of O_3, the increase in CH_4 has contributed to an increase in ozone throughout the background

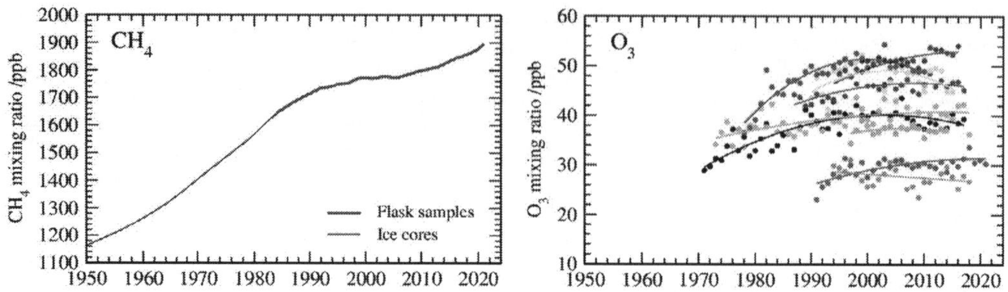

FIGURE 10.10 Trends in surface mixing ratios of CH_4 (left) and O_3 (right) at remote monitoring stations, revealing a rapid increase in CH_4 and an increase in O_3 that has slowed over the past decade. CH_4 measurements are from the NOAA GML network (Lan et al., 2022) and O_3 measurements are from the TOAR dataset (Schröder et al., 2021).

troposphere, as shown in Fig. 10.9. The increase in O_3 since 1980 exceeds 1.5 ppb in continental northern midlatitudes in summertime, when O_3 in polluted regions is at its greatest, and thus contributes to regional O_3 exceedances. The effect in wintertime is smaller, reflecting less efficient production, but increases over tropical marine regions still exceed 1.0 ppb.

Global anthropogenic emissions of the main O_3 precursor, NO_x, have increased strongly over the past century, particularly between 1950 and 2000, although this has slowed in many regions over the past decade due to emission controls. This increase has led to increases in O_3 production both close to emission sources and in the background atmosphere. The tropospheric burden of ozone has increased by about 15% since 1950, and remote monitoring stations at northern midlatitudes record an increase in surface O_3, although the magnitude of this differs by location as shown in Fig. 10.10. Prior to 2000, observations show an increase in surface O_3 at most locations, averaging about 0.6 ppb $year^{-1}$, reflecting the wider increase in tropospheric O_3 abundance, but this growth has slowed in the past decade, and surface O_3 has started to decrease in some parts of the world, most notably in summertime in North America and Europe (Gaudel et al., 2018). Model studies suggest that current annual global mean surface ozone mixing ratios average 28–30 ppbv, and that this is about 8 ppb higher than that in the preindustrial period (Turnock et al., 2019). However, the mean abundance over the continents is greater than that over the oceans, and has increased by a greater margin, as much as a factor of two, under the influence of local precursor sources.

Future changes in global surface O_3 remain highly uncertain, but depend strongly on future precursor emissions as well as on natural changes in emissions and photochemistry associated with changes in climate. Fig. 10.11 shows the expected changes in annual mean surface O_3 over different continents associated with the IPCC's Shared Socioeconomic Pathway (SSP) storylines for regional and global development. The pathway focussed on sustainability and climate change mitigation measures (SSP1) shows strong decreases in ozone globally, while those characterized by rapid economic development or regional rivalry (SSP3 and SSP5) show increases. While these surface changes reflect global changes in the tropospheric abundance of ozone, there are also substantial regional differences in some scenarios that reflect differing development pathways in different parts of the world.

The global burden of CO has increased by about 50% since preindustrial times, but since 2000 has started to decrease at a rate of about 1% $year^{-1}$ (Szopa et al., 2021). This likely reflects reductions in global anthropogenic emissions due to better regulation of combustion sources, although the production of CO from atmospheric oxidation of CH_4 and VOC may have increased. The increase in atmospheric NO_x and decrease in CO is thought to be responsible for an increase in tropospheric OH between 1980 and 2014 of about 9%. This is expected to have increased atmospheric oxidation, reducing the lifetime of CH_4, VOC and many other pollutants, although there is limited observational evidence yet to support these changes in photochemistry.

10.5.2 Regional trends

Trends in regional air quality depend strongly on regional economic development pathways and the effectiveness of emission control policies. In general, $PM_{2.5}$ and NO_2 concentrations are now dropping in North America, Europe and Japan, but increasing in the Middle East, Africa and southern Asia. Urban O_3 shows similar changes, with reductions in the highest concentrations in midlatitude regions and increases in equatorial regions. We provide a brief summary of

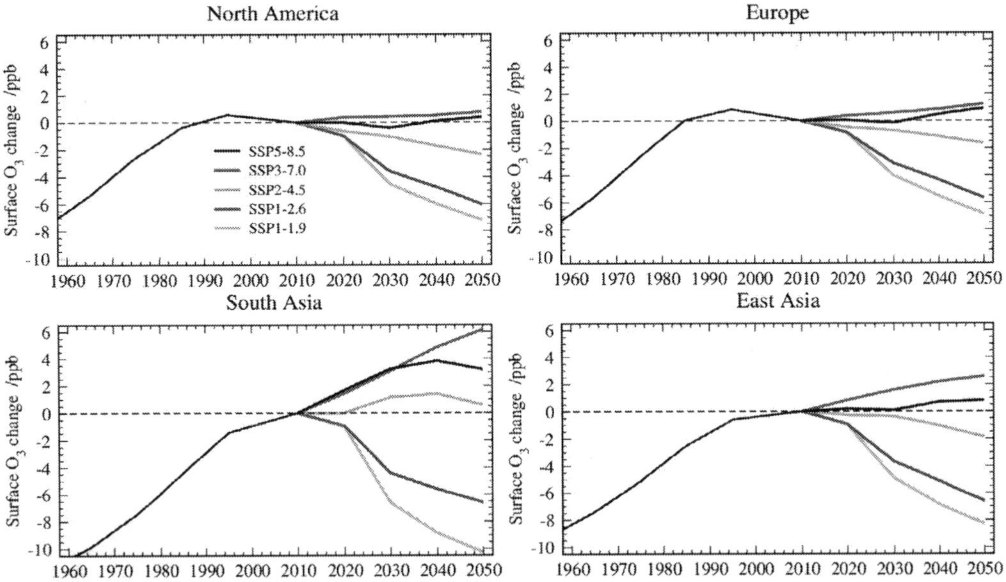

FIGURE 10.11 Regional average changes in surface ozone compared to present-day (2010) conditions based on historical and possible future precursor emission changes along the SSP scenarios. These estimates are based on simulations from 14 global chemistry transport models using the approach described in Wild et al. (2012).

regional trends here, but refer the reader to Sicard et al. (2023) for a more in-depth analysis of observed urban air quality changes over the 2000–19 period.

Air quality over North America has shown substantial improvement over the last 30 years, particularly in the United States where emission control policies have been implemented effectively. Under the Clean Air Act, there have been strict controls on major emission sources, including measures to control industrial pollution, enforcement of vehicle fuel standards, and limits on the use of coal for energy generation. Concentrations of key criteria pollutants have dropped by large margins, and over the last 30 years 1-hour averaged SO_2 has dropped by more than 90%, 8-hour averaged CO by 75%, 1-hour averaged NO_2 by 54%, and 24-hour averaged PM_{10} by about 30%, see Fig. 10.12. The sharp reductions in SO_2 and CO principally reflect the shift away from coal combustion, while reductions in NO_2 and $PM_{2.5}$ have been less rapid as these are strongly influenced by vehicle usage and other combustion sources. However, since 2000 annual mean surface NO_2 concentrations have dropped by 55% and annual mean $PM_{2.5}$ by about 40%. Average concentrations for these pollutants all now lie well below the thresholds set by the US National Ambient Air Quality Standards (NAAQS). These declines have occurred despite increases from some natural pollution sources, in particular from wildfires. The annual area of land in the US burned by fires has increased by a factor of three since 1990, contributing substantially to NO_x, CO and $PM_{2.5}$ in major fire years such as 2017 and 2020, and this is believed to have been responsible for an increase in annual mean $PM_{2.5}$ over the western US of as much as 5 $\mu g\, m^{-3}$ over the last decade (Childs et al., 2022). More limited improvements in air quality have been seen for daytime O_3, for which 8-hour average concentrations have dropped by only about 20% since 1990 and now just meet the national air quality standards of 70 ppb. This reflects the wider, regional nature of ozone pollution and substantial background concentrations, but also the greater difficulty of mitigating secondary pollutants. However, the total number of days with either $PM_{2.5}$ or O_3 reaching levels classified as "Unhealthy for Sensitive Groups" over 35 major cities has dropped by a factor of three since 2000. Los Angeles faces the greatest challenge, with more than 100 days per year still exceeding healthy limits, while East Coast cities such as New York see lower pollutant levels, with only 15–20 days per year now exceeding these limits.

Over Europe there have also been notable reductions in emissions of major pollutants supported by comprehensive legislation particularly targeting the energy, industry and transport sectors. Between 2005 and 2020 emissions of SO_2 were reduced by 80%, NO_2 reduced by 48%, VOC and $PM_{2.5}$ both by 30% and NH_3 reduced by 9%. However, reductions in pollutant concentrations are somewhat smaller than over the US, with $PM_{2.5}$ dropping 20%–25% between

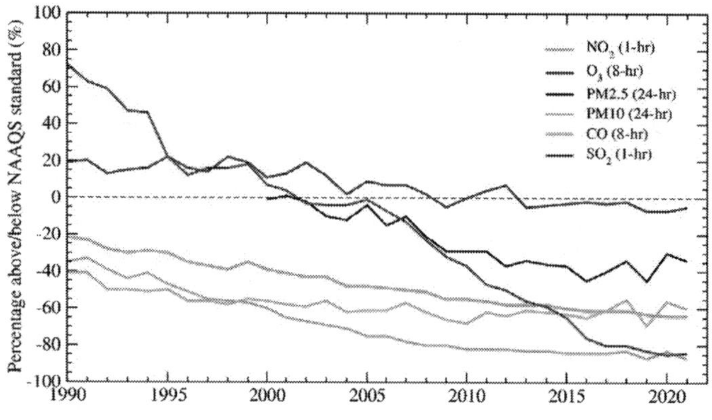

FIGURE 10.12 Annual average air pollutant concentrations in the US relative to the most recent National Ambient Air Quality Standards (NAAQS) from the US Environmental Protection Agency showing a rapid decline in most pollutants since 1990 (US EPA, 2022). *Data from the US Environmental Protection Agency (US EPA) (2022).*

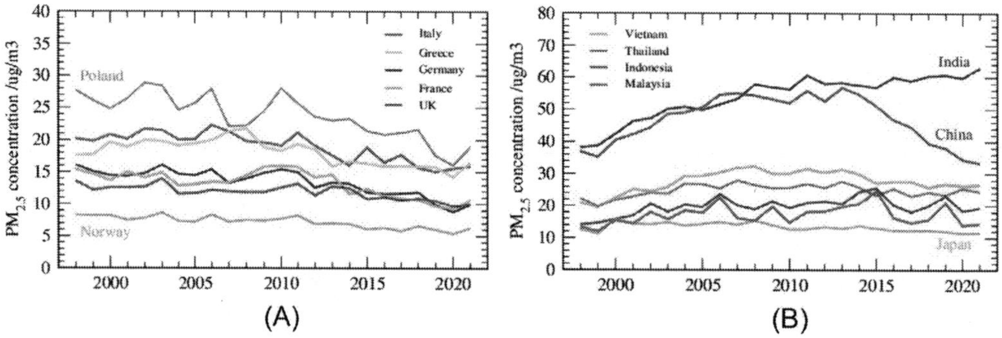

FIGURE 10.13 Estimated trends in annual mean population-weighted $PM_{2.5}$ concentrations over selected countries in (A) Europe and (B) South and East Asia derived by combining satellite and surface measurements with model simulations. Note the different scales used in the two panels. *Data from van Donkelaar et al. (2021).*

2000 and 2019. There is substantial variation in population-weighted $PM_{2.5}$ concentrations across Europe, with relatively low concentrations in Scandinavia and higher concentrations in Mediterranean countries, see Fig. 10.13, although eastern European countries also show higher concentrations reflecting weaker regulation and a greater reliance on coal. Over Europe as a whole, the proportion of urban residents exposed to annual $PM_{2.5}$ levels above 25 $\mu g\,m^{-3}$ fell from about 20% in 2006 to below 1% in 2019, marking a substantial improvement (EEA, 2022). Over the same period, the proportion exposed to PM_{10} levels above 50 $\mu g\,m^{-3}$ for more than 35 days per year dropped from 44% to 11%. However, 97% of urban residents are still exposed to annual $PM_{2.5}$ levels above the WHO's new 5 $\mu g\,m^{-3}$ standard, highlighting further progress to be made.

For surface ozone over Europe, there was no apparent trend in annual mean maximum daily 8-hr O_3 between 2003 and 2012. Nevertheless, there has been a significant decrease in the 95th percentile of O_3 concentrations, and an increase in the fifth percentile, suggesting a reduction in the contribution of local photochemistry associated with lower regional NO_x emissions. There has also been a drop in the proportion of the urban population exposed to 8-hour O_3 above the European threshold of 120 $\mu g\,m^{-3}$ for 25 days per year from about 35% to about 15% between 2000 and 2020. However, there is a large variability in exceedances that reflects meteorological conditions, and exceedances for O_3 were unusually high in 2003, 2006, 2015, and 2018 due to the summer heatwaves in these years.

Further improvement in air quality in Europe presents a continuing challenge as the largest and most easily regulated sources are effectively controlled. There have been a number of setbacks, including support for diesel vehicles in the 2000s to reduce CO_2 emissions, which enhanced urban emissions of NO_x and $PM_{2.5}$, hindering the attainment of air quality objectives. While regulation is now addressing this issue, the slowing rate of improvement in air quality reflects a shift from the dominance of major vehicular, industrial and power generation sources to less easily regulated sources including domestic heating and cooking, tyre and brake wear, and agricultural emissions. While these sources are often smaller in magnitude, addressing them presents a tougher challenge as many involve personal behavioral choices as well as government regulation.

Air pollution in China has shown more dramatic changes over the past 30 years, reflecting very rapid economic development and industrialization since the 1990s with little emissions control but with increasingly tight regulation over the last decade. A rapid increase in demand for energy drove a large increase in SO_2 emissions from coal combustion, but regulations on the use of scrubber technology have gradually brought this under control, and SO_2 concentrations peaked in 2006 and have since fallen rapidly. Satellite measurements of the tropospheric NO_2 column show an increase of almost 300% between 1996 and 2011–12, when NO_2 peaked, followed by a rapid reduction of 40% by 2017 (Georgoulias et al., 2019). Similarly, $PM_{2.5}$ concentrations increased rapidly across China until 2012 but then dropped by 33% between 2013 and 2017, and subsequently a further 10%, see Fig. 10.13. This reflects weak enforcement of environmental regulations until 2013, when a particularly bad pollution episode in North China in January 2013 brought $PM_{2.5}$ levels above 500 $\mu g\, m^{-3}$ for an extended period. This event, sometimes referred to as the "airpocalypse," catalyzed the establishment of a new national monitoring network of more than 1600 stations and the implementation of strict air quality legislation under the Air Pollution Prevention and Control Action Plan. This plan set specific air pollution reduction targets for the North China Plain and the Yangtze and Pearl River Deltas, regions that include Beijing, Shanghai and Guangzhou respectively. The legislation was strictly enforced and was successful in rapidly reducing urban $PM_{2.5}$ and NO_2 concentrations in all three regions over the five-year period from 2013 to 2017. Annual mean $PM_{2.5}$ in Beijing was reduced from 89.5 to 58 $\mu g\, m^{-3}$, a drop of 35%. While the absolute reductions were smaller elsewhere, they were accompanied by drops in other pollutants, most notably SO_2 which fell by 57% on a national scale reflecting a shift away from local coal combustion. Subsequent legislation has expanded to encompass a larger number of cities across the country and to tighten emission controls further, with an aim of all cities meeting the national $PM_{2.5}$ standard of 35 $\mu g\, m^{-3}$. This has now been achieved in southern China, and parts of central and eastern China are close to the standard, but concentrations in the heavily populated parts of northern China remain above this, as shown in Fig. 10.14. Despite this notable success, concentrations in all regions remain well above the WHO guideline for 24-hour $PM_{2.5}$ of 5 $\mu g\, m^{-3}$.

Reductions in NO_x emissions in China associated with shifts in energy generation and tighter fuel standards have not been accompanied by reductions in VOC emissions. One consequence of this is that the photochemical environment in urban areas has changed so that oxidant production is more efficient. This has led to a substantial increase in daytime O_3 in summer in many major cities, with the highest concentrations increasing by 20% across major cities and by almost 25% in the Beijing region, see Fig. 10.14. This increase in summertime O_3 presents a new and important challenge for urban air quality in China, and new policies tackling VOCs and other O_3 precursors will be needed to address this (Wang et al., 2022).

Other countries in South and East Asia have seen less dramatic changes in air pollutant concentrations, as shown in Fig. 10.13. Key pollutants such as NO_2, O_3, and $PM_{2.5}$ are continuing to increase in many parts of Southern Asia, although typically at a slower rate than over the 2000–10 period. $PM_{2.5}$ concentrations remain high in urban regions in India and Pakistan, and air quality here is now substantially worse than in China, as seen in Fig. 10.6. Satellite

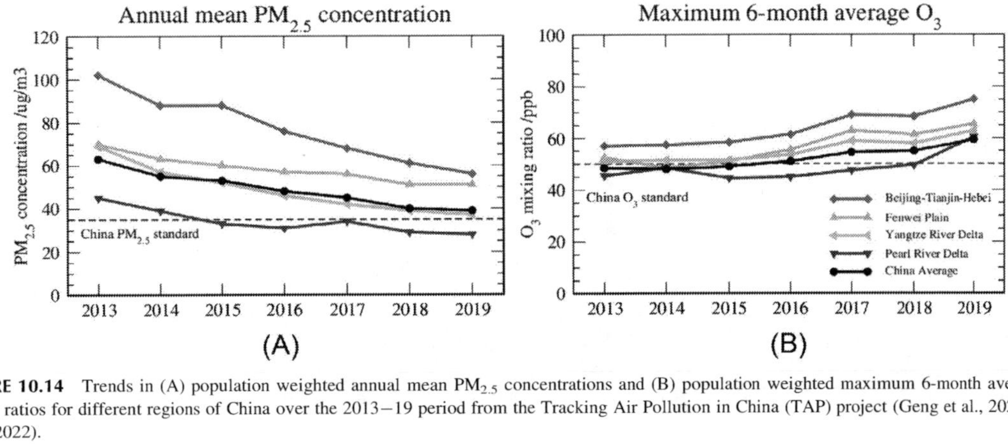

FIGURE 10.14 Trends in (A) population weighted annual mean $PM_{2.5}$ concentrations and (B) population weighted maximum 6-month average O_3 mixing ratios for different regions of China over the 2013–19 period from the Tracking Air Pollution in China (TAP) project (Geng et al., 2021; Xiao et al., 2022).

measurements show an increase in regional NO_2 columns of about one third over the past two decades, and this is associated with increasing urban ozone, particularly in the hot, premonsoon season in April and May. Tighter vehicle fuel standards have been implemented in India in recent years, but the benefits of these have been outweighed by an increase in vehicle usage, with a doubling of vehicle numbers between 2011 and 2019. The principal sources to address are now from the residential and industrial sectors, although waste burning and agricultural sources also present major challenges. It is likely that the Indian subcontinent will follow a similar trajectory to east Asia in bringing down $PM_{2.5}$, albeit at a slower rate, and major cities in India are already starting to experience increased ozone.

Other parts of the world have typically seen less dramatic changes in air quality, although the lack of long-term measurement networks in many regions makes identification of trends difficult. $PM_{2.5}$ concentrations in Southeast Asian countries have generally risen between 2000 and 2010, as shown in Fig. 10.13, but subsequently started to decline. These trends are partly masked by substantial interannual variability, particularly over Indonesia and Malaysia where the major haze events due to forest fires in 2015 and 2019 stand out clearly in annual average $PM_{2.5}$ concentrations. Global reanalysis studies suggest that air quality is deteriorating in the Middle East and in sub-Saharan Africa, with increases in $PM_{2.5}$ and NO_2 concentrations associated with economic development and urbanization, and substantial increases in O_3 concentrations in Africa (Sicard et al., 2023). However, a clearer assessment of these trends requires the establishment of reliable long-term measurement networks across these regions.

Worked Example 10.4: What role do natural emissions play in regional air quality trends?

Over most regions, the trends in air quality principally reflect changes in anthropogenic emissions due to industrial and economic development, population changes and/or emission controls. However, in cleaner regions, or regions where anthropogenic sources are more effectively controlled, natural emissions may play a larger role. Changes in climate may influence emissions from soil and vegetation, lead to an increase in wildfires, and alter transport patterns; these effects may be particularly prevalent in Tropical regions influenced by major climate phenomena such as El Nino. Changes in land use (e.g., afforestation, urbanization) and agricultural practices may also alter natural sources at a local and regional level. It is important to be aware of these natural influences when setting long-term regional air quality goals.

10.6 Contributions to local air quality

This chapter has focussed on regional and global air quality and explored differences in regional air quality across the world. However, very poor air quality is often highly localized, reflecting the presence of specific small-scale emission sources such as industrial facilities or transportation corridors, and these sources build upon the wider urban and regional background. The pollution experienced at a particular location is often viewed conceptually as the superposition of contributions from regional, urban and local sources (Lenschow et al., 2001). The regional contribution represents the relatively well-mixed background from many sources across an extended area that may be chemically aged and upon which meteorological processes have acted, see Fig. 10.15. The urban increment is present over cities, reflecting the combined effect of sources over the metropolitan region transported and mixed over urban scales. The local increment reflects the contribution from specific sources such as factories or roads as they disperse from the point of emission over short distances. While this approach to source apportionment is simplistic, neglecting the interactions

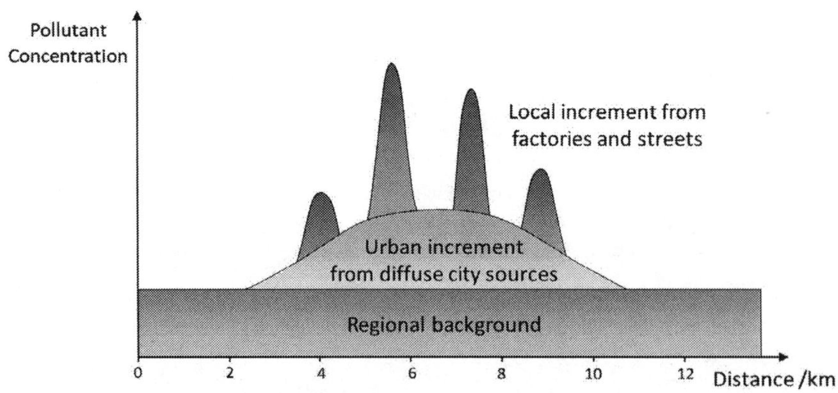

FIGURE 10.15 Conceptual diagram of the contribution of local, urban and regional background sources to pollutant concentrations experienced in different locations across a city.

that may occur between pollutants from different sources, it provides a useful method to characterize the spatial variation in pollutant concentrations.

The importance of the urban and local increments depends strongly on the lifetime of the pollutant under consideration and on the spatial heterogeneity of sources. Concentrations of short-lived VOC or NO from car exhausts may be high by the roadside, but low elsewhere as they are rapidly reduced by dispersion and oxidation. In contrast, longer-lived pollutants such as $PM_{2.5}$ are typically better mixed, and the urban and regional contributions generally dominate except very close to the largest sources. The situation is more complex for secondary oxidants such as ozone, where dispersed sources of precursors lead to net production, but intense local sources of NO associated with roads or industrial facilities may deplete ozone through chemical titration. Under these situations the local and urban pollutant increments may be negative, and precursors from urban regions are often transported out of the source region before substantial production occurs. This highlights the important role that regional background concentrations may play in influencing local air quality.

10.7 Useful links

https://aqicn.org/map/world/—Map of real-time air quality indices from monitoring sites around the world showing measurements of a range of pollutants ($PM_{2.5}$, PM_{10}, O_3, NO_2, SO_2, and CO) from national monitoring networks.

https://www.iqair.com/air-quality-map—Dynamic and interactive real-time map of surface $PM_{2.5}$ based on national monitoring networks.

https://worldview.earthdata.nasa.gov/—NASA Worldview provides historic and near real-time satellite imagery of fires, dust storms, volcanic eruptions, meteorological features and a range of atmospheric pollutants.

https://www.who.int/data/gho/data/themes/air-pollution/who-air-quality-database—The World Heath Organization maintain a database of annual mean pollutant concentrations as part of their commitment to the Sustainable Development Goal indicator 11.6.2, Air quality in cities.

https://www.eea.europa.eu/data-and-maps/dashboards/air-quality-statistics—Air quality trends and statistics over Europe.

http://tapdata.org.cn—The Tracking Air Pollution in China project combines observations and model data to provide past trends and near real-time air pollutant concentrations across China.

10.8 Questions

1. Compare and contrast the distributions of NO_2 and O_3 across North America and Europe. Why do the distributions of the two pollutants look so different in each region?
2. Initial policies to control urban air pollution often lead to reductions in $PM_{2.5}$ and NO_2 but to increases in O_3. Explain why this is, and suggest approaches to mitigate this problem.
3. Observations of regional surface O_3 are starting to show decreases at northern midlatitudes but are continuing to increase in many tropical regions. Suggest reasons for these changes.
4. Methane is not usually classified as a hazardous air pollutant. How would you explain to a policymaker that controlling its emissions would benefit air quality?
5. Discuss the importance of long-range transport of pollutants for air quality in your region.

Solutions to the end-of-chapter questions will be available at: https://www.elsevier.com/books-and-journals/book-companion/9780128225912.

References

AIRS Science Team/Joao Teixeira (2013). Monthly Standard Physical Retrieval (AIRS-only) 1 degree x 1 degree V006, Greenbelt, MD, USA, Goddard Earth Sciences Data and Information Services Center (GES DISC) 2013. Available from https://doi.org/10.5067/Aqua/AIRS/DATA321 AIRS/Aqua L3.

Akimoto, H. (2003). Global air quality and pollution. *Science*, *302*, 1716–1719. Available from https://doi.org/10.1126/science.1092666.

Childs, M., Li, J., Wen, J., Heft-Neal, S., Driscoll, A., Wang, S., Gould, C., Qiu, M., Burney, J., & Burke, M. (2022). Daily local-level estimates of ambient wildfire smoke $PM_{2.5}$ for the contiguous US. *Environmental Science & Technology*, *56*, 13607–13621. Available from https://doi.org/10.1021/acs.est.2c02934.

van Donkelaar, A., Hammer, M., Bindle, L., Bauer, M., Brook, J., Garay, M., Hsu, N., Kalashinoka, L., Sayer, A., & Martin, R. (2021). Monthly global estimates of fine particulate matter and their uncertainty. *Environmental Science & Technology, 55*(22), 15287–15300. Available from https://doi.org/10.1021/acs.est.1c05309.

EEA (2022). *Exceedance of air quality standards in Europe*. <https://www.eea.europa.eu/ims/exceedance-of-air-quality-standards>.

Fiore, A., Dentener, F., Wild, O., Cuvelier, C., Schultz, M. G., Hess, P., Textor, C., & Zuber, A. (2009). Multi-model estimates of intercontinental source-receptor relationships for ozone pollution. *Journal of Geophysical Research, 114*(D04301). Available from https://doi.org/10.1029/2008JD010816.

Fiore, A., West, J., Horowitz, L., Naik, V., & Schwarzkopf, M. (2008). Characterizing the tropospheric ozone response to methane emission controls and the benefits to climate and air quality. *Journal of Geophysical Research, 113*, D08307. Available from https://doi.org/10.1029/2007JD009162.

Fischer, E., Hsu, N., Jaffe, D., Jeong, M.-J., & Gong, S. (2009). A decade of dust: Asian dust and springtime aerosol load in the U.S. Pacific Northwest. *Geophysical Research Letters, 36*(3), L03821. Available from https://doi.org/10.1029/2008GL036467.

Gaudel, A., Cooper, O., Ancellet, G., Barret, B., Boyndard, A., Burrows, J., Clerbaux, C., Zeng, G., & Ziemke, J. (2018). Tropospheric ozone assessment report: Present-day distribution and trends of tropospheric ozone relevant to climate and global atmospheric chemistry model evaluation. *Elementa, 6*(39). Available from https://doi.org/10.1525/elementa.291.

van Geffen, J., Eskes, H., Boersma, K., & Veefkind, J. (2022). KNMI TROPOMI ATBD of the total and tropospheric NO_2 data products, Report S5P-KNMI-L2-0005-RP 2022. De Bilt, The Netherlands.

Geng, G., Xiao, Q., Liu, S., Liu, X., Cheng, J., Zheng, Y., Xue, T., Tong, D., Zheng, B., Peng, Y., Huang, X., He, K., & Zhang, Q. (2021). Tracking air pollution in China: Near real-time $PM_{2.5}$ retrievals from multisource data fusion. *Environmental Science & Technology, 55*(17), 12106–12115. Available from https://doi.org/10.1021/acs.est.1c01863.

Georgoulias, A., van der A, R., Stamnes, P., Boersma, K., & Eskes, H. (2019). Trends and trend reversal detection in 2 decades of tropospheric NO_2 satellite observations. *Atmospheric Chemistry Physics, 19*, 6269–6294. Available from https://doi.org/10.5194/acp-19-6269-2019.

Gouveia, N., Kephart, J., Dronova, I., McClure, L., Granados, J., Betancourt, R., O'Ryan, A. C., Texcalac-Sangrador, J., Martinez-Folgar, K., Rodriguez, D., & Diez-Roux, A. (2021). Ambient fine particulate matter in Latin American cities: Levels, population exposure, and associated urban factors. *Science of the Total Environment, 772*. Available from https://doi.org/10.1016/j.scitotenv.2021.145035.

Guo, S., Hu, M., Zamora, M., Peng, J., Shang, D., Zheng, J., Du, Z., Wu, Z., Shao, M., Zeng, L., Molina, M., & Zhang, R. (2014). Elucidating severe urban haze formation in China. *Proceedings of the National Academy of Sciences, 111*(49), 17373–17378. Available from https://doi.org/10.1073/pnas.1419604111.

HTAP (2010). Hemispheric transport of air pollution 2010. *Part A: Ozone and Particulate Matter, 2010*. United Nations, New York and Geneva. Available at https://doi.org/10.18356/2c908168-en.

Lan, X., Dlugokencky, E.J., Mund, J., Crotwell, A., Crotwell, M., Moglia, E., Madronich, M., Neff, D., & Thonig, K. (2022). *Atmospheric methane dry air mole fractions from the NOAA GML carbon cycle cooperative global air sampling network, 1983-2021, version: 2022-11-21*. <https://doi.org/10.15138/VNCZ-M766>.

Lenschow, P., Abraham, H.-J., Kutzner, K., Lutz, M., Preuss, J.-D., & Reichenbacher, W. (2001). Some ideas about the sources of PM10. *Atmospheric Environment, 35*(Supp. 1), S23–S33. Available from https://doi.org/10.1016/S1352-2310(01)00122-4.

Lin, Y., Lin, C., Lin, P., Engling, G., Lin, Y., Lan, Y., Chang, C., Kuo, T., Hsu, W., & Ting, C. (2013). Influence of Southeast Asian biomass burning on ozone and carbon monoxide over subtropical Taiwan. *Atmospheric Environment, 64*, 358–365. Available from https://doi.org/10.1016/j.atmosenv.2012.09.050.

Liu, M., Westphal, D., Wang, S., Shimizu, A., Sugimoto, N., Zhou, J., & Chen, Y. (2003). A high-resolution numerical study of the Asian dust storms of April 2001. *Journal of Geophysical Research, 108*, 8653. Available from https://doi.org/10.1029/2002JD003178.

Monks, P., Archibald, A., Colette, A., Cooper, O., Coyle, M., Derwent, R., Fowler, D., Granier, C., Law, K., Mills, G., Stevenson, D., Tarasova, O., Thouret, V., von Schneidemesser, E., Sommariva, R., Wild, O., & Williams, M. (2015). Tropospheric ozone and its precursors from the urban to the global scale from air quality to short-lived climate forcer. *Atmospheric Chemistry Physics, 15*, 8889–8973. Available from https://doi.org/10.5194/acp-15-8889-2015.

NASA Worldview (2022). <https://worldview.earthdata.nasa.gov/>.

NASA Earth Observations (2022). <https://neo.gsfc.nasa.gov/>.

NASA/LEARC/SD/ASDC. (2000). *NASA Langley Atmospheric Science Data Center DAAC MOPITT CO gridded monthly means (near and thermal infrared radiances), V008*, 2000. Available from https://doi.org/10.5067/TERRA/MOPITT/MOP03JM_L3.008.

Nawaz, M., & Henze, D. (2020). Premature deaths in Brazil associated with long-term exposure to $PM_{2.5}$ from Amazon fires between 2016 and 2019. *GeoHealth, 4*. Available from https://doi.org/10.1029/2020GH000268.

Pozzer, A., Schultz, M., & Helmig, D. (2020). Impact of U.S. Oil and natural gas emission increases on surface ozone is most pronounced in the Central United States. *Environmental Science & Technology, 54*(19), 12423–12433. Available from https://doi.org/10.1021/acs.est.9b06983.

Saunois, M., Stavert, A., Poulter, B., Bousquet, P., Canadell, J., Jackson, R., Raymond, P., & Zhuang, Q. (2020). The global methane budget 2000–2017. *Earth Systems Science Data, 12*, 1561–1623. Available from https://doi.org/10.5194/essd-12-1561-2020.

Schmale, J., Arnold, S., Law, K., Thorp, T., Anenberg, S., Simpson, W., Mao, J., & Pratt, K. (2018). Local Arctic air pollution: A neglected but serious problem. *Earth's Future, 6*, 1385–1412. Available from https://doi.org/10.1029/2018EF000952.

Schmidt, A., Ostro, B., Carslaw, K., & Simmons, A. (2011). Excess mortality in Europe following a future Laki-style Icelandic eruption. *PNAS, 108*(38), 15710–15715. Available from https://doi.org/10.1073/pnas.110856910.

Schröder, S., Schultz, M., Selke, N., Sun, J., Ahring, J., Mozaffari, A., Romberg, M., Epp, E., Lensing, M., Apweiler, S., Leufen, L., Betancourt, C., Hagemeier, B., & Rajveer, S. (2021). FZ-Juelich B2SHARE TOAR Data Infrastructure 2021. Available from https://doi.org/10.34730/4d9a287dec0b42f1aa6d244de8f19eb3.

Seguel, R., Gallardo, L., & Fleming, Z. (2020). Two decades of ozone standard exceedances in Santiago de Chile. *Air Qual Atmospheric Health*, 593–605. Available from https://doi.org/10.1007/s11869-020-00822-w.

Sicard, P., Agathokleous, E., Anenberg, S., De Marco, A., Paoletti, E., & Calatayud, V. (2023). Trends in urban air pollution over the last two decades: A global perspective. *Science Total Environment, 858*. Available from https://doi.org/10.1016/j.scitotenv.2022.160064.

Solberg, S., Hov, Ø., Søvde, A., Isaksen, I., Coddeville, P., De Backer, H., Forster, C., Orsolini, Y., & Uhse, K. (2008). European surface ozone in the extreme summer 2003. *Journal of Geophysical Research, 113*, D07307. Available from https://doi.org/10.1029/2007JD009098.

Sun, Y., Wang, Z., Wild, O., Xu, W., Chen, C., Fu, P., Du, W., Zhou, L., & Worsnop, D. (2016). "APEC blue": Secondary aerosol reductions from emission controls in Beijing. *Scientific Reports*, *6*(20668). Available from https://doi.org/10.1038/srep20668.

Szopa, S., Naik, V., Adhikary, B., Artaxo, P., Berntsen, T., Collins, W. D., Fuzzi, S., Gallardo, L., Kiendler-Scharr, A., Klimont, Z., Liao, H., Unger, N., & Zanis, P. (2021). In V. Masson-Delmotte, P. Zhai, A. Pirani, S. L. Connors, C. Péan Péan, S. Berger, N. Caud, Y. Chen, L. Goldfarb, M. I. Gomis, M. Huang, K. Leitzell, E. Lonnoy, J. B. R. Matthews, T. K. Maycock, T. Waterfield, O. Yelekci, R. Yu, & B. Zhou (Eds.), *Chapter 6: Short lived climate forcers*. Cambridge University Press. Available from https://doi.org/10.1017/9781009157896.008.

Tagaris, E., Stergiou, I., & Sotiropoulou, R. (2017). Impact of shipping emissions on ozone levels over Europe: assessing the relative importance of the Standard Nomenclature for Air Pollution (SNAP) categories. *Environmental Science Pollution Research.*, *24*, 14903–14909. Available from https://doi.org/10.1007/s11356-017-9046-x.

Turnock, S., Wild, O., Sellar, A., & O'Connor, F. (2019). 300 years of tropospheric ozone changes using CMIP6 scenarios with a parameterised approach. *Atmospheric Environment*, *213*, 686–698. Available from https://doi.org/10.1016/j.atmosenv.2019.07.001.

US EPA (2022). *Our nation's air: Trends through 2021*. <https://gispub.epa.gov/air/trendsreport/2022/>.

Wang, T., Xue, L., Feng, Z., Dai, J., Zhang, Y., & Tan, Y. (2022). Ground-level ozone pollution in China: A synthesis of recent findings on influencing factors and impacts. *Environmental Research Letters.*, *17*. Available from https://doi.org/10.1088/1748-9326/ac69fe.

Wild, O., Fiore, A., Shindell, D., Doherty, R., Collins, W., Dentener, F., Schultz, M., MacKenzie, I., Zeng, G., Hess, P., Keating, T., & Zuber, A. (2012). Modelling future changes in surface ozone: a parameterized approach. *Atmospheric Chemistry Physics*, *12*, 2037–2054. Available from https://doi.org/10.5194/acp-12-2037-2012.

Wild, O., Prather, M., Akimoto, H., Sundet, J., Isaksen, I., Crawford, J., Davis, D., Avery, M., Kondo, Y., Sachse, G., & Sandholm, S. (2003). Chemical transport model ozone simulations for spring 2001 over the western Pacific: Regional ozone production and its global impacts. *Journal of Geophysical Research*, *109*, D15S02. Available from https://doi.org/10.1029/2003JD004041.

Xiao, Q., Geng, G., Xue, T., Liu, S., Cai, C., He, K., & Zhang, Q. (2022). Tracking $PM_{2.5}$ and O_3 pollution and the related health burden in China 2013–2020. *Environmental Science & Technology*, *56*(11), 6922–6932. Available from https://doi.org/10.1021/acs.est.1c04548.

Xu, L., Crounse, J., Allen, H., Wennberg, P., Bourgeois, I., Brown, S., Campuzano-Jost, P., Coggon, M., & Yokelson, R. (2021). Ozone chemistry in western U.S. wildfire plumes. *Science Advances*, *7*(50). Available from https://doi.org/10.1126/sciadv.abl3648.

Young, P., Naik, V., Fiore, A., Gaudel, A., Guo, J., Lin, M., Neu, J., Parrish, D., & Zeng, G. (2018). Tropospheric ozone assessment report: Assessment of global-scale model performance for global and regional ozone distributions, variability, and trends. *Elementa Science Anth*, *6*(10). Available from https://doi.org/10.1525/elementa.265.

Chapter 11

Air quality and climate interactions

Fan Wang[1], Meng Gao[1] and Gregory R. Carmichael[2]

[1]*Department of Geography, Hong Kong Baptist University, Hong Kong SAR, P.R. China,* [2]*Department of Chemical and Biochemical Engineering, The University of Iowa, Iowa City, IA, United States*

11.1 Introduction

Air quality and climate are significant public concerns that greatly impact human health, vegetation growth, and overall environmental safety. The intricate relationship between climate and air quality is well-established (Fiore et al., 2012). Climate factors can influence long-distance transport, chemical processes, and local circulation, thereby affecting air pollution levels (Jacob & Winner, 2009). Additionally, air pollutants and their precursors interact with solar radiation, disturbing the planet's energy budget and resulting in climate change. The accumulation of air pollutants and their precursors under specific weather and topography conditions primarily contributes to air pollution (Akimoto, 2003). Emissions can be categorized into natural and anthropogenic sources. Natural emissions arise from processes such as volcanic eruptions, forest fires, vegetation release, soil weathering, and sea spray bursts. Anthropogenic emissions originate from human activities like the combustion of fossil fuels, industry, transportation, and agricultural practices (Guenther et al., 2000).

Within the atmosphere, air pollutants, particularly aerosols and certain absorbing gases, impede shortwave radiation (SWD) from reaching the surface through scattering, reflecting, and absorbing mechanisms (Satheesh & Moorthy, 2005). Consequently, these alterations in the surface energy balance impact near-ground meteorological variables such as temperature, wind speed, and relative humidity. Surface energy changes also affect the development of vertical convection and turbulence, thereby modifying planetary boundary layer (PBL) and cloud formation. Referred to as the direct effect, this phenomenon directly affects the radiation balance of the land-atmosphere system (Li et al., 2017). Aerosols also act as cloud condensation nuclei (CCN) and ice nuclei (IN), altering cloud lifetimes and optical properties, thereby exerting an indirect radiative forcing on the global climate (Fan et al., 2016). Furthermore, aerosols have a semidirect effect on clouds (Johnson et al., 2004). Certain aerosols, like dust and black carbon (BC), absorb solar radiation and re-emit thermal radiation into the surrounding atmosphere. This process can heat the atmosphere and cloud masses, leading to the evaporation of cloud droplets, reduced cloud volume, shortened cloud lifetime, and decreased cloud albedo. Additionally, clouds can bring aerosols back to the ground through precipitation, thereby complicating the influence of aerosols on clouds.

The interactions between air quality, weather, and climate can amplify the adverse effects of these factors (Fuzzi et al., 2015). Consequently, there is a growing interest in implementing measures that can effectively address both air pollution and climate change concurrently. For example, replacing fossil fuels with renewable energy such as solar, wind, and geothermal power can reduce greenhouse gas (GHGs) emissions and minimize primary emissions of air pollutants such as particles, CO, SO_2, and NO_2. Cities, as the largest anthropogenic source of GHGs and air pollutants, play a pivotal role in future air pollution and climate change (Baklanov et al., 2016). The urban heat island (UHI), caused by contrasting thermal properties with surrounding rural areas, exacerbates heat stress and worsens urban air quality. Numerous studies have suggested increasing green spaces within cities and implementing roof and wall strategies to alleviate urban heat stress. However, these measures have implications for air quality, as decreasing urban temperatures can alter the chemical reaction processes and dispersion of air pollutants (Getter & Rowe, 2006).

11.2 Processes and feedback

11.2.1 Radiation processes and feedback

Photolysis is the process of chemical decomposition induced by natural or artificial light in a chemical mixture. It fundamentally affects the production of secondary pollutants including ozone (O_3) and secondary aerosols

Air Quality. DOI: https://doi.org/10.1016/B978-0-12-822591-2.00011-1

(Oum et al., 1998; Wu et al., 2020). Ozone is a secondary air component in the atmosphere. It is generated by complex photochemical reactions. Stratospheric ozone is generated from O_2 by the reaction powered by solar ultraviolet radiation. Tropospheric ozone forms when heat and sunlight trigger chemical reactions between nitrogen oxides (NO_x) and volatile organic compounds (VOCs). Photolysis of nitrates can generate hydroxyl (OH) radicals, NO_2 radicals, and nitrite ions, thereby facilitating the oxidation of dissolved SO_2 in particles, leading to sulfate production. The rate of photolysis is influenced by the intensity of solar radiation. Several studies have shown that O_3 concentrations tend to be higher in clear sky conditions during summer because of intense solar radiation (Vautard et al., 2005; Vieno et al., 2010; Vinod et al., 2002).

Radiation is also influenced by air pollutants through direct effects. Air pollutants directly regulate radiation through absorption and scattering mechanisms. Shortwave solar radiation is the primary energy source that regulates weather and climate (Charlson et al., 1991, 1992). After solar radiation enters the atmosphere, aerosols reflect or absorb a significant portion of SWD (Andrews et al., 2006; Ramanathan et al., 2001). The process of reflecting energy back to space is known as scattering, while absorption refers to the process of energy being absorbed. Aerosols, which are tiny solid particles or liquid droplets suspended in the air, are emitted directly or formed from various sources. Certain types of aerosols can also absorb solar radiation, with BC being the most significant absorber, followed by mineral dust and some organic carbon components. These absorbing aerosols cut down the amount of SWD reaching the ground and contribute to atmospheric heating, consequently affecting longwave radiation processes (Ding et al., 2016). Ozone, as an air pollutant, also absorbs solar radiation. The majority of ozone is concentrated in the stratosphere, approximately 9–18 miles (15–30 km) above the ground, where it absorbs most of the ultraviolet (UV) radiation from the Sun.

Example 1:

The initial aerosol optical depth (AOD) is 0.2 and the aerosol single scattering albedo is 0.9. Calculate the original aerosol radiative forcing (ARF) and the change in radiative forcing caused by aerosols due to a 20% increase in the AOD, assuming all other parameters remain constant. Given the solar constant is 1361 $W\,m^{-2}$ and Earth's albedo is 0.3.

Answer:

Step 1: Calculate the initial radiative forcing using the given parameters:

Radiative Forcing (Initial) = AOD × Solar Constant × (1 − Earth's Albedo) × (1 − Aerosol Single Scattering Albedo) = 0.2 × 1361 × (1−0.3) × (1−0.9) = 19.064 $W\,m^{-2}$

Step 2: Calculate the new AOD after the 20% increase:

New AOD = Initial AOD + (Initial AOD × 20%) = 0.24

Step 3: Calculate the new radiative forcing using the updated AOD:

Radiative Forcing (New) = New AOD × Solar Constant × (1 − Earth's Albedo) × (1 − Aerosol Single Scattering Albedo) = 0.24 × 1361 × (1−0.3) × (1−0.9) = 22.877 $W\,m^{-2}$

Step 4: Calculate the change in radiative forcing:

Change in Radiative Forcing = Radiative Forcing (New) − Radiative Forcing (Initial) = 22.877−19.064 = 3.813 $W\,m^{-2}$

11.2.2 Microphysical processes and feedback

Aerosol indirect effects refer to the influence of aerosols acting as CCN or IN on diverse cloud characteristics, which ultimately have indirect impacts on the climate system (see Fig. 11.1; Kuniyal and Guleria, 2019). These effects can be categorized into two main types (Lohmann & Feichter, 2005).

The first type is known as the Twomey effect. When the concentration of aerosols elevates, it results in an enhancement in the number concentration of cloud particles and a decline in their radius. Consequently, this change in cloud particle characteristics increases the cloud albedo, which is associated with the absorption properties and optical thickness of the aerosols when the liquid water content of the cloud remains constant.

The second type is commonly referred to as the Albrecht effect or cloud life-cycle effect. An increase in anthropogenic aerosols results in a reduction in the radius of cloud particles. This, in turn, hampers precipitation processes and alters the cloud life cycle. In addition to these indirect effects, aerosols can also have a semidirect effect on clouds. For instance, soot and BC aerosols possess strong absorption capabilities for solar radiation and emit thermal radiation into the surrounding atmosphere. This heating effect can cause the evaporation of cloud droplets, leading to reduced cloud volume, shortened cloud lifetime, and a decrease in the average albedo of clouds. Conversely, clouds have the ability to remove aerosols from the atmosphere through precipitation, further influencing the influence of aerosols on clouds.

The complexity of these indirect effects poses challenges in thoroughly studying the impacts of aerosols on various cloud properties including macroscopic characteristics (e.g., cloud area, cloud thickness, liquid water paths), microscopic properties (e.g., number concentration and size distribution of droplet and ice crystal) and precipitation effects.

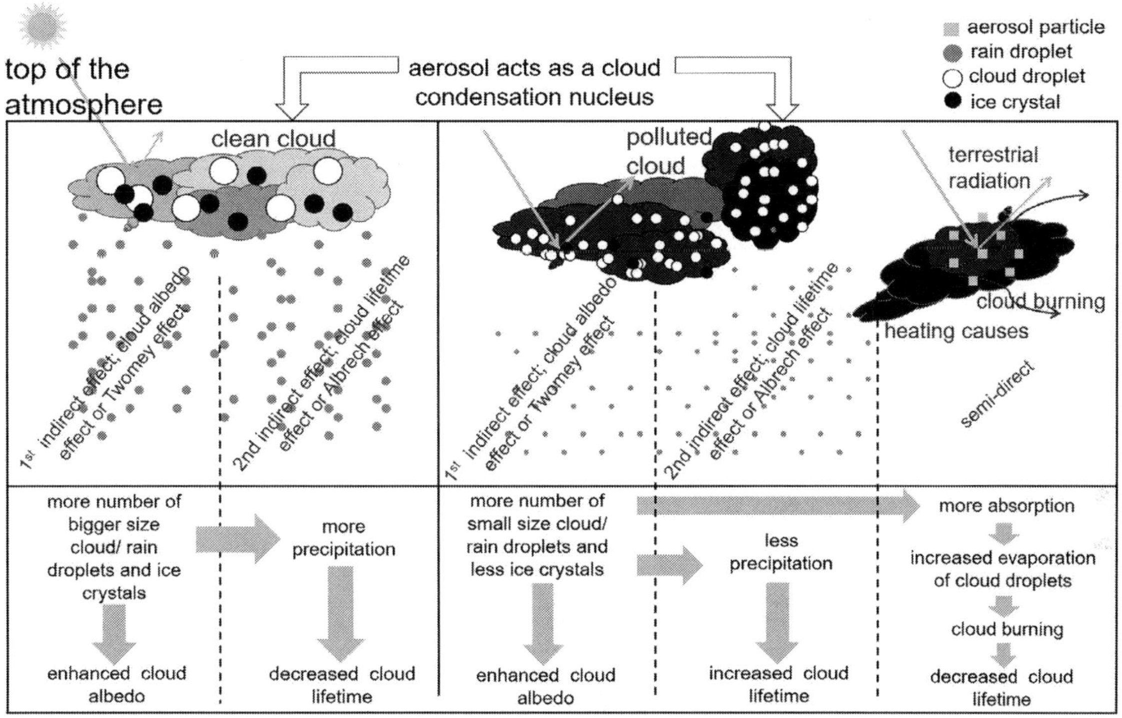

FIGURE 11.1 **Schematic diagram of the aerosol indirect effects.** Aerosols are emitted from natural sources and act as CCN and also give rise to the formation of clean clouds. The clean clouds contain larger cloud drops. The large concentrations of natural and anthropogenic aerosols increase cloud droplet number concentration and hence formed polluted clouds. The polluted clouds contain a greater number of smaller cloud droplets. *From Kuniyal, J.C. & Guleria, R.P. (2019). The current state of aerosol-radiation interactions: A mini review.* Journal of Aerosol Science, 130, *45–54. https://doi.org/10.1016/j.jaerosci.2018.12.010.*

The research in this field has not fully addressed the dynamics, thermal, cloud microphysical processes, and chemical processes involved. Moreover, different types of clouds exhibit distinct macroscopic characteristics, especially in the case of stratiform and convective clouds. These cloud types exhibit significant differences in properties and physical processes. Consequently, the impacts of interactions between aerosol and clouds on climate remain highly uncertain due to these complexities and research gaps.

11.2.3 Physical processes and feedback

Meteorological conditions play a crucial role in influencing air quality by regulating dilution, dispersion, transport, and transformation of atmospheric pollutants, which impact the distribution and concentration of pollutants (Xing et al., 2011). Temperature is a fundamental factor that significantly influences air quality by simultaneously affecting chemical reactions, emissions, and dispersion. Increased temperature generally enhances the efficiency of natural emissions, such as biogenic volatile organic compounds (BVOCs). Temperature variations also influence vegetation transpiration, which, in turn, affects the rate of dry deposition of pollutants. Moreover, temperature has a notable impact on chemical reaction rates. Horizontal and vertical temperature changes particularly affect pollutant diffusion. Horizontal temperature gradients influence the development of horizontal atmospheric convection, while vertical temperature anomalies can trigger atmospheric inversions that hinder the vertical transport of air pollutants. Surface air temperature is also a crucial factor in determining the structure of the thermal boundary layer. The evolution of the boundary layer directly influences the vertical mixing of pollutants, which subsequently impacts near-surface pollutant concentrations. It is worth noting that air pollutants, particularly aerosols, can also have feedback effects on meteorological conditions (Ding et al., 2016). Aerosols, through aerosol radiative effects (ARE), modify the surface radiation balance, leading to potential impacts on UHI intensity. ARE leads to a decline in the downward SWD reaching the ground. It also decreases the sensible heat (SH) flux and contributes to a decrease in the height of the PBL (PBLH). These effects exacerbate the severity of haze events (see Fig. 11.2; Zhang et al., 2022).

The UHI is an urban area that experiences higher temperatures than surrounding rural regions. The formation of UHI is attributed to the evolution of natural pervious land to impervious pavement, buildings, and other hard surfaces. Additionally, human activities in cities contribute to the emission of significant anthropogenic heat (Oke, 1973, 1982). The UHI phenomenon creates a temperature gradient between urban and rural regions, resulting in the circulation of UHI winds blowing from rural to urban regions. This circulation pattern creates unfavorable conditions for the diffusion of air pollutants and tends to worsen air quality within cities. Conversely, air pollutants can also influence the intensity of UHI. Urban areas, typically being centers of pollution, tend to have higher aerosol concentrations compared to rural areas. In such cases, aerosols may reduce the amount of downward SWD, leading to a more pronounced decrease in air temperatures in urban areas during daytime (Cao et al., 2016). However, absorbing aerosols, like BC, possess the ability to both absorb and emit radiation. This characteristic results in an elevation of the longwave radiation energy received by urban surfaces, thereby intensifying the UHI effect. Additionally, light-absorbing particles contribute to atmospheric warming, enhance the stability of the PBL, and diminish diurnal temperature variations over urban regions, thereby amplifying UHI, especially during nighttime.

11.3 Modeling approaches

11.3.1 Online and coupled models

The modeling approach serves as a fundamental method for analyzing the interactions among air pollution, weather, and climate, and it is a significant contributor to the progress made in this field (Grell & Baklanov, 2011). This approach utilizes mathematical tools to describe atmospheric physical and chemical processes. By employing mathematical methods, models simulate horizontal and vertical distribution and variation of meteorology and air quality. They aim to replicate the processes of pollutant transport, reaction, and removal in the atmosphere. Modeling is an essential technique for analyzing the spatial and temporal evolution of atmospheric pollution, investigating underlying mechanisms, identifying sources, and examining the interactions between climate and air pollution.

Initially, air chemistry models and weather models developed as separate disciplines due to their complexity and the limited availability of computational resources. As a result, early models were offline and uncoupled. In an offline model, the chemical transport model is driven separately using meteorological data obtained from observations,

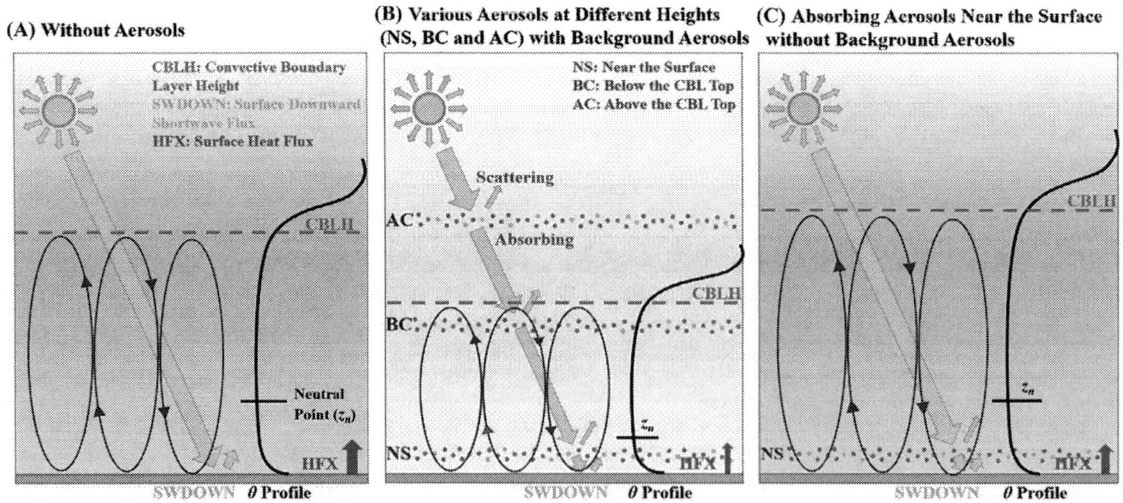

FIGURE 11.2 A schematic figure depicting the aerosol–boundary-layer feedback. Aerosol cooling effect by attenuating incident solar radiation and surface heat flux exceeded its warming effect by reheating the atmosphere layer with absorbed shortwave radiation, and eventually declined net heating rate, which inhibited CBL development, lowered mixed-layer potential temperature and stabilized atmospheric stratification. *From Zhang, X., Cai, C., Hu, X.M., Gao, L., Xu, X., Hu, J. & Chen, H. (2022). Aerosols consistently suppress the convective boundary layer development.* Atmospheric Research, 269. *https://doi.org/10.1016/j.atmosres.2022.106032.*

analyzed meteorological data, or forecasted meteorological data from meteorological models. On the other hand, in an online model, the chemical transport model receives real-time output data directly from a meteorological model through a model interface. Online models have evolved into online coupled models, where the meteorological and chemical components share the same grid without spatial and temporal interpolations. Fully online coupled models allow for two-way interactions, as all 3D meteorological variables and air pollutant concentrations are available at each time step. However, due to computational and time constraints, fully online coupled models are not as prevalent as offline models. The third generation of models, such as the Weather Research and Forecasting model coupled with Chemistry (WRF-Chem), represents fully online coupled meteorological-chemical models. WRF-Chem's major advantage lies in its complete temporal and spatial resolution coupling between the meteorological and chemical transport modes, enabling true online feedback. This feature has led to the widespread adoption of WRF-Chem in studies focusing on air pollution, weather, and climate interactions. Online coupled models, capable of considering the interactions between weather, climate, and air quality, represent the mainstream in current model development.

However, the current coupled models face limitations in spatial resolution due to computational constraints. Subgrid parameterization schemes are employed to account for calculations below the grid resolution scale. As a result, there is no unified model that can address all spatial scales. Different models need to be developed to handle specific spatial scales, but these models can be coupled with each other as complementary tools to address practical problems (see Fig. 11.3; Zhang et al., 2012).

11.3.2 Earth system approaches

The Earth system model (ESM) is a crucial tool for comprehending the mechanisms behind past climate and environmental changes and predicting potential scenarios for future global changes. The ESM evolved from its predecessor, the climate system model (CSM). The traditional CSM primarily incorporated the atmospheric circulation system, land surface physical system, ocean circulation, and sea ice system. However, as research on global climate change intensified from the end of the last century to the beginning of this century, the ESM was developed and expanded its scope to encompass the Earth's surface ecological and environmental systems, including terrestrial and marine ecosystems, atmospheric chemistry, aerosols, and more.

At present, the ESM is a mathematical model that accounts for the physical, chemical, and biological processes occurring in each of the Earth's spheres, as well as the material and energy exchanges between them. These processes are solved using numerical calculation methods and compiled into a comprehensive computational program on a large scale (see Fig. 11.4). By utilizing the ESM, researchers are now capable of simulating interactions between ocean ecosystems and GHGs, understanding the climate effects of atmospheric chemicals like ozone and dust, investigating the carbon cycle across the atmosphere, ocean, and surface, and exploring the impacts of GHGs on the upper atmosphere.

Example 2:

How can ESM be employed to address air pollution and climate change issues?

Answer:

1. **Simulating air pollutant transport and transformation:** ESMs can simulate the transport and transformation of air pollutants in the atmosphere, capturing the complex interactions between emissions, atmospheric chemistry, meteorology, and land surface processes. By incorporating emissions inventories and atmospheric chemistry modules, ESMs can help identify the sources, pathways, and fate of air pollutants, aiding in the development of effective mitigation strategies.
2. **Assessing air quality impacts:** ESMs can be used to assess the impacts of air pollution on air quality and human health. By simulating pollutant concentrations, ESMs can provide information on the spatial and temporal distribution of pollutants, helping to identify areas of high pollution levels and vulnerable populations. This information can guide policy-making and the implementation of targeted measures to improve air quality and protect human health.
3. **Investigating feedback mechanisms:** ESMs enable the study of feedback mechanisms between air pollution and climate change. They can capture the interactions between atmospheric composition, aerosols, radiation, and climate dynamics. ESMs can help assess how air pollutants, such as aerosols or GHGs, affect climate variables, including temperature, precipitation, and atmospheric circulation, and how climate change influences air quality.
4. **Evaluating emission reduction scenarios:** ESMs can be employed to evaluate the effectiveness of emission reduction scenarios in mitigating both air pollution and climate change. By simulating different emission scenarios, ESMs can estimate the impacts of policy measures or technological interventions on air pollutant concentrations, GHG concentrations, and climate variables. This information is crucial for guiding decision-making and identifying the most effective strategies for achieving both clean air and climate goals.

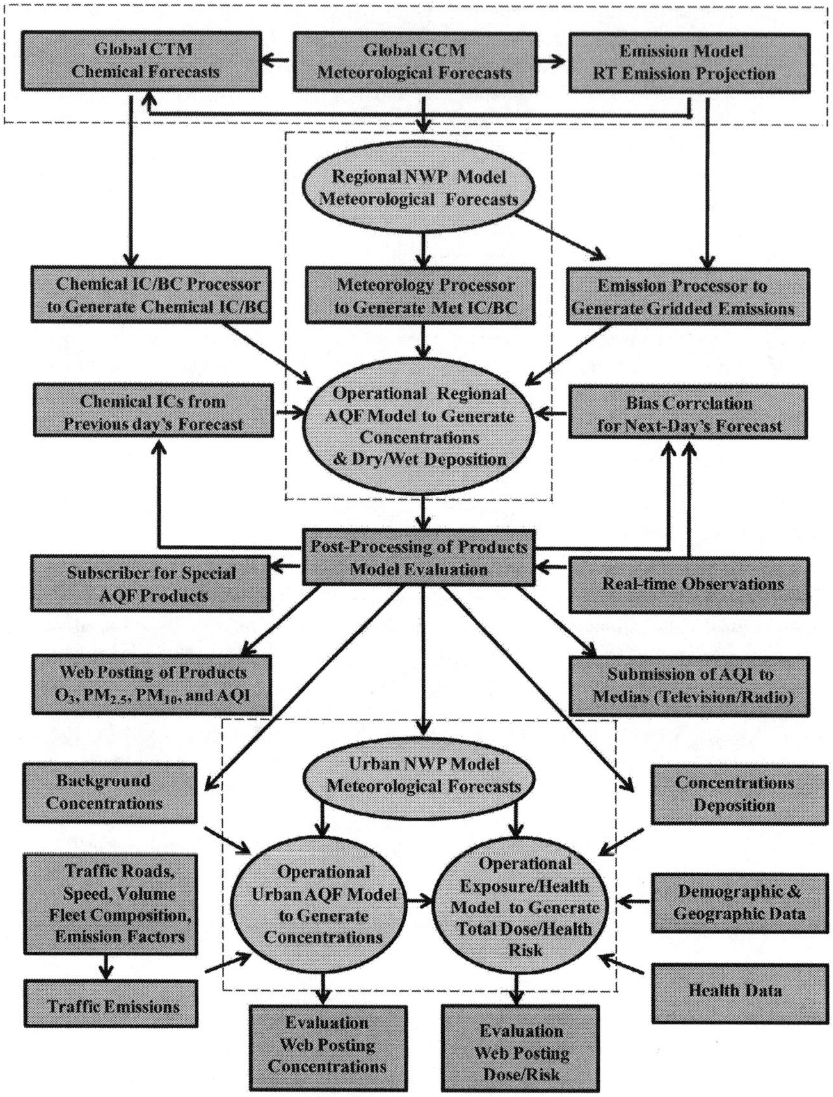

FIGURE 11.3 Anautomated real-time air quality forecasting system from global to urban scales. Global, regional, and urban scales model systems are illustrated as offline-coupled meteorology-chemistry models. It is important to note that these systems have the capability to be implemented as online-coupled models as well. *From Zhang, Y., Bocquet, M., Mallet, V., Seigneur, C. & Baklanov, A. (2012). Real-time air quality forecasting, part I: History, techniques, and current status.* Atmospheric Environment, 60, *632–655. https://doi.org/10.1016/j.atmosenv.2012.06.031.*

5. **Studying climate-air pollution co-benefits:** ESMs allow the assessment of co-benefits between climate change mitigation and air pollution control measures. By exploring the synergies and trade-offs between reducing GHG emissions and air pollutant emissions, ESMs can provide insights into strategies that simultaneously address both challenges. This integrated approach can lead to more sustainable and efficient solutions, promoting cleaner air and mitigating climate change.
6. **Enhancing policy-making and international cooperation:** ESMs provide valuable information for policy-makers, facilitating evidence-based decision-making and supporting the development of air quality and climate policies. ESMs contribute to international cooperation by providing a common framework for modeling and evaluating the impacts of air pollution and climate change at regional and global scales. They facilitate the exchange of scientific knowledge and enable collaboration in addressing these interconnected challenges.

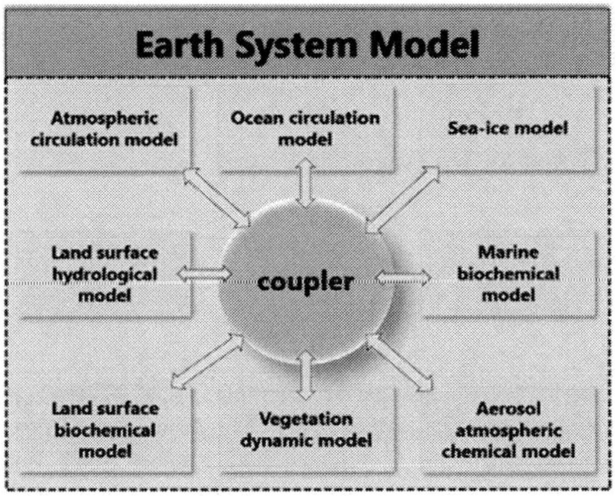

FIGURE 11.4 A schematic figure of the structure of Earth system model (ESM). The ESM is a mathematical model that accounts for the physical, chemical, and biological processes occurring in each of the Earth's spheres, as well as the material and energy exchanges between them.

11.3.3 Predictions and projections

The development of models at various scales has greatly supported weather, climate, and air quality predictions. For regional weather and air quality forecasts, mesoscale meteorological and chemical models are widely employed. ESMs play a crucial role in projecting future climate and air quality scenarios. The community Earth system model (CESM) is one extensively used ESM. It was established and operated by the National Center for Atmospheric Research. CESM has garnered prominence for its inclusion in reports published by the Intergovernmental Panel on Climate Change (IPCC). By employing ESMs, researchers can assess future climate and air quality outcomes based on different emission scenarios for GHGs and pollutants. This allows scientists to understand how changes in ocean and atmospheric circulation will impact regional climates in the coming decades, considering the interactions between climate and air pollution. Additionally, ESMs enable the investigation of how climate change and aerosol feedback will affect the frequency and intensity of tropical cyclones and other extreme events. These predictions and projections contribute to the development of relevant emission reduction measures and climate policies.

11.4 Impacts of air quality on weather

11.4.1 Impacts of air quality on weather processes

Air quality has direct and indirect impacts on various weather conditions, such as monsoon circulation, heat waves, drought, extreme precipitation, and tornadoes (Kinney, 2008). Aerosols, as a prominent air pollutant, play a significant role in influencing weather conditions. They scatter, reflect, and absorb shortwave radiation (SWD), thereby reducing the amount of downwelling solar radiation that reaches the ground. This results in a lower surface sensible heat flux and a decrease in near-ground air temperature, affecting other meteorological variables. Apart from cooling down the surface, absorbing aerosols can also contribute to atmospheric warming. The impacts of aerosols on variables such as 2 m air temperature, 2 m water vapor, and 10 m wind speed have been extensively studied at urban and regional scales. Topic 3 of the Model Inter-Comparison Study for Asia (MICS-Asia) focused on the evaluation of aerosol pollution events in the North China Plain using online coupled air quality models. Additionally, it aimed to assess the significance of aerosol radiative feedback. The study revealed reductions in near-surface temperature ranging from 0.3°C to 1.6°C and decreases in 2 m water vapor ranging from 0.07 to 0.29 $g\,kg^{-1}$ during severe haze days in the Beijing–Tianjin–Hebei region (Gao et al., 2020).

Aerosols have the capacity to impact precipitation through two main mechanisms. Firstly, suspended particles in the atmosphere can exert radiative effects, altering the distribution of solar radiation and consequently influencing atmospheric heating, circulation patterns, and precipitation patterns (Creamean et al., 2013; Li et al., 2011; Tao et al., 2012). Secondly, aerosols can interfere with and modify the formation of clouds and precipitation, thereby affecting the availability and characteristics of CCN and IN, ultimately influencing the formation of clouds and rainfall (Khain et al., 2008). The direct ARE results in a reduction of the solar radiation reaching the ground, resulting in a decrease in surface temperatures. This can enhance atmospheric stability and decline convection potential, creating unfavorable conditions for cloud formation.

Apart from surface cooling, absorbing aerosols can also warm the atmosphere. Atmospheric warming can decrease the number of low-level clouds by enhancing the evaporation of cloud droplets. On the other hand, warming can trigger updrafts that enhance water vapor convergence at lower elevations, thereby increasing precipitation. The potential warming resulting from increased precipitation can trigger feedback processes within the large-scale circulation system. These feedback processes can further amplify the initial response of aerosols to warming, ultimately resulting in intensified rainfall. In addition, aerosols have the capability to raise the amount of CCN, which in turn increases cloud volume and decreases coalescence and collision rates, resulting in a decrease in precipitation. However, when there is an upward movement in moist and warm air, the reduction in coalescence and collision of cloud droplets can result in the production of supercooled droplets. Latent heat release during freezing and deep melting at high altitudes enhances upward heat transport in polluted clouds and energizes deep convection. Consequently, aerosols can contribute to enhanced convection. The influences of aerosols on rainfall can vary depending on large-scale environmental conditions and dynamic feedback processes, resulting in positive, negative, or mixed impacts.

Aerosols also influence tropical cyclones and tornadoes (Rosenfeld & Bell, 2011; Rosenfeld et al., 2012). Extreme weather conditions are influenced by both dynamic and thermodynamic conditions in the atmosphere. Aerosols have the ability to modulate the thermodynamic state of the atmosphere, thereby affecting these events. The size of tropical cyclones is largely determined by sea surface temperature (SST), as the release of latent heat from the ocean serves as the most important energy source of these storms. On the other hand, aerosols, acting as CCN or IN, exert influence on the intensities of storms by regulating adiabatic heating and the thermodynamic structure of the atmosphere. Tornadoes are commonly associated with supercell storms, where the updraft continues for extended periods and is accompanied by a shifting downdraft and rainfall shaft caused by significant wind shear. In contrast, multicell storms exhibit short-lived cloud growth and the replacement of updrafts with downdrafts. Aerosols transit a multicell storm into a supercell storm through two mechanisms. Firstly, aerosols can delay the onset of rainfall in the upward air and cause rainfall in farther downwind regions. Consequently, the descending cold air must travel a longer distance upwind before undercutting the updraft and terminating it. Secondly, aerosols weaken coalescence, reducing the amount of precipitation particles. These particles have the potential to grow larger, leading to increased hail formation due to reduced competition for cloud water. Although the precipitation intensity remains unchanged, a decrease in the number of smaller hydrometeors affects several factors, including the overall surface area, evaporation amount, and subsequent cooling. This reduction in negative buoyancy of the air weakens its tendency to propagate upwind and undermine the updraft, ultimately resulting in the termination of the updraft.

Monsoons are large-scale wind systems characterized by significant seasonal changes in wind direction due to temperature differences between continents and adjacent oceans. In the past, air pollution and monsoon research were regarded as distinct issues, with little overlap or integration between the two fields of study. However, recent studies have highlighted the intrinsic linkages between these two phenomena and the need to study them jointly (Lau et al., 2008, 2009; Li et al., 2016). An intriguing observation is that major monsoon systems around the world are predominantly found in regions that often encounter significant episodes of aerosol pollution. These regions include areas with high levels of pollution events in East and South Asia, as well as regions affected by dust storms and biomass burning in Africa. Aerosols can influence the onset, evolution, intensity, and rainfall patterns of monsoons through radiative forcing and microphysical effects. On a continental scale, aerosols play a role in diminishing surface insolation and weakening the land-ocean thermal gradient, thereby impeding the formation of monsoons. On a local scale, the radiative effects of aerosols bring about changes in the stability of thermodynamic and convective potential of the lower atmosphere. Consequently, these changes lead to declined temperatures, enhanced atmospheric stability, and suppressed atmospheric circulations. Aerosols also have the capacity to modify the thermodynamic conditions of the air, which governs cloud formation, convection, and precipitation, by acting as CCN or IN. Notably, absorbing aerosols like BC and desert dust in monsoon regions in Asia regulate the evolution of the monsoon system. They can initiate dynamical feedback processes that intensify the early monsoon and influence its subsequent evolution.

Ultrafine particles (UFPs), acting as CCN, play a significant role in changing rainfall patterns. Recent observations of weather extremes, such as floods and droughts, might be attributed to changing anthropogenic UFP emissions. Studies have reported that UFPs substantially affect cloud physics and, consequently, on weather (Junkermann & Hacker, 2018). Aircraft measurements have confirmed an increase in particle number emissions despite a decrease in coarse fine dust concentration, attributing it to the combustion of fossil fuels in exhaust gas cleaning systems.

11.4.2 Impacts of air quality on numerical weather prediction

Numerical Weather Prediction (NWP) is an essential tool for weather forecasting, and its accuracy significantly determines the quality of forecast results. In recent years, aerosols have been highlighted as one of the largest contributions to uncertainty in NWP. Studies have shown that aerosols can introduce biases in temperature forecasts influenced by atmospheric aerosols. For example, (Huang & Ding, 2021) compared observations with the Global Forecast System prediction and found a monthly-average of $\pm$ 1.5°C bias in the 24-hour forecast of air temperature because of high aerosol loadings.

Copernicus Atmosphere Monitoring Service (CAMS) operates as part of the Copernicus Earth observation program initiated by the European Union. The service is managed by the European Centre for Medium-Range Weather Forecasts (ECMWF) and is responsible for the continuous monitoring of air quality. It achieves this through the utilization of satellite and ground-based observations, as well as advanced numerical models. The global CAMS forecasting system, integrated into the ECMWF's Integrated Forecasting System, includes prognostic fields for aerosols and O_3 in its radiation scheme to account for their interactions with shortwave and longwave radiative transfer.

To specify the influence of prognostic dust on meteorological forecasts, particularly 2 m temperature predictions, a comparison is made with a control forecast that employs an aerosol climatology derived from the CAMS re-analysis. Notably, it was demonstrated that the transport of Saharan dust to Europe in spring 2022 resulted in a cooling effect of up to 4 K in the 2 m temperature forecast. This improvement in forecast accuracy was evident when compared to other observations (Flemming et al., 2022).

The Working Group on Numerical Experimentation (WGNE) focuses on the development of atmospheric circulation models for prediction and studies of weather and climate across various timescales. It conducted the WGNE Subseasonal to Seasonal Global Atmosphere Watch (WGNE-S2S-GAW) Aerosol Project to assess the impact of aerosols on NWP and subseasonal to seasonal (S2S) predictions. The findings from the project demonstrated that dust aerosols had discernible impacts on surface temperature and winds, despite the synoptic conditions not being significantly influenced by the radiative forcing of predicted aerosols (Colarco, 2016; Rémy et al., 2015).

11.5 Impacts of air quality on climate

Air quality and climate are closely interconnected due to the emission of chemical species that contribute to air pollution and climate change. These emissions often occur simultaneously, leading to changes in both air quality and climate. Short-lived climate pollutants (SLCPs) are a group of compounds with relatively short atmospheric lifetimes, typically less than two decades. They include aerosols (such as sulfates, nitrates, carbon aerosols, and mineral dust), as well as reactive gases (such as methane, O_3, NO_x, CO, VOCs, SO_2, and ammonia).

SLCPs interact with meteorological variables and solar radiation, influencing climate by either warming or cooling the Earth's surface. Warming SLCPs, such as ozone and methane, act as GHGs that absorb energy and contribute to warming. On the other hand, cooling SLCPs primarily consist of aerosol particles, such as sulfates, nitrates, and organic aerosols, which reflect incoming SWD and cool down the ground.

The IPCC's Sixth Assessment Report (AR6) highlighted the effects of changes in aerosol loadings since 1850 on global mean surface temperature (Masson-Delmotte et al., 2021). The report found a 0.66°C $\pm$ 0.51°C decrease in the ensemble-mean global average surface temperature over the recent past (1995–2014), with larger temperature decreases in the Northern Hemisphere (0.97°C $\pm$ 0.54°C) compared to the Southern Hemisphere (0.34°C $\pm$ 0.2°C). Negative responses of zonal-mean temperature were observed at all latitudes. The values become more negative with increasing latitude, with the most significant decrease observed around northern polar latitudes (around 2.7°C).

SLCPs can also affect precipitation and other climate variables. For instance, the deposition of aerosols on snow and ice surfaces may reduce surface albedo, resulting in enhanced absorption of solar radiation and positive radiative forcing. Furthermore, the deposition of aerosols and their dissolved nutrients in terrestrial and marine ecosystems can alter biogeochemical processes, further influencing climate dynamics.

Overall, the complex interactions between SLCPs, air quality, and climate underscore the need to address both air pollution and climate change as interconnected challenges. By mitigating SLCP emissions, it is possible to simultaneously improve air quality and reduce the rate of near-term climate warming, providing co-benefits for human health, ecosystems, and the climate system as a whole.

11.6 Impacts of climate on air quality

11.6.1 Impacts of future climates on air quality, human health and food security

Air pollution is influenced by both high emissions and unfavorable weather conditions. Temperature, precipitation, relative humidity, and wind play crucial roles in the generation, transport, and accumulation of air pollutants. Future climate change can significantly impact air quality through changes in natural emissions, meteorological conditions, and chemical reactions.

Numerous studies have shown that air pollution, including extreme events, is likely to increase in many densely populated areas under a warming climate. Specific regions can exhibit heightened sensitivity to significant feedback from natural sources of ozone and aerosols, including wildfires, dust, and biological precursors. Additionally, variations in chemical and depositional sinks can also contribute to the overall feedback dynamics in these areas. For instance, southern Europe may experience heightened sensitivity to climate change due to increases in BVOCs emissions and decreases in dry deposition associated with plant responses to heat stress.

Global and regional model projections have been used to assess the response of O_3 and $PM_{2.5}$ to future climate change. The IPCC's AR6 report indicates that a warmer climate is expected to increase surface O_3 levels regionally, with a magnitude ranging from 0.2 to 2 parts per billion per degree Celsius (ppb/°C) in regions with high O_3 precursor emissions. However, the projected warming is only anticipated to cause an approximately 0.21 $\mu g\,m^{-1}$ increase (in 2100 for the RCP8.5 scenario) in global mean $PM_{2.5}$ concentrations.

These climate-induced air quality penalties can have significant implications for public health. Studies have estimated the additional mortalities associated with increased $PM_{2.5}$ and O_3 exposure due to a population-weighted average concentration increase of 3% and 4% respectively. For instance, one study reported that such concentration increases would lead to an additional 12,100 deaths $year^{-1}$ from $PM_{2.5}$ exposure and 8900 deaths per year from O_3 exposure (Hong et al., 2019). Another study suggested that reducing the global average temperature by about 0.5°C by 2050 could help avoid 0.7–4.7 million premature deaths annually from ambient air pollution and increase annual crop yields by 30–135 million metric tons through O_3 reductions (Shindell et al., 2012).

These results highlight the benefits of addressing climate change and air pollution together to mitigate the adverse impacts on public health and the environment. Implementing measures to cut down GHG and air pollutant emissions can yield substantial co-benefits by improving air quality, protecting public health, and enhancing agricultural productivity.

11.6.2 Changes in emissions, biogenic, and fires

Climate plays a crucial role in regulating air quality, particularly in influencing source emissions (Lu et al., 2019; Tegen & Schepanski, 2018). Natural emissions are significantly influenced by climate factors, including ambient temperature and solar radiation. Plants emit large amounts of volatile compounds such as NO, CO, and nonmethane VOCs. BVOCs contribute significantly to the production of tropospheric O_3 and secondary aerosols. Strong solar radiation, often accompanied by high ambient temperature, triggers various physiological, biochemical, and molecular responses in organisms. It enhances vegetation transpiration and accelerates BVOC emission rates (Fig. 11.5). However, these conditions can also lead to cell and tissue damage, resulting in mass vegetation loss. The impact of these stresses on VOC emissions can have additive or antagonistic effects, amplifying or suppressing plant responses.

Aerosols originate from various natural sources, including mineral dust released from desert surfaces, marine aerosols emitted by oceans, and carbonaceous aerosols produced by wildfires. The characteristics of these aerosols are influenced by both surface properties and atmospheric conditions. For instance, dust emissions from bare land are influenced by soil moisture, which, in turn, is highly affected by the intensity of solar radiation. The impacts of climate on wildfires are strong but complex, as climate determines vegetation's primary productivity and, consequently, the supply and availability of biofuel. High temperatures and drought-induced by intense solar radiation increase the risk of

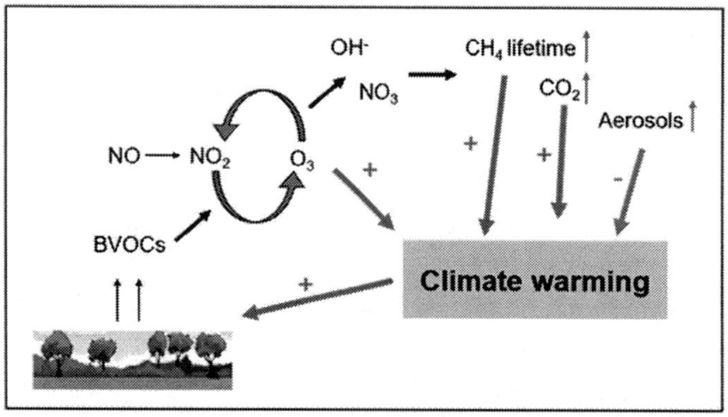

FIGURE 11.5 Effects of increased biogenic volatile organic compound (BVOC) emissions on atmospheric chemistry and climate. Climate warming enhances vegetation transpiration and accelerates BVOC emission rates, regulating chemical reactions of greenhouse gases and aerosols, which will have feedback on the climate itself.

wildfires and enhance the emission of natural aerosols. Additionally, artificial structures like buildings and streets can also emit VOCs, particularly under high temperatures and strong solar radiation conditions.

Example 3:

Wildfires have become one major natural disaster because of global warming. How do wildfires influence air quality and what are their long-term effects on weather and climate?

Answer:

Wildfires have significant impacts on air quality, with the release of pollutants such as particulate matter (PM), carbon monoxide, nitrogen oxides, and VOCs leading to poor air quality, reduced visibility, and health risks. Smoke particles, including fine PM ($PM_{2.5}$), can be transported over long distances, affecting air quality far from the fire source and contributing to respiratory health issues. Wildfires also influence atmospheric chemistry, leading to the formation of secondary pollutants like ozone and secondary organic aerosols. In terms of climate, wildfires can create localized convective processes, impacting weather patterns, and can release large amounts of carbon dioxide and other GHGs, contributing to climate change. The loss of vegetation cover and carbon release from burning vegetation can further affect surface albedo, carbon storage, and ecosystem dynamics, influencing long-term climate regulation. Monitoring and understanding these impacts are vital for effective wildfire management, air quality control, and climate change mitigation.

11.6.3 Co-occurrences of pollution and heat waves

Heat extremes and air pollution often occur simultaneously due to shared atmospheric patterns, such as slow-moving high-pressure systems. High air temperatures, strong solar insolation, small rainfall, and weak wind, create a favorable environment for the regional accumulation of pollutants and heat.

During heat extremes, multiple interactions and feedback mechanisms further exacerbate air pollution. For example, the accelerated emissions of BVOCs contribute to the generation of secondary organic aerosols and O_3. Increased energy demands for air-conditioning lead to elevated anthropogenic emissions. Additionally, heat-related drought can increase dust emissions.

Evidence suggests that the health impacts of the combined effects of heat extremes and air pollution can be more severe than the sum of their individual effects. This is of particular concern under the backdrop of global warming. Climate-driven shifts in meteorological conditions, such as heatwaves or stagnation events, are anticipated to enhance the occurrence of severe air pollution episodes in heavily polluted regions. These changes in weather patterns create conditions that are more conducive to the development and intensification of extreme levels of air pollution.

Numerous studies worldwide have reported increased co-occurrences of pollution and heat waves during the early decades of this century. Machine-learning models and numerical predictions indicate that under climate change, the probability of these co-occurrences will continue to increase, especially in densely populated regions. (Lelieveld et al., 2014; Xu et al., 2020).

11.7 Applying an understanding of these interactions for a better future

Combining strategies to address air pollution and global warming can yield significant advantages in mitigating the adverse impacts of interactions between weather, climate, and air pollution (Harlan & Ruddell, 2011). Many measures taken to tackle climate change also bring co-benefits in terms of improving air quality. For instance, the combustion of fossil fuels, a major contributor to both air pollutants and GHGs, can be addressed through technological solutions that simultaneously reduce emissions of air pollutants such as PM, sulfur dioxide, and nitrous oxide, as well as CO_2, by transitioning to renewable energy sources. Studies indicate that reducing GHG emissions not only helps combat climate change but also has the potential to prevent millions of premature deaths by reducing air pollution-related health risks (West et al., 2013).

Modeling results have shown that maintaining a global temperature increase within 2°C would yield significant air quality benefits. In the context of the United States, the reduction of GHG emissions leading to improved air quality has the potential to avert an estimated 4.5 million premature deaths, prevent 3.5 million hospitalizations and emergency room visits, and result in a decrease of 300 million lost workdays (Shindell et al., 2021). Another study provides additional support to the idea that the benefits of enhancing air quality outweigh the costs associated with mitigating climate change (Fang et al., 2013). Through a comparison of future scenarios with rigorous climate policies to those without, researchers found that the reduction of GHG emissions would effectively prevent a significant number of premature deaths attributable to air pollution on a global scale. The estimated health benefits resulting from this approach range from \$50 to \$380 ton^{-1} of CO_2 reduced. These benefits surpass the marginal costs of GHG mitigation in 2030 and 2050 and are within the lower cost range in 2100 (Barbier, 2014).

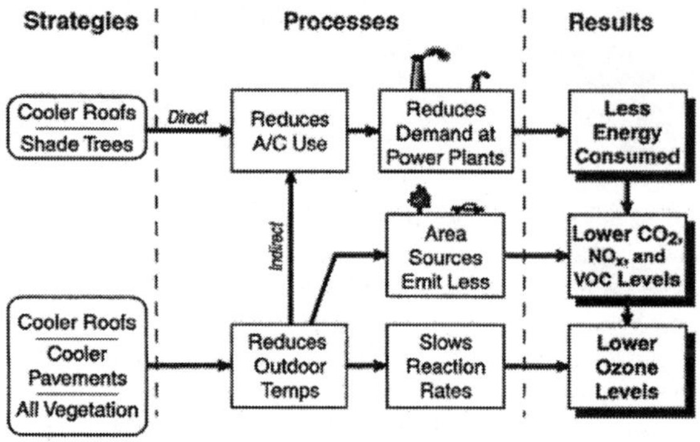

FIGURE 11.6 Methodology to analyze the impact of shade trees, cool roofs, and cool pavements on energy use and air quality. Cool surfaces (cool roofs and cool pavements) and urban trees can have a substantial effect on urban air temperature and, hence, can reduce cooling energy use and smog. *From Akbari, H, Pomerantz, M & Taha, H. (2001). Cool surfaces and shade trees to reduce energy use and improve air quality in urban areas.* Solar Energy, 70*(3), 295–310. https://doi.org/10.1016/s0038-092x(00)00089-x.*

The city is a major source of anthropogenic air pollutants and heat, and is also the region most affected by air quality and climate interactions. The concept of sustainable cities is gaining traction. Measures, like increasing the area of urban green space and using a range of roofing measures (i.e., cool roofs and green roofs), can be taken within cities to mitigate the penalties of such interactions (Fig. 11.6; Akbari et al., 2001). Cool roof and green roof strategies are feasible measures that have been formally introduced in some major cities in recent decades. Cool roofs use high albedo materials to reflect more sunlight back into the space, thereby absorbing less heat than standard roofs, while green roofs use vegetation to cool the built environment through evaporation. Both cool roofs and green roofs reduce air temperatures near surfaces and cut the energy consumption of air conditioning systems. Both field experiments and numerical model simulations have shown that the use of cool and green roofs can reduce maximum ambient temperatures by 1°C–5°C and reduce energy consumption by 10%–50% (Yang et al., 2015). The presence of green spaces in urban areas, ranging from individual green walls to expansive urban forests, has been linked to various benefits such as alleviating heat stress, mitigating UHIs, and reducing air pollution (Száraz, 2014). However, it is important to consider specific factors when planting trees in urban street canyons, such as wind direction, street orientation, and building height, as these factors can create conditions that potentially hinder the dispersion of air pollutants within the canyon, leading to their accumulation. Consequently, the advantages of green spaces in street canyons are site-specific and require careful consideration. Additionally, the emission of BVOCs from vegetation should be evaluated, particularly in areas with high traffic volumes.

11.8 Questions

1. What are the primary sources of air pollution and how do they impact air quality and climate?
2. How does air quality differ between urban and rural areas, and what are the factors contributing to these differences?
3. How do GHG emissions contribute to both climate change and air pollution?
4. What are some effective strategies and technologies for improving air quality and mitigating climate change simultaneously?
5. Aerosol optical thickness can be calculated by combining extinction coefficient and aerosol layer thickness. The extinction coefficient (σ) can be calculated using the following equation: σAerosol optical thickness can be calculated by combining the extinction coefficient and aerosol layer thickness. The extinction coefficient (σ) can be calculated using the following equation: $\sigma = \pi \times \text{particle size}^2 \times \text{number concentration}$. Assume that the thickness of the aerosol layer is 5 km, the particle size is 0.1 mm, and the number concentration is $5000\,\text{cm}^{-3}$ Calculate the optical thickness of the aerosol layer.
6. The ARF due to a mixture of two aerosol species can be calculated using the equation: $\text{ARF} = \sum(\alpha_i \times \Delta R_i)$, where α_i is the aerosol radiative efficiency and ΔR_i is the change in radiative flux. If the aerosol radiative efficiency α_1 is $0.3\ \text{W m}^{-2}$ and α_2 is $0.5\ \text{W m}^{-2}$, and the changes in radiative flux ΔR_1 and ΔR_2 are -1 and $-2\ \text{W m}^{-2}$, respectively, what is the ARF?
7. The direct radiative effect (DRE) of an aerosol species is given by the equation: $\text{DRE} = \alpha \times C$, where α is the aerosol radiative efficiency and C is the aerosol concentration. If the aerosol radiative efficiency α is $0.3\ \text{W m}^{-3}$ and the aerosol concentration C is 1000 particles cm^{-3}, what is the DRE?

Solutions to the end-of-chapter questions will be available at: https://www.elsevier.com/books-and-journals/book-companion/9780128225912.

References

Akbari, H., Pomerantz, M., & Taha, H. (2001). Cool surfaces and shade trees to reduce energy use and improve air quality in urban areas. *Solar Energy*, *70*(3), 295–310.

Akimoto, H. (2003). Global air quality and pollution. *Science (New York, N.Y.)*, *302*, 1716–1719.

Andrews, E., Sheridan, P. J., Fiebig, M., McComiskey, A., Ogren, J. A., Arnott, P., Covert, D., Elleman, R., Gasparini, R., Collins, D., & Jonsson, H. (2006). Comparison of methods for deriving aerosol asymmetry parameter. *Journal of Geophysical Research: Atmospheres*, 111.

Baklanov, A., Molina, L. T., & Gauss, M. (2016). Megacities, air quality and climate. *Atmospheric Environment*, *126*, 235–249.

Barbier, E. B. (2014). Climate change mitigation policies and poverty. *Wiley Interdisciplinary Reviews: Climate Change*, *5*, 483–491.

Cao, C., Lee, X., Liu, S., Schultz, N., Xiao, W., Zhang, M., & Zhao, L. (2016). Urban heat islands in China enhanced by haze pollution. *Nature Communication*, *7*, 12509.

Charlson, R. J., Langner, J., Rodhe, H., Leovy, C. B., & Warren, S. G. (1991). Perturbation of the northern hemisphere radiative balance by backscattering from anthropogenic sulfate aerosols. *Tellus A: Dynamic Meteorology and Oceanography, 43*, 152–163.

Charlson, R. J., Schwartz, S. E., Hales, J.M., Cess, R. D., Coakley Jr, J. A., Hansen, J. E., & Hofmann, D. J. (1992). Climate forcing by anthropogenic aerosols. *Science, 255*, 423–430.

Colarco, P. (2016). Aerosol complexity and implications for predictability and short-term forecasting. In *ECMWF annual seminar 2016*,

Creamean, J. M., Suski, K. J., Rosenfeld, D., Cazorla, A., Demott, P. J., Sullivan, R. C., White, A. B., Ralph, F. M., Minnis, P., Comstock, J. M., Tomlinson, J. M., & Prather, K. A. (2013). Dust and biological aerosols from the Sahara and Asia influence precipitation in the western US. *Science (New York, N.Y.)*, *339*, 1572–1578.

Ding, A. J., Huang, X., Nie, W., Sun, J. N., Kerminen, V. M., Petäjä, T., Su, H., Cheng, Y. F., Yang, X. Q., Wang, M. H., Chi, X. G., Wang, J. P., Virkkula, A., Guo, W. D., Yuan, J., Wang, S. Y., Zhang, R. J., Wu, Y. F., Song, Y., ... Fu, C. B. (2016). Enhanced haze pollution by black carbon in megacities in China. *Geophysical Research Letters*, *43*, 2873–2879.

Fan, J., Wang, Y., Rosenfeld, D., & Liu, X. (2016). Review of aerosol–cloud interactions: Mechanisms, significance, and challenges. *Journal of the Atmospheric Sciences*, *73*, 4221–4252.

Fang, Y., Mauzerall, D. L., Liu, J., Fiore, A. M., & Horowitz, L. W. (2013). Impacts of 21st century climate change on global air pollution-related premature mortality. *Climatic Change*, *121*, 239–253.

Fiore, A. M., Naik, V., Spracklen, D. V., Steiner, A., Unger, N., Prather, M., Bergmann, D., Cameron-Smith, P. J., Cionni, I., Collins, W. J., Dalsoren, S., Eyring, V., Folberth, G. A., Ginoux, P., Horowitz, L. W., Josse, B., Lamarque, J. F., MacKenzie, I. A., Nagashima, T., ... Zeng, G. (2012). Global air quality and climate. *Chemical Society Reviews*, *41*, 6663–6683.

Flemming, J., Remy, S., Inness, A., Ades, M., Parrington, M., Garrigues, S., Hogan, R., Haiden, T. & Engelen, R. (2022). The impact of the dust transport events over Europe in March 2022 on the weather forecast with the ECMWF model. In *Copernicus meetings*.

Fuzzi, S., Baltensperger, U., Carslaw, K., Decesari, S., van der Gon, H. D., Facchini, M. C., Fowler, D., Koren, I., Langford, B., Lohmann, U., Nemitz, E., Pandis, S., Riipinen, I., Rudich, Y., Schaap, M., Slowik, J. G., Spracklen, D. V., Vignati, E., Wild, M., ... Gilardoni, S. (2015). Particulate matter, air quality and climate: Lessons learned and future needs. *Atmospheric Chemistry and Physics*, *15*, 8217–8299.

Gao, M., Han, Z., Tao, Z., Li, J., Kang, J. E., Huang, K., Dong, X., Zhuang, B., Li, S., Ge, B., Wu, Q., Lee, H. J., Kim, C. H., Fu, J. S., Wang, T., Chin, M., Li, M., Woo, J. H., Zhang, Q., ... Carmichael, G. R. (2020). Air quality and climate change, Topic 3 of the model inter-comparison study for Asia Phase III (MICS-Asia III)—Part 2: Aerosol radiative effects and aerosol feedbacks. *Atmospheric Chemistry and Physics*, *20*, 1147–1161.

Getter, K. L., & Rowe, D. B. (2006). The role of extensive green roofs in sustainable development. *HortScience: A Publication of the American Society for Horticultural Science*, *41*, 1276–1285.

Grell, G., & Baklanov, A. (2011). Integrated modeling for forecasting weather and air quality: A call for fully coupled approaches. *Atmospheric Environment*, *45*, 6845–6851.

Guenther, A., Geron, C., Pierce, T., Lamb, B., Harley, P., & Fall, R. (2000). Natural emissions of non-methane volatile organic compounds, carbon monoxide, and oxides of nitrogen from North America. *Atmospheric Environment*, *34*, 2205–2230.

Harlan, S. L., & Ruddell, D. M. (2011). Climate change and health in cities: Impacts of heat and air pollution and potential co-benefits from mitigation and adaptation. *Current Opinion in Environmental Sustainability*, *3*, 126–134.

Hong, C., Zhang, Q., Zhang, Y., Davis, S. J., Tong, D., Zheng, Y., Liu, Z., Guan, D., He, K., & Schellnhuber, H. J. (2019). Impacts of climate change on future air quality and human health in China. *Proceedings of the National Academy of Sciences*, *116*, 17193–17200.

Huang, X., & Ding, A. (2021). Aerosol as a critical factor causing forecast biases of air temperature in global numerical weather prediction models. *Science Bulletin*, *66*, 1917–1924.

Jacob, D. J., & Winner, D. A. (2009). Effect of climate change on air quality. *Atmospheric Environment*, *43*, 51–63.

Johnson, B., Shine, K., & Forster, P. (2004). The semi-direct aerosol effect: Impact of absorbing aerosols on marine stratocumulus. *Quarterly Journal of the Royal Meteorological Society*, *130*, 1407–1422.

Junkermann, W., & Hacker, J. M. (2018). Ultrafine particles in the lower troposphere: major sources, invisible plumes, and meteorological transport processes. *Bulletin of the American Meteorological Society*, *99*, 2587–2602.

Khain, A., Benmoshe, N., & Pokrovsky, A. (2008). Factors determining the impact of aerosols on surface precipitation from clouds: An attempt at classification. *Journal of the Atmospheric Sciences*, *65*, 1721–1748.

Kinney, P. L. (2008). Climate change, air quality, and human health. *American Journal of Preventive Medicine*, *35*, 459–467.

Kuniyal, J. C., & Guleria, R. P. (2019). The current state of aerosol-radiation interactions: A mini review. *Journal of Aerosol Science*, *130*, 45–54.

Lau, K.-M., Ramanathan, V., Wu, G.-X., Li, Z., Tsay, S. C., Hsu, C., Sikka, R., Holben, B., Lu, D., Tartari, G., Chin, M., Koudelova, P., Chen, H., Ma, Y., Huang, J., Taniguchi, K., & Zhang, R. (2008). The joint aerosol–monsoon experiment: A new challenge for monsoon climate research. *Bulletin of the American Meteorological Society*, *89*, 369–384.

Lau, W. K., Kim, K.-M., Hsu, C. N., & Holben, B. N. (2009). Possible influences of air pollution, dust-and sandstorms on the Indian monsoon. *World Meteorological Organization (WMO) Bulletin, 58*, 22.

Lelieveld, J., Hadjinicolaou, P., Kostopoulou, E., Giannakopoulos, C., Pozzer, A., Tanarhte, M., & Tyrlis, E. (2014). Model projected heat extremes and air pollution in the eastern Mediterranean and Middle East in the twenty-first century. *Regional Environmental Change, 14*, 1937–1949.

Li, Z., Lau, W. M., Ramanathan, V., Wu, G., Ding, Y., Manoj, M., Liu, J., Qian, Y., Li, J., Zhou, T., Fan, J., Rosenfeld, D., Ming, Y., Wang, Y., Huang, J., Wang, B., Xu, X., Lee, S. -S., Cribb, M., … Brasseur, G. P. (2016). Aerosol and monsoon climate interactions over Asia. *Reviews of Geophysics, 54*, 866–929.

Li, Z., Niu, F., Fan, J., Liu, Y., Rosenfeld, D., & Ding, Y. (2011). Long-term impacts of aerosols on the vertical development of clouds and precipitation. *Nature Geoscience, 4*, 888–894.

Li, Z., Guo, J., Ding, A., Liao, H., Liu, J., Sun, Y., Wang, T., Xue, H., Zhang, H., & Zhu, B. (2017). Aerosol and boundary-layer interactions and impact on air quality. *National Science Review, 4*, 810–833.

Lohmann, U., & Feichter, J. (2005). Global indirect aerosol effects: A review. *Atmospheric Chemistry and Physics, 5*, 715–737.

Lu, X., Zhang, L., & Shen, L. (2019). Meteorology and climate influences on tropospheric ozone: A review of natural sources, chemistry, and transport patterns. *Current Pollution Reports, 5*, 238–260.

Masson-Delmotte, V., Zhai, P., Pirani, A., Connors, S. L., Péan, C., Berger, S., Caud, N., Chen, Y., Goldfarb, L., Gomis, M. I., & Huang, M. (2021). Climate change 2021: The physical science basis. In *Contribution of working group I to the sixth assessment report of the intergovernmental panel on climate change*, 2.

Oke, T. R. (1973). City size and the urban heat island. *Atmospheric Environment (1967), 7*, 769–779.

Oke, T. R. (1982). The energetic basis of the urban heat island. *Quarterly Journal of the Royal Meteorological Society, 108*, 1–24.

Oum, K. W., Lakin, M., Dehaan, D. O., Brauers, T., & Finlayson-Pitts, B. J. (1998). Formation of molecular chlorine from the photolysis of ozone and aqueous sea-salt particles. *Science (New York, N.Y.), 279*, 74–76.

Ramanathan, V. C. P. J., Crutzen, P. J., Kiehl, J. T., & Rosenfeld, D. (2001). Aerosols, climate, and the hydrological cycle. *Science, 294*, 2119–2124.

Rémy, S., Benedetti, A., Bozzo, A., Haiden, T., Jones, L., Razinger, M., Flemming, J., Engelen, R., Peuch, V., & Thepaut, J. (2015). Feedbacks of dust and boundary layer meteorology during a dust storm in the eastern Mediterranean. *Atmospheric Chemistry and Physics, 15*, 12909–12933.

Rosenfeld, D., & Bell, T. L. (2011). Why do tornados and hailstorms rest on weekends? *Journal of Geophysical Research: Atmospheres*, 116.

Rosenfeld, D., Woodley, W. L., Khain, A., Cotton, W. R., Carrió, G., Ginis, I., & Golden, J. H. (2012). Aerosol effects on microstructure and intensity of tropical cyclones. *Bulletin of the American Meteorological Society, 93*, 987–1001.

Satheesh, S., & Moorthy, K. K. (2005). Radiative effects of natural aerosols: A review. *Atmospheric Environment, 39*, 2089–2110.

Shindell, D., Kuylenstierna, J. C. I., Vignati, E., Van Dingenen, R., Amann, M., Klimont, Z., Anenberg, S. C., Muller, N., Janssens-Maenhout, G., Raes, F., Schwartz, J., Faluvegi, G., Pozzoli, L., Kupiainen, K., Höglund-Isaksson, L., Emberson, L., Streets, D., Ramanathan, V., Hicks, K., … Fowler, D. (2012). Simultaneously mitigating near-term climate change and improving human health and food security. *Science (New York, N.Y.), 335*, 183–189.

Shindell, D., Ru, M., Zhang, Y., Seltzer, K., Faluvegi, G., Nazarenko, L., Schmidt, G. A., Parsons, L., Challapalli, A., Yang, L., & Glick, A. (2021). Temporal and spatial distribution of health, labor, and crop benefits of climate change mitigation in the United States. *Proceedings of the National Academy of Sciences, 118*(46), e2104061118.

Száraz, L. R. (2014). The impact of urban green spaces on climate and air quality in cities. *Geographical Locality Studies, 2*, 326–354.

Tao, W. K., Chen, J. P., Li, Z., Wang, C., & Zhang, C. (2012). Impact of aerosols on convective clouds and precipitation. *Reviews of Geophysics*, 50.

Tegen, I., & Schepanski, K. (2018). Climate feedback on aerosol emission and atmospheric concentrations. *Current Climate Change Reports, 4*, 1–10.

Vautard, R., Honore, C., Beekmann, M., & Rouil, L. (2005). Simulation of ozone during the August 2003 heat wave and emission control scenarios. *Atmospheric Environment, 39*, 2957–2967.

Vieno, M., Dore, A., Stevenson, D. S., Doherty, R., Heal, M. R., Reis, S., Hallsworth, S., Tarrason, L., Wind, P., Fowler, D., Simpson, D., & Sutton, M. A. (2010). Modelling surface ozone during the 2003 heat-wave in the UK. *Atmospheric Chemistry and Physics, 10*, 7963–7978.

Vinod, V., Varghese, S., & Kuriakose, B. (2002). Degradation behaviour of natural rubber—aluminium powder composites: effect of heat, ozone and high energy radiation. *Polymer Degradation and Stability, 75*, 405–412.

West, J. J., Smith, S. J., Silva, R. A., Naik, V., Zhang, Y., Adelman, Z., Fry, M. M., Anenberg, S., Horowitz, L. W., & Lamarque, J.-F. (2013). Co-benefits of mitigating global greenhouse gas emissions for future air quality and human health. *Nature climate change, 3*, 885–889.

Wu, J., Bei, N., Hu, B., Liu, S., Wang, Y., Shen, Z., Li, X., Liu, L., Wang, R., Liu, Z., Cao, J., Tie, X., Molina, L. T., & Li, G. (2020). Aerosol—photolysis interaction reduces particulate matter during wintertime haze events. *Proceedings of the National Academy of Sciences, 117*, 9755–9761.

Xing, J., Zhang, Y., Wang, S., Liu, X., Cheng, S., Zhang, Q., Chen, Y., Streets, D. G., Jang, C., Hao, J., & Wang, W. (2011). Modeling study on the air quality impacts from emission reductions and atypical meteorological conditions during the 2008 Beijing Olympics. *Atmospheric Environment, 45*, 1786–1798.

Xu, Y., Wu, X., Kumar, R., Barth, M., Diao, C., Gao, M., Lin, L., Jones, B., & Meehl, G. A. (2020). Substantial increase in the joint occurrence and human exposure of heatwave and high-PM hazards over South Asia in the mid-21st century. *AGU Advances, 1*, e2019AV000103.

Yang, J., Wang, Z.-H., & Kaloush, K. E. (2015). Environmental impacts of reflective materials: Is high albedo a 'silver bullet' for mitigating urban heat island? *Renewable and Sustainable Energy Reviews, 47*, 830–843.

Zhang, X., Cai, C., Hu, X. M., Gao, L., Xu, X., Hu, J., & Chen, H. (2022). Aerosols consistently suppress the convective boundary layer development. *Atmospheric Research, 269*, 106032.

Zhang, Y., Bocquet, M., Mallet, V., Seigneur, C., & Baklanov, A. (2012). Real-time air quality forecasting, part I: History, techniques, and current status. *Atmospheric Environment, 60*, 632–655.

Chapter 12

Indoor air quality

Prashant Kumar, Hamid Omidvarborna, Soma Sekhara Rao Kolluru, Abidemi Kuye, Yasemin Guler, Leonardo Yoshiaki Kamigauti and Huai-Wen Wu

Faculty of Engineering and Physical Sciences, Global Centre for Clean Air Research (GCARE), School of Sustainability, Civil and Environmental Engineering, University of Surrey, Guildford, United Kingdom

12.1 Introduction

An indoor environment is an enclosed space in which the quality of indoor air has major implications on the health of its occupants. Poor indoor air quality (IAQ) can generally cause headaches, dizziness, eye irritation, allergies and fatigue during short-term exposure and lung diseases such as asthma, chronic obstructive pulmonary disease (COPD) and lung cancer during long-term exposures. However, people exposed to indoor air pollutants (IAP) in different indoor environments do not usually have the same health response. For example, children are particularly more vulnerable to poor IAQ. Because their lungs are still developing and their airways are smaller, any inflammation can narrow down their airways and make breathing difficult. Likewise, adults over 65 suffering from lung/cardiovascular diseases are also among the highest risk as compared with healthy adults.

The indoor environment can be characterized as static or mobile (Section 12.2). The static environment includes IAQ in stationary spaces such as buildings, classrooms, offices, hospitals, and underground stations. While the mobile environment includes IAQ in mobile spaces such as cars, buses, trains, and ships. Both environments have different characteristics such as different types of IAP (Section 12.3), interactions (Section 12.4), exposure levels and health impacts (Section 12.6) that are affected by the outdoor environment in different ways. A holistic consideration of all these diverse factors is necessary to evaluate the reliable indoor health exposure assessment.

To further support the health assessment process, various analytical models are used. The modelling approaches have evolved over time. These models can help predict indoor air pollution events by using the contributing factors obtained from indoor monitoring. However, the regulations for indoor air pollution are still lacking. More efforts are needed by environmental legislators and policymakers to regulate IAPs similar to outdoor air pollutants.

The aim of this section is to provide an overview of various factors affecting the IAQ and the associated challenges in its modelling and to assess, mitigate, and predict exposure to IAP. It starts by introducing the different types of indoor environments (Section 12.2) and the sources of air pollutants that affect the IAQ (Section 12.3). The processes and interactions involved are discussed in Section 12.4. The efforts to monitor IAQ levels are discussed in Section 12.5 with the aim of understanding individual exposure levels (Section 12.6). The methods to model the IAQ are discussed in Section 12.7. Section 12.8 concludes with a few case studies summarising the methodologies and instruments used to gain an understanding of practically implementing the above steps in real-world applications. Lastly, the challenges associated with the legislation concerning the IAQ are discussed in Section 12.9.

12.2 Type of indoor microenvironments

Indoor environments contribute significantly to human exposure to air pollutants because people spend most of their time indoors. According to the US Environmental Protection Agency (USEPA), indoor air can be two to five times more polluted than outdoor air. This is because enclosed areas allow pollutants to accumulate more than in open spaces. Therefore, understanding different indoor spaces and their specifications with respect to exposure is necessary. We discussed the static and mobile microenvironments in Section 12.1. The most common indoor microenvironments can be broadly considered as homes, schools, offices, hospitals, and transport. Below, we discuss their characteristics.

Air Quality. DOI: https://doi.org/10.1016/B978-0-12-822591-2.00012-3

12.2.1 Homes

There are several sources that impact IAQ in residential buildings. Among them, the primary causes of indoor pollution are the activities/sources that release particles or gases into the air. Regular activities such as cooking, heating via open fires, cleaning, smoking, and outdoor air infiltration, which is the case in homes, are known as the main activities that threaten IAQ. These activities could include emissions from combustion sources (e.g., burning of fossil fuel-based products, such as natural gas or kerosene; frying oil, biomass materials such as wood or coal, tobacco products, and any other combustible materials), interior building and furniture materials (e.g., such as furniture made of pressed wood products, household cleaning and maintenance, asbestos, domestic heating and cooling systems), and infiltration from outdoor sources such as vehicular emissions, pesticides, radon, and so on, which are discussed in Section 12.3. Also, elevated temperature/humidity and poor ventilation can break thermal comfort and increase pollutant concentrations due to insufficient clean outdoor air available to dilute.

12.2.2 Schools

The school represents another indoor microenvironment where children and personnel spend ample time. On average, school children spend close to 6 hours or more in the school environment on school days. Children in nursery and primary schools are more vulnerable to air pollution compared with healthy adults because they have faster breathing rates and their lungs and immune systems are still developing. It has been reported that children attending schools close to high-traffic density roads have higher exposure to vehicular emissions which can result in childhood asthma (Perez et al., 2013).

The pollution level in the classrooms depends on the building characteristics, design, and maintenance. Major sources of IAP in schools include cleaning products, air fresheners, paints used for artwork, carpets, and outdoor air infiltration. Kumar et al. (2021a) studied the impact of traffic congestion on children's exposure to $PM_{2.5}$ concentrations during peak hours with the intention of assessing the levels of potential harm and used the findings to influence drop-off/pick-up styles. The research showed that the average concentration and the pupil's levels of exposure to $PM_{2.5}$ using low-cost sensors were below the average daily concentration recommended by the World Health Organization (WHO), but the frequent exposure to the peak $PM_{2.5}$ concentration during the drop-off hours over the school academic year could be potentially harmful. (Kumar et al., 2020b) also provided guidelines for mitigating children's exposure to traffic pollution in and around schools, which include general guidelines and targeted recommendations to school children, parents, and schools.

12.2.3 Offices

An office is a critical microenvironment because office workers spend over a third of their waking hours (8/24 hours) on weekdays (Schweizer et al., 2007). On average, office employees work 40 to 60 hours a week. As a result, the IAQ in offices has a significant impact on the total human exposure level. Over the last decades, the occupational health problems of office workers have increased, with symptoms generally linked to exposure to IAP and poor building ventilation systems. The common sources of IAP in offices are cleaning products, perfumes or sprays, carpet, upholstery, furniture, and laser printers, which are linked to VOCs, PMs, ultrafine particles and O_3 pollutants. Morawska et al. (2017) reported that outdoor air infiltration is one of the major sources of air pollution in offices, with internal sources being insignificant. Earlier Zhu et al. (2015) showed that indoor air pollution in offices may cause a serious health risk for office workers; an example is the emission of ultrafine particles by laser printers that can easily enter the human lungs and bloodstream.

12.2.4 Hospitals

The quality of indoor air in hospitals is a key risk factor for the health of patients, staff, and visitors who use the facility. Several research studies have examined the quantities, sources, and properties of bioaerosol in hospitals because infection is a common occurrence there. Pollutants in hospitals can be classified into biological and nonbiological sources. Examples of biological pollutants are bacterial and fungal; nonbiological pollutants include CO, CO_2, O_3, total volatile organic compounds (TVOCs), formaldehyde (HCHO), $PM_{2.5}$, PM_{10}, etc. (see Section 12.3). There are different activities that can impact IAQ in hospitals. For example, detergents, sanitizers, and cleaning solutions, which are used to reduce the risk of infection in hospitals can increase the levels of TVOCs. Additionally, the use of HCHO solution during treatment can pollute the air with its high concentration.

The hospitals have different units such as the intensive care units (ICU), accident and emergency units, radiology units, pharmacies, theater rooms, and reception areas. Previous research studies have shown that IAQ parameters and distributions can differ among units because medical activities and materials used vary across the different work areas.

Likewise, the ventilation system used in the different work areas affects the IAQ. There are two types of ventilation systems used in hospitals, which are mechanical and natural ventilation systems. The transmission and dissemination of air pollutants in hospitals might be impacted by air conditioning. Jung et al. (2015) showed the distribution of IAP across different hospital units and their potential relationships with different types of air conditioning systems. The study showed that human activities and air conditioning types in various units have an impact on IAQ. The central air conditioning system in hospitals efficiently eliminated aerosol contaminants as compared to noncentral air conditioning hospitals. Furthermore, the pharmacy and ward units have higher TVOC, and CO_2 concentrations compared with other units with no significant differences in the average concentrations of CO, O_3, HCHO, $PM_{2.5}$, PM_{10}, bacteria, or fungi among different units. Section 12.6.4 explains indoor sources and IAQ exposure in hospitals and their health impacts.

12.2.5 Transport microenvironments

People spend commuting time in vehicles to move from one place to another. Therefore, in-vehicle air pollution has become a major public health concern in urban areas because of the increase in transportation time due to high traffic volume and heavily populated public transport during peak hours. Examples include buses, underground and overground trains, cruise ships, and so on. Poor in-vehicle air quality is due to emissions from infiltration from the ambient air (contaminated with exhaust emissions), fuel leakage vapor, and the vehicle interior materials. Air pollutants such as PM, VOC, CO, and NO_x are reported as the primary pollutants detected in-vehicle microenvironments. Recently, due to the emergence of COVID-19, the nexus between in-car aerosol concentrations, ventilation, and the risk of respiratory infection during typical journeys was studied by Kumar et al. (2021a). The study advised that driving with windows open significantly contributes to daily traffic exposure in the morning peak hours. However, driving with windows shut and air conditioning on recirculation mode could substantially reduce exposure to $PM_{2.5}$ and particle number concentrations (PNCs). However, the CO_2 level could rapidly increase due to exhalation during the windows shut for a longer duration and increasing the probability of COVID-19 transmission with time spent on journeys. Fresh air supply by periodically opening windows and efficient facemasks were reported to substantially reduce the respiratory infection probability in-vehicle microenvironments.

12.3 Indoor air pollutant sources

Table 12.1 summarizes the diverse range of indoor environmental conditions, indoor and infiltrated outdoor air pollutants, their sources, and associated health risks. IAQ has become of concern in recent years due to the rapid construction of airtight and energy-efficient buildings. IAP are usually generated from human activities, such as cooking, biomass burners, cleaning materials and solvents, smoke, and natural pollutants such as allergens, dampness, mold, and radon; their levels are mostly linked to lifestyle and how the indoor air is refreshed or purified. Indoor air in such enclosed environments can become stagnant and IAP can elevate instantly. Poor IAQ can cause several symptoms and respiratory diseases, which are associated with the type of pollutants, intensity, and duration of emission. Nowadays, IAQ problems or sick building syndrome (SBS) are not primarily blamed on VOCs. The indoor environmental conditions, exhaled CO_2 level, ingressed outdoor pollutants, and indoor-sourced pollutants impact the IAQ notably. For example, increases in the level of indoor air temperature, RH, and reduced ventilation rate could make a suitable situation to experience damp conditions, resulting in mold and bacterial growth. Lack of proper ventilation, especially during cold seasons, could also facilitate the spread of disease-causing microbes or viruses, such as seasonal flu or coronavirus as the transmission rates are highest indoors. Besides, the burning of fossil fuels or biomass for cooking and heating produces a significant number of toxic gaseous and particulate emissions (Kumar et al., 2022). Furthermore, building components, such as furniture, generate some types of pollutants by themselves. As summarized in Table 12.1, all these pollutants could endanger IAQ. Although the presence of all these pollutants in one indoor microenvironment is unlikely and therefore familiarity with these sources and their associated health impacts is essential to occupants.

Example 12.1: Name two primary sources of IAP? What are the main sources of $PM_{2.5}$, CO_2, and TVOC indoors?

Solution: The indoor environment is impacted by infiltrated outdoor air pollutants as well as indoor activities by residents, such as cooking, burning biomass fuels, cleaning solvents, smoke, and natural pollutants such as allergens, dampness, mold, and radon.

- $PM_{2.5}$: Tobacco smoke, combustion activities.
- CO_2: Combustion activities (e.g., vehicle exhaust), metabolic activity (exhalation).
- TVOC: Personal care products (hair spray, perfumes, etc.), cleaning agents, paints, furniture.

TABLE 12.1 The comprehensive indoor, generated indoor, and ingressed outdoor pollutants that affect IAQ with their associated health consequences.

Environmental parameters/ Pollutants	Major indoor (I) and outdoor (O) sources	The main associated health problems with no specific order	References
Temperature/RH	Poor ventilation, HVAC failure, cooking activities, take a shower/bath	Fatigue, headache, fever	WHO (2009)
CO_2	I: Combustion activities (fossil fuels/ biomass in stoves, boilers, heaters, etc.), metabolic activity due to occupancy, ingress of vehicular emissions (from indoor car parking), lack of fresh air supply	Headache, sleepiness, poor concentration, loss of attention, nasal congestion, wheezing	USEPA (2022)
CO	I: Tobacco smoke, combustion activities O: Traffic and industrial emissions	Low-birth weight, fatigue, potential in perinatal deaths, chest pain in people with heart disease, headaches, dizziness, confusion, nausea	USEPA (2022), WHO (2010)
VOCs	I: Personal care products (hair spray, perfumes, etc.), tobacco smoke, cleaning agents and disinfectants, paints, glues and solvents, air fresheners, candles and furniture. O: Traffic emissions, contaminated subsurface (contaminated soil and groundwater)	Allergic reactions on the skin, visual disorders, memory impairments, possible damage to the central nervous system, SBS, etc.	USEPA (2022)
NO_2	I: Vehicular emissions (from indoor car parking or outdoors), combustion activities (fuel burning) O: Traffic and industrial emissions	Exacerbation of asthma, reduced lung function in kids, respiratory infection diseases, wheeze	USEPA (2022), WHO (2010)
HCHO	I: Used in the manufacture of building materials, household fabrics/products, paints, and furniture (made from formaldehyde-containing resins)	Rhino conjunctivitis, asthma, irritation of the skin, eyes, nose and throat	USEPA (2022), WHO (2010)
SO_2	I: Combustion activities O: Combustion coal and other fuels, industries	Bronchial activity, wheeze	Seow et al. (2016)
Polycyclic aromatic hydrocarbons (PAHs)	I: Fuel/biomass/coal combustion O: Mobile sources (vehicle exhaust, evaporation from tanks), industries, natural sources (volcanic eruptions, wildfire, etc.), agricultural sources (waste/residual burning, pesticides, etc.)	Breast/lung cancer, childhood cancers, reduced lung function, exacerbation of asthma, and increased rates of obstructive lung diseases and cardiovascular diseases	WHO (2010), WHO (2021)
Ozone	I: Electrostatic air cleaners, photocopiers, and laser printers O: Secondary photochemical reactions	Airway irritation, lung damage, pneumonia, bronchitis, aggravate asthma, wheeze	USEPA (2022)
Radon	I: Soil building concentration materials (e.g., stone, concrete, etc.)	Risk of lung cancer, breathing problems	USEPA (2022), WHO (2010)
$PM_{2.5}$ /PM_{10}	I: Tobacco smoke, combustion activities, dust re-suspension due to indoor movements O: Traffic and industrial emissions	Exacerbation of asthma, respiratory infections diseases, exacerbation of COPD, chronic bronchitis, sore throat	USEPA (2022)

(Continued)

TABLE 12.1 (Continued)

Environmental parameters/ Pollutants	Major indoor (I) and outdoor (O) sources	The main associated health problems with no specific order	References
Biological contaminants	I: Encompass any biological agents transmitted by the airborne route such as bacteria, molds, mildew, viruses, fungal spores, foodstuffs, animal dander and cat saliva, house dust, mites, cockroaches, and pollen sourced from indoor occupants, animals, and soil and plants O: Pollens, outdoor air	Trigger symptoms of allergy, asthma, allergic rhinitis, atopic eczema, eye/ throat/nose irritation, sinus and other respiratory problems, fever, infectious diseases, chronic respiratory diseases, wheezing, sensitization	Baldacci et al. (2015), Hospodsky et al. (2012), USEPA (2022), WHO (1990, 2009)
Pesticides	I: Termiticides, insecticides, rodenticides, fungicides, disinfectants, and herbicides, building materials (carpet, textiles, and cushioned furniture) O: Infiltration of disinfectants and herbicides that are used to control pests (insects and rodents, bacteria, fungi, and other organisms).	Eye/nose/throat irritation, damage to central nervous system and kidney, increased risk of cancer	USEPA (2022)
Asbestos	I: Insulation, building materials, packaging, coating, and fire-retardant materials	Cancers (e.g., mesothelioma), pleural thickening, pleural plaques and asbestosis	USEPA (2022)
Heavy metals	I: Paints, firearms, dust, soil, radiators, consumer products, incense burning, tobacco smoke O: Soil	Memory/hearing loss, damage to the nervous system in new-borns, high blood pressure, kidney/heart disease, reduced fertility, hyperactivity or loss of consciousness, respiratory illnesses, cardiovascular deaths	Komarnicki (2005), Madany et al. (1994), Rashed (2008), USEPA (2022)
Persistent organic pollutants (POPs)	O: Pest and disease control in agricultural activities, crop production, and manufacturing/industrial processes, such as dioxins, DDT, etc.	Potency to affect reproduction, promote cancer, development of metabolic diseases like type 2 diabetes and obesity	Amoatey et al. (2020), Ruzzin (2012)

Notes: The health problems could vary depending upon the concentration level and exposure period.
All the references are cited in the far right column of the table.

12.4 Processes and interactions of indoor air pollutant

12.4.1 Indoor air chemistry

As explained in Section 12.3, the IAP are influenced by many indoor and outdoor factors, including particulate matter (PM), various chemical organic/inorganic compounds and biological pollutants. These factors play important roles in human health indoors due to dermal exposure, air inhalation and dust ingestion (Abbatt & Wang, 2020), and could threaten human lives (Bekö et al., 2020). Therefore, it is essential to understand the indoor air chemistry of the pollutants. Fig. 12.1 illustrates a typical residential building, where the major release of chemical compounds to the environment is occurring through: (i) cooking (e.g., fatty acids, triglycerides, terpenes, proteinaceous material, ethanol); (ii) cleaning (e.g., chlorinated molecules, ammonia, acetic acid, terpenes); (iii) building materials, upholstery, and furnishings (e.g., formaldehyde, alcohols, small carboxylic acids, terpenes); (iv) human (e.g., siloxanes, ammonia, amino acids, squalene, triglycerides, fatty acids, isoprene, lactic acid); (v) consumer products (e.g., phthalates); and (vi) microbes (e.g., volatile organic compounds or VOCs) (Abbatt & Wang, 2020).

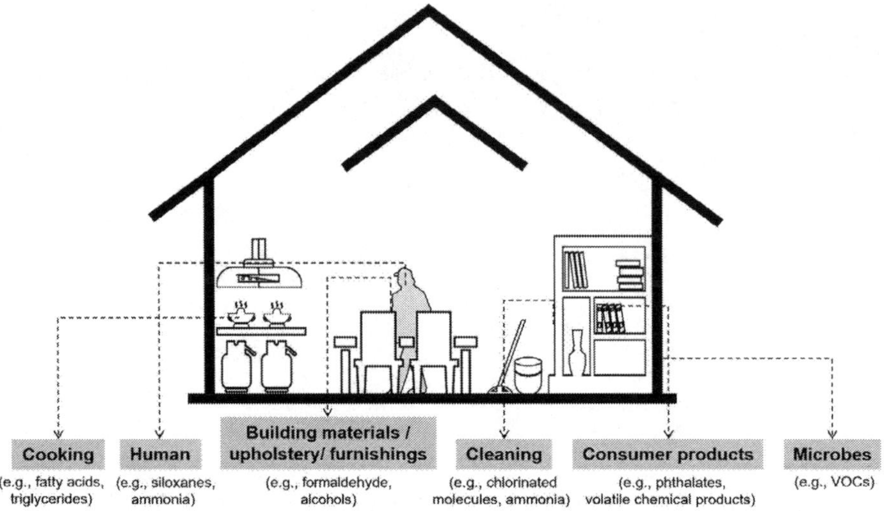

FIGURE 12.1 A schematic diagram shows the major sources of reactive chemicals indoors.

In addition, there are three important types of indoor chemistry that should not be neglected:

Indoor gas phase chemistry: The concentration of any pollutants in a room depends on the balance between the sources and sinks (Weschler & Carslaw, 2018). To affect the indoor environment, the time scales of the reactions between gas phase pollutants must compete with air exchange (Weschler & Shields, 2000). To date, many studies have focused on the reactions between oxidants (O_3 and OH radicals) and VOCs, which produce thousands of complex and often multifunctional products in the gas phase.

Indoor surface chemistry: Indoor surface reactions often occur at the interface, including air/particle and air/surface films (Morrison, 2008). The time constraints of air exchange rates on gas phase reactions in the room do not apply to surface reactions, except for those involving airborne particles. Moreover, surface chemistry is more important in indoor areas because of a lack of time constraints, combined with high surface-to-volume ratios ($2-4\ m^2\,m^{-3}$) (Weschler & Carslaw, 2018).

Indoor photochemistry: There is a common misconception that no photolytic reaction takes place indoors due to the absence of direct sunlight. In fact, light can enter the indoor environment through doors and windows, as well as through interior lighting (Weschler & Carslaw, 2018). Photolysis therefore still takes place indoors, but relatively slowly compared with outdoors.

12.4.2 Controlling the transformation

Chemical transformation is usually described by detailed kinetic reaction mechanisms that may incorporate hundreds of species and thousands of reaction steps (Turányi, 2016). Controlling indoor air chemistry is different from outdoors due to less direct sunlight or rain, less extreme temperature, a higher surface-to-volume ratio, and higher concentrations of organic compounds (Weschler & Carslaw, 2018). For example, in outdoor conditions, O_3 has a 97% chance of reacting with NO, while it has a 1% chance of reacting with VOCs, a 1% chance of depositing on surfaces and a 1% chance of photolysis. However, in a typical indoor condition, O_3 has a higher than 40% chance of reacting with NO, a lower than 40% chance of surface deposition, a 20% chance of removal by air exchange and a 1% chance of reacting with VOCs. Therefore, it is evident that the photolysis of O_3 is usually negligible indoors (Weschler & Carslaw, 2018). This identifies the following priorities indoors with respect to the outdoors (Bekö et al., 2020):

The potential of gas-phase chemistry indoors relies on (i) photolysis; (ii) the relative importance of OH, NO_3, Cl, and O_3 as oxidants, as well as their dominant conditions and formation pathways; (iii) the relationship between different reaction pathways and the formation of secondary products; and (iv) chemical reactions and ventilation rate. The abundance of organic trace components in indoor air is driven by (i) the effects of indoor solid-phase and liquid-phase processes, such as indoor liquid-phase reactivity, including acid-base chemistry; and (ii) parametric heterogeneous processes on indoor surfaces and real environments (e.g., HONO and aldehydes).

To control the negative transformation of pollutants in indoor microenvironments, it is essential to consider the impacts of moisture, occupants, and buildings on indoor chemistry.

Impact of moisture on indoor chemistry: At an RH of 65%, water is a significant fraction (approximately 30%) of the airborne particles. When the RH changes, so does the presence and thickness of the water surface film, the water content of the particles in the air, and the amount of water absorbed into the porous material (Ho et al., 1974; Leygraf et al., 2000). Inorganic acids (e.g., hydrochloric, nitric acid), organic acids (e.g., acetic, formic, lactic acid), inorganic bases (e.g., ammonia) and organic bases (e.g., amines, nicotine) are partitioned between the gas phase and the surface/airborne particles in water (Weschler & Carslaw, 2018).

Impact of occupants on indoor chemistry: Occupants emit skin oils and shed their skin oil-rich dander at a very rapid rate (Wisthaler & Weschler, 2010). The ozone/skin oil oxidation chemistry produces several gas phase products, including 6-methyl-5-heptene-2-one (6-MHO), acetone, 4-oxopentanal (4-OPA) and geranyl acetone. Less volatile chemicals are also produced by these reactions and remain on the skin, clothing, and surfaces, such as succinic, levulinic, suberic and adipic acids. In addition, breath is an important source of reactive indoor chemicals, including nitric oxide (NO), isoprene and ammonia (Weschler and Carslaw, 2018).

Impact of buildings on indoor chemistry: Building ventilation rate is a key factor affecting indoor air chemistry. In naturally ventilated buildings, there is a need to measure temperature, air exchange rates and relative humidity to get a proper understanding of the chemical composition. Basements, wall cavities, attics and crawl spaces should not be overlooked. In mechanically ventilated buildings, heating, ventilation, and air conditioning systems are often used. It is, therefore, necessary to know when the system operates, the air exchange rate, the fraction of recirculated air and outdoor air, humidification and dehumidification, the filtration efficiency, and the temperature throughout the day.

Further, understanding household demands and providing suggestions to change their unhealthy habitats are crucial to controlling IAP. The quantitative relationship between health outcomes (e.g., morbidity risk), indoor air concentrations and cooking styles (e.g., fuel use) need to be clearly understood and transformed into policy decisions and general recommendations for households (Larson & Rosen, 2002).

The indoor environment changes over time with the occupants' activities. We should fully understand indoor chemical transformation so that we can avoid transforming relatively environmentally benign chemicals (e.g., NaClO and HCl in cleaning products) into ones that cause malodour, irritation, and adverse health consequences (e.g., Cl_2). This is crucial to ensure that building occupants and other contents are protected from accidental exposure to hazardous chemicals (Weschler & Carslaw, 2018).

12.5 Indoor air quality monitoring

A variety of air quality monitoring technologies are used for monitoring air quality such as government-operated stationary monitors, known as reference-grade federal reference method (FRM) or federal equivalent method (FEM) monitors, satellite-based remote sensors, and mobile air quality monitors. Governments and local agencies jointly manage networks of air quality monitors for alerting, policymaking, environmental legislation, and other purposes. As shown in Table 12.2, these technologies offer resolution, precision, and accuracy. However, this is a costly process, because such equipment requires expensive maintenance, calibration and supporting infrastructure to operate. Additionally, the other downside of this process is the trouble in detecting widespread, localized, and IAPs.

As the public demand for IAQ monitoring grows, the rapid rise of low-cost sensing technologies started to meet those demands. In other words, with more affordable monitors, the creation of a dense and widespread network is becoming possible. Furthermore, air quality sensors can be deployed anywhere inside both static and mobile environments (Section 12.2), considering their size and requirements for operation. With such flexibility, it is possible to pinpoint localized sources and pollution hotspots, which can help to understand community exposure in finer grids. Table 12.2 summarizes some of the regulatory and advanced low-cost technologies that can serve the purpose of IAQ monitoring. Table 12.3 demonstrates the schematic diagram of each technology in brief. Low-cost sensors use different technologies to measure the concentration of air pollutants. These could be lasers to estimate the number/size of particles carried by sample air through a chamber or solid-state sensors to measure the amount of adsorbed gas on the surface. The sensors generally use specific algorithms, which may vary from one developer to another, to convert raw data into meaningful concentration values. The temperature, RH, and interferences of other gases may be considered in the development of the algorithms with the aim of improving the performance of sensors over time in different environments.

TABLE 12.2 Examples of the most common air quality monitoring technologies serve the purpose of both regulatory and nonregulatory applications (Kebabian, et al., 2005; Omidvarborna et al., 2021; Stedman et al. 1972; Yi et al., 2015).

Group	Technology	Description	Pollutant	Reported resolution/ range	Approximate cost
Regulatory					
PM	Beta attenuation	Measuring the absorption of beta radiation by solid particles extracted from the sampled air-flow.	PM_{10} and $PM_{2.5}$	$0.1-1\ \mu g\,m^{-3}$ /$0-1500\ \mu g\,m^{-3}$	$$$$$
CO	IR gas filter correlation	It is an advanced form of IR sensing technology, which is ideal in applications with interferent background gases, delivering accurate, sensitive measurement results.	CO	0.1 ppb/ Up to 1000 ppm	$$$$$
NO_2	Cavity attenuated phase shift spectroscopy	In this technology, a modulated broadband incoherent light source (a 430-nm LED) is coupled to an optically resonant cavity formed by two high-reflectivity mirrors. The presence of NO_2 in the cell causes a phase shift in the signal received by a photodetector that is proportional to the NO_2 concentration.	NO_2	0.1 ppb/ Up to 1 ppm	$$$$$
O_3	Chemiluminescence detection	The principle of the detector in detecting O_3 is based on the chemiluminescent reaction $O_3 + NO \rightarrow O_2 + NO_2 + h_v$. The main features are - Selective response to the measured gas, - Sensitivity down to the ppb range - Linearity in concentration (factor of 10^6) - High temporal resolution (~ l sec)	O_3	0.1 ppb/ Up to 2 ppm	$$$$$
Nonregulatory					
PM	Light scattering	Measuring the intensity of light scattered from particles through a focused light. This intensity is proportional to the number of particles in the chamber. The mass of the particles is derived by the assumption of an average density.	PNC, PM_1, $PM_{2.5}$ and PM_{10}	NR/ up to $2000\ \mu g\,m^{-3}$	$$$-$$$$
CO_2	Nondispersive infrared	An NDIR sensor works by an infrared (IR) lamp directing waves of light through a tube filled with an air sample. As the IR light passes through the tube, the CO_2 molecules absorb the specific band of the IR light, while allowing other wavelengths to pass through. The wavelength absorbed by CO_2 molecules is equal to CO_2 concentration.	CO_2	1 ppm/ 0–5000 ppm	$$-$$$
CO, NO_2	Electrochemical sensor	Measure electrical output from a chemical reaction proportional to the CO level within the sensor.	CO, NO_2, O_3	CO: 10 ppb/ Up to 1000 ppm NO_2: 10 ppb/Up to 20 ppm O_3: 5 ppb/ Up to 5 ppm	$$$-$$$$*

CO, NO_2, O_3	Solid-State sensor	Solid-state sensors or metal oxide (MOS) sensors include a sensing element that is heated to 250°C–350°C. The pollutant reacts with the oxide coating, causing a change in resistance between the two electrodes, which is then measured as the pollutant concentration. This type of sensor has a limited working range of 1–50 ppm.	CO, NO_2, O_3, H_2S	CO: NR/ Up to 2000 ppm NO_2: NR/ Up to 10 ppm O_3: 1 ppb/ Up to 1000 ppm		$$-$$$$*
VOCs	Photoionization detectors	It uses high-energy photons (ultraviolet (UV) light in the energy scale of 10 eV) for ionization of gaseous molecules. The resulting ions are proportional to the concentration of the pollutant.	VOCs	1 ppb/ 1 ppb to 50 ppm		$$-$$$

Note: All the references are cited as a part of the table caption.
NR: Not reported. *Higher prices are usually for precise and low-concentration level sensors.

TABLE 12.3 Schematic diagram of different air quality monitoring techniques. Details of each technique can be found under their reference link (see references).

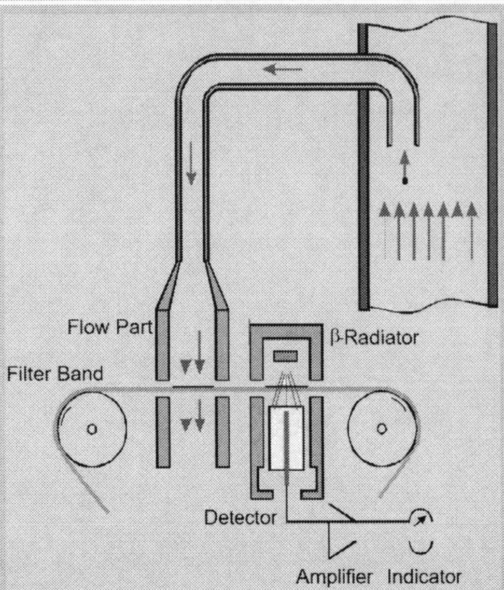

Beta attenuation monitoring technology for PM measurement (Environmental, 2011)

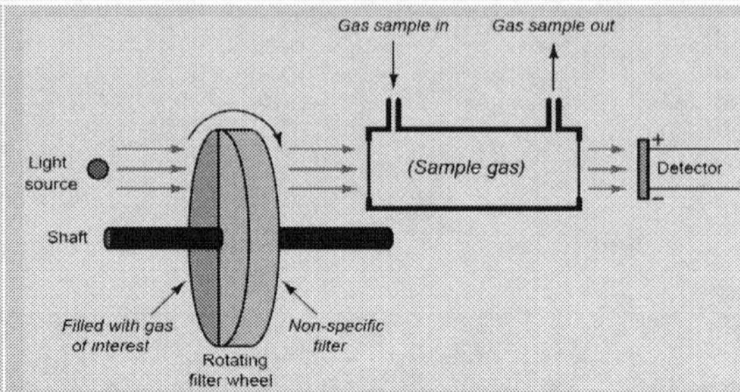

Gas-filter correlation spectroscopy (Tools, 2018)

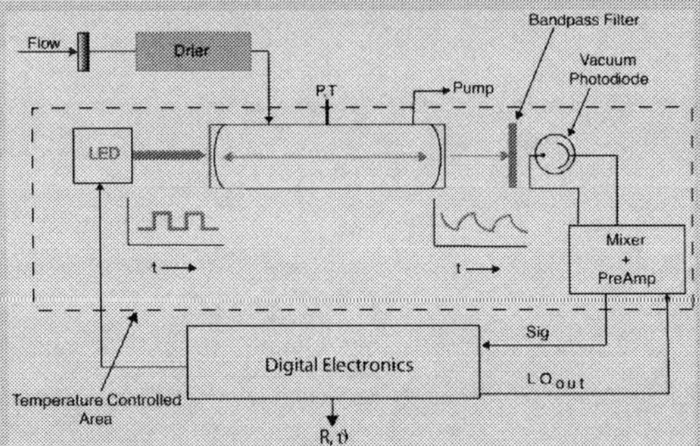

Cavity-attenuated phase shift spectroscopy (Freedman et al., 2019)

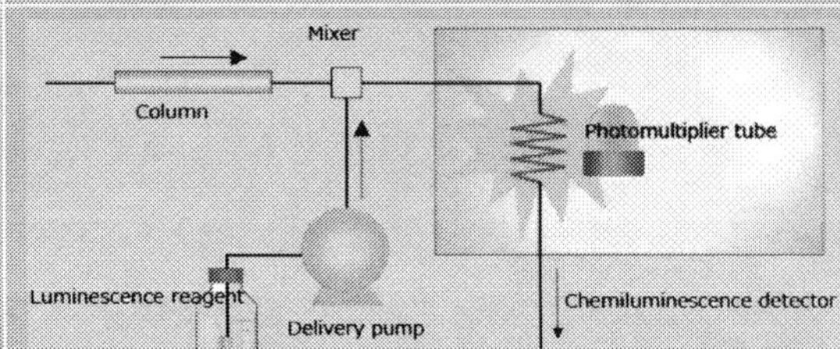

Chemiluminescence detection (Shimadzu, 2022)

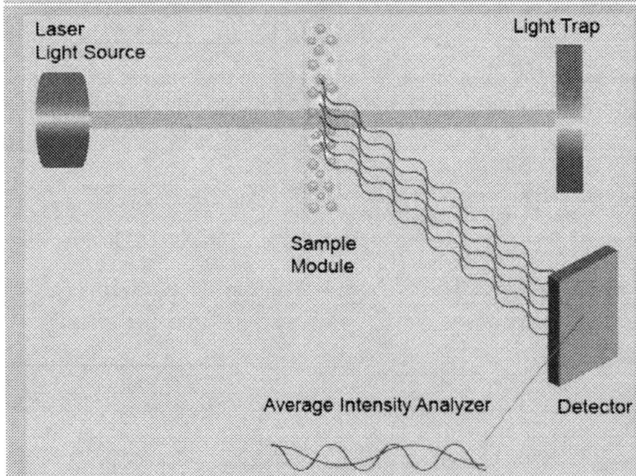

Light scattering sensing technology (PTL, 2022)

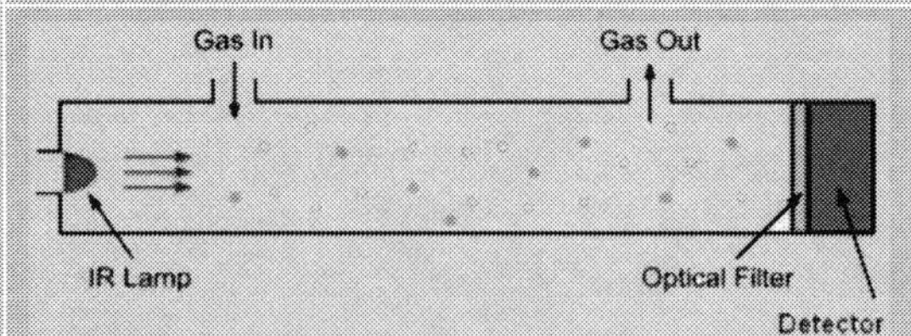

Non-dispersive infrared sensor (CO_2 Meter, 2022)

(*Continued*)

TABLE 12.3 (Continued)

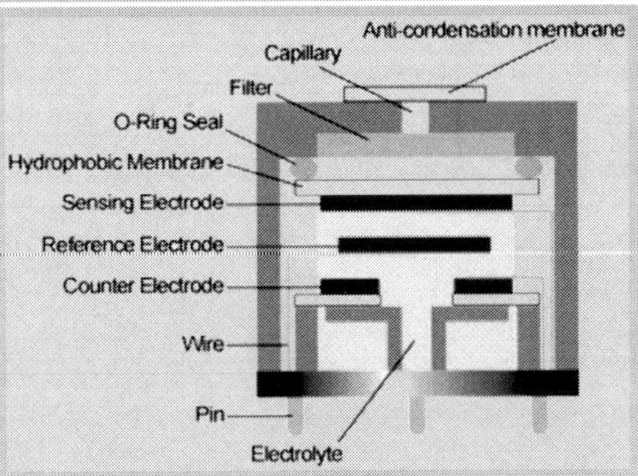

Electrochemical gas sensor (Membrapor, 2022)

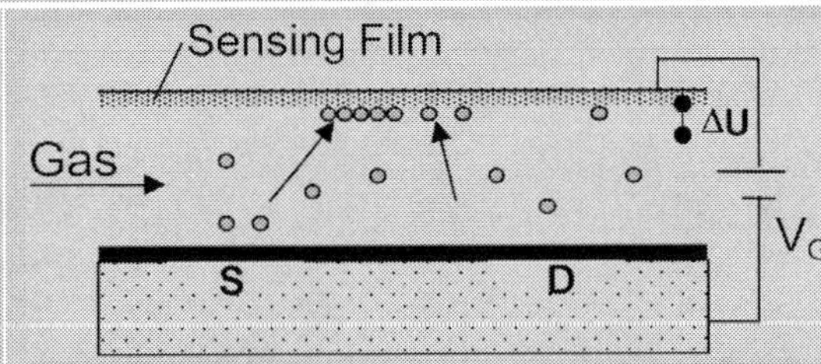

Solid state gas sensors (Moos, et al., 2009)

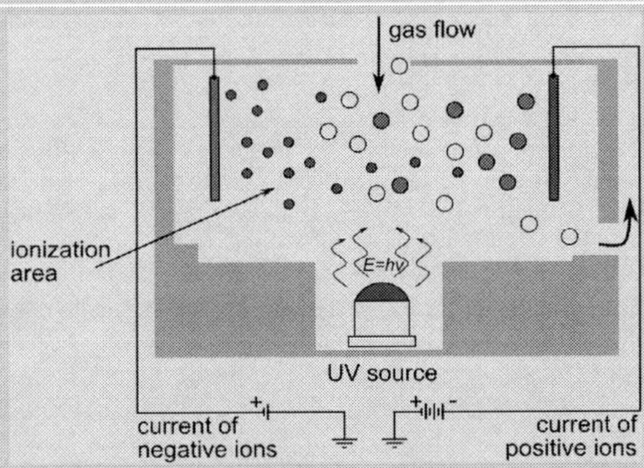

Photoionization detector (Zimmer, et al., 2015)

Example 12.2.: Why is low-cost sensing technology taking over traditional monitoring systems, especially for indoor applications? Name advanced sensing methods to measure $PM_{2.5}$ /PM_{10}, CO_2, CO, NO, and TVOCs.

Solution: The traditional monitoring systems (e.g., FRM or FEM monitors, satellite-based remote sensors, and mobile air quality monitors) offer resolution, precision, and accuracy, while they are expensive to operate and require continuous maintenance and calibration. Therefore, low-cost sensing technologies, which are more affordable, smaller in size, easy to operate, etc. are welcomed. The advanced sensing methods are:

- Light scattering technology, for example, $PM_{2.5}$/PM_{10}.
- Nondispersive infrared sensors, for example, CO_2.
- Electrochemical sensors, for example, CO and NO_2.
- Solid-state sensor, for example, CO, NO_2, and O_3.
- Photoionization detectors, for example, TVOCs.

There is a trade-off between using high-fidelity regulatory-grade air quality equipment against low-fidelity and low-cost air quality sensors. It provides a challenge in ensuring the quality and appropriate interpretation of measured air pollution data by the sensors. Hence, low-cost sensors are usually utilized for relative comparison in certain applications to raise awareness among citizens and public engagement, for example, citizen science activities to improve their understanding of environmental/public health concerns (Kumar et al., 2020; Mahajan et al., 2020).

12.6 Indoor air quality exposure

IAQ monitors, as discussed in Section 12.5, usually measure pollutant concentrations in terms of either ppm/ppb or $\mu g\ m^{-3}$. However, to estimate the personal exposure to each pollutant, apart from the concentration level, breathing rate, and the exposure duration of the individuals in the polluted environment need to be considered. The individual exposures to pollutants can be estimated using Eq. (12.1):

$$\text{Inhaled exposures }(\mu g) = \text{indoor concentrations }\left(\mu g\ m^{-3}\right) \times \text{inhalation rate }\left(m^3\ min^{-1}\right) \times \text{time duration (min)} \quad (12.1)$$

Here, the inhalation rate is the rate at which breathing occurs. The inhalation rate varies with age, gender, and level of physical activity. This section will discuss the various levels of exposure in different indoor environments presented in the reviewed literature.

12.6.1 Indoor air quality exposure in homes

As mentioned earlier, people spend most of their time in their homes. A study was conducted from November 2016 to October 2017 in 32 residences in Tianjin, China (25 homes with natural ventilation and seven homes with mechanical ventilation) to determine the relationship between IAQ and occupant SBS symptoms (Sun et al., 2019). SBS describes situations in which indoor occupants experience unhealthy conditions indoors mostly due to the building's dampness, improper temperature and relative humidity, ventilation, illumination levels, noise, air quality etc. (Ghaffarianhoseini et al., 2018). They assessed pollutants including TVOCs, formaldehyde, $PM_{2.5}$, ultrafine particles, and ozone, as well as environmental characteristics like ventilation rate. Lower ventilation rates inside the residences resulted in higher levels of formaldehyde ($0.10 \pm 0.06\ mg\ m^{-3}$) and TVOCs ($586 \pm 732\ \mu g\ m^{-3}$) that caused the symptoms of SBS according to the researchers. The reported formaldehyde value was comparable to IAQ standards (WHO, 2010). During the initial COVID-19 outbreak, another IAQ research was undertaken on four types of legacy housing in Madrid, Spain (Domínguez-Amarillo et al., 2020). Before and during the lockdown, the researchers measured $PM_{2.5}$, NO_2, TVOCs, and CO_2 in all the legacy houses considered. During the lockdown and as compared to before the lockdown, the mean daily indoor $PM_{2.5}$ levels increased by 12% (from $14\ \mu g\ m^{-3}$ to $17\ \mu g\ m^{-3}$) and the mean TVOC concentration increased by 2.75-times from 5550 ppb to 15,285 ppb. They reported that during the COVID-19 outbreak, pollution levels rose due to a lack of proper ventilation and intensive use of cleaning agents within the dwellings. In Lithuania, Kaunelienė et al. (2016) evaluated IAP in 11 recently constructed low-energy dwellings. Despite low exchange rates in most buildings, formaldehyde levels ($3.3-52.3\ \mu g\ m^{-3}$) were found 5.2 times greater than the national limit of $10\ \mu g\ m^{-3}$. The levels of several VOCs such as B4 p-xylene ($214.5\ \mu g\ m^{-3}$), o-xylene ($328.9\ \mu g\ m^{-3}$) and ethylbenzene ($521.7\ \mu g\ m^{-3}$) were found to exceed daily indoor limits of Lithuania. They reported that the values of VOCs increased soon following the installation of the interior surfaces and furniture in some buildings.

12.6.2 Indoor air quality exposure in schools

Schools are one of the public places that attract many exposure assessment studies. A study was undertaken in 16 schools in a city in the mid-Atlantic region between 2015 and 2017 to determine the indoor concentrations of $PM_{2.5}$, NO_2, CO, and CO_2 (Majd et al., 2019). The levels of $PM_{2.5}$ ($5.2\ \mu g\ m^{-3}$) and CO (0.5 ppm) were below WHO daily standards, whereas NO_2 (12.6 ppb) and CO_2 (750 ppm) concentrations were above the WHO and ASHRAE (American Society of Heating, Refrigerating, and Air-conditioning Engineers) daily norms. NO_2 and CO concentrations varied seasonally, with higher levels reported during the winter months due to increased combustion from heating sources and decreased atmospheric mixing during the winters. Another study was conducted in Cassino, Italy, at three public schools (two urban and one suburban) (Buonanno et al., 2013), where they measured UFP (in terms of PNC) and BC. They discovered that PNC levels ranged from 1.95×10^4 to $3.49 \times 10^{4,}$ with higher levels in urban schools than in suburban schools. Similarly, urban schools had greater amounts of BC, with the highest median value of $10\ \mu g\ m^{-3}$. Traffic emissions are identified as the main source of PNC and BC in urban schools. In another comprehensive study, which lasted around 9 days, 25 nurseries and eight primary schools in urban and rural Northern Portugal were monitored for indoor pollutants such as PM_1, $PM_{2.5}$, PM_{10}, total suspended particles (TSP), CO, CO_2, NO_2, O_3, HCOH, and TVOCs, as well as meteorological and comfort parameters (Branco et al., 2019). According to the study, students experienced thermal discomfort and insufficient humidity in 60% and 44% of the classrooms, respectively. They were also exposed to higher amounts of $PM_{2.5}$ and CO_2 in urban classrooms, which could lose focus and attention towards effective learning.

12.6.3 Indoor air quality exposure in offices

The IAQ inside offices and other workplaces are important not only for employees' comfort but also for visitors. The Occupational Safety and Health Administration (OSHA) provides guidelines addressing IAQ in the most common workplaces in the US. The IAQ guidelines are typically related to temperature, humidity, lack of outside air ventilation, or smoking (OSHA, 2015). A study was conducted during the spring season in Poland to investigate the microbiological content of air-conditioned and naturally ventilated offices. *Bacillus*, *Staphylococcus xylosus*, *Macrococcus equipercicus*, and *Micrococcus luteus D* are among the 12 distinct bacteria species found in the study that were mainly sourced from indoor air and of human origin (Brągoszewska et al., 2018). The average concentration of viable culturable count of bacteria in the indoor offices with air conditioning was 541 CFU m^{-3} (colony forming units per cubic metre), which was below the Polish threshold standards of 1000 CFU m^{-3} for offices. In another project named OFFICEAIR, supported by European Nations (EU), the IAQ of 37 office buildings in the summer and 35 office buildings in the winter in eight European nations, including Italy, Finland, Greece, Spain, France, the Netherlands, Hungary, and Portugal was assessed for $PM_{2.5}$, O_3, NO_2, 12 VOCs (including D-limonene, α-pinene and benzene), and seven aldehydes (including formaldehyde) (Mandin et al., 2017). They claimed that the amounts of D-limonene and α-pinene found in their samples were significantly greater than those found in other office building investigations. They discovered that D-limonene, α-pinene, NO_2, and benzene levels were greater in the winter due to high emissions from indoor heating sources and high atmospheric stability outdoors (lower wind speeds and lower mixing of air) which impacts the indoor air. On the other hand, O_3 and formaldehyde levels were higher in the summer due to increased indoor emissions of formaldehyde due to higher temperatures and higher ozone production outdoors (due to increased photochemical production) which infiltrated indoors. In another study, IAQ assessments were conducted in the Painting and Printmaking department of a Turkish university to quantify pollutants such as O_3, NO_2, and 29 VOCs. The concentration of benzene in painting workshops was doubled as compared to offices (5.34 $\mu g\,m^{-3}$ vs 2.50 $\mu g\,m^{-3}$). Overall, the study found that painting and stained-glass workshops had the highest levels of VOCs compared to offices. From the exposure point of view, the study also anticipated that the faculty and workshop technicians would be at risk for cancer throughout their lives (Can et al., 2015).

12.6.4 Indoor air quality exposure in hospitals

IAQ in hospitals is also critical not only for minimizing exposure to typical air pollutants but also to further ensure airborne infection control and efficient care, now more than ever. A study was conducted in two oncology units of hospitals located in Italy and Spain to measure TVOCs, $PM_{2.5}$ and CO_2 using low-cost sensors (Palmisani et al., 2021). The temporal profiles of the pollutants displayed higher concentration peaks during the weekdays which are characterized by higher human occupancy. Another study attempted to determine the levels of CO_2, $PM_{2.5}$, and NO_2 levels in the Hospital's reception area, otorhinolaryngology outpatient clinic, and orthopedic clinic, respectively (Lee et al., 2020). They discovered that pollutant concentrations were higher in the otorhinolaryngology outpatient clinic than in other regions and that IAQ was worse during the winter. The authors speculate that several medical procedures and heavy usage of oronasal sprays might have contributed to higher concentrations in the otorhinolaryngology clinic. Moreover, the increased number of incoming patients and the higher outdoor concentrations typically observed during the winter season might have contributed to the elevated indoor concentrations during this period. In another study performed in Turkey, pollutants such as PM_1, $PM_{2.5}$, CO_2, CO, temperature, and humidity were examined in eight different sites throughout a hospital. The study found that emergency services had the highest PM_1 and $PM_{2.5}$ concentrations, while pediatric clinics had the highest CO_2 levels. These higher concentrations were thought to be the result of a large number of patients and caregivers in the pediatric clinic as well as the busy and packed environment in the emergency services (Babaoglu et al., 2019). To analyze the impact of COVID-19 on IAQ, a study was conducted in two mechanically ventilated classrooms in Spain during the winter season. As a part of the study, the CO_2 concentrations were measured before the pandemic and after the pandemic. They reported that weekly mean CO_2 levels decreased by 300 ppm during the pandemic while using hybrid (mechanical and natural) ventilation systems. Also, they observed that weekly mean CO_2 levels decreased by 400 ppm when the schools were naturally ventilated (Alonso et al., 2021).

12.6.5 Indoor air quality exposure in kitchens

Assessing IAQ in kitchens is most important, as a significant proportion of the population is exposed to various pollutants generated from kitchens, especially in low- and middle-income countries. To evaluate the in-kitchen exposures to $PM_{2.5}$ and PM_{10}, a study was conducted in 60 low-income homes across 12 cities in different countries: Dhaka (Bangladesh); Chennai (India);

Nanjing (China); Medellín (Colombia); São Paulo (Brazil); Cairo (Egypt); Sulaymaniyah (Iraq); Addis Ababa (Ethiopia); Akure (Nigeria); Blantyre (Malawi); Dar-es-Salaam (Tanzania) and Nairobi (Kenya) (Kumar et al., 2022). The study reported that the type of fuels used, volume of the kitchen, cooking types and kitchen ventilation were the most prominent factors affecting in-kitchen exposures. The highest $PM_{2.5}$ ($185 \pm 48\ \mu g\ m^{-3}$) and PM_{10} ($220 \pm 58\ \mu g\ m^{-3}$) concentrations were found in Dhaka due to extensive frying, prolonged cooking, and smaller kitchen volumes. $PM_{2.5}$ concentrations due to charcoal-based cooking increased by approximately 3.1-times compared with LPG. Dual ventilation (mechanical and natural) in Chennai, Cairo and Sulaymaniyah reduced average in-kitchen $PM_{2.5}$ and PM_{10} by 2.3- and 1.8-times compared with natural ventilation (open doors) in Addis Ababa, Dar-es-Salam, and Nairobi. The lowest exposure doses were observed in South American homes and the highest in Asian homes. The study concluded that increased cooking exhaust extraction, cleaner fuels, awareness of improved cooking practices and minimizing passive occupancy in kitchens as the measures to mitigate harmful cooking emissions. Another study was conducted in rural households in Northern India to investigate the IAQ due to biomass burning used as fuel for kitchens (Sharma & Jain, 2019). In this study, to estimate the levels of PM (PM_{10}, $PM_{2.5}$, and PM_1), CO and CO_2, a comparative assessment has been made between traditional cookstoves (TCS) and improved cookstoves (ICS) combined with kitchen characteristics. They reported that using ICS, IAQ improved in kitchen areas resulting in a lower concentration of PM_{10}, $PM_{2.5}$, PM_1 and CO by 21%–62%, 20%–80%, 24%–87% and 19%–93%, respectively. They also observed that the worst IAQ was found in enclosed kitchens compared to semienclosed and open kitchens. The study concluded that the use of ICS combined with efficient kitchen designs can significantly improve the overall IAQ. A study was conducted in charcoal-fueled kitchens of Mbarara municipality in Western Uganda to evaluate $PM_{2.5}$ and CO levels (Nakora et al., 2020). They observed that the mean levels of $PM_{2.5}$ were $449\ \mu g\ m^{-3}$ in the wet season and $526\ \mu g\ m^{-3}$ in the dry season followed by CO levels of 41.52 ppm. They reported that pollutant levels exceeded the WHO limit values of $24\ \mu g\ m^{-3}$ for $PM_{2.5}$ and 6.340 ppm for CO. The study concluded that kitchen size and ventilation were found to significantly influence pollution levels.

12.6.6 Indoor air quality exposure in transport microenvironments

The transportation sector in various ways generates different pollutants that can impact in-vehicle air quality levels. A study was undertaken in Hong Kong to quantify $PM_{2.5}$, UFP, BC, and CO in various public transportation microenvironments such as liquefied petroleum gas (LPG) and diesel buses, and train to systematically address the inherent heterogeneity in personal exposures to pollutant concentrations in diverse microenvironments (Yang et al., 2015). According to the study, traffic-related air pollutants such as UFP, BC, and CO had greater fluctuations among different modes of transportation, indicating heterogeneity. According to the findings, the pollutants showed unique patterns, with greater CO concentrations within LPG buses and BC and UFP concentrations in diesel bus cabins, implying vehicle self-pollution and infiltration of external pollutants during travel. Another assessment was carried out in Guildford, a typical UK town, to examine personal PM_{10}, $PM_{2.5}$, and PM_1 exposure levels in two microenvironments (bus and car) during the morning peak, afternoon off-peak, and evening peak hours (Kumar et al., 2018). The highest PM_{10} concentrations were reported in buses due to infiltration of outdoor pollutants into the buses while frequent door openings and accumulation of particles, while fine particles ($PM_{2.5}$ and PM_1) concentrations were found in cars due to infiltration of the fine particles through the air-conditioning filtration systems. Another study in Hangzhou, China, investigated the sources of polybrominated diphenyl ethers (PBDEs) in air and dust samples obtained from subways, hybrid, and electric buses (Jin et al., 2022). PBDEs are a type of chemicals that are used in certain products to reduce the chance of being caught on fire. Subways exhibited $407.6\ pg\ m^{-3}$ and $925.9\ ng\ g^{-1}$ of PBDEs in the air and dust samples, electric buses showed $747.5\ pg\ m^{-3}$ and $1160.1\ ng\ g^{-1}$, and hybrid buses had $625.4\ pg\ m^{-3}$ and $1200.6\ ng\ g^{-1}$, respectively. The study found that the overall concentrations in subways were lower than in buses, and the interior materials such as cushion fibers were the main sources of the pollutants. In a different study, exposure to $PM_{2.5}$ and PM_{10} were measured in cars across 10 global cities, including Dhaka (Bangladesh), Chennai (India), Guangzhou (China), Medellín (Colombia), São Paulo (Brazil), Cairo (Egypt), Sulaymaniyah (Iraq), Addis Ababa (Ethiopia), Blantyre (Malawi), and Dar-es-Salaam (Tanzania) (Kumar et al., 2021b). Measurements were made on predetermined routes during morning, off-peak, and evening peak times with the windows open and closed (fan on in recirculation mode). Study results showed that measurements under the open window scenario had the highest $PM_{2.5}$ and PM_{10} concentrations. In most cities, exposure to $PM_{2.5}$ and PM_{10} was much higher during morning peak hours. In-car $PM_{2.5}$ levels were up to 80% lower with recirculation than with the windows open. Furthermore, in-car $PM_{2.5}$ exposures were disproportionately higher for drivers in the least affluent cities. They finally concluded that lower car speeds and temporally longer journeys were the major determinants of higher exposures. In another similar study, the relationship between in-car $PM_{2.5}$ and PM_{10} exposures and socioeconomic indicators [i.e., road density, fuel prices, city-specific gross domestic product (GDP), etc.] was assessed in the same ten cities discussed earlier (Kumar et al., 2021c). Higher $PM_{2.5}$ concentrations and inhaled doses were observed in hotspots across all city routes. For instance, the amount of time spent at the hotspots in Addis Ababa and Guangzhou, which accounted for 26% and 28% of the overall trip time, respectively, caused 54% and 56% of the total $PM_{2.5}$ inhaled dose. Except for Guangzhou, every city demonstrated a reduction in the percentage length of hotspots as GDP increased. The study concluded that lower GDP was found to be associated with higher economic losses due to health burdens caused by air pollution in most cities, indicating a socioeconomic discrepancy.

12.7 Indoor air quality modelling

The IAQ monitoring provides a snapshot of the reality in those selected moments. However, questions emerge from the monitoring data in the form of what-if and why. For instance, one may identify a high concentration of NO_2 inside a school, but to mitigate such incidents, it is important to know why the concentration is so high. To address such questions, certain models are developed for source location and identification, which will be discussed later in this section. After the causes are identified through modelling, mitigation methods can be selected. To assist this process, another model can be used to generate scenarios where the mitigation methods can be compared. Finally, after the mitigation method is successfully employed, one may want to use the model to forecast the concentrations of NO_2 in the environment for future planning. Therefore, models are fundamental tools to work with IAQ and basic knowledge about them is needed not only for its application but also in the study planning. In this section, two main approaches are used for IAQ modelling: analytical and machine learning (ML). Analytical models are directly based on physics, while ML is based on data variability. These two approaches with their newly shaped structures are discussed next.

12.7.1 Analytical models

Most analytical models use mass conservation equations looking at the physical system as a whole. Normally, a bulk volume of space (a virtual box) containing the sources of a pollutant and people is determined (e.g., room), and the concentration of the pollutant, emission sources and sinks are calculated. This is commonly denominated as a box model. This type of model has limited precision, but it is less prone to numeric errors as compared with the other types of models. The simplicity of box models also helps with analysis and troubleshooting and makes simplifications and improving computational costs possible. The generic mass conservation model used in the box model can be summarized in Eq. (12.2):

$$\text{Concentration }(t_n) = \text{Concentration }(t_{n-1}) + \Delta\text{Concentration }(\text{Sources}, \text{Sinks}, dt) \tag{12.2}$$

Here, the pollutant concentration in determining time is calculated by the sum of the previous concentration and the variation of the concentration in the time interval, which is a function of the source's emissions and the sink's absorptions. It is noteworthy that the use of this equation is not limited to the concentration as an incognito variable, but also can be used to estimate the sources and sinks. The definition of the sources and sinks depends on the pollutant. For instance, CO_2 is produced by respiration, so it is a function of not only the outdoor concentrations but also the number of people inside and the physical activity they perform. Meanwhile, O_3 rarely has indoor sources, making it entirely a function of the outdoors. Therefore, the equation used for the modelling can normally be found in the format of the ratio between the indoor and outdoor concentrations [Eq. (12.3)–(12.6)].

$$\Delta\text{Ozone} = (P.a)\text{Ozone}_{\text{out}} - (a + K)\text{Ozone}_{\text{in}} + \Delta S \tag{12.3}$$

$$0 = (1.a)\text{Ozone}_{\text{out}} - (a + K)\text{Ozone}_{\text{in}} \tag{12.4}$$

$$(a + K)\text{Ozone}_{\text{in}} = (a)\text{Ozone}_{\text{out}} \tag{12.5}$$

$$\frac{\text{Ozone}_{\text{in}}}{\text{Ozone}_{\text{out}}} = \frac{a}{a + K} \tag{12.6}$$

where P is the value of the outdoor ozone that penetrates the building, a is the air exchange rate (h^{-1}), K is the surface removal rate (h^{-1}), and S is the indoor sources of ozone. ΔS can be neglected. The value of P is approximately 1 and the variation of ozone concentration is very small for a short time [Eqs. (12.4) and (12.5)]. Using such approximations, the ratio of ozone inside and outside of the building can be calculated by the air exchange rate and surface removal rate only [Eq. (12.6)].

The same theory applies to NO_2. In the case of highly reactive compounds, such as VOCs, indoor chemical reactions may be necessary to account for. Those compounds generate challenges in the models because their reactions can be very sensitive to the timestep of the model. Fig. 12.2 illustrates the impact of the timestep on the reactions. System (A)

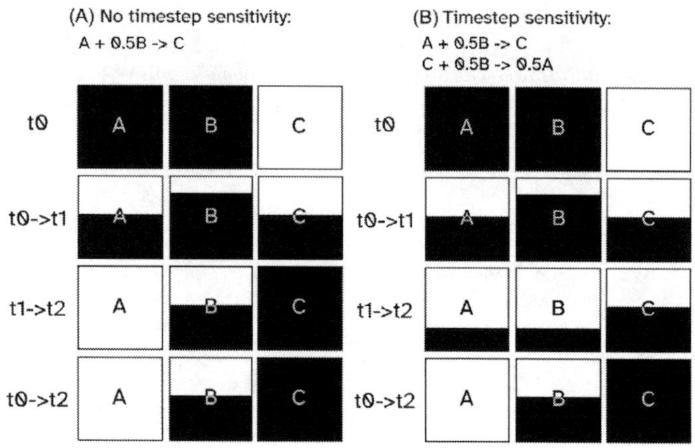

FIGURE 12.2 Representation of a chemical system with (A) and without (B) timestep sensitivity.

uses one equation with three compounds (A, B, and C), and system (B) uses two equations with the same compounds. The squares represent the quantity of each compound in the referred timestep. The timestep is indicated as t# with # being the step number. The lines present four states: the first is the initial quantity at t_0, the second is the quantity calculated in the next step t_1, and the third is the subsequent step t_2 calculated from the t_1. The last line represents the result derived from calculating t_2 directly from t_0, which is different from the intermediate step t_1 in a system sensitive to the timestep.

The sources of indoor pollutants were discussed in Section 12.3. The sinks depend on the pollutant, but ventilation is a major factor in all cases. Many heating, ventilation, and air conditioning (HVAC) units have their own PM filters, so they must be considered in the calculations. In addition, the deposition of PM on interior surfaces is an important sink. The deposition coefficient is a function of the particle size as described by Riley et al. (2002), with a minimum of 0.00001 second^{-1} for PM between 0.10 and 0.20 μm, maximums of 0.002–0.003 second^{-1} for PM with 10 μm and a theoretical 0.003 per second for PM with 0.001 μm aerodynamic diameter. In the case of ozone, surface removal is a relevant sink due to its high reactivity. However, the rate depends on many variables such as the age of the surface material, the airflow, and the relative humidity. It makes the ozone concentration extremely difficult to forecast with precision without laboratory measurements. Lai et al. (2015) investigated 12 studies about ozone removal rates in real conditions. They found a median removal rate of 4.13 hour^{-1}, with the 10th percentile of 2.80 hour^{-1} and the 90th percentile of 5.40 hour^{-1}. Another form of representation of the removal rates is the deposition velocity (v), that is the flux of the deposition (F, deposited mass per area per time) per concentration of the pollutant. Grøntoft and& Raychaudhuri, (2004) compiled the deposition velocities for ozone, NO_2 and SO_2 from various sources as listed in Table 12.4.

Surfaces can act as sources and/or sinks for VOC pollutants, depending on their material and structure. For instance, wood, sealants, carpet, and adhesives can act as VOC sources, meanwhile, vinyl flooring, polyurethane foam, styrene – butadiene rubber carpet backing, and other polymeric substances can act as sources or sinks, depending on the relative magnitudes of the gas- and material-phase concentrations. Analytical equations can model the behavior of simple single-layer materials (e.g., linoleum) like the ones developed by Kumar and Little (2003). However, multilayered surfaces (e.g., stacked wallpapers) with abrupt changes in composition and porous materials demand additional efforts in numerical modelling. Liu et al. (2013) made a systematic review of the subject with a comprehensive analysis of single- and multilayer models for VOCs.

The second and well-implemented approach of the analytical models is computational fluid dynamics (CFD) modelling. In this approach, the air velocity and its derivatives are calculated inside a determined volume, subdivided by a tri-dimensional grid. It is the most applicable approach used in available software. Unlike the analytical models, it does not assume homogeneous pollutant concentration inside the modelling domain. CFD models can use either Eulerian or Lagrangian approaches. The Eulerian approach, which is the most common approach, uses a room as a point of reference, calculating the derivatives of each cell on the grid in the timestep and adding it to the current state. Its computational cost is directly dependent on the number of cells in the rooms, and, consequently, its resolution. The Lagrangian approach starts with a selected cell in a grid and follows its changes over time. Lagrangian models demand more computational power, however, they can trace back the origin of the pollutants in a specific region with precision. However, this level of precision is not necessary in most of the modelling cases, making Lagrangian models more applicable in the presence of fast transport (e.g., vents with high-speed airflows).

TABLE 12.4 Deposition velocities of ozone, NO_2, and SO_2 for various materials. Compilation by (Grøntoft & Raychaudhuri, 2004).

Material	Ozone removal rate ($cm\,s^{-1}$)	NO_2 removal rate ($cm\,s^{-1}$)	SO_2 removal rate
Lacquered ash	0.0007 ± 0.0008[a]	–	–
Oiled ash	0.003 ± 0.0007[a]	–	–
Oiled beech parquet	0.0078 ± 0.0027[a]	–	–
Plywood	0.030[b]	–	–
Lacquered plywood	–	0.00639[c]	–
Softwood	–	–	0.024[d]
Hardwood	–	–	0.048[d]
Painted wood	–	0.003[c]	–
Wallpaper (50%–60% RH)	4% *,[e]	–	0.015[f]
Sandstone	–	0.02[g]	–
Used brick	–	0.059[h]	–
Aluminum (30%–32% RH)	0.000698[i]	0.0006[j]	–
Aluminum (50%–60% RH)	–	0.002[j]	0.002[k]
Aluminum (87%–90% RH)	0.0294[l]	–	0.01[k]
Painted stainless steel	–	0.00083[c]	–
Linen	0.0095[b]	–	–
Cotton bathmat (41% RH)	–	0.0211[c]	–
Cotton bathmat (67% RH)	–	0.055[c]	–
Cotton terry cloth	–	0.0047[h]	–
Cotton muslin	0.109[b]	–	–
Cotton/polyester bedspread	–	0.0386[h]	–
Wool (lamb)	0.107[b]	–	–
Wool/polyester fabric	–	0.0047[h]	–
Wool carpet (50%–55% RH)	–	0.0919[c] /0.0856[h]	0.021[m]
Linoleum (50%–55% RH)	0.007 ± 0.004[a]	–	0.003[m]
Nylon	0.032[b]	–	–
Nylon carpet (50%–55% RH)	0.032 ± 0.0043[a,**]	–	0.007[m]
Painted gypsum board (urethane modified alkyd binding agent)	0.03 ± 0.0052[a]	–	–
Painted gypsum board (acrylic binding agent)	0.042 ± 0.0023[a]	–	–
Vinylchloride treated floor-sheet (45% RH)	–	0.0022[c]	–
Vinylchloride treated floor-sheet (70% RH)	–	0.00278[c]	–
Plastic tile	–	0.00333[c]	–
Plasterboard (44% RH)	–	0.0169[c]	–
Plasterboard (67% RH)	–	0.0272[c]	–

The percentage of ozone deposited in the test section.
**With latex backing.*
[a]*Klenø et al. (2001).*
[b]*Sabersky et al (1973).*
[c]*Miyazaki (1984).*
[d]*Spedding (1970).*
[e]*Reiss et al. (1994).*
[f]*Spedding and Rowlands (1970).*
[g]*Behlen et al. (1996).*
[h]*Spicer et al. (1989).*
[i]*Cox and Penkett (1972).*
[j]*Leaderer et al. (1986).*
[k]*Spedding (1972a).*
[l]*Mueller et al. (1973).*
[m]*Spedding (1972b).*

12.7.2 Machine learning

ML is an old field of study with a recent uprising due to the technological advances of the last few years. ML models for IAQ studies normally focus on the IAQ prediction based on a set of inputs (e.g., pollutant concentration, temperature, relative humidity, and air velocity). ML is a statistical approach to modelling, where a target variable is defined by a function of known variables (called features, Fig. 12.3A). The function is completely naive to the underlying equations in the system, being completely dependent on the features' variability and information. They are generic and adaptative, capable of approximating the analytical functions using a set of parameters and an error function that is minimized using part of the data (called training set). The advantage of ML is the capacity to model unaccounted relationships and even hidden variables in some cases. The five main problems in ML modelling are (i) the lack of interpretability, as the final function may not express its underlying mechanisms intuitively; (ii) the range of the data, because the function may not properly learn the behaviors outside the range of the training set; (iii) the need of a huge training set; (iv) the high computational cost (with computational time between hours and days, depending on the model and case); and (v) the overfitting. Overfitting is the situation where any algorithm interprets the bias and errors of the training dataset as valid information and incorporates it into the model (Fig. 12.3B). It leads to the loss of generality in the model, as the errors and biases of the training dataset are different from other datasets. The concept of "algorithm bias" derives from this concern, which is extremely important to anyone who uses ML to comprehend. Kordzadeh and Ghasemaghaei (2022) systematically reviewed this concept. There is no definitive solution to overfitting, however many algorithms try to cope with diverse cases. Ultimately, the main factor in the quality of the model is the quality of the dataset.

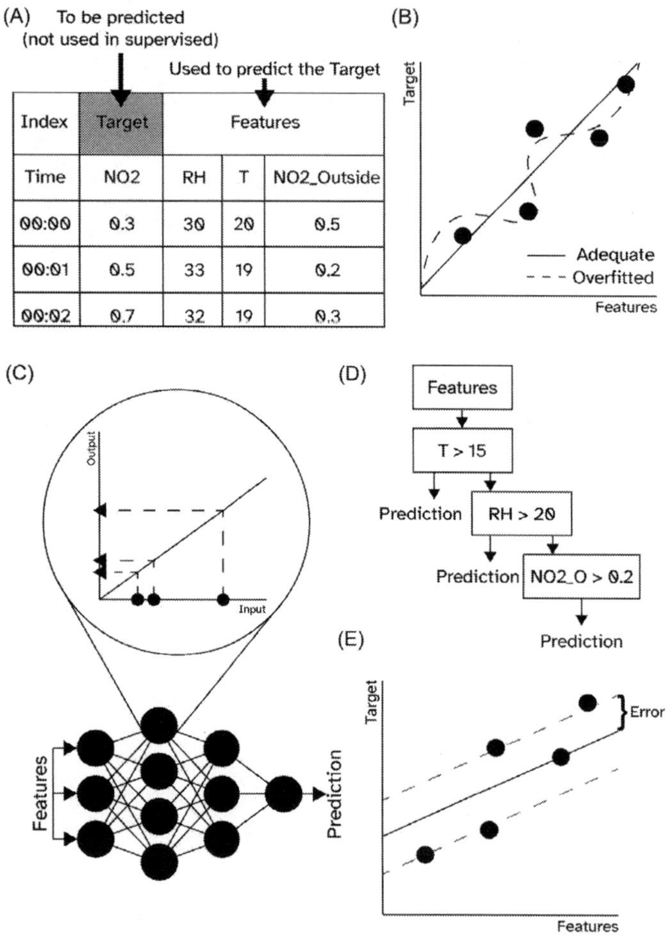

FIGURE 12.3 Representations of (A) target (label) and features; (B) overfitting, (C) Artificial Neural Networks; (D) Decision Trees; and (E) Support Vector Machines in two dimensions.

Models based on ML follow two main categories, supervised and unsupervised. Supervised models use a training set composed of explanatory variables (called features) and the target variable (called label in case of classification models, or target in regression models, explained below). A basic and well-known example of supervised models is linear regression. The parameters of a linear regression model can be calculated analytically, however, they can be estimated using ML algorithms too. To do so, initial random guesses are made for the parameters and an error term is added to the equation. Then, an iterative algorithm (e.g., least squares) is implemented to minimize the error, which leads to an approximation of real parameters. This process is more computationally efficient than analytical calculations.

ML tries to solve two main tasks, classification, and regression. Classification is the attribution of classes to points in a dataset based on its features, while regression is the attribution of a value to points based on its features. Statistically, the main difference between classification and regression is the final value being a discrete label variable in classification and a continuous variable in the regression. For instance, the concentration of a pollutant in a room can be classified as nonharmful and harmful, or its value of concentration in ppm can be estimated by regression. Another difference is the lack of ordinance in the classification output, that is, the classes do not necessarily have a relative value. For example, the dataset can have some outliers, so a model can be developed to flag the data points as "valid," "outlier," and "peak event," and none of them is numerically greater or lesser than the other.

Regression models strongly depend on the chosen equations, available explanatory variables, and functions to minimize the error. Each choice comes with a set of underlying statistical assumptions. One of the most important assumptions is the sufficiency of the explanatory variables. Unfortunately, real-world studies cannot monitor all variables in the system. Therefore, regressions will always present disturbances. In the cases when disturbances do not overcome the signal of the target variables, and its distribution is random around the modelled function, the regression may be appropriate. However, in pollution data, it is expected that the variance of the disturbances will be correlated to the pollutant concentration. Hence, it violates the assumption of heteroscedasticity of ordinary linear regressions, making some statistical analyses, specifically t-tests and F-tests, inapplicable (Milionis & Davies, 1994). Another fundamental factor to be considered is the use of the pollutant concentration in the past in the model (autocorrelated variable). It is a common practice because an autocorrelated variable can carry unaccounted information from the real system. However, similar to the disturbances' variance problem mentioned above, it makes t-tests and F-tests inapplicable.

A solution to those problems is the use of stochastic models such as autoregressive integrated moving averages (MAs) or ARIMA (Box & Jenkins, 1970). Stochastic models use random variables in their algorithms. The ARIMA model uses a combination of two patterns of autocorrelation, autoregression (AR) and stochastic MAs to estimate the pollutant concentration. In an AR model, the pollutant concentration can be described as the concentrations in previous times multiplied by a parameter ϕ. The stochastic MA describes the pollutant concentration using a window of previous data for the model. This window considers the previous values as random shocks in the average. Stochastic models need fewer assumptions due to the random nature of the error, making them more adequate in IAQ studies than ordinary regressions (Milionis & Davies, 1994).

The main family of supervised algorithms used nowadays in IAQ applications is artificial neural networks (ANN). ANN is composed of a sequential arrangement of dozens (or more, depending on the computational power and data availability) of multiple regressions (Fig. 12.3C). This allows an information flow inside the algorithm that processes different aspects of the data without explicit instruction. Other types of noteworthy supervised ML algorithms are random forests (RFs) and support vector machines (SVM) families. RF algorithms are based on the ensemble of many decision trees. Decision trees are algorithms that predict the target values by a set of conditions based on the features (Fig. 12.3D). For instance, a tree can assert the concentration of PM in a room as $12\,\mu g\,m^{-3}$ if the temperature is greater than 20°C, then, if the relative humidity is below 0.2%, then if the PM outside is below $15\,\mu g\,m^{-3}$, and so on. The number of conditions, variables, and thresholds are defined in the training process. SVM algorithms calculate hyperplanes to divide the data into target categories given an error threshold. The equivalent in two dimensions is to find the best line that can separate the categories by the given features. In IAQ, the categories can be thresholds of pollutant concentration. The use of SVM algorithms in regression problems (e.g., predicting the pollutant concentration using other variables) is similar to linear regression, but with an error threshold in the calculation (Fig. 12.3E).

Unsupervised models are used in different applications compared to supervised models. One of the applications is dimensionality reduction, that is, reducing the number of features of the data by combining the information of similar ones in a new set of variables. Unsupervised algorithms range from principal component analysis to ANN called autoencoders. For instance, the model developed by Loy-Benitez et al. (2020) uses a type of autoencoder to impute missing data in IAQ data inside subway metro systems. However, supervised ANN models are more often used in IAQ studies. Another important application of unsupervised models is clustering, i.e., the grouping of similar observations. Unsupervised techniques can be used to improve the supervised algorithms, filtering useless information from the dataset, and reducing the complexity. They can also be used as a basis for source apportionment analysis.

Example 12.3.: Which aspects should one consider when choosing between analytical and ML models?

Solution: *Data availability*—Analytical models require emission data from sources and sinks in the case of a study aiming to forecast the IAQ, meanwhile, ML models can forecast using just the concentration history.

Computational resources—The most complex ML models can demand a lot of computational power, especially in the use of ANN, which requires GPU power, instead of CPU. Analytical models can be more accessible, especially if the model is not a CFD.

Regulatory needs—In the case where the study needs to follow regulatory guidelines, it is recommended to utilize models recommended by regulatory agencies as they may need to follow some principles for example, prioritize overestimation to improve reliability. Nowadays, models based on ML are not as common as analytical ones in regulatory studies.

Human interpretability—Many models based on ML can provide no information about its intern processes. It can be a hindrance on occasions, for example, when debugging is required and when it is necessary to build trust in a non-specialized audience.

Other aspects can be necessary as well depending on the application.

12.7.3 Examples of indoor air quality models

Two good examples of accessible operational IAQ models are the EPA's Simulation Tool Kit for IAQ and Inhalation Exposure (IAQX) and CONTAM (Guo, 2000a, 2000b). The IAQX is free software developed by the EPA's National Risk Management Research Laboratory. The first well-supported and validated version of the IAQX was released in 2000 when the most advanced techniques were not included in its development. It is designed to analyze the impact of pollutant sources and sinks, ventilation, and air cleaners. It calculates the pollutant concentration and personal exposure as a function of time and can derive useful analysis from its calculation. The IAQX uses an analytical mass-balance chemical model, dividing the building into room-size air zones assuming that the air in each zone is well mixed. The latest IAQX tool (version 1.1) is currently available on EPA's website, air research section.

The CONTAM is a complex open-source model supported by the US National Institute of Standards and Technology (NIST) and is designed to determine the airflow rates inside and outside the building, contaminant concentrations, and personal exposures (Dols & Polidoro, 2020). CONTAM is used not only to assess various environment variables over time (e.g., ventilation rates, distribution of ventilation, and infiltration rates) but also to predict pollutant concentrations, estimate personal exposure based on occupancy patterns, and evaluate intervention scenarios. In principle, the CONTAM model uses a one-dimensional convection-diffusion flow algorithm to simulate ducts and long hallways as a compromise between the room-scale mass balance and the CFD. CONTAM also uses a Lagrangian model for high-speed flows in ducts and automatically changes to Eulerian in low-speed scenarios.

A pair of ML studies (Sofuoglu, 2008; Xie et al., 2009) used the data from the EPA's Building Assessment Survey and Evaluation (BASE) study to develop a model with the aim to predict IAQ. The BASE database combined the data from 100 random office buildings from 10 regions in the United States. This variety in the input data made the ML model generic in cost of precision. However, most ML models need to be tailor-made to some degree as indoor environments can be radically different. An example of this is the study of Elbayoumi et al. (2015), where the authors developed an ANN to predict PM concentrations inside schools using three consecutive days of measurements in each site. For those, the main input variables are the target pollutant concentration, indoor temperature, relative humidity, wind speed, outdoor concentrations, and ventilation rates. By now, few articles have reviewed the implementation of ML models in research-based IAQ applications (Omidvarborna et al., 2021; Wei et al., 2019). They concluded that ML algorithms are powerful tools in IAQ problems, however, there are few studies about feature and hyperparameter selection, performance metrics, and model comparisons. Thus, the current research in ML applied in IAQ still lacks a solid base to develop.

12.8 Case studies

12.8.1 Kitchens

Kumar et al. (2022) conducted a study to assess $PM_{2.5}$ and PM_{10} exposures in kitchens of 60 low-income homes across four continents and 12 cities, including Dhaka (Bangladesh), Chennai (India), Nanjing (China), Medellín (Colombia),

São Paulo (Brazil), Cairo (Egypt), Sulaymaniyah (Iraq), Addis Ababa (Ethiopia), Akure (Nigeria), Blantyre (Malawi), Dar-es-Salaam (Tanzania) and Nairobi (Kenya). The primary goals were to establish household exposure profiles across a range of low-income houses and to specify the key determinants influencing in-kitchen exposure to fine and coarse particles. In each city, five low-income homes were examined for one week of continuous air quality monitoring of their kitchens. Apart from $PM_{2.5}$ and PM_{10} monitoring, the qualitative information of the building and occupants (i.e., building location, apartment type, kitchen type, ventilation conditions, floor area, kitchen volume, window size, door dimensions, and monitoring location) and the outdoor surrounding area through the surveys were carried out by the field researchers. The occupant survey considered factors such as the number of kitchen occupants during cooking, type of cookstoves and fuel used, time and duration of cooking, type of cooking carried out, types of cuisines, and status of natural and mechanical ventilation during cooking. To confirm that findings across in-kitchen exposure settings in the examined homes in all cities were comparable, the authors checked whether or not the following criteria were met by all selected homes: (i) residents belonged to low-income families, (ii) homes were either ground or first floor, (iii) all measurements were made in kitchens, (iv) monitors were placed at the breathing height (1.5 m above ground level) and approximately 1.5 m from the cooking, (v) identical measuring equipment was used, (vi) the same duration of data collection, (vii) the same number of homes in all cities, and (viii) whether cooking occurred on a daily basis or not. The factory-calibrated Aeroqual Series 500 Portable Air Quality monitor with a PM sensor was used to measure $PM_{2.5}$ and PM_{10} concentrations in each house across all cities. Measurement campaigns were conducted between March and October 2021 (for 24 hours in 7 days continuously) in each home. One-minute measurement intervals of $PM_{2.5}$ and PM_{10} were collected in 60 homes, adding up to 420 days (10,080 hours) across all cities. The mean highest $PM_{2.5}$ ($185 \pm 48\ \mu g\ m^{-3}$) and PM_{10} ($220 \pm 58\ \mu g\ m^{-3}$) concentrations were recorded in Dhaka due to small kitchen volume, extensive frying and prolonged cooking compared with the lowest values from Medellín ($PM_{2.5}$: $10 \pm 3\ \mu g\ m^{-3}$; PM_{10}: $14 \pm 2\ \mu g\ m^{-3}$). Exposure profiles of kitchen occupants showed that the fuel, volume of the kitchen, type of cooking and ventilation rates were the most prominent factors affecting in-kitchen exposure. Dual ventilation (mechanical and natural) in Chennai, Cairo and Sulaymaniyah reduced average in-kitchen $PM_{2.5}$ and PM_{10} exposure by 2.3- and 1.8-times compared with natural ventilation (open doors) in Addis Ababa, Dar-es-Salam, and Nairobi. The study reported that smaller-volume kitchens had higher cooking exposures compared with their larger-volume counterparts. Potential exposure doses were highest for Asian homes, followed by African, Middle Eastern and South American homes. The study concluded that increased cooking exhaust extraction, cleaner fuels, awareness of improving cooking practices and minimizing passive occupancy in kitchens as the necessary measures to mitigate harmful cooking emissions in kitchens.

12.8.2 Schools

A study was conducted in two urban schools in Gliwice, Poland to determine the levels of both indoor and outdoor VOCs (Benzene, Toluene, Ethylbenzene, M, P-Xylene, O-Xylene, Styrene, sec-Butylbenzene, *n*-Butylbenzene, and Naphthalene) and PM ($PM_{2.5}$ and PM_{10}) during the winter in younger and older children's classrooms (Mainka et al., 2015). The school for younger children was situated in a residential neighborhood next to an air quality station on the southwest side of the city center. The school for older students was situated close to a busy street. The study's main goal was to assess the various pollutant concentrations in indoor settings. In each school, the measurements were taken for ten days. The sampling location was chosen to be away from the main door between 0.8 and 1.0 meters above the ground, which represents the normal child's breathing height. The SKC air sampling pump (model 224-PCMTX8) with a flow rate of $6\ m^3\ h^{-1}$ was utilized to measure VOCs. The samples were taken using stainless steel tube samplers from Perkin Elmer. Thermal desorber TurboMatrix 100 linked to a gas chromatograph Clarus 500 (PerkinElmer, Inc., Waltham, MA, USA) with a flame ionization detector (FID) was used to analyze the samples of VOCs. PM fractions were measured using an Atmoservice PNS-15 with a flow rate of $2.3\ m^3\ h^{-1}$. Among all VOCs, benzene had the highest concentration ($1.63\ \mu g\ m^{-3}$ in schools for younger children, $2.93\ \mu g\ m^{-3}$ in schools for older children). However, the benzene concentration did not exceed the annual ambient air quality standard of $5\ \mu g\ m^{-3}$. The measured PM_{10} concentrations were $118\ \mu g\ m^{-3}$ in the school for older students and $68\ \mu g\ m^{-3}$ in the school for younger students. Additionally, the older children's school had higher $PM_{2.5}$ levels ($71\ \mu g\ m^{-3}$) than the younger children's school ($41\ \mu g\ m^{-3}$). They concluded that indoor sources combined with outdoor infiltration of pollutants were the major cause of higher PM levels in schools.

12.8.3 Offices

A study was conducted in Kunming, Changsha, and Tianjin, China, to examine the interior air quality of office buildings and how it relates to the interplay between employees' moods and health. The study's main objectives were to (1) investigate the IAQ on different floors of multi-story office buildings through on-site measurements of the CO_2, formaldehyde, and TVOC concentrations; (2) to estimate the indoor environment quality and people's perceptions of offices; (3) conduct pollutant risk assessments and compare differences at the city level; (4) provide quantifications for IAQ in offices and suggestions for improving the indoor environment. According to the China Energy Efficiency Testing of Public Buildings Standard, three levels of sampling sites were taken into consideration to characterise the IAQ in relation to building height, including low-level, medium-level, and high-level. The sampling locations within the building were between 1.2 and 1.5 m above the ground. Additionally, at least two office spaces and four sampling locations were established for each floor level, with the number of sampling locations selected based on the size of the office space. According to the regulation, a 30-m-distance outdoor sampling location in front of the office building was considered. A questionnaire survey was also carried out to find out how people felt about IAQ in the office building. To measure the pollutant levels, the study utilized instruments such as the Formaldehyde 400 for formaldehyde, the PGM-7240 ppb RAE for TVOCs, and the Telaire 7001 for CO_2 concentrations. More than 900 questionnaires were gathered from the offices in these cities. In the study, it was found that formaldehyde concentrations were highest in Tianjin (0.05 $mg\,m^{-3}$) followed by Kunming (0.04 $mg\,m^{-3}$), and Changsha (0.01 $mg\,m^{-3}$). TVOC concentrations followed the order of Kunming (0.44 $mg\,m^{-3}$) > Tianjin (0.20 $mg\,m^{-3}$) > Changsha (0.04 $mg\,m^{-3}$). The highest CO_2 concentrations were found in Tianjin (671 ppm) and the lowest in Kunming (513 ppm). It was discovered that the air quality in Changsha's offices was generally better. They noticed that pollutant concentrations varied depending on the time of day, with the morning and afternoon having the poorest air quality. They also noticed that outdoor and indoor pollutant concentrations were strongly correlated. The study found that using outdoor fresh air with filters and purifiers can help lower interior pollution (Cheng et al., 2022).

12.8.4 Hospitals

A study was undertaken at a busy and large Iranian teaching hospital to evaluate the factors affecting IAQ and their impact on respiratory health. The goal of this cross-sectional study was to find out what factors affected the hospital staff's respiratory functions. The study included 228 hospital employees as a case group (those with more than one year of experience), including 86 nurses, 37 paramedics, 33 cleaning staff, 19 surgical technicians, 20 medical laboratory technicians, and 33 other workers (Veysi et al., 2019). People were chosen from each case group if they had not suffered from respiratory illnesses. A control group of 228 office workers was also chosen, with the inclusion criterion of having more than a year of work experience and being free of respiratory ailments. The researchers employed the American Thoracic Society's standard respiratory questionnaire to investigate respiratory disorders such as phlegm, wheezing, acute cough, and shortness of breath. Several parameters were measured with the instruments as part of the investigation. The aerosol monitoring equipment model GT-31 from SIBATA, Japan, was used to assess PM_{10} concentrations. A Phocheck Tiger VOCs detector from Ion Science was used to measure the VOCs. Trypticase soy agar and Sabouraud dextrose agar were used to examine the bacterial and fungal cultures present. Bioaerosol concentrations were determined using The National Institute for Occupational Safety and Health (NIOSH) technique 0800. These characteristics were measured in three rounds at the hospital, according to the personnel's shifts (morning, afternoon, and evening). Similarly, measurements were conducted at the hospital's office buildings. To assess the data, several statistical approaches were utilized, including the Chi-Square test, T-test, and logistic regression. The concentration of bioaerosols such as bacteria was the highest in hospital wards (80.77 ± 169.48 $cfu\,m^{-3}$) compared to offices (19.35 ± 30.07 $cfu\,m^{-3}$). Similarly, the highest concentration of fungus was observed in hospital wards (29.97 ± 55.15 $cfu\,m^{-3}$). For PM_{10} and VOCs concentrations, the highest concentrations were found in hospital wards (28.18 ± 38.7 $\mu g\,m^{-3}$, 455 ± 389.5 ppb) compared to offices (10.02 ± 22.61 $\mu g\,m^{-3}$, 38.7 ± 120.7 ppb), respectively. The overall concentrations of pollutants were higher in hospital wards than in offices, according to the authors. They also discovered that hospital employees had greater phlegmatic coughs and wheezing. The investigation determined that the higher concentrations inside the hospital were caused by a breakdown of the ventilation systems within the wards.

12.8.5 Transport microenvironment

A study was carried out to assess the personal exposure levels to PM (PM_{10}, $PM_{2.5}$, and PM_1) in two transport microenvironments (bus and car) and compare with the personal exposures while cycling and walking in a predefined route in a typical town Guildford, United Kingdom (Kumar et al., 2018). To achieve the broader goal of quantification of personal exposures to PM fractions in these microenvironments, the study was based on clearly defined objectives: (a) to compare the different PM fractions among all the modes and for different times of the day (morning peak: 08:30–11:00 hours, mid-day off-peak: 13:00–15:00 hours, and evening peak:17:00–19:00 hours); (b) to evaluate various levels of respiratory deposition doses (RDDs) of PM fractions during peak and off-peak hours, and (c) to identify the factors influencing the temporal variability of PM fractions in different transport microenvironments. The PM concentration in the study was measured using a portable environmental dust monitor (GRIMM, model 1.108), which works on the principle of light scattering. Measurements were carried out during February—April 2015, in which a total of 142 trips were made, of which 30 trips by bus, 52 trips by car and 30 each by cycle and walk. The study reported that the mean PM_{10} concentrations were the highest on the bus ($90 \pm 63\ \mu g\ m^{-3}$), followed by walking ($63 \pm 76\ \mu g\ m^{-3}$), on the car ($23 \pm 9\ \mu g\ m^{-3}$) and the lowest by cycling ($14 \pm 17\ \mu g\ m^{-3}$). However, for fine particles ($PM_{2.5}$ and PM_1), the highest concentration was obtained for the car and the lowest concentration was for the cyclist. For PM (PM_{10}, $PM_{2.5}$, and PM_1) concentrations during different periods of the day, the highest concentrations were found during the evening peak periods for bus journeys, while the highest PM concentrations were found during morning peak periods for car journeys. RDD, which is a function of physical activity and particle size distribution, for coarse particles was the highest for the walking mode ($56 \pm 14\ \mu g\ h^{-1}$) and the lowest for the car mode ($1.2 \pm 0.3\ \mu g\ h^{-1}$). However, the highest RDD for fine particles was obtained for walking ($5.5 \pm 0.3\ \mu g\ h^{-1}$) and cycling ($5.1 \pm 1.2\ \mu g\ h^{-1}$) modes followed by bus ($4.1 \pm 0.7\ \mu g\ h^{-1}$) and car ($2.0 \pm 0.2\ \mu g\ h^{-1}$) modes. The study concluded that the particle concentration and RDD of coarse particles were the lowest for the car mode. The car mode displayed the lowest RDD despite higher concentrations of fine particles. Due to the mode of operation of car journeys with closed windows and filtration through the air conditioning system, the car journeys exhibited lower concentrations of coarse particles. Physical activities of the car passengers were relatively modest compared to cycling or walking mode, which makes the rank ordering of RDD different from the particle concentrations. The study concluded that RDD needs to be considered in addition to particle concentrations for curtailing the adverse health effects.

Example 12.4.: According to the reviewed case studies, what are the reported pollutants and their contributors in common indoor spaces, such as transport microenvironments, hospitals, schools, and offices?

Solution: *Transport microenvironments:* PM (PM_{10}, $PM_{2.5}$, and PM_1) from on-road vehicles such as buses and cars.

Hospitals: PM_{10} (dust resuspension probably due to people movement/activity), VOCs (probably due to building materials, hospital furniture, cleaning agents, disinfectants, etc.) and bioaerosols (such as bacteria, molds, and pollen that sourced from indoor occupants, animals, and soil and plants outdoor).
Schools: VOCs (such as benzene, toluene, ethylbenzene, M,P-xylene, O-xylene, styrene, sec-butylbenzene, n-butylbenzene, and naphthalene) and PM ($PM_{2.5}$ and PM_{10}) mainly sourced from nearby traffic.
Offices: CO_2 (from exhalation), formaldehyde and TVOC from the building materials, office furniture, cleaning agents, disinfectants, etc.

12.9 Policy and regulatory considerations

Exposure to IAP has been a big concern in recent decades (Klepeis et al., 2001). As discussed in Section 12.3, IAQ is disrupted by air pollutants of both indoor and outdoor origins. Although the major source of air pollution indoors is emissions from indoor activities, the infiltration of outdoor air pollutants through open windows and doors and supplying air ventilation systems are the other important contributors (Shrestha et al., 2019). Therefore, it is expected that reducing emissions of pollutants from the outdoors should result in improving IAQ levels.

Governments and agencies issue regulations and standards to keep air pollution levels under control. The available guidelines are summarized in Table 12.5. As shown in this table, regulations and standards are used for air quality control in several leading countries worldwide. Some countries have developed guidelines for both ambient and indoor air contaminants; however, regulations on IAQ are still not available in some countries like the European Union (EU) and Australia (Abdul-Wahab et al., 2015). In the United States (US), the Environmental Protection Agency (EPA) has set threshold limits for common ambient air pollutants under the Clean Air Act. Although EPA helps to protect IAQ in different ways, unlike ambient air quality it does not regulate IAQ (USEPA, 2022).

Similarly, in the United Kingdom (UK), concentrations of key pollutants in outdoor air are regulated by the Air Quality Standards Regulations 2010, the Air Quality Standards (Wales) Regulations 2010, the Air Quality Standards (Northern Ireland) 2010 and the Air Quality Standards (Scotland) Regulations 2010 (DEFRA, 2022). However, with assistance from the government and authorities, legislation, and threshold limits for regulating IAQ have not been set yet. In the absence of current limits and regulations for indoor air in the UK, governmental authorities recommend using the WHO Air Quality Guidelines (Shrubsole et al., 2019).

TABLE 12.5 Indoor and outdoor air quality guidelines and standards in the leading countries.

Country	Current policies and standards	
	Outdoor air quality	Indoor air quality
US	The Clean Air Act 1970[a] (National Ambient Air Quality Standards (NAAQS)	Standards for Ventilation and Indoor Air Quality ANSI/ASHRAE Standard 62.1–2019
UK	Air Quality Standards Regulations 2010[b] Air Quality Standards (Wales) Regulations 2010[c] Air Quality Standards (Northern Ireland) 2010[d] Air Quality Standards (Scotland) Regulations 2010[e]	UK Guidelines for Volatile Organic Compounds in Indoor Spaces
	The National Emission Ceilings Regulations 2018	
EU	Directive 2004/107/EC, 15 December 2004 relating to arsenic, cadmium, mercury, nickel and polycyclic aromatic hydrocarbons in ambient air	Not Available
	Directive 2008/50/EC, 21 May 2008 on ambient air quality and cleaner air for Europe	
	Directive (EU) 2015/1480, 28 August 2015 amending several annexes to Directives 2004/107/EC and 2008/50/EC of the European Parliament and of the Council laying down the rules concerning reference methods, data validation and location of sampling points for the assessment of ambient air quality (Text with EEA relevance)	
China	Ambient Air Quality Standard GB 3095-2012	IAQ Standard GB/T 18883–2002
Australia	National Environment Protection Measure for Ambient Air Quality (Air NEPM)	Not Available
Canada	Canadian Ambient Air Quality Standards (CAAQS)	Residential IAQ Guidelines
Worldwide	WHO Global Air Quality Guidelines (AQGs) 2021	WHO AQGs 2021

[a]The legal authority for federal programs. Also, The Clean Air Act requires EPA to set National Ambient Air Quality Standards (NAAQS) for six common air pollutants.
[b]The Secretary of State for Environment, Food and Rural Affairs has the responsibility for meeting the limited values in England.
[c]Responsibility for meeting air quality limit values is devolved to the national administrations in Wales. Department for Food and Rural Affairs (DEFRA) coordinates assessment and air quality plans for the UK as a whole.
[d]Responsibility for meeting air quality limit values is devolved to the national administrations in Northern Ireland. DEFRA coordinates assessment and air quality plans for the UK as a whole.
[e]Responsibility for meeting air quality limit values is devolved to the national administrations in Scotland. DEFRA coordinates assessment and air quality plans for the UK as a whole.

Two sets of IAQ-relevant standards have been established in China that are directly related to IAQ (Bai et al., 2003). In contrast with China, there are no specific regulative controls for IAQ in Australia in the current situation (see Table 12.5. The Australian perspective on the available US standards is practised in public buildings in Australia (Dingle & Murray, 1993).

Apart from the regulations and standards, the WHO has recently updated its guideline in 2021, which is more stringent than its previous one for both indoor and ambient air quality in 2005. The WHO guideline, as shown in Table 12.6, recommends the air quality levels for six pollutants which have evidence of the most health effects from exposure (Zhao et al., 2022).

Consequently, if there is a concern about outdoor air quality, IAQ should get even more attention due to spending time indoors. Measurements of pollutant concentrations and thermal comfort are required for IAQ regulations and methodologies for the evaluation and management of indoor air (Kumar et al., 2016). Also, the COVID-19 pandemic has changed people's lifestyles, and they have begun to spend more time indoors (Ravindra et al., 2022). Since the beginning of the pandemic, international institutions and agencies have initiated and updated consistently air quality guidelines and standards to meet requirements for the management of IAQ. However, it seems that there is a regulatory gap for a generalized standard (Zhang et al., 2020).

TABLE 12.6 The air quality guide levels recommended by WHO (2021b).

Pollutant ($\mu g\, m^{-3}$)	Averaging time	Interim target				Air quality guide level
		1	2	3	4	
$PM_{2.5}$	Annual	35	25	15	10	5
	24-hour[a]	75	50	37.5	25	15
PM_{10}	Annual	70	50	30	20	15
	24-hour[a]	150	100	75	50	45
O_3	Peak season[b]	100	70	–	–	60
	8-hour[a]	160	120	–	–	100
NO_2	Annual	40	30	20	–	10
	24-hour[a]	120	50	–	–	25
SO_2	24-hour[a]	125	50	–	–	40
CO	24-hour[a]	7	–	–	–	4

μg = microgram
[a]*99th percentile (i.e., 34 exceedance days per year).*
[b]*Average of daily maximum 8-hour mean O_3 concentration in the six consecutive months with the highest six-month running-average O_3 concentration.*

Example 12.5.: In the air quality guidelines published by the World Health Organization in 2021, what limit values are recommended for which air pollutants? Why is IAP largely unregulated?

Solution: The WHO guideline published in 2021 recommends the air quality levels for six pollutants: $PM_{2.5}$, PM_{10}, O_3, NO_2, SO_2, and CO which could impact IAQ.

IAP comes from a variety of indoor and infiltrated outdoor sources. Activities such as smoking cigarettes or burning fossil fuels (e.g., natural gas) or biomass (e.g., woods) for heating or cooking purposes, are somehow regulated by installing CO sensors or banning smoking in public places. However, the guidelines set by federal and international agencies have not translated much into action or they are not enforced on the ground by now. Therefore, additional efforts are needed to make this happen.

12.10 Summary and conclusions

Exposure to IAP is linked to significant health impacts. The source characteristics of various indoor microenvironments differ which makes the understanding of IAP exposure challenging. This section focused on various indoor environments such as homes, schools, hospitals, offices, and transport microenvironments, where individuals spend most of their daily time. In each microenvironment, the sources of IAP and suggested mitigation steps were elaborated. We summarized environmental parameters and IAP, common indoor and outdoor sources, and associated health impacts.

IAP originates mainly from indoor human activities, building components, and the ingress of polluted outdoor air. Due to the confined indoor microenvironments, surfaces, and occupants, indoor activities are important factors in affecting indoor chemical processes. Therefore, indoor gas phase chemistry, indoor surface chemistry, and indoor photochemistry could contribute to the formation of secondary pollutants. We summarized the key factors controlling indoor chemical transformation (e.g., humidity, occupants, and buildings).

To understand the emissions from indoor sources, monitoring is required. We summarized the IAQ monitoring methods that can also allow us to better understand indoor exposure levels. There are several technologies available that offer resolution, precision, and accuracy. Despite the cost associated with some of those technologies, we also discussed alternative solutions in the form of low-cost sensors that are affordable and give reasonable performance.

Later, IAQ modelling tools are introduced to address questions that emerge from the IAQ monitoring data. Among the reviewed ML algorithms, ANN-based algorithms have been widely incorporated in predicting IAP. Several case

FIGURE 12.4 Schematic diagram for Question 12.5.

studies have summarized an example of combining various monitoring methods, instruments, and exposure levels in different microenvironments. The section concludes with a review of available policies and regulatory considerations in maintaining the IAQ and highlighting the gaps.

12.11 Questions

1. Question 12.1. What is IAQ and why is this important?
2. Question 12.2. How can good IAQ be achieved?
3. Question 12.3. Are there any symptoms that suggest that the air quality of an indoor environment is bad?
4. Question 12.4. A 30-year-old individual has a habit of washing dishes daily for 30 minutes in a kitchen. The kitchen has a window which is facing a city road. The $PM_{2.5}$ mean concentration levels in the kitchen were found to be 50 $\mu g\, m^{-3}$ during that period. Calculate the total exposure levels of the individual during that period.
5. Question 12.5. Calculate the O_3 concentration in a square office (16 m^2 with 3 m height) with large half opened windows, acrylic painted gypsum board ceiling and walls (3), a table ($0.8 \times 2.5 \times 0.5$ m) and a side table ($0.6 \times 0.6 \times 0.2$ m) made of plywood, two chairs made of plywood arms (two curved stripes of $1.5 \times 0.05 \times 0.1$ m $chair^{-1}$) and linen seats (two $0.5 \times 0.3 \times 0.05$ blocks per chair), a nylon rug (2×0.6 m), and lacquered ash floor Fig. 12.4. The external O_3 concentration is 5 ppb. The air in the room is completely exchanged in two hours.

Solutions to the end-of-chapter questions will be available at: https://www.elsevier.com/books-and-journals/book-companion/9780128225912.

References

Abbatt, J. P., & Wang, C. (2020). The atmospheric chemistry of indoor environments. *Environmental Science: Processes & Impacts*, *22*(1), 25–48.

Abdul-Wahab, S. A., En, S. C. F., Elkamel, A., Ahmadi, L., & Yetilmezsoy, K. (2015). A review of standards and guidelines set by international bodies for the parameters of indoor air quality. *Atmospheric Pollution Research*, *6*(5), 751–767.

Alonso, A., Llanos, J., Escandón, R., & Sendra, J. (2021). Effects of the Covid-19 pandemic on indoor air quality and thermal comfort of primary schools in winter in a Mediterranean climate. *Sustainability.*, *13*(5), 2699.

Amoatey, P., Al-Mayahi, A., Omidvarborna, H., Baawain, M., & Sulaiman, H. (2020). Occupational exposure to pesticides and associated health effects among greenhouse farm workers. *Environmental Science and Pollution Research*, *27*, 22251–22270.

Babaoglu, U., Sezgin, F., & Yag, F. (2019). Sick building symptoms among hospital workers associated with indoor air quality and personal factors. *Indoor and Built Environment*, *29*(5), 645–655.

Bai, Z., Wang, Z., Zhu, T., & (Jim) Zhang, J. (2003). Developing indoor air quality related standards in China. *Journal of Asian Architecture and Building Engineering*, *2*, 55–60.

Baldacci, S., Maio, S., Cerrai, S., Sarno, G., Baïz, N., Simoni, M., Annesi-Maesano, I., & Viegi, G. (2015). Allergy and asthma: Effects of the exposure to particulate matter and biological allergens. *Respiratory Medicine*, *109*(9), 1089–1104.

Behlen, A., Wittenburg, C., Steiger, M., & Dannecker, W., 1996. Dry deposition of NO, NO_2, HONO, HNO_3, and Pan on historical building stones. In 8th International Congress on Deterioration and Conservation of Stone, Berlin, 30 Sept.-4 Oct. 1996: Proceedings (377-385).

Bekö, G., Carslaw, N., Fauser, P., Kauneliene, V., Nehr, S., Phillips, G., Saraga, D., Schoemaecker, C., Wierzbicka, A., & Querol, X. (2020). The past, present, and future of indoor air chemistry. *Indoor Air*, *30*(3), 373–376.

Box, G. E. P., & Jenkins, G. M. (1970). *Time series analysis forecasting and control*. Holden-Day.

Branco, P. T. B. S., Alvim-Ferraz, M. C. M., Martins, F. G., & Sousa, S. I. V. (2019). Quantifying indoor air quality determinants in urban and rural nursery and primary schools. *Environmental Research*, *176*, 108534.

Brągoszewska, E., Biedroń, I., Kozielska, B., & Pastuszka, J., Quality Atmosphere & Health. Microbiological indoor air quality in an office building in Gliwice, Poland: Analysis of the case study. *11* (2018), 729–740. http://dx.doi.org/10.1007/s11869-018-0579-z

Buonanno, G., Fuoco, F. C., & Morawska, L. (2013). L. Stabile, Airborne particle concentrations at schools measured at different spatial scales. *Atmospheric Environment*, *67*, 38–45.

Can, E., Üzmez, Ö. Ö., Döğeroğlu, T., & Gaga, E. (2015). Indoor air quality assessment in painting and printmaking department of a fine arts faculty building. *Atmospheric Pollution Research*, *6*(6), 1035–1045.

Cheng, Z., Lei, N., Bu, Z., Sun, H., Li, B., & Lin, B. (2022). Investigations of indoor air quality for office buildings in different climate zones of China by subjective survey and field measurement. *Building and Environment*, *214*, 108899.

CO_2 Meter. (2022). *How does an NDIR CO_2 sensor work?*. [Online] Available at: https://www.co2meter.com/blogs/news/how-does-an-ndir-co2-sensor-work [Accessed 25 September 2022].

Cox, R. A., & Penkett, S. A. (1972). Effect of relative humidity on the disappearance of ozone and sulphur dioxide in contained systems. *Atmospheric Environment*, *6*, 365–368.

DEFRA 2022 UK Air Quality Policy Context https://uk-air.defra.gov.uk/air-pollution/uk-eu-policy-context 2022.

Dingle, P., & Murray, F. (1993). Control and regulation of indoor air: An Australian perspective. *Indoor and Built Environment*, *2*(4), 217–220.

Dols W. S., & Polidoro B. (2020). National Institute of Standards and Technology CONTAM User Guide and Program Documentation Version 3.4. <https://doi.org/10.6028/NIST.TN.1887r1>.

Domínguez-Amarillo, S., Fernández-Agüera, J., Cesteros-García, S., & González-Lezcano, R. A. (2020). Bad air can also kill: Residential indoor air quality and pollutant exposure risk during the COVID-19 crisis. *International Journal of Environmental Research and Public Health*, *17*(19), 7183.

Elbayoumi, M., Ramli, N. A., & Md Yusof, N. F. F. (2015). Development and comparison of regression models and feedforward backpropagation neural network models to predict seasonal indoorPM2.5–10 and PM2.5 concentrations in naturally ventilated schools. *Atmospheric Pollution Research*, *6*(6), 1013–1023.

Environmental, C. (2011). *Particulate matter monitoring technologies and detection principles*. [Online] Available at: https://slideplayer.com/slide/5264610/ [Accessed 26 October 2022].

Freedman, A., Onasch, T., Kebabian, P., & Massoli, P. (2019). *Cavity attenuated phase shift (CAPS)-based detection of gas phase species and aerosols*. IEEE.

Ghaffarianhoseini, A., AlWaer, H., Omrany, H., Ghaffarianhoseini, A., Alalouch, C., Clements-Croome, D., & Tookey, J. (2018). Sick building syndrome: are we doing enough? *Architectural Science Review*, *61*(3), 99–121.

Grøntoft, T., & Raychaudhuri, M. (2004). Compilation of tables of surface deposition velocities for O3, NO2 and SO2 to a range of indoor surfaces. *Atmospheric Environment*, *38*(4), 533–544.

Guo Z. (2000a). Research Triangle Park, NC US Environmental Protection Agency Simulation tool kit for indoor air quality and inhalation exposure (IAQX) Version 1.0 user's guide. <https://nepis.epa.gov/Adobe/PDF/P1000A0G.pdf>.

Guo, Z. (2000b). Development of a Windows-based indoor air quality simulation software package. *Environmental Modelling & Software*, *15* (4) 403–410. https://doi.org/10.1016/S1364-8152(00)00020-7

Ho., Hidy, G., & Govan, R. (1974). Microwave measurements of the liquid water content of atmospheric aerosols. *Journal of Applied Meteorology*, *13*(8), 871–879.

Hospodsky, D., Qian, J., Nazaroff, W., Naomichi, Y., Bibby, K., Rismani-Yazdi, H., & Peccia, J. (2012). Human occupancy as a source of indoor airborne bacteria. *PLOS ONE*, *7*(4), e34867.

Jin, M., Ye, N., Lu, Z., Zhang, S., Zhou, S., & He, J. (2022). Pollution characteristics and source identification of PBDEs in public transport microenvironments. *Science of The Total Environment*, *820*, 153159.

Jung, C. C., Wu, P. C., Tseng, C. H., & Su, H. J. (2015). Indoor air quality varies with ventilation types and working areas in hospitals. *Building and Environment*, *85*, 190–195.

Kaunelienė, V., Prasauskas, T., Krugly, E., Stasiulaitienė, I., Čiužas, D., Šeduikytė, L., & Martuzevičius, D. (2016). Indoor air quality in low energy residential buildings in Lithuania. *Building and Environment*, *108*, 63–72.

Kebabian, P. L., Herndon, S. C., & Freedman, A. (2005). Detection of nitrogen dioxide by cavity attenuated phase shift spectroscopy. *Analytical Chemistry*, *77*, 724–728.

Klepeis, N. E., Nelson, W. C., Ott, W. R., Robinson, J. P., Tsang, A. M., Switzer, P., Behar, J. V., Hern, S. C., & Engelmann, W. H. (2001). The National Human Activity Pattern Survey (NHAPS): A resource for assessing exposure to environmental pollutants. *Journal of Exposure Science & Environmental Epidemiology*, *11*, 231–252.

Klenø, J. G., Clausen, P. A., Weschler, C. J., & Wolkoff, P. (2001). Determination of ozone removal rates by selected building products using the FLEC emission cell. *Environmental Science & Technology*, *35*, 2548–2553.

Komarnicki, G. J. (2005). Lead and cadmium in indoor air and the urban environment. *Environmental Pollution*, *136*(1), 47–61.

Kordzadeh, N., & Ghasemaghaei, M. (2022). Algorithmic bias: Review, synthesis, and future research directions. *European Journal of Information Systems*, *31*(3), 388–409.

Kumar, D., & Little, J. C. (2003). Single-layer model to predict the source/sink behavior of diffusion-controlled building materials. *Environmental Science & Technology*, *37*(17), 3821–3827.

Kumar, P., Hama, S., Abbass, R. A., Nogueira, T., Brand, V. S., Abhijith, K.V., de Fatima Andrade, M., Asfaw, A., Aziz, K. H., Cao, S. J., & El-Gendy, A. (2021c). Potential health risks due to in-car aerosol exposure across ten global cities. *Environment International*, *155*, 106688. https://doi.org/10.1016/j.envint.2021.106688

Kumar, P., Hama, S., Abbass, R. A., Nogueira, T., Brand, V. S., Wu, H. W., Abulude, F. O., Adelodun, A. A., Anand., de Fatima Andrade, M., & Apondo, W. (2022). In-kitchen aerosol exposure in twelve cities across the globe. *Environment International*, *162*, 107155.

Kumar, P., Hama, S., Nogueira, T., Abbass, R. A., Brand, V. S., de Fatima Andrade, M., Asfaw, A., Aziz, K. H., Cao, S. J., El-Gendy, A., & Islam, S. (2021b). In-car particulate matter exposure across ten global cities. *Science of the Total Environment*, *750*, 141395. https://doi.org/10.1016/j.scitotenv.2020.141395

Kumar, P., Omidvarborna, H., Barwise, Y., & Tiwari, A. (2020b). Guildford mitigating exposure to traffic pollution in and around schools. 10.5281/zenodo.3754131. https://zenodo.org/record/3754131#.Y2wb4XbMKUk

Kumar, P., Omidvarborna, H., Pilla, F., Lewin, N. (2020). A primary school driven initiative to influence commuting style for dropping-off and picking-up of pupils. *Science of the Total Environment*, *727*, 138360. https://doi.org/10.1016/j.scitotenv.2020.138360

Kumar, P., Omidvarborna, H., Tiwari, A., & Morawska, L. (2021a). The nexus between in-car aerosol concentrations, ventilation and the risk of respiratory infection. *Environment International*, *157*, 106814. https://doi.org/10.1016/j.envint.2021.106814

Kumar, P., Rivas, I., Singh, A. P., Ganesh, V. J., Ananya, M., & Frey, H. C. (2018). Dynamics of coarse and fine particle exposure in transport microenvironments. *NPJ Climate and Atmospheric Science*, *1*, 1–12.

Kumar, P., Skouloudis, A. N., Bell, M., Viana, M., Carotta, M. C., Biskos, G., & Morawska, L. (2016). Real-time sensors for indoor air monitoring and challenges ahead in deploying them to urban buildings. *Science of the Total Environment*, *560*, 150–159.

Lai, D., Karava, P., & Chen, Q. (2015). Study of outdoor ozone penetration into buildings through ventilation and infiltration. *Building and Environment*, *93*, 112–118.

Larson, B. A., & Rosen, S. (2002). Understanding household demand for indoor air pollution control in developing countries. *Social Science & Medicine*, *55*(4), 571–584.

Leaderer, B.P., Renes, S., Bluyssen, P., & Van De Loo, H., 1986, June. Chamber studies of NO_2, SO_2, and RSP deposition rates indoors. In the 79th Annual Meeting Air Pollution Control Association.

Lee, H. J., Lee, K. H., & Kim, D. K. (2020). Evaluation and comparison of the indoor air quality in different areas of the hospital. *Medicine*, *99*(52), 23942.

Leygraf, C., Wallinder, I., Tidblad, J., & Graedel, T. (2000). *Atmospheric Corrosion*. Wiley-Interscience.

Liu, Z., Ye, W., & Little, J. C. (2013). Predicting emissions of volatile and semivolatile organic compounds from building materials: A review. *Building and Environment*, *64*, 7–25.

Loy-Benitez, J., Heo, S., & Yoo, C. (2020). Imputing missing indoor air quality data via variational convolutional autoencoders: Implications for ventilation management of subway metro systems. *Building and Environment*, *182*, 107–135.

Madany, I. M., Akhter, M. S., & Al Jowder, O. A. (1994). The correlations between heavy metals in residential indoor dust and outdoor street dust in Bahrain, 1994 *Environment international*, *20*(4), 483–492.

Mahajan, S., Kumar, P., Pinto, J. A., Riccetti, A., Schaaf, K., Camprodon, G., Smari, V., Passani, A., & Forino, G. (2020). A citizen science approach for enhancing public understanding of air pollution. *Sustainable Cities and Society*, *52*, 101800.

Mainka, A., Brągoszewska, E., Kozielska, B., Pastuszka, J. S., & Zajusz-Zubek, E. (2015). Indoor air quality in urban nursery schools in Gliwice, Poland: Analysis of the case study. *Atmospheric Pollution Research*, *6*(6), 1098–1104.

Majd, E., Curriero, F., Berman, J., Connolly, F., Leaf, P., Rule, A., Green, T., Clemons-Erby, D., & Gummerson, C. (2019). Indoor air quality in inner-city schools and its associations with building characteristics and environmental factors. *Environmental Research*, *170*, 83–91.

Mandin, C., Trantallidi, M., Cattaneo, A., Canha, N., Mihucz, V. G., Szigeti, T., Mabilia, R., Perreca, E., Spinazzè, A., Fossati, S., & De Kluizenaar, Y. (2017). Assessment of indoor air quality in office buildings across Europe–The OFFICAIR study. *Science of the Total Environment*, *579*, 169–178.

Membrapor. (2022). Electrochemical gas sensors. [Online] Available at: https://www.membrapor.ch/electrochemical-gas-sensors/ [Accessed 25 September 2022].

Milionis, A. E., & Davies, T. D. (1994). Regression and stochastic models for air pollution—I. Review, comments and suggestions. *Atmospheric Environment*, *28*, 2801–2810.

Miyazaki, T. (1984). Adsorption characteristics of NOx by several kinds of interior materials. *Indoor Air: Chemical Characterization and Personal Exposure*, *4*, 103–110.

Moos, R., Sahner, K., Fleischer, M., Guth, U., Barsan, N., & Weimar, U. (2009). Solid state gas sensor research in Germany–a status report. *Sensors*, *9*(6), 4323–4365.

Morawska, L., Ayoko, G. A., Bae, G. N., Buonanno, G., Chao, C. Y. H., Clifford, S., Fu, S. C., Hänninen, O., He, C., Isaxon, C., & Mazaheri, M. (2017). Airborne particles in indoor environment of homes, schools, offices and aged care facilities: The main routes of exposure. *Environment International*, *108*, 75–83.

Morrison, G. (2008). Interfacial chemistry in indoor environments. *Environmental Science & Technology*, *42*, 3495–3499.

Mueller, F. X., Loeb, L., & Mapes, W. H. (1973). Decomposition rates of ozone in living areas. *Environmental Science & Technology*, *7*, 342–346.

Nakora, N., Byamugisha, D., & Birungi, G. (2020). Indoor air quality in rural Southwestern Uganda: particulate matter, heavy metals and carbon monoxide in kitchens using charcoal fuel in Mbarara Municipality. *SN Applied Sciences*, *2*, 1–16.

Omidvarborna, H., Kumar, P., Hayward, J., Gupta, M., & Nascimento, E. G. S. (2021). Low-cost air quality sensing towards smart homes. *Atmosphere*, *12*, 453.

OSHA. (2015). *Indoor Air Quality in Commercial and Institutional buildings*. Maroon Ebooks.

Palmisani, J., Di Gilio, A., Viana, M., de Gennaro, G., & Ferro, A. (2021). Indoor air quality evaluation in oncology units at two European hospitals: Low-cost sensors for TVOCs, $PM_{2.5}$ and CO_2 real-time monitoring. *Building and Environment*, *205*, 108237.

Perez, L., Declercq, C., Iñiguez, C., Aguilera, I., Badaloni, C., Ballester, F., Bouland, C., Chanel, O., Cirarda, F. B., Forastiere, F., & Forsberg, B. (2013). Chronic burden of near-roadway traffic pollution in 10 European cities (APHEKOM network). *European Respiratory Journal*, *42*(3), 594–605.

PTL. (2022). Dynamic light scattering (DLS). [Online] Available at: https://www.particletechlabs.com/analytical-testing/particle-size-distribution-analyses/dynamic-light-scattering [Accessed 25 September 2022].

Rashed, M. N. (2008). Total and extractable heavy metals in indoor, outdoor and street dust from Aswan City, Egypt. *Clean–Soil, Air, Water*, *36*, 850–857.

Ravindra, K., Singh, T., Vardhan, S., Shrivastava, A., Singh, S., Kumar, P., & Mor, S. (2022). COVID-19 pandemic: What can we learn for better air quality and human health? *Journal of Infection and Public Health*, *15*, 187–198.

Reiss, R., Ryan, P. B., & Koutrakis, P. (1994). Modeling ozone deposition onto indoor residential surfaces. *Environmental Science & Technology*, *28*, 504–513.

Riley, W. J., McKone, T. E., Lai, A. C., & Nazaroff, W. W. (2002). Indoor particulate matter of outdoor origin: Importance of size-dependent removal mechanisms. *Environmental Science & Technology*, *36*, 200–207.

Ruzzin, J. (2012). Public health concern behind the exposure to persistent organic pollutants and the risk of metabolic diseases. *BMC Public Health*, *12*, 1–8.

Sabersky, R. H., Sinema, D. A., & Shair, F. H. (1973). Concentrations, decay rates, and removal of ozone and their relation to establishing clean indoor air. *Environmental Science & Technology*, *7*, 347–353.

Schweizer, C., Edwards, R. D., Bayer-Oglesby, L., Gauderman, W. J., Ilacqua, V., Juhani Jantunen, M., Lai, H. K., Nieuwenhuijsen, M., & Künzli, N. (2007). Indoor time–- microenvironment–activity patterns in seven regions of Europe. *Journal of Exposure Science & Environmental Epidemiology*, *17*(12), 170–181.

Seow, W. J., Downward, G. S., Wei, H., Rothman, N., Reiss, B., Xu, J., Bassig, B. A., Li, J., He, J., Hosgood, H. D., & Wu, G. (2016). Indoor concentrations of nitrogen dioxide and sulfur dioxide from burning solid fuels for cooking and heating in Yunnan Prov. *Indoor Air*, *26*, 776–783.

Sharma, D., & Jain, S. (2019). Impact of intervention of biomass cookstove technologies and kitchen characteristics on indoor air quality and human exposure in rural settings of India. *Environment International*, *123*, 240–255.

Shimadzu. (2022). About chemiluminescence detection. [Online] Available at: https://www.shimadzu.com/an/service-support/technical-support/analysis-basics/basic/chemiluminescence_detection.html [Accessed 25 September 2022].

Shrestha, P., Humphrey, J., Carlton, E., Adgate, J., Barton, K., Root, E., & Miller, S. (2019). Impact of outdoor air pollution on indoor air quality in low-income homes during wildfire seasons. *International Journal of Environmental Research and Public Health*, *16*, 3535.

Shrubsole, C., Dimitroulopoulou, S., Foxall, K., Gadeberg, B., & Doutsi, A. (2019). IAQ guidelines for selected volatile organic compounds (VOCs) in the UK. *Building and Environment*, *165*, 106382.

Spedding, D. J., & Rowlands, R. P. (1970). Sorption of sulphur dioxide by indoor surfaces I. Wallpapers. *Journal of Applied Chemistry*, *20*, 143–146.

Spedding, D. J. (1970). Sorption of sulphur dioxide by indoor surfaces II. Wood. *Journal of Applied Chemistry*, *20*, 226–228.

Spedding, D. J. (1972a). Kinetic study of the uptake of sulphur dioxide by aluminium. *British Corrosion Journal*, *7*, 281–284.

Spedding, D. J. (1972b). The sorption of sulphur dioxide by indoor surfaces. IV. Flooring materials. *Journal of Applied Chemistry and Biotechnology*, *22*, 1–8.

Spicer, C. W., Coutant, R. W., Ward, G. F., Joseph, D. W., Gaynor, A. J., & Billick, I. H. (1989). Rates and mechanisms of NO2 removal from indoor air by residential materials. *Environment International*, *15*, 643–654.

Stedman, D. H., Daby, E. E., Stuhl, F., & Niki, H. (1972). Analysis of ozone and nitric oxide by a chemiluminescent method in laboratory and atmospheric studies of photochemical smog. *Journal of the Air Pollution Control Association*, *22*, 260–263.

Sofuoglu, S. C. (2008). Application of artificial neural networks to predict prevalence of building-related symptoms in office buildings. *Building and Environment*, *43*, 1121–1126.

Sun, Y., Hou, J., Cheng, R., Sheng, Y., Zhang, X., & Sundell, J. (2019). Indoor air quality, ventilation and their associations with sick building syndrome in Chinese homes. *Energy and Buildings*, *197*, 112–119.

Tools, I. (2018). Gas filter correlation (GFC) spectroscopy. [Online] Available at: https://instrumentationtools.com/gas-filter-correlation-gfc-spectroscopy/ [Accessed 25 September 2022].

USEPA (2022). *Introduction to indoor air quality*. <https://www.epa.gov/indoor-air-quality-iaq/introduction-indoor-air-quality>.

Veysi, R., Heibati, B., Jahangiri, M., Kumar, P., Latif, M. T., & Karimi, A. (2019). Indoor air quality-induced respiratory symptoms of a hospital staff in Iran. *Environmental Monitoring and Assessment*, *191*, 1–8.

Wei, W., Ramalho, O., Malingre, L., Sivanantham, S., Little, J. C., & Mandin, C. (2019). Machine learning and statistical models for predicting indoor air quality. *Indoor Air*, *29*, 704–726.

Weschler, C., & Carslaw, N. (2018). Indoor chemistry. *Environmental Science & Technology*, *52*, 2419–2428.

Weschler, C., & Shields, H. (2000). The influence of ventilation on reactions among indoor pollutants: Modeling and experimental observations. *Indoor Air*, *10*, 92–100.

WHO (2021a). Human health effects of polycyclic aromatic hydrocarbons as ambient air pollutants. <https://apps.who.int/iris/bitstream/handle/10665/350636/9789289056533-eng.pdf?sequence=1&isAllowed=y>.

WHO (1990). *Rautavaara WHO indoor air quality: Biological contaminants.*

WHO (2009). WHO guidelines for indoor air quality: Dampness and mould.

WHO (2010). WHO guidelines for indoor air quality: Selected pollutants.

Wisthaler, A., & Weschler, C. (2010). Reactions of ozone with human skin lipids: Sources of carbonyls, dicarbonyls, and hydroxycarbonyls in indoor air. *Proceedings of the National Academy of Sciences*, *107*, 6568–6575.

Xie, H., Ma, F., & Bai, Q. (2009). *IEEE prediction of indoor air quality using artificial neural networks*, *2*, 414–418.

Yang, F., Kaul, D., Wong, K. C., Westerdahl, D., Sun, L., Ho, K. F., Tian, L., Brimblecombe, P., & Ning, Z. (2015). Heterogeneity of passenger exposure to air pollutants in public transport microenvironments. *Atmospheric Environment*, *109*, 42–51.

Yi, W. Y., Lo, K. M., Mak, T., Leung, K. S., Leung, Y., & Meng, M. L. (2015). A survey of wireless sensor network based air pollution monitoring systems. *Sensors*, *15*, 31392–31427.

Zhang, H., Srinivasan, R., & Systematic, A. (2020). Review of air quality sensors, guidelines, and measurement studies for indoor air quality management. *Sustainability.*, *12*, 9045.

Zhao, B., Shi, S., & Ji, J. (2022). The WHO air quality guidelines 2021 promote great challenge for indoor air. *Science of The Total Environment*, *827*, 154376.

Zhu, Y., Yang, L., Meng, C., Yuan, Q., Yan, C., Dong, C., Sui, X., Yao, L., Yang, F., Lu, Y., & Wang, W. (2015). Indoor/outdoor relationships and diurnal/nocturnal variations in water-soluble ion and PAH concentrations in the atmospheric PM2.5 of a business office area in Jinan, a heavily polluted city in China. *Atmospheric Research*, *153*, 276–285.

Zimmer, C., Kallis, K., & Giebel, F. (2015). Micro-structured electron accelerator for the mobile gas ionisation sensor technology. *Journal of Sensors and Sensor Systems*, *4*, 151–157.

Chapter 13

Quantitative data analysis methods for air quality prediction

Pushp Raj Tiwari[1,2,3] **and Saurabh Kumar**[2]

[1]*Centre for Atmospheric and Climate Physics Research, University of Hertfordshire, Hatfield, United Kingdom,* [2]*Centre for Climate Change Research (C3R), University of Hertfordshire, Hatfield, United Kingdom,* [3]*Centre for Future Society Research, University of Hertfordshire, Hatfield, United Kingdom*

13.1 Introduction

The science of statistics deals with the gathering, organizing and analysis of data as well as the inferences that can be made about the entire population from samples. For this, a relevant statistical test must be used, together with an acceptable study design and study sample selection. To properly design an epidemiological study or a clinical trial, one must have a solid understanding of statistics. Data analysis techniques are proving to be a crucial tool for evaluating the efficacy of the air quality model. There are two methods for evaluating the quality of the air: quantitative measures and qualitative analyses. The quantitative technique quantifies the quantities of air pollutants, including particulate matters (PMs), carbon monoxide (CO), alkanes, VOCs, secondary organic aerosols, and other flammable and hazardous gases (NO_2, SO_2) in a specific site under specified temperature, humidity, and wind speed. These measurements are heavily influenced by and rely on the location of the pollutant sources, the time of measurement, the pollutant and operator work schedules, the dispersion time, vehicular traffic, etc. For the purpose of predicting air pollution, numerous techniques have been developed. Deterministic models, conventional statistical techniques, and methods based on artificial intelligence (AI) can be used to classify these basically into three groups. At various three-dimensional scales, deterministic models are capable of simulating atmospheric chemical and physical processes. The atmospheric dispersion modeling system model, the weather research and forecasting model combined with chemistry model, and the community multiscale air quality modeling system model are examples of representative deterministic models. From conventional multiple linear regression techniques, statistical techniques have been created. Traditional regression techniques and spatial statistical techniques have been blended into a sophisticated analysis method thanks to the advancement of computer science and the ongoing invention and enhancement of statistical prediction techniques. In recent years, AI techniques have arisen. These techniques, which have data adaptability and data-driving at their core, include machine learning techniques like artificial neural networks (ANNs), deep learning, and support vector machines (SVMs). Numerous works have demonstrated that traditional statistical approaches are less accurate than AI technology (Taheri & Sodoudi, 2016). These techniques for forecasting air pollution have been examined in a number of research. For instance, precise statistical methods for the forecasting of air pollution were introduced by Bai et al. (2018). They highlighted the fundamental aspects of these methods but paid less attention to their benefits, drawbacks, and application-related impacts. In their analysis of various research articles on machine learning. Rybarczyk and Zalakeviciute (2018) found that while neural networks and SVMs were frequently used for prediction applications, integrated learning and regression were chosen for estimation applications. For this reason, it is required to conduct a quantitative analytical assessment of a substantial number of studies concerning air pollution prediction in in order to assist researchers in this field in more rapidly and precisely identifying the research objective.

Air Quality. DOI: https://doi.org/10.1016/B978-0-12-822591-2.00013-5

13.2 Data analysis methods and tools

The most important stage in predicting climate change is data analysis. Through a variety of methods or methodologies, its main goal is to improve understanding of the atmosphere and how it interacts with the oceans, cryosphere, and the terrestrial surface. These include mathematical data modeling, diagnostic analyses, and empirical investigations on climate data.

We can examine data using particular tools or processes known as data analysis methods. Depending on whether the data is quantitative or qualitative, they can be divided into two major types. Data with a quantitative component can be expressed numerically. Because of this, a significant portion of quantitative data analysis involves applying statistical techniques to identify unbiased patterns in the data. In general, since the associated procedures for data analysis do not allow for interpretation, this form of data is chosen.

Any data that cannot be stated in numbers is considered qualitative data. It frequently rests on beliefs or experiences that are hard to measure or impossible to define. As a result, qualitative data approaches are far less common and typically less trustworthy for reaching scientific results.

13.2.1 Statistical tools in analytical method validation

To show the analytical method's validity, a statistical analysis of the data gathered during a method validation should be carried out. The computation of the mean (or average), standard deviation, relative standard deviation, confidence intervals, and regression analysis, as well as the application of regression and correlation analysis, are the major parameters utilized for the interpretation of the results of the analytical technique validation. Statistical computing tools like SPSS, R, and Minitab are frequently used to carry out these calculations.

The objective of statistical analysis is to provide an insight into the method characteristic under investigation by summarizing a set of data. The basic parameters utilized for interpreting analytical technique validation results are the mean (or average), standard deviation, relative standard deviation, confidence intervals, and regression analysis, and the primary tools are the F-test, t-test, regression and correlation analysis, and so on. The following is a description of the statistical tools:

Mean: The most fundamental and often used statistic is the mean or average of a set of data. By summing together all the data points and dividing the result by the number of samples, the mean is calculated. It is typically denoted by $\overline{X}$ and is computed using the following formula:

$$\overline{X} = \frac{\sum X_i}{n} = \frac{X_1 + X_2 + \ldots + X_n}{n} \tag{13.1}$$

where *Xi* are individual values and n is the number of individual data points.

Standard deviation: The difference between a set's mean and its individual values is used to calculate a data set's standard deviation, which is a measurement of the dispersion of values within the sample set. Standard deviation is computed using the following formula:

$$S = \sqrt{\frac{\sum_i (X_i - \overline{X})^2}{(n-1)}} = \sqrt{\frac{(X_1 - \overline{X})^2 + (X_2 - \overline{X})^2 + \ldots + (X_n - \overline{X})^2}{(n-1)}} \tag{13.2}$$

where *Xi* is individual value, $\overline{X}$ is the sample mean, and n is the number of individual data points.

Relative standard deviation: The standard deviation of the sample set multiplied by 100% is divided by the average of the sample set to produce the relative standard deviation. A percentage is used to represent the relative standard deviation. In general, %RSD is used to represent the acceptance criterion for data's accuracy, precision, and recurrence.

$$\%\mathrm{RSD} = \frac{s}{\overline{X}} \times 100\% \tag{13.3}$$

Confidence interval: To show how reliable an estimate is, confidence intervals are utilized. Confidence intervals set boundaries around the sample mean in order to forecast the range of the true population of the mean. The prediction is often predicated on a 95% probability. The sample standard deviation and sample mean have an impact on the confidence interval. Confidence interval for:

$$\mu = \overline{X} \pm \frac{zs}{\sqrt{n}} \tag{13.4}$$

where s is the sample deviation, $\overline{X}$ is the sample mean, n is the number of individual data points, and z is constant obtained from statistical tables for z. The value of z depends on the confidence level listed in statistical tables for z. For 95%, z is 1.96. For small samples, z can be replaced by t-value obtained from the Student's t-distribution tables. The value of t corresponds to $n-1$. For example, for

$n = 6$,
mean = 101.82,
standard deviation = 1.2,
95% confidence interval = 0.49.

So, this calculated confidence interval indicates that the range of the true population of the mean is between 101.33 and 102.31.

In short, the confidence interval can be used in a variety of ways to assess the magnitude of uncertainties (confidence) in analytical work or measurements, including single determinations made as part of routine tasks, specific calibrations, determinations for which there are no prior data, etc.

Regression analysis: To determine whether test findings have a linear connection, regression analysis is utilized. Typically, a linear relationship is assessed across the entire analytical procedure. Plotting data received from analysis of solutions made at various concentration levels on a graph is a common method of investigation. By applying a linear equation to the data, linear regression analyses the relationship between two variables. A linear regression line has an equation of the form $Y = b_1{*}X + b_0$, where X is the independent variable and Y is the dependent variable. The slope of the line is b_1 and b_0 is the intercept (the value of y when $x = 0$). Obtaining a line between the points to reduce the deviations of the points from the projected line is the statistical process for finding the "best-fitting" straight line. The principle of least squares is the name for the goodness of fit test's best-fit criterion. In mathematics, the line that minimizes the sum of squares of the differences between observed and anticipated values of Y is said to be the best-fit line. Examining the residuals (the differences between the fitted line and the observed values) after a regression model has been fitted to a set of data enables testing the accuracy of the linear relationship assumption. Any potential nonlinear relationship between the variables can be shown by plotting the residuals on the y-axis versus the independent variable on the x-axis, which may prompt the investigation of outliers. The coefficient of determination (r^2) and linear correlation coefficient are two more significant computations that are typically provided (r). The amount of variation that the model is able to explain is indicated by its coefficient of determination (r^2). The ideal value of r^2 is 1, which denotes a zero error. Between the values that were predicted and those that were actually observed is a correlation, or r. This will have a value between -1 and $+1$; the closer it is to 1, the better the correlation. Therefore, extra care should be needed while analyzing the correlation coefficient since any data that forms a straight line will yield a high correlation coefficient. Not merely the correlation or outcomes, but also additional statistical analysis is advised to obtain estimates of systemic errors. For instance, in method comparison studies, the results would show linear correlation and have a high correlation coefficient, even though there would be a difference between the two methods. If one method consistently produces higher results than the other method, the results would also be different.

The hypothesis tests: The purpose of hypothesis testing is to determine whether the experimental results support a particular theoretical concept. The null hypothesis, represented by H_0, assumes equality between the two elements or sequence of elements. The second stage entails calculating the variance between various characteristics. The probability P that this deviation will occur if H_0 is true is determined in the third step. The required conclusions are drawn in the fourth phase, which is: H_0 is acknowledged as credible if P is big; but, if P is little, the deviation is incompatible with H_0. The level of confidence or significance level is a fixed value limit for P that is used to decide whether P is large or small (often, we choose $P = 0.95$ as the level of confidence and $a = 0.05$ as the significance level). Four scenarios are plausible:

1. Acceptance of H_0 true.
2. Rejecting true H_0: first error species (α). α: first error species (α).
3. Acceptance false H_0: second error species (β). β: second error species (β).
4. Rejecting false H_0.

Student's t-test: Student's t-test is used to test the null hypothesis that there is no difference between the means of the two groups. It is used in three circumstances:

To test if a sample mean (as an estimate of a population mean) differs significantly from a given population mean (this is a one-sample t-test) The formula for one sample t-test is

$$t = \frac{X - \mu}{SE} \tag{13.5}$$

where X is the sample mean, μ is the population mean, and SE is the standard error of mean.

To test if the population means estimated by two independent samples differ significantly (the unpaired t-test). The formula for unpaired t-test is

$$t = \frac{X_1 - X_2}{SE_{x_1 - x_2}} \tag{13.6}$$

where $x_1 - x_2$ is the difference between the means of the two groups and SE denotes the standard error of the differences.

Fisher's test: If there is a significant correlation between two categorical variables, Fisher's Exact Test is performed to determine it. When one or more of the cell counts in a table are lower than five, it is sometimes used as a potential substitute for the Chi-Square Test of Independence. The following null and alternative hypotheses are used with Fisher's Exact Test:

H_0 (null hypothesis): The two variables are independent.

H_1 (alternative hypothesis): The two variables are not independent.

The Fisher's Exact Test's one-tailed p-value is determined as follows:

$$P = \frac{(a+b)!(c+d)!(a+c)!(b+d)!}{(a!b!c!d!n!)} \tag{13.7}$$

Example: The Environmental science class is divided into six random pupils, and each is tested on their understanding of atmospheric phenomenon. The instructor expects the students to be able to get a test score of greater than 70. Scores for the six pupils are 62, 92, 75, 68, 83, and 95. Can the Professor say with 90% certainty that the average test result for the class will be higher than 70?

Solution: We'll approach this problem step-by-step.

Step1: Null hypothesis: H_0: $\mu = 70$; alternative hypothesis: H_a: $\mu > 70$.

Step 2: To compute the sample mean and standard deviation.

$$X = \frac{62 + 92 + 75 + 68 + 83 + 95}{6} = \frac{475}{6} = 79.17$$

and standard deviation $SD = 13.17$

Step 3: To calculate the t-value, we can use Eq. (13.5) as

$$t = \frac{X - \mu}{SE}$$

where X is the sample mean, μ is the population mean, and SE is the standard error of mean standard error of mean (SE) can be calculated as

$$\mathrm{SE} = \frac{SD}{\sqrt{n}},$$

Therefore $t = \frac{X-\mu}{SE} = \frac{79.17 - 70}{\frac{13.17}{\sqrt{6}}} = \frac{9.17}{5.38} = 1.71$.

The computed t-value of 1.71 will be compared to the critical value in the t-table in order to test the hypothesis. The null hypothesis can be disproved if the computed t-value exceeds the critical t-value from the table. A level of confidence of 90% corresponds to an alpha level of 0.10. The given problem has 6−1 (=5) degrees of freedom. The t-table value for $t_{.10,5}$ is 1.476 (Appendix 1). The computed t value of 1.71 is more than the critical value in the table, which allows the null hypothesis to be disproved and provides the tutor with proof that the class mean on the atmospheric phenomenon would be at least 70. Since this is a one-tailed test, the alpha level is not divided by two.

Analysis of variation (ANOVA): Three or more groups cannot be compared using the Student's t-test. ANOVA is used to determine whether there are any substantial differences between the means of two or more groups. Within-group variability and between-group variability are the two variances that are examined in an ANOVA. The variation that cannot be accounted for by the study design is known as the within-group variability (error variance). It is focused on random variations found in our samples. However, our approach is what led to the between-group (or effect variance) variance. The F-test is used to compare these two variance values. A simplified formula for the F statistic is

$$F = \frac{MS_b}{MS_w} \tag{13.8}$$

where Ms_b is the mean squares between the groups and Ms_w is the mean squares within groups.

Monte Carlo Simulation: The Monte Carlo method, commonly referred to as Monte Carlo simulation, uses computers to create models of potential outcomes and display their probability distributions. It takes into account a variety of potential outcomes before attempting to determine how likely each one is to materialize. Data analysts utilize it to conduct an advanced risk analysis that aids in projecting future occurrences and guiding decision-making.

Wilcoxon's signed rank test: The sign test has a significant drawback in that it just uses the + or − signs and leaves out the quantitative information of the input data. The relative sizes are also taken into account by Wilcoxon's signed rank test, which gives the test greater statistical power by looking at both the observed values and how they compare to θ_0 as well. The same as with the sign test, the sample is discarded if any observed values are equal to the reference value θ_0. The Wilcoxon rank sum test evaluates the difference in the rank sums after ranking all of the data points in order, computing the rank sum for each sample, and then comparing them.

Mann–Whitney test: It is used to determine whether observations in one sample generally tend to be larger than observations in the other or, alternatively, to test the null hypothesis that the median of the two samples is equal.

Mann–Whitney test compares all data (x_i) belonging to the X group and all data (y_i) belonging to the Y group and calculates the probability of x_i being greater than y_i: $P(x_i > y_i)$. The null hypothesis states that $P\ (x_i > y_i) = P(x_i < y_i) = 1/2$ while the alternative hypothesis states that $P(x_i > y_i) \neq 1/2$.

The Mann–Kendall (MK) test: If a trend is present in time series data, it can be identified using the MK Trend Test (Mann, 1945; Kendall, 1975). The MK Trend Test is sometimes also named as Kendall's tau test or MK test. Because it is a nonparametric test, no underlying assumptions regarding the normality of the data are made.

The hypotheses for the test are as follows:

H_0 (null hypothesis): The data do not show any trend.

H_A (alternative hypothesis): The data show a trend. (This trend might be positive, negative and neutral in nature.)

The presence of a trend in the time series data is statistically supported if the test's p-value is less than a predetermined level of significance (popular choices include 0.10, 0.05, and 0.01).

For the analysis of API time series, recognizing significant patterns and identifying changes in a trend over time are crucial. A monotonic trend, either increasing or decreasing, can be detected statistically using the MK test on a time series.

The analysis should validate the following before running the test:

1. There is no seasonal data collection (i.e., data collected only during summer or winter period). This is because if the obtained data exhibits an alternating rising or downward trend, the test will not be effective.
2. The obtained data shouldn't contain any kind of covariate (a factor that could affect your data that is shown).
3. Each time period should have a single data point. In order to conduct the analysis, if we have numerous points, utilize the median value.

The MK trend test is computed using the equation as follows:

$$S_{mk} = \sum_{i=1}^{n-1} \sum_{j=i+1}^{n} \operatorname{sgn}(x_j - x_i) \tag{13.9}$$

where n is the length of data, x_i and x_j is the two generic sequential data, and $\operatorname{sgn}(x_j - x_i)$ can be written as

$$\operatorname{sgn}(x) = \begin{cases} 1: & x_j - x_i > 0 \\ 0: & x_j - x_i = 0 \\ -1: & x_j - x_i < 0 \end{cases}$$

The test statistic S_{mk} is asymptotically normally distributed, and under the null hypothesis it has an expectation $E(S_{mk}) = 0$ and variance $V(S_{mk})$ is given by

$$V(S_{mk}) = \frac{n(n-1)(2n+5)}{18} \tag{13.10}$$

in the case of that there is no tie in the data (Mann, 1945; Kendall, 1948). Otherwise, it can be given by

$$V(S_{mk}) = \frac{[n(n-1)(2n+5) - \sum_j t_j(t_j-1)(2t_j+5)]}{n} \tag{13.11}$$

where t is the number of data points in the jth tied group. For this test, the statistics Z can be calculated as

$$Z = \begin{cases} \dfrac{s+1}{\sqrt{V(S_{\text{mk}})}}: & S>0 \\ 0: & S=0 \\ -1: & S<0 \end{cases}$$

13.2.2 Data mining and machine learning for air pollution

Computing science's data mining and machine learning techniques offer a wide range of affordable and efficient solutions that have proven effective for similar issues in other fields. The large datasets involved in climate and air pollution modeling are complicated, making it challenging to examine them using classic epidemiology or environmental health models. New techniques such as data mining and machine learning are therefore necessary to improve our comprehension of the data analysis. Data mining is a computational technique that is frequently used to examine huge databases, spot patterns, extract useful information, and forecast the results of upcoming or ambiguous events. Methods from a variety of computing fields, such as AI, statistics, mathematics, machine learning, and database systems, are used in this approach. Understanding the root cause of the issue and providing accurate policy recommendations could result from identifying the sources and impacts of air pollution. Researchers working in the field of climate and air quality now have access to more data acquisition resources thanks to recent advancements in sensor technologies and big data, including machine learning (Reid et al., 2015), neural network (NN) models, radial base functions, SVMs fuzzy models (Wang & Song, 2018), gray models (Wu et al., 2018), and feedforward NN models (Dunea et al., 2015; Guo et al., 2020). A hybrid computational intelligence system for combined machine learning has been proposed by Bougoudis et al. (2016). In an effort to determine the root cause of pollutants, the approach is used to find correlations between air pollution levels and weather patterns. To study the problem, they gather information from the larger Attica region. They use ensemble approaches that incorporate Random Forests and ANNs. They assert that this strategy increases accuracy, but the feed-forward neural network is unable to correctly forecast continuous values.

13.2.2.1 Learning algorithms

Decision Tree (DT) Regression: DTs refer to the nonparametric supervised learning procedure used for regression and classification/categorization. With the aid of straightforward decision rules generated from the key characteristics of the data, the DTs' main goal is to provide a prediction model for the values of the outcome variable. Classification and regression trees (CARTs) are employed for the quantitative result variable(s), but they do not compute the sets of decision rules (Breiman et al., 1984). Binary trees are created by the CART utilizing the threshold and traits that produce the most information at each node.

Random Forest (RF) Regression: RF is one of the most widely used ensemble learning approaches based on decision tree predictors. RF is a straightforward, effective, and understandable technique. The trees are bagged in the first phase, and then the tree is split in the second step using the random split selection algorithm, which is applied to each node of the algorithm with a subset of the characteristics only. The key benefits of RF were its ability to do both regression and classification tasks as well as its ability to generate accurate predictions and results that are simple to understand. Regression models are constructed using RF and SVR to predict the nitrogen oxide (NO_x) levels in Italy and the Beijing air quality index (AQI). The outcomes demonstrate that RF fared better in the NO_x concentration forecast in an Italian city. Liu et al. (2017) reported successful forecasting of AQI in Beijing using support vector regression (SVR). RF was also successfully utilized to anticipate the PM10 concentrations in Blagoevgrad, Bulgaria (Ivanov et al., 2018).

Gradient Boosting Regression: The GB is a concept used to describe the generalization of boosting to any differentiable loss function. It represents a practical and accurate method that may be applied to both regression and classification issues (Ridgeway, 2007). Among the advantages of GB is that no pre-processing of the dataset is necessary, and missing values are handled automatically. It is not necessary to impute data. The feature selection is automated, and the significance of each variable in the models has been determined. The ecology and web search ranking are two areas where GB has found practical applications. Recently, GB has been used to analyze the spread of the COVID-19 epidemic using air pollution data in Quito, Ecuador, with the best predictive performance in congested regions (Rybarczyk & Zalakeviciute, 2021).

SVR: SVR, a machine learning technique, creates hyperplanes to divide data into distinct classes and is typically applied to data analysis with a categorical output variable. SVR, a type of regression analysis, is applied in the situation of the continuous numeric output variable. Vapnik and his colleagues first introduced a version of SVM for regression known as SVR (Boningari & Smirniotis, 2016; Boser et al., 1992; Cortes & Vapnik, 1995; Drucker et al., 1997; Guyon et al., 1993; Schölkopf et al., 1995, 1996; Vapnik et al., 1997). The collection of training data in SVR contains predictor variables as well as observed response values. For each training point x, the objective is to develop a function $f(x)$ that deviates from y_n (sample labels) by a value no larger than bias ε, or to remain as flat as possible. SVR is also referred to as tube regression for this reason. Fig. 13.1 depicts its schematic diagram.

The solution of SVR is given by

$$f(x) = \sum_{n=1}^{N} \left(a_n - a_n^*\right)\left(x_n^T x\right) + b \tag{13.12}$$

where a_n and a_n* are the introduced Lagrange multipliers, **x** is representing the input feature vector, b is the distance parameter. Some regression issues are too complex to be fully explained by a linear model. In this case we generally use a nonlinear SVR by using a nonlinear kernel function $K(x_1, x_2) = < \phi(x\phi) = < \phi(x_1), \phi(x\ \phi), \phi(x_2) >$ instead of dot product $x_n{}^T x$. Here where $\phi(x)$ maps x to a high dimensional space. As a result, the ultimate solution to nonlinear SVR can be expressed as ϕx. Here where $\phi(x)$ maps x to a high dimensional space. As a result, the ultimate solution to nonlinear SVR can be expressed as

$$f(x) = \sum_{n=1}^{N} \left(a_n - a_n^*\right) K(x, x_n) + b \tag{13.13}$$

ANN Multi-Layer Perceptron Regression: The input layer, hidden layer, and output layer are the three layers of an ANN. Although each ANN only has one input and one output, they may also contain zero, one, or more hidden layers. The architecture of ANNs can be categorized into a wide range of styles, including single-layer, multilayer, feed-forward, and recurrent networks. In an ANN, each input neuron and bias carries an assigned weight. The net input is applied across an activation function to determine the output. The weights are then modified when the output is compared to the objective. There are many different types of activation functions, including Sigmoidal, Bipolar, and Binary step functions. A multilayer perceptron network is an implementation of the perceptron network with one or more hidden layers. A feed-forward network is one that uses several perceptrons. In general, it is made up of one input layer, one or more hidden layers, and one output layer.

13.3 Statistical trend and cluster analysis

Cluster analysis (CA): Through the use of CA, we are able to visualize the geographical structure of observational and simulation data, which entails a spatial classification of grid cells or station sites with similar climatic characteristics. The goal of CA is to divide data into groups whose identities are unknown beforehand. In most cases, the number of groups into which the data should be categorized is not known in advance while performing a CA. Instead, the groups are defined and membership is determined by how similar and different individual observations are to one another.

Instead of being an inferential tool, CA is largely an exploratory tool for data analysis. The technique will construct groups and assign group memberships at various degrees of aggregation given a sample of data vectors x defining the

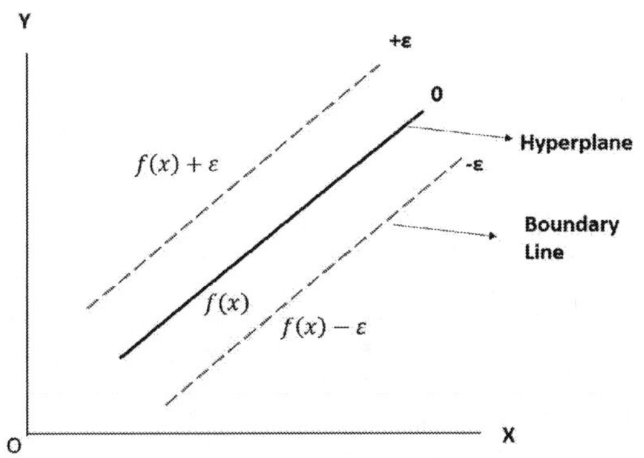

FIGURE 13.1 Schematic diagram of support vector regression.

rows of a $n \times K$ data matrix [X]. Future observations cannot be assigned membership according to the procedure's rules. A CA, on the other hand, might highlight groupings in the data that might otherwise go unnoticed, potentially leading to an empirically meaningful stratification of the data or aiding in the suggestion of a physical basis for the observed structure in the data. For instance, geopotential height data have been subjected to CA in an effort to distinguish between several atmospheric flow regimes (e.g., Cheng & Wallace, 1993). Data are sorted using the CA approach, or more precisely, hierarchical CA. With this technique, objects are grouped progressively based on how similar their attributes are. The ideal number of clusters can be graphically estimated using a dendrogram. As a result, hierarchically or nonhierarchically ordered clusters are generated. Overall, it can be claimed that CA is an effective method for categorizing and streamlining the behavior of environmental pollutants in a certain study area.

Generally. in CA, a data set is divided into a number of observational groups that differ from one another in terms of the usual group values of the variables. The goal is to reduce within-group variance while maximizing between-group variance. CA is a categorization technique that allows the use of any number of variables to categorize the sample's constituents. Large amounts of unlabeled input data are used in the unsupervised clustering algorithm approach, which groups the data into groups based on metrics of similarity. By assessing the outlook of our accessible dataset, we can choose the optimal option. There various type of clustering methods such as: Hierarchical Clustering, Partitioning Clustering, Density-Based Clustering, DBSCAN (Density-Based Spatial Clustering of Applications with Noise), Grid-Based Clustering, Ordering Points to Identify Clustering Structure (OPTICS), Hierarchical Density-Based Spatial Clustering of Applications with Noise (HDBSCAN), Fuzzy Clustering, PAM (Partitioning Around Medoids), K-means clustering etc.

13.3.1 K-means clustering

The most well-known clustering algorithm is undoubtedly K-means. Many beginning data science and machine learning courses cover it. It is simple to comprehend and use. A clustering approach that divides data into K groups is the K-means clustering algorithm. With its ability to quickly and effectively cluster large amounts of data, the K-means clustering technique is more well-known. The K-means algorithm's ultimate clustering results' quality is dependent on the initial centroids' random selection. For different runs of the same input data in the original K-means algorithm, distinct clusters are discovered. The original data points are arranged in the enhanced K-means clustering method according to the distance between the centroid and the data points. K equal sets are formed from the sorted data points. The initial centroids for each set are determined by using its middle points (Ramzi et al., 2015).

13.3.2 Hierarchical clustering

Depending on the hierarchy, Hierarchical clustering algorithms generate a tree-like structure in which newly generated clusters are produced utilizing previously formed clusters and are classified as bottom-up approach (agglomerative) and top-down approach (divisive). The following is a typical implementation of an agglomerative hierarchical clustering algorithm:

Stage I: The first step is to treat each observation as an initial cluster.

Stage II: This stage involves computing the distances between clusters.

Stage III: In this step two clusters with the shortest distance are joined to form a single cluster. Then, the distance matrix is recalculated to reflect this merging procedure.

Stage IV: Continue using Steps II and III until there is just one cluster that contains all observations.

An output from a hierarchical clustering technique is a dendrogram, which is a two-dimensional tree-like structure that represents the order of nested clusters. The structure also shows the separation between each fusion (or division). When the dendrogram is cut at the required level, a number of disconnected groupings are produced (or clusters).

13.4 Interpolation, extrapolation methods

The techniques of Interpolation and Extrapolation are used to identify the analytic function $f(x)$ that passes through the provided set of data points and to utilize that function to calculate the desired value at x. These methods form the cornerstone of numerous commonly used applications in science, economics, and engineering. Due to practical constraints, most experimental findings are based on a subset of observations that are discrete in character and thus only applicable under certain circumstances. Values must be found for intermediate conditions as well, though, in order to find a generic or approximative function. In this situation, extrapolation and interpolation are crucial. They are actually widely utilized in real-world settings, such as data modeling, which may be used to predict weather, stock market movements, and many other things. The temperature is measured at the weather data centre, for instance, every hour of the day.

However, we need to make temperature predictions for unknowable values, such as every 30 minutes or the following day. Interpolation is the process of predicting these unidentified intermediate values within the data range, while extrapolation is the process of estimating data points outside the data range.

Interpolation and extrapolation may sound like sophisticated scientific concepts, but they are actually basic mathematical and statistical processes.

13.4.1 Linear interpolation technique

Two data points with the coordinates given as (x_0, y_0) and (x_1, y_1) respectively. The equation of a straight line, which depicts a straight line drawn through these two data points, can be used to find a function that reflects these data points. The equation of a straight line is given by

$$f(x) = y = mx + c \tag{13.14}$$

Here, m represents the gradient of the line and c represents the y-intercept of the equation, which is the value of y at $x = 0$, and are given by the following formula:

$$m = \frac{(y_1 - y_0)}{(x_1 - x_0)}$$

$$c = y_1 - mx_1$$

After substituting the values of m and c and rearranging, the interpolating function $f(x)$ can be written as:

$$f(x) = y = \frac{y_1 - y_0}{x_1 - x_0}(x - x_0) + y_0$$

This straight line equation is basically a polynomial of degree 1. Fig. 13.2 is an illustration of linear interpolation.

13.4.2 Piecewise interpolation technique

Another method for polynomial interpolation is piecewise interpolation. Piecewise and linear interpolation are very similar. The quantity of data points is the only distinction between this technique and the preceding one. Instead of fitting a single function to all of the data points, piecewise interpolation fits different functions to cover different portions of the data range. Consider N data points and each pair of successive data points define the interval and for each interval, a specific function $f_k(x)$ is assigned, where $k = N - 1$ and represents the number of functions which is needed for N data points. Find the interval in which the desired value of x lies, then interpolate using that interval's appropriate function. Piecewise interpolation technique is also of first degree. A straight line function that represents the linearly interpolated estimate for y is written as:

$$y = f_k(x) = y_k + \frac{x - x_k}{x_{k+1} - x_k}(y_{k+1} - y_k) \tag{13.15}$$

The piecewise linear interpolation technique is shown in Fig. 13.3 along with the interpolated data points. The function it offers is continuous, but it cannot be differentiable since $f_k'(x) \neq f_{k+1}'(x)$.

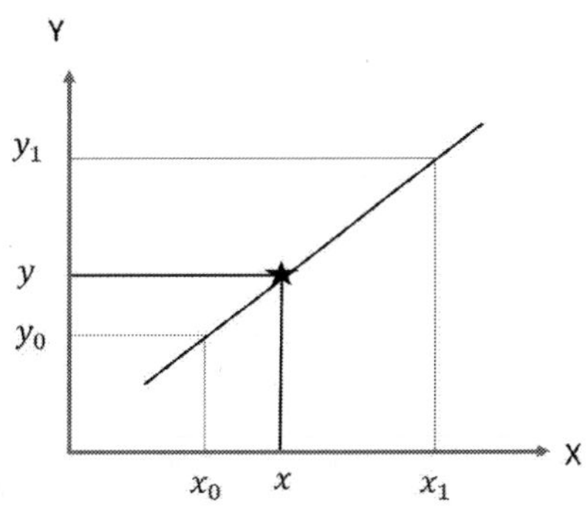

FIGURE 13.2 Schematic diagram of linear interpolation.

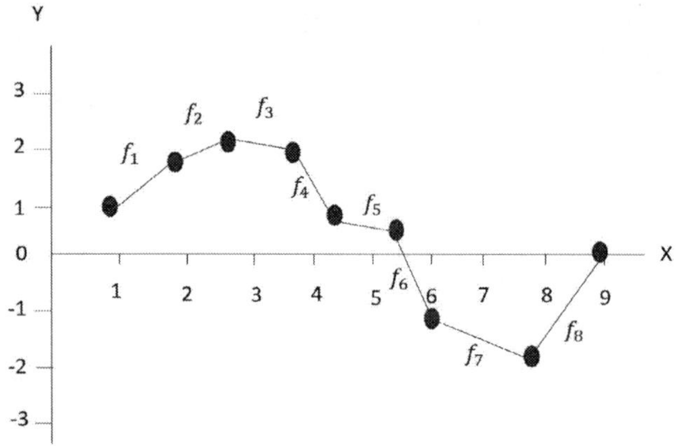

FIGURE 13.3 Interpolation of data points using piecewise linear interpolation.

13.4.3 Nearest neighbor interpolation technique

According to this technique, the value of an interpolated point is set to the value of the closest existing data point. A succession of plateaus, or zero-order polynomials, can be conceived of as what the interpolation appears to be.

Let us consider two successive data points as x_k and x_{k+1}. In this method first we find the mid value using these data points. Then, values of x which is smaller than the mid value treated as the value of of y_k and values greater than the mid value corresponds to the value of y_{k+1}. The formula for the function $f_k(x)$, which defines the nearest neighbor interpolation method is written as

$$f_k(x) = \begin{cases} y_k, x \leq \dfrac{(x_k + x_{k+1})}{2} \\ y_{k+1}, x > \dfrac{(x_k + x_{k+1})}{2} \end{cases} \tag{13.16}$$

According to the $f_k(x)$ function mentioned above, there is a discontinuity at the midpoint of the interval. Fig. 13.4 illustrates this discontinuity by a dashed line that passes through the midpoint and moves halfway from one data point to the next. Consequently, there is a leap between the two data points. The only breaks are at the halfway points. Compared to piecewise interpolation, it is quick and simple to implement.

13.4.4 Polynomial interpolation technique

The polynomial interpolation of degree 1 that is generated from two data points is also known as linear interpolation. A polynomial quadratic equation of degree 2 is created for 3 data points. The quadratic equation has the form $y = ax^2 + bx + c$, where a, b, and c are the coefficients. The coefficient values must be discovered in order to determine which quadratic function will pass through each of these data points. Similarly, a polynomial of degree $N-1$ is required for N points. The function $f(x)$ can be written as

$$f(x) = c_0 + c_1x + c_0x^2 + \ldots + c_{N-1}x^{N-1} \tag{13.17}$$

When using polynomial interpolation, the number of data points increases along with the number of parameters and equations, which leads to more complicated mathematical equations. For the polynomial interpolating N data points, a more useful version called Lagrange form exists and is denoted as

$$f(x) = \sum_{k=1}^{N} \left(\prod_{\substack{j=1 \\ j \neq k}}^{N} \frac{x - x_j}{x_k - x_j} \right) y_k \tag{13.18}$$

Above equation of Lagranges form can be treated as sum of N different polynomials. Every polynomial of degree $(N-1)$ and is weighted by the measurement values y_k for $k = 1, \ldots, N$. This approach has certain drawbacks because it

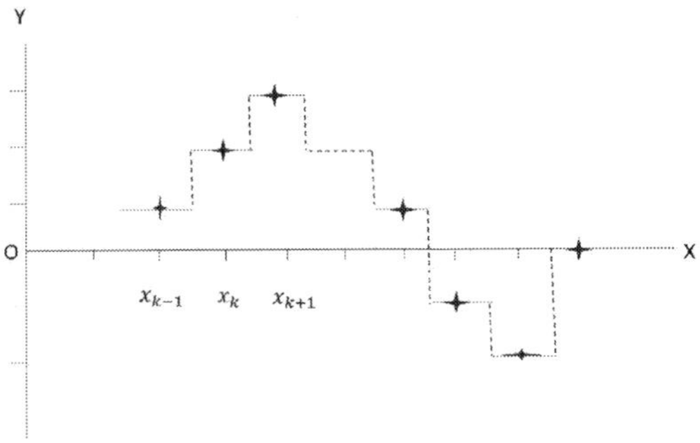

FIGURE 13.4 The nearest neighbor method is used to interpolate the data points.

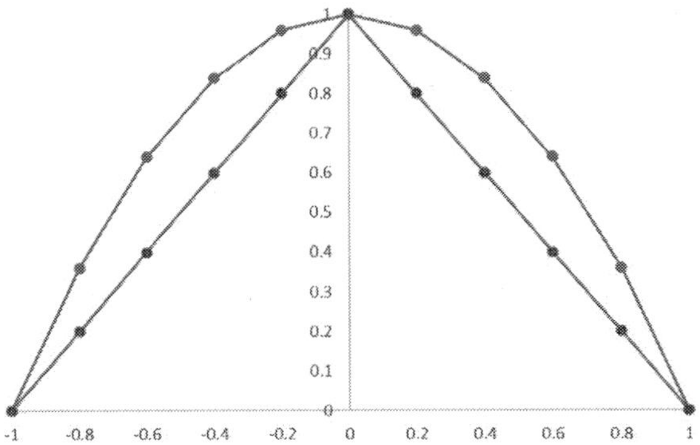

FIGURE 13.5 Piecewise linear approximation (continuous green graph) and quadratic polynomial $p(x) = 1 - x^2$.

becomes more computationally expensive to use as more points are added. Additionally, higher order polynomials may cause round off errors, which will cause them to vary from our target function and possibly cause oscillations in our interpolated function.

If additional interpolation points (x_i, y_i) are employed, piecewise linear functions approximate smooth functions better. By reducing the subinterval lengths and so increasing the number of subintervals, we can improve the accuracy of the piecewise linear approximation. From the next example we come to the conclusion that piecewise linear function approximations are simple to compute. Links, however, are visible in piecewise linear approximation (Fig. 13.5). Consequently, it may take a lot of subintervals to find a piecewise linear approximation with good precision.

Example: To approximate a function on the interval $[-1, 1]$. The function values $y_i = f(x_i)$ given as $x_1 = -1$, $x_2 = 0$, $x_3 = 1$, and $y_1 = y_3 = 0$, $y_2 = 1$.

Solution: To approximate $f(x)$ by a linear function on each subinterval $[x_1, x_2]$ and $[x_2, x_3]$.

By using Lagranges form Eq. (13.18) we can write

$$p(x) = y_1 \frac{(x - x_2)}{(x_1 - x_2)} + y_2 \frac{(x - x_1)}{(x_2 - x_1)} = x + 1, \ -1 \leq x \leq 0, \ \leq, \ -1 \leq x \leq 0, \ \leq, \ -1 \leq x \leq 0,$$

$$p(x) = y_2 \frac{(x - x_3)}{(x_2 - x_3)} + y_3 \frac{(x - x_2)}{(x_3 - x_2)} = 1 - x, 0 \leq x \leq 1. \leq, 0 \leq x \leq 1. \leq, 0 \leq x \leq 1$$

(Fig. 13.5). Piecewise linear approximation (continuous green graph) and quadratic polynomial $p(x) = 1 - x^2$ (blue graph).

13.4.5 Cubic spline interpolation

Round-off error and oscillations in the $(N - 1)$th order polynomial might produce inaccurate results. To avoid this error piecewise lower order polynomials to subsets of data points is applied. Spline functions are the name for these polynomials that connect such subset of data points. For N data points, $(i = 1, \ldots, N)$ there are $N - 1$ intervals, and each interval has its own spline function $s_i(x)$. The number of data points used in each spline function depends on the order of the spline function. Splines are the simplest representations that appear smooth and don't have higher order polynomial difficulties. Because a cubic polynomial has continuous first and second derivatives, cubic spline interpolation is one of the smoothest methods of interpolation and is sufficiently smooth piecewise.

The ith interval cubic spline function can be written as

$$S_i(x) = a_i + b_i(x - x_i) + c_i(x - x_i)^2 + d_i(x - x_i)^3 \tag{13.19}$$

13.4.6 Bilinear interpolation technique

This method's fundamental principle is to interpolate in one direction before interpolating in the other direction that is still left. Therefore, it is surrounded by four points—two in one direction and two in the opposite way. As a result, the problem is separated into various combinations of linear subproblems. Bilinear interpolation on the grid square is the most straightforward interpolation in 2D.

Four data points are $Q_{11} = (x_1, y_1)$, $Q_{12} = (x_1, y_2)$, $Q_{21} = (x_2, y_1)$, and $Q_{22} = (x_2, y_2)$ as shown in the above (Fig. 13.6)

Interpolate in x-direction is given by

$$f(x, y_1) \approx \frac{x_2 - x}{x_2 - x_1} f(Q_{11}) + \frac{x - x_1}{x_2 - x_1} f(Q_{21}) \tag{13.20}$$

$$f(x, y_2) \approx \frac{x_2 - x}{x_2 - x_1} f(Q_{12}) + \frac{x - x_1}{x_2 - x_1} f(Q_{22}) \tag{13.21}$$

And in the Y-direction

$$f(x, y) \approx \frac{y_2 - y}{y_2 - y_1} f(x, y_1) + \frac{y - y_1}{y_2 - y_1} f(x, y_2) \tag{13.22}$$

Example: Consider a function $f(x, y)$ which takes on the values as in Table 13.1 and that in between f is obtained by bilinear interpolation. Calculate the value of $f(1/4, 1/2)$.

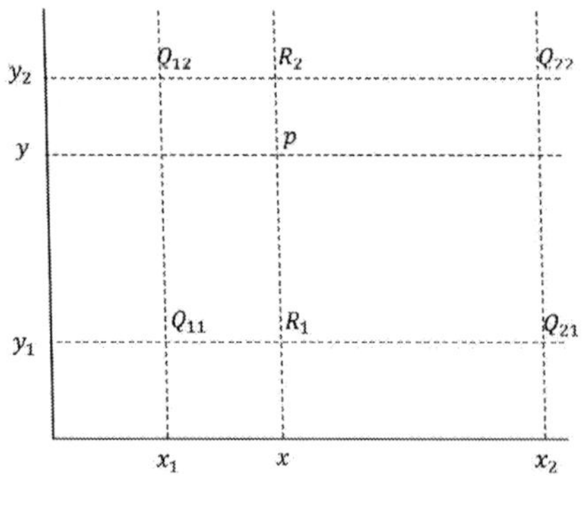

FIGURE 13.6 Working principle of bilinear interpolation.

TABLE 13.1 Function $f(x, y)$ values.

x	y	f(x, y)
0	0	1
0	1	−3
1	0	5
1	1	−11

Solution: we first interpolate in x to find:

$$f\left(\frac{1}{4},0\right)=\frac{3}{4}f(0,0)+\frac{1}{4}f(1,0)=2$$

$$f\left(\frac{1}{4},1\right)=\frac{3}{4}f(0,1)+\frac{1}{4}f(1,1)=-5$$

Now we interpolate in y:

$$f\left(\frac{1}{4},\frac{1}{2}\right)=\frac{1}{2}f\left(\frac{1}{4},0\right)+\frac{1}{2}f\left(\frac{1}{4},1\right)=-\frac{3}{2}$$

13.4.7 Types of extrapolation

Polynomial Extrapolation: It is possible to construct a polynomial curve using all of the available data or data close to the endpoints. Lagrange interpolation or Newton's system of finite series, which supplies the data, are frequently used in this procedure. Using the corresponding endpoints and the final polynomial, the data is extrapolated.

Conic Extrapolation: Five locations closer to the end of the given (i.e., known) data can be used to construct a conic section. A circle or an ellipse will cause the conic section to curl back on itself. However, since the curve is relative to the X-axis, neither a parabola nor a hyperbola will reverse. Software is mostly utilized to implement this method, which extrapolates using conic section templates.

French Curve Extrapolation: Any distribution that tends to be exponential but has accelerating or decelerating factors is suited for French curve extrapolation.

Extrapolation using Fast Fourier Transform: A mathematical method called Fast Fourier Transform uses convoluted data. Extrapolation can be done quickly by using the Fast Fourier Transform.

13.5 Graphical representation, percentiles, when to use and when not to use

Graphs are frequently used to visually represent data relationships. A chart, which is another word for graph, is a graphical depiction of the data. In a chart, the symbols that represent the data include bars in a bar chart, lines in a line chart, or slices in a pie chart. Graphs make it possible to examine the causal link between two variables. Graphs are useful for quantifying how much one variable changes when another variable changes by a specific amount.

Bar Graph: Bar graph display data using rectangular bars with heights or lengths proportional to the values they represent. The length of the bar allows the reader to compare the quantity with ease. In general bar graph uses category variables on the x-axis and numerical values on the y-axis. The plotting of the bars in a bar graph can be done either horizontally or vertically. There are three sorts of bar graphs: simple, neighboring, and cumulative. Multiple bar graphs are another name for the last two. A simple bar graph can only use a maximum of two variables i.e. one category and one quantitative, whereas a multiple bar diagram can only use a maximum of three i.e. two categorical and one quantitative. When comparing data from two or more separate datasets, a researcher may find it advantageous to use multiple bar graphs.

Pie Chart: A pie chart is a graph that has a circular and is segmented into sections. The sectors' arc lengths match the numerical value they stand for in a linear fashion. Only categorical data are used with it.

Histogram: Histogram generally used to represent the frequency distribution of the data. The frequency distribution of a continuous variable is shown as a histogram, whose areas are proportional to the corresponding frequencies. Both a histogram and a bar graph are composed of rectangular bars, which makes them extremely comparable. There is no space between any two bars in the histogram. In contrast to bar graphs, which require at least two variables, including one quantitative and one categorical variable, histograms are used to examine the normal distribution of continuous data and only contain a single continuous variable. Histograms are not used to display categorical variables. The same function as histograms is accomplished by frequency polygons, which are especially useful for comparing different data sets. For showing cumulative frequency distributions, frequency polygons are a useful option. In a histogram, the frequency polygon is created by connecting the midpoints of the rectangular bars tops. In contrast to a one-dimensional bar graph, a histogram is a two-dimensional plot. In a histogram, each rectangle's area displays the frequency, but in a bar graph, the height does so and the width is irrelevant.

Box Plot: A set of numerical data is shown in this style of chart by the quartiles that make up the data. It is a straightforward method for showing how our data are structured. It makes it very simple to compare data features across categories. Box plots are used to display the distributions of numerical data values, particularly when comparing them across various groups. This gives high-level information at a glance and provide details like the symmetry, skew, variance, and outliers of a set of data. It is simple to determine where the primary concentration of the data is and to compare different groups. Box plot is also known as Whisker plot. A box plot provides a five-number summary of a dataset. Minimum—It is the lowest value in the dataset, excluding outliers. First Quartile (Q1)—25% of the data falls below the First (lower) Quartile. The median (Q2) represents the dataset's midpoint. Half of the values are below it and half are above it. Third Quartile (Q3)—75% of the data falls below the Third (Upper) Quartile. Maximum—It is the highest value in the dataset, excluding outliers (Fig. 13.7).

The Inter Quartile Range (IQR) refers to the data contained within the box (50% of the data). IQR is computed as IQR = Q3 − Q1. The data points below and above the lower and upper boundary are considered outliers. The formula used to determine the lower and upper limit are Lower Limit = Q1−1.5 × IQR and Upper Limit = Q3 + 1.5 × IQR. The numbers that fall between the lower and upper boundaries are used to calculate the lowest and maximum values, with values that are outside of these ranges being regarded as outliers.

With the help of box plots, we can easily figure out the mean value of the data, how spread the data is, and if the data is skewed or not (skewness) (Fig. 13.8). The median gives the mean value of the data. The data is said to be normally distributed if the Median is in the center of the box. In this case the whiskers are almost the same on both ends. Positive Skew (Right Skew) exists when the Median is nearer to the first Quartile and the whisker at the lower end is smaller. Data is said to have a negative Skew (left Skew) if the Median is nearer to the third Quartile and the whisker at the top end is smaller.

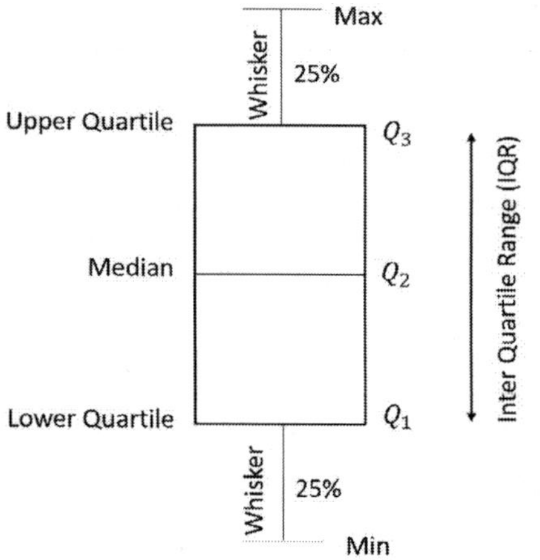

FIGURE 13.7 Schematic diagram of box plot.

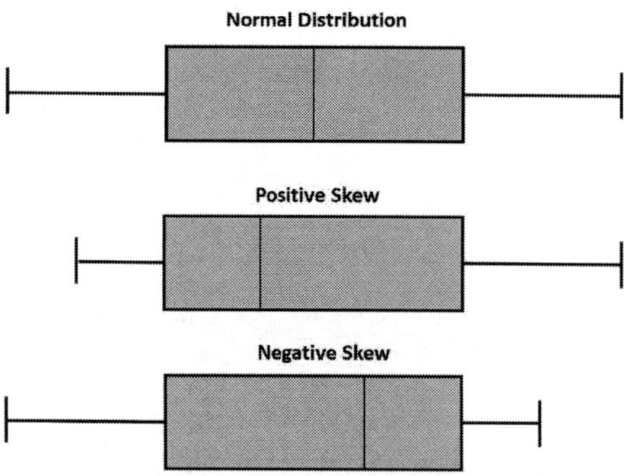

FIGURE 13.8 Distribution and skewness of box plot.

Scatter Plot: A scatter plot, sometimes referred to as a scatter diagram, is a graph in which the values of two quantitative variables are drawn along two axes, with the arrangement of the resulting points indicating any relationship among variables given a set of data. Using scatter plots, we can visualize how much one variable has an impact on another. The term "correlation" refers to the connection between two variables. Positive correlation can be defined as a scatter plot with rising values for both variables. It is possible to describe a scatter plot as having a negative correlation if one variable is increasing while another is declining. It is referred to as having no correlation if there is no trend.

To appropriately implement statistical tools, it is necessary to understand an useful statistical notion called data type. In order to evaluate data sets using the proper statistical techniques, a proper understanding of data types is required. The use of various statistical diagrams to show data set summaries and findings relies on our goals and the sorts of data we are using. We should utilize relevant illustrations for our data sets since they help viewers understand the synopsis and conclusions quickly and readily.

13.6 Uncertainty analysis and comparison with air quality standards

According to Morgan and Henrion (1990), uncertainty analysis is the process of computing the overall amount of uncertainty that is introduced into the output as a result of quantifiable uncertainty in the model's inputs and outputs, as well as the relative significance of these uncertainties. The sum of model uncertainty, variability, and uncertainty regarding the input data can thus be used to establish total model uncertainty. It's possible that errors or omissions in a model's representation of the dynamics and chemistry of the atmosphere, incommensurability, the use of numerical solution techniques, and the choice of the modeling domain and grid structure are all causes of model formulation uncertainty.

On the basis of emissions, observational data, meteorological, chemistry, and model resolution, input data uncertainties are defined. Stochastic anthropogenic and atmospheric processes are referred to as variability. It adds to the previously noted uncertainties, particularly those related to emissions estimation and depictions of chemistry and meteorology. Through the use of data quality indicators that represent a model's capacity to imitate real-world occurrences, it is possible to compare observations and model predictions to assess the overall level of model uncertainty. Applications of such indicators allow to identify model constraints and provide a support for model intercomparison, notwithstanding the difficulty of defining quantitative quality indicators for model evaluation.

Three primary elements—scientific, statistical, and operational—are included in the evaluation of air quality models (Chang & Hanna, 2004). A detailed analysis of the model's algorithms, physics, presumptions, and codes is done for the purpose of determining how accurate, effective, and sensitive they are in a scientific analysis. Usually, this task necessitates thorough model comprehension. The model's predictions are compared to the observations for statistical evaluation to determine how well they match. The user-friendliness of the model (user's manual, user interface, etc.) is primarily taken into account by the operational evaluation component. In this work, we'll concentrate mostly on the estimation of overall uncertainty as well as the evaluation of the scientific and statistical model.

13.6.1 Guidelines for modeling uncertainty estimation

As a metric of acceptability to ensure good model performance and trustworthy modeling findings for decision makers, the current European legislation establishes modeling quality objectives. However, a practical implementation of these requirements and an interpretation of the results of an uncertainty analysis based on the suggested methodology are imprecise and, in some situations, difficult for nonexpert users to understand. For the scientific community, developing a standardized process for evaluating uncertainty remains a challenge. However, based on the compilation completed and here presented, it is possible to define some guidelines to estimate the overall model uncertainties for a specific modeling application.

There are three application levels that can be tailored to end users' demands, objectives, and capabilities. Observed and simulated pollutant concentrations are represented graphically in the first level of analysis, which is a straightforward qualitative analysis (e.g., time series plots, contour plots, frequency distributions, scatter plots). The second level includes the estimation of overall model uncertainty based on statistical parameters and FWD settlements. Investigating the contributions of the various elements of the overall model uncertainty is required in order to enhance model performance. As a result, the third level describes the overall model uncertainty by estimating its various components. As a first step, the variability is addressed.

Since stochastic perturbations cannot be eliminated, their impact on the overall model uncertainty needs to be studied independently. The method for calculating how much the stochastic processes contribute to the overall model uncertainty includes (i) a spectrum analysis of the air quality monitoring data, (ii) removing the stochastic short-term variation from the original data before the model is validated, and (iii) quantifying variability.

Utilizing cyclical patterns of data is the emphasis of spectrum analysis. A complex time series with cyclical components will be broken down into a few fundamental sinusoidal functions with specific wavelengths for the analysis's goal of separating the many phenomena that are present in the time series. One of the methods that can be applied for data spectrum analysis is fast Fourier transformation. Using a moving average filter, it is possible to separate the air quality data into deterministic and stochastic components. The difference between the original and filtered data is referred to as the stochastic component.

In terms of the variation coefficient, the variability is finally expressed as

$$VC = \frac{\sigma}{\overline{C}} \times 100\% \quad (13.23)$$

where is $\vec{C}$ the average concentration and σ is the standard deviation of the differences between the original and filtered series. Data should be log converted before analysis since the statistical distribution of the data deviates from a normal distribution (Borrego et al., 2006).

13.7 Data quality metrics/objectives

Data quality objectives (DQOs) provide direction for articulating problems and identifying the choices that must be made. DQO defines the quality criteria for the decisions, which results in the quality requirements for the data, and then creates a justifiable sampling and evaluation plan. This methodical planning procedure can contribute to the gathering of "excellent" environmental data to support regulatory bodies in decision-making. The DQO process is a technique for organizing data collection tasks. It offers a framework for harmonizing choice uncertainty with the resources at hand. All large projects involving the acquisition of data must follow the DQO methodology. The national ambient air quality standard adopted by the governing bodies collected the data from all sort of sources according to DQO process. The Ambient Air Quality Monitoring Program's statistics are utilized to make highly specific judgments that may have an economic impact on the region they cover. The DQO Planning Process produces qualitative and quantitative statements. In addition to defining the best types of data to gather and the ideal environments in which to do so, this section also clarifies the study's objectives and specifies the acceptable limits for probable decision errors.

13.7.1 Population uncertainties

Representativeness is the most crucial factor in determining the quality of any ambient air monitoring network. A population feature, a parameter fluctuation at a sampling point, a process state, or a condition are all described in terms of how accurately and precisely data represent them. Representativeness can be impacted by population uncertainty, the spatial and temporal components of error, and other factors. The choice of suitable boundary conditions (the monitoring region and the sampling time period/frequency) which the decision will apply and the creation of an adequate statistical sampling design can be used to minimize these uncertainty. The following elements are necessary to ensure the acquisition of a representative air quality sample:

1. deciding on a network size that is in line with the monitoring goals and finding representative sampling locations;
2. detecting and recording the physical and meteorological restrictions that are placed on the sampling sites due to local terrain, emission sources, land access, and other factors;
3. choosing frequency and schedules for sample that are in line with the goals of the monitoring.

13.8 Recent geographic information system spatial exposure techniques/maps and examination of potential hotspots and implications for exposure reduction

The term "geographic information system" (GIS) refers to a computer-based tool for mapping and analyzing geographical phenomena and events that take place on Earth. The unique visualization and spatial analytical advantages provided by maps are integrated with conventional database operations like query and statistical analysis by GIS technology. Due to its unique capabilities that set it apart from other information systems, GIS are useful for a variety of public and commercial companies for explaining events, forecasting results, and strategizing. In general, a GIS offers tools for data collection, management, processing, and analysis, as well as the display of findings in both visual and narrative form, with a focus on maintaining and interpreting the intrinsic properties of spatial data. GISs are distinguished by their capacity to incorporate spatial data, manage it, analyze it, and respond to spatial issues.

Air quality modeling in relation to transportation pollution has only recently begun to use GIS (Bruckman et al., 1992; USEPA, 1998). To provide a more straightforward and adaptable technique of modeling source-receptor connections, a variety of GIS-based methodologies have thus been developed recently (Briggs, 2005; Jerrett et al., 2005). GIS Systems support three fundamental types of data to construct a geographical data model.

1. **Vector data:** Events or Points: Pattern expressed as a set of points in space; Lines: Pattern expressed as a network, and Polygons: Pattern expressed as a set of analytical units with clearly defined closed bounds.
2. **Raster data:** Data from grid cells.
3. **Image:** Satellite photos and images.

13.8.1 Land use regression models

As a result of the considerable geographical and temporal variability of urban air pollution levels numerous monitoring sites are necessary to quantify concentrations as precisely as possible and to reduce exposure misclassification. Land use regression (LUR) models have been popular in community health research as a method for capturing the smaller-scale heterogeneity as GIS's ability to offer land use data has grown. These statistical models regress observed pollutant quantities at sample locations (the dependent variable) on geographic factors such traffic volume, proximity to roads, harbors, and industrial sources; population and housing density; and geographic latitude and longitude etc. The parameter estimates generated from the regression model are then used to predict air pollution levels for any place, such as individual buildings (Jerrett et al., 2007; Ross et al., 2006; Vienneau et al., 2010). LUR modeling is also endorsed by the Health Effects Institute panel (Health Effects Institute, 2010) as a more precise tool for exposure analysis in a recently completed critical evaluation on exposure and health effects.

Using the surrounding land use, transportation, physical geography, and population variables, LUR is a potential method for forecasting pollutant concentrations for a specific place. Compared to dispersion modeling, LUR's key advantages are its low cost and relatively simple inputs as well as its empirical structure (such as the choice of an appropriate buffer size) for regression mapping (Jerrett et al., 2005). The final model performance depends on the choice of spatial covariates at the right impact distances such as distance to road, industrial land use, etc.

LUR models have a few drawbacks despite their benefits of being straightforward, practical, and frequently utilized by academics. They include the need for precise monitoring data and at many different locations, particularly in highly industrialized urban regions with a variety of emission sources. The gathering of monitoring data in these areas is costly and time-consuming. In addition to it, the recommended buffer distances for one urban region may not be the best option for another because to variations in geography, meteorology, land use, traffic, and population mix between metropolitan areas. LUR using GIS have certain limitations:

1. The majority of the time, these models cannot be applied in different metropolitan settings.
2. They are unable to establish a relationship between particular emission sources and concentrations to create pollution mitigation policy.
3. They are normally not created to address issues of air pollution that involve many pollutants.

13.9 Data visualization approaches of model outputs

The data generated by the air quality models is substantial. This data needs to be visualized in a way that consumers can easily interpret in order to give useful insight into how the atmosphere functions. The visualization system must be able to manage large quantities of spatial data. Different file formats, such as netcdf, grid, ascii, etc., must be supported due to the variable nature of the data source.

The support of all current aspects should be ensured through a system for air pollution data exploration (three spatial, one temporal, multiple thematic and possible scenarios). Geo-referenced coordinates are necessary for the visual presentation and interpretation of air pollution data. A suitable software solution should contain a variety of visualization techniques. For the purposes under consideration, the most widely used shading techniques, texture mapping tools, isolines and isosurfaces generation, slicing functions, vector field displays, flow visualization, color editing and reclassification tools, glyph representations, volume rendering, and animation are required.

The National Center for Atmospheric Research, in association with U.C. Davis and Ohio State University, developed the software platform known as Visualization and Analysis Platform for Ocean, Atmosphere, and Solar Researchers (VAPOR). It can create graphics and videos using very huge mesh-based datasets, such as wind speed and other two- and three-dimensional physical fields (Gruchalla et al., 2009). Although VAPOR's input file format is VDF, it also enables conversion from other file types like NetCDF, particularly those produced by weather research and forecasting models. Although it is free software, it is not open source.

HYSPLIT:

HYSPLIT is a comprehensive system for simulating sophisticated transport, dispersion, chemical transformation, and deposition simulations as well as basic air parcel trajectories (HYSPLIT – Air Resources Laboratory (noaa.gov)). HYSPLIT is frequently used to pinpoint local pollution sources and propagation routes for the spread of air pollution (Koracin et al., 2011; Stein et al., 2015; McGowan & Clark, 2008). HYSPLIT makes an effort to reconstruct the paths of numerous air parcels starting from a certain location (i.e., receptor) for each timestamp using meteorological data. A value for the estimated concentration is given to each of these trajectories. There is no temporal information given, but these trajectories can be further grouped to find potential trends or to create a heatmap as a static image (Wang et al., 2009). In the field of atmospheric sciences, HYSPLIT is still one of the most frequently used models for air transport and dispersion. A back trajectory analysis is frequently used to identify the source of air masses and construct source-receptor associations.

NCVIEW: Ncview is a visual file viewer for files in the netCDF format. Normally, we can use ncview to quickly and simply push a button to view your netCDF files. Simple data movies, viewing in other dimensions, looking at the real data values, changing color mappings, inverting the data, and other options are available. It operates using X11, R4 or above on UNIX platforms. We can tell whether it's worthwhile to spend your time using a comprehensive analysis package by quickly skimming over a netCDF file using ncview. Carbon monoxide (CO) spatial average concentration at 00 UTC over South Asia region using NCVIEW tool is shown in Fig. 13.9.

NCL: A free interpreted language developed exclusively for processing and visualizing scientific data, the National Center for Atmospheric Research (NCAR) Command Language (NCL) is a creation of the Computational & Information Systems Laboratory at the NCAR and is supported by the National Science Foundation (https://www.ncl.ucar.edu/). NCL provides strong file input and output capabilities. It can read and write data in netCDF-3, netCDF-4 classic, netCDF-4, HDF4, binary, and ASCII formats. It can read HDF-EOS2, HDF-EOS5, GRIB1, GRIB2, and OGR files (shapefiles, MapInfo, GMT). Hundreds of graphics resources in NCL make it simple to create and customize high-quality visuals.

Panoply: It is also a user-friendly tool for data visualization. Panoply is a cross-platform programme developed by Goddard institute for space studies (GISS) personnel. It can be used on desktop computers running Windows, Mac OS X, Linux, and other operating systems. A compatible Java 11 (or later version) JRE or JDK must have been installed on your computer in order to use Panoply. Spatial average concentration of O3 over South East Asia region is presented using panoply plotting tool is shown in Fig. 13.10.

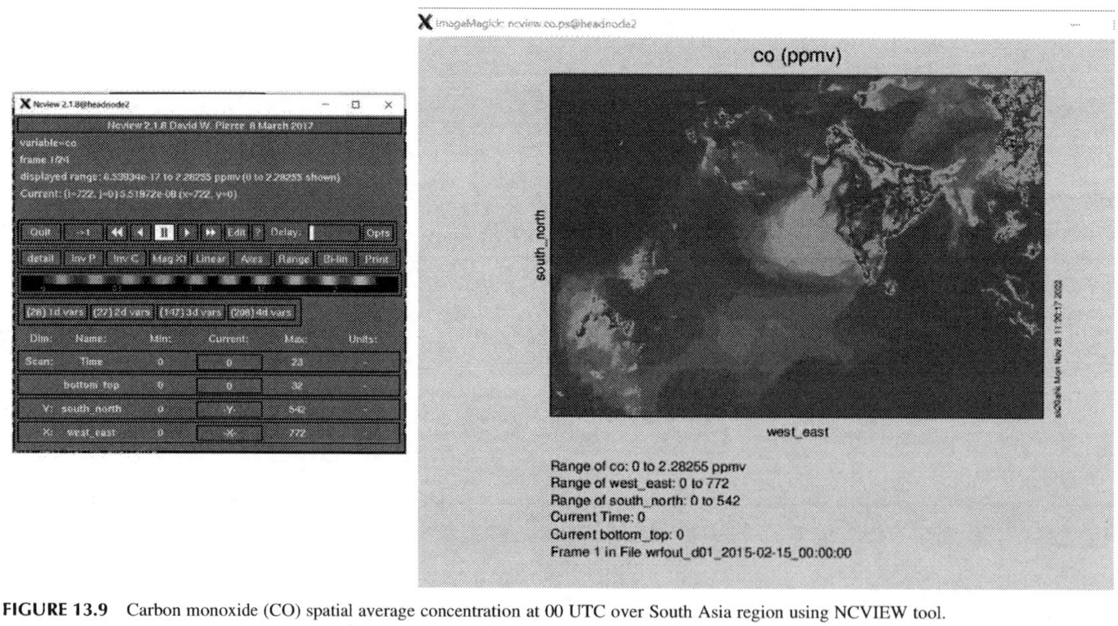

FIGURE 13.9 Carbon monoxide (CO) spatial average concentration at 00 UTC over South Asia region using NCVIEW tool.

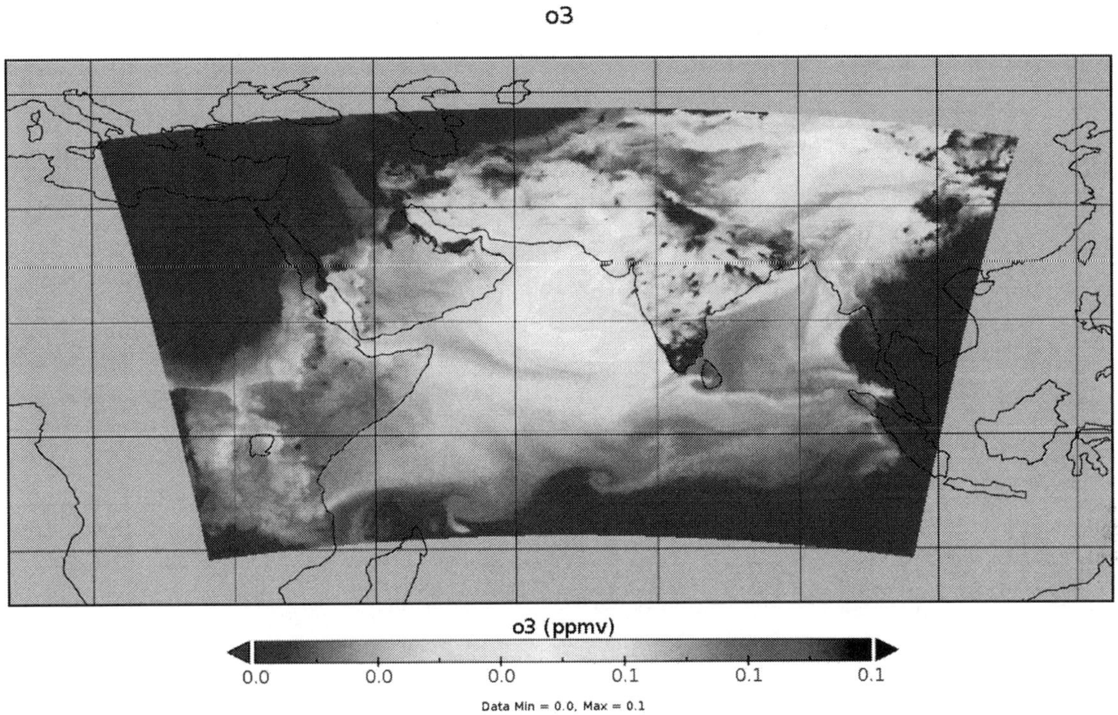

FIGURE 13.10 Spatial average concentration of O3 over South East Asia using panoply plotting tool.

13.10 Questions

1. How many neighboring cells does Bilinear Interpolation use when interpolating a value? (a) 3 (b) 4 (c) 8 (d) 16.
2. When using bilinear interpolation, what are the two underlying assumptions?
3. Bilinear Interpolation can only be performed with a grid based system (True/False).
4. Bilinear Interpolation uses a distance weighted average when interpolating an unknown value (True/False).
5. Explain the distinction between mean squared error and mean absolute deviation as forecast accuracy measures.
6. Explain the differences between the paired and two-sample t-tests.
7. Each year, an exam is given. The test has a mean score of 140 and a standard deviation of 20. What was test score if z-score is 1.50?
8. What is P-value? How it is significant for a statistical test?
9. Show that the constant k that minimizes the expected value of the squared error is the mean value of the target variable.
10. The null and alternative hypotheses for a one-way ANOVA test are...

$$H_0: \mu_1 = \mu_2 = \ldots\ldots = \mu_k$$

$$H_a: \text{Notallmeansareequal.}$$

Assume that the null hypothesis is incorrect in practise. Does this imply that there is no common mean among the populations? What does it mean if not?

11. Find a polynomial of degree 3 or less that is unique and agrees with the given data in Table 13.2 using Lagrange interpolation.

Solutions to the end-of-chapter questions will be available at: https://www.elsevier.com/books-and-journals/book-companion/9780128225912.

TABLE 13.2 Data provided for Lagrange interpolation.

i	x_i	y_i
0	−1	3
1	0	−4
2	1	5
3	2	−6

References

Bai, L., Wang, J., Ma, X., & Lu, H. (2018). Air pollution forecasts: An overview. *Int. J. Environ. Res. Public Health, 15*(4).

Boningari, T., & Smirniotis, P. G. (2016). Impact of nitrogen oxides on the environment and human health: Mn-based materials for the NOx abatement. *Current Opinion in Chemical Engineering, 13*, 133–141. Available from https://doi.org/10.1016/j.coche.2016.09.004, http://www.elsevier.com/wps/find/journaldescription.cws_home/725837/description#description.

Borrego, C., Miranda, A. I., Costa, A. M., Monteiro, A., Martins, H., Ferreira, J., et al. (2006). Air4EU — M2: uncertainties of models & monitoring. Final report. AMBQA-12/2006. Milestone Report M2 of Air4EU Project (SSPI-CT-2003- 503596). Aveiro, Portugal: Universidade de Aveiro.

Boser B. E., Guyon I. M., & Vapnik V. N. (1992). A training algorithm for optimal margin classifiers. In: Haussler D. (Ed.), *Proceedings of the Annual Conference on Computational Learning Theory* (pp. 144–152). ACM Press, Pittsburgh, PA.

Bougoudis, I., Demertzis, K., & Iliadis, L. (2016). HISYCOL a hybrid computational intelligence system for combined machine learning: The case of air pollution modeling in Athens. *Neural Computing. Applications., 27*(5), 1191–1206.

Breiman, L., Friedman, J., Olshen, R., & Stone, C. (1984). *Classification and Regression Trees* (vol. 37, pp. 237–251). Belmont, CA, USA: Wadsworth Int. Group, no. 15.

Briggs, D. (2005). The role of Gis: Coping with space (and time) in air pollution exposure assessment. *Journal of Toxicology and Environmental Health, Part A, 68*(13-14), 1243–1261. Available from https://doi.org/10.1080/15287390590936094.

Bruckman, L., Dickson, R.J., & Wilkosen, J.G. (1992). The use of GIS software in the development of emissions inventories and emission modelling. In *Proceedings of the air and waste management association*. Pittsberg, PA (USA).

Chang, J. C., & Hanna, S. R. (2004). Air quality model performance evaluation. *Meteorology and Atmospheric Physics, 87*(1-3), 167–196. Available from https://doi.org/10.1007/s00703-003-0070-7.

Cheng, X., & Wallace, J. M. (1993). Cluster analysis of the Northern Hemisphere wintertime 500-hPa height field. *Journal of Atmospheric Science, 50*, 2674–2696. Available from https://doi.org/10.1175/1520-0469(1993)050,2674:CAOTNH.2.0.CO;2.

Cortes, C., & Vapnik, V. (1995). Support vector networks. *Machine Learning, 20*, 273–297.

Drucker, H., Surges, C. J. C., Kaufman, L., Smola, A., & Vapnik, V. (1997). Neural information processing systems foundation United States support vector regression machines. *Advances in Neural Information Processing Systems*, 155–161.

Dunea, D., Pohoata, A., & Iordache, S. (2015). Using wavelet–feedforward neural networks to improve air pollution forecasting in urban environments. *Environmental Monitoring and Assessment, 187*, 477. Available from https://doi.org/10.1007/s10661-015-4697-x.

Gruchalla, K., Rast, M., Bradley, E., Clyne, J., & Mininni, P. (2009). Visualization-driven structural and statistical analysis of turbulent flows. In *Lecture notes in computer science (including subseries Lecture notes in artificial intelligence and lecture notes in bioinformatics)* (Vol. 5772, pp. 321–332). United States 10.1007/978-3-642-03915-7_28.

Guo, Q., He, Z., Li, S., Li, X., Meng, J., Hou, Z., Liu, J., & Chen, Y. (2020). Air pollution forecasting using artificial and wavelet neural networks with meteorological conditions. *Aerosol and Air Quality Research, 20*(6), 1429–1439. Available from https://doi.org/10.4209/aaqr.2020.03.0097, https://aaqr.org/articles/aaqr-20-03-oa-0097.pdf.

Guyon, I., Boser, B., & Vapnik, V. (1993). Automatic capacity tuning of very large VC-dimension classifiers. In S. J. Hanson, J. D. Cowan, & C. L. Giles (Eds.), Advances in Neural Information Processing Systems *5* (pp. 147–155). Morgan Kaufmann Publishers.

Health Effects Institute (2010). Traffic-related air pollution: A critical review of the literature on emissions, exposure, and health effects. Special Report #17, 2010-01-12.

Ivanov, A., Voynikova, D., Stoimenova, M., Gocheva-Ilieva, S., & Iliev, I. (2018). Random forests models of particulate matter PM10: A case study. In *AIP conference proceedings*. American Institute of Physics Inc., Bulgaria <http://scitation.aip.org/content/aip/proceeding/aipcp> 10.1063/1.5064879.

Jerrett, M., Arain, A., Kanaroglou, P., Beckerman, B., Potoglou, D., Sahsuvaroglu, T., Morrison, J., & Giovis, C. (2005). A review and evaluation of intraurban air pollution exposure models. *Journal of Exposure Analysis and Environmental Epidemiology, 15*(2), 185–204. Available from https://doi.org/10.1038/sj.jea.7500388.

Jerrett, M., Arain, M. A., Kanaroglou, P., Beckerman, B., Crouse, D., Gilbert, N. L., Brook, J. R., Finkelstein, N., & Finkelstein, M. M. (2007). Modeling the intraurban variability of ambient traffic pollution in Toronto, Canada. *Journal of Toxicology and Environmental Health, Part A, 70* (3-4), 200–212. Available from https://doi.org/10.1080/15287390600883018.

Kendall, M. G. (1948). The advanced theory of statistics, 4th edn, vol 1.

Kendall, M. G. (1975). Rank correlation methods. Griffin.

Koracin, D., Vellore, R., Lowenthal, D. H., Watson, J. G., Koracin, J., McCord, T., DuBois, D. W., Chen, L. W. A., Kumar, N., Knipping, E. M., et al. (2011). Regional source identification using lagrangian stochastic particle dispersion and hysplit backward-trajectory models. *Journal of the Air & Waste Management Association*, *61*(6), 660–672.

Liu, B. C., Binaykia, A., Chang, P. C., Tiwari, M. K., & Tsao, C. C. (2017). Urban air quality forecasting based on multidimensional collaborative support vector regression (SVR): A case study of Beijing-Tianjin-Shijiazhuang. *PLoS One*, *12*(7). Available from https://doi.org/10.1371/journal.pone.0179763, http://journals.plos.org/plosone/article/file?id = 10.1371/journal.pone.0179763&type = printable.

Mann, B. (1945). Nonparametric tests against trend. *Econom J Econom Soc*, *13*, 245–259.

McGowan, H., & Clark, A. (2008). Identification of dust transport pathways from lake eyre, australia using hysplit. *Atmospheric Environment*, *42*(29), 6915–6925.

Morgan, M. G., & Henrion, M. (1990). *Uncertainty: A Guide to Dealing with Uncertainty in Quantitative Risk and Policy Analysis*. Cambridge Univ. Press.

Ramzi, A. H., Dimishkieh, M., & MehediMasud. (2015). *An enhanced k-means clustering algorithm for pattern discovery in healthcare data. International Journal of Distributed Sensor Networks Volume*. Hindawi Publishing Corporation.

Reid, C. E., Jerrett, M., Petersen, M. L., Pfister, G. G., Morefield, P. E., Tager, I. B., Raffuse, S. M., & Balmes, J. R. (2015). Spatiotemporal prediction of fine particulate matter during the 2008 Northern California wildfires using machine learning. *Environmental Science and Technology*, *49* (6), 3887–3896. Available from https://doi.org/10.1021/es505846r, http://pubs.acs.org/journal/esthag.

Ridgeway, G. (2007). Generalized boosted models: A guide to the GBM package. R Package Vignette.

Ross, Z., English, P. B., Scalf, R., Gunier, R., Smorodinsky, S., Wall, S., & Jerrett, M. (2006). Nitrogen dioxide prediction in Southern California using land use regression modeling: potential for environmental health analyses. *Journal of Expos Science Environment. Epidemiology.*, *16*, 106–114.

Rybarczyk, Y., & Zalakeviciute, R. (2018). Machine learning approaches for outdoor air quality modelling: A systematic review. *Applied Sciences*, *8* (12). Available from https://doi.org/10.3390/app8122570.

Rybarczyk, Y., & Zalakeviciute, R. (2021). Assessing the COVID-19 impact on air quality: A machine learning approach. *Geophysical Research Letters*, *48*, e2020GL091202.

Schölkopf, B., Burges, C., & Vapnik, V. (1995). Extracting support data for a given task. In Fayyad U.M. and Uthurusamy R. (Eds.), *Proceedings, first international conference on knowledge discovery & data mining*. Menlo Park, AAAI Press.

Schölkopf, B., Burges, C., & Vapnik, V. (1996). Incorporating invariances in support vector learning machines. In C. von der Malsburg, W. von Seelen, J. C. Vorbrüggen, & B. Sendhoff (Eds.), *Artificial neural networks ICANN'96* (Vol. 1112, pp. 47–52). Berlin: Springer Lecture Notes in Computer Science.

Stein, A. F., Draxler, R. R., Rolph, G. D., Stunder, B. J. B., Cohen, M. D., & Ngan, F. (2015). NOAA's HYSPLIT atmospheric transport and dispersion modeling system. *Bulletin of the American Meteorological Society*, *96*(12), 2059–2077. Available from https://doi.org/10.1175/bams-d-14-00110.1.

Taheri, S. H., & Sodoudi, S. (2016). Statistical modeling approaches for PM10 prediction in urban areas; A Review of 21st-CenturyStudies. *Atmosphere*, *7*, 15.

USEPA (1998). A GIS based modal model of automobile exhaust emissions. USEPA Report No. EPA-600/-98 -097, Research Triangle Park, NC (USA).

Vapnik, V., Golowich, S., & Smola, A. (1997). Support vector method for function approximation, regression estimation, and signal processing. In M. C. Mozer, M. I. Jordan, & T. Petsche (Eds.), Advances in Neural Information Processing Systems 9 (pp. 281–287). Cambridge, MA: MIT Press.

Vienneau, D., de Hoogh, K., Beelen, R., Fischer, P., Hoek, G., & Briggs, D. (2010). Comparison of land-use regression models between Great Britain and the Netherlands. *Atmospheric Environment*, *44*(5), 688–696. Available from https://doi.org/10.1016/j.atmosenv.2009.11.016.

Wang, J., & Song, G. (2018). A deep spatial-temporal ensemble model for air quality prediction. *Neurocomputing*, *314*, 198–206.

Wang, Y., Zhang, X., & Draxler, R. R. (2009). Trajstat: GIS-based software that uses various trajectory statistical analysis methods to identify potential sources from long-term air pollution measurement data. *Environmental Modelling and Software*, *24*(8), 938–939.

Wu, L., Li, N., & Yang, Y. (2018). Prediction of air quality indicators for the Beijing-Tianjin-Hebei region. *Journal of Cleanear Production*. Available from https://doi.org/10.1016/j.jclepro.2018.06.068.

Chapter 14

Air quality and health impacts

Ulas Im[1], Zorana Jovanovic Andersen[2] and Camilla Geels[1]

[1]Department of Environmental Science, Aarhus University, Denmark, [2]Department of Public Health, Section of Environmental Health, University of Copenhagen, Copenhagen, Denmark

14.1 Introduction

Air pollution is a major environmental stressor responsible for over 6.7 million deaths globally in 2019, due to ischemic cardiovascular disease, stroke, diabetes, chronic obstructive respiratory diseases (COPD), acute lower respiratory infections (ALRI), lung cancer, and neonatal deaths (European Environment Agency (EEA), 2021; Health Effects Institute, 2020). Over 300,000 of these occur in Europe. Besides leading to premature mortality, long-term exposure to air pollution causes every year globally millions of new cases of these diseases as well as exacerbations of disease in chronic lung and heart patients, in terms of symptoms, need for medication, doctor visits, hospitalizations, sick days, and loss of quality of life. Because of the links with multiple diseases and the ubiquitous nature of the exposure, air pollution ranked as the fourth leading risk factor for morbidity and mortality in 2019, surpassed only by high blood pressure, tobacco use, and poor diet (Abbafati et al., 2020). Still, the total burden of air pollution is not yet fully mapped. As we are learning more about the effects of air pollution on new health outcomes such as neurocognitive development in children, neurodegenerative diseases including dementia and Parkinson's disease, mental health, cancers other than lung, etc., overall estimates of disease burden caused by air pollution are still rising. Decreasing air pollution levels in Europe, Canada and the US, over the last decades, are a result of successful air pollution legislation and/or the implementation of air quality standards (e.g., legal limit values) leading to the introduction of cleaner fuels, regulation of industrial emissions, and the implementation of pollution control technologies. This development has generated interest in the effects of air pollution at low levels, below legal limit values in these parts of the world. This research is important to help evaluate whether the current legislations in the USA, Europe, Canada, UK, etc. adequately protect human health, and whether revisions of current legislation are needed. On the other hand, in the other parts of the world, such as in Southeast Asia, Northern Africa, the Middle East, and Latin America, where air pollution levels are generally much higher, there is a general lack of regulation and monitoring programs, presenting major obstacles to conduct research on health effects in these areas and translate research into successful policies. Health impact assessment is an important tool used to quantify the public health impacts of the air pollution levels a population is exposed to.

In this chapter, we introduce the air pollution epidemiology and health impact assessment tools used to quantify the health effects of air pollution. We give a short history of the air pollution epidemiology field and explain the main study designs used to examine the health effects of exposure to air pollution, give an overview over the health effects of air pollution, introduce WHO 2021 Air Quality Guidelines and its main take-home messages, as well as EU legislation on air quality. Finally, we introduce and explain how HIA is used to estimate the burden of air pollution, with several case studies and examples.

14.2 Air pollution epidemiology

Air pollution epidemiology is concerned with the health effects related to exposure to air pollution and is typically divided into two main areas: (1) short-term studies - studies of health effects related to acute or short-term exposures to air pollution, lasting over several hours, days or weeks and (2) long-term studies - studies of health effects of chronic or long-term exposures to air pollution, lasting several months, years, decades, or an entire lifetime. Short-term studies give answer to the question of whether short-term exposure to high levels of air pollution (on days with high levels of

Air Quality. DOI: https://doi.org/10.1016/B978-0-12-822591-2.00014-7

air pollution in winter months, several hours of exposure in heavy traffic, etc.) can trigger adverse effects, such as an exacerbation of an existing disease, hospitalization, death, etc. Typical health effects from short-term or acute exposure to air pollution include cough, headache, irritation of the eyes, nose, and throat, allergic reactions, asthma symptom aggravation, wheezing, and complication of symptoms chronic respiratory and cardiovascular disease patients, which can be serious and lead to hospitalization or even death. Long-term air pollution studies give an answer to the question of whether everyday exposure to air pollution accumulated over years, decades, or a lifetime, can result in the development of a certain disease or premature death. Typical health effects studied with respect to long-term or chronic exposure to air pollution are the incidence of chronic diseases, such as asthma in children, and adults, incidence of COPD, cardiovascular disease, stroke, diabetes, lung cancer, breast cancer, pneumonia, or death. Here we will give a brief overview of the history of air pollution epidemiology, and introduce different types of studies used in air pollution epidemiology, with major examples.

14.2.1 Short history of air pollution epidemiology

A common feature for all environmental exposures is that data are observed, and usually involve low-level exposures of the general public, which are difficult to measure and difficult to link to disease. Thus, early realizations of the adverse effects of environmental pollutants typically came from "natural experiments," or disasters, where extremely high levels of a pollutant led to a dramatic increase in mortality or morbidity. Similarly, early realizations that air pollution may have adverse effects on human health came from short and intense air pollution episodes, where stagnant and cold weather conditions mixed with emissions from the combustion of fossil fuel (wood, coal) and industry, created sharp increases in the concentration of air pollutants, leading to detrimental health consequences, in terms of the unusually high number of deaths, hospital admissions, etc. The first documented air pollution episode occurred in the Meuse Valley in Belgium, in 1930, where over four days in early December, thick fog due to heavy industrial emissions from steelworks, zinc smelters, glass manufacturers, fertilizer and explosive plants, developed around Liege in the valley of the river Meuse, wherein area with around 35,000 inhabitants, more than sixty people died, ten times more than expected, raising considerable media attention. Another similar episode 20 years later, became the landmark in air pollution epidemiology, because of its scale and since it provided data to researchers for the first time to perform analyses of the association between air pollution and mortality. This was "The Great London Smog" of 1952, the most famous air pollution episode, also known as "The Killer Fog." The "Killer Fog" was created when unusually cold and calm weather trapped the heavy, motionless, thick layer of smoky, dusty fumes from the coal stoves and factories in the London Thames basin (Fig. 14.1) (Logan, 1953). This severe air pollution event affected London for 4 days (5–9th December), during which the heavy smog caused major disruption in traffic by reducing visibility, penetrating indoor areas, and causing adverse health effects. The episode led to a fivefold increase in SO_2 levels (several thousands of $\mu g\,m^{-3}$) and a subsequent threefold increase in the number of deaths above the expected, to over 4000 (Wilkins, 1954). Reanalysis of the London Fog data discovered that the air pollution episode contributed to higher death rates several months later, resulting in 12,000 deaths (Bell & Davis, 2001). A similar air pollution episode occurred on October 27–31, 1948, in Donora, Pennsylvania, USA, a steel mill town on the Monongahela River with a population of 1400, where 20 people died and over 6000 suffered from respiratory problems. These air pollution episodes of the 1940s to 1950s have made an important impact on science, public perception of air pollution, and government regulation, and led to some important legislation on air pollution.

The pollution levels from the 1940s and 1950s episodes decreased steadily in the 1960s and 1970s in Western Europe and the USA, mainly due to effective legislation that led to a dramatic reduction in the use of coal and the introduction of central heating in cities. Along with economic development, traffic-related air pollution gained increasing importance in the 1970s and 1908s, but still, adverse effects were considered unlikely at air pollution concentrations seen at this time. In the 1980s, a few air pollution episodes in Western Europe were caused by long-range transported air pollution from Eastern Europe. Some of these episodes in 1985 led to an increase in mortality and hospital admissions in Germany (Wichmann, 2004). The health effects of this short episode were however difficult to detect since the concentrations of SO_2 and PM were in hundreds, instead of thousands of $\mu g\,m^{-3}$, which were seen in the early episodes of the 1930s to 1950s. Studies that evaluated air pollution disasters or episodes present examples of the earliest and methodologically simplest epidemiological studies, which typically compared mortality before, during and after air pollution episodes. These studies still have major contemporary relevance in the areas of the world with high and rising air pollution levels such as China and India, which experience some of the highest air pollution levels in the world, and where major cities frequently experience air pollution episodes as seen in Europe and the USA in the 1930s and 1940s.

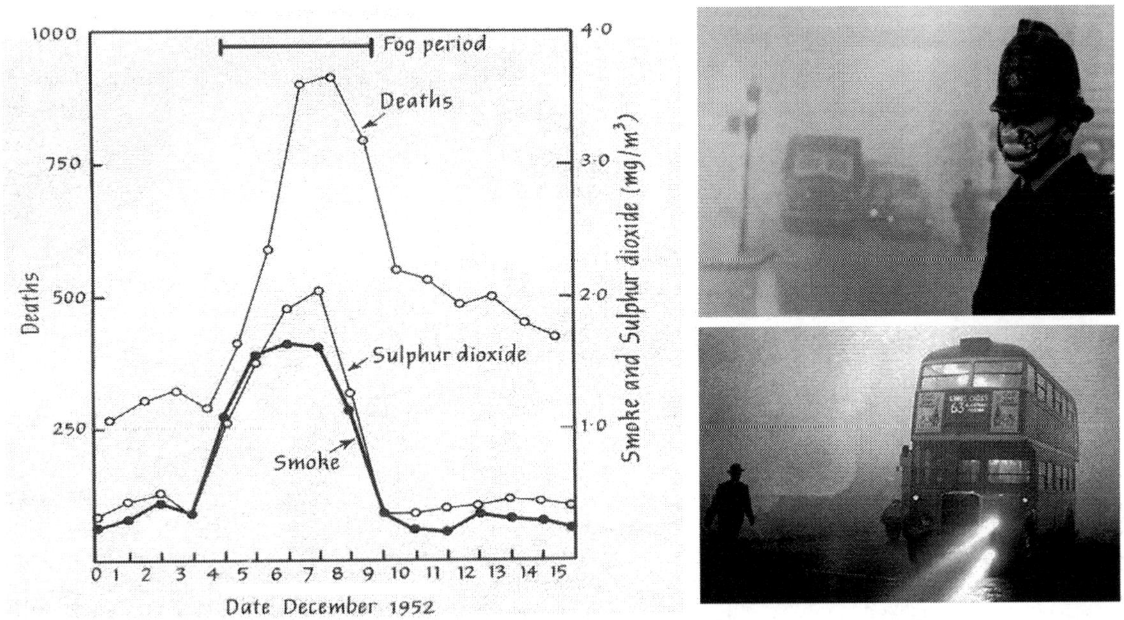

FIGURE 14.1 The Great London Smog episode of 1952 resulted in 12,000 deaths. *Based on Wilkins, E. T. (1954). Air pollution aspects of the London fog of December 1952. Quarterly Journal of the Royal Meteorological Society, 80 (344), 267–271, https://doi.org/10.1002/qj.49708034420.*

14.2.2 Epidemiological designs for studying short-term health effects of air pollution

The study design used to study the health effects of short-term, acute exposures to air pollution is called a time-series study. Short-term or time-series air pollution study design examines time-series of day-to-day variations in air pollution over long periods of time as determinants of day-to-day variations in mortality, hospital admissions, and other public health indicators. The tool to analyze time-series data was introduced by Schwartz in 1990 (Schwartz & Marcus, 1990), leading to an exponential increase in studies of short-term effects of air pollution. The Poisson generalized additive model (GAM) time-series model became the standard tool as it could easily utilize routinely collected data from air pollution monitors, hospital, and death registries. In these types of studies, typically based in a single city, time-series (typically several years) data of daily variations in air pollution levels (for example, 24-h mean pollutant concentrations measured at a central monitor) are associated with daily variations in health outcomes (daily count of deaths or hospitalizations in that city), and adjusted for population-level confounders, such as weather (temperature, humidity), season, day of the week, influenza epidemics, public holidays, pollen counts, etc. In these models, it is important to account for other unmeasured confounders such as improvements in medical care, changes in population size, trends in the occurrence of major diseases, population habit changes with respect to lifestyle, drinking and smoking rates, etc. These unmeasured time trends are modeled by calendar time.

Another design used in analyzing time-series data is a case-crossover design where air pollution concentration on the day a person died or was hospitalized is compared to other weekdays within the same month when the person did not experience the health outcome. The data are then analyzed by a conditional logistic regression model. This design is attractive and has gained popularity over time-series GAM approach, as it is a case-only design, where persons are used as their own control, and because it provides personal risk estimates. In an example from a study on short-term exposure to air pollution and childhood asthma in Copenhagen (Fig. 14.2), a time-series of daily mean concentrations of NO_2 ($\mu g\,m^{-3}$) from a central background monitor in Copenhagen from 2001 until 2009, was linked to daily number of asthma hospitalizations in children aged 18 years or younger, in nine Copenhagen hospitals, during the same period. In this study, the authors found that for each 6 ppb increase in mean NO_2 levels over 5 days, there was a 10% increase in risk of childhood asthma hospitalization the following day (Iskandar et al., 2012).

By the late 1990s, abundant single-site time-series studies have established that air pollution levels, even at much lower levels than that seen during air pollution episodes, were associated with increased rates of mortality and

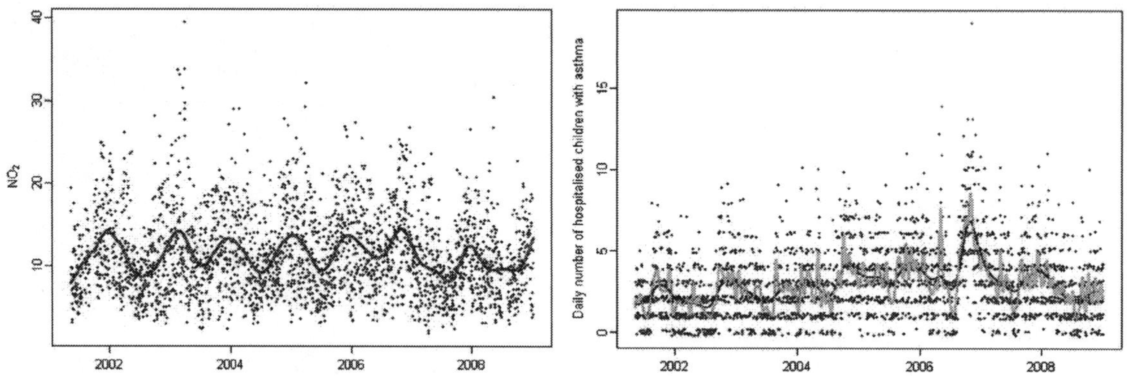

FIGURE 14.2 Time-series of daily mean levels of NO_2 from background monitor (*left*) and number of hospital admissions for asthma in children (0−18 years old) in Copenhagen between 2001 and 2009.

morbidity in cities in the USA and Europe. More recently, time-series studies in China and other sites in South-East Asia with some of the highest air pollution levels globally, have been an invaluable tool for showing health effects related to air pollution exposure in this part of the world and for pushing forward public awareness and political action for stricter regulation and cleaner air. Furthermore, the application of a time-series model has been extended by a hierarchy, allowing for analyses of data from many sites under a common framework, giving pooled estimates. Two such large collaborative multicity studies, APHEA (Air Pollution and Health: a European Approach) (Katsouyanni et al., 1995) in Europe and NMMAPS (National Mortality, Morbidity and Air Pollution Studies) (Samet, Dominici, Zeger, et al., 2000; Samet, Zeger, et al., 2000) in the USA have provided many new insights that were especially useful to policymakers. The APHEA was extended to the APHEA-2 study, which included 43 million people from 29 European cities, followed over 5 years in the mid-1990s for mortality (Katsouyanni et al., 2001). Similarly, the NMMAPS mortality study included the 20 largest cities in the USA, with 50 million people followed from 1984 to 1994 (Samet, Dominici, Curriero, et al., 2000). The two studies have provided remarkably similar results with respect to all-cause mortality, where the combined effect for Europe and the USA was respectively 0.6% (95% CI:0.4−0.8)and 0.5% (0.1−0.9) increase in daily mortality for each 10 $\mu g\,m^{-3}$ increase in $PM_{10.}$ Although these relative rates are small, the burden of disease attributable to air pollution is substantial considering the large populations exposed to air pollution and the large number of persons to whom the relative rates of mortality and morbidity apply. The results from NMMAPS had a profound impact on policymakers and played a central role in the USA Environmental Protection Agency's (EPA) development of National Air Quality Standards (NAAQS) for the six "criteria" pollutants defined by EPA (O_3, PM, CO, lead (Pb), SO_2, NO_{2}). The most recent multicity study, the Multi-Country Multi-City (MCC) Collaborative Research Network is an international global collaboration of research teams working on a program aiming to produce epidemiological evidence on the associations between air pollution, climate, and health. The MCC project collected data on air pollution and mortality from cities all around the world, and in a recent publication based on data from 652 cities found 0.44% (95% CI: 0.39−0.43) and 0.68% (95% CI: 0.59−0.77) increase in daily all-cause mortality for each 10 $\mu g\,m^{-3}$ increase in 2-day moving average of PM_{10} and $PM_{2.5}$, respectively (Liu et al., 2019).

14.2.3 Epidemiological designs for studying long-term health effects of air pollution

While the studies above indicate that short-term exposure to elevated air pollution levels is associated with short-term changes in cardiorespiratory health, they do not tell us how much life is shortened due to air pollution, how does air pollution affect long-term mortality rates in a given population, or whether air pollution can lead to the development of a chronic disease. Studies of chronic- or long-term exposure to air pollution evaluate the effects of typically low or moderate exposure to air pollution (in Western Europe and USA, although emerging studies from China examine the effects of high levels of air pollution) that persist over long periods of time, as well as cumulative effects of repeated exposure to substantially elevated levels of air pollution. This can be done in cohort or case-control studies.

One of the earliest and the most famous long-term air pollution studies is the Harvard Six-Cities study, a prospective cohort study of 8111 adults from six American cities, which compared mortality rates for each city with mean air pollution levels in each city. This is an example of an ecological study, as there are no individual-level data available on air pollution exposure or outcome: mean air pollution for an entire city was assigned for all inhabitants, and mortality rate ratios were compared between cities. The study reported the highest mortality in the cities with the highest particle levels, with a linear dose-response relationship, detecting 26% higher mortality in the most (Steubenville, Ohio) as compared to the least (Portage, Wisconsin) polluted city. Recent updated reanalysis of the Six-Cities Study until 2009 has shown that despite decreases in air pollution levels in the USA to levels well below the USA annual standard of $12\,\mu g\,m^{-3}$, the estimated effects of $PM_{2.5}$ did not change over time, suggesting stable toxicity of $PM_{2.5}$ and no safe threshold (Lepeule et al., 2012).

A typical long-term exposure study is a cohort study, where air pollution models are used to predict air pollution levels at the residence of the cohort participants. As an example, in the Danish Nurse Cohort, individual concentrations of NO_2, $PM_{2.5}$, PM_{10}, were estimated at each nurse's residence (28,731 nurses), as an annual mean from 1990 until 2014. A Cox regression model was used to estimate whether between-person differences in air pollution levels explain an individual's risk of developing type 2 diabetes (Fig. 14.3) (https://pubmed.ncbi.nlm.nih.gov/26989812/). Nurses were followed up from the cohort baseline (recruitment) in 1993 or 1999 for up to 20 years, to examine whether nurses who live in areas with high pollution had a higher risk of developing type 2 diabetes than those living in areas with low air pollution. Authors found that each $3\,\mu g\,m^{-3}$ increase in $PM_{2.5}$ was associated with 11% increase in the risk of developing type 2 diabetes. In this type of studies, it is very important to consider relevant confounders, such as smoking, physical activity, body mass index, alcohol use, diet, etc., which are typically available in the cohort studies, and what makes these types of studies the gold standard in air pollution epidemiology. Another important feature of cohort studies is the temporality between exposure and effect, that exposure to air pollution comes before the health effect of interest is observed.

Mortality is one of the most studied long-term exposure outcomes. If lifelong exposure to air pollution contributes to an increased risk of a number of chronic and infectious diseases throughout lifetime, then the ultimate consequence is that air pollution exposure shortens one's life, leading to premature death. A number of cohort studies have evaluated the association between long-term exposure to air pollution and mortality, typically all-cause natural mortality (where death due to external causes, such as accidents, suicides, etc. is excluded, as these are not believed to be plausibly related to air pollution). Cause-specific mortality has also been evaluated, including cardiovascular mortality, respiratory mortality, lung cancer mortality, and more recently diabetes mortality. This substantial amount of evidence has shown that long-term exposure to air pollution increases the risk of premature mortality. Most studies on long-term exposure to air pollution and mortality are based on well-known cohort studies from the USA, Canada and Europe and are summarized in the recent metaanalyses by Chen and Hoek, which found an 8% excess risk in all-cause mortality per $10\,\mu g\,m^{-3}$ increase in $PM_{2.5}$ (Chen & Hoek, 2020), a notable 30% increase from the earlier estimate from 2013 metaanalyses of 6% (Hoek et al., 2013).

Several large multicenter studies from Europe have examined mortality, documenting even stronger effects of air pollution in Europe. The recent Effects of Low-Level Air Pollution: A Study in Europe (ELAPSE) study, which is an

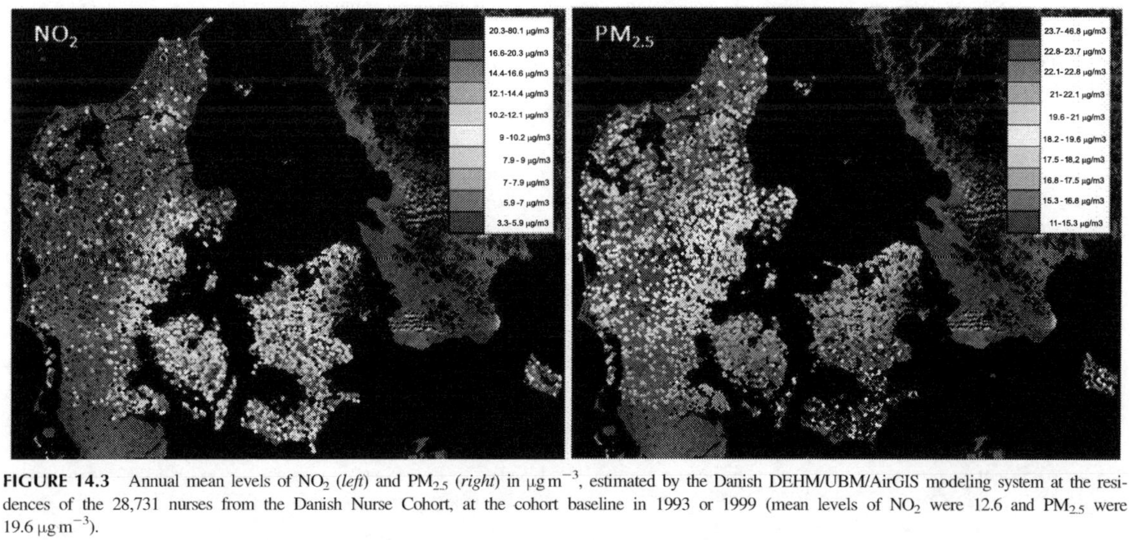

FIGURE 14.3 Annual mean levels of NO_2 (*left*) and $PM_{2.5}$ (*right*) in $\mu g\,m^{-3}$, estimated by the Danish DEHM/UBM/AirGIS modeling system at the residences of the 28,731 nurses from the Danish Nurse Cohort, at the cohort baseline in 1993 or 1999 (mean levels of NO_2 were 12.6 and $PM_{2.5}$ were $19.6\,\mu g\,m^{-3}$).

extension of ESCAPE study, in a pooled analysis of over 325,000 persons from eight cohorts from six EU countries found an even stronger effect of 13% increased risk of all-cause natural mortality for each 5 $\mu g\, m^{-3}$ increase in $PM_{2.5}$ (Strak et al., 2021).

More recently, the latest developments in modeling of air pollution with a combination of measurements with satellite data have facilitated modeling of air pollution exposure estimates for entire countries. This led to a new approach, the use of large nationwide administrative data in the study of air pollution health effects, allowing for large statistical power, but lacking data on lifestyle risk factors (smoking, body mass index, physical activity), which may be important confounders of the association between air pollution and a health outcome. An example is a large US Medicare mortality study with over 60 million US citizens (Di et al., 2017) which found a 7.3% increase in all-cause mortality for each 10 $\mu g\, m^{-3}$ increase in $PM_{2.5}$, and even stronger effects of 13.6% at air pollution levels below 12 $\mu g\, m^{-3}$ of $PM_{2.5}$, which is the US EPA's limit value for $PM_{2.5.}$ Another recent example is the European ELAPSE study which has included combined estimates from analyses in six EU countries (Stafoggia et al., 2022).

Another type of design used to study the association between long-term exposure to air pollution and health is a case-control study, which is typically used in studies of rare diseases, such as Parkinson's disease (PD) (Ritz et al., 2016), etc. where cohort studies would not have sufficient power to detect associations. In these types of studies, cases, for example, all patients with PD, would be matched with multiple controls who did not develop PD, and air pollution exposure would be estimated at residences of cases and controls at the time of the study or historically, to estimate risk of development of PD due to air pollution (Ritz et al., 2016).

Panel or longitudinal studies with data on repeated measurements of a health outcome are a valuable design in showing how changes in air pollution can contribute in changes in a health outcome over time. A landmark study from Swiss Study on Air Pollution And Lung Disease In Adults, (SAPALDIA) cohort with repeated measurements of lung function, associated improvements in air pollution levels with attenuated lung function decline (Downs et al., 2007) and reduced rates of respiratory symptoms, providing important clues about the dynamics behind long-term air pollution exposures and development of chronic respiratory disease, and giving strong support to inference on causality.

14.2.4 Experimental designs in air pollution epidemiology

Although environmental epidemiology mainly relies on observed data that have been typically collected for purposes other than studies of air pollution health effects, experimental designs provide invaluable insights into the health effects related to air pollution and strengthen the evidence for causality between air pollution and adverse health effects. Experimental designs include controlled experimental exposure studies used in examining short-term effects of air pollution, often in real-world exposure settings, where, health outcome collected immediately after exposure is compared in subjects exposed to high and low air pollution levels. An example includes the Oxford Street study where 60 adult volunteers with mild or moderate asthma spent 2 hours walking along one of London's busiest roads, Oxford Street, and through a nearby park with low air pollution levels, Hyde Park while undertaking air pollution measurements with portable sensors, and providing lung function measurements before, during and after each walk (McCreanor et al., 2007). The study found that in adults with asthma, walking for 2 hours at a leisurely pace along Oxford Street, with diesel-powered vehicles as the primary air pollution source, resulted in a significant reduction in lung function. This study provides important support to the results of cohort and cross-sectional studies that traffic-related air pollution worsens lung function. A similar experimental design was used in a study with 28 volunteers cycling in two exposure settings (quiet square without traffic and on a pedestrian bridge above a busy highway) in Barcelona to evaluate whether the benefits of exercise in a polluted environment outweigh the adverse effects of added exposure to air pollution during exercise, due to increased inhalation rates (Kubesch et al., 2015). The authors found that cycling decreased systolic blood pressure in both low and high air pollution exposure levels, documenting that the short-term benefits of exercise seem to outweigh the harms due to air pollution exposure.

Natural experimental studies take advantage of an unusual event resulting in extremely high or low air pollution levels that provide unique opportunities for studying how sudden changes in air pollution levels, whether it be increase or decrease, impact health. The Great London Smog Example is an example of a natural experiment study. Another example includes the study of a steel mill closure in Utah Valley over a year period in 1987–88, which documented that 3 times lower levels of PM_{10} during the steel mill closure period resulted in two to three times lower hospital admission rates for respiratory disease in children, as compared to the period before and after the steel mill's closure (Pope, 1989). A similar approach was used in the studies evaluating the health effects of traffic control measures during the 1996 Olympic Games in Atlanta in 1996 (Peel et al., 2010), the effect of the coal ban in Dublin in 1990 on mortality rates (Clancy et al., 2002) among others.

Worked example 14.1

Air pollution is one of the major risk factors for disease burden (mortality and morbidity) globally. Which of the following is correct:

1. Air pollution is the leading (#1) risk factor for morbidity and mortality.
2. Air pollution is fourth leading risk factor for morbidity and mortality after blood pressure, tobacco use, and poor diet.
3. Air pollution is sixth leading risk factor for morbidity and mortality after blood pressure, tobacco use, physical inactivity, alcohol consumption, and poor diet.
4. Air pollution is eighth leading risk factor for morbidity and mortality after blood pressure, tobacco use, physical inactivity, alcohol consumption, poor diet, obesity, and poor sleep.

Solution to worked example 14.1

Answer is (b): Air pollution is fourth leading risk factor for morbidity and mortality after blood pressure, tobacco use, and poor diet. This is because everyone is exposed to air pollution levels that are harmful to health (99% of the global population) and because air pollution is linked to so many major chronic and infectious diseases, that are major causes of death globally (cardiovascular and respiratory diseases, lung cancer, etc.).

Worked example 14.2

If we want to determine whether air pollution is a risk factor for chronic kidney disease (CKD), which of the following designs would we apply?

1. Time-series study of daily short-exposure to air pollution and daily number of hospital admissions from CKD.
2. Natural experiment study comparing CKD prevalence before and after residential coal burning ban in Krakow, Poland.
3. Controlled experimental study comparing kidney function before and after exposure to diesel in a chamber.
4. Cohort study of long-term exposure to air pollution and incidence of CKD.

Solution to worked example 14.2

Answer is (d): Cohort study of long-term exposure to air pollution and incidence of CKD, as these types of design, comparing long-term exposure contrasts between people, prospectively, with exposure assigned/measured before onset of the disease, is the best design to evaluate whether a certain exposure is a risk factor for a given disease. It is important to note that it is important to account for other important confounders of air pollution associated with CKD.

14.3 Health effects of air pollution

In this section, we give an overview over major health effects of air pollution in adults, children, and pregnant women.

14.3.1 Air pollution and the lung

The respiratory tract is the first portal of entry for air pollutants and consequently, the effects of air pollution on the respiratory system have been studied the earliest and most extensively. Air pollution affects lungs already before birth, and continuing throughout the entire lifetime. Typical biologic effects of air pollution include a suppressed immune system, inflammation and oxidative stress effects, impaired lung growth in children 2017, lung function decrements in children and adults, and carcinogenic effects. Thereby, long-term exposure to air pollution in healthy people increases risk of development of a range of respiratory diseases, such as asthma in children (Khreis et al., 2017) and adults (Liu et al., 2021a, 2021b), chronic obstructive pulmonary disease (Liu et al., 2021a, 2021b), acute lower respiratory infections (Grigg, 2018), and lung cancer (Hvidtfeldt et al., 2021), and it increases the risk of premature death due to respiratory disease (Chen & Hoek, 2020; Stafoggia et al., 2022). Furthermore, air pollution presents a substantial burden in the daily life of respiratory disease patients, where exposure to short-term peaks in air pollution can trigger exacerbations of manifest disease, such as asthma attacks, increased use of relief medication, emergency room visits, hospitalizations, and even death. Finally, people with respiratory disease are more susceptible to developing other diseases which have been linked to air pollution, such as cardiovascular diseases, diabetes, neurodegenerative diseases, etc. (Health Effects Institute, 2020).

As everyone is exposed to varying levels of air pollution, and the prevalence of potentially impacted health outcomes is high—the burden of air pollution on respiratory diseases is huge. The traffic-related air pollutant nitrogen dioxide (NO_2) has been estimated to be associated with 4 million or 13% of all new asthma cases in children, globally every year (Achakulwisut et al., 2019). Similarly, 40% of COPD deaths, 30% of ALRIs deaths, and 19% of all lung cancer deaths globally have been estimated to be attributable to total air pollution (Health Effects Institute, 2020). A recent coroner's ruling in a landmark case in London, UK, acknowledging air pollution as a primary cause in the death of Ella Kissi-Debrah, a 6-year-old severe asthma patient, has major significance in recognizing the severe health consequences of air pollution. The ruling stated that Ella's death in February 2013 was caused by acute respiratory failure, severe asthma, and air pollution exposure to traffic-related NO_2 and particulate matter (PM) pollution around her home, which exceeded the WHO guidelines (World Health Organization, 2021). The ruling also stated that the failure to reduce pollution levels to legal limits, as well as the failure to provide her mother with information about the potential for air pollution to exacerbate asthma, possibly contributed to Ella's death. This is the first ruling of its kind globally, that has referred to air pollution as a public health emergency. It will likely play a role in increasing pressure on governments to tackle unlawful levels of air pollution, as well as on respiratory clinicians to better inform their patients on harms related to air pollution.

14.3.2 Air pollution and the heart

A wealth of evidence links air pollution to increased cardiovascular disease morbidity and mortality, which have been summarized in several important position papers on air pollution and cardiovascular disease by the American Heart Association from 2010 (Brook et al., 2010; Kaufman et al., 2020) and by the European Society of Cardiology, (Newby et al., 2015). These statements by the major global medical societies published in 2010 and 2015, marked the point in time when air pollution was recognized as a risk factor for cardiovascular disease.

Air pollution exposure to both short- and long-term exposure to air pollution is linked to increased risk of cardiovascular diseases. Several recent meta-analyses assessing the impact of short-term exposures to $PM_{2.5}$ (per $10\,\mu g\,m^{-3}$ increase during the prior few hours-to-days) have been published. In 34 studies, short-term $PM_{2.5}$ exposure was linked to triggering of the acute myocardial infarction by 2.5% (Mustafić et al., 2012), hospitalization or death from heart failure (2.1%; 35 studies) (Shah et al., 2013), stroke (1.1%; 94 studies) (Shah et al., 2015) and arrhythmia (1.5%; 23 studies) (Song et al., 2016), and similar risks were also reported for short-term exposures to $NO_{2.}$ While these relative risks are modest, short-term exposures to $PM_{2.5}$ account for up to 5% (population-attributable fraction) of myocardial infarctions worldwide because hundreds of millions of people are continuously impacted (Nawrot et al., 2011).

Longer-term exposures over several years appear to pose amplified risks over and above the acute risks. A meta-analysis reported for that each $10\,\mu g\,m^{-3}$ increase in $PM_{2.5}$ leads to an 11% increase in cardiovascular mortality, roughly 5- to 10-fold the risks following acute exposures (Chen & Hoek, 2020). Several large European studies published after this metaanalyses confirmed the association between long-term exposure to air pollution and cardiovascular disease mortality (Stafoggia et al., 2022; Strak et al., 2021). Long-term exposure to air pollution is also linked to the incidence of myocardial infarction (Zhu et al., 2021), stroke (Yuan et al., 2019), heart failure (Jia et al., 2023), and atrial fibrillation (Chen et al., 2021).

14.3.3 Air pollution and metabolic diseases

A plausible biological mechanism of air pollution promoting diabetes was provided by Sun et al. (2009) showing that exposure to particulate air pollution caused increased blood glucose, inflammation in adipose tissue, and insulin resistance in high-fat-diet mice. This led to a surge in epidemiological studies on long-term exposure to air pollution and type 2 diabetes, and there is nowhere is sufficient evidence to establish a causal relationship between air pollution and type 2 diabetes, incidence (Yang et al., 2020). Globally, it has been estimated that air pollution is responsible for 20% of diabetes related deaths (Health Effects Institute, 2020).

14.3.4 Air Pollution and cancer

Outdoor air pollution was, in 2013, classified as carcinogenic to humans by the International Agency on Research for Cancer (IARC) (Loomis et al., 2013), which sparked interest in research of air pollution as a risk factor for a number of cancer types. The association between air pollution and lung cancer is well established. Meta-analyses from 2014 found a 9% and 10% increase in the risk of lung cancer for each $10\,\mu g\,m^{-3}$ increase in $PM_{2.5}$ (Hamra et al., 2014) and NO_2 (Hamra et al., 2015), respectively. An important study of 17 cohorts participating in the European Studies on Chronic Air Pollution Effects (ESCAPE) project from 2016 found an 18% increase in lung cancer incidence per $5\,\mu g\,m^{-3}$ increase in $PM_{2.5}$ (Raaschou-Nielsen et al., 2013), which was reproduced in a recent study based on seven cohorts from low air pollution areas in Western Europe (including two Danish cohorts) in the ELAPSE project (13% per 5 $\mu g\,m^{-3}$) (Hvidtfeldt et al., 2021). Finally, lung cancer mortality has also been linked to air pollution, with recent metaanalyses showing a 12% increased risk per 10 $\mu g\,m^{-3}$ increase in $PM_{2.5}$ (Chen & Hoek, 2020). The association between air pollution and cancers other than lung is more sparse. The largest tudy on air pollution and cancer by Turner et al. from 2017 was based on the American Cancer Prevention Study II (CPS-II), which linked $PM_{2.5}$ and NO_2 to mortality from all types of nonlung cancers in 623,048 subjects, and reported associations with kidney, bladder, and colorectal cancer death (Turner et al., 2017). Another important contribution was made by the European ESCAPE and ELAPSE projects. Besides the study on lung cancer incidence, they also presented associations of $PM_{2.5}$ and NO_2 with breast (Andersen et al., 2017; Hvidtfeldt, Chen, et al., 2022), liver (Pedersen et al., 2017; So et al., 2021), stomach (Nagel et al., 2018), kidney (Hvidtfeldt, Taj, et al., 2022; Raaschou-Nielsen et al., 2017), bladder (Chen et al., 2022; Pedersen et al., 2018) or brain (Andersen et al., 2018) cancer incidence. Recent metaanalyses on air pollution and gastrointenstinal cancers reported associations with liver and colorectal, but not with esophageal, gastric, and pancreatic cancers (Pritchett et al., 2022).

14.3.5 Air pollution and the brain

Particles can reach the brain after inhalation into the lungs, via circulation crossing the blood-brain barrier, or via nasal passage and olfactory neurons, inducing neuroinflammation and oxidative stress (Heusinkveld et al., 2016). A groundbreaking study from 2002 showed that the brains of dogs from polluted areas of Mexico City accumulated nickel and vanadium, and had enhanced oxidative and DNA damage in olfactory bulbs, frontal cortex, and hippocampus (Calderón-Garcidueñas et al., 2003). Chronic inflammation and Alzheimer's disease (AD)-like pathology were later confirmed in postmortem studies of human brains, showing infiltration of monocytes, microglia activation, elevated pro-inflammatory markers, increased Aβ deposition (hallmark disease protein of AD), BBB damage, endothelial cell activation, and brain lesions (Calderón-Garcidueñas et al., 2004). A recent study showed accumulation of magnetite nanoparticles in the human brain (Maher et al., 2016). Magnetite is a toxic mineral producing free radicals that have been associated with AD and memory loss. All these studies provide convincing biological plausibility for a link between air pollution and neurodegenerative diseases, such as dementia, AD, Parkinson's disease (PD) and possibly psychological disorders.

14.3.6 Air pollution and neurodegenerative diseases

Dementia has been studied the most with respect to air pollution and recent meta-analyses showed significant associations between $PM_{2.5}$ and dementia incidence (Wilker et al., 2023). A number of studies have also examined whether air pollution can lead to the development of PD or PD mortality, with rather mixed results (Kasdagli et al., 2019). Several more recent papers based on the large US, Canadian and European populations have however detected an association with PD incidence or mortality (Cole-Hunter et al., 2022; Kwon et al., 2024; Shi et al., 2020).

14.3.7 Air pollution and mental health

Recent research suggested that both, short- and long-term exposure to air pollution can worsen, exacerbate existing psychological disorders, or lead to the development of new disease, with most studies focusing on depression, major depressive disorders, anxiety or suicide (Braithwaite et al., 2019). Several studies found a strong association between daily and weekly variations of air pollution and depressive behaviors (Lim et al., 2012; Szyszkowicz et al., 2009). Long-term exposure to air pollution has also been linked to an increased risk of developing depression in a few studies, including a large South Korean study (Kim et al., 2016), the American Nurses' Health Study (Kioumourtzoglou et al., 2017), and large Roman longitudinal study (Nobile et al., 2023), which also linked air pollution to anxiety and schizophrenia. Long-term exposure to air pollution has been linked to suicide or overall mortality from psychiatric disorders (Andersen et al., 2022; Min et al., 2018; So et al., 2022); (b) Evidence is still sparse, but suggestive of associations between air pollution with depression, and suicide, while more evidence is needed on anxiety, and other psychological outcomes.

In summary, air pollution has been linked to several adverse health outcomes in adults as illustrated in. Although everyone is exposed to air pollution, not everyone is equally affected, and several groups are especially vulnerable to the adverse effects of air pollution. Elderly subjects are especially vulnerable to effects of air pollution, related to aging and natural decline in several functions (lung function, cognitive function, etc.) as well as to several chronic diseases and co-morbidities related to aging, which make them more vulnerable to the effects of air pollution. We also describe in more detail the effects of air pollution on two other sensitive groups, children, and pregnant women.

14.3.8 Air pollution and children health

Children are more susceptible to the adverse effects of ambient air pollution than adults due to several factors: their lungs, brains, and immune systems are under development, and toxicants from ambient air can interfere with multiple biological processes throughout life, but the impact on health is likely to be worse when exposure occurs early in life and the expected lifespan ahead is longer; children inhale more air per unit of body weight than adults; children are more active and spent more time outdoors, and therefore breathe in more air pollution. As mentioned above, there is compelling evidence that exposure to ambient air pollution damages the health of children in numerous ways, as illustrated in Fig. 14.4. There is growing and robust evidence that both short-term and long-term exposure to ambient air pollution contribute to poor respiratory health in children and that it is biological plausible that long-term exposure to ambient air pollution not only worsens respiratory health but also increases the risk of developing of asthma (Khreis et al., 2017). In addition to asthma, exposure to ambient air pollution has been associated with a wide range of respiratory outcomes in children ranging from mild and transient to severe symptoms, admissions to hospitals, emergency room visits and mortality related to acute lower respiratory infections like bronchiolitis (King et al., 2018) and pneumonia (Fuertes et al., 2014; Nhung et al., 2017) and reduced lung function (Gehring et al., 2013) in addition to asthma. Early-life exposure to $PM_{2.5}$ has also been associated with sensitization to both aero- and food allergens in childhood (Bowatte et al., 2015). Finally, there is some evidence that early life exposure to traffic-related air pollution is associated with eczema and hay fever.

Number of other outcomes have been studied in children, but evidence is still sparse and/or mixed for inference on causal associations. Link between air pollution and childhood cancers has been explored in a number of studies. Leukemia and lymphoma are the most common in children, accounting for almost half of all childhood cancers, followed by brain tumors. The literature on air pollution and childhood cancers is limited in scope because most studies examined only leukemias, brain tumors, or all childhood cancer types combined, and few had sufficient sample sizes to stratify by cancer subtypes or estimate associations with rarer tumors. Acute leukemia is the most frequent childhood cancer that has been extensively studied with respect to air pollution. A number of systematic reviews have evaluated the association between traffic-related air pollution and acute childhood leukemia indicating association (Filippini et al., 2019). The 2016 IARC-WHO review of air pollution noted that weak associations with childhood leukemia, especially acute lymphoblastic leukemia (ALL) could not be ruled out based on epidemiological studies, defining these associations as "suggestive" but "inconsistent" (International Agency for Research on Cancer (IARC), 2004). Few studies have examined the relationship between air pollution and childhood brain tumors, but evidence remains inconclusive (Zumel-Marne et al., 2019).

Air pollution and neurodevelopment in children is another important topic with increasing evidence in recent years. Autism spectrum disorder (ASD) and attention-deficit/hyperactivity disorder (ADHD) are neurodevelopmental disorders associated with broad functional impairments that can substantially affect children's and families' quality of life. ASD and ADHD cover a wide range of conditions, which are usually identified in childhood and for the majority persist into

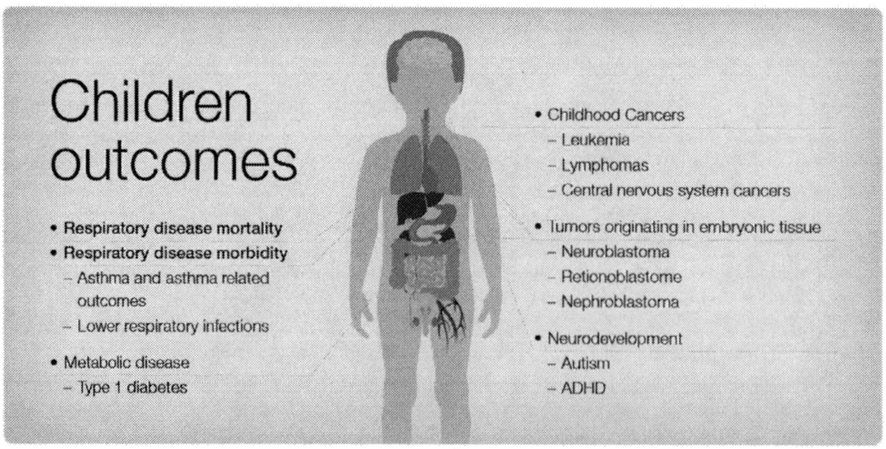

FIGURE 14.4 Overview of health effects related to air pollution exposure in children (outcomes for which evidence is sufficient to establish causal association are listed in bold text, while outcomes that are currently being studied, but evidence is still not sufficient to establish whether association is causal, are listed in nonbolded text).

adulthood and are characterized by difficulties in attention, behavior, social interactions, communication and language. Recent reviews have concluded that there is an association between prenatal exposure to $PM_{2.5}$ and ASD in newborns (Dutheil et al., 2021), as well as prenatal and postnatal exposures and ASD in childhood (Imbriani et al., 2021). Less evidence exists on air pollution and ADHD, most of which relied on teacher- or parent-reported symptoms of ADHD. This body of literature presents contradicting results (Rosi et al., 2023).

14.3.9 Air pollution and pregnancy

Air pollution exposure during pregnancy can lead to complications that can affect adversely the health of both, the mother and the child (Fig. 14.5). In the mother, exposure to air pollution can lead to pregnancy-induced hypertensive disorders, including gestational hypertension and preeclampsia (onset of high blood pressure in the pregnancy and often also protein in the urine), as established by metaanalyses of 17 studies from 2014 (Pedersen et al., 2014). More sparse and mixed evidence exists for gestational diabetes mellitus, with a metaanalyses from 2020 showing significant associations based on 11 studies (Hu et al., 2020). Very modest, but suggestive evidence has shown an increased risk of miscarriage, also called spontaneous abortions, or early fetal loss (Kioumourtzoglou et al., 2019). To conclude, there is strong evidence for an association between air pollution and hypertensive pregnancy disorders, including preeclampsia and gestational hypertension, while more studies on air pollution and gestational diabetes mellitus and miscarriage are needed.

A growing body of research provides evidence of an association between maternal exposure to air pollution during pregnancy (prenatal or pregnancy exposure) and a range of adverse birth outcomes, including miscarriage, stillbirth, preterm birth, decreased birth weight, and/or other indicators of intrauterine growth restriction (IUGR) such as low birth weight, small for gestational age (SGA), and congenital anomalies. Meta-analyses have shown association between prenatal exposure to air pollution and low birth weight (https://pubmed.ncbi.nlm.nih.gov/35569625/), which is now considered well established. Epidemiological evidence on air pollution and preterm birth (birth occurring before 37 weeks of gestation) is more mixed and less convincing than that for IUGR. The lower rates of preterm birth among women who were pregnant around the time of the closure of the steel mill in Utah Valley as compared among those who were pregnant before or after the closure, provide some of the early evidence supportive of an association between air pollution and preterm birth (Parker et al., 2008).

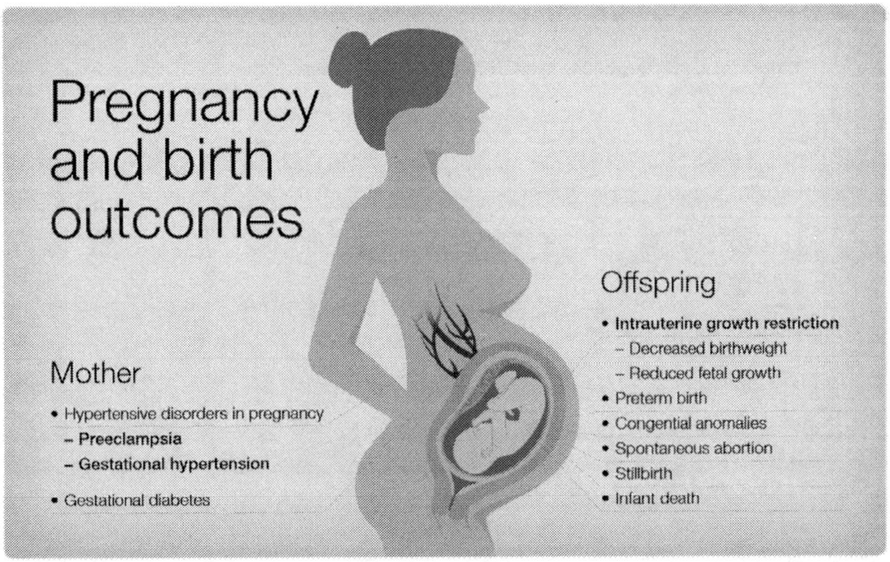

FIGURE 14.5 Overview of health effects related to air pollution exposure during pregnancy in mother and on birth outcomes (outcomes for which evidence is sufficient to establish causal association are listed in bold text, while outcomes that are currently being studied, but evidence is still not sufficient to establish whether association is causal, are listed in nonbolded text)

Stillbirth refers most commonly to an infant born with no sign of life at or after 28 weeks of gestation. A review from 2021 concluded that there is likely link between air pollution and stillbirth (Zhang et al., 2021).

Worked example 14.3:

Air pollution is causally linked to mortality from all of these diseases except:

1. Stroke
2. Chronic obstructive pulmonary disease (COPD)
3. Malaria
4. Ischemic heart disease

Solution to worked example 14.3

Answer is c): Malaria. Malaria is a disease caused by a parasite spread to humans through the bites of infected mosquitoes.

14.4 European Union air quality legislation and who air quality guidelines

European Union policy on air quality includes ambient air quality directives that set air quality standards to provide protection from excessive pollution concentrations in the entire Union and thereby protect the health of European citizens. The EU's Ambient Air Quality Directives (AAQD) (2008/50/EC Directive on Ambient Air Quality and Cleaner Air for Europe and 2004/107/EC Directive on heavy metals and polycyclic aromatic hydrocarbons in ambient air) set legally binding pollutant concentration thresholds, so-called limit values that shall not be exceeded in a given period of time (Table 14.1). For the assessment of the health-related scientific basis, the EU relies on the World Health Organization (WHO) and its Air Quality Guidelines on outdoor air pollution. However, while the 2005 WHO Guidelines already recommended that annual average concentrations of $PM_{2.5}$ should not exceed 10 $\mu g\,m^{-3}$ and annual average concentration for NO_2 should not exceed 40 $\mu g\,m^{-3}$, the currently still effective limit values of 2008 were set at much higher 25 and 40 $\mu g\,m^{-3}$, respectively. Therefore, while following the 2005 WHO recommendations for NO_2, the high limit value for $PM_{2.5}$ already refutes the EUs overarching aim of comprehensive protection.

In September 2021, the WHO has updated its 2005 recommendations, providing a comprehensive synthesis of the research on the health effects of air pollution, including extensive amount of high-quality evidence that has been published during last 15 years (World Health Organization, 2021). The new WHO 2021 Air Quality Guidelines make a historically bold statement, by setting strikingly lower recommendations than those set in 2005, with greatest changes for annual average concentrations of $PM_{2.5}$ set at 5 $\mu g\,m^{-3}$ and for NO_2 set at 10 $\mu g\,m^{-3}$ (Table 14.1). The new Guidelines also highlight the linear exposure-response relationship of air pollution with mortality and other major health effects, down to the lowest observable concentrations. Most importantly, this implies that any improvements in air quality will

TABLE 14.1 Comparison of limit values set by the 2008 EU Ambient Air Quality Directive and WHO Air Quality Guidelines 2005 and 2021 for several major pollutants.

		EU 2008 air quality directives	WHO 2005 air quality guidelines	WHO 2021 air quality guidelines
Pollutant	Averaging period	Objective and legal nature: concentration	Concentrations	Concentration
$PM_{2.5}$	Annual	Limit value: 25 $\mu g\,m^{-3}$	10 $\mu g\,m^{-3}$	5 $\mu g\,m^{-3}$
$PM_{2.5}$	Daily (24-h)	Limit value: -	25 $\mu g\,m^{-3}$	15 $\mu g\,m^{-3}$
PM_{10}	Annual	Limit value: 40 $\mu g\,m^{-3}$	20 $\mu g\,m^{-3}$	15 $\mu g\,m^{-3}$
PM_{10}	Daily (24-h)	Limit value: 50 $\mu g\,m^{-3}$	50 $\mu g\,m^{-3}$	45 $\mu g\,m^{-3}$
NO_2	Annual	Limit value: 40 $\mu g\,m^{-3}$	40 $\mu g\,m^{-3}$	10 $\mu g\,m^{-3}$
NO_2	Daily (24-h)	Limit value: 50 $\mu g\,m^{-3}$	–	25 $\mu g\,m^{-3}$

Source: Air Quality Directives (2008/50/EC, 2004/107EC); WHO (2006), Air Quality Guidelines: Global Update (2005); WHO (2021); and Air Quality Guidelines: Global Update (2021).

result in health benefits, even at concentrations well below current or future limit values (Chen & Hoek, 2020). The new Guidelines also point to increasing evidence for adverse health effects of air pollution well and far beyond those on the cardiorespiratory system, acknowledging new associations with diabetes, neonatal deaths, and increasing research on cognitive development, neurodegenerative diseases, cancers other than lung, and even psychological disorders. The main messages of the updated WHO Air Quality Guidelines are clear and alarming - there are no levels of air pollution below which air pollution is safe for human health.

14.5 Heath impact assessment models

Based on the improved understanding of the link between exposure to air pollution and negative health effects, there has been a growing interest in developments of methods for estimating the health effects across, for example, a national population. That has led to the development of Health Impact Assessment (HIA) models (Anenberg et al., 2016). These models (often also referred to as tools) integrates a number of different data and can be used to assess the total health burden of the current air pollution levels at a given spatial scale, for example, at regional to national or even urban scale. They can typically also be used to evaluate the impact of a given air pollution measure in terms of the expected health benefit. By answering "what if" questions, they can hence support the development of new policies and regulations related to air pollution. Some of the HIA models also include economic data whereby the economic impact can be assessed. These costs are typically referred to as the external costs that can be attributed to exposure to air pollution. In the following, we will go through the different steps that are part of a HIA and give examples of applications of HIA models.

14.5.1 The impact pathway approach

In the early 1990s a collaboration between the European Commission and the US Department of Energy initiated a research project to develop a method to assess the environmental externalities of energy use. The background for the project was a growing recognition and understanding of the fact that energy use (for example a power plant) caused damages to the surrounding environment, for example, as negative impacts on human health, damage on buildings and crops and/or to natural ecosystems. The cost related to these damages was not included and hence reflected in the market price of energy and therefore referred to as external or indirect cost (Holland & Berry, 1995). To be able to make more comprehensive and fair evaluations and comparisons of different technologies, a common methodology to estimate external costs was needed. External Costs of Energy (ExternE), the European part of the project, led to the development of the Impact Pathway Approach that is the basis for most HIA models today. The philosophy behind this approach is to estimate the environmental benefits and costs by following the pathway from the sources via transport and transformation in air (or soil/water) to the health impacts, before being expressed in monetary benefits and costs (Bickel & Friedrich, 2005). The Impact Pathway Approach hence includes a number of different steps as illustrated in Fig. 14.6. The first two steps cover the emissions and the resulting air pollution concentrations. Depending on the case to be studied, this can be the emission from a single power plant or, for example, a full set of gridded emissions representing the main emission sectors in a given area. This is then used as input to chemistry-transport models whereby gridded air pollution concentrations are obtained. These are then combined with population data (distribution and age profiles) in order to estimate the exposure. By applying concentration exposure-response functions (described further below) and background health data the mortality (# people dying prematurely) as well as a range of morbidity endpoints (including, e.g., # hospitalizations) can be estimated. Finally, economic valuations related to each of these cases are applied to estimate the resulting external costs. This can, for example, include the cost of a bronchodilator in relation to asthma or the cost related to hospital-based health care.

The setup in terms of included area covered, air pollutants and health endpoints can vary from model to model (Hassan Bhat et al., 2021). Most models include effects associated to both long- and short-term exposure to $PM_{2.5}$ (in some cases also PM components like black carbon or heavy metals), NO_2, SO_2, and O_3, which typically are based on annual mean concentrations and daily/hourly maximum concentrations at the surface. A number of the existing HIA models. for example, the German EcoSense model (https://www.ier.uni-stuttgart.de/en/research/models/ecosense/) and the Danish EVA model (Brandt et al., 2013a, 2013b) include the full pathway. This makes it possible to make sector-wise simulations with the chemistry-transport models, whereby the cost can be divided into emission sectors and components. This leads to the so-called unit costs, reflecting the health cost for emitting, for example, a kg NH_3 in the agricultural sector. A similar division can be made for regions whereby the cost of emission within or outside, for example, a country/city can be estimated, see Fig. 14.7. Other HIA models are developed to be used more widely, for

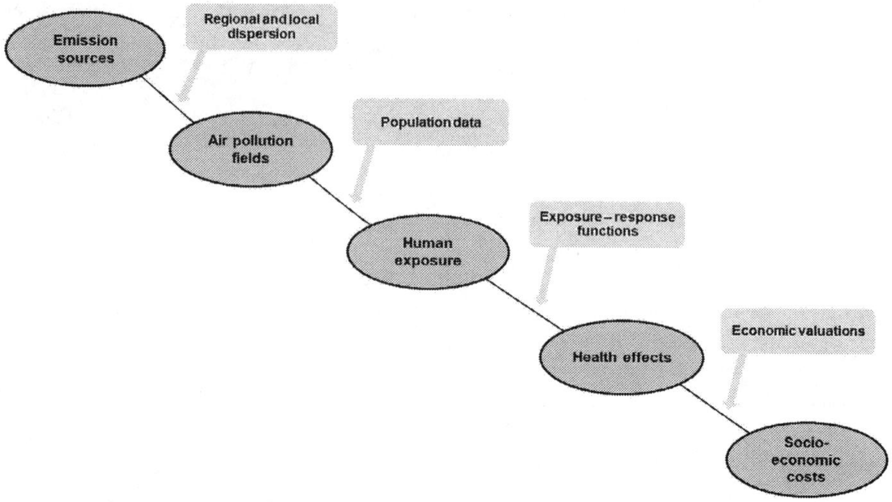

FIGURE 14.6 **Illustration of the impact pathway approach based on the economic valuation of air (EVA) pollution system** (Brandt et al., 2013b). Overview of the impact pathway approach within the EVA system. The diagram outlines the systematic process of tracing the effects of air pollution on the human health, leading to the quantification of its economic impact. The Impact Pathway Approach provides a structured framework for evaluating the costs associated with air pollution, facilitating informed decision-making for policymakers and stakeholders.

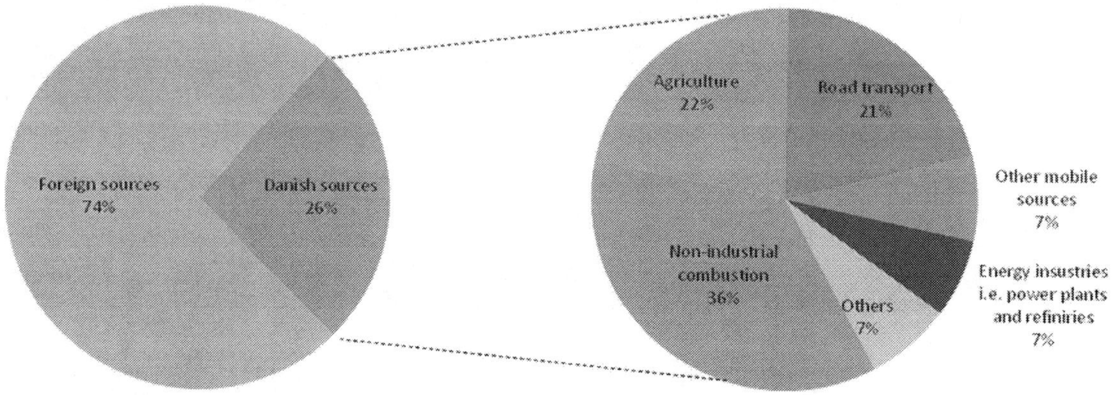

FIGURE 14.7 Contribution to annual premature deaths in Denmark from Danish and foreign sources (*left*), and from Danish sources (*right*). The right side highlights the top five categories of Danish sources. The remaining sources are grouped as "Others." The figures gives an example of how a HIA model can be used to assess the contribution from international and national sources to the total annual number of premature deaths in Denmark (*left*). Through sectorwise model simulations, the contribution from the main sectors can also be estimate (*right*) (Ellermann et al., 2023). *From Ellermann, T., Nordstrøm, C., Brandt, J., Christensen, J., Ketzel, M., Massling, A., Bossi, R., Frohn, L. M., Geels, C., Solvang Jensen, S., Nielsen, O-K, Winther, M., Bech Poulsen, M., Monies, C., Bjært Sørensen, M., Skou Andersen M., & Sigsgaard, T. (2023). Luftkvalitet 2021. Status for den nationale luftkvalitetsovervågning. Aarhus Universitet, DCE – Nationalt Center for Miljø og Energi, 148 s. – Videnskabelig rapport nr. 533. http://dce2.au.dk/pub/SR533.pdf.*

example, by policymakers and health experts at the local/national/regional level and are therefore allowing the user to input values for the air pollution, for example, in the form of annual mean $PM_{2.5}$ concentration for a specific urban area based on observations. The AirQ $^{+}$ developed by WHO and BenMAP—CE developed by the US Environmental Protection Agency are examples of such tools that are widely used across the world (Sacks et al., 2020). They also allow the user to make experiments with different types of exposure-response functions, whereby the user can get an understanding of some of the uncertainty related to this kind of assessment (Lehtomäki et al., 2020).

14.5.2 Risk functions used in health impact assessment models

The risk functions derived from the epidemiological studies described above are used as input in HIA models, together with surface concentrations of the particular pollutants and population information, which is broken down into several age categories as the response to the exposure of a particular pollutant depends on the age of the person exposed. This information is brought together in two different approaches: linear and nonlinear response functions. The linear response functions assume a linear relationship between the change in concentration of a particular pollutant and the health endpoint. However, several studies showed that while this assumption holds in certain regions, for example, Pope, (2000); the joint World Health Organization/UNECE Task Force on Health (EU, 2004; Watkiss et al., 2005), the relationship is rather nonlinear in other regions (e.g., Samoli et al., 2005). Using one of the approaches in a particular region may end up with under or overestimation of the health impact.

The linear all-cause mortality, which gives the total impact from different causes such as heart or lung diseases is shown in Eq. (14.1)

$$R = \alpha \times \delta c \times P \tag{14.1}$$

where R is the response of the mortality rate or the years of life lost (in cases or days), δ δ_c denotes the pollutant concentration, P denotes the affected share of the population, and α an empirically determined constant for the particular health outcome or exposure-response function (ERF). The applied ERFs can originate from particular studies in literature or for example those recommended by WHO, for example, (Héroux et al., 2015), which are based on a comprehensive review and only pollutant/health outcome pairs supported by sufficient evidence for a causal association are included in the recommendations. The underlying studies are larger meta-studies like (Hoek et al., 2013) that include studies from Europe, the US and Asia, where data are most extensive and up to date.

A recent example of the nonlinear approach is the Global Exposure Mortality Model (GEMM) (Burnett et al., 2018), where relative risks (RR) are calculated through the hazard ratios functions. These functions are based on 41 cohort studies from 16 countries and model the association between $PM_{2.5}$ and nonaccidental mortality. Hence, the methodology of Burnett et al. (2018) as implemented in Lelieveld et al. (2019) was followed for estimating the RR as in Eq. (14.2):

$$RR = \exp \theta \log[(z/\alpha) + 1]/(1 + \exp - (z - \mu)/\nu), \text{ where } z = \max(0, \text{PM2.5} - 2.4\,\mu\text{g m}^{-3}) \tag{14.2}$$

where θ, α, μ, and ν are variables obtained from z is the $PM_{2.5}$ concentration. Due to lack of long-term knowledge about the shape of the concentration–mortality association below 2.4 $\mu g\,m^{-3}$, concentrations below this threshold value were taken from cohort studies where subjects were exposed to relatively low concentrations.

This method can calculate cause-specific mortality due to exposure to ambient $PM_{2.5}$, including lung cancer (LC), chronic obstructive pulmonary disease (COPD), cerebrovascular disease (CEV), ischemic heart disease (IHD) and lower respiratory infection (LRI). The category of non-accidental diseases, which was defined as NCD + LRI; and the so-called "other NCD," defined as the subtraction of the above categories to NCD + LRI, were added when using GEMM. RRs and hence premature mortality have been estimated for each pathology and different ages: 25–29, 30–34, 35–39, 40–44, 45–49, 50–54, 55–59, 60–64, 65–69, 70–74, 75–79, +80 and all ages.

Worked example 14.4

Estimating the number of premature deaths related to $PM_{2.5}$ exposure in two different countries.

The total present-day populations of countries A and B are 5 million, with both countries having 50% of their population being above 30 years old. In 2050, the population of country A is expected to increase to 7 million and its above-30 population will reach 70%. The total and the above-30 fraction of the country B population is expected to stay at present-day levels.

Country A has a present-day annual mean $PM_{2.5}$ concentration of 15 $\mu g\,m^{-3}$, while country B has 10 $\mu g\,m^{-3}$. Country A is very ambitious in its air quality objectives and aims to decrease its annual mean $PM_{2.5}$ concentrations by 35% in 2050 compared to today, while country B plans to follow the business-as-usual emission pathway, which will lead to an increase in 2050 $PM_{2.5}$ concentrations by 5% compared to present-day.

Calculate the number of $PM_{2.5}$-related mortality cases in the two countries at present day and in 2050. Assume a linear relationship between the concentration and the number of premature deaths and that the exposure response function (ERF) is equal to 1.23E-3 chronic YOLL μg^{-1} $PM_{2.5}$ m^{-3} in both countries. This should be applied for the adult share of the population (above the age of 30 years). It can also be assumed that a premature death is equivalent to 11 YOLL (Years of Life Loss).

Solution for the present day:

Following the equation for linear all-cause mortality in Eq. (14.1), the number of premature deaths can be calculated for the two countries.

Country A: R = ERF × pollutant concentration × affected share of the population = 1.23E-3 YOLL μg^{-1} $PM_{2.5}$ m^{-3} × 15 μg $PM_{2.5}$ m^{-3} × (5000000 × 0.50) = 46125 YOLL.

The number of YOLL is then converted to number of premature deaths by dividing with 11. So, for country A the number of premature deaths associated with $PM_{2.5}$ is 46125/11 = 4193 per year.

Country B: R = 1.23E-3 YOLL μg^{-1} $PM_{2.5}$ m^{-3} × 10 μg $PM_{2.5}$ m^{-3} × (5000000 × 0.50) = 30750 YOLL. Equivalent to 2697 premature deaths per year.

Solution for the 2050 including the changes in population and $PM_{2.5}$ concentration:

Country A: R = 1.23E-3 YOLL μg^{-1} $PM_{2.5}$ m^{-3} × (15 × (1−0.35)) μg $PM_{2.5}$ m^{-3} × (7000000 × 0.70) = 58763 YOLL, equivalent to 5342 premature deaths per year.

Country B: R = 1.23E-3 YOLL μg^{-1} $PM_{2.5}$ m^{-3} × (10 × (1 + 0.05)) μg $PM_{2.5}$ m^{-3} × (5000000 × 0.50) = 32288 YOLL, equivalent to 2831 premature deaths per year.

14.6 Global and regional trends in mortality and morbidity

For the first time in 2013, WHO acknowledged air pollution as the largest environmental risk factor for the global population health. WHO data show that almost all the global population (99%) breathe air that exceeds WHO guideline limits, containing high levels of pollutants, with low- and middle-income countries suffering from the highest exposures. Many of the drivers of air pollution (i.e. combustion of fossil fuels) are also sources of greenhouse gas emissions. Policies to reduce air pollution, therefore, offer a win-win strategy for both climate and health, lowering the burden of disease attributable to air pollution, as well as contributing to the near- and long-term mitigation of climate change.

In 2014, the Health Effects Institute (HEI) initiated the Global Burden of Disease from Major Air Pollution Sources (GBD MAPS) project. Using the GBD framework, the burden of disease attributable to major ambient air pollution sources in countries around the world has been estimated. The GBD 2015 study (Cohen et al., 2017) estimated global population-weighted mean concentrations of $PM_{2.5}$ and O_3 at an approximate 11 km × 11 km resolution with satellite-based estimates, chemical transport models, and ground-level measurements. Using integrated exposure−response functions for each cause of death, the study estimated the relative risk of mortality from ischemic heart disease, cerebrovascular disease, chronic obstructive pulmonary disease, lung cancer, and lower respiratory infections from epidemiological studies using non-linear exposure−response functions spanning the global range of exposure.

GBD 2015 (Cohen et al., 2017) showed that ambient $PM_{2.5}$ was the fifth-ranking mortality risk factor in 2015. Exposure to $PM_{2.5}$ caused 4.2 million (3.7−4.8 million) deaths in 2015, representing 7·6% of total global deaths (). 59% of these premature deaths have been estimated to take place in east and south Asia. Deaths attributable to ambient $PM_{2.5}$ increased from 3.5 million (3 million to 4 million) in 1990 to 4.2 million (3.7−4.8 million) in 2015. Exposure to ozone caused an additional 254,000 (97,000−422,000) deaths due to chronic obstructive pulmonary disease in 2015 (Fig. 14.8).

According to the GBD 2015, deaths attributable to long-term exposure to $PM_{2.5}$ in 2015 varied substantially among countries (Fig. 14.9). South and east Asia contributed 59% of the 4.2 million global deaths attributable to ambient $PM_{2.5}$ in 2015 [1.36 million deaths (1.19−1.56 million) in south Asia and 1.14 million deaths (0.97−1.31 million) in east Asia]. In World Bank high-income countries, exposure to ambient $PM_{2.5}$ contributed to 4.3% of total deaths in 2015 versus 9.0% in upper-middle-income, 8.7% in lower-middle-income, and 4.9% in low-income countries. These differences in attributable mortality mostly reflect the fraction of total deaths from cardiovascular disease among countries.

Ambient air pollution contributed substantially to the global burden of disease in 2015, which increased over the past 25 years, due to population ageing, changes in non-communicable disease rates, and increasing air pollution in low-income and middle-income countries (Fig. 14.9). Modest reductions in burden will occur in the most polluted countries unless $PM_{2.5}$ values are decreased substantially, but there is potential for substantial health benefits from exposure reduction.

The Global Burden of Disease Study 2019 (GBD, 2019), coordinated by the Institute for Health Metrics and Evaluation (IHME), estimated the burden of diseases, injuries, and risk factors for 204 countries and territories and selected subnational locations. This dataset contains air pollution exposure estimates for ozone pollution, ambient particulate matter pollution, and household air pollution by year from 1990 to 2019 (Fig. 14.10). According to GBD

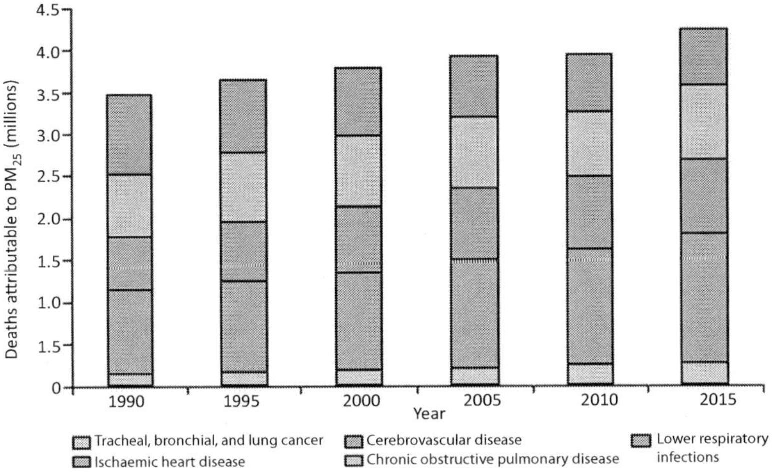

FIGURE 14.8 Global deaths attributable to ambient particulate matter pollution by year and cause. *From Cohen, A. J., Brauer, M., Burnett, R., Anderson, H Ross, Frostad, J., Estep, K., Balakrishnan, K., Brunekreef, B., Dandona, L., Dandona, R., Feigin, V., Freedman, G., Hubbell, B., Jobling, A., Kan, H., Knibbs, L., Liu, Y., Martin, R., . . . H Forouzanfar, M. (2017). Estimates and 25-year trends of the global burden of disease attributable to ambient air pollution: an analysis of data from the Global Burden of Diseases Study 2015. Lancet, 389, 1907–1918. http://doi.org/10.1016/S0140-6736(17)30505.*

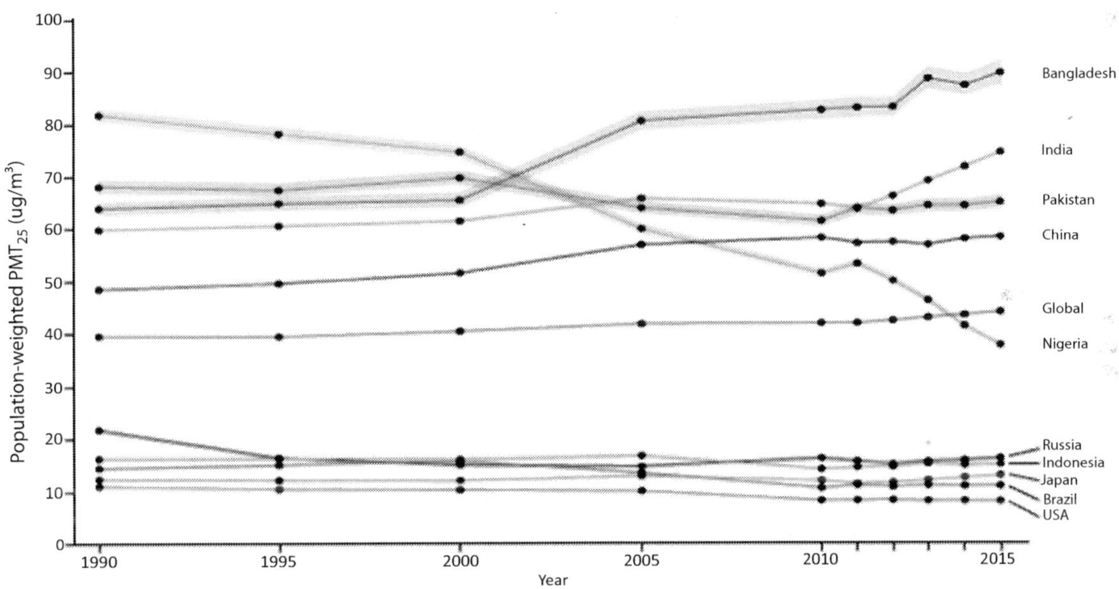

FIGURE 14.9 Trends in population-weighted mean concentrations of particle mass with aerodynamic diameter less than 2.5 μm. *From Cohen, A. J., Brauer, M., Burnett, R., Anderson, H Ross, Frostad, J., Estep, K., Balakrishnan, K., Brunekreef, B., Dandona, L., Dandona, R., Feigin, V., Freedman, G., Hubbell, B., Jobling, A., Kan, H., Knibbs, L., Liu, Y., Martin, R., . . . H Forouzanfar, M. (2017). Estimates and 25-year trends of the global burden of disease attributable to ambient air pollution: an analysis of data from the Global Burden of Diseases Study 2015.* Lancet, 389, *1907–1918. http://doi.org/10.1016/S0140-6736(17)30505.*

(2019), combustion of fossil fuels (coal, oil, and natural gas) contributed to an estimated one million deaths globally (27.3% of all mortality); 800,000 of those deaths were in South Asia or East Asia (32.5% of deaths in those regions).

Recently, Lelieveld et al. (2019) and (Im et al., 2022a, 2022b), using the Global Exposure Mortality Model (GEMM) (Burnett et al., 2018), estimated more than 8 million $PM_{2.5}$ and O_3-related premature deaths, globally in 2015, doubling up the GDB (2021) estimates. The GEMM model calculates relative risks (RR) through the hazard ratio

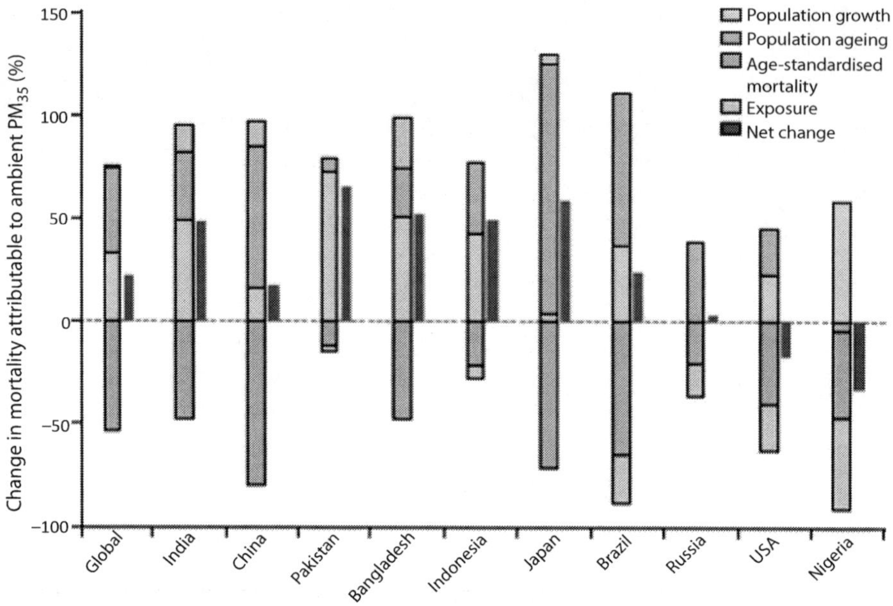

FIGURE 14.10 Changes in mortality attributable to ambient particulate matter pollution according to population-level determinants by country from 1990 to 2015. *From Cohen, A. J., Brauer, M., Burnett, R., Anderson, H Ross, Frostad, J., Estep, K., Balakrishnan, K., Brunekreef, B., Dandona, L., Dandona, R., Feigin, V., Freedman, G., Hubbell, B., Jobling, A., Kan, H., Knibbs, L., Liu, Y., Martin, R., ... H Forouzanfar, M. (2017). Estimates and 25-year trends of the global burden of disease attributable to ambient air pollution: an analysis of data from the Global Burden of Diseases Study 2015.* Lancet, 389, *1907–1918. http://doi.org/10.1016/S0140-6736(17)30505.*

functions, which are based on 41 cohort studies from 16 countries, including Asian cohorts. The inclusion of the Asian cohorts led to much larger risk ratios compared to earlier studies including those from GBD, which led to the factor 2 higher mortality estimates in these studies.

14.7 Evolution of air pollution-related premature mortality in the future

How the exposure and mortality change in the future depends on the changes in emissions and the surface concentrations of air pollutants change, as well as population and the age distribution. Using a business-as-usual scenario, Lelieveld et al. (2015) estimated that in 2050 outdoor air pollution will likely roughly double compared to the 3.3 million global premature deaths in 2010. This change in mortality was due to both change in emissions and the population. Similarly, Silva et al. (2016), using future projections of emissions and population, calculated that the global mortality burden of O_3 will increase from 382,000 deaths in 2000 to between 1.09 and 2.36 million in 2100. They also estimated that the global mortality burden of $PM_{2.5}$ will decrease from 1.70 million deaths in 2000 to between 0.95 and 1.55 million deaths in 2100.

Recently, (Im et al., 2022a, 2022b), assuming only emission changes, without a change in the population, estimated that globally, $PM_{2.5}$ -related mortality is projected to decrease in the high and middle-of-the-road emission mitigation scenarios by up to 60% and 30%, respectively, while it will increase by up to 25% in low mitigation scenarios (Fig. 14.11). The long-term O_3-related COPD deaths follow the same trend as the $PM_{2.5}$-related mortality. Similarly, O_3-related global premature mortality decreases by up to 20% and 2%, in the high and the middle-of-the-road mitigation scenarios, respectively, while the low mitigation scenario leads to a 10% increase. Overall, the total premature mortality due to exposure to both $PM_{2.5}$ and O_3 is projected to decrease by up to 55% in the high mitigation scenario, and by 25% in the middle-of-the-road scenario, and increase by up to 22% in the low mitigation scenario. The largest reductions in both $PM_{2.5}$ and O_3-related premature mortality are projected to take place in the Asian region, where emissions are highest.

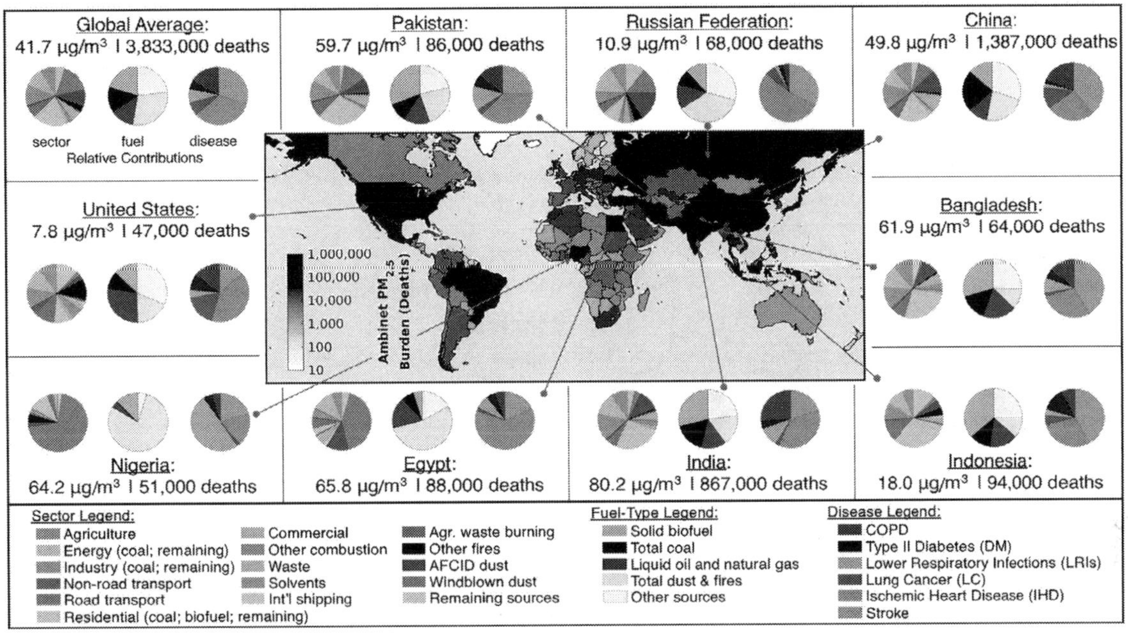

FIGURE 14.11 Ambient $PM_{2.5}$ burden, population-weighted mean $PM_{2.5}$ mass, and fractional source sector, fuel, and disease contributions globally and for the nine countries with the largest burden in 2017. *From GBD (2019).*

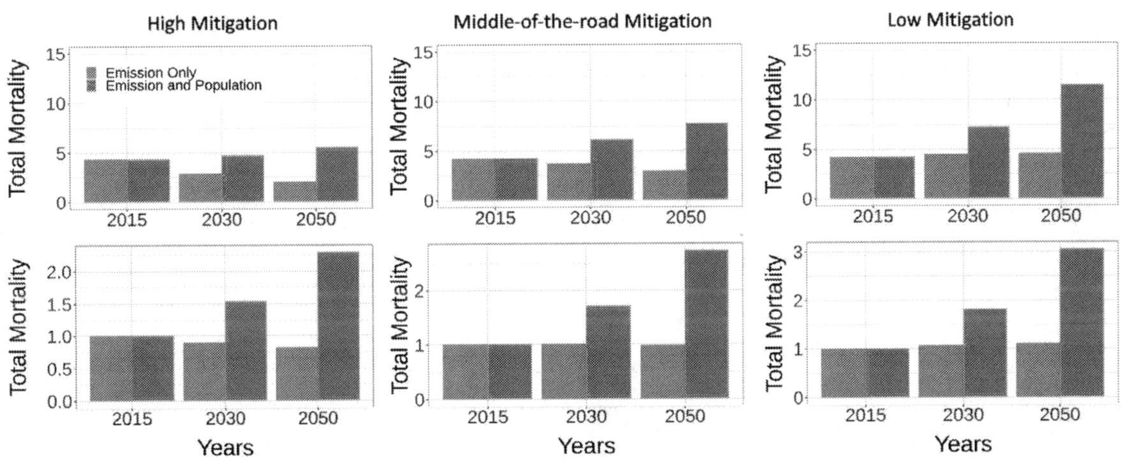

FIGURE 14.12 Present-day (2015) and future (2030 and 2050) global $PM_{2.5}$ and O_3-related premature mortalities in future scenarios under fixed 2015 (impact of emission changes only) and changing population (impact of both emission and population changes). The high mitigation refers to the Shared Socio-economic Pathways (SSP) and Representative Concentration Pathways (RCP) used in the Coupled Model Intercomparison Project Sixth-Phase (CMIP6). Source: *Im, U., Bauer, S. E. Frohn, L. M., Geels, C., Tsigaridis, K., & Brandt, J. (2022a). Present-day and future PM2.5 and O3-related global and regional premature mortality in the EVAv6.0 health impact assessment model.* Environment International, 216 *(4), 114702. Im, U., Bauer, S. E. Frohn, L. M., Geels, C., Tsigaridis, K., & Brandt, J. (2022b). Present-day and future PM2.5 and O3-related global and regional premature mortality in the Evav6.0 health impact assessment model.* SSRN, *Denmark SSRN. https://www.ssrn.com/index.cfm/en/. https://doi.org/10.2139/ssrn.4094648.*

Additional calculations in Im et al. (2022a, 2022b), which consider population changes in the future (Fig. 14.12) show that despite the positive impacts of emission reductions on premature mortality, the increasing and aging population leads to increased $PM_{2.5}$-related premature mortality of up to 13% under the high mitigation scenario, and a factor of 1.8 under the middle-of-the-road scenario. The low mitigation scenario led to more than a factor of 2 increase in the $PM_{2.5}$-related mortality. Similarly, the increasing and aging populations on top of these increased surface O_3 levels leads to an increase in O_3-related mortality by a factor of 1.3–2.1. The results clearly show that the population exposed to the surface $PM_{2.5}$ and O_3 is more important than the level of these pollutants, which is an important challenge for the future.

Although the population change has a much larger impact globally on $PM_{2.5}$ and O_3-related premature mortality, there are large regional differences (Fig. 14.12). In Europe and North America, reductions of up to 70% and 80% are calculated, respectively, while SSP2–4.5 gives smaller reductions of up to 45%. O_3-related premature mortality in Europe and North America are projected to increase by 18% (SSP1–2.6) to 75% (SSP3–7.0). Other regions experience similar changes in response to the changing emissions and changing populations, except for the Asian region, where changes are much larger. The increase in $PM_{2.5}$-related mortality in southeast Asia is estimated to be up to 85%, while in East Asia, the mortality is projected to decrease by 15%.

Fig. 14.13 presents the difference between 2050 and 2015 total ($PM_{2.5} + O_3$) premature mortalities for the different SSPs in one high-pollution region (Asia) and one with relatively low pollution levels (Europe), also having different populations and age distributions. The figure clearly shows that in both high and middle-of-the-road scenarios, emission reductions are projected to lead to decreases in the number of premature deaths, while in the low-mitigation scenario, mortality in Asia is projected to continue to increase due to increasing emissions. In Europe, the reduction in emissions even in the low mitigation scenario leads to a reduction of the number of premature deaths. The figure also shows that even though the population increases and ages in Europe, emission reductions take over in Europe, leading to a decrease in the number of premature deaths in the high and the middle-of-the-road scenarios. On the other hand, the emission reductions cannot compensate for the increasing and especially aging population in the low mitigation scenario. In Asia, however, emission reductions alone are not enough to decrease the number of deaths, except for in high mitigation scenario (Fig. 14.14).

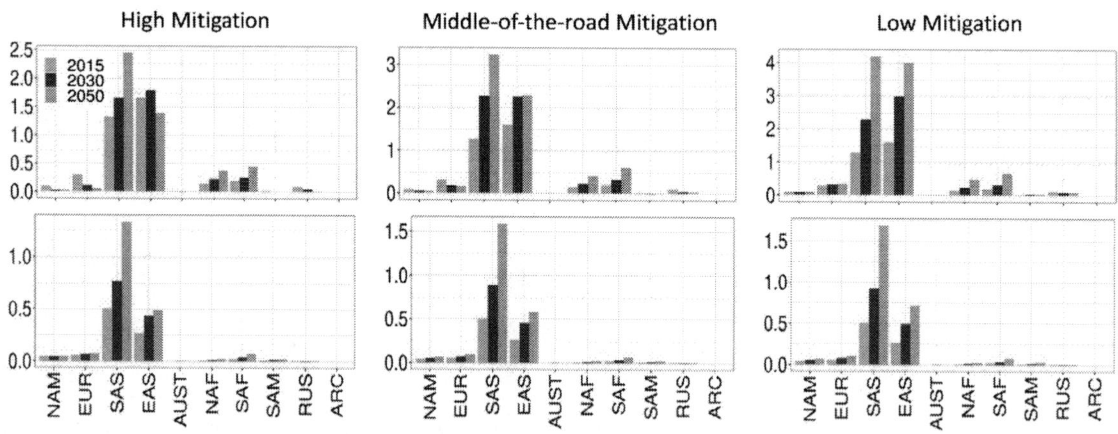

FIGURE 14.13 Present-day (2015) and future (2030 and 2050) regional $PM_{2.5}$ and O_3-related premature mortalities in different SSP scenarios under changing population. The regions are defined as: NAM: North America, EUR: Europe, SAS: Southeast Asia, EAS: East Asia, AUST: Australia, NAF: North Africa, SAF: South Africa, SAM: South America, RUS: Russia, ARC: Arctic. Source*: Im, U., Bauer, S. E. Frohn, L. M., Geels, C., Tsigaridis, K., & Brandt, J. (2022a). Present-day and future PM2.5 and O3-related global and regional premature mortality in the EVAv6.0 health impact assessment model.* Environment International, 216 *(4), 114702. Im, U., Bauer, S. E. Frohn, L. M., Geels, C., Tsigaridis, K., & Brandt, J. (2022b). Present-day and future Pm2.5 and O3-related global and regional premature mortality in the Evav6.0 health impact assessment model.* SSRN, Denmark *SSRN. https://www.ssrn.com/index.cfm/en/. https://doi.org/10.2139/ssrn.4094648.*

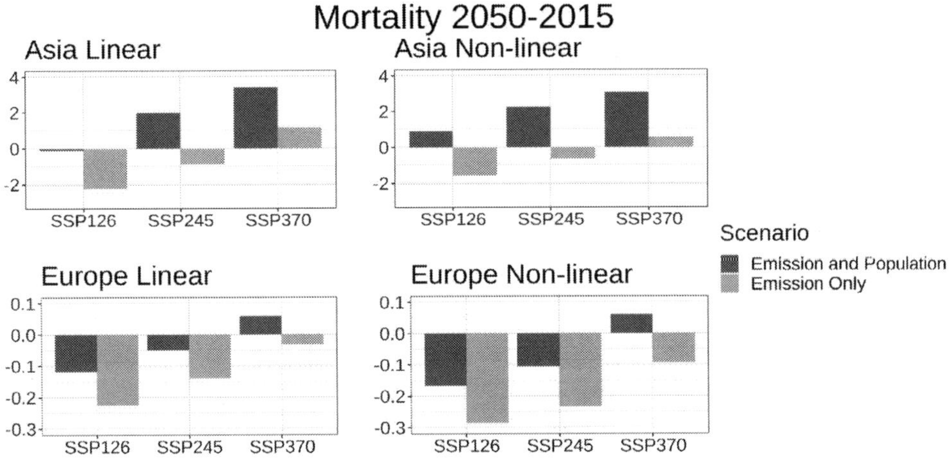

FIGURE 14.14 Impacts of emissions only and changes in both emissions and population in 2050 compared to 2015 in different SSPs for Asia (*top panel*) and Europe (*lower panel*), estimated using linear (*left panel*) and non-linear (*right panel*) ERFs. Source*: Im, U., Bauer, S. E. Frohn, L. M., Geels, C., Tsigaridis, K., & Brandt, J. (2022a). Present-day and future PM2.5 and O3-related global and regional premature mortality in the EVAv6.0 health impact assessment model. Environment International*, 216 *(4), 114702. Im, U., Bauer, S. E. Frohn, L.M., Geels, C., Tsigaridis, K., & Brandt, J. (2022b). Present-day and future Pm2.5 and O3-related global and regional premature mortality in the Evav6.0 health impact assessment model. SSRN, Denmark SSRN. https://www.ssrn.com/index.cfm/en/. https://doi.org/10.2139/ssrn.4094648.*

14.8 Questions

14.1 How many people die prematurely from air pollution globally every year?

1. 650,000
2. 1.2 million
3. 4.5 million
4. 6.7 million

Further discuss how this number compares with mortality due to other factors such as smoking and how this number changes regionally.

14.2 Why is air pollution one of the major risk factor for mortality and morbidity globally?

14.3 What is the difference between long- and short-term exposure to air pollution?

14.4 Based on the Worked Example 14.4, discuss why some regions experience an increase in air pollution-related mortality despite reductions in air pollution concentrations.

14.5 Discuss potential measures to reduce air pollution-related health impacts in an urban setting in your country.

14.6 Why are international agreements and/or recommendations like, for example, from the WHO important in order to mitigate adverse health impacts of air pollution?

Solutions to the end-of-chapter questions will be available at: https://www.elsevier.com/books-and-journals/book-companion/9780128225912.

References

Abbafati, C., Machado, D. B., Cislaghi, B., Salman, O. M., Karanikolos, M., McKee, M., Abbas, K. M., Brady, O. J., Larson, H. J., Trias-Llimós, S., Cummins, S., Langan, S. M., Sartorius, B., Hafiz, A., Jenabi, E., Mohammad Gholi Mezerji, N., Borzouei, S., Azarian, G., Khazaei, S., … Zhu, C. (2020). Global burden of 87 risk factors in 204 countries and territories, 1990–2019: a systematic analysis for the Global Burden of Disease Study 2019. *The Lancet*, *396*(10258), 1223–1249. Available from https://doi.org/10.1016/S0140-6736(20)30752-2.

Achakulwisut, P., Brauer, M., Hystad, P., & Anenberg, S. C. (2019). Global, national, and urban burdens of paediatric asthma incidence attributable to ambient NO_2 pollution: Estimates from global datasets. *The Lancet Planetary Health*, *3*(4), e166–e178. Available from https://doi.org/10.1016/S2542-5196(19)30046-4, https://www.journals.elsevier.com/the-lancet-planetary-health.

Andersen, Z. J., Pedersen, M., Weinmayr, G., Stafoggia, M., Galassi, C., Jørgensen, J. T., Sommar, J. N., Forsberg, B., Olsson, D., Oftedal, B., Aasvang, G. M., Schwarze, P., Pyko, A., Pershagen, G., Korek, M., Faire, U. D., ÖÖstenson, C. G., Fratiglioni, L., Eriksen, K. T., ... Raaschou-Nielsen, O. (2018). Long-term exposure to ambient air pollution and incidence of brain tumor: The European Study of Cohorts for Air Pollution Effects (ESCAPE). *Neuro-Oncology*, *20*(3), 420–432. Available from https://doi.org/10.1093/neuonc/nox163, http://neuro-oncology.oxfordjournals.org/current.dtl.

Andersen, Z. J., Stafoggia, M., Weinmayr, G., Pedersen, M., Galassi, C., Jørgensen, J. T., Oudin, A., Forsberg, B., Olsson, D., Oftedal, B., Aasvang, G. M., Aamodt, G., Pyko, A., Pershagen, G., Korek, M., de Faire, U., Pedersen, N. L., Östenson, C. G., Fratiglioni, L., ... Raaschou-Nielsen, O. (2017). Long-term exposure to ambient air pollution and incidence of postmenopausal breast cancer in 15 European cohorts within the ESCAPE project. *Environmental Health Perspectives*. Available from https://doi.org/10.1289/EHP1742.

Andersen, Z. J., Zhang, J., Jørgensen, J. T., Samoli, E., Liu, S., Chen, J., Strak, M., Wolf, K., Weinmayr, G., Rodopolou, S., Remfry, E., de Hoogh, K., Bellander, T., Brandt, J., Concin, H., Zitt, E., Fecht, D., Forastiere, F., Gulliver, J., ... Lim, Y. H. (2022). Long-term exposure to air pollution and mortality from dementia, psychiatric disorders, and suicide in a large pooled European cohort: ELAPSE study. *Environment International*, *170*. Available from https://doi.org/10.1016/j.envint.2022.107581, http://www.elsevier.com/locate/envint.

Anenberg, S. C., Belova, A., Brandt, J., Fann, N., Greco, S., Guttikunda, S., Heroux, M. E., Hurley, F., Krzyzanowski, M., Medina, S., & Miller, B. (2016). Survey of ambient air pollution health risk assessment tools. *Risk Analysis*, *36*(9), 1718–1736. Available from https://doi.org/10.1111/risa.12540.

Bell, M. L., & Davis, D. L. (2001). Reassessment of the Lethal London Fog of 1952: Novel indicators of acute and chronic consequences of acute exposure to air pollution. *Environmental Health Perspectives*, *109*(3). Available from https://doi.org/10.2307/3434786.

Bickel, P., & Friedrich, R. (Eds) (2005). ExternE – Externalities of Energy, Methodology 2005 Update, EUR 21951. Available from https://doi.org/10.18419/opus-11923.

Bowatte, G., Lodge, C., Lowe, A. J., Erbas, B., Perret, J., Abramson, M. J., Matheson, M., & Dharmage, S. C. (2015). The influence of childhood traffic-related air pollution exposure on asthma, allergy and sensitization: A systematic review and a meta-analysis of birth cohort studies. *Allergy: European Journal of Allergy and Clinical Immunology*, *70*(3), 245–256. Available from https://doi.org/10.1111/all.12561, http://onlinelibrary.wiley.com/journal/10.1111/(ISSN)1398-9995.

Braithwaite, I., Zhang, S., Kirkbride, J. B., Osborn, D. P. J., & Hayes, J. F. (2019). Air pollution (particulate matter) exposure and associations with depression, anxiety, bipolar, psychosis and suicide risk: A systematic review and meta-analysis. *Environmental Health Perspectives*, *127*(12), 126002. Available from https://doi.org/10.1289/EHP4595, https://ehp.niehs.nih.gov/doi/pdf/10.1289/EHP4595.

Brandt, J., Silver, J. D., Christensen, J. H., Andersen, M. S., Bønløkke, J. H., Sigsgaard, T., Geels, C., Gross, A., Hansen, A. B., Hansen, K. M., Hedegaard, G. B., Kaas, E., & Frohn, L. M. (2013a). Assessment of past, present and future health-cost externalities of air pollution in Europe and the contribution from international ship traffic using the EVA model system. *Atmospheric Chemistry and Physics*, *13*(15), 7747–7764. Available from https://doi.org/10.5194/acp-13-7747-2013.

Brandt, J., Silver, J. D., Christensen, J. H., Andersen, M. S., Bønløkke, J. H., Sigsgaard, T., Geels, C., Gross, A., Hansen, A. B., Hansen, K. M., Hedegaard, G. B., Kaas, E., & Frohn, L. M. (2013b). Contribution from the ten major emission sectors in Europe and Denmark to the health-cost externalities of air pollution using the EVA model system – An integrated modelling approach. *Atmospheric Chemistry and Physics*, *13*(15), 7725–7746. Available from https://doi.org/10.5194/acp-13-7725-2013.

Brook, R. D., Rajagopalan, S., Pope, C. A., Brook, J. R., Bhatnagar, A., Diez-Roux, A. v, Holguin, F., Hong, Y., Luepker, R. v, Mittleman, M. A., Peters, A., Siscovick, D., Smith, S. C., Whitsel, L., & Kaufman, J. D. (2010). Particulate matter air pollution and cardiovascular disease: An update to the scientific statement from the American Heart Association. *Circulation*, *121*(21), 2331–2378. Available from https://doi.org/10.1161/CIR.0B013E3181DBECE1.

Burnett, R., Chena, H., Szyszkowicza, M., Fann, N., Hubbell, B., Pope, C. A., III, Apte, J. S., Brauer, M., Cohen, A., Weichenthal, S., Coggins, J., Di, Q., Brunekreef, B., Frostad, J., Lim, S. S., Kan, H., Walker, K. D., Thurston, G. D., Hayes, R. B., ... Spadar, J. V. (2018). Global estimates of mortality associated with longterm exposure to outdoor fine particulate matter. *Proceedings of the National Academy of Sciences*, *38*(115), 9592–9597. Available from https://doi.org/10.1073/pnas.1803222115.

Calderón-Garcidueñas, L., Maronpot, R. R., Torres-Jardon, R., Henríquez-Roldán, C., Schoonhoven, R., Acuña-Ayala, H., Villarreal-Calderón, A., Nakamura, J., Fernando, R., Reed, W., Azzarelli, B., & Swenberg, J. A. (2003). DNA damage in nasal and brain tissues of canines exposed to air pollutants is associated with evidence of chronic brain inflammation and neurodegeneration. *Toxicologic Pathology*, *31*(5), 524–538. Available from https://doi.org/10.1080/01926230390226645, http://tpx.sagepub.com/archive/.

Calderón-Garcidueñas, L., Reed, W., Maronpot, R. R., Henríquez-Roldáná, C., Delgado-Chavez, R., Calderón-Garcidueñas, A., Dragustinovis, I., Franco-Lira, M., Aragón-Flores, M., Solt, A. C., Altenburg, M., Torres-Jardón, R., & Swenberg, J. A. (2004). Brain inflammation and Alzheimer's-like pathology in individuals exposed to severe air pollution. *Toxicologic Pathology*, *32*(6), 650–658. Available from https://doi.org/10.1080/01926230490520232.

Chen, J., & Hoek, G. (2020). Long-term exposure to PM and all-cause and cause-specific mortality: A systematic review and meta-analysis, In *Environment International*. Available from https://doi.org/10.1016/j.envint.2020.105974.

Chen, J., Rodopoulou, S., Strak, M., de Hoogh, K., Taj, T., Poulsen, A. H., Andersen, Z. J., Bellander, T., Brandt, J., Zitt, E., Fecht, D., Forastiere, F., Gulliver, J., Hertel, O., Hoffmann, B., Hvidtfeldt, U. A., Verschuren, W. M. M., Jørgensen, J. T., Katsouyanni, K., ... Hoek, G. (2022). Long-term exposure to ambient air pollution and bladder cancer incidence in a pooled European cohort: the ELAPSE project. *British Journal of Cancer*, *126*(10), 1499–1507. Available from https://doi.org/10.1038/s41416-022-01735-4, http://www.nature.com/bjc/index.html.

Chen, M., Zhao, J., Zhuo, C., & Zheng, L. (2021). The association between ambient air pollution and atrial fibrillation. *International Heart Journal*, *62*(2), 290–297. Available from https://doi.org/10.1536/ihj.20-523.

Clancy, L., Goodman, P., Sinclair, H., & Dockery, D. W. (2002). Effect of air-pollution control on death rates in Dublin, Ireland: An intervention study. *Lancet*, *360*(9341), 1210–1214. Available from https://doi.org/10.1016/S0140-6736(02)11281-5, http://www.journals.elsevier.com/the-lancet/.

Cohen, A. J., Brauer, M., Burnett, R., Anderson, H., Ross., Frostad, J., Estep, K., Balakrishnan, K., Brunekreef, B., Dandona, L., Dandona, R., Feigin, V., Freedman, G., Hubbell, B., Jobling, A., Kan, H., Knibbs, L., Liu, Y., Martin, R., ... H Forouzanfar, M. (2017). Estimates and 25-year trends of the global burden of disease attributable to ambient air pollution: an analysis of data from the Global Burden of Diseases Study 2015. *Lancet*, *389*, 1907–1918. Available from https://doi.org/10.1016/S0140-6736(17)30505-6.

Cole-Hunter, T., Zhang, J., So, R., Samoli, E., Liu, S., Chen, J., Strak, M., Wolf, K., Weinmayr, G., Rodopolou, S., Remfry, E., de Hoogh, K., Bellander, T., Brandt, J., Concin, H., Zitt, E., Fecht, D., Forastiere, F., Gulliver, J., ... Andersen, Z. J. (2022). Long-term air pollution exposure and Parkinson's disease mortality in a large pooled European cohort: An ELAPSE study. *Environment International*, *171*, 107667. Available from https://doi.org/10.1016/J.ENVINT.2022.107667.

Di, Q., Wang, Y., Zanobetti, A., Wang, Y., Koutrakis, P., Choirat, C., Dominici, F., & Schwartz, J. D. (2017). Air pollution and mortality in the medicare population. *New England Journal of Medicine*, *376*(26), 2513–2522. Available from https://doi.org/10.1056/NEJMoa1702747, http://www.nejm.org/doi/pdf/10.1056/NEJMoa1702747.

Downs, S. H., Schindler, C., Liu, L. J. S., Keidel, D., Bayer-Oglesby, L., Brutsche, M. H., Gerbase, M. W., Keller, R., Künzli, N., Leuenberger, P., Probst-Hensch, N. M., Tschopp, J. M., Zellweger, J. P., Rochat, T., Schwartz, J., Ackermann-Liebrich, U., Rochat, T., Ackermann-Liebrich, U., Gaspoz, J. M., ... Nilly, R. (2007). Reduced exposure to PM10 and attenuated age-related decline in lung function. *New England Journal of Medicine*, *357*(23), 2338–2347. Available from https://doi.org/10.1056/NEJMoa073625, http://content.nejm.org/cgi/reprint/357/23/2338.pdf.

Dutheil, F., Comptour, A., Morlon, R., Mermillod, M., Pereira, B., Baker, J. S., Charkhabi, M., Clinchamps, M., & Bourdel, N. (2021). Autism spectrum disorder and air pollution: A systematic review and meta-analysis. *Environmental Pollution (Barking, Essex 1987)*, *278*. Available from https://doi.org/10.1016/J.ENVPOL.2021.116856.

Ellermann, T., Nordstrøm, C., Brandt, J., Christensen, J., Ketzel, M., Massling, A., Bossi, R., Frohn, L.M., Geels, C., Solvang Jensen, S., Nielsen, O.-K., Winther, M., Bech Poulsen, M., Monies, C., Bjært Sørensen, M., Skou Andersen M., & Sigsgaard, T. (2023). Luftkvalitet 2021. Status for den nationale luftkvalitetsovervågning. Aarhus Universitet, DCE – Nationalt Center for Miljø og Energi, 148 s. – Videnskabelig rapportnr. 533. http://dce2.au.dk/pub/SR533.pdf.

EU. (2004). Modelling and assessment of the health impact of particulate matter and ozone, Economic commission for Europe, Executive body for the convention on long-range transboundary air pollution, Working group on effects, twenty-third session, Geneva, 1–3 September 2004.

European Environment Agency (EEA). (2021). Air quality in Europe 2022. In *EEA Report. Reportno.*

Filippini, T., Hatch, E. E., Rothman, K. J., Heck, J. E., Park, A. S., Crippa, A., Orsini, N., & Vinceti, M. (2019). Association between outdoor air pollution and childhood leukemia: A systematic review and dose–response meta-analysis. *Environmental Health Perspectives*, *127*(4). Available from https://doi.org/10.1289/EHP4381, https://ehp.niehs.nih.gov/doi/pdf/10.1289/EHP4381.

Fuertes, E., MacIntyre, E., Agius, R., Beelen, R., Brunekreef, B., Bucci, S., Cesaroni, G., Cirach, M., Cyrys, J., Forastiere, F., Gehring, U., Gruzieva, O., Hoffmann, B., Jedynska, A., Keuken, M., Klümper, C., Kooter, I., Korek, M., Krämer, U., ... Heinrich, J. (2014). Associations between particulate matter elements and early-life pneumonia in seven birth cohorts: Results from the ESCAPE and TRANSPHORM projects. *International Journal of Hygiene and Environmental Health*, *217*(8), 819–829. Available from https://doi.org/10.1016/j.ijheh.2014.05.004.

Gehring, U., Gruzieva, O., Agius, R. M., Beelen, R., Custovic, A., Cyrys, J., Eeftens, M., Flexeder, C., Fuertes, E., Heinrich, J., Hoffmann, B., de Jongste, J. C., Kerkhof, M., Klümper, C., Korek, M., Mölter, A., Schultz, E. S., Simpson, A., Sugiri, D., ... Brunekreef, B. (2013). Air pollution exposure and lung function in children: The ESCAPE project. *Environmental Health Perspectives*, *121*(11-12), 1357–1364. Available from https://doi.org/10.1289/ehp.1306770, http://ehp.niehs.nih.gov/wp-content/uploads/121/11-12/ehp.1306770.pdf, Netherlands.

Global Burden of Disease Collaborative Network (GBD) (2021). Global Burden of Disease Study 2019 (GBD 2019) Air Pollution Exposure Estimates 1990-2019. Seattle, United States of America: Institute for Health Metrics and Evaluation (IHME). Available from https://doi.org/10.6069/70JS-NC54.

Grigg, J. (2018). Air pollution and respiratory infection: An emerging and troubling association. *American Journal of Respiratory and Critical Care Medicine*, *198*(6), 700–701. Available from https://doi.org/10.1164/rccm.201804-0614ed.

Hamra, G. B., Guha, N., Cohen, A., Laden, F., Raaschou-Nielsen, O., Samet, J. M., Vineis, P., Forastiere, F., Saldiva, P., Yorifuji, T., & Loomis, D. (2014). Outdoor particulate matter exposure and lung cancer: A systematic review and meta-analysis. *Environmental Health Perspectives*, *122*(9), 906–911. Available from https://doi.org/10.1289/ehp.1408092, http://ehp.niehs.nih.gov/wp-content/uploads/122/9/ehp.1408092.pdf.

Hamra, G. B., Laden, F., Cohen, A. J., Raaschou-Nielsen, O., Brauer, M., & Loomis, D. (2015). Lung cancer and exposure to nitrogen dioxide and traffic: A systematic review and meta-analysis. *Environmental Health Perspectives*, *123*(11), 1107–1112. Available from https://doi.org/10.1289/ehp.1408882, http://ehp.niehs.nih.gov/wp-content/uploads/123/11/ehp.1408882.alt.pdf.

Hansen, A. B., Ravnskjær, L., Loft, S., Andersen, K. K., Bräuner, E. V., Baastrup, R., Yao, C., Ketzel, M., Becker, T., Brandt, J., Hertel, O., & Andersen, Z. J. (2016). Long-term exposure to fine particulate matter and incidence of diabetes in the Danish Nurse Cohort. *Environment International*, *91*, 243–250. Available from https://doi.org/10.1016/j.envint.2016.02.036.

Hassan Bhat, T., Jiawen, G., & Farzaneh, H. (2021). Air pollution health risk assessment (AP-HRA), principles and applications. *International Journal of Environmental Research and Public Health*, *18*(4). Available from https://doi.org/10.3390/ijerph18041935.

Health Effects Institute. (2020). *State of Global Air 2020. Special Report.* https://www.stateofglobalair.org/.

Heusinkveld, H. J., Wahle, T., Campbell, A., Westerink, R. H. S., Tran, L., Johnston, H., Stone, V., Cassee, F. R., & Schins, R. P. F. (2016). Neurodegenerative and neurological disorders by small inhaled particles. *Neurotoxicology, 56*, 94–106. Available from https://doi.org/10.1016/j.neuro.2016.07.007, http://www.elsevier.com/inca/publications/store/6/2/1/3/5/5.

Hoek, G., Krishnan, R. M., Beelen, R., Peters, A., Ostro, B., Brunekreef, B., & Kaufman, J. D. (2013). Long-term air pollution exposure and cardio-respiratory mortality: A review. *Environmental Health: A Global Access Science Source, 12*(1). Available from https://doi.org/10.1186/1476-069X-12-43.

Holland, M., & Berry, J. (1995). ExternE: externalities of energy. Vol. 1: summary. Retrieved from CEC.

Hu, C. Y., Gao, X., Fang, Y., Jiang, W., Huang, K., Hua, X. G., Yang, X. J., Chen, H. B., Jiang, Z. X., & Zhang, X. J. (2020). Human epidemiological evidence about the association between air pollution exposure and gestational diabetes mellitus: Systematic review and meta-analysis. *Environmental Research, 180*. Available from https://doi.org/10.1016/j.envres.2019.108843, http://www.elsevier.com/inca/publications/store/6/2/2/8/2/1/index.htt.

Hvidtfeldt, U. A., Chen, J., Rodopoulou, S., Strak, M., de Hoogh, K., Andersen, Z. J., Bellander, T., Brandt, J., Fecht, D., Forastiere, F., Gulliver, J., Hertel, O., Hoffmann, B. H., Katsouyanni, K., Ketzel, M., Brynedal, B., Leander, K., Ljungman, P. L. S., Magnusson, P. K. E., ... Raaschou-Nielsen, O. (2022). Breast cancer incidence in relation to long-term low-level exposure to air pollution in the ELAPSE pooled cohort. *Cancer Epidemiology, Biomarkers & Prevention: A Publication of the American Association for Cancer Research, Cosponsored by the American Society of Preventive Oncology*, OF1–OF9. Available from https://doi.org/10.1158/1055-9965.EPI-22-0720.

Hvidtfeldt, U. A., Severi, G., Andersen, Z. J., Atkinson, R., Bauwelinck, M., Bellander, T., Boutron-Ruault, M. C., Brandt, J., Brunekreef, B., Cesaroni, G., Chen, J., Concin, H., Forastiere, F., van Gils, C. H., Gulliver, J., Hertel, O., Hoek, G., Hoffmann, B., de Hoogh, K., ... Fecht, D. (2021). Long-term low-level ambient air pollution exposure and risk of lung cancer – A pooled analysis of 7 European cohorts. *Environment International, 146*. Available from https://doi.org/10.1016/j.envint.2020.106249, http://www.elsevier.com/locate/envint.

Hvidtfeldt, U. A., Taj, T., Chen, J., Rodopoulou, S., Strak, M., de Hoogh, K., Andersen, Z. J., Bellander, T., Brandt, J., Fecht, D., Forastiere, F., Gulliver, J., Hertel, O., Hoffmann, B., Jørgensen, J. T., Katsouyanni, K., Ketzel, M., Lager, A., Leander, K., ... Raaschou-Nielsen, O. (2022). Long term exposure to air pollution and kidney parenchyma cancer – Effects of low-level air pollution: A Study in Europe (ELAPSE). *Environmental Research, 215*. Available from https://doi.org/10.1016/j.envres.2022.114385.

Héroux, M. E., Anderson, H. R., Atkinson, R., Brunekreef, B., Cohen, A., Forastiere, F., Hurley, F., Katsouyanni, K., Krewski, D., Krzyzanowski, M., Künzli, N., Mills, I., Querol, X., Ostro, B., & Walton, H. (2015). Quantifying the health impacts of ambient air pollutants: Recommendations of a WHO/Europe project. *International Journal of Public Health, 60*(5), 619–627. Available from https://doi.org/10.1007/s00038-015-0690-y, http://www.springerlink.com/content/1661-8564/.

Imbriani, G., Panico, A., Grassi, T., Idolo, A., Serio, F., Bagordo, F., De Filippis, G., De Giorgi, D., Antonucci, G., Piscitelli, P., Colangelo, M., Peccarisi, L., Tumolo, M. R., De Masi, R., Miani, A., & De Donno, A. (2021). Early-life exposure to environmental air pollution and autism spectrum disorder: A review of available evidence. *International Journal of Environmental Research and Public Health, 18*(3), 1–24. Available from https://doi.org/10.3390/ijerph18031204, https://www.mdpi.com/1660-4601/18/3/1204/pdf.

Im, U., Bauer, S. E., Frohn, L. M., Geels, C., Tsigaridis, K., & Brandt, J. (2022a). Present-day and future PM2.5 and O3-related global and regional premature mortality in the EVAv6.0 health impact assessment model. *Environment International, 216*(4), 114702.

Im, U., Bauer, S. E., Frohn, L. M., Geels, C., Tsigaridis, K., & Brandt, J. (2022b). *Present-day and future Pm2.5 and O3-related global and regional premature mortality in the Evav6.0 health impact assessment model. SSRN*. Denmark: SSRN. Available from https://www.ssrn.com/index.cfm/en/, 10.2139/ssrn.4094648.

International Agency for Research on Cancer (IARC). (2004). IARC Working Group on the Evaluation of Carcinogenic Risks to Humans. Tobacco smoke and involuntary smoking. *IARC Working Group on the Evaluation of Carcinogenic Risks to Humans, 83*(June), 8–11.

Iskandar, A., Andersen, Z. J., Bønnelykke, K., Ellermann, T., Andersen, K. K., & Bisgaard, H. (2012). Coarse and fine particles but not ultrafine particles in urban air trigger hospital admission for asthma in children. *Thorax, 67*(3), 252–257. Available from https://doi.org/10.1136/thoraxjnl-2011-200324, http://thorax.bmj.com/content/67/3/252.full.pdf + html.

Jia, Y., Lin, Z., He, Z., Li, C., Zhang, Y., Wang, J., Liu, F., Li, J., Huang, K., Cao, J., Gong, X., Lu, X., & Chen, S. (2023). Effect of air pollution on heart failure: Systematic review and meta-analysis. *Environmental Health Perspectives, 131*(7). Available from https://doi.org/10.1289/EHP11506, https://ehp.niehs.nih.gov/doi/epdf/10.1289/EHP11506.

Kasdagli, M. I., Katsouyanni, K., Dimakopoulou, K., & Samoli, E. (2019). Air pollution and Parkinson's disease: A systematic review and meta-analysis up to 2018. *International Journal of Hygiene and Environmental Health, 222*(3), 402–409. Available from https://doi.org/10.1016/j.ijheh.2018.12.006, http://www.urbanfischer.de/journals/intjhyg.

Katsouyanni, K., Touloumi, G., Samoli, E., Gryparis, A., Le Tertre, A., Monopolis, Y., Rossi, G., Zmirou, D., Ballester, F., Boumghar, A., Anderson, H. R., Wojtyniak, B., Paldy, A., Braunstein, R., Pekkanen, J., Schindler, C., & Schwartz, J. (2001). Confounding and effect modification in the short-term effects of ambient particles on total mortality: Results from 29 European cities within the APHEA2 project. *Epidemiology (Cambridge, Mass.), 12*(5), 521–531. Available from https://doi.org/10.1097/00001648-200109000-00011.

Katsouyanni, K., Zmirou, D., Spix, C., Sunyer, J., Schouten, J. P., Ponka, A., Anderson, H. R., Le Moullec, Y., Wojtyniak, B., Vigotti, M. A., & Bacharova, L. (1995). Short-term effects of air pollution on health: A European approach using epidemiological time-series data. The APHEA project: Background, objectives, design. *European Respiratory Journal, 8*(6), 1030–1038.

Kaufman, J. D., Elkind, M. S. V., Bhatnagar, A., Koehler, K., Balmes, J. R., Sidney, S., Burroughs Peña, M. S., Dockery, D. W., Hou, L., Brook, R. D., Laden, F., Rajagopalan, S., Kendrick, K. B., Turner, J. R., Cohen, A. J., Mittleman, M. A., Siscovick, D., & Thurston, G. D. (2020). Guidance to reduce the cardiovascular burden of ambient air pollutants: A policy statement from the American Heart Association. *Circulation, 142*(23), E432–E447. Available from https://doi.org/10.1161/CIR.0000000000000930, http://circ.ahajournals.org.

Khreis, H., Kelly, C., Tate, J., Parslow, R., Lucas, K., & Nieuwenhuijsen, M. (2017). Exposure to traffic-related air pollution and risk of development of childhood asthma: A systematic review and meta-analysis. *Environment International*, *100*, 1–31. Available from https://doi.org/10.1016/j.envint.2016.11.012.

Kim, K. N., Lim, Y. H., Bae, H. J., Kim, M., Jung, K., & Hong, Y. C. (2016). Long-term fine particulate matter exposure and major depressive disorder in a community-based urban cohort. *Environmental Health Perspectives*, *124*(10), 1547–1553. Available from https://doi.org/10.1289/EHP192, http://ehp.niehs.nih.gov/wp-content/uploads/124/10/EHP192.alt.pdf.

King, C., Kirkham, J., Hawcutt, D., & Sinha, I. (2018). The effect of outdoor air pollution on the risk of hospitalisation for bronchiolitis in infants: A systematic review. *PeerJ*, *6*(8). Available from https://doi.org/10.7717/peerj.5352.

Kioumourtzoglou, M. A., Power, M. C., Hart, J. E., Okereke, O. I., Coull, B. A., Laden, F., & Weisskopf, M. G. (2017). The association between air pollution and onset of depression among middle-aged and older women. *American Journal of Epidemiology*, *185*(9), 801–809. Available from https://doi.org/10.1093/aje/kww163, http://aje.oxfordjournals.org/.

Kioumourtzoglou, M. A., Raz, R., Wilson, A., Fluss, R., Nirel, R., Broday, D. M., Yuval., Hacker, M. R., McElrath, T. F., Grotto, I., Koutrakis, P., & Weisskopf, M. G. (2019). Traffic-related air pollution and pregnancy loss. *Epidemiology (Cambridge, Mass.)*, *30*(1), 4–10. Available from https://doi.org/10.1097/EDE.0000000000000918, http://journals.lww.com/epidem/pages/default.aspx.

Kubesch, N., De Nazelle, A., Guerra, S., Westerdahl, D., Martinez, D., Bouso, L., Carrasco-Turigas, G., Hoffmann, B., & Nieuwenhuijsen, M. J. (2015). Arterial blood pressure responses to short-term exposure to low and high traffic-related air pollution with and without moderate physical activity. *European Journal of Preventive Cardiology*, *22*(5), 548–557. Available from https://doi.org/10.1177/2047487314555602, https://academic.oup.com/eurjpc.

Kwon, D., Paul, K. C., Yu, Y., Zhang, K., Folle, A. D., Wu, J., Bronstein, J. M., & Ritz, B. (2024). Traffic-related air pollution and Parkinson's disease in central California. *Environmental Research*, *240*. Available from https://doi.org/10.1016/j.envres.2023.117434, https://www.sciencedirect.com/journal/environmental-research.

Lehtomäki, H., Geels, C., Brandt, J., Rao, S., Yaramenka, K., Åström, S., Andersen, M. S., Frohn, L. M., Im, U., & Hänninen, O. (2020). Deaths attributable to air pollution in Nordic countries: Disparities in the estimates. *Atmosphere*, *11*, 467. Available from https://doi.org/10.3390/atmos11050467.

Lelieveld, J., Evans, J., Fnais, M., et al. (2015). The contribution of outdoor air pollution sources to premature mortality on a global scale. *Nature*. Available from https://doi.org/10.1038/nature15371.

Lelieveld, J., Klingmüller, K., Pozzer, A., Pöschl, U., Fnais, M., Daiber, A., & Münzel, T. (2019). Cardiovascular disease burden from ambient air pollution in Europe reassessed using novel hazard ratio functions. *European Heart Journal*. Available from https://doi.org/10.1093/eurheartj/ehz135.

Lepeule, J., Laden, F., Dockery, D., & Schwartz, J. (2012). Chronic exposure to fine particles and mortality: An Extended follow-up of the harvard six cities study from 1974 to 2009. *Environmental Health Perspectives*, *120*(7), 965–970. Available from https://doi.org/10.1289/ehp.1104660.

Lim, Y. H., Kim, H., Kim, J. H., Bae, S., Park, H. Y., & Hong, Y. C. (2012). Air pollution and symptoms of depression in elderly adults. *Environmental Health Perspectives*, *120*(7), 1023–1028. Available from https://doi.org/10.1289/ehp.1104100, http://ehp.niehs.nih.gov/article/fetchObjectAttachment.action?uri = info%3Adoi%2F10.1289%2Fehp.1104100&representation = PDF.

Liu, C., Chen, R., Sera, F., Vicedo-Cabrera, A. M., Guo, Y., Tong, S., Coelho, M. S. Z. S., Saldiva, P. H. N., Lavigne, E., Matus, P., Ortega, N. V., Garcia, S. O., Pascal, M., Stafoggia, M., Scortichini, M., Hashizume, M., Honda, Y., Hurtado-Díaz, M., Cruz, J., … Kan, H. (2019). Ambient particulate air pollution and daily mortality in 652 cities. *New England Journal of Medicine*, *381*(8), 705–715. Available from https://doi.org/10.1056/NEJMoa1817364, http://www.nejm.org/medical-index.

Liu, S., Jørgensen, J. T., Ljungman, P., Pershagen, G., Bellander, T., Leander, K., Magnusson, P. K. E., Rizzuto, D., Hvidtfeldt, U. A., Raaschou-Nielsen, O., Wolf, K., Hoffmann, B., Brunekreef, B., Strak, M., Chen, J., Mehta, A., Atkinson, R. W., Bauwelinck, M., Varraso, R., … Andersen, Z. J. (2021a). Long-term exposure to low-level air pollution and incidence of asthma: The ELAPSE project. *European Respiratory Journal*, *57*(6). Available from https://doi.org/10.1183/13993003.03099-2020.

Liu, S., Jørgensen, J. T., Ljungman, P., Pershagen, G., Bellander, T., Leander, K., Magnusson, P. K. E., Rizzuto, D., Hvidtfeldt, U. A., Raaschou-Nielsen, O., Wolf, K., Hoffmann, B., Brunekreef, B., Strak, M., Chen, J., Mehta, A., Atkinson, R. W., Bauwelinck, M., Varraso, R., … Andersen, Z. J. (2021b). Long-term exposure to low-level air pollution and incidence of chronic obstructive pulmonary disease: The ELAPSE project. *Environment International*, *146*. Available from https://doi.org/10.1016/J.ENVINT.2020.106267.

Logan, W. P. D. (1953). MORTALITY IN THE LONDON FOG INCIDENT, 1952. *The Lancet*, *261*(6755), 336–338. Available from https://doi.org/10.1016/S0140-6736(53)91012-5.

Loomis, D., Grosse, Y., Lauby-Secretan, B., Ghissassi, F. E., Bouvard, V., Benbrahim-Tallaa, L., Guha, N., Baan, R., Mattock, H., & Straif, K. (2013). The carcinogenicity of outdoor air pollution. *The Lancet Oncology*, *14*(13), 1262–1263. Available from https://doi.org/10.1016/S1470-2045(13)70487-X.

Maher, B. A., Ahmed, I. A. M., Karloukovski, V., MacLaren, D. A., Foulds, P. G., Allsop, D., Mann, D. M. A., Torres-Jardón, R., & Calderon-Garciduenas, L. (2016). Magnetite pollution nanoparticles in the human brain. *Proceedings of the National Academy of Sciences of the United States of America*, *113*(39), 10797–10801. Available from https://doi.org/10.1073/pnas.1605941113, http://www.pnas.org/content/113/39/10797.full.pdf.

McCreanor, J., Cullinan, P., Nieuwenhuijsen, M. J., Stewart-Evans, J., Malliarou, E., Jarup, L., Harrington, R., Svartengren, M., Han, I. K., Ohman-Strickland, P., Kian, F. C., & Zhang, J. (2007). Respiratory effects of exposure to diesel traffic in persons with asthma. *New England Journal of Medicine*, *357*(23), 2348–2358. Available from https://doi.org/10.1056/NEJMoa071535, http://content.nejm.org/cgi/reprint/357/23/2348.pdf.

Min, Jy, Kim, H. J., & Min, Kb (2018). Long-term exposure to air pollution and the risk of suicide death: A population-based cohort study. *Science of the Total Environment*, *628-629*, 573–579. Available from https://doi.org/10.1016/j.scitotenv.2018.02.011, http://www.elsevier.com/locate/scitotenv.

Mustafić, H., Jabre, P., Caussin, C., Murad, M. H., Escolano, S., Tafflet, M., Périer, M. C., Marijon, E., Vernerey, D., Empana, J. P., & Jouven, X. (2012). Main air pollutants and myocardial infarction: A systematic review and meta-analysis. *JAMA - Journal of the American Medical Association, 307*(7), 713–721. Available from https://doi.org/10.1001/jama.2012.126, http://jama.ama-assn.org/content/307/7/713.full.pdf + html.

Nagel, G., Stafoggia, M., Pedersen, M., Andersen, Z. J., Galassi, C., Munkenast, J., Jaensch, A., Sommar, J., Forsberg, B., Olsson, D., Oftedal, B., Krog, N. H., Aamodt, G., Pyko, A., Pershagen, G., Korek, M., De Faire, U., Pedersen, N. L., ÖÖstenson, C. G., ... Weinmayr, G. (2018). Air pollution and incidence of cancers of the stomach and the upper aerodigestive tract in the European Study of Cohorts for Air Pollution Effects (ESCAPE). *International Journal of Cancer, 143*(7), 1632–1643. Available from https://doi.org/10.1002/ijc.31564, http://onlinelibrary.wiley.com/journal/10.1002/(ISSN)1097-0215.

Nawrot, T. S., Perez, L., Künzli, N., Munters, E., & Nemery, B. (2011). Public health importance of triggers of myocardial infarction: A comparative risk assessment. *The Lancet, 377*(9767), 732–740. Available from https://doi.org/10.1016/S0140-6736(10)62296-9, http://www.journals.elsevier.com/the-lancet/.

Newby, D. E., Mannucci, P. M., Tell, G. S., Baccarelli, A. A., Brook, R. D., Donaldson, K., Forastiere, F., Franchini, M., Franco, O. H., Graham, I., Hoek, G., Hoffmann, B., Hoylaerts, M. F., Künzli, N., Mills, N., Pekkanen, J., Peters, A., Piepoli, M. F., Rajagopalan, S., & Storey, R. F. (2015). Expert position paper on air pollution and cardiovascular disease. *European Heart Journal, 36*(2), 83–93. Available from https://doi.org/10.1093/eurheartj/ehu458, http://eurheartj.oxfordjournals.org/.

Nhung, N. T. T., Amini, H., Schindler, C., Kutlar Joss, M., Dien, T. M., Probst-Hensch, N., Perez, L., & Künzli, N. (2017). Short-term association between ambient air pollution and pneumonia in children: A systematic review and meta-analysis of time-series and case-crossover studies. *Environmental Pollution, 230*, 1000–1008. Available from https://doi.org/10.1016/j.envpol.2017.07.063, https://www.journals.elsevier.com/environmental-pollution.

Nobile, F., Forastiere, A., Michelozzi, P., Forastiere, F., & Stafoggia, M. (2023). Long-term exposure to air pollution and incidence of mental disorders. A large longitudinal cohort study of adults within an urban area. *Environment International, 181*, 108302. Available from https://doi.org/10.1016/j.envint.2023.108302.

Nyadanu, S. D., Dunne, J., Tessema, G. A., Mullins, B., Kumi-Boateng, B., Lee Bell, M., Duko, B., & Pereira, G. (2022). Prenatal exposure to ambient air pollution and adverse birth outcomes: An umbrella review of 36 systematic reviews and meta-analyses. *Environmental Pollution, 306*, 119465. Available from https://doi.org/10.1016/j.envpol.2022.119465.

Parker, J. D., Mendola, P., & Woodruff, T. J. (2008). Preterm birth after the utah valley steel mill closure: A natural experiment. *Epidemiology (Cambridge, Mass.), 19*(6), 820–823. Available from https://doi.org/10.1097/EDE.0b013e3181883d5d.

Pedersen, M., Andersen, Z. J., Stafoggia, M., Weinmayr, G., Galassi, C., Sørensen, M., Eriksen, K. T., Tjønneland, A., Loft, S., Jaensch, A., Nagel, G., Concin, H., Tsai, M. Y., Grioni, S., Marcon, A., Krogh, V., Ricceri, F., Sacerdote, C., Ranzi, A., ... Raaschou-Nielsen, O. (2017). Ambient air pollution and primary liver cancer incidence in four European cohorts within the ESCAPE project. *Environmental Research*. Available from https://doi.org/10.1016/j.envres.2017.01.006.

Pedersen, M., Stafoggia, M., Weinmayr, G., Andersen, Z. J., Galassi, C., Sommar, J., Forsberg, B., Olsson, D., Oftedal, B., Krog, N. H., Aamodt, G., Pyko, A., Pershagen, G., Korek, M., De Faire, U., Pedersen, N. L., ÖÖstenson, C. G., Fratiglioni, L., Sørensen, M., ... Raaschou-Nielsen, O. (2018). Is there an association between ambient air pollution and bladder cancer incidence? Analysis of 15 European Cohorts. *European Urology Focus, 4*(1), 113–120. Available from https://doi.org/10.1016/j.euf.2016.11.008, http://www.journals.elsevier.com/european-urology-focus.

Pedersen, M., Stayner, L., Slama, R., Sørensen, M., Figueras, F., Nieuwenhuijsen, M. J., Raaschou-Nielsen, O., & Dadvand, P. (2014). Ambient air pollution and pregnancy-induced hypertensive disorders: A systematic review and meta-analysis. *Hypertension, 64*(3), 494–500. Available from https://doi.org/10.1161/HYPERTENSIONAHA.114.03545, http://hyper.ahajournals.org/.

Peel, J. L., Klein, M., Flanders, W. D., Mulholland, J. A., & Tolbert, P. E.HEI Health Review Committee. (2010). Impact of improved air quality during the 1996 Summer Olympic Games in Atlanta on multiple cardiovascular and respiratory outcomes. *Research Report (Health Effects Institute)* (148), 3–2333.

Pope, C. A. (1989). Respiratory disease associated with community air pollution and a steel mill, Utah Valley. *American Journal of Public Health, 79* (5), 623–628. Available from https://doi.org/10.2105/ajph.79.5.623.

Pope, C. A., 3rd (2000). Epidemiology of fine particulate air pollution and human health: Biologic mechanisms and who's at risk? *Environmental Health Perspectives, 108*(Suppl 4), 713–723. Available from. Available from https://doi.org/10.1289/ehp.108-1637679.

Pritchett, N., Spangler, E. C., Gray, G. M., Livinski, A. A., Sampson, J. N., Dawsey, S. M., & Jones, R. R. (2022). Exposure to outdoor particulate matter air pollution and risk of gastrointestinal cancers in adults: A systematic review and meta-analysis of epidemiologic evidence. *Environmental Health Perspectives, 130*(3), 36001. Available from https://doi.org/10.1289/EHP9620, https://ehp.niehs.nih.gov/doi/pdf/10.1289/EHP9620.

Raaschou-Nielsen, O., Andersen, Z. J., Beelen, R., Samoli, E., Stafoggia, M., Weinmayr, G., Hoffmann, B., Fischer, P., Nieuwenhuijsen, M. J., Brunekreef, B., Xun, W. W., Katsouyanni, K., Dimakopoulou, K., Sommar, J., Forsberg, B., Modig, L., Oudin, A., Oftedal, B., Schwarze, P. E., ... Hoek, G. (2013). Air pollution and lung cancer incidence in 17 European cohorts: Prospective analyses from the European Study of Cohorts for Air Pollution Effects (ESCAPE). *The Lancet Oncology*. Available from https://doi.org/10.1016/S1470-2045(13)70279-1.

Raaschou-Nielsen, O., Pedersen, M., Stafoggia, M., Weinmayr, G., Andersen, Z. J., Galassi, C., Sommar, J., Forsberg, B., Olsson, D., Oftedal, B., Krog, N. H., Aasvang, G. M., Pyko, A., Pershagen, G., Korek, M., De Faire, U., Pedersen, N. L., ÖÖstenson, C. G., Fratiglioni, L., ... Hoek, G. (2017). Outdoor air pollution and risk for kidney parenchyma cancer in 14 European cohorts. *International Journal of Cancer, 140*(7), 1528–1537. Available from https://doi.org/10.1002/ijc.30587, http://onlinelibrary.wiley.com/journal/10.1002/(ISSN)1097-0215.

Ritz, B., Lee, P. C., Hansen, J., Lassen, C. F., Ketzel, M., Sørensen, M., & Raaschou-Nielsen, O. (2016). Traffic-related air pollution and Parkinson's disease in Denmark: A case−control study. *Environmental Health Perspectives, 124*(3), 351−356. Available from https://doi.org/10.1289/ehp.1409313, http://ehp.niehs.nih.gov/wp-content/uploads/124/3/ehp.1409313.alt.pdf.

Rosi, E., Crippa, A., Pozzi, M., De Francesco, S., Fioravanti, M., Mauri, M., Molteni, M., Morello, L., Tosti, L., Metruccio, F., Clementi, E., & Nobile, M. (2023). Exposure to environmental pollutants and attention-deficit/hyperactivity disorder: An overview of systematic reviews and meta-analyses. *Environmental Science and Pollution Research, 30*(52), 111676−111692. Available from https://doi.org/10.1007/s11356-023-30173-9, https://www.springer.com/journal/11356.

Sacks, J. D., Fann, N., Gumy, S., Kim, I., Ruggeri, G., & Mudu, P. (2020). Quantifying the public health benefits of reducing air pollution: Critically assessing the features and capabilities of WHO's AirQ + and U.S. EPA's environmental benefits mapping and analysis program-community edition (BenMAP-CE). *Atmosphere, 11*(5), 516. Available from https://doi.org/10.3390/atmos11050516, https://res.mdpi.com/d_attachment/atmosphere/atmosphere-11-00516/article_deploy/atmosphere-11-00516-v2.pdf.

Samet, J. M., Dominici, F., Curriero, F. C., Coursac, I., & Zeger, S. L. (2000). Fine particulate air pollution and mortality in 20 U.S. cities, 1987-1994. *New England Journal of Medicine, 343*(24), 1742−1749. Available from https://doi.org/10.1056/NEJM200012143432401.

Samet, J. M., Dominici, F., Zeger, S. L., Schwartz, J., & Dockery, D. W. (2000). The National Morbidity, Mortality, and Air Pollution Study. Part I: Methods and methodologic issues. *Research Report (Health Effects Institute)* (94), 5−1484.

Samet, J. M., Zeger, S. L., Dominici, F., Curriero, F., Coursac, I., Dockery, D. W., Schwartz, J., & Zanobetti, A. (2000). The national morbidity, mortality, and air pollution study. Part II: Morbidity and mortality from air pollution in the United States. *Research Report (Health Effects Institute), 94*(2), 5−79.

Schwartz, J., & Marcus, A. (1990). Mortality and air pollution J London: A time series analysis. *American Journal of Epidemiology, 131*(1), 185−194. Available from https://doi.org/10.1093/oxfordjournals.aje.a115473.

Shah, A. S. V., Langrish, J. P., Nair, H., McAllister, D. A., Hunter, A. L., Donaldson, K., Newby, D. E., & Mills, N. L. (2013). Global association of air pollution and heart failure: A systematic review and meta-analysis. *The Lancet, 382*(9897), 1039−1048. Available from https://doi.org/10.1016/S0140-6736(13)60898-3, http://www.journals.elsevier.com/the-lancet/.

Shah, A. S. V., Lee, K. K., McAllister, D. A., Hunter, A., Nair, H., Whiteley, W., Langrish, J. P., Newby, D. E., & Mills, N. L. (2015). Short term exposure to air pollution and stroke: Systematic review and meta-analysis. *BMJ (Online), 350*. Available from https://doi.org/10.1136/BMJ.h1295, http://www.bmj.com/content/bmj/350/bmj.h1295.full.pdf.

Shi, L., Wu, X., Danesh Yazdi, M., Braun, D., Abu Awad, Y., Wei, Y., Liu, P., Di, Q., Wang, Y., Schwartz, J., Dominici, F., Kioumourtzoglou, M. A., & Zanobetti, A. (2020). Long-term effects of PM2·5 on neurological disorders in the American Medicare population: a longitudinal cohort study. *The Lancet Planetary Health, 4*(12), e557−e565. Available from https://doi.org/10.1016/S2542-5196(20)30227-8, https://www.journals.elsevier.com/the-lancet-planetary-health.

Silva, R. A., West, J. J., Lamarque, J. F., Shindell, D. T., Collins, W. J., Dalsoren, S., Faluvegi, G., Folberth, G., Horowitz, L. W., Nagashima, T., Naik, V., Rumbold, S. T., Sudo, K., Takemura, T., Bergmann, D., Cameron-Smith, P., Cionni, I., Doherty, R. M., Eyring, V., . . . Zengast, G. (2016). The effect of future ambient air pollution on human premature mortality to 2100 using output from the ACCMIP model ensemble. *Atmospheric Chemistry and Physics, 16*(15), 9847−9862. Available from https://doi.org/10.5194/acp-16-9847-2016, http://www.atmos-chem-phys.net/volumes_and_issues.html.

Song, X., Liu, Y., Hu, Y., Zhao, X., Tian, J., Ding, G., & Wang, S. (2016). Short-term exposure to air pollution and cardiac arrhythmia: A meta-analysis and systematic review. *International Journal of Environmental Research and Public Health, 13*(7). Available from https://doi.org/10.3390/ijerph13070642.

So, R., Andersen, Z. J., Chen, J., Stafoggia, M., de Hoogh, K., Katsouyanni, K., Vienneau, D., Rodopoulou, S., Samoli, E., Lim, Y. H., Jørgensen, J. T., Amini, H., Cole-Hunter, T., Mahmood Taghavi Shahri, S., Maric, M., Bergmann, M., Liu, S., Azam, S., Loft, S., . . . Mehta, A. J. (2022). Long-term exposure to air pollution and mortality in a Danish nationwide administrative cohort study: Beyond mortality from cardiopulmonary disease and lung cancer. *Environment International, 164*. Available from https://doi.org/10.1016/j.envint.2022.107241, http://www.elsevier.com/locate/envint.

So, R., Chen, J., Mehta, A. J., Liu, S., Strak, M., Wolf, K., Hvidtfeldt, U. A., Rodopoulou, S., Stafoggia, M., Klompmaker, J. O., Samoli, E., Raaschou-Nielsen, O., Atkinson, R., Bauwelinck, M., Bellander, T., Boutron-Ruault, M. C., Brandt, J., Brunekreef, B., Cesaroni, G., . . . Andersen, Z. J. (2021). Long-term exposure to air pollution and liver cancer incidence in six European cohorts. *International Journal of Cancer, 149*(11), 1887−1897. Available from https://doi.org/10.1002/ijc.33743, http://onlinelibrary.wiley.com/journal/10.1002/(ISSN)1097-0215.

Samoli, E., Analitis, A., Touloumi, G., Schwartz, J., Anderson, H. R., Sunyer, J., Bisanti, L., Zmirou, D., Vonk, J. M., Pekkanen, J., Goodman, P., Paldy, A., Schindler, C., & Kaysouyanni, K. (2005). Estimating the exposure−response relationships between particulate matter and mortality within the APHEA multicity project. *Environmental Health Perspectives, 113*, 88−95.

Stafoggia, M., Oftedal, B., Chen, J., Rodopoulou, S., Renzi, M., Atkinson, R. W., Bauwelinck, M., Klompmaker, J. O., Mehta, A., Vienneau, D., Andersen, Z. J., Bellander, T., Brandt, J., Cesaroni, G., de Hoogh, K., Fecht, D., Gulliver, J., Hertel, O., Hoffmann, B., . . . Janssen, N. A. H. (2022). Long-term exposure to low ambient air pollution concentrations and mortality among 28 million people: results from seven large European cohorts within the ELAPSE project. *The Lancet Planetary Health, 6*(1), e9−e18. Available from https://doi.org/10.1016/S2542-5196(21)00277-1, https://www.journals.elsevier.com/the-lancet-planetary-health.

Strak, M., Weinmayr, G., Rodopoulou, S., Chen, J., De Hoogh, K., Andersen, Z. J., Atkinson, R., Bauwelinck, M., Bekkevold, T., Bellander, T., Boutron-Ruault, M. C., Brandt, J., Cesaroni, G., Concin, H., Fecht, D., Forastiere, F., Gulliver, J., Hertel, O., Hoffmann, B., . . . Samoli, E. (2021). Long term exposure to low level air pollution and mortality in eight European cohorts within the ELAPSE project: Pooled analysis. *The BMJ (Clinical Research Ed.), 374*. Available from https://doi.org/10.1136/bmj.n1904, http://www.bmj.com/.

Sun, Q., Yue, P., Deiuliis, J. A., Lumeng, C. N., Kampfrath, T., Mikolaj, M. B., Cai, Y., Ostrowski, M. C., Lu, B., Parthasarathy, S., Brook, R. D., Moffatt-Bruce, S. D., Chen, L. C., & Rajagopalan, S. (2009). Ambient air pollution exaggerates adipose inflammation and insulin resistance in a mouse model of diet-induced obesity. *Circulation, 119*(4), 538−546. Available from https://doi.org/10.1161/CIRCULATIONAHA.108.799015.

Szyszkowicz, M., Rowe, B., & Colman, I. (2009). Air pollution and daily emergency department visits for depression. *International Journal of Occupational Medicine and Environmental Health, 22*(4). Available from https://doi.org/10.2478/v10001-009-0031-6.

Turner, M. C., Krewski, D., Ryan Diver, W., Arden Pope, C., Burnett, R. T., Jerrett, M., Marshall, J. D., & Gapstur, S. M. (2017). Ambient air pollution and cancer mortality in the cancer prevention study II. *Environmental Health Perspectives, 125*(8). Available from https://doi.org/10.1289/EHP1249, https://ehp.niehs.nih.gov/wp-content/uploads/2017/08/EHP1249.alt_.pdf.

Watkiss, P., Pye, S., & Holland, M. (2005). Cafe CBA: Baseline analysis 2000 to 2020. Service contract for carrying out cost-benefit analysis of air quality related issues, in Particular in the Clean Air for Europe (Cafe) Programme. Available from http://ec.europa.eu/environment/archives/cafe/activities/pdf/cba_baseline_results2000_2020.pdf (last access: 24 May 2017).

Wichmann, H. E. (2004). What can we learn today from the Central European smog episode of 1985 (and earlier episodes)? *International Journal of Hygiene and Environmental Health, 207*(6), 505–520. Available from https://doi.org/10.1078/1438-4639-00322, http://www.urbanfischer.de/journals/intjhyg.

Wilker, E. H., Osman, M., & Weisskopf, M. G. (2023). Ambient air pollution and clinical dementia: systematic review and meta-analysis. *BMJ (Clinical Research ed.), 381*, e071620. Available from https://doi.org/10.1136/bmj-2022-071620.

Wilkins, E. T. (1954). Air pollution aspects of the London fog of December 1952. *Quarterly Journal of the Royal Meteorological Society, 80*(344), 267–271. Available from https://doi.org/10.1002/qj.49708034420.

World Health Organization. (2006). Air quality guidelines for particulate matter, ozone, nitrogen dioxide and sulfur dioxide.

World Health Organization. (2021). *WHO global air quality guidelines. Particulate matter (PM2.5 and PM10), ozone, nitrogen dioxide, sulfur dioxide and carbon monoxide*.

Yang, M., Cheng, H., Shen, C., Liu, J., Zhang, H., Cao, J., & Ding, R. (2020). Effects of long-term exposure to air pollution on the incidence of type 2 diabetes mellitus: a meta-analysis of cohort studies. *Environmental Science and Pollution Research, 27*(1), 798–811. Available from https://doi.org/10.1007/s11356-019-06824-1, https://link.springer.com/journal/11356.

Yuan, S., Wang, J., Jiang, Q., He, Z., Huang, Y., Li, Z., Cai, L., & Cao, S. (2019). Long-term exposure to PM2.5 and stroke: A systematic review and meta-analysis of cohort studies. *Environmental Research, 177*, 108587. Available from https://doi.org/10.1016/J.ENVRES.2019.108587.

Zhang, H., Zhang, X., Wang, Q., Xu, Y., Feng, Y., Yu, Z., & Huang, C. (2021). Ambient air pollution and stillbirth: An updated systematic review and meta-analysis of epidemiological studies. *Environmental Pollution (Barking, Essex: 1987), 278*, 116752. Available from https://doi.org/10.1016/j.envpol.2021.116752.

Zhu, W., Cai, J., Hu, Y., Zhang, H., Han, X., Zheng, H., & Wu, J. (2021). Long-term exposure to fine particulate matter relates with incident myocardial infarction (MI) risks and post-MI mortality: A meta-analysis. *Chemosphere, 267*, 128903. Available from https://doi.org/10.1016/j.chemosphere.2020.128903.

Zumel-Marne, A., Castano-Vinyals, G., Kundi, M., Alguacil, J., & Cardis, E. (2019). Environmental factors and the risk of brain tumours in young people: A systematic review. *Neuroepidemiology, 53*(3-4), 121–141. Available from https://doi.org/10.1159/000500601.

Chapter 15

Air quality and the environmental impacts

Tuukka Petäjä, Ilona Ylivinkka, Tom Kokkonen, Pauliina Schiestl-Aalto, Veli-Matti Kerminen, Jaana Bäck and Markku Kulmala

Institute for Atmospheric and Earth System Research, University of Helsinki, Helsinki, Finland

15.1 Terrestrial–biosphere interactions

About 30% of the global land area is covered by forests (Fig. 15.1). The largest forests in area covered are in the tropics (45% of world's forested areas) and in the boreal region (27%), followed by temperate (16%) and subtropical forests (11%) (FAO & UNEP, 2020). The tropical and temperate forests are rich in different species, whereas boreal forests are dominated by few, mainly conifer tree species (*Abies, Larix, Picea*, and *Pinus*), some broadleaved trees (*Alnus, Betula, Populus*, and *Salix*) and low shrubs, which are adapted to the short growing season, cold winter temperatures and snow accumulation. For example, the conical shape of evergreen boreal conifers minimizes branch damages by reducing snow accumulation, and the small surface area of needles and their waxy coating is beneficial against cold wind (FAO & UNEP, 2020; Jõgiste et al., 2008).

Forests and the atmosphere are in a continuous interaction via different gas and energy fluxes. The canopy takes up CO_2 and releases water, stems transport water and nutrients from the soil, and all living cells consume energy releasing CO_2 in respiration. Life in forests is based on photosynthesis, which creates a flux of CO_2 from the atmosphere into the vegetation, making forests a carbon sink.

In photosynthesis energy from solar radiation is stored as sugars, which are formed from water, taken up by roots, and CO_2, taken into the leaves through stomatal pores. These sugars are consumed in respiration to release the energy for growth and maintenance of living cells. As a result, some of the CO_2 is returned to the atmosphere. In the boreal region, the forest ecosystem takes up about 4700 g CO_2 m^{-2} in photosynthesis and releases 3600 g CO_2 m^{-2} in foliage, stem, ground vegetation and root respiration as well as in decomposition in the soil (Ilvesniemi et al., 2009; Kolari et al., 2006, 2009). Hence, about 20% of the carbon fixed in photosynthesis is stored in the forest.

Fig. 15.2 provides a detailed view of the carbon cycle determined at the Station for Measuring Ecosystem–Atmosphere Relations (SMEAR II, Hari & Kulmala, 2005) in Hyytiälä, Southern Finland. The site presents a Scots pine (*Pinus sylvestris*) dominated boreal environment with a stand age of around 60 years. The largest sink component is photosynthesis of the biosphere composed of 4300 g CO_2 m^{-2} sink to the trees and 400 g CO_2 m^{-2} to the ground vegetation (Ilvesniemi et al., 2009; Kolari et al., 2006, 2009). The emissions of CO_2 from the root respiration and decomposition of the litter accumulates up to 2000 g CO_2 m^{-2}, while the above-ground respiration consists of foliage (900 g CO_2 m^{-2}), wood structures (400 g CO_2 m^{-2}) and ground vegetation (300 g CO_2 m^{-2}) respiration. The net annual carbon budget of the forest is ca. 1100 g CO_2 m^{-2} (Fig. 15.2) varying yearly depending on environmental conditions.

Fig. 15.3 describes the parameters that are important for the carbon cycle of a single pine tree. The rate of photosynthesis depends on the level of photosynthetically active radiation (PAR), atmospheric CO_2 concentration, atmospheric humidity and availability of water at the root zone as well as temperature since plant metabolism and enzyme activation are highly temperature dependent.

Especially in the boreal region, the seasonal cycle of light and temperature regulate the annual growth of vegetation (Hänninen & Tanino, 2011). The active growing season is interrupted by decreasing daylight hours in autumn leading to growth cessation and, possibly along with low temperatures, to dormancy period (Fig. 15.4). The dormancy release requires chilling accumulation while bud burst occurs along with warming temperatures, and the longing photoperiod starts a new active growing season.

Air Quality. DOI: https://doi.org/10.1016/B978-0-12-822591-2.00015-9

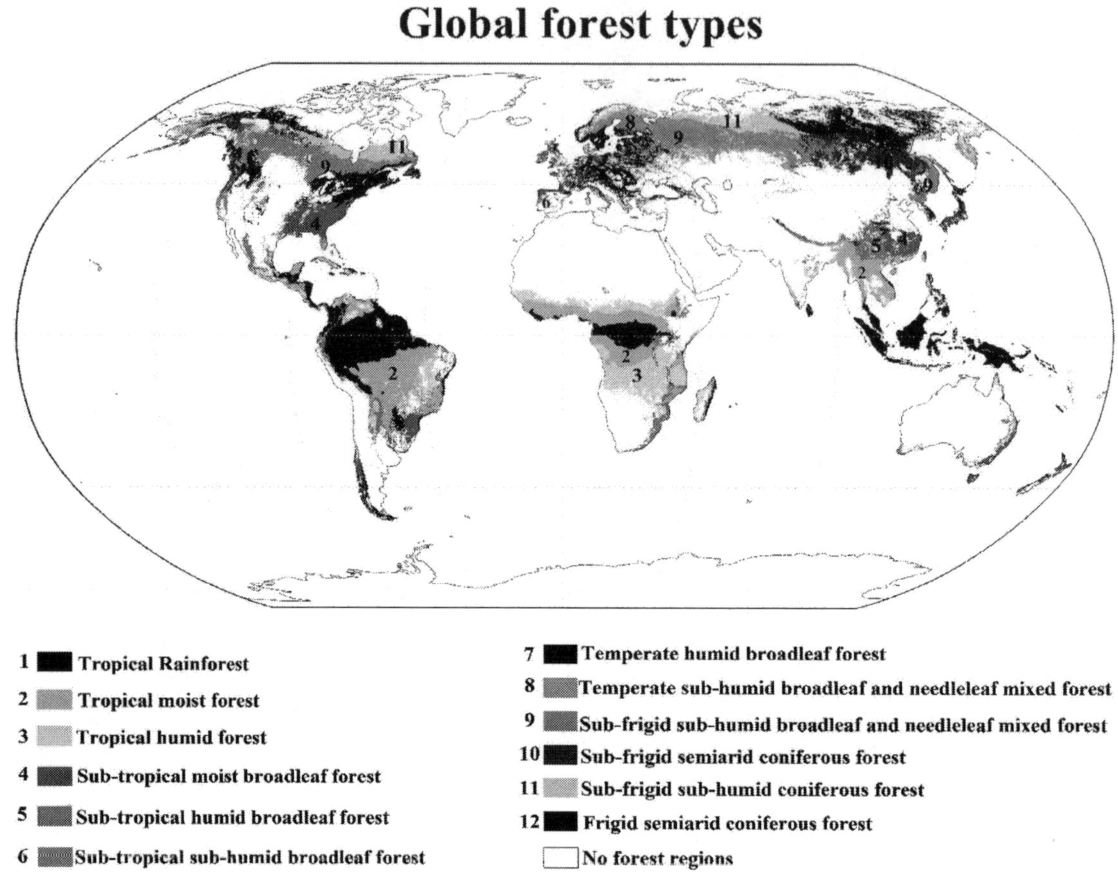

FIGURE 15.1 The global distribution of forests by climatic and vegetation domain.

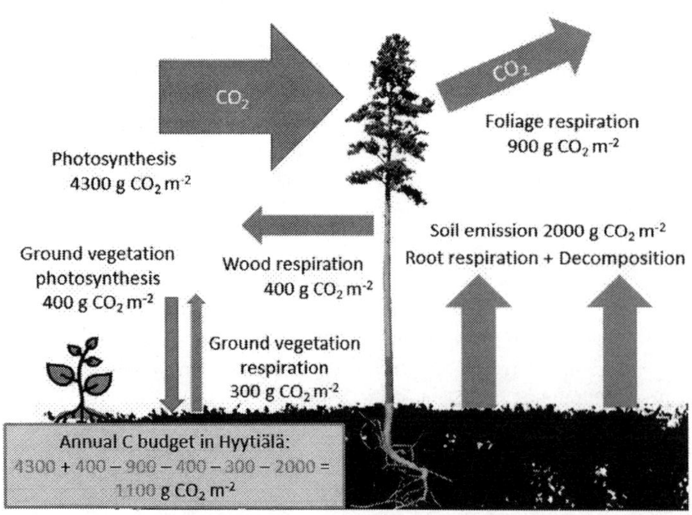

FIGURE 15.2 Carbon cycle and annual budget in the boreal pine forest. *Photograph of the pine by Juho Aalto and the root by Evie S. on Unsplash.*

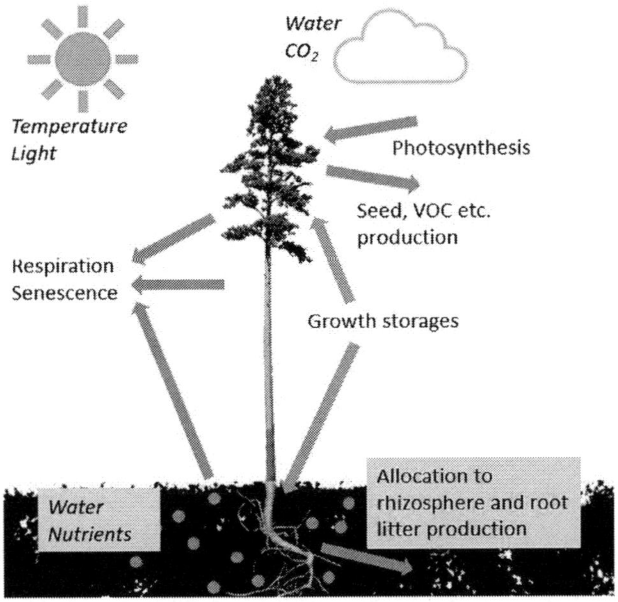

FIGURE 15.3 Components of the carbon cycle of a tree Carbon gain (green arrows) and loss (orange arrows) and governing environmental factors (in italics). *Photograph of the pine by Juho Aalto and the root by Evie S. on Unsplash.*

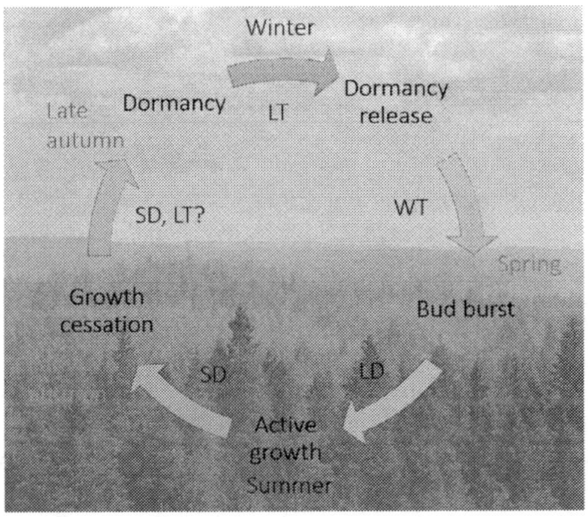

FIGURE 15.4 Annual growing cycle of boreal forests. LD/SD = long/short day, LT/WT = low/warm temperature (Singh et al., 2017). *Adapted from Singh, R.K., Svystun, T., AlDahmash, B., Jönsson, A.M. and Bhalerao, R.P. (2017), Photoperiod- and temperature-mediated control of phenology in trees—A molecular perspective. New Phytolygist, 213, 511–524. https://doi.org/10.1111/nph.14346. Photograph by Juho Aalto.*

The seasonal cycle leads to the loss of carbon through leaf senescence as the dead leaves decompose. Additionally, roots and microbes in the soil exchange carbon (Fig. 15.3). Trees feed carbon to the microbes through roots while microbes transport nutrients to the tree. Hence, the symbiosis is beneficial for both. Some of the stored carbon is used for seed production and lost via emissions of biogenic volatile organic compounds (BVOCs), which are a side product of photosynthesis. The difference between the input and output of carbon is stored in the growth of the tree, either above ground (stem, branch, foliage growth) or below ground (root growth).

Worked example 1: Carbon exchange

Carbon exchange of a forest consists of gross primary production (GPP; i.e., photosynthesis), total ecosystem respiration (TER), and net ecosystem exchange (NEE), which is the difference between respiration and photosynthesis: NEE = TER−GPP. Look at the following figures of NEE, GPP, TER, air temperature, and PAR from the SMEAR II

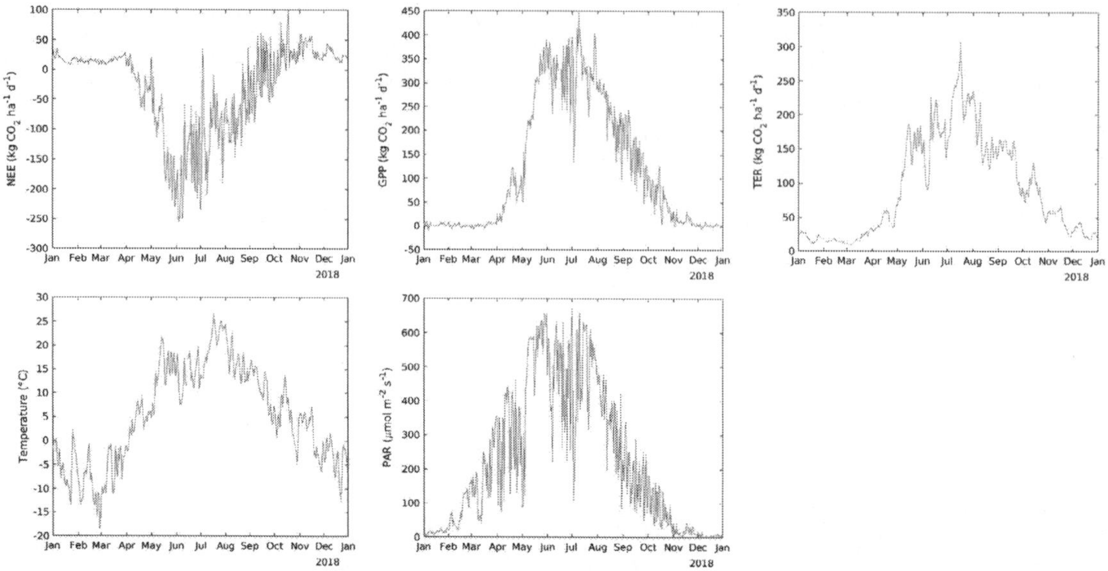

FIGURE 15.5 Annual cycle of net ecosystem exchange, gross primary production, TER, air temperature, and PAR in 2018 at the SMEAR II site in southern Finland. *From Ilona Ylivinkka. Data from smart-SMEAR database. https://smear.avaa.csc.fi/.*

site in southern Finland. Explain when the forest is a carbon sink and when a source. Explain what it means when NEE equals zero. Why respiration is high in summer and low in winter? (Fig. 15.5).

The forest is a carbon sink when NEE is negative. It means that photosynthesis exceeds ecosystem respiration. The figure shows that the forest around SMEAR II is carbon sink approximately from April to September. That covers roughly the period when the daily average temperature is above freezing and PAR is more than 200 μmol m^{-2} s^{-1}. In wintertime, forest is a source of carbon, because beneath the humus layer soil temperature seldom decreases to freezing temperatures and thus, soil decomposition continues during winter creating a small but existing CO_2 source, but the level of photosynthesis is zero or very low. On a yearly scale, a mature forest is a carbon sink as photosynthesis exceeds ecosystem respiration.

NEE is zero when photosynthesis and respiration are in balance. For example, there is some level of photosynthesis in October, but the level of respiration is higher in autumn compared to spring due to higher air temperature, and hence the forest if carbon source / neutral in October.

Both anthropogenic respiration by plants and heterotrophic respiration by, for example, soil microbes, are temperature dependent, and therefore there is more respiration in summer compared to winter. Moreover, trees respire when they use energy from sugars for metabolism; they are active in summer and hence use energy, leading to higher levels of respiration in summer than in winter.

15.1.1 Biogenic volatile organic compounds

A fraction of the carbon fixed in photosynthesis is allocated to the synthesis of BVOCs. They are a variety of hydrocarbons, including alkanes, alkenes, alcohols, aldehydes, ethers, esters, carboxylic acids and terpenoids (isoprenoids). They are emitted by all plants both from aerial and below-ground parts, though foliage is the largest source of BVOCs (Fig. 15.6). Additionally, BVOCs are released into the atmosphere from decaying plant material, and some compounds may be deposited on vegetation or by microbes (Fig. 15.6). The emissions have strong seasonal variation, and the proportions of different emission sources may be altered (Mäki et al., 2019).

The most emitted BVOCs are terpenoids, which are comprised of C_5 units, often called isoprene units. The simplest terpenoid is isoprene (C_5H_8), which covers about 50% of the global BVOC emissions (Guenther et al., 2012). In mass, the global isoprene emissions are 0.5 Pg (500 Tg or 5×10^{14} g) yr^{-1}. Monoterpenes, in turn, are comprised of two C_5 units ($C_{10}H_{16}$; 15% of global BVOC emissions) and sesquiterpenes three C_5 units ($C_{15}H_{24}$; 3% of global BVOC

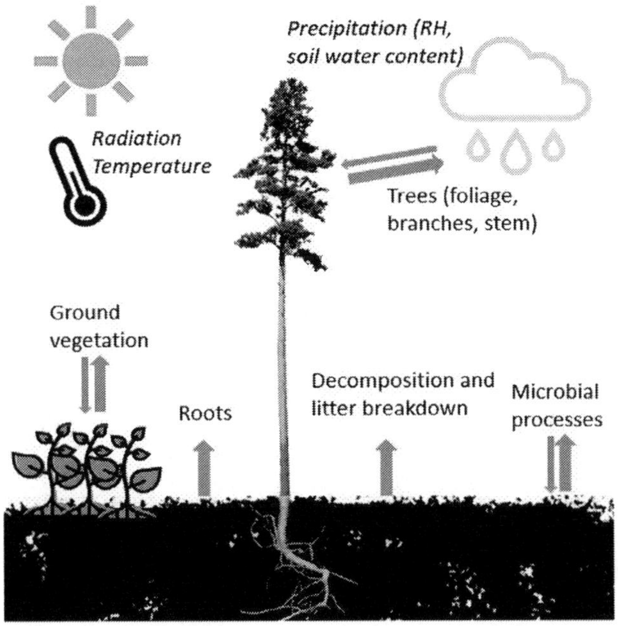

FIGURE 15.6 Monoterpene exchange in a conifer stand. *Photograph of the pine by Juho Aalto and the root by Evie S. on Unsplash.*

emissions). Monoterpenes and sesquiterpenes are actually groups of several isomeric compounds that is, compounds with the same chemical formula but with a different chemical structure.

Different plant species have unique emission profiles which are further modified by genotypes, environmental factors and growing conditions, leading to the emission of tens of thousands of different BVOC compounds. Conifers and broadleaved trees are generally strong BVOC emitters while the emission rates from grasses and crops are lower (Guenther et al., 2012). Plants can roughly be categorized into isoprene and monoterpene emitters. The dominance of emitting either isoprene or monoterpenes may stem from different ecological strategies of plants since the isoprene emitters have usually higher photosynthetic capacity and specific leaf area while the lifespan of leaves is shorter compared to monoterpene emitters. The trade-off may be related to the competing substrate availability of isoprene and monoterpene synthesis (Harrison et al., 2013). Many tropical tree species are isoprene emitters whereas boreal forest trees are commonly monoterpene emitters.

In normal conditions, plants allocate about 0.05%–0.5% of the carbon fixed in photosynthesis to the synthesis of BVOCs, but the share can increase significantly when plants are stressed. BVOCs are plant scents and odors meant for pollination, communication, and defense. They have several functions for plants, and many functions are still unknown to us. Constitutively emitted BVOCs are primary metabolites, meaning that they are essential for growth and reproduction, whereas induced emissions are secondary metabolites meant for improving the fitness of plants for instance in stressed conditions. The functions of induced BVOCs are discussed in more detail in Section 15.2.

The synthesis of BVOCs is regulated by substrate availability in short time scales and by transcriptional dynamics in longer time scales. The BVOCs may be emitted directly after their synthesis (de novo emissions) or be stored in special storage pools, such as resin ducts, and be released into the atmosphere on longer time scales. Due to the substrate availability, de novo emissions are coupled with photosynthesis, while stored BVOCs may be emitted continuously. Commonly, mono- and sesquiterpenes from conifers are emitted from storage pools, although de novo emissions are also possible (e.g., Taipale et al., 2011). Broadleaved trees typically have only de novo emissions since they lack effective storage.

The emissions of BVOCs depend exponentially on temperature due to various reasons (e.g., Aalto et al., 2015, Bäck et al., 2012, Guenther et al., 1993, Guenther et al., 1995). Firstlyenzyme activation, which controls both photosynthesis and synthesis of BVOCs, is highly temperature dependent. Secondly, substrates needed in BVOC synthesis are obtained from photosynthesis and are thus also temperature dependent. Thirdly, vapor pressure and diffusion, which control the vaporization and fluxes of compounds in the leaves, are increased at higher temperatures. The higher the vapor pressure is, the easier it is for the compound to escape from plant tissue. Vapor pressure of a compound depends on the temperature and concentration of the compound in plant tissue. Moreover, the synthesized BVOCs must cross several cellular barriers, at minimum cell membrane, cell wall and cuticle, which regulate the emissions. From the plant surface, the compounds escape either through cuticles or through stomata. Compounds such as methanol, are more strongly bound by stomatal openness whereas, for example, isoprene can easily escape from tissues due to a large diffusion gradient into the atmosphere. Hence, isoprene cannot be stored by any plant and is always emitted de novo. In extremely high temperatures (typically >45°C) the exponential temperature dependence of BVOC emissions breaks down as enzymes are damaged by the heat. BVOC emissions can also be light dependent, through photosynthesis providing substrates for the BVOC synthesis.

Continuous, long-term observations on Scot's pine needle and shoot BVOC emissions have, however, revealed that the conventional view on temperature and light as the only emission-controlling factors is likely to be too simplified. For example, new foliar biomass growth has been found to be a significant source of many BVOCs (Aalto et al., 2014). Also, the photosynthetic recovery in spring, especially the period measured at SMEAR II site when light use efficiency was still low after cold winter season, there was a transient increase in both monoterpene synthesis and emissions (Aalto et al., 2015).

Due to the role of BVOCs in atmospheric chemistry, ozone formation, aerosol formation and boundary layer processes, emission estimates are needed as input in tropospheric chemistry and climate models. The emissions can be modeled by combining information on vegetation type and coverage with emission algorithms describing emissions from certain plant types (e.g., Guenther et al., 2012). They formulate a mathematical relationship between environmental conditions and BVOC emissions without the requirement of detailed information about physicochemical and biochemical theory on a plant's BVOC emission mechanisms. However, there are also process-based models describing the emission mechanisms in detail, but they are not currently used in large-scale emission inventories due to their high computational expenses.

The empirical models describe the temperature and/or light dependence of BVOC emissions. Commonly the emission models rely on the concept of emission potential, also known as standard emission factor or normalized emission, which explains the BVOC emissions at standard temperature and light conditions (usually at a temperature of 30°C and light intensity at 1000 μmol m^{-2} s^{-1}. Different sampling conditions can then be easily compared using emission potentials. Widely used models were developed by Guenther et al. in the nineties (Guenther et al., 1995) that have provided a core for a global emission model MEGAN (Model of Emissions of Gases and Aerosols from Nature; Guenther et al., 2012).

Worked example 2: Temperature dependency of monoterpene emissions from the biosphere using *Larix cajanderi* trees as an example

The traditional monoterpene emission algorithm (pool algorithm, Guenther et al., 1993) assumes that the emission is exponentially dependent on ambient temperature. In practice, this depicts the unhindered evaporation of monoterpenes from storage pools.

The total emission (E) can be described with the equation:

$$E = E_{\text{pool}}\exp[\beta(T - T_S)] \tag{15.1}$$

where E is the measured emission in the units of $\mu g\ g_{dw}^{-1}\ h^{-1}$ (micrograms of monoterpenes per grams of dry weight of biomass per hour) and E_{pool} ($\mu g\ g_{dw}^{-1}\ h^{-1}$) is the emission in standardized temperature (T_s), which is typically 30°C. The T is the measured temperature of the emission chamber. b is the temperature dependency of the emission, typically having value of 0.09°C^{-1}. Values are based on experimental data from *Larix cajanderi* tree emission measurements presented in Kajos et al. (2013).

Plot the temperature dependency of the emissions from 15°C to 35°C in July.

The emissions are given by Eq. (15.1) using the temperatures between 15°C and 35°C. Kajos et al. (2013) give emission potentials for summer months (from June to August) for two trees. The emission potentials in July differ between tree A (1.5 $\mu g\ g_{dw}^{-1}\ h^{-1}$) and tree B (7.4 $\mu g\ g_{dw}^{-1}\ h^{-1}$), and hence they are both used separately for calculating July monoterpene emissions from *L. cajanderi* trees.

15.1.2 Feedbacks between the biogenic volatile organic compounds and aerosol particles

Although volatile hydrocarbons are also emitted from anthropogenic sources, globally the biogenic emissions dominate by one order of magnitude, totaling approximately 1000 Tg yr^{-1} (Guenther et al., 2012). The atmospheric concentrations of individual BVOCs are in the order of ppb (parts per billion) or ppt (parts per trillion), but they are highly reactive and are thus rapidly oxidized in the atmosphere by ozone and hydroxyl and nitrate radicals (see Chapter 4). The oxidation products have lower volatilities (Ehn et al., 2014) and have therefore a higher tendency to nucleate or condense onto the particulate phase. Organic low-volatility vapors can, along with other low-volatility vapors such as sulfuric acid, ammonia and amines, participate in new particle formation (NPF; Kulmala et al., 2014a). New aerosol particles are formed from condensing vapors. If the formed particles continue growth reaching about 100 nm in diameter, they can contribute to the radiative balance of the Earth directly by interacting with light and indirectly by changing cloud properties and cloudiness in general (see Chapters 4 and 11).

BVOCs have variable ability to form low volatility vapors and therefore contribute to NPF. For example, mono- and sesquiterpenes emitted from the boreal forests have been shown to be efficient in forming lower volatility compounds since they can undergo rapid autoxidation forming highly oxygenated molecules (HOMs) that are likely to condense and participate in aerosol formation and growth processes (Ehn et al., 2014). Isoprene, in turn, can even suppress NPF when found in high concentrations. This may be caused by the production of molecules with 14–15 carbon atoms in termination reaction compared to molecules with 19–20 carbon atoms in pure monoterpene oxidation, leading to reduced nucleation and growth rates (Heinritzi et al., 2020). Hence, although most of the BVOCs are emitted by tropical forests and boreal forests form a relatively small source of BVOCs, the emission profiles from these two ecosystems are rather different (Artaxo et al., 2022), and the emissions from boreal forests have a high atmospheric impact in aerosol formation perspective.

Besides limiting the amount of incoming solar radiation, radiation scattered from aerosol particles and clouds also increases the diffuse fraction of global radiation reaching the surface. Diffuse radiation, coming from all directions, is beneficial for photosynthesis compared to direct sunlight because in this case radiation is distributed more evenly inside the canopy (Gu et al., 2002). Additionally, leaves angled away from direct sunlight receive higher levels of radiation, and light saturation of photosynthesis in the upper canopy occurs less frequently. Light saturation can commonly take place during clear-sky days, meaning that photosynthesis has reached its maximum and increasing the level of sunlight cannot enhance photosynthesis any further. Moreover, usually under diffuse radiation air temperature is slightly lower, leading to reduced respiration, and thus increased net carbon uptake. The optimal diffuse fraction of global radiation for photosynthesis is about 0.5 in the boreal environment (Ezhova et al., 2018). Therefore, the maximal ecosystem-scale photosynthesis is not reached at clear-sky and clean atmospheric conditions but rather in the presence of fair-weather clouds and relatively high aerosol load. In these conditions, photosynthesis can increase up to 30%. However, it should be noted that when the fraction of diffuse radiation is high, the overall level of incoming solar radiation tends to be reduced, and hence restricts photosynthetic activity. Consequently, maximal photosynthesis consequently occurs when the observed level of global radiation is high, and the fraction of diffuse radiation is about 0.5.

Since photosynthetic activity depends on CO_2 concentration, the described processes form a negative feedback mechanism between ecosystems and the atmosphere (Fig. 15.7; Kulmala et al., 2020, 2014b; Yli-Juuti et al., 2021). The

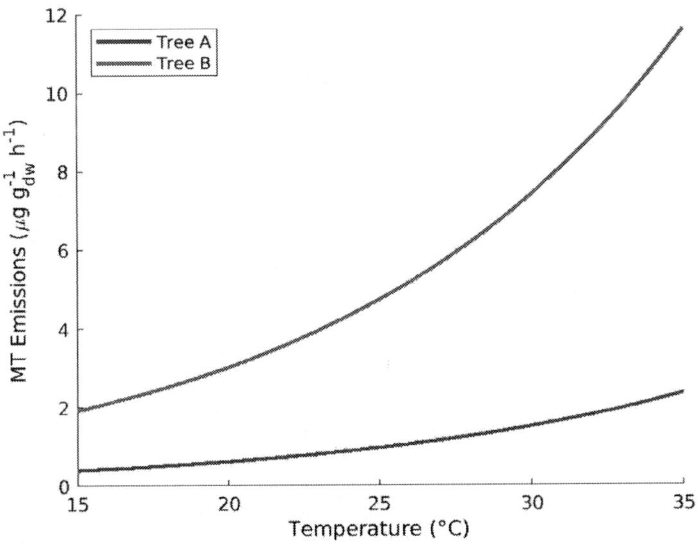

FIGURE 15.7 Temperature dependency of monoterpene emissions of two *L. cajanderi* trees. *From Ilona Ylivinkka. Emission data from Kajos, M.K., Hakola, H., Holst, T., Nieminen, T., Tarvainen, V., Maximov, T., Petäjä, T., Arneth, A., & Rinne, J. (2013). Terpenoid emissions from fully grown east Siberian Larix cajanderi trees. Biogeosciences, 10, 4705–4719, https://doi.org/10.5194/bg-10-4705-2013.*

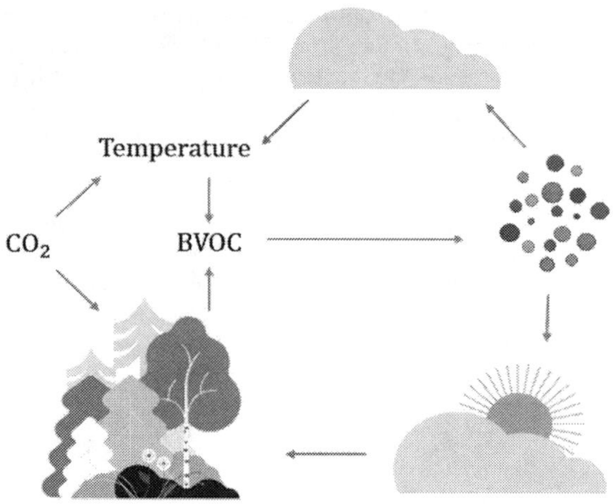

FIGURE 15.8 Illustration of feedback mechanisms between ecosystems and the atmosphere. Red arrows indicate increase and blue arrows decrease. Feedback connecting the elevated CO_2 concentration, ambient temperature, biogenic volatile organic compound emissions and aerosol population influencing properties of clouds and radiative transfer of solar radiation through the atmosphere. *From Ilona Ylivinkka.*

increasing CO_2 concentration due to anthropogenic emissions boosts photosynthesis, enhancing also BVOC emissions. Increasing CO_2 concentration additionally raises air temperature, growing the emission rates of BVOCs further. The elevated concentrations of BVOCs can lead to higher concentrations of aerosol particles, which can backscatter solar radiation and act as cloud condensation nuclei (CCN), modifying cloud properties, and cloudiness, and thereby cooling the climate (Petäjä et al., 2022). Furthermore, the increased scattering from clouds and aerosol particles can enhance photosynthesis, closing the feedback loop (Fig. 15.8). These perturbations where a change in one variable leads to subsequent changes in other processes, either amplifying (positive feedback) or dampening (negative feedback) the original change, are called feedbacks.

In the changing climate, photosynthesis is strengthened at high latitudes by lengthening the growing season and migration of woody plant species both northward and upslope on hills, allowing prominent changes to the local vegetation diversity (Sizov et al., 2021). These changes are enabled by global warming, which proceeds faster at high latitudes compared to the global average (IPCC, 2021; Rantanen et al., 2022).

Forests are additionally a source of ice nuclei, which function similarly to CCN. Homogeneous ice nucleation requires temperatures below $-36°C$, but in the presence of ice nuclei crystallization can take place also at higher temperatures. Several biogenic components from forests can act as ice nuclei, including bacteria, viruses, fungal spores, algae, pollen, biological crusts and lichens, and other plant or animal fragments and particles from biomass burning (DeMott et al., 2010; Després et al., 2012). The role of secondary biogenic aerosol particles in ice nuclei concentrations is not yet determined. However, the ice processes in the atmospheric are highly relevant as most of the precipitation that reaches the surface originates through ice nucleation processes.

15.2 Air pollution impacts the ecosystems, vegetation and aquatic ecosystems and vice versa

On one hand, air pollution does damage to the vegetation due to oxidative stress caused by ozone and acidic aerosol particles (acid rain). On the other hand, the vegetation influences air pollution by contributing to ozone formation via the oxidation of biogenic volatile organics in the air. Therefore, air pollution and air quality are relevant in the context of the well-being of the biosphere as well as of human health. Consequently, legislation related to emissions and concentrations has been revised and strengthened. Everything surrounding us is affected by the air quality—and vice versa—and the issues are not restricted to humans.

BVOCs have multiple functions for plants, and therefore their synthesis and emissions are highly regulated by plant species, different stresses, and environmental conditions (Fig. 15.9). Plant stressors are divided into biotic and abiotic stressors. The biotic stressors are, for example, fungal infections and herbivore infestations whereas the abiotic stressors include adverse growing conditions, such as drought, heat waves, cold spells and air pollution. The induced emissions act as plant defense against stresses, mitigating the damages. The induced BVOCs example, during a herbivore infestation are toxic or act as repellents for the attacking species, or they can be a signal to nearby parasites and predators as

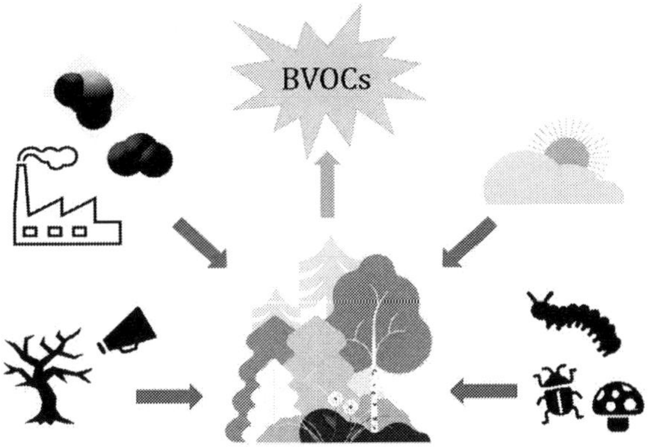

FIGURE 15.9 Biogenic volatile organic compound (BVOC) emissions from forests are largely modified by different biotic and abiotic stress Processes and activities influencing BVOC emissions from the biosphere. *From Ilona Ylivinkka.*

well as to the nearby trees. The induced emissions can thus diminish herbivory when the birds come to feed on the herbivores, or the neighboring trees also start producing repellents. This way, the BVOC emissions are part of the communication between the trees and their surroundings.

The Plant stress factors can both increase and decrease constitutively emitted BVOCs or induce the synthesis of particular stress compounds (Faiola & Taipale, 2020; Peñuelas & Staudt, 2010). Additionally, in conditions of multiple successive or cooccurring stresses, the induced BVOC emissions may be different than in a case when the same stresses act individually. The plants can be acclimated to the stress, or the stress tolerance may decline due to the other stresses (Niinemets, 2010). Plants' responses to the stresses are variable and can depend on species and age of the plant, hence a complete understanding of the stress responses is difficult to obtain. Moreover, timing and severity of the stresses also affect the induction of BVOCs (Niinemets, 2010). For example, mature leaves may have higher stress tolerance compared to young leaves because they have already accumulated secondary metabolites. Also, a cold spell in late autumn may not be harmful for boreal forest trees, since they have started growth cessation and prepared themselves for the dormancy period, whereas a cold spell in spring may interrupt shoot and needle growth and thus, photosynthesis.

Climate change and increased anthropogenic activities cause alterations in CO_2 concentration and temperature as well as in, for example, O_3 concentration, precipitation patterns and nutrient availability, modifying the growing conditions of vegetation in many ways. Hence, in the future the occurrence of simultaneous stresses is likely to increase and can result in increased emission rates of BVOCs to the atmosphere (Holopainen et al., 2018; Li et al., 2019; Räisänen et al., 2008).

The elevated CO_2 concentration can lead to a decrease in isoprene emissions because at high CO_2 concentrations, isoprene synthesis is uncoupled from photosynthesis. The effect on monoterpene emissions is more complex since the size of storages may be changed at high CO_2 concentrations, altering the source-sink balance, but de novo emissions may be diminished similarly to isoprene (Calfapietra et al., 2013b; Holopainen et al., 2018).

The elevated temperatures in turn increase BVOC emissions exponentially, but trees emitting especially isoprene have even shown increased thermotolerance under high temperatures. Isoprene synthesis can function as a tool to lose excess energy, and thus avoid overheating (Possell & Loreto, 2013). However, high temperatures are often associated with low precipitation, and thus drought. In drought conditions, trees tend to close the stomata to avoid extensive loss of water, which also restricts the CO_2 intake and the level of photosynthesis. Hence, upon severe and prolonged drought the BVOC emissions of plants may decrease.

Several weather extremes, including for instance strong winds, hailstorms and flooding, may also become more frequent in the future (IPCC, 2021). Strong winds and hailstorms cause mechanical damage to plants, releasing bursts of BVOCs from broken cells (Possell & Loreto, 2013). Flooding, in turn, causes anoxic conditions in the root zone, leading to the formation of ethanol. Ethanol is then transported in the tree to the leaves and emitted into the atmosphere, causing strong emissions of ethanol and acetaldehyde (Kreuzwieser & Rennenberg, 2013).

Progressing global warming influences the probability of occurrence and severity of different biotic stresses. Similar to plant species, many herbivores and pathogen species may spread to areas where they did not thrive previously (Lehmann et al., 2020). For example, insect egg mortality during wintertime typically requires periods with very low temperatures (e.g., under $-36°C$), which may not be reached as often in the future, creating propitious conditions for insect herbivores and pathogens. Also, the northward and upslope spread of suitable host plants as well as monoculture farming and forest management create favorable circumstances for many biotic stressors to thrive and influence the wellbeing of the biosphere.

Other stresses such as drought or air pollution can expose trees to further biotic stresses as their fitness is already diminished (Holopainen et al., 2013). Trees grown under high SO_2 and NO_x concentrations as well as heavy metal pollution have been particularly associated with leaf-mining and phloem feeding herbivores. Heavy metal exposure is additionally often visible as a decreased photosynthetic activity, growth and even abnormalities in the tree structures. Legislation on air pollution aiming for reductions of SO_2 and NO_x emissions from anthropogenic industrial processes has already decreased plant damage caused by depositing sulfur and nitrous compounds in North America and Europe (Holopainen et al., 2013). However, elsewhere rapid industrialization may result in regional or local pollution hotspots.

Ozone and other reactive oxygen species can lead to oxidative stress in plants causing cell damage. BVOCs, especially terpenoids with one or several double bonds, are highly reactive, and can therefore act as antioxidants protecting leaf tissues and cell membranes. However, photochemistry of BVOCs can also lead to the production of ozone in NO_x-rich environments, such as in urban areas. Hence, BVOC emissions can amplify oxidative stress and worsen air quality in cities (see Chapter 4).

As underlined above, temperatures influence BVOC emissions, and the emissions increase exponentially as a function of temperature. In urban areas, local air temperatures are commonly higher compared to the surrounding environments due to anthropogenic heating in the cities. The phenomenon is called urban heat island. Influenced by higher temperatures, the urban trees and urban greenery emit more BVOCs than the corresponding trees outside the city, which can further influence ozone concentrations and contribute to regional air pollution. To solve the issue, bio-engineering tests have already been made to cultivate nonemitting species for cities to ease the pollution exposure due to BVOC-involved photochemistry.

Tropospheric ozone is a significant greenhouse gas, but BVOCs can additionally affect the atmospheric lifetime of other strong greenhouse gases, such as methane. Because BVOCs are highly reactive, they can consume oxidants from the atmosphere, prolonging the lifetime of other greenhouse gases, and thus causing positive radiative forcing (Boy et al., 2022).

Increased concentrations of atmospheric oxidants, particularly tropospheric ozone, have been noticed to be harmful to plants also due to loss of inter and intra-plant as well as multitrophic connection (Holopainen et al., 2013). It means that the emitted BVOCs react more rapidly, and hence they do not travel as long distances in the atmosphere as previously. The oxidized compounds are different from the original compounds—at least they have additional oxygen atom(s) but potentially also other atoms attached to them—and hence their chemical properties are altered after oxidation. Therefore, the signals aimed for internal or plant–plant communication, or for attraction of pollinators, seed dispersers or herbivore parasites and predators may be lost in atmospheric processing. The transport distances may have dropped from kilometers to a couple of hundred meters. Particularly difficult the situation is for pollinator species specialized in certain flowers. They might not be able to find these flowers if the scents are not transmitted long enough in the atmosphere for them to find their way from flower to flower. Some species might be able to adapt to the oxidized scents although due to anthropogenic activities and increased emissions of VOC the world has become more fragrant hindering BVOC-based communication even further.

A direct consequence of the droughts caused by climate change on the biosphere is an increased risk of forest fires. The forest fires emit large amounts of gaseous and particulate pollution, depending on the stages of the fire. Smoldering fires tend to emit higher amounts of carbon monoxide and black carbon aerosol whereas open fires are "cleaner" producing a higher fraction of CO_2. The emissions from the smoldering fires tend to linger closer to the surface whereas the high-intensity fires can even initiate cloud formation around the fires (Pyrocumulus). Annually, 3% of the Earth's land area burns causing matter and energy exchange between the atmosphere and land surface. In some areas, forest fires are a natural part of the forest renewal process although land-use changes and prolonged drought periods due to changing climate have increased the fire risk in temperate regions. Fire ignition is most often caused by humans and lightning has only a secondary role.

15.3 Effect of air pollution on urban energy and water balances and boundary layer dynamics

Haze formation, together with atmospheric boundary layer processes, form a feedback mechanism. Increased aerosol concentration in the boundary layer modifies local urban climate by attenuating the incoming solar radiation reducing the surface energy availability and heat fluxes (Gao et al., 2020; Kajino et al., 2017). This decreases the near-surface air temperature (Ding et al., 2013; Wang et al., 2014) and furthermore the vertical turbulent heat fluxes and mixing and the daytime boundary layer height (BLH) (Ding et al., 2016; Liu et al., 2018; Petäjä et al., 2016; Wang et al., 2018), but increases incoming longwave radiation emissions from the polluted atmosphere (Petäjä et al., 2016; Tang et al., 2016). In addition, increased absorption of radiation by the polluted atmosphere changes the vertical temperature profile, leading to more stable conditions. With reduced turbulence and mixing, the BLH is consequently lower. The reduced BLH constrains diverse emissions into a smaller and smaller air volume, further enhancing the accumulation of pollutants within the boundary layer during haze episodes (Petäjä et al., 2016). This vicious cycle can deteriorate the strongest haze episodes even further and prolong their existence until the synoptic weather changes and pushes the highly polluted urban air elsewhere. Formation of secondary aerosol particles (Chu et al., 2019; Kerminen et al., 2018; Kulmala et al., 2004) can contribute to the initial steps of haze formation (Chu et al., 2020; Du et al., 2021; Kulmala et al., 2017, 2022; Lin et al., 2021; Wang et al., 2020).

During this kind of haze episodes, solar heat stored on urban surfaces during the daytime is reduced, weakening the nocturnal heat emissions and the urban heat island (UHI, a temperature difference between the urban and surrounding rural area) and therefore also the night time BLH. Additionally, calm synoptic scale winds enable UHI-induced circulation, also called the urban heat dome flow, which drives the transport of pollutants within urban regions (Fan et al., 2017b; Zhang et al., 2014). The urban heat dome flow includes a convergent inflow to the city in the lower atmosphere, an upward flow caused by turbulent plumes in the urban area, a divergent outflow at higher altitudes, and a dome-shaped flow field at the boundary between the inversion layer and the outflow region. This circulation, together with a low-level entrainment flow of cleaner air from rural areas, increases ventilation, which improves urban air quality (Fan et al., 2017b).

Unfortunately, weak synoptic scale winds that enable UHI circulation, are favorable also for haze formation (Zhong et al., 2018), which decreases UHI magnitude, and thus restricts UHI circulation and ventilation in urban areas. As a summary, the feedback associated with the BLH and particulate pollution will reduce vertical mixing and the overall weakening of the UHI will decrease the urban heat dome flow, inhibiting the ventilation of urban areas (Miao et al., 2015; Sang et al., 2000).

With less energy available at the surface, due to the attenuated incoming solar radiation, also evaporation may be reduced, modifying other water balance terms (Kokkonen et al., 2019). In addition, haze can increase the condensation nuclei number concentrations and therefore precipitation in the form of drizzle. The higher surface runoff rates due to the modified water balance and increased precipitation may increase pollutant loads in urban water bodies, by flushing pollutants from the contaminated surfaces.

However, the linkage between increased aerosol concentrations and the urban hydrological cycle is not yet completely understood due to the lack of detailed studies at the local scale, despite its potential contribution to the deterioration of the aquatic environment. Poor air quality is the result of several factors, including pollutant emissions, atmospheric transport, atmospheric chemistry, and meteorological conditions. Therefore, the accuracy and spatial coverage of meteorological variables are essential to be able to study the effects of pollutants on the local hydrological cycle correctly. The lack of these studies is partly because of a lack of high-resolution meteorological and/or hydrological observations for detailed analyses and modeling from regions with heavy air pollution episodes. Global reanalysis products could provide the essential variables to enable modeling where necessary observations are unavailable or have too coarse a temporal resolution (Kokkonen et al., 2018). Several reanalysis products are available, but many of them have not been properly evaluated in highly polluted environments, especially in urban regions.

Several urban land surface models with different levels of complexity have been developed to examine the urban land-surface interaction (Grimmond et al., 2010; 2011). These models range from highly abstract models used only to guide research, to more advanced ones used for predicting and analyzing real-world phenomena as well as simulating the urban climate in various urban development scenarios. One of the first models was introduced by Myrup (1969) to study urban energy balance, which represented the urban environment as an extensive homogeneous surface. After that many urban land surface models have been developed, from simple single-layer urban canopy models (e.g., Kusaka et al., 2001) to multilayer urban canopy models (e.g., Kondo et al., 2005) and for example Community Land Model−Urban (CLM-Urban; Oleson et al., 2008) that has an unique purpose to simulate local urban climate as an operational component of a global climate model. Usually, these models are designed to depict the urban energy balance of an idealized urban canopy comprised of various urban facets but originally excluded the effect of vegetation. While the exclusion of the vegetation might be a reasonable simplification for a model design in central downtown locations of major cities, in suburban areas vegetation influences urban latent heat flux substantially (Best & Grimmond, 2016; Grimmond & Oke, 1986; Mitchell et al., 2001, 2003; Oke et al., 2017). The first international urban model comparison experiment, evaluating 32 urban land surface schemes, concludes also that the models with a representation of vegetation performed much better in simulating latent and sensible heat fluxes than the models neglecting it (Grimmond et al., 2011) even though urban land surface models tend to underestimate vegetative evaporation. Therefore, the inclusion of the description of vegetation in models plays a crucial role in understanding energy and water balances in urban areas. Thus, the newer models [e.g., Surface Urban Energy and Water Balance Scheme (SUEWS; Järvi et al., 2011) and SURFEX (Masson et al., 2013)] have been specifically designed to incorporate the description of the vegetation and many earlier models have subsequently been modified to included vegetation afterwards.

Also, a variety of urban hydrological models have been developed to study the water balance and flow within urban areas (Peña-Guzman et al., 2017). Notable examples of these models are EPANET (Rossman, 1993), Aquacycle (Mitchell et al., 2001) and Storm Water Management Model (SWMM; Rossman, 2015). These models primarily focus on water distribution in cities and the design of pipe networks. For example, the EPANET is commonly used to simulate the quality and transportation of potable water in the distribution systems. The Aquacycle is frequently utilized in studies concerning the reuse of storm water and wastewater in urban areas, as it allows the separation of storm water and wastewater as the source of supply in nonpotable water applications. The main application of SWMM has been in studies related to pipe network design and dynamic flood routing, since it includes schemes for determining the flow depth of water in each pipe of the study area and routing of above-ground runoff along with pumps, regulators, pipes and channels. However, the description of evaporation is often insufficient in these models, which is needed when examining the urban water balance and the effect of air pollution on the hydrological cycle.

As a summary, air pollution has dramatic impacts on urban energy balance by influencing the incoming solar radiation and therefore surface heating. In extreme haze events, vertical mixing is limited which causes further deterioration in the air quality. The air pollutants can be water-soluble or washed away by precipitation from the air or from urban surfaces. This adds to the pollution levels in the urban water systems.

Worked example 3: Boundary layer dynamics in an urban environment.

Based on the information given by Kokkonen et al. (2019), calculate how much the incoming solar radiation would be attenuated in Beijing on June 1, 2023 at 12:00 due to air pollution when the aerosol optical depth (AOD) at 440 nm is increasing from 0.1 (fairly clean air) to 3 (very hazy conditions). You can use online calculators for the solar zenith angle.

Figure 1 in Kokkonen et al. (2019) shows that the equation for the relationship for incoming solar radiation and AOD for summer months at 12:00 is

$$K_{\downarrow}/\varphi_s = -0.091\text{AOD} + 0.693,$$

where, $K_{\downarrow}$ is the incoming solar radiation, AOD is the AOD value, and φ_s is the clear sky radiation which is defined as

$$\varphi_s = I_{sc} \times \cos\theta_z,$$

where, I_{sc} is the solar constant (1367 W m^{-2}) and θ_z is the solar zenith angle. Therefore,

$$K_{\downarrow} = (-0.091\text{AOD} + 0.693)(I_{sc} \times \cos\theta_z)$$

The cosine of solar zenith angle ($\cos\theta_z$) given by the NOAA solar position calculator (https://gml.noaa.gov/grad/solcalc/azel.html) is 0.9506.

Therefore, substituting AOD = 0.1 and AOD = 3 to the equations gives us:

$$K_{\downarrow}(\text{AOD} = 0.1) \approx 888.7 \text{ W m}^{-2}$$

$$K_{\downarrow}(\text{AOD} = 3) \approx 545.8 \text{ W m}^{-2}$$

The attenuation of incoming solar radiation due to increasing air pollution is $\Delta K_{\downarrow} \approx 343 \text{ W m}^{-2}$.

15.4 Impacts of urban structures and buildings on air quality

The air quality can vary substantially in different parts of the city. This is partly due to differences in local spot emissions (Apte et al., 2017; Zhou et al., 2017), traffic conditions (Amato et al., 2016; Huang et al., 2014; Samoli et al., 2019) and because of urban structures (Fig. 15.10) on the ventilation conditions in street canyons (Fu et al., 2017; Wolf et al., 2020). Street-level pollutants in an urban environment are dispersed and mixed into the upper layers by the wind in the street canyons. However, urban morphological characteristics can affect near-surface ventilation conditions and therefore also pollutant concentrations at a pedestrian level (Gal & Unger, 2009; Ng et al., 2011). In addition, good

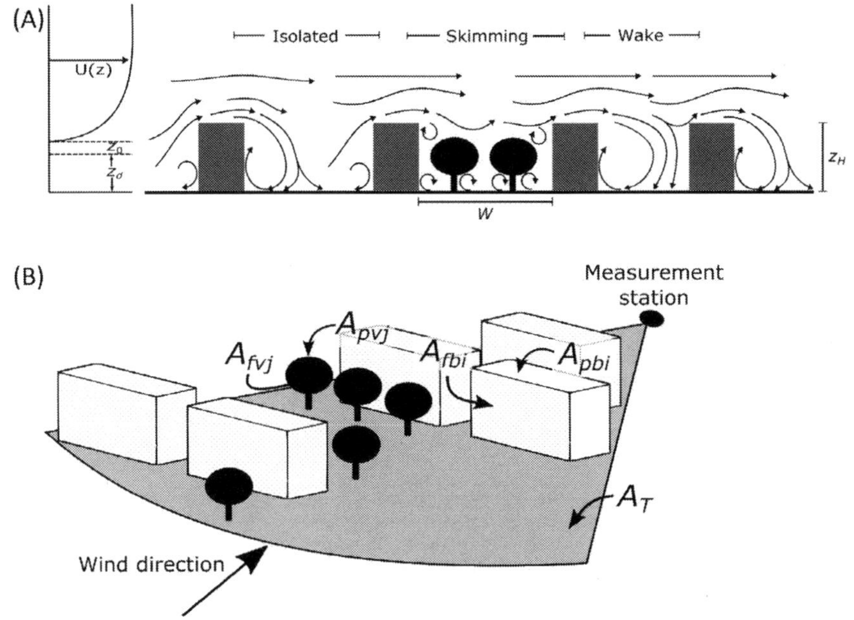

FIGURE 15.10 Processes influencing street canyon ventilation. Schematic illustration of (A) street canyon ventilation and the effect of street canyon aspect ratio and vegetation to ventilation conditions and (B) urban surfaces that affect the ventilation in street level and are used in the morphometric analyses. A_T is the total surface area, $A_{fv\ j}$ and $A_{fb\ i}$ are frontal area of individual tree or building, respectively, and $A_{pv\ j}$ and $A_{pb\ i}$ are plan area of individual tree or building, respectively. *From Kokkonen, T.V., Xie, Y., Paasonen, P., Gani, S., Jiang, L., Wang, B., Zhou, D., Qin, W., Nie, W., Kerminen, V.-M., Petäjä, T., Sun, J., Kulmala, M., & Ding, A. (2021). The effect of urban morphological characteristics on the spatial variation of PM2.5 air quality in downtown Nanjing. Environmental Science: Atmospheres, 1(7), 481–497. https://doi.org/10.1039/D1EA00035G. Creative Commons Attribution 3.0 Unported Licence.*

ventilation conditions in cities are beneficial not only in terms of air quality but also in terms of restricting the spreading of airborne transmitted diseases like COVID-19 (Morawska & Milton, 2020, Ng, 2009).

In recent decades, the effect of urban spatial morphological characteristics has gained more attention in urban climate research. The local climate zones proposed by Stewart and Oke (2012) is one of the essential studies emphasizing the importance of these characteristics in terms of urban climate. In recent years, there have been several studies, for example, using wind tunnels (Sharma et al., 2005), focusing on idealized street canyons and individual roughness elements (Xia et al., 2014), or by modeling (Kadaverugu et al., 2019). Edussuriya et al. (2011), studied the effect of 21 different morphological characteristics on the spatial variation of urban air quality. However, only six of the variables (complete aspect ratio, occlusivity, roughness height, zero-plane displacement height, total building volume/number of buildings, and standard deviation of building height) were found to vary significantly at the district level and therefore also responsible for the spatial variation of air quality within the urban environment. Some studies have found that the height-normalized roughness length was better representing the effect of surface roughness on the $PM_{2.5}$ air quality (e.g., Kokkonen et al., 2021). Many of these characteristics are widely used in estimating the near-surface ventilation conditions in urban environments, but for example, occlusivity requires very detailed data on buildings, which are not often available. Here we focus on the most used urban characteristics that are relatively easy to define using datasets that are commonly available for cities.

15.4.1 Roughness length

Sum of the roughness length (z_0) and zero-plane displacement height (z_d) define the height after which the boundary layer wind follows the logarithmic wind profile ($U(z)$; Fig. 15.10). The roughness length (z_0) has been identified as one of the most important urban morphological characteristics in terms of near-surface air quality, for example, by Edussuriya et al. (2014) and Kokkonen et al. (2021). There are several different methods to calculate the roughness length if proper micrometeorological observations are not available. Grimmond and Oke (1999) performed a rather comprehensive review and evaluation of nine different most used methods to estimate z_0. They found that most morphometric method formulas predicted reasonable values for small to medium densities, but some of the methods failed at medium to large densities. One of the most widely used morphometric formulas to estimate the roughness length is the MacDonald et al. (1998) as it functions rather well across the full range of urban surface morphometry typically found in cities (Grimmond & Oke, 1999). The height normalized roughness length based on the MacDonald et al. (1998) formula is

$$\frac{z_0}{z_H} = \left(1 - \frac{z_d}{z_H}\right)\exp\left\{-\left[0.5\beta\frac{C_d}{\kappa^2}\left(1-\frac{z_d}{z_H}\right)\lambda_F\right]^{-0.5}\right\} \quad (15.2)$$

where C_d is a drag coefficient for buildings = 1.2, κ is von Kármán's constant = 0.4, β is a drag correction coefficient, λ_F is the frontal area index, z_H is the mean building height, and z_d is zero-plane displacement height, which in MacDonald et al. (1998) is defined as

$$\frac{z_d}{z_H} = 1 + \alpha^{-\lambda_P}(\lambda_P - 1) \quad (15.3)$$

where α is an empirical coefficient and λ_P is the plan area fraction of buildings. The values for the empirical coefficients (α and β) have been estimated using wind tunnel experiments and the recommended values for staggered arrays of cubes, which is a fairly good assumption for urban areas, are $\alpha = 4.43$ and $\beta = 1.0$.

The standard deviation of building height has been found to increase the vertical turbulent flux rates (Kurppa et al., 2018) and therefore also to improve the near-surface air quality (Kurppa et al., 2018; Zhang et al., 2021).

Worked example 4: Aerodynamic roughness parameters

Calculate the roughness length for an urban area with a surface area of 1 km^2. The number of buildings in the area is 800. You can assume that the buildings are cube-shaped with a building height and width of 20 m. The buildings are aligned perpendicular to the wind direction.

Answer: Roughness length can be calculated in several ways (e.g., Grimmond & Oke, 1999). Here we are using the MacDonald et al. (1998) method which was given in the text. Then the height normalized roughness length is

$$\frac{z_0}{z_H} = \left(1 - \frac{z_d}{z_H}\right)\exp\left\{-\left[0.5\beta\frac{C_d}{\kappa^2}\left(1-\frac{z_d}{z_H}\right)\lambda_F\right]^{-0.5}\right\}$$

In the text it was given that C_d is a drag coefficient for buildings = 1.2, κ is von Kármán's constant = 0.4, β is a drag correction coefficient, λ_F is the frontal area index, z_H is the mean building height, and z_d is zero-plane displacement height, which in MacDonald et al. (1998) is defined as

$$\frac{z_d}{z_H} = 1 + \alpha^{-\lambda_P}(\lambda_P - 1),$$

where α is an empirical coefficient and λ_P is the plan area fraction of buildings. The values for the empirical coefficients (α and β) for staggered arrays of cubes were given in the text and they are $\alpha = 4.43$ and $\beta = 1.0$.

The plan area fraction is calculated as the fraction of the area of the buildings compared to the total area.

Therefore,

$$\lambda_P = \frac{800 \cdot (0.02 \cdot 0.02)}{1} = 0.32.$$

Thus, the height normalized zero-plane displacement height is

$$\frac{z_d}{z_H} = 1 + 4.43^{-0.32}(0.32 - 1) \approx 0.58$$

As the buildings are cube-shaped and the frontal area index is calculated as the fraction of the total vertical area of buildings perpendicular to wind compared to the total surface area, it is the same value as the plan area fraction (thus, $\lambda_F = 0.32$).

Then we can calculate the roughness length.

$$z_0 = (1 - 0.58)\exp\left\{-\left[0.5 \cdot 1\frac{1.2}{0.4^2}(1-0.58) \cdot 0.32\right]^{-0.5}\right\} \cdot 20 \approx 2\text{m}$$

15.4.2 Street canyon aspect ratio

Street canyon aspect ratio (λ_S) is the ratio of the mean building height to the width of the street (z_H/W). The street canyon aspect ratio is directly describing the effectiveness of the ventilation of the street canyon. The lower the ratio is, the better the ventilation of the street canyon is. When the wind is perpendicular to the street canyon, with $\lambda_S < 0.3$ the street canyon flows are isolated and the canyon is fairly well ventilated, while with approximately λ_S $0.3 < \lambda_S < 1.0$ there can occur wake effect flow and with $\lambda_S > 1.0$ there can be skimming flow with a substantially reduced ventilation of the street canyon (Fig. 15.10). The free flow area of the street canyon can be reduced also with obstacles, such as urban trees, within the street canyon, which can cause the skimming flow even if $\lambda_S < 1.0$ (Fig. 15.10a). The street canyon aspect ratio has been shown to be a very important factor in determining the ventilation conditions of the street canyons and therefore also the near-surface $PM_{2.5}$ air quality in many wind tunnel and modeling studies (Li et al., 2006; Voordeckers et al., 2021). However, these studies have usually focused on idealized street canyons without trees present. The ratio is calculated typically only accounting for the buildings and not the vegetation in the street canyon. Therefore, the street canyon aspect ratio alone does not represent very good ventilation conditions in real urban environments where trees are commonly present in street canyons (Gromke & Ruck, 2012; Kokkonen et al., 2021) and will reduce the freely ventilated space between the buildings.

15.4.3 Urban vegetation

The effect of trees on air quality in urban environments is still a hot research topic with controversial results. It is generally accepted that urban trees can purify air by absorbing gaseous pollutants and by deposition of particulate matter (Janhäll, 2015), so planting urban trees has been understood as an effective air pollution mitigation technique in cities. However, based on recent studies, trees can actually deteriorate the air quality on a local scale based on their influence on the flow dynamics and atmospheric chemistry. Trees are porous and flexible and can therefore absorb momentum and reduce the turbulent intensity and the velocity of the wind flow, even during the winter when there are no leaves, and therefore trees can deteriorate the air quality locally (Di Sabatino et al., 2015; Jeanjean et al., 2017; Karttunen et al., 2020; Vos et al., 2013; Xing & Brimblecombe, 2020). Furthermore, certain tree species can release significant amounts of BVOC (Calfapietra et al., 2013a; Sections 15.1 and 15.2), contributing to ozone formation in the atmosphere. However, there are still considerable uncertainties in the net impact of trees on air quality in different urban areas, for example in urban parks and street canyons.

Several studies have examined the effect of urban street trees on the near-surface wind velocities and $PM_{2.5}$ concentrations. The wind flow has been shown to decrease due to the street trees up to 79% during the leaf-on period and 39% during the leaf-off period when compared to rooftop observations (Di Sabatino et al., 2015). The increase in pedestrian level $PM_{2.5}$ concentration due to the decreased ventilation has been shown to be 7%–42% (Di Sabatino et al., 2015; Jeanjean et al., 2017; Karttunen et al., 2020; Kokkonen et al., 2021). Even though there is deposition of $PM_{2.5}$ on the trees, the aerodynamic effect decreasing the ventilation and accumulating pollutants under the canopy increases the particulate pollution in the street canyons much more than the cleaning effect by deposition cleaning the air.

In addition, even if trees can purify air by absorbing gaseous pollutants, at the same time certain tree species can release significant amounts of BVOC emissions (Calfapietra et al., 2013a). The emissions of isoprene contribute to the formation of near-surface ozone in urban areas, and monoterpenes and sesquiterpenes can increase $PM_{2.5}$ and PM_{10} concentrations (Churkina et al., 2017).

Due to the cleaning effect of vegetation, planting urban trees has often been seen as a solution to improve urban air quality (Jayasooriya et al., 2017) and in urban design, the green city idea with a lot of urban trees has been thought to be one with pure air. However, planting trees in street canyons in order to purify air in urban areas should be carefully reconsidered, since it tends to deteriorate the air quality locally. Especially, the use of urban trees for alleviating a local pollution hotspot should be avoided, since it would most likely just accumulate the emissions from the hotspot and make the local pollution even worse.

However, in the more open urban areas, such as urban parks, sports fields and parking lots, the trees have been shown to clean the air. The pollutant removal by the trees has even been shown to be more effective with higher concentrations of particulate pollution (Kokkonen et al., 2021; Tallis et al., 2011), and therefore it could work well, especially in highly polluted urban environments. Thus, urban planners should include large open areas within the cities with a high fraction of vegetation, for example, urban parks with trees and low vegetation (e.g., shrubs and grass). In addition, low vegetation (e.g., hedges), that do not capture the street level pollutants under the canopy, planted between the road traffic and pedestrian lanes in the street canyons have been shown to improve air quality in the pedestrian lanes and therefore protect the pedestrians from pollutants coming from the road (Abhijith et al., 2017).

Urban trees have an important role in controlling human thermal comfort in cities (Hami et al., 2019; Taleghani, 2018), particularly in cities that suffer from high temperatures during the summertime. In addition, urban trees have also been shown to improve the well-being of the citizens (van den Berg et al., 2015). Therefore, a substantial number of trees should be planted also in the street canyons to protect pedestrians from the heat and to make the urban areas more pleasant and liveable for the citizens. In order to minimize the deterioration effect on air quality, special attention and further research should be given to the choice of tree species, tree placement and tree heights. For example, variation in the tree height has been shown to be effective in increasing turbulence, compared to homogenous canopy height, therefore leading to a smaller deterioration of the ventilation and air quality at the pedestrian level (Karttunen et al., 2020). In addition, it is important to choose the correct tree species that can grow sustainably in urban environments, would have smaller BVOC emissions and would benefit the urban biodiversity in order to design and build healthier and more livable cities.

15.5 Air pollutant and climate implications

An increasing number of people live in urban environments (United Nations, 2014; Keith et al., 2023) and are exposed to cities' local climate (see Chapter 11). Urban areas commonly have high concentrations of atmospheric pollutants, increased heat stress, decreased evaporation and increased risk of flooding compared to surrounding natural areas. All these phenomena influence human well-being and one of the main aims of urban planning is to minimize their negative effects. Only with research-based knowledge, urban planners can design sustainable cities that can effectively adapt and mitigate the problems related to urban climate (Baklanov et al., 2018).

Exposure to atmospheric pollution causes up to 4 million premature deaths per year worldwide (Cohen et al., 2017; Lelieveld et al., 2015). In addition to health problems, air pollution is connected to climate change (Molina & Molina, 2004). The interactions span from local, regional and global scales. The consequences of air pollution in local and regional scale include negative impacts to the ecosystem due to deposition of acidic compounds, such as SO_2. Photochemical oxidation of the anthropogenic VOCs contributes to the formation of ozone and consequent oxidative stress to the vegetation, which can lead to elevated biogenic VOC emissions. Together with elevated primary emissions from anthropogenic activities to higher aerosol number and mass concentrations altering visibility and attenuation of incoming solar radiation. Therefore, there is a reduction of surface energy availability and turbulent heat fluxes (Kajino et al., 2017; Petäjä et al., 2016). With less energy available at the surface also evaporation decreases. The increasing atmospheric stability due to absorption of solar radiation by the aerosols can also lead to an increased risk of heavy rainfall increasing substantially the risk of flooding (Fan et al., 2015). For example, Ding et al. (2013) showed that the anthropogenic air pollution influenced regional precipitation patterns. In addition, even small rainfall totals flush pollutants from the air and from contaminated surfaces to surface runoff water (Yufen et al., 2008). Therefore, it is important to understand the impact of air quality on the hydrological cycle, and in urban scales, on runoff in particular. In larger spatial scales, these impacts expand to continental scale via aerosol-boundary layer – monsoon interactions (Ding et al. 2021).

In recent years the air quality in urban areas has been deteriorating and especially in developing countries the air pollution is a significant problem (Kan et al., 2012; Kulmala, 2015; Mayer, 1999; Molina and Molina, 2004). With continuing economic growth, the air pollution is estimated to increase even further (Zhao et al, 2013) with controlling air quality policies influencing the trends in the concentrations (Wang et al., 2017; Kuula et al., 2022). Overall, there are significant geographical differences depending on regional background concentrations and the pace of industrialization, urbanization and pollution control (Wang et al., 2017; Lorelei de Jesus et al. 2020; Ganguly et al. 2020). Even though important advances have been made in the past, new studies are needed to increase understanding of interactions between the air quality and regional and global climate and the hydrological cycle (Baklanov et al, 2018) particularly in the urban gigacity environments (Kulmala et al., 2021).

15.6 Questions

Question 1: Carbon exchange

Carbon exchange of a forest is typically measured in units $\mu mol\ m^{-2}\ s^{-1}$ (micromoles of CO_2 per square metre per second). Change the unit to $kg\ C^{-1}\ ha^{-1}\ day^{-1}$ (kilograms of C per hectare per day).

Question 2: Biogenic emissions versus anthropogenic emissions

A politician announced in the 1980s that the trees pollute more than cars. True or false? Take a stand and defend your argument based on the material provided in this chapter and discuss it with a fellow student.

Question 3: Biogenic emissions in a changing climate
Following the Solved Example 1, estimate in percentage, how much do the emissions from a single branch of *Larix cajanderi* tree increase, if the average temperature of a typical Siberian July of 19°C rises by 4°C?

Question 4: Biogenic emissions from Siberian Taiga
By how much would the BVOC emissions increase from Siberian Taiga in the next 50 years (in $Tg\,year^{-1}$), if the average temperature in this area is expected to increase by up to 4°C? The temperatures in the Arctic and boreal environment are expected to increase faster than the global mean (Rantanen et al., 2022). Discuss the assumptions, limitations, and consequences. To get background data for your numerical estimate, you may refer to the studies by Kajos et al. (2013), Shubuya et al. (2004), and Delcourt and Veraverbeke (2022).

Question 5: Biogenic secondary aerosol formation over boreal forests, the relative importance of different BVOCs
Let us consider the North-Western region of the boreal forest environment, covering essentially the land areas of Finland, Sweden and Norway.

In the forest air, what is the relative importance of isoprene, monoterpenes and sesquiterpes as groups of BVOC emitted by the forest? See Hellén et al. (2018) in order to make this estimate.

Question 6: Biogenic secondary aerosol formation and oxidation by ozone
The typical time scale for biogenic secondary organic aerosol formation (BSOA) in the boreal forest environment is about 2–3 days (Petäjä et al., 2022). Within this time scale, is it correct to assume that the dominant fraction of emitted monoterpes have reacted with ozone? see Hellén et al. (2018) and Atkinson and Arey (2003) for the relevant mix of monoterpenes and reaction rates, respectively. You may assume an ozone concentration of 30 ppb (about 7×10^{11} molecules cm^{-3}).

Question 7: Biogenic secondary aerosol formation and production of low volatility vapors
After ozone oxidation, the subsequent oxidation steps leading to very low-volatile compounds that will condense into preexisting aerosol particles to form BSOA are thought to be fast, and the lifetime of these compounds in the gas phase against BSOA formation is equal to the inverse of the condensation sink (CS). Considering typical values of CS in this boreal forest environment (see Neefjes et al., 2022), can we approximate that the dominant fraction of low-volatile oxidation products of monoterpenes will form BSOA within the same time scale of 2–3 days?

Question 8: Biogenic secondary aerosol formation, accumulated mass concentration
The mole fraction of BVOC that will eventually form BSOA via oxidation and subsequent condensation is the called the molar BSOA yield, and typical molar yields estimated for monoterpene oxidation vary from about 2% to values in excess of 10% in the literature (see Liao et al., 2014; Jokinen et al., 2015). Assuming the typical mixing layer height of 1 km over the boreal forest environment during the summer time, an ambient temperature of 10°C and BSOA molar yield of 5%, how large BSOA concentration (in $\mu g\,m^{-3}$) will form in the forest air during 2 days? Take BVOC emissions using the Eq. (15.1) and setting $\beta = 0.09°C^{-1}$ and $T_s = 30°C$, and by assuming that $E_{pool} = 1.5\,\mu g\,g_{dw}^{-1}\,h^{-1}$ and $g_{dw} = 1000\,g\,m^{-2}$. You may further assume that no BSOA is lost from the air via deposition and that the molecular weight of monoterpenes and its condensable reaction products are equal to 136 and 400 $g\,mol^{-1}$, respectively.

Question 9: Biogenic secondary aerosol formation at a higher temperature
Expansion from question 7, how large would the BSOA concentration would be at an ambient temperature of 25°C? How does the estimated mass concentrations compare with measured organic aerosol concentrations at the SMEAR II station in Hyytiälä (see Neefjes et al., 2022)?

Question 10: Biogenic secondary aerosol formation, contribution to CCN concentrations
The climatic effects of BSOA particles over the boreal forest environment depends essentially on whether these particles are able to grow into sizes large enough to act as cloud condensation nuclei (CCN), and especially on how large this CCN concentration is. At an ambient temperature of 10°C similar to Q7, what would be the BSOA particle concentration (in units particles cm^{-3}) after 2 days of BSOA formation if assuming that all BSOA particles have grown to the size of 100 nm and that the density of these particles is 1 $g\,cm^{-3}$? How would this concentration change if we assume sizes of 50 nm or 150 nm? How do these numbers compare with typical concentrations of Aitken and accumulation mode particles (potential CCN) measured in this boreal forest environment during the summer (see Neefjes et al., 2022)?

Question 11: Aspect ratio of street canyons
Calculate the street canyon aspect ratio (λ_S) for street canyon that has

1. building height of 20 m and street width of 100 m
2. building height of 20 m and street width of 50 m
3. building height of 100 m and street width of 80 m

Question 12: Flow patterns in the street canyons
Sketch the flow patterns in the street canyons in Q11, when the wind is coming perpendicular to the street.

Question 13: Logarithmic wind profile

The logarithmic wind profile illustrated in Fig. 15.10a describes the vertical wind profile above the roughness elements. It is widely used for example in pollutant dispersion models to estimate the vertical wind conditions. The logarithmic wind profile for urban areas under neutral atmospheric conditions can be described as $U(h_1) = (u^* / \kappa) \ln ((h_1 - z_d)/z_0)$, where U is the wind speed at a given height, u^* is the friction velocity, κ is the von Kármán constant, h_1 is the height, z_d is the zero-plane displacement height, and z_0 is the roughness length.

1. Calculate the wind speed at height of 30, 50, and 100 m when the observed wind at 25 m height is U(25 m) = 5.0 m s^{-1}. You can use the values for z_0 and z_d calculated in the Worked example 4 from this chapter. Assume neutral atmospheric conditions.
2. The wind speed is typically measured at standard height of 10 m. Consider what problems this might cause in dense high-rise urban areas.

Solutions to the end-of-chapter questions will be available at: https://www.elsevier.com/books-and-journals/book-companion/9780128225912.

References

Aalto, J., Kolari, P., Hari, P., Kerminen, V. M., Schiestl-Aalto, P., Aaltonen, H., Levula, J., Siivola, E., Kulmala, M., & Bäck, J. (2014). New foliage growth is a significant, unaccounted source for volatiles in boreal evergreen forests. *Biogeosciences*, *11*(5), 1331–1344.

Aalto, J., Porcar-Castell, A., Atherton, J., Kolari, P., Pohja, T., Hari, P., Nikinmaa, E., Petäjä, T., & Bäck, J. (2015). Onset of photosynthesis in spring speeds up monoterpene synthesis and leads to emission bursts. *Plant, Cell & Environment*, *38*(11), 2299–2312.

Abhijith, K. V., Kumar, P., Gallagher, J., McNabola, A., Baldauf, R., Pilla, F., Broderick, B., Di Sabatino, S., & Pulvirenti, B. (2017). Air pollution abatement performances of green infrastructure in open road and built-up street canyon environments—A review. *Atmospheric Environment*, *162*, 71–86. Available from https://doi.org/10.1016/j.atmosenv.2017.05.014.

Amato, F., Favez, O., Pandolfi, M., Alastuey, A., Querol, X., Moukhtar, S., Bruge, B., Verlhac, S., Orza, J. A. G., Bonnaire, N., Le Priol, T., Petit, J. F., & Sciare, J. (2016). Traffic induced particle resuspension in Paris: Emission factors and source contributions. *Atmospheric Environment*, *129*, 114–124. Available from https://doi.org/10.1016/j.atmosenv.2016.01.022.

Apte, J. S., Messier, K. P., Gani, S., Brauer, M., Kirchstetter, T. W., Lunden, M. M., Marshall, J. D., Portier, C. J., Vermeulen, R. C. H., & Hamburg, S. P. (2017). High-resolution air pollution mapping with google street view cars: Exploiting big data. *Environmental Science & Technology*, *51*(12), 6999–7008. Available from https://doi.org/10.1021/acs.est.7b00891.

Artaxo, P., Hansson, H.-C., Andreae, M. O., Bäck, J., Gomez Alves, E., Barbosa, H. M. J., Bender, F., Boursoukidis, E., Carbone, S., Chi, J., Decesari, S., Despés, V. R., Ditas, F., Ezhova, E., Fuzzi, S., Hasselquist, N. J., Heintzenberg, J., Holanda, B. A., Guenther, A., ... Kesselmeier, J. (2022). Tropical and boreal forest–Atmosphere interactions, a review. *Tellus B: Chemical and Physical Meteorology*, *74*, 24–163. Available from https://doi.org/10.16993/tellusb.34.

Atkinson, R., & Arey, J. (2003). Gas-phase tropospheric chemistry of biogenic volatile organic compounds: a review. *Atmos. Environ.*, *37*(Suppl. 2), S197–S219.

Baklanov, A., Grimmond, C. S. B., Carlson, D., Terblanche, D., Tang, X., Bouchet, V., Lee, B., Langendijk, G., Kolli, R. K., & Hovsepyan, A. (2018). From urban meteorology, climate and environment research to integrated city services. *Urban Climate*, *23*, 330–341. Available from https://doi.org/10.1016/j.uclim.2017.05.004.

Best, M. J., & Grimmond, C. S. B. (2016). Modeling the partitioning of turbulent fluxes at urban sites with varying vegetation cover. *Journal of Hydrometeorology*, *17*(10), 2537–2553. Available from https://doi.org/10.1175/Jhm-D-15-0126.1.

Boy, M., Zhou, P., Kurtén, T., Chen, D., Xavier, C., Clusius, P., Roldin, P., Baykara, M., Pichelstorfer, L., Foreback, B., Bäck, J., Petäjä, T., Makkonen, R., Kerminen, V.-M., Pihlatie, M., Aalto, J., & Kulmala, M. (2022). Positive feedback mechanism between biogenic volatile organic compounds and the methane lifetime in future climates. *NPJ Climate and Atmospheric Science*, *5*. Available from https://doi.org/10.1038/s41612-022-00292-0.

Bäck, J., Aalto, J., Henriksson, M., Hakola, H., He, Q., & Boy, M. (2012). Chemodiversity of a Scots pine stand and implications for terpene air concentrations. *Biogeosciences*, *9*, 689–702. Available from https://doi.org/10.5194/bg-9-689-2012.

Calfapietra, C., Fares, S., Manes, F., Morani, A., Sgrigna, G., & Loreto, F. (2013). Role of biogenic volatile organic compounds (BVOC) emitted by urban trees on ozone concentration in cities: A review. *Environmental Pollution*, *183*, 71–80. Available from https://doi.org/10.1016/j.envpol.2013.03.012.

Calfapietra, C., Pallozzi, E., Lusini, I., & Velikova, V. (2013). Modification of BVOC emissions by changes in atmospheric [CO_2] and air pollution. In Ü. Niinemets, & R. K. Monson (Eds.), *Biology, controls and models of tree volatile organiccompound emissions* (pp. 253–284). Dordrecht: Springer Netherlands.

Chu, B., Dada, L., Liu, Y. C., Yao, L., Wang, Y., Du, W., Cai, J., Dällenbach, K., Bianchi, F., Chen, X., Simonen, P., Zhou, Y., Fu, Y., Yin, R., Li, H., He, X.-C., Feng, Z., Yan, C., Kangasluoma, J., … Kulmala, M. (2020). Particle growth with photochemical age in new particle formation and haze events in the winter of Beijing, China. *Science and Total Environment*, *753*, 142207.

Chu, B., Kerminen, V.-M., Bianchi, F., Yan, C., Petäjä, T., & Kulmala, M. (2019). Atmospheric new particle formation in China. *Atmosphere Chemistry and Physics.*, *19*, 115–138.

Churkina, G., Kuik, F., Bonn, B., Lauer, A., Grote, R., Tomiak, K., & Butler, T. M. (2017). Effect of VOC emissions from vegetation on air quality in Berlin during a heatwave. *Environmental Science & Technology*, *51*(11), 6120–6130. Available from https://doi.org/10.1021/acs.est.6b06514.

Cohen, A. J., Brauer, M., Burnett, R., Anderson, H. R., Frostad, J., Estep, K., Balakrishnan, K., Brunekreef, B., Dandona, L., Dandona, R., Feigin, V., Freedman, G., Hubbell, B., Jobling, A., Kan, H., Knibbs, L., Liu, Y., Martin, R., Morawska, L., Pope, A. D., III., Shin, H., Straif, K., Shaddick, G., Thomas, M., van Dingenen, R., van Donkelaar, A., Vos, T., Murray, C. J. L., & Forouzanfar, M. H. (2017). Estimates and 25-year trends of the global burden of disease attributable to ambient air pollution: an analysis of data from the Global Burden of Diseases Study 2015. *The Lancet*, *389*(10082), 1907–1918. Available from https://doi.org/10.1016/S0140-6736(17)30505-6.

Delcourt, C. J. F., & Veraverbeke, S. (2022). Allometric equations and wood density parameters for estimating aboveground and woody debris biomass in Cajander larch (Larix cajanderi) forests of northeast Siberia. *Biogeosciences*, *19*, 4499–4520. Available from https://doi.org/10.5194/bg-19-4499-2022.

DeMott, P. J., Prenni, A. J., Liu, X., Kreidenweis, S. M., Petters, M. D., Twohy, C. H., Richardson, M. S., Eidhammera, T., & Rogers, D. (2010). Predicting global atmospheric ice nuclei distributions and their impacts on climate. *Proceedings of the National Academy of Sciences*, *107*(25), 11217–11222.

Després, V., Huffman, J. A., Burrows, S. M., Hoose, C., Safatov, A., Buryak, G., Fröhlich-Nowoisky, J., Elbert, W., Andreae, M. O., Pöschl, U., & Jaenicke, R. (2012). Primary biological aerosol particles in the atmosphere: A review. *Tellus B: Chemical and Physical Meteorology*, *64*(1), 15598.

Di Sabatino, S., Buccolieri, R., Pappaccogli, G., & Leo, L. S. (2015). The effects of trees on micrometeorology in a real street canyon: Consequences for local air quality. *International Journal of Environment and Pollution*, *58*(1-2), 100–111. Available from https://doi.org/10.1504/Ijep.2015.076587.

Ding, K., Huang, X., Ding, A., Wang, M., Su, H., Kerminen, V.-M., Petäjä, T., Tan, Z., Wang, Z., Zhou, D., Sun, J., Liao, H., Wang, H., Carslaw, K., Wood, R., Zuidema, P., Rosenfeld, D., Kulmala, M., Fu, C., Pöschl, U., Cheng, Y., & Andreae, M. O. (2021). Aerosol-boundary-layer-monsoon interactions amplify semi-direct effect of biomass smoke on low cloud formation in Southeast Asia. *Nature Communications*, *12*, 6416. Available from https://doi.org/10.1038/s41467-021-26728-4.

Ding, A. J., Fu, C. B., Yang, X. Q., Sun, J. N., Petäjä, T., Kerminen, V.-M., Wang, T., Xie, Y., Herrmann, E., Zheng, L. F., Nie, W., Liu, Q., Wei, X. L., & Kulmala, M. (2013). Intense atmospheric pollution modifies weather: A case of mixed biomass burning with fossil fuel combustion pollution in eastern China. *Atmospheric Chemistry and Physics*, *13*(20), 10545–10554. Available from https://doi.org/10.5194/acp-13-10545-2013.

Ding, A. J., Huang, X., Nie, W., Sun, J. N., Kerminen, V. M., Petäjä, T., Su, H., Cheng, Y. F., Yang, X. Q., Wang, M. H., Chi, X. G., Wang, J. P., Virkkula, A., Guo, W. D., Yuan, J., Wang, S. Y., Zhang, R. J., Wu, Y. F., Song, Y., … Fu, C. B. (2016). Enhanced haze pollution by black carbon in megacities in China. *Geophysical Research Letters*, *43*(6), 2873–2879. Available from https://doi.org/10.1002/2016gl067745.

Du, W., Dada, L., Zhao, J., Chen, X., Daellenbach, K. R., Xie, C., Wang, W., He, Y., Cai, J., Yao, L., Zhang, Y., Wang, Q., Xu, W., Wang, Y., Tang, G., Cheng, X., Kokkonen, T. V., Zhou, W., Yan, C., … Sun, Y. (2021). A 3D study on the amplification of regional haze and particle growth by local emissions. *NPJ Climate Atmosprere Science*, *4*, 4. Available from https://doi.org/10.1038/s41612-020-00156-5.

Edussuriya, P., Chan, A., & Malvin, A. (2014). Urban morphology and air quality in dense residential environments: Correlations between morphological parameters and air pollution at street-level. *Journal of Engineering Science and Technology*, *9*(1), 64–80.

Edussuriya, P., Chan, A., & Ye, A. (2011). Urban morphology and air quality in dense residential environments in Hong Kong. Part I: District-level analysis. *Atmospheric Environment*, *45*(27), 4789–4803. Available from https://doi.org/10.1016/j.atmosenv.2009.07.061.

Ehn, M., Thornton, J. A., Kleist, E., Sipilä, M., Junninen, H., Pullinen, I., Springer, M., Rubach, F., Tillmann, R., Lee, B., Lopez-Hilfiker, F., Andres, S., Acir, I.-H., Rissanen, M., Jokinen, T., Schobesberger, S., Kangasluoma, J., Kontkanen, J., Nieminen, T., … Mentel, T. F. (2014). A large source of low-volatility secondary organic aerosol. *Nature*, *506*(7489), 476–479.

Ezhova, E., Ylivinkka, I., Kuusk, J., Komsaare, K., Vana, M., Krasnova, A., Noe, S., Arshinov, M., Belan, B., Park, S.-B., Lavrič, J. V., Heimann, M., Petäjä, T., Vesala, T., Mammarella, I., Kolari, P., Bäck, J., Rannik, Ü., Kerminen, V.-M., & Kulmala, M. (2018). Direct effect of aerosols on solar radiation and gross primary production in boreal and hemiboreal forests. *Atmospheric Chemistry and Physics*, *18*(24), 17863–17881.

Faiola, C., & Taipale, D. (2020). Impact of insect herbivory on plant stress volatile emissions from trees: A synthesis of quantitative measurements and recommendations for future research. *Atmospheric Environment: X*, *5*, 100060.

Fan, J., Rosenfeld, D., Yang, Y., Zhao, C., Leung, L. R., & Li, Z. (2015). Substantial contribution of anthropogenic air pollution to catastrophic floods in Southwest China. *Geophysical Research Letters*, *42*, 6066–6075. Available from https://doi.org/10.1002/2015GL064479.

Fan, Y. F., Hunt, J. C. R., & Li, Y. G. (2017a). Buoyancy and turbulence-driven atmospheric circulation over urban areas. *Journal of Environmental Sciences-China*, *59*, 63–71. Available from https://doi.org/10.1016/j.jes.2017.01.009.

Fan, Y. F., Li, Y. G., Bejan, A., Wang, Y., & Yang, X. Y. (2017b). Horizontal extent of the urban heat dome flow. *Scientific Reports*, *7*, 11681. Available from https://doi.org/10.1038/s41598-017-09917-4.

FAO & UNEP (2020). *The state of the world's forests 2020. Forests, biodiversity and people. Rome.* https://doi.org/10.4060/ca8642en.

Fisher, S., Bellinger, D. C., Cropper, M., Kumar, P., Binagwaho, A., Koudenoukpo, J. B., Park, Y., Taghian, G., & Landrigan, P. J. (2021). Air pollution and development in Africa: impacts on health, the economy, and human capital. *The Lancet*, *5*(10), e681–e688. Available from https://doi.org/10.1016/S2542-5196(21)00201-1.

Fu, X. W., Liu, J. F., Ban-Weiss, G. A., Zhang, J. C., Huang, X., Ouyang, B., Popoola, O., & Tao, S. (2017). Effects of canyon geometry on the distribution of traffic-related air pollution in a large urban area: Implications of a multicanyon air pollution dispersion model. *Atmospheric Environment*, *165*, 111–121. Available from https://doi.org/10.1016/j.atmosenv.2017.06.031.

Gal, T., & Unger, J. (2009). Detection of ventilation paths using high-resolution roughness parameter mapping in a large urban area. *Building and Environment*, *44*(1), 198–206. Available from https://doi.org/10.1016/j.buildenv.2008.02.008.

Ganguly, T., Selvaraj, K. L., & Guttikunda, S. K. (2020). National Clean Air Programme (NCAP) for Indian cities: Review and outlook of clean air action plans. *Atmospheric Environment: X*, *8*100096. Available from https://doi.org/10.1016/j.aeaoa.2020.100096.

Gao, M., Han, Z. W., Tao, Z. N., Li, J. W., Kang, J. E., Huang, K., Dong, X. Y., Zhuang, B. L., Li, S., Ge, B. Z., Wu, Q. Z., Lee, H. J., Kim, C. H., Fu, J. S., Wang, T. J., Chin, M., Li, M., Woo, J. H., Zhang, Q., … Carmichael, G. R. (2020). Air quality and climate change, topic 3 of the model inter-comparison study for Asia phase III (MICS-Asia III)—Part 2: Aerosol radiative effects and aerosol feedbacks. *Atmospheric Chemistry and Physics*, *20*(2), 1147–1161. Available from https://doi.org/10.5194/acp-20-1147-2020.

Grimmond, C. S. B., Blackett, M., Best, M. J., Baik, J. J., Belcher, S. E., Beringer, J., Bohnenstengel, S. I., Calmet, I., Chen, F., Coutts, A., Dandou, A., Fortuniak, K., Gouvea, M. L., Hamdi, R., Hendry, M., Kanda, M., Kawai, T., Kawamoto, Y., Kondo, H., … Zhang, N. (2011). Initial results from phase 2 of the international urban energy balance model comparison. *International Journal of Climatology*, *31*(2), 244–272. Available from https://doi.org/10.1002/joc.2227.

Grimmond, C. S. B., Blackett, M., Best, M. J., Barlow, J., Baik, J. J., Belcher, S. E., Bohnenstengel, S. I., Calmet, I., Chen, F., Dandou, A., Fortuniak, K., Gouvea, M. L., Hamdi, R., Hendry, M., Kawai, T., Kawamoto, Y., Kondo, H., Krayenhoff, E. S., Lee, S. H., … Zhang, N. (2010). The international urban energy balance models comparison project: First results from phase 1. *Journal of Applied Meteorology and Climatology*, *49*(6), 1268–1292. Available from https://doi.org/10.1175/2010jamc2354.1.

Grimmond, C. S. B., & Oke, T. R. (1986). Urban water-balance.2. Results from a suburb of vancouver, British-Columbia. *Water Resources Research*, *22*(10), 1404–1412. Available from https://doi.org/10.1029/WR022i010p01404.

Grimmond, C. S. B., & Oke, T. R. (1999). Aerodynamic properties of urban areas derived, from analysis of surface form. *Journal of Applied Meteorology*, *38*(9), 1262–1292. Available from https://doi.org/10.1175/1520-0450(1999)038 < 1262:Apouad > 2.0.Co;2.

Gromke, C., & Ruck, B. (2012). Pollutant concentrations in street canyons of different aspect ratio with avenues of trees for various wind directions. *Boundary-Layer Meteorology*, *144*(1), 41–64. Available from https://doi.org/10.1007/s10546-012-9703-z.

Gu, L., Baldocchi, D., Verma, S. B., Black, T., Vesala, T., Falge, E. M., & Dowty, P. R. (2002). Advantages of diffuse radiation for terrestrial ecosystem productivity. *Journal of Geophysical Research: Atmospheres*, *107*(D6), ACL 2-1–ACL 2-23.

Guenther, A., Hewitt, C. N., Erickson, D., Fall, R., Geron, C., Graedel, T., Harley, T., Klinger, L., Lerday, M., McKay, W. A., Pierce, T., Scholes, B., Steinbrecher, R., Tallamraju, R., Taylor, J., & Zimmerman, P. (1995). A global model of natural volatile organic compound emissions. *Journal of Geophysical Research*, *100*, 8873–8892.

Guenther, A. B., Jiang, X., Heald, C. L., Sakulyanontvittaya, T., Duhl, T. A., Emmons, L. K., & Wang, X. (2012). The model of emissions of gases and aerosols from nature version 2.1 (MEGAN2. 1): An extended and updated framework for modeling biogenic emissions. *Geoscientific Model Development*, *5*(6), 1471–1492.

Guenther, A. B., Zimmerman, P. R., Harley, P. C., Monson, R. K., & Fall, R. (1993). Isoprene and monoterpene emission rate variability: Model evaluations and sensitivity analyses. *journal of geophysical research*, *98*, 12609–12617. Available from https://doi.org/10.1029/93JD00527.

Hami, A., Abdi, B., Zarehaghi, D., & Bin Maulan, S. (2019). Assessing the thermal comfort effects of green spaces: A systematic review of methods, parameters, and plants' attributes. *Sustainable Cities and Society*, *49*, 101634. Available from https://doi.org/10.1016/j.scs.2019.101634.

Hari, P., & Kulmala, M. (2005). Station for measuring ecosystem–Atmosphere relations (SMEAR II). *Boreal Environment Research*, *10*, 315–322.

Harrison, S. P., Morfopoulos, C., Dani, K. G. S., Prentice, I. C., Arneth, A., Atwell, B. J., Barkley, M. P., Leihman, M. R., Loreto, F., Medlyn, B. E., Niinemets, Ü., Possell, M., Peñuelas, J., & Wright, I. J. (2013). Volatile isoprenoid emissions from plastid to planet. *New Phytologist*, *197*, 49–57. Available from https://doi.org/10.1111/nph.12021.

Heinritzi, M., Dada, L., Simon, M., Stolzenburg, D., Wagner, A. C., Fischer, L., Ahonen, L. R., Amanatidis, S., Baalbaki, R., Baccarini, A., Bauer, P. S., Baumgartner, B., Bianchi, F., Brilke, S., Chen, D., Chiu, R., Dias, A., Dommen, J., Duplissy, J., … Curtius, J. (2020).). Molecular understanding of the suppression of new-particle formation by isoprene. *Atmospheric Chemistry and Physics*, *20*(20), 11809–11821.

Hellén, H., Praplan, A. P., Tykkä, T., Ylivinkka, I., Vakkari, V., Bäck, J., Petäjä, T., Kulmala, M., & Hakola, H. (2018). Long-term measurements of volatile organic compounds highlight the importance of sesquiterpenes for the chemistry of a boreal forest. *Atmos. Chem. Phys.*, *18*, 13839–13863.

Holopainen, J. K., Nerg, A. M., & Blande, J. D. (2013). Multitrophic signalling in polluted atmospheres. In Ü. Niinemets, & R. Monson (Eds.), Biology, controls and models of tree volatile organic compound emissions. *Tree Physiology* (5). Dordrecht: Springer. Available from https://doi.org/10.1007/978-94-007-6606-8_11.

Holopainen, J. K., Virjamo, V., Ghimire, R. P., Blande, J. D., Julkunen-Tiitto, R., & Kivimäenpää, M. (2018). Climate change effects on secondary compounds of forest trees in the northern Hemisphere. *Frontiers in Plant Science*, *9*, 1445.

Huang, R. J., Zhang, Y. L., Bozzetti, C., Ho, K. F., Cao, J. J., Han, Y. M., Daellenbach, K. R., Slowik, J. G., Platt, S. M., Canonaco, F., Zotter, P., Wolf, R., Pieber, S. M., Bruns, E. A., Crippa, M., Ciarelli, G., Piazzalunga, A., Schwikowski, M., Abbaszade, G., … Prevot, A. S. H. (2014). High secondary aerosol contribution to particulate pollution during haze events in China. *Nature*, *514*(7521), 218–222. Available from https://doi.org/10.1038/nature13774.

Hänninen, H., & Tanino, K. (2011). Tree seasonality in a warming climate. *Trends in Plant Science*, *16*(8), 412−416.

Ilvesniemi, H., Levula, J., Ojansuu, R., Kolari, P., Kulmala, L., Pumpanen, J., Launiainen, S., Vesala, T., & Nikinmaa, E. (2009). Long-term measurements of the carbon balance of a boreal Scots pine dominated forest ecosystem. *Boreal Environment Research*, *14*, 731−753.

IPCC. (2021). *Climate change 2021: The physical science basis. contribution of working group I to the sixth assessment report of the intergovernmental panel on climate change*. Cambridge, United Kingdom and New York, NY, USA: Cambridge University Press.

Janhäll, S. (2015). Review on urban vegetation and particle air pollution—Deposition and dispersion. *Atmospheric Environment*, *105*, 130−137. Available from https://doi.org/10.1016/j.atmosenv.2015.01.052.

Jayasooriya, V. M., Ng, A. W. M., Muthukumaran, S., & Perera, B. J. C. (2017). Green infrastructure practices for improvement of urban air quality. *Urban Forestry & Urban Greening*, *21*, 34−47. Available from https://doi.org/10.1016/j.ufug.2016.11.007.

Jeanjean, A. P. R., Buccolieri, R., Eddy, J., Monks, P. S., & Leigh, R. J. (2017). Air quality affected by trees in real street canyons: The case of Marylebone neighbourhood in central London. *Urban Forestry & Urban Greening*, *22*, 41−53. Available from https://doi.org/10.1016/j.ufug.2017.01.009.

Järvi, L., Grimmond, C. S. B., & Christen, A. (2011). The surface urban energy and water balance scheme (SUEWS): Evaluation in Los Angeles and vancouver. *Journal of Hydrology*, *411*(3−4), 219−237. Available from https://doi.org/10.1016/j.jhydrol.2011.10.001.

Jögiste, K., Kulmala, L., Bäck, J., & Hari, P. (2008). Boreal zone. In P. Hari, & L. Kulmala (Eds.), *Boreal Forest and Climate Change. Advances in Global Change Research* (34). Dordrecht: Springer.

Jokinen, T., Berndt, T., Makkonen, R., Kerminen, V.-M., Junninen, H., Paasonen, P., Stratmann, F., Herrmann, H., Guenther, A., Worsnop, D. R., Kulmala, M., Ehn, M., & Sipilä, M. (2015). Production of extremely low-volatile organic compounds from biogenic emissions: measured yields and atmospheric implications. *Proc. Nat. Acad. Sci.*, *112*, 7123−7128.

Kadaverugu, R., Sharma, A., Matli, C., & Biniwale, R. (2019). High resolution urban air quality modeling by coupling CFD and mesoscale models: A review. *Asia-Pacific Journal of Atmospheric Sciences*, *55*(4), 539−556. Available from https://doi.org/10.1007/s13143-019-00110-3.

Kajino, M., Ueda, H., Han, Z. W., Rei, K. D., Inomata, Y., & Kaku, H. (2017). Synergy between air pollution and urban meteorological changes through aerosol-radiation-diffusion feedback. A case study of Beijing in January 2013. *Atmospheric Environment*, *171*, 98−110. Available from https://doi.org/10.10164/j.atmosenv.2017.10.018.

Kajos, M. K., Hakola, H., Holst, T., Nieminen, T., Tarvainen, V., Maximov, T., Petäjä, T., Arneth, A., & Rinne, J. (2013). Terpenoid emissions from fully grown east Siberian *Larix cajanderi* trees. *Biogeosciences*, *10*, 4705−4719. Available from https://doi.org/10.5194/bg-10-4705-2013.

Kan, H., Chen, R., & Tong, S. (2012). Ambient air pollution, climate change, and population health in China. *Environment International*, *42*, 10−19. Available from https://doi.org/10.1016/j.envint.2011.03.003.

Karttunen, S., Kurppa, M., Auvinen, M., Hellsten, A., & Järvi, L. (2020). Large-eddy simulation of the optimal street-tree layout for pedestrian-level aerosol particle concentrations—A case study from a city-boulevard. *Atmospheric Environment-X*, *6*100073. Available from https://doi.org/10.1016/j.aeaoa.2020.100073.

Keith, M., Birch, E., Bouchoud, N. J. A., Cardama, M., Cobbett, W., Cohen, M., Elmqvist, T., Espey, J., Hajer, M., Martmann, G., Matsumoto, T., Parnell, S., Revi, A., Roberts, D. C., Saiz, E., Schwanen, T., Seto, K. C., Tuts, R., & wan der Pütten, M. (2023). A new urban narrative for sustainable development. *Nature Sustainability*, *6*, 115−117. Available from https://doi.org/10.1038/s41893-022-00979-5.

Kerminen, V.-M., Chen, X., Vakkari, V., Petäjä, T., Kulmala, M., & Bianchi, F. (2018). Atmospheric new particle formation and growth: Review of field observations. *Environmental Research Letters*, *13*, 103003.

Kokkonen, T. V., Grimmond, C. S. B., Räty, O., Ward, H. C., Christen, A., Oke, T. R., Kotthaus, S., & Järvi, L. (2018). Sensitivity of surface urban energy and water balance scheme (SUEWS) to downscaling of reanalysis forcing data. *Urban Climate*, *23*, 36−52. Available from https://doi.org/10.1016/j.uclim.2017.05.001.

Kokkonen, T. V., Grimmond, S., Murto, S., Liu, H. Z., Sundström, A. M., & Järvi, L. (2019). Simulation of the radiative effect of haze on the urban hydrological cycle using reanalysis data in Beijing. *Atmospheric Chemistry and Physics*, *19*(10), 7001−7017. Available from https://doi.org/10.5194/acp-19-7001-2019.

Kokkonen, T. V., Xie, Y., Paasonen, P., Gani, S., Jiang, L., Wang, B., Zhou, D., Qin, W., Nie, W., Kerminen, V.-M., Petäjä, T., Sun, J., Kulmala, M., & Ding, A. (2021). The effect of urban morphological characteristics on the spatial variation of $PM_{2.5}$ air quality in downtown Nanjing. *Environmental Science: Atmospheres*, *1*(7), 481−497. Available from https://doi.org/10.1039/D1EA00035G.

Kolari, P., Kulmala, L., Pumpanen, J., Launiainen, S., Ilvesniemi, H., Hari, P., & Nikinmaa, E. (2009). CO_2 exchange and component CO_2 fluxes of a boreal Scots pine forest. *Boreal Environment Research*, *14*, 761−783.

Kolari, P., Pumpanen, J., Kulmala, L., Ilvesniemi, H., Nikinmaa, E., Grönholm, T., & Hari, P. (2006). Forest floor vegetation plays an important role in photosynthetic production of boreal forests. *Forest Ecology and Management*, *221*, 241−248.

Kondo, H., Genchi, Y., Kikegawa, Y., Ohashi, Y., Yoshikado, H., & Komiyama, H. (2005). Development of a multi-layer urban canopy model for the analysis of energy consumption in a big city: structure of the urban canopy model and its basic performance. *Boundary-Layer Meteorology*, *116*, 395−421. Available from https://doi.org/10.1007/s10546-005-0905-5.

Kreuzwieser, J., & Rennenberg, H. (2013). Flooding-driven emissions from trees. In Ü. Niinemets, & R. Monson (Eds.), Biology, Controls and Models of Tree Volatile Organic Compound Emissions. *Tree Physiology* (5). Dordrecht: Springer. Available from https://doi.org/10.1007/978-94-007-6606-8_9.

Kulmala, M., Pekkanen, J., Kokkonen, T. V., Paatero, S., Petäjä, T., Kerminen, V.-M., & Ding, A. J. (2021). Opinion: Gigacity: a source of problems or the new way to sustainable development. *Atmospheric Chemistry and Physics*, *21*, 8313−8322. Available from https://doi.org/10.5194/acp-21-8313-2021.

Kulmala, M. (2015). Atmospheric chemistry: China's choking cocktail. *Nature*, *526*, 497−499. Available from https://doi.org/10.1038/526497a.

Kuula, J., Timonen, H., Niemi, J. V., Manninen, H., Rönkkö, T., Hussein, T., Fung, P. L., Tarkoma, S., Laakso, M., Saukko, E., Ovaska, A., Kulmala, M., Karppinen, A., Johansson, L., & Petäjä, T. (2022). Opinion: Insights into updating Ambient Air Quality Directive 2008/50/EC. *Atmospheric Chemistry and Physics*, *22*, 4801–4808. Available from https://doi.org/10.5194/acp-22-4801-2022.

Kulmala, M., Cai, R., Stolzenburg, D., Zhou, Y., Dada, L., Guo, Y., Yan, C., Petäjä, T., Jiang, J., & Kerminen, V.-M. (2022). The contribution of new particle formation and subsequent growth to haze formation. *Environental Science Atmospheres*, *2*, 352–361. Available from https://doi.org/10.1039/D1EA00096A.

Kulmala, M., Ezhova, E., Kalliokoski, T., Noe, S., Vesala, T., Lohila, A., Liski, J., Makkonen, R., Bäck, J., Petäjä, T., & Kerminen, V.-M. (2020). CarbonSink + . *Boreal Environment Research.*, *25*, 145–159.

Kulmala, M., Kerminen, V.-M., Petäjä, T., Ding, A. J., & Wang, L. (2017). Atmospheric gas-to-particle conversion: Why NPF events are observed in megacities? *Faraday Discussions*, *200*, 271–288. Available from https://doi.org/10.1039/c6fd00257a.

Kulmala, M., Petäjä, T., Ehn, M., Thornton, J., Sipilä, M., Worsnop, D. R., & Kerminen, V.-M. (2014). Chemistry of atmospheric nucleation: On the recent advances on precursor characterization and atmospheric cluster composition in connection with atmospheric new particle formation. *Annual Review of Physical Chemistry*, *65*, 21–37.

Kulmala, M., Nieminen, T., Nikandrova, A., Lehtipalo, K., Manninen, H. E., Kajos, M. K., Kolari, P., Lauri, A., Petäjä, T., Krejci, R., Hansson, H.-C., Swietlicki, E., Lindroth, A., Christensen, T. R., Arneth, A., Hari, P., Bäck, J., Vesala, T., & Kerminen, V.-M. (2014). CO_2-induced terrestrial climate feedback mechanism: From carbon sink to aerosol source and back. *Boreal Environment Research*, *19*(suppl.B)), 122–131.

Kulmala, M., Vehkamäki, H., Petäjä, T., Dal Maso, M., Lauri, A., Kerminen, V.-M., Birmili, W., & McMurry, P. H. (2004). Formation and growth rates of ultrafine atmospheric particles: A review of observations. *Journal of Aerosol Science*, *35*, 143–176.

Kurppa, M., Hellsten, A., Auvinen, M., Raasch, S., Vesala, T., & Järvi, L. (2018). Ventilation and air quality in city blocks using large-eddy simulationurban planning perspective. *Atmosphere*, *9*(2), 65. Available from https://doi.org/10.3390/atmos9020065.

Kusaka, H., Kondo, H., Kikegawa, Y., & Kimura, F. (2001). A simple single-layer urban canopy model for atmospheric models: Comparison with multilayer and slab models. *Bound.-Layer Meteorology*, *101*(3), 329–358.

Lehmann, P., Ammunét, T., Barton, M., Battisti, A., Eigenbrode, S. D., Jepsen, J. U., Kalinkat, G., Neuvonen, S., Niemelä, P., Terblanche, J. S., Økland, B., & Björkman, C. (2020). Complex responses of global insect pests to climate warming. *Frontiers in Ecology and the Environment*, *18*(3), 141–150.

Lelieveld, J., Evans, J. S., Fnais, M., Giannadaki, D., & Pozzer, A. (2015). The contribution of outdoor air pollution sources to premature mortality on a global scale. *Nature*, *525*, 367–371. Available from https://doi.org/10.1038/nature15371.

Liao, L., Kerminen, V.-M., Boy, M., Kulmala, M., & Dal Maso, M. (2014). Temperature influence on the natural aerosol budget over boreal forests. *Atmos. Chem. Phys.*, *14*, 8295–8308.

Li, T., Holst, T., Michelsen, A., & Rinnan, R. (2019). Amplification of plant volatile defence against insect herbivory in a warming Arctic tundra. *Nature Plants*, *5*(6), 568–574.

Li, X. X., Liu, C. H., Leung, D. Y. C., & Lam, K. M. (2006). Recent progress in CFD modelling of wind field and pollutant transport in street canyons. *Atmospheric Environment*, *40*(29), 5640–5658. Available from https://doi.org/10.1016/j.atmosenv.2006.04.055.

Lin, Z., Wang, Y., Zheng, F., Zhou, Y., Guo, Y., Feng, Z., Li, C., Zhang, Y., Hakala, S., Chan, T., Yan, C., Daellenbach, K., Chu, B., Dada, L., Kangasluoma, J., Yao, L., Fan, X., Du, W., Cai, J., ... Kulmala, M. (2021). Rapid mass growth and enhanced light extinction of atmospheric aerosols during the heating season haze episodes in Beijing revealed by aerosol-chemistry-boundary layer interaction. *Atmospheric Chemistry.and Physics*, *21*, 12173–12187. Available from https://doi.org/10.5194/acp-21-12173-2021.

Liu, Q., Jia, X., Quan, J., Li, J., Li, X., Wu, Y., Chen, D., Wang, Z., & Liu, Y. (2018). New positive feedback mechanism between boundary layer meteorology and secondary aerosol formation during severe haze events. *Scientific Reports*, *8*(1), 6095. Available from https://doi.org/10.1038/s41598-018-24366-3.

Lorelei de Jesus, A., Thompson, H., Knibbs, L. D., Kowalski, M., Cyrys, J., Niemi, J. V., Kousa, A., Timonen, H., Luoma, K., Petäjä, T., Beddows, D., Harrison, R. M., Hopke, P., & Morawska, L. (2020). Long-term trends in PM2.5 mass and particle number concentrations in urban air: the impacts of mitigation measures and changing climates. *Environmental Pollution*, *263*114500. Available from https://doi.org/10.1016/j.envpol.2020.114500, Avaiable from.

MacDonald, R. W., Griffiths, R. F., & Hall, D. J. (1998). An improved method for the estimation of surface roughness of obstacle arrays. *Atmospheric Environment*, *32*(11), 1857–1864. Available from https://doi.org/10.1016/S1352-2310(97)00403-2.

Masson, V., Le Moigne, P., Martin, E., Faroux, S., Alias, A., Alkama, R., Belamari, S., Barbu, A., Boone, A., Bouyssel, F., Brousseau, P., Brun, E., Calvet, J.-C., Carrer, D., Decharme, B., Delire, C., Donier, S., Essaouini, K., Gibelin, A.-L., ... Voldoire, A. (2013). The SURFEXv7.2 land and ocean surface platform for coupled or offline simulation of earth surface variables and fluxes. *Geoscientific Model Develoment.*, *6*(4), 929–960.

Mayer, H. (1999). Air pollution in cities. *Atmospheric Environment*, *33*(24-25), 4029–4037. Available from https://doi.org/10.1016/S1352-2310(99)00144-2.

Molina, M. J., & Molina, L. T. (2004). Megacities and Atmospheric Pollution. *Journal of the Air & Waste Management Association*, *54*(6), 644–680. Available from https://doi.org/10.1080/10473289.2004.10470936.

Miao, Y., Liu, S., Zheng, Y., Wang, S., Chen, B., Zheng, H., & Zhao, J. (2015). Numerical study of the effects of local atmospheric circulations on a pollution event over Beijing–Tianjin–Hebei, China. *Journal of Environmental Sciences*, *30*, 9–20. Available from https://doi.org/10.1016/j.jes.2014.08.025.

Mitchell, V., Mein, R., & McMahon, T. (2001). Modelling the urban water cycle. *Environmental Modelling Software*, *16*(7), 615–629.

Mitchell, V. G., McMahon, T. A., & Mein, R. G. (2003). Components of the total water balance of an urban catchment. *Environmental Management*, *32*(6), 735–746. Available from https://doi.org/10.1007/s00267-003-2062-2.

Morawska, L., & Milton, D. K. (2020). It is time to address airborne transmission of coronavirus disease 2019 (COVID-19). *Clinical Infectious Diseases*. Available from https://doi.org/10.1093/cid/ciaa939.

Myrup, L. O. (1969). A numerical model of the urban heat island. *Journal of Applied Meteorology*, *8*(6), 908–918.

Mäki, M., Aalto, J., Hellén, H., Pihlatie, M., & Bäck, J. (2019). Interannual and seasonal dynamics of volatile organic compound fluxes from the boreal forest floor. *Frontiers in plant science*, *10*(191), 1–13.

Neefjes, I., Laapas, M., Liu, Y., Medus, E., Miettunen, E., Ahonen, L., Quelever, L., Aalto, J., Bäck, J., Kerminen, V.-M., Lampilahti, J., Luoma, K., Mäki, M., Mammarella, I., Petäjä, T., Räty, M., Sarnela, N., Ylivinkka, I., Hakala, S., Kulmala, M., Nieminen, T., & Lintunen, A. (2022). 25 years of atmospheric and ecosystem measurements in a boreal forest – Seasonal variation and responses to warm and dry years. *Boreal Env. Res*, *27*, 1–31.

Ng, E. (2009). Policies and technical guidelines for urban planning of high-density cities—Air ventilation assessment (AVA) of Hong Kong. *Building and Environment*, *44*(7), 1478–1488. Available from https://doi.org/10.1016/j.buildenv.2008.06.013.

Ng, E., Yuan, C., Chen, L., Ren, C., & Fung, J. C. H. (2011). Improving the wind environment in high-density cities by understanding urban morphology and surface roughness: A study in Hong Kong. *Landscape and Urban Planning*, *101*(1), 59–74. Available from https://doi.org/10.1016/j.landurbplan.2011.01.004.

Niinemets, Ü. (2010). Responses of forest trees to single and multiple environmental stresses from seedlings to mature plants: Past stress history, stress interactions, tolerance and acclimation. *Forest Ecology and Management*, *260*(10), 1623–1639.

Oke, T. R., Mills, G., Christen, A., & Voogt, J. A. (2017). *Urban climates*. Cambridge, UK: Cambridge University Press.

Oleson, K. W., Bonan, G. B., Feddema, J., Vertenstein, M., & Grimmond, C. S. B. (2008). An urban parameterization for a global climate model. Part I: Formulation and evaluation for two cities. *J. Appl. Meteorol. Clim.*, *47*(4), 1038–1060.

Petäjä, T., Järvi, L., Kerminen, V. M., Ding, A. J., Sun, J. N., Nie, W., Kujansuu, J., Virkkula, A., Yang, X. Q., Fu, C. B., Zilitinkevich, S., & Kulmala, M. (2016). Enhanced air pollution via aerosol-boundary layer feedback in China. *Scientific Reports*, *6*18998. Available from https://doi.org/10.1038/srep18998.

Petäjä, T., Tabakova, K., Manninen, A., Ezhova, E., O'Connor, E., Moisseev, D., Sinclair, V. A., Backman, J., Levula, J., Luoma, K., Virkkula, A., Paramonov, M., Räty, M., Äijälä, M., Heikkinen, L., Ehn, M., Sipilä, M., Yli-Juuti, T., Virtanen, A., ... Kerminen, V.-M. (2022). Influence of biogenic emissions from boreal forests on aerosol–Cloud interactions. *Nature Geoscience 15*, 42–47. Available from https://doi.org/10.1038/s41561-021-00876-0.

Peña-Guzman, C. A., Melgarejo, J., Prats, D., Torres, A., & Martinez, S. (2017). Urban water cycle simulation/management models: A review. *Water*, *9*(4), 285. Available from https://doi.org/10.3390/w9040285.

Peñuelas, J., & Staudt, M. (2010). BVOCs and global change. *Trends in plant science*, *15*(3), 133–144.

Possell, M., & Loreto, F. (2013). The role of volatile organic compounds in plant resistance to abiotic stresses: Responses and mechanisms. In Ü. Niinemets, & R. K. Monson (Eds.), *Biology, controls and models of tree volatile organic compound emissions* (pp. 209–235). Dordrecht: Springer Netherlands.

Rantanen, M., Karpechko, A. Y., Lipponen, A., Nordling, K., Hyvärinen, O., Ruosteenoja, K., Vihma, T., & Laaksonen, A. (2022). The arctic has warmed nearly four times faster than the globe since 1979. *Communications Earth & Environment*, *3*(1), 1–10.

Rossman, L. (1993). *EPANET water quality model*. Washington, D.C.: US Environmental Protection Agency, EPA/600/A-93/226 (NTIS PB93236552).

Rossman, L. A. (2015). *Storm water management model user's manual version 5*. Cincinnati, OH: US Environmental Protection Agency, *1*. EPA/600/R-14/413b.

Räisänen, T., Ryyppö, A., & Kellomäki, S. (2008). Effects of elevated CO_2 and temperature on monoterpene emission of Scots pine (*Pinus sylvestris L.*). *Atmospheric Environment*, *42*(18), 4160–4171.

Samoli, E., Stergiopoulou, A., Santana, P., Rodopoulou, S., Mitsakou, C., Dimitroulopoulou, C., Bauwelinck, M., de Hoogh, K., Costa, C., Mari-Dell'Olmo, M., Corman, D., Vardoulakis, S., Katsouyanni, K., & Consortium, E.-H. (2019). Spatial variability in air pollution exposure in relation to socioeconomic indicators in nine European metropolitan areas: A study on environmental inequality. *Environmental Pollution*, *249*, 345–353. Available from https://doi.org/10.1016/j.envpol.2019.03.050.

Sang, J. G., Liu, H. P., Liu, H. Z., & Zhang, Z. K. (2000). Observational and numerical studies of wintertime urban boundary layer. *Journal of Wind Engineering and Industrial Aerodynamics*, *87*(2-3), 243–258. Available from https://doi.org/10.1016/S0167-6105(00)00040-4.

Sharma, N., Chaudhry, K. K., & Rao, C. V. C. (2005).). Air pollution dispersion studies through environmental wind tunnel (EWT) investigations: A review. *Journal of Scientific & Industrial Research*, *64*(8), 549–559, <Go to ISI>://WOS:000231079000001.

Shubuya, M., Saito, H., Sawamoto, T., Hatano, R., Yajima, T., Takahashi, K., Cha, J., Isaev, A., & Maximov, T. (2004). Time trend in aboveground biomass, net primary production, and carbon storage of natural larix gmelinii stands in eastern Siberia. *Eurasian Journal of Forest Research*, *7*(2), 67–74.

Singh, R. K., Svystun, T., AlDahmash, B., Jönsson, A. M., & Bhalerao, R. P. (2017). Photoperiod- and temperature-mediated control of phenology in trees–A molecular perspective. *New Phytolygist*, *213*, 511–524. Available from https://doi.org/10.1111/nph.14346.

Sizov, O., Ezhova, E., Tsymbarovich, P., Soromotin, A., Prihod'ko, N., Petäjä, T., Zilitinkevich, S., Kulmala, M., Bäck, J., & Köster, K. (2021). Fire and vegetation dynamics in northwest Siberia during the last 60 years based on high-resolution remote sensing. *Biogeosciences*, *18*(1), 207–228.

Stewart, I. D., & Oke, T. R. (2012). Local climate zones for urban temperature studies. *Bulletin of the American Meteorological Society*, *93*(12), 1879–1900. Available from https://doi.org/10.1175/Bams-D-11-00019.1.

Taipale, R., Kajos, M. K., Patokoski, J., Rantala, P., Ruuskanen, T. M., & Rinne, J. (2011). Role of de novo biosynthesis in ecosystem scale monoterpene emissions from a boreal Scots pine forest. *Biogeosciences*, *8*(8), 2247–2255.

Taleghani, M. (2018). Outdoor thermal comfort by different heat mitigation strategies–A review. *Renewable & Sustainable Energy Reviews*, *81*, 2011–2018. Available from https://doi.org/10.1016/j.rser.2017.06.010.

Tallis, M., Taylor, G., Sinnett, D., & Freer-Smith, P. (2011). Estimating the removal of atmospheric particulate pollution by the urban tree canopy of London, under current and future environments. *Landscape and Urban Planning*, *103*(2), 129–138. Available from https://doi.org/10.1016/j.landurbplan.2011.07.003.

Tang, G. Q., Zhang, J. Q., Zhu, X. W., Song, T., Munkel, C., Hu, B., Schafer, K., Liu, Z. R., Zhang, J. K., Wang, L. L., Xin, J. Y., Suppan, P., & Wang, Y. S. (2016). Mixing layer height and its implications for air pollution over Beijing, China. *Atmospheric Chemistry and Physics*, *16*(4), 2459–2475. Available from https://doi.org/10.5194/acp-16-2459-2016.

United Nations (2014), World urbanization prospects: The 2014 revision - highlights, Statistical Papers - United Nations (Ser. A), Population and Vital Statistics Report, UN, New York, https://doi.org/10.18356/527e5125-en.

van den Berg, M., Wendel-Vos, W., van Poppel, M., Kemper, H., van Mechelen, W., & Maas, J. (2015). Health benefits of green spaces in the living environment: A systematic review of epidemiological studies. *Urban Forestry & Urban Greening*, *14*(4), 806–816. Available from https://doi.org/10.1016/j.ufug.2015.07.008.

Voordeckers, D., Lauriks, T., Denys, S., Billen, P., Tytgat, T., & Van Acker, M. (2021). Guidelines for passive control of traffic-related air pollution in street canyons: An overview for urban planning. *Landscape and Urban Planning*, *207*103980. Available from https://doi.org/10.1016/j.landurbplan.2020.103980.

Vos, P. E. J., Maiheu, B., Vankerkom, J., & Janssen, S. (2013). Improving local air quality in cities: To tree or not to tree? *Environmental Pollution*, *183*, 113–122. Available from https://doi.org/10.1016/j.envpol.2012.10.021.

Wang, J., Zhao, B., Wang, S., Yang, F., Xing, J., Morawska, L., Ding, A., Kulmala, M., Kerminen, V.-M., Kujansuu, J., Wang, Z., Ding, D., Zhang, X., Wang, H., Tian, M., Petäjä, T., Jiang, J., & Hao, J. (2017). Particulate matter pollution over China and the effects of control measures. *Science of the Total Environment*, *584-585*, 426–447. Available from https://doi.org/10.1016/j.scitotenv.2017.01.027.

Wang, L. L., Zhang, N., Liu, Z. R., Sun, Y., Ji, D. S., & Wang, Y. S. (2014). The influence of climate factors, meteorological conditions, and boundary-layer structure on severe haze pollution in the Beijing–Tianjin–Hebei region during january 2013. *Advances in Meteorology*. Available from https://doi.org/10.1155/2014/685971.

Wang, Y. H., Yu, M., Wang, Y., Tang, G., Song, T., Zhou, P., Liu, Z., Hu, B., Ji, D. S., Wang, L., Zhu, X., Yan, C., Ehn, M., Gao, W. K., Pan, Y. P., Xin, J. Y., Sung, Y., Kerminen, V.-M., Kulmala, M., & Petäjä, T. (2020). Rapid formation of intense haze episode via aerosol-boundary layer feedback in Beijing. *Atmospheric Chemistry and Physics*, *10*, 45–53.

Wang, Z. L., Huang, X., & Ding, A. J. (2018). Dome effect of black carbon and its key influencing factors: A one-dimensional modelling study. *Atmospheric Chemistry and Physics*, *18*(4), 2821–2834. Available from https://doi.org/10.5194/acp-18-2821-2018.

Wolf, T., Pettersson, L. H., & Esau, I. (2020). A very high-resolution assessment and modelling of urban air quality. *Atmospheric Chemistry and Physics*, *20*(2), 625–647. Available from https://doi.org/10.5194/acp-20-625-2020.

Xia, Q., Niu, J. L., & Liu, X. P. (2014). Dispersion of air pollutants around buildings: A review of past studies and their methodologies. *Indoor and Built Environment*, *23*(2), 201–224. Available from https://doi.org/10.1177/1420326x12464585.

Xing, Y., & Brimblecombe, P. (2020). Trees and parks as "the lungs of cities". *Urban Forestry & Urban Greening*, *48*126552. Available from https://doi.org/10.1016/j.ufug.2019.126552.

Yli-Juuti, T., Mielonen, T., Heikkinen, L., Arola, A., Ehn, M., Isokääntä, S., Keskinen, H.-M., Kulmala, M., Laakso, A., Lipponen, A., Luoma, K., Mikkonen, S., Nieminen, T., Paasonen, P., Petäjä, T., Romakkaniemi, S., Tonttila, J., Kokkola, H., & Virtanen, A. (2021). Significance of the organic aerosol driven climate feedback in the boreal area. *Nature Communications*, *12*(1), 1–9.

Yufen, R., Xiaoke, W., Zhiyun, O., Hua, Z., Xiaonan, D., & Hong, M. (2008). Stormwater Runoff Quality from Different Surfaces in an Urban Catchment in Beijing, China. *Water Environ. Res.*, *80*, 719–724,. Available from https://doi.org/10.2175/106143008X276660.

Zhao, B., Wang, S. X., Liu, H., Xu, J. Y., Fu, K., Klimont, Z., Hao, J. M., He, K. B., Cofala, J., & Amann, M. (2013). NOx emissions in China: historical trends and future perspectives,. *Atmospheric Chemistry and Physics*, *13*, 9869–9897. Available from https://doi.org/10.5194/acp-13-9869-2013.

Zhang, N., Wang, X. Y., & Peng, Z. (2014). Large-eddy simulation of mesoscale circulations forced by inhomogeneous urban heat island. *Boundary-Layer Meteorology*, *151*(1), 179–194. Available from https://doi.org/10.1007/s10546-013-9879-x.

Zhang, Y. X., Ye, X. P., Wang, S. B., He, X. J., Dong, L. Y., Zhang, N., Wang, H. K., Wang, Z. R., Ma, Y., Wang, L., Chi, X. G., Ding, A. J., Yao, M. Z., Li, Y. P., Li, Q. L., Zhang, L., & Xiao, Y. L. (2021). Large-eddy simulation of traffic-related air pollution at a very high resolution in a mega-city: evaluation against mobile sensors and insights for influencing factors. *Atmospheric Chemistry and Physics*, *21*(4), 2917–2929. Available from https://doi.org/10.5194/acp-21-2917-2021.

Zhong, J. T., Zhang, X. Y., Dong, Y. S., Wang, Y. Q., Liu, C., Wang, J. Z., Zhang, Y. M., & Che, H. C. (2018). Feedback effects of boundary-layer meteorological factors on cumulative explosive growth of $PM_{2.5}$ during winter heavy pollution episodes in Beijing from 2013 to 2016. *Atmospheric Chemistry and Physics*, *18*(1), 247–258. Available from https://doi.org/10.5194/acp-18-247-2018.

Zhou, Y. D., Zhao, Y., Mao, P., Zhang, Q., Zhang, J., Qiu, L. P., & Yang, Y. (2017). Development of a high-resolution emission inventory and its evaluation and application through air quality modeling for Jiangsu Province, China. *Atmospheric Chemistry and Physics*, *17*(1), 211–233. Available from https://doi.org/10.5194/acp-17-211-2017.

Chapter 16

Air quality management and policy applications

Jo Barnes[1], Enda Hayes[1] and Anil Namdeo[2]

[1]University of the West of England, Bristol, United Kingdom, [2]Department of Geography and Environmental Sciences, Northumbria University, Newcastle upon Tyne, United Kingdom

16.1 Introduction

This chapter introduces the role of air quality management and presents policies to reduce emissions and concentrations of air pollutants, illustrated with case studies from around the world.

16.1.1 The role of air quality management

There are numerous sources of air pollution, both natural and anthropogenic (see Chapter 3), that affect our health (see Chapter 14), and our ecosystems (see Chapter 15), damage our built environments and contribute to climate change (see Chapter 11). There is limited scope to be able to control the effects of natural sources, for example, SO_2 and PM_{10} from volcanoes, but, through emissions controls on polluting activities and management of pollutant concentrations in ambient (outdoor) air, we can help to reduce the impacts of the anthropogenic contribution. Although human exposure to air pollution is also affected by indoor air pollution (see Chapter 12), air quality management, to date, has focused on addressing the outdoor environment.

The types of pollutants and the extent of the problem vary across different cities, regions and countries due to the specific sources affecting those areas, the local topography and the unique meteorological conditions that affect formation and dispersal. In some low- and middle-income countries (LMICs), cities, such as Delhi and Mexico City, experience emissions from road traffic, industry, waste burning, and construction, inter alia, which are compounded by the physical terrain limiting dispersal of pollutants, and meteorology that creates inversions, trapping pollution close to the ground. In high-income countries (HICs), for example the United Kingdom, road traffic has become the primary issue as industrial and other sources have largely been addressed (or at least moved away from populous areas), and typically pollutant concentrations are significantly lower than in LMICs. However, as policies to address road traffic sources in HICs take effect, other pollutant sources are coming to the fore. The case studies presented within this chapter provide insights from these different perspectives.

Chapter 3 introduces the concept of top-down and bottom-up emissions inventories at different scales. Understanding the relative contributions of emissions from the various sources (including background sources), and their relationship to the concentrations of pollutants in ambient air, is key to targeted air quality management. As well as emissions, we also need to assess the concentrations of different pollutants in specific areas through modeling (see Chapter 8) and monitoring (see Chapter 9). These concentrations can then be compared to standards set by respective governments.

Worked example 16.1: Explain the difference between emissions and concentrations? Why is it important to manage both emissions and concentrations?

Solution: Pollutants leaving a source, such as a vehicle exhaust or chimney, are known as emissions (usually measured by weight over time, example, tonnes per annum), whilst pollutants in ambient air are referred to as concentrations [measured by volume, example, micrograms per cubic meter ($\mu g\,m^{-3}$), or parts per billion (ppb)/million (ppm)]. There is not a linear relationship between emissions and concentrations due to the effects of meteorology and

Air Quality. DOI: https://doi.org/10.1016/B978-0-12-822591-2.00016-0

atmospheric chemistry. Whilst tackling air pollution at source is the most effective way to reduce emissions, managing exposure to air pollution requires an understanding of how these strategies affect concentrations of pollutants spatially and temporally.

16.1.2 Setting air quality standards

Chapter 14 introduces the World Health Organization (WHO) air quality guidelines (AQGs), which form the basis for most countries' air quality standards to protect public health. The WHO AQGs set levels for various pollutants based on a review of the latest epidemiological evidence, and hence have been iteratively updated since their first publication in 2005, most recently in September 2021 (WHO, 2021). There may be different averaging periods for the guideline levels applicable to pollutants to reflect the specific health effects associated with short-term and long-term exposure. In most cases the 2021 AQGs are considerably stricter than the 2005 WHO AQGs.

Whilst the WHO guidelines provide an evidence-based reference for policymakers, they are not intended to be adopted wholesale as legislative standards. This is a recognition that countries will need to balance the risks and costs associated with setting thresholds and that these will differ depending on their economic status, political climate and public opinion (WHO, 2017). For instance, setting a relatively high threshold standard would be more easily achievable and hence lower cost, but would incur a higher risk to public health (and associated health costs); a lower threshold would potentially protect more of the population but would have greater economic impacts (Fig. 16.1). In many countries, background concentrations of some pollutants, that is pollutants transported from other areas, will exceed the WHO AQGs. Reducing air pollution to concentrations at which the majority of the population will be protected is therefore an ambition, rather than a realistic goal, in most urban areas at present.

Examples of air quality standards are presented for the case study areas in this chapter.

Worked example 16.2: Why is it important to manage ambient air quality?

Solution: Air pollution has a significant impact on public health, on climate change and on the built and natural environments. The WHO has estimated that ambient air pollution alone caused an estimated 4.2 million premature deaths in 2019, and it is a major contributor to a wide range of other noncommunicable diseases, including cardiovascular disease, respiratory diseases and cancers. Many of the sources of air pollution are also sources of greenhouse gas emissions, although not all climate change policies are good for air quality, for example, solid fuel burning. Deposition of air pollutants on vulnerable ecosystems can result in eutrophication and acidification. Acid deposition can also corrode buildings and monuments, whilst black smoke can stain them. Pollutants generated outside can also penetrate indoor environments. The scale of air pollution requires governments to regulate and reduce sources to protect vulnerable people, ecosystems and infrastructure.

16.1.2.1 Air quality indices

To communicate levels of air pollution in different areas, many governments use air quality indices (AQIs) rather than reporting specific pollutant concentrations or exceedances of standards (Priti & Kumar, 2022). AQIs are often used as the basis for health alert messaging to protect the most vulnerable members of the public from acute, short-term episodes of elevated concentrations that might trigger asthma attacks or cardiac arrest. They are not therefore reflective of more chronic exposure over longer periods, or of localized pollution hotspots. AQIs are often used to represent a range of pollutants, simplified to a color-coded scale with alert thresholds linked to specific health advice. Given the different concentration thresholds experienced across the world, and the diversity of methodologies used to calculate them, these

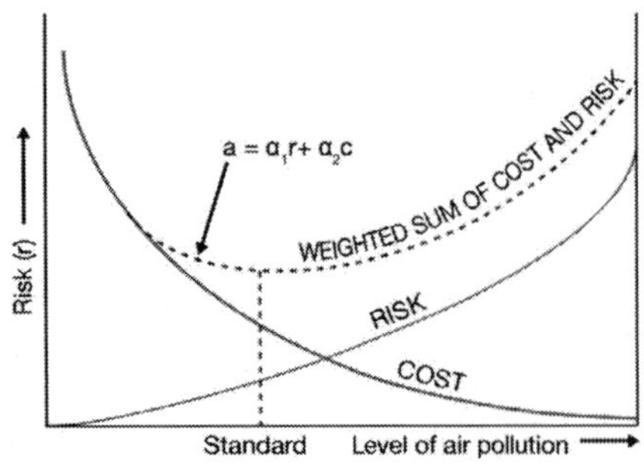

FIGURE 16.1 Derivation of ambient air quality standards. Key: α_1 and α_2 are weightings on risk and cost, respectively. Source*: From Evolution of WHO air quality guidelines: Past, present and future (Originally Suess & Craxford (1976). The manual on urban air quality management).*

country-specific AQIs are not comparable. To that end, unofficial AQIs (e.g., https://waqi.info/ and https://air-quality.com/) have also been developed that may incorporate official and unofficial measurement data standardized to a common scale, such as the US AQI. The lack of consistency between unofficial AQIs, however, can also be problematic for accurately communicating air quality.

Worked example 16.3: How do concentrations of air pollutants relate to AQIs?

Solution: AQIs normally relate to short-term averaging periods of concentrations as they are used to communicate episodic air pollution events. They do not therefore capture the long-term concentrations or associated health effects. The bands of AQIs usually relate to a range of concentrations aligned to health evidence, with the highest forecast pollutant concentration driving the overall AQI reading.

16.1.3 Air quality management interventions

Where ambient concentrations of air pollution exceed the standards set in that area, governments have a range of potential options at their disposal. These will typically fall within the categories of constraints ("sticks") or incentives ("carrots"). Constraint measures may include road charging schemes, which restrict access to areas, or charge drivers for entering, based on the emission standards of the vehicles (e.g., Low Emission or Clean Air Zones). Incentives may be the provision of cycle paths and infrastructure to encourage active travel. In most situations, a combination of constraint and incentive measures will be required to address the air quality issue in that area.

Worked example 16.4: What other "carrots" or "sticks" can policy makers use to control air pollution?

Solution: Examples of "carrot policies," or incentives include tax incentives for purchasing of cleaner vehicles or subsidies for heat pumps. Examples of "sticks" or constraints are fines for failed roadside emissions tests or banning solid fuel-burning appliances and fuels.

16.1.4 Challenges of assessing improvements in air quality

In practice, it is extremely difficult to attribute the implementation of specific management actions to measured reductions in air pollutants. This is due in part to the complex nonlinear relationship between emissions and concentrations, the variable effects of meteorology and seasonality (see Chapter 5), the confounding effects of other factors (e.g., highway maintenance), the implementation of combinations of interventions, and the frequent absence of data pre/postintervention against which to evaluate. Some attempts have been made to account for the meteorological variability through applying "de-weathering" tools (e.g., https://github.com/davidcarslaw/deweather and Shi et al. (2022)), which, provided the other variables can be controlled, can enable comparisons of observed concentrations before and after the intervention more accurately. These approaches can also prove more accurate than more traditional air quality modeling (see Chapter 8) or statistical approaches (see Chapter 13) (Shi et al., 2022).

16.1.5 Integrated air quality management

In introducing policies to improve air quality, policymakers should be aware of the potential for unintended consequences. For example, on other policy areas, such as carbon reduction targets, and on individuals and households, particularly those living in low socioeconomic areas. It is equally as important that other policy areas do not impinge on air quality or the ability to implement measures to improve air quality. Effective air quality policies must therefore be developed across sectors and departments to ensure a holistic, integrated approach. One such approach is presented in the C40 Cities Framework for integrating air quality management and climate action planning (C40 Cities Climate Leadership Group, 2022).

16.1.6 Multilevel governance

Given the boundaryless nature of air pollution (see Chapter 10), it is vital that authorities work together at all tiers of governance to ensure a coherent and consistent set of integrated policies that address pollutant sources and effects in a coordinated way. The Forum for International Cooperation on Air Pollution was established in 2019 under the UNECE Convention on Long-Range Transboundary Air Pollution, to promote and facilitate worldwide collaboration and cooperation on air pollution (Forum for International Cooperation on Air Pollution, n.d.). Within countries, however, there is also a need for central, state and municipal governments to share the responsibility for reducing air pollution effectively and efficiently, as policies introduced at each level can have both positive and negative impacts on emissions, concentrations and exposure, and can affect the success of respective mitigation strategies.

Worked example 16.5: Why is multilevel governance important in managing air pollution?

Solution: Different levels of government will have variable levels of control over sources of air pollution and the ability to protect those that are exposed. For example, local sources, such as local road traffic or small industrial sites, will be best managed by local authorities, whilst large-scale or widespread sources, such as trunk roads or large infrastructure projects, will normally be the responsibility of regional or national governments. It is important to ensure coordination of air quality policies across all tiers of governance so that measures are implemented most effectively and efficiently, and so that exposure in one area is not negatively affected by transboundary sources of pollution from another area.

16.2 Case studies

The following case studies illustrate the evolution of air quality management and policy in different contexts throughout the world, specifically in the United Kingdom, India, South Africa, and Mexico City.

16.2.1 United Kingdom (London and Bristol)

The development of air quality management in the UK has a relatively long history as scientific understanding of pollutants and how to detect them, relative contributions of emissions from changing sources, and evidence of epidemiological and ecological effects have evolved. As a result, the UK has provided a useful reference point for air quality management in other locations.

16.2.1.1 Historical context for current legislation

During the Industrial Revolution in the UK, the use of coal for industry and domestic heating led to extremely poor air quality in towns and cities as sulphur dioxide (SO_2) and particulate matter combined with meteorological conditions to create dense sulphurous smogs. One particularly significant event, known as the Great Smog of London, in December 1952, led to more than 4000 excess deaths in less than a week, and up to 12,000 overall (Brimblecombe, 2017; Corton, 2015). In response, the UK Government introduced the Clean Air Act in 1956 (amended 1968 and 1993), which brought in the first legislation to control both domestic and industrial emission sources (Clean Air Act 1956, 1956).

However, as these domestic and industrial sources of air pollution declined, the near exponential increase in road traffic that began in the 1940s has created a new source (Department for Transport, 2022). Petrol and diesel road traffic have become the main sources of nitrogen oxides (NO_x) in many urban areas, particularly close to busy roads. They are also a source of particulate matter, especially from nonexhaust emissions (brake and tyre wear and resuspended dusts), which can additionally be generated by electric vehicles (EVs). Other forms of transport, for example, diesel trains, shipping and aviation are also a source of NO_x, SO_2 and particulate matter.

16.2.1.2 Legislation and air quality thresholds

In the UK, legislation on air quality has come from two directions, almost simultaneously: the UK government and the European Union.

16.2.1.2.1 United Kingdom air quality legislation

This section sets out the evolution of air quality legislation in the UK focusing on emissions control and management of pollutant concentrations.

16.2.1.2.1.1 Clean Air Act 1956 As previously mentioned, one of the first pieces of national legislation in the UK was the Clean Air Act 1956 (Clean Air Act 1956). The Clean Air Act of 1956 introduced a number of key strategies to reduce emissions of SO_2 and particulate matter from domestic burning and industrial sources, including:

- Moving power stations out of cities, thereby separating source from receptors;
- Legislating for higher chimneys to dispel emissions;
- Allowing local authorities to create Smoke Control Areas to control the production of emissions in urban areas;
- Enabling grants for households to switch from coal to (relatively) cleaner gas or electric for cooking and heating.

16.2.1.2.1.2 Environmental Protection Act 1990 In addition to the Clean Air Acts, the Environmental Protection Act 1990 (Environmental Protection Act 1990, §Part III Statutory Nuisances and Clean Air, 1990) also introduced a statutory nuisance regime in England, Wales and Scotland. The Act specifies 11 different categories of statutory nuisance, which include:

- smoke emitted from premises so as to be prejudicial to health or a nuisance;
- fumes or gases emitted from premises so as to be prejudicial to health or a nuisance;
- any dust, steam, smell or other effluvia arising on industrial, trade or business premises and being prejudicial to health or a nuisance.

Under section 79 of the Environmental Protection Act, a local authority must take "such steps as are reasonably practicable" to investigate a statutory nuisance complaint. Local authorities also have a duty to inspect their areas to detect whether a nuisance exists or is likely to occur or recur[1].

16.2.1.2.1.3 Environment Act (Part IV) 1995 The measures implemented under the Clean Air Act 1956 (and its subsequent amendments, 1968 and 1993), significantly helped to improve air quality in urban areas, thereby protecting public exposure in the most densely populated zones. However, by the 1990s it was recognized that controlling *emissions* was only part of the picture, and that managing *concentrations* of pollutants in ambient air was required to better understand and reduce the exposure levels of the public.

In 1995, the UK government introduced the Environment Act (Part IV) 1995 (Environment Act 1995, § Part IV Air Quality, 1995). The Environment Act 1995 defined responsibilities for national and local government to reduce air pollution concentrations. This required the government to produce a National Air Quality Strategy and Air Quality Regulations that set out the national air quality objectives (AQOs), which national and local government were required to work to achieve. The local government responsibilities introduced by the Environment Act 1995 are known as local air quality management (LAQM).

LAQM gives statutory responsibilities for local government to:

- Review and assess ambient air quality against the national AQOs.
- Declare air quality management areas (AQMAs) where there is a risk of these AQOs being exceeded.
- Prepare air quality action plans (AQAPs) and implement measures therein to work towards achievement of the AQOs.

The purpose of LAQM is to identify and manage local pollution hotspots in areas of public exposure to complement national actions as set out in the Air Quality Strategy.

As a devolved responsibility, the national AQOs were set out in separate regulations for each UK nation[2] and hence there are subtle differences between them (Table 16.1). As per the Regulations, the AQOs apply at locations which are situated outside of buildings or other natural or artificial structures above or below ground; and at which members of the public are regularly present.

Worked example 16.6: Why are there different AQO thresholds set for different averaging periods for some pollutants?

Solution: The different averaging periods relate to the periods of exposure to pollutants. Short term exposure (e.g., over periods of 1 hour or 24 hours) to high concentrations of pollutants can cause acute health effects, whilst long-term exposure (e.g., over periods of a year) to relatively lower concentrations of pollutants can lead to chronic health effects. Evidence has shown that some pollutants can cause both acute and chronic health effects, hence there are multiple thresholds set for these pollutants relating to different average exposure periods to protect the public from short-term episodes of air pollution as well as longer-term exposure.

1. A similar regime applies in Northern Ireland under the Clean Neighborhoods and Environment Act (Northern Ireland) 2011. (The EPA as it applies in Scotland is slightly amended by the Public Health etc. (Scotland) Act 2008.)

2. England: Air Quality (England) Regulations 2000 (SI 928), as amended by the Air Quality (England) (Amendment) Regulations 2002 (SI 3043); Wales: Air Quality (Wales) Regulations 2000, No. 1940 (Wales 138), as amended by the Air Quality (Wales) (Amendment) Regulations 2002, No. 3182 (Wales 298); Scotland: Air Quality (Scotland) Regulations 2000 (Scottish SI 2000 No. 97), as amended by the Air Quality (Scotland) (Amendment) Regulations 2002 (Scottish SI 2002 No. 297); Northern Ireland: Air Quality Regulations (Northern Ireland) 2003, Statutory Rules of Northern Ireland 2003, No. 342.

TABLE 16.1 UK air quality objectives.

Pollutant	Objective	Averaging period	Obligation
Nitrogen dioxide (NO_2)	200 $\mu g\,m^{-3}$ not to be exceeded more than 18 times a year	1-h mean	All local authorities
	40 $\mu g\,m^{-3}$	Annual mean	All local authorities
Particulate matter (PM_{10})	50 $\mu g\,m^{-3}$ not to be exceeded more than 35 times a year	24-h mean	All local authorities
	50 $\mu g\,m^{-3}$ not to be exceeded more than 7 times a year	24-h mean	Scotland only
	40 $\mu g\,m^{-3}$	Annual mean	All local authorities
	18 $\mu g\,m^{-3}$	Annual mean	Scotland only
Particulate matter ($PM_{2.5}$)	Work towards reducing emissions/concentrations of fine particulate matter ($PM_{2.5}$) 20 $\mu g\,m^{-3}$	Annual mean	England only (encouraged in Wales)
	10 $\mu g\,m^{-3}$	Annual mean	Scotland only
Sulphur dioxide (SO_2)	266 $\mu g\,m^{-3}$ not to be exceeded more than 35 times a year	15-min mean	All local authorities
	350 $\mu g\,m^{-3}$ not to be exceeded more than 24 times a year	1-h mean	All local authorities
	125 $\mu g\,m^{-3}$ not to be exceeded more than 3 times a year	24-h mean	All local authorities
Benzene (C_6H_6)	16.25 $\mu g\,m^{-3}$	Running annual mean	All local authorities
	5 $\mu g\,m^{-3}$	Annual mean	England and Wales only
	3.25 $\mu g\,m^{-3}$	Running annual mean	Scotland and Northern Ireland only
1,3-Butadiene (C_4H_6)	2.25 $\mu g\,m^{-3}$	Running annual mean	All local authorities
Carbon monoxide (CO)	10 $mg\,m^{-3}$	Maximum daily running 8-h mean	England, Wales and Northern Ireland only
	10 $mg\,m^{-3}$	Running 8-h mean	Scotland only
Lead (Pb)	0.25 $\mu g\,m^{-3}$	Annual mean	All local authorities

Source: Adapted from Defra (2022c). Local air quality management technical guidance (TG22). Table 1.1 https://laqm.defra.gov.uk/wp-content/uploads/2022/08/LAQM-TG22-August-22-v1.0.pdf. Defra (2022a). Local air quality management policy guidance (PG22). https://laqm.defra.gov.uk/guidance/.

16.2.1.2.1.4 Environment Act 2021 In November 2021, the 1995 Environment Act was amended with the 2021 Environment Act, which applies in England (Environment Act 2021, 2021). The 2021 Environment Act (Schedule 11) sets out the following amendments to the LAQM framework for local authorities in England:

- An update of the 2007 Air Quality Strategy within 12 months (by May 2023), with 5-yearly updates subsequently and a duty for the Secretary of State to report annually against AQOs and standards in England;
- Clarification and expansion of 'relevant public authorities' who should have regard for the Strategy (who may be designated by the Secretary of State);
- Identification of sources causing breaches of the AQOs or standards by local authorities in England and a duty for air quality partners (neighboring authorities, Environment Agency) responsible to cooperate;
- Requirement for English local authorities' AQAPs to include scheduled measures to secure and maintain achievement of AQOs and standards;
- Specific roles for (upper tier) county councils and combined authorities to support local authorities in England.

The 2021 Environment Act (Schedule 12) also replaces fines for noncompliance with Smoke Control Areas under the Clean Air Act 1993 with civil penalty notices. Buying and selling unapproved solid fuels for use in Smoke Control Areas will also be fined. Amendments to LAQM and the Clean Air Act came into force from 1 May 2022.

Under Part 1 of the 2021 Environment Act, two new $PM_{2.5}$ concentration targets were introduced in 2023: a reduced annual mean $PM_{2.5}$ threshold (10 $\mu g\,m^{-3}$ annual mean concentration $PM_{2.5}$ nationwide by 2040, with an interim target of 12 $\mu g\, m^{-3}$ by January 2028) and a long-term target (35% reduction in average population exposure by 2040, with an interim target of a 22% reduction by January 2028, both compared to a 2018 baseline) (The Environmental Targets Fine Particulate Matter England Regulations 2022). These are national targets applicable in England, for which local authorities will have some responsibility to help achieve, although they are not required to assess $PM_{2.5}$ in their LAQM reports (Air quality strategy: Framework for local authority delivery, n.d).

16.2.1.2.2 European Union Air Quality Legislation in the United Kingdom

Parallel to UK legislation on air quality has been the development of similar legislation on emissions sources and ambient concentrations of key pollutants in the EU. As a member state of the EU (from 1973 to 2020), the UK was required to implement these EU Directives and transpose them into UK legislation.

Although the EU Directives ceased to have effect in the UK following its exit from the EU, the EU Withdrawal Act 2018 (European Union Withdrawal Act 2018) ensured that EU environmental law would continue under the UK transpositions of the Directives. In 2023, UK transposition of EU legislation on air quality and emissions control was reviewed as part of the Retained EU Law (Revocation and Reform) Act 2023 (Retained EU Law, 2023), by which the government revoked and modified certain EU legislation following the UK's departure from the European Union.

16.2.1.2.2.1 Environmental Permitting Regulations The Environment Agency, Natural Resources Wales, the Scottish Environment Protection Agency (SEPA) and the Northern Ireland Environment Agency (NIEA) are responsible for Part A(1) activities (Schedule 1 to the Environmental Permitting Regulations (England and Wales) 2010[3]/Annex 1 to the Industrial Emissions Directive (IED) (Directive 2010/75/EU)). This covers industrial processes like refineries, food and drink factories intensive farming activities (e.g., large-scale chicken farms), certain waste activities like disposing of waste to landfill, hazardous waste treatment and waste incineration.

Under the Environmental Permitting (England and Wales) Regulations 2016 (S.I. 2016/1154) (subsequently the Environmental Permitting (England and Wales) (Amendment) (EU Exit) Regulations 2019), local authorities (or sometimes Port Authorities) are responsible for:

- Part A(2) installations regulated under integrated pollution prevention and control, known as (Local Authority Industrial Pollution Prevention and Control (LA-IPPC)). Part A(2) processes include glass and ceramics manufacture, galvanizing, rendering, animal carcass incineration, etc.
- Part B installations (Local Authority Pollution Prevention and Control (LAPPC)) for emissions to air only (except for some combustion activities and generation of waste wood, where Medium Combustion Plant Directive (MCPD) or specified generator regulations apply). Part B processes include quarrying, cement works, crematoria, dry cleaners, maggot breeding, etc.

16.2.1.2.2.2 National Emission Ceilings Regulations 2018 The UK National Emission Ceilings Regulations 2018 transposed the revised National Emission reduction Commitments Directive (NECD) 2016/2284/EU on July 1, 2018, requiring the Secretary of State to prepare and implement a national air pollution control program (NAPCP). The 2019 NAPCP, revised February 9, 2023, sets out how the UK would meet the legally binding 2020 and 2030 emission reduction commitments (ERCs). Under the Retained EU Law (Revocation and Reform) Act 2023, specific provisions of the National Emissions Ceilings Regulations 2018 (SI 2018/129) and Decision 2018/1522, relating to the NAPCP, were removed on December 31, 2023 to be included within the Environmental Improvement Plan reporting in England (Smith and Boulton, 2024).

16.2.1.2.2.3 Air Quality Standards Regulations 2010 (England) The Ambient Air Quality Directive (AAQD) 2008/50/EC was transposed into UK legislation on 11 June 2010 by the Air Quality Standards Regulations 2010 (England) (The Air Quality Standards Regulations 2010, 2023), and equivalent legislation in the Devolved Administrations. In the UK, responsibility for meeting air quality limit values set in the AAQD is devolved to the national administrations in Scotland, Wales and Northern Ireland. The Secretary of State for Environment, Food and Rural Affairs has responsibility for meeting the limit values in England. The Department for Environment, Food and Rural Affairs (Defra) co-ordinates assessment and air quality plans for the UK as a whole. For compliance assessment reporting purposes, the UK is divided into 43 zones and agglomerations defined by population density.

3. N.B. The Environmental Permitting (England and Wales) Regulations 2016 (S.I. 2016/1154) (now The Environmental Permitting (England and Wales) (Amendment) (EU Exit) Regulations 2019) (consolidate and replace the EPR 2010 (S.I. 2010/675)).

To assess air quality against the limit values, Defra report:

- Monitored concentrations from the Automatic Urban and Rural Network (AURN) and the UK Urban NO_2 Network (UUNN) monitoring sites;
- Modeled concentrations using the Pollution Climate Mapping model.

Annual compliance assessment reports are available online: https://uk-air.defra.gov.uk/library/annualreport/index.

16.2.1.2.3 Relationship between air quality limit values and objectives

The air quality limit values, under the Air Quality Standards Regulations arising from the AAQD, and the national AQOs, under each of the devolved administrations' Air Quality Regulations arising from the Environment Act 1995 (Part IV) (Table 16.1), are similar in that they refer to many of the same pollutants and share many of the same averaging periods and threshold values. There are, however, spatial differences in where each of them applies. Whilst both sets of regulations require there to be "relevant exposure" of members of the public, the guidance given to local authorities (Defra, 2022a) in the assessment of AQOs is more specific than that which the AAQD, and therefore the Air Quality Standards Regulations, confers on the national devolved administrations. Thus there are instances within local authority areas where UK government modeled limit value exceedances have been identified outside of local authority designated AQMAs, and exceedance areas within AQMAs that do not constitute breaches of the limit values (Barnes, Hayes, Chatterton, & Longhurst, 2018). This confusing situation was compounded by the ruling of the UK High Court (Clientearth No.3, 2018) which established an unprecedented legal requirement for local authorities with areas in breach of the limit values to achieve them "in the shortest time possible" through the implementation of Clean Air Zones or equivalent measures (ClientEarth, 2019).

16.2.1.2.4 United Kingdom daily air quality index

The UK government publishes a Daily Air Quality Index (DAQI) (https://uk-air.defra.gov.uk/air-pollution/daqi) which translates short-term average concentrations of five pollutants (NO_2, SO_2, O_3, $PM_{2.5}$, and PM_{10}) into an index of 1 to 10, divided into four bands (Low, Moderate, High and Very High). The purpose of the DAQI is to provide health advice to protect people who may be susceptible to the effects of high levels experienced in air pollution episodes. Five-day forecast maps based on the DAQI are published on the government's UK Air website (https://uk-air.defra.gov.uk/forecasting/) Table 16.2.

TABLE 16.2 UK daily air quality index (Defra).

Air pollution banding	Value	Accompanying health messages for at-risk individuals	Accompanying health messages for the general population
Low	1–3	Enjoy your usual outdoor activities.	Enjoy your usual outdoor activities.
Moderate	4–6	Adults and children with lung problems, and adults with heart problems, who experience symptoms, should consider reducing strenuous physical activity, particularly outdoors.	Enjoy your usual outdoor activities.
High	7–9	Adults and children with lung problems, and adults with heart problems, should reduce strenuous physical exertion, particularly outdoors, and particularly if they experience symptoms. People with asthma may find they need to use their reliever inhaler more often. Older people should also reduce physical exertion.	Anyone experiencing discomfort such as sore eyes, cough or sore throat should consider reducing activity, particularly outdoors.
Very high	10	Adults and children with lung problems, adults with heart problems, and older people, should avoid strenuous physical activity. People with asthma may find they need to use their reliever inhaler more often.	Reduce physical exertion, particularly outdoors, especially if you experience symptoms such as cough or sore throat.

Source: From https://uk-air.defra.gov.uk/air-pollution/daqi.

16.2.1.3 Responsibilities of different tiers of governance

In the UK, air quality is a devolved responsibility, which means that each of the nations of England, Scotland, Wales and Northern Ireland have responsibility for managing air quality, and for adopting their own air quality legislation. They also have responsibility for achieving the limit values as set out in the respective nations' Air Quality Standards Regulations derived from the AAQD 2008/50/EC, and the permitting of large-scale industrial processes via their respective environmental agencies. Within each devolved administration there are also local government responsibilities in fulfilment of their statutory LAQM duties, certain industrial emissions permitting, and in responding to directions from the relevant national government. In London boroughs, the Mayor of London has responsibility for overseeing LAQM.

16.2.1.4 Specific measures at different spatial/temporal scales

16.2.1.4.1 London

In London, the Mayor has established a number of road charging schemes that cover different geographical areas and with different restrictions (Greater London Authority, 2022b, 2022a). The Congestion Charge area covers Central London only, and applies to almost all vehicles but only during certain daytime hours; the Low Emission Zone (LEZ) (launched in February 2008) covers almost all of Greater London and applies to only the most polluting heavy vehicles (lorries, buses and some vans, etc.); and the Ultra-Low Emission Zone (ULEZ), originally established in April 2019 to cover the same area as the Congestion Charge, but operating 24 hours a day throughout the year, applies to older, more polluting, light vehicles (cars, motorbikes and smaller vans). The ULEZ was expanded to include the area up to the North and South circular roads in October 2021, and it was further extended to cover the same area as the Low Emission Zone (i.e., all London boroughs) on August 29, 2023 (Transport for London, n.d.).

Whilst some assessments suggest that the London LEZ and ULEZ alone appeared to have delivered little or no improvement in vehicle emissions (Beevers et al., 2016), ambient concentrations (Ellison et al., 2013; Font et al., 2019; Ma et al., 2021) and child health (Mudway et al., 2019; Wood et al., 2015), the implementation period coincided with a period during which it was identified that real-world driving emissions were higher than anticipated based on the "type approval" emissions tests that motor manufacturers were required to achieve for new vehicles (see Section 16.2.1.5). Other studies have modeled significant positive long-term health impacts (Webber et al., 2020) and observed reductions in concentrations of NO_2 (Greater London Authority, 2020, 2023, 2025; Hajmohammadi & Heydecker, 2022) from the ULEZ and its 2021 expansion.

16.2.1.4.2 Bristol

As a local authority, Bristol City Council (BCC) fulfils its statutory LAQM duties. It has had an AQMA in place since 2001 for exceedences of the AQOs for PM_{10} (24-hour mean) and NO_2 (annual mean, and more recently 1-hour mean) due to road traffic emissions. The AQAP has been integrated into the regional Joint Local Transport Plan and Bristol Transport Strategy and is complemented by the One City Plan and Climate Emergency Action Plan (Bristol City Council, 2022). In November 2022, Bristol also launched a Clean Air Zone to support national government in achieving the NO_2 annual mean limit value. Clean Air Zones (CAZs) are local authority defined areas in which the most polluting vehicles are charged for entering. As prescribed in the government's Clean Air Zone Framework, restrictions may apply to all vehicles (CAZ D) or just commercial vehicles (CAZ C) or further subsets, and target petrol light duty vehicles below Euro standard 4 and diesel light duty vehicles below Euro standard 6 (Defra, 2022b). The Bristol CAZ D (charging non-compliant private and commercial vehicles) covers a small area of the city center with the intention of reducing the number of polluting vehicles entering the city. In addition, BCC is encouraging and facilitating a modal shift by providing safe, convenient, and reliable alternatives to car use, alongside facilitating a shift towards cleaner vehicles. Like many UK cities, Bristol also has a Smoke Control Area, and residents are restricted from burning unauthorized fuels or using nonexempt appliances. Despite this, domestic solid fuel burning is increasing in the city, contributing to emissions of PM_{10} and $PM_{2.5}$.

16.2.1.5 Performance of past and current management plans/strategies

Since the early 1990s, new vehicles in Europe have had to pass type approval tests and achieve increasingly stringent emissions standards (known as Euro standards) to support national emissions reduction plans. In the 2000s, it became apparent that roadside concentrations of NO_2 were not falling at expected rates based on the forecast impacts of these emissions controls (Carslaw, Beevers, Tate, Westmoreland, & Williams, 2011). Investigation revealed that the emissions tests were not sufficiently reflective of on-street driving and, as a result, real-world emissions of diesel cars were much higher than expected (O'Driscoll, Stettler, Molden, Oxley, & ApSimon, 2018). At the same time, the US EPA in

California found that Volkswagen diesel cars had been illegally employing "defeat devices" to pass type approval emissions tests in America. These same devices were also found to be in use in European tests, and discrepancies were not exclusive to VW models (Arbour, 2022; Ayala, 2018). The combination of these factors resulted in NO_x emissions from diesel cars being higher in use than permitted. In the UK, this was further compounded by the government's climate change policy that gave tax incentives to drivers of diesel cars due to their lower CO_2 emissions per kilometer relative to petrol vehicles. This policy led to a significant growth in the sale of diesel vehicles, which in turn led to an increase in concentrations of roadside NO_2. This failure of emissions control processes coupled with incompatible carbon reduction policies therefore resulted in higher air pollution and consequent morbidity and mortality. Although emissions testing is now required to represent typical driving behavior, there is a legacy of higher polluting early Euro 6 vehicles on the roads, which are permitted to enter LEZs and CAZs (Demuynck, Bosteels, Bunar, & Spitta, 2020; Fuller, 2019).

National air quality policy has also received repeated criticism for its lack of ambition and action in achieving the NO_2 limit values (House of Commons Environment, Food and Rural Affairs, Environmental Audit, Health and Social Care, and Transport Committees, 2018), resulting in legal action from ClientEarth (ClientEarth, 2019) and the European Commission. On May 17, 2018 the European Commission referred France, Germany, and the UK to the Court of Justice of the EU for failure to respect limit values for NO_2 and for failing to take appropriate measures to keep exceedance periods as short as possible. Action was brought on October 23, 2018—European Commission v United Kingdom of Great Britain and Northern Ireland (Case C-664/18) and the EU ruling, published on March 4, 2021, found that the UK had "systematically and persistently" exceeded the annual (and hourly in London) mean limit value for NO_2, and ruled that the arguments put forward by the UK cannot justify the long delay to achieve compliance, given the requirement to ensure that the period of exceedance is kept as short as possible (C-664/18, 2021).

A number of studies have also critiqued LAQM (Barnes et al., 2014; Brunt et al., 2016; Longhurst et al., 2011), highlighting a bureaucratic focus on the diagnosis of exceedance areas over implementation of Action Plans, as well as barriers to improving air quality, that is, resource limitations, absence of local authority powers, and a lack of intra-governmental co-operation and inter-governmental co-ordination.

16.2.1.6 Future policy directions

The role of England local authorities in managing air quality is being more closely aligned with the national targets, for example, through the use of CAZs and the realignment of regional air quality zones with local government boundaries in order to achieve rapid compliance (Defra, 2023). This may include a legal duty on local authorities to take action to reduce $PM_{2.5}$ emissions if voluntary action is not considered to be sufficient (Defra, 2023).

Clean air is a strategic goal of the UK, as stated in Defra's 25-Year Environment Plan (Defra 2018, 2023), which it intends to achieve through meeting its national emission reduction commitments, ending the sale of new conventional petrol and diesel cars and vans, and maintaining continuous improvement in industrial emissions. Further action is also planned to reduce domestic burning and reduce ammonia from agricultural sources. Seven headline indicators relevant to this goal (and to mitigating and adapting to climate change) are reported against annually in the Environmental Improvement Plan with an overview provided via the Outcome Indicator Framework: https://oifdata.defra.gov.uk/1/. In Scotland, implementation of their second Clean Air for Scotland Strategy is underway (https://www.gov.scot/publications/cleaner-air-scotland-2-towards-better-place-everyone/); in Wales, the Environment (Air Quality and Soundscapes) (Wales) Act 2024 has been published (https://www.legislation.gov.uk/asc/2024/2) as outlined in the 10-year 2020 Clean Air Plan for Wales: Healthy Air, Healthy Wales; and a consultation on Northern Ireland's draft Clean Air Strategy was undertaken in 2020 (https://niopa.qub.ac.uk/bitstream/NIOPA/11575/1/Clean%20Air%20Strategy%20for%20NI%20-%20Public%20Discussion%20Doc.pdf).

The UK is also a founding signatory of the Convention on Long Range Transboundary Air Pollution and is leading the Task Force for International Cooperation on Air Pollution, alongside Sweden, to promote international collaboration towards preventing and reducing air pollution to improve air quality globally.

16.2.2 India (Delhi)

As a rapidly developing LMIC, India has faced air quality issues arising from economic growth and its expanding population, particularly inward migration to urban areas. Management of multiple complex and often unregulated pollutant sources in close proximity to highly populous areas, coupled with limited available resource and additional demands, presents national and state regulators with significant challenges.

16.2.2.1 Historical context for current legislation

The air pollution levels in India are among the highest in the world, posing a significant threat to the country's health and economy. Most of India's 1.4 billion people are exposed to unhealthy levels of air pollutants emanating from multiple sources. In 2017, around 97% of the country's population were exposed to particulate matter ($PM_{2.5}$) (Narain, O'Hagan, Kshirsagar, & Skoufias, 2020) in excess of the value specified by the World Health Organization (WHO) 2005 air quality guidelines, thereby making them vulnerable to serious health risks. Based on the concentrations of $PM_{2.5}$ emissions, India is consistently ranked among the most polluted countries in the world. The 2017 Global Burden of Disease Study (Balakrishnan et al., 2019) estimated that 1.24 million deaths (12.5% of the total) in India in 2017 were attributable to air pollution. This included 0.67 million from ambient PM and 0.48 million from household air pollution. India contributed 18.1% of the global population but had 26.2% of the global air pollution DALYs (disability-adjusted life years) in 2017. The Lancet Study (Balakrishnan et al., 2019) highlighted that the health impacts of air pollution also represent a heavy cost to India's economy estimating that the lost labor income due to fatal illness from $PM_{2.5}$ pollution in 2017 was in the range of \$30–78 billion. This is equal to about 0.3–0.9 percent of the country's GDP.

Over half of $PM_{2.5}$ emissions in India are "secondary," with precursor emissions from agriculture, industry, power plants and, households, and transport. This secondary form spreads farther and wider than primary $PM_{2.5}$ and travels across states and cities, and crosses jurisdictional borders.

Delhi, the national capital of India, is one of the cities with nonattainment areas where national ambient air quality standards (NAAQS) are not met. In the National Capital Region (NCR), extreme air pollution events occur routinely in winter months bringing attention from national and international media often with expressions such as "Delhi Airpocalypse" or "Death by breath" to describe the scale of the issue. The air pollution episodes bring about a range of reactionary emergency measures. While these episodes are short-term and extreme, it must be stressed that air quality in Delhi is poor all year round (De Vito et al., 2018).

Delhi is among the most polluted cities in the world. Of the top 10 most polluted cities in the world list released by WHO, nine cities are from India, and among them, Delhi remains at sixth position. In 2019 alone, air pollution caused over 17,000 premature deaths and an economic loss of US\$1207 million in Delhi (Pandey et al., 2021). In recent years, severe pollution episodes associated with exceptionally high levels of $PM_{2.5}$ concentrations (more than $600\ \mu g\ m^{-3}$) were found to occur more frequently in the NCR during the postmonsoon winter period.

The main sources of air pollution in Delhi include .vehicle exhaust; heavy industry including power generation; small-scale industries like brick kilns; suspended dust from roads due to vehicle movement; construction activities; open waste burning; combustion of fuels for cooking, lighting, and heating; and in-situ power generation via diesel generator sets. The variety of emission sources, the city's landlocked geographical location and unfavorable meteorological conditions are found to be drivers of high pollution days. In addition to the contributions from the local emission sources, open crop residue burning in the north-western states of India is thought to be significantly responsible for some of the extreme air pollution episodes during the postmonsoon period.

16.2.2.2 Legislation and air quality thresholds

Air quality management in India has evolved along with broader issues of environmental protection and the Indian constitutional and political framework. In 1972, the Indian constitution was amended to include environmental protection. The link between environmental protection and the right to life was substantiated throughout the years by the Supreme Court. Indeed, in key court cases [e.g., but not exclusively in the M.C. Mehta v. Union of India)] the interpretation of the Indian Constitution that sees life—and therefore, public health—as dependent on a healthy and pollution-free environment is made explicit by the Supreme Court (De Vito et al., 2018).

Since 1972, key pieces of environmental legislation have been adopted, including the Water Act (1974); the Air Pollution Act (1981); the Environment Act (1986), and the National Environment Policy (2006). The National Environment Policy (2006) represented the first strategic initiative for environmental·protection aimed at integrating environmental protection with economic and social development, and that explicitly considered issues of intergenerational and intragenerational equity, the Environment Pollution (Prevention and Control) Authority (EPCA) and the National Green Tribunal (2010) (De Vito et al., 2018).

Many policies and programmes have been implemented in India to address the issue of air pollution. Like any other policy measure, the success and efficacy of these programmes has been contingent upon collaboration and coordination across various stakeholders.

One of the most recent policies launched to tackle air pollution is the National Clean Air Program (Ministry of Environment, Forest and Climate Change, 2019). The NCAP is envisioned as a comprehensive strategy to provide the

State and Union governments with a blueprint to combat air pollution. The NCAP has set a time-bounded goal for improving air quality across the country, with a focus on around 102 "nonattainment" cities where air pollution standards are not being met. The NCAP provides cities an overall framework for developing air quality management plans, with guidance on policies across a range of sectors. It set a target of reducing key air pollutants PM_{10} and $PM_{2.5}$ by 20%−30% by 2024, with 2017 as the base year. The NCAP plans to create a pan-India air quality monitoring network and generate citizen awareness. It has been observed that without the legal mandate, the NCAP will not be able to implement the action plan. Roy, Chandra, and Ratho (2020) note that the primary reason for noncompliance is lack of appropriate monitoring and inspection elements and have suggested that training be given to personnel and regular inspection drives conducted.

16.2.2.2.1 National ambient air quality standards (NAAQS)

The Air (Prevention and Control of Pollution) Act (1981) entrusted the Central Pollution Control Board (CPCB) and several State Pollution Control Boards (SPCBs) as agencies responsible for air quality standard creation and monitoring. CPCB and SPCBs fall under the control of the Ministry of Environment, Forest and Climate Change. With the goal of providing for the prevention, control, and abatement of air pollution, the first ambient air quality standards were adopted in 1982 by CPCB, and revised in 1994 and 2009.

Table 16.3 shows the Indian NAAQS introduced in 2009 (CPCB, 2009).

To check whether NAAQS are being met the CPCB initiated the National Air Quality Monitoring Program (NAMP) in 1984 with seven stations. By 2021, the number of stations was increased to 804 covering 344 cities/towns/villages in 28 states and six Union territories. The NAMP monitors three pollutants NO_2, SO_2, and PM_{10}. Other parameters like $PM_{2.5}$, CO, NH_3, Pb, O_3, benzene, benzo(a)pyrene, arsenic and nickel are being monitored at selected locations and are slowly being added to the monitoring network under NAMP. If an area fails to meet the NAAQS, then it is classified as nonattainment area where adverse health effects are expected. The NCAP was formulated to tackle air pollution in cities, specifically to reduce particulate matter ($PM_{2.5}$) pollution by 20%−30% from 2017 levels by 2024. But the nature of air pollution means there are limitations to a policy that focuses on urban areas (CPCB, 2021).

A comparison of NAAQS in major Asian countries including China, Japan, India, and South Korea with the WHO (2021) AQGs suggests there is still much effort to be made to improve air quality (Ouyang et al., 2022). Fig. 16.1 compares WHO AQG 2021 with NAAQS of China, India, Japan and South Korea. Achieving the stricter WHO (2021) AQGs is considerably more challenging for developing countries, as they are struggling to meet even the 2005 AQGs. The results obtained during COVID-19 period indicate that even with an unprecedented lockdown (i.e., dramatic reduction in emissions), most cities still struggled to achieve AQGs for several pollutants including O_3 (due to reduction in NO_3 in the atmosphere) and $PM_{2.5}$ (due to excess biomass burning) (Venter, Aunan, Chowdhury, & Lelieveld, 2021). It has been suggested that developing countries need to adopt a stepwise approach by utilizing the interim targets of WHO AQGs as stepping-stones (Ouyang et al., 2022).

Worked example 16.7: How do background sources affect some countries' ability to achieve the WHO AQGs?

Solution: In countries like India, the background concentration of PM from sources such as wildfires, sea salt, or sand/dust storms is already higher than the 2021 AQGs levels, making it challenging to achieve the guideline levels. For example, the background $PM_{2.5}$ level at Mumbai, India, was found to be $33\,\mu g\,m^{-3}$ ($\pm 7\,\mu g\,m^{-3}$) (Beig et al., 2020), which is more than 6 times higher than the 2021 AQG (Fig. 16.2).

16.2.2.2.2 Indian air quality index (IND-AQI)

The AQI transforms weighted values of individual air pollution-related parameters into a single number. In India, this has been based on a maximum subindex approach using five parameters that is, suspended particulate matter (SPM), SO_2, CO, PM_{10}, and NO_2 (Sharma, Maheshwari, Sengupta, & Shukla, 2003). However, the calculated AQI was always dominated by subindex of SPM due to lack of data availability for other pollutants. The Indian Institute of Tropical Meteorology (IITM), Pune has evolved an AQI, which provides a subindex for PM_{10}, $PM_{2.5}$, O_3, NO_2, and CO (Beig, Ghude, & Deshpande, 2010) air quality in terms of very unhealthy, very poor, poor (unhealthy for sensitive groups), moderate and good. The revised CPCB Air Quality Standards (for 12 pollutants) resulted in the development of a National AQI to strengthen the air quality information dissemination system for greater public awareness and participation in air quality management. The Indian Air Quality Index (IND-AQI) is based on human exposure and health effects and may not be applicable to ecologically sensitive areas (CPCB, 2014). To present status of air quality and its effect on human health, description categories have been adopted for IND-AQI (Table 16.4).

TABLE 16.3 India national air quality standards.

Revised national ambient air quality standards (NAAQS) (NAAQS Notification dated November. 2009)

S. no.	Pollutants	Time weighted average	Concentration in ambient air		Methods of measurement
			Industrial, residential, rural and other areas	Ecologically sensitive area (notified by central government	
1	Sulphur dioxide (SO_2) $\mu g\,m^{-3}$	Annual[a]	50	20	1. Improved West and Gaeke 2. Fluorescence
		24 h[b]	80	80	
2	Nitrogen dioxide (NO_2) $\mu g\,m^{-3}$	Annual[a]	40	30	1. Modified Jacob & Hochheiser 2. Chemiluminescence
		24 h[b]	80	80	
3	Particulate matter (size <10 $\mu g\,m^{-3}$)	Annual[a]	60	60	1. Gravimetric 2. TEOM 3. Beta attenuation
		24 h[b]	100	100	
4	Particulate matter (size <2.5 μm) or $PM_{2.5}$ $\mu g\,m^{-3}$	Annual[a]	40	40	1. Gravimetric 2. TEOM 3. Beta attenuation
		24 h[b]	60	60	
5	Ozone (O_3) $\mu g\,m^{-3}$	8 h[b]	100	100	1. UV photometric 2. Chemiluminescence 3. Chemical method
		1 h[b]	180	180	
6	Lead (Pb) $\mu g\,m^{-3}$	Annual[a]	0.50	0.50	1. AAS/ICP Method after sampling using EPM 2000 or equivalent filter paper 2. EO-XRF using Teflon filter
		24 h[b]	1.0	1.0	
7	Carbon monoxide (CO) $mg\,m^{-3}$	8 h[b]	02	02	Non dispersive infra red (NDIR) spectroscopy
		1 h[b]	04	04	
8	Ammonia (NH_3) $\mu g\,m^{-3}$	Annual[a]	100	100	1. Chemiluminescence 2. Indophenol blue method
		24 h[b]	400	400	
9	Benzene (C_6H_6) $\mu g\,m^{-3}$	Annual[a]	05	05	1. Gas chromatography based continuous analyzer 2. Adsorption and Desorption followed GC analysis
10	Benzo(a)Pyrene (BaP) particulate phase only, $ng\,m^{-3}$	Annual[a]	01	01	Solvent extraction followed by HPLC/GC analysis
11	Arsenic (As) $ng\,m^{-3}$	Annual[a]	06	06	AAS/ICP after sampling on EPM 2000 or equivalent filter paper
12	Nickel (Ni) $ng\,m^{-3}$	Annual[a]	20	20	AAS/ICP after sampling on EPM 2000 or equivalent filter paper

Note: Whenever and wherever monitoring results in two consecutive days of monitoring exceed the limits specified above for the respective category, it shall be considered adequate reason to institute regular or continuous monitoring and further investigation.
[a]*Annual Arithmetic mean of minimum 104 measurements in a year at a particular site taken twice a week 24 hourly at uniform interval.*
[b]*24 hourly 08 hourly or 01 hourly monitored values, as applicable shall be compiled with 98% of the time in a year. 2% the time, they may exceed the limits but not on two consecutive days of monitoring.*
Source: https://cpcb.nic.in/upload/NAAQS_2019.pdf.

16.2.2.3 Responsibilities of different tiers of governance

The air pollution challenge in India is inherently multisectoral and multijurisdictional. In India, the central government formulates environmental policies and legislation which are then implemented at state levels by state governments.

Several policy measures have been taken by the Ministry of Environment, Forest and Climate Change, Government of India, to tackle the adverse effects of air pollution in the short and long term. The introduction of

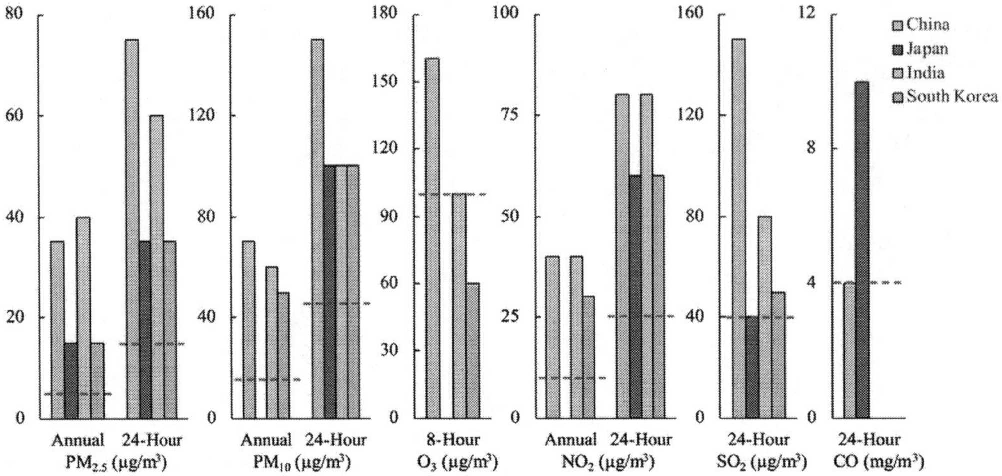

FIGURE 16.2 Comparison of national ambient air quality standards of China (blue), Japan (purple), India (green), and South Korea (yellow) with the air quality guidelines (red line). Source: *From Ouyang, H., Tang, X., Kumar, R., Zhang, R., Brasseur, G., Churchill, B., Alam, M., Kan, H., Liao, H., Zhu, T., Chan, E.Y.Y., Sokhi, R., Yuan, J., Baklanov, A., Chen, J., & Patdu, M.K. (2022). Toward better and healthier air quality implementation of WHO 2021. Global air quality guidelines in Asia.* Bulletin of the American Meteorological Society, 103*(7), E1696–E1703. https://doi.org/10.1175/BAMS-D-22-0040.1 © American Meteorological Society. Used with permission.*

TABLE 16.4 IND-air quality index.

Air quality index category	Air quality index range	Possible health impacts
Good	0–50	Minimal impact
Satisfactory	51–100	Minor breathing discomfort to sensitive people
Moderate	101–200	Breathing discomfort to the people with lungs, asthma and heart diseases
Poor	201–300	Breathing discomfort to most people on prolonged exposure
Very poor	301–400	Respiratory illness on prolonged exposure
Severe	401–500	Affects healthy people and seriously impacts those with existing disease

Source: Adapted from National air quality index (https://cpcb.gov.in).

odd-even license-plate based driving restrictions in Delhi whereby vehicles were restricted from use on alternate days, the government's decision to adopt compressed natural gas (CNG) as an alternative fuel to petrol and diesel, and improvements in fuel and vehicle quality to lower emissions are some of the measures adopted to reduce air pollution (Gurjar, 2021).

In November 2016, the Supreme Court instructed the government to produce and implement a Graded Response Action Plan (GRAP) to limit the cases of severe pollution in Delhi NCR. In January 2017 the GRAP was notified by the Ministry of Environment, Forest and Climate Change. The GRAP includes a set of reactive and increasingly tougher measures based on the level of air pollution in the city. The actions suggested for each level are cumulative and add up to the level of emergency.

Recognizing the need for concerted cross-jurisdiction and airshed level action and coordination, India's Parliament established the Commission for Air Quality Management in National Capital Region and Adjoining Areas Bill 2021 (Ministry of Environment, Forest and Climate Change, Government of India, The Commission for Air Quality Management in National Capital Region and Adjoining Areas Bill, 2021). This includes Haryana, Punjab, Rajasthan, and Uttar Pradesh, states adjoining the National Capital Territory of Delhi, where any source of pollution may cause adverse impact on air quality in the NCR. The Bill also dissolved the Environment Pollution Prevention and Control Authority established in the NCR in 1998. The functions of CAQM include: (1) coordinating actions by adjoining state governments, (2) planning and executing plans to prevent and control air pollution in NCR, (3) providing a framework for identifying air pollutants, (4) conducting research and development through networking with technical institutions, (5) training and creating a special workforce to deal with issues related to air pollution, and (6) preparing action plans such as increasing plantation and addressing stubble burning (Ministry of Environment, Forest and Climate Change, Government of India, The Commission for Air Quality Management in National Capital Region and Adjoining Areas Bill, 2021).

16.2.2.4 Specific measures at different spatial/temporal scales

Controlling the air quality in the NCR has emerged as a challenging task because of complex interactions between local/regional emissions and typical atmospheric conditions. Specific interventions have been introduced in the NCR. For the transport sector, for instance, some of these measures include the use of EVs as a mode of public transportation, the use of bioethanol as fuel, and the construction of multilevel car parking facilities to tackle congestion. To tackle the problem of open burning of garbage and household wastes, door-to-door collection of segregated wastes has been introduced and several compost pits have been established.

The GRAP has been developed in NCR to advise the government to take necessary measures to prevent severe air quality episodes in the capital city. GRAP is a set of emergency measures that are imposed to prevent further deterioration of air quality once it reaches a certain threshold in the Delhi-NCR region. Approved by the Supreme Court in 2016 and notified in 2017, the plan was formulated after several meetings that the Environment Pollution (Prevention and Control) Authority (EPCA) held with state government representatives and experts. The result was a plan that institutionalized measures to be taken when air quality deteriorates.

Stage 1 of GRAP is activated when the AQI is in the "poor" category (201—300), Stage 2 is when it's in the "very poor" category (301—400), Stage 3 is when the AQI is the "Severe" category (401—450) and finally Stage 4 is when it rises to the "Severe + " category (more than 450). GRAP is incremental in nature and thus, when the air quality dips from "poor" to "very poor," measures listed under both sections have to be followed.

For example, on December 15, 2022 when the AQI in NCR dipped to the "severe" category, Stage 4 of GRAP was set in motion. A central panel directed authorities to prohibit the use of diesel four-wheelers that are not BS-VI compliant, as well as the entry of trucks in Delhi. Petrol cars were allowed to continue to operate as usual. Additionally, all industries in the NCR running on fuels apart from those approved for the NCR were to be shut, except for dairy units and those manufacturing medical equipment or drugs. Construction and demolition activities, including those on linear public projects like highways, roads, flyovers, pipelines and power lines, which were previously allowed to continue, were also banned (Fig. 16.3).

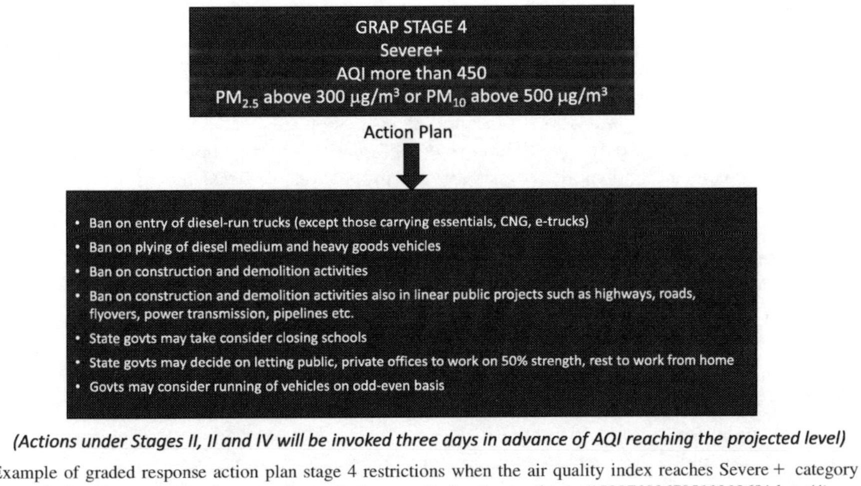

FIGURE 16.3 Example of graded response action plan stage 4 restrictions when the air quality index reaches Severe + category in Delhi national capital region. Source: *Indian Express Twitter account (https://twitter.com/IndianExpress/status/1588719267851091968/photo/4).*

NCR schools have regularly been closed during winter episodes under the provision of GRAP Stage 4. The implementation of GRAP measures is complicated by the fact that multiple actors are involved in the process and have different responsibilities.

Another set of interventions include the provision of liquefied petroleum gas (LPG) to the (mostly poorer) households who often use biomass as fuel for cooking and heating. The Government of India (GoI) in 2016 launched Pradhan Mantri Ujjwala Yojana (Government of India, Ministry of Petroleum and Natural Gas, 2016), which aimed to provide 80 million "below poverty level" households with LPG by 2019. Under the program, about 75% of the planned LPG connections had been dispersed by January 2019 (Dabadge, Sreenivas, & Josey, 2018). GoI also initiated the Deen Dayal Upadhyay Grameen Jyoti Yojana in 2014, which promises to electrify all villages in India, thus eradicating emissions from kerosene lighting.

In Delhi NCR, more stringent measures and large-scale shifts in practice have already been undertaken. These include switching the public transport system to CNG in 2001 (Narain & Krupnick, 2007) and the 2018 implementation of Bharat Stage VI emissions standards to curb vehicular emission in this region. In 2014, the Delhi government restricted the entry of heavy goods vehicles from 7 am to 10 pm and in 2016, the government banned the use of diesel vehicles older than 10 years and registration of diesel vehicles >2000 cc capacity. A total ban on open waste burning was also introduced in 2015.

16.2.2.5 Performance of past and current management plans/strategies

The GRAP measures to be undertaken under "severe" air pollution conditions have been implemented during winter episodes since 2017. The efficacy of each of the enforced actions remains largely unclear. By design these actions might prove to be helpful to address critical episodes of high pollution concentrations, however, they do not address the issue of year-round exceedances of the annual means which would also require more cooperation between multiple authorities, and even multiple states. The GRAP thus falls short of a comprehensive, long-term mitigation strategy for Delhi. There are many issues with GRAP's implementation, and with the adequacy of the proposed measures against the scale of the challenge.

Performance of the odd-even car trial, conducted in Delhi for two periods January 1–15 and April 15–30, 2016, was evaluated by Kumar et al. (2017). The authors found that although the scheme reduced the hourly average net $PM_{2.5}$ and PM_{10} by up to 74% during most of the day (after 1100 hours) compared with the corresponding hours during the previous year, daily average $PM_{2.5}$ and PM_{10} were up to 3 times higher during the trial periods when compared with the pre–trial days. The persistence of overnight emissions from heavy goods vehicles into the morning hours (0800–1100 hours) meant the scheme was less effective at that time. The major setback to the odd-even car policy during winters can be attributed to the stable atmosphere (winds are not strong enough to disperse the $PM_{2.5}$) and the failure to control the $PM_{2.5}$ from outside the periphery of the city (Gulia, Mittal, & Khare, 2018).

Narain and Krupnick (2007) found that the conversion of buses from diesel to CNG has helped to reduce PM_{10}, CO, and SO_2 concentrations in the city and has not led to the recent increase in NO_2. At the same time, however, the conversion of three-wheelers from petrol to CNG has not had the same benefit, possibly because of poor technology. Narain and Krupnick (2007) state that another policy measure that appears to have had a positive impact on air quality is the reduction in the sulfur content of diesel and petrol. This has led to a decrease in SO_2 levels and, consequently, PM_{10} concentrations because of the conversion of SO_2 to sulfates ($PM_{2.5}$). Goyal and Sidhartha (2003) found that the introduction of the CNG bus fleet had resulted in the reduction of suspended particulate matter and other pollutants by nearly 10%. Some of these gains from fuel switching and fuel-quality improvements are, however, being negated by the increase in number of vehicles and more specifically the proportion of diesel-fueled cars, which is leading to an increase in PM_{10} and NO_2 levels (Narain & Krupnick, 2007).

To tackle the air pollution caused by burning crop-residues, the Punjab State Government relied heavily on awareness campaigns and arranging 42,342 new subsidized crop residue management (CRM) machines in 2022, even though the farm unions continued to insist on financial assistance to clear their fields. The Punjab Government claims to have provided over 122,000 CRM machines to farmers in the last few years (As season ends, Punjab farm fires lowest in 11 years, 2022).

16.2.2.6 Future policy directions

The management of air pollution sources, particularly those which affect territories over thousands of kilometers such as stubble burning, require high levels of coordination and cooperation. Now, scientists are looking to a different approach (airshed air quality management) to tackling the country's toxic air, based on the recognition that pollution transcends national and state borders (Centre for Science and Environment, 2022). The idea of airsheds is not new. The Canadian province of British Columbia and California in the USA have both used an airshed approach to tackle air pollution. The idea is that all towns and cities that lie within an airshed need to coordinate with each other for an effective solution (Government of British Columbia, 2017). The region of the Indo-Gangetic plain, for example, may be considered as one airshed. The northern and central subcontinents (Pakistani and Indian Punjab and much of NCR) are a subset of this broader airshed. The towns in this airshed don't have any control over the pollution in their jurisdiction. That is why an airshed approach as air-quality management is required. It will also require a legal policy to ensure that cities take responsibility (Ganguly, Selvaraj, & Guttikunda, 2020).

Green sectors such as clean energy and e-mobility are likely to be the driving force for developing and implementing tangible solutions to improve air quality (Sinha & Chandra, 2020).

Business and industry can significantly accelerate momentum towards achieving India's air quality standards by providing the necessary investment, technical expertise and large-scale coordination. IKEA is a case in point. IKEA launched a collection in 2019 that used rice straws (crop residue) as raw material. Farmers in Northwest India burn approximately 39 million tonnes of rice straw residue on their farms every year in order to clear the field for the next harvest—these activities contribute to the formation of smog that envelops Delhi during the stubble burning season (Singh et al., 2022). IKEA's innovative solution showcases the potential of repurposing a waste product into an economically viable resource that reduces air pollution.

It has been felt by many observers that more still needs to be done to further reduce levels of air pollution in Delhi NCR. The existing measures need to be strengthened and scaled up. Governmental efforts alone are not enough. Participation of the community is crucial to make a palpable effect in the reduction of pollution. Delhi Government, in 2020, launched the Green Delhi app to enable users to file complaints related to pollution causing activities. With the launch of the mobile application, the Delhi government plans to involve citizens in the fight against air pollution. The application is a part of the Delhi government's anti air pollution campaign "Yudh Pradushan ke Virudh" (War Against Pollution) launched in October 2020. The statement said that "a government cannot bring any change without the support of its citizens."

Another approach could be to set up an ombudsman body of eminent citizens as an independent public grievance redressal mechanism to maintain citizen and government engagement and transparency. Establishing a public information and feedback mechanism between citizens and government before plans are implemented, could ensure a successful collaboration with ownership of the problem and solutions.

16.2.3 Republic of South Africa

This section explores air quality management in the Republic of South Africa. The current air quality management process is relatively young and has evolved from the National Environmental Management: Air Quality Act, 2004 (Act No. 39 of 2004) (Republic of South Africa, National Environment Management: Air Quality Act 39 of 2004, 2004). This section provides an overview of the key tools used in South Africa within the context of their specific air pollution and societal challenges.

16.2.3.1 Historical context for current air quality legislation

The approach to air pollution control in South Africa originated from the Atmospheric Pollution Prevention Act, 1965 (Republic of South Africa, 1965) (APPA) and the Bill of Rights in the Constitution of the Republic of South Africa, 1996. The APPA was regarded as an inadequate legislative tool to efficiently and effectively address modern air quality challenges due to its lack of proactive management; a perceived bias towards industry through the application of a "best practicable means" strategy; an inadequate emissions control and penalty system; and a general weakness in the legal enforcement of the Act (Barnard, 1999; Scott, 2005). Consequently, the National Environmental Management Act, 1998, (NEMA) was established which set out the regulatory tools for cooperative environmental governance and the principles for decision-making. One of these tools was the Republic of South Africa, National Environmental Management: Air Quality Act, 2004 (Act No. 39 of 2004) (AQA) which set out the requirements for the practical mechanisms, systems, procedures, norms and standards to allow South Africa to achieve compliance with ambient air quality standards and to meet international agreements and obligations (see Table 16.5). The AQA introduced international best practice including the devolution of responsibility to local government and effects-based management (Naiker, Diab, Zunckel, & Hayes, 2012).

TABLE 16.5 Key air quality management tools established by the Air Quality Act.

Section 7(1) of Air Quality Act	Control of emissions from point and nonpoint sources
	Air quality monitoring
	Air quality management planning
	Air quality information management
	Any other matter which the Minister considers necessary
Section 7(2) of Air Quality Act	**Opportunities for public participation**
	Public access to air quality information
	Prevention of air pollution and degradation of air quality
	Reduction of discharges likely to impair air quality, including the reduction of air pollution at source
	The promotion of efficient and effective air quality management
	Effective air quality monitoring
	Regular reporting on air quality
	Compliance with South Africa's obligations in terms of international agreements

Source: Adapted from the Republic of South Africa, National environment management: Air quality act 39 of 2004. (2004). https://www.gov.za/sites/default/files/gcis_document/201409/a39-04.pdf.

16.2.3.2 Legislation and air quality thresholds

16.2.3.2.1 National framework for air quality

The National Framework for Air Quality Management, established in 2007 as a requirement of the AQA, set the foundations for the air quality management regime implemented in South Africa today (Hanekom, 2018). As required by the AQA, the National Framework has been reviewed and updated in 2012 and 2017. These five-year milestones have allowed the air quality management regime to evolve from fundamental capacity building projects in 2007 and 2012, to more ambitious regulatory activities since 2017. In the two decades since the AQA, there have been several significant achievements, including the establishment of national ambient air quality standards; the declaration of three national priority areas and the development of related management plans; the publication and continual upgrading of the South African Air Quality Information System (SAAQIS); the establishment of the South African Atmospheric Emission Licensing and Inventory Portal (SAAELIP); and the publication of annual National Air Quality Officer's Reports for context and accountability.

16.2.3.2.2 National ambient air quality standards

In addition to the air quality management mechanisms outlined in Sections 7(1) and 7(2) of the AQA, it also introduced ambient air quality standards (Naiker et al., 2012). Table 16.6 sets out the national ambient air quality standards for South Africa as established by the AQA.

16.2.3.3 Responsibilities of different tiers of governance

Like the approach in the United Kingdom (see Section 16.2.1), air quality management is devolved to provincial and municipal governments. However, the process required the spheres of government to work cooperatively. There is a positive working relationship between the different municipalities and the province but there is notable variation in government capacity and capabilities to fulfill its duties. National and provincial governments provided strategic direction, guidance and oversight to support municipal governments who have a responsibility to implement the technical management processes (e.g., air quality monitoring) and control measures (Naiker et al., 2012).

TABLE 16.6 Ambient air quality standards for South Africa.

	Averaging period	Concentration	Number of exceedances	Compliance date
Sulphur dioxide (SO_2)	10-m	500 $\mu g\,m^{-3}$	526	Immediate
	1-h	350 $\mu g\,m^{-3}$	88	Immediate
	24-h	125 $\mu g\,m^{-3}$	4	Immediate
	Annual	50 $\mu g\,m^{-3}$	N/A	Immediate
Nitrogen dioxide (NO_2)	1-h	200 $\mu g\,m^{-3}$	88	Immediate
	Annual	40 $\mu g\,m^{-3}$	N/A	Immediate
Particulate matter (PM_{10})	24-h	120 $\mu g\,m^{-3}$	4	Immediate – 31/12/2014
	24-h	75 $\mu g\,m^{-3}$	4	01/01/2015
	Annual	50 $\mu g\,m^{-3}$	N/A	Immediate – 31/12/2014
	Annual	40 $\mu g\,m^{-3}$	N/A	01/01/2015
Ozone (O_3)	8-h	120 $\mu g\,m^{-3}$	11	Immediate
Benzene (C_6H_6)	Annual	10 $\mu g\,m^{-3}$	N/A	Immediate – 31/12/2014
	Annual	5 $\mu g\,m^{-3}$	N/A	01/01/2015
Lead (Pb)	Annual	0.5 $\mu g\,m^{-3}$	N/A	Immediate
Carbon monoxide (CO)	1-h	30 $mg\,m^{-3}$	88	Immediate
	8-h	10 $mg\,m^{-3}$	11	Immediate

Source: Adapted from Republic of South Africa, National environment management: Air quality act 39 of 2004. (2004). https://www.gov.za/sites/default/files/gcis_document/201409/a39-04.pdf.

1. National Government assigns air quality management to the Department of Environment, Forestry and Fisheries (DEFF). This department is responsible for the development and implementation of national air quality policies, the air quality standards, coordination of monitoring programmes and establishing technical and policy guidance for the two lower tiers of government.
2. Provincial Governments work with DEFF to identify national air quality priority areas, develop and implement related Air Quality Management Plans, apply the air quality standards and support air quality monitoring networks.
3. Municipal Governments work with provincial and national governments to implement air quality management plans, maintain monitoring networks, and raise awareness about air quality through public engagement activities.

16.2.3.3.1 South African air quality information system

Air quality in South Africa is monitored through a strategic network of more than 135 continuous analyzers operated by the Government and 50 privately owned analyzers (accurate as of February 2023). However, the data capture from these stations is variable, for example, in 2020 only 37 analyzers met the minimum data threshold of $>=75\%$ for national reporting, 63 did not meet the threshold and 37 were not operational (Republic of South Africa, 2021).

Worked example 16.8: Why are there difficulties in maintaining good data capture from monitoring stations in South Africa?

Solution: The reasons for this poor performance include resourcing (finance, personnel and infrastructure), supply chain challenges, power supply, security and technical capacity (Republic of South Africa, 2021).

The information from this network is disseminated to the public via the SAAQIS (https://saaqis.environment.gov.za/) which is described as *"a 'one-stop-shop' for all air quality information, from monitoring to legislation, as well as notices, guidelines and contact information of air quality officials in different jurisdictions across the country"* (Gwaze & Mashele, 2018). The SAAQIS also reports a South African specific AQI derived from six criteria pollutants (i.e., PM_{10}, $PM_{2.5}$, CO, O_3, SO_2, and NO_2) based on National Ambient Air Quality Standards. The SAAQIS also provides educational information on pollutants such as a description of what the different pollutions are, the sources of these pollutants and the related health impacts (Table 16.7).

TABLE 16.7 Air quality indicators published by South African air quality information system.

Air quality indicator		Advice for at-risk individuals	Advice for general population
Descriptor	Low/Good	Enjoy your usual outdoor activities	Enjoy your usual outdoor activities
Score	1–3		
Color	Green		
Descriptor	Moderate	Individuals who experience symptoms should consider reducing strenuous physical activity, especially outdoors	Enjoy your usual outdoor activities
Score	4–5		
Color	Yellow		
Descriptor	High/ Unhealthy	Reduce strenuous physical exertion, especially outdoors. People with asthma may find they need their inhaler more. Older people should reduce physical exertion.	Anyone experiencing discomfort (e.g., sore eyes, cough, sore throat) should consider reducing activity especially outdoors.
Score	6–7		
Color	Orange		
Descriptor	Very high/ Very unhealthy	Avoid strenuous activity outdoors. People with asthma may find they need their inhaler more.	Reduce physical exertion especially outdoors and especially if you are experiencing symptoms (e.g., sore eyes, cough, sore throat)
Score	8–9		
Color	Red		
Descriptor	Hazardous	Avoid strenuous activity outdoors. People with asthma may find they need their inhaler more.	Reduce physical exertion especially outdoors and especially if you are experiencing symptoms (e.g., sore eyes, cough, sore throat)
Score	10		
Color	Purple		

Source: Adapted from the South African air quality information system (SAAQIS) (https://saaqis.environment.gov.za/).

16.2.3.3.2 South African atmospheric emission licensing

The South African National Atmospheric Emission Licensing process was established in 1998 and is the main process for the regulation of industrial and commercial activities. The National Atmospheric Emissions Inventory System (NAEIS) is facilitated by an online platform, the South African Atmospheric Emission Licensing and Inventory Portal (SAAELIP) (https://saaelip.environment.gov.za/SAAELIP/home/), which acts as a repository for Atmospheric Emissions Licenses (AEL) and reporting. The system sets out the criteria for obtaining an AEL including the registration and monitoring requirements for any activities that require a license. It sets emission standards for industries, regulates compliance with these standards through inspections and audits of activities and, if required, the imposition of fines and other penalties for noncompliance. The National Air Quality Officer's Annual Report (Republic of South Africa, 2022) indicated that reporting by the regulated sector had improved from 76% in 2015 to >90% in 2021. There is also an ambition for the development and refinement of the system to include emission inventories from nonindustrial sources to determine a more robust total emission picture for the country.

16.2.3.3.3 National priority areas and air quality management plans

Three national priority areas have been declared to date—(1) Vaal Triangle Airshed declared in 2006, (2) Highveld declared in 2007 and (3) Waterberg-Bojanala declared in 2015. Additionally, there are seven provinces, seven metropolitan and 34 district authorities with action plans under implementation. Following the declaration of the national priority areas, the responsible authority developed an Air Quality Management Plan (AQMP) which, as defined by the AQA, sets out the measures and resources needed to implement an Air Quality Strategy or to achieve the air quality standards in the priority area. The AQMP provide the framework for implementing departments and industry to include air quality management in business planning to ensure effective implementation and monitoring.

16.2.3.3.3.1 Vaal Triangle and Highveld air quality management plan The Gauteng Province includes the Cities of Johannesburg and Pretoria and has two air quality priority areas [Vaal Triangle airshed priority area (VTAPA) and Highveld priority area (HPA)]. Gauteng Province has three Metropolitan Municipalities (City of Tshwane, City of Johannesburg, City of Ekurhuleni) and two District Municipalities (Sedibeng and West Rand).

The VTAPA was declared in April 2006 and amended in August 2007. The first priority area declared in South Africa, the VTAPA has complex air pollution challenges due to the high density of commercial, agricultural, residential, industrial (including a coal-fired power plant) and mining activities (Scorgie, Kneen, & Burger, 2003). The VTAPA is heavily populated with large high-density informal residential settlements, where coal and wood-burning are typical (Moreoane, Mukwevho, & Burger, 2021). Based on limited monitoring data at the time of declaration, the VTAPA was established primarily due to concerns with PM_{10}, with SO_2 and O_3 also problematic. The VTAPA AQMP (National Environmental Management: Air Quality Act, 2004 (Act No. 39 of 2004) Vaal Triangle Air-Shed Priority Area Air Quality Management Plan, 2009) identified a suite of objectives with related activities, timeline and indicators for success for various sectors:

1. Minimize the biomass burning impacts.
2. Reduce air pollution concentrations in domestic fuel-burning areas and to promote affordable alternative energy sources (e.g., energy-efficient stoves).
3. Ensure Iron, Steel and Ferroalloy industries achieve compliance with national emission limits and ambient air quality standards.
4. Minimize fugitive dust and gaseous emissions from mining operations (e.g., materials handling operations, crushing and screening, chemically treated haul roads, control of wind erosion etc.).
5. Use best available techniques in the petrochemical sector to minimize emissions from waste dumps, stockpiles combustion processes.
6. Reduce emissions in the vicinity of power generation plants.
7. Achieve acceptable local air quality near small industries (e.g., fuel-burning appliances at schools, hospitals, stand-by generators etc).
8. Minimize emissions from waste burning and uncontrolled burning in landfill sites.
9. Enhance governance at a National, Provincial and Local level through the prioritization of air quality in government, better finances and resources and capacity building.
10. Ensure readily available air quality information by prioritization of ambient air quality monitoring and the standardization of monitoring practices and accreditation.

The HPA was declared in November 2007. Like the Vaal Triangle, the HPA has a high concentration of industrial and nonindustrial sources. As the HPA cuts across multiple municipality and provincial boundaries, the national government acts as the lead authority in the development of the AQMP. Within the HPA, approximately half of the annual emissions of PM_{10} is attributed to opencast mine haulage roads, 17% from metallurgical industry, and 12% from power generation. Power generation accounted for 73% of NO_x and 82% of SO_2 emissions. The HPA was declared because ambient air quality standards for SO_2, PM_{10} and O_3 were exceeded in nine extensive locations (although not all pollutants were of concern in all locations). The HPA AQMP (Republic of South Africa, Highveld Priority Area Air Quality Management Plan, 2012) sets out seven key goals with related objectives, activities, timeline and indicators for success. The seven HPA goals are as follows:

1. By 2015, organizational capacity in government is optimized to efficiently and effectively maintain, monitor and enforce compliance with ambient air quality standards.
2. By 2020, industrial emissions are equitably reduced to achieve compliance with ambient air quality standards and dust fallout limit values.
3. By 2020, air quality in all low-income settlements is in full compliance with ambient air quality standards.
4. By 2020, all vehicles comply with the requirements of the National Vehicle Emission Strategy.
5. By 2020, a measurable increase in awareness and knowledge of air quality exists.
6. By 2020, biomass burning and agricultural emissions will be 30% less than current.
7. By 2020, emissions from waste management are 40% less than current.

16.2.3.3.3.2 Waterberg–Bojanala air quality management plan The Waterberg-Bojanala Area (WBPA), located in the north-east of South Africa, was declared in June 2012. The WBPA cuts across the Limpopo and North-West Provinces including the district municipalities of Waterberg and Bojanala Platinum. The region has extensive mining activities including notable coalfields and is strategically important to South Africa due to its coal-fired power plants. The WBPA AQMP (National Environmental Management: Air Quality Act: Waterberg Bojanala Priority Area Air Quality Management Plan, 2015) reported that 99.9% of SO_2 and 87% of NO_x emissions are from industry, while 74% and 24% of PM_{10} emissions came from mining and industrial activities respectively. Based on limited monitoring data and extensive modeling study, the WBPA was established primarily due to concerns with SO_2, NO_2 and PM_{10}. The WBPA AQMP (2015) sets out five key goals with related objectives, activities, timeline and indicators for success. The five WBPA goals are as follows:

1. Address shortcomings in cooperative governance in the WBPA to improve local structures, mechanisms and capacity to implement the AQMP.
2. Improve the systems and tools such as emission inventories, ambient monitoring and modeling, and enforcement.
3. Compliance with national ambient air quality standards through emission control and reduction across all sectors.
4. Ensure sound science and research are used to establish a health baseline and prioritizes emission reduction interventions.
5. Improve communication and current levels of knowledge of air quality amongst stakeholders.

16.2.3.4 Performance of past and current management plans/strategies

The Air Quality Act of 2004 is considered a significant success as it provides a legal framework for air quality management in South Africa and underpins the development and implementation of various air quality management tools over the last two decades including the National Framework for Air Quality Management, SAAQIS, NAEIS and national and local action plans. However, the data indicates that measured concentrations, especially for PM_{10}, remain a concern. Tshehla and Wright (2019) provided a review of air quality management since the introduction of the legislation to tackle air pollution in South Africa. They concluded that many of the strategic objectives of the AQA are still outstanding:

1. While national and provincial action plans are in place or under development, the decentralization of air quality management is struggling with many district authorities without an action plan as required under the Municipal Systems Act No. 32 of 2000.
2. Moreoane et al. (2021) identified that action plans needed better funding mechanisms to assist in implementing intervention strategies. This could be through a system of environmental pollution taxes/levies for polluting industries in line with the polluter-pays principle.
3. There is a lack of source apportionment of air pollution across all three spheres of government. This lack of evidence impacts the ability of the responsible authority to effectively target interventions and mitigation strategies.
4. There is a lack of enforcement, possibly related to a lack of capacity and capabilities, to scrutinize industry compliance reports, chase noncompliance, and check consistency and adherence to guidance within the regulatory authority. This has been recognized by the national government who are implementing a suite of training programmes across all aspects of air quality management.
5. South Africa has invested in air quality monitoring systems to collect and report data on pollution levels across the country. Section 8(b) of the Air Quality Act states that each municipality should have a minimum of one monitoring station. Tshehla and Wright (2019) report that there are only 121 government-managed stations across 278 municipalities. Additionally, there are concerns about the data quality from many stations. The significant gaps in the monitoring network and the upskilling of staff are a priority for the national government.
6. There is a need for a detailed cost-benefit analysis to determine the impact of air quality on the economy.
7. There is a need to close the gap between science and policy. Additionally, there is a need to address the limited public participation in air quality management.

16.2.3.5 Current state of air quality in South Africa

The combination of multiple emission sources, complex topography and meteorology, and the social inequalities that exist in South African society make the management of air pollution and its public health impacts challenging for the province and the country. Determining the effectiveness of different measures spatially and temporally is challenging with limited measure-specific data available. Many of the measures implemented in South Africa to date have been primarily focussed on establishing and refining appropriate air quality management tools and regulating and reducing emissions from industrial pollution. The State of Environment Report uses a National Air Quality Indictor (NAQI) as a measure of the efficacy of policy interventions (Republic of South Africa, 2022). The NAQI is based on the normalized ratio of the annual averages of PM_{10} and SO_2 measured at the air quality monitoring stations to determine long-term

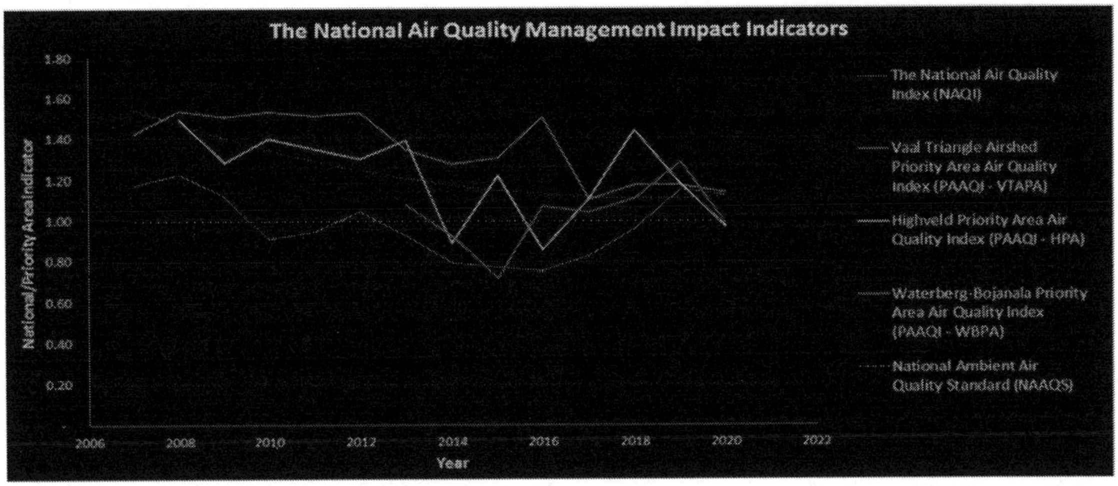

FIGURE 16.4 National Air Quality Indicator values from 2007 to 2020. Annual trends in National Air Quality Indicator values from 2007 to 2020 in four districts of South Africa against the National Ambient Air Quality Standard. Source: *From National air quality officer's annual report on air quality management in the republic of South Africa, South African National government. (2022). (Original work published 2022).*

trends (Fig. 16.4). The NAQI shows continued annual exceedances but the levels of PM_{10} in the VTAPA and HPA are improving due to specific measures such as national emissions standards, cleaner fuel regulations and the priority area action plans.

The National Air Quality Officer's Annual Report (Republic of South Africa, 2022), indicated that SO_2 concentrations are comfortably below the air quality standard but PM_{10} continues to be a notable concern with concentrations close to or exceeding the air quality standard for the last 10–15 years. Moving forward, there is a recognized need to continue with national and local actions and explore the cobenefits of low carbon interventions. Additionally, there is a need to widen the perspective of air quality management to focus on other sources, especially road transport and domestic solid fuel burning, and the role of people and their behavior in the generation of pollution.

16.2.3.6 Future policy directions

Across South Africa, particulate matter remains the pollutant of most concern due to the variety of sources, health impacts and the unique climate conditions in the country. National policy programmes have made notable progress in establishing the mechanisms and tools for air quality management but limited progress in pollution reduction. Priorities include:

1. Continue uptake of clean technologies and fuels.
2. Enhancement of the air quality monitoring network density and data quality.
3. Strengthening environmental emission standards and regulation of compliance.
4. Raising public awareness of their role in generating pollution and their exposure to pollution and the subsequent health implications.
5. Capacity development across all spheres of government for technical assessment, licensing and monitoring.
6. Maximizing the ancillary benefits or comanagement of air pollution and carbon emissions.

Despite progress made in recent years, several challenges and emerging issues need to be considered in future legislation and management approaches such as illegal burning, illegal industrial operators, addressing road transport emissions, domestic solid fuel burning for heating and cooking and indoor air quality. A coordinated effort from multiple stakeholders such as government, industry, civil society, and the public will be required to address these.

16.2.4 Mexico (Mexico City)

This section focuses on Mexico, more specifically the Mexico City Metropolitan Area (MCMA), which includes Mexico City and contiguous municipalities. MCMA is one of the world's largest megacities with an estimated 20 million inhabitants living on the dried bed of the elevated lake Texcoco and its surroundings. The inland basin is at an altitude of 2240 meters above sea level and is surrounded on three sides by mountains and volcanoes. It has over 40,000 industries and >5.5 million vehicles that consume >49 million liters of petroleum fuels per day (Calderón-Garcidueñas et al., 2015). The MCMA geographic setting and climatological characteristics along with the low residential mobility allow for the opportunity to study chronic health effects associated with sustained year-long air pollutant exposures (Calderón-Garcidueñas et al., 2015).

Air pollution levels in MCMA frequently exceed the WHO Guidelines (López-Feldman et al., 2021). Mexico City has some of the worst air quality in the Western Hemisphere, with particulate levels (PM_{10}) that are three to four times higher than New York, Los Angeles, São Paulo, or Buenos Aires (Davis, 2017). Vehicular emissions are major contributors to air pollution in MCMA, reflected by a vehicular fleet of 5.7 million. Air pollution in MCMA is also influenced by its topography and high-altitude location.

16.2.4.1 Historical context for current legislation

During the 20th century, Mexico City experienced a huge increase in population and urbanization as it attracted migrants from other parts of the country and industrialization stimulated economic growth. Population growth, increasing motorization and industrial activities, and a constrained geographical basin and intense solar radiation combined to cause intense air quality problems of both primary and secondary pollutants (Molina et al., 2019).

Mexico government recognized air pollution as a social and public health problem and began implementing air pollution control strategies at the end of the 1980s. Mexico City was once the most polluted mega-city worldwide, however, there has been a significant achievement in reducing air pollution exposures to the population after the implementation of several policies, over more than three decades. Significant progress has been made since the 1990s in improving the air quality of MCMA by implementing successive programmes that combined regulatory actions with the support of technology and scientific research. Despite air quality improvements, current $PM_{2.5}$ and O_3 levels in Mexico City are still above air quality standards.

In Mexico, the right to clean air and a healthy environment is supported by different levels of Mexican legislation: (a) Constitution: Article 4 recognizes the right to a healthy environment, (b) the recent Constitution of Mexico City: Article 16 guarantees the right to a healthy environment and requires the development of public policies for the protection of the environment, including the atmosphere (Molina et al., 2019).

16.2.4.2 Legislation and air quality thresholds

Air quality legislation has been in place in Mexico since the 1970s. The current Air Quality Mexican Official Standards defined by the Secretary of Health, regulate seven criteria pollutants by setting maximum allowable concentrations (Table 16.8). In 1994, the Mexican government published the first ambient air quality standards for ozone, carbon monoxide, nitrogen dioxide, sulphur dioxide, lead and particulate matter. The standards have been reviewed periodically resulting in updated standards for ozone and particulate matter in 2014. The air quality standards published by the government are almost identical to those of the United States of America. Current air quality standards can be viewed at INECC's website https://sinaica.inecc.gob.mx/pags/nomsIAS.php and also at https://www.transportpolicy.net/standard/mexico-air-quality-standards/.

The Official Mexican Standards (NOM-172-SEMARNAT-2019) have been set to establish the guidelines for obtaining and communicating the Air Quality and Health Risk Index (Indice de Calidad del Aire), in order to report in a clear, timely and continuous manner the status of air quality, probable damage to health it causes and the measures that can be taken to reduce exposure.

The Air and Health Index was introduced to raise public awareness of air pollution. The National Air Quality Information System, SINAICA (https://sinaica.inecc.gob.mx/index.php) provides the latest data on Air and Health Index across Mexico. The information at SINAICA comes from Air Quality Monitoring Systems, SMCA, which are managed by different governments (national, state and municipal). SINAICA also provides recommendations to the general public for different Air Quality and Health Index categories (Table 16.9).

16.2.4.3 Responsibilities of different tiers of governance

The Secretariat of Environment (SEDEMA) is responsible for Mexico City's environmental programs, including air quality management and climate action plans. Similarly, the environment secretariats of the State of Mexico and the State of Hidalgo manage the environmental programs of their respective states. In 1996, the Metropolitan Environmental Commission (CAM) was created to coordinate governance of metropolitan environmental problems (Molina et al., 2019). In 2013, CAM was replaced by the Megalopolis Environmental Commission (CAMe), covering

TABLE 16.8 Official Mexican standards (NOM-172-SEMARNAT-2019) *N.B. ppm = parts per million.*

Pollutant [final rule NOM]		Standard	Averaging time	Form
Ozone [NOM-020-SSA1–2014]		0.095 ppm	1-h	Not to be exceeded over a period of one year
		0.070 ppm	8-h	75th percentile, over a period of one year
Carbon monoxide [NOM-021-SSA1–1993]		11.00 ppm	8-h	Not to be exceeded more than once per year
Sulphur dioxide [NOM-022-SSA1–1993]		0.110 ppm	24-h	Not to be exceeded more than once per year
		0.025 ppm	Annual	Not to be exceeded
Nitrogen dioxide [NOM-023-SSA1–1993]		0.210 ppm	1-h	Not to be exceeded more than once per year
Particle pollution [NOM-025-SSA1–2014]	PM_{10}	75 $\mu g\,m^{-3}$	24-h	75th percentile of each hour concentration per day, averaged over a period of one year
		40 $\mu g\,m^{-3}$	Annual	75th percentile of each 24-h concentration per trimester, averaged over one year (at least 3 trimesters)
	$PM_{2.5}$	45 $\mu g\,m^{-3}$	24-h	75th percentile of each hour concentration per day, averaged over a period of one year
		12 $\mu g\,m^{-3}$	Annual	75th percentile of each 24-h concentration per trimester, averaged over one year (at least 3 trimesters)
Lead [NOM-026-SSA1–1993]		1.5 $\mu g\,m^{-3}$	3 months	Not to be exceeded

Source: From NOM-172-SEMARNAT-2019.

TABLE 16.9 Recommendations for various categories of air and health index, Mexico.

Air quality	Risk level	Recommendations	
		Sensitive groups	**The whole population**
Good	Low	The risk is considered to be minimal or nil	Enjoy outdoor activities
Acceptable	Moderate	Consider reducing vigorous physical activities outdoors	Enjoy outdoor activities
Bad	High	Avoid physical activities (both moderate and vigorous) outdoors	Reduce vigorous physical activities outdoors
Very bad	Very high	Do not do outdoor activities. See a doctor if respiratory or cardiac symptoms occur	Avoid moderate and vigorous physical activities outdoors.
Extremely bad	Extremely high	Stay indoors. See a doctor if respiratory or cardiac symptoms occur	

Source: From https://sinaica.inecc.gob.mx/recomendaciones.html.

Mexico City and the five surrounding states (Puebla, Tlaxcala, Morelos, Hidalgo, and Mexico) (Molina et al., 2019). Mexico's Ecology and Climate Change Institute (INECC) monitors air quality nationwide to allow for more effective enforcement.

The Mexican government has created four Comprehensive Programs Against Air Pollution since 1990, including strategies for transport, planning, environmental education and reforestation.

The first air quality management plan was developed for Mexico City as part of an extensive technical cooperation of the Federal District authorities and several international organizations.

Subsequently, a series of Proaire comprehensive air pollution control plans were released by the CAM, which comprises the Federal District and State of Mexico environment agencies and the federal environmental authority (the Secretariat for the Environment and Natural Resources, SEMARNAT) (Fernández-Bremauntz, 2008). The current program, ProAire IV contains 81 measures and 116 actions across eight themes: reduction of energy consumption, cleaner energy, promotion of public transport, emission control, environmental education, citizen participation, green areas, and scientific research and strengthening of health protection (Cárdenas Gonzáles, 2018).

Air quality management plans for the cities of Guadalajara, Monterrey, Ciudad Juarez, Tijuana, and Mexicali were also developed by the National Institute of Ecology and the respective local authorities between 1997 and 2000. A plan to improve air quality in the industrial area of Salamanca was also agreed between SEMARNAT and Guanajuato authorities, introducing control measures for regional refinery and power generation facilities and a pollution-episodes-plan to address extreme peaks of suphfur dioxide (Fernández-Bremauntz, 2008).

16.2.4.4 Specific measures at different spatial/temporal scales

Both the Mexican Government and the citizens of Mexico City have recognized air pollution as a major environmental and social concern since the mid-1980s. In the 1990s, comprehensive air quality management programmes were developed and implemented. Specific actions included the removal of lead from gasoline and the implementation of catalytic converters in automobiles; reduction of sulfur content in diesel transportation fuel; substitution of fuel oil in industry and power plants with natural gas; and reformulation of liquified petroleum gas used for heating and cooking. The government also strengthened the vehicle inspection and maintenance program; vehicles were required to be inspected in a centralized system and more frequent inspections were required for higher emitting vehicles as incentives to promote fleet turnover and help ensure the proper maintenance of vehicles.

Mexico City has one of the oldest and best-known driving restriction policies. *Hoy No Circula* (HNC) (No-Driving Day) program was introduced in 1989, and it bans most drivers from using their vehicles one weekday per week on the basis of the last digit of the vehicle's license plate. For example, vehicles with a license plate ending in "5" or "6" cannot be used on Mondays. The restrictions are in place week days between 5:00 am and 10:00 pm and affect both residential and commercial vehicles. When first imposed in 1989, the restrictions applied to 2.3 million vehicles, or 460,000 vehicles per day. Compliance with the program is nearly universal. City police vigorously enforce the restrictions and vehicles violating the ban are easy to spot. Drivers caught violating the rules must pay a hefty fine and their vehicles are impounded for 48 hours (Davis, 2017). The HNC program was expanded to include Saturdays in July 2008. The primary rationale for the expansion was air quality, with the Mexico City government pointing to a steady recent increase in Saturday air pollution levels.

Since 2015, the city has run a pilot diesel particulate filter retrofit program which is effective in reducing particulate emissions. Automatic number plate recognition (ANPR) systems are used in Mexico City to enforce speed limits on sensitive parts of the urban network (International Transport Forum, 2017). The penalties for contravening the restriction in Mexico City are mostly enforced by police with fines.

In addition to the HNC program, at a local scale, a number of programmes such as the Clean Transport Program, the Auto Regulation Program, the Vehicle Substitution Program and the Program for Modernizing the Federal Road Transport Fleet have been developed in Mexico City to promote fleet improvements and renewal for diesel heavy-duty diesel vehicles.

16.2.4.5 Performance of past and current management plans/strategies

Hoy No Circula was evaluated by Davis (2008). The study found that across pollutants there is no evidence that the restrictions have improved air quality. Evidence from additional sources (Eskeland & Feyzioglu, 1997; Gallego, Montero, & Salas, 2013) indicates that the restrictions led to an increase in the total number of vehicles in circulation as well as a change in composition toward high-emissions vehicles. The expansion of *Hoy No Circula* to Saturdays was evaluated by Davis (2017). The study also found that there is little evidence that the program expansion improved air quality. Across eight major pollutants, the program expansion had virtually no discernible effect on pollution levels.

Worked example 16.9: Why did the *Hoy No Circula* strategy fail to improve air quality? How could this be redressed?

Solution: Data from subway, bus, and light rail ridership showed no evidence that the expansion was successful in getting drivers to switch to lower-emitting forms of transportation. In addition, Mexico City has one of the largest taxi fleets in the world. Taxis are available everywhere in the city and are relatively cheap. For a driver who is prohibited from using their vehicle one Saturday per month, taking a taxi is an easy, convenient alternative (Davis, 2017). To help solve this problem, regulations on taxi fleet emissions could reduce air pollution from this source, however this would not tackle traffic congestion. Incentives for drivers to use alternative modes, including public transport, coupled with an effective awareness-raising campaign, could help to motivate a switch away from private vehicles and taxis.

Nevertheless, as a result of these regulatory actions, combined with technology change, concentrations of criteria pollutants have been decreasing over the past decade, despite the continuing increase in population and economic activity. However, Mexico City's levels of particulate matter and ozone still exceed NOM and WHO standards.

In the MCMA, although the Megalopolis Environmental Commission was established to coordinate the regional air quality programs of Mexico City with the surrounding municipalities of five states, the different jurisdictions and capacities are major barriers to policy implementation.

A recent study (Muñoz-Pizza et al., 2022) illustrates differences across cities in the technical capacity to produce evidence, and document barriers shared to incorporate evidence into air quality management – including institutional capacity, interest groups, legal frameworks, and limited knowledge transfer. The study highlighted that the opportunities that increase the integration of evidence into air quality policies include follow-up programs, guaranteed research funding, contextualizing the topics to be investigated as a public problem, and assigning a crucial role to research so that formal and informal links among decision-makers and scientists are formed and strengthened.

16.2.4.6 Future policy directions

Mexico City has been working on improving air quality for a number of years. Much progress has been made in tackling air-pollution problems through comprehensive air-quality management programmes based on scientific, technical, social and political considerations. Yet continuing pressure from an increasing urban population and the desire of people to have a better quality of life causes an ongoing need to improve air quality.

The original PROAIRE program recognized that people must be involved in any solution and included various formal and informal programs to inform people about the problem and invite them to action. It was recognized that a cultural change was needed to modify the society-city-environment relation. However, the public participation and engagement has been low so far. This needs to be strengthened in any future policy and program.

Pilot studies in Mexico have catalyzed a national program to identify and control lead exposure in children and pregnant women (Téllez-Rojo et al., 2019).

A National Air Quality Strategy, called Estrategia Nacional de Calidad del Aire (ENCA), is being designed in Mexico City for the year 2030 with objectives and strategic lines for prevention, control and mitigation of air pollution (Secretaría de Medio Ambiente y Recursos Naturales, n.d.).

16.3 Summary

The case studies presented in this chapter illustrate commonalities and diversities in terms of the air pollution challenges faced in different locations, and in the management plans and strategies adopted by governments in response. In all cases, the process of air quality management has continued to evolve as the understanding of the sources and effects increases and the effectiveness of measures is evaluated. Although these case studies present only a few key examples of air quality management, it is hoped that they provide an indication of the basic concepts, and of the opportunities and challenges that governments face, that may be applicable to other areas in the world, especially in countries that do not currently have air quality management regimes in place.

16.3.1 Dual approach: emissions and concentrations

One common theme in addressing air quality is the dual-pronged approach of emissions control from specific source sectors, as identified in national and local emission inventories, and the management of ambient concentrations of a range of pollutants, as determined through monitoring and modeling studies. Whilst tackling air pollution at source is the most effective way to reduce emissions, managing exposure to air pollution requires an understanding of how these strategies affect concentrations of pollutants spatially and temporally.

16.3.2 Setting health-based standards

The 2021 WHO Air Quality Guidelines represent the most recent epidemiological evidence, however, 99% of the global population is exposed to concentrations exceeding these thresholds. Many governments, following the WHO's recommendation, have therefore adopted their own air quality standards for certain pollutants that take into account local priorities and risk management. Typically LMICs are subject to higher pollutant concentrations due to the combination of contributing sources arising from rapid population and economic growth. The costs and practical challenges of reducing concentrations in these areas to WHO AQG levels may be incompatible with priorities to grow local economies and raise standards of living of the populace. One of the biggest challenges governments face is how to decouple economic growth from emissions generation and achieve sustainable development.

16.3.3 Using air quality indices

Many countries use national AQIs, normally to communicate short-term health impacts of air pollution episodes to vulnerable groups and the wider populations. As with the different air quality standards set in different countries, governments use slightly different AQI methodologies, descriptors and bandings and are reflective of the levels of pollution typically experienced in those locations. For example, the UK Daily AQI uses bandings from 1–10 (Low to Very High) based on the highest short-term concentration of five pollutants using different averaging periods (hourly mean NO_2, 15-minute mean SO_2, 8-hourly mean O_3, and 24-hourly mean PM_{10} and $PM_{2.5}$). The Indian AQI uses a range of 1–500 (Good to Severe) based on the highest 24-hourly averages (PM_{10}, $PM_{2.5}$, NO_2, SO_2, NH_3, and Pb) and 8-hourly means (CO, O_3). In the UK the "Very High" category for the $PM_{2.5}$ component is 71 $\mu g\,m^{-3}$ or more, whereas in India, the "Severe" $PM_{2.5}$ subindex equates to 250 + $\mu g\,m^{-3}$. A $PM_{2.5}$ subindex of 60 $\mu g\,m^{-3}$ would be classed as "Satisfactory" in the Indian AQI, but in the UK the same concentration would be classified as "High." The health impact at this $PM_{2.5}$ concentration is clearly the same regardless of which country you are in, however, the norm in many Indian cities is so much higher than that typically experienced in the UK. In India, where the AQI thresholds also act as triggers for the implementation of GRAP measures, adopting UK DAQI equivalent descriptor thresholds would effectively grind the local economy to a halt. In addition, some platforms have been developed using other AQIs (e.g., https://www.aqi.in/, https://waqi.info/) which, although can help to harmonize across different locations, can also be confusing when compared to countries' official AQIs.

16.3.4 Developing action plans

Another commonality across the case studies is the use of exceedances of air quality standards as a trigger for the development of action plans to remediate. For example, LAQM AQAPs in the UK, the NCAP in India, Delhi NCR's short-term GRAP, and AQMPs in South Africa. Whilst pragmatic in terms of targeting resources where it is most required, this approach means that members of the public, and ecosystems, must be subject to damaging levels of pollution before any action can be taken to resolve the situation. In the case of long-term exceedances, the development and subsequent implementation of action plans can take years, during which public health is continuously undermined. For short-term exceedances, although the implementation of preprepared short-term action plans may be immediate, the impact may take longer to manifest and the consequences of exposure to such high levels of air pollution will be fatal for many. Adoption of air quality strategies, for example in UK local authorities that are not in breach of AQOs, can be a useful way to work towards reducing pollutant levels before they become a problem. However, recognizing that vulnerable people may be susceptible to the effects of air pollution even below so-called health-based standards, it would seem to be a more equitable approach to pursue continual exposure reduction, with particular attention to those at high risk.

16.3.5 Integrated policy approach

Whilst in many cases AQAPs and strategies are standalone documents, these are sometimes integrated into other related policy area plans, that is, climate action plans, transport plans and spatial development plans. Even where plans are targeted specifically at improving air quality, these are most effective when developed with input from other policy leads to ensure that common goals (win-wins) and potential conflicts (trade-offs) are identified. Furthermore, the actions that are required to reduce air pollution are dependent on these other policy leads, as government departments that undertake an assessment of air quality do not have the means to be able to directly influence the sources and causes of the exceedances measured.

16.3.6 Multilevel governance

As demonstrated in the case studies in this chapter, responsibility for the assessment and management of air quality usually requires multilevel governance. National governments generally set standards (e.g., National AQOs in the UK, operate national scale monitoring networks and develop national strategies to tackle common or large-scale sources. State or municipal governments are then tasked with adopting these strategies, identifying local hotspots of air pollution and enacting plans to address these. This approach adopts a principle of subsidiarity which should ensure cost-effective solutions, however, there can be tensions where there are political differences between national and state level, as in Delhi, and where national policies run counter to local objectives, example, adopting tax incentives for diesel vehicles in the UK (Barnes et al., 2014, 2018).

16.3.7 Challenges

Transboundary pollution can pose a challenge to the achievement of air quality standards, particularly in the case of the production of secondary $PM_{2.5}$, which effectively elevates background concentrations, rendering local plans ineffective. Managing secondary pollution requires management of the precursor pollutants; for example, management of O_3 concentrations requires the reduction of both NO_x and VOCs. Natural sources, although usually discounted in reporting air quality exceedances, can also elevate concentrations and catalyze production of secondary pollutants, whilst topography can hamper the dispersion of air pollution and add to the challenges of management. As noted in the India case study, adopting an airshed approach may be an effective way to manage this complexity.

One of the greatest challenges for policymakers is the reliance on behavior change. For example, encouraging people to leave their cars at home and walk, cycle or catch the bus requires a change of mindset and depends on the availability of these alternative means. Policy measures that restrict traffic, such as Low Traffic Neighborhoods or School Streets, or charging zones, such as London's ULEZ expansion and Clean Air Zones in some UK local authority areas, have met forceful opposition. Even those members of the public who are supportive of such measures in principle, may not consider the alternatives to private car use workable, or even desirable, for them, particularly where they have already spent money purchasing, insuring and taxing a vehicle. Engendering behavior change therefore requires policymakers to facilitate, by making the desired behavior cheaper, easier and more convenient than the status quo, and as equitably as possible to avoid entrenching systemic health disparities. Involving people in cocreating or codeveloping air quality policy, example, via citizens' assemblies and community sensing projects, can be an effective way to both inform and empower publics, and to achieve solutions that have public support and fit their ways of living.

16.3.8 Future considerations

As we have seen in the case studies presented, air quality management is ever-evolving as new issues come to the fore. Here we highlight a few examples of areas where it is likely that future policy will be required. (Chapter 17 addresses the issue of future challenges more fully.)

16.3.8.1 Indoor air pollution

In many LMICs, indoor pollution has reduced relative to outdoor pollution as traditional practices of domestic biomass burning, for cooking, heating and lighting, have been replaced with lower emission fuels, such as LPG, as in India. In HICs, like the UK, however, the reverse is true. As sources of outdoor pollution have reduced through the successful implementation of policy measures, indoor air quality has gone unregulated and increases in wood-burning for esthetic or economic reasons, has increased. This trend is particularly concerning in urban areas where this contributes to concentrations of outdoor $PM_{2.5}$, however, there is also the consideration of the impact that this is having on indoor concentrations. There are multiple sources of indoor air pollutants, apart from heating. For example, cooking, cleaning, refurbishments and mold can all contribute varying amounts of pollution, which, without adequate ventilation, can lead to significant concentrations in indoor environments. Increasing attention is being given to understanding and addressing indoor pollution sources to protect public health. (Further information on indoor air pollution can be found in Chapter 12.)

16.3.8.2 Emerging sources

As well as the increases in solid fuel burning and awareness of indoor sources, there are a number of other new or potential pollutant sources that policymakers will need to consider. Wildfires, for instance, are increasing in frequency and ferocity as climate change leads to drought-like conditions. As well as their obvious immediate threat to life, they are also responsible for extremely high concentrations of PM in populated areas, often for extended periods of time. Similarly, desertification can lead to dust storms with widespread impacts.

Non-exhaust emissions, from brake and tyre-wear and resuspended road dust, have been identified as a significant contributor of PM and metals (e.g., Cu and Zn) and in some places now exceed exhaust emissions (Air Quality Expert

Group, 2019; Fussell et al., 2022). As even EVs contribute to nonexhaust emissions, policymakers must ensure that they are not simply substituting fossil-fueled vehicles like-for-like.

Airborne microplastics are a novel source that is proving to be ubiquitous, as they are shed from synthetic fabrics and road tyres. Research is exploring their potential toxicity to determine whether this is another area that will need to come under regulation.

16.4 Questions

1. How does assessment of air quality concentrations relate to exposure of receptors and health effects?
2. What types of emission controls on road vehicles have been implemented and how effective have they been at reducing pollutants?
3. Why is important to consider policies to reduce air pollution and to achieve net zero emissions together, and what are the obstacles in doing so?
4. What are some of the challenges that governments face in implementing policies that require individuals and households to change their behavior?
5. Describe the relationship between economic growth and air pollution in high income countries and low- and middle-income countries.

Solutions to the end-of-chapter questions will be available at: https://www.elsevier.com/books-and-journals/book-companion/9780128225912.

References

Air Quality Expert Group. (2019). Non-Exhaust Emissions from Road Traffic, 1907101151_20190709_Non_Exhaust_Emissions_typeset_Final.pdf. <https://uk-air.defra.gov.uk/assets/documents/reports/cat09/1907101151_20190709_Non_Exhaust_Emissions_typeset_Final.pdf>.

Arbour, M.-E. (2022). The Volkswagen scandal at the CJEU: Defeat devices between the conformity guarantee and environment law. *European Journal of Risk Regulation, 13*(4), 670–681. Available from https://doi.org/10.1017/err.2022.32.

As season ends, Punjab farm fires lowest in 11 years. (2022). *The Times of India*. <https://timesofindia.indiatimes.com/city/chandigarh/as-season-ends-punjab-farm-fires-lowest-in-11-years/articleshow/95838783.cms>.

Ayala, A. (2018). In recognition of this special journal issue on real-world emissions on CARB's leadership and central role in uncovering and resolving the environmental violations from Volkswagen's use of illegal diesel defeat devices. *Science of The Total Environment, 642*, 1441–1443. Available from https://doi.org/10.1016/j.scitotenv.2018.06.235.

Balakrishnan, K., Dey, S., Gupta, T., Dhaliwal, R. S., Brauer, M., Cohen, A. J., Stanaway, J. D., Beig, G., Joshi, T. K., Aggarwal, A. N., Sabde, Y., Sadhu, H., Frostad, J., Causey, K., Godwin, W., Shukla, D. K., Kumar, G. A., Varghese, C. M., Muraleedharan, P., & Dandona, L. (2019). The impact of air pollution on deaths, disease burden, and life expectancy across the states of India: The global burden of disease study 2017. *The Lancet Planetary Health, 3*(1), e26–e39. Available from https://doi.org/10.1016/S2542-5196(18)30261-4.

Barnard, D. (1999). *Environmental law for all: A practical guide for the business community, the planning professions, environmentalists, and lawyers*. Impact Books. Available from https://openlibrary.org/books/OL6826882M/Environmental_law_for_all.

Barnes, J. H., Hayes, E. T., Chatterton, T. J., & Longhurst, J. W. S. (2014). Air quality action planning: Why do barriers to remediation in local air quality management remain? *Journal of Environmental Planning and Management, 57*(5), 5. Available from https://doi.org/10.1080/09640568.2012.762573.

Barnes, J. H., Hayes, E. T., Chatterton, T. J., & Longhurst, J. W. S. (2018). Policy disconnect: A critical review of UK air quality policy in relation to EU and LAQM responsibilities over the last 20 years. *Environmental Science and Policy*, 85. Available from https://doi.org/10.1016/j.envsci.2018.03.024.

Beevers, S. D., Carslaw, D. C., Dajnak, D., Stewart, G. B., Williams, M. L., Fussell, J. C., & Kelly, F. J. (2016). *Traffic management strategies for emissions reduction: Recent experience in London*, 27–39. Available from https://doi.org/10.2147/EECT.S69858.

Beig, G., Bano, S., Sahu, S. K., Anand, V., Korhale, N., Rathod, A., Yadav, R., Mangaraj, P., Murthy, B. S., Singh, S., Latha, R., & Shinde, R. (2020). COVID-19 and environmental -weather markers: Unfolding baseline levels and veracity of linkages in tropical India. *Environmental Research, 191*, 110121. Available from https://doi.org/10.1016/j.envres.2020.110121.

Beig, G., Ghude, S. D., & Deshpande, A. (2010). Scientific evaluation of air quality standards and defining air quality index for India. Indian Institute of Tropical Meteorology, Pune. India Ministry of Earth Sciences, Govt. of India. Available from: https://citeseerx.ist.psu.edu/document?repid=rep1&type=pdf&doi=4f7d28f727afa813d93fea5306886668758b914f.

Brimblecombe, P. (2017). *Air pollution episodes* (6). World Scientific Publishing Co. Pte. Ltd. Available from https://doi.org/10.1142/q0098#t = toc.

Bristol City Council. (2022). Air quality annual status report 2022. <https://www.bristol.gov.uk/files/documents/5074-air-quality-annual-status-report-2022/file>.

Brunt, H., Barnes, J., Longhurst, J. W. S., Scally, G., & Hayes, E. (2016). Local air quality management policy and practice in the UK: The case for greater public health integration and engagement. *Environmental Science and Policy*, 58. Available from https://doi.org/10.1016/j.envsci.2016.01.009.

C40 Cities Climate Leadership Group. (2022). Clean air, healthy planet: A framework for integrating air quality management and climate action planning. <https://www.c40knowledgehub.org/s/article/Clean-air-healthy-planet-A-framework-for-integrating-air-quality-management-and-climate-action-planning?language = en_US>.

Calderón-Garcidueñas, L., Kulesza, R. J., Doty, R. L., D'Angiulli, A., & Torres-Jardón, R. (2015). Megacities air pollution problems: Mexico city metropolitan area critical issues on the central nervous system pediatric impact. *Environmental Research*, *137*, 157–169. Available from https://doi.org/10.1016/j.envres.2014.12.012.

Cárdenas Gonzáles, B. (2018). PROGRAM TO IMPROVE AIR QUALITY IN MEXICO CITY - PROAIRE. In: Sawin, E., McCauley, S., Edberg, S., Mwaura, G. & José Gutiérrez, M., (2018). Multisolving at the Intersection of Health and Climate: Lessons from Success Stories, https://img.climateinteractive.org/2018/01/Multisolving-at-the-Intersection-of-Health-and-Climate.pdf

Carslaw, D. C., Beevers, S. D., Tate, J. E., Westmoreland, E. J., & Williams, M. L. (2011). Recent evidence concerning higher NO_x emissions from passenger cars and light duty vehicles. *Atmospheric Environment*, *45*(39), 7053–7063. Available from https://doi.org/10.1016/j.atmosenv.2011.09.063.

Centre for Science and Environment. (2022). Managing regional air quality—Need for a framework. <https://www.cseindia.org/managing-regional-air-quality-need-for-a-framework-11205>.

Clean Air Act 1956, (1956). <https://www.legislation.gov.uk/ukpga/Eliz2/4-5/52/enacted>.

ClientEarth. (2019). What do ClientEarth's legal cases mean for local authority plans to deliver nitrogen dioxide compliance in England and Wales? <https://www.clientearth.org/media/m1borg0p/what-do-clientearths-legal-cases-mean-for-local-authority-plans-to-deliver-nitrogen-dioxide-compliance-in-england-and-wales-ce-en.pdf>.

Clientearth No.3, R (on the application of) v Secretary of state for environment, food and rural affairs & Ors [2018] EWHC 315 (Admin), (EWHC (Admin) 21 February 2018). <http://www.bailii.org/ew/cases/EWHC/Admin/2018/315.html>.

Corton, C. L. (2015). *London fog: The biography*. Belknap Press.

CPCB. (2009). National ambient air quality standards. <https://cpcb.nic.in/uploads/National_Ambient_Air_Quality_Standards.pdf>.

CPCB. (2014). National air quality index. Final report. <https://app.cpcbccr.com/ccr_docs/FINAL-REPORT_AQI_.pdf>.

CPCB. (2021, September 16). About NAMP. <https://cpcb.nic.in/about-namp/>.

C-664/18 (CJEU) (2021, March 4). <https://eur-lex.europa.eu/legal-content/EN/TXT/HTML/?uri = CELEX:62018CJ0664>.

Dabadge, A., Sreenivas, A. and Josey, A. (2018) What has the Pradhan Mantri Ujjwala Yojana achieved so far? Economic & Political Weekly. LIII (20).

Davis, L. W. (2008). The effect of driving restrictions on air quality in Mexico City. *Journal of Political Economy*, *116*(1), 38–81. Available from https://doi.org/10.1086/529398.

Davis, L. (2017). Saturday driving restrictions fail to improve air quality in Mexico City. *Scientific Reports*, *7*, 41652. Available from https://doi.org/10.1038/srep41652.

Defra. (2018, January 11). 25 Year environment plan. GOV.UK. <https://www.gov.uk/government/publications/25-year-environment-plan>.

Defra. (2022a). Local air quality management policy guidance (PG22). <https://laqm.defra.gov.uk/guidance/>.

Defra. (2022b, October 6). Clean air zone framework. GOV.UK. <https://www.gov.uk/government/publications/air-quality-clean-air-zone-framework-for-england/clean-air-zone-framework>.

Defra (2022c). Local air quality management technical guidance (TG22). Table 1.1 <https://laqm.defra.gov.uk/wp-content/uploads/2022/08/LAQM-TG22-August-22-v1.0.pdf>.

Defra. (2023). Air quality strategy: framework for local authority delivery GOV.UK. 28 April 2023 [online]. Available from: https://www.gov.uk/government/publications/the-air-quality-strategy-for-england/air-quality-strategy-framework-for-local-authority-delivery [Accessed 16 May 2023].

Defra. (2023). Environmental improvement plan. <https://www.gov.uk/government/publications/environmental-improvement-plan>.

Demuynck, J., Bosteels, D., Bunar, F., & Spitta, J. (2020). Diesel passenger car with ultra-low NOx emissions in real driving conditions. *MTZ Worldwide*, *81*(1), 40–43. Available from https://doi.org/10.1007/s38313-019-0151-8.

Department for Transport. (2022). Annual road traffic estimates 2021 [Statistical Release]. <https://assets.publishing.service.gov.uk/government/uploads/system/uploads/attachment_data/file/1107056/road-traffic-estimates-in-great-britain-2021.pdf>.

De Vito, L., Chatterton, T., Namdeo, A., Nagendra, S., Gulia, S., Goyal, S., Bell, M., Goodman, P., Longhurst, J., Hayes, E., Kumar, R., Sethi, V., Sengupta, B., Ramadurai, S. B. G., Majumder, S., Menon, J. S., Turamari, M. N., & Barnes, J. (2018). *Air Pollution in Delhi: A Review of Past and Current Policy Approaches*, *230*, 441–451. Available from https://doi.org/10.2495/AIR180411.

Ellison, R. B., Greaves, S. P., & Hensher, D. A. (2013). Five years of London's low emission zone: Effects on vehicle fleet composition and air quality. *Transportation Research Part D: Transport and Environment*, *23*, 25–33. Available from https://doi.org/10.1016/j.trd.2013.03.010.

Environment Act (2021). <https://www.legislation.gov.uk/ukpga/2021/30/introduction/enacted>.

Environment Act 1995, § Part IV Air Quality (1995). <https://www.legislation.gov.uk/ukpga/1995/25/part/IV>.

Environmental Protection Act 1990, § Part III Statutory Nuisances and Clean Air (1990). <https://www.legislation.gov.uk/ukpga/1990/43/part/III>.

Eskeland, G. S., & Feyzioglu, T. (1997). Rationing can backfire: The 'Day without a Car' in Mexico City. *The World Bank Economic Review*, *11*(3), 383–408. Available from https://www.jstor.org/stable/3990252.

European Union (Withdrawal) Act 2018. Retrieved 29 May 2023, from <https://www.legislation.gov.uk/ukpga/2018/16/contents/enacted>.

Fernández-Bremauntz, A. (2008). Air quality management in Mexico. *Journal of Toxicology and Environmental Health, Part A, 71*(1), 56–62. Available from https://doi.org/10.1080/15287390701557909.

Font, A., Guiseppin, L., Blangiardo, M., Ghersi, V., & Fuller, G. W. (2019). A tale of two cities: Is air pollution improving in Paris and London. *Environmental Pollution, 249*, 1–12. Available from https://doi.org/10.1016/j.envpol.2019.01.040.

Forum for International Cooperation on Air Pollution. (n.d.). <https://www.naturvardsverket.se/en/environmental-work/international-cooperation/multilateral-cooperation/forum-for-international-cooperation-on-air-pollution/> (accessed 14.12.22).

Fuller, G. (2019). *The invisible killer: the rising global threat of air pollution - and how we can fight back.* Melville House UK.

Fussell, J. C., Franklin, M., Green, D. C., Gustafsson, M., Harrison, R. M., Hicks, W., Kelly, F. J., Kishta, F., Miller, M. R., Mudway, I. S., Oroumiyeh, F., Selley, L., Wang, M., & Zhu, Y. (2022). A review of road traffic-derived non-exhaust particles: Emissions, physicochemical characteristics, health risks, and mitigation measures. *Environmental Science & Technology, 56*(11), 6813–6835. Available from https://doi.org/10.1021/acs.est.2c01072.

Gallego, F., Montero, J.-P., & Salas, C. (2013). The effect of transport policies on car use: Evidence from Latin American cities. *Journal of Public Economics, 107*, 47–62. Available from https://doi.org/10.1016/j.jpubeco.2013.08.007.

Ganguly, T., Selvaraj, K. L., & Guttikunda, S. K. (2020). National Clean Air Programme (NCAP) for Indian cities: Review and outlook of clean air action plans. *Atmospheric Environment: X, 8*, 100096. Available from https://doi.org/10.1016/j.aeaoa.2020.100096.

Government of British Columbia. (2017). B.C. Airsheds—Province of British Columbia. <https://www2.gov.bc.ca/gov/content/environment/air-land-water/air/air-quality-management/airsheds?keyword = 2016&keyword = unit> (accessed 01.07.23).

Government of India, Ministry of Petroleum and Natural Gas. (2016, May 1). Pradhan Mantri UjjwalaYojana—Scheme guidelines. <https://www.pmuy.gov.in/about.html>.

Goyal, P., & Sidhartha. (2003). Present scenario of air quality in Delhi: A case study of CNG implementation. *Atmospheric Environment, 37*(38), 5423–5431. Available from https://doi.org/10.1016/j.atmosenv.2003.09.005.

Greater London Authority. (2020). Central London ultra low emission zone – Ten month report. <https://www.london.gov.uk/sites/default/files/ulez_-ten_month_evaluation_report_23_april_2020.pdf>.

Greater London Authority. (2022a). London air quality map. LGOV. <https://www.london.gov.uk/programmes-and-strategies/environment-and-climate-change/pollution-and-air-quality/london-air-quality-map>.

Greater London Authority. (2022b). The Mayor's ultra low emission zone (ULEZ) for London. LGOV. <https://www.london.gov.uk/programmes-strategies/environment-and-climate-change/pollution-and-air-quality/mayors-ultra-low-emission-zone-ulez-london>.

Greater London Authority. (2023). Inner London ultra low emission zone expansion one year report. London City Hall. <https://www.london.gov.uk/programmes-strategies/environment-and-climate-change/environment-and-climate-change-publications/inner-london-ultra-low-emission-zone-expansion-one-year-report>.

Greater London Authority (2025). London-wide ultra low emission zone one year report. London City Hall [online]. Available from <https://www.london.gov.uk/programmes-strategies/environment-and-climate-change/environment-and-climate-change-publications/london-wide-ultra-low-emission-zone-one-year-report>. (Accessed 30 March 2025).

Gulia, S., Mittal, A., & Khare, M. (2018). Quantitative evaluation of source interventions for urban air quality improvement—A case study of Delhi city. *Atmospheric Pollution Research, 9*(3), 577–583. Available from https://doi.org/10.1016/j.apr.2017.12.003.

Gurjar, B.R. (2021, April 5). Air pollution in India: Major issues and challenges. <https://www.teriin.org/article/air-pollution-india-major-issues-and-challenges>.

Gwaze, P., & Mashele, S.H. (2018). South African air quality information system (SAAQIS) mobile application tool: Bringing real time state of air quality to South Africans. Clean Air Journal, 28 (1). Available from https://doi.org/10.17159/2410-972X/2018/v28n1a1.

Hajmohammadi, H., & Heydecker, B. (2022). Evaluation of air quality effects of the London ultra-low emission zone by state-space modelling. *Atmospheric Pollution Research, 13*(8). Available from https://doi.org/10.1016/j.apr.2022.101514.

Hanekom, D.A. (2018). The 2017 National Framework for Air Quality Management in the Republic of South Africa.

House of Commons Environment, Food and Rural Affairs, Environmental Audit, Health and Social Care, and Transport Committees. (2018). Improving air quality, Fourth Report of the Environment, Food and Rural Affairs Committee, Fourth Report of the Environmental Audit Committee, Third Report of the Health and Social Care Committee, and Second Report of the Transport Committee of Session 2017–19.

International Transport Forum. (2017). Strategies for Mitigating Air Pollution in Mexico City: International best practice, OECD. https://www.itf-oecd.org/sites/default/files/docs/air-pollution-mitigation-strategy-mexico-city.pdf.

Kumar, P., Gulia, S., Harrison, R. M., & Khare, M. (2017). The influence of odd–even car trial on fine and coarse particles in Delhi. *Environmental Pollution, 225*, 20–30. Available from https://doi.org/10.1016/j.envpol.2017.03.017.

Longhurst, J. W. S., Olowoporoku, A. O., Barnes, J., Chatterton, T. J., Hayes, E. T., & Irwin, J. G. (2011). Process efficiency and outcome effectiveness in the United Kingdom's local air quality management regime. *International Journal of Sustainable Development and Planning, 6*(3). Available from https://doi.org/10.2495/SDP-V6-N3-286-298.

López-Feldman, A., Heres, D., & Marquez-Padilla, F. (2021). Air pollution exposure and COVID-19: A look at mortality in Mexico City using individual-level data. *The Science of the Total Environment, 756*, 143929. Available from https://doi.org/10.1016/j.scitotenv.2020.143929.

Ma, L., Graham, D. J., & Stettler, M. E. J. (2021). Has the ultra low emission zone in London improved air quality. *Environmental Research Letters, 16*(12), 124001. Available from https://doi.org/10.1088/1748-9326/ac30c1.

Ministry of Environment, Forest and Climate Change. (2019). National Clean Air Programme (NCAP). <https://moef.gov.in/wp-content/uploads/2019/05/NCAP_Report.pdf>.

Ministry of Environment, Forest and Climate Change, Government of India, The Commission for Air Quality Management in National Capital Region and Adjoining Areas Bill (2021). Testimony of ministry of environment, forest and climate change, government of India. <https://prsindia.org/billtrack/the-commission-for-air-quality-management-in-national-capital-region-and-adjoining-areas-bill-2021>.

Molina, V., Retama., & Zavala. (2019). Experience from integrated air quality management in the Mexico City Metropolitan Area and Singapore. *Atmosphere*, *10*(9), 512. Available from https://doi.org/10.3390/atmos10090512.

Moreoane, L., Mukwevho, P., & Burger, R. (2021). The quality of the first and second Vaal Triangle airshed priority area (VTAPA) air quality management plans. *Clean Air Journal*, *31*(2). Available from https://doi.org/10.17159/caj/2020/31/2.12178.

Mudway, I. S., Dundas, I., Wood, H. E., Marlin, N., Jamaludin, J. B., Bremner, S. A., Cross, L., Grieve, A., Nanzer, A., Barratt, B. M., Beevers, S., Dajnak, D., Fuller, G. W., Font, A., Colligan, G., Sheikh, A., Walton, R., Grigg, J., Kelly, F. J., & Griffiths, C. J. (2019). Impact of London's low emission zone on air quality and children's respiratory health: A sequential annual cross-sectional study. *The Lancet Public Health*, *4*(1), e28–e40. Available from https://doi.org/10.1016/S2468-2667(18)30202-0.

Muñoz-Pizza, D. M., Villada-Canela, M., Rivera-Castañeda, P., Osornio-Vargas, Á., Martínez-Cruz, A. L., & Texcalac-Sangrador, J. L. (2022). Barriers and opportunities to incorporate scientific evidence into air quality management in Mexico: A stakeholders' perspective. *Environmental Science & Policy*, *129*, 87–95.

Naiker, Y., Diab, R. D., Zunckel, M., & Hayes, E. T. (2012). Introduction of local air quality management in South Africa: Overview and challenges. *Environmental Science & Policy*, *17*, 62–71. Available from https://doi.org/10.1016/j.envsci.2011.11.009.

Narain, U., & Krupnick, A. (2007). The impact of Delhi's CNG program on air quality (SSRN scholarly paper No. 969727). Available from https://doi.org/10.2139/ssrn.969727.

Narain, U., O'Hagan, R.G., Kshirsagar, V., & Skoufias, E. (2020, August 4). In India, air quality has been improving despite the COVID-19 lockdown. <https://blogs.worldbank.org/endpovertyinsouthasia/india-air-quality-has-been-improving-despite-covid-19-lockdown>.

O'Driscoll, R., Stettler, M. E., Molden, N., Oxley, T., & ApSimon, H. M. (2018). Real world CO_2 and NO_x emissions from 149 Euro 5 and 6 diesel, gasoline and hybrid passenger cars. *The Science of the Total Environment*, *621*, 282–290. Available from https://doi.org/10.1016/j.scitotenv.2017.11.271.

Ouyang, H., Tang, X., Kumar, R., Zhang, R., Brasseur, G., Churchill, B., Alam, M., Kan, H., Liao, H., Zhu, T., Chan, E. Y. Y., Sokhi, R., Yuan, J., Baklanov, A., Chen, J., & Patdu, M. K. (2022). Toward better and healthier air quality: Implementation of WHO 2021 global air quality guidelines in Asia. *Bulletin of the American Meteorological Society*, *103*(7), E1696–E1703. Available from https://doi.org/10.1175/BAMS-D-22-0040.1.

Pandey, A., Brauer, M., Cropper, M. L., Balakrishnan, K., Mathur, P., Dey, S., Turkgulu, B., Kumar, G. A., Khare, M., Beig, G., Gupta, T., Krishnankutty, R. P., Causey, K., Cohen, A. J., Bhargava, S., Aggarwal, A. N., Agrawal, A., Awasthi, S., Bennitt, F., & Dandona, L. (2021). Health and economic impact of air pollution in the states of India: The global burden of disease study 2019. *The Lancet Planetary Health*, *5*(1), 1. Available from https://doi.org/10.1016/S2542-5196(20)30298-9.

Priti, K., & Kumar, P. (2022). A critical evaluation of air quality index models (1960–2021. *Environmental Monitoring and Assessment*, *194*(5), 5. Available from https://doi.org/10.1007/s10661-022-09896-8.

Republic of South Africa. (1965). Atmospheric pollution prevention act 45 of 1965. <https://www.up.ac.za/media/shared/600/LAS%20Legislation/air-pollution-prevention-act_17-april-1965.zp53523.pdf>.

Republic of South Africa, Highveld Priority Area Air Quality Management Plan, (2012). <https://www.gov.za/sites/default/files/gcis_document/201409/35072144.pdf>.

Republic of South Africa, National Environment Management: Air Quality Act 39 of 2004, (2004). <https://www.gov.za/documents/national-environment-management-air-quality-act>.

Republic of South Africa, National Environmental Management: Air Quality Act, 2004 {ActNo. 39 of 2004) VAAL Triangle air-shed priority area alr quality management plan (2009). <https://www.gov.za/sites/default/files/gcis_document/201409/32263613.pdf>.

Republic of South Africa, National Environmental Management: Air Quality Act: Waterberg Bojanala Priority Area Air Quality Management Plan, (2015). <https://www.gov.za/sites/default/files/gcis_document/201512/39489gon1207re.pdf>.

Republic of South Africa. (2021, October 4). State of air report and national priority areas air quality indicators. 15th air quality governance Lekgotlas, Session 1,2, Gauteng. <https://saaqis.environment.gov.za/Lekgotla%20Proceedings/2021/1.2%202020%20State%20of%20Air%20Report%20and%20NAQI%20-%20Final.pdf>.

Republic of South Africa. (2022). 2021 National air quality officer's annual report on air quality management in the republic of South Africa, South African National government. <https://saaqis.environment.gov.za/>.

Retained EU Law (Revocation and Reform) Act—Government Bill—UK Parliament. (2023). <https://bills.parliament.uk/bills/3340> (accessed 14.08.24).

Roy, A., Chandra, T., & Ratho, A. (2020). Finding solutions to air pollution in India: The role of policy, finance, and communities. ORF. <https://www.orfonline.org/research/finding-solutions-to-air-pollution-in-india-the-role-of-policy-finance-and-communities-74311/>.

Schedule of retained EU law. (2023). GOV.UK. <https://www.gov.uk/government/publications/schedule-of-retained-eu-law> (accessed 16.05.23).

Scorgie, Y., Kneen, M. A., & Burger, L. W. (2003). Air pollution in the Vaal Triangle - quantifying source contributions and identifying cost-effective solutions. *Clean Air Journal*, *13*(2), 2. Available from https://doi.org/10.17159/caj/2003/13/2.7152.

Scott, G. (2005). *National environmental management: Air quality act, entry into effect*. Durban, South Africa: National Association for Clean Air Symposium.

Secretaría de Medio Ambiente y Recursos Naturales. (n.d.). Estrategia nacional de calidad del aire (ENCA). gob.mx. <http://www.gob.mx/semarnat/acciones-y-programas/estrategia-nacional-de-calidad-del-aire-enca-153921> (accessed 18.05.23).

Sharma, M., Maheshwari, M., Sengupta, B., & Shukla, B. P. (2003). Design of a website for dissemination of air quality index in India. *Environmental Modelling & Software*, *18*(5), 405–411. Available from https://doi.org/10.1016/S1364-8152(03)00003-3.

Shi, Z., Liu, B., Cheng, K., Elliot, R., Cole, M., & Bryson, J. R. (2022). *Quantifying the impact of clean air policy interventions for air quality management*. Available from https://doi.org/10.25500/EPAPERS.BHAM.00004040.

Singh, R., Kandpal, B., Garg, R., Vasudeva, T., Shrma, Y., & Kamal, M. A. (2022). Challenges associated with air pollution in sustainable city development: case of New Delhi and New York. *American Journal of Civil Engineering and Architecture*, *10*(3), 116–125.

Sinha, J., & Chandra, T. (2020). Getting to the green frontier faster: The case for a green frontier superfund. *ORF*. Available from https://www.orfonline.org/research/the-case-for-a-green-frontier-superfund/.

Smith, L., & Boulton, P. (2024). Air quality: policies, proposals and concerns, House of Commons Library. https://commonslibrary.parliament.uk/research-briefings/cbp-9600/.

Téllez-Rojo, M. M. et al. (2019). Reporte nacional de niveles de plomo en sangre y uso de barro vidriado en población infantil vulnerable. Salud Pública de México [online]. 61 (6, nov-dic), p. 787. https://www.saludpublica.mx/index.php/spm/article/view/10555.

The Air Quality Standards Regulations 2010. (2023). Retrieved 29 May, from <https://www.legislation.gov.uk/uksi/2010/1001/contents/made>.

The Environmental Targets (Fine Particulate Matter) (England) Regulations 2022. (n.d.) <https://www.legislation.gov.uk/ukdsi/2022/9780348242959> (accessed 16.05.23).

Transport for London. (n.d.). ULEZ expansion 2023. Transport for London. <https://www.tfl.gov.uk/modes/driving/ultra-low-emission-zone/ulez-expansion-2023> (accessed 14.12.23).

Tshehla, C., & Wright, C. Y. (2019). 15 Years after the national environmental management air quality act: Is legislation failing to reduce air pollution in South Africa? *South African Journal of Science*, *115*(9–10), 1–4. Available from https://doi.org/10.17159/sajs.2019/6100.

Venter, Z. S., Aunan, K., Chowdhury, S., & Lelieveld, J. (2021). Air pollution declines during COVID-19 lockdowns mitigate the global health burden. *Environmental Research*, *192*. Available from https://doi.org/10.1016/j.envres.2020.110403, Scopus.

Webber, L., Xu, M., & Graff, H. (2020). Modelling the long-term health impacts of changing exposure to NO_2 and $PM_{2.5}$ in London. HealthLumen Limited. Available from https://www.london.gov.uk/sites/default/files/modelling_the_long-term_health_impacts_of_changing_exposure_to_no2_and_pm2.5_in_london_final_250220_-4.pdf.

WHO. (2017). Evolution of WHO air quality guidelines: Past, present and future. WHO regional office for Europe. <https://www.euro.who.int/__data/assets/pdf_file/0019/331660/Evolution-air-quality.pdf>.

WHO. (2021). WHO global air quality guidelines: Particulate matter ($PM_{2.5}$ and PM_{10}), ozone, nitrogen dioxide, sulphur dioxide and carbon monoxide. <https://www.who.int/publications-detail-redirect/9789240034228>.

Wood, H. E., Marlin, N., Mudway, I. S., Bremner, S. A., Cross, L., Dundas, I., Grieve, A., Grigg, J., Jamaludin, J. B., Kelly, F. J., Lee, T., Sheikh, A., Walton, R., & Griffiths, C. J. (2015). Effects of air pollution and the introduction of the london low emission zone on the prevalence of respiratory and allergic symptoms in schoolchildren in east London: A sequential cross-sectional study. *PLoS One*, *10*(8), e0109121. Available from https://doi.org/10.1371/journal.pone.0109121.

Chapter 17

Air quality in cities around the world

Ranjeet S. Sokhi*
Centre for Climate Change Research (C3R), School of Physics, Engineering and Computer Science, University of Hertfordshire, Hatfield, United Kingdom

Subchapter 17.1

17.1 Introduction

Improving air quality in cities is a worldwide challenge. Many studies are reporting that air pollution, along with climate change, is one of the greatest environmental risks facing the world (e.g., Lelieveld et al., 2023; Shaddick et al., 2020; Sokhi et al., 2022; Southerland et al., 2022). It is estimated that most of the world's population (90%) is exposed to air quality that does not meet World Health Organization (WHO) guidelines (WHO, 2016). This is especially true when local air quality is compared to the latest WHO air quality guidelines (WHO, 2021). The main drivers for poor air quality are dependent on regional factors including air pollution emissions, prevailing meteorology and increasingly on climate change. Road traffic remains a major contributor to air pollutants such as NO_2 and $PM_{2.5}$ in most cities but in many parts of the world, air pollution is exacerbated by increasing urbanization and rapid industrial expansion as well as growing human density which have led to many of the most polluted cities being in developing regions.

Cities in developing regions have specific challenges with vehicle ownership driven by economic development, and urban population growth along with significant sand and dust intrusion. Resuspended particles from roads, adding to emissions from construction activities, unregulated waste burning and industry are common sources that lead to poor air quality. Other contributions that can significantly deteriorate air quality include long-range transport from surrounding regions and distances further afield including wildfire smoke and also from agricultural burning that affects many cities in all regions.

Increases in urban populations bring many challenges unless adequate infrastructure is in place. An important mechanism to limit additional air pollution resulting from urbanization, is to have integrated responsive urban services (Grimmond et al., 2020) to ensure environmental measures are put in place to handle waste and emissions along with other urban services such as clean water and sanitation and traffic and road management. These integrated urban services are essential to reduce the adverse impacts on air quality in cities and reduce inequalities.

In addition to anthropogenic emission sources, there is a greater appreciation of natural emissions that can exacerbate air quality through episodic air pollution events. These include wildfires, dust and sandstorms and biogenic emissions from vegetation.

Studies on health impacts from exposure to air pollution have largely been based on concentration-response functions from more developed regions. It is not yet clear how the health impacts may change for air pollution with a different composition profile relative to previous epidemiological studies (e.g., a global study by Southerland et al., 2022). In addition, population age and health distribution are changing in many cities with people living longer. The current understanding is that an older population is likely to be more susceptible to adverse impacts from air pollution, even if levels are low with greater uptake of cleaner fuels (e.g., through road fleet electrification).

Emissions of air pollutants create a hazard that leads to poor air quality, but meteorology plays a large role in influencing the variability of the concentrations throughout the day and night and also seasonally. This variability is influenced by factors such as wind speed, air temperature, boundary and mixing layer height, the occurrence of thermal

*Case study contributors.

Air Quality. DOI: https://doi.org/10.1016/B978-0-12-822591-2.00017-2

inversions as well as synoptic meteorology which can result in high concentrations over several days and even weeks, for example, summertime photochemical smog and wintertime episodes.

There are complex mechanisms by which climate change can influence weather and atmospheric composition both of which affect air quality either directly or indirectly through chemical transformations (IPCC, 2021). There is already growing evidence that higher temperatures resulting from climate change are leading to increases in the frequency of wildfires which are affecting concentrations of $PM_{2.5}$ and O_3 (UNEP, 2022).

17.1.1 Scope of chapter

This chapter provides an overview of air quality trends over the last one to two decades for selected major cities across the world. The location of the cities covers all the main regions and is shown in Fig. 17.1.1. The main characteristics of the cities, emission sources and challenges are summarized in Table 17.1.1. The selected cities represent different emissions and source profiles and weather climatology as well as a mixed range of challenges. One clear message that comes across when examining the status of air quality in these cities, is that while there are common challenges affecting all cities, there are many that are specific to the regions where they are located.

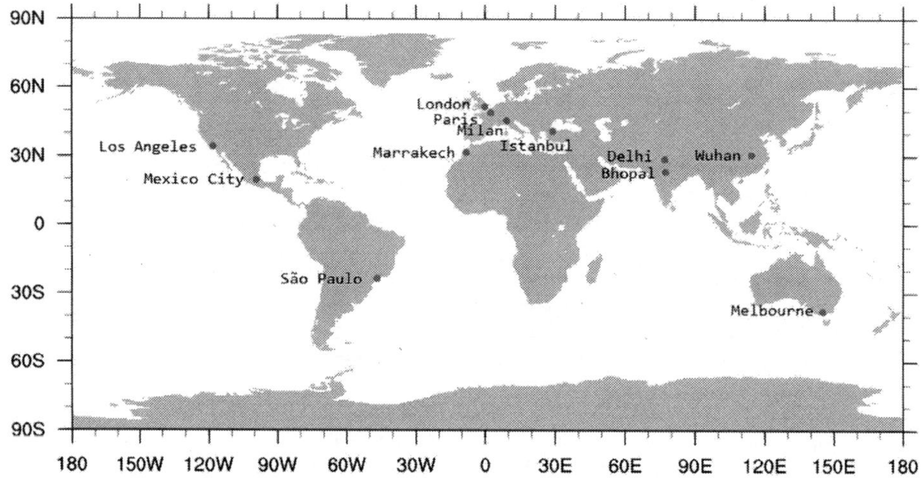

FIGURE 17.1.1 Location of cities for which air quality is examined in this chapter.

TABLE 17.1.1 Main characteristics, sources and air quality challenges facing selected cities. Population statistics for 2024 have been taken from World Population Review, https://worldpopulationreview.com/.

Name of city	Country	Geography and climatology	Population (year)	Main air pollution sources	Main air quality challenges
Bhopal	India	Located in central India, area of about 465 km^2, elevation of 550 m above mean sea level, with a warm, humid subtropical climate	2.4 million (Census, 2011), 2.62 million (2024)	Industry, domestic, road transport, agriculture	Traffic growth, limited air pollution monitoring, regulate open waste-burning, increase cleaner fuels
Delhi	India	Capital city of India is located in the Indo-Gangetic Plains, area of nearly 1500 km^2 and about 215 m above sea level, with monsoon-influenced humid subtropical climate	32 million (Census, 2011), 33.81 million (2024)	Road transport, nonexhaust, residential, industrial, power sectors, construction industry, transboundary contributions	Increasing vehicle density, need for electric vehicle infrastructure, need shift from fossil fuels to renewable energy, improve waste management techniques, increase public transportation and natured-based solution

(Continued)

TABLE 17.1.1 (Continued)

Name of city	Country	Geography and climatology	Population (year)	Main air pollution sources	Main air quality challenges
Wuhan	China	Located in Central China, area of nearly 8500 km^2 and elevation about 20 m above sea level, with humid subtropical climate	8.85 million (2024)	Industry, residential, transport, power, agricultural contributing to haze and ozone pollution, regional transport of air pollution from North China Plain (NCP)	Addressing air complex pollution of $PM_{2.5}$ and O_3, need measures to reduce NO_x and VOCs, need reduction of regional air pollution
Melbourne	Australia	Australia's second city is located in the state of Victoria, southeast Australia, area of nearly 10,000 km^2 and elevation of about 31 m above sea level, with a warm temperate oceanic climate	5.32 million (2024)	Road vehicles, domestic wood heaters, power stations, ecological and landscape burns. Bushfires and dust storms	Population growth, susceptibility to air pollution impacts for the growing elderly population, climate change driven increase in intensity of bushfires, and changes in meteorology and weather
Marrakech	Morocco	Fourth largest city is located in central Morocco, with an area of 230 km^2 and elevation of 470 m above sea level with a hot, dry climate	1.07 million (2024)	Residential combustion, road transport, industry activities	Rapidly industrialization and urbanization, traffic growth, increase in electricity production and energy consumption, lack of local evaluation of exposure and health impacts
Los Angeles	USA	Second most populated city in the United States, it is located in southern California with Santa Monica mountains to the west and San Fernando Valley to the north, with a mean elevation of 93 m above sea level and has a Mediterranean type climate	3.80 million for the city and 9.61 million of LA county (2024)	Transport (road, aircraft, trains, ships, commercial boats), off-road vehicles, fuel combustion from stationary sources and other processes (e.g., residential fuel combustion, windblown dust and construction activities)	Emphasis shifting from transportation sector to volatile chemical products, regional transport of ozone and increase in wildfires
Mexico City Metropolitan Area	Mexico	Mexico City is the capital of Mexico, located on an elevated basin at an altitude of 2240 m above mean sea level, with a subtropical highland climatology. The metropolitan area includes contiguous municipalities of the states of Mexico and Hidalgo.	22.51 million (2024)	Fossil fuels, mobile sources, solvents use, industrial activities	Ozone trends are showing an increase, while decrease of PM has stalled, emissions from consumer and personal care products and solvent use are exacerbating ozone levels, seasonal agricultural and forest fires
São Paulo	Brazil	This most populous city in Brazil is located near the southeast coast, area of about 8000 km^2 at an elevation of 755 m above sea level, with humid subtropical climate	22.81 million (2024)	Mainly vehicular emissions, with seasonal wild fire smoke, fossil fuel combustion, use of ethanol maybe worsening photochemical episodes	Stricter emissions control needed, regional control needed for smoke intrusion, improved understanding of VOC/NO_x regime needed to reduce O_3, vehicle maintenance scheme is needed, phase out gross-emitters, prioritize nonexhaust and resuspension sources, improve local emission inventories

(Continued)

TABLE 17.1.1 (Continued)

Name of city	Country	Geography and climatology	Population (year)	Main air pollution sources	Main air quality challenges
Istanbul	Turkey	Financial capital of Turkey is located where European and Asian continents meet, average elevation of 30 m above sea level, with climate of mix of Mediterranean Climate and temperate zone	16.05 million (2024)	Road traffic, airport emissions, shipping emissions, Industry, power generation, residential heating, waste and agricultural emissions, solvents, natural dust	Impacts from increases in population, economic growth and transportation (road transport, maritime and air travel)
London	United Kingdom	Capital of the UK is located on the river Thames, south east England, Greater London has an average elevation of 42 m, with a temperate marine climate	9.75 million (2024)	Road transport, construction industry, industrial and commercial power, aviation, industrial processes, domestic heating, resuspended dust, wood burning, commercial cooking	Need for local and regional policies, nonroad-transport sources, construction industry emissions, indoor and other nonambient sources, wood burning, commercial cooking, aviation industry and river transportation, regional air pollution, ultrafine particles
Milan	Italy	Milan is the capital of Lombardy region and is located north of the Po River in northern Italy, has an average elevation of about 120 m, with a mid-latitude, four-season humid subtropical climate	3.16 million (2024)	Road traffic, wood combustion for residential heating, agriculture, industrial production processes and solvent	Control of precursors of secondary air pollutants (O_3 and $PM_{2.5}$), need for regional measures to control secondary species, reduce biomass burning and ammonia emissions, increase public transport availability and move to cleaner fuels, change public perception e.g., to reduce wood burning
Paris	France	Capital of France is located in northern central France, on the river Seine, has an average elevation of about 35 m with temperate marine *climate*	11.28 million (2024)	Road traffic, freight transport and the residential and tertiary sectors, aviation, energy production, agriculture	Equitable implementation of the Low Emission Zone (LEZ), strengthening control of mobility and travel, and more durable urban logistics, e.g., cleaner vehicles

References

Grimmond, S., Bouchet, V., Molina, L. T., Baklanov, A., Tan, J., Schlünzen, K. H., Mills, G., Golding, B., Masson, V., Ren, C., Voogt, J., Miao, S., Lean, H., Heusinkveld, B., Hovespyan, A., Teruggi, G., Parrish, P., & Joe, P. (2020). Integrated urban hydrometeorological, climate and environmental services: concept, methodology and key messages. *Urban Clim*, *33*, 100623. Available from https://doi.org/10.1016/j.uclim.2020.100623.

IPCC. (2021). In V. Masson-Delmotte, P. Zhai, A. Pirani, S. L. Connors, C. Péan, S. Berger, N. Caud, Y. Chen, L. Goldfarb, M. I. Gomis, M. Huang, K. Leitzell, E. Lonnoy, J. B. R. Matthews, T. K. Maycock, T. Waterfield, O. Yelekçi, R. Yu, & B. Zhou (Eds.), *Climate change 2021: The physical science basis. Contribution of working group I to the sixth assessment report of the intergovernmental panel on climate change*. Cambridge University Press. Available from https://www.ipcc.ch/report/ar6/wg1/#FullReport.

Lelieveld, J., Haines, A., Burnett, R., Tonne, C., Klingmüller, K., Münzel, T., & Pozzer, A. (2023). Air pollution deaths attributable to fossil fuels: Observational and modelling study. *BMJ (Clinical Research ed.)*, *383*, e077784. Available from https://doi.org/10.1136/bmj-2023-077784.

Southerland, V. A., Brauer, M., Mohegh, A., Hammer, M. S., van Donkelaar, A., Martin, R. V., Apte, J. S., & Anenberg, S. C. (2022). Global urban temporal trends in fine particulate matter ($PM_{2.5}$) and attributable health burdens: estimates from global datasets. *Lancet Planet Health*, *6*, e139–e146. Available from https://doi.org/10.1016/S2542-5196(21)00350-8.

Shaddick, G., Thomas, M. L., Mudu, P., Ruggeri, G., & Gumy, S. (2020). Half the world's population are exposed to increasing air pollution. *NPJ Climate and Atmospheric Science*, *3*, 23. Available from https://doi.org/10.1038/s41612-020-0124-2.

Sokhi, R. S., Moussiopoulos, N., Baklanov, A., Bartzis, J., Coll, I., Finardi, S., Rainer Friedrich., Geels, C., Grönholm, T., Halenka, T., Ketzel, M., Maragkidou, A., Matthias, V., Moldanova, J., Ntziachristos, L., Schäfer, K., Suppan, P., Tsegas, G., Carmichael, G., Franco, V., Hanna, S., Jalkanen, J.-P., Velders, G. J. M., & Kukkonen, J. (2022). Advances in air quality research – current and emerging challenges. *Atmospheric Chemistry &. Physics*, *22*, 4615–4703. Available from https://doi.org/10.5194/acp-22-4615-2022.

UNEP (2022). Spreading like wildfire – The rising threat of extraordinary landscape fires. A UNEP Rapid Response Assessment. United Nations Environment Programme, Nairobi, 2022.

WHO. (2016). *Ambient air pollution: A global assessment of exposure and burden of disease* (p. 131) Geneva, Switzerland: World Health Organization. Available from https://apps.who.int/iris/handle/10665/250141.

WHO (2021). WHO global air quality guidelines. Particulate matter ($PM_{2.5}$ and PM_{10}), ozone, nitrogen dioxide, sulfur dioxide and carbon monoxide, 2021, World Health Organization, electronic version, ISBN 978-92-4-003422-8. https://apps.who.int/iris/handle/10665/345329.

Subchapter 17.2

Air quality case study on the central Indian city of Bhopal, India

Arushi Sharma[1] and Chandra Venkataraman[1,2]

[1]*Interdisciplinary Program in Climate Studies, Indian Institute of Technology Bombay, Mumbai, Maharashtra, India,* [2]*Department of Chemical Engineering, Indian Institute of Technology Bombay, Mumbai, Maharashtra, India*

17.2.1 Background

Bhopal is the capital city of the central Indian state of Madhya Pradesh, known as the "city of lakes," and is considered one of the greenest cities in India. It is situated at 77° 25' E longitude and 23° 25' N latitude and has an altitude of approximately 550 m above mean sea level with an approximate area of 436 km^2, accommodating a population of approximately 25 lakhs (that is 2.5 million, projected for 2022) with 2.37 million being the last Census (2011) population. It has a strong economic base with numerous large and medium-scale industries running on and outskirts of the city. It has recently been chosen as one of the initial 20 cities selected for development as a smart city in India under the government's flagship Smart Cities Mission.

The urbanization rate of this city has remained consistent at around 80% over the past few decades, indicating a stable level of urban development. This rapid urbanization resulted in many challenges, such as insufficient core infrastructure of sanitation and waste management, unplanned industrialization, water scarcity, poor road and highway conditions, high growth of vehicular population, and a proliferation of major slum populations associated with it. The rise in air quality degradation problem in the city also needs urgent attention. Studies have shown an increase in traffic across the city, leading to traffic congestion in several areas during peak hours (Mehta and Mehta Associates, 2006). The major sources of air pollution in the old city area in Bhopal include tourist inflow, vehicular density, roadside dust, and burning of coal and fuel wood on a large scale. Resuspended crustal dust and biomass-burning aerosols transport from central and northern India increase $PM_{2.5}$ mass concentration in this city (Kumar & Raman, 2020).

While this city strives for urbanization, efforts have been made to protect its green areas and reduce air pollution. The government has taken various measures under the "Air plan" to control air pollution, like regular checking of vehicular emissions and issuing of Pollution under control certificates, good traffic management redirection of traffic movement to avoid congestion, promotion and operationalization of E-rickshaw, regular cleaning of road dust, development of green belts, regulations on construction activities as per Bhopal Municipal Corporation, restriction on open burning of municipal solid waste, biomass, plastic, and horticulture waste and installation and operation of air pollution control devices to limit industrial emissions (http://www.mppcb.mp.gov.in/proc/Action%20Plan2019/Action_Plan_Non_attainment_Bhopal.pdf). While these efforts are making a positive impact, it is still essential to understand the air pollution sources, their trends and subsequent health impacts to support adequate policy-driven solutions to mitigate the air pollution problem in Bhopal.

17.2.2 Major air pollution sources and emissions

The regional emission inventory of India, SMoG-India (Pandey et al., 2014; Sadavarte & Venkataraman, 2014) identifies annual contributions from the five major source-sector of $PM_{2.5}$ emissions in Bhopal (shown in Fig. 17.2.1). The primary sector contributing to $PM_{2.5}$ emissions is Industry, accounting for approximately 50% of emissions in 2019 (refer to Fig. 17.2.1). Following this, the domestic sector, involving the use of biomass fuel for cooking, lighting, and heating, contributes 30%, while vehicular transport contributes 12%, and agricultural combustion contributes 7%. It is noteworthy that the emission sources influencing this central Indian city differ from those in other cities across the country, such as the northern city of Delhi, where elevated pollution levels have been observed. In Delhi, major contributors to emissions include road transport, residential sources, agricultural combustion, and power plants.

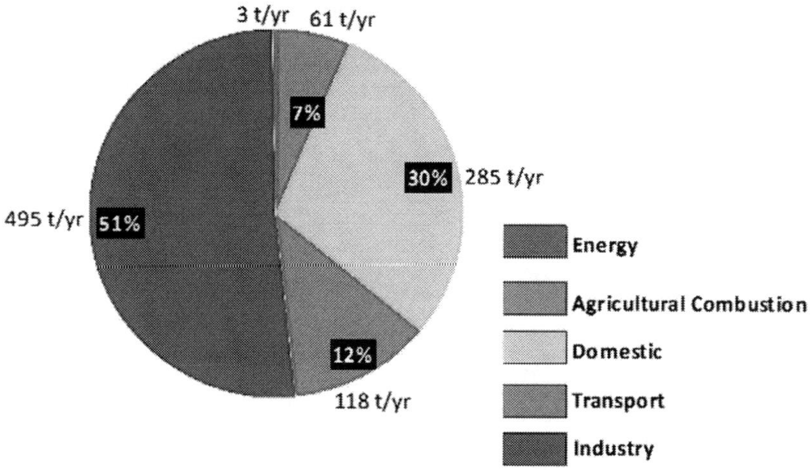

FIGURE 17.2.1 Source-sector contribution of $PM_{2.5}$ emissions obtained from SmoG-India emission inventory for Bhopal for 2019. The black highlights emphasize the percentage contributions, with industry being the primary contributor and energy being a minor source. The specific magnitudes are expressed in tons per year ($t\,yr^{-1}$).

17.2.3 Trends of key air pollutant concentrations

The annual surface $PM_{2.5}$ concentrations data obtained from the Central Pollution Control Board (CPCB), which is regulated by the Ministry of Environment and Forests (https://cpcb.nic.in/air-quality-data/ and https://cpcb.nic.in/namp-data/) was analyzed from 2014 to 2021 for which it has been recorded and is available (see Fig. 17.2.2). Over these several years, the annual surface $PM_{2.5}$ concentration has shown mean value of $48\,\mu g\,m^{-3}$, with a maximum of $143\,\mu g\,m^{-3}$ and a minimum of $19\,\mu g\,m^{-3}$.

17.2.4 Comparison to WHO AQG (2021) and national standards

The updated air quality guidelines (AQG, 2021) state that annual average concentrations of $PM_{2.5}$ should not exceed $5\,\mu g\,m^{-3}$. The annual average $PM_{2.5}$ concentration from 2014 to 2021 in Bhopal on the other hand was $48\,\mu g\,m^{-3}$, which is approximately 10 times the WHO annual air quality guideline value and 1.2 times India's annual average standard for $PM_{2.5}$ of $40\,\mu g\,m^{-3}$. Various Indian cities like Bhopal experience challenging meteorological and climatic conditions with the added challenge of topography, high pollution episodes and urban heat island effect which calls for achievable regional-based air quality standards for various Indian cities. To achieve the prescribed air quality standards, an informed and customized approach is necessary depending on the region's background pollution level and primary sources and will differ for different cities of India to control their air pollution emissions.

17.2.5 Health impacts

People with heart and lung diseases, older adults and children are at greater risk from the presence of particles in the air. High exposure to annual surface $PM_{2.5}$ concentrations of approximately 50 μg m-3 in Bhopal has been estimated to lead to approximately 1300 premature deaths (Maheshwarkar & Sunder Raman, 2021). A study conducted on school-going children in Bhopal revealed long-term exposure to high ambient air pollution in early childhood associated with higher respiratory impedance in children (De, 2020). Traffic-related air pollution exposure can increase respiratory morbidity and cause the highest risk of breathlessness in shopkeepers working in heavily trafficked roadside shops with poor air conditioning in Bhopal (De et al., 2019). Given the growing evidence, long-term exposures to high $PM_{2.5}$ concentrations will cause an increased risk of cardiorespiratory symptoms in the general population in Bhopal city.

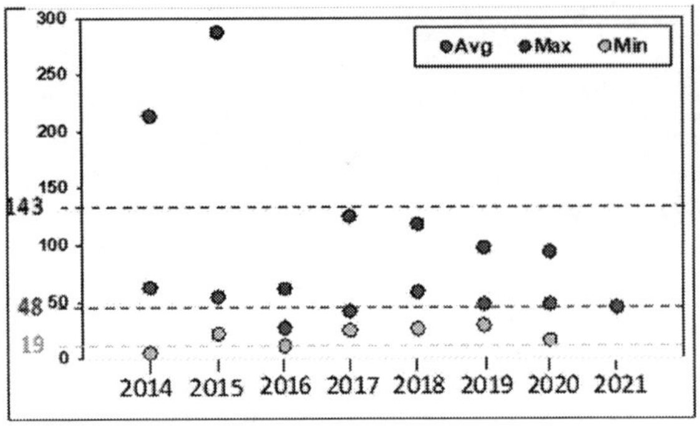

FIGURE 17.2.2 Annual surface $PM_{2.5}$ observations ($\mu g\, m^{-3}$) obtained from CPCB observations (2014–21) for Bhopal. *From https://cpcb.nic.in/air-quality-data/ and https://cpcb.nic.in/namp-data/.*

17.2.6 Remaining challenges to improving air quality

Since the last few years, the city has seen extraordinary growth in the development of roads/highways, residential/commercial hubs, industries, parks and maintenance of drainage systems, but unfortunately, there are only six air pollution monitoring stations. According to The Air Pollution Knowledge Assessment city program, Bhopal should have at least 18 Continuous Air Monitoring Stations (CAMS) for efficient reporting for all the criteria pollutants. There is a need for stricter regulations for the combustion of fossil fuel and industrial processes than the standard prescribed RSPM (Respirable Suspended Particulate Matter) of 120 $\mu g\, m^{-3}$ and biomass gasifiers for informal industries located in the Mandideep, Govindpura and Bagroda industrial areas. A shift from biomass to cleaner fuel like LPG or electricity for cooking and heating purposes will improve air quality, along with establishing CNG/LPG filling stations. Using cleaner vehicular fuels like electricity and CNG, reducing traffic congestion by widening urban roads in the old city area and developing an efficient public transport system along the corridor can help reduce traffic emissions. Also, road infrastructure improvement can help reduce on-road dust resuspension in Bhopal. Stricter regulations are also required to regulate open waste-burning activity dispersed across the city to avoid high pollution seasonal episodes.

References

Census (2011). Bhopal Population 2024, https://www.census2011.co.in/census/city/302-bhopal.html. (accessed 02.02.24).

De, S., Kushwah, G. D. S., Dharwey, D., & Shanmugasundaram, D. (2019). Respiratory morbidity of roadside shopkeepers exposed to traffic-related air pollution in Bhopal, India. *Journal of Health and Pollution*, *9*(21).

De, S. (2020). Long-term ambient air pollution exposure and respiratory impedance in children: A cross-sectional study. *Respiratory Medicine*, *170*, 105795.

Kumar, S., & Raman, R. S. (2020). Source apportionment of fine particulate matter over a National Park in Central India. *Science of the Total Environment*, *720*, 137511.

Maheshwarkar, P., & Sunder Raman, R. (2021). Population exposure across central India to PM2. 5 derived using remotely sensed products in a three-stage statistical model. *Scientific reports*, *11*(1), 1–13.

Mehta and Mehta Associates. (2006). *City development plan*. Bhopal: JNNURM.

Pandey, A., Sadavarte, P., Rao, A. B., & Venkataraman, C. (2014). Trends in multi-pollutant emissions from a technology-linked inventory for India: II. Residential, agricultural and informal industry sectors. *Atmospheric Environment*, *99*, 341–352.

Sadavarte, P., & Venkataraman, C. (2014). Trends in multi-pollutant emissions from a technology-linked inventory for India: I. Industry and transport sectors. *Atmospheric Environment*, *99*, 353–364.

Subchapter 17.3

Air quality trend and the impact of air quality over Megacity Delhi

Khaiwal Ravindra[1], Akshi Goyal[2], Vikas Singh[3] and Suman Mor[2]

[1]Department of Community Medicine and School of Public Health, Post Graduate Institute of Medical Education and Research (PGIMER), Chandigarh, India, [2]Department of Environment Studies, Panjab University, Chandigarh, India, [3]National Atmospheric Research Laboratory (NARL), Gadanki, Andhra Pradesh, India

17.3.1 Background

Air pollutants are considered a major risk factor for premature mortality worldwide (Cohen et al., 2017). Increasing urbanization and rapid industrial expansion have accelerated the concentration of air pollutants in the atmosphere. The highest concentrations of pollutants were found in urban areas due to their fast expansion and growing human density. In India, notably, the megacity Delhi is reported to be among the most polluted cities.

Delhi is an expanding city with decreasing air quality conditions. The issue of air pollution worsened in the 1990s as car growth outpaced economic development and population growth.

Significant increases in vehicle numbers, dust from road and construction operations, burning of municipal solid waste, and industrial emissions are important contributors to these dangerously high pollution levels. Numerous studies have demonstrated, however, that Delhi's location in the Indo-Gangetic Plains and the transboundary air pollution from neighboring states make it a funnel for pollutants to concentrate (Gurjar et al., 2016). Moreover, lower ambient temperature during the winter season with inversion condition, shallower mixing depth, and increased fuel consumption exacerbate air pollution issues.

In order to find the solution for better air quality, Delhi has unveiled many initiatives. A complete ban on BS-IV diesel cars, 15-year-old petrol and 10-year-old diesel cars, mandatory Pollution Under Control Certificates, closure of industries in residential areas, Green Delhi app to address the complaints registered by the residents, installing smog towers, vertical gardening on the roadside are some of the measures taken by the government. Not only these, but the Delhi government have also encouraged the neighboring states to check stubble burning, use bio-decomposer for crop residue increase cleaner fuel and use CNG-powered vehicles. The governmental campaigns included: Swachh Bharat Abhiyan, antiopen burning, and anti-road dust campaign, and also imposed a Graded Response Action Plan (GRAP) III action plan in the National Clean Air Program (NCAP) to enhance the air quality. The proposed individual initiatives to mitigate air pollution include using public transport, better waste management, car-pooling, more CNG vehicles, solar power, planting more trees, and spreading awareness.

Regardless of these many initiatives taken by the government, there has not been a significant effect on the air quality of Delhi. Consequently, further governmental effort is needed to focus more on finding innovative local solutions to address the current air pollution challenges.

17.3.2 Major air pollution sources and emissions

Globally, Megacity Delhi is the fifth largest populous, whereas it is the third largest city in India. Delhi's population grew from 11 million to approximately 32 million from 2011 to 2023 (Census of India, 2021). Similarly, the number of vehicle-registered with the Transport Department has risen from 2.5 million to about 14 million in the last decade (Open Government Data (OGD) Platform, 2022). Emission inventory data has shown that vehicular emissions are the primary contributors to the poor air quality in Delhi. Apart from this, residential, industrial, power sectors, construction and demolition and tertiary sectors are significant factors leading to an increase in the concentration of various air pollutants, as shown in Fig. 17.3.1. During winter months, a divergent source of air pollution, that is, burning of crop residue in the neighboring states situated in the Indo-gangetic plain, undesirably affects Delhi's air quality. According to Lawrence et al. (2016) and Singh et al. (2020) the nonexhaust emission sources would account for the majority of the

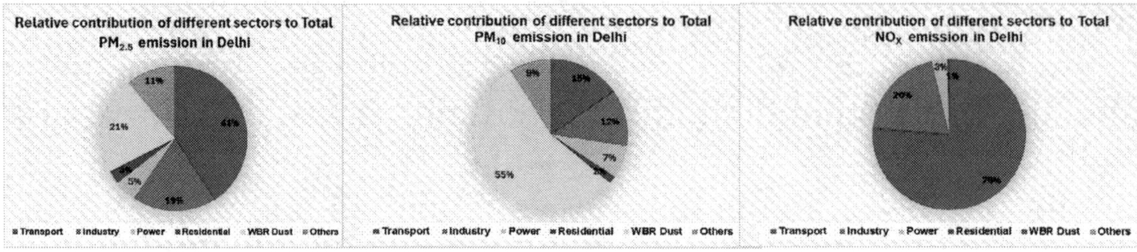

FIGURE 17.3.1 Relative contribution of different sectors in PM_{10}, $PM_{2.5}$, and NO_x emission in Delhi (SAFAR - India, 2018).

total PM_{10} emissions of Delhi. Therefore, the major focus should be given to the source apportionment and control of nonexhaust emission sources along with the exhaust emission sources.

17.3.3 Trends of key air pollutant concentrations

Numerous studies have reported the increasing trend of major air pollutants in Delhi, such as particulate pollutants (PM_{10} and $PM_{2.5}$), gaseous pollutants [oxides of nitrogen (NO_x), oxides of sulfur (SO_x), carbon monoxide (CO), and ozone (O_3)] and toxic volatile organic compounds (VOCs). In the early 1990s, Delhi was one of the top five megacities in terms of SO_2 emissions. The total SO_2 emissions in Delhi increased from 45 Gg in 1990 to 81 Gg in 1995 majorly due to vehicular pollution and power plant emissions, but a declining trend was seen after 2000 due to various policy initiatives, including desulfurizing fossil fuels. In 2010, about 37.40 Gg of SO_2 emissions were reported from all sources in Delhi due to continuous abatement policies and efforts.

The annual average of PMexceeded the NAAQS standard; the levels of PM rose from $325\,\mu g\,m^{-3}$ in 2005 to $525\,\mu g\,m^{-3}$ in 2012 in Delhi (Gurjar et al., 2016) and they also predicted an increase in the future. Due to high vehicular activity, a steep rise in NO_x and O_3 was also seen during summers while an increase in SO_2 value was witnessed in 2018 due to small-scale industries, diesel vehicles, crop residue burning and unplanned activities such as tandoor. Fig. 17.3.2 shows the time-series analysis of the annual average concentration of major air pollutants at a single monitoring station in Delhi, that is, the ITO site, from 2010 to 2022. This data was acquired from the Central Pollution Control Board's official website, commonly known as CPCB (https://airquality.cpcb.gov.in/ccr/#/caaqm-dashboard-all/caaqm-landing). The trend analysis shows daily maximum hourly CO averages, 8-hour O_3 averages, and 24-hour mean SO_2, whereas annual averages are reported for NO_x, PM_{10} and $PM_{2.5}$.In 2020, a significant decline in PM and NO_x concentration was observed, which was found to be associated with lockdown measures to restrict the spread of COVID-19 disease (Mahato et al., 2020; Vadrevu et al., 2020).

17.3.4 Comparison to WHO AQG (2021) and national standards and health impacts

According to the Global Burden of Diseases, air pollution is the fifth leading cause of premature mortality globally. While WHO reported that 7 million people die each year from air pollution, and 9 out of 10 people breathe air with excessive levels of contaminants. Short-term and long-term exposure to air pollutants is associated with major human health issues such as eye irritation, asthma, pneumonia, lung infections, chronic obstructive pulmonary disease (COPD), and other cardiovascular and neurological problems.

Delhi has poor air quality, and airborne concentrations of key air pollutants frequently exceed the National Ambient Air Quality Standards (NAAQS) established by CPCB. The permissible limits by WHO AQG (2021) are shown in a brown dotted line, whereas Indian NAAQS are shown in a blue dotted line (Fig. 17.3.2).

In the last 12 years, Delhi has seen to exceed both WHO and NAAQS prescribed annual average levels of NO_x, PM_{10} as well as $PM_{2.5}$. However, the missing values of ozone and sulfur dioxide (Fig. 17.3.2) were due to instrument failure. An increase in the concentration of air pollutants has not only resulted in Delhi's poor air quality but also has detrimental effects on the built infrastructure, local atmospheric visibility, human health, ecosystems, historical landmarks and global climate. The trend shows short-term improvements in pollutant concentration due to several policy initiatives taken by the government to improve air quality. However, compliance with NAAQS and WHO standards is still a major challenge for the city.

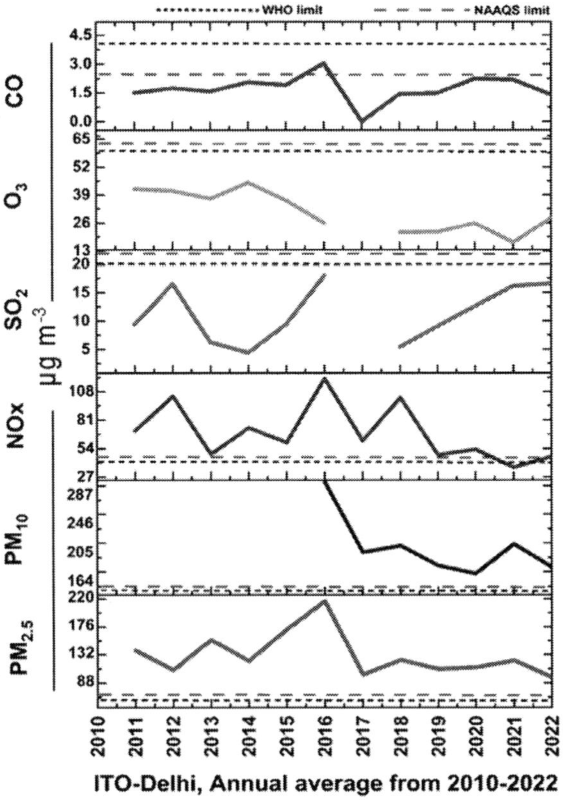

FIGURE 17.3.2 Trend analysis of the annual average concentration (μg m^{-3}) of major air pollutants at ITO monitoring station of Delhi from 2010 to 2022. WHO AQG (2021) air quality guidelines are shown in a brown dotted line and the Indian NAAQS standards are shown in a blue dotted line.

17.3.5 Remaining challenges to improving air quality

Bad air quality and public health are a continual concern for the megacity which accounts for 30% of the North-India population. The government took several interventions to enhance the air quality of Delhi. Switching commercial vehicles from petrol/diesel to CNG was a practical solution, but the impact was nullified due to increasing vehicle density. The odd-even scheme was a stop-gap solution for Delhi, and citizens' compliance was one of the major barriers to achieving the defined goals. Therefore, permanent long-term control measures with solid institutional and governmental support are necessary instead of aiming at temporary solutions.

India has already made progress in the electric vehicles (EVs) market with the Delhi EV policy 2020. However, with increased EV manufacturing, availability and battery charging station infrastructure still need to be built to match the increasing demand. To ease the adoption of EVs among the public, incentives under various schemes and policies are required. Moreover, in 2016, the rooftop solar policy was introduced with a target of about 2 GW by 2025 and Delhi has been steadily gaining in the field of solar power. Avoiding potentially catastrophic air pollution and climate change impacts requires an urgent shift from fossil fuels to renewable energy.

In addition to technology options for reducing air pollution, urban vegetation is recognized as one of the ameliorative methods to eliminate pollutants primarily through dry deposition. Furthermore, implementing improved waste management techniques, providing better public transportation options, and vertical gardening on roadsides by the government are some crucial measures.

References

Census of India (2021). <https://censusindia.gov.in/census.website/> (accessed 06.02.24).

Cohen, A. J., Brauer, M., Burnett, R., Anderson, H. R., Frostad, J., Estep, K., Balakrishnan, K., Brunekreef, B., Dandona, L., Dandona, R., & Feigin, V. (2017). Estimates and 25-year trends of the global burden of disease attributable to ambient air pollution: an analysis of data from the Global Burden of Diseases Study 2015. *The Lancet*, *389*(10082), 1907–1918.

Gurjar, B. R., Ravindra, K., & Nagpure, A. S. (2016). Air pollution trends over Indian megacities and their local-to-global implications. *Atmospheric Environment*, *142*, 475–495.

Lawrence, S., Sokhi, R., & Ravindra, K. (2016). Quantification of vehicle fleet PM10 particulate matter emission factors from exhaust and non-exhaust sources using tunnel measurement techniques. *Environmental Pollution*, *210*, 419–428.

Mahato, S., Pal, S., & Ghosh, K. G. (2020). Effect of lockdown amid COVID-19 pandemic on air quality of the megacity Delhi, India. *Science of the total environment*, *730*, 139086.

Open Government Data (OGD) Platform (2022). <https://data.gov.in/catalog/total-number-registered-motor-vehicles-india> (accessed 06.02.23).

Singh, V., Biswal, A., Kesarkar, A. P., Mor, S., & Ravindra, K. (2020). High resolution vehicular PM10 emissions over megacity Delhi: Relative contributions of exhaust and non-exhaust sources. *Science of the Total Environment*, *699*, 134273.

SAFAR, India (2018). SAFAR-high resolution emission inventory of mega city Delhi for 2018, Special Scientific Report, SAFAR-Delhi-2018-A, ISSN: 0252-1075, Publ.- IITM-Pune, Ministry of Earth Sciences (Govt. of India), 2018. <http://safar.tropmet.res.in/source.pdf>.

Vadrevu, K. P., Eaturu, A., Biswas, S., Lasko, K., Sahu, S., Garg, J. K., & Justice, C. (2020). Spatial and temporal variations of air pollution over 41 cities of India during the COVID-19 lockdown period. *Scientific Reports*, *10*(1), 16574.

Further reading

Ravindra, K., Sidhu, M. K., Mor, S., John, S., & Pyne, S. (2016). Air pollution in India: Bridging the gap between science and policy. *Journal of Hazardous, Toxic, and Radioactive Waste*, *20*(4), A4015003.

Subchapter 17.4

Air quality in Wuhan, China

Shaofei Kong and Huang Zheng
Department of Atmospheric Sciences, School of Environmental Studies, China University of Geosciences, Wuhan, P.R. China

17.4.1 Background

Wuhan is the largest megacity in central China with over 13 million inhabitants (as of 2022). It is on the east margin of Twain-Hu Basin, which is surrounded by several mountains. It is an important transport hub and has experienced rapid development through growth in real estate and new technology industries such as optoelectronics, biological science and environmental science. Air pollution is a long-standing issue affecting Wuhan characterized by haze and ozone pollution.

Since 2005, the Wuhan Municipal Ecology and Environmental Bureau released the Bulletin of Ecology and Environment. The bulletin only reported the annual concentration of sulfur dioxide (SO_2), nitrogen oxides (NO_x), and inhalable particulate matter (PM_{10}). Since the implication of the Air Pollution Prevention and Control Action Plan, nine national automatic ambient air monitoring stations were installed in 2013. According to the Ambient air quality standards (GB3095-2012), hourly concentrations of three additional air pollutants: carbon monoxide (CO), ozone (O_3), and fine particulate matter ($PM_{2.5}$) were reported. With the improvement in ambient air monitoring ability, more number and types of stations were installed in Wuhan (Fig. 17.4.1). Currently, there are 11 automatic monitoring stations operated and maintained by Hubei province. These stations mainly focus on the monitoring of ambient CO, NO_x, O_3, PM_{10}, $PM_{2.5}$, and SO_2. In addition to these regular air pollutant monitoring stations, there is one station in the port near the Yangtze River, two stations in the industrial park, and four stations near the street. According to monitoring requirements, additional air pollutants such as volatile organic carbon (VOC) and trace element were monitored in the industrial park and port sites,

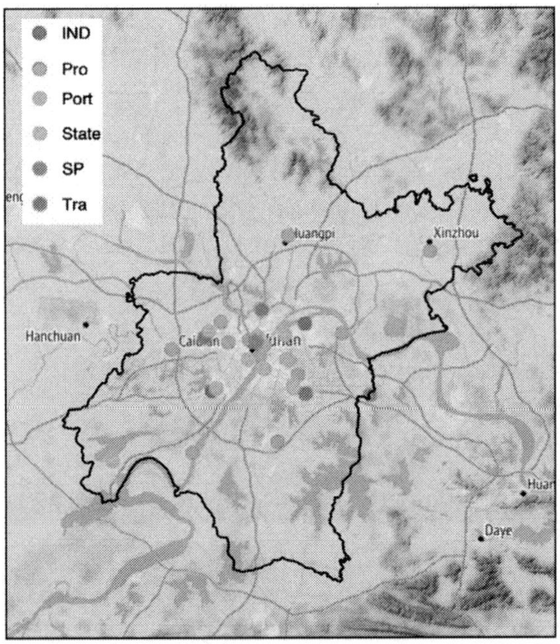

FIGURE 17.4.1 Location of Wuhan (black boundary) and the distribution of different types of ambient air monitoring stations including industrial (ND), port (Port), the station operated and maintained by State (State), Hubei province (Pro), superstation (SP), and transportation (Tra).

respectively. Two super stations which can monitor the chemical speciation of $PM_{2.5}$ (carbonaceous matter, water-soluble ions, trace elements), VOC and aerosol extinction coefficient were distributed in the urban and suburban areas in Wuhan. It should be noted that there was a revision of reference temperature from 273 to 298 K since September 1, 2018, and the air pollutant concentrations before this point were converted and reported here.

17.4.2 Major air pollutant emission sources

Fig. 17.4.2 shows the emission inventory of key air pollutants in Wuhan from 2008 to 2017, which is extracted from the Multi-resolution Emission Inventory model for Climate and air pollution research. Emissions from industrial sectors contribute the most to CO, NO_x, PM_{10}, $PM_{2.5}$, SO_2, and VOC in Wuhan. Except for SO_2 and VOC, the emissions of these pollutants peaked around 2010 and have shown continuous decreases since then. The emissions of SO_2, however, peaked in 2008 and decreased until 2017. Compared to other pollutants, the emission of VOC has shown a flat, gradual increase in recent years, which needs further attention. Despite the reductions of air pollutant emissions from all sectors, the industrial sector still contributes most to air pollutant emissions. An exception is NO_x, which is equally contributed by the industrial and transportation sectors since 2014 in Wuhan. With the growing population of vehicles, the number of cars has been more than 4 million since 2021 in Wuhan which indicates that further emission reduction should take priority in the industrial and transport sectors in Wuhan.

17.4.3 Trends of key air pollutants

As a result of emission reduction, some air pollutant concentrations in Wuhan have decreased significantly (Fig. 17.4.3). The CO decreased by 5.6% from 0.89 $mg\,m^{-3}$ in 2013 to 0.84 $mg\,m^{-3}$ in 2022 with a reduction of 0.01 $mg\,m^{-3}\,year^{-1}$. NO_2 also showed a reduction with annual concentrations decreasing from 38.7 $\mu g\,m^{-3}$ in 2013 to 30.6 $\mu g\,m^{-3}$ in 2022. Ambient PM_{10} in Wuhan shows increasing from 2013 (64.2 $\mu g\,m^{-3}$) to 2015 (89.7 $\mu g\,m^{-3}$) and then decreases to 52.8 $\mu g\,m^{-3}$ in 2022. The overall reduction rate of PM_{10} is $-3.66\,\mu g\,m^{-3}\,year^{-1}$ between 2013 and 2022. $PM_{2.5}$ has shown a significant decrease in Wuhan with $-4.42\,\mu g\,m^{-3}\,year^{-1}$ at 0.001 confidence level. Compared to the $PM_{2.5}$ level in 2013 (72.3 $\mu g\,m^{-3}$), $PM_{2.5}$ in 2022 (31.8 $\mu g\,m^{-3}$) decreased by 127%. Contrary to the decreasing trends of other air pollutants, O_3 has shown an increase in Wuhan with the daily maximum 8 h average (MDA8) O_3 by 3.27 $\mu g\,m^{-3}\,year^{-1}$. As a result, the annual MDA8 O_3 increased from 74.7 $\mu g\,m^{-3}$ in 2013 to 93.5 $\mu g\,m^{-3}$ in 2022.

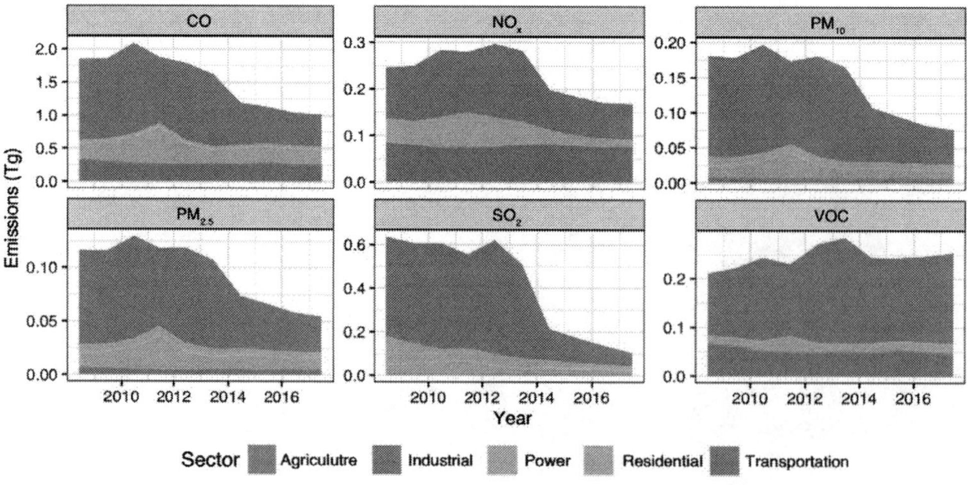

FIGURE 17.4.2 Time series of air pollutant emissions from different anthropogenic sectors between 2008 and 2017 in Wuhan, China. Data were extracted from Zheng et al. (2018).

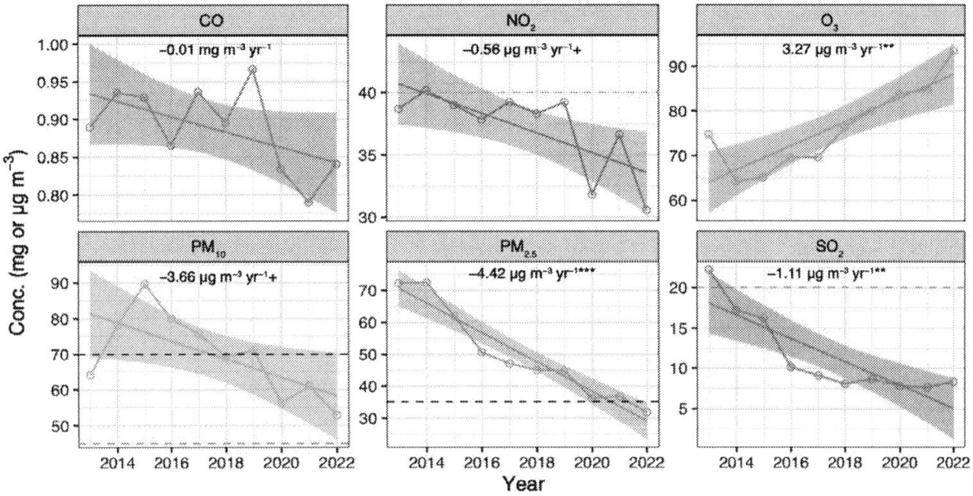

FIGURE 17.4.3 Time series of annual concentrations of air pollutants in Wuhan from 2013 to 2022. The duration of data was from 2013-1-18 to 2022-11-30. The texts in the figures represent the trends of air pollutants and the +, **, and *** represent the trends that are significant at 0.1, 0.01, and 0.001 confidence levels respectively. The black and green dashed horizontal lines represent the grand II and grade I of the air quality guideline of China.

17.4.4 Comparison to WHO AQG (2021) and Chinese national standards

The concentrations of air pollutants in the air quality guidelines (AQG) of Chinese air quality standards (GB3095-2012) are generally higher than the latest AQG released by WHO in 2021. Compared to other air pollutants, the annual CO and SO_2 concentration in Wuhan meets the AQG of China and WHO. The ambient NO_2 in Wuhan meets the Grade II of the standards ($< 40\ \mu g\, m^{-3}$) in most years except for 2014. However, the annual NO_2 concentration is higher than the AQG ($10\ \mu g\, m^{-3}$) of WHO. The annual PM_{10} and $PM_{2.5}$ meet the Grade II of AQG of China in 2018 and 2022, respectively. According to the AQG of WHO, the annual PM_{10} and $PM_{2.5}$ in Wuhan just meet the interim target I (PM_{10}: $70\ \mu g\, m^{-3}$, $PM_{2.5}$: $35\ \mu g\, m^{-3}$), and there is a long journey to meet the higher targets in AQG of WHO (PM_{10}: $15\ \mu g\, m^{-3}$, $PM_{2.5}$: $5\ \mu g\, m^{-3}$).

17.4.5 Health impacts

Due to the improvement of air quality in China, the adverse impacts on public health has reduced. For instance, the $PM_{2.5}$-attributable excess deaths in 2017 was 1,975,400 and the excess deaths would have been 2,380,700 if there were no emission control measures. Therefore, the avoided excess deaths were 405,300 with the strengthen industrial emission standards contributing the most (157,200) (Zhang et al., 2019).

17.4.6 Challenges to improving air quality

Despite the achievements of air quality control in China, there are still challenges in further improving air quality: (1) the complex pollution of $PM_{2.5}$ and O_3 in Wuhan. As shown in Fig. 17.4.3, on the one hand, despite the reduction in air pollution emissions, the annual $PM_{2.5}$ concentration just meets Grade II of AQG of China and further reduction would be tougher. On the other hand, the ozone concentration in Wuhan and China has been increasing in recent years probably due to the $PM_{2.5}$ reduction which has been examined by Li et al. (2019). Due to the co-precursors of $PM_{2.5}$ and O_3, the same study by Li et al. (2019) suggests that the complex pollution could be synchronously solved by aggressive reduction of NO_x and VOC emissions; (2) regional transport of air pollutants from North China Plain (NCP) is another key issue of air quality control in Wuhan. It was estimated that the regional transport from NCP and the basin terrain of Wuhan and its surrounding area can increase the $PM_{2.5}$ concentration by 43.5%. The joint control of gaseous precursor emissions in upwind regions is essential to prevent heavy air pollution in Wuhan (Hu et al., 2023).

References

Hu, W., Zhao, Y., Zhao, T., Bai, Y., Zhao, C., Kong, S., Chen, L., Du, Q., Zheng, H., Lu, W., Liu, W., & Sun, X. (2023). Aggravated chemical production of aerosols by regional transport and basin terrain in a heavy $PM_{2.5}$ pollution episode over central China. *Atmospheric Environment (Oxford, England: 1994)*, *294*, 119489.

Li, K., Jacob, D. J., Liao, H., Zhu, J., Shah, V., Shen, L., Bates, K. H., Zhang, Q., & Zhai, S. (2019). A two-pollutant strategy for improving ozone and particulate air quality in China. *Nature Geoscience*, *12*, 906–910.

Zhang, Q., Zheng, Y., Tong, D., Shao, M., Wang, S., Zhang, Y., Xu, X., Wang, J., He, H., Liu, W., Ding, Y., Lei, Y., Li, J., Wang, Z., Zhang, X., Wang, Y., Cheng, J., Liu, Y., Shi, Q., Yan, L., Geng, G., Hong, C., Li, M., Liu, F., Zheng, B., Cao, J., Ding, A., Gao, J., Fu, Q., Huo, J., Liu, B., Liu, Z., Yang, F., He, K., & Hao, J. (2019). Drivers of improved $PM_{2.5}$ air quality in China from 2013 to 2017. *Proceedings of the National Academy of Science USA*, *116*, 24463–24469.

Zheng, B., Tong, D., Li, M., Liu, F., Hong, C., Geng, G., Li, H., Li, X., Peng, L., Qi, J., Yan, L., Zhang, Y., Zhao, H., Zheng, Y., He, K., & Zhang, Q. (2018). Trends in China's anthropogenic emissions since 2010 as the consequence of clean air actions. *Atmospheric Chemistry &. Physics*, *18*, 14095–14111.

Subchapter 17.5

Air quality in Melbourne, Australia

Jason Choi[1], Timothy Chaston[1] and Nigel Tapper[2]
[1]*Environment Protection Authority Victoria, Australia,* [2]*School of Earth, Atmosphere and Environment, Monash University, Australia*

17.5.1 Background

The city of Melbourne is situated in the state of Victoria on the southeast coast of Australia at 37.8°S, 145.0°E and has a warm temperate oceanic climate (Sturman & Tapper, 2006). With a population of 4,976,157 (Australian Bureau of Statistics, 2020), Melbourne is Australia's 2nd largest city (total population of Australia 25,890,800) and it comprises 75.5% of the total Victorian population (Australian Bureau of Statistics, 2022). Melbourne is the 79th largest city in the world by population. Overall, Melbourne's air pollution has decreased since monitoring began in 1979 and is considered low by international standards. However, this improvement is not consistent across the metropolitan area and air pollution, even at low concentrations, can have environmental and human health impacts (Hanigan et al., 2019). Of the criteria air pollutants, $PM_{2.5}$, O_3, and NO_2 are of greatest concern.

Whilst emissions of pollutants are the main driver of ambient concentrations, meteorology can also be a significant factor. During certain times of the year, especially in late summer and autumn, atmospheric conditions can trap air pollution above Melbourne. This was first observed in the late 1970s and is known as the Melbourne or Spillane Eddy, named after Kevin Spillane, who first wrote about it (Spillane, 1978). The Spillane Eddy is a gyre that forms under stable anticyclonic conditions and recirculates air over the city. It is often associated with the formation of high levels of O_3 during the summer months.

17.5.2 Major air pollution sources and emissions

According to Environment Protection Authority of Victoria (EPA)'s emissions inventory (Environment Protection Authority Victoria, 2022), major natural sources of pollution include bushfires, dust storms and biogenic emissions from vegetation. Major anthropogenic sources include motor vehicles, domestic wood heaters, power stations, and ecological and landscape burns. Industrial sources, such as oil refineries and port activities, significantly impact local environments. Some locations, such as Brooklyn in the inner west of Melbourne, are impacted by dust from nearby industry. Victoria's power is mainly generated by the combustion of brown coal in the Latrobe Valley, about 130 km (about 81 million) East of Melbourne. Those emissions contribute little to air pollution in Melbourne.

17.5.3 Air pollution trends since 2000

The Victorian air pollution monitoring network is operated by the EPA and data has been collected since 1979. EPA has continually improved data capture, quality, and spatial and temporal coverage of the network. This includes improvements to capture hourly $PM_{2.5}$ measurements since 2014. Prior to this, $PM_{2.5}$ data was recorded using a 24 hour gravimetric method run every three days. Other improvements include the use of low-cost air pollution sensors to increase coverage of regional Victoria and have enabled monitoring in locations where a full monitoring station could not be installed.

Since 2000, CO (Fig. 17.5.1A), and NO_2 (Fig. 17.5.1B) concentrations have decreased, in part due to improved vehicle emissions standards and controls. In contrast, ambient $PM_{2.5}$, PM_{10} (Fig. 17.5.1C), O_3 (Fig. 17.5.1D) and SO_2 concentrations have remained constant on average and are strongly influenced by large-scale events, such as the major bushfires of 2019/2020. That event markedly increased annual average concentrations of $PM_{2.5}$ and PM_{10} across the entire state.

17.5.4 Compliance with regulatory and international standards and guidelines

In Melbourne, air pollution is assessed against the Environment Reference Standard (ERS), which generally adopts the standards in the Australian National Environment Protection (Ambient Air Quality) Measure (NEPM) but includes more stringent guidance for O_3 and SO_2. Reviews of PM_{10} and $PM_{2.5}$ standards are planned for 2025. When compared

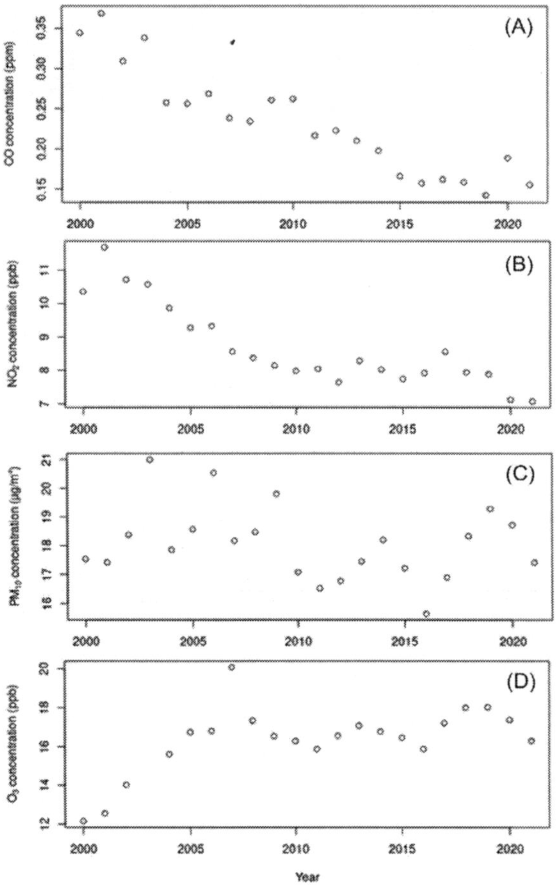

FIGURE 17.5.1 Annual average one-hour CO, NO_2, PM_{10}, and O_3 concentrations from EPA monitoring stations from 2000 to 2021.

against the WHO Air Quality Guidelines (AQG), the standards for PM_{10} and $PM_{2.5}$ are similar to the stage 4 interim targets, but overall, the AQG concentrations are lower. Compliance against the AQG and ERS for CO and SO_2 is generally achieved. For NO_2 there is compliance with the ERS but not the AQG. For $PM_{2.5}$, PM_{10} and O_3, ambient concentrations at times were higher than the standards and guidelines in the ERS and AQG for all years.

17.5.5 Estimate of published health impacts

Despite comparatively moderate levels of air pollution in Melbourne Australia, considerable health impacts have been demonstrated. For example, the mortality burden of $PM_{2.5}$ in Victoria was estimated at 9366 years of life lost (YLL) annually (95% confidence intervals: 6,103, 12,421) based on chronic exposures to $PM_{2.5}$ over recent decades (Hanigan et al., 2020). Using the value of a statistical life year, that study estimated a premature mortality cost of about \$1.5B annually (Australian dollars and thereafter for this section) in Victoria (Hanigan et al., 2020). Using a similar WHO recommended health impact assessment approach, we estimate that chronic NO_2 exposures in Melbourne add 7086 YLL. In total, the mortality cost attributable to $PM_{2.5}$ and NO_2 therefore exceeds \$2.5B annually.

In addition to sustained anthropogenic air pollution, Victorians are exposed intermittently to landscape fire smoke (LFS) haze, leading to additional premature deaths, hospital admissions for cardiovascular and respiratory diseases and emergency department visits for asthma episodes (Johnston et al., 2021). In a national study, the costs of these outcomes were estimated for each of the years 2000–20, and in Victoria, LFS-related health costs were about \$200M during the 2006/07 fire season and approximately $500M during the 2019/20 season. The annual health cost estimates of LFS during the intervening years ranged from $10 to $100M (Johnston et al., 2021).

If the energy transition away from fossil fuels can be achieved in Melbourne, landscape fires will become the main source of air pollution. In our comparison of chronic air pollution exposure scenarios in inner Melbourne, the mortality burden of $PM_{2.5}$ would be more than 50% greater under sustained conditions of 2020 (mega fire season) than under those of 2021 (baseline landscape fire activity), indicating the degree to which LFS impacts annual average air pollutant concentrations. The LFS of 2020 represented a much larger portion of $PM_{2.5}$ exposures in all other parts of Victoria, where anthropogenic sources are fewer and flammable landscapes are nearer.

17.5.6 Challenges to improving air quality

Since 2011 Melbourne's population is estimated to have grown by 19.4%. If this trend was to continue, there would likely be increases in anthropogenic emissions and increases in air pollution, though this would not be linear compared with population growth. In addition, as with many major economies, there is a trend of increasing population age; this will mean that future populations may be more susceptible to impacts from air pollution.

Climate change is also expected to influence future air quality, this may be due to a drying climate which may result in longer seasons of more intense bushfires, as well as changes in Melbourne's meteorology which could alter the development, formation, chemical transformation, transport and deposition of various air pollutants.

References

Australian Bureau of Statistics (2020). Regional population, 2018-19 | Australian Bureau of Statistics. [online] <http://www.abs.gov.au>. <https://www.abs.gov.au/statistics/people/population/regional-population/latest-release>.

Australian Bureau of Statistics (2022). National, state and territory population, December 2019, Australian Bureau of Statistics. [online] <http://www.abs.gov.au. Available at: https://www.abs.gov.au/statistics/people/population/national-state-and-territory-population/latest-release>.

Environment Protection Authority Victoria. (2022). 2016 emissions inventory report | Environment Protection Authority Victoria. [online] https://www.epa.vic.gov.au/about-epa/publications/2028 (accessed December 12, 2022).

Hanigan, I. C., Rolfe, M. I., Knibbs, L. D., Salimi, F., Cowie, C. T., Heyworth, J., Marks, G. B., Guo, Y., Cope, M., Bauman, A., Jalaludin, B., & Morgan, G. G. (2019). All-cause mortality and long-term exposure to low level air pollution in the '45 and up study' cohort, Sydney, Australia, 2006-2015. *Environment International*, *126*, 762–770. Available from https://doi.org/10.1016/j.envint.2019.02.044, Epub 2019 Mar 15. PMID: 30878871.

Hanigan, I. C., Broome, R. A., Chaston, T. B., Cope, M., Dennekamp, M., Heyworth, J. S., Heathcote, K., Horsley, J. A., Jalaludin, B., Jegasothy, E., Johnston, F. H., Knibbs, L. D., Pereira, G., Vardoulakis, S., Vander Hoorn, S., & Morgan, G. G. (2020). Avoidable mortality attributable to anthropogenic fine particulate matter (PM2.5) in Australia. *International Journal of Environmental Research and Public Health*, *18*(1), 254. Available from https://doi.org/10.3390/ijerph18010254, PMID: 33396338; PMCID: PMC7795118.

Johnston, F. H., Borchers-Arriagada, N., Morgan, G. G., et al. (2021). Unprecedented health costs of smoke-related PM2.5 from the 2019–20 Australian megafires. *Nature Sustainable*, *4*, 42–47. Available from https://doi.org/10.1038/s41893-020-00610-5.

Spillane, K. T. (1978). Atmospheric characteristics on high oxidant days in Melbourne. *Clean Air*, *12*, 50–56.

Sturman, A. P., & Tapper, N. J. (2006). The weather and climate of Australia and New Zealand (2nd ed., 541 pp.). Melbourne VIC, Australia: Oxford University Press.

Subchapter 17.6

Air quality in Marrakech, Morocco

Kenza Khomsi[1] **and Imane Sekmoudi**[2]

[1]*Mohammed 6 University of Sciences and Health, Casablanca, Morocco,* [2]*Independent Researcher, Casablanca, Morocco*

17.6.1 Background

Marrakech is the fourth largest city and first tourist destination in Morocco. It is located in the central-south of Morocco in the plain of Haouz. It is the capital city of the Marrakech-Safi region. The city covers an area of 230 km^2. It represents a uniform topography of moderate altitude. It is a uniform plain sloping gently at a rate of about 8% oriented from the South-East to the North-West over 180 km. The reliefs of the city consist mainly of hills and the average altitude is about 465 m above sea level (Inchaouh et al., 2017). According to the last population census of 2014, the city of Marrakech totals 1,330,468 inhabitants, including 980,548 inhabitants in urban areas and 349,920 inhabitants in rural areas. The density is estimated at 0.41 km $inhabitant^{-1}$ and 507 inhabitant km^{-2} in the prefecture of Marrakech (Haut Commissariat au Plan, 2018).

Marrakech has a continental climate which is influenced by the very high altitudes of the High Atlas. The arid and semiarid character dominates. Average annual temperatures are of 20°C and ranges between −7°C and 46.6°C as observed over several years (Inchaouh et al., 2017). Average rainfall is 281 mm $year^{-1}$. Winters in Marrakech are often cold at night and in the morning while summers are generally hot (El Hassar, 2021). The relative humidity is 73% in January and 33% in July and can even vanish during this month when dry winds blow. The winds are generally calm flowing in west and northwest directions (Inchaouh et al., 2017). Between 1941 and 2010, the hot season for the city was found between April and October and the city is exposed to the drying winds of the Chergui and Sirocco type from the East and the South (Khomsi et al., 2020).

The economy of Marrakech-Safi region is based mainly on agriculture and livestock, tourism, crafts (45% of national crafts) and industry that is mainly composed of agro-industry, chemical and para-chemicals. Some previous studies focused on the assessment of air pollution in the city of Marrakech. Over the period 2009 and 2012, (Inchaouh et al., 2017) showed that Marrakech is subject to photochemical pollution mainly due to its geographical location and climatology (strong sunshine). Between 2009 and 2012, the O_3 seasonal profile shows that the ozone levels exhibited strong seasonality, with a seasonal maximum occurring during spring and minimum levels in winter (Inchaouh et al., 2017).

In this short survey, we assess air quality in the city of Marrakech between 2003 and 2021. Trends in nitrogen dioxide (NO_2), particulate matter (PM_{10}), O_3, carbon monoxide (CO) and sulfur dioxide (SO_2) are studied using Sen's Slope (SS) and Man Kendall test (Hamed & Ramachandra Rao, 1998; Kamruzzaman et al., 2018). These statistical approaches allow trends of the rate of change in variables like air pollutant concentrations to be quantified. These pollutants were chosen because their surveillance is mandatory according to the Moroccan regulation. Unfortunately, observed data for the studied pollutants with a long temporal coverage were not accessible. Thus, for the purpose of this work, data from the Atmosphere Data Store (ADS) of the Copernicus Atmosphere Monitoring Service (CAMS)[1] were used. Global and regional PM_{10} CAMS Reanalysis (CAMSRA) have been examined against the average daily PM_{10} concentrations collected from six fixed air quality stations in 2016, including Marrakech from Morocco (Sekmoudi et al., 2021). The authors concluded that the CAMSRA global analysis data can be used to compute climatology, study trends, evaluate models, benchmark other reanalysis, or serve as boundary conditions for past periods. They also highlighted that the reanalysis data are closer to observed data mainly for the stations in the south, such as Marrakech, which are impacted by natural particulate matter pollution.

1. Atmosphere Data Store (ADS) of the Copernicus Atmosphere Monitoring Service.

17.6.2 Major air pollution sources and emissions

The city of Marrakech is the main industrial center of the region and contributes 60% to its job creation. Moreover, the city possesses a structured road network which plays a key role in its economic and social development. It has an international airport: Marrakech Menara which is located 6 km to the south. These diverse activities in Marrakech either industrial or traffic expose the city to atmospheric pollution. In the work reported by Saidi et al. (2023), the authors have used anthropogenic emission inventory provided by the Moroccan Ministry of Environment, Mines, and Sustainable Energy that contains annual fluxes, for 2013, of anthropogenic air pollutants over the area. These authors showed that the main source of PM_{10} and $PM_{2.5}$ emissions is residential/tertiary combustion (75% and 83%) followed by road transport (12% and 13%, respectively). Road transport contributes almost 50% of NO_x emissions and another 31% comes from agriculture. The residential/tertiary sector makes a contribution of 13%. The industrial activity in the region leads to high emission fluxes of SO_x.

17.6.3 Trends of key air pollutant concentrations

Figs 17.6.1 and 17.6.2 show respectively the temporal evolution of the quantities of total columns (TC) of SO_2, NO_2, O_3, and CO, and the concentration of PM_{10}, between 2003 and 2021, in the city of Marrakech. According to Fig. 17.6.1A, the average annual TC-SO_2 over the 19 years was found to be 2.229×10^{-6} kg m^{-2}. The minimum and maximum annual

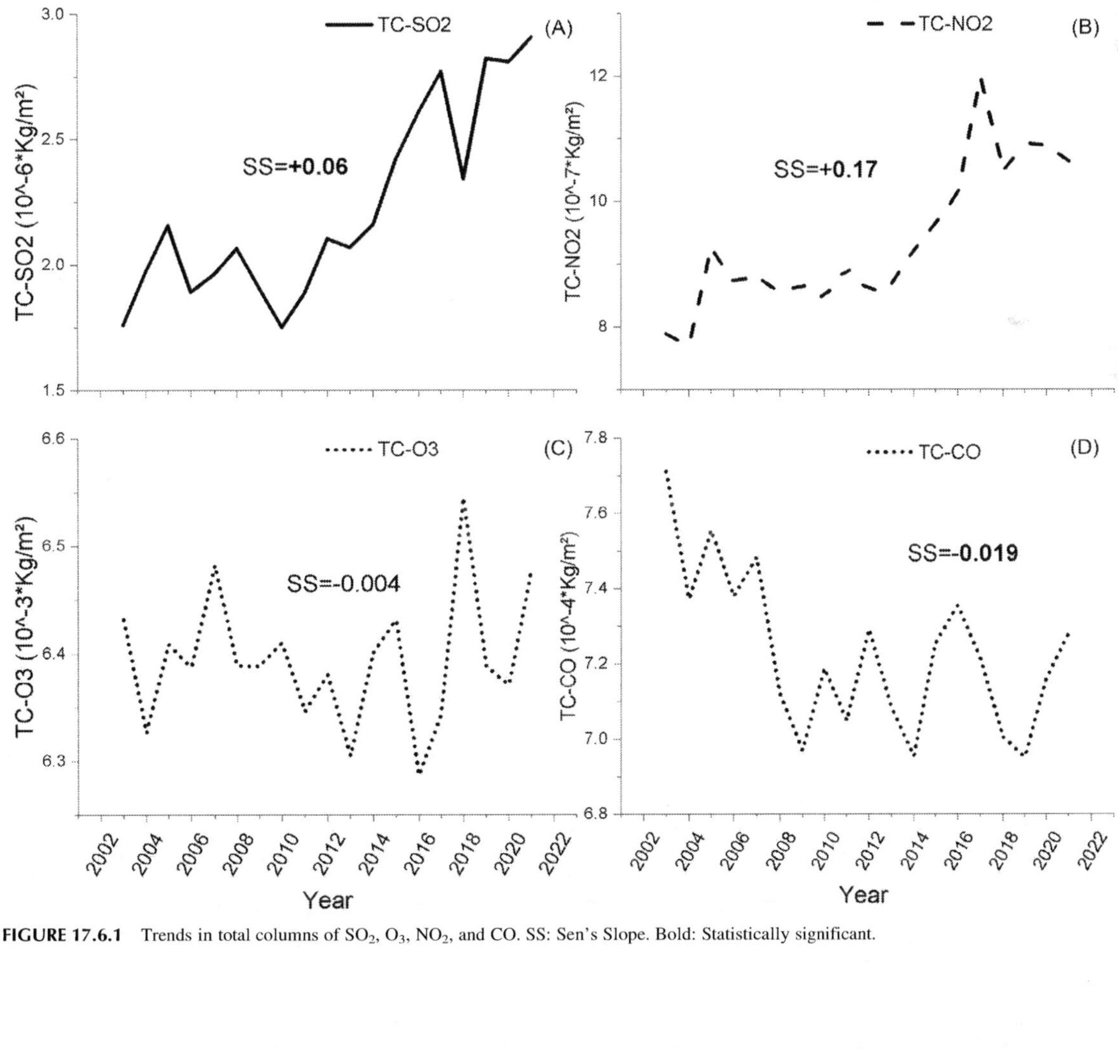

FIGURE 17.6.1 Trends in total columns of SO_2, O_3, NO_2, and CO. SS: Sen's Slope. Bold: Statistically significant.

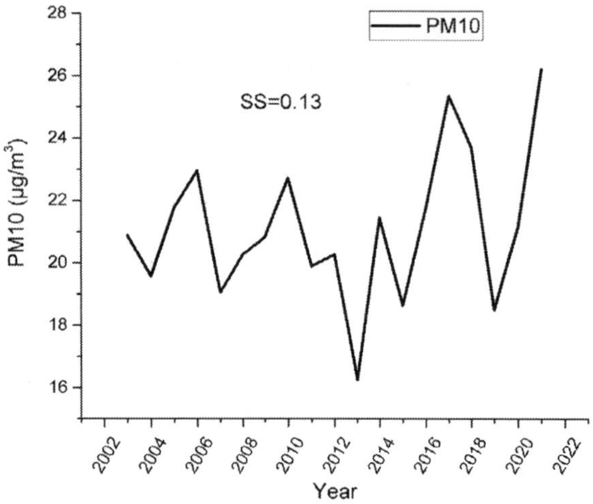

FIGURE 17.6.2 Trends in PM_{10} concentrations, SS: Sen's Slope

quantities were 1.752×10^{-6} kg m^{-2} and 2.906×10^{-6} kg m^{-2}, respectively observed in 2010 and 2021. The figure shows a statistically significant increasing with an SS of the order of 0.06 kg m^{-2}year^{-1}. Regarding TC-NO_2, Fig. 17.6.1B shows a significant upward trend during the study period with an SS = +0.17 kg m^{-2}year^{-1}. For this period, the average annual TC-NO_2 amount was found to be 9.37×10^{-7} kg m^{-2}, and the minimum and maximum annual amounts were 7.695×10^{-7} kg m^{-2} and 11.98×10^{-7} kg m^{-2}, respectively for the years 2004 and 2017. On the other hand, the quantity of TC-O_3 in Marrakech varies from 6.286×10^{-3} kg m^{-2} in 2016 to 6.545×10^{-3} kg m^{-2} in 2018 with an average of 6.395×10^{-3} kg m^{-2}. The SS was negative with a value of −0.004 kg m^{-2}year^{-1} which represents a decreasing trend (Fig. 17.6.1C). Similarly, for the TC-CO in Marrakech, Fig. 17.6.1D shows a statistically significant decrease with an SS of −0.019 kg m^{-2}year^{-1}. The quantity of TC-CO varies from 6.951×10^{-4} kg m^{-2} in 2019 to 7.712×10^{-4} kg m^{-2} in 2003 with an average of 7.230×10^{-4} kg m^{-2}. Fig. 17.6.2 shows the temporal evolution of PM_{10} concentrations in Marrakech from 2003 to 2021. We notice minimum and maximum annual concentrations of 16.3 and 26.3 μg m^{-3}, respectively for 2013 and 2021. The average concentration is of around 21.11 μg m^{-3}. The SS was positive with a statistically significant value of 0.13 μg m^{-3}year^{-1}, this represents evidence of an upward trend. All the average concentrations of PM_{10} recorded during the study period exceed the updated WHO guidelines for 2021 (15 μg m^{-3}).

17.6.4 Health Impacts

The number of annual deaths due to air pollution by ambient particulate matter has increased by 36% (6014 premature deaths in Morocco in 2013) (Roy, 2016). Deaths from household air pollution continued to increase by 18% (953 premature deaths in Morocco in 2013). The estimated economic cost of premature deaths from air pollution was approximately: \$215 billion for ambient particulate matter and \$232 billion for household air pollution (Roy, 2016). $PM_{2.5}$ fine particulate exposure resulted in an estimated 5450 premature deaths in 2014 among adults, from ischemic heart disease, stroke, lung cancer, and chronic obstructive pulmonary disease, and in children, following acute infections of the lower respiratory tract. The groups most vulnerable to premature death from air pollution are adults over 55 due to ischemic heart disease and children under 5 due to acute lower respiratory tract infections (Croitoru & Sarraf, 2018).

17.6.5 Challenges to improving air quality

Nowadays the city of Marrakech is raising the same challenges as many other major cities in the world. One issue would be: "how will development occur in a rapidly industrializing and urbanizing society with reduced adverse effects on air quality and thus on human health?". Answering this question implies seeking the availability of the tools that can help to address air pollution challenges mainly at the urban scale. These start from the availability of the appropriate infrastructure and the adequate strategy for data accessibility in order to monitor air quality till the implementation of the appropriate policies to improve decision-making in this field. Indeed, being an important economic center in Morocco, Marrakech is maintaining its position by multiplying its activities, further developing its economic sectors, and creating employment and opportunities for social development. In particular, the city continues to face challenges such as:

- The presence of a large number of industrial establishments, some of which are emitters of atmospheric pollutants;
- Emissions of atmospheric pollutants resulting from electricity production and energy consumption;
- The large car fleet in continuous growth; and
- The lack of local evaluating exposure-response functions to assess the real impact of air quality on human health.

References

Croitoru, L., & Sarraf, M. (2018). How much does environmental degradation cost? The case of Morocco. *Journal of Environmental Protection (Irvine,. Calif)*. Available from https://doi.org/10.4236/jep.2018.93017.

El Hassar, N. (2021). Une Approche de Deep Learning pour la Prévision de la Qualité de l'air à Marrakech. Université Mohammed V Faculté des Sciences de Rabat Master de Statistique et Économétrie.

Hamed, K. H., & Ramachandra Rao, A. (1998). A modified Mann-Kendall trend test for autocorrelated data. *Journal of Hydrology*, *204*, 182–196. Available from https://doi.org/10.1016/S0022-1694(97)00125-X.

Haut Commissariat au Plan (2018). Monographie Régionale : Région de Marrakech-Safi.

Inchaouh, M., Tahiri, M., Johra, B., El., & Abboubi, R. (2017). State of ambient air quality in Marrakech city (MOROCCO) over the period 2009-2012. *International Journal of Geomate*. Available from https://doi.org/10.21660/2017.29.1254.

Kamruzzaman, M., Rahman, A. T. M. S., Ahmed, M. S., Kabir, M. E., Mazumder, Q. H., Rahman, M. S., & Jahan, C. S. (2018). Spatio-temporal analysis of climatic variables in the western part of Bangladesh. *Environmental Development Sustainable*. Available from https://doi.org/10.1007/s10668-016-9872-x.

Khomsi, K., Najmi, H., Chelhaoui, Y., & Souhaili, Z. (2020). The contribution of large-scale atmospheric patterns to pollution with PM10: The new Saharan Oscillation. *Aerosol and Air Quality Research*, *20*(5), 1038–1047. Available from https://doi.org/10.4209/aaqr.2019.08.0401.

Roy, R. (2016). *The cost of air pollution in Africa*. OECD Development Centre Working Paper, OECD Publishing, Paris.

Saidi, L., Valari, M., & Ouarzazi, J. (2023). Air quality modeling in the city of Marrakech, Morocco using a local anthropogenic emission inventory. *Atmospheric Environment (Oxford, England: 1994)*, *293*, 119445. Available from https://doi.org/10.1016/j.atmosenv.2022.119445.

Sekmoudi, I., Khomsi, K., Faieq, S., & Idrissi, L. (2021). Assessment of global and regional PM10 CAMSRA data: Comparison to observed data in Morocco. *Environmental Science Pollution Research*. Available from https://doi.org/10.1007/s11356-021-12783-3.

Subchapter 17.7

Air quality in Los Angeles, California, United States

Somayyeh Arghavani[1], Ummugulsum Alyuz[1], Ranjeet S. Sokhi[1] and Brian McDonald[2]
[1]Centre for Climate Change Research (C3R), School of Physics, Engineering and Computer Science, University of Hertfordshire, Hatfield, United Kingdom, [2]NOAA Chemical Sciences Laboratory, Boulder, CO, United States

17.7.1 Background

The city of Los Angeles is located in Southern California and has a population of about 3.8 million (estimated in 2022) with Los Angeles County having a population of around 9.7 million (United States Census Bureau, 2024). It is a densely urbanized area bordered by the Santa Monica Mountains to the west and the San Fernando Valley to the north. In terms of air quality, particulate matter (PM), nitrogen oxides (NO_x), and ozone (O_3) are the main air pollutants of interest in this area (Jury, 2020). Los Angeles (LA) is in the South Coast Air Basin (Basin), and air quality is managed by the South Coast Air Quality Management District (SCAQMD), which has been at the forefront of air quality improvement policies since 1974 when the first air pollution control program was established. SCAQMD controls emissions primarily from stationary sources such as large power plants and refineries to local gas stations, using a vast network of air quality monitors at 40 permanent sites and 10 additional sites near the large refineries. Despite the continuous increase in population and on-road vehicles in this area, air quality has significantly improved. The maximum level of ozone is now less than one-quarter of its concentration in the 1950s (SCAQMD, 2024). According to the Air Quality Index (AQI) reports, in 2023, LA County experienced 46 unhealthy days for sensitive groups, which was less than in 2022 (65 days), and 26 unhealthy days compared to 28 days in 2022 (EPA, 2024a).

Apart from anthropogenic emissions, pacific high-pressure cell, mountains, and concave coastline affect the wind pattern and air quality in LA (Jury, 2020). In addition, wildfires are one of the most important factors that affect LA ambient air quality air by emitting significant amounts of smoke, which is composed of different pollutants such as PM (including $PM_{2.5}$ and smaller particles), carbon dioxide (CO_2), carbon monoxide (CO), nitrogen oxides (NO_x) and volatile organic compounds (VOCs) (Enayati Ahangar et al., 2021). Due to climate change, the number and intensity of wildfires are also expected to increase (Williams et al., 2019).

17.7.2 Major air pollutant emission sources

The main emission sources for NO_x, PM_{10}, and $PM_{2.5}$ in LA County in 2020 are outlined in Fig. 17.7.1. Nonroad mobile sources, including aircraft, trains, ships, commercial boats, and off-road vehicles, accounted for 41% of NO_x emissions. On-road motor vehicles contributed 38%, followed by fuel combustion from stationary sources at 13%, and

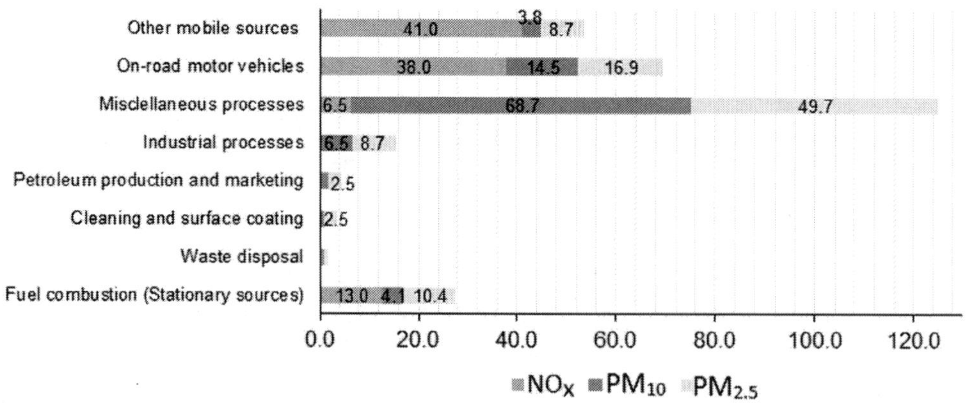

FIGURE 17.7.1 Percentage contributions of sectors to NO_x, PM_{10}, and $PM_{2.5}$ emissions in 2020 in LA County (CARB, 2024a).

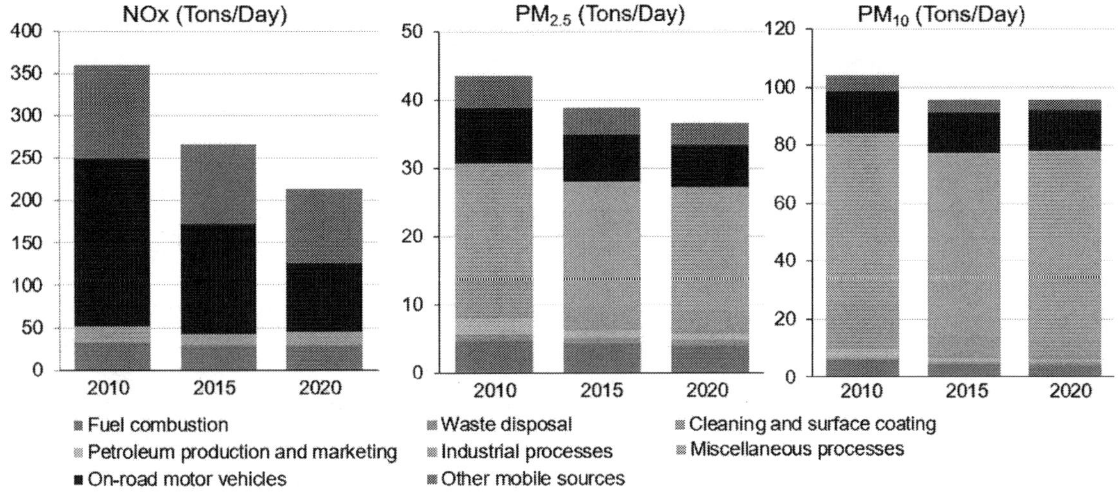

FIGURE 17.7.2 Estimated annual average anthropogenic emissions by sectors in LA County for 2010, 2015, and 2020 (CARB, 2024a).

miscellaneous processes (such as residential fuel combustion, windblown dust and construction activities) at 6.5%. For PM_{10}, miscellaneous processes were the primary source at 68.7%, followed by on-road mobile sources at 14.5%, industrial processes at 6.5%, and fuel combustion at 4.1%. Similarly, for $PM_{2.5}$ emissions, miscellaneous processes led at 49.7%, followed by on-road mobile sources at 16.9%, fuel combustion at 10.4%, and industrial processes at 8.7% (the percentages were calculated from the data obtained from CARB, 2024a).

Annual reduction trends for NO_x, $PM_{2.5}$ and PM_{10} emissions are shown in Fig. 17.7.2 from various sectors in LA County between 2010 and 2020. The most substantial reduction in NO_x emissions was observed in on-road motor vehicles (59%) and nonroad mobile sources (40%) sectors during this period. For $PM_{2.5}$, a total reduction of 16% is estimated from 2010 to 2020. Petroleum production and marketing showed the highest reduction trend (62%), followed by industrial processes (38%), nonroad mobile sources (33%), on-road motor vehicles (23%) and fuel combustion (17%), with a slight increase of 4 in the miscellaneous processes sector. Regarding PM_{10} emissions, the total reduction is estimated at 8%. The most significant reductions for PM_{10} were estimated in industrial processes (61%), petroleum production and marketing (51%), nonroad mobile sources (33%), waste disposal (33%) and fuel combustion (30%) sectors, while a slight increase of 11% was reported in the miscellaneous processes sector (the percentages were calculated from the data obtained from CARB, 2024a).

17.7.3 Trends of key air pollutants

Fig. 17.7.3 displays the annual average concentrations of NO_2, O_3, $PM_{2.5}$ and PM_{10} in LA County from 2010 to 2023. The largest reduction occurred in NO_2 concentration, with a decrease of 24.63 (approximately 41%) between 2010 and 2023, mainly due to on-road and nonroad mobile sources. Similarly, $PM_{2.5}$ concentration decreased by approximately 22% (10.98 $\mu g\,m^{-3}$) during the same period. A slight decrease of 2.6 (approximately 11%) was observed in PM_{10} concentration between 2010 to 2023. In the case of O_3, however, maximum daily 8-hour average (MDA8) values showed a slight increasing trend. In 2022, the annual average of MDA8 in LA County was approximately 124.5 $\mu g\,m^{-3}$, representing about 7% increase compared to 2011 (116 $\mu g\,m^{-3}$) (EPA, 2024b).

17.7.4 Comparison to WHO AQG (2021) and US National Air Quality Standards

In 2015, the US EPA updated the National Ambient Air Quality Standard (NAAQS) for O_3 to 70 ppb, approximately 137 $\mu g\,m^{-3}$ for an 8-hour average. This standard was also adopted in the California Ambient Air Quality Standard (CAAQS). For $PM_{2.5}$, CAAQS for an annual average of 12 $\mu g\,m^{-3}$ whereas, the equivalent NAAQS, has recently been lowered to 9 $\mu g\,m^{-3}$ (https://seetheair.org/2024/02/08/us-epa-2024/#more-19259). The CAAQS standards for the annual mean of PM_{10} is 20 and 50 $\mu g\,m^{-3}$ for a 24-hour average. In the case of PM_{10}, 24-hour average NAAQS is 150 $\mu g\,m^{-3}$

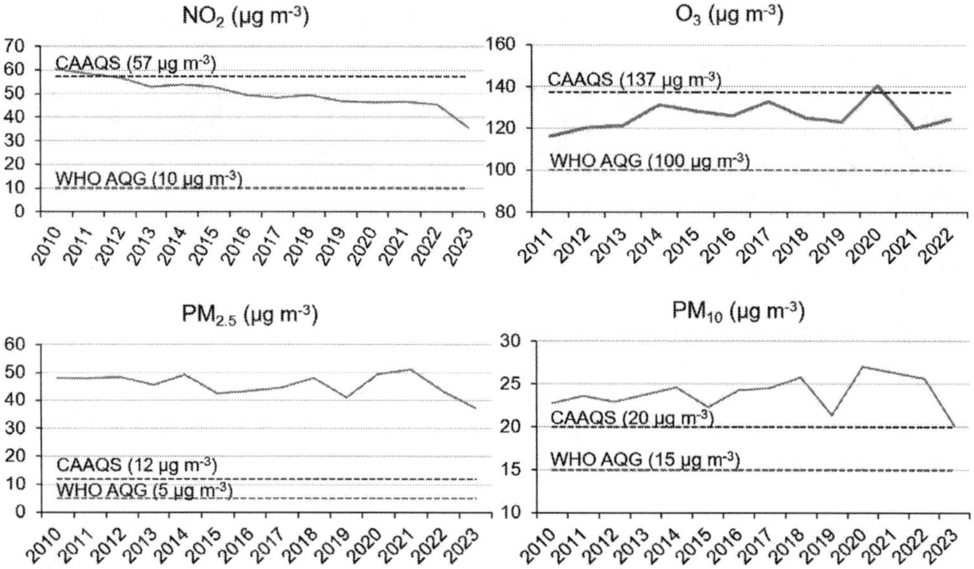

FIGURE 17.7.3 Annual averages of NO_2, O_3, $PM_{2.5}$ and PM_{10} concentrations observed between 2010 to 2023 and maximum daily 8-hour average (MDA8) for O_3 for 2011 to 2022. Measurement data has been extracted from 14 stations in LA County (EPA, 2024b). The dashed horizontal black line shows the California Ambient Air Quality Standards while the red line shows the WHO 2021 Air Quality Guidelines.

Additionally, the annual average standards for NO_2 are 101.3 $\mu g\,m^{-3}$ (53 ppb) and 57.3 $\mu g\,m^{-3}$ (30 ppb) for NAAQS and CAAQS, respectively (CARB, 2024b).

In 2020, one or more stations in the Basin exceeded the NAAQSs on 181 days, including 8-hour ozone (157 days), 24-hour $PM_{2.5}$ (39 days), PM_{10} (3 days), and NO_2 (1 day). Nine of the ten stations that exceeded the O_3 limit (100 days), were located within the Basin, including LA County $PM_{2.5}$ concentrations that have failed to meet the 2012 annual national standard of 12 $\mu g\,m^{-3}$ from 2010 to 2023. In 2020, 16 stations in the Basin including LA County, recorded one or more $PM_{2.5}$ daily average concentrations exceeding the national standard of 35 $\mu g\,m^{-3}$. The concentration of NO_2 in LA County exceeded the short-term (1 hour) national standard of approximately 191.2 $\mu g\,m^{-3}$ only once, at one site (AQMP, 2022). The annual average of NO_2 concentrations in LA County remained below the CAAQS annual average of approximately 57.3 $\mu g\,m^{-3}$ since 2012 (see Fig. 17.7.3).

17.7.5 Health impacts

Studies show that a 50 $\mu g\,m^{-3}$ increase in PM_{10} can lead to a 6% increase in cardiovascular and a 10% increase in respiratory disease admissions among individuals aged 65 years or older (Los Angeles city planning, 2019). Similarly, a 10 $\mu g\,m^{-3}$ increase in fine particles correlates with a 4% rise in total mortality (Pope et al., 2002). Long-term exposure to $PM_{2.5}$ is associated with a threefold higher risk of mortality compared to the national cohort (Jerrett et al., 2005).

Research in LA County has linked monthly NO_2, $PM_{2.5}$, and CO concentrations with asthma-related hospitalizations (Delamater et al., 2012). Additionally, exposure to $PM_{2.5}$ has been linked to low birth weight and decreased verbal learning in middle-aged and older adults (Coker et al., 2016; Gatto et al., 2014; Wilhelm et al., 2012). Studies indicate that a 20-ppb increase in ozone concentration is associated with an 83% increase in illness-related absence rates among children in Southern California (Gilliland et al., 2001).

To assess the health impacts of NO_x emission reduction on O_3 and $PM_{2.5}$ concentrations in LA County, a numerical study using the CMAQ model and BenMAP-CE was conducted. The study found that reducing NO_x emissions by 88% from 2008 levels could decrease $PM_{2.5}$ concentrations and respiratory hospital admissions by 17 cases in LA County. This reduction could also lead to a 79.6% decrease in hospital admissions and prevent 60 cases of asthma due to reduced O_3 concentrations (Stewart et al., 2017).

17.7.6 Challenges for improving air quality

The reduction in transportation emissions in the United States has shifted human exposure from the transportation sector to volatile chemical products (VCPs, such as personal care products, printing inks and cleaning agents). This shift highlights the evolution of urban air pollution sources (McDonald et al., 2018). Additionally, the increase in biogenic VOC emissions from urban green programmes in LA could offset reductions achieved through anthropogenic VOC emissions. An option to mitigate these effects is to utilize low VOC emission vegetation in green space development (Gu et al., 2021).

Analysis of annual trends in a broader range of air pollutant species including ultrafine particulates, NO_x, CO, elemental carbon (EC), organic carbon (OC), and O_3, in LA from 2007 to 2023 reveals an overall decreasing trend. However, when comparing the changes during two sub periods (2007–15 and 2016–23), a more significant reduction in air pollutant levels was observed between 2007 and 2015 possibly indicating higher effectiveness of air quality improvement policies in the earlier period. Continuing challenges in controlling air pollution in LA County also include transported ozone from the Pacific Ocean and increasing impacts of wildfires (Badami et al., 2024; Cooper et al., 2015; Enayati Ahangar et al., 2021; Shi et al., 2019).

References

AQMP. (2022). *2022 Air quality management plan*. South Coast Air Quality Management District. <http://www.aqmd.gov/home/air-quality/clean-air-plans/air-quality-mgt-plan>.

Badami, M. M., Tohidi, R., & Sioutas, C. (2024). Los Angeles Basin's air quality transformation: A long-term investigation on the impacts of PM regulations on the trends of ultrafine particles and co-pollutants. *Journal of Aerosol Science*, *176*, 106316. Available from https://doi.org/10.1016/j.jaerosci.2023.106316.

CARB (2024a). California Air Resources Board. CEPAM2019v1.03 Emission Projection Data. <https://ww2.arb.ca.gov/applications/emissions-user-defined-query>.

CARB. (2024b). California Air Resources Board. California Ambient Air Quality Standards. <https://ww2.arb.ca.gov/resources/california-ambient-air-quality-standards>.

Coker, E., Liverani, S., Ghosh, J. K., Jerrett, M., Beckerman, B., Li, A., Ritz, B., & Molitor, J. (2016). Multi-pollutant exposure profiles associated with term low birth weight in Los Angeles County. *Environment International*, *91*, 1–13. Available from https://doi.org/10.1016/j.envint.2016.02.011.

Cooper, O. R., Langford, A. O., Parrish, D. D., & Fahey, D. W. (2015). Atmosphere. Challenges of a lowered U.S. ozone standard. *Science (New York, N.Y.)*, *348*(6239), 1096–1097. Available from https://doi.org/10.1126/science.aaa5748.

Delamater, P. L., Finley, A. O., & Banerjee, S. (2012). An analysis of asthma hospitalizations, air pollution, and weather conditions in Los Angeles County, California. *The Science of the Total Environment*, *425*, 110–118.

Enayati Ahangar, F., Pakbin, P., Hasheminassab, S., Epstein, S. A., Li, X., Polidori, A., et al. (2021). Long-term trends of PM2.5 and its carbon content in the South Coast Air Basin: A focus on the impact of wildfires. *Atmospheric Environment*, *255*, 118431.

EPA. (2024a). Air quality index report. United States Environmental Protection Agency. <https://www.epa.gov/outdoor-air-quality-data/air-quality-index-report>.

EPA. (2024b). Air data: Air quality data collected at outdoor monitors across the US. United States Environmental Protection Agency. <https://www.epa.gov/outdoor-air-quality-data>.

Gatto, N. M., Henderson, V. W., Hodis, H. N., St, John, J. A., Lurmann, F., Chen, J. C., & Mack, W. J. (2014). Components of air pollution and cognitive function in middle-aged and older adults in Los Angeles. *Neurotoxicology*, *40*, 1–7.

Gilliland, F. D., Berhane, K., Rappaport, E. B., Thomas, D. C., Avol, E., Gauderman, W. J., London, S. J., Margolis, H. G., McConnell, R., Islam, K. T., & Peters, J. M. (2001). The effects of ambient air pollution on school absenteeism due to respiratory illnesses. *Epidemiology (Cambridge, Mass.)*, *12*(1), 43–54.

Gu, S., Guenther, A., & Faiola, C. (2021). Effects of Anthropogenic and biogenic volatile organic compounds on Los Angeles air quality. *Environmental Science & Technology*, *55*, 12191–12201.

Jerrett, M., Burnett, R. T., Ma, R., Pope, C. A., III, Krewski, D., Newbold, K. B., Thurston, G., Shi, Y., Finkelstein, N., Calle, E. E., & Thun, M. J. (2005). Spatial analysis of air pollution and mortality in Los Angeles. *Epidemiology (Cambridge, Mass.)*, *16*, 727–736.

Jury, M. R. (2020). Meteorology of air pollution in Los Angeles. *Atmospheric Pollution Research*, *11*(7), 1226–1237.

Los Angeles City Planning (2019). City of Los Angeles, Department of City Planning. Air quality and health effects report. <https://planning.lacity.gov/odocument/e1a00fbf-6134-4fa9-b6fd-54eee631effb/City_of_LA_-_Air_Quality_and_Health_Effects_and_Attachments.pdf>.

McDonald, B. C., Gouw, J. A., Gilman, J. B., Jathar, S. H., et al. (2018). Volatile chemical products emerging as largest petrochemical source of urban organic emissions. *Science (New York, N.Y.)*, *359*, 760–764.

Pope, C. A., III, Burnett, R. T., Thun, M. J., Calle, E. E., Krewski, D., Ito, K., & Thurston, G. D. (2002). Lung cancer, cardiopulmonary mortality, and long-term exposure to fine particulate air pollution. *JAMA: the Journal of the American Medical Association*, *287*(9), 1132–1141.

SCAQMD. (2024). The South Coast air quality management district. <https://www.aqmd.gov/nav/about/>.

Shi, H. R., Jiang, Z., Zhao, B., Li, Z. J., Chen, Y., Gu, Y., Jiang, J. H., Lee, M., Liou, K. N., Neu, J. L., Payne, V. H., Su, H., Wang, Y., Witek, M., & Worden, J. (2019). Modeling study of the air quality impact of record-breaking Southern California wildfires in December 2017. *Journal of Geophysics Reseaerch-Atmosphere*, *124*, 6554–6570.

Stewart, D. R., Saunders, E., Perea, R. A., Fitzgerald, R., Campbell, D. E., & Stockwell, W. R. (2017). Linking air quality and human health effects models: An application to the Los Angeles Air Basin. *Environmental Health Insights*, *1*, 1–13.

United State Census Bureau. (2024). <https://www.census.gov/> (accessed 09.02.24).

Wilhelm, M., Ghosh, J. K., Su, J., Cockburn, M., Jerrett, M., & Ritz, B. (2012). Traffic-related air toxics and term low birth weight in Los Angeles County, California. *Environmental Health Perspectives*, *120*(1), 132–138.

Williams, A. P., Abatzoglou, J. T., Gershunov, A., Guzman-Morales, J., Bishop, D. A., Balch, J. K., et al. (2019). Observed impacts of anthropogenic climate change on wildfire in California. *Earth's Future*, *7*(8), 892–910.

Subchapter 17.8

Air quality in the Mexico City Metropolitan Area, Mexico

Luisa T. Molina and Armando Retama
Molina Center for Energy and the Environment, Boston, MA, United States

17.8.1 Background

With 21.8 million inhabitants in 2020 (Secretaría de Economía, 2023), the Mexico City Metropolitan Area (MCMA) is one of the largest megacities of the world and the most populous metropolitan area in North America. About 9.2 million inhabitants live in Mexico City, the capital of Mexico, and 12.6 million live in 59 contiguous municipalities of the State of Mexico and one municipality of the State of Hidalgo (INEGI, 2023). The MCMA lies in an elevated basin at an altitude of 2240 m above mean sea level and is surrounded on three sides by mountains and volcanoes. The MCMA has undergone massive transformation in urbanization and demographics throughout its history. Rapid population growth, uncontrolled urban expansion, increased energy consumption and motorization, a constrained basin surrounded by mountains at high altitude and intense solar radiation combined to cause serious air pollution problems for the metropolitan area in the 1980s. The automatic air-quality monitoring network, established in the late 1980s, revealed high concentrations of all criteria pollutants, placing Mexico City air pollution problems among the worst in the world.

Since the 1990s, the Mexican government has made significant progress in improving the air quality of the MCMA by developing and implementing successive comprehensive air quality improvement programs (ProAire, Spanish acronym) that combined regulatory actions with technological change based on scientific, technical, social, and political considerations. Vehicle emissions reduction focused on adapting advanced technology and improved fuel quality that had been successfully implemented in other countries, including removal of lead from gasoline, mandatory use of catalytic converters, reduction of sulfur content in diesel fuel, reinforcement of vehicle inspection and maintenance, and mandatory "no driving day" rule. Emissions reduction measures for industries and commercial establishments included closing an oil refinery, relocating high-polluting industries outside the valley, substituting fuel oil in industry and power plants with natural gas, and reformulating liquefied petroleum gas for cooking and water heating (Molina et al., 2019).

One of the actions taken by the environmental authorities in the MCMA is the collaboration with national and international scientific and technical experts, including integrated assessment of the air quality in the MCMA and intensive field measurement campaigns (Molina & Molina, 2002; Molina et al., 2010). Scientific research has played an important role in providing the Mexican environmental authorities with an essential understanding of the emission sources of pollutant species and their transport and transformation in the atmosphere, their effects on human health and the environment, and the identification of effective emission reduction strategies. The key findings and policy implications from these collaborative studies have been incorporated in the design of ProAire.

17.8.2 Major air pollution sources and emissions

Fig. 17.8.1 shows the emission inventory (EI-MCMA) for the year 2018 reported by the Secretariat of Environment (SEDEMA, Spanish acronym). Since the early 1990s, SEDEMA has published the EI-MCMA every two years, and the most recent version also includes emissions of toxic pollutants, greenhouse gases (expressed as CO_2eq), and black carbon (BC) (http://www.aire.cdmx.gob.mx). Currently, the main emission sources are the use of fossil fuels, mobile sources, solvents used in businesses and services, and industrial activities. Area sources contribute to high emissions of PM_{10}, $PM_{2.5}$, NMVOC, NH_3, SO_2, and air toxics. SEDEMA has made a concerted effort to improve the inventory of VOC emissions as it is one of the major uncertainties in modeling O_3 formation; one of the main challenges is the unregulated informal sector.

17.8.3 Trends of key pollutant concentrations

After three decades of implementing comprehensive air quality improvement programmes, Mexico City has significantly reduced the concentrations of all criteria pollutants. Fig. 17.8.2 shows the air quality trends for the period

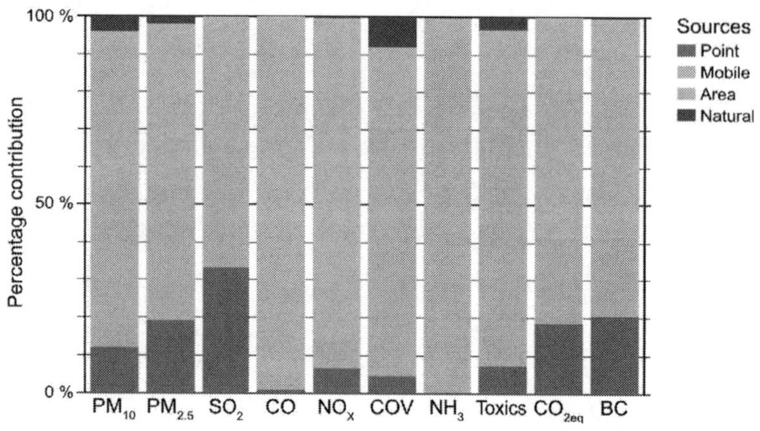

FIGURE 17.8.1 Emission Inventory for the year 2018 for Mexico City Metropolitan Area. *Mexico City Secretariat of the Environment, http://www.aire.cdmx.gob.mx.*

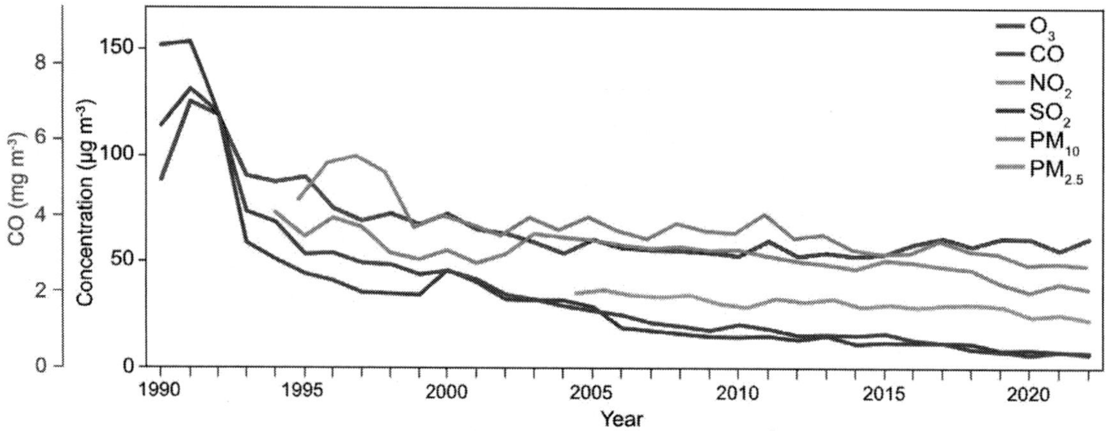

FIGURE 17.8.2 The 1990–2022 time series represents the average of the annual averages of monitoring stations that had more than 75% hourly values for each year, for each air pollutant. Mass concentrations (mg m^{-3} and μg m^{-3}) are reported at 298 K and 1 atm pressure. *From Mexico City Secretariat of the Environment, http://www.aire.cdmx.gob.mx.*

1990 to 2022 reported by the Mexico City Air Quality Monitoring System (SIMAT, Spanish acronym). Air quality in Mexico City improved significantly during the 1990s with drastic reductions in all pollutants. However, in the last 20 years, the rate of decline has decreased and in the case of O_3, there is an emerging upward trend (Fig. 17.8.3).

17.8.4 Comparison to WHO AQG (2021) and Mexican National Standards

The time series illustrated in Fig. 17.8.3 shows the maximum values for each day of the hourly averages of NO_2 and CO, the daily maximum values of the 8-hour averages for O_3, and the maximum values of the 24-hour average for SO_2. For PM_{10} and $PM_{2.5}$, the median values of the annual averages of all monitoring stations are shown. In the case of $PM_{2.5}$ monitoring started in the last quarter of 2003.

Fig. 17.8.3 shows annual means of PM for two conditions: firstly, for local pressure and temperature conditions, as reported in Mexico City, and secondly, adjusted to the 298 K and 1 atm condition for comparison with WHO air quality guidelines. The intention for presenting both values is to show the impact of the adjustment made for the high elevation of Mexico City on the concentrations. As Mexico City is at an average altitude of 2240 m above sea level, when the

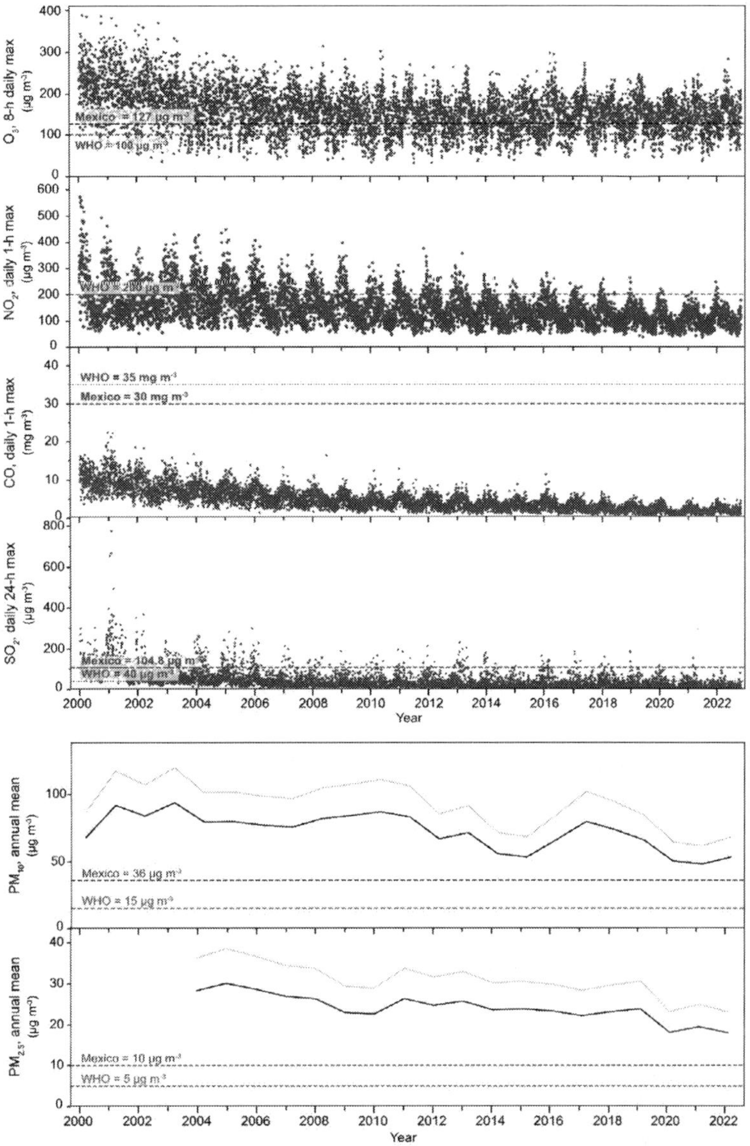

FIGURE 17.8.3 Air quality trends based on monitoring data for the Mexico City Metropolitan Area between January 2000 to October 2022 for gaseous air pollutants are presented for different daily average metrics. The daily values of the indicators were calculated for comparison with the WHO guideline values (blue dashed line) and the Mexican air quality standard (red dashed line). In all cases, except O_3, the concentrations decreased in 2020 due to the COVID-19 pandemic. In the case of PM, concentrations are shown as annual means for local pressure and temperature conditions (black line), as reported in Mexico City, and adjusted to the 298 K and 1 atm condition (gray line) for WHO comparison. *From Mexico City Secretariat of the Environment, http://www.aire.cdmx.gob.mx.*

adjustment is made, the concentrations are higher by approximately 28% which implies that for Mexico City the WHO guideline values are much stricter compared to cities that are at lower altitudes close to sea level. This is an aspect that deserves the attention of the WHO since many cities in Latin America are located at altitudes more than 1000 m above sea level.

17.8.5 Health impacts

The improvements in air quality since the 1990s suggest that important health benefits have been achieved in terms of reducing chronic exposure. A recent collaborative study between SEDEMA and the Harvard School of Public Health estimated that 22,500 premature deaths were avoided during the period of 1990 to 2015 due to air quality improvement (SEDEMA-HSPH, 2016). The ProAire programs also included the Atmospheric Environmental Contingency Program (PCAA, Spanish acronym) that aim to alert the public during severe episodes and implement actions to alleviate pollution levels. It is based on the concentrations of O_3, $PM_{2.5}$, and PM_{10} reported by SIMAT and meteorological forecast. Since its initiation in 1986, the program had been updated several times with respect to the pollutant concentrations used as contingency thresholds, implementation mechanisms, and actions.

17.8.6 Remaining challenges to improving air quality

Despite significant progress in cleaning the air, there are remaining challenges. As shown in Fig. 17.8.3, concentration trends of O_3 and fine particles have stalled since around 2010. Mexico City has a robust air quality monitoring network and an emissions inventory updated biennially that provide the data needed for daily air quality forecasting and support air quality management. Nevertheless, most of the current air quality mitigation strategies in the MCMA are still based on scientific findings from the MILAGRO field campaign that took place in 2006 (Molina et al., 2010). A recent study shows that there is an overall reduction in the VOC-OH reactivity during the morning hours in urban areas with large spatial variability, implying a large spatial variability in O_3 production (Zavala et al., 2020). While alkanes (from leakage and unburned LPG used for cooking and water heating) are still key contributors to VOC-OH reactivity, the contribution from aromatic and alkene species has decreased, consistent with the reduction of VOCs from mobile sources. Changes in ozone production suggest that increases in the relative contributions from highly oxygenated volatile chemical products, such as consumer and personal care and solvent use, are responsible for sustained high O_3 levels in recent years and strongly suggest a need for new measurements of radical budgets, expanding VOCs measurements to include oxygenated species and using the data to support modeling studies in the design of new air quality improvement programs.

The MCMA facing additional challenges from regional contributions. Urban expansion of the MCMA has produced the megalopolis, consisting of Mexico City and contiguous municipalities of six surrounding states. The Megalopolis Environmental Commission (CAMe, https://www.gob.mx/comisionambiental) was created in 2013 to coordinate regional policies and programs; however, the different administrative and legislative jurisdictions and the available resources have created an ongoing challenge. Except for the MCMA, there is limited air quality monitoring and air pollution studies in the other states, making it difficult to evaluate the regional air quality and the impacts of pollutants in the region. Burning of regional biomass is a major contributor of fine particles to the MCMA during air pollution episodes. During the dry season, agricultural and forest fires in the surrounding areas are frequent, severely impacting the air quality.

To design new air quality policies, firstly, an updated and locally generated scientific information on the impact of atmospheric chemistry on changes in emission profiles resulting from new regulatory measures and advanced technology will be needed. Secondly, information on the changes in urban micrometeorology resulting from a growing urbanization and climate change will also be needed.

References

INEGI (2023) México en cifras, Ciudad de México. <https://www.inegi.org.mx/app/areasgeograficas/?ag = 09#collapse-Resumen> (accessed 05.02.24).

Molina, L. T., & Molina, M. J. (2002). *Air quality in the Mexico Megacity: An integrated assessment* (p. 384) Dordrecht, The Netherlands: Kluwer Academic Publishers.

Molina, L. T., et al. (2010). An overview of the MILAGRO 2006 Campaign: Mexico City emissions and their transport and transformation. *Atmospheric Chemistry & Physics*, *10*, 8697–8760.

Molina, L. T., Velasco, E., Retama, A., & Zavala, M. (2019). Experience from integrated air quality management in the Mexico City Metropolitan Area and Singapore. *Atmosphere*, *10*(9), 512.

Secretaría de Economía (2023, November). Data México, Valle de México. <https://www.economia.gob.mx/datamexico/es/profile/geo/valle-de-mexico#> (accessed 02.02.24).

SEDEMA-HSPH (2016). Análisis histórico de los beneficios para la salud asociados a una mejor calidad del aire en la Ciudad de México (CDMX) entre 1990 y 2015. <http://www.data.sedema.cdmx.gob.mx/beneficios-en-salud-por-la-mejora-de-la-calidad-del-aire/descargas/analisis-espanol.pdf> (accessed 20.11.22).

Zavala, M., Brune, W. H., Velasco, E., Retama, A., Cruz-Alavez, L. A., & Molina, L. T. (2020). Changes in ozone production and VOC reactivity in the atmosphere of the Mexico City Metropolitan Area. *Atmospheric Environment*, *238*, 117747.

Subchapter 17.9

Air quality in São Paulo, Brazil

Admir Créso Targino, Patricia Krecl and Gabriel Yoshikazu Oukawa
Federal University of Technology (UTFPR), Graduate Program in Environmental Engineering, Londrina, Brazil

17.9.1 Background

There are currently five megacities in Latin America: Bogotá, Buenos Aires, Mexico City, Rio de Janeiro and São Paulo. With 21 million inhabitants in 39 municipalities in 2022, the Metropolitan Area of São Paulo (MASP) houses 10.2% of Brazil's population and is the largest urban agglomeration in the Southern Hemisphere (IBGE, 2023). MASP lies on an elevated plateau (at an altitude of about 755 m), has a humid subtropical climate, with dry winters and hot summers (Cwa in the Köppen-Geiger classification), an annual mean air temperature of 20.4 °C and an annual mean cumulative rainfall of 1658 mm (INMET, 2021).

The MASP's population has expanded about 3.5 times since 1970, establishing its position as a major hub for services, finance, trade and industries. As a result, the air quality in the MASP is dominated by the intense local anthropogenic activities, such as the use of fossil fuels in on-road vehicles and industrial processes (Pereira et al., 2017), and by the sporadic advection of remote air pollution plumes (de Arruda Moreira et al., 2021; Targino et al., 2019). Since the 1970s, the Environmental Company of São Paulo State (CETESB) has been monitoring the air quality in the MASP, playing a key role in assessing air pollution trends and streamlining measures for improving the air quality. The MASP currently has 29 automatic air quality monitoring stations distributed across about 7950 km^2.

17.9.2 Major air pollution sources

The MASP's air quality is largely influenced by vehicular emissions, which in 2019 released 113, 48, and 1.2 Gg of CO, NO_x, and PM, respectively (CETESB, 2020). Restricting vehicular movement in the MASP can result in substantial reductions in primary emissions, such as NO_x, as demonstrated during both the COVID-19 lockdown in 2020 and the truck drivers' strike in 2018 (Krecl et al., 2020). Backyard and waste burning, food establishment activities, construction work and dust from poorly maintained roads also contribute to the deterioration of the air quality. Air pollution is generally worst from July to September when rain is less frequent and thermal inversions prevail. In this period, the city is frequently affected by the outflow of wildfire smoke transported from central and northern Brazil by the South American low-level jet, causing manifold increases in gaseous and particulate concentrations. The intrusion of smoke contributes between 40 and 70 $\mu g\, m^{-3}$ of the total O_3 in the MASP, with the largest influence in the dry period (Targino et al., 2019).

The vehicular fleet of the MASP has 7.3 million units, of which 81% are light-duty vehicles (LDV) and 13% are motorcycles that run on ethanol, gasohol (a blend of gasoline with up to 27.5% v/v of anhydrous ethanol) or any mixture between these two. The remainder is made up of 3% LDV and 3% heavy-duty vehicles (HDVs) that run on diesel with an 8% biodiesel blend. The use of ethanol in Brazil has been advocated as an important source of renewable energy and a viable path to independence from fossil fuels. Since 2005, ethanol sales have grown substantially in the MASP and now makeup about 50% of all fuel sold. The widespread use of ethanol has been disputed as a benefit for reducing air pollution in some cities, since it may boost certain photochemical cycles. However, it is still unclear whether this is the case for the MASP, despite its O_3 formation regime being VOC limited (Andrade et al., 2017; Nogueira et al., 2017).

17.9.3 Air pollution trends

Sulfur dioxide concentrations have been steadily reducing over the years, mainly as a result of the control over fixed sources and the reduction in the S-content of fuels, both industrial and automotive. Brazil has been regulating vehicular emissions since 1987 through the Motor Vehicle Air Pollution Control Program, with a succession of increasingly more stringent stages, and reductions of S-content in diesel fuels. Since 2013, automotive diesel has been sold with S-content

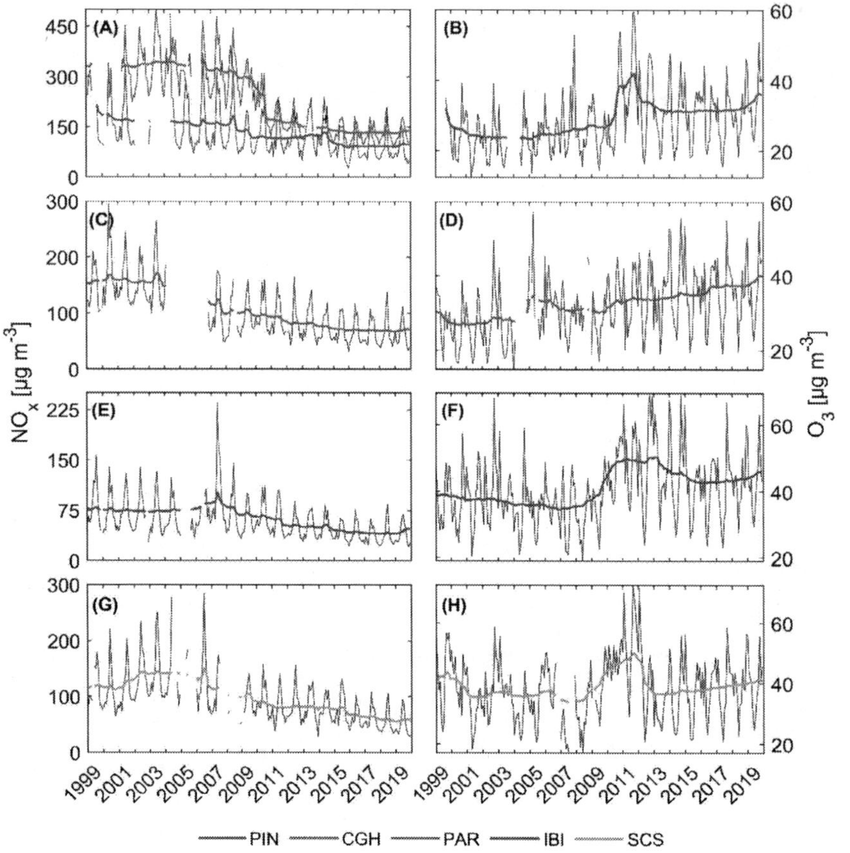

FIGURE 17.9.1 Monthly mean NOx and O_3 concentrations for several sites in the MASP (black lines) in the period 1999–2019, together with meteorologically normalized concentrations (colored lines).

of 500 and 10 mg kg^{-1}, the latter being mandatory for vehicles manufactured starting in 2012. The maximum S-content in gasoline was set at 50 mg kg^{-1} in 2014, replacing the previous limit of 800 mg kg^{-1}. The concentrations of traffic-related pollutants in the MASP (PM_{10}, CO and NO_x) decreased over the period 2000–2013 (Pérez-Martínez et al., 2015), which has been attributed to improvements in vehicle emission technology and the quality of the fuels currently used. The NO_x and O_3 monthly means as well as their meteorologically normalized trends are shown in Fig. 17.9.1 for selected sites in the period 1999–2019. Mean annual NO_x concentrations decreased between 44% and 59%, with a steeper decrease at the site Congonhas (CGH) in 2010. This occurred as a response to restrictions on trucks (diesel-fueled) around CGH, diverted to the newly opened southern sector of the ring road. The intervention demonstrates clearly how local traffic management strategies can complement national or local government mandates to limit vehicle emissions and reduce primary air pollutant concentrations. The O_3 trends showed mixed signals, but mostly upward trends over the last couple of years, underscoring the challenges in controlling O_3. Because the O_3 photochemical regime in the MASP is VOC limited, O_3 formation will be more sensitive to variations in VOC emissions than to variations in NO_x emissions, aside from weather conditions and a changing climate.

17.9.4 Ambient air pollution standards

São Paulo State reviewed its ambient air quality standards (AAQS) in 2013, also introducing thresholds for $PM_{2.5}$. The new AAQS are being implemented progressively, and is currently at the second interim target (MI2, in force since January 1, 2022). Most stations in the MASP in the year 2019 complied with the state annual PM_{10}, $PM_{2.5}$ and NO_2 AAQS (35, 17, and 50 µg m^{-3}, respectively), however none met the 2021 WHO guidelines (15, 5, and 10 µg m^{-3}, respectively) (Fig. 17.9.2B–D). Thus, considerable effort is required to reduce these pollutants below the WHO

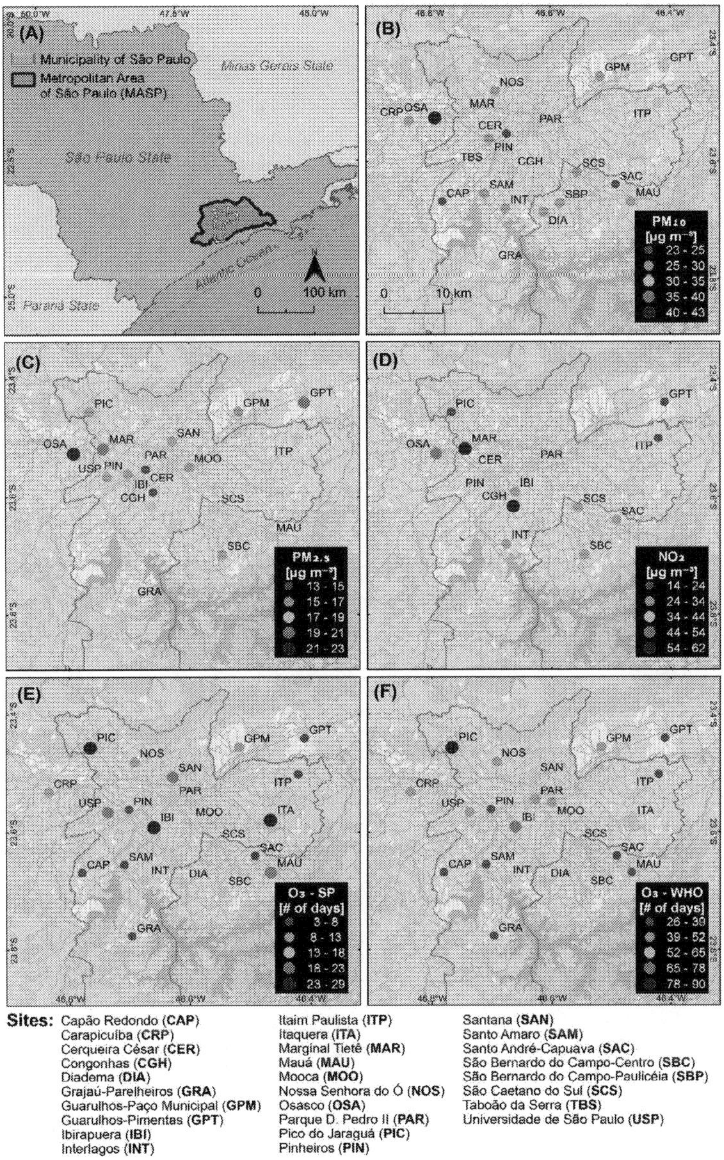

FIGURE 17.9.2 (A) Locations of São Paulo State, MASP and the municipality of São Paulo. For the year 2019, mean annual concentrations of (B) PM_{10}, (C) $PM_{2.5}$ and (D) NO_2, together with exceedance days of 8-hour average O_3 considering (E) São Paulo State AAQS (target MI2, 130 $\mu g\,m^{-3}$) and (F) 2021 WHO guidelines (100 $\mu g\,m^{-3}$).

thresholds, and particularly for the stations located next to busy motorways, such as Osasco (OSA), Marginal Tietê (MAR, 10,700 vehicles $hour^{-1}$, afternoon rush hour) and CGH (8670 vehicles $hour^{-1}$, afternoon rush hour). The challenge imposed by the new WHO guidelines is also highlighted by the number of days exceeding the 8-hour mean O_3 concentration, when compared to the state AAQS (target MI2, 130 $\mu g\,m^{-3}$) (Fig. 17.9.2E and F). In the case of SO_2 and CO, no exceedances were observed in 2019 either for the state AAQS or the WHO thresholds (daily mean of 40 $\mu g\,m^{-3}$ for SO_2 and 8-hour mean of 10 $mg\,m^{-3}$ for CO).

17.9.5 Remaining challenges to improving air quality

The continued nonattainment of O_3, $PM_{2.5}$, PM_{10} and NO_2 in some neighborhoods of the MASP may pose long-term difficulties for air quality management, as meeting the state AAQS or the WHO guidelines will require tighter local measures. Moreover, an integrated geographical approach is required to combat intrusions of remote biomass smoke that contributes to extreme air quality episodes. In terms of O_3 and secondary PM formation, the speciation of VOCs and inorganic species is imperative and would help elucidate which precursors contribute the most to O_3 and PM formation and how they compete with one another. Answering these questions may assist in designing strategies to combat specific precursors while avoiding unintended effects of favouring the formation of other species.

There is currently no regular vehicle maintenance and inspection program in place in the MASP, which results in the existence of gross emitters. Phasing out the gross emitters is required to accelerate the decrease in NO_x and PM concentrations. Retrofitting the obsolete diesel-fueled HDV fleet gradually with diesel particle filters until replacement with newer units is essential for the reduction of particulates. As cleaner fuels and new vehicular technologies have been successfully introduced in the MASP, the characterization of resuspended, tire and wear PM should be prioritized as their relative contribution will increase. Despite industrial emissions and activity data being stored by environmental institutions, access to the information is either fragile or often disrupted by data outages. However, these data are essential for feeding into local industrial emission inventories used by air quality models. Open data policies should be put into place to facilitate a wide range of actions to improve air quality, public health, and general environmental health.

References

Andrade, M., de, F., Kumar, P., de Freitas, E. D., Ynoue, R. Y., Martins, J., Martins, L. D., Nogueira, T., Perez-Martinez, P., de Miranda, R. M., Albuquerque, T., Gonçalves, F. L. T., Oyama, B., & Zhang, Y. (2017). Air quality in the megacity of São Paulo: Evolution over the last 30 years and future perspectives. *Atmospheric Environment (Oxford, England: 1994)*, *159*, 66–82. Available from https://doi.org/10.1016/j.atmosenv.2017.03.051.

CETESB (2020). Qualidade do ar no Estado de São Paulo 2019. <https://cetesb.sp.gov.br/ar/publicacoes-relatorios> (accessed 26.11.22).

de Arruda Moreira, G., da Silva Andrade, I., Cacheffo, A., da Silva Lopes, F. J., Calzavara Yoshida, A., Gomes, A. A., da Silva, J. J., & Landulfo, E. (2021). Influence of a biomass-burning event in PM2.5 concentration and air quality: A case study in the metropolitan area of São Paulo. *Sensors*, *21*, 425. Available from https://doi.org/10.3390/s21020425.

IBGE (2023). Censo Demográfico de 2022 (in Portuguese). <https://censo2022.ibge.gov.br/panorama/> (accessed 10.12.23).

INMET (2021). Normais climatológicas do Brasil 1991-2020 (in Portuguese). <https://portal.inmet.gov.br/normais> (accessed 26.11.22).

Krecl, P., Targino, A. C., Oukawa, G. Y., & Cassino Junior, R. P. (2020). Drop in urban air pollution from COVID-19 pandemic: Policy implications for the megacity of São Paulo. *Environmental Pollution (Barking, Essex: 1987)*, *265*, 114883. Available from https://doi.org/10.1016/j.envpol.2020.114883.

Nogueira, T., Dominutti, P. A., Fornaro, A., & de Fatima Andrade, M. (2017). Seasonal trends of formaldehyde and acetaldehyde in the megacity of São Paulo. *Atmosphere (Basel)*, *8*. Available from https://doi.org/10.3390/atmos8080144.

Pereira, G. M., Teinilä, K., Custódio, D., Santos, A. G., Xian, H., Hillamo, R., Alves, C. A., De Andrade, J. B., Da Rocha, G. O., Kumar, P., Balasubramanian, R., Andrade, M. D. F., & Vasconcellos, P. D. C. (2017). Particulate pollutants in the Brazilian city of Saõ Paulo: 1-year investigation for the chemical composition and source apportionment. *Atmospheric Chemistry &. Physics*, *17*, 11943–11969. Available from https://doi.org/10.5194/acp-17-11943-2017.

Pérez-Martínez, P. J., Andrade, M. F., & Miranda, R. M. (2015). Traffic-related air quality trends in São Paulo, Brazil. *Journal of Geophysics Research Atmosphere.*, *120*, 6290–6304. Available from https://doi.org/10.1038/175238c0.

Targino, A. C., Harrison, R. M., Krecl, P., Glantz, P., de Lima, C. H., & Beddows, D. (2019). Surface ozone climatology of South Eastern Brazil and the impact of biomass burning events. *Journal of Environmental Management*, *252*, 109645. Available from https://doi.org/10.1016/j.jenvman.2019.109645.

Subchapter 17.10

Air quality in Istanbul, Turkey

Alper Unal[1], Yusuf Aydin[1], Goksel Demir[2] and Ali Deniz[1]

[1]Department of Climate and Marine Sciences, Eurasia Institute of Earth Sciences, Istanbul Technical University, Maslak, Istanbul, Turkey, [2]Hamidiye Faculty of Sciences, Department of Occupational Health, and Safety, University of Health Sciences Turkey, Uskudar, Istanbul, Turkey

17.10.1 Background

Istanbul is located at latitude 41° 0' 54'' N and longitude 28° 58' 46'' E where European and Asian continents meet connecting the Black Sea in the north and the Sea of Marmara in the south. Istanbul with an official population of almost 16 million (TURKSTAT, 2022) is the world's 13th most populated city and one of Europe's largest agglomerations. It has a population density of 2523 people km^{-2} and covers an area of 5343 km^2. The historical city was built on seven hills and has the highest elevation over 500 m on the Asian side with an average elevation of 30 m. Istanbul has always been the financial, economic, and cultural capital of Turkey with a current GDP per capita around US$13.000 and the share of city's GDP in Turkey's economy is over 30 percent (TURKSTAT, 2022). Istanbul's climate can be classified as a mix of Mediterranean Climate and temperate zone. Summers in Istanbul are warm, humid, and dry, while winters are cold and windy. Throughout the year, the temperature typically ranges from 3°C to 30°C. In winter prevailing wind direction is north-easterly and south-westerly, while it is northerly in summer.

17.10.2 Current air quality challenges and emission sources

Istanbul has been facing an air pollution problem since the 1980s when rapid urbanization and industrialization with an increase in vehicle ownership has started. The population of the city has increased four times while the urban area has increased almost twice. During this period car ownership per thousand people has increased tenfold (from 30 in the 1980s to 312 in 2022). This has created major traffic congestion in the city making it the worst traffic in the world (time lost to congestion is 142 hours $year^{-1}$ as estimated from traffic navigation data) (Tomtom, 2022).

In Istanbul there are two major airports, Istanbul Airport (14th busiest airport in the world with 37 million passengers per year) on the European side, and Sabiha Gökçen Airport on the Asian side. Bosphorus strait is the fifth busiest strait in the world with over 40.000 ships carrying 630 million gross tonnages annually. Istanbul has eight Organized Industrial Zones textile industry accounts for approximately 37% of all industrial activities in Istanbul, followed by the metal industry (30%) and the chemical industry (21%). According to the latest official emissions inventory as reported to the European Monitoring and Evaluation Program (EMEP, 2022), Industry (including power generation), residential heating, and transportation (road transport, aviation, shipping and off-road) make almost 100% of all emission sources for NO_x, SO_X, and CO while it is 80% for PM_{10}. For NH_3, the highest contributors are waste and agricultural emissions while solvents make the highest contribution to NMVOC emissions (Fig. 17.10.1). Saharan dust is another major source for Istanbul and its vicinity. There are episodes where PM_{10} concentrations reach as high as 200 $\mu g\,m^{-3}$ (Kabatas et al., 2018).

17.10.3 Air quality measurements and comparisons with WHO guidelines

Istanbul has an extensive air quality monitoring network. Number of stations for PM_{10}, $PM_{2.5}$, CO, NO_2, SO_2, and O_3 are 37, 5, 20, 12, 11, and 25 respectively. Although some of these stations have been operated since the 1990s, data after 2008 has been reliable. A summary of these data (as annual averages) is given in Fig. 17.10.2 PM_{10} values are over the WHO 2021 Guideline of 15 $\mu g\,m^{-3}$ for the entire 2008–2020 period. There has been a decreasing trend in PM_{10} starting in 2013 and the rate of decrease increased with the pandemic years of 2019 and 2020, with an overall reduction of almost 30%. $PM_{2.5}$ measurements were started in 2014 and between 2014 and 2020, although $PM_{2.5}$ values are always over the WHO (2021) Air Quality Guideline of 5 $\mu g\,m^{-3}$, there is a 24% reduction. Another parameter that shows a reduction is NO_2. Between 2008 and 2020, although NO_2 values were much higher compared to the WHO Guideline of 10 $\mu g\,m^{-3}$, over 50% reduction has been observed. Ozone peak period averages (as estimated from daily

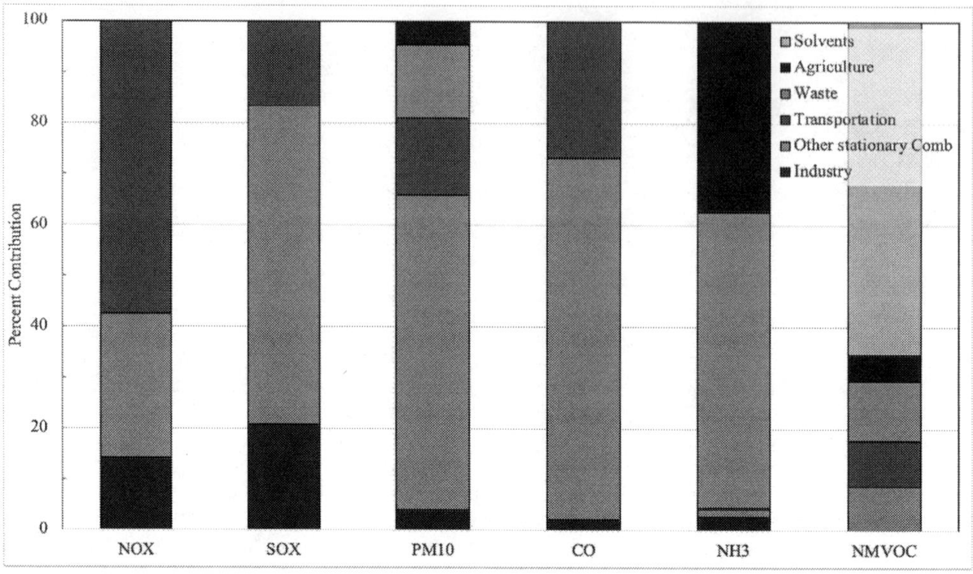

FIGURE 17.10.1 Sectoral contributions for each pollutant inventory for Istanbul.

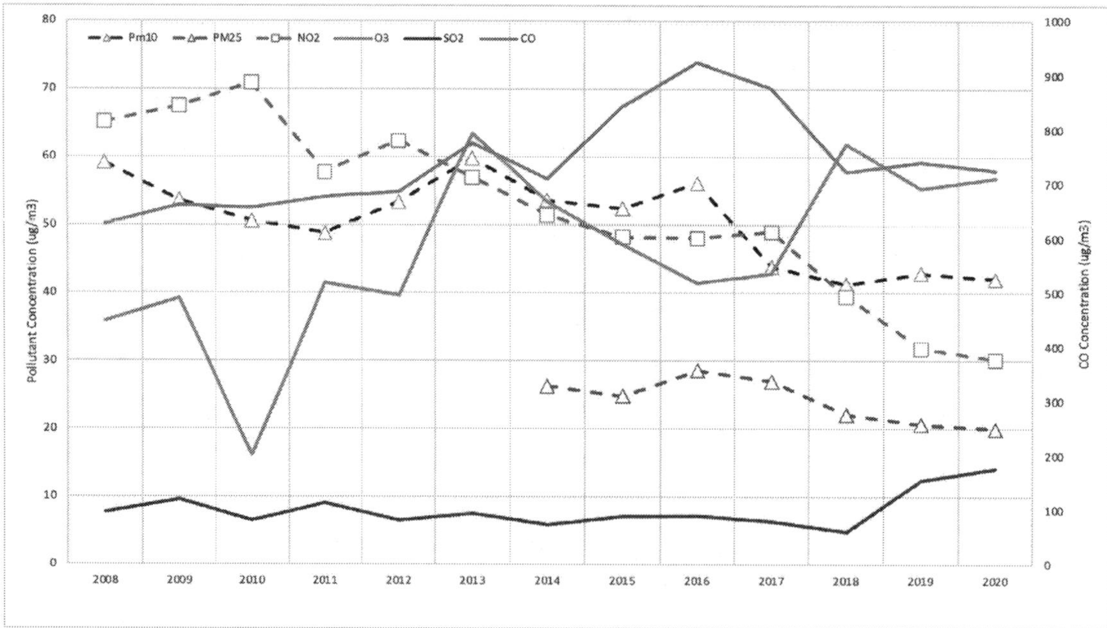

FIGURE 17.10.2 Temporal change in Pollutant Concentration in Istanbul (RTMEUCC, 2022) [Note that *y*-axis for CO is given on the right-hand side].

8 hours rolling average) are between 54 and 83 $\mu g\,m^{-3}$ which are close to or over the WHO guideline of 60 $\mu g\,m^{-3}$. Years 2018, 2019, and 2020 show a significant increase in ozone values. The most significant change is observed with SO_2 data. Although annual averages are lower than the WHO guideline of 20 $\mu g\,m^{-3}$, as seen in the figure, up to 2018 there is a significant decrease in the number of exceedance days, 2019 and 2020 show a significant jump to values close

to the suggested limit. This is highly correlated with natural gas prices which showed a 40% increase in 2018 and 80% increase in 2019 as compared to 2017 which suggests an increase in coal for residential heating. While annual averages for CO are much lower than the WHO guideline of 4000 $\mu g\, m^{-3}$, there is 16% increase between 2008 and 2020.

17.10.4 Future challenges

It can be concluded that increased control over industrial emissions, switching to natural gas for residential heating an increase in vehicle technology and an increase in the use of public transport, as well as limits on using high sulfur levels for ships have contributed to significant reductions in almost all pollutant levels in Istanbul. However, the impact of these changes is either being neutralized or even pollution levels tend to increase due to an increase in population, increased economy and transportation (including not just road transport but also maritime and air travel). These combined with increased dust transportation (due to changes in meteorological conditions) as well as an increase in the use of coal for residential heating have the potential to reverse the changes and bring high pollution episodes of the 1990s.

References

EMEP website <https://www.emep.int/> (accessed 20.11.22).

Kabatas, B., Pierce, R. B., Unal, A., Rogal, M. J., & Lenzen, A. (2018). April 2008 Saharan dust event: Its contribution to PM 10 concentrations over the Anatolian Peninsula and relation with synoptic conditions. *Science of The Total Environment*, *633*, 317–328, ISSN: 0048-9697.

Republic of Türkiye Ministry of Environment, Urbanization, and Climate Change (RTMEUCC) website, https://www.havaizleme.gov.tr/. (accessed 20.11.22).

Tomtom, Traffic Index <https://www.tomtom.com/traffic-index/ranking/> (accessed 20.11.22)

TURKSTAT, Turkish Statistical Institute's website <https://data.tuik.gov.tr> (accessed 20.11.22)

Subchapter 17.11

Air quality in Greater London, United Kingdom

Somayyeh Arghavani, Ummugulsum Alyuz and Ranjeet S. Sokhi

Centre for Climate Change Research (C3R), School of Physics, Engineering and Computer Science, University of Hertfordshire, Hatfield, United Kingdom

17.11.1 Background

London, the capital of the UK, was home to almost 8.8 million people in 2021 (ONS, 2022), with a projected population of 9.81 million by 2043 (ONS, 2020). The term 'Greater London' encompasses the 32 boroughs of London in addition to the City of London, which serves as the inner financial center rather than a borough itself, and is commonly called the 'square mile'. The 12 boroughs surrounding the City of London are collectively referred to as 'inner London'.

Air pollution has been a long-standing challenge in London. Historically, coal burning was the primary contributor to London's air pollution until the late 20th century marked by the infamous smog episodes of the 1950s and 1960s (GLA, 2022). Over the last couple of decades, vehicle traffic and construction have emerged as significant sources of air pollution in London. Exposure to air pollution is estimated to be responsible for 3600 to 4100 deaths in Greater London attributable to anthropogenic $PM_{2.5}$ and NO_2 (Dajnak et al., 2019).

In 1956, the UK introduced its first Clean Air Act in response to growing concerns about air pollution. In 1961, the world's first coordinated national air pollution monitoring network, named the National Survey, was established to monitor black smoke and sulfur dioxide at around 1200 sites across the UK, including London (DEFRA, 2007). National survey measurements provided valuable data to understand the extent and impact of air pollution on public health and the environment. The modern London Air Quality Network (LAQN) was formed in 1993 to coordinate and improve air pollution monitoring in London. Currently, LAQN comprises about 100 monitoring sites in most of London's boroughs, providing essential data for air quality management (LAQN, 2023).

Since the 1950s, the profile of air pollution sources in London has undergone significant changes. Major polluting industries have relocated beyond the city limits, accompanied by increased efforts to adopt cleaner fuels (Fuller, 2018). Environmental regulations have been strengthened with the introduction of national air quality standards and European Union limit values. Additionally, the major clean air zone interventions were implemented in London for complying with the UK and European Union legislations, and to improve London's air, such as the Low Emission Zone (LEZ) and Ultra-Low Emission Zone (ULEZ).

The introduction of the LEZ in 2008 aimed to improve air quality in Greater London by reducing PM_{10} emissions from heavily polluting road vehicles, particularly larger and heavier diesel vehicles, and positively impacting other pollutants like NO_x (Transport for London, 2008). Between 2008 and 2013, the LEZ directly contributed to a 20% decrease in PM_{10} emissions and indirectly to a 27% reduction in $PM_{2.5}$ emissions, and a 25% decrease in NO_x emissions from all sources (GLA, 2021). Despite these efforts, compliance with legal limits for PM_{10} and NO_2, as well as with WHO recommended targets for $PM_{2.5}$, prompted the introduction of the ULEZ in April 2019 for central London (applicable to all vehicles that do not meet the Euro 4 and Euro 6 vehicle standards), and progressively expanded to cover all of Greater London since the end of August 2023. However, despite these efforts, key air quality metrics still fail to meet WHO air quality guideline (AQG) values (GLA, 2020) and even the stricter 2021 AQG values (WHO, 2021).

17.11.2 Major air pollutant emission sources

The percentage contributions by sectors to NO_x, PM_{10}, and $PM_{2.5}$ emissions in Greater London for the year 2019 are illustrated in Fig. 17.11.1, utilizing data sourced from the London Atmospheric Emissions Inventory (LAEI) (GLA, 2023). The figure highlights that road transport is the largest contributor to NO_x emissions in London, accounting for 43%, followed by industrial/commercial and power sectors, aviation, industrial processes, and domestic heating. In terms of particulate matter (PM which comprises PM_{10}, and $PM_{2.5}$), the construction industry (30%) and road transport (27%) sectors account for the largest part of the PM emissions in London. Other significant sources of PM include resuspended dust, wood burning, and commercial cooking.

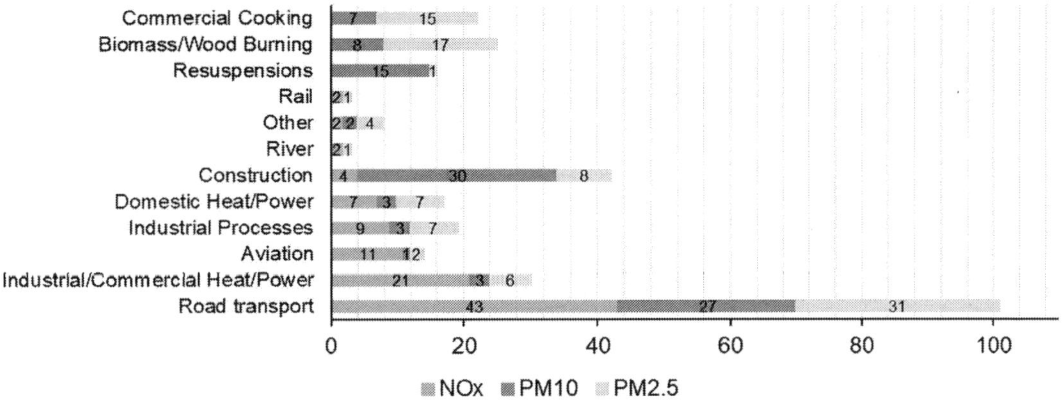

FIGURE 17.11.1 Percentage contributions of sectors to NO_x, PM_{10} and $PM_{2.5}$ emissions in 2019 in Greater London (GLA, 2023). The numbers in the bars indicate the percentage contributions of each sector for each pollutant.

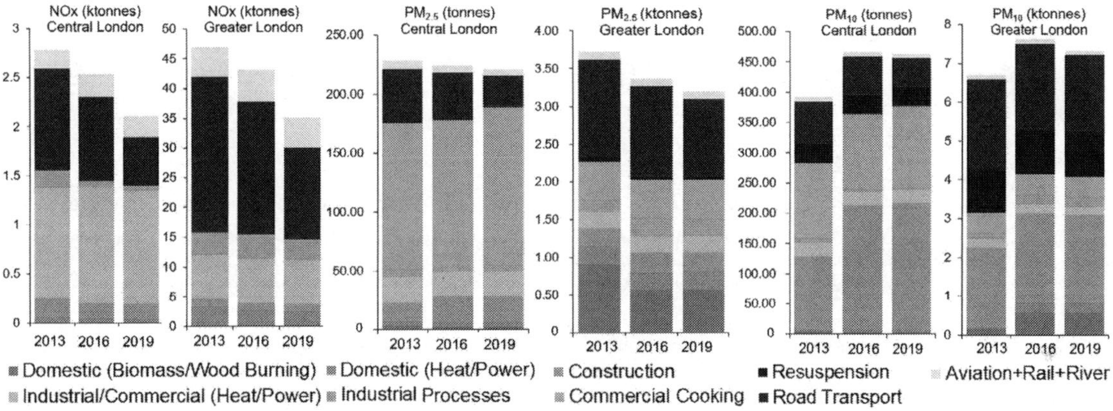

FIGURE 17.11.2 The trend of contributions of sectors to NO_x, $PM_{2.5}$ and PM_{10} emissions in Greater London and Central London for 2013, 2016 and 2019. *From GLA (Greater London Authority) (2023). London Atmospheric Emissions Inventory (LAEI) 2019 Update, Greater London Authority. London: Data source. <https://data.london.gov.uk/dataset/london-atmospheric-emissions-inventory–laei–2019>.*

Fig. 17.11.2 illustrates the annual reduction trends in NO_x, $PM_{2.5}$, and PM_{10} emissions from various source sectors in Greater London and central London between 2013 and 2019, utilizing data from the GLA (2023). According to Fig. 17.11.2, road transport and industrial/commercial (heat/power) sectors significantly contribute to NO_x emissions in both Greater London and Central London, while road transport emissions exhibit a decreasing trend in both areas. These notable reductions in NO_x are attributed to the operation of the Low Emission Zone (LEZ) and Ultra Low Emission Zone (ULEZ).

Regarding $PM_{2.5}$, the commercial cooking sector is dominant in Central London, showing an increase in 2019, whereas road transport dominates in Greater London, along with domestic biomass wood burning and commercial cooking. The construction sector is the primary contributor to PM_{10} emissions (mainly the coarse fraction) in both Central and Greater London, with growing economic activity, the trend significantly increases in Central London. Commercial cooking remains dominant in central London for $PM_{2.5}$ as well. In Greater London, road transport is the second dominant sector, with a decreasing trend observed in 2019. Resuspended particles also contribute significantly to the coarse fraction and, consequently, to PM_{10}, but with markedly less importance in Central London due to controlled road traffic.

17.11.3 Trends of key air pollutants

Fig. 17.11.3 illustrates annual changes in observed nitrogen dioxide (NO_2), $PM_{2.5}$, PM_{10} and maximum daily 8-hour values of ozone (O_3) concentrations between 2010 and 2023 at each roadside and urban background sites in Greater

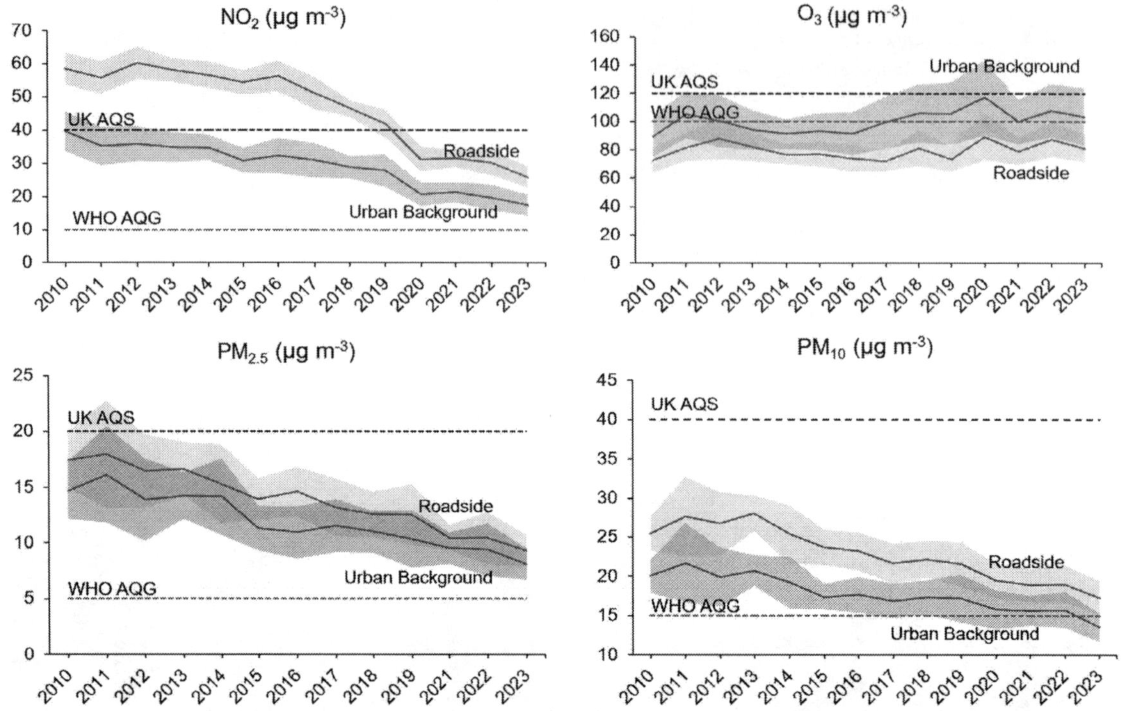

FIGURE 17.11.3 Annual averages of NO_2, maximum daily 8-hour average (MDA8) of O_3, $PM_{2.5}$ and PM_{10} concentrations observed in roadside and urban background sites in Greater London between 2010 and 2023 with shaded 95% confidence intervals (data from LAQN, 2023). The dashed horizontal black line shows UK Air Quality Standards while the red one shows the standard from WHO 2021 Air Quality Guidelines.

London. There is a decreasing trend in NO_2, $PM_{2.5}$, and PM_{10} concentrations, with a slight downward trend in NO_2 concentrations noticeable between 2010 and 2017, followed by a more rapid reduction in 2020, particularly evident at roadside sites. Since about half of the $PM_{2.5}$ originates from sources outside London it is evident that the reduction in particulate matter exhibits a shallower trend compared to NO_2. Specifically, the decrease in $PM_{2.5}$ concentrations between 2010 and 2023 is $8.34\ \mu g\ m^{-3}$ (47%) at roadside sites, compared to $5.66\ \mu g\ m^{-3}$ (40%) at urban background sites. Likewise, the reduction in PM_{10} concentrations was around $9\ \mu g\ m^{-3}$ (34%) at roadsides and approximately $7\ \mu g\ m^{-3}$ at urban background sites.

In the case of O_3, there was only a slight decrease observed until 2017, followed by an increase thereafter. Possible explanations for this ozone trend include the UK heatwave event in 2018 and the exceptionally sunny and dry summer with high radiation intensity (Kendon et al., 2019; Zhang & Stevenson, 2022). After 2017, reductions in NO_x and VOC concentrations were observed, contributing to an increase in O_3 concentration. The data presented in Fig. 17.11.3, as well as the percentage reductions mentioned in the text, were calculated using data sourced from LAQN (2023).

17.11.4 Comparison of London Observations to WHO AQG (2021) and UK National Standards

The limit values set by the WHO and the UK are also indicated in Fig. 17.11.3 for each of the air pollutants. The UK national standards are largely derived from those of the European Union (EU). According to the data presented in Fig. 17.11.3, both roadside and urban background stations surpass the WHO limits and even exceed the higher UK limits for NO_2, $PM_{2.5}$, and PM_{10}.

In 2019, all monitoring sites across London exceeded the World Health Organization (WHO) annual average guidelines (WHO, 2005) for both NO_2 and $PM_{2.5}$. Out of 119 NO_x monitoring sites in London, 48 sites failed to meet the UK annual mean air quality limit of $40\ \mu g\ m^{-3}$. However, there was a notable decrease in the number of Londoners living in areas exceeding the NO_2 UK annual mean legal limit, dropping from two million in 2016 to 161,000 in 2019. Despite this improvement, none of the major roads in London met the WHO annual mean air quality guideline (WHO, 2005) for NO_2 in 2019. Nevertheless, 84% of the major roads in London met the UK NO_2 annual mean legal limit, compared to 46% in 2016. While all sites in London achieved the UK standard limit of 20 $\mu g\,m^{-3}$ for $PM_{2.5}$, they still did not meet the updated WHO limits in 2021 (WHO, 2021) of 5 $\mu g\,m^{-3}$ (GLA, 2023; London Assembly, 2022).

17.11.5 Health impacts, estimate of published health impacts

Air pollution is linked to increased health risks, including the stunted growth of lungs in children and the worsening of chronic illnesses, such as asthma, lung and heart disease, and dementia in adults. It was estimated that in 2019, 61,800 to 70,200 life years were lost in Greater London, which was related to anthropogenic $PM_{2.5}$ and NO_2 (Dajnak et al., 2019). Many health impact-related studies have focussed on London. $PM_{2.5}$ exposures have been associated with slower fetal growth rate and low birth weight of infants (Smith et al., 2017). Studies have shown that even short-term exposure to NO_2 and PM can lead to an increase in the daily number of respiratory consultations and inhaler prescriptions for all ages, especially in children (Ashworth et al., 2021). Similar health-related studies in London confirmed the negative consequences of air pollution on lung volumes of primary school-aged children (Mudway et al., 2019), respiratory health (Evangelopoulos et al., 2021), asthma (Evangelopoulos et al., 2022), blood pressure (Karamanos et al., 2023) and childhood mental health problems (Roberts et al., 2019). A study by Webber et al. (2020) reports that by 2050 air pollution control measures in London save almost 300,000 London habitants from air pollution-related diseases which translates to a cost saving for the National Health Service (NHS) and social care system of around £5 billion.

17.11.6 Challenges for improving air quality in London

The current challenges to improve air quality further are vastly different to those encountered during the period of the London smog in the 1950s. Some previous policies have actually been detrimental to efforts to reduce air pollution emissions, for example, moving from petrol to diesel vehicles increased NO_x and fine particulate emissions in the 1990s (Graham & Glaister, 2002). Although the increase in the use of public transport has been a core emission-reduction policy in London in recent years, other local and regional policies are needed for an effective air quality improvement strategy. For example, emissions from nonroad-transport sources are showing an increasing trend. In 2019, construction emissions were responsible for 30 percent of total PM_{10} emissions in London (GLA, 2023). In terms of people's exposure, indoor and other nonambient air quality is also crucial (Jung et al., 2018). Saunders et al., 2019, for example, have shown that $PM_{2.5}$ concentrations in London underground trains are 18 times higher than street level and the chemical composition of these PM is different to the PM found in ambient air. Strategies to control wood burning, commercial cooking, aviation industry and river transportation also need to be strengthened as they are important sources of NO_x, $PM_{2.5}$ and PM_{10} in both inner and outer London. Studies such as (Hama et al., 2018) have shown the importance of contributions from outside the urban environment especially for $PM_{2.5}$ and O_3 and hence require action at a regional, national, and international level to manage these sources (GLA, 2020). Another issue is the control of ultrafine particulates (diameter less than 100 nm) which have significant health impacts as they penetrate deep into the lung, bloodstream, and even the brain (Li et al., 2017; Schraufnagel, 2020).

References

Ashworth, M., Analitis, A., Whitney, D., et al. (2021). Spatio-temporal associations of air pollutant concentrations, GP respiratory consultations and respiratory inhaler prescriptions: A 5-year study of primary care in the borough of Lambeth, South London. *Environmental Health, 20*, 54.

Dajnak, D., Evangelopoulos, D., Kitwiroon, N., Beevers, S., & Walton, H. (2019). London health burden of current air pollution and future health benefits of mayoral air quality policies. *Environmental Research Group: Imperial College London*, 2019.

Evangelopoulos, D., Chatzidiakou, L., Walton, H., Katsouyanni, K., Kelly, F. J., Quint, J. K., et al. (2021). Personal exposure to air pollution and respiratory health of COPD patients in London. *European Respiratory Journal, 58*, 2003432.

Evangelopoulos, D., Dajnak, D., Wood, D., Fecht, D., & Walton, H. (2022). *Health impact assessment of current and past air pollution on asthma in London*. Imperial College London.

Fuller, G. (2018). *Invisible killer: The rising global threat of air pollution - and how we can fight back*. Melville House UK.

GLA (Greater London Authority) (2020). Mayer of London, Air quality in London 2016-2020, London Environment Strategy: Air Quality Impact Evaluation. <https://www.london.gov.uk/sites/default/files/air_quality_in_london_2016-2020_october2020final.pdf>.

GLA (Greater London Authority) (2021). London Low Emission Zone – Six Month Report, September 2021. London. Published by Greater London Authority. <http://www.london.gov.uk>.

GLA (Greater London Authority) (2022). 70 years since the great London smog, 1952 air quality in a modern context [Report]. <https://www.london.gov.uk/sites/default/files/2022-12/70th%20anniversary%20of%20the%20Great%20London%20Smog.pdf>.

GLA (Greater London Authority) (2023). London Atmospheric Emissions Inventory (LAEI) 2019 Update, Greater London Authority. London: Data source. <https://data.london.gov.uk/dataset/london-atmospheric-emissions-inventory–laei–2019>.

Graham, D., & Glaister, S. (2002). Evaluating the impact of local air quality policies on atmospheric pollution levels: Recent trends in NOx and particulates across the UK. *Journal of Environmental Management*, *64*(3), 247–257.

Hama, S. M. L., Mørck, T. A., & Shaw, J. (2018). Investigating the effect of regional climate change on air quality in the UK using regional climate modeling techniques. *Science of The Total Environment*, *615*, 495–505.

Jung, K. H., Bernabé, K., Moors, K., Yan, B., & Hong, Y. C. (2018). Ambient air pollution and type 2 diabetes: A systematic review of epidemiologic research. *Current Environmental Health Reports*, *5*(4), 411–424.

Karamanos, A., Lu, Y., Mudway, I. S., Ayis, S., Kelly, F. J., Beevers, S. D., et al. (2023). Associations between air pollutants and blood pressure in an ethnically diverse cohort of adolescents in London, England. *PLoS One*, *18*(2), 1–18.

Kendon, M., Mccarthy, M., Jevrejeva, S., Matthews, A., & Legg, T. (2019). State of the UK climate 2018. *International Journal of Climatology*, *39*, 1–55.

LAQN (London Air Quality Network) (2023). London Air Quality Network monitoring data. <https://www.londonair.org.uk/LondonAir/Default.aspx>.

Li, Y., Lane, K. J., Corlin, L., Patton, A. P., Durant, J. L., Thanikachalam, M., Woodin, M., Wang, M., & Brugge, D. (2017). Association of long-term near-highway exposure to ultrafine particles with cardiovascular diseases, diabetes, and hypertension. *International Journal of Environmental Research and Public Health*, *14*(5), 461. Available from https://doi.org/10.3390/ijerph14050461.

London Assembly. (2022). Mayer of London, 70 years since the great London smog, 1952 air quality in a modern context. <https://www.london.gov.uk/programmes-strategies/environment-and-climate-change/environment-and-climate-change-publications/70-years-great-london-smog>.

Mudway, I. S., Dundas, I., Wood, H. E., Marlin, N., Jamaludin, J. B., Bremner, S. A., et al. (2019). Impact of London's low emission zone on air quality and children's respiratory health: a sequential annual cross-sectional study. *Lancet Public Health*, *4*(1), 28–40.

ONS (Office for National Statistics) (2020). Household projections for England: 2018-based.

ONS (Office for National Statistics) (2022). Population and household estimates, England and Wales: Census 2021 [Statistical bulletin, unrounded data].

Peters, A., & Wichmann, H. E. (2015). Traffic-related particulate matter and cardiovascular risk: The role of ultrafine particles. *International Journal of Hygiene and Environmental Health*, *218*(1), 275–282.

Roberts, S., Arseneault, L., Barratt, B., Beevers, S., Danese, A., Odgers, C. L., et al. (2019). Exploration of NO_2 and $PM_{2.5}$ air pollution and mental health problems using high-resolution data in London-based children from a UK longitudinal cohort study. *Psychiatry Research*, *272*(8-17).

Schraufnagel, D. E. (2020). The health effects of ultrafine particles. *Molecular Medicine – Nature*, *52*(3), 311–317. Available from https://doi.org/10.1038/s12276-020-0403-3.

Smith, R. B., Fecht, D., Gulliver, J., Beevers, S. D., Dajnak, D., Blangiardo, M., et al. (2017). Impact of London's road traffic air and noise pollution on birth weight: retrospective populnation-based cohort study. *BMJ (Clinical Research ed.)*, *359*, 5299.

Transport for London. (2008). London Low Emission Zone Impacts Monitoring Baseline Report, July 2008. London.

UK DEFRA (2007). Air pollution in the UK: 2006, Section 3.2: Brief History of Monitoring in the UK, Report number: AEA/ENV/R/2544A, Department for Environment, Food and Rural Affairs, Crown Copyright.

Webber, L., Xu, M., & Graff, H. (2020). Modelling the long-term health impacts of changing exposure to NO2 and $PM_{2.5}$ in London. Transport for London and the Greater London Authority.

World Health Organization. (2005). WHO air quality guidelines for particulate matter, ozone, nitrogen dioxide and sulfur dioxide: Global update 2005.

World Health Organization. (2021). WHO global air quality guidelines: Particulate matter (PM2.5 and PM10), ozone, nitrogen dioxide, sulfur dioxide and carbon monoxide. World Health Organization. <https://iris.who.int/handle/10665/345329>. License: CC BY-NC-SA 3.0 IGO.

Zhang, Ch, & Stevenson, D. (2022). Characteristic changes of ozone and its precursors in London during COVID-19 lockdown and the ozone surge reason analysis. *Atmospheric Environment*, *273*, 118980.

Subchapter 17.12

Air quality and atmospheric pollutants composition in Milan

Sandro Finardi[1], Guido Lanzani[2], Anna Di Leo[2] and Paola Radice[1]
[1]ARIANET, Milan, [2]ARPA Lombardia (Environmental Protection Agency of Lombardia Region), Milan

17.12.1 Background

Milan is the capital of Lombardy, the most populous and richest region of Italy. With about 10M inhabitants (year 2021), and a Gross Domestic Product (GDP) reaching 442298M€ (year 2022), Lombardy hosts 16.9% of the Italian population and produces more than 22% of the total Italian GDP (ISTAT, 2022), it is home to 25% of the country's industrial production, 11% of the agricultural production, hosting over 26% of livestock (ISTAT, 2022; Pretolani & Rama, 2023). Milan Municipality, with 1.3M inhabitants, and a density of 7560 inhabitant km^{-2}, is the central portion of Greater Milan agglomeration that, following the EU air quality directive 2008/50/EC, has been identified for air quality management purposes as an area including about 3.6M inhabitants, nearly uniform in terms of population density, industrial activities and traffic concentration. Lombardy region records 63.1 cars/100 inhabitants, and for the Milan municipality, the ownership is 50.5 cars/100 inhabitants (ACI, 2023).

17.12.2 Major air pollution sources and emissions

The Milan conurbation is exposed to substantial emission loads, which are summarized in Table 17.12.1 and subdivided among activities macro-sectors.

Emission inventory data (ARPA Lombardia, 2022a) show the significant contribution of nonindustrial combustion to primary $PM_{2.5}$ and PM_{10} emissions (28% and 23%, respectively), which provides almost the same mass contribution of road traffic (34% and 40%) and is attributable to wood combustion for residential heating, resulting the dominant contribution during wintertime. Although only a small percentage of houses is heated with wood (about 7% in the whole Lombardy), PM emission factors for wood-burning appliances are 100–1000 higher than those for natural gas-fueled boilers. Noticeable is the dominant contribution of agriculture to ammonia emission, 97% at the regional level, 58% in Milan Agglomeration, mainly due to livestock farming, while the second larger share in the town is due to road traffic (11%). As expected, road transport is the main contributor to NO_x emissions (67%) in the urbanized area. It should be noted that the relevance of agriculture and residential heating contribution to ammonia and PM emissions grows at the regional level when suburban and rural areas are included. The different industrial production processes (including mechanical engineering, metal, chemical and textile industries) also provide contributions to primary PM, NO_x, VOC emissions that cannot be neglected. The use of solvents is the major source of (light) VOCs (67%).

17.12.3 Trends of key air pollutant concentrations

The atmospheric circulation of the Po Valley is characterized by the strong modification of synoptic flow due to the high mountains (Alps and Apennines chains) that surround the valley on three sides. The local atmospheric circulation features, dominated by calms and weak winds, create unfavorable meteorological conditions for the dispersion of air pollutants. Furthermore, persisting anticyclonic conditions with temperature inversions are frequently recorded in the region causing winter air pollution episodes (Kukkonen et al. 2005). Due to the concentration of urban and industrial emissions and the adverse meteorological conditions affecting the region, Milan suffers air quality conditions worse than those experienced by many other European cities like, for example, Paris and London, even if per GDP and per capita emissions in Lombardy are comparable to those of the most developed European Countries. Nevertheless, the long-term trends of the main air pollutants show significant improvement in the air quality achieved over the whole Lombardy region and inside the Milan urban area.

Fig. 17.12.1 shows the decreasing trend for NO_2 and $PM_{2.5}$ annual mean concentrations, that continue to the present period showing that the relevant EU air quality limits (Directive 2008/50/EC) were met from 2014 for NO_2 and from 2018 for $PM_{2.5}$ over most of the Lombardy territory (75 percentile of monitoring stations), with the exclusion of some

TABLE 17.12.1 Total yearly emissions within the Milan Agglomeration in 2021.

	SO_2	NO_x	VOC	CH_4	CO	CO_2	N_2O	NH_3	PM2.5	PM10	TSP
	$t\,year^{-1}$	$t\,year^{-1}$	$t\,year^{-1}$	$t\,year^{-1}$	$t\,year^{-1}$	$kt\,year^{-1}$	$t\,year^{-1}$	$t\,year^{-1}$	$t\,year^{-1}$	$t\,year^{-1}$	$t\,year^{-1}$
Combustion in energy and transformation industry	6	794	75	163	326	738	4	1	13	13	13
Nonindustrial combustion plants	134	3,111	1,307	561	6,809	4,706	111	84	597	610	631
Combustion in manufacturing industry	797	1,585	680	45	648	1,651	20	11	205	241	280
Production Processes	44	101	1,517	9	615	97	3	1	60	90	139
Fossil fuels production and distribution			2,536	8,863							
Solvent and other product use	0	62	27,601		65			4	222	243	365
Road Transport	12	13,348	4,908	334	18,629	5,067	171	219	720	1,054	1,424
Other mobile sources and machinery	27	481	63	1	292	77	2	0	16	16	16
Waste treatment and disposal	45	384	160	4,371	72	323	157	27	2	2	2
Agriculture		28	937	2,174			98	1,121	1	3	6
Other sources and sinks	5	94	1,336	160	1,988	-32	3	469	302	334	376
Total	1,072	19,987	41,120	16,680	29,445	12,626	569	1,937	2,137	2,608	3,253

Source: ARPA Lombardia INEMAR emission inventory; ARPA Lombardia (2022a).

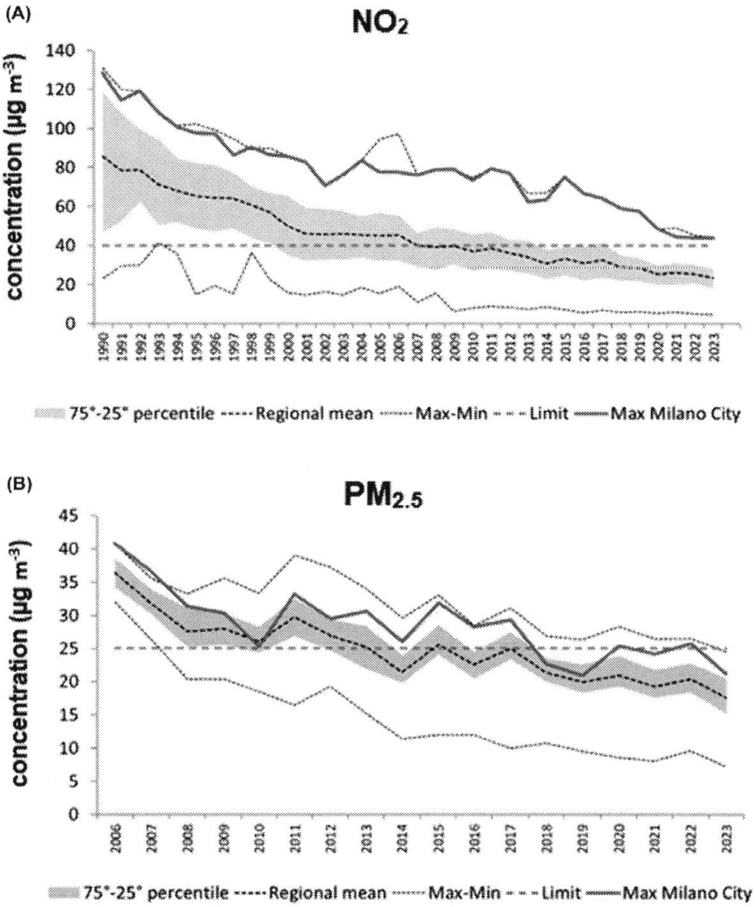

FIGURE 17.12.1 NO_2 annual mean concentrations from 1990 to 2023 (A) and $PM_{2.5}$ annual mean concentrations from 2006 to 2023 (B) observed in Lombardy (min/max, black dotted lines; interquartile range, green shaded area; and mean value, black dashed line) and Milan (maximum value recorded inside the city, red solid line). EU air quality limits are indicated by dashed red lines.

urban locations. This analysis does not consider the newly adopted directive (PE-CONS, 2024) that has introduced tighter air quality standards for a better alignment with the WHO 2021 guidelines. To assess its impact in the Italian context, the new limit values are presently exceeded by a major fraction of the largest Italian cities for NO_2, $PM_{2.5}$ and PM_{10} (EEA, 2024). The maximum annual mean recorded within Milan agglomeration for NO_2 concentrations is very often coincident with the maximum over the whole Lombardy, indicating the impact of urban and suburban traffic, while the same indicator for $PM_{2.5}$ is often lower than the maximum over the whole region, due to the impact of nonurban sources like wood burning and to the contribution of secondary aerosols (ARPA Lombardia, 2022).

17.12.4 Comparison with WHO AQG (2021) and national standards

The present air quality conditions in the Milan agglomeration (ARPA Lombardia, 2022b) still do not meet the EU limits for the annual mean concentration for NO_2 ($40\,\mu g\,m^{-3}$), for the number of days (35) exceeding PM10 daily average concentration limit ($50\,\mu g\,m^{-3}$), and for the target value for health protection of O_3 (no more than 25 days exceeding $120\,\mu g\,m^{-3}$ for the daily maximum of the 8-hour running mean concentration). It should be noted that the ozone air quality standard is not met over whole Lombardy region and doesn't show a clear long-term trend correlated with local precursors NO_x and VOCs concentrations trend.

The decreasing trend of major air pollutants, together with the successful story of SO_2 pollution that is now reduced to concentrations near the detection limit, show that it is possible to improve air quality even in adverse geographic and climatological conditions. This objective of further improvements is proving to be more difficult than in the past because air pollution in Milan is not only of primary origin but is contributed by secondary substances formed in the atmosphere by chemical reactions such as ozone and a fraction of PM. In this context, the new WHO air quality guidelines carry a new challenge for Milan and for the whole Lombardy region, where the guidelines values for NO_2, PM_{10} and $PM_{2.5}$ are presently not attained over most of the territory and a large reduction of presently measured concentrations would be necessary to reach the proposed targets. It is very important to reduce air pollutant concentrations for their recognized health impact (see, e.g., Carugno et al., 2016; Consonni et al., 2018). PM_{10}, $PM_{2.5}$ and ozone, because of their total or partial secondary origin, have concentration levels that are rather uniform over large portions of the territory, making air quality monitor representative of the population exposure. Gariazzo et al. (2020) used monitoring network data and different modeling techniques to estimate Milan population-weighted exposure that showed the highest among the major Italian conurbations for both NO_2 and $PM_{2.5}$, due to population density and concentration levels.

17.12.5 Remaining challenges to improving air quality

In order to reduce NO_2 concentrations, road traffic remains the main emission sector to target. In the case of particulate matter, additional considerations are needed because of its chemical composition, for which continuous analysis is available in Milan since 2013 (Fig. 17.12.2A). The long-term average composition highlights the importance of secondary inorganic (ammonium nitrate and ammonium sulfate) and organic compounds (part of the organic carbon) showing

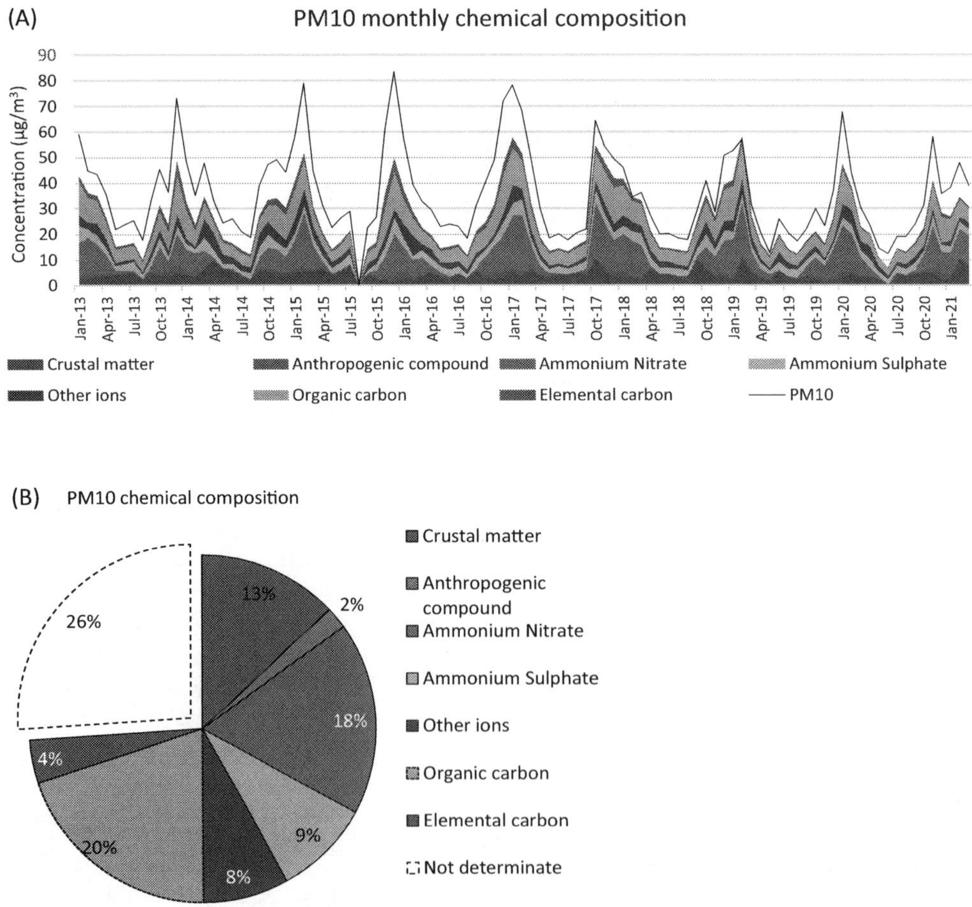

FIGURE 17.12.2 PM10 chemical composition measured in Milan city Pascal station from 2013 to 2021. Monthly mean time series (A) and mean composition recorded during the whole period (B).

an overall contribution of nearly 50% of the total PM_{10} mass (Amato et al., 2016). The secondary component is even more important for $PM_{2.5}$, specially during high air pollution episodes (Larsen et al., 2012). The monthly mean chemical composition time series shows its seasonal variation and, in particular, the growth of ammonium nitrate and organic carbon fractions during wintertime (Fig. 17.12.2B).

It should be noted that although concentrations of ammonium nitrate have been quite stable in the last decade, elemental carbon has been decreasing constantly during the last years, especially because of the introduction of the particulate filter on diesel vehicles. The aerosol composition suggests the need to control the emissions of precursors that lead to secondary aerosol components over the space scale that characterizes atmospheric chemistry reactions. To reduce O_3 and PM concentrations, local actions can give only small or even no results, while actions at the scale of the whole Po Valley basin or larger would give the desired benefits. These measures must reduce primary PM emissions and gaseous precursors such as NO_x, NH_3 and VOCs for particulate matter, and NO_x and VOCs for ozone. In addition to transport and industrial sectors, which remain major contributors to NO_x emissions at urban locations, further control measures are needed to reduce biomass burning and ammonia emissions, having agriculture as the most significant source within the Po Valley.

Considering that EU air quality limit values were not met, Regione Lombardia (2018) has implemented an air quality plan, updated in 2018 with the revision of the planned actions and introduction of new ones. The process has been shared with the other Po Valley regions (Emilia Romagna, Piemonte and Veneto) signing an agreement in June 2017 (Regione Lombardia, 2017). The items covered by this agreement are related to the more critical emissions: NO_x from diesel cars, wood burning and agricultural emissions. In particular, new limitations were added for the Low Emission Zones of Lombardia, a timetable with the progressive ban of the oldest wood stoves was introduced, and more restrictive actions were introduced regarding the spreading of manure in the breeding sector (such as the requirements of direct injection in the field). Full implementation of the planned actions is expected to produce compliance with the EU limit values, even if the secondary nature of PM, the nonlinearity of atmospheric chemistry processes, and the continental background of ozone and PM components impact increase the uncertainty of the results.

The new WHO guidelines, and the consequent recently approved revision of the EU air quality directive, introduce the need for a renewed strategy that should significantly increase the ambition of emission control measures to be planned and coordinated from local to continental scale taking into account even the impacts from the changing climate. The possible impact of transport emission reduction at urban locations in particular for NO_2, has been clearly demonstrated by the air quality impact of COVID-19 lockdown measures during the year 2020 (see, e.g., Sokhi et al., 2021). These results foster the need to increase public transport availability, to favor alternative mobility, for example, offering safe cycle routes and related infrastructures, with the final aim of reducing the use of private cars. In parallel the phasing out of internal combustion engines in favor of electrical vehicles should be promoted together with the reduction of commuting needs through adequate planning of urban structure development and smart working implementation. The limited impact on PM concentrations obtained during the lockdown measures, on the other hand, confirms the need to consider the impact of long-range transport of biomass-burning products, ozone and secondary PM precursors characterized by very low local emissions as sulfur compounds. To achieve these goals, appropriate investments should be made where needed, integrating local measures with regional to continental scale intervention and policies to overtake regional and national administrative limits that are not enough to effectively reduce air pollution, due to the growing relevance of air pollutants of secondary origin like ozone and a relevant fraction of PM. In order to implement measures that carry economic and social costs, it is important to have the support of public opinion for the needed lifestyle changes. For example, changing the public perception that "wood is good," because it has been used from ancient times, and that "agriculture is natural" by itself, being aware that intensive modern agriculture is not better than a normal industrial activity. On the other hand, a change in the behavior should be pursed, considering that also the application of the best technical available measures in all the sectors could be not enough to achieve WHO targets in the Po Valley context.

References

ACI, Annuario Statistico (2023). (in Italian). <https://www.aci.it/laci/studi-e-ricerche/dati-e-statistiche/annuario-statistico/annuario-statistico-2023.html>.

Amato, F., Alastuey, A., Karanasiou, A., Lucarelli, F., Nava, S., Calzolai, G., Severi, M., Becagli, S., Gianelle, V., Colombi, C., Alves, C., Custódio, D., Nunes, T., Cerqueira, M., Pio, C., Eleftheriadis, K., Diapouli, E., Reche, C., Minguillón, M., Manousakas, M., Maggos, T., Vratolis, S., Harrison, R., & Querol, X. (2016). AIRUSE-LIFEC: a harmonized PM speciation and source apportionment in five southern European cities. *Atmospheric Chemistry & Physics*, *16*, 3289–3309. Available from https://doi.org/10.5194/acp-16-3289-2016, http://www.atmos-chemphys.net/16/3289/2016/.

ARPA Lombardia (2022a). INEMAR2021 – Inventario regionale delle emissioni – anno 2021.<https://www.inemar.eu/; http://inemar.arpalombardia.it/inemar/webdata/main.seam>.

ARPA Lombardia (2022b). Rapporto annuale sulla qualità dell'aria – anno 2021. <https://www.arpalombardia.it/Pages/Aria/Relazioni-e-valutazioni.aspx>.

ARPA Lombardia (2022). Rapporto sulla qualità dell'aria - Città Metropolitana di Milano. <https://www.arpalombardia.it/media/knzfx5rm/rqa_-mi_2022.pdf>.

Carugno, M., Consonni, D., Randi, G., Catelan, D., Grisotto, L., et al. (2016). Air pollution exposure, cause-specific deaths and hospitalizations in a highly polluted Italian region. *Environmental Research*, *147*, 415–424. Available from https://doi.org/10.1016/j.envres.2016.03.003.

Consonni, D., Carugno, M., De Matteis, S., Nordio, F., Randi, G., Bazzano, M., Caporaso, N. E., Tucker, M. A., Bertazzi, P. A., Pesatori, A. C., Lubin, J. H., & Landi, M. T. (2018). Outdoor particulate matter (PM10) exposure and lung cancer risk in the EAGLE study. *PLoS One*, *13*(9), e0203539. Available from https://doi.org/10.1371/journal.pone.0203539.

EEA (2024). Italy – air pollution country fact sheet. <https://www.eea.europa.eu/en/analysis/maps-and-charts/italy-air-pollution-country-2023-country-fact-sheets> last visited on 04/11/2024.

Gariazzo, C., Carlino, G., Silibello, C., Renzi, M., Finardi, S., Pepe, N., Radice, P., Forastiere, F., Michelozzi, P., Viegi, G., et al. (2020). A multi-city air pollution population exposure study: Combined use of chemical-transport and random-Forest models with dynamic population data. *The Science of the Total Environment*, *2020*(724), 138102. Available from https://doi.org/10.1016/j.scitotenv.2020.138102.

ISTAT (2022). Istituto Nazionale di Statistica, data available from the access portal. <http://dati.istat.it/>.

Kukkonen, J., Pohjola, M., Sokhi, S., Luhana, L., Kitwiroon, N., Fragkou, L., Rantamäki, M., Berge, E., Ødegaard, V., Slørdal, H., Denby, B., & Finardi, S. (2005). Analysis and evaluation of selected local-scale PM10 air pollution episodes in four European cities. *Atmospheric Environment*, *39*, 2759–2773. Available from https://doi.org/10.1016/j.atmosenv.2004.09.090.

Larsen, B. R., Gilardoni, S., Stenström, K., Niedzialek, J., Jimenez, J., & Belis, C. A. (2012). Sources for PM air pollution in the Po Plain, Italy: II. Probabilistic uncertainty characterization and sensitivity analysis of secondary and primary sources. *Atmospheric Environment*, *50*(2012), 203–213.

PE-CONS. (2024). Directive of the European Parliament and of the Council on Ambient Air Quality and Cleaner Air for Europe (recast), PE-CONS 88/24, <https://data.consilium.europa.eu/doc/document/PE-88-2024-INIT/en/pdf> last visited on 04/11/2024.

Pretolani, R., & Rama, D. (2023). Il sistema agroalimentare della Lombardia. Rapporto 2022 (in italian). <https://www.regione.lombardia.it/wps/portal/istituzionale/HP/servizi-e-informazioni/imprese/imprese-agricole/ricerca-e-statistiche-in-agricoltura>.

Regione Lombardia (2017). Accordo di bacino padano per il miglioramento della qualità dell'aria. <http://www.regione.lombardia.it/wps/portal/istituzionale/HP/DettaglioRedazionale/servizi-e-informazioni/cittadini/Tutela-ambientale/Qualita-dell-aria/nuovo-accordo-programma-miglioramento-qualita-aria/nuovo-accordo-programma-miglioramento-qualita-aria>.

Regione Lombardia (2018). Aggiornamento del piano regionale degli interventi per la qualità dell'aria. <https://www.regione.lombardia.it/wps/wcm/connect/fb99bca7-07c6-4af3-bdaa-e1f89a6982c4/Allegato + 1 + Aggiornamento + PRIA + 2018.pdf?MOD = AJPERES&CACHEID = ROOTWORKSPACE-fb99bca7-07c6-4af3-bdaa-e1f89a6982c4-mkqDO3S>.

Sokhi, R. S., Singh, V., Querol, X., Finardi, S., Targino, A. C., Andrade, M. F., Pavlovic, R., Garland, R. M., Massagué, J., Kong, S., Baklanov, A., Kong, S., Baklanov, A., Ren, L., Tarasova, O., Carmichael, G., Peuch, V. H., Anand, V., Arbilla, G., Badali, K., Beig, G., . . . Zavala, M. (2021). A global observational analysis to understand changes in air quality during exceptionally low anthropogenic emission conditions. *Environment International*, *157*, 106818. Available from https://doi.org/10.1016/j.cnvint.2021.106818.

Subchapter 17.13

Air quality in the Paris metropolis area

Isabelle Coll

Univ Paris Est Creteil and Université Paris Cité, CNRS, LISA, Créteil, France

17.13.1 Background

Île-de-France is the region that hosts the Great Paris Metropolis (GPM), an administrative territory of about 30 km wide comprising the city of Paris and its first urban ring (https://www.apur.org/dataviz/portraits-metropole-grand-paris-donnees/). Concentrated on 7% of the regional territory, the metropolis was home to 7.1 million inhabitants in 2019 (60% of the Ile-de-France population). Although most of the urban dynamic is linked to this urban ring (+1% of inhabitants per year over the last decade), Paris remains a very densely populated area (20,000 inhabitants km^{-2}) which, together with its neighboring municipalities, forms a very attractive territory in terms of employment, shops and cultural and leisure areas. Public transport is unevenly distributed across the territory. It has a very concentric structure that aims essentially to ensure access to the dense urban area from outlying zones, which are less well served. Thus, the public transport accessibility indicator[2] is 92% in Paris, compared with only 17% to 65% in the inner suburbs, and even less outside the metropolitan area. Similarly, the road and highway network is structured in a star pattern and converges on a ring road that distributes the accesses to Paris. Since the year 2000, the region has transcribed in the territorial planning documents its ambition to reduce the use of private cars and to promote public transport as well as soft and alternative forms of mobility through a reflection on urban planning, transport and functional mix of the city (Chazal & Nouailhat, 2021). These efforts have already led to a significant increase in the number of trips made by public transport in the city (+2 million since 2000, i.e., +33% increase) and the emergence of active mobility. Despite this, one third of daily trips are still made by car in Île-de-France, almost half of which (6 million) are between Paris and the inner suburbs, even though these areas are well served by public transport.

17.13.2 Major air pollution sources and emissions

Air quality problems in Greater Paris are largely related to the organization of activities in the area, the high demand for transport and the omnipresence of road traffic in the urban space (AIRPARIF, 2022c). The GPM is a zone with a high density of activities that emit atmospheric pollutants, including first and foremost individual mobility, freight transport and the residential and tertiary sectors - which emit a decisive share of NO_2 (60%), PM_{10} (50%) and $PM_{2.5}$ (75%) - but also shops and industries, as well as the airport and energy sectors. Finally, agricultural activities located immediately on the outskirts of the metropolis (farm machinery, land spreading, etc.) complete this spatially contrasted panorama of sources in the area. Background air levels of primary pollutants increase logically with proximity to the Parisian center, but those related to traffic also show—as in all large cities—marked gradients around the main roads.

17.13.3 Trends of key air pollutant concentrations

Since the 2000s, continuous technological action to reduce emissions, efforts to rethink the territory and mobility, and energy management plans have led to a significant decrease in air pollutant emissions (AIRPARIF, 2022a, 2022c). For example, nitrogen oxide emissions have decreased by 45% from 2005 to 2018, which represents more than a 40% reduction in road emissions and nearly a 25% reduction in residential emissions. For PM_{10}, the decrease is 34% over the same period, that is, −55% and −30%, respectively for road transport and the residential sector. This evolution is illustrated on Fig. 17.13.1 which shows a progressive decrease in concentration levels, from 30% (background) to 40% (traffic) for NO_2 and from 35% (background) to 40% (traffic) for PM_{10}. In addition, the areas with very degraded air quality (right tail of the boxplots) tend to disappear. Over the same period, the $PM_{2.5}$ fraction (not emitted by resuspension) has benefited the most from technological advances in the road sector, with a 65% reduction between 2005 and 2018. The annual averages measured on the network decreased by 35% (background) and 50% (traffic).

2. Percentage of the territory's population living less than 500 m from a metro or tram, and less than 1000 m from a train.

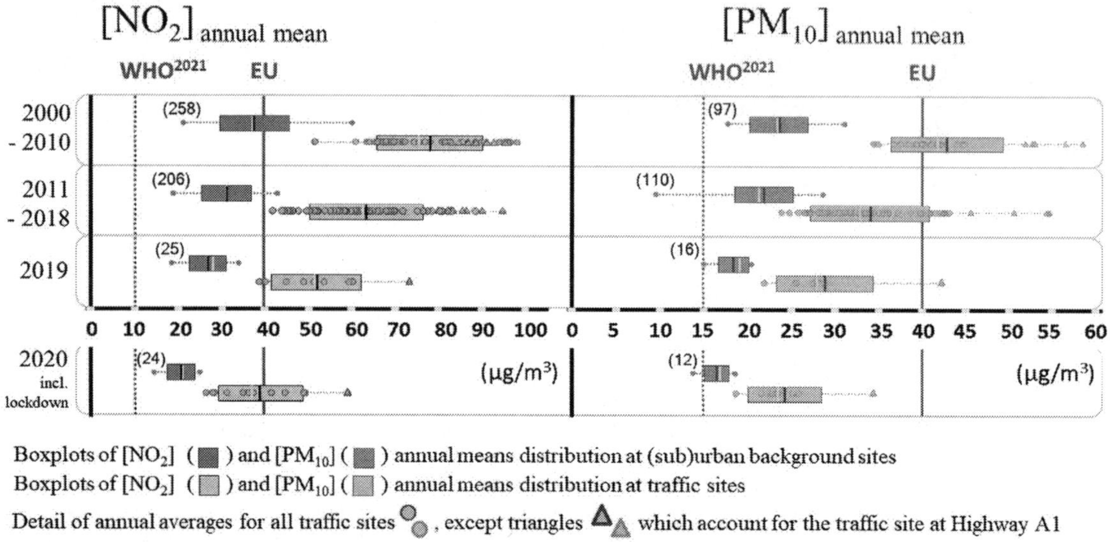

FIGURE 17.13.1 Boxplots of NO_2 (red, left) and PM_{10} (blue, right) annual averages (in $\mu g\ m^{-3}$) observed on all the urban background (dark colors) and traffic (light colors) sites of the AIRPARIF network along different time periods between 2000 and 2020. The boxplots show the distributions of the min, max, 25th and 75th percentile, median (yellow) and mean annual averages of the relevant sites on the period of interest. The number shown for each background boxplot is the number of data considered. Data are plotted over the boxplots for traffic sites as they are not numerous and not homogeneously distributed. The highway A1 site is always shown as triangles. The vertical lines represent the effective EU regulatory thresholds and WHO 2021 air quality guideline (AQG) values. Figure produced from AIRPARIF data available on the EEA data portal. *From https://discomap.eea.europa.eu/App/AirQualityStatistics/index.html?.*

Concerning ozone, as in many cities, an increase in background levels was observed due to the decrease in nitrogen oxide (NO_x) emissions. In the Parisian urban background, the current average ozone level (increasing since 2016) is now 46 $\mu g\ m^{-3}$, that is, 2.3 times the 1992 level. At the same time, monitoring has been greatly reduced for several pollutants. Carbon monoxide concentrations have dropped significantly over the past 25 years in the Ile-de-France region (− 90%), due to technological improvements in car fuels. The maximum concentrations currently reach 2 $mg\ m^{-3}$, which does not require continuous monitoring. Similarly, the average annual SO_2 levels have fallen below the detection limit (5 $\mu g\ m^{-3}$) in 2019.

17.13.4 Comparison to WHO AQG (2021) and national standards, and health impacts

As can be seen in Fig. 17.13.1, there have been no exceedances of the annual regulatory threshold for PM_{10} (40 $\mu g\ m^{-3}$) in urban background areas in the last 20 years. Even better, the quality of these environments has improved significantly as shown in Fig. 17.13.2 C and D. In road proximity, on the other hand, the situation is only recently on the way to compliance, since 2019 is the first year for which only one station remains with values exceeding 40 $\mu g\ m^{-3}$. On a daily scale, the improvement is also noticeable: exceedances of the daily threshold (50 $\mu g\ m^{-3}$) occurred for 40% of the length of the regional road network in 2007 whereas it has been less than 1% since 2016. However, the roads still affected by exceedances are located on the ring road and along the major Parisian boulevards, where they constitute a chronic exposure point for a very large number of drivers and city dwellers. For $PM_{2.5}$, current average concentrations in the whole urban background atmosphere are close to 12 $\mu g\ m^{-3}$, and 18 $\mu g\ m^{-3}$ near traffic. No citizen is exposed to concentrations that exceed the annual threshold of 25 $\mu g\ m^{-3}$. The improvement is slower for NO_2. Although the regional background level is around 25 $\mu g\ m^{-3}$ (see the improvement from Fig. 17.13.2A and B), well below the annual regulatory threshold (40 $\mu g\ m^{-3}$), the situation near traffic shows a considerable amount of inertia with more than 10,000 km of roads - mainly located in the Paris agglomeration - still exceeding the annual threshold in 2019. The map in Fig. 17.13.2B shows a very contrasted air quality with a quasisystematic exceedance of the regulatory thresholds on the road transport network in the metropolis.

The concern for air quality near roads is also related to specific particulate matter fractions, whose health impacts are being explored. Black carbon, notably, is a relevant tracer of car combustion and its levels have shown to reach 5 to

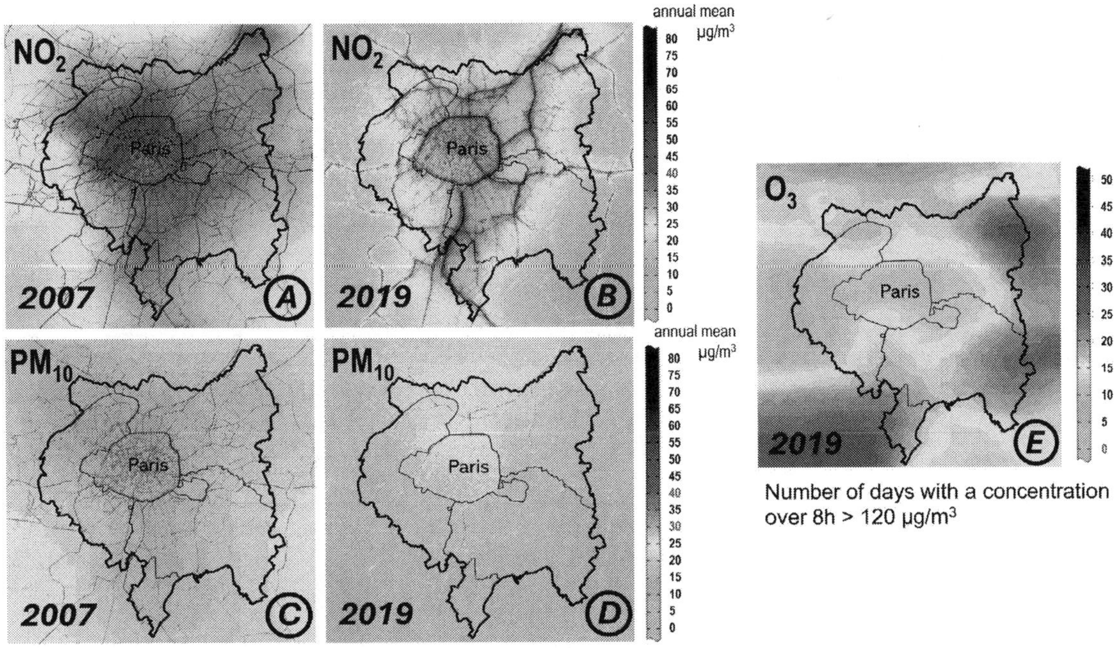

FIGURE 17.13.2 Maps of annual average concentrations of NO_2 (A and B) and PM_{10} (C and D) for the year 2007 and 2019 for the Paris territory. Panel (E) shows number of days when the target O_3 value was exceeded in 2019. The thick black line is the administrative contour of the Great Paris metropolis. *From AIRPARIF, https://www.airparif.fr/surveiller-la-pollution/bilans-et-cartes-annuels-de-pollution.*

10 times the urban background level ($1-2\ \mu g\ m^{-3}$) close to heavy traffic sites. Similarly, along a series of field campaigns carried out in the Paris region by the air quality monitoring network AIRPARIF in 2021, a gradient was observed in the number of ultrafine particles (diameter less than 100 nm) between the urban background and the roads, with a factor ranging from 2 to 5 depending on the traffic intensity. These are major elements in the exploration of the risk of exposure to pollutants in metropolises. In addition to this chronic pollution, there are occasional events created by very low dispersive meteorological conditions, such as low wind conditions, low mixing heights and thermal inversions (AIRPARIF, 2020).

In Ile-de-France, winter pollution episodes that occur during very cold weather are associated with a clear predominance of particles from wood combustion and PM_{10} levels that exceed the daily threshold. In spring, the combination of favorable anticyclonic conditions and the fertilizer applications leads to the emergence of particulate episodes in which more than half of the particulate ammonium nitrate comes from local agricultural activities. Over the period of January to November 2022, two days of winter episode (daily maximum in PM_{10}: 74 μgm^{-3}), and three days of spring episode were recorded (daily maximum in PM_{10}: 61 $\mu g\ m^{-3}$).

Regarding ozone (AIRPARIF, 2022b), the EU target value of $120\ \mu g\ m^{-3}$ averaged over 8 hour (maximum daily eight-hour mean) has been exceeded more than 25 days $year^{-1}$ (on average over 3 years) in suburban and rural areas of Ile-de-France in recent years, which had not happened since 2010 (see Fig. 17.13.2E). As for the trend on O_3 summer peaks, it is yet not very clear due to the small perspective we have on this phenomenon, and due to high interannual meteorological variability. In addition, despite an increase in exceedances of the $180\ \mu g\ m^{-3}$ hourly threshold since 2017 (over the summer of 2022, six days of exceedance were observed, with an hourly maximum of $218\ \mu g\ m^{-3}$), this tendency could be countered by the continued reduction in ozone precursors emissions. Indeed, it has already been observed that at equivalent temperatures, ozone production in the Ile-de-France is currently not as intense as in the 2000s, which could indicate a change in its dynamics of formation.

The Regional Health Observatory estimated in 2018 that 6,600 premature deaths could be avoided in the Paris metropolis by complying with the regulatory thresholds (Host et al., 2018). But the challenge is of course to go beyond, especially in view of the current noncompliance with WHO guidelines. While in 2019 only a minor part of the inhabitants of Ile-de-France remained exposed to exceedances of the WHO guidelines for particulate matter (this phenomenon was even accentuated by the restrictions of activity related to the COVID-19 pandemic in 2020), the 2021 revision of the WHO air quality guidelines has given new ambitious objectives to public actors. In 2021, the percentage of Ile-de-France residents exposed to exceedances of the thresholds recommended by the WHO was thus 100% (O_3), 95% (NO_2), 100% ($PM_{2.5}$) and 80% (PM_{10}).

17.13.5 Remaining challenges to improving air quality

The GPM's air quality ambitions are now largely based on integrated action plans that consider land-use planning, the organization of mobility and the supply of urban services. Thus, in addition to the renovation of buildings, changes in practices and technological developments related to the energy transition (from which air quality will also benefit), the policy implemented under both the master plan of the Ile-de-France region and the Atmosphere Protection Plan relies on the gradual implementation of a metropolitan Low Emission Zone (LEZ) initiated in 2015. The Regional Health Observatory estimated that it could, for example, significantly reduce the occurrence of childhood asthma (−4% in an optimistic scenario). In particular, implementing the LEZ beyond the Paris perimeter would ensure a fairer social distribution of benefits, with a higher proportion of avoidable cases among the least-favoured young populations. The greatest difficulty, however, lies in the feasibility and operationalization of this scheme, which must be designed with social equity in mind, and which takes place in a context of health, energy and economic crisis. As far as traffic is concerned, the search for more durable urban logistics (such as cleaner vehicle fleets, integration of warehouses into planning schemes, management of delivery parking) is also an urgent and major challenge. Finally, although they showed significant one-off effects of wood heating on particulate pollution, the 2020 lockdown phases, were associated with significant improvement in air quality (see the 2020 data in Fig. 17.13.1) which demonstrated the importance of mobility control for air quality in Île-de-France. In this frame, post-COVID-19 changes in mobility practices (telecommuting, cycling, shortening commuting distances) could provide additional leverage for air quality improvement.

References

AIRPARIF: Bilan de la qualité de l'air, Année 2019, Surveillance et information en Île-de-France (2020). <https://www.airparif.fr/sites/default/files/documents/2020-06/bilan-2019_0.pdf>.

AIRPARIF: Bilan de la qualité de l'air, Année 2021, Surveillance et information en Île-de-France (2022a). <https://www.airparif.fr/sites/default/files/document_publication/BilanQA_IDF_2021.pdf>.

AIRPARIF (2022b). Ozone - Etat des connaissances en Île-de-France. <https://www.airparif.fr/sites/default/files/pdf/Note_O3.pdf>.

AIRPARIF (2022c). Surveillance et information sur la qualité de l'air, Bilan Métropole du Grand Paris – 2021. <https://www.airparif.fr/sites/default/files/document_publication/Bilan_MGP_2021.pdf>.

Chazal, P., & Nouailhat, V. (2021). Évolution des mobilités dans le Grand Paris, tendances historiques, évolutions en cours et émergentes, Atelier Parisien d'URbanisme (APUR). <https://www.apur.org/sites/default/files/evolution_mobilites_grand_paris.pdf?token = uM1LFlmo>.

Host, S., Saunal, A., Honoré, C., Joly, F., Le Tertre, A., & Medina, S. (2018). Bénéfices sanitaires attendus d'une zone à faibles émissions: évaluation quantitative d'impact sanitaire prospective pour l'agglomération parisienne, Paris, Observatoire régional de santé Île-de-France: 106. <http://www.ors-idf.org/nos-travaux/publications/benefices-sanitaires-attendus-dune-zone-a-faible-emissions.html>.

Subchapter 17.14

Final remarks—toward cleaner air for all

The topic of air quality is more than just a scientific discipline; air quality has a direct impact on our lives and health and on the environment in which we live. Reflecting on the contents of the book, the reader will appreciate the complex nature of the processes linking emissions, meteorology, chemistry and climate that influence air quality. Going beyond processes and interactions, this book has introduced some of the main approaches to measure and predict air pollution which are critical to understand and quantify impacts on human health and on ecosystems. Similarly, these approaches form the basis for developing and testing air quality management strategies and policies.

The main characteristics of the atmosphere and its behavior relevant to air quality have been introduced in Chapter 2. Remarkably, this thin layer of gases that we call the atmosphere, serves multiple purposes from providing the oxygen that we need to live, protecting us from harmful radiation and creating the weather that nurtures our planet and all life on it. Chapters 2 and 3 introduced that anthropogenic emissions of air pollution are the main driver for poor air quality. However, natural processes also play a critical role in influencing atmospheric composition through exchange with the Earth's surface which critically serves as sinks for species from the air. There are many other interactions between the atmosphere and the rest of the natural environment including natural emissions (e.g., from soil and vegetation) that have a large influence on the quality of air and illustrate the linkages between the different components of the Earth as a system.

These interlinkages involve different scales and processes. Chapter 4 discussed the role of chemical transformations which often involve reactions in gas-phase and gas-particle phase. These reactions occur between anthropogenic and naturally emitted species which play a crucial role both at urban and regional scales for secondary air pollutants (e.g., long-range transport of ozone and aerosols). For aerosols, scavenging and deposition (dry and wet deposition), and microphysical processes (e.g., aerosol-cloud interactions) affect their concentration and size distributions.

Dynamical processes, introduced in Chapter 5, govern meteorological transport and have a major influence on air pollution, its concentration and spatial and temporal distributions. These operate from the smallest micro, meso and synoptic scales with connections with global scale atmospheric circulations. As explained in Chapter 6, nearer the surface, atmospheric boundary layer processes largely govern diurnal and seasonal variations of air pollution concentrations which, along with emission variabilities and faster chemical reactions (e.g., those involving ozone and nitrogen oxides), directly determine our exposure.

Chapters 7 and 8 examined many of the above considerations and illustrated the complex nature of air quality prediction which includes parameterization of many of the processes discussed above. Although air quality models can differ considerably in their complexity and assumptions, their common features include emissions processing, meteorological driver (e.g., incorporating treatment for turbulence, advection transport, mixing and stability), physical and chemical transformations and geographical characteristics (e.g., terrain and street geometry). When using theses models, it is therefore important to understand their limitations and range of applicability to make sensible interpretation of their outputs. Model applications can be specific such as for isolated point source emissions or for streets and cities as well as for regional and continental assessments. Depending on their application, models are critical to quantify air pollutant concentrations necessary for estimating exposure to air pollution, to test the efficacy of control measures and policies, including for compliance with regulation, and for basic research.

Measurement of air pollution, discussed in Chapter 9, is equally important as modeling. In fact, they are often used in combination, for example, for model development and evaluation and when measurements are not available. Monitoring networks, supported by quality assurance and quality control processes, have now become a necessary tool to manage air quality in many regions of the world although there are still large areas where measurements are lacking (e.g., in many parts of Africa and other regions).

While air pollution is acutely experienced every day by large sections of the population in towns and cities across the world, regional and global contributions can have a major influence on local air quality. As discussed in Chapter 10, these influences can differ in terms of their contribution and nature for different parts of the world. Long-range transport of air pollution, as an important example, can affect local air quality beyond continental scales which implies that controlling and improving air quality is not necessarily within the responsibility of a single region.

On a global scale, the relationships between climate and air quality are now well recognized as discussed in Chapter 11. The intricate interactions, however, influence air pollution along its full chain from emissions to receptor impacts depending on the specific species. Climate affects air quality through many routes including via chemical, dynamical and microphysical processes and local and regional circulation and transport mechanisms. In addition to changes due to temperature, relative humidity and precipitation patterns, air pollutant species such as aerosols and their precursors interact to affect solar radiation and perturb the radiation budget of the planet. Synergistic effects that combine direct air pollution exposure and the consequences of a changing climate (e.g., heat stress) can exacerbate health impacts. Feedback interactions are another layer of complexity, for example, with climate change affecting boundary layer processes and emissions which in turn affects climate.

When considering exposure to air pollution, Chapter 12 discussed the importance of the time we spend in indoor environments and how sources and conditions of these enclosed spaces can be even more important than ambient pollution adding to our health burden. Risks from indoor air pollution, however, can vary significantly with many factors including the nature of sources, age of individuals and their prevailing health condition. The type of enclosed environment will also significantly affect exposure to air pollution (e.g., fixed or mobile spaces) which implies that a more integrated approach is needed to health exposure assessment.

Quantitative analysis, discussed in Chapter 13, of air pollution data, either from measurements or from models, requires increasingly sophisticated approaches which comprise a set of tools for air quality scientists. These tools are routinely used to analyze measurements and model outputs for air quality management and research purposes. As explained in earlier chapters, air quality data can vary in complexity and extent depending on the type of measurement instrument or type of mathematical model used. Tools for analyzing air quality data are equally diverse in terms of their capabilities and applications and include statistical, visualization, and now increasingly artificial intelligence techniques.

The main motivation for air quality research is to understand and develop approaches to reduce the impact on health. There is increasingly ample evidence that air pollution is responsible for millions of premature deaths globally and for exacerbating many medical conditions, especially for vulnerable populations. Chapter 14 has introduced health impact assessment and air pollution epidemiology along with providing an overview of the health effects of air pollution and main methods for estimating the health burden from air pollution.

The impact of air pollution goes beyond human health as it also has a major impact on our surrounding environment as explained in Chapter 15. The atmosphere is inextricably connected to the terrestrial biosphere, such as through natural emissions and their influence on air quality. However, the terrestrial environment is equally complex with different soil and vegetation types, land use practices (e.g., urban and agriculture) and prevailing meteorological conditions. These interactions with the atmosphere can have major influences on the concentration of secondary air pollutants including aerosols and gaseous air pollutants (e.g., ozone).

As the impact of air pollution has been progressively appreciated, air quality management practices have been developed to curtain these impacts. Chapter 16 has introduced some of the main approaches to air quality management and policies for reducing air pollution including through region-specific case studies and providing insights into their local and regional challenges.

Chapter 17 also focuses on case studies centered on major cities which face disparate challenges ranging from growing urban populations and vehicle ownership, changing emissions distributions, and prevailing weather patterns to the higher frequency of wildfires and natural hazards from sand and dust storms. Not only do these case studies illustrate the pressures from poor air quality around the world, they also serve to highlight the progress that has been achieved. We hope that this book has provided the reader not only with the foundations of air quality science but also with a wider appreciation of the intricacies of our environment and our own role in determining the future of our world.

Bibliography

There are many books available that are relevant to the subject of air quality science. The following are a small selection that the reader may find useful and will complement this book.

Air Pollution - Concepts, Theory, and Applications
Christian Seigneur. Cambridge University Press, 2019. Online ISBN 9781108674614.

Air Quality
Wayne T. Davis, Joshua S. Fu and Thad Godish. CRC Press, 2023. ISBN 9780367705237.

An Introduction to Atmospheric Physics
David G. Andrews. Cambridge University Press, 2012. Online ISBN 9780511800788.

Atmospheric Chemistry and Physics: From Air Pollution to Climate Change
Spyros N. Pandis and John H. Seinfeld. Wiley, 2016. ISBN 9781118947401.

Atmospheric Science - An Introductory Survey
John M. Wallace and Peter V. Hobbs. Elsevier and Academic Press, 2006. ISBN 9780127329512.

Handbook of Air Quality
Edited by Frieda Bush. States Academic Press, 2023. ISBN 9781639897544.

Indoor Air Quality: The Latest Sampling and Analytical Methods
Kathleen Hess-Kosa. CRC Press, 2020. ISBN 9780367656775.

Mesoscale Modelling for Meteorological and Air Pollution Applications
Edited by Ranjeet S. Sokhi, Alexander Baklanov and K. Heinke Schlünzen. Anthem Press, 2018. ISBN 9781783088263.

Principles of Air Quality Management
Roger D. Griffin. CRC Press, 2020. ISBN 9780367577803.

Physics of the Atmosphere and Climate
Salby L Murry. Cambridge University Press, 2012. Online ISBN 9781139005265.

Training Materials and Best Practices for Chemical Weather/Air Quality Forecasting
Editors by Yang Zhang and Alexander Baklanov. Published by the World Meteorological Organization (WMO), reference ETR-26, 2020.

Issues in Environmental Science and Technology - Indoor Air Pollution
Edited by R M Harrison and R E Hester Hardback, Royal Society of Chemistry, 2019. Hardback ISBN 9781788015141.

Air Pollution: Science, Engineering and Management Fundamentals
Mukesh Khare, Prateek Sharma, Sri Harsha Kota, Sumanth Chinthala. CRC Press, 2024. Hardback ISBN-13 9780367750527.

Index

Note: Page numbers followed by "*f*" indicate figures, "*t*" indicate tables, and "*b*" indicate boxes.

F

G

H

I

N

O

P

T

X

Y

Z

Made in the USA
Monee, IL
13 September 2026

63138396R00330